AF333812

BIOLOGICAL REACTIVE INTERMEDIATES VI

ADVANCES IN EXPERIMENTAL MEDICINE AND BIOLOGY

Recent Volumes in this Series

BIOLOGICAL REACTIVE INTERMEDIATES VI

Chemical and Biological Mechanisms in Susceptibility to and Prevention of Environmental Diseases

Edited by

Patrick M. Dansette

Université René Descartes
Paris, France

Robert Snyder

Rutgers University and The Environmental & Occupational Health Sciences Institute
Piscataway, New Jersey

Marcel Delaforge

CNRS
Paris, France

G. Gordon Gibson

University of Surrey
Guildford, Surrey, England

Helmut Greim

Institute of Toxicology and Environmental Health
The Technical University of Munich
Munich, Germany

David J. Jollow

Medical University of South Carolina
Charleston, South Carolina

Terrence J. Monks

University of Texas
Austin, Texas

I. Glenn Sipes

University of Arizona
Tucson, Arizona

Kluwer Academic / Plenum Publishers
New York, Boston, Dordrecht, London, Moscow

Library of Congress Cataloging-in-Publication Data

Biological reactive intermediates VI: chemical and biological mechanisms in
susceptibility to and prevention fo environmental disease/edited by Patrick M. Dansette
... [et al.].
     p. ; cm. — (Advances in experimental medicine and biology; v. 500)
    Includes bibliographical references and index.
    ISBN 0-306-46659-7
     1. Biochemical toxicology—Congresses. 2. Environmental toxicology—Congresses. 3.
Pollutants—Structure-activity relationship—Congresses. I. Title: Biological reactive
intermediates six. II. Title: Biological reactive intermediates 6. III. Dansette, Patrick M.
IV. International Symposium on Biological Reactive Intermediates (6th: 2000:
Université René Descartes) V. Series.
    [DNLM: 1. Toxicology—methods—Congresses. 2. Biotransformation—Congresses. 3.
Environmental Illness—etiology—Congresses. 4. Environmental Illness—prevention &
control—Congresses. 5. Environmental Pollutants—toxicity—Congresses. 6.
Structure-Activity Relationship—Congresses. QV 602 B6155 2001]
    RA1219.5 .B576 2001
    615.9′02—dc21

    2001038407

Proceedings of the International Symposia on Biological Reactive Intermediates VI, held July 16–20, 2000
held at the Université René Descartes, Paris, France

ISBN 0-306-46659-7

©2001 Kluwer Academic / Plenum Publishers, New York
233 Spring Street, New York, N.Y. 10013

http://www.wkap.nl/

10 9 8 7 6 5 4 3 2 1

A C.I.P. record for this book is available from the Library of Congress

PREFACE

Historically we have separated the disciplines of Chemistry and Biochemistry by recognizing that the distinguishing characteristic of Biochemistry is the catalysis of reactions by enzymes. Enzymes permit metabolic reactions which would otherwise require extremes of temperature, pressure or pH, often associated with Chemistry, to proceed under ambient conditions of the body. Under some conditions chemical reactions occur *in vivo* in which products of enzymatic reactions proceed to undergo further reactions non-enzymatically with cellular macromolecules. The results can often be seen as toxic or carcinogenic responses. The chemicals that initiate these reactions are termed "biological reactive intermediates."

The International Symposia on Biological Reactive Intermediates (BRI) began in 1975 at the University of Turku, Finland and have since convened at the University of Surrey, Guildford, The United Kingdom (1980), the University of Maryland, College Park, Maryland (1985), the University of Arizona, Tucson, Arizona (1990), the GSF Forschungszentrum and Technical University of Munich (1995) and, most recently, at the Université René Descartes, Paris, France (2000).

The Symposium was organized by an International Planning Committee co-chaired by P. Dansette (Paris, France) and T.J. Monks (Austin, Texas). The committee included: P.H. Beaune (Paris, France), M. Delaforge (Saclay, France), G.P. Gervasi (Pisa, Italy), G.G. Gibson (Guildford, UK), H. Greim (Munich, Germany), D.J. Jollow (Charleston, South Carolina), P. Moldeus (Sodertalje, Sweden), I.G. Sipes (Tucson, Arizona), R. Snyder (Piscataway, New Jersey), and P.J. van Bladderen (Zeist, The Netherlands). They were assisted by an International Scientific Program Advisory Committee which included: T.J. Monks (chr.) (Austin, Texas), M.W. Anders (Rochester, New York), J. Bolton (Chicago, Illinois), F. DeMatteis (Turin, Italy), D.L. Laskin (Piscataway, New Jersey), A. Puga (Cincinnati, Ohio), D. Williams (Corvallis, Oregon) and G. Yost (Salt Lake City, Utah).

BRI VI convened at the Université René Descartes, July 16–20, 2000, and was attended by approximately 300 participants. There were 67 invited papers presented and 159 volunteer presentations in the form of posters. Reports on these presentations are the substance of this volume.

The organizers are indebted to the following institutional sponsors for support of the meeting: the Université René Descartes- Paris V, UFR Biomedicale- Centre Universitaire des Saints -Peres, Centre National de la Recherche Scientifique, Department de Chemie, Programme Physique et Chemie di Vivant, Centre D'Etudes Atomiques- Department Sciences du Vivant, the European Commission, the U.S. Environmental Protection Agency, National Institute of Environmental Health Sciences, Center for Disease Control and Prevention (Agency for Toxic Substances and Disease Registry), the U.S. Department of Energy, Rutgers, The State University of New Jersey, Rutgers College of Pharmacy, The University of Medicine and Dentistry of New Jersey/Robert Wood Johnson Medical School, the Environmental and Occupational Health Sciences Institute, the Medical University of South Carolina, the Center for Molecular and Cellular Toxicology, University of Texas, Austin, Texas, and the University of Arizona.

The organizing committee also recognizes the following commercial sponsors: Allied Signal, Inc., Astra-Zeneca, Aventis Pharma, BASF AG, Bayer AG, Biopredic, BYK Gulden, Eastman Kodak Company, Fournier, Hoescht Marion Roussel, Hoffmann La Roche, SA, Lipha, SA, Mc Neil Specialty Products Company, Merck & Company, Inc., Merck KGaA, Nestle, SA, Novartis Pharma, Procter and Gamble, Upjohn, Phoenix International Life Sciences, Rohm and Haas Company, Smith, Kline Beecham, Schering AG, Schering Plough Research Institute, SKW Trostberg AG, Solvay Pharma GmbH, and Totalfina-Elf. Without the help of our loyal sponsors BRI VI would not have been possible.

The chairs of each session play a key role in keeping the symposium on time and moving forward with appropriate pace. Many of the session chairs were either speakers or members of the organizing or program committees. Others were invited solely to act as session chair people. The organizers wish to thank R.W. Estabrook (Dallas, TX), N.P.E. Vermeulen (Amsterdam, The Netherlands), D.M. Jerina (Bethesda, MD). J. Gorrod (Colchester, UK), K. Netter (Marburg, Germany), J.J. Kocsis (Philadelphia, PA), D.J. Reed (Corvallis, OR), H. Vainio (Stockholm, Sweden), and E. Dybing (Oslo, Norway) for their important work as session chairmen.

Professor Sten Orrenius of the Karolinska Institute in Stockholm was the guest of honor and keynote speaker at the symposium. Professor Orrenius is a world renowned scholar who has made significant contributions to our understanding of the mechanism of cell death. For his participation in BRI VI, and in recognition of his many scientific contributions, the University of Paris has awarded him a doctorate, *honorus causa*.

The effort that goes into mounting a major international symposium requires much of the time between these meetings which are held at five-year intervals. The planning and program committees begin quite early. A venue must be chosen along with a local chair person who agrees to accept responsibility for local arrangements. We were delighted when Dr. Patrick Dansette agreed to act as local chairman and to take on this onerous task. He and his associates in Paris performed splendidly. We must also thank the conference secretariat, Michèle Centonze Conseil, for creating an atmosphere which permitted the scientific program to go forward with success, while Michèle Centonze and her staff managed the meeting in all of its aspects with efficiency, cordiality, and aplomb. For funds to be raised and correspondence via every means available utilized requires the efforts of many people. We wish to thank the staff of the Université René Descartes, and the other contributing universities for donating their time and making the effort needed to insure success of the symposium. Special thanks are conveyed to Bernadine Chmielowicz for coordinating the effort at the Environmental and Occupational Health Sciences Institute.

The success of BRI VI is directly related to the interest and enthusiasm of the contributing speakers, presenters, and authors. Over the past 25 years their interest and efforts have grown. This meeting was still in progress when many of the participants began asking about plans for the next BRI. The science remains at the cutting edge of basic research in Toxicology. The advent of genomics and proteomics have led to their adoption by toxicologists as important tools for the study of mechanistic toxicology. In addition, we have begun to ask serious questions about the utilization of data from studies of biological reactive intermediates in the risk assessment process. As a result the symposium is relevant to the most basic aspects of Toxicology as well as at the practical level of risk analysis.

Although there were many new faces of young investigators at BRI VI, many of the participants have attended these meetings from the beginning. We interpret that to mean that the science is excellent, and that the spirit of scientific comradeship and collegiality remains strong. The participants will not soon forget the stimulating scientific discussions held against the background of the glorious days in summertime Paris.

Patrick M. Dansette *G. Gordon Gibson* *Terrence J. Monks*
Robert Snyder *Helmut Greim* *I. Glenn Sipes*
Marcel Delaforge *David J. Jollow*

CONTENTS

SHORT COMMUNICATIONS

SESSION III

SHORT COMMUNICATIONS

SESSION IV

SHORT COMMUNICATIONS

SESSION VII

SHORT COMMUNICATIONS

SESSION VIII

SHORT COMMUNICATIONS

SESSION IX

WORKSHOP ON UTILIZATION OF COVALENT BINDING DATA IN RISK ASSESSMENT

SESSION I—COVALENT BINDING TO PROTEINS

SESSION II—COVALENT BINDING TO DNA

xv

BIOLOGICAL REACTIVE INTERMEDIATES AND MECHANISMS OF CELL DEATH

John D. Robertson, Joya Chandra, Vladimir Gogvadze, and Sten Orrenius[1]

Division of Toxicology
Institute of Environmental Medicine
Karolinska Institutet
Box 210
SE-171 77 Stockholm
Sweden

INTRODUCTION

Apoptosis and necrosis are two modes of cell death with distinct morphological and biochemical features. Apoptosis is an active process characterized by cell shrinkage, nuclear and cytoplasmic condensation, chromatin fragmentation and phagocytosis. In contrast, necrosis is a passive form of cell death associated with inflammation resulting from cellular and organelle swelling, rupture of the plasma membrane and spilling of cellular contents into the intercellular milieu. Lethal levels of biological reactive intermediates may trigger either apoptotic or necrotic cell death, depending on the cell type and severity of insult. Further, effectuation of the apoptotic death program requires maintenance of a sufficient intracellular energy level and of a redox state compatible with caspase activity. Thus, ATP depletion or severe oxidative stress may redirect otherwise apoptotic cell death to necrosis.

In the current paradigm for apoptotic cell death, the activity of a family of caspases (cysteine proteases with a stringent requirement for Asp in the P_1 position of the substrate) orchestrates the multiple downstream events that comprise apoptosis. In TNF- and CD95-mediated apoptosis the proteolytic cascade is believed to be triggered directly by pro-caspase-8 binding to the activated plasma membrane receptor complex. In other forms of

---

[1] To whom correspondence should be addressed. Fax: +46 8 32 90 41. Email: Sten.Orrenius@imm.ki.se

apoptosis, the mechanisms of activation of the proteolytic cascade are less well established but may involve other proteases, *e.g.* granzyme B and calpain, or factors released from the mitochondria. Recently, the possibility that cytochrome *c* may serve to activate dormant caspases in the cytosol, and thereby to trigger or amplify the apoptotic process, has attracted considerable attention.

Lethal oxidative stress may cause apoptotic or necrotic cell death, depending on the severity of the insult. While it is yet unclear how an oxidative stimulus can activate the caspase cascade, there is emerging evidence that this may occur by either upregulation of the CD95 system or stimulation of cytochrome *c* release from mitochondria. Although oxidative stress can induce apoptosis, this is slow to develop and usually occurs some time after the original source of the oxidative stress has decayed. During the acute phase of oxidative stress, cells appear to be partially resistant to a secondary induction of apoptosis. There is accumulating evidence that this inhibition operates via redox inactivation of caspases, and that caspase activity may be modulated by fluctuations in intracellular redox state. Our recent studies with menadione, dithiocarbamates and hydrogen peroxide demonstrate the toxicological significance of this mechanism, which may also apply to modulation of caspase activity by nitric oxide or alterations in the glutathione redox state under physiological conditions.

Finally, another mechanism by which reactive intermediates can modulate cell death is by induction of heat shock proteins (Hsps). Thus, increased expression of Hsp27 and Hsp72 in thermotolerant T-lymphocytes protects the cells from apoptosis induced by several triggers. Whereas Hsp27 appears to exert its antiapoptotic effect by preventing the release of mitochondrial cytochrome *c*, Hsp72 interferes with the subsequent activation of caspases.

From these and other studies it is clear that biological reactive intermediates can not only determine cell fate but can also modulate the mode of cell death. This modulation may be mediated by their effect on components of the apoptotic machinery.

ROS, MITOCHONDRIA, AND CYTOCHROME *c* RELEASE

Intracellular accumulation of reactive oxygen species (ROS), such as superoxide anion, hydrogen peroxide, singlet oxygen, hydroxyl radical, and peroxy radical, can arise from toxic insults or as a result of normal metabolic processes. These species may perturb the cell's natural antioxidant defense systems, resulting in damage to all of the major classes of biological macromolecules, including nucleic acids, proteins and lipids. Normally, depletion of intracellular glutathione (GSH) precedes cytotoxicity, which is associated with perturbations of Na^+ and Ca^{2+} ion homeostasis, mitochondrial dysfunction and ATP depletion.[1] Morphologically, formation of cell surface blebs is a characteristic feature of oxidative cell damage and is caused by effects of both thiol oxidation and elevated intracellular Ca^{2+} concentration on microfilament structure.[2]

Severe oxidative damage leads to ATP depletion and necrotic cell death, whereas modest oxidative stress can kill cells by apoptosis. Several mechanisms for ROS-mediated induction of apoptosis have been proposed, including upregulation of the CD95 system and ROS/Ca^{2+}-induced mitochondrial swelling and cytochrome *c* release.[3]

Both *in vitro* and *in vivo* studies have demonstrated the release of cytochrome *c* from mitochondria during apoptosis,[4,5] although the mechanism controlling this event is unknown. Once in the cytosol, cytochrome *c* interacts with apoptotic protease activating factor-1 (Apaf-1) and pro-caspase-9 forming the apoptosome complex (discussed below).[6] The result is the cleavage and activation of pro-caspase-9 and other pro-caspases, which are responsible for the executive stages of apoptotic cell death (Figure 1).

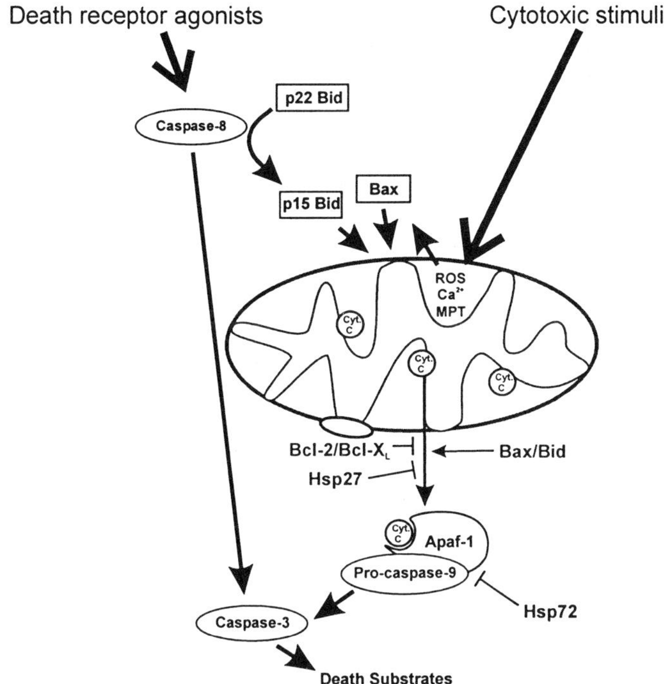

Figure 1. A proposed model of the relationships (described in text) among mitochondria, caspases, Hsps and Bcl-2-related proteins involved in regulating apoptotic cell death.

Cytochrome c is a component of the mitochondrial respiratory chain and, as such, is involved in the formation of the mitochondrial membrane potential; the primary driving force for ATP production. Specifically, it supports respiration by shuttling electrons between complexes III (bc1) and IV (cytochrome oxidase). Apo-cytochrome c is encoded by a nuclear gene and synthesized on ribosomes in the cytoplasm. It is imported into the mitochondrial intermembrane space in an unfolded configuration where it receives its heme group.[7] Covalent attachment of this heme group stimulates a conformational change, and holo-cytochrome c subsequently assumes its functional role as a component of the electron transport chain. It is important to note that only holo-cytochrome c, and not apo-cytochrome c, is able to stimulate pro-caspase-9 activation.[4] Moreover, observations made by our laboratory indicate that both reduced and oxidized cytochrome c are able to activate pro-caspase-9 with similar efficiencies.[8]

As mentioned, the precise molecular mechanism(s) controlling cytochrome c release from mitochondria in the presence of a pro-apoptotic stimulus is (are) not clear, although several models have been proposed, including one that is mitochondrial permeability transition (MPT)-dependent and another that is MPT-independent. It is well known that Ca^{2+} is obligatory for MPT induction, although the sensitivity of MPT to Ca^{2+} can be greatly increased by a depletion of adenine nucleotides, an elevated level of inorganic phosphate, or oxidative stress.[9] When phosphate or hydrogen peroxide are added to mitochondria, these organelles swell, become leaky, and their membrane potential ($\Delta\psi_m$) drops. The induction of MPT, resulting in the opening of the permeability transition pore

(PTP), stimulates the release of cytochrome *c*, and this effect can be blocked by the MPT inhibitor cyclosporin A. The PTP consists of both inner and outer mitochondrial membrane proteins, such as adenine nucleotide translocator (ANT) and voltage-dependent anion channel (VDAC), respectively, and is formed at contact sites between these two membranes.[10] An extension of this model was recently proposed by Tsujimoto's group, who demonstrated the ability of recombinant Bax and Bak, two pro-apoptotic Bcl-2 family proteins, to hasten the opening of VDAC in liposomes and induce changes in $\Delta\psi_m$, whereas the anti-apoptotic Bcl-X$_L$ was able to bind and close VDAC directly.[11] A subsequent study reported the ability of other pro-apoptotic Bcl-2 family members, *i.e.*, Bid and Bik, to stimulate apoptosis in a fashion different from that of Bax and Bak.[12] Specifically, Bid and Bik stimulated cytochrome *c* release without inducing changes in $\Delta\psi_m$ or interacting directly with VDAC. These data supported the existence of MPT-independent modes of cytochrome *c* release. Further evidence of an MPT-independent mechanism was reported by Martinou *et al.*[13], who demonstrated that mitochondria of NGF-deprived sympathetic neurons undergoing apoptosis released cytochrome *c* prior to any changes in $\Delta\psi_m$. Moreover, recent evidence from our own work indicates that under certain conditions the topoisomerase II inhibitor etoposide stimulates cytochrome *c* release without inducing a complete collapse of $\Delta\psi_m$ and MPT (J.D. Robertson, V. Gogvadze, B. Zhivotovsky, and S. Orrenius, unpublished data) (Figure 1).

It is important to note that cytochrome *c* release is often a common feature of cells undergoing either apoptosis or necrosis.[14] Specifically, we demonstrated a similar degree of cytochrome *c* release in response to an apoptosis-inducing dose of CD95 or a necrosis-inducing dose of menadione. In this respect, the release of cytochrome *c*, by itself, would seem to be an insufficient estimation of the apoptotic status of a cell. Instead, this form of cell death is more accurately distinguished by the involvement of caspases as described below.

CASPASES

Considerable interest has surrounded a growing family of cysteine-aspartate proteases (caspases) that are intimately involved in apoptotic cell signaling and share sequence similarity with the *Caenorhabditis elegans* cell death protease (Ced-3).[15-17] The precise way in which caspases execute cell death is not known largely because of the variety of intracellular protein substrates for these enzymes that have been identified and because of their ability to translocate across subcellular compartments upon activation.[18,19]

Since the discovery of interleukin-1β-converting enzyme (ICE or caspase-1), thirteen additional caspases have been identified.[15,16] All caspases are present constitutively and synthesized in inactive, precursor forms (30-50 kDa) that must be proteolytically cleaved in order to be activated.[18] Each pro-caspase consists of a pro-domain, a large (~20 kDa) and a small (~10 kDa) subunit.[20] Cleavage and subsequent heterodimerization of the larger and smaller subunits results in caspase activation. An active caspase can subsequently activate additional pro-caspases like itself and/or different pro-caspases.[21] Caspases are commonly divided into two groups based upon the primary structure of their NH$_2$-terminal prodomain: initiator (caspase-2, -8, -9 and −10) and effector (caspase-3, -6 and −7) enzymes. Initiator pro-caspases are activated with the help of adaptor molecules, *e.g.*, Apaf-1 and FADD/MORT1, which bring these proteases into close proximity, permitting auto-processing. Once activated, initiator caspases are responsible for cleaving and activating effector pro-caspases. Effector caspases, in turn, cleave various proteins leading to morphological and biochemical features characteristic of apoptosis.[22]

Different initiator pro-caspases signal via distinct pathways. For instance, pro-caspase-9 is involved in apoptosis induced by cytotoxic stimuli, whereas pro-caspase-8 is recruited during death receptor activation (Figure 1). Moreover, and as mentioned previously, initiator pro-caspases require the assistance of adaptor molecules to become activated. Pro-caspase-9 requires the cytosolic adaptor molecule Apaf-1. In the presence of different chemical stimuli, mitochondria release specific proteins into the cytosol. Among these molecules are cytochrome c, apoptosis inducing factor and adenylate kinase-2.[23-25] Once in the cytosol, cytochrome c has been shown to activate Apaf-1, which then oligomerizes, binds and activates pro-caspase-9 in the presence of dATP.[6,26] Active caspase-9 can signal downstream resulting in the activation of effector pro-caspase-3 and -7. Cleavage of pro-caspase-9 is inhibitable by the caspase inhibitor z-VAD.fmk, whereas the release of cytochrome c is unaffected by caspase inhibitors in most models of chemically-induced apoptosis.[27]

Although pro-caspase-8 is primarily associated with receptor-mediated apoptosis, it bears mentioning because of its ability to activate all known caspases *in vitro* and because of a report indicating that p22 Bid, a caspase-8 substrate, was cleaved during chemical-induced apoptosis.[27,28] Pro-caspase-8 activation requires the cytoplasmic adaptor molecule FADD/MORT1, which is essential for CD95(Fas/APO-1)-mediated forms of apoptosis.[29] Traditionally, when Fas ligand binds to its receptor, FADD/MORT1 is recruited to the receptor through a death effector domain (DED). This allows binding of pro-caspase-8, auto-processing and activation.[30-32] Once activated, caspase-8 can signal downstream activating the effector pro-caspase-3, seemingly bypassing pro-caspase-9 and mitochondria. However, recent studies demonstrate the ability of mitochondria to significantly amplify Fas-mediated apoptosis by releasing cytochrome c.[33,34] Considerable evidence indicates that p22 Bid (a pro-apoptotic Bcl-2 family protein) is cleaved by active caspase-8, yielding a p15 truncated Bid (tBid) product. tBid, the active form of this protein, was shown to target mitochondria and stimulate the release of cytochrome c.[35,36]

REDOX MODULATION OF CASPASES

Caspases contain an active site cysteine nucleophile[37] that is prone to oxidation or thiol alkylation.[38] It is, therefore, not surprising that the activity of caspases is optimal

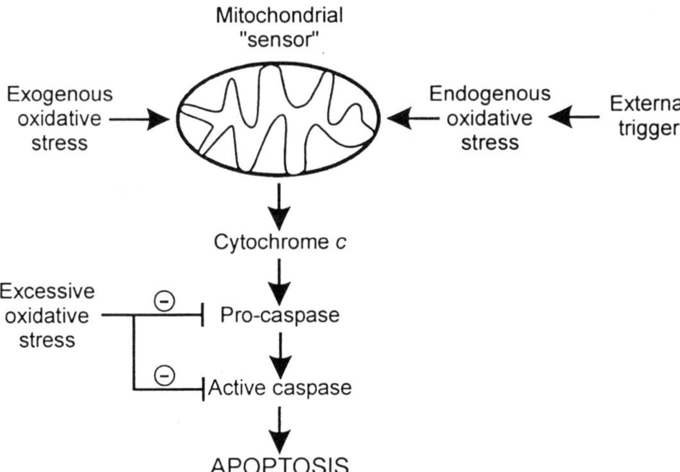

Figure 2. Schematic representation of ROS-mediated induction and inhibition of caspase activity as described in text.

under reducing environments. Any deviation from such reducing conditions, as would be the case in the presence of an oxidative stress, could be detrimental to caspases and render them inactive (Figure 2). It was demonstrated by our group that hydrogen peroxide suppresses both the activation of pro-caspases and activity of mature caspases, possibly through modulation of the redox status of the cell and the oxidation of cysteine residues within caspases (Figure 2).[39] Furthermore, the oxidation of dithiocarbamates to thiuram disulfides has been shown to inhibit apoptosis[38] through disulfide formation at the active site cysteine of caspases and inhibition of their enzymatic activity.[40]

Indirect effects on caspases can also be a component of toxicity in certain systems. As mentioned previously, we demonstrated that menadione could not only induce necrosis in HepG2 cells, but that exposure of CD95-treated cells to menadione switched the mode of cell death to necrosis.[14] Menadione, like other redox-active quinones, is known to have cytotoxic effects, which are mediated through oxidative stress and alterations in cellular Ca^{2+} homeostasis. Metabolism of menadione leads to the generation of ROS[1] and to the oxidation of reduced glutathione to glutathione disulfide.[41] Pre-incubation of apoptotic cells with catalase, but not with Cu, Zn-SOD, reduced ROS levels and reversed the inhibitory effect of menadione on caspase activity. Taken together, these results suggest that the inhibitory effects of menadione are due to the production of hydrogen peroxide and subsequent inactivation of caspases and not to a direct effect of menadione on caspase thiol group(s).

While apoptosis and necrosis can occur simultaneously within a cell population, studies suggest that intracellular energy levels are important determinants of the mode of cell death.[42,43] Leist *et al.*[43] demonstrated that different aspects of the cell death program could be halted by manipulating intracellular levels of ATP. Specifically, apoptosis induced by established pro-apoptotic stimuli, *e.g.,* staurosporine or anti-CD95 agonistic antibody, could be switched to necrosis when ATP levels were reduced; an effect believed to be linked to reduced caspase activity and/or formation of the apoptosome complex.[6]

HEAT SHOCK PROTEINS

Exposure of cells to transient non-lethal hyperthermia causes the activation of a cellular stress response and induces a state of acquired thermotolerance, which renders cells resistant to subsequent lethal insults.[44] This phenomenon is highly conserved and is now recognized as a general protective strategy of cells against many other stress-inducing agents, including heavy metals, radiation, nitric oxide, and other oxidants. Accumulation of a set of highly conserved proteins, called heat shock proteins (Hsps) correlates with the thermotolerant phenotype.[45,46] The major Hsps of mammalian cells include proteins of 110, 90, 70, 60, 40, 32 and 27 kDa.[45,46]

Thermotolerance has been linked to inhibition of apoptosis by a number of studies.[47] Similarly, constitutive overexpression of Hsps yielded a comparable anti-apoptotic effect.[47,48] A number of Hsps, in particular Hsp32 (heme oxygenase) and Hsp27, were shown to be induced and to protect against oxidative stress.[48,49] The inducible form of heme oxygenase (HO-1) was recently reported to protect cells against oxidant-induced cell death by regulating the intracellular redox active iron.[50] Hsp27, on the other hand, appears to inhibit apoptosis by a different mechanism. It was suggested that the small Hsps (Hsp27, αB-crystallin) regulate apoptosis by maintaining the redox equilibrium of the cell.[48] In this regard, it was shown that overexpression of Hsp27 or αB-crystallin, can inhibit oxidant-induced cellular damage due to a decrease in the intracellular level of ROS in a glutathione-dependent manner.[51] This effect may be mediated by glucose-6-phosphate dehydrogenase (G6PD), since Hsp27-overexpressing cells have increased levels of this enzyme.[52] A separate study showed that overexpression of G6PD bolsters intracellular glutathione levels, and protects cells against H_2O_2 and TNF-induced apoptosis.[53]

Recent results from our laboratory showed that Hsp72 and Hsp27 were maximally induced at 42°C in heat-treated Jurkat T-lymphocytes, and neither Hsp was induced at temperatures beyond 42°C, despite heat shock factor (HSF) activation. Moreover, cells underwent apoptosis at 44°C and necrosis at 46°C, whereas thermotolerant cells, in which Hsp27 and Hsp72 were elevated, were resistant to apoptosis, as determined by cytochrome *c* release and caspase activation. Hsp27 and Hsp72 were not able to rescue cells from death at temperatures beyond 44°C. This experiment allowed us to identify the temperature at which Hsps and thermotolerance are maximally induced, as well as temperatures that cause different forms of cell death.[54] A subsequent study carried out in our laboratory demonstrated that Hsp27, and to a far less extent Hsp72, associated with mitochondria during thermotolerance, suggesting that Hsp27 plays a larger role in regulating mitochondrial changes, *e.g.*, cytochrome *c* release, than does Hsp72 during apoptosis. This was further supported by evidence showing that inhibition of Hsp27 induction during thermotolerance in cells transfected with *hsp27* antisense potentiated mitochondrial cytochrome *c* release following exposure to various apoptosis stimuli, including H_2O_2 (A. Samali, J.D. Robertson, and Sten Orrenius, unpublished data). Additionally, *in vitro* studies revealed that recombinant Hsp72 more efficiently blocked cytochrome *c*-mediated caspase activation, than did recombinant Hsp27. Taken together, these results suggest that Hsp27 exerts its anti-apoptotic effect by preventing the release of cytochrome c, whereas Hsp72 interferes with the apoptosomal activation of caspases (Figure 1).

CONCLUDING REMARKS

Various paradigms integrating biological reactive intermediates and cell death are evolving as pathways become better characterized. Depending upon the severity of the insult, lethal oxidative stress can stimulate necrotic or apoptotic cell death. A central theme among different forms of apoptosis induced by biological reactive intermediates is the involvement of mitochondria. Specifically, it seems that the release of cytochrome *c*, if not a universal event, occurs in many ROS-mediated apoptotic models. Once released, cytochrome *c* is able to activate Apaf-1, which oligomerizes and recruits pro-caspase-9 in the presence of dATP. Active caspase-9, in turn, cleaves and activates effector pro-caspases, which are responsible for carrying out biochemical and morphological changes characteristic of apoptosis. Recent evidence indicates that the initiation of apoptotic cell death following a lethal insult depends heavily on sufficient levels of ATP and a redox state compatible with caspase activation; two phenomena that can be influenced by Hsps. Taken together, our current understanding of the relationship between reactive intermediates and cell death suggests that these molecules are capable of directly stimulating cell death, as well as redirecting apoptosis to necrosis by altering specific steps within the apoptotic program.

ACKNOWLEDGMENTS

This work was supported by grants from the Swedish Medical Research Council (03X-2471). J.D.R. is supported by a Visiting Scientist grant from The Swedish Foundation for International Cooperation in Research and Higher Education (STINT) and V.G. by a grant from the Royal Swedish Academy of Sciences.

REFERENCES

1. H. Thor, M.T. Smith, P. Hartzell, G. Bellomo, S. Jewell, and S. Orrenius, The metabolism of menadione (2-methyl-1,4-naphthoquinone) by isolated hepatocytes. A study of the implications of oxidative stress in intact cells, *J. Biol. Chem.* 257:12419 (1982).
2. S. Jewell, G. Bellomo, H. Thor, S. Orrenius, and M.T. Smith, Bleb formation in hepatocytes during drug metabolism is caused by disturbances in thiol and calcium ion homeostasis, *Science* 217:1257 (1982).
3. H. Stridh, M. Kimland, D.P. Jones, S. Orrenius, and M.B. Hampton, Cytochrome c release and caspase activation in hydrogen peroxide- and tributyltin-induced apoptosis, *FEBS Lett.* 429:351 (1998).
4. J. Yang, X. Liu, K. Bhalla, C.N. Kim, A.M. Ibrado, J. Cai, T.-I. Pen, D.P. Jones, and X. Wang, Prevention of apoptosis by Bcl-2: release of cytochrome c from mitochondria blocked, *Science* 275:1129 (1997).
5. R.M. Kluck, E. Bossy-Wetzel, D.R. Green, and D.D. Newmeyer, The release of cytochrome c from mitochondria: a primary site for Bcl-2 regulation of apoptosis, *Science* 275:1132 (1997).
6. P. Li, D. Nijhawan, I. Budihardjo, S.M. Srinivasula, M. Ahmad, E.S. Alnemri, and X. Wang, Cytochrome c and dATP dependent formation of Apaf-1/caspase-9 complex initiate an apoptotic protease cascade, *Cell* 91:479 (1997).
7. M. Crompton, The mitochondrial permeability transition pore and its role in cell death, *Biochem. J.* 341:233 (1999).
8. M.B. Hampton, B. Zhivotovsky, A.F. Slater, D.H. Burgess, and S. Orrenius, Importance of the redox state of cytochrome c during caspase activation in cytosolic extracts, *Biochem. J.* 329:95 (1998).
9. A.P. Halestrap, Interaction between oxidative stress and calcium overload on mitochondrial function, in: *Mitochondria: DNA, Proteins and Disease.* V. Darley-Usmar and A.H.V. Schapira, eds., Portland Press, London (1994).
10. M. Crompton, The mitochondrial permeability transition pore and its role in cell death, *Biochem. J.* 341:233 (1999).
11. S. Shimizu, M. Narita, and Y. Tsujimoto, Bcl-2 family proteins regulate the release of apoptogenic cytochrome c by the mitochondrial channel VDAC, Nature 399:483 (1999).
12. S. Shimizu and Y. Tsujimoto, Proapoptotic BH3-only Bcl-2 family members induced cytochrome c release, but not mitochondrial membrane potential loss, and do not directly modulate voltage-dependent anion channel activity, *Proc. Natl. Acad. Sci. USA* 97:577 (2000).
13. I. Martinou, S. Desagher, R. Eskes, B. Antonsson, E. Andre, S. Fakan, and J.C. Martinou, The release of cytochrome c from mitochondria during apoptosis of NGF-deprived sympathetic neurons is a reversible event, *J. Cell. Biol.* 144:883 (1999).
14. A. Samali, H. Nordgren, B. Zhivotovsky, E. Peterson, and S. Orrenius, A comparative study of apoptosis and necrosis in HepG2 cells: Oxidant-induced caspase inactivation leads to necrosis, *Biochem. Biophys. Res. Commun.* 255:6 (1999).
15. E.S. Alnemri, D.J. Livingston, D.W. Nicholson, G. Salvesen, N.A. Thornberry, W.W. Wong, and J. Yuan, Human ICE/CED-3 protease nomenclature, *Cell* 87:171 (1996).
16. N.A. Thornberry and Y. Lazebnik, Caspases: enemies within, *Science* 28:1312 (1998).
17. A. Samali, B. Zhivotovsky, D.P. Jones, S. Nagata, and S. Orrenius, Apoptosis: cell death defined by caspase activation, *Cell Death Diff.* 6:495 (1998).
18. D.W. Nicholson and N.A. Thornberry, Caspases: killer proteases, *Trends Biochem. Sci.* 22:299 (1997).
19. B. Zhivotovsky, A. Samali, A. Gahm, and S. Orrenius, Caspases: their intracellular localization and translocation during apoptosis, *Cell Death Differ.* 6:644 (1999).
20. M. Whyte and G. Evan, (1995) Apoptosis. The last cut is the deepest, *Nature* 376:17 (1995).
21. E.A. Slee, M.T. Harte, R.M. Kluck, B.B. Wolf, C.A. Casiano, D.D. Newmeyer, H.G. Wang, J.C. Reed, D.W. Nicholson, E.S. Alnemri, D.R. Green, and S.J. Martin, Ordering the cytochrome c-initiated caspase cascade: hierarchical activation of caspases-2, -3, -6, -7, -8, and -10 in a caspase-9-dependent manner, *J. Cell Biol.* 144:281 (1999).
22. J.D. Robertson, S. Orrenius, and B. Zhivotovsky, Review: Nuclear events in apoptosis, *J. Struct. Biol.* 129:346 (2000).
23. X. Liu, C.N. Kim, J. Yang, R. Jemmerson, and X. Wang, Induction of apoptotic program in cell-free extracts: requirement for dATP and cytochrome c, *Cell* 86:147 (1996).
24. S.A. Susin, H.K. Lorenzo, N. Zamzami, I. Marzo, B.E. Snow, G.M. Brothers, J. Mangion, E. Jacotot, P. Costantini, M. Loeffler, N. Larochette, D.R. Goodlett, R. Aebersold, D.P. Siderovski, J.M. Penninger, and G. Kroemer, Molecular characterization of mitochondrial apoptosis-inducing factor, *Nature* 397:441 (1999).

25. C. Köhler, A. Gahm, T. Noma, A. Nakazawa, S. Orrenius, and B. Zhivotovsky, Release of adenylate kinase 2 from the mitochondrial intermembrane space during apoptosis, *FEBS Lett.* 447:10 (1999).

26. H. Zou, W.J. Henzel, X. Liu, A. Lutschg, and X. Wang, Apaf-1, a human protein homologous to C. elegans CED-4, participates in cytochrome c-dependent activation of caspase-3, *Cell* 90:405 (1997).

27. X.M. Sun, M. MacFarlane, J. Zhuang, B.B. Wolf, D.R. Green, and G.M. Cohen, Distinct caspase cascades are initiated in receptor-mediated and chemical-induced apoptosis, *J. Biol. Chem.* 274:5053 (1999).

28. S.M. Srinivasula, M. Ahmad, T. Fernandes-Alnemri, G. Litwack, and E.S. Alnemri, Molecular ordering of the Fas-apoptotic pathway: the Fas/APO-1 protease Mch5 is a CrmA-inhibitable protease that activates multiple Ced-3/ICE-like cysteine proteases, *Proc. Natl. Acad. Sci. U S A* 93:14486 (1996).

29. A. Strasser and K. Newton, FADD/MORT1, a signal transducer that can promote cell death or cell growth, *Int. J. Biochem. Cell Biol.* 31:533 (1999).

30. M. Muzio, A.M. Chinnaiyan, F.C. Kischkel, K. O'Rourke, A. Shevchenko, J. Ni, C. Scaffidi, J.D. Bretz, M. Zhang, R. Gentz, M. Mann, P.H. Krammer, M.E. Peter, and V.M. Dixit, FLICE, a novel FADD-homologous ICE/CED-3-like protease, is recruited to the CD95 (Fas/APO-1) death--inducing signaling complex, *Cell* 85:817 (1996).

31. M.P. Boldin, T.M. Goncharov, Y.V. Goltsev, and D. Wallach, Involvement of MACH, a novel MORT1/FADD-interacting protease, in Fas/APO-1- and TNF receptor-induced cell death. *Cell* 85:803 (1996).

32. J.P. Medema, C. Scaffidi, F.C. Kischkel, A. Shevchenko, M. Mann, P.H. Krammer, and M.E. Peter FLICE is activated by association with the CD95 death-inducing signaling complex (DISC), *EMBO J.* 16:2794 (1997).

33. T. Kuwana, J.J. Smith, M. Muzio, V. Dixit, D.D. Newmeyer, and S. Kornbluth, Apoptosis induction by caspase-8 is amplified through the mitochondrial release of cytochrome c, *J. Biol. Chem.* 273:16589 (1998).

34. E. Bossy-Wetzel and D.R. Green, Caspases induce cytochrome c release from mitochondria by activating cytosolic factors, *J. Biol. Chem.* 274:17484 (1999).

35. H. Li, H. Zhu, C.J. Xu, and J. Yuan, Cleavage of BID by caspase 8 mediates the mitochondrial damage in the Fas pathway of apoptosis, *Cell* 94:491 (1998).

36. X. Luo, I. Budihardjo, H. Zou, C. Slaughter, and X. Wang, Bid, a Bcl2 interacting protein, mediates cytochrome c release from mitochondria in response to activation of cell surface death receptors, *Cell* 94:481 (1998).

37. E.S. Alnemri, D.J. Livingston, D.W. Nicholson, G. Salvesen, N.A. Thornberry, W.W. Wong, and J. Yuan, Human ICE/CED-3 protease nomenclature, *Cell* 87:171 (1996).

38. C.S. Nobel, D.H. Burgess, B. Zhivotovsky, M.J. Burkitt, S. Orrenius, and A.F. Slater, Mechanism of dithiocarbamate inhibition of apoptosis: thiol oxidation by dithiocarbamate disulfides directly inhibits processing of the caspase-3 proenzyme, *Chem. Res. Toxicol.* 10:636 (1997).

39. M.B. Hampton and S. Orrenius, Dual regulation of caspase activity by hydrogen peroxide: implications for apoptosis, *FEBS Lett.* 414:552 (1997).

40. C.S. Nobel, M. Kimland, D.W. Nicholson, S. Orrenius, A.F. Slater, Disulfiram is a potent inhibitor of proteases of the caspase family, *Chem. Res. Toxicol.* 10:1319 (1997).

41. D. Di Monte, G. Bellomo, H. Thor, P. Nicotera, S. Orrenius, Menadione-induced cytotoxicity is associated with protein thiol oxidation and alteration in intracellular Ca2+ homeostasis, *Arch. Biochem. Biophys.* 235:343 (1984).

42. P. Nicotera, M. Leist, and E. Ferrando-May, Intracellular ATP, a switch in the decision between apoptosis and necrosis, *Toxicol. Lett.* 102-103:139 (1998).

43. M. Leist, B. Single, A.F. Castoldi, S. Kuhnle, P. Nicotera, Intracellular ATP concentration: a switch deciding between apoptosis and necrosis, *J. Exp. Med.* 185:1481 (1997).

44. G.C. Li and G.M. Hahn, A proposed operational model of thermotolerance based on effects of nutrients and the initial treatment temperature, *Cancer Res.* 40:4501 (1980).

45. S. Lindquist and E.A. Craig, The heat-shock proteins, *Annu. Rev. Genet.* 22:631 (1988).

46. P.L. Moseley, Heat shock proteins and heat adaptation of the whole organism, *J. Appl. Physiol.* 83:1413 (1997).

47. A. Samali and T.G. Cotter, Heat shock proteins increase resistance to apoptosis, *Exp. Cell. Res.* 223:163 (1996).

48. A.P. Arrigo, Small stress proteins: chaperones that act as regulators of intracellular redox state and programmed cell death, *Biol. Chem.* 379:19 (1998).

49. R. Foresti, P. Sarathchandra, J.E. Clark, C.J. Green, and R. Motterlini, Peroxynitrite induces heme oxygenase-1 in vascular endothelial cells: a link to apoptosis, *Biochem. J.* 339:729 (1999).

50. C.D. Ferris, S.R. Jaffrey, A. Sawa, M. Takahashi, S.D. Brady, R.K. Barrow, S.A. Tysoe, H. Wolosker, D.E. Baranano, S. Dore, K.D. Poss, and S.H. Snyder, Haem oxygenase-1 prevents cell death by regulating cellular iron, *Nat. Cell Biol.* 1:152 (1999).

51. P. Mehlen, X. Preville, P. Chareyron, J. Briolay, R. Klemenz, A.P. Arrigo, Constitutive expression of human hsp27, Drosophila hsp27, or human alpha B-crystallin confers resistance to TNF- and oxidative stress-induced cytotoxicity in stably transfected murine L929 fibroblasts, *J. Immunol.* 154:363 (1995).

52. X. Preville, F. Salvemini, S. Giraud, S. Chaufour, C. Paul, G. Stepien, M.V. Ursini, and A.P. Arrigo, Mammalian small stress proteins protect against oxidative stress through their ability to increase glucose-6-phosphate dehydrogenase activity and by maintaining optimal cellular detoxifying machinery, *Exp. Cell Res.* 247:61 (1999).

53. F. Salvemini, A. Franze, A. Iervolino, S. Filosa, S. Salzano, and M.V. Ursini, Enhanced glutathione levels and oxidoresistance mediated by increased glucose-6-phosphate dehydrogenase expression, *J. Biol. Chem.* 274:2750 (1999).

54. A. Samali, C.I. Holmberg, L. Sistonen, and S. Orrenius, Thermotolerance and cell death are distinct cellular responses to stress: Dependence on heat shock proteins, *FEBS Lett.* 461:306 (1999).

STRUCTURE ACTIVITY RELATIONSHIPS FOR THE CHEMICAL BEHAVIOUR AND TOXICITY OF ELECTROPHILIC QUINONES / QUINONE METHIDES

Ivonne M.C.M. Rietjens[1,2,3], Hanem M. Awad[1], Marelle G. Boersma[1], Marlou L.P.S. van Iersel[3], Jacques Vervoort[1] and Peter J. Van Bladeren[2,3]

1) Laboratory of Biochemistry, Wageningen University, Dreijenlaan 3, 6703 HA Wageningen, The Netherlands
2) Division of Toxicology, Wageningen University, Tuinlaan 5, 6703 HE Wageningen, The Netherlands.
3) TNO/WU Centre for Food Toxicology, PO Box 8000, 6700 EA Wageningen, The Netherlands

INTRODUCTION

Electrophilic quinones and quinone methides from a variety of natural and synthetic compounds have been classified as likely toxic reactive metabolites. This includes the quinones / quinone methides of catechol-type metabolites from estrogens and polycyclic aromatic hydrocarbons[1,2,3]. Metabolic activation of estrogens to redox active and/or electrophilic metabolites has been proposed as one of the mechanisms responsible for the link between estrogen exposure and the risk of developing cancer[1,2]. Especially catechol (ortho-diol)-type of metabolites resulting from cytochrome P450 catalysed hydroxylation of estrogens (Figure 1) may be involved.

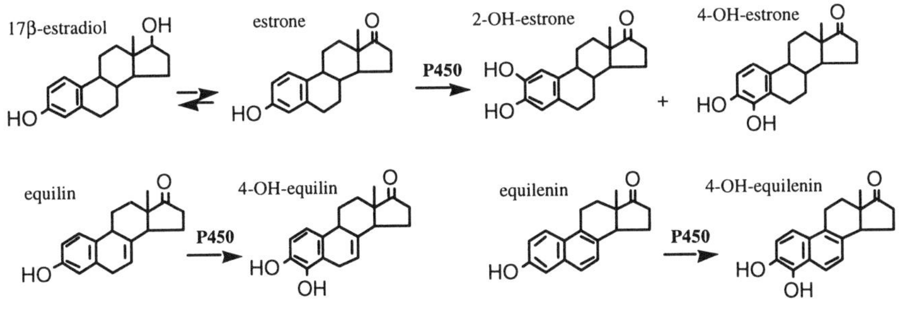

Figure 1. Metabolism of endogenous and equine estrogens to catechol metabolites by cytochromes P450[1,2].

Biological Reactive Intermediates VI, Edited by Dansette *et al.*
Kluwer Academic / Plenum Publishers, 2001

The involvement of catechol-type metabolites has also been outlined to play a role in the metabolic activation of polycyclic aromatic hydrocarbons[2,3]. In addition to the conversion of the dihydrodiol metabolites to diol-epoxides by cytochromes P450, the conversion of the dihydrodiol metabolites of polycyclic aromatic hydrocarbons by dihydrodiol dehydrogenase may result in formation of reactive catechol-type metabolites (Figure 2). These catechol-type metabolites are suggested to contribute to the carcinogenicity and toxicity of the aromatic hydrocarbons.

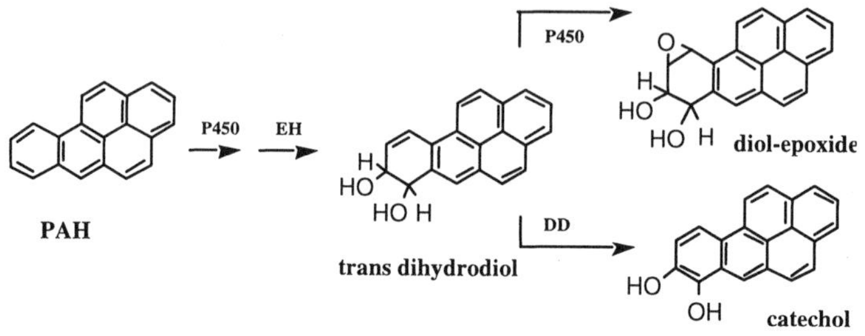

Figure 2. Metabolic activation of polycyclic hydrocarbons to catechol metabolites[2,3].

Figure 3 presents the mechanism behind the redox chemistry and alkylating toxicity of catechol-type metabolites from estrogens and polycyclic aromatic hydrocarbons. Redox cycling between the catechol and its quinone generates reactive oxygen species able of damaging the DNA. In addition, oxidation of the catechols to quinones and their isomeric quinone methides (see below) generates potent electrophiles that could alkylate DNA.

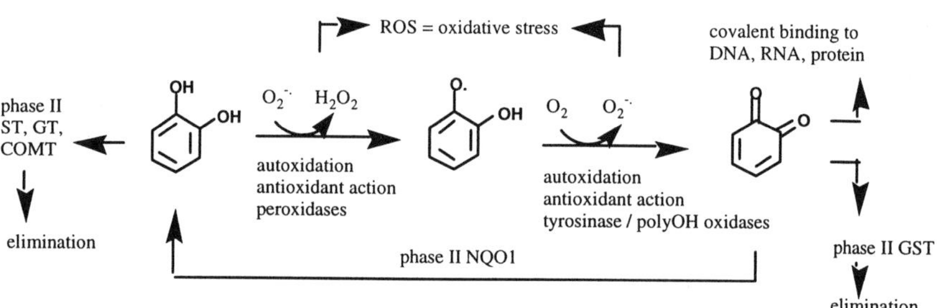

Figure 3. Schematic presentation of the redox chemistry and alkylating toxicity of catechol-type metabolites from estrogens and polycyclic aromatic hydrocarbons.

Taking this toxic pro-oxidative behaviour of catechol metabolites into account it is of interest that many flavonoids already contain this catechol-type structural element without the need for metabolic activation. This especially holds for flavonoids like for example quercetin, luteolin, taxifolin and fisetin containing a 3',4'-dihydroxy structural element (Figure 4). For these 3',4'-dihydroxyflavonoids their pro-oxidative quinone / quinone methide chemistry is especially of interest because of their increasing use as functional food ingredients and/or food supplements.

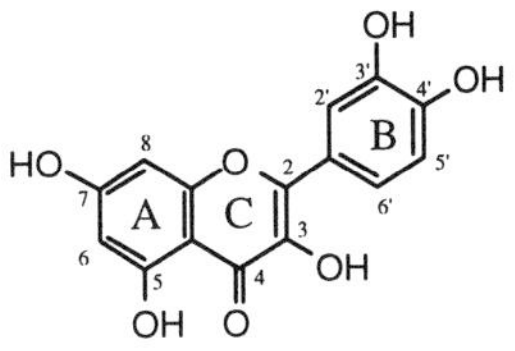

	3-OH	5-OH	7-OH	C2=C3
quercetin	+	+	+	+
taxifolin	+	+	+	-
luteolin	-	+	+	+
fisetin	+	-	+	+

Figure 4. Schematic presentation of the structure of quercetin and of other 3',4'-dihydroxyflavonoids.

Natural polyphenols like flavonoids and isoflavonoids and their corresponding glycosides are important constituents of fruits, vegetables, nuts, seeds, tea, olive oil and red wine[3,4]. On average our daily western diet contains about 1 gram of polyphenols. The anti-oxidative properties of these compounds are often claimed to be responsible for the protective effects of these food components against cardiovascular disease, certain forms of cancer and/or photosensitivity diseases[4-10]. In addition, it has been proposed that accumulation of oxidative damage is an important contributor to not only pathological conditions but also to the ageing process. Therefore, beneficial health effects in ageing have also been related to antioxidant action[7-10]. These studies provide the basis for the present rapidly increasing interest for the use of antioxidants as functional food ingredients. As a result increased human exposure to polyphenolic flavonoid-type antioxidants can be expected in the near future. In shops and at the internet, food and food supplements based on (iso)flavonoids as functional ingredients are intensively marketed. This, although scientific data supporting the health claims as well as data allowing a balanced risk-evaluation are lacking.

With respect to possible pro-oxidant toxicity it is of interest to notice that the mutagenic properties of the flavonoid quercetin in a variety of bacterial and mammalian mutagenicity tests, has been related to its quinone / quinone methide chemistry (Figure 5).

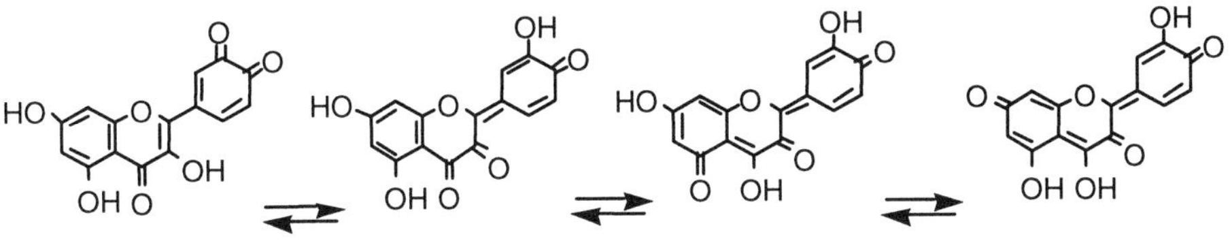

Figure 5. Quinone / quinone methide isomerisation of quercetin.

The objective of our present investigations is to perform structure activity studies on the chemical behaviour and toxicity of 3',4'-dihydroxyflavonoids with special emphasis on the nature of the quinone / quinone methide type metabolites formed. Using the GSH trapping method, HPLC, LC-MS, MALDI-TOF, ^{1}H NMR and quantum mechanical computer calculations the quinone / quinone methide chemistry of a series of 3',4'-dihydroxyflavonoids could be characterised. The results provide insight in structure activity relationships for the chemical behaviour and pro-oxidative toxicity of these electrophilic quinone / quinone methide metabolites. The results obtained also illustrate that this quinone / quinone methide chemistry is far from straight forward.

MATERIALS AND METHODS

Mutagenicity and toxicity
Data for mutagenicity and toxicity of the 3',4'-dihydroxyflavonoids were taken from the literature as indicated.

Detection of flavonoid quinone / quinone methide glutathione adducts
Quinone / quinone methides from the different flavonoids were made using either tyrosinase or horseradish peroxidase / H_2O_2 and incubation conditions as previously described[12,13]. Using the GSH trapping method, HPLC, LC-MS, MALDI-TOF and ^{1}H NMR the quinone / quinone methide chemistry of a series of 3',4'-dihydroxyflavonoids could be characterised. Experimental details of these methods can be found in our recent manuscripts[12,13].

Quantum mechanical computer calculations
Quantum mechanical calculations were carried out on a Silicon Graphics Indigo2 workstation using Spartan 5.0 (Wavefunction Inc.). Density functional theory (DFT) calculations were carried out using the B3LYP/6-311G**//HF/3-21G* method as implemented in the Gaussian 94 program[12]. In addition, the semi-empirical AM1 Hamiltonian was used for the present study.

RESULTS AND DISCUSSION

Structure mutagenicity relationships for flavonoids

Extensive literature data exist on the mutagenicity of flavonoids including the model compounds of the present study. Table 1 presents a selection of data for the 3',4'-dihydroxyflavonoids and some of their derivatives[11].

Table 1. Mutagenicity of selected flavonoids in *Salmonella typhimurium* strain TA98 in the presence of S9 mix. Data taken from literature[11]

Flavonoid	characteristics	nanomoles/plate	mutants/plate
quercetin	3,5,7,3',4'OH, C2=C3	55	705
		83	1055
		166	2000
luteolin	lacks 3-OH	166	0
		1660	6
fisetin	lacks 5-OH	166	9
		1660	26
taxifolin	lacks C2=C3	166	12
		1660	119
5-*O*-methylquercetin	conjugated 5-OH	166	19
		1660	65
3-*O*-methylquercetin	conjugated 3-OH	166	7
		1660	2

These data indicate that structural features which are essential for mutagenic activity are a flavonoid ring structure with:

1) a free hydroxyl group at the 3 position,
2) a double bond at the 2,3 position,
3) a keto group at the 4 position, and
4) a structure which permits the proton of the 3-hydroxyl group to tautomerise to a 3-keto moiety.
5) Furthermore, especially compounds with free hydroxyl groups at the 3' and 4' positions of the B-ring exhibit mutagenic activity already without requirement for metabolic activation. When free hydroxyl groups are not present in the B-ring a metabolic activating system is required for induction of the mutagenicity[11].

Based on these requirements a metabolic pathway for flavonoid activation to DNA-reactive species was proposed. For quercetin, this pathway includes enzymatic and/or chemical oxidation of quercetin to quercetin *ortho*-quinone, followed by isomerisation of the *ortho*-quinone to quinone methides (Figure 5). These quinone methides are suggested to be the active alkylating DNA-reactive intermediates. The essential characteristic of this hypothesis is that the mutagenic activity requires the ability of the B ring to oxidise to a quinoid intermediate and of the proton of the 3-hydroxyl group to rearrange to give a 3-keto isomer, which requires a double bond at the 2-3 position. These requirements are in line with the requirements derived for mutagenic activity of the flavonoids. This also explains why conjugation of the 3-hydroxyl moiety leading to 3-methoxy- or 3-glucuronyl flavones eliminates the mutagenicity. Comparison of the mutagenicity of quercetin to that of fisetin also illustrates the potentiating effect of a 5-hydroxyl group. This observation is related to the fact that the 5-hydroxyl moiety makes a strong hydrogen bond with the 4-keto group, thereby preventing and/or weakening the hydrogen bonding of the 3-hydroxyl with the 4-keto group. This weakening of the 3-OH 4-keto hydrogen bond increases possibilities for the essential tautomerisation of the 3-hydroxyl group to give the quinone methides. Thus, conjugation of the 5-hydroxyl (as in 5-O-methylquercetin) or the absence of the 5-hydroxyl (fisetin) reduces mutagenicity. Altogether the quinone / quinone methide chemistry of flavonoids appears to play an important role in their possible pro-oxidative and/or mutagenic action. This supports the need for investigation of the quinone / quinone methide chemistry and metabolism of this series of interesting functional food ingredients.

Structure activity relationships for flavonoid quinone / quinone methide chemistry as derived from glutathione adduct formation

Since the high reactivity of quinones and quinone methides is known to hamper their direct identification, the *ortho*-quinone / quinone methides were trapped by glutathione, and the glutathione adducts formed were separated by HPLC and identified by MALDI-TOF, LC-MS and ^{1}H NMR. Recently this *ortho*-quinone trapping method appeared to be an accurate method for quantification of *ortho*-quinone metabolites[12,14-16]. As an example Figure 6 presents the HPLC chromatogram of the incubation of quercetin with tyrosinase in the presence of glutathione (GSH). Two major metabolites are observed. In MALDI-TOF analysis these metabolites have a m/z of 608.2, which is identical to the mass expected for a protonated mono-glutathionyl quercetin adduct. ^{1}H NMR spectra of these metabolites were recorded and the representative part of the spectra, showing the region representing the ^{1}H chemical shifts of the aromatic protons, are presented in Figure 6 as well. For comparison Figure 6d shows the same region of the spectrum of quercetin itself.

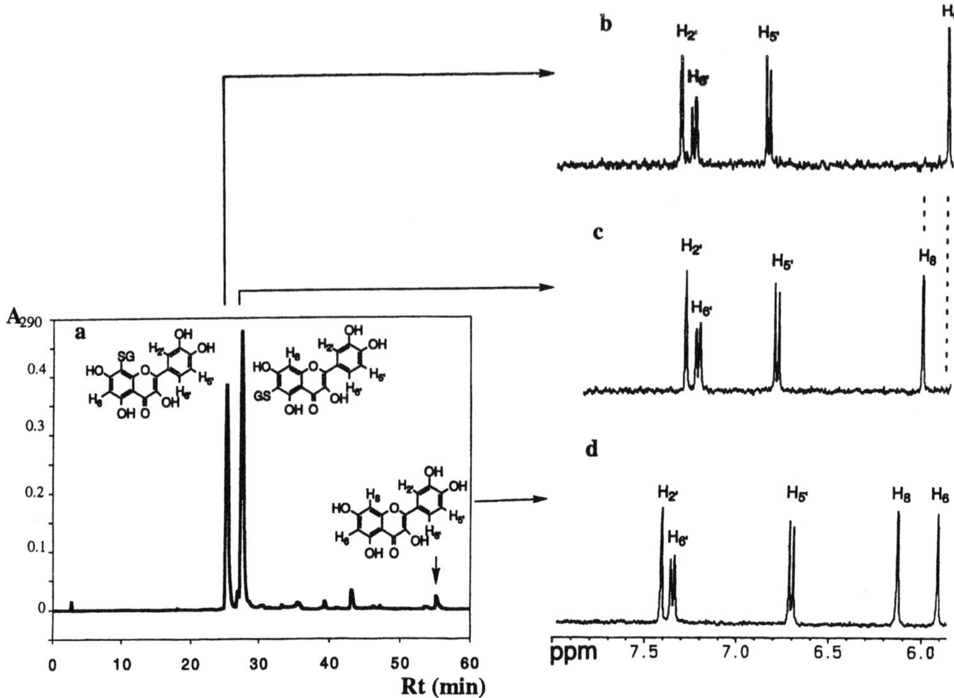

Rt (min)

Figure 6. HPLC pattern (a) and (b+c) the aromatic part of the ^{1}H NMR spectra of the two major metabolites formed upon incubation of quercetin with tyrosinase in the presence of GSH. For comparison the aromatic region of the ^{1}H NMR spectrum of quercetin is also presented (d).

The spectrum of quercetin (Figure 6d) shows the ^{1}H resonances of H6 (at 5.91 ppm, doublet with ${}^4J_{H6\text{-}H8}$ = 1.9 Hz), of H8 (at 6.13 ppm, doublet with ${}^4J_{H8\text{-}H6}$ = 1.9 Hz), of H5' (at 6.71 ppm, doublet with ${}^3J_{H5'\text{-}H6'}$ = 8.5 Hz), of H6' (at 7.35 ppm, double doublet with ${}^3J_{H6'\text{-}H5'}$ = 8.5 Hz and ${}^4J_{H6'\text{-}H2'}$ = 2.1 Hz) and of H2' (at 7.41 ppm, doublet with ${}^4J_{H2'\text{-}H6'}$ = 2.1 Hz). This identification and NMR pattern of quercetin is in line with literature data[17]. Comparison of the aromatic region of the ^{1}H NMR spectra of the two glutathionyl-quercetin adducts in Figure 6 to that of quercetin reveals the loss of respectively the H6 and H8 proton signals and of the ${}^4J$ coupling of 1.9 Hz between the H6 and H8 proton. Both metabolites still contain the characteristic ^{1}H NMR pattern of the H2', H5' and H6', the protons in the B ring of quercetin. This identifies the two products formed upon incubation of quercetin with tyrosinase and GSH as 6-glutathionyl-quercetin and 8-glutathionyl quercetin[12]. In incubations with horseradish peroxidase instead of tyrosinase, catalysing one- instead of two-electron substrate oxidations, similar results were obtained[13]. This implies that upon oxidation of quercetin to its quinone, glutathione adduct formation occurs in the A-ring instead of in the B-ring where the *ortho*-quinone moiety is expected to be formed upon oxidation of its catechol-like motif. This regioselectivity of the glutathione conjugation can best be explained by taking into consideration the quinone methides I and/or II from quercetin (Figure 5).

Similar experiments were performed for other 3',4'-dihydroxyflavonoid model compounds, the results of which will be described in detail elsewhere (Awad *et al.* in preparation). Table 2 summarises the regioselectivity of ring adduct formation observed for the different 3',4'-dihydroxyflavonoid model compounds.

Table 2. Regioselectivity of the glutathione addition of the flavonoid quinones / quinone methides

Flavonoid	characteristics	regioselectivity: conjugation at
quercetin	3,5,7,3',4'OH, C2=C3	ring A
luteolin	lacks 3-OH	ring B
fisetin	lacks 5-OH	ring C
taxifolin	lacks C2=C3	ring B
3,3',4' trihydroxyflavone	lacks 5-OH and 7-OH	ring B

With luteolin, which lacks the C3-OH but still contains the C2=C3 double bond, glutathione adduct formation was observed preferentially at the B ring. This could be concluded from the clear presence of the ^{1}H NMR characteristic of the A-ring protons H6 and H8 in the luteolin-glutathione adducts formed and isolated on HPLC. Detailed analysis of these results will be published separately (Awad *et al.* in preparation). Clearly, elimination of the 3-OH, preventing quinone methide isomerisation, restricts the glutathione addition to the B-ring and the *ortho*-quinone isomer of oxidized luteolin.

With taxifolin, which lacks the C2=C3 double bond, the formation of B-ring glutathionyl taxifolin adducts is observed. As for the luteolin adducts this could be derived from detailed ^{1}H NMR analysis showing the presence of all protons of the A and C ring of taxifolin in the purified adducts. Although saturation of the C2=C3 double bond does not eliminate quinone methide chemistry of taxifolin *ortho-* quinone (Figure 7), its conjugation with glutathione is dominated by the *ortho*-quinone isomer.

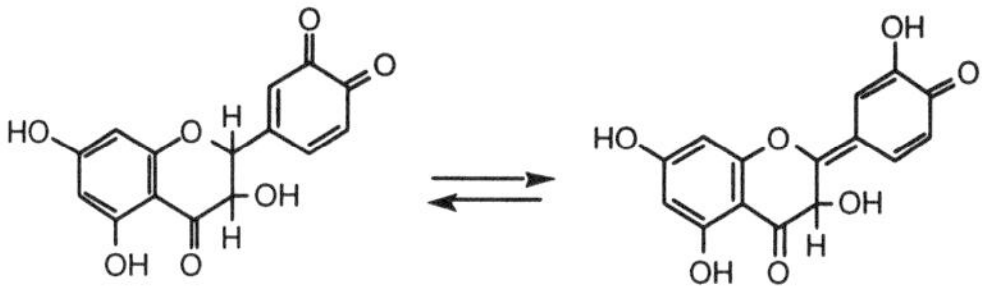

Figure 7. Quinone / quinone methide isomerisation of taxifolin.

Results with 3,3'4'-trihydroxyflavone were in line with the results obtained with luteolin and taxifolin. Again preferential formation of only B-ring glutathionyl adducts was observed.

For fisetin, the presence of the C2=C3 double bond, the C3-OH but also the C7-OH would, in theory, allow the *ortho*-quinone to extend to the quinone methides III and I (Figure 5), introducing electrophilic reactivity and possibilities for conjugate formation similar to those of quercetin. However, fisetin behaved different from all other flavonoid model compounds investigated, and, surprisingly, showed the formation of an adduct still containing all six aromatic protons, i.e. showing the ^{1}H NMR resonances of H6, H8 and H5 (A-ring) and of H2', H5' and H6' (B-ring). This indicates that glutathione adduct formation must proceed by adduct formation in ring C. Adduct formation of flavonoid quinones / quinone methides in the C-ring may be in line with the reactivity of quercetin *ortho*-quinone with OH⁻ described to occur upon protonation of the quinone methide followed by an attack at C2 (Figure 8) and, ultimately, resulting in reordering and further reactivity of the C-ring[17]. Thus, the reactivity of the C2 position in a flavonoid quinone for electrophilic attack has been reported before.

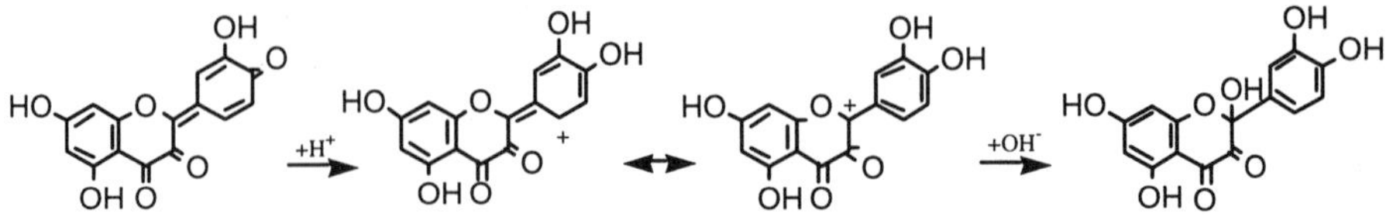

Figure 8. Nucleophilic attack of OH⁻ at C2 of quercetin quinone as described in the literature[17].

Altogether the results presented indicate that the quinone / quinone methide chemistry of the 3',4'-dihydroxyflavonoids investigated is far from straight forward.

AM1 calculated relative stability and electrophilic reactivity of the various quinones / quinone methides.

In a previous study[12] we performed density functional theory (DFT) calculations using the B3LYP/6-311G**//HF/3-21G* to investigate the chemical characteristics possibly underlying the preferential regioselectivity of glutathione conjugation for quercetin quinone / quinone methide. The results obtained indicated that especially the electrophilic reactivity of the most reactive quercetin quinone / quinone methide isomers appeared to dominate the regioselectivity and reactivity of the quinone / quinone methide in adduct formation with GSH. Thus, the preferential GSH conjugation of the quinone / quinone methide of quercetin in the A-ring instead of in the B-ring could be explained from the high electrophilicity of the quinone methides I and II (Figure 5) apparently shifting the equilibria in favour of the GSH adducts of these most reactive alkylating isomers. The parameter used to quantify the relative electrophilic reactivity of the various quinone / quinone methides was the calculated energy of the lowest unoccupied molecular orbital (E(LUMO)) of the isomers of interest. Table 3 presents these E(LUMO) values for the model compounds of the present study calculated by semi-empirical AM1 calculations. For quercetin the results obtained are in line with those reported previously, and indicate the quinone methides I and II to be the most reactive electrophiles.

For taxifolin the data obtained (Table 3) support the experimental observation that the alkylating reactivity of the quinone / quinone methide is dominated by the *ortho*-quinone which, for taxifolin, is slightly more reactive and also more stable than the quinone methide.

For fisetin the results of the calculations are similar to those for quercetin showing fisetin quinone methide I to be the isomer with the highest electrophilicity. Thus, isomers theoretically expected for fisetin are adducts at the C6 and C8 position of the fisetin quinone methide I. This is not what is observed. Apparently, the role of the intramolecular hydrogen bridges in the fisetin quinone are underestimated by the calculations. And, although being most reactive, the quinone methide I in spite of its high electrophilicity does not dominate the regioselectivity of glutathione conjugation. This may be in line with the observed loss in mutagenicity going from quercetin to fisetin, and could be due to the fact that in the absence of the 5-OH the hydrogen bond between the 4-keto and the 3-OH seriously hampers isomerisation to the quinone methide.

Table 3. AM1 calculated relative stability and electrophilic reactivity of the quinone and quinone methides of the flavonoid model compounds. Energetic stability is presented relative to the most stable isomer for each flavonoid, electrophilic reactivity is indicated as the energy of the lowest unoccupied molecular orbital (E(LUMO)). Numbering of the quinone methides is as indicated in Figure 5.

flavonoid	Relative stability (kcal/mol)	E(LUMO) in eV
quercetin *ortho*-quinone	0	-1.96
quercetin quinone methide I	21.8	-2.72
quercetin quinone methide II	21.0	-2.65
quercetin quinone methide III	1.3	-2.20
fisetin *ortho*-quinone	0	-1.88
fisetin quinone methide I	17.6	-2.63
fisetin quinone methide III	0.9	-2.12
luteolin *ortho*-quinone	-	-2.14
taxifolin *ortho*-quinone	0	-1.68
taxifolin quinone methide	0.3	-1.60

Pro-oxidant toxicity and biotransformation of flavonoids

The pro-oxidant toxicity of the flavonoids was investigated in HL-60 cells and revealed that in addition to the pro-oxidant effects and their ease of oxidation especially an increase in polyphenol lipophilicity increases their cytotoxicity[18]. The polyphenols have to pass the cell membrane and enter the cells before exerting their toxic effects. Once within the cell several metabolic factors may also influence the pro-oxidant toxicity of the flavonoid model compounds (Figure 3). Conjugation of the flavonoids by cellular enzymes like *O*-methyl transferases and glucuronyl transferases was shown to occur efficiently[19-22]. Clearly the regioselectivity of this conjugation will determine the extent to which pro-oxidative quinone / quinone methide chemistry will be influenced. ^{1}H NMR appears to be an efficient tool for identification of this exact regioselectivity of flavonoid conjugation since modification of an OH moiety, for example by glucuronidation, was shown to shift the ^{1}H NMR chemical shift value of the hydrogen atoms positioned *ortho* with respect to this hydroxyl moiety by 0.3 to 0.4 ppm (unpublished results).

Once oxidised to the reactive quinone, enzymes like glutathione transferase but also quinone reductase (NADPH: quinone oxidoredutase or DT-diaphorase), may affect the ultimate fate and toxicity of the compounds. Especially of interest is the observation that in CHO cells genetically enriched in quinone reductase, the toxicity of quercetin is enhanced[7]. This, in spite of the fact that quinone reductase has been shown and is well known for its ability to reduce the toxicity of quinones[23,24]. This increased toxicity of quercetin for quinone reductase enriched cells, together with the swift and unique isomerisation of its *ortho*-quinone to the quinone methides, now raises the question of the substrate behaviour of these quinone methides towards quinone detoxifying enzymes like quinone reductase and glutathione S-transferases. This and the possible role of quinone methides in quercetin mediated pro-oxidative toxicity is presently under investigation

CONCLUDING REMARKS

The results of the present study, in which the quinone / quinone methide chemistry of a series of 3',4'-dihydroxyflavonoids was investigated indicate that this quinone / quinone methide chemistry of 3',4'-dihydroxyflavonoids is far from straight forward. This conclusion is in line with the conclusion of *Bolton et al.*[1] who, on the basis of experiments with estrogen *ortho*-quinones / quinone methides stated that the *in vivo* fate of these estrogen *ortho*-quinones may be much more complex than currently realised.

REFERENCES

1. J.L. Bolton, E. Pisha, Z. Fagan and S. Qiu, Role of quinoids in estrogen carcinogenesis, *Chem. Res. Toxicol.* 11:1113(1998).
2. J.L. Bolton, Trush M.A., T.M. Penning, G. Dryhurst and T.J. Monks, Role of quinones in Toxicology. *Chem. Res. Toxicol.* 13:135 (2000).
3. T.M. Penning, M.E. Burczynski, C.F. Hung, K.D. McCoull, N.T. Palackal and L.S. Tsuruda, Dihydrodiol dehydrogenases and polycyclic aromatic hydrocarbon activation: generation of reactive and redox-active quinones, *Chem. Res. Toxicol.* 12:1 (1999).
4. E. Middleton and C. Kandaswami, The impact of plant flavonoids on mammalian biology: implications for immunity, inflammation and cancer. In: The Flavonoids, J.B. Harborne, ed., Chapman & Hall, London (1994) 619-652
5. C.A. Rice-Evans, N.J. Miller and G. Paganga, Structure-antioxidant activity relationships of flavonoids and phenolic acids. *Free Rad. Biol. Med.* 20: 933 (1996).
6. A.T. Diplock, J.-L. Charleux, G. Crozier-Will, F.J. Kok, C. Rice-Evans, M. Roberfroid, W. Stahl and J. Vina-Ribes, Functional food science and defense against reactive oxidative species. *Br. J. Nutr.* 80: S77 (1998).
7. D. Metodieva, A.K. Jaiswal, N. Cenas, E. Dickancaite, J. Segura-Aguilar, Quercetin may act as a cytotoxic prooxidant after its metabolic activation to semiquinone and quinoidal product, *Free Rad. Biol. Med.* 26: 107 (1999).
8. G. Block, S. Patterson and A. Subar. Fruit, vegetables and cancer prevention: a review of the epidemiological evidence. *Nutr. Cancer* 19: 1 (1992).
9. A.T. Diplock, The Leon Goldberg memorial lecture: Antioxidants and disease prevention. *Fd. Chem. Toxicol.* 34: 1013 (1996).
10. G.S. Omenn, Micronutrients (vitamins and minerals) as cancer-preventive agents. *IARC Sci. Publ.* 139: 33 (1997).
11. J.T. MacGregor and L. Jurd, Mutagenicity of plant flavonoids: structural requirements for mutagenic activity in Salmonella typhimurium. *Mut. Res.* 54: 297 (1978).
12. M.G. Boersma, J. Vervoort, H. Szymusiak, K. Lemanska, B. Tyrakowska, N. Cenas, J. Segura-Aguilar and I.M.C.M. Rietjens, Regioselectivity and reversibility of the glutathione conjugation of quercetin quinone methide, *Chem. Res. Toxicol.* 13:185 (2000).
13. H.M. Awad, M.G. Boersma, J. Vervoort and I.M.C.M. Rietjens, Horseradish peroxidase mediated oxidation of quercetin to quercetin quinone/quinone methides.. *Arch. Biochem. Biophys.* 378: 224 (2000).
14. M. Butterworth, S.S. Lau and T.J. Monks, 17β-estradiol metabolism by hamster hepatic microsomes: comparison of catechol estrogen O-methylation with catechol estrogen oxidation and glutathione conjugation. *Chem. Res. Toxicol.* 9; 793 (1996).
15. S.L. Iverson, L. Shen, N. Anlar and J.L. Bolton. Bioactivation of estrone and its catechol metabolites to quinoid-glutathione conjugates in rat liver microsomes. *Chem. Res. Toxicol.* 9:492 (1996).

16. M. Butterworth, S.S. Lau and T.J. Monks. Formation of catechol estrogen glutathione conjugates and γ-glutamyl transpeptidase-dependent nephrotoxicity of 17β-estradiol in the golden Syrian hamster. *Carcinogenesis* 18:561 (1997).

17. L. Viborg-Jorgensen, C. Cornett, U. Justesen, L.H. Skibsted and L.O. Dragsted. Two-electron electrochemical oxidation of quercetin and kaempferol changes only the flavonoid C-ring. *Free Rad. Res.* 29: 339 (1998).

18. E. Sergediene, K. Jonsson, H. Szymusiak, B. Tyrakowska, I.M.C.M. Rietjens and N. Cenas, Prooxidant toxicity of polyphenolic antioxidants to HL-60 cells: description of quantitative structure-activity relationships. *FEBS Lett.* 462: 392 (1999).

19. K. Shimoi, H. Okada, M. Furugori, T. Goda, S. Takase, M. Suzuki, Y. Hara, H. Yamamoto and N. Kinae. Intestinal absorption of luteolin and luteolin 7-O-beta-glucoside in rats and humans. *FEBS Lett.* 438: 220 (1998).

20. B.T. Zhu, E.L. Ezell and J.G. Liehr. Catechol-*O*-methyltransferase-catalyzed rapid O-methylation of mutagenic flavonoids. *J. Biol. Chem.* 269: 292 (1994).

21. J.P. Spencer, G. Chowrimootoo, R. Choudhury, E.S. Debnam, S.K. Srai and C. Rice-Evans. The small intestine can both absorb and glucuronidate luminal flavonoids. *FEBS Lett.* 458: 224 (1999).

22. G. Williamson, A.J. Day, G.W. Plumb and D. Couteau, Human metabolic pathways of dietary flavonoids and cinnamates. *Biochem. Soc. Transact.* 28: 16 (2000).

23. J.M. Karczewski, J.G.P. Peters and J. Noordhoek. Quinone toxicity in DT-diaphorase-efficient and -deficient colon carcinoma cell lines. *Biochem. Pharmacol.* 57:27 (1999).

24. G. Powis, P.Y. Gadaska, A. Gallegos, K. Sherrill and D. Goodman. Over-expression of DT-diaphorase in transfected NIH 3T3 cells does not lead to increased anticancer quinone drug sensitivity: A questionable role for the enzyme as a target for bioreductively activated anticancer drugs. *Anticancer Res.* 15: 1141 (1995).

USE OF STRUCTURE-ACTIVITY RELATIONSHIPS FOR PROBING BIOCHEMICAL MECHANISMS: GLUTATHIONE TRANSFERASE ZETA CONJUGATION OF HALOACIDS

Paul D. Swartz and Ann M. Richard

US Environmental Protection Agency
Office of Research and Development
National Health and Environmental Effects Research Lab
Research Triangle Park, NC 27111 USA

INTRODUCTION

A structure-activity relationship (SAR) represents an empirical means for generalizing chemical information relative to biological activity, rationalizing existing data, and, ideally, providing an effective linkage between theory and experiment. In addition, by virtue of its implicit relationship to one or more rate determining step(s) of a biological interaction, an SAR can serve as a valuable tool for inquiry into the molecular mechanisms that underlie empirical observation. Such efforts are most productively guided and constrained by plausible hypotheses, some molecular-level resolution of the problem, and some knowledge of the probable chemistry and biochemical activation pathways of structurally similar chemicals. It follows that information pertaining to biologically reactive intermediates and their formation is highly relevant to, and can be potentially inferred from successful SAR models.

Two previously published computational studies pertaining to glutathione-mediated bioactivation of haloalkenes have provided useful illustration of these concepts, highlighting the ability of molecular modeling and SAR approaches to complement experimental observation and probe chemical mechanisms. In the first study, computational methods were used to evaluate the relative plausibility of two proposed bioactivation pathways for a series of 2,2-dihalo-1,1-difluoroethane-1-thiolate (2,2-dihalo-DFET) intermediates formed from glutathione-mediated bioactivation of the corresponding 1,1-dihalo-2,2-difluoroethenes (Shim and Richard, 1997). Comparison of calculated reaction path profiles (i.e. activation barriers and product energies) indicated that the brominated 2,2-dihalo-DFETs were significantly more likely to form a thiirane rather than a thionoacyl fluoride intermediate, a preference that offered a clear mechanism-based rationale for the distinct metabolites and mutagenicity of the bromine-containing 2,2-dihalo-DFETs. In a second modeling study, quantum mechanical calculations were used to compute stereoisomeric product outcomes of the trichloroethylene-glutathione conjugation (Shim and Richard, 1999). Lacking details of the enzyme interaction, general knowledge pertaining to nucleophilic vinylic substitution (S_NV) reactions was used to inform and constrain the modeling problem. Calculated transition state activation barriers for a model CH_3S-TCE conjugation system were used to infer product isomer outcomes and produced results in good agreement with experiment, suggesting that detailed knowledge of the local chemistry of the thiolate interaction was sufficient to rationalize product outcome. Both of these computational studies relied, for their success, upon the availability of knowledge of probable reaction outcomes, and experimental data pertaining to biochemical activation pathways to anchor and validate the modeling results.

In this paper, we present results of a current SAR modeling study of glutathione transferase zeta (GSTZ) mediated glutathione conjugation for a series of α-haloacetic and halopropionic acids. Dichloroacetic acid (DCA) is a prominent environmental contaminant formed as a by-product of water disinfection that has been found to be a hepatocarcinogen in rodents (Bull et al., 1990; DeAngelo et al., 1996), driving concern for potential adverse human health effects of this and other haloacids in treated water. The GSTZ bioactivation pathway has been implicated in hepatic biotransformation of DCA to glyoxylic acid (Tong et al., 1998a, Cornett et al., 1999), a major metabolite possibly involved in toxicity. DCA and other α-haloacetic acids have been shown to be efficient substrates for GSTZ, which catalyzes conversion of the haloacids to glutathione conjugates and glyoxylate (Tong et al., 1998a,b). To investigate the structural aspects of the GSTZ substrate specificity, Tong et al. (1998b) measured relative rates of glutathione conjugate or glyoxylate formation for a series of dihaloacetic acids and α-halopropionic acids. These activities ranged over 3-4 orders of magnitude and indicated sensitive structural specificity in enzyme function. Other investigators have further probed the nature and function of the GSTZ enzyme, revealing an identical correspondence of GSTZ with a recently characterized enzyme in the tyrosine reduction pathway, i.e. MAA isomerase (Fernández-Cañón and Peñalva, 1998). Structural investigations have identified hydrophilic residues in the GSTZ binding domain potentially involved in catalytic function (Board et al., 1997); however, the 3-dimensional structure and nature of the binding domain and mechanism of catalytic enhancement of GSTZ are unknown at the present time. Finally, recent work of Anders and coworkers (Anderson et al., 1999; Tzeng et al., 2000) have provided evidence that DCA and other non-fluorinated haloacids partially inhibit their own metabolism and deplete GSTZ by covalent modification of the enzyme, introducing a complicating factor in interpreting experimentally measured rates of conjugation.

In the first part of the present analysis, we employed an empirical SAR approach for identifying mechanistically relevant molecular parameters that correlate with GSTZ substrate activity. In addition to physico-chemical properties, such as pKa and log(octanol/water) partition coefficients, we computed reaction energies (energies of products minus reactants) for S_N1 dehalogenation and S_N2 methylsulfide halogen displacement as approximate models for glutathione conjugation reaction (Figure 1). In the second part of this study, we performed a computational experiment to test the hypothesis that GSTZ catalytic function might be enhanced by polarization of the acid carbonyl bond, e.g. through H-bonding interactions in the active site. For this purpose, we simulated a carbonyl bond stretch for each haloacid and assessed its effect on carbon-halogen bond distances and bond orders, i.e. changes that could facilitate the dehalogenation process. We demonstrate how consistencies in the pattern of these changes, with both the empirical SAR results and measured biological activities, offer a means to explore possible chemical mechanisms for GSTZ catalytic function.

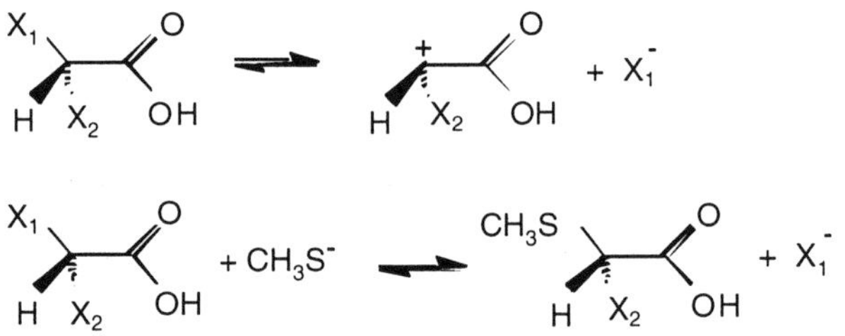

Figure 1. Modeled reactions of dehalogenation (S_N1) or methylsulfide halogen displacement (S_N2) for the dihaloacetic acids.

METHODS

All energies, electronic properties and molecular geometries were computed using the SPARTAN molecular modeling program (Wavefunction, Inc., Irvine, CA, ver. 5.1.1 for

UNIX) on a Silicon Graphics OCTANE workstation. Initial geometry optimization was performed at the semi-empirical AM1 level, followed by full geometry optimization and calculation of electronic properties using density functional theory (DFT) as implemented within SPARTAN at the pBP/DN** basis set level, a level found in to be suitable for treatment of molecules containing bromine and larger atoms. Calculations involving forced stretches of either the carbonyl or the carbon-halogen bond were carried out by imposing a distance constraint on the bond of interest and allowing the remainder of the molecule to fully relax. Single point HF/3-21G** calculations of the DFT geometries were required to compute bond orders within SPARTAN since this property option was unavailable for DFT calculations. For calculations modeling explicit H-bonding to the carbonyl (using methanol, ethanol and phenol), we constrained the O--H distance to a value of 1.0 and the C=O--H angle to 120 degrees to simulate a maximal H-bonding interaction such as would be found in a protein or in solution. Acid-dissociation constants (pKa) were computed by the SPARC method (Karickhoff, 1991), and log(octanol/water) partition coefficients (a.k.a. log $K_{o/w}$ or logP) were computed by the CLOGP method (ver. 2.5, Biobyte Corp., Pomona, CA). Statistical analysis was performed using SYSTAT (ver. 5.2.1 for Macintosh, SPSS Inc, Chicago, IL).

RESULTS

Table 1 presents a listing of the 11 haloacids, experimental activities, and computed properties considered for the present SAR analysis, with measured substrate enzyme activities obtained from the Tong et al. (1998b) study. Three haloacids from that study were excluded from the present analysis: Monofluoroacetic acid was the sole monohaloacetic acid analyzed and had very weak enzyme activity. 3,3-dichloropropionic acid was the sole β-haloacid analyzed, has a different elimination mechanism than the

Table 1. GSTZ activities and calculated physico-chemical properties of 11 α-haloacids.

Haloacid		Activity[a]	Log(A)[b]	ΔE(X⁻ loss)[c]	ΔE(MeS⁻ conj)[d]
bromofluoroacetic acid	BFA	5028	3.70	202.14 (Br)[e]	-36.00 (Br)[e]
chlorofluoroacetic acid	CFA	3883	3.59	210.02 (Cl)[e]	-28.12 (Cl)[e]
difluoroacetic acid	DFA	ND	-1.0	257.45	19.31
bromochloroacetic acid	BCA	1411	3.15	181.28 (Br)[e]	-40.76 (Br)[e]
dichloroacetic acid	DCA	1038	3.02	190.02	-23.02
2,2-dichloropropionic acid	22DCPA	244	2.39	176.41	-22.46
dibromoacetic acid	DBA	155	2.19	189.79	-15.79
2-iodopropionic acid	2-IPA	2532	3.40	166.63	-47.00
2-bromopropionic acid	2-BPA	2142	3.33	176.48	-37.76
2-chloropropionic acid	2-CPA	1655	3.22	183.77	-30.45
2-fluoropropionic acid	2-FPA	ND	-1.0	225.86	11.64

[a] GSTZ substrate activities, measuring production of glutathione conjugation or glyoxylate, were obtained from Tong et al. (1998), in units of [nmol min⁻¹(mg purified protein)⁻¹], where ND=no detectable activity.

[b] Log(A) = Log(Activity), where compounds with no detectable activity were assigned values of -1.0.

[c] ΔE(X⁻ loss) is halide dissociation energy (XA –> X⁻ + A⁺) in kcal/mol, calculated within SPARTAN at the DFT(pBP/DN**) level of theory.

[d] ΔE(MeS⁻ conj) is conjugation energy (XA + MeS⁻ –> X⁻ + MeSA) in kcal/mol, calculated within SPARTAN at the DFT(pBP/DN**) level of theory.

[e] Reported energies correspond to reaction involving loss of the preferred halide, indicated in parentheses.

α-haloacids, and had no measured enzyme activity. Finally, different reported enzyme activities for stereoisomers *R*- and *S*-2-chloropropionic acid reported by Tong et al. (1998b), which clearly implicate stereospecific interactions in enzyme function, were not modeled due to lack of stereospecific information on the three-dimensional structure of the GSTZ binding domain. Hence, only activities of the racemic (*R,S*) forms of the propionic acids were considered in the present analysis.

Initially, a Pearson matrix of r values was computed to examine single parameter correlations and intercorrelations for substrate enzyme activity, i.e. log(A) in Table 1, and a variety of calculated parameters, including pKa, CLOGP, $\Delta E(X^- loss)$ and $\Delta E(MeS^- conj)$. CLOGP and pKa [values not shown] poorly correlated with log(A), yielding r values of 0.56 and 0.03, respectively. The reaction energies for halide loss, $\Delta E(X^- loss)$, and methyl sulfide displacement, $\Delta E(MeS^- conj)$, display similar trends in Table 1 and were reasonably intercorrelated, with r=0.86. However, in relation to log(A), $\Delta E(X^- loss)$ gave an intermediate correlation of r=0.78 (r^2=0.61), whereas $\Delta E(MeS^- conj)$ yielded a much higher correlation of r=0.95 (r^2=0.90), by far the best correlation of any single parameter examined. A plot of this correlation, presented in Figure 2, shows its ability to clearly distinguish actives from inactives; however, it is also obvious that the inactives exert a large leverage upon the overall regression, as evidenced by the n=9 regression fit and significantly lower r=0.71 (r^2=0.50) for the set of 9 haloacids when the 2 inactives are excluded. This lowered correlation for $\Delta E(MeS^- conj)$ remains significantly higher than for any other parameter correlation with log(A) on this subset of 9 chemicals, the next highest being r=0.29 for $\Delta E(X^- loss)$. To test the reasonableness of assigning an arbitrarily low value of log(A)=-1.0 to the two inactives, DFA and 2-FPA, we used the n=9 regression equation to predict log(A) values for these two compounds, yielding predictions of 1.27 and 1.55, respectively. This ability of the n=9 regression to extrapolate well beyond its range of modeled activity to predict very low activities for DFA and 2-FPA increases confidence in its validity. Furthermore, overall regression statistics for the n=11 data set were found to vary little (r=0.92 to 0.95) upon varying the inactive log(A) assignment from +1.0 to -2.0.

The overall regression statistic of r=0.95 for log(A) vs. $\Delta E(MeS^- conj)$ represents the most successful attempt to relate all 11 haloacids according to the same property metric. Another measure of success of this parameter correlation, however, is its ability to reproduce the correct relative orderings of activities within chemically meaningful subclasses of the data set. A comparison of experimental vs. predicted activities for the 11

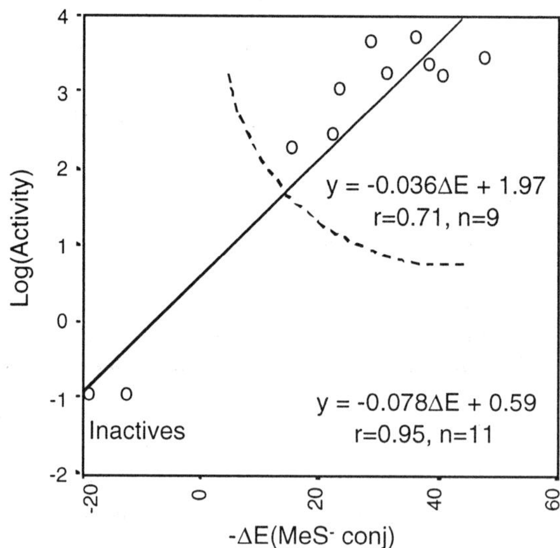

Figure 2. Regression plot of log(GSTZ activity) vs. $\Delta E(MeS^- conj)$ for 11 haloacids.

haloacids arranged in 3 subclasses is represented in Figure 3 for: 1) fluorine-containing dihaloacetic acids, 2) nonfluorinated dihaloacetic and propionic acids, and 3) mono α-halopropionic acids. Recently published results of Anders and coworkers (Anderson et al., 1999; Tzeng et al., 2000) indicating that the non-fluorinated dihaloacetic acids inhibit their own metabolism through covalent modification of the GSTZ enzyme, whereas the fluorine-containing dihaloacetic acids did not, provided the motivation for considering these two groups separately. Figure 3 clearly indicates that, in spite of the less impressive r=0.71 correlation for the n=9 set, the relative orderings of activities within each of the three subgroups is accurately reproduced by the n=11 ΔE(MeS⁻ conj) regression predictions. Particularly note-worthy is the apparent ability of ΔE(MeS⁻ conj) to capture the salient chemistry to predict correctly the ordering of activities for BCA>DCA>DBA.

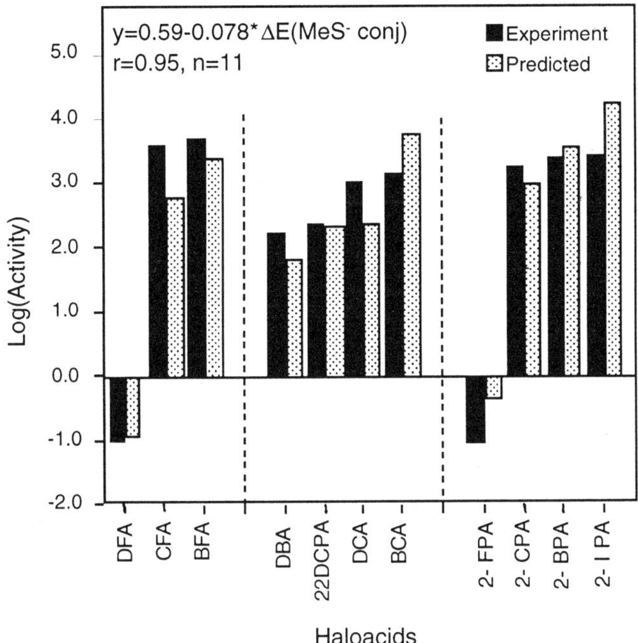

Figure 3. Comparison of experimental vs. predicted values of log(GSTZ substrate activity), with predicted values based on n=11 regression in ΔE(MeS⁻ conj).

In the second part of this study, we performed a series of calculations on the 11 haloacids listed in Table 1 to assess the affect of an artificially induced carbonyl stretch on the carbon-halide bond(s). Questions we wished to address were: Would such a stretch, which simulates a possible hydrogen bonding interaction within the active site of the GSTZ, weaken the bond of the leaving halide, thereby facilitating the conjugation reaction? Which halogen in a mixed haloacid would be marked as the leaving halide, and would this agree with the thermodynamically predicted results? And, finally, would the induced bond order changes reflect the pattern of activities observed in Table 1 and Figure 3?

Computed values of the change in bond order of the carbon-[leaving halogen] bond, i.e. $\Delta BO(C\text{-}X_1)$, as a function of a 0.5Å carbonyl bond stretch from the minimum energy starting geometry, i.e. $\Delta d(C\text{=}O)=0.5$Å, are presented in the bar graph in Figure 4, where more negative values of $\Delta BO(C\text{-}X_1)$ correspond to greater weakening of the $C\text{-}X_1$ bond. Comparison of Figures 3 and 4 indicates that $\Delta BO(C\text{-}X_1)$ values successfully reproduce most of the trends observed for substrate enzyme activities within the 3 subclasses of haloacids. Only the DBA value within the set of non-fluorinated dihaloacids does not fit the decreasing pattern precisely.

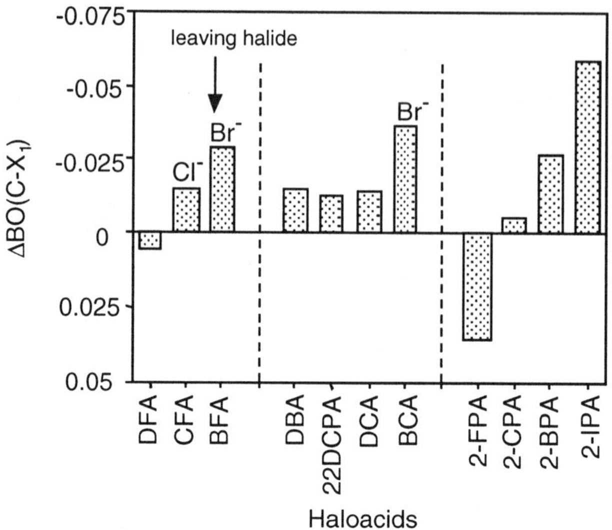

Figure 4. Computed values of the change in bond order of the carbon-[leaving halogen] bond, i.e. $\Delta BO(C\text{-}X_1)$, corresponding to a 0.5Å stretch of the carbonyl bond.

For each of the haloacids, in addition to bond order changes in the carbon-[leaving halogen] bond, the carbonyl stretch calculations produced changes throughout the molecule that were consistent with potentiation for halide displacement, including: 1) increasing lability of one (in the case of mixed haloacids) or both halogens at the $\alpha\text{-}C_2$ carbon; 2) increased positive charge on the C_2 carbon corresponding to an increasingly electrophilic center; 3) increased distance of the carbon-[leaving halogen] (i.e. $C\text{-}X_1$) bond and, in the case of mixed haloacids, corresponding decreased distance and increased bond order of the carbon-[retained halogen] bond (i.e. $C\text{-}X_2$). These changes are represented for BCA in Figure 5. For the mixed haloacids, the preferred leaving halogen followed the order Br>Cl>F (Figure 4), results that are in agreement with the thermodynamic reaction energy predictions in Table 1. The total molecule energy changes associated with the 0.5Å carbonyl stretch at the DFT(pBP/DN**) level were on the order of 70-80 kcal/mol. Additional calculations modeling explicit hydrogen bonding (of methanol, ethanol and phenol) to the carbonyl were performed for BCA (results not shown). Although these

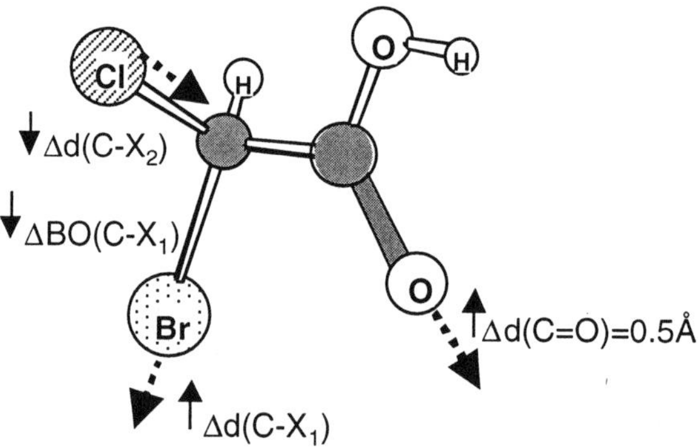

Figure 5. Coordinated molecular property changes for bromochloroacetic acid (BCA), and other haloacids, resulting from a simulated 0.5Å stretch in the carbonyl bond distance.

calculations produced smaller changes in bond orders and distances than resulted from the 0.5Å carbonyl stretch calculations, i.e. C-X$_1$ distance changes corresponding to approximately a 10 kcal/mol change in total molecule energy, they confirmed each of the trends represented in Figure 5.

DISCUSSION

Since the relatively recent discovery and characterization of the GSTZ family of enzymes (Board et al., 1997; Fernández-Cañón and Peñalva, 1998), experimental investigation has characterized many aspects of GSTZ function, particularly in relation to metabolism of the haloacids (Tong et al., 1998a,b; Anderson et al., 1999; Cornett et al., 1999; Blackburn et al., 2000; Tzeng et al., 2000). Published experimental results corresponding to a wide range GSTZ substrate activities for series of closely related haloacetic and halopropionic acids (Tong et al., 1998b) motivated the present SAR analysis and computational investigation. The nature of such data, presumed to be narrowly focused from both a chemical and biological mechanism standpoint, provides an opportunity for developing an SAR model with the potential to shed light on underlying chemical reactivity mechanisms driving the biochemical transformation.

The results of our empirical SAR analysis, summarized in Figures 2 and 3, identified a compelling association of a single calculated property with GSTZ activity for a set of 11 haloacetic and halopropionic acids, i.e. ΔE(MeS$^-$ conj) -- the reaction energy corresponding to methyl sulfide displacement of a halogen on the α-carbon. The successful overall correlation of this parameter with enzyme activity (r=0.95), its ability to clearly separate active and inactive acids (Figure 2), its ability to extrapolate from the set of 9 active acids to predict very low activity for the 2 inactive acids, and its ability to reproduce the most prominent patterns of relative activity among the most closely related haloacids represented in the 3 subclasses in Figure 3, all provide support for the validity of this parameter association. The success of ΔE(MeS$^-$ conj) over ΔE(X$^-$ loss) as a superior correlate to activity also distinguishes the two proposed reaction scenarios in Figure 1, implying that S$_N$2 sulfide displacement of a halogen is an important driver of the conjugation reaction and determinant of relative substrate enzyme activities. Furthermore, the finding that relative product energies correlate well with these enzyme activities suggests a thermodynamically controlled reaction with a late transition state resembling products.

It was mentioned earlier that a motivation for considering the fluorine-containing dihaloacids separately from the remaining dihaloacids in examining relative enzyme activity trends (as in Figure 3) came from the reported observations of Anders and coworkers (Anderson et al., 1999; Tzeng et al., 2000) that only the non-fluorinated dihaloacetic acids inhibit their own metabolism through covalent modification of the GSTZ enzyme. Note that both BFA and CFA have significantly greater experimental enzyme activities than the corresponding non-fluorinated dihaloacids, BCA, DCA and DBA (Table 1 and Figure 3). For DCA, the inhibition of GSTZ activity *in vivo* has been reported to be as high as 79% over a 24 treatment period (Anderson et al., 1999). Clearly, this introduces uncertainty in the reported substrate enzyme activities used as the basis of the present SAR analysis. However, it is perhaps reasonable to assume that if no competitive inhibition of GSTZ were occurring, the experimental activities of the non-fluorinated haloacids in Figure 3 would be higher in relation to the fluorinated dihaloacids than reported. This trend would be consistent with the errors in the SAR predictions, which appear to overestimate the measured activities of the non-fluorinated dihaloacids in relation to the fluorinated haloacids.

The experimentally measured substrate enzyme activities listed in Table 1, coupled with the structural and reactivity properties of the haloacids, have the potential for yielding some insight into reactivity mechanisms of the newly discovered GSTZ enzymes. A more complete understanding, of course, requires knowledge of the GSTZ enzyme itself, its 3 dimensional structure and stereospecific substrate binding requirements, mechanism of catalytic enhancement, and glutathione conjugation reaction mechanism. Although this level of understanding of GSTZ function is presently unavailable, much is known of GST catalytic function, in general, that can help guide hypothesis generation and constrain a computational modeling study. It is generally accepted that a major role of all GSTs in catalyzing glutathione conjugation of weak electrophiles is in stabilizing the more reactive thiolate form of glutathione, making this thiolate sterically accessible to an electrophilic substrate, and desolvating the thiolate and/or the electrophile within the hydrophilic binding domain of the enzyme (Armstrong, 1991). GSTs, including GSTZ, have strong sequence homology in the region where glutathione binding and, presumably, thiolate stabilization occur (Armstrong, 1991; Board et al., 1997). However, a major distinguishing feature of GSTZ in comparison to other more well known GSTs, and a contributing factor in its relatively late discovery, is its limited activity with classic GST substrates, such as 1-chloro-2,4-dinitrobenzene and its rather unusual affinity for α-haloacid substrates (Board et al., 1997; Tong et al., 1998a,b).

Board et al. (1997) proposed, from sequence analysis and homology modeling of the N-terminal domain of GST1-1, possible involvement of a serine (Ser-14) or other polar residue (such as tyrosine) in catalytic function. The likelihood of one or more H-bonding-capable polar residues in the active site, coupled with the apparent requirement of an acid moiety in GSTZ substrates, led us to propose that polarization of the acid carbonyl bond might be directly involved in catalytic enhancement of glutathione conjugation with the α-haloacids. A maximal bond polarization effect was approximately modeled, computationally, by a simulated 0.5Å stretch of the carbonyl bond for each of the α-haloacids. What we observed in all cases was that the simulated bond stretch caused molecular changes (i.e. changes in bond distances, bond orders, and charge on the nucleophilic α–carbon center) entirely consistent with facilitated halide loss, i.e. changes that marked the beginning of the halide displacement reaction. The further observation that the changes of bond orders for the various α-haloacids followed the enzyme activity trends represented in Figure 3 surprisingly well provided additional support for the carbonyl bond polarization hypothesis.

CONCLUSIONS

The results of our empirical SAR analysis of GSTZ substrate activity, combined with the computational simulation of the carbonyl bond polarization, have produced a remarkably consistent picture of chemical reactivity considerations potentially important to catalytic function and glutathione conjugation outcome. The carbonyl bond polarization simulates potentiation of the α-haloacid by the GSTZ protein at the earliest stage of the conjugation reaction, i.e. increasing the electrophilic character of the α–carbon and weakening the carbon-[leaving halogen] bond, whereas calculated reaction energies of product formation, $\Delta E(MeS^- \text{ conj})$, are modeling a thermodynamically controlled reaction outcome. The consistency in the pattern of activities predicted by both of these calculated properties in relation to measured substrate enzyme activities provide the most compelling argument for their relevance to enzyme function.

The SAR and computational approaches of the present study have involved major simplifications and approximations in treatment of a complex enzymatic biotransformation process. However, we have illustrated a modeling approach which attempts to make best use of limited existing knowledge and experimental data to infer and evaluate plausible mechanisms for GSTZ conjugation. The results of the present study should serve as a useful foundation and starting point for further experimental and theoretical investigation.

Acknowledgments

The authors wish to thank to Dr. Prof. M. W. Anders for helpful suggestions. At the time of this study, P. D. S. held a post doctoral fellowship in the UNC Curriculum of Toxicology, funded by the EPA/UNC Toxicology Research Program, Training Agreement T901915. This manuscript has been reviewed by the National Health and Environmental Effects Research Laboratory, U. S. Environmental Protection Agency and approved for publication. Approval does not signify that the contents necessarily reflect the views and policies of the Agency, nor does mention of trade names or commercial products constitute endorsement or recommendation for use.

REFERENCES

Anderson, W. B., Board, P. G. Gargano, B., and Anders, M. W., 1999, Inactivation of glutathione transferase zeta by dichloroacetic acid and other fluorine-lacking α-haloalkanoic acids, *Chem. Res. Toxicol.* **12**:1144-1149.

Armstrong, R. N., 1991, Glutathione *S*-transferases: reaction mechanisms, structure, and function, *Chem. Res. Toxicol.* **4**:131-140.

Blackburn, A. C., Tzeng, H. F., Anders, M. W., and Board, P. G., 2000, Discovery of a functional polymorphism in human glutathione transferase zeta by expressed sequence tag database analysis, *Pharmacogenetics* **10**:49-57.

Board, P. G., Baker, R. T., Chelvanayagam, G., and Jermiin, L. S., 1997, Zeta, a novel class of glutathione transferases in a range of species from plants to humans, *Biochem. J.* **328**:929-935.

Bull, R. J., Sanchez, I. M., Nelson, M. A., Larson, J. L., and Lansing, A. J., 1990, Liver tumor induction in B6C3F1 mice by dichloroacetate and trichloroacetate, *Toxicol.* **63**:341-365.

Cornett, R., James, M. O., Henderson, G. N., Cheung, J., Shroads, A. L., and Stacpoole, P. W., 1999, Inhibition of glutathione *S*-transferase ζ and tyrosine metabolism by dichloroacetate: A potential unifying mechanism for its altered biotransformation and toxicity, *Biochem. Biophys. Res. Comm.* **262**:752-756.

DeAngelo, A. B., Daniel, F. B., Most, B. M., and Olson, G. R., 1996, The carcinogenicity of dichloroacetic acid in the male Fischer 344 rat, *Toxicol.* **114**:207-221.

Fernández-Cañón, J. M. and Peñalva, M. A., 1998, Characterization of a fungal maleylacetoacetate isomerase gene and identification of its human homologue, *J. Biol. Chem.* **273**:329-337.

Karickhoff, S. W., McDaniel, V. K., Melton, C., Vellino, A. N., Nute, D. E., and Carriera, L. A., 1991, Predicting chemical reactivity by computer, *Environ. Toxicol. Chem.* **10**:1405-1416.

Shim, J.-Y. and Richard, A. M., 1997, Theoretical evaluation of two plausible routes for bioactivation of S-(1,1-difluoro-2,2-dihaloethyl)-L-cysteine conjugates: thiirane vs. thionoacyl fluoride pathway, *Chem. Res. Toxicol.* **10**:103-110.

Shim, J.-Y. and Richard, A. M., 1999, Theoretical study of the nucleophilic vinylic substitution reaction of trichloroethylene (TCE) + CH_3S^- as a model for glutathione conjugation of TCE, *Chem. Res. Toxicol.* **12**:308-316.

Tong, Z., Board, P. G., and Anders, M. W., 1998a, Glutathione transferase zeta catalyses the oxygenation of the carcinogen dichloroacetic acid to glyoxylic acid, *Biochem. J.* **331**:371-374.

Tong, Z., Board, P. G., and Anders, M. W., 1998b, Glutathione transferase zeta-catalyzed biotransformation of dichloroacetic acid and other alpha-haloacids, *Chem. Res. Toxicol.* **11**:1332-1338.

Tzeng, H.F., Blackburn, A. C., Board, P. G., and Anders, M.W., 2000, Polymorphism- and species-dependent inactivation of glutathione transferase zeta by dichloroacetate, *Chem. Res. Toxicol.* **13**:231-236.

STRUCTURE TOXICITY RELATIONSHIPS - HOW USEFUL ARE THEY IN PREDICTING TOXICITIES OF NEW DRUGS?

Sidney D. Nelson

School of Pharmacy
University of Washington
Box 357631
Seattle, WA 98195-7631 U.S.A.

1. INTRODUCTION

Several new drugs have either been removed from the market soon after their release, or been flagged for careful monitoring and restricted use because of severe toxic effects. Although in some cases the parent drug structure may be responsible for the observed toxicities, in most cases reactive metabolites are formed that interact with cellular macromolecules to stress cells directly, and/or more commonly, yield immunogenic products. Since it is the structures of drugs and their metabolites that dictate reactivity, and not the metabolic pathways themselves, this chapter will focus on what can be empirically learned from comparisons of structures associated with toxic effects and similar structures that are less toxic.

Throughout the discussion of these structural comparisons, the reader should keep in mind the fact that there are several factors other than structure that can influence risk of toxicity. Some of these other determinants of toxicity include dose of the drug and fraction of the dose converted to toxic metabolite(s) which is dependent on concentrations of enzymes, cofactors, and other cellular constituents involved in catalyzing toxic metabolite formation and detoxication. These concentrations, in turn, are affected by genetics and a host of environmental factors including diet and other drugs as inducers, inhibitors, etc. In many cases, the largely unknown factors that determine immune response to macromolecular adducts of a drug or its metabolites adds a major layer of complexity to toxic responses. Despite the complexity, a few substructures have surfaced that appear to carry greater potential for risk of toxicity than others, and these will be summarized by reviewing some drugs that have either failed to be marketed, or were removed from the market or were restricted in use due to toxic effects.

2. STRUCTURES IN DRUGS ASSOCIATED WITH TOXIC EFFECTS

This section will focus on structures that for the most part have a recorded history of causing toxic effects via reactive metabolite formation. Reports are from the medical and biomedical literature and not from case reports listed in the FDA database which would require additional examination and verification.

Biological Reactive Intermediates VI, Edited by Dansette *et al.*
Kluwer Academic / Plenum Publishers, 2001

2.1. Bromfenac

Bromfenac (Figure 1), a nonsteroidal anti-inflammatory drug (NSAID), was withdrawn in 1998, less than a year after its introduction, because of 20 reports of serious hepatotoxicity (Friedman *et al.,* 1999). Based on several structural moieties present in bromfenac, it is not surprising that the drug caused some tissue toxic effects.

Halogenated aromatic compounds have been associated with toxicity, and bromobenzene is a well-known hepatotoxin (Brodie *et al.,* 1971). Although the bromophenyl ring in bromfenac is different than bromobenzene itself because it is part of a bromophenone structure, another *p*-bromophenyl substituted drug, ebrotidine (Figure 1), has also been found to be hepatotoxic whereas its hydrogen analog, famotidine, is not (Andrade *et al.,* 1999). Arene oxides of the bromophenyl ring have been implicated as the electrophilic reactive intermediates most responsible for hepatotoxicity (Rombach and Hanzlik, 1999).

Bromfenac also contains an aniline ring, and most aniline-containing drugs have been associated with toxic effects. Hydroxylamine and nitroso oxidation products of simple anilines have been known to cause methemoglobinemia and hemolysis of red blood cells (Kiese, 1966), and there is strong evidence to suggest these same reactive metabolites can be formed from most aniline drugs (e.g., sulfonamides), and are responsible for several autoimmune related toxic effects (Rubin, 1989; Uetrecht, 1990; Cribb and Spielberg, 1992; Pirmohamed *et al.,* 1996; Bluhm *et al.,* 1999; Hess *et al.,* 1999).

Finally, bromfenac is an arylacetic acid, and more arylacetic and arylpropionic acid NSAIDs have been discontinued because of toxicity than any other class of drugs (Bakke *et al.,* 1984; Bakke *et al.,* 1994). Although acyl glucuronides are formed from this class of drugs that can react with proteins either through trans-acylation or by acyl migration and Schiff-base formation (Spahn-Langguth and Benet, 1992), structural features in addition to the arylacetic and arylpropionic acid moieties likely are important in causing toxicity.

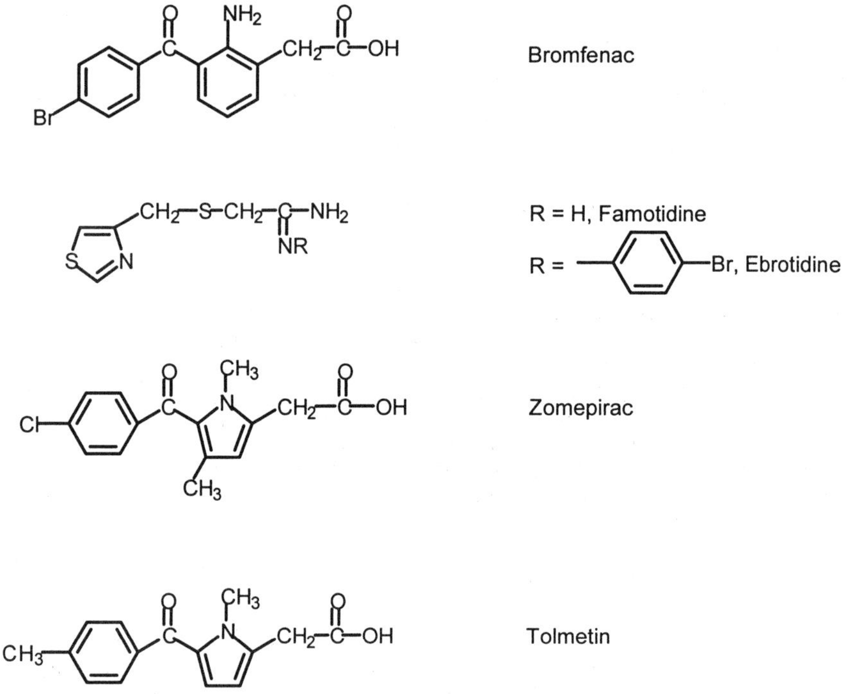

Bromfenac

R = H, Famotidine

R = Br, Ebrotidine

Zomepirac

Tolmetin

Figure 1. Structures of drugs discussed that may relate to toxic effects caused by Bromfenac.

For example, zomepirac and tolmetin are analogs with similar structures (Figure 1), and they both form acyl glucuronides that bind approximately the same extent to plasma proteins with similar elimination rates (Benet *et al.*, 1993). However, zomepirac was removed from the market in 1983 due to a high incidence of anaphylactic reactions (Samuel, 1981; U.S. Congress Proceedings, 1982), and it also caused renal failure in patients (Warren, S. E. and Mosley, C., 1983; Clive, D. M. and Stoff, J. F., 1984). Tolmetin has caused similar toxicities, but the incidence seems to be far less (Chatterjee, 1981; Brekza and Novey, 1985; Tietjen, 1989; Shaw and Anderson, 1991). Structurally, it would be interesting to know if the *p*-chlorobenzoyl group in zomepirac vs. *p*-methylbenzoyl group in tolmetin and/or the addition of a methyl group to the pyrrole ring in zomepirac leads to the differences in apparent incidence of toxic reactions associated with zomepirac.

The hypothesis that reactive metabolites other than those generated from acyl glucuronides are involved in the immunogenic toxic effects of NSAIDs is supported by recent investigations of protein adduct formation from the NSAID diclofenac (Figure 2) which show that both a quinone imine oxidation product and acyl glucuronide form reactive metabolites that bind to liver proteins (Kretz-Rommel and Boelsterli, 1993; Hargus *et al.*, 1994; Tang *et al.*, 1999; Shen *et al.*, 1999), and may be involved in hepatotoxicity associated with the drug (Banks *et al.*, 1995). Thus, with the multiple substructures that can form reactive metabolites in bromfenac, one or more of them was likely involved in causing the hepatotoxicities observed.

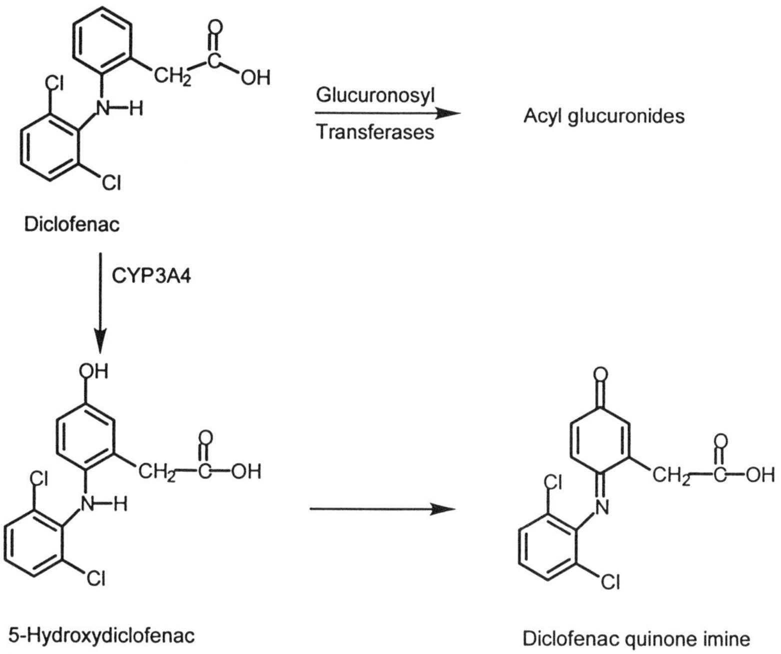

Figure 2. Formation of two different reactive metabolites of diclofenac.

2.2. Tenidap and Other Thiophene-Containing Drugs

Tenidap (Figure 3) was developed as an NSAID with cytokine inhibitory properties, and was marketed for a short time in Europe but not released in the U.S. because of several cases of hepatic and renal injury that occurred in patients in Europe and patients in clinical trials in Japan (SCRIP, 1995). Tenidap has structural similarities to tienilic acid (Figure 3) which was removed from use after many cases and some fatalities due to hepatic injury (Zimmerman *et al.*, 1984). Mechanistic studies have implicated a reactive thiophene oxidative metabolite in the immune-mediated toxicity (Beaune *et al.*, 1987), though it is not known if it is a thiophene epoxide or a thiophene sulfoxide metabolite of a positional isomer that causes direct hepatic toxicity in animals (Bonierbale *et al.*, 1999). In any case, the thiophene ring in tenidap is oxidized extensively in a manner similar to tienilic acid (Fouda *et al.*, 1997).

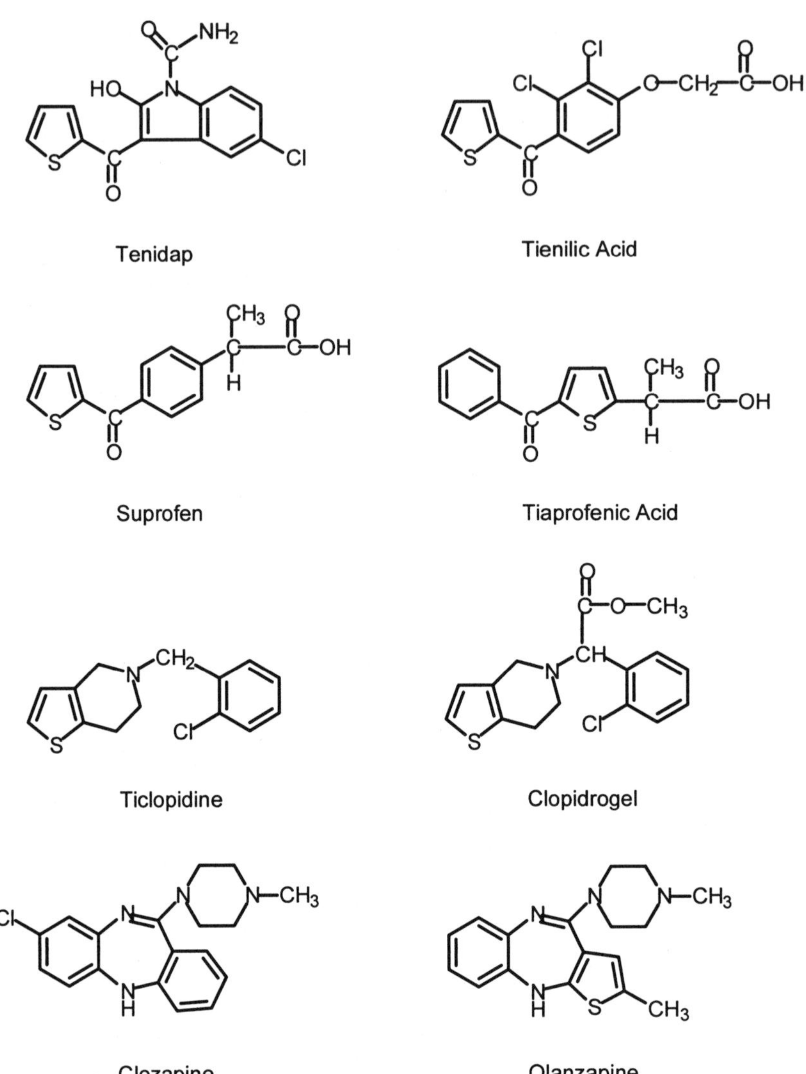

Figure 3. Structures of some thiophene-containing drugs and analogs.

Several other drugs that contain thiophene rings have either been removed from the market because of severe toxic effects, or must be carefully monitored when used clinically. Suprofen, a 2-arylpropionic acid NSAID (Figure 3) was removed from the market because of serious renal toxicity (Bakke *et al.*, 1994), and its positional isomer, tiaprofenic acid, has also been associated with several cases of interstitial cystitis (Gheyi *et al.*, 1999). Although the role of reactive metabolite formation from the thiophene ring in renal toxicity caused by suprofen and tiaprofenic acid is not known, the incidence of renal injury appears to be less when the thiophene ring is replaced by the isosteric phenyl ring as in ketoprofen.

Ticlopidine (Figure 3) is an antiplatelet drug whose use must be carefully monitored because of severe hematological abnormalities (Ono *et al.*, 1991; Tsatalas *et al.*, 1995; Bennett *et al.*, 1998). A mechanism for metabolic activation of the thiophene ring of ticlopidine by myeloperoxidase has been proposed (Liu and Uetrecht, 2000). Though it is not known if the mechanism is responsible for the adverse effects caused by ticlopidine, a recently introduced analog, clopidrogel (Figure 3), contains the same thiophene ring, and reports of thrombotic thrombocytic purpura associated with its use have begun to appear (Bennett *et al.*, 2000). It will be of interest to see if the incidence of hematogical toxicities is less with clopidrogel than ticlopidine, since it has a built in ester group as a "soft-spot" for metabolism not present in ticlopidine.

Olanzapine (Figure 3) is an interesting example of how the incorporation of a substituted thiophene into a structure may improve its safety profile. Clozapine (Figure 3) is a dibenzodiazepine antipsychotic drug with restricted use because it causes a high incidence of agranulocytosis (Alvir and Lieberman, 1994).

A hypothetical mechanism for the formation of reactive metabolites of clozapine involves myeloperoxidase-mediated oxidation of the chlorophenylazepine piperidine system to a conjugated electrophilic imminium ion (Uetrecht, 1990; Fischer *et al.*, 1991; Maggs *et al.*, 1995) which may generate antimyeloperoxidase antibodies (Jaunkalns *et al.*, 1992). The same reaction could occur with olanzapine, but the thiophene group would be more easily oxidized, and the products of this oxidation are apparently not reactive electrophiles (Gardner *et al.*, 1998). In fact the methyl group on the 2 (benzylic-like) position of the thiophene ring is the ultimate metabolic site rather than the thiophene ring itself (Kassahun *et al.*, 1997). This factor may only in part play a role, since olanzapine normally requires a dose one-tenth that of clozapine and when given in higher doses has caused agranulocytosis (Naumann *et al.*, 1999; Benedetti *et al.*, 1999).

2.3. Flutamide and Other Nitroaromatic Drugs

Flutamide and nilutamide (Figure 4) were the first two androgen receptor antagonists used primarily in the treatment of prostate cancer. However, patients taking flutamide must be monitored for development of hepatotoxicity, and those taking nilutamide for pneumotoxicity and hepatotoxicity because of several case reports of severe toxic reactions in patients (Wysowski and Fourcroy, 1996; Andrade *et al.*, 1999; Pfitzenmeyer *et al.*, 1992; Gometz *et al.*, 1992). These toxicities are similar to those observed with the nitroheteroaromatic antibacterial drug, nitrofurantoin (Holmberg *et al.*, 1980; Reinhart *et al.*, 1992). Mechanisms of toxicity caused by nitrofurantoin have focused on reduction of the aryl nitro group to anion radicals and redox cycling, and/or further reduction to an electrophilic aryl hydroxylamine or aryl nitroso compound (Boyd *et al.*, 1979; Silva *et al.*, 1993). Evidence has been provided for similar reactions with nilutamide coupled with effects on the mitochondrial respiratory chain (Berson *et al.*, 1994). Interestingly, bicalutamide (Figure 4), in which the nitro group is replaced by a cyano group, has been associated with only one case of hepatotoxicity that may have resulted from prior treatment with flutamide (Sarosdy, 1999).

Other relatively new drugs that have nitro aromatic structures (Figure 4) are the COMT inhibitor, tolcapone, that has caused hepatotoxicity (Assal *et al.*, 1998), and the COX-2 selective inhibitor, nimesulide that has caused renal failure (Peruzzi *et al.*, 1999; Balasubramaniam *et al.*, 2000).

2.4. Trovofloxacin and Other Quinolone Antibiotics

Trovofloxacin (Figure 5) was severely restricted in its use in 1999 because of more than 100 cases of hepatotoxicity, fourteen with acute liver failure, associated with its use (Chen *et al.*, 2000). The only other quinolone antibiotic known to cause so many cases of tissue organ damage is temafloxacin (Figure 5), which was discontinued in 1992 because of fever, jaundice, renal dysfunction, liver dysfunction, coagulopathy and hemolytic anemia (Blum *et al.*, 1994). Both trovofloxacin and temafloxacin contain a difluorophenyl moiety not present in other quinolone antibiotics, though this structure is present in other drugs (e.g. diflunisal) not associated with such severe toxic effects. However, this structure may influence extent of acyl glucuronide formation or reactivity of the quinolone ring. It is known that trovofloxacin is extensively glucuronidated in humans (Dalvie *et al.*, 1997). Alternatively, oxidation of the novel bicyclic aminopiperidine ring may play a role in hepatotoxicity caused by trovofloxacin, though it is not known if reactive metabolite formation is involved in the toxicities caused by either trovofloxacin or temafloxacin. It is noteworthy that the only other pyridinoquinolone, enoxacin, has an isomeric piperazine ring, and contains an *N*-ethyl group rather than the *N*-difluorophenyl group. No cases of hepatotoxicity have been reported with enoxacin.

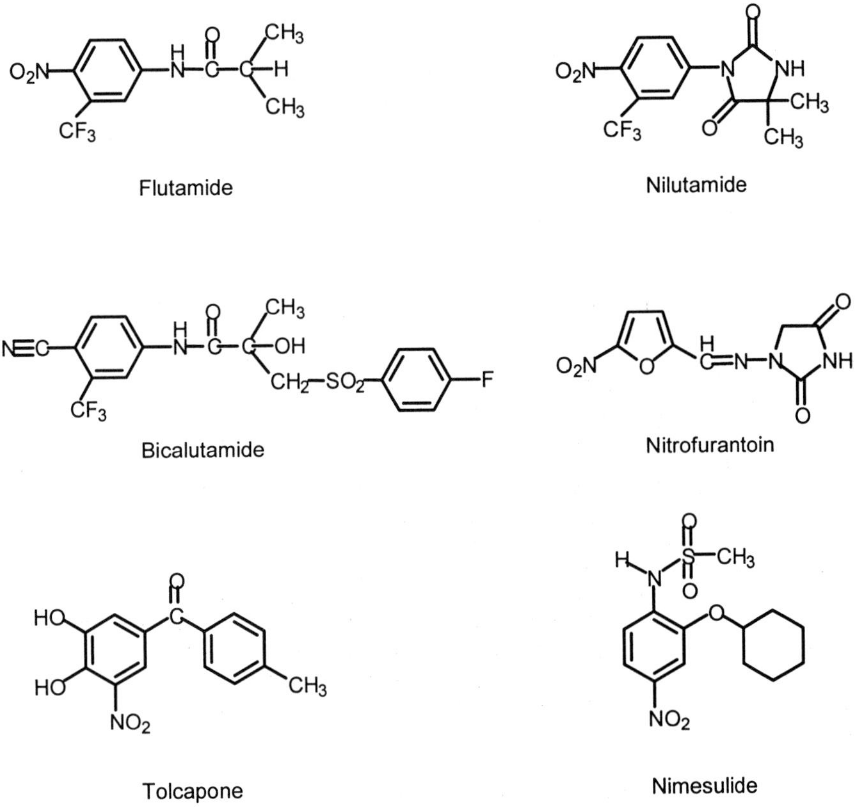

Figure 4. Structures of nitroaromatic drugs and analogs.

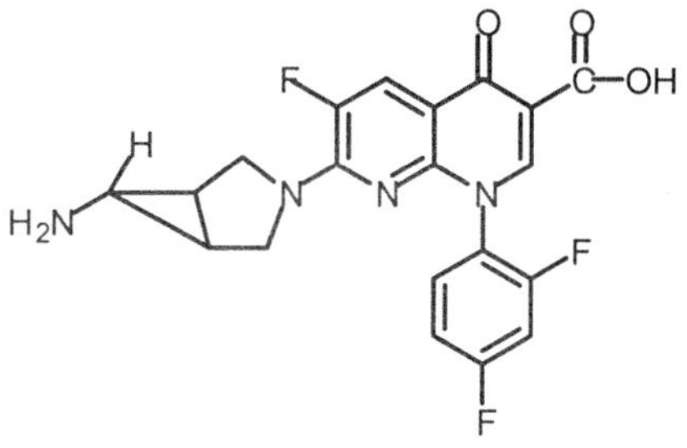

Trovofloxacin

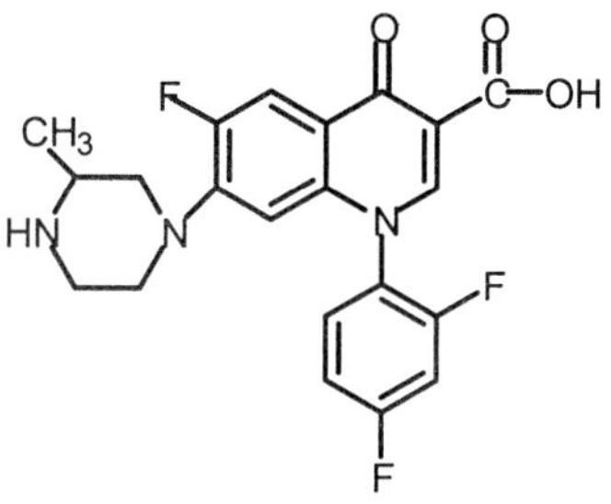

Temafloxacin

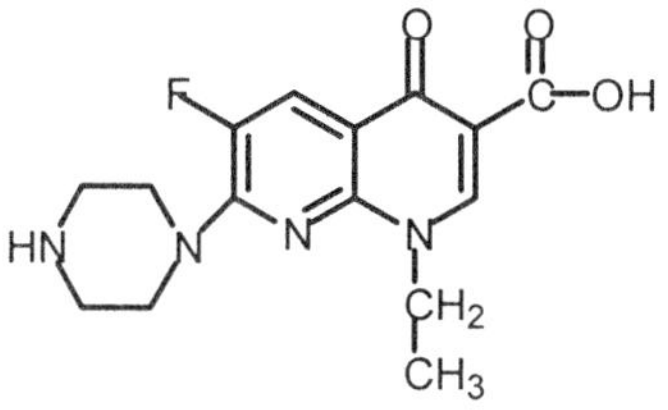

Enoxacin

Figure 5. Structures of trovofloxacin and structurally related quinolone antibiotics.

2.5. Troglitazone and Other Thiazolidinediones

Troglitazone (Figure 6) was the first of a new class of oral antidiabetic agents that enhance sensitivity to insulin of cells in several tissues. It was withdrawn from the market in March 2000 because of several cases of idiosyncratic hepatotoxicity, including some deaths of patients (The Pink Sheet, 2000). Two newer glitazones, rosiglitazone and pioglitazone (Figure 6) appear to be safer, though two case reports of hepatotoxicity after rosiglitazone therapy have recently appeared (Forman *et al.*, 2000; Al-Salman *et al.*, 2000). If the toxicity is dose-related with the glitazone class of drugs, the incidence of hepatotoxicity caused by rosiglitazone and pioglitazone would be expected to be much less than with troglitazone inasmuch as the concentrations of the latter two are approximately 10-100 times less than with troglitazone.

The mechanism of toxicity is unknown. However, troglitazone is known to be metabolized to a quinone (Yamazaki *et al.*, 1999), a reaction that is not possible with either rosiglitazone and pioglitazone. Alternatively, since patients on rosiglitazone have developed hepatotoxicity, the thiazolidinedione structure may be metabolized to reactive intermediates (W. Chen and T. Baillie, personal communications). Such studies with this new drug structure will be interesting and important to studies of structure-toxicity relationships.

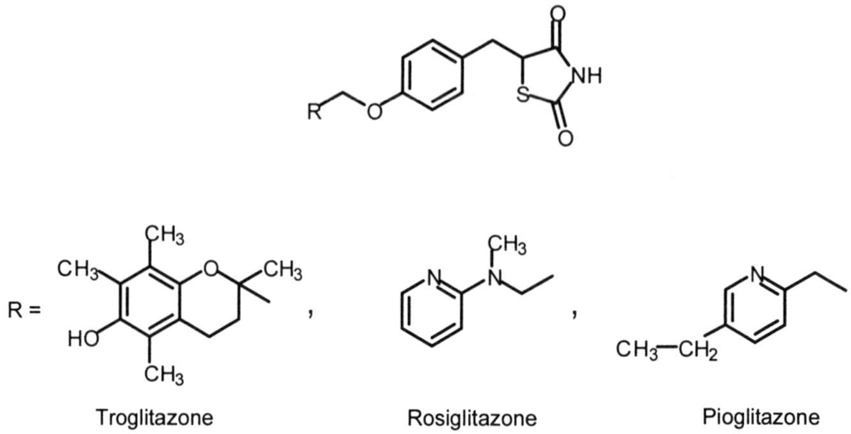

Figure 6. Structures of glitazone drugs.

3. SUMMARY

This chapter provides just a few newer examples of structural moieties found in drugs that have been associated with reactive metabolite formation and toxicities. For a discussion of several other structures in drugs that undergo metabolic activation to reactive intermediates, the reader is directed to previous volumes in this series and other chapters in this book, as well as a previous condensed review (Nelson, 1982). Since that review, some new knowledge allows us to better predict that some structural moieties are more likely than others to form drug reactive metabolites that may be involved in causing toxic effects in humans. For example, most aniline-, thiophene-, and nitroaromatic-containing drugs have had a relatively high incidence of adverse effects, and it would be prudent in the drug discovery process to avoid these substructures if possible. However, as illustrated by the case of olanzapine, these structures may be important for potent activity, and could therefore be beneficial in some cases.

The glitazones represent a new class of drugs with a unique thiazolidinedione structure. This raises two important points. First, it demonstrates how limited our knowledge base is in regard to structure toxicity relationships when new structures are introduced. Our approaches must be very empirical and are far from quantitative for the reasons outlined in the introduction. Secondly, the glitazones point out the importance of benefit/risk considerations. This was a new structural class of drugs with a unique spectrum of action that is very beneficial in the treatment of a major disease. Despite some suspected risk of toxicity, based on early trials, troglitazone was approved for use with careful monitoring. This author believes that was the right decision, as was the decision to withdraw the drug when the risk became unacceptable, especially with the introduction of safer alternatives. If this were just another NSAID (e.g., bromfenac), there would be little reason for approval.

In summary, as I pointed out previously (Nelson, 1982), with our limited knowledge of structure toxicity relationships, we can only make reasonable judgments as to risk assessment of a new drug in humans, and hope that we neither release a dangerous chemical entity nor, as importantly, abort an effective one.

REFERENCES

Al-Salman, J., Arjomand, H., Kemp, D.G., and Mittal, M., 2000, Hepatocellular injury in a patient receiving rosiglitazone. A case report, *Ann. Intern. Med.* **132**:121-124.

Alvir, M.J., and Lieberman, J.A., 1994, A reevaluation of the clinical characteristics of clozapine-induced agranulocytosis in light of the United States Experience, *J. Clin. Psychopharmacol.* **14**:87-89.

Andrade, R.J., Lucena, M.I., Martin-Vivaldi, R., Fernandez, M.C., Nogueras, F., Pelaez, G., Gomez-Outes, A., Garcia-Escaño, M.D., Bellot, U., Hervás, A., Cárdenas, F., Bermudez, F., Romero, M., and Salmerón, J., 1999, Acute liver injury associated with ebrotidine, a new H_2-receptor antagonist, *J. Hepatology* **31**:641-646.

Andrade, R.J., Lucena, M.I., Fernández, M.C., Suárez, F., Montero, J.L., Fraga, E., and Hidalgo, F., 1999, Fulminant liver failure associated with flutamide therapy for hirsutism, *Lancet* **353**:983.

Assal, F., Spahr, L., Hadengue, A., Rubbici-Brandt, L., and Burkhard, P.R., 1998, Tolcapone and fulminant hepatitis, *Lancet* **352**:958.

Bakke, O.M., Wardell, W. M., and Lasagna, L., 1984, Drug discontinuations in the United Kingdom and the United States, 1964 to 1983: Issues of safety, *Clin. Pharmacol. Ther.* **35**:559-567.

Bakke, O.M., Manocchia, M., de Abayo, F., Kaitin, K.I., and Lasagna, L., 1994, Drug safety discontinuations in the United Kingdom, the United States, and Spain from 1974 through 1993: A regulatory perspective, *Clin. Pharmacol. Ther.* **58**:108-117.

Balasubramaniam, J., 2000, Nimesulide and neonatal renal failure, *Lancet* **355**:575.

Banks, A.T., Zimmerman, H.J., Ishak, K.G., and Harter, J.G., 1995, Diclofenac-associated hepatotoxicity: Analysis of 180 cases reported to the Food and Drug Administration as adverse reactions, *Hepatologoy* **22**:820-827.

Beaune, P., Dansette, P.M., Mansuy, D., Finck, M., Amar, C., Leroux, J.P., and Homberg, J.C., 1987, Human anti-endoplasmic reticulum autoantibodies appearing in a drug-induced hepatitis are directed against a human liver cytochrome P-450 that hydroxylates the drug, *Proc. Natl. Acad. Sci. U.S.A.,* **84**:551-555.

Benedetti, F., Cavallaro, R., and Smeraldi, E., 1999, Olanzapine-induced neutropenia after clozapine-induced neutropenia, *Lancet* **354**:567.

Benet, L.Z., Spahn-Langguth, H., Iwakawa, S., Volland, C., Mizuma, T., Mayer, S., Mutschler, E., and Lin, E.T., 1993, Predictability of the covalent binding of acidic drugs in man, *Life Sci.* **53**:141-146.

Bennett, C.L., Weinberg, P.D., Rozenberg-Ben-Dror, K., Yarnold, P.R., Kwaan, H.C., and Green, D., 1998, Thrombotic thrombocytopenic purpura associated with ticlopidine: A review of 60 cases, *Ann. Intern. Med.* **128**:541-544.

Bennett, C.L., Connors, J.M., Carwile, J.M., Moake, J.L., Bell, W.R., Tarantolo, S.R., McCarthy, L.J., Sarode, R., Hatfield, A.J., Feldman, M.D., Davidson, C.J., and Tsai, H.-M., 2000, Thrombotic thrombocytopenic purpura associated with clopidrogel, *N. Engl. J. Med.,* **342**:1773-1777.

Berson, A., Schmets, L., Fisch, C., Fau, D., Wolf, C., Fromenty, B., Deschamps, D., and Pessayre, D., 1994, Inhibition by nilutamide of the mitochondrial respiratory chain and ATP formation. Possible contribution to the adverse effects of this antiandrogen, *J. Pharmacol. Exp. Ther.* **270**:167-176.

Bluhm, R.E., Adedoyin, A., McCarver, D.G., and Branch, R.A., 1999, Development of dapsone toxicity in patients with inflammatory dermatoses: Activity of acetylation and hydroxylation as risk factors, *Clin. Pharmacol. Ther.* **65**:598-605.

Blum, M.D., Graham, D.J., and McCloskey, C.A., 1994, Temafloxacin syndrome, *Clin. Infect. Dis.* **18**:946-950.

Bonierbale, E., Valadon, P., Pous, C., Desfosses, B., Dansette, P.M., and Mansuy, D., 1999, Opposite behaviors of reactive metabolites of tienilic acid and its isomer toward liver proteins: Use of specific anti-tienilic acid-protein adduct antibodies and the possible relationship with different hepatotoxic effects of the two compounds, *Chem. Res. Toxicol.* **12**:286-296.

Boyd, M.R., Catignani, G.L., Sasame, H.A., Mitchell, J.R., and Stiko, A.W., 1979, Acute pulmonary injury in rats by nitrofurantoin and modification by vitamin E, dietary fat and oxygen, *Am. Rev. Respir. Dis.,* **120**:93.

Bretza, J.A. and Novey, H.S., 1985, Anaphylactoid reactions to tolmetin after interrupted dosage, *Western J. Med.* **143**:55-59.

Brodie, B.B., Reid, W.D., Cho, A.K., Sipes, G., Krishna, G., and Gillette, J.R., 1971, Possible mechanism of liver necrosis caused by aromatic organic compounds, *Proc. Natl. Acad. Sci. U.S.A.* **68**:160-164.

Chatterjee, G.P., 1981, Nephrotic syndrome with Tolmetin, *J. Amer. Med. Assoc.* **246**:1589.

Chen, H.J.L., Bloch, K.J., and MacLean, J.A., 2000, Acute eosinophilic hepatitis from trovofloxacin, *N. Engl. J. Med.,* **342**:359-360.

Clive, D.M., and Stoff, J.F., 1984, Renal syndromes associated with non-steroidal anti-inflammatory drugs, *N. Engl. J. Med.*, **310:**563-572.

Cribb, A.E., and Spielberg, S.P., 1992, Sulfamethoxazole is metabolized to the hydroxylamine in humans, *Clin. Pharmacol. Ther.*, **51:**522-526.

Dalvie, D.K., Khosla, N., and Vincent, J., 1997, Excretion and metabolism of trovofloxacin in humans, *Drug Metab. Dispos.* **25:**423-427.

Fischer, V., Haar, J.A., Greiner, L., Lloyd, R.V., and Mason, R.P., 1991, Possible role of free radical formation in clozapine (clozaril)-induced agranulocytosis, *Mol. Pharmacol.* **40:**846-853.

Forman, L.M., Simmons, D.A., and Diamond, R.H., 2000, Hepatic failure in a patient taking rosiglitazone, *Ann. Intern. Med.* **132:**118-121.

Fouda, H.G., Avery, M.J., Dalvie, D., Falkner, F.C., Melvin, L.S., and Ronfeld, R.A., 1997, Disposition and metabolism of tenidap in the rat, *Drug Metab. Dispos.* **25:**140-148.

Friedman, M.A., Woodcock, J., Lumpkin, M.M., Shuren, J.E., Hass, A.E., and Thompson, L.J., 1999, The safety of newly approved medicines: Do recent market removals mean there is a problem?, *J. Amer. Med. Assoc.* **281:**1728-1734.

Gardner, I., Zahid, N., MacCrimmon, D., and Uetrecht, J.P., 1998, Comparison of the oxidation of clozapine and olanzapine to active metabolites and the toxicity of these metabolites to human leukocytes, *Mol. Pharmacol.* **53:**991-998.

Gheyi, S.K., Robertson, A., and Atkinson, P.M., 1999, Severe interstitial cystitis caused by tiaprofenic acid, *J. Royal Soc. Med.* **92:**17.

Gometz, J.L., Dupont, A., Cusan, L., Tremblay, M., and Labrie, F., 1992, Simultaneous liver and lung toxicity related to the nonsteroidal antiandrogen nilutamide (Anandron): A case report, *Am. J. Med.* **92:**563-566).

Hargus, S.J., Amouzedeh, H.R., Pumford, N.R., Myers, T.J., McCoy, S.C., and Pohl, L.R., 1994, Metabolic activation and immunochemical localization of liver protein adducts of the nonsteroidal anti-inflammatory drug diclofenac, *Chem. Res. Toxicol.* **7:**575-582.

Hess, D.A., Sisson, M.E., Suria, H., Wijsman, J., Puvanesasingham, R., Madrenas, J., and Reider, M.J., 1999, Cytotoxicity of sulfonamide reactive metabolites: Apoptosis and selective toxicity of CD8+ cells by the hydroxylamine of sulfamethoxazole, *FASEB J.* **13:**1688-1698.

Holmberg, L., Roman, G., Bottiger, L.E., Eriksson, B., Spross, R., and Wessling, A., 1980, Adverse reactions to nitrofurantoin, *Am. J. Med.* **69:**733-738.

Jaunkalns, R., Shear, N.H., Sokoluk, B., Gardner, D., Claas, F., and Uetrecht, J.P., 1992, Antimyeloperoxidase antibodies and adverse reactions to clozapine, *Lancet* **339:**1611-1612.

Kassahun, K., Mattiuz, E., Nyhart, E., Obermeyer, B., Gillespie, T., Murphy, A., Goodwin, R.M., Tupper, T., Callaghan, J.T., and Lemberger, L., 1997, Disposition and biotransformation of the antipsychotic agent olanzapine in humans, *Drug Metab. Dispos.* **25:**81-93.

Kretz-Rommel, A., and Boelsterli, U.A., 1993, Diclofenac covalent protein binding is dependent on acyl glucuronide fomation and is inversely related to P450-mediated acute cell injury in cultured rat hepatocytes, *Toxicol. Appl. Pharmacol.* **120:**155-161.

Liu, X.C., and Uetrecht, J.P., 2000, Metabolism of ticlopidine by activated neutrophils: Implications for ticlopidine-induced agranulocytosis, *Drug Metab. Dispos.* **28:**726-730.

Maggs, J.L., Williams, D., Pirmohamed, M., and Park, B.K., 1995, The metabolic formation of reactive intermediates from clozapine, a drug associated with agranulocytosis in man, *J. Pharmacol. Exp. Ther.* **275:**1463-1475.

Naumann, R., Felber, W., Heilemann, H., and Reuster, T., 1999, Olanzapine-induced agranulocytosis, *Lancet* **354:**566-567.

Nelson, S.D., 1982, Metabolic activation and drug toxicity, *J. Med. Chem.* **25:**753-765.

Ono, K., Kurohara, K., Yoshihara, M., Shimamoto, Y., and Yamaguchi, M., 1991, Agranulocytosis caused by ticlopidine and its mechanism, *Am. J. Hematol.* **37:**239-242.

Peruzzi, L., Gianoglio, B., Porcellini, M.G., and Coppo, R., 1999, Neonatal end-stage renal failure associated with maternal ingestion of cyclo-oxygenase-type-2 selective inhibitor nimesulide as tocolytic, *Lancet* **354:**1615.

Pfitzenmeyer, P., Feucher, P., and Piard, F., 1992, Nilutamide pneumonitis: A report of eight patients, *Thorax* **47:**622-627.

Pirmohamed, M., Madden, S., and Bark, B.K., 1996, Idiosyncratic drug reactions: Metabolic bioactivation as a pathogenic mechanism, *Clin. Pharmacokin.* **31:**215-230.

Reinhart, H.H., Reinhart, E., Korlipara, P., and Peleman, R., 1992, Combined nitrofurantoin toxicity to liver and lung, *Gastroenterology,* **102:**1396-1399.

Rombach, E.M., and Hanzlik, R.P., 1999, Detection of adducts of bromobenzene 3,4-oxide with rat liver microsomal protein sulfhydryl groups using specific antibodies, *Chem. Res. Toxicol.* **12:**159-163.

Rubin, R.L., 1989, Autoimmune reactions induced by procainamide and hydralazine, in: *Autoimmunity and Toxicology* (M.E. Kammüller, N. Bloksma, and W. Seinen, eds.), Elsevier Science Publishers, New York, pp. 119-150.

Samuel, S.A., 1981, Apparent anaphylactic reaction to zomepirac (Zomax), *N. Engl. J. Med.* **304:**978.

Sarosdy, M.F., 1999, Which is the optimal antiandrogen for use in combined androgen blockade of advanced prostate cancer? The transition from a first- to second-generation antiandrogen, *Anti-cancer Drugs* **10:**791-796.

SCRIP, No. 2073, October 31, 1995, p. 21.

Shaw, G.R. and Anderson, R.W., 1991, Multisystem failure and hepatic microvesicular fatty metamorphosis associated with tolmetin ingestion, *Arch. Pathol. Lab. Med.* **115:**818-821.

Shen, S., Marchick, M.R., Davis, M.R., Doss, G.A., and Pohl, L.R., 1999, Metabolic activation of diclofenac by human cytochrome P450 3A4: Role of 5-hydroxydiclofenac, *Chem. Res. Toxicol.* **12:**214-222.

Silva, J.M., Khan, S., O'Brien, P.W., 1993, Molecular mechanisms of nitrofurantoin-induced hepatocyte toxicity in aerobic versus hypoxic conditions, *Arch. Biochem. Biophys.* **305:**362-369.

Spahn-Langguth, H., and Benet, L.Z., 1992, Acyl glucuronides revisited: Is the glucuronidation process a toxification as well as detoxification mechanism?, *Drug Metab. Rev.* **24:**5-48.

Tang, W., Stearns, R.A., Wang, R.W., Chiu, S.H.L., and Baillie, T.A., 1999, Roles of human hepatic cytochrome P450s 2C9 and 3A4 in the metabolic activation of diclofenac, *Chem. Res. Toxicol.* **12:**192-199.

The Pink Sheet, March 27, 2000, p.4.

Tietjen, D.P., 1989, Recurrence and specificity of nephrotic syndrome due to tolmetin, *Am. J. Med.* **87:**354-355.

Tsatalas, C., Chalkia, P., Garyfallos, A., Kakoulidis, I., and Xanthakis, I., 1995, Ticlopidine-induced aplastic anemia: Case report and review of the literature, *Clin. Drug Invest.* **9:**127-130.

Uetrecht, J., 1990, Drug metabolism by leukocytes and its role in drug-induced lupus and other idiosyncratic drug reactions, *Crit. Rev. Toxicol.,* **20:**213-235.

U.S. Congress, 1982, Committee on Government Operations, Intergovernment Relations and Human Resources: FDA's regulation of Zomax, Proceedings of the Ninety-eighth Congress, Washington, D.C.

Warren, S.E., and Mosley, C., 1983, Renal failure and tubular dysfunction due to zomepirac therapy, *J. Amer. Med. Assoc.* **249:**396-398.

Wysowski, D., and Fourcroy, J., 1996, Flutamide hepatotoxicity, *J. Urol.* **155:**209-212.

Yamazaki, H., Shibata, A., Suzuki, M., Nakajima, M., Shimada, N., Guengerich, F.P., and Yokoi, T., 1999, Oxidation of troglitazone to a quinone-type metabolite catalyzed by cytochrome P-450 2C8 and P-450 3A4 in human liver microsomes, *Drug Metab. Dispos.* **27:**1260-1266.

Zimmerman, H.J., Lewis, J.L., Ishak, K.G., and Maddrey, W.C., 1984, Ticrynafen-associated hepatic injury: Analysis of 340 cases, *Hepatology* **4:**315-323.

BIOLOGICAL REACTIVE INTERMEDIATES IN DRUG DISCOVERY AND DEVELOPMENT

A Perspective from the Pharmaceutical Industry

Thomas A. Baillie and Kelem Kassahun

Department of Drug Metabolism
Merck Research Laboratories, WP75A-303
West Point, PA 19486 USA

The detection of chemically-reactive, electrophilic metabolites poses a particular problem in the discovery and development of drug candidates in pharmaceutical research, inasmuch as it is not possible to accurately predict the likely toxicological consequences of these intermediates in animal safety studies or in clinical trials. Advances in analytical instrumentation (notably liquid chromatography-tandem mass spectrometry [LC-MS/MS]) have facilitated the detection of reactive intermediates through the identification of the glutathione (GSH) adducts to which they normally give rise, while the increased use of radiolabeled tracers in drug development permits an early assessment to be made of the propensity of a drug candidate to undergo covalent binding to cellular macromolecules. Unfortunately, these advances in analytical methods for the detection and characterization of reactive drug metabolites have far outstripped our understanding of the mechanisms of foreign compound-mediated toxicities at the molecular level, and of the role of both covalent binding and oxidative stress in the cascade of events that lead ultimately to cellular injury or immune-mediated toxicities. In light of these uncertainties, it seems reasonable to argue that one should attempt to minimize, through structural modification, the extent to which a drug candidate undergoes metabolism to reactive intermediates. Therefore, it becomes imperative to have close collaboration between Drug Metabolism scientists and their counterparts in Medicinal Chemistry during both the discovery and early development phases.

Research over the past three decades in the field of biological reactive intermediates has yielded a wealth of information on the functional groups which may be converted by either Phase I or Phase II enzymes to electrophilic metabolites, and some examples of these are given in Figure 1. Terminal olefins and acetylenes, typified by allylisopropylacetamide (AIA) and biphenylacetylene, respectively, are known to serve as mechanism-based inhibitors of cytochrome P-450 enzymes, as are certain dihydropyridine derivatives, such as DDC (Figure 1) (1). The hepatotoxic properties of both furosemide and tienilic acid are believed to be associated with metabolic activation of their furan and thiophene ring systems, respectively, while the simple *N*-formyl compound, *N*-methylformamide (NMF), undergoes oxidative biotransformation to methyl isocyanate, a very reactive carbamoylating species (2). Trifluoroacetylation of the *N*-terminal amino acid residues of certain proteins resulting from exposure to a reactive metabolite of halothane is thought to be the causative event in the immune-mediated hepatotoxicity of the inhalation anesthetic halothane (3). Thus, while the xenobiotics depicted in Figure 1 represent a group of diverse chemical structures, their toxic properties are believed, in each case, to be a consequence of oxidative biotransformation of the indicated functional groups. While it would be impractical for medicinal chemists to avoid the use of these functionalities in the design of new chemical entities, their presence nevertheless should be considered as a "structural alert" for the drug metabolism scientist, in that studies can be conducted at an early stage to determine whether the compound in question undergoes metabolism at that site to generate an electrophilic, potentially toxic, intermediate.

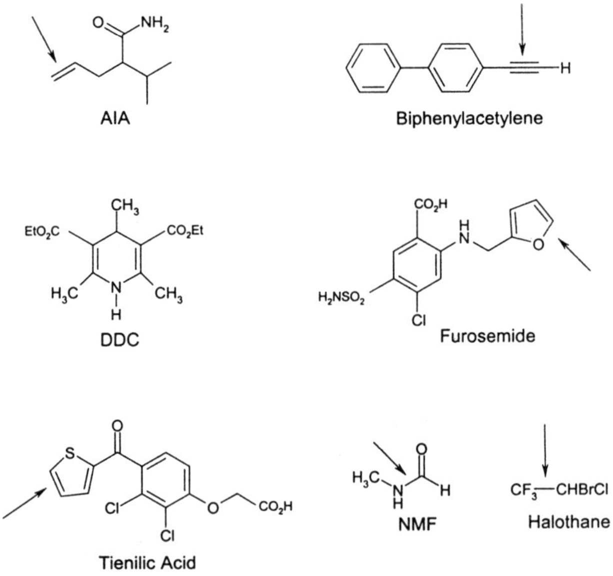

AIA

Biphenylacetylene

DDC

Furosemide

Tienilic Acid

NMF

Halothane

Figure 1. Prototypical substrates for metabolic activation to reactive electrophiles. The functional group involved in each case is indicated by an arrow.

Despite the rather extensive literature on the mechanisms by which drugs and other foreign compounds undergo metabolic activation, it is likely that numerous functional groups which have not hitherto been recognized as precursors to reactive intermediates also can undergo bioactivation. Support for this notion derives from recent studies carried out in our laboratories (4) on the metabolic fate of the orally-active antidiabetic agent troglitazone (Rezulin), a member of the thiazolidinedione (TZD) family of PPAR-γ agonist insulin sensitizers. Troglitazone was approved by the FDA in January, 1997, for the treatment of non-insulin-dependent diabetes mellitus, and proved to be a useful therapeutic agent for this indication. However, extended clinical use of troglitazone was associated with rare instances of serious hepatic injury and, as of March, 2000, 90 cases of liver failure had been reported to the FDA, 63 of which proved fatal (5). With the

46

advent of apparently safer TZD derivatives, troglitazone was withdrawn from the U.S. market in March, 2000. The mechanism of troglitazone-induced hepatotoxicity remains unknown, although it is possible that reactive metabolites of the drug may play a key role in the process. Evidence in support of the formation of a reactive metabolite(s) from troglitazone was obtained from *in vitro* studies on the effects of the drug on human liver microsomal testosterone 6β-hydroxylase activity, a marker for CYP3A4. As shown in Figure 2, in the presence of NADPH but without preincubation, troglitazone weakly inhibited the metabolism of testosterone in a pooled human liver microsomal preparation, whereas after a 30-min preincubation period in the presence of NADPH, troglitazone reduced markedly the subject activity, with an apparent IC_{50} of approximately 10 μM. These findings suggested that troglitazone undergoes CYP3A-mediated bioactivation in human liver.

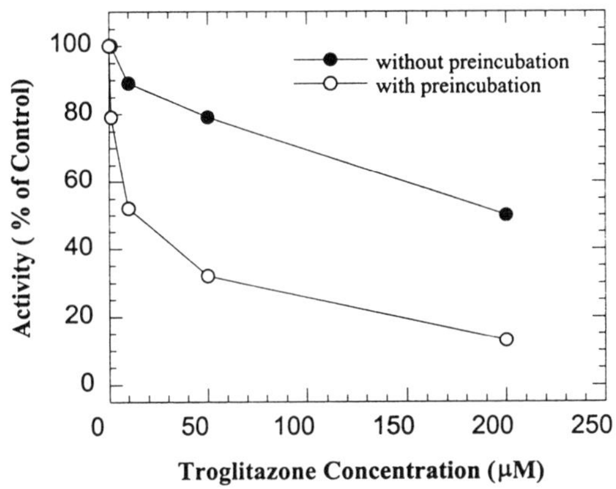

Preincubation-dependent inhibition of CYP3A activity (testosterone 6β-hydroxylase) in pooled human liver microsomal preparations incubated with troglitazone for 30 min at 37°C. [Testosterone] = 250 μM.

Figure 2. Effects of troglitazone on human liver microsomal CYP3A4 activity *in vitro*.

These findings were followed up by trapping experiments, in which troglitazone was incubated with human liver microsomal preparations fortified with GSH. Analysis of the products by LC-MS/MS revealed the formation of several GSH conjugates, which were detected on the basis of their characteristic fragmentation upon collisionally-activated decomposition (CID), namely the loss of the elements of pyroglutamic acid (129 Da) from the parent (MH^+) ion (6). The MS/MS spectrum of one such adduct (M1), which is reproduced in Figure 3, indicated that the TZD ring of troglitazone had undergone cleavage, with expulsion of the elements of CO and concomitant formation of a mixed disulfide with GSH. This interpretation was supported by ^{1}H NMR analysis of an isolated specimen and, subsequently, by comparison with an authentic sample prepared by synthesis.

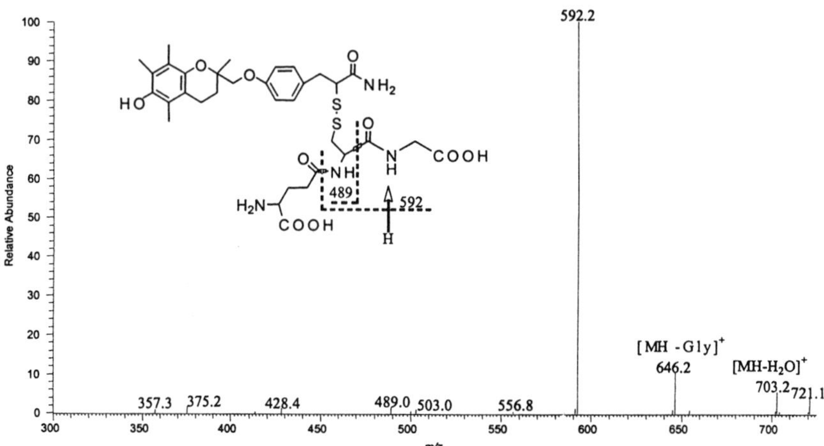

Product ion spectrum obtained by CID of the parent ion (MH$^+$) of GSH conjugate M1 at *m/z* 721. The origin of characteristic ions are as indicated.

Figure 3. MS/MS Spectrum of mixed disulfide troglitazone-GSH conjugate M1.

Characterization of two additional products (M2 and M3) of troglitazone metabolism in the above *in vitro* preparation led to their tentative identification as carbamate thioester derivatives, in which GSH appears to have added to a reactive isocyanate intermediate of TZD ring opening. The proposed structures of conjugates M2 and M3 are depicted in Figure 4, while the likely metabolic origins of these adducts are shown in Figure 5. Collectively, the results of this study suggest that troglitazone undergoes a previously-unrecognized CYP3A-catalyzed reaction leading to a an unstable TZD-*S*-oxide, which collapses spontaneously to a ring-opened product bearing two electrophilic functionalities, *viz.* a sulfenic acid and an α-ketoisocyanate. Capture of these two reactive moieties by GSH, attended in one case by hydrolysis and decarboxylation of the isocyanate, provides a logical route to the observed GSH conjugates M1 and M2, while M3 may represent a product of dehydration of M2. Although alternative pathways may be considered for the formation of these GSH conjugates of troglitazone, it is evident that the drug does undergo metabolic activation in human liver microsomes, and it may be surmised that the intermediates thus formed likely will covalently modify cellular proteins. The significance of these observations with respect to the hepatotoxicity of troglitazone remains to be established.

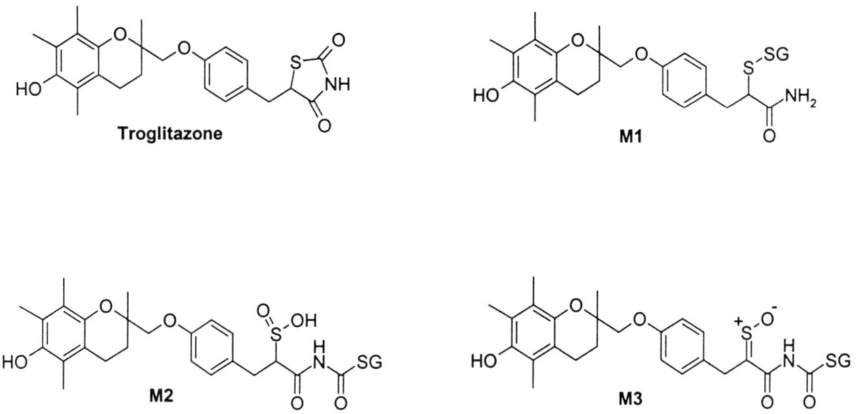

Figure 4. Structures of GSH conjugates of troglitazone (M1-M3) formed by TZD ring cleavage.

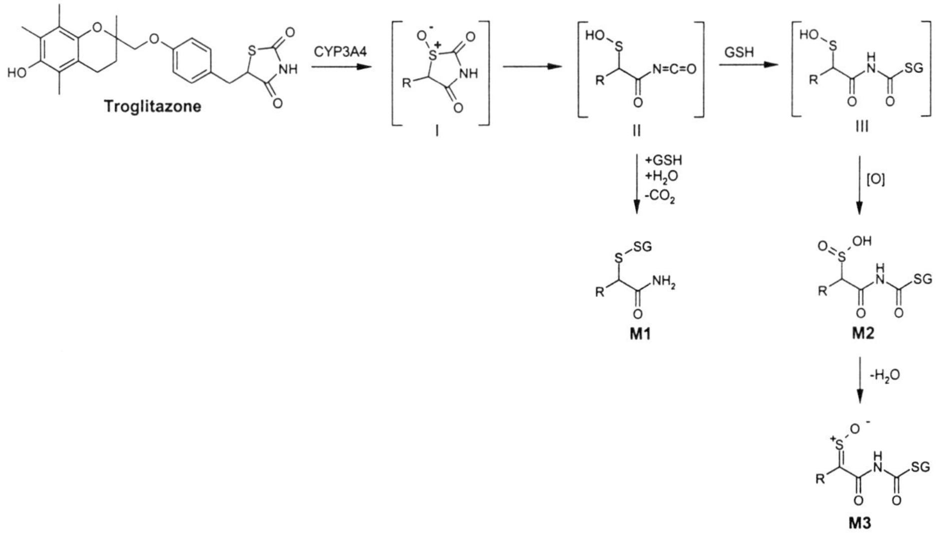

Figure 5. Proposed mechanism for bioactivation of troglitazone *via* TZD ring cleavage.

The example cited above illustrates the point that many novel functional groups present in drug candidates may be subject to metabolic activation, and that the products of these reactions may be trapped *in vitro* by nucleophiles such as GSH and detected relatively readily by LC-MS/MS techniques. The relevance of the *in vitro* findings to the situation *in vivo* then may be addressed by studies with intact animals, where GSH adducts may be detected in samples of bile. Subsequent experiments with a radiolabeled analog of the compound in question will reveal the propensity of the reactive intermediate(s) to covalently modify cellular proteins, *in vitro* and *in vivo*. The key issue in such investigations, however, relates to how best to exploit this type information in the context of drug discovery and development? Questions which are posed frequently, and to which, unfortunately, there are no clear-cut answers as yet, include the following: What level (if any) of covalent binding to proteins in liver (or other potential target organs for toxicity) might be acceptable in a drug candidate taken into development? Which protein targets are critical to the viability of the cell (7), and which of these are likely to be adducted by a specific reactive intermediate? Might there be a dose threshold for toxicity of a given reactive intermediate, and what are the factors that determine this value? How can we forecast the risk of idiosyncratic toxicity mediated by the human immune system when presented with potentially immunogenic drug-protein conjugates?

As stated above, it is evident that advances in analytical methods have greatly facilitated the indirect detection of biological reactive intermediates. However, although the formation of a reactive intermediate is an undesirable feature of any drug candidate, the phenomenon needs to be viewed as only one component of the overall risk assessment. With prototype drugs for novel pharmacological targets or therapeutic indications, a higher risk may be acceptable in order to establish "proof of concept." In the case of "back-up" compounds, on the other hand, structural diversity to minimize reactive metabolite formation should be a specific objective of the program. Certainly, there are examples where reactive metabolites are not toxic (8), and therefore the findings of appropriate safety assessment studies in animal models remain critically important in establishing the potential for many forms of toxicity. It should be recognized, however, that idiosyncratic reactions in humans mediated by the covalent binding of reactive drug metabolites to proteins cannot be predicted due to the lack of suitable animal models, and this limitation represents perhaps the most serious concern for all candidate drugs which are subject to metabolic activation (9). The overall goal, therefore, is to make reasonable judgments as to the risks in humans in the hope that we neither release a new drug that is toxic nor, as importantly, abort an effective agent for the treatment of human disease.

Acknowledgments

The authors would like to acknowledge the contributions of their colleagues at Merck Research Laboratories to the studies on troglitazone metabolism summarized in this article: Drs. Paul Pearson, Wei Tang, Kwan Leung, Charles Elmore, Dennis Dean and

George Doss, Ms. Regina Wang and Mr. Ian McIntosh. We also thank Dr. Kathleen Schultz with assistance in manuscript preparation.

References

1. Ortiz de Montellano, P. R. (1975). In "Cytochrome P-450. Structure, Mechanism and Biochemistry," 2nd edition., edited by P. R. Ortiz de Montellano, Plenum Press, New York, pp. 245-303.

2. Pearson, P. G., Threadgill, M. D., Howald, W. N., and Baillie, T. A. (1988). Applications of tandem mass spectrometry to the characterization of derivatized glutathione conjugates. Studies with S-(N-methylcarbamoyl)glutathione, a metabolite of the antineoplastic agent N-methylformamide. *Biomed. Environ. Mass Spectrom.,* **16**, 51-56.

3. Thomassen, D., Martin, B. M., Martin, J. L., Pumford, N. R., and Pohl, L. R. (1990). Characterization of a halothane induced trifluoroacetylated 100 kDa neoantigen that is related to a glucose-regulated protein. *FASEB J.,* **4**, A599.

4. Kassahun, K., Pearson, P.G., Tang, W., McIntosh, I., Leung, K., Elmore, C., Dean, D., Doss, G. and Baillie, T. A. (2000). Studies on the metabolism of troglitazone to reactive intermediates *in vitro* and *in vivo*. Evidence for novel biotransformation pathways involving quininomethide formation and thiazolidinedione ring scission – submitted.

5. Kohlroser, J., Mathai, J., Reichheld, J., Banner, B. F. and Bonkovsky, H. L. (2000). Hepatotoxicity due to troglitazone: report of two cases and review of adverse events reported to the United States Food and Drug Administration. *Am. J. Gastroenterol,* **95**, 272-276.

6. Baillie, T. A., and Davis, M.R. (1993). Mass spectrometry in the analysis of glutathione conjugates. *Biol. Mass Spectrom.,* **22**, 319-325.

7. Pumford, N. R., and Halmes, N. C. (2000). Protein targets of xenobiotic reactive intermediates. *Annu. Rev. Pharmacol. Toxicol.,* **37**, 91-117.

8. Nelson, S. D., and Pearson, P. G. (1990). Covalent and noncovalent interactions in acute lethal cell injury caused by chemicals. *Annu. Rev. Pharmacol. Toxicol.,* **30**, 169-195.

9. Uetrecht, J. P. (1999). New concepts in immunology relevant to idiosyncratic drug reactions: the "danger hypothesis" and innate immune system. *Chem. Res. Toxicol.,* **12**, 387-395.

BIOACTIVATION OF TOXICANTS BY CYTOCHROME P450-MEDIATED DEHYDROGENATION MECHANISMS

Garold S. Yost

Department of Pharmacology and Toxicology
30 South 2000 East, Room 201
University of Utah
Salt Lake City, UT 84112-5820

INTRODUCTION

Normal cytochrome P450-mediated metabolism of toxicants proceeds through an electron or hydrogen atom abstraction mechanism that generally produces hydroxyl radical equivalents bound to prosthetic heme iron and eventually leads to oxygenated products through a hydroxyl rebound mechanism (Ortiz de Montellano, 1995). In a growing number of examples, certain P450 enzymes initiate oxidation of toxicants through the first step of one-electron abstraction (or hydrogen atom abstraction), but subsequently catalyze a second-electron oxidation that leads to dehydrogenated (desaturated) products. Many of these products are highly reactive electrophiles that initiate toxicities through binding to proteins and/or DNA (Yost, 1997; Guengerich and Kim, 1991; Lewis *et al.*, 1996; Han *et al.*, 1990). The precise chemical environments of the active sites of the enzymes that direct selective dehydrogenation, rather than hydroxylation, are not known. Several of the enzymes that catalyze dehydrogenation of toxicants are selectively expressed in respiratory tissues (Pelkonen and Raunio, 1997; Mace *et al.*, 1998), and much of our work (Thornton-Manning *et al.*, 1996; Lanza *et al.*, 1999) has addressed the mechanisms of dehydrogenation by several human lung-expressed enzymes such as CYP2F1 (Nhamburo *et al.*, 1990) and CYP4B1 (Nhamburo *et al.*, 1989).

The precise mechanisms that control the production of dehydrogenated intermediates by selective P450 enzymes, and the mechanisms that are responsible for cell death after alkylation of critical protein and/or DNA targets have not been adequately elucidated. A working hypothesis of our studies is that the spatial and electronic parameters of these enzymes must direct two-electron oxidation rather than oxygenation through facilitated electron transport from ferryl heme and/or proton relay mechanisms. Presented below are several examples of dehydrogenation reactions that are catalyzed by P450 enzymes, including recent work in our laboratory that has evaluated the formation of a dehydrogenated

Biological Reactive Intermediates VI, Edited by Dansette *et al.*
Kluwer Academic / Plenum Publishers, 2001

product of 3-methylindole (Skiles and Yost, 1996; Thornton-Manning *et al.*, 1996; Lanza *et al.*, 1999), and the formation of adducts of this product with thiols (Skordos *et al.*, 1998a) and with deoxynucleosides (Regal *et al.*, 1999).

DEHYDROGENATION OF ALKANES

Oxidation of substrates that are simple hydrocarbons would appear to be the most difficult type of dehydrogenation reactions because of the relatively strong "unactivated" C-H bond that must be broken, and because intermediate radicals or radical cations are not resonance stabilized. Resonance stabilization of intermediates in the dehydrogenation of typical aromatic substrates such as acetaminophen, butylated hydroxytoluene, and 3-methylindole would be extensive. Thus, it is surprising that hydrocarbons are substrates for this process.

However, the formation of alkenes from alkyl groups of substrates such as lauric acid can be catalyzed by P450 enzymes. The CYP4B1 enzyme from rabbits possess a unique propensity to dehydrogenate lauric acid (Guan *et al*, 1998) and valproic acid (Rettie *et al.*, 1995) with only a modest preference for formation of the hydroxylated metabolites (Figure 1). Dehydrogenation of valproic acid occurs through selective hydrogen abstraction at the omega-1 methylene position, as demonstrated by elegant intramolecular deuterium isotope effects (Rettie *et al.*, 1995), which is presumably followed by a second hydrogen abstraction at the methyl carbon.

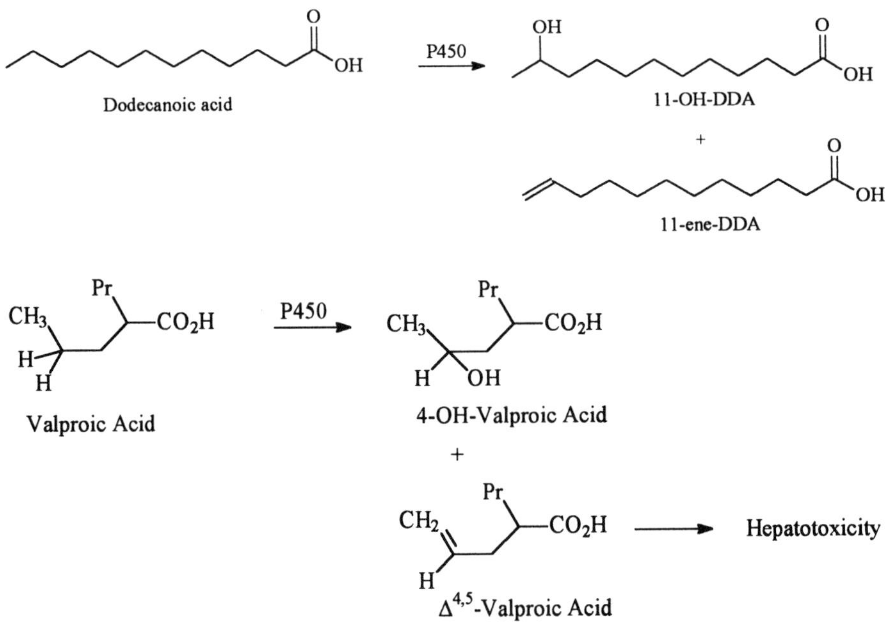

Figure 1. Dehydrogenation of a fatty acid and valproic acid that illustrate dehydrogenation of hydrocarbons by cytochrome P450 enzymes. Dehydrogenation of valproic acid leads to production of the putative toxic alkene intermediate. (Pr = propyl)

DEHYDROGENATION OF TOXICANTS

There are a number of examples of substrates that are bioactivated to reactive, toxic, electrophilic intermediates by dehydrogenation: acetaminophen, valproic acid (Figure 1) (Rettie *et al.*, 1995), 4-hydroxytamoxifen (Figure 2) (Fan *et al.*, 2000), urethane (Figure 3) (Guengerich and Kim, 1991; Forkert and Lee, 1997), butylated hydroxytoluene (Figure 3) (Bolton and Thompson, 1991; Lewis *et al.*, 1996), *N*-methylformamide (Han *et al.*, 1990), and 3-methylindole (Figure 5) (Yost, 1997; Skiles and Yost, 1996). The reactive intermediates that are produced are generally conjugated alkenes, but some are formed from phenolic substrates and are called "quinone methides," such as that formed by oxidation of 4-hydroxytamoxifen. These intermediates bind avidly to nucleophilic sites such as cysteine thiols of proteins or the primary amines of nucleotides like the N^2-position of deoxyguanosine.

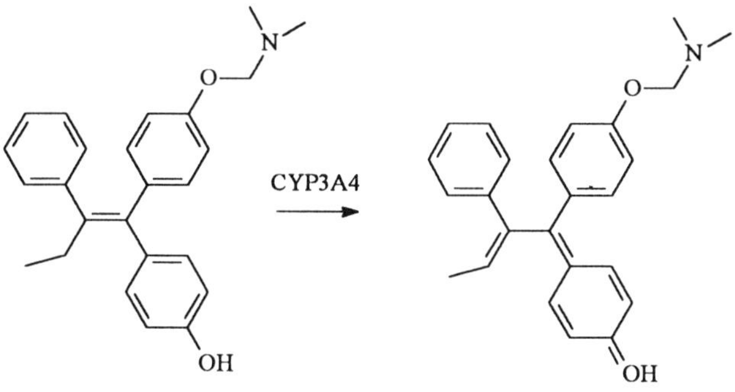

Figure 2. Dehydrogenation of 4-hydroxytamoxifen to its quinone methide is efficiently catalyzed by the cytochrome P450 3A4 enzyme.

The bioactivation of toxicants in lung tissues has been shown to be mediated in several cases by dehydrogenation of these chemicals (Figure 3). These bioactivation events are generally catalyzed by P450 enzymes that are often selectively expressed in lung tissues (Pelkonen and Raunio, 1997; Mace *et al.*, 1998).

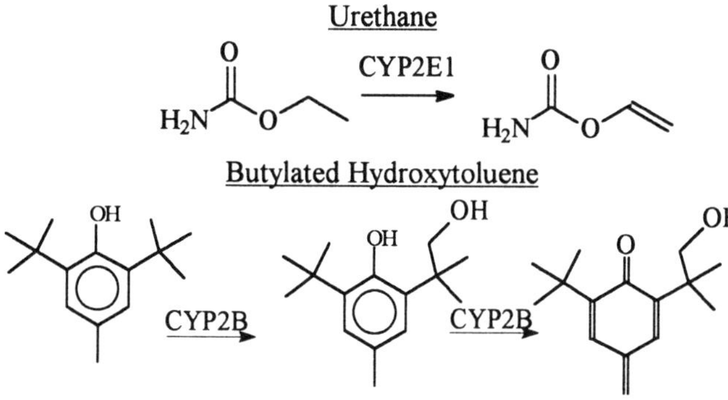

Figure 3. Dehydrogenation of lung toxicants by the specific cytochrome P450 enzymes that have been implicated in the bioactivation of each toxicant. The ethylene ether group of bioactivated urethane is oxygenated by CYP2E1 to the carcinogenic epoxide.

A lung carcinogen that is bioactivated by dehydrogenation is the simple carbamate ethyl ether, urethane, by P450 2E1 to the alkene (Guengerich and Kim, 1991) that is oxidized to the epoxide, and this metabolite appears to be the ultimate lung carcinogen. Bioactivation of butylated hydroxytoluene by a lung CYP2B enzyme, through successive hydroxylation and dehydrogenation steps, produces a quinone methide that is responsible for lung damage (Bolton and Thompson, 1991) and may be involved in the tumor promoting activity of this chemical. 3-Methylindole (discussed below) is a prototypical example of a pneumotoxicant that is bioactivated by dehydrogenation.

An interesting example of dehydrogenation of the heterocyclic xanthine, furafylline, by cytochrome P450 1A2 has been described (Racha *et al.*, 1998). This elegant work showed that the dehydrogenated product, an imidazomethide (Figure 4), efficiently inactivated the enzyme with a partition ratio of hydroxylation to inactivation of approximately 5:1. High intramolecular isotope effects for carbinol formation ($k_H/k_D > 9$) have led to the conclusion that hydrogen atom abstraction (Higgins *et al.*, 1998) from the substrate was probably the initial oxidative step in the formation of a common methyl radical intermediate which then lost a second hydrogen atom to form the dehydrogenated electrophile. Hydrolysis of the dehydrogenated intermediate, rather than hydroxyl rebound from the methyl radical, produced the bulk (70-80%) of the carbinol. There are significant correlations between this work and the studies with 3-methylindole regarding the mechanisms of dehydrogenation by P450 enzymes.

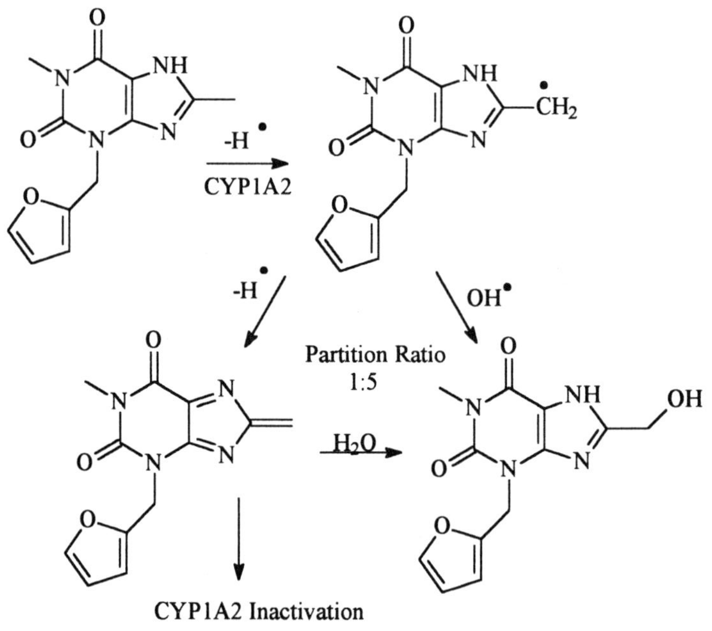

Figure 4. Dehydrogenation of furafylline by cytochrome P450 1A2 is mediated by two one-electron oxidations. Formation of the dehydrogenated product leads to inhibition of 1A2 through mechanism-based inactivation. The partition ratio between hydroxylation and dehydrogenation is 5:1.

DEHYDROGENATION OF 3-METHYLINDOLE

Significant work in our laboratory has focused on the mechanisms of dehydrogenation and toxicity of 3-methylindole (3MI), a pneumotoxicant that selectively destroys lung bronchiolar epithelial and olfactory epithelial cells (Yost, 1997). Members of the 2F subfamily of P450 enzymes are selectively expressed in lung epithelial cells and they bioactivate 3MI to the dehydrogenated intermediate, 3-methyleneindolenine, without formation of hydroxylated products (Wang *et al.*, 1998; Lanza *et al.*, 1999). Studies with goat lung microsomes (Wang *et al.*, 1998) have indicated that the goat CYP4B2 enzyme also efficiently catalyzes the production of the dehydrogenated intermediate. The selective expression of several P450 enzymes in human lung tissues (Pelkonen and Raunio, 1997; Mace *et al.*, 1998; Anttila *et al.*, 1997), and efficient bioactivation of pneumotoxicants and lung carcinogens by these enzymes, lead to the intriguing possibility that humans are susceptible to injury by these chemicals. Figure 5 depicts the alternate pathways of dehydrogenation and oxygenation of 3MI (Skiles and Yost, 1996; Skordos *et al.*, 1998a, b) to produce electrophilic intermediates that can be trapped with nucleophilic thiols, or as seen below, with nucleosides.

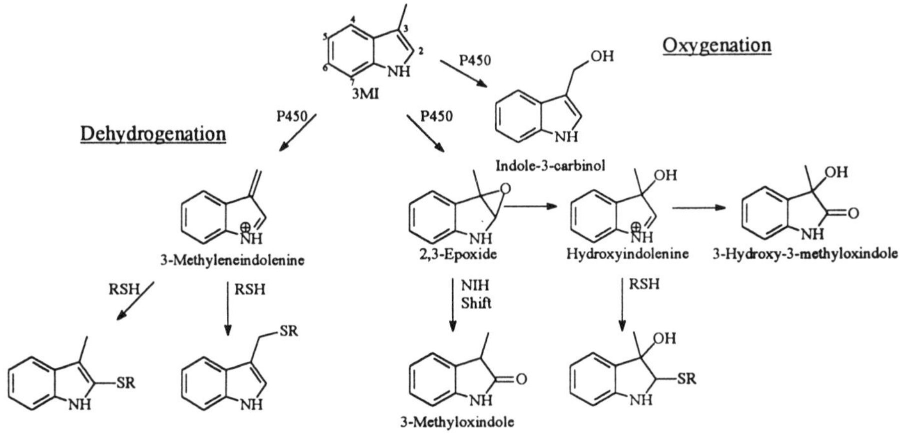

Figure 5. Competition between dehydrogenation and oxygenation of 3-methylindole illustrates the production of reactive electrophiles that can bind nucleophilic thiols (RSH), or detoxification products like indole-3-carbinol or 3-methyloxindole.

The efficiency of the formation of the dehydrogenated product by members of the 2F subfamily leads to the conclusion that this substrate is chemically primed for this type of transformation, rather than the normal oxygenation process that is generally encountered for most substrates. In order to evaluate the hypothesis that the dehydrogenation of 3MI was a "substrate-dependent" and not an "enzyme-dependent" transformation, we evaluated the ability of other common P450 enzymes to catalyze the dehydrogenation and/or oxygenation of this substrate.

The human recombinant CYP1A1, CYP1A2, CYP1B1, and CYP2E1 enzymes that were expressed in the lymphoblast microsomal matrix from Gentest Corp. (Woburn, MA) were incubated with 3MI under normal conditions (Lanza, *et al.*, 1999) for linear product formation at various concentrations of substrate. The products were identified by HPLC and rates of product formation were analyzed to determine apparent K_M and V_{max} values. The relative efficiencies of each enzyme for formation of the dehydrogenation product, 3-methyleneindolenine (trapped as its N-acetylcysteine thioether adduct), indole-3-carbinol, and 3-methyloxindole are reported in Table 1.

The most striking finding from this study was that CYP2E1 produced exclusively the oxindole product without any detectable formation of either of the products derived from oxidation of the methyl position of 3MI, i.e. 3-methyleneindolenine or indole-3-carbinol. This result was exactly the converse of the results obtained by incubation of 3MI with either the CYP2F1 or CYP2F3 enzymes, where only the dehydrogenated product was produced. The other enzymes were reasonably efficient at producing both dehydrogenated and oxygenated products. Interestingly, the CYP1B1 enzyme was very adept at making the carbinol but no dehydrogenated product could be detected. Thus, these studies proved that the dehydrogenation of 3MI was not "substrate-dependent." Rather, specific cytochrome P450 enzymes catalyze the formation of either the dehydrogenated product, or the oxygenated products, or all of these products. Thus, the active sites of the enzymes must have electronic or steric environments that direct these divergent pathways towards alternate routes. One of the more fascinating aspects of this work is the finding that the enzymes that are expressed predominately in the lung, i.e. 2F and 4B P450 enzymes (Pelkonen and Raunio, 1997), exclusively form what appears to be the most pneumotoxic product, the dehydrogenated intermediate of 3MI.

Table 1. Dehydrogenation vs. oxygenation of 3-methyindole is highly enzyme dependent.

P450 Enzymes	3-Methyleneindolenine[a]	Indole-3-carbinol[a]	3-Methyloxindole[a]
CYP2F1	72	n.d.	n.d.
CYP2F3	1.6	n.d.	n.d.
CYP1A1	0.7	49	5
CYP1A2	24	111	91
CYP1B1	n.d.	171	5
CYP2E1	n.d.	n.d.	100

n.d. = none detected

[a]Efficiency of product formation expressed as V_{max}/K_M (nmol product/nmol P450 x min/μM)

Nucleophilic thiols are excellent trapping agents for 3-methyleneindolenine, but a possibility exists that the bases of DNA could also be alkylated by this reactive intermediate and produce mutagenic and carcinogenic events. In order to evaluate the possibility that this intermediate would form adducts with DNA, we incubated indole-3-carbinol with each of the four deoxynucleosides under acidic conditions where the carbinol loses water to spontaneously form the methylene imine. These studies demonstrated (Regal *et al.*, 1999) that adducts of 2'-deoxyguanosine (dG), 2'-deoxyadenosine (dA), and 2'-deoxycytosine were formed. Adducts of thymidine could not be detected. Analysis by LC/MS (Figure 6) and NMR showed that the adducts formed a covalent bond between the exocyclic amine of each deoxynucleoside and the exocyclic methylene of dehydrogenated 3MI.

When either 3MI or deuterium labeled 3MI [3-(^{2}H$_3$)-methylindole] were incubated with goat lung microsomes that were supplemented with dA, adducts were identified from the separate incubations. These adducts differed by two mass units for the fragments (Figure 6) that were derived from 3MI that had lost one atom of deuterium during the bioactivation of the substrate to its dehydrogenated product. Thus, these studies demonstrated that 3-methyleneindolenine, formed by

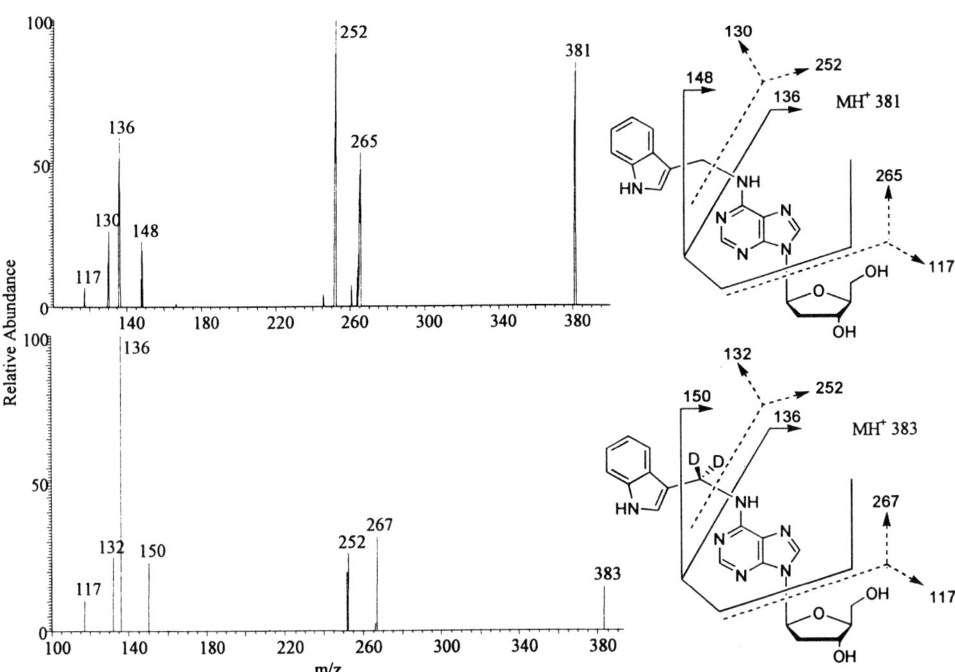

Figure 6. Mass spectra of deoxyadenosine adducts, formed by incubation with 3MI (upper panel) or 3-(^{2}H$_3$)-methylindole (lower panel).

P450-mediated dehydrogenation of 3MI, reacted efficiently with deoxynucleosides to form stable adducts. Studies are ongoing to evaluate the hypothesis that production of nucleoside adducts may indicate that 3MI is mutagenic and/or carcinogenic.

CONCLUSIONS

The studies presented in this manuscript have attempted to provide an overview of the process of dehydrogenation to produce reactive intermediates. However, the mechanisms employed by these P450 enzymes in the production of dehydrogenated metabolites (reactive intermediates) have not been established. The vast majority of the substrates that have been studied to date are toxicants that are bioactivated to electrophilic reactive intermediates. Thus, the inclusion of discussions of this bioactivation pathway in this issue of Biological Reactive Intermediates VI is appropriate. As the list of interesting and relevant substrates of the dehydrogenation pathway grows, substantial mechanistic knowledge is continuously gained about this fascinating process. However, precise descriptions of the catalytic environment of the enzymes that will permit predictions of likely substrates are not available yet. When the mechanisms responsible for the production of dehydrogenated reactive intermediates by P450 enzymes have been established, scientists should be able to envisage the scope of appropriate substrates and possible toxic intermediates.

ACKNOWLEDGMENTS

The scientific contributions of Gary Skiles (Merck and Co., West Point, PA) and Kelly Regal (University of Michigan, Ann Arbor, MI) to the deoxynucleoside adduct studies; and Diane Lanza, Kon Skordos, Janice Thornton-Manning, and Huifen Wang (University of Utah) to the dehydrogenation of 3MI by specific P450 enzymes are gratefully acknowledged. The author acknowledges the generous gift of several human cytochrome P450 enzymes, expressed in lymphoblast cell microsomes, by Dr. Charles Crespi (Gentest Corp., Woburn, MA). This research was supported by National Institutes of Health grants HL13645 and HL60143 from the National Heart, Lung, and Blood Institute.

REFERENCES

Anttila, S., Hukkanen, J., Hakkola, J., Stjernvall, T., Beaune, P., Edwards, R. J., Boobis, A. R., Pelkonen, O., and Raunio, H., 1997, Expression and localization of CYP3A4 and CYP3A5 in human lung. *Am. J. Respir. Cell Mol. Biol.* **16**:242-249.

Bolton, J. L., and Thompson, J. A., 1991, Oxidation of butylated hydroxytoluene to toxic metabolites. Factors influencing hydroxylation and quinone methide formation by hepatic and pulmonary microsomes, *Drug Metabol. Dispos.* **19**:467-472.

Fan, P. W., Zhang, F., and Bolton, J. L., 2000, 4-Hydroxylated metabolites of the antiestrogens tamoxifen and toremifene are metabolized to unusually stable quinone methides, *Chem. Res. Toxicol.* **13**:45-52.

Forkert, P.-G., and Lee, R. P., 1997, Metabolism of ethyl carbamate by pulmonary cytochrome P450 and carboxylesterase isozymes: Involvement of CYP2E1 and hydrolase A[1], *Toxicol. Appl. Pharmacol.* **146**:245-254.

Guan, Y., Fisher, M. B., Lang, D. H., Zheng, Y.-M., Koop, D. R., and Rettie, A. E., 1998, Cytochrome P450-dependent desaturation of lauric acid: isoform selectivity and mechanism of formation of 11-dodecenoic acid, *Chem.-Biol. Inter.* **110**:103-121.

Guengerich, F. P., and Kim, D.-H., 1991, Enzymatic oxidation of ethyl carbamate to vinyl carbamate and its role as an intermediate in the formation of 1,N^6-ethenoadenosine, *Chem. Res. Toxicol.* **4**:413-421.

Han, D. H., Pearson, P. G., Baillie, T. A., Dayal, R., Tsang, L. H., and Gescher, A., 1990, Chemical synthesis and cytotoxic properties of *N*-alkylcarbamic acid thioesters, metabolites of hepatotoxic formamides, *Chem. Res. Toxicol.* **3**:118-124.

Higgins, L., Bennett, G.A., Shimoji, M., and Jones, J. P., 1998, Evaluation of cytochrome P450 mechanism and kinetics using kinetic deuterium isotope effects, *Biochem.* **37**:7039-7046.

Lanza, D. L., Code, E., Crespi, C. L., Gonzalez, F. J., and Yost, G. S., 1999, Specific dehydrogenation of 3-methylindole and epoxidation of naphthalene by CYP2F1 expressed in human lymphoblastoid cells. *Drug Metab. Dispos.* **27**:798-803.

Lewis, M. A., Yoerg, D. G., Bolton, J. L., and Thompson, J. A., 1996, Alkylation of 2'-deoxynucleosides and DNA by quinone methides derived from 2,6-di-*tert*-butyl-4-methylphenol, *Chem. Res. Toxicol.* **9**:1368-1374.

Mace, K., Bowman, E. D., Vautravers, P., Shields, P. G., Harris, C. C., and Pfeifer, A. M. A., 1998, Characterisation of xenobiotic-metabolising enzyme expression in human bronchial mucosa and peripheral lung tissues, *Eur. J. Cancer*, **34**:914-920.

Nhamburo, P. T., Gonzalez, F. J., McBride, O. W., Gelboin, H. V., and Kimura, S., 1989, Identification of a new P-450 expressed in human lung: Complete cDNA sequence, cDNA-directed expression and chromosome mapping, *Biochemistry*, **28**: 8060-8066.

Nhamburo P. T., Kimura S., McBride O. W., Kozak C. A., Gelboin H., and Gonzalez F. J., 1990, The human CYP2F gene subfamily: identification of a cDNA encoding a new cytochrome P450, cDNA-directed expression, and chromosome mapping, *Biochemistry* **29**:5491-5499.

Ortiz de Montellano, P. R., 1995, Oxygen activation and reactivity in *Cytochrome P450: Structure, Mechanism and Biochemistry* (2nd Edition, P. R. Ortiz de Montellano, ed.) Plenum Press, New York, 245-303.

Pelkonen, O. and Raunio, H., 1997, Metabolic activation of toxins: tissue-specific expression and metabolism in target organs, *Environ. Health Perspec.* **105**:767-774.

Racha, J. K., Rettie, A. E., and Kunze, K. L., 1998, Mechanism-based inactivation of human cytochrome P450 1A2 by furafylline: detection of a 1:1 adduct to protein and evidence for the formation of a novel imidazomethide intermediate, *Biochem.* **37**:7407-7419.

Regal, K. A., Laws, G., Yost, G. S., Yuan, C., and Skiles, G. L., 1999, Formation and detection of 3-methylindole DNA adducts, *ISSX Proceedings*, 9th North American ISSX Meeting, **15**:223.

Rettie, A. E., Sheffels, P. R., Korzekwa, K. R., Gonzalez, F. J., Philpot, R. M., and Baillie, T. A., 1995, CYP4 isozyme specificity and the relationship between ω-hydroxylation and terminal desaturation of valproic acid, *Biochem.* **34**:7889-7895.

Skiles, G. L., and Yost, G. S., 1996, Mechanistic studies on the cytochrome P450-catalyzed dehydrogenation of 3-methylindole, *Chem. Res. Toxicol.* **9**:291-297.

Skordos, K. W., Laycock, J. D., and Yost, G. S., 1998a, Thioether adducts of a new imine reactive intermediate of the pneumotoxin 3-methylindole, *Chem. Res. Toxicol.* **11**:1326-1331.

Skordos, K. W., Skiles, G. L., Laycock, J. D., Lanza, D. L., and Yost, G. S., 1998b, Evidence supporting the formation of 2, 3-epoxy-3-methylindoline: a reactive intermediate of the pneumotoxin 3-methylindole, *Chem. Res. Toxicol* **11**:741-749.

Thornton-Manning, J. R., Appleton, M. L., Gonzalez, F. J. and Yost, G. S., 1996, Metabolism of 3-methylindole by vaccinia-expressed P450 enzymes: correlation of 3-methyleneindolenine formation and protein-binding, *J. Pharmacol. Exp. Ther.* **276:**21-29.

Wang, H., Lanza, D. L., and Yost, G. S., 1998, Cloning and expression of CYP2F3, a cytochrome P450 that bioactivates the selective pneumotoxins 3-methylindole and naphthalene, *Arch. Biochem. Biophys.* **349:**329-340.

Yost, G. S., 1997, Selected, nontherapeutic agents in *Comprehensive Toxicology,* (Sipes, I. G., McQueen, C. A., and Gandolfi, A. J., eds), Vol. 8, *Toxicology of the Respiratory System* (Roth, R. A., ed.) Elsevier Science, New York, 591-610.

NEW ASPECTS OF DNA ADDUCT FORMATION BY THE CARCINOGENS CROTONALDEHYDE AND ACETALDEHYDE

Stephen S. Hecht, Edward J. McIntee, Guang Cheng, Yongli Shi, Peter W. Villalta, and Mingyao Wang

University of Minnesota Cancer Center
Minneapolis, MN 55455

INTRODUCTION

Crotonaldehyde and acetaldehyde are mutagenic in *S. typhimurium* and other systems used for detection of genetic damage (IARC, 1985; 1995; 1999). Crotonaldehyde induces altered liver cell foci, neoplastic nodules, and hepatocellular carcinoma upon oral administration to F344 rats (Chung et al, 1986; IARC, 1995). Acetaldehyde causes tumors of the respiratory tract in rats and hamsters upon exposure by inhalation (IARC, 1985; 1999). Crotonaldehyde and acetaldehyde are commonly detected in cigarette smoke, mobile source emissions and other products of thermal degradation (IARC, 1985; 1995; 1999). Their concentrations in cigarette smoke are far higher, while their carcinogenic activities are considerably lower, than those of polycyclic hydrocarbons, aromatic amines and N-nitrosamines, which are considered to be important carcinogens in cigarette smoke (IARC, 1985; 1986; 1995; 1999). Crotonaldehyde and acetaldehyde are also important endogenous compounds. Crotonaldehyde is a product of lipid peroxidation while acetaldehyde is the principal metabolite of ethanol.

Crotonaldehyde forms adducts with DNA and deoxyguanosine (dG)(Chung et al, 1999; Eder et al, 1999). The exocyclic 1, N^2-propano-dG adducts **2** and **3** (Figure 1) are the major

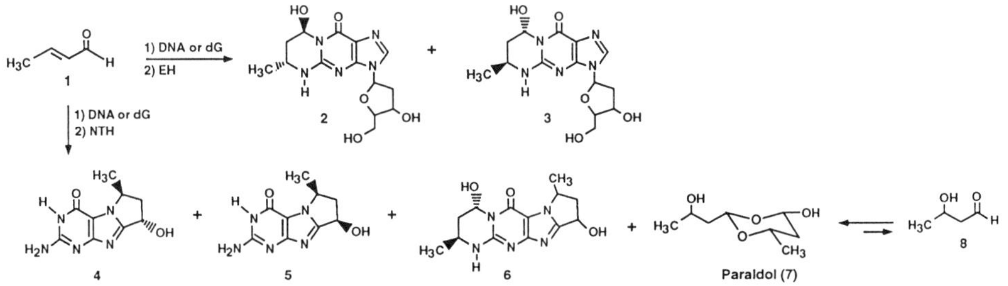

Figure 1. Summary of products formed in the reaction of crotonaldehyde (**1**) with DNA or dG, according to previous studies. DNA reactions were followed by enzymatic hydrolysis (EH) or neutral thermal hydrolysis (NTH). Other isomers of **2**, **3**, and **6** have been reported (Chung et al, 1999; Eder et al, 1999).

products characterized to date in reactions of crotonaldehyde with DNA. Adducts **4-6** have also been detected in reactions with dG. Using ^{32}P-postlabelling, Chung and Nath demonstrated that adducts **2** and **3** are present in DNA of various tissues from untreated laboratory rodents as well as humans (Chung et al, 1999). Acetaldehyde reacts with the exocyclic amino group of dG and DNA to form an unstable Schiff base, N^2-ethylidene-dG, which can be stabilized by reduction, producing N^2-ethyl-dG (Vaca et al, 1995; Fang and Vaca, 1995). This adduct has been detected in the DNA of peripheral white blood cells of alcohol abusers and in human buccal cells exposed to acetaldehyde (Fang and Vaca, 1997; Vaca et al, 1998).

Recently, we demonstrated that enzymatic, acid, or neutral thermal hydrolysis of DNA that had been reacted with crotonaldehyde produced 2-(2-hydroxypropyl)-4-hydroxy-6-methyl-1,3-dioxane (paraldol, 7, Figure 1), the dimer of 3-hydroxybutanal (8) (Wang et al, 1998). Levels of adducts releasing 7 were considerably higher than those of adducts 2-6, suggesting that they could play a role in the DNA damaging properties of crotonaldehyde. Paraldol-releasing DNA adducts of crotonaldehyde could arise, in part, by reaction of paraldol itself with DNA. Since paraldol can also be formed from acetaldehyde via aldol condensation to 3-hydroxybutanal, we investigated acetaldehyde DNA adduct formation to determine if any similar products were formed.

RESULTS AND DISCUSSION

Crotonaldehyde was allowed to react with DNA. The DNA was extensively purified to remove unreacted crotonaldehyde. It was then hydrolyzed enzymatically and analyzed by HPLC (Figure 2). The numbered peaks were not present in control incubations lacking crotonaldehyde, and released paraldol upon acid hydrolysis. In this study, we characterized peaks 7-9.

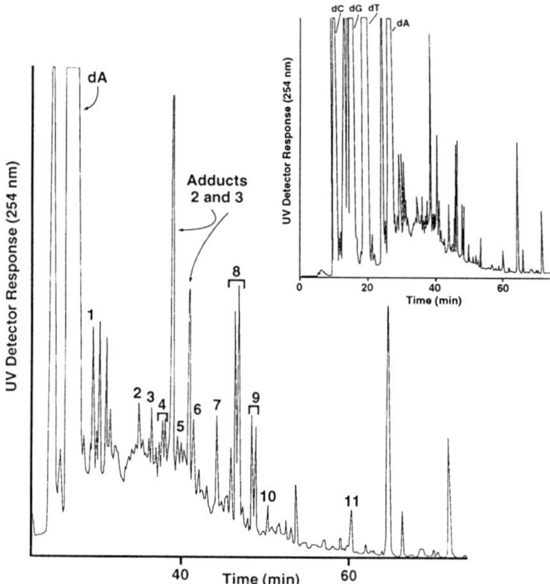

Figure 2. Chromatogram obtained upon HPLC analysis of an enzymatic hydrolysate of DNA that had been reacted with crotonaldehyde. Peaks 1-11 were not present in hydrolysates of untreated DNA. Each of these peaks released paraldol upon hydrolysis. Adducts 2 and 3 are shown in Figure 1. Inset: entire chromatogram.

Peaks 8 and 9 each consisted of at least two isomers. They were characterized as diastereomers of N^2-[2-(2-hydroxypropyl)-6-methyl-1,3-dioxane-4-yl]deoxyguanosine (N^2-paraldol-dG) (12, Figure 3) by their ^{1}H-NMR, mass, and UV spectra, and by reaction of paraldol with dG. NOESY spectra indicated that the peak 8 isomers have all equatorial substituents in the dioxane ring while peak 9 isomers have an axial methyl group. N^2-paraldol-dG was also produced upon reaction of paraldol with DNA, and of crotonaldehyde with dG. There are 16 possible stereoisomers of N^2-paraldol-dG. Peaks 8 and 9 comprise at least four, and possibly eight, of these isomers. Preliminary LC-MS data indicate that small amounts of eight additional isomers elute earlier than peaks 8 and 9.

Peak 7 consisted of at least 4 isomers, 2 major and 2 minor. They were characterized as diastereomers of N^2-[2-(2-hydroxypropyl)-6-methyl-1,3-dioxane-4-yl]deoxyguanylyl-(5'-3')-thymidine (N^2-paraldol-dG-(5'-3')-thymidine, **13**, Figure 3) by their ^{1}H-NMR, mass, and UV spectra, and by reaction of paraldol with DNA. Further evidence was obtained by hydrolysis with snake venom phosphodiesterase which produced N^2-paraldol-dG-5'-monophosphate and thymidine, while hydrolysis with spleen phosphodiesterase gave N^2-paraldol-dG and thymidine-3'-monophosphate.

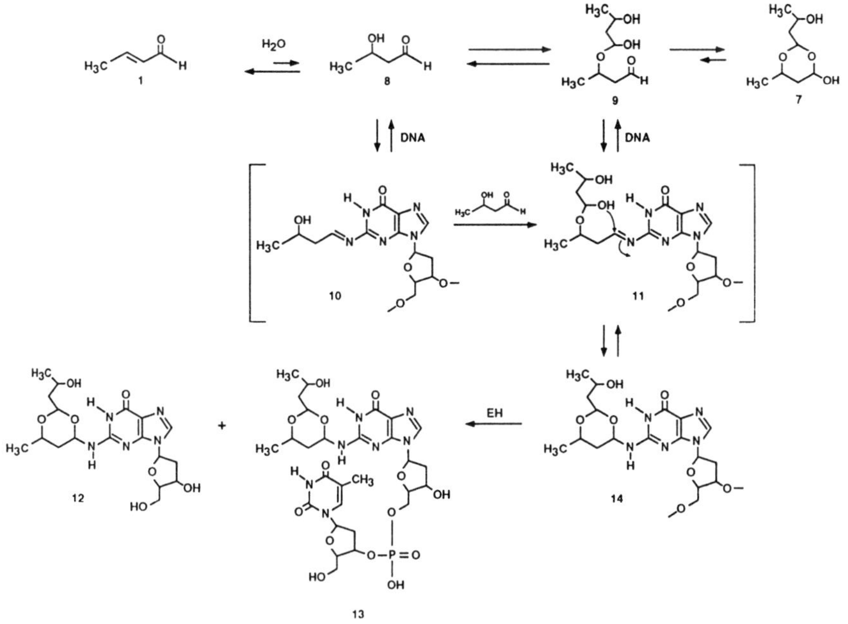

Figure 3. Proposed mechanisms of formation of crotonaldehyde-derived DNA adducts **12** and **13**.

As some of the other peaks in Figure 2 might have resulted from reactions of paraldol with other DNA constituents, we examined the reaction of paraldol with various nucleosides and nucleotides. Products were found only in the reaction of paraldol with dG, dG-3'-mono-phosphate, and dG-5'-monophosphate. Reactions with deoxyadenosine, deoxycytidine, thymidine, thymidine-3'-monophosphate, and thymidine-5'-monophosphate did not produce detectable products.

Levels of the N^2-paraldol-dG adducts **12** and **13** in DNA that had been reacted with crotonaldehdye were similar to those of Michael addition adducts **2** and **3**. However, total paraldol-releasing adducts were 13-76 times greater than those of adducts **2** and **3**. Adducts **12** and **13** constitute less than 10% of total paraldol-releasing adducts. Enzymatic hydrolysis of DNA that has been reacted with crotonaldehyde causes extensive release of paraldol itself, and it is our hypothesis, supported by unpublished data, that a Schiff base such as **10** or **11** (Figure 3) is the source of most of the released paraldol. The results of these studies, together with those of Chung and co-workers (1999) on the presence of adducts **2** and **3** in human DNA, suggest that the actual levels of crotonaldehyde-derived DNA adducts in human DNA are substantially higher than previously recognized.

Possible mechanisms of formation of adducts **12** and **13** are outlined in Figure 3. Aqueous solutions of crotonaldehyde develop 3-hydroxybutanal (**8**) which can react with DNA to form Schiff base **10**. Reaction of **10** with another molecule of 3-hydroxybutanal would give **11** which can cyclize to **14**. The latter is converted to **12** and **13** by enzyme hydrolysis. Alternatively,

3-hydroxybutanal (**8**) dimerizes to paraldol (**7**) via intermediate **9**. This can react with DNA to form Schiff base **11**, which then cyclizes to **14**. Reaction of paraldol with DNA to produce adducts **12** and **13** supports this mechanism.

The relative roles of adducts such as **2-6** versus N^2-paraldol-dG adducts **12** and **13** in crotonaldehyde mutagenesis and tumorigenesis is of course unclear at this point. Site specific mutagenesis studies with $1,N^2$-propano-dG, an unsubstituted analogue of adducts **2** and **3**, have been reported (Moriya et al, 1999). This adduct is highly mutagenic in *E. coli*, directing the incorporation of dAMP, resulting in dG→dT transversions. Frameshift mutations have also been observed. However, in simian kidney (COS7) cells, $1,N^2$-propano-dG was far less mutagenic, with favored incorporation of dC. In a shuttle vector system, crotonaldehyde induced a variety of mutations including dG→dT transversions, dG→dA transitions, and tandem base substitutions (Kawanishi et al, 1998). Our results clearly demonstrate that, in contrast to adducts **2** and **3**, N^2-paraldol-dG adducts are resistant to enzymatic hydrolysis, indicating that they may cause significant steric perturbations in DNA. This suggests that they may also be resistant to DNA repair. A comparison of the persistence of adducts **2** and **3** versus N^2-paraldol-dG in cells or rodents treated with crotonaldehyde would be of interest.

We began our studies of acetaldehyde-DNA reactions by investigating the reaction of acetaldehyde with dG. HPLC analysis of this reaction mixture produced the chromatogram illustrated in Figure 4. The major peak eluting at 31 min has a half life of less than 5 min at 37° C. It was identified as N^2-ethylidene-dG (**16**, Figure 5) by its UV, by treatment with 2,4-dinitrophenylhydrazine reagent which gave acetaldehyde-2,4-dinitrophenylhydrazone, and by treatment with NaBH₃CN, which gave N^2-ethyl-dG (**18**, Figure 5). The latter was identified by comparison of its UV spectrum, MS, and HPLC retention time to those of a standard. Acetaldehyde was then allowed to react with calf thymus DNA. When the DNA was treated with NaBH₃CN prior to enzyme hydrolysis, the major adduct peak was identified as N^2-ethyl-dG by comparison of its UV spectrum and retention time to those of a synthetic standard. When the NaBH₃CN step was omitted prior to enzymatic hydrolysis, the chromatogram illustrated in Figure 6 was obtained. Owing to its instability, N^2-ethylidene-dG was not detected in these hydrolysates. However, we did observe peaks 1-6, as in the reactions with dG illustrated in Figure 4.

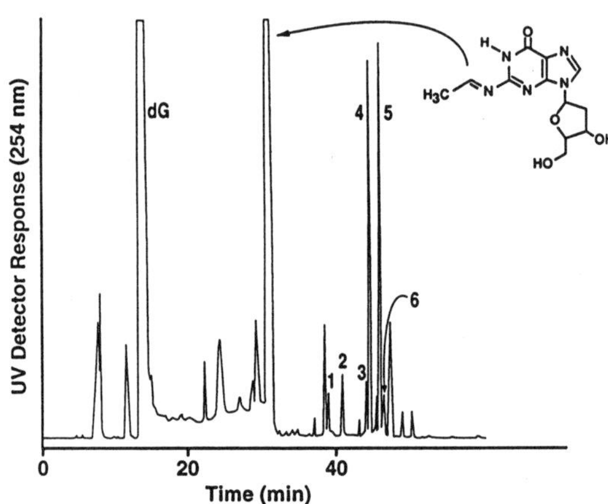

Figure 4. Chromatogram obtained upon HPLC analysis of the products of reaction of acetaldehyde with dG. The identities of peaks 1-6 are: 1 and 2, **2/3** (Figure 5); 3, **22** (Figure 5); 4-6, at least 3 diastereomers of **21** (Figure 5).

Peaks 1 and 2 were identified as diastereomers of 3-(2-deoxyribos-1-yl)-5,6,7,8-tetrahydro-8-hydroxy-6-methylpyrimido[1,2-*a*]purine-10(3*H*)one (**2** and **3**, Figure 1 and   Figure 5) by their spectral properties and by comparison to standards prepared by reaction of crotonaldehyde with dG.   These products are common to both the acetaldehyde- and crotonaldehyde-DNA reactions, but the diastereomeric ratio of the products is different.   The ratio of peak 2 to peak 1 was 5 to 1 in the acetaldehyde-DNA reaction (Figure 6), in contrast to the 2 to 1 ratio of the two peaks obtained upon reaction of crotonaldehyde with DNA (Figure 2).   These 1,N^2-propano-dG adducts were also produced in the reaction of acetaldehyde with dG, and the ratio of peak 2 to peak 1 was 1.7 to 1 (Figure 4).

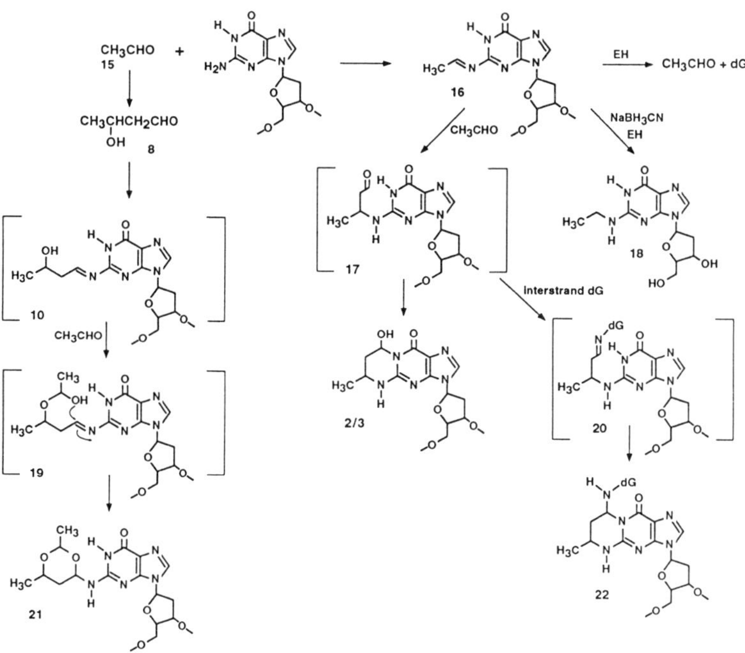

Figure 5.   Product formation in the reaction of acetaldehyde with DNA.   With the exception of **18**, adducts are shown as the structures in DNA in order to conserve space.   In the text, they are identified at the nucleoside level.   EH: enzyme hydrolysis

Peak 3 of Figure 6 was identified as the cross-linked adduct **22** (Figure 5), based on its UV, MS, and ^{1}H-NMR spectra.   Further confirmation of the structure of adduct **22** was obtained by hydrolysis with acid, which produces the corresponding guanine derivative, identified by MS and UV.   This cross-linked adduct was observed in reactions of acetaldehyde with double stranded DNA, but not with single stranded DNA or poly-G.   Therefore, it is an interstrand cross-link.

Peaks 4-6 of Figure 6 were identified as diastereomers of N^2-(2,6-dimethyl-1,3-dioxan-4-yl)dG (**21**, Figure 5) by their UV, mass, and ^{1}H-NMR spectra, and by conversion to a mixture of the 2,4-dinitrophenylhydrazones of acetaldehyde, crotonaldehyde, and 3-hydroxybutanal upon treatment with 2,4-dinitrophenylhydrazine reagent.

In this study we have characterized three new types of stable DNA adducts of acetaldehyde (Figure 5) - the 1,N^2-propano-dG adduct **2/3**, the N^2-dimethyldioxane-dG adduct **21**, and the cross-link adduct **22**.   While these adducts are formed in substantially lower yield than the major adduct of acetaldehyde, N^2-ethylidene-dG (**16**), they are stable at the nucleoside level, in contrast to **16**, and may be more stable in DNA.

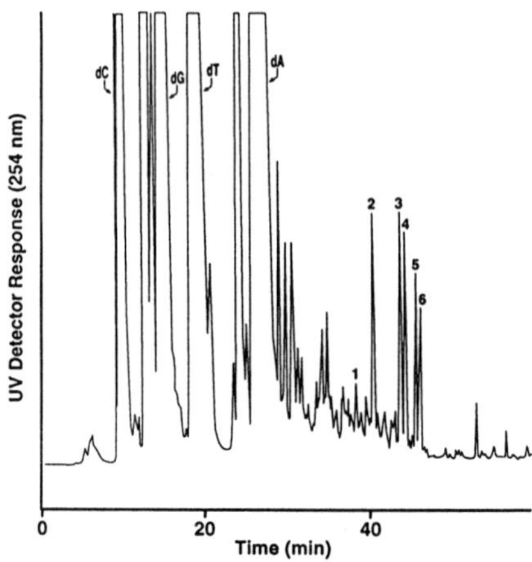

Figure 6. Chromatogram obtained upon HPLC analysis of an enzymatic hydrolysate of DNA that had been allowed to react with acetaldehyde. The identities of peaks 1-6 are: 1 and 2, **2/3** (Figure 5); 3, **22** (Figure 5); 4-6, at least 3 diastereomers of **21** (Figure 5).

The 1,N^2-propano-dG adducts **2** and **3** are also formed in reactions of crotonaldehyde with dG and DNA, as described above. Crotonaldehyde can be produced in aqueous solutions of acetaldehyde by aldol condensation to 3-hydroxybutanal (**8**) followed by dehydration (Baigrie et al, 1985). The crotonaldehyde generated in this way could then react with dG or DNA to give adducts **2** and **3**. However, our data indicate that this is not the mechanism by which adducts **2** and **3** are formed from acetaldehyde. In the reaction of acetaldehyde with DNA, the ratio of peak 2 to peak 1 (Figure 6), which are the two diastereomers- adducts **2** and **3**- is 5 to 1, whereas the corresponding ratio in the reaction of crotonaldehyde with DNA is 1-2 to 1 (Figure 2 and Chung et al, 1999). Therefore, we propose the mechanism illustrated in Figure 5 for the formation of adducts **2** and **3**. The major DNA adduct of acetaldehyde, N^2-ethylidene-dG (**16**), reacts with a second molecule of acetaldehyde to give intermediate **17**. The transition state for this reaction is diastereomeric, and the approach of the second molecule of acetaldehyde from one direction is clearly favored, producing one diastereomer of **17** in excess over the other. Vaca et al (1995) isolated two diastereomers of N^2-(4-hydroxybut-2-yl)dG from reactions of dG with acetaldehyde, after reduction with NaBH$_4$, and proposed a similar mechanism of formation. Although their conditions were different from those employed here, their products may have been produced from **2** and **3**, as NaBH$_4$ reduction of these adducts is known to give N^2-(4-hydroxybut-2-yl)dG (Chung and Hecht, 1983).

Intermediate **17** of Figure 5 is also the most likely precursor to the cross-linked adduct **22**. Reaction of **17** with dG in the opposite strand would produce Schiff base **20**, which then reacts with N-1 to give **22**. Our results demonstrate that predominantly one diastereomer of **22** is formed in the acetaldehyde-DNA reaction. On standing for 30 days at 4°C, this diastereomer is gradually converted back to the major diastereomer of **2/3** which we have observed in the reactions of acetaldehyde with DNA and dG. Therefore, it is highly probable that **22** and **2/3** are formed via the same intermediate (**17**) as illustrated in Figure 5. Although we have not detected the cross-link adduct **22** in reactions of crotonaldehyde with DNA, it is likely that it is present.

Acetaldehyde is known to produce interstrand cross-links in DNA (IARC, 1985; 1999). It also causes cross-links in human lymphocytes *in vitro* without metabolic activation. DNA-protein cross-links have been observed as well. Adduct **22** represents the first structural characterization of a DNA cross-link resulting from reaction with acetaldehyde. The availability of this structure should facilitate future studies on mechanisms of mutagenesis by acetaldehyde involving cross-linking. Matsuda et al, 1998 have identified specific tandem GG to TT mutations induced by acetaldehyde and have attributed these to intrastrand crosslinking between adjacent guanine bases, although the structure of the crosslink was not determined. We did not

68

observe adduct **22** in single stranded DNA or polyG, but we did not extensively investigate the possibility of intrastrand formation of **22**, and it is possible that this would occur under other conditions.

The aldol condensation product 3-hydroxybutanal (**8**) is clearly the initial source of the N^2-dimethyldioxane-dG adducts **21**. This can react with the exocyclic amino group of dG producing Schiff base **10**. Reaction of **10** with another molecule of acetaldehyde would produce intermediate **19**, which then cyclizes giving **21**. We did not investigate the presence of intermediate **10** in this study. Vaca et al (1995) did obtain evidence for **10** in the reaction of acetaldehyde with dG by carrying out $NaBH_4$ reduction yielding N^2-(3-hydroxybut-1-yl)dG. However, it should be noted that $NaBH_4$ reduction of **21** would also be expected to produce this product. An alternate pathway to **21** would involve initial reaction of 3-hydroxybutanal with acetaldehyde producing 2,4-dimethyl-1,3-dioxane-6-ol. This could then react with DNA, analogous to the reaction of paraldol with DNA. We did not detect N^2-paraldol-dG adducts in the acetaldehyde-DNA reactions, probably because the concentration of paraldol is low in the presence of excess acetaldehyde.

In summary, the results of this study provide new information on DNA adduct formation by crotonaldehyde and acetaldehyde. These carcinogens are ubiquitous in the human environment and exposures can be exceptionally high, for example, through cigarette smoking or alcohol consumption. The paraldol-releasing crotonaldehyde DNA adducts are formed in quantities considerably higher than the Michael addition adducts that have been previously detected in human DNA, suggesting that crotonaldehyde-DNA modifications may be considerably higher than previously suspected. Although the new acetaldehyde adducts described here are formed in quantities less than 10% of that of N^2-ethylidene-dG, and require more than one molecule of acetaldehyde per molecule of dG, levels of acetaldehyde generated *in vivo* could be sufficient for this to occur. As an example, Tuma et al (1996) have detected protein adducts resulting from two molecules of acetaldehyde and one molecule of malondialdehyde in the liver of rats treated with ethanol.

ACKNOWLEDGMENTS This study was supported by Grant No. CA-85702 from the National Cancer Institute. Stephen S. Hecht is an American Cancer Society Research Professor, supported by ACS Grant RP-00-138.

REFERENCES

Baigrie, L. M., Cox, R. A., Slebocka-Tilk, H., Tencer, M., and Tidwell, T. T., 1985, Acid-catalyzed enolization and aldol condensation of acetaldehyde. *J. Chem. Soc.* **107:** 3640-3645.

Chung, F.-L. and Hecht, S. S., 1983, Formation of cyclic 1,N^2-adducts by reaction of deoxy-guanosine with α-acetoxy-*N*-nitrosopyrrolidine, 4-(carbethoxynitrosamino)butanal, or crotonaldehyde. *Cancer Res.* **43:** 1230-1235.

Chung, F.-L., Tanaka, T., and Hecht, S. S., 1986, Induction of liver tumors in F344 rats by crotonaldehyde. *Cancer Res.* **46:** 1285-1289.

Chung, F.-L., Zhang, L., Ocando, J. E., and Nath, R. G., 1999, Role of 1,N^2-propanodeoxy-guanosine adducts as endogenous DNA lesions in rodents and humans. in: *Exocyclic DNA Adducts in Mutagenesis and Carcinogenesis* (Singer, B. and Bartsch, H., Eds.), IARC Scientific Publ. 150, International Agency for Research on Cancer, Lyon, France, pp. 45-54.

Eder, E., Schuler, D., and Budiawan, 1999, Cancer risk assessment for crotonaldehyde and 2-hexenal: an approach. in: *Exocyclic DNA Adducts in Mutagenesis and Carcinogenesis* (Singer, B. and Bartsch, H., Eds.), IARC Scientific Publ. 150, International Agency for Research on Cancer, Lyon, France, pp. 219-232.

Fang, J.-L. and Vaca, C. E., 1995, Development of a ^{32}P-postlabelling method for the analysis of adducts arising through the reaction of acetaldehyde with 2'-deoxyguanosine-3'- mono-phosphate and DNA. *Carcinogenesis* **16:** 2177-2185

Fang, J.-L. and Vaca, C. E.,1997, Detection of DNA adducts of acetaldehyde in peripheral white blood cells of alcohol abusers. *Carcinogenesis* **18:** 627-632.

International Agency for Research on Cancer, 1985, Allyl compounds, aldehydes, epoxides and peroxides. in: *Monographs on the Evolution of the Carcinogenic Risk of Chemicals to Humans*, Vol. 36, IARC, Lyon, France, pp. 101-132.

International Agency for Research on Cancer, 1986, Tobacco smoking. in: *Monographs on the Evaluation of the Carcinogenic Risk of Chemicals to Humans*, Vol. 38, IARC, Lyon, France, pp. 312-314.

International Agency for Research on Cancer, 1995, Dry cleaning, some chlorinated solvents and other industrial chemicals. in: *Monographs on the Evaluation of Carcinogenic Risks to Humans,* Vol. 63, IARC, Lyon, France, pp. 373-391.

International Agency for Research on Cancer, 1999, Re-evaluation of some organic chemicals, hydrazine and hydrogen peroxide (part two). in: *Monographs on the Evaluation of the Carcinogenic Risk of Chemicals to Humans,* Vol. 71, IARC, Lyon, France, pp. 319-335.

Kawanishi, M., Matsuda, T., Sasaki, G., Yagi, T., Matsui, S., and Takebe, H., 1998, A spectrum of mutations induced by crotonaldehyde in shuttle vector plasmids propagated in human cells. *Carcinogenesis* **19:** 69-72.

Matsuda, T., Kawanishi, M., Yagi, T., Matsui, S., and Takebe, H., 1998, Specific tandem GG to TT base substitutions induced by acetaldehyde are due to intra-strand crosslinks between adjacent guanine bases. *Nucleic Acids Res.* **26:** 1769-1774.

Moriya, M., Pandya, G. A., Johnson, F., and Grollman, A. P., 1999, Cellular response to exocyclic DNA adducts, in: *Exocyclic DNA Adducts in Mutagenesis and Carcinogenesis* (Singer, B. and Bartsch, H., Eds.), IARC Scientific Publ. 150, International Agency for Research on Cancer, Lyon, France, pp. 263-270.

Tuma, D. J., Thiele, G. M., Xu, D., Klassen, L. W., and Sorrell, M. F, 1996, Acetaldehyde and malondialdehyde react together to generate distinct protein adducts in the liver during long-term ethanol administration. *Hepatol.* **23**: 872-880.

Vaca, C. E., Fang, J.-L., and Schweda, E. K. H., 1995, Studies of the reaction of acetaldehyde with deoxynucleosides. *Chem.-Biol. Interact.* **98:** 51-67.

Vaca, C. E., Nilsson, J. A., Fang, J. L., and Grafstrom, R. C., 1998, Formation of DNA adducts in human buccal epithetial cells exposed to acetaldehyde and methylglyoxal *in vitro*. *Chem-Biol Interact.* **108:** 197-208.

Wang, M., Upadhyaya, P., Dinh, T. T., Bonilla, L. E., and Hecht, S. S., 1998, Lactols in hydrolysates of DNA reacted with α-acetoxy-*N*-nitrosopyrrolidine and crotonaldehyde. *Chem. Res. Toxicol.* **11:** 1567-1573.

MECHANISMS OF OVOTOXICITY INDUCED BY ENVIRONMENTAL CHEMICALS: 4-VINYLCYCLOHEXENE DIEPOXIDE AS A MODEL CHEMICAL

Patricia B. Hoyer,[1] Ellen A. Cannady,[2] Nicole A. Kroeger,[1] I. Glenn Sipes[2]

[1]Department of Physiology
[2]Department of Pharmacology and Toxicology
Southwest Environmental Health Sciences Center
The University of Arizona
Tucson AZ 85724 USA

INTRODUCTION

Females are born with a finite number of undeveloped, primordial follicles. Environmental chemicals that destroy oocytes contained in these follicles can produce premature ovarian failure (early menopause in women) because once destroyed, they cannot be replaced. Menopause is known to be associated with an increased incidence of a variety of health problems such as osteoporosis, cardiovascular disease, and ovarian cancer. Exposure of women to chemicals that are ovotoxic in laboratory animals (chemotherapeutic agents, contaminants of cigarette smoke) is known to be associated with early menopause. Therefore, an overall understanding of mechanisms involved in chemical-induced ovotoxicity is of general health-related concern (Hoyer and Sipes, 1996).

One chemical that destroys oocytes contained in small ovarian follicles, is the industrial chemical, 4-vinylcyclohexene (VCH, Smith et al., 1990a). VCH and its diepoxide metabolite, VCD, are produced during the manufacture of rubber tires, flame retardants, insecticides, plasticizers, and antioxidants. In 1986 the National Toxicology Program (NTP, 1986) reported that following two years of dosing by oral gavage with VCH, female mice, but not rats, developed benign mixed ovarian tumors, and rare ovarian neoplasms. Tumorigenesis was preceded by a decrease in oocyte-containing follicles in mice after 13 weeks exposure. Earlier studies had determined that the diepoxide of VCH, 4-vinylcyclohexene diepoxide (VCD), is mutagenic in bacteria (Simmon and Baden, 1980; Voogd et al, 1981). Shorter term dosing studies (30d) determined that VCH directly targets the mouse ovary and destroys oocytes contained in small (primordial, primary, and secondary) pre-antral follicles (Smith et al., 1990). This oocyte loss was extensive enough that premature ovarian failure resulted within a year (Hooser et al., 1994).

Biological Reactive Intermediates VI, Edited by Dansette *et al.*
Kluwer Academic / Plenum Publishers, 2001

SPECIES SPECIFICITY

Direct evidence of ovotoxicity produced by VCH and its epoxide metabolites came from estimates of follicle loss by histological counting of oocytes (Figure 1). VCH caused a significant reduction in the number of oocytes contained in small pre-antral follicles in immature female B6C3F$_1$ mice, but not in Fischer-344 rats, following 30 days of daily dosing via the intraperitoneal route. Because VCH can be metabolized to two possible monoepoxides, 1,2-vinylcyclohexene monoepoxide (1,2-VCME) and 7,8-vinylcyclohexene monoepoxide (7,8-VCME), and ultimately to VCD, the ability of these epoxides to cause follicular loss was also determined in rats and mice. The monoepoxides were more potent than VCH and less potent than VCD, in terms of follicle loss. VCH only produced ovotoxicity in mice; however, VCME and VCD were effective in both mice and rats, with mice displaying a greater sensitivity. The results suggested that mice have a greater capacity to form the epoxides. In a detailed structure-activity study, strong evidence was provided that VCD is the ultimate ovarian toxicant (Doerr et al., 1995). Analogues of VCH that can only form monoepoxides (vinylcyclohexane, ethylcyclohexene, and cyclohexene) were not found to cause follicular loss when administered to mice for 30 days. However, VCH analogues that are or can form diepoxides (butadiene monoepoxide and diepoxide, and isoprene) produced a significant destruction of oocyte-containing small follicles in mice. It therefore, appears that metabolism to VCD is required for the destruction of follicles that is observed with VCH in mice.

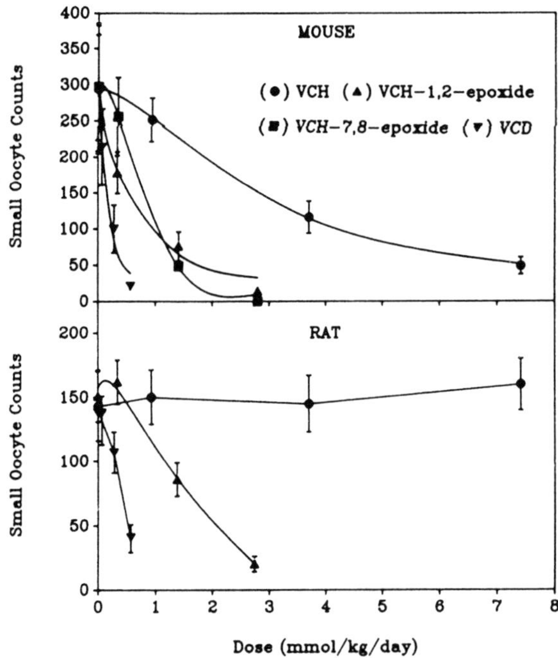

Figure 1. Comparison of the dose-response relationship in the reduction in small oocyte counts in the ovaries of rats and mice treated i.p. with VCH, or VCH epoxides for 30 days. The standard deviations are indicated by bars; n=4-10 animals/group (Reprinted with permission, *Toxicology and Applied Pharmacology*; Smith et al., 1990a)

OVOTOXICITY IN RATS

Based on the evidence that VCD is ovotoxic in rats, most mechanistic studies have focused on VCD-induced ovotoxicity in Fischer-344 female rats. A substantial amount of follicle loss was initially observed following dosing of rats with VCD for 30 days. To identify specific events involved in the induction of this loss, it was important to identify a time course for the onset of ovotoxicity (Kao et al., 1999). Figure 2 shows a summary of the effect of VCD-dosing on primordial follicles in rats following various days of daily dosing, between 2 and 30d. Ovarian sections were evaluated for follicle loss, as well as atretic (unhealthy) appearance, and the combined height of the bar on each day provides an estimate, relative to control animals, of the total follicle damage that has been caused by VCD at that time. The first appearance of VCD-induced atretic follicles was observed following the final dose on day 10, but a significant reduction in follicle numbers did not occur until day 12. These results demonstrate that follicular damage and eventual follicle loss occur as a fairly gradual process, with a similar percentage displaying damage (atretic in appearance) after each subsequent dose (Figure 2). Using this approach, examination of the earliest onset of ovotoxicity has determined that VCD-induced follicle loss results from direct targeting of the smallest pre-antral follicles (primordial and primary), and requires repeated daily dosing (10d; Springer et al., 1996a; Borman et al., 1999).

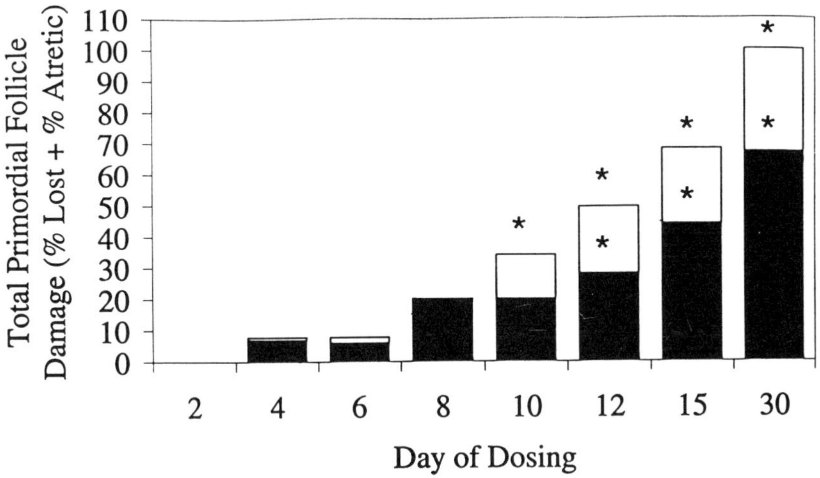

Figure 2. Summary of total ovotoxicity following daily dosing of rats with VCD. Female F344 rats were dosed daily (80mg/kg; i.p.) for the indicated days. Four hours following the final dose, ovaries were collected and histologically evaluated. Data are expressed as percent loss (number of control − number of VCD/number of control X 100; solid portion) and percent atretic (percent VCD − percent control; open portion) relative to control. Total bar height reflects total follicle damage (ovotoxicity) at each time point in response to daily dosing. n=5-10 animals per group, * above each portion, p<0.05, significant effect of VCD.

Having determined the time course of the onset of follicular loss, it was of interest to identify cellular and molecular mechanism(s) involved. To obtain mechanistic information related to VCD-induced ovotoxicity, it was important to develop an approach for making *in vitro* ovarian measurements. The dynamic heterogeneity of ovarian structures has been well documented (Hirschfield, 1991). Primordial and primary follicles are the smallest structures, and are relatively inactive   (Hirshfield and Schmidt, 1987). The physiological targets of VCH and VCD are oocytes contained in the smallest and most immature stages of development (primordial and primary follicles). Thus, if measurements were to be made in preparations from whole ovaries (also containing many larger structures), specific responses in those smallest compartments of ovarian function would be obscured. In order to investigate mechanisms of VCD-induced ovotoxicity at the cellular and molecular level, a technique for isolation of highly purified fractions of small pre-antral follicles (25-100 µm) was developed in rats (Flaws et al., 1994). Additionally, during the isolation, a fraction containing larger, growing pre-antral follicles (100-250 µm) is also obtained. This provides comparative results in non-targeted follicles. More recent studies in rats have utilized this *in vitro* method and taken an integrated morphological, biochemical and molecular approach to investigate generalized mechanisms of chemical-induced ovotoxicity, using VCD as a model.

OVARIAN METABOLISM

The ovary contains enzymes responsible for biotransformation and detoxification of many xenobiotics. Both the rat and mouse ovary contain epoxide hydrolase, glutathione-S-transferase, and cytochromes P450 which metabolize known ovarian toxicants (Bengtsson et al., 1983; Heinrichs and Juchau, 1980; Mukhtar et al, 1978a,b). Metabolism within the ovary could occur within selected structures, causing a compartmentalization of exposure that may affect toxicity within certain classes of follicles.

Bioactivation

In mice, hepatic metabolism of VCH has been well investigated (Smith et al, 1990b; Doerr et al., 1995). Evidence has been provided for hepatic epoxidation of VCH by the cytochrome P450 enzymes, specifically the cyp 2a and cyp 2b isoforms. Repeated daily dosing of mice with VCH also induced these isoforms, and thus its own metabolism (Doerr-Stevens et al., 1999). However, whether specific ovarian compartments in mice also express the enzyme isoforms capable of direct bioactivation of VCH to VCD, is not known. It is plausible that metabolism within the ovary could more directly expose the target population of follicles to ovotoxic effects of these chemicals. Thus, investigation into the ability of mouse ovaries to convert VCH to VCD has been undertaken. Preliminary studies have evaluated the presence of the same cyp 450 isoforms thought to be involved in the hepatic metabolism of VCH in isolated ovarian compartments. Constitutive expression of mRNA encoding *cyp* 2a, *cyp* 2b, and *cyp* 2e1 has been observed in the mouse ovary. Repeated daily dosing with VCH or VCD also affects expression of mRNA encoding the isoforms under consideration in the ovary. Table 1 provides a summary of the initial identification of expression of mRNA encoding these *cyp* P450 isoforms in select ovarian compartments, small pre-antral (Fraction 1; 25-100 µm, primordial, primary and secondary), large pre-antral (Fraction 2; 100-250 µm, growing), and large antral (>250 µm) follicles, as well as interstitial tissue. Cyp 2e1 protein has also been evaluated in the ovary by confocal microscopy. Interestingly, there was extensive distribution of the protein in interstitial tissue, suggesting that this non-follicular ovarian compartment could play an important role in ovarian metabolism of xenobiotics.

Table 1. Effect of VCH and VCD on expression of mRNA encoding cytochromes P450.[a]

Ovarian Fraction	*cyp* 2e1	*cyp* 2b9/10	*cyp* 2a4/5
Fraction 1 follicles	↓ VCH/VCD	—	↓ VCH
Fraction 2 follicles	↓ VCH/VCD	↓ VCD	↑ VCD
Antral follicles	↓ VCH/VCD	—	↑ VCH/ ↓ VCD
Interstitial cells	↓ VCH/VCD	↓ VCD	↓ VCH

[a]Ovarian fractions were isolated from d42 B6C3F₁ mice dosed daily (15 d; i.p.) with control, VCH (800 mg/kg), or VCD (80 mg/kg). Total RNA was prepared and analyzed by RT-PCR to amplify the cytochrome P450 isoforms of interest (*cyp* 2e1, *cyp* 2b9/10, *cyp* 2a4/5). Amplified PCR products were separated on a 1.5% agarose gel and visualized by ethidium bromide staining. ↓ = decreased; ↑ = increased; — = no change.

Detoxification

Although VCD produced follicular loss in rats as well as mice, lower doses were more effective in mice. Subsequently, it was shown that the mouse, as compared with the rat, has a reduced capacity to convert VCD to its inactive tetrol derivative (Salyers et al., 1993). Similar results have been reported for epoxides of butadiene (Csanady et al., 1992). Therefore, mice appear to be deficient in a major detoxification pathway for this class of compounds, and the greater susceptibility of mice compared to rats relates to both an enhanced formation and reduced detoxification of the ovotoxic epoxides. Because of this, studies involving detoxification of VCD were performed in rats. The major enzymatic pathways for detoxification of xenobiotic epoxides are hydration to corresponding diols (catalyzed by microsomal epoxide hydrolase, EH), and conjugation with glutathione (catalyzed by glutathione-S-transferase, GST).

Table 2. Metabolism of VCD to tetrol in isolated rat ovarian follicles and liver cells.[a]

	p mol tetrol produced/µg protein
Fraction 1 follicles 25-100 µm diameter	17 ± 3*
Fraction 2 follicles 100-250 µm diameter	30 ± 1
Liver cells	26 ± 2

* $p < 0.05$, different from liver cells

[a]Fraction 1 and Fraction 2 follicles were isolated, and liver cells were prepared from undosed F344 female rats (age d28). Tissues were incubated (1 h, 37°C) with [¹⁴C]VCD (76µM). Following incubation, media content of [¹⁴C]tetrol was measured by HPLC. (data adapted from Flaws et al., 1994).

A role for EH conversion of VCD to 4-(1,2-dihydroxy)ethyl-1,2-dihydroxycyclohexane, the inactive tetrol metabolite, was demonstrated in isolated rat ovarian follicles (Table 2). *In vitro* experiments were performed using isolated follicles collected from undosed rats to identify the direct ability of the ovary to metabolize [¹⁴C]VCD. Following *in vitro* incubation, the smallest follicles (fraction 1) displayed a lower capacity to convert VCD to the tetrol than did larger pre-antral follicles (fraction 2) or liver cells. These results provide evidence that the rat ovary can directly detoxify VCD to the tetrol, but that, compared with non-target tissues, diminished capacity may reside in those smallest follicles that are physiologically targeted. Further evidence for follicular EH activity participating in the response to VCD was obtained in small follicles from rats that had been dosed daily with VCD for 10 days (Figure 3). The day 10 time point had been determined to be the earliest day of dosing in which VCD-induced follicle damage can be detected, yet no significant follicle loss has occurred (Springer et al., 1996a). Compared with controls, VCD dosing caused an increase in expression of mRNA encoding EH in small pre-antral follicles from rats. Conversely, there was no effect of VCD on expression of EH in liver, and a decrease was observed in larger pre-antral follicles (fraction 2). Therefore, these findings provided evidence for an early protective effect (enhanced detoxification) induced by VCD specifically in the target population of small follicles. This observation also suggests an ability of ovarian compartments to respond selectively in metabolizing xenobiotic chemicals to which they may be exposed. Additionally, the collective information presented here demonstrates that the ovary may play a direct role in the metabolism of VCH and VCD in the mouse as well as the rat.

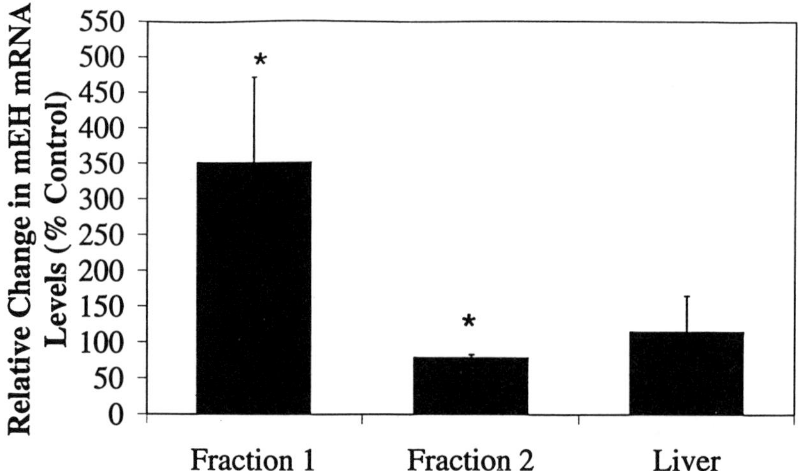

Figure 3. Effect of VCD dosing on EH mRNA levels. RNA was prepared from isolated fraction 1 (25-100μm) or fraction 2 (100-250 μm) ovarian follicles, or liver cells collected from rats dosed daily for 10d with VCD (80mg/kg;i.p.). Total RNA was analyzed by RT-PCR, agarose gel electrophoresis, and autoradiography. Quantitative data were normalized to RP-L19 mRNA. Data are expressed as effect of VCD treatment versus vehicle control. *$p<0.05$; VCD different from control; n=3 (modified from Springer et al., 1996b)

MOLECULAR EFFECTS

It has been determined by morphological evaluation that VCD-induced follicle destruction is via physiological cell death, apoptosis (Springer et al., 1996a). Molecular studies also support this conclusion. Expression of mRNA encoding *bax*, a cell death enhancer gene, was increased in small pre-antral follicles isolated from rats dosed with VCD for 10 days, the time of earliest evidence of impending follicle loss (Figure 4).

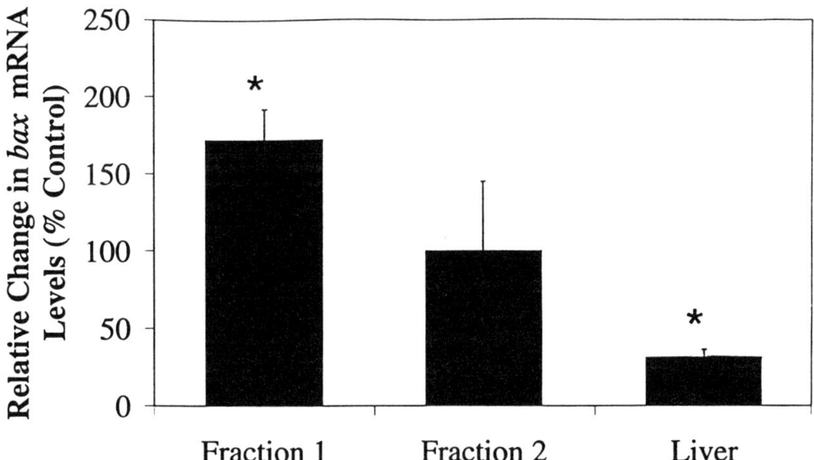

Figure 4. Effect of VCD dosing on *bax* mRNA levels. RNA was prepared from isolated fraction 1 (25-100 µm) or fraction 2 (100-250 µm) ovarian follicles, or liver cells collected from rats dosed i.p. daily for 10 days with VCD (80mg/kg). Total RNA was analyzed by RT-PCR, agarose gel electrophoresis and autoradiography. Quantitative data were normalized to RP-L19 mRNA. Data are expressed as effect of VCD treatment versus vehicle control. *$p<0.05$; VCD different from control; n=3 (modified from Springer et al., 1996b)

The effect was specific for follicle fraction 1 (which contains those follicles specifically targeted by VCD), and was not seen in large, pre-antral follicles or liver. mRNA encoding *bax* in liver, a non-target tissue, was decreased compared with control. The reason for this is not known, however, it does provide support that VCD is not initiating apoptosis in the liver. To investigate whether the increase in expression of *bax* in small follicles on day 10 is consistent with the observed pattern of follicle destruction, these measurements were also made following a single dose of VCD (Borman et al., 1999). Unlike small follicles after 10 daily doses, when *bax* was increased 73% above control, mRNA for *bax* was significantly reduced after only a single dose of VCD. This finding was underscored 15 days after this single dose of VCD, when ovaries from VCD-treated rats contained an increased number of small follicles as compared with matched controls. On the other hand, after 15 days of repeated daily doses of VCD, mRNA for *bax* was increased 230% above controls, and follicle loss was extensive. Thus, these results support that Bax is integrally involved in the pathway of induction of follicular apoptosis caused by VCD because the extent of down-regulation or up-regulation of *bax* correlates well with protection against atresia or with accelerated follicle loss.

Bax protein is known to have a mitochondrial site of action. In addition to increased expression of *bax*, other cellular components known to be associated with a mitochondrial pathway of apoptosis have been identified in small follicles collected from rats dosed with VCD for 15 days. Increased cytochrome c diffusion into the cytosolic compartment of small follicles was observed using confocal microscopy (unpublished data). This is likely in response to mitochondrial membrane damage caused by Bax protein (Wolter et al.,

1997). Activation of caspase-3, a proteolytic enzyme associated with apoptosis (Alnemri et al., 1996), has also been measured by western blotting in small follicles collected from VCD-dosed rats (unpublished data). Thus, evidence for a mechanism of activation of the VCD-induced apoptotic pathway in small follicles appears to involve Bax, cytochrome c, and caspase-3. The focus of our mechanistic studies is now on upstream events that initiate the Bax-regulated cascade.

In summary, VCD-induced ovotoxicity in rats results from activation of intracellular events that are generally associated with activation of a *Bax/Bcl*-mediated apoptotic pathway in other tissue types. Additionally, this ovotoxicity requires repeated exposure. Thus, this physiological form of toxicity is 'silent' because it mimics normal cellular atresia, and would go unnoticed in exposed individuals that are affected. As a result, chronic exposure in women to low levels of this chemical may represent a risk of early menopause. The results of these studies can also serve as an approach for predicting potential damage that may be caused in women exposed to a variety of other environmental chemicals that are known to have ovotoxic effects.

ACKNOWLEDGEMENTS

This work has been supported by funding from the March of Dimes, Arizona Disease Control Research Commission (9809), Chemical Manufacturers' Association, and National Institutes of Environmental Health Services (Center Grant ES06694, R01ES08979, R0 ES09246).

REFERENCES

E.S. Alnemri, D.J. Livingston, D.W. Nicholson, G. Salvesen, N.A. Thornberry, W.W. Wong, and J. Yuan, Human ICE/CED-3 protease nomenclature. *Cell*, 87:171 (1996).

M. Bengtsson, J. Montelius, L. Mankowitz, and J. Rydstrom, Metabolism of polycyclic aromatic hydrocarbons in the rat ovary. *Bioch. Pharmacol.* 32:129 (1983).

S.M. Borman, B.J. VanDePol, S.W. Kao, K.E. Thompson, I.G. Sipes, and P.B.Hoyer, A single dose of the ovotoxicant 4-vinylcyclohexene diepoxide is protective in rat primary ovarian follicles. *Toxicol. Appl. Pharmacol.* 158:244 (1999).

G.A. Csanady, F.P. Guengerich, J.A. Bond, Comparison of the biotransformation of 1,3-butadiene and its metabolite, butadiene monoepoxide by hepatic and pulmonary tissues from humans, rats and mice. *Carcinogenesis* 13:1143 (1992).

J.K. Doerr, S.B. Hooser, B.J. Smith, and I.G. Sipes, Ovarian toxicity of 4-vinylcyclohexene and related olefins in B6C3F1 mice: Role of diepoxides. *Chem. Res. Toxicol.* 8:963 (1995).

J.K. Doerr-Stevens, J. Liu, G.J. Stevens, J.C. Kraner, S.M. Fontaine, J.R. Halpert, and I.G. Sipes, Induction of cytochrome P-450 enzymes after repeated exposure to 4-vinylcyclohexene in B6C3F1 mice. *Drug Metab. Disp* 27(2):281-287 (1999).

J.A. Flaws, K.L. Salyers, I.G. Sipes, and P.B. Hoyer, Reduced ability of rat pre-antral ovarian follicles to metabolize 4-vinyl-1-cyclohexene diepoxide in vitro. *Tox. Appl. Pharmacol.* 126:286 (1994).

W.L. Heinrichs, and M.R. Juchau, Extrahepatic drug metabolism: The gonads. In: Extrahepatic Metabolism of Drugs and Other Foreign Compounds, TE Gram (ed), SP Medical and Scientific Books, New York, pp. 319 (1980).

A.N. Hirshfield, Development of follicles in the mammalian ovary. *Int. Rev. Cytol.* 124 (1991).

A.N. Hirshfield, and W.A. Schmidt, Kinetic aspects of follicular development in the rat. In: Regulation of Ovarian and Testicular Function. ed. VB Mahesh, DS Dindsa, NY Plenum Press New York pp. 211-237 (1987).

S.B. Hooser, D.P. Douds, D.G. DeMerell, P.B. Hoyer, and I.G. Sipes, Long-term ovarian and gonadotropin changes in mice exposed to 4-vinylcyclohexene *Reprod. Toxicol.* 8:315 (1994).

P.B. Hoyer, and I.G. Sipes, Assessment of follicle destruction in chemical-induced ovarian toxicity. *Ann. Rev. Pharmacol. Toxicol.* 36:307-331 (1996).

S.W. Kao, I.G. Sipes, and P.B. Hoyer, Early effects of ovotoxicity induced by 4-vinylcyclohexene diepoxide in rats and mice. *Reprod. Toxicol.* 13:67 (1999).

H. Mukhtar, R.M. Philpot, and J.B. Bend, The postnatal development of microsomal epoxide hydrase, cytosolic glutathione S-transferase, and mitochondrial and microsomal cytochrome P-450 in adrenals and ovaries of female rats. *Drug Metab. Disp.* 6:577 (1978a).

H. Mukhtar, R.M. Philpot, and J.R. Bend, Metabolizing enzyme activities and cytochrome P-450 content of rat ovaries during pregnancy. *Bioch. Bioph. Res. Comm.* 81:89 (1978b).

National Toxicology Program, Toxicology and carcinogenesis studies of 4-vinylcyclohexene in F344/N rats and B6C3F1 mice. *NTP Tec. Rep. No. 303* US Dept. Health Hum. Serv. Public Health Serv. Natl. Inst. Health. Public Inf. Natl. Toxicol. Program, Research Triangle Park NC. (1986).

K.L. Salyers, W. Zheng, and I.G. Sipes, Disposition and toxicokinetics of 4-vinyl-1-cyclohexene diepoxide in female F344 rats and B6C3F1 mice. *ISSX Proc.* 4:Abstract 169 (1993).

Simmon VF nd Baden JM (1980). Mutagenic activity of vinyl compounds and derived epoxides. *Mutat. Res.* 78:227-231

B.J. Smith, D.R. Mattison, and I.G. Sipes, The role of epoxidation in 4-vinylcyclohexene-induced ovarian toxicity. *Toxicol. Appl. Pharmacol.* 105:372 (1990a).

B.J. Smith, I.G. Sipes, J.C. Stevens, and J.R. Halpert, The biochemical basis for the species difference in hepatic microsomal 4-vinylcyclohexene epoxidation between female mice and rats. *Carcinogenesis* 11:1951 (1990b).

L.N. Springer, M.E. McAsey, J.A. Flaws, J.L. Tilly, I.G. Sipes, and P.B. Hoyer, Involvement of apoptosis in 4-vinylcyclohexene diepoxide-induced ovotoxicity in rats. *Toxicol. Appl. Pharmacol.* 139:394 (1996a).

L.N. Springer, J.L. Tilly, I.G. Sipes, and P.B. Hoyer, Enhanced expression of *bax* in small preantral follicles during 4-vinylcyclohexene diepoxide-induced ovotoxicity in the rat. *Toxicol. Appl. Pharmacol.* 139:402 (1996b).

C.E Voogd, J.J. VanderStel, and J.A. Jacobs, The mutagenic action of aliphatic epoxides. *Mutat. Res.* 89:269 (1981).

K.G. Wolter, Y.T. Hsu, C.L. Smith, A. Nechushtan, X.G. Xi, and Y.J. Youle, Movement of Bax from the Cytosol to Mitochondria during Apoptosis. *J. Cell Biol.* 139: 1281 (1997).

MUTAGENICITY AND CARCINOGENICITY OF BIOLOGICAL REACTIVE INTERMEDIATE'S DERIVED FROM A "NON-GENOTOXIC" CARCINOGEN

Serrine S. Lau,[1] Hae-Seong Yoon,[1] Sonal K. Patel,[1] Jeffrey I. Everitt,[2] Cheryl L. Walker,[3] and Terrence J. Monks[1]

[1]Center for Molecular and Cellular Toxicology, Division of Pharmacology and Toxicology, College of Pharmacy, University of Texas at Austin, Austin, Texas 78712-1074; [2]Chemical Industry Institute for Toxicology, Research Triangle Park, North Carolina 27709-2233; [3]The University of Texas M.D. Anderson Cancer Center, Science Park – Research Division, Smithville, Texas 78957

INTRODUCTION

Hydroquinone (HQ) is used as a developer in the photographic industry, as an antioxidant in the rubber industry, and as an intermediate in the manufacturing of food antioxidants. HQ is also an important metabolite of benzene (1,2). HQ has been identified in relatively high concentrations in the smoke of unfiltered cigarettes (up to 155 μg per cigarette) (3) and was chosen for study by the National Cancer Institute because it is produced in large quantities, humans are frequently exposed to it, and there is little adequate carcinogenicity data on it (4). Smoking unfiltered cigarettes and long-term cigarette smoking ($\geq$ 30 years) correlate with a high incidence of renal cell carcinoma in men (5-7). Although the basis for the increased incidence of renal tumors in cigarette smokers is not known, cigarette smoke contains high concentrations of oxidants and free radicals, the principal radical in the tar phase being the 1,4-benzoquinone/hydroquinone redox couple (8).

Although HQ is generally not mutagenic in short-term bacterial mutagenicity assays (9,10), and no mutagenic activity has been found in mouse cells *in vivo* (11), HQ causes base-pair changes in the TA1535 *Salmonella* test strain (12) and is mutagenic in oxidant-sensitive (TA104 and TA2637) *Salmonella* test strains (13), consistent with the mutagenicity of 1,4-benzoquinone in several Ames bacterial test strains (14). HQ is also clastogenic and induces sister chomatid exchange (4,12,15), catalyzes the *in vitro* formation of 8-oxo-deoxyguanosine (16,17) and causes single-strand DNA breaks in isolated hepatocytes (18). In addition, HQ causes renal tubular cell degeneration in the renal cortex of male Fischer 344 rats, and at high doses, markedly increases the number of tubular cell adenomas (4). Renal cell tumors also arise in animals exposed to HQ, but only at nephrotoxic doses and only in male rats (4,19) and mice (19). HQ also acts as a tumor promoter dependent on the target organ and on the initiation protocol used (20). Neither the mechanism of HQ-mediated nephrocarcinogenicity in male rats, nor the basis for the species and sex differences are known.

The acute nephrotoxicity of HQ is dependent upon the activity of γ-glutamyl transpeptidase (γ-GT), indicating that the toxicity of HQ is dependent upon the formation of metabolites that are substrates for this enzyme (21). Consistent with this view, glutathione (GSH) conjugates of HQ, in particular 2,3,5-*tris*-(glutathion-S-yl)HQ (TGHQ), are potent nephrotoxicants (21,22), and HQ is metabolized *in vivo* to GSH conjugates in amounts sufficient to support their role in the acute nephrotoxic effects of HQ (23).

Spontaneous renal Cell carcinoma (RCC) is rare in rats, occurring in most strains with a frequency of <0.05% (24). However, Eker rats (25) carry a single autosomal mutation that predisposes them to the development of spontaneous renal cell tumors at a high incidence. A germline insertion of an endogenous retrovirus in the tuberous sclerosis 2 (*Tsc-2*) tumor suppressor gene is responsible for the predisposing Eker mutation (26,27). The *Tsc-2* gene has been primarily associated with the development of RCC in rats (28-30), although humans with tuberous sclerosis are at increased risk for the development of benign and malignant renal tumors (28,31). In rats carrying the Eker mutation, preneoplastic lesions in the renal tubules begin to appear at about 2-3 months of age, and by the age of 1 yr, the incidence of renal cell tumors in gene carriers approaches 100%. Most renal cell tumors in Eker rats originate from the renal proximal tubules and are histologically similar to renal tumors in humans (28,32).

After exposure to known renal carcinogens, the number of renal tumors in susceptible Eker rats (i.e., those carrying a germline *Tsc-2* mutation) is greatly increased (33,34), those carrying the mutation (Tsc-2$^{Ek/+}$) being more than 70-fold more susceptible to the induction of RCC than their homozygous wild-type (Tsc-2$^{+/+}$) littermates (33). Loss of the second allele of this gene as a somatic event leads to the development of RCC in these animals, supporting Knudson's two- hit hypothesis for the loss of tumor suppressor gene function in tumorigenesis (24). Thus, the Eker rat model system offers a unique opportunity to investigate mechanisms of chemical-induced renal carcinogenesis in kidney epithelial cells that are predisposed to tumor development. We therefore used Eker rats to test the hypothesis that a quantitatively minor metabolite of HQ plays an important, if not essential, role in "nongenotoxic" carcinogen-mediated carcinogenesis (34).

RESULTS AND DISCUSSION

<u>Cell Proliferation Following TGHQ Treatment in the Eker Rat</u>
Increased cell proliferation in response to TGHQ (2.5 µmol/kg for 4 mo. followed by 3.5 µmol/kg for additional 6 mo.; i.p.; 5 days/week) was evident after 4 months treatment and was predominantly located in a band-like region of the OSOM extending to the medullary rays (Figure 1A), as previously reported for acute HQ exposure (21). The labeling index was increased 19-fold in kidneys of Eker rats treated with TGHQ compared with saline-treated controls. Figure 1B shows characteristic BrdU immunostaining, which was localized to the regions of toxicity. Exfoliated necrotic cells were evident within dilated tubules lined by cuboidal epithelium, and many of the cells were in S phase and were strongly immunoreactive (Figure 1B arrowhead). In contrast, kidney sections from control animals showed a distinct absence of cells in S phase (Figure 1C).

The correlation appears clear between TGHQ-induced nephrotoxicity and nephrocarcinogenicity. Acute toxicity after both HQ and TGHQ exposure occurs in the outer stripe of the outer medulla (OSOM) of the kidney and progresses with time along the medullary rays (21). This region is the site of the vast majority of preneoplastic and neoplastic lesions observed in TGHQ-exposed kidneys. The site-selectivity of this toxicity may be a consequence of the susceptibility of this area to oxidative stress and of the high concentrations of γ-GT in the brush-border membrane of proximal tubular cells (35). Pretreatment of rats with acivicin to inhibit γ-GT completely prevents HQ-mediated nephrotoxicity (21), implying that metabolism of HQ to metabolites requiring processing by this enzyme is a prerequisite for toxicity (22). The site-selective toxicity is also reflected in the rapid excretion of γ-GT into urine, in the absence of gross kidney dysfunction or accumulation of blood urea nitrogen (21). Moreover, reactive electrophilic metabolites of TGHQ become covalently adducted to proteins in the same region of the kidney that ultimately give rise to tumors (36). Thus, the proteins necessary for both the transport and bioactivation of TGHQ (37) reside at the site of the initial tissue injury and cell proliferation.

<u>Histological Changes Induced by TGHQ in the Eker Rat</u>
TGHQ-treated rats (10 months treatment) developed numerous toxic tubular dysplasias (Figure 1D) of a form not observed in age-matched untreated Eker rats. Toxic tubules differ from the normal proliferative lesions of the Eker rat in that they exhibit a thick circumscribing peritubular fibrosis surrounding a dilated tubular profile. These preneoplastic lesions were found as early as 4 months of TGHQ treatment, and are believed to represent early transformation within tubules undergoing regeneration in response to

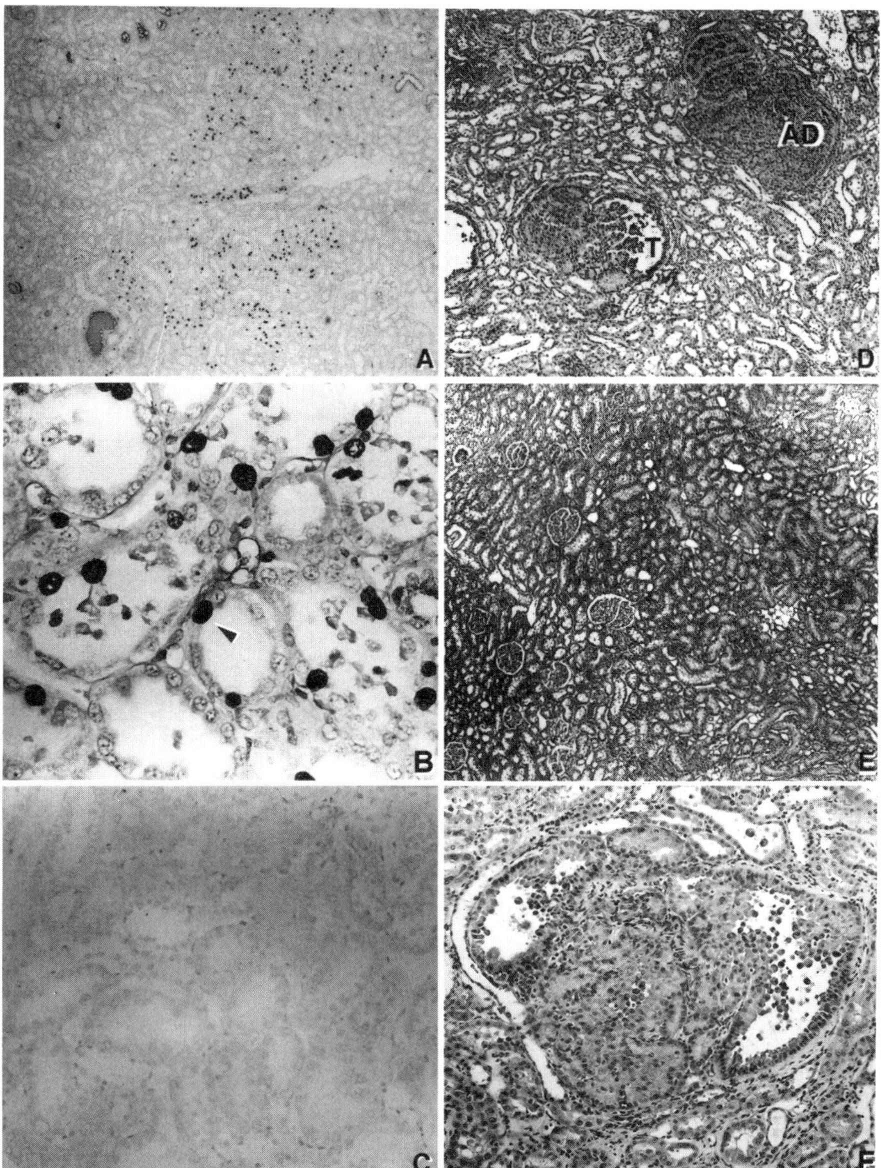

Fig. 1. Cell proliferation determined by BrdU immunostaining. (A-C) Kidney slices from Eker rats treated with 2.5 μmol/kg TGHQ i.p. for 4 months (**A**, 60x; **B**, 300x) or saline (**C**, 300x). Arrowhead indictes BrdU positive nucleus (B). Kidney slices from Eker rats treated with 2.5 μmol/kg TGHQ i.p. for 4 months followed by 3.5 μmol/kg TGHQ for 6 months (**D**, 150x H&E) or with saline (**E**, 150x, H&E). AD, adenoma; T, tubular dysplasia. (**F**) Neoplastic development within tubular dysplasias (300x, H&E). From reference 34.

toxic injury. However, while microscopic examination of kidney tissues obtained from animals as early as 4 months of TGHQ exposure revealed increases in the numbers of basophilic dysplasias. The number of renal cell tumors was not significantly greater at this early time, most probably because 16 weeks is insufficient time for TGHQ-induced preneoplastic lesions to progress to frank renal tumors. However, following 10 months of TGHQ treatment, rats exhibited a significant increase in renal tumors. In contrast, in control 12-month-old Eker rats (Figure 1E) the typical appearance of the cortical medullary junction exhibited only a few small basophilic atypical tubules. Figure 1D depicts the renal parenchyma from a TGHQ-treated rat at the junction of the OSOM, with a solid adenoma and an adenoma arising within the region of tubular dysplasia (T). Figure 1F shows, in higher magnification, the neoplastic development within tubular dysplasias from another 12-month-old TGHQ-treated rat. Papillary tumor development was also evident within dilated tubular profiles (Figure 1F).

In addition to microscopic lesions, numerous gross lesions were observed in TGHQ treated rats (10 mo). The most common form of spontaneous renal cell tumors in the Eker rat is an adenoma of the basophilic type, arising from the proximal tubular epithelium. However eosinophilic lesions, which may originate in the more distal segments of the nephron (38), are often observed in these rats as well. Numerous lesions of both types were found in rats treated with TGHQ for 10 months. Interestingly, the increase in the number of lesions (preneoplasia and adenoma) was mostly due to an increase in basophilic lesions, suggesting that these tumors arise from renal epithelial cells of the P2/P3 segment of the renal nephron, which is the region of acute toxicity in TGHQ-treated rats. The numbers of basophilic dysplasias, adenomas and RCCs increased 6-, 7- and 10-fold respectively, in Eker rats after 10 months treatment with TGHQ. Most of the renal cell tumors observed in the TGHQ-exposed animals were in the region of TGHQ-induced acute renal injury. The existence of basophilic and eosinophilic subtypes of the preneoplastic renal tubular dysplasias and renal cell tumors has been previously noted in Eker rats. These cytologic variants occur in Eker rats spontaneously and after chemical carcinogen exposure (31,33). A unique feature of the study reported here was the finding of neoplastic proliferations arising within the so-called toxic tubular dysplasias. These dysplastic lesions do not occur spontaneously in Eker rats and are therefore considered to be specifically associated with TGHQ treatment. The anatomical distribution of these lesions in the OSOM, the target region for TGHQ-induced toxicity, supports this idea. Furthermore, most of the lesions induced by TGHQ were of the basophilic type, consistent with the cell type of origin being epithelial cells of the P2/P3 segment of the renal nephron.

The ratios of renal cell adenomas to dysplasias were identical (1:4) in control and treated Eker rats, suggesting that TGHQ acts at an early stage of carcinogenesis to induce preneoplastic lesions, rather than by simply promoting the progression of pre-existing lesions to adenomas. This may be a result of synergism between the clastogenic and DNA-damaging effects of TGHQ and cell proliferation induced in response to nephrotoxicity. That incubation of the *supF* gene with TGHQ followed by transfection and replication in either human AD293 or bacterial cells significantly increases the mutation frequency (39) is consistent with this view. Although the mutation spectrum induced by TGHQ *in vitro* in short-term assays is consistent with the participation of hydroxyl radicals in quinone-thioether-mediated DNA damage and cytotoxicity, several mutations occur with a frequency indicative of the involvement of additional mutagenic species, most likely quinone-thioether DNA adducts (39). TGHQ-mediated tumorigenesis may therefore involve both genotoxic and non-genotoxic mechanisms.

<u>Loss of heterozygosity of the *Tsc-2* gene</u>
Tumor formation in susceptible Eker rats exposed to TGHQ presumably involves two events: inheriting a germline mutation of the *Tsc-2* gene and acquiring another mutation in the wild-type *Tsc-2* allele, either spontaneously or as a result of carcinogen exposure (24). The resulting loss of function of the remaining normal allele of this gene is a rate-limiting event for the development of preneoplastic lesions and, ultimately, tumor formation. Using PCR primers that specifically amplify mutant and normal alleles of the *Tsc-2* gene, loss of the normal *Tsc-2* allele was examined in tumors from TGHQ-treated rats. Using the new technique of HPLC-based DNA fragment analysis (Figure 2, bottom) (40), and traditional PCR methodologies (Figure 2, top) loss of heterozygosity at the *Tsc-2* locus was observed in 12/12 tumors analyzed from 12-month-old TGHQ-treated Eker rats. In all cases, lack of amplification of the wild-type *Tsc-2* allele was apparent, as indicated by the absence of the

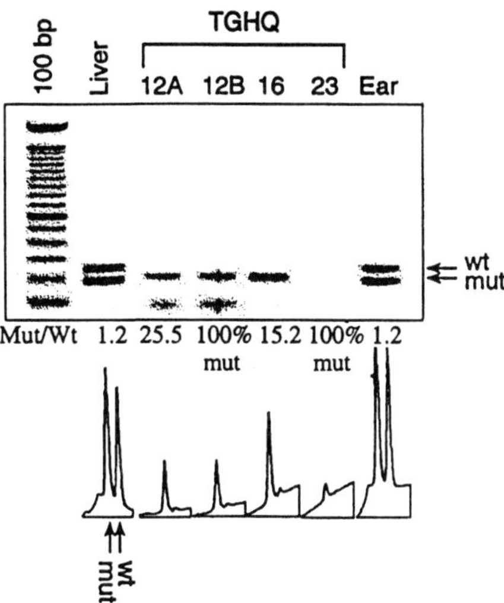

Figure 2   Loss of heterozygosity of the Tsc-2 gene. Top panel represents analysis of the wild type and mutant allelle of the Tsc-2 gene on a 1 % agarose gel visualized by ethidium bromide staining. DNA from liver and ear are included as non-target organ controls. The bottom panel represents quantitative determination of the Tsc-2 allelles by HPLC analysis. Data adapted from reference 34.

wild-type PCR product from the chromatograms and as visualized by traditional gel electrophoresis, consistent with a role for *Tsc-2* as a tumor suppressor gene whose function is lost during TGHQ-mediated carcinogenesis. Loss of the normal allele of the *Tsc-2* tumor suppressor gene in TGHQ-associated tumors suggests that loss of tuberin function is involved in tumor development, but the precise cellular and molecular mechanism by which TGHQ induces renal tumors in the Eker rat is not known. Importantly, using laser capture micro-dissection technology we showed that loss of the *Tsc-2* gene occurred within the toxic tubules. Tuberin is widely expressed in most adult tissues (41) and exhibits GTPase activating protein (GAP) activity toward at least two proteins involved in signal transduction, Rap-1a and Rab5 (42,43). GAP proteins such as Rap1GAP act as regulators of cell signaling by catalyzing the hydrolysis of GTP to GDP. The homology between tuberin and Rap1GAP predicts that tuberin might be a potential regulator of Rap1. In fact, tuberin exerts weak but specific GAP activity toward Rap1 *in vitro* (42), and tuberin and Rap1 colocalize *in vivo* to the Golgi apparatus (44). The Eker germline insertion in the *Tsc-2* gene results in premature termination of tuberin synthesis upstream of the catalytic domain, leading to the loss of the Rap1GAP-like domain. Loss of tuberin expression in cells would therefore maintain Rap1 in the active GTP-bound state, contributing to the development of RCC. Rab5 is involved in regulating endocytosis (45), which is critical for normal functioning of renal epithelial cells. Proteins such as polypeptide growth factors are reabsorbed and/or degraded by luminal endocytosis at the apical surface of epithelial cells of the proximal tubule. The GTPase activity of Rab5, which is regulated by tuberin, modulates endosome fusion (46-49), and loss of *Tsc-2* gene function is associated with disrupted endocytosis (43). However, the role of this disruption in tumor pathogenesis requires further investigation.

It has been proposed that certain chemicals may induce carcinogenesis by mechanisms involving cytotoxicity followed by sustained regenerative hyperplasia and, ultimately, tumor formation, and HQ has been proposed to belong to such a class of nongenotoxic carcinogens (9-11). Both HQ and TGHQ are acutely nephrotoxic, producing a proliferative response subsequent to tissue damage (21). Although it is unlikely that proliferation per se initiates tumorigenesis, additional rounds of DNA synthesis associated with compensatory cell proliferation in the kidney could increase the opportunity for sustaining genetic alterations relevant for tumor initiation. In this regard, the ability of HQ and its metabolites to act as clastogens, inducing both numerical and structural chromosome alterations, is well established (4,11,15,50-52). TGHQ-induced loss of heterozygosity of the normal allele of the *Tsc-2* gene is consistent with mechanisms of gene loss associated with periods of high mitotic activity, such as chromosome nondisjunction.

<u>Cell transformation by HQ or TGHQ treatment</u>
A single treatment with HQ (25 or 50 µM) or TGHQ (100 or 300 µM) for 4 h induced cell transformation in primary renal epithelial cells carrying a mutation in the Tsc-2 gene (Tsc-$2^{EK/+}$). In cell lines derived from transformed colonies, LOH at the Tsc-2 locus was observed resulting in loss of tuberin expression. Colonies transformed by TGHQ were able to give rise to cell lines that formed tumors when injected into nude mice. The tumor mass was composed of cells exhibiting a well differentiated epithelial phenotype, growing in small nests and separated by thin septa of stromal connective tissue. Although the neoplastic epithelial cells exhibited a considerable degree of atypia, they also exhibited a clear tendency to reorganize into well-developed tubular structures, resembling kidney tubules. In contrast, treatment of Tsc-$2^{+/+}$ renal epithelial cells with HQ or TGHQ did not result in cell transformation. The results indicate that TGHQ can induce neoplastic transformation, and that the Tsc-2 tumor suppressor gene is a target for quinol-thioether-induced renal carcinogenesis. These data suggest that TGHQ, a metabolite of HQ, is mutagenic (39), and is potentially a genotoxic carcinogen.

Cell transformation is indispensable for carcinogenesis, which is a multi-stage process generally requiring multiple genetic alterations (53,54). Following TGHQ-induced cell transformation, the remaining wild-type allele of the Tsc-2 gene was lost, with the concomitant loss of tuberin expression. It appears that the preexisting mutation in one allele of the Tsc-2 gene in Eker rat cells is a prerequisite for transformation of these cells by HQ and TGHQ, since Tsc-$2^{+/+}$ cells were refractory to transformation by these compounds at the doses tested. As a result of the presence of the Tsc-2 mutation, only a single genetic event at the wild-type Tsc-2 allele would be required to inactivate this tumor suppressor gene in Tsc-$2^{EK/+}$ epithelial cells. However, in Tsc-$2^{+/+}$ cells, two genetic events would be necessary to inactivate both alleles of this gene. The increased susceptibility of Tsc-$2^{EK/+}$ cells to transformation by HQ and TGHQ and the LOH at the Tsc-2 gene locus suggest that the Tsc-2 gene is a target for TGHQ-mediated transformation. Moreover, TGHQ may be acting at the initiation stage of the carcinogenic process by inactivating the remaining wild-type Tsc-2 allele in Tsc-$2^{EK/+}$ cells. In this *in vitro* transformation assay, transformation is measured as the ability to induce preneoplastic cells, identified in this assay as transformed colonies with enhanced growth potential (55). These in vitro data are in excellent agreement with the above *in vivo* findings.

As noted above, tuberin possesses GAP (GTPase activating protein) activity toward Rap1, thereby negatively regulating Rap1, a small Ras-related GTP-binding protein (42). Tuberin is also involved in cell cycle regulation (56,57) and suppresses cell proliferation and tumorigenicity (58,59). Although the precise physiological role of tuberin is not yet clear, it may function as a regulator of many signaling pathways, including the cell cycle. Loss of tuberin contributes to disturbances in signaling pathways, which lead to cell transformation as well as tumor development (34). Consistent with this view, MAPK$^{p42/p44}$ is constitutively activated in all three QT-RRE cell line derived from TGHQ transformed epithelial cells. Interestingly, the Rap 1 and MAPK signaling pathways are linked through B-Raf (60-62), suggesting that loss of tuberin function as a negative regulator of Rap1 is causally related to high MAPK$^{p42/p44}$ activity in QT-RRE cell lines.

The mechanisms responsible for LOH at the Tsc-2 locus in TGHQ transformed cells are not clear. Cytogenetic analysis of the cell lines revealed no indication of monosomy of chromosome 10 in the transformed cell lines, suggesting that mechanisms other than chromosome loss are responsible for loss of the wild-type Tsc-2 allele. However, TGHQ can generate ROS (63) and produce oxidative damage (17) and alkylating species (36), which could play a role in TGHQ-induced transformation. Similar studies showed that HQ was capable of transforming BALBc/3T3 cells and that the transformation frequency was significantly increased by the tumor promoting agent, TPA (64). HQ also caused an increase in transformation frequency of Syrian hamster embryo (SHE) cells (15). Moreover, analysis of the mutation spectra of HQ in the *supF* gene showed that deletion of one cytosine was a unique HQ-induced mutation (64). We have recently shown that TGHQ increases the mutation frequency of the *supF* gene replicated in human AD293 cells and E. coli MBL50 cells, and that base substitutions are the most common type of mutations (39).

In summary, we have demonstrated that TGHQ, a potent, albeit quantitatively minor, nephrotoxic metabolite of HQ, produced renal tumors in an animal model of RCC. Preneoplastic and neoplastic lesions developed within the OSOM in regions of nephrotoxicity and high rates of compensatory cell proliferation. These data underscore the importance of obtaining adequate data on the mechanism of action of potential human carcinogens, as well as their metabolites, before classifying these compounds, particularly those considered to be nongenotoxic carcinogens. Such a designation has important implications for human risk assessment and should be considered carefully in the absence of data on a compounds' metabolic profile and mechanism of action in target organs relevant to carcinogenicity. HQ and TGHQ are capable of transforming primary rat kidney epithelial cells derived from Eker Tsc-2$^{EK/+}$ rats. The remaining normal allele of the Tsc-2 gene appears to be a target for TGHQ-mediated transformation. Transformed colonies from TGHQ-derived transformants exhibit a variety of chromosomal aberrations and are tumorigenic when injected into nude mice. The combined use of genetically susceptible renal epithelial cells from the Eker rat with an *in vitro* transformation assay specific for this cell type appears to provide a powerful approach for unveiling the potential activity of weak carcinogens. Such *in vitro* data can provide valuable support for interpretation of *in vivo* data.

ACKNOWLEDGEMENTS

This work was supported by grants GM 39338 to Dr. Serrine Lau and CA 63613 to Dr. Cheryl Walker. We also acknowledge the support of NIEHS Center Grant ES 07784 and thank Drs. Irma Gimenez-Conti, (Director of the Histology and Tissue Processing Core) and Dennis Johnston (Director of the Biostatistics and Data Processing Core) for their contributions to this work.

REFERENCES

1. Proteus, J.W. and Williams R.T. (1949) The metabolism of benzene. II. The isolation of phenol, catechol, quinol and hydroxyquinol from the etheral sulphate fraction of the urine of rabbits receiving benzene orally. *Biochem*, **44**, 56-61.
2. Parke, D.V. and Williams R.T. (1953) Studies in detoxication. 54. The metabolism of benzene. (a) The formation of phenylglucuronide and phenylsulphuric acid from [^{14}C]benzene. (b) The metabolism of [^{14}C]phenol. *Biochem J*, **55**, 337-340.
3. Ishiguro, S., Saugawara H., Kusama M., Yano S., Shimojima N. and Sugawara S. (1976) Glass capillary column gas chromatographic analysis of tobacco and cellulose cigarette smoke. I. Acidic fractions. *Sci. Pap. Cent. Res. Inst. Jpn. Tob. Salt Public Corp.*, **118**, 207-211.
4. Kari, F.W., Bucher J., Eustis S.L., Haseman J.K. and Huff J.E. (1992) Toxicity and carcinogenicity of hydroquinone in F344/N rats and B6C3F1 mice. *Food Chem Toxicol*, **30**, 737-47.
5. Muscat, J.E., Hoffmann D. and Wynder E.L. (1995) The epidemiology of renal cell carcinoma. A second look. *Cancer*, **75**, 2552-7.
6. Tavani, A. and La Vecchia C. (1997) Epidemiology of renal-cell carcinoma. *J Nephrol*, **10**, 93-106.
7. Randerath, E. and Randerath K. (1993) Monitoring tobacco smoke-induced DNA damage by 32P-postlabelling. *IARC Sci Publ*, **124**, 305-14.
8. Church, D.F. and Pryor W.A. (1985) Free-radical chemistry of cigarette smoke and its toxicological implications. *Environ Health Perspect*, **64**, 111-26.
9. Florin, I., Rutberg L., Curvall M. and Enzell C.R. (1980) Screening of tobacco smoke constituents for mutagenicity using the Ames' test. *Toxicology*, **15**, 219-232.
10. Sakai, M., Yoshida D. and Mizusaki S. (1985) Mutagenicity of polycyclic aromatic hydrocarbons and quinones on Salmonella typhimurium TA97. *Mutat Res*, **156**, 61-7.

11.	Gocke, E., Wild D., Eckhardt K. and King M.T. (1983) Mutagenicity studies with the mouse spot test. *Mutat Res*, **117**, 201-12.

12.	Gocke, E., King M.T., Eckhardt K. and Wild D. (1981) Mutagenicity of cosmetics ingredients licensed by the European Communities. *Mutat Res*, **90**, 91-109.

13.	Hakura, A., Tsutsui Y., Mochida H., Sugihara Y., Mikami T. and Sagami F. (1996) Mutagenicity of dihydroxybenzenes and dihydroxynaphthalenes for Ames Salmonella tester strains. *Mutat Res*, **371**, 293-9.

14.	Hakura, A., Mochida H., Tsutsui Y. and Yamatsu K. (1995) Mutagenicity of benzoquinones for Ames Salmonella tester strains. *Mutat Res*, **347**, 37-43.

15.	Tsutsui, T., Hayashi N., Maizumi H., Huff J. and Barrett J.C. (1997) Benzene-, catechol-, hydroquinone- and phenol-induced cell transformation, gene mutations, chromosome aberrations, aneuploidy, sister chromatid exchanges and unscheduled DNA synthesis in Syrian hamster embryo cells. *Mutat Res*, **373**, 113-23.

16.	Leanderson, P. and Tagesson C. (1990) Cigarette smoke-induced DNA-damage: role of hydroquinone and catechol in the formation of the oxidative DNA-adduct, 8-hydroxydeoxyguanosine. *Chem Biol Interact*, **75**, 71-81.

17.	Lau, S.S., Peters M.M., Kleiner H.E., Canales P.L. and Monks T.J. (1996) Linking the metabolism of hydroquinone to its nephrotoxicity and nephrocarcinogenicity. *Adv Exp Med Biol*, **387**, 267-73.

18.	Walles, S.A. (1992) Mechanisms of DNA damage induced in rat hepatocytes by quinones. *Cancer Lett*, **63**, 47-52.

19.	Shibata, M.A., Hirose M., Tanaka H., Asakawa E., Shirai T. and Ito N. (1991) Induction of renal cell tumors in rats and mice, and enhancement of hepatocellular tumor development in mice after long-term hydroquinone treatment. *Jpn J Cancer Res*, **82**, 1211-9.

20.	Yamaguchi, S., Hirose M., Fukushima S., Hasegawa R. and Ito N. (1989) Modification by catechol and resorcinol of upper digestive tract carcinogenesis in rats treated with methyl-N-amylnitrosamine. *Cancer Res*, **49**, 6015-8.

21.	Peters, M.M., Jones T.W., Monks T.J. and Lau S.S. (1997) Cytotoxicity and cell-proliferation induced by the nephrocarcinogen hydroquinone and its nephrotoxic metabolite 2,3,5-(tris-glutathion-S- yl)hydroquinone. *Carcinogenesis*, **18**, 2393-401.

22.	Lau, S.S., Hill B.A., Highet R.J. and Monks T.J. (1988) Sequential oxidation and glutathione addition to 1,4-benzoquinone: correlation of toxicity with increased glutathione substitution. *Mol Pharmacol*, **34**, 829-36.

23.	Hill, B.A., Kleiner H.E., Ryan E.A., Dulik D.M., Monks T.J. and Lau S.S. (1993) Identification of multi-S-substituted conjugates of hydroquinone by HPLC-coulometric electrode array analysis and mass spectroscopy. *Chem Res Toxicol*, **6**, 459-69.

24.	Yeung, R.S., Xiao G.H., Everitt J.I., Jin F. and Walker C.L. (1995) Allelic loss at the tuberous sclerosis 2 locus in spontaneous tumors in the Eker rat. *Mol Carcinog*, **14**, 28-36.

25.	Eker, R. and Mossige J. (1961) A dominant gene for renal adenomas in the rat. *Nature*, **189**, 858-859.

26.	Yeung, R.S., Xiao G.H., Jin F., Lee W.C., Testa J.R. and Knudson A.G. (1994) Predisposition to renal carcinoma in the Eker rat is determined by germ- line mutation of the tuberous sclerosis 2 (TSC2) gene. *Proc Natl Acad Sci U S A*, **91**, 11413-11416.

27.	Hino, O., Kobayashi E., Hirayama Y., Kobayashi T., Kubo Y., Tsuchiya H., Kikuchi Y. and Mitani H. (1995) Molecular genetic basis of renal carcinogenesis in the Eker rat model of tuberous sclerosis (Tsc2). *Mol Carcinog*, **14**, 23-7.

28.	Walker, C. (1998) Molecular genetics of renal carcinogenesis. *Toxicologic Pathology*, **26**, 113-20.

29.	Satake, N., Urakami S., Hirayama Y., Izumi K. and Hino O. (1998) Biallelic mutations of the Tsc2 gene in chemically induced rat renal cell carcinoma. *Int J Cancer*, **77**, 895-900.

30.	Toyokuni, S., Okada K., Kondo S., Nishioka H., Tanaka T., Nishiyama Y., Hino O. and Hiai H. (1998) Development of high-grade renal cell carcinomas in rats independently of somatic mutations in the Tsc2 and VHL tumor suppressor genes. *Jpn J Cancer Res*, **89**, 814-20.

31.	Green, A.J., Smith M. and Yates J.R. (1994) Loss of heterozygosity on chromosome 16p13.3 in hamartomas from tuberous sclerosis patients. *Nat Genet*, **6**, 193-6.

32. Everitt, J.I., Goldsworthy T.L., Wolf D.C. and Walker C.L. (1992) Hereditary renal cell carcinoma in the Eker rat: a rodent familial cancer syndrome. *J Urol*, **148**, 1932-6.

33. Walker, C., Goldsworthy T.L., Wolf D.C. and Everitt J. (1992) Predisposition to renal cell carcinoma due to alteration of a cancer susceptibility gene. *Science*, **255**, 1693-5.

34. Lau, S.S., Monks, T.J., Everitt, J. I., Kleymenova, E., and Walker, C. Carcinogenicity of a nephrotoxic metabolite of the "nongenotoxic" carcinogen hyroquinone. Chem. Res. Toxicol., in press.

35. Monks, T.J. and Lau S.S. (1994) Glutathione conjugate-mediated toxicities. In FC, K. (ed.), *Conjugation-Deconjugation Reactions in Drug Metabolism and Toxicity*. Springer-Verlag, Berlin, pp. 459-510.

36. Kleiner, H.E., Jones T.W., Monks T.J. and Lau S.S. (1998) Immunochemical analysis of quinol-thioether-derived covalent protein adducts in rodent species sensitive and resistant to quinol-thioether- mediated nephrotoxicity. *Chem Res Toxicol*, **11**, 1291-300.

37. Hughey, R.P., Rankin B.B., Elce J.S. and Curthoys N.P. (1978) Specificity of a particulate rat renal peptidase and its localization along with other enzymes of mercapturic acid synthesis. *Arch Biochem Biophys*, **186**, 211-7.

38. Wolf, D.C., Whiteley H.E. and Everitt J.I. (1995) Preneoplastic and neoplastic lesions of rat hereditary renal cell tumors express markers of proximal and distal nephron. *Vet Pathol*, **32**, 379-86.

39. Jeong, J.K., Wogan G.N., Lau S.S. and Monks T.J. (1999) Quinol-glutathione conjugate-induced mutation spectra in the supF gene replicated in human AD293 cells and bacterial MBL50 cells. *Cancer Res*, **59**, 3641-5.

40. Kleymenova, E., Muga S., Fischer S. and Walker C.L. (in press) Application of high-pressure liquid chromatography-based analysis of DNA fragments to molecular carcinogenesis. *Molecular Carcinogenesis*.

41. Wienecke, R., Maize J.C., Jr., Reed J.A., de Gunzburg J., Yeung R.S. and DeClue J.E. (1997) Expression of the TSC2 product tuberin and its target Rap1 in normal human tissues. *Am J Pathol*, **150**, 43-50.

42. Wienecke, R., Konig A. and DeClue J.E. (1995) Identification of tuberin, the tuberous sclerosis-2 product. Tuberin possesses specific Rap1GAP activity. *J Biol Chem*, **270**, 16409-14.

43. Xiao, G.H., Shoarinejad F., Jin F., Golemis E.A. and Yeung R.S. (1997) The tuberous sclerosis 2 gene product, tuberin, functions as a Rab5 GTPase activating protein (GAP) in modulating endocytosis. *J Biol Chem*, **272**, 6097-100.

44. Wienecke, R., Maize J.C., Jr., Shoarinejad F., Vass W.C., Reed J., Bonifacino J.S., Resau J.H., de Gunzburg J., Yeung R.S. and DeClue J.E. (1996) Co-localization of the TSC2 product tuberin with its target Rap1 in the Golgi apparatus. *Oncogene*, **13**, 913-23.

45. Field, H., Farjah M., Pal A., Gull K. and Field M.C. (1998) Complexity of trypanosomatid endocytosis pathways revealed by Rab4 and Rab5 isoforms in Trypanosoma brucei. *J Biol Chem*, **273**, 32102-10.

46. Barbieri, M.A., Hoffenberg S., Roberts R., Mukhopadhyay A., Pomrehn A., Dickey B.F. and Stahl P.D. (1998) Evidence for a symmetrical requirement for Rab5-GTP in in vitro endosome-endosome fusion. *J Biol Chem*, **273**, 25850-5.

47. Gournier, H., Stenmark H., Rybin V., Lippe R. and Zerial M. (1998) Two distinct effectors of the small GTPase Rab5 cooperate in endocytic membrane fusion. *Embo J*, **17**, 1930-40.

48. Simonsen, A., Lippe R., Christoforidis S., Gaullier J.M., Brech A., Callaghan J., Toh B.H., Murphy C., Zerial M. and Stenmark H. (1998) EEA1 links PI(3)K function to Rab5 regulation of endosome fusion. *Nature*, **394**, 494-8.

49. Christoforidis, S., McBride H.M., Burgoyne R.D. and Zerial M. (1999) The Rab5 effector EEA1 is a core component of endosome docking. *Nature*, **397**, 621-5.

50. Eastmond, D.A., Rupa D.S. and Hasegawa L.S. (1994) Detection of hyperdiploidy and chromosome breakage in interphase human lymphocytes following exposure to the benzene metabolite hydroquinone using multicolor fluorescence in situ hybridization with DNA probes. *Mutat Res*, **322**, 9-20.

51.   Stillman, W.S., Varella-Garcia M., Gruntmeir J.J. and Irons R.D. (1997) The benzene metabolite, hydroquinone, induces dose-dependent hypoploidy in a human cell line. *Leukemia*, **11**, 1540-5.

52.   Rupa, D.S., Schuler M. and Eastmond D.A. (1997) Detection of hyperdiploidy and breakage affecting the 1cen-1q12 region of cultured interphase human lymphocytes treated with various genotoxic agents. *Environ Mol Mutagen*, **29**, 161-7.

53.   Yamasaki, H and Mironov, N. Genomic instability in multistage carcinogenesis. Toxicol. Lett., *112*: 251-256, 2000.

54.   Pitot, H.C. The role of receptors in multistage carcinogenesis. Toxicol. Lett., *77*: 55-61, 1995.

55.   Walker, C. and Ginsler, J. Development of a quantitative in vitro transformation assay for kidney epithelial cells. Carcinogenesis, *13*: 25-32, 1992.

56.   Soucek, T., Pusch, O., Wienecke, R., DeClue, J.E., and Hengstschlager, M. Role of Tuberous Sclerosis gene-2 product in cell cycle control. J. Biol. Chem., *272*: 29301-29308, 1997.

57.   Soucek, T., Yeung, R.S., and Hengstschlager, M. Inactivation of the cyclin-dependent kinase inhibitor p27 upon loss of the tuberous sclerosis complex gene-2. Proc. Natl. Acad. Sci. USA, *95*: 15653-15658, 1998.

58.   Jin, F., Wienecke, R., Xiao, G.H., Maize, J.R.Jr., Declue, J.E., and Yeung, R.S. Suppression of tumorigenicity by the wild-type tuberous sclerosis 2 (Tsc2) gene and its C-terminal region. Proc. Natl. Acad. Sci. USA, *93*: 9154-9159, 1996.

59.   Orimoto, K., Tsuchiya, H., Kobayashi, T., Matsuda, T., and Hino, O. Suppression of the neoplastic phenotype by replacement of the Tsc2 gene in Eker rat renal carcinoma cells. Biochem. Biophys. Res. Communi., *219*: 70-75, 1996.

60.   Ohtsuka, T., Shimizu, K., Yamamori, B., Kuroda, S., and Takai, Y. Activation of brain B-Raf protein kinase by Rap1B small GTP-binding protein. J. Biol. Chem., 271: 1258-1261, 1996.

61.   Wan, Y. and Huang, X.Y. Analysis of the Gs/mitogen-activated protein kinase pathway in mutant S49 cells. J. Biol. Chem., *273*: 14533-14537, 1998

62.   York, R.D., Yao, H., Dillon, T., Ellig, C.L., Eckert, S.P., McCleskey, E.W., and Stork, P.J.S. Rap1 mediates sustained MAP kinase activation induced by nerve growth factor. Nature, *392*: 622-626, 1998.

63.   Towndrow, K.M., Mertens, J.J.W.M., Jeong, J.K., Weber, T.J., Monks, T.J., and Lau, S.S. Stress- and growth-related gene expression are independent of chemical-induced prostaglandin E_2 synthesis in renal epithelial cells. Chem. Res. Toxicol., *13*: 111-117, 2000.

64.   Joseph, P., Klein-Szanto, A.J.P., and Jaiswal, A.K. Hydroquinones cause specific mutations and lead to cellular transformation and in vivo tumorigenesis. Brit. J. of Cancer, *78*: 312-320, 1998.

REACTIVE METABOLITES OF 1,3-BUTADIENE: DNA AND HEMOGLOBIN ADDUCT FORMATION AND POTENTIAL ROLES IN CARCINOGENICITY

Adnan A. Elfarra, Thomas S. Moll, Renee J. Krause, Raymond A. Kemper, and Rebecca R. Selzer

University of Wisconsin - Madison
2015 Linden Drive - West
Madison, WI 53706

INTRODUCTION

1,3-Butadiene (BD), a petrochemical widely used in the manufacture of synthetic rubber and plastics, has recently been added to the list of chemicals *Known To Be Human Carcinogens* based upon epidemiological and mechanistic data indicating a causal relationship between occupational exposure to BD and excess mortality from lymphatic and/or hematopoietic cancers (U.S. Department of Health and Human Services, 2000). Long-term BD inhalation studies in mice and rats have also been associated with genotoxicity and carcinogenicity, with mice being much more sensitive to BD-induced carcinogenicity than rats (Melnick *et al.*, 1990; Owen *et al.*, 1987). Because BD toxicities are believed to be mediated by reactions of BD metabolites with nucleophilic sites on macromolecules, studies in our laboratory have focused on the characterization of potential metabolic pathways of BD bioactivation, in terms of the enzymes involved, the metabolites formed, and the kinetics of the reactions. Bioactivation reactions of several primary and secondary BD metabolites were also examined. In addition, the reactions of butadiene monoxide (BMO), a primary metabolite of BD, with macromolecules were characterized.

The objective of this review is to summarize and discuss our data on the metabolic pathways of BD bioactivation, and the *in vitro* reactions of BMO with nucleosides, calf-thymus DNA, and mouse and rat hemoglobin. A more detailed review of earlier studies has been published previously (Elfarra *et al.*, 1996).

CYTOCHROME P450-MEDIATED BIOACTIVATION REACTIONS

The primary oxidation products of BD by cytochrome P450s (Fig. 1) were shown to be BMO, and 3-butenal, which tautomerizes to produce the known carcinogen, 2-butenal (crotonaldehyde, CA; Elfarra *et al.*, 1991; Duescher and Elfarra, 1993). The amounts of CA formed in mouse liver, lung, and kidney microsomes were 2-5% of BMO amounts detected

Biological Reactive Intermediates VI, Edited by Dansette *et al.*
Kluwer Academic / Plenum Publishers, 2001

in these microsomes (Sharer *et al.*, 1992). BMO can be oxidized by cytochrome P450s to yield *meso*- and (±)-diepoxybutane (Fig. 1; Krause and Elfarra, 1997). Because the P450-dependent oxidation products of BD are known to cause genotoxicity and carcinogenicity in rodents, the P450 enzymes that catalyze BD (2E1, 2A6, 2B6, 2D6, 1A2, 4B1) and BMO (2E1, 2A6, 2C9, 4B1) oxidations in human and rodent tissues have been implicated in the mechanisms of BD-induced toxicities (Duescher and Elfarra, 1994a, 1994b; Krause and Elfarra, 1997; Krause *et al.*, 1997a, 1999). Thus, species, sex, and tissue differences in the expression of these P450s may play roles in differences in susceptibility.

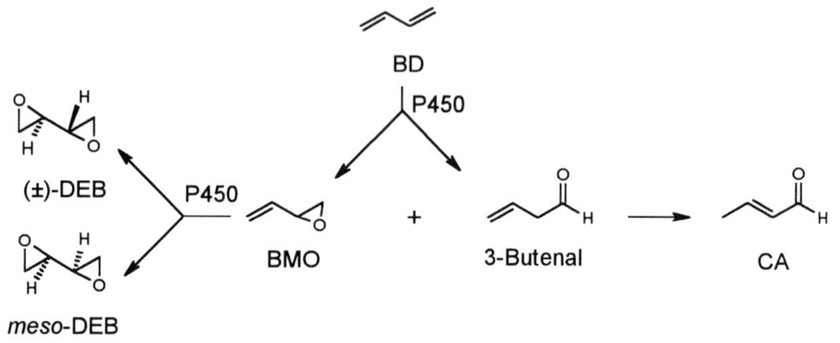

Figure 1. Scheme of 1,3-butadiene (BD) bioactivation to butadiene monoxide (BMO), 3-butenal, crotonaldehyde (CA), and diepoxybutane (DEB).

BMO is metabolized to 3-butene-1,2-diol (BDD) by both microsomal and cytosolic epoxide hydrolases, but rats given BMO or BDD excreted only a small fraction (<5%) of the BMO or BDD dose as BDD and BDD sulfate or glucuronide conjugates into urine within 24 h after BMO or BDD administration (Krause *et al.*, 1997; Kemper *et al.*, 1998). *In vivo* treatment of mice with BDD depleted hepatic and renal non-protein thiols and inactivated hepatic microsomal *p*-nitrophenol hydroxylase activity, a 2E1 marker. These results and additional data provided evidence for BDD metabolism by cytochrome P450s and alcohol dehydrogenases to yield three reactive metabolites, hydroxymethylvinylketone (HMVK), 2-hydroxy-3-butenal (HBAL), and epoxybutanediol (EBD; Fig. 2; Kemper and Elfarra, 1996; Kemper *et al.*, 1997; 1998). EBD could also be formed by further metabolism of *meso*- and (±)-DEB by epoxide hydrolases (Krause and Elfarra, 1997).

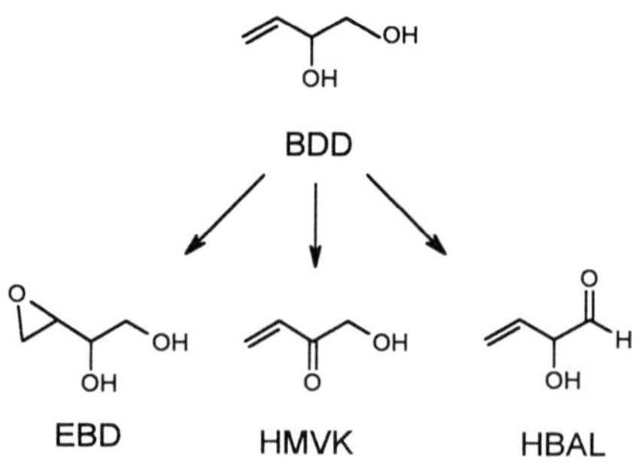

Figure 2. Bioactivation pathways for 3-butene-1,2-diol (BDD) by cytochrome P450s and/or alcohol dehydrogenases; EBD, epoxybutanediol; HMVK, hydroxymethylvinylketone; HBAL, 2-hydroxy-3-butenal.

When coordinated metabolism of BMO was examined in freshly isolated mouse and rat hepatocytes (Kemper and Elfarra, 1999; Elfarra *et al.*, 2000), the hepatocytes from both species were found to catalyze BMO oxidation to *meso-* and ($\pm$)-DEB, BMO hydrolysis to BDD, and BMO conjugation with GSH to form GSH conjugates (GSBMO). The latter reaction has been previously characterized both *in vitro* and *in vivo* (Sharer *et al.*, 1992; Sharer and Elfarra, 1992; Elfarra *et al.*, 1995). Metabolite area under the curve (AUC) exhibited dependence on BMO concentration and incubation time. At low BMO concentrations (5 or 25 μM), the observed BMO activation/detoxication ratios (obtained by dividing the AUC for total DEB by the summed AUC values for BDD and GSBMO) with the mouse hepatocytes were nearly 40-fold higher than the ratios observed with the rat hepatocytes.

MYELOPEROXIDASE-MEDIATED BIOACTIVATION REACTIONS

BD can also be oxidized by myeloperoxidase (MPO), an enzyme present in polymorphonuclear leukocytes and bone marrow (Duescher and Elfarra 1992a). The products formed (Fig. 3) were dependent on the KCl concentration in the incubation; in the absence of KCl, BD was oxidized to BMO and CA, but in the presence of KCl, 1-chloro-2-hydroxy-3-butene (CHB) was also formed and became the major product at KCl concentrations higher than 50 mM. Detailed kinetic and mechanistic investigations provided evidence for BD oxidation by MPO by two mechanisms: the first is direct oxygen transfer from the hemoprotein to BD to yield BMO and CA, and the second is the reaction of BD with HOCl, the oxidation product of chloride ion by MPO. MPO did not catalyze BMO oxidation to DEB, and the rate of direct BD oxidation by MPO was much higher than the rate measured with styrene (Duescher and Elfarra, 1995), even though styrene was readily metabolized to the expected chlorohydrin by MPO in the presence of KCl (Duescher and Elfarra, 1992b). These results are consistent with a restrictive active site that would allow access to BD but not BMO or bulky olefins for direct oxidation. While these results suggest a role for MPO in BD activation and toxicity, further investigation is needed to clarify this potential role.

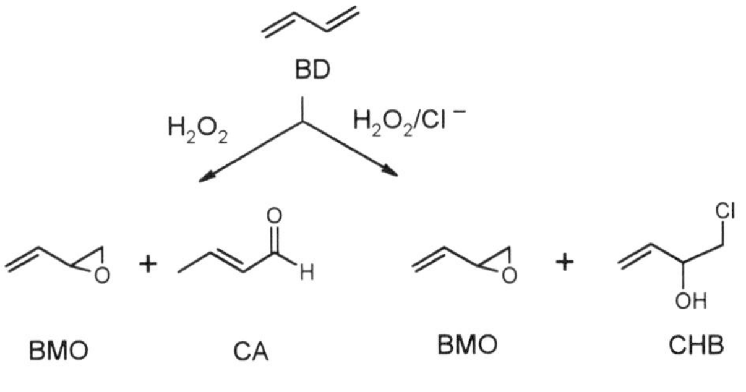

Figure 3. Scheme of 1,3-butadiene (BD) bioactivation by myeloperoxidase in the presence and absence of chloride ion; BMO, butadiene monoxide; CA, crotonaldehyde; CHB, 1-chloro-2-hydroxy-3-butene.

REACTIONS OF BMO WITH NUCLEOSIDES AND DNA

We have characterized 29 BMO adducts formed from the four DNA bases based on their UV, ^{1}H NMR, and fast atom bombardment mass spectra (Selzer and Elfarra, 1996a; 1996b; 1997a; 1997b). These adducts consist of 15 pairs of diasteromeric regioisomers of BMO (Figs. 4-7). Initial characterization of the purine adducts was done with nucleoside

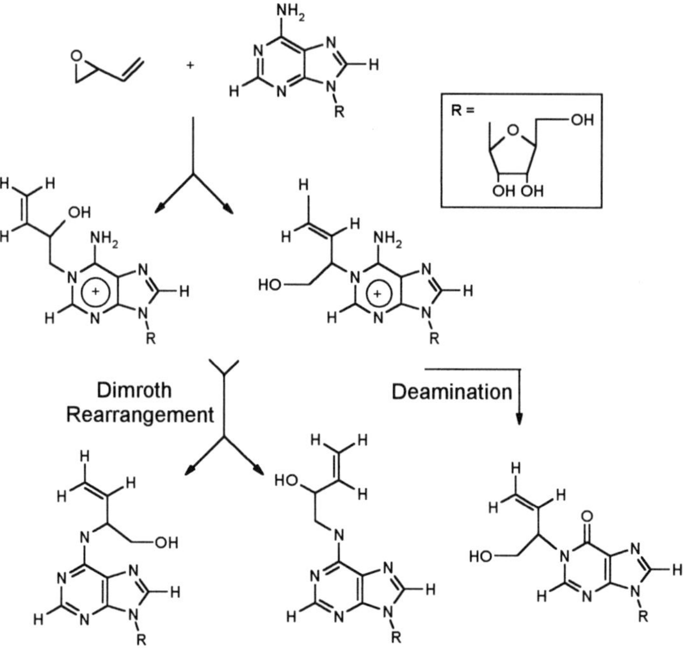

Figure 4. Reaction of butadiene monoxide with adenosine.

bases; however, additional studies with the deoxynucleoside bases showed that the reactivity was similar regardless of the sugar moiety.

The N-1- and N^2-guanosine, N^6-adenosine, N-1-inosine, N-3-deoxyuridine, and N-3-thymidine adducts of BMO were stable under *in vitro* physiological conditions. Other adducts underwent depurination, deamination, and rearrangement reactions. The N-7-guanosine adducts depurinated to the corresponding N-7-guanine adducts, with half-lives of 50-90 h. The N-1 adducts of adenosine, the initial products of the reaction of BMO with adenosine, underwent Dimroth rearrangement to the corresponding N^6-adenosine adducts (Fig. 4). In a competing reaction, the N-1-adenosine regioisomer containing a terminal hydroxyl group deaminated to corresponding N-1-inosine adducts. The N-3-deoxycytidine

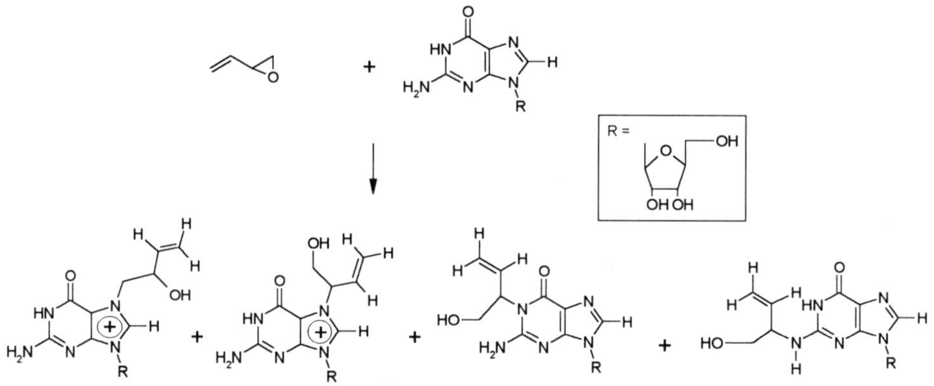

Figure 5. Reaction of butadiene monoxide with guanosine.

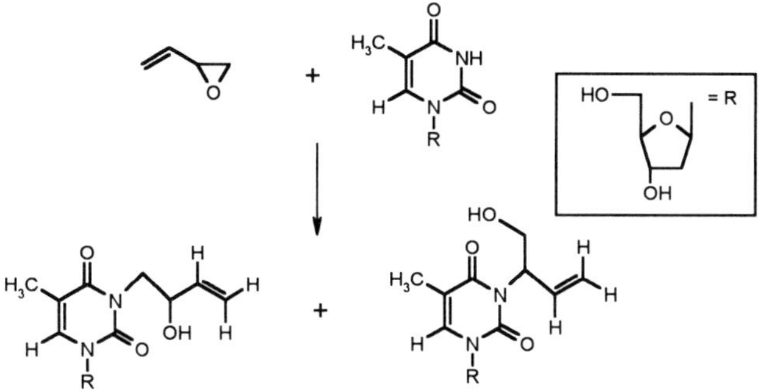

Figure 6. Reaction of butadiene monoxide with thymidine.

adducts, with half-lives ranging from 0-2.5 h, deaminated to the corresponding N-3-deoxy-uridine adducts, whereas the N^2-deoxycytidine adducts dealkylated with half-lives of 11 h (Fig. 7).

Pseudo-first-order rate constants for BMO adduct formation were calculated from the *in vitro* reactions of nucleosides with an excess of BMO. Comparison of the rate constants (Table 1) demonstrates variable reactivity depending upon the individual nucleoside and its site of modification.

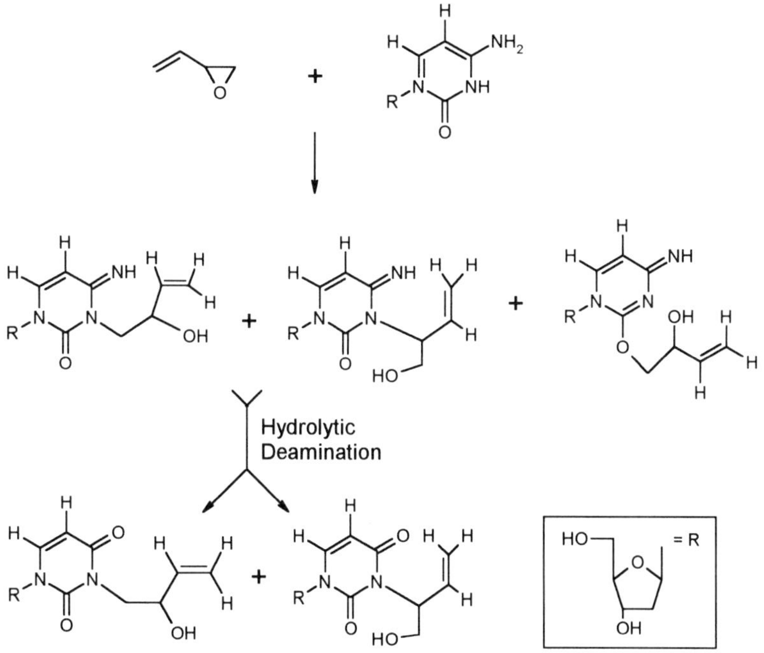

Figure 7. Reaction of butadiene monoxide with cytidine.

When BMO was reacted with single- (ss) or double-stranded (ds) calf thymus DNA, several guanine, adenine, deoxyadenosine and deoxyuridine adducts were detected after enzymatic hydrolysis of the DNA (Fig. 8), suggesting the involvement of multiple DNA

Table 1. Pseudo-First-Order Rate Constants of the Reaction of BMO with Nucleosides.

Adduct	Pseudo-first-order rate constant (h^{-1})[a]
N-7-(1-hydroxy-3-buten-2-yl)guanosine	2.58×10^{-2}
N-7-(2-hydroxy-3-buten-1-yl)guanosine	2.58×10^{-2}
N^2-(1-hydroxy-3-buten-2-yl)guanosine	4.00×10^{-3}
N-1-(1-hydroxy-3-buten-2-yl)guanosine	3.38×10^{-3}
N^6-(1-hydroxy-3-buten-2-yl)adenosine	3.38×10^{-3}
N^6-(2-hydroxy-3-buten-1-yl)adenosine	9.68×10^{-3}
N-1-(1-hydroxy-3-buten-2-yl)inosine	4.10×10^{-3}
N-3-(1-hydroxy-3-buten-2-yl)deoxyuridine	9.71×10^{-4}
N-3-(2-hydroxy-3-buten-1-yl)deoxyuridine	6.19×10^{-3}
N-3-(2-hydroxy-3-buten-1-yl)deoxycytidine	6.40×10^{-3}
O^2-(2-hydroxy-3-buten-1-yl)deoxycytidine	1.35×10^{-3}
N-3-(1-hydroxy-3-buten-2-yl)thymidine	1.50×10^{-4}
N-3-(2-hydroxy-3-buten-1-yl)thymidine	1.49×10^{-3}

[a]Data were taken from Selzer and Elfarra, 1996a, 1996b, 1997a, and 1997b. Pseudo-first-order rate constants were similar for diastereomers and are reported as the averages of rate constants calculated for the two diastereomers.

adducts in BMO-induced genotoxicity (Selzer and Elfarra, 1999). Consistent with the pseudo-first-order constants (Table 1), the primary products were regioisomeric *N*-7-guanine adducts (Table 2). Regioisomeric *N*-3-adenine adducts, which depurinated from the DNA more rapidly than the *N*-7-guanine adducts, were also formed. Regioisomeric N^6-deoxyadenosine adducts were detected; these adducts are apparently formed by the Dimroth rearrangement of the corresponding *N*-1-deoxyadenosine adducts produced after the release of the *N*-1-alkylated nucleosides by enzymatic hydrolysis. *N*-3-(2-Hydroxy-3-buten-1-yl)-deoxyuridine adducts, deamination products of the corresponding deoxycytidine adducts, were also detected. Rates of formation of the *N*-3-deoxyuridine and N^6-deoxyadenosine adducts were 10-20-fold higher in ssDNA compared to dsDNA. The *N*-7-guanine adducts increased only slightly, presumably because the *N*-7 guanine position is not involved in hydrogen bonding in dsDNA.

Table 2. Quantitation of adducts formed in the reaction of butadiene monoxide (BMO) with calf thymus ssDNA and dsDNA.

Adduct	25 mM BMO		100 mM BMO		500 mM BMO	
	ssDNA	dsDNA	ssDNA	dsDNA	ssDNA	dsDNA
I	1/220[a]	1/280	1/61	1/95	1/20	1/28
II	1/187	1/251	1/45	1/101	1/15	1/30
III	1/2107	n.d.	1/565	1/7951	1/129	1/1833
IV	1/1683	n.d.	1/455	1/5973	1/106	1/1050
V	1/1796	n.d.	1/480	1/9500	1/105	1/2166

[a]The BMO-DNA adduct concentrations were determined using HPLC. Adduct levels were expressed as adducts per unmodified base. Structures of adducts I-V can be found in Figure 8. n.d. = not detectable

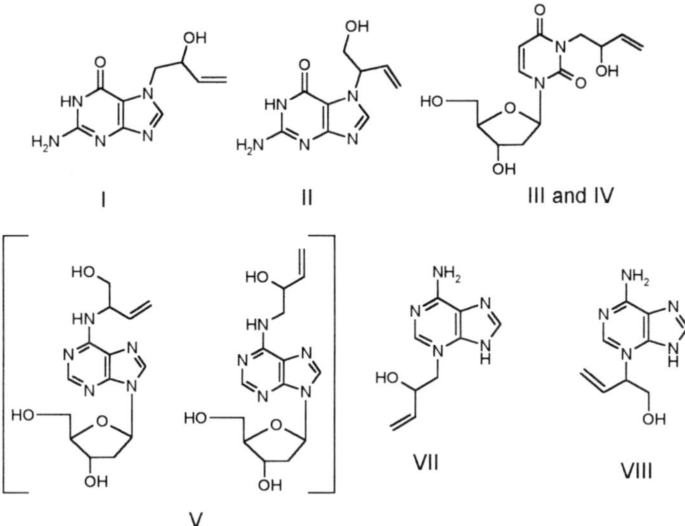

Figure 8. Characterized DNA adducts of butadiene monoxide. I, *N*-7-(2-hydroxy-3-buten-1-yl)guanine; II, *N*-7-(1-hydroxy-3-buten-2-yl)guanine; III and IV, *N*-3-(2-hydroxy-3-buten-1-yl)deoxyuridine; V, N^6-deoxyadenosine, both regioisomers; VII, *N*-3-(2-hydroxy-3-buten-1-yl)adenine; VIII, *N*-3-(1-hydroxy-3-buten-2-yl)adenine.

REACTIONS OF BMO WITH HEMOGLOBIN

As biomarkers of human exposure to BD, BMO adducts at the *N*-terminal valine residues of hemoglobin have been measured (Osterman-Golkar, *et. al.*, 1996). However, mechanistic details concerning the relative reactivity, regioselectivity, and stereoselectivity of the reaction of BMO with the *N*-terminal valine of hemoglobin were lacking. Therefore, we studied the reaction *in vitro* using intact erythrocytes from rats and mice (Moll and Elfarra, 1999). The reaction of BMO with the *N*-terminal valine residue of both rat and mouse hemoglobin produced mostly C-1 adducts (Figs. 9-10). However, the reaction with rat hemoglobin produced a higher ratio of C-1:C-2 adducts in comparison with the reaction with mouse hemoglobin. These results and the finding that the rates obtained with rat hemoglobin were slower than the rates obtained with mouse hemoglobin demonstrate species differences in this reaction, indicating the importance of measuring all adducts when comparing the relative rates of adduct formation among different species.

Because the *N*-terminal valine adducts may not be the first to be formed on the α- and β-globin chains, and may not be the most prevalent adducts since there are only four *N*-terminal valine residues per hemoglobin molecule, new electrospray- and liquid chromatography-mass spectrometry methods were developed to characterize the reaction of BMO with mouse hemoglobin. The new methods allowed detection of up to ten BMO adducts on each of the α- and β-globin chain. This suggests the utility of our approach to develop new biomonitoring methods that are likely to be more sensitive than the methods used to determine adducts at the *N*-terminal valine residues of hemoglobin (Moll and Elfarra, 2000).

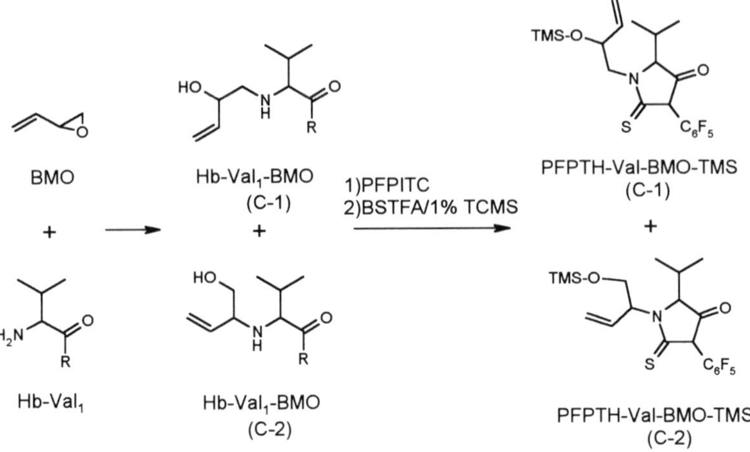

Figure 9. Scheme of butadiene monoxide reaction with the N-terminal valine of mouse and rat hemoglobin (Hb-Val$_1$); PFPITC, pentafluorophenyl isothiocyanate; BSTFA, bis(trimethylsilyl)trifluoroacetamide; TCMS, trimethylchlorosilane; PFPTH, pentafluorophenylthiohydantoin; TMS, trimethylsilyl.

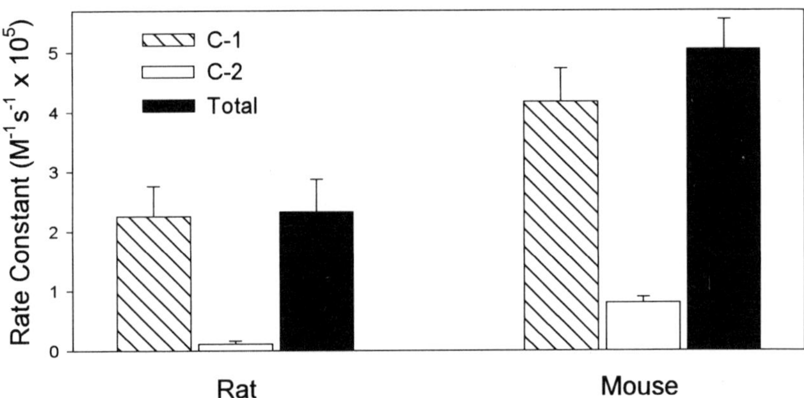

Figure 10. Kinetic parameters (second-order rate constants) from the in vitro reaction of mouse and rat erythrocytes with butadiene monoxide. Values represent the mean ± S.D. (n = 6). For chemical structures of C-1 and C-2, see figure 9.

CONCLUSIONS

The studies described above provide evidence for the formation of several primary and secondary reactive metabolites of BD *in vitro*. These metabolites may play a role in BD-induced toxicity. These studies also illustrate the potential of the isolated hepatocyte model for estimating flux through competing metabolic pathways and predicting *in vivo* metabolism and carcinogenicity of BMO, the primary reactive metabolite of BD. The hepatocyte model could be useful in assessing species differences in BD bioactivation and may help predict the extent to which many of the BD reactive metabolites are formed *in vivo*. Development of highly sensitive biomarker assays for the various reactive metabolites, along with metabolic and toxicity studies using MPO and/or P450 knock-out mice will enhance our understanding of the roles of various metabolic pathways in BD-induced genotoxicity and carcinogenicity. Because each reactive metabolite of BD is likely to react with multiple sites on DNA, and various adducts are likely to exhibit different rates of formation and different stabilities, more studies are needed to assess the relative roles of specific DNA adducts in BD-induced genotoxicity and carcinogenicity.

ACKNOWLEDGEMENTS

The studies conducted in the authors' laboratory were supported by NIEHS grant ES06841; T.S.M. was supported by NRSA fellowship ES05835 from NIEHS; R.R.S. was supported by an NSF graduate research fellowship.

REFERENCES

Duescher, R.J., and Elfarra, A.A. (1992a) 1,3-Butadiene oxidation by human myeloperoxidase: Role of chloride ion in catalysis of divergent pathways. *J. Biol. Chem.*, **267**, 19859-19865.

Duescher, R.J., and Elfarra, A.A. (1992b). Chlorohydrins are novel metabolites of 1,3-butadiene, styrene, and cyclohexene formed by human myeloperoxidase. *Toxicologist* **12**, 63.

Duescher, R.J., and Elfarra, A.A. (1993). Chloroperoxidase-mediated oxidation of 1,3-butadiene to 3-butenal, a crotonaldehyde precursor. *Chem. Res. Toxicol.*, **6**, 669-673.

Duescher, R.J., and Elfarra, A.A. (1994a) Human liver microsomes are efficient catalysts of 1,3-butadiene oxidation. Evidence for major roles by cytochromes P450 2A6 and 2E1. *Arch. Biochem. Biophys.*, **311**, 342-349.

Duescher, R.J., and Elfarra, A.A. (1994b) Sex-related differences in mouse lung and kidney microsomal oxidation of 1,3-butadiene. Evidence for the involvement of multiple P450 enzymes. *Toxicologist*, **14**, 49.

Duescher, R.J., and Elfarra, A.A. (1995) Effects of structural modifications on direct alkene oxidation by human myeloperoxidase. *FASEB J.*, **9**, A369.

Elfarra, A.A., Duescher, R.J., and Pasch, C. (1991) Mechanisms of 1,3-butadiene oxidation to butadiene monoxide and crotonaldehyde by mouse liver microsomes and chloroperoxidase. *Arch. Biochem. Biophys.* **286**, 244-251.

Elfarra, A.A., Sharer,J.E., and Duescher, R.J. (1995) Synthesis and characterization of *N*-acetyl-L-cysteine *S*-conjugates of butadiene monoxide and their detection and quantitation in urine of rats and mice given butadiene monoxide. *Chem. Res. Toxicol.*, **8**, 68-76.

Elfarra, A.A., Krause, R.J., and Selzer, R.R. (1996) Biochemistry of 1,3-butadiene metabolism and its relevance to 1,3-butadiene-induced carcinogenicity. *Toxicology,* **113**, 23-30.

Elfarra, A.A., Krause, R.J., and Kemper, R.A. (2000). Cellular and molecular basis for species, sex and tissue differences in 1,3-butadiene metabolism. Abstracts of the International Symposium on "Evaluation of Butadiene, Isoprene & Chloroprene Health Risks," London, England, 12-14 September.

Kemper, R.A., and Elfarra, A.A. (1996) Oxidation of 3-butene-1,2-diol by alcohol dehydrogenase. *Chem. Res. Toxicol.,* **9**, 1127-1134.

Kemper, R.A., and Elfarra, A.A. (1999) Metabolism of butadiene monoepoxide by freshly isolated hepatocytes from mice and rats. *Toxicol. Sci. (suppl.)* **48**, 225.

Kemper, R.A., Krause, R.J., and Elfarra, A.A. (1997) Oxidation of 3-butene-1,2-diol by mouse, rat, and human liver microsomes. *Fund. Appl. Toxicol. (suppl.)* **36**, 317.

Kemper, R.A., Elfarra, A.A., and Myers, S.R. (1998) Metabolism of 3-butene-1,2-diol in B6C3F1 Mice: Evidence for involvement of alcohol dehydrogenase and cytochrome P-450. *Drug Metab. Dispos.,* **26**, 914-920.

Krause, R.J., and Elfarra, A.A. (1997) Oxidation of butadiene monoxide to *meso-* and (±)-diepoxybutane by cDNA-expressed human P450s and by mouse, rat, and human liver microsomes: Evidence for preferential hydration of *meso*-DEB in rat and human liver microsomes. *Arch. Biochem. Biophys.,* **337**, 176-184.

Krause, R.J. Philpot, R.M., and Elfarra, A.A. (1997a). Role of P450 4B1 in 1,3-butadiene and butadiene monoxide oxidations in mouse tissues. *Fund. Appl. Toxicol. (suppl.)* **36**, 133.

Krause, R.J., Sharer, J.E., and Elfarra, A.A. (1997b) Epoxide hydrolase-dependent metabolism of butadiene monoxide to yield 3-butene-1,2-diol in mouse, rat, and human liver. *Drug Metab. Dispos.,* **25**, 1013-1015.

Krause, R.J., Philpot, R.M., and Elfarra, A.A. (1999). Role of cytochrome P450 4B1 in 1,3-butadiene oxidation in lung microsomes of humans, rats, and rabbits. *Toxicol. Sci. (suppl.)* **48**, 411.

Melnick, R.L., Huff, J.E., Chou, J.B., and Miller, R.A. (1990) Carcinogenicity of 1,3-butadiene in C57BL/6 x C3H F1 mice at low exposure concentrations. *Cancer Res.,* **50**, 6592-6599.

Moll, T.S., and Elfarra, A.A. (1999) Characterization of the reactivity, regioselectivity and stereoselectivity of the reactions of butadiene monoxide with valinamide and the N-terminal valine of mouse and rat hemoglobin. *Chem. Res. Toxicol.,* **12**, 679-689.

Moll, T.S., and Elfarra, A.A. (2000) A comprehensive structural analysis of hemoglobin adducts formed after *in vitro* exposure of erythrocytes to butadiene monoxide. Abstracts of the International Symposium on "Evaluation of Butadiene, Isoprene & Chloroprene Health Risks," London, England, September 12-14.

Osterman-Golkar, S., Peltonen, K., Anttinen-Klemetti, T., Landin, H.H., Zorec, V., and Sorsa, M. (1996) Haemoglobin adducts as biomarkers of occupational exposure to 1,3-butadiene. *Mutagenesis* **11**, 145-149.

Owen, P.E., Glaister, J.R., Gaunt, I.F., and Oullinger, D.H. (1987) Inhalation toxicity studies with 1,3-butadiene. III. Two year toxicity/carcinogenicity studies in rats. *Am. Ind. Hyg. Assoc.,* **48**, 407-413.

Selzer, R.R., and Elfarra, A.A. (1996a) Synthesis and biochemical characterization of N^1-, N^2-, and N^7-guanosine adducts of butadiene monoxide. *Chem. Res. Toxicol.,* **9**, 126-132 .

Selzer, R.R., and Elfarra, A.A. (1996b) Characterization of N^1- and N^6-adenosine adducts and N^1-inosine adducts formed by the reaction of butadiene monoxide with adenosine. Evidence for the N^1-adenosine adducts as major initial products. *Chem. Res. Toxicol.,* **9**, 875-881.

Selzer, R.R., and Elfarra, A.A. (1997a) Chemical modification of deoxycytidine at different sites yields adducts of different stabilities. Characterization of N^3- and O^2-deoxycytidine and N^3-deoxyuridine adducts of butadiene monoxide. *Arch. Biochem. Biophys.* **343**, 63-72.

Selzer, R.R., and Elfarra, A.A. (1997b) Characterization of four N-3-thymidine adducts formed *in vitro* by the reaction of thymidine and butadiene monoxide. *Carcinogenesis,* **18**, 1993-1998.

Selzer, R.R., and Elfarra, A.A. (1999) *In vitro* reactions of butadiene monoxide with single and double stranded DNA: Characterization and quantitation of several purine and pyrimidine adducts. *Carcinogenesis,* **20**, 285-292.

Sharer, J.E., and Elfarra, A.A. (1992) *S*-(2-Hyroxy-3-buten-1-yl)glutathione and *S*-(1-hydroxy-3-buten-2-yl)glutathione are *in vivo* metabolites of butadiene monoxide. Detection and quantitation in bile. *Chem. Res. Toxicol.* **5,** 787-790.

Sharer, J.E., Duescher, R.J., and Elfarra, A.A. (1992) Species and tissue differences in the microsomal oxidation of 1,3-butadiene and the glutathione conjugation of butadiene monoxide in mice and rats. Possible role in 1,3-butadiene-induced toxicity. *Drug Metab. Dispos.,* **20,** 658-664.

U.S Department of Health and Human Services, Public Health Service, National Toxicology Program (2000). The Ninth Report on Carcinogens.

CHEMISTRY AND BIOLOGICAL ACTIVITY OF NOVEL SELENIUM-CONTAINING COMPOUNDS

Jan N.M. Commandeur, Martijn Rooseboom, and Nico P.E. Vermeulen

Division of Molecular Toxicology, Leiden/Amsterdam Center for Drug Research, FEW, Vrije Universiteit, de Boelelaan 1083, 1081 HV Amsterdam

INTRODUCTION

A large number of animal studies and epidemiological studies which have been performed in the last decades has revealed that a number of selenium (Se)-containing compounds possess chemopreventive activity. In a meta-analysis from a number of studies comparing the significance of serum Se, retinol, beta-carotene and vitamin E, Se emerged as the factor with the most consistent protective effect (Comstock et al., 1992). On the other hand, geographic analysis and several prospective and case-control studies showed that people with low blood Se had an increased risk of cancer. Recently, a placebo-controlled double-blind study involving 1312 patients with a history in skin cancer showed a much lower prevalence of developing and dying from lung, colon or prostate cancer in the group receiving supplementation of 200 µg Se per day (Clark et al., 1996). A selenized yeast-extract was given to the people of this study. Relative risk of cancer incidence in lung, colon and prostate was reduced to 0.54 (P=0.04), 0.37 (P=0.002) and 0.42 (P=0.03), respectively. The Se-compounds responsible for the chemopreventive effect of selenized yeast still are not completely characterized yet. Selenomethionine accounts for 20% of the Se-containing compounds (Bird et al., 1997). Other compounds identified include selenocystine, Se-methylselenocysteine and selenoethionine, representing approximately 20%. However, still 40-50% of the total Se is unidentified.

In the present paper, the chemistry of Se-compounds and their interactions with biomolecules will be briefly reviewed. For more comprehensive reviews on these subjects, see Ip (1998) and Ganther (1999). Subsequently, some recent results will be presented on the bioactivation mechanisms of novel selenocysteine (SeCys)-conjugates.

CHEMICAL FORMS OF SELENIUM AND ANTICARCINOGENIC ACTIVITY

The chemopreventive activity of several individual Se-compounds as well as Se-enriched diet has now been well established in many animal cancer models (Ip, 1998; Ganther, 1999). More than 90% of these chemoprevention studies on selenium have been carried out with selenomethionine and sodium selenite. Both of these compounds are non-organ specific chemopreventive agents as they suppress tumorigenesis in a variety of

Biological Reactive Intermediates VI, Edited by Dansette *et al.*
Kluwer Academic / Plenum Publishers, 2001

organs, such as liver, skin, pancreas, esophagus, colon, mammary gland and a few other sites. The disposition of selenomethionine and sodium selenite has been the subject of several studies and is summarized in Figure 1. It has been demonstrated that the Se of selenite and selenomethionine to some extent becomes incorporated in selenoproteins as selenocysteine through a unique co-translational mechanism. With both compounds hydrogen selenide is a key metabolite that is subsequently activated to selenophosphate. Whether incorporation in selenoproteins contributes to the chemopreventive activity, however, is still questionable since activity of several selenoproteins, such as glutathione peroxidase and type I deiodinase, is already maximal at normal dietary levels of Se which are not chemopreventive. Results from several in vivo studies suggest that a monomethylated Se-species is critical for the chemopreventive activity of these compounds.

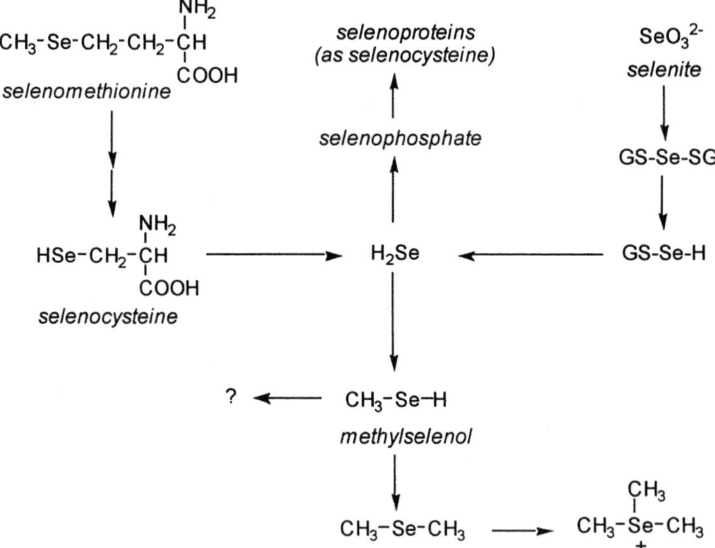

Figure 1. Biotransformation of selenite and selenomethionine to chemopreventive methylated selenium compounds.

Alternative precursors for generating monomethylated Se, such as selenobetaine, Se-methylselenocysteine and methylselenocyanate, Fig. 2, all appeared to be more efficacious than either selenite or selenomethionine in cancer chemoprotection in the range of 1-3 ppm (Ip and Ganther, 1990; Ip and Ganther, 1992; Ip et al., 1994). Recently, methyl seleninic acid also appeared to be equipotent as chemopreventive agent in vivo, when compared to Se-methylselenocysteine (Ip et al., 2000). Se-methylselenocysteine is believed to be converted to methylselenol directly via a beta-lyase reaction. Methyl seleninic acid is rapidly and non-enzymatically reduced by GSH to form methyl selenenic acid and methyl selenol, subsequently, Fig. 2. In an in vitro study using mammary hyperplastic epithelial cells methylseleninic acid appeared to be 10 times more active than Se-methylselenocysteine, which may be explained by a limiting beta-lyase activity in the cell lines used. Both compounds were able to induce apoptosis in cell lines with wild-type or non-functional p53 and these effects were not attributable to DNA damage (Ip et al., 2000).

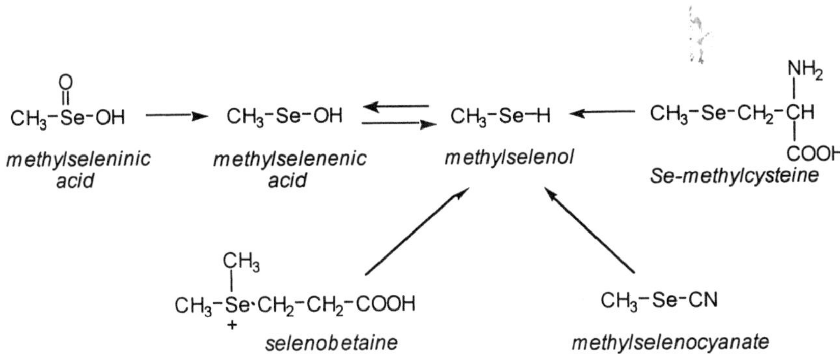

Figure 2. Alternative precursors for chemopreventive monomethylated selenium-compounds.

A concern with compounds forming monomethylated Se-metabolites is the close proximity between their effective dose and the toxic range. The chemopreventive index (defined as the ratio MTD/ED$_{50}$), for several Se-compounds is: selenomethionine, 1.0; sodium selenite, 1.3; Se-methylselenocysteine, 2.0; methylselenocyanate, 2.0. In an attempt to develop compounds with higher chemopreventive index several aliphatic and aromatic selenocyanates have been synthesized and tested. The benzyl-type compounds, benzylselenocyanate and phenylene-bis(methylene)selenocyanates, showed potent anticarcinogenic activity in the DMBA-induced mammary tumor model with an improved chemopreventive indices of 2.5 and 4.0, respectively (Ip et al., 1994). In a series of aliphatic selenocyanates the chemopreventive potency increased with increasing length of the carbon side chain (Ip et al., 1995). These results indicate that the chemopreventive activity is not restricted to monomethylated Se-compounds, but that other Se-substituents are allowed as well. This is confirmed by the recent observation that Se-allylselenocysteine and Se-propylselenocysteine showed higher chemopreventive activity when compared to Se-methylselenocysteine (Ip et al., 1999). This means that novel Se-compounds may be designed with optimal chemopreventive activity and maximal tolerance.

MECHANISM OF ACTION OF LOW MOLECULAR WEIGHT SELENIUM COMPOUNDS

Because activity of selenoproteins is already maximal at normal dietary Se-levels (0.1 ppm), it is likely that the chemopreventive activity of Se-compounds is mediated by low molecular weight Se-compounds. The biological activity of these Se-compounds on the one hand may be explained by their excellent ability to inactivate oxidants like hydrogen peroxide and peroxynitrite (Sies, 1998), or by their ability to modify protein thiol-homeostasis.

In his recent review, Ganther classified the reactions by which Se-compounds might modify proteins into four different types of reactions (Ganther, 1999):

• *Type 1* reactions involve the formation of selenotrisulfide bonds (S-Se-S) by reaction of selenite with proteins with cysteine clusters.

• *Type 2* reactions involve formation of selenylsulfide bonds (Se-S) and may involve reactions of selenenic acids with protein thiols-groups.

• *Type 3* reactions involve the catalysis of protein disulfide (S-S) bond formation or its reversal. Disulfide bond formation may result from reaction of selenenic acid with a protein dithiol moiety: selenylsulfide bond formed is subsequently attacked by the second thiol-group, releasing the Se as a selenol compound. Reduction of disulfides by GSH is known to be strongly catalysed by selenol-compounds. A selenol-group has a pKa-value of 5, implicating that under physiological conditions the Se-atom is strongly nucleophilic and able to open disulfides and form a selenylsulfide derivate. A selenylsulfide undergoes facile reaction with GSH forming a GSH-protein mixed disulfide thereby releasing the catalytic selenol-group again. The GSH-protein mixed disulfide subsequently can again be reduced by a thiol-disulfide oxidoreductases or protein disulfide isomerase.

• *Type 4* reactions involve the formation of diselenide bonds (Se-Se) by reaction of the selenocysteine-residue of selenoproteins with selenenic acid. It has been speculated that the selenocysteine of mammalian thioredoxin reductase may be a target for methylselenenic acid to give a stable diselenide that is inactive and is resistant to reduction by cellular disulfide reductases. The gradual formation of the inactive diselenide is suggested as a mechanism to explain the declining levels of thioredoxin reductase activity observed upon prolonged Se-supplemention (Ganther, 1999). Protein diselenides appear to be more difficult to reduce than protein disulfide as demonstrated previously with thioredoxin disulfide and its diselenide analog (Muller et al., 1994). Whereas the disulfide was reduced by 1% 2-mercaptoethanol within minutes, the diselenide form was not reduced under the same conditions.

As an alternative mode of action of Se-compounds, interference with the Zn-homeostases has been proposed recently (Jacob et al., 1999). Over 200 proteins are known to contain zinc-specific inhibitory sites where zinc is complexed with high affinity to cysteine-clusters. The enzyme caspase-3 which plays an important role in apoptosis is inhibited by zinc that has a binding constant of 1.5 nM (Perry et al., 1997). As a possible explanation for Se-induced apoptosis, release of this inhibitory zinc by complexing to selenolates was proposed. The higher affinity of selenolate for Zn could result in release of the cysteinylthiolate-bound zinc and activate the enzyme (Ganther, 1999). More recently, however, it was shown that zinc-release from metallothionein by benzeneselenol was dramatically enhanded by addition of t-butylhydroperoxide (Jacob et al., 1999). This may indicate that selenenic acids also may interfere with zinc-homeostatis by formation of proteinselenylsulfides (*Type 2*) or by oxidation of protein-thiols (*Type 3*).

NOVEL SELENOCYSTEINE CONJUGATES AS POTENTIAL KIDNEY-SELECTIVE PRODRUGS

As indicated above, three naturally occuring SeCys-conjugates, Se-methylselenocysteine, Se-allylselenocysteine and Se-propylselenocysteine, have been shown to be relatively potent chemopreventive agents in chemically-induced tumor models (Ip et al., 1999). Se-Allylselenocysteine, acheiving an efficiency of close to 90% inhibition of methylnitrosoureau-induced mammary tumors in rats when added at 2 ppm level in the diet, even appears to be the most active chemopreventive Se-compound identified as yet. As a mechanism of action of these compounds a beta-elimination reaction is proposed, leading to the formation of selenol-compounds. The enzymes involved in these beta-elimination reactions, however, have not been characterized yet.

Cysteine S-conjugates, the sulfur-analogues of SeCys-conjugates are also metabolised via beta-elimination reactions forming ammonia, pyruvate and corresponding thiol-compounds. In case of cysteine S-conjugates, beta-elimination activity is highest in kidney whereas lower activity was observed in liver. Three mammalian beta-lyase enzymes have now been characterized, all of which appear to be cytosolic pyridoxal-5'-phosphate (PLP)-dependent enzymes (Cooper, 1998). In kidney, glutamine transaminase K (GTK) has been identified as the major cytosolic cysteine conjugate beta-lyase. This 90 kD protein appeared to be identical to kynurenine aminotransferase (KAT). In addition a high MW (Mr 330.000) protein was demonstrated with slightly different substrate selectivity. The hepatic beta-lyase is distinct from the renal forms and after cloning appeared to be identical to kynureninase.

Cysteine S-cojugates of halogenated alkenes are beta-eliminated at a high rate by the renal beta-lyase enzymes to form ammonia, pyruvate and highly reactive thiol-compounds (Commandeur et al., 1995). The combination of active renal uptake and high renal beta-lyase activity results in a selective nephrotoxicity when administered to rodents. In an attempt to develop kidney-selective antitumor-agents Elfarra et al. demonstrated that this targeting concept can also be applied to target the antitumor agents mercaptopurine and thioguanine to the kidney by administration of S-(purinyl)-L-cysteine and S-(guanin-6-yl)-L-cysteine, respectively (Elfarra et al., 1993 and 1995). However a limitation of this prodrug-concept is the fact that only significant beta-elimination occurs with strongly electron-withdrawing S-substituents.

In 1996, a series of 20 novel SeCys-conjugates were synthesized in order to evaluate whether the same targeting concept can be applied to target antitumor selenol-compounds to the kidney (Andreadou et al., 1996a). Interestingly, all selenocysteine conjugates tested showed very high beta-elimination activities in kidney cytosol, when measured by formation of pyruvate. Activities were more than 50-fold higher when compared to their sulfur-analogues. Electron-withdrawing substituents did not appear to be required for high beta-lyase activities anymore, which may be explained by a better leaving-group ability of selenols when compared to thiols.

Subsequent studies have been dedicated to the identification of the proteins involved in beta-elimination of the novel SeCys-conjugates. By using highly purified rat renal cysteine S-conjugate/GTK, we recently demonstrated that this protein is extremely active in beta-elimination of SeCys-conjugates (Commandeur et al., 2000). Next to beta-elimination these conjugates are also actively transaminated by GTK which in fact appeared to be the major pathway. When correlating specific activities of purified GTK with those observed in rat renal cytosol, however, a relatively poor correlation was observed. For example, para-substituted Se-phenylselenocysteine conjugates showed extremely low activity with GTK, whereas in cytosol a significant beta-elimination activity was observed. Several additional experiments indicated that the poor correlation observed is explained by the involvement of multiple enzymes in the pyruvate formation from SeCys-conjugates.

Firstly, rat renal cytosol was fractionated by anion exchange chromatography and fractions were screened for beta-lyase activity and cross-reactivity to polyclonal antibody to GTK (anti-GTK) (Commandeur et al., 2000). Two fractions displaying high beta-lyase activity showed the presence of cross-reactive proteins to anti-GTK. Other fractions displaying significant pyruvate formation from SeCys-conjugates did not contain cross-reactive proteins, indicating involvement of additional proteins.

By using aminooxyacetic acid (AOAA), an extremely potent and aselective inhibitor of PLP-dependent enzymes, it was tested whether non-PLP-dependent enzymes may contribute to cytosolic pyruvate formation from SeCys-conjugates (Andreadou et al., 1996b). It appeared that for several SeCys-conjugates, AOAA inhibited less than 40% of pyruvate formation strongly suggesting involvement of non-PLP-enzymes. As a possible explanation, selenoxidation followed by syn-elimination to a selenenic acid and 2-aminoacrylic acid was proposed, in analogy with a described synthetical pathway to introduce double bonds in compounds by oxidising phenylselenides by hydrogen peroxide (Jones et al., 1970). 2-Aminoacrylic acid is the same compound form by PLP-dependent beta-lyase and is rapidly hydrolysed to ammonia and pyruvic acid.

To test the hypothesis whether selenoxidation of SeCys-conjugates can lead to syn-elimination, experiments were performed using purified flavin-containing mono-oxygenase (FMO)-enzymes from human (Rooseboom et al., 2001a). FMO's are known to catalyse oxygenation of S-benzylcysteine (SBC) leading to chemically stable S-benzylcysteine sulfoxide (Duescher et al., 1994) and are also extremely active in selenoxidation of various organoselenocompounds (Chen and Ziegler, 1994). Upon incubation of Se-benzylselenocysteine (SeBC) in presence of human FMO1 and FMO3, indeed significant time-dependent pyruvate formation was observed, in contrast to parallel incubations with SBC. In incubations of SBC a stable sulfoxide was observed by HPLC-analysis, whereas an analogous selenoxide of SeBC, could not be identified. When incubating at different concentrations of SeBC pyruvic acid formation showed Michaelis-Menten kinetics, allowing determining enzyme kinetic parameters k_{cat} and K_m. Human FMO1 appeared to be more active (K_m 198 µM and k_{cat} 6.16 min^{-1}) when compared to human FMO3 (K_m 462 µM and k_{cat} 3.75 min^{-1}). In the study of Duescher (1994) rabbit FMO1 was more active than FMO3 with SBC as substrate. In presence of thiol-compounds GSH or thiocholine FMO-catalysed formation of pyruvate from SeBC was strongly reduced (Rooseboom et al., 2001a), which may be explained by the efficient non-enzymic reduction of selenoxides by thiol compounds,thus regenerating SeBC (Assman et al., 1998).

Because FMO's are microsomal enzymes, they cannot explain the non-PLP-dependent pyruvate formation in rat renal cytosol. As an alternative mechanism of selenoxidation, the involvement of L-amino acid oxidase (L-AAO) was considered. L-AAO is known to catalyse the oxidative deamination of L-amino acids, forming the keto acid form, ammonia and hydrogen peroxide. The hydrogen peroxide formed may oxygenate SeCys-conjugates to their corresponding selenoxides, which then is expected to decompose to a selenenic acid, ammonia and pyruvate. To test this hypothesis SeCys-conjugates and Cys-conjugates were incubated in presence of commercialy available snake venom L-AAO and screened for formation of pyruvate and hydrogen peroxide. Indeed, with all SeCys-conjugates tested significant formation of both pyruvate and hydrogen peroxide (Rooseboom et al., 2001b). However, when catalase was added to the incubation, pyruvate formation was unaffected although hydrogen peroxide could not be detected anymore, implying that pyruvate formation was independent of hydrogen peroxide formation. As an explanation for these observations a common enzyme-substrate complex may be proposed that can undergo two independent pathways of product formation. A similar observation has been described in case of biotransformation of beta-chloro-D-alanine by D-amino acid oxidase (Walsh et al., 1971). At 100% oxygen concentration beta-chloro-D-alanine undergoes oxidative deamination, forming 3-chloropyruvate, ammonia and hydrogen peroxide. At low oxygen concentration the enzyme produces chloride, pyruvate and ammonia, indication change to beta-elimination reaction. We propose that a similar situation may apply to the combination

SeCys-conjugate/L-AAO. This mechanism implies that L-AAO like PLP-dependent beta-lyases can produce ammonia, pyruvate and selenol.

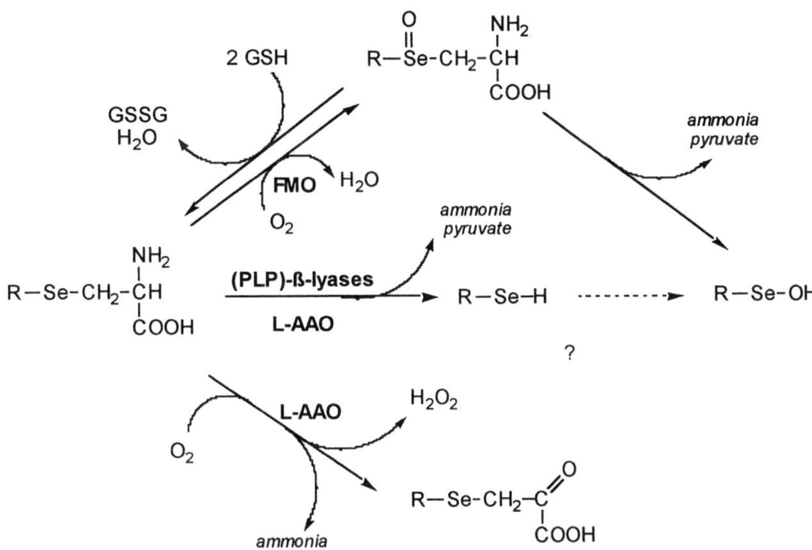

Figure 3. Identified enzymes active in the bioactivation of selenocysteine Se-conjugates.

In conclusion, as summarised in Figure 3, SeCys-conjugates can be activated by at least three different classes of enzymes, FMO, PLP-dependent beta-lyases and L-amino acid oxidases. During biotransformation by these enzymes at least four biologically active products may be formed: selenol-compounds, selenenic acids, selenoxides and hydrogen peroxide. Which of these products are involved in the potent chemoprevention observed with SeCys-conjugates remains to be established.

REFERENCES

Andreadou, I., Menge, W. P. B., Commandeur, J. N. M., Worthington, E. A., and Vermeulen, N. P. E., 1996a, Synthesis of novel Se-substituted selenocysteine derivatives as potential kidney selective prodrugs of biologically active selenol compounds: Evaluation of kinetics of beta-elimination reactions in rat renal cytosol.e. 39: 2040-2046.

Andreadou, I., Van De Water, B., Commandeur, J. N. M., Nagelkerke, F. J., and Vermeulen, N. P. E., 1996b, Comparative cytotoxicity of 14 novel selenocysteine Se-conjugates in rat renal proximal tubular cells. *Toxicol. Appl. Pharmacol.* 141: 278-287.

Assmann, A., Briviba K., and Sies, H., 1998, Reduction of methionine selenoxide to selenomethionine by glutathione. *Arch. Biochem. Biophys.* 349: 201-203.

Bird, S.M., Uden, P.C., Tyson, J.F., Block, E. and Denoyer, E., 1997, Speciation of selenoaminoacids and organoselenium compounds in selenium-enriched yeast using high-performance liquid chromatography-inductively coupled plasma mass spectrometry. *J.Anal.Atom.Spectrom.* 12: 785-788.

Chen, G. -P., and Ziegler, D. M., 1994, Liver microsomes and flavin-containing monooxygenase catalyzed oxidation of organic selenium compounds. *Arch. Biochem. Biophys.* 312: 566-572.

Clark, L.C., Combs, G.F., Turnbull, B.W., Slate, E.H., Chalker, D.K., Chow, L.S., Davis, L.S., Glover, R.A., Graham, G.F., Gross, E.G., Krongrad, A., Lesher, J.L., Park, K., Sanders, B.B., Smith, C.L. and Taylor,

R., 1996, Effects of selenium supplemention for cancer prevention in patients with carcinoma of the skin. *J.Am.Med.Assoc.* 276: 1957-1985.

Commandeur, J. N. M., Andreadou, I., Rooseboom, M., Out, M., de Leur, L. J., Groot, E., and Vermeulen, N. P. E., 2000, Bioactivation of selenocysteine Se-conjugates by a highly purified rat renal cysteine conjugate beta-lyase/glutamine transaminase K. *J. Pharmacol. Exp. Ther.* 294: 753-761.

Commandeur, J.N.M., Stijntjes, G.J. and Vermeulen, N.P.E., 1995, Enzymes and transport systems involved in the formation and disposition of glutathione S-conjugates. Role in bioactivation and detoxication mechanisms of xenobiotics. *Pharmacol. Rev.* 47: 271-330.

Comstock, G.W., Bush, T.L. and Herzlsouer, K., 1992, Serum retinol, beta-carotene, vitamin E and selenium as related to subsequent cancer of specific sites. *Am.J.Epidemiol.* 135: 115-121.

Cooper, A.J.L., 1998, Mechanisms of cysteine S-conjugate beta-lyases. *Adv.Enzymol.* 72: 199-238.

Duescher, R. J., Lawton, M. P., Philpot, R. M., and Elfarra, A. A., 1994, Flavin-containing monooxygenase (FMO)-dependent metabolism of methionine and evidence for FMO3 being the major FMO involved in methionine sulfoxidation in rabbit liver and kidney microsomes. *J. Biol. Chem.* 269: 17525-17530.

Elfarra, A. A., Renee, J. D., Hwang, I. Y., Sicuri, A. R., and Nelson, J.A., 1995, Targeting 6-thioguanine to the kidney with S-(guanin-6-yl)-L-cysteine. *J. Pharmacol. Exp.Ther.* 274: 1298-1304.

Elfarra, A.A. and Hwang, I.Y., 1993, Targeting of 6-mercaptopurine to the kidneys. Metabolism and kidney-selectivity of S-(6-purinyl)-L-cysteine analogs in rats. *Drug Metab. Dispos.*, 21: 841-845.

Ganther, H. E., 1999, Selenium metabolism, selenoproteins and mechanisms of cancer prevention: complexities with thioredoxin reductase *Carcinogenesis* 20: 1657-1666.

Ip, C., Vadhanavikit, S. and Ganther, H., 1995, Cancer chemoprevention by aliphatic selenocyanates: effect of chain length on inhibition of mammary tumors and DMBA adducts. *Carcinogenesis* 16: 35-38.

Ip, C., El-Bayoumy, K., Upadhyaya, P., Ganther, H., Vadhanavikit, S. and Thompson, H., 1994, Comparative effect of inorganic and organic selenocyanate derivates in mammary cancer chemoprevention. *Carcinogenesis* 15: 187-192.

Ip, C., Zhu, Z., Thompson, H. J., Lisk, D., and Ganther, H. E., 1999, Chemoprevention of mammary cancer with Se-allylselenocysteine and other selenoamino acids in the rat. *Anticancer Res.* 19, 2875-2880.

Ip, C., Thompson, H.J., Zhu, Z. and Ganther, H.E., 2000, In vitro and in vivo studies of methyl seleninic acid: evidence that a monomethylated selenium metabolite is critical for cancer chemoprevention. *Cancer Res.* 60: 2882-2886.

Ip, C. and Ganther, H., 1990, Activity of methylated forms of selenium in cancer prevention. *Cancer Res.* 50: 1206-1211.

Ip, C. and Ganther, H.E., 1992, Comparison of selenium and sulfur analogs in cancer prevention. *Carcinogenesis* 13: 1167-1170.

Ip, C. and Ganther, H., 1992, Relationship between the chemical form of selenium, critical metabolites, and cancer prevention. In: *Cancer Chemoprevention* (Wattenberg et al., eds.), pp. 479-488. CRC Press, Boca Raton, FL.

Ip, C., 1998, Lessons from basic research in selenium and cancer prevention. *J. Nutr.* 128:1845-1854.

Jacob, C., Maret, W., and Vallee, B. L., 1999, Selenium redox biochemistry of zinc-sulfur coordination sites in proteins and enzymes. *Proc. Natl. Acad. Sci. USA* 96: 1910-1914.

Jones, D.N., Mundy, D. and Whitehouse, R.D., 1970, Steroidal selenoxides diastereomeric at selenium: syn-elimination, absolute configuration, and optial rotary dispersion characteristics. *J.Chem.Soc.Chem.Commun.* 86.

Muller, S., Senn, H., Gsell, B., Vetter, W., Baron, C. and Bock, A., 1994, The formation of diselenide bridges in proteins by incorporation of selenocysteine residues: biosynthesis and characterization of (Se)2-thioredoxin. *Biochemistry* 33: 3404-3412.

Perry, D. K., Smyth, M. J., Stennicke, H. R., Salvesen, G. S., Duriez, P., Poirier, G. G., and Hannun, Y. A., 1997, Zinc is a potent inhibitor of the apoptotic protease, caspase-3. A novel target for zinc in the inhibition of apoptosis. *J. Biol. Chem.* 272: 18530-18533.

Rooseboom, M., Commandeur, J.N.M., Floor, G., Rettie, A and Vermeulen, N.P.E., 2001a, Selenoxidation by flavin-containing monooxygenases as a novel pathway for beta-elimination of selenocysteine Se-conjugates. *Chem.Res.Toxicol.* 14:127–134.

Rooseboom, M., Vermeulen, N.P.E., Van Hemert, N., and Commandeur, J.N.M., 2001b, Bioactivation of chemopreventive selenocysteine Se-conjugates and related amino acids by amino acid oxidases. Novel route of metabolism of selenoamino acids. *Chem. Res. Toxicol.* 14: 996–1005.

Sies, H., Klotz, L.O., Sharov, V.S., Assmann, A., Briviba, K., 1998, Protection against peroxynitrite by selenoproteins. *Z.Naturforsch.* (C) 53: 228-232.

Walsh, C.T., Schonbrunn, A. and Abeles, R.H., 1971, Studies on the mechanism of action of D-amino acid oxidase. Evidence for removal of substrate a-hydrogen as a proton. *J.Biol.Chem.* 246: 6855-66.

FORMATION AND FATE OF REACTIVE INTERMEDIATES OF HALOALKANES, HALOALKENES, AND α-HALOACIDS

M. W. Anders

Department of Pharmacology and Physiology
University of Rochester Medical Center
601 Elmwood Avenue, Box 711
Rochester, NY 14642

INTRODUCTION

Haloalkanes, haloalkenes, and α-haloacids are important industrial chemicals and environmental contaminants. For example, 1,2-dibromoethane and 1,2-dibromo-3-chloro-propane were formerly used as fumigants and nematocides, trichloroethylene is a common environmental contaminant, and dichloroacetate (DCA), which is produced during the chlorination of drinking water, is present in finished drinking water supplies in the U.S. Many haloalkanes, haloalkenes, and α-haloacids are toxic, and some are rodent or suspected human carcinogens. The toxicity of these chemicals is associated with their bioactivation to reactive intermediates by the cytochromes P450 or glutathione transferases (GSTs).

This review will focus on the glutathione- and GST-catalyzed biotransformation of haloalkanes, haloalkenes, and α-haloacids. A common feature of the glutathione-dependent bioactivation of these compounds is the formation of sulfur-containing metabolites that may undergo chemical conversion to reactive intermediates or enzymatic processing to other metabolites are enzymatically converted to reactive intermediates. In all cases, the biological effects are associated with reactive-intermediate formation.

HALOALKANES

The GSTs, particularly theta-class GSTs (GSTT1-1), catalyze the biotransformation of a range of haloalkanes. Dichloromethane (methylene chloride) is an important solvent used in many industrial processes. Dichloromethane and other dihalomethanes undergo GST-dependent biotransformation to formaldehyde (Hashmi et al., 1994). The reaction mechanism involves the GSTT1-1-catalyzed displacement of chloride from dichloromethane (Figure 1, **1**) to give *S*-(chloromethyl)glutathione (Figure 1, **2**). Hydrolysis of *S*-(chloromethyl)glutathione **2** gives *S*-

(hydroxymethyl)glutathione (Figure 1, **3**), which is the hemithioacetal of formaldehyde and glutathione (Figure 1). Hemithioacetal **3** is in equilibrium with glutathione and formaldehyde (Figure 1, **4**). The observed carcinogenicity of dichloromethane in mice is associated with its GSTT1-1-dependent metabolism (Sherratt et al., 1997). Although dichloromethane is not mutagenic in most in vitro assay systems, it is mutagenic in the Ames test with *Salmonella typhimurium* TA1535 transfected with GSTT1-1 (Thier et al., 1993).

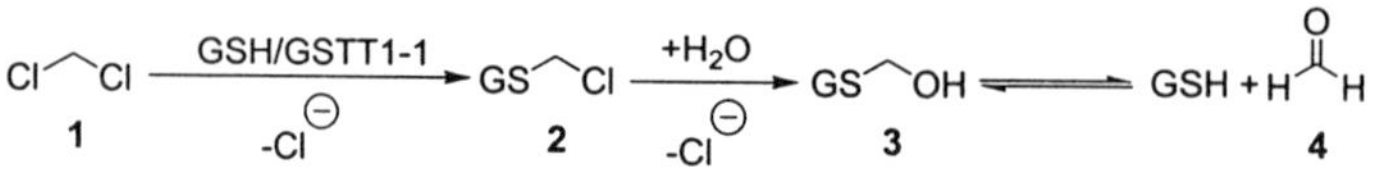

Figure 1. Glutathione transferase-dependent biotransformation of dichloromethane. **1**, dichloromethane; **2**, *S*-(chloromethyl)glutathione; **3**, *S*-(hydroxymethyl)glutathione; **4**, formaldehyde; GSH, glutathione; GSTT1-1, theta-class glutathione transferase.

A role for dichloromethane-derived formaldehyde, which leads to the formation of protein-DNA crosslinks, in the carcinogenicity of dichloromethane has been demonstrated (Casanova et al., 1992). Other studies implicate the reactive intermediate *S*-(chloromethyl)glutathione in the mutagenicity of dichloromethane (Gisi et al., 1999).

1,2-Dihaloalkanes (vicinal dihaloalkanes or ethylene halides) are an important group of chemicals. 1,2-Dibromoethane, 1,2-dichloroethane, 1,2-dibromo-3-chloropropane are examples of commercially important 1,2-dihaloalkanes. 1,2-Dibromoethane, for example, has seen considerable use as a lead scavenger in gasoline and as a fumigant. 1,2-Dibromo-3-chloropropane was formerly used extensively as a soil nematocide. Significantly, many 1,2-dihaloalkanes are mutagenic in the Ames test and some are rodent carcinogens, which has lead to restrictions in their commercial applications. The mutagenicity and, perhaps, carcinogenicity of this group of chemicals is associated with GST-dependent biotransformation to reactive intermediates. 1,2-Dihaloethanes (Figure 2, **1**) undergo GST-catalyzed biotransformation to *S*-(2-haloethyl)glutathiones (half-sulfur mustards) (Figure 2, **2**), which undergo intramolecular reactions to yield thiiranium (episulfonium) ions (Figure 2, **3**). The thiiranium ions thus formed may react with nucleophilic sites in DNA to give the adduct *S*-[2-N^7-guanyl)ethyl]glutathione (Ozawa and Guengerich, 1983) (Figure 2, **4**).

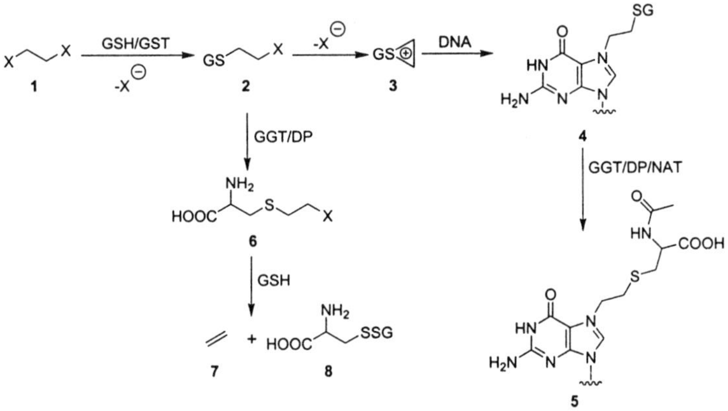

Figure 2. Glutathione transferase-dependent biotransformation of 1,2-dihaloethanes. **1**, 1,2-dihaloethane; **2**, *S*-(2-haloethyl)glutathione; **3**, thiiranium ion; **4**, *S*-[2-N^7-guanyl)ethyl]glutathione; **5**, *S*-[2-N^7-guanyl)ethyl]-*N*-acetyl-L-cysteine; **6**, *S*-(2-haloethyl)-L-cysteine; **7**, ethylene; **8**, cysteinyl glutathione disulfide; GSH, glutathione; GST, glutathione transferase; GGT, γ-glutamyltransferase; DP, dipeptidases; NAT, *N*-acetyltransferase.

The mercapturate S-[2-N^7-guanyl)ethyl]-N-acetyl-L-cysteine (Figure 2, **5**) is excreted in the urine of rats given 1,2-dibromoethane (Kim and Guengerich, 1989). The cysteine analog (Figure 2, **6**) of the intermediate half-sulfur mustard formed from 1,2-dichloroethane may react with glutathione to yield the novel metabolite ethylene (Figure 2, **7**) and cysteinyl glutathione disulfide (Livesey et al., 1982).

Evidence for the glutathione-dependent formation of a thiiranium ion from 1,2-dibromo-3-chloropropane has been presented (Pearson et al., 1990). Although glutathione conjugate formation is important in the mutagenicity of 1,2-dibromo-3-chloropropane, both conjugate formation and cytochrome P450-dependent oxidation appear to play roles (Holme et al., 1989).

HALOALKENES

Haloalkenes constitute an important group of toxic chemicals that are widely used in commerce and industry, and workplace exposure and environmental contamination have accompanied this use. Also, haloalkenes may be formed by the degradation of anesthetic agents. Many haloalkenes are selective nephrotoxins, and this selectivity is associated with bioactivation via the cysteine conjugate β-lyase pathway. The β-lyase pathway is a multiorgan, multistep pathway: the first step in the β-lyase pathway is the GST-catalyzed formation of S-(haloalkyl)- or S-(haloalkenyl)glutathione conjugates. For several haloalkenes, this pathway is catalyzed by the hepatic microsomal GST. The glutathione S-conjugates are hydrolyzed to the corresponding cysteine S-conjugates by γ-glutamyltransferase and the dipeptidases cysteinylglycine dipeptidase and aminopeptidase M; these hydrolytic steps may take place in the bile duct, intestine, or kidney. The cysteine S-conjugates thus formed are actively transported and concentrated in the kidney by amino acid and organic anion transporters. In the kidney, the cysteine S-conjugates are bioactivated by mitochondrial and cytosolic β-lyases to afford reactive intermediates. A recent review about the β-lyase-dependent bioactivation of haloalkenes has appeared (Anders and Dekant, 1998).

The β-lyase pathway has been implicated in the nephrotoxicity of a range of 1,1-dichloroethylenes and 1,1-difluoroethylenes, including chlorotrifluoroethylene, tetrafluoro-ethylene, hexafluoropropene, trichloroethylene, tetrachloroethylene, hexachlorobutadiene, 2-bromo-2-chloro-1,1-difluoroethylene, and 2-(fluoromethoxy)-1,1,3,3,3-pentafluoro-1-propene (Compound A).

The first step in the β-lyase-dependent bioactivation of haloalkenes is the GST-catalyzed formation of glutathione conjugates. With 1,1-dichloroalkenes, a GST-catalyzed addition-elimination reaction gives S-(1-chloroalkenyl)glutathione conjugates, whereas most 1,1-difluoroalkenes undergo an addition reaction to give S-(1,1-difluoroalkyl)glutathione conjugates. Some 1,1-difluoroalkenes, *e.g.*, hexafluoropropene and 2-(fluoromethoxy)-1,1,3,3,3-pentafluoro-1-propene, undergo both addition and addition-elimination reactions.

The β-lyase-dependent bioactivation of haloalkenes will be illustrated with 2-(fluoro-methoxy)-1,1,3,3,3-pentafluoro-1-propene (Compound A; Figure 3, **2**). Compound A is a degradation product of the anesthetic sevoflurane (Figure 3, **1**). Carbon dioxide absorbents, which contain strong bases, catalyze the elimination of HF from sevoflurane in the anesthetic circuit to give Compound A. Hence, human subjects anesthetized with sevoflurane are exposed to low concentrations of Compound A, which has lead to considerable discussion and controversy about the possible risk to humans of anesthetization with sevoflurane. Compound A is nephrotoxic in rats and produces renal proximal tubular damage similar to that produced by other haloalkenes (Gonsowski et al., 1994; Gonsowski et al., 1994).

Metabolite analysis supports a role for the glutathione- and β-lyase-dependent bioactivation of Compound A. When Compound A is given to rats, the Compound A-derived glutathione conjugates S-[2-(fluoromethoxy)-1,3,3,3-tetrafluoro-1-propenyl]glutathione (Figure 3, **3**) and S-[2-(fluoromethoxy)-1,1,3,3,3-pentafluoropropyl]glutathione (Figure 3, **4**) are excreted in the bile, and the corresponding mercapturates S-[2-(fluoromethoxy)-1,3,3,3-tetrafluoro-1-propenyl]-N-acetyl-L-cysteine (Figure 3, **7**) and S-[2-(fluoromethoxy)-1,1,3,3,3-pentafluoropropyl]-N-acetyl-L-cysteine (Figure 3, **8**) are excreted in urine (Jin et al., 1996). The Compound A-derived cysteine S-conjugates S-[2-(fluoromethoxy)-1,3,3,3-tetrafluoro-1-propenyl]-L-cysteine (Figure 3, **5**) and S-[2-(fluoromethoxy)-1,1,3,3,3-pentafluoropropyl]-L-cysteine (Figure 3, **6**) are substrates for rat, nonhuman primate, and human renal β-lyases, although activity with human renal tissue is much lower than that with rat renal tissue (Iyer and Anders, 1996). Also, cysteine S-conjugates **5** and **6** are biotransformed in vitro to 2-(fluoromethoxy)-3,3,3-trifluoropropanoate (Figure 3, **9**) (Iyer and Anders, 1997), which is also excreted in the urine of rats given Compound A (Kharasch et al., 1999) and in the urine of human subjects anesthetized with sevoflurane (Iyer et al., 1998). In addition to acid **9**, ^{19}F nuclear magnetic resonance studies showed the presence of mercapturates **7** and **8** in the urine of human subjects anesthetized with sevoflurane (Iyer et al., 1998). These data show that Compound A undergoes β-lyase-dependent bioactivation in both rats and humans. The studies on the fate of Compound A in human subjects are important because they constitute the first demonstration of the β-lyase pathway in humans exposed to a haloalkene (Iyer et al., 1998).

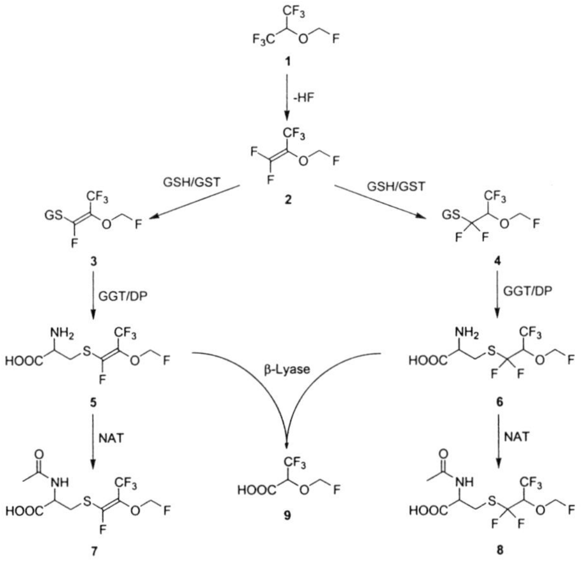

Figure 3. Glutathione- and β-lyase-dependent bioactivation of Compound A. **1**, sevoflurane; **2**, 2-(fluoromethoxy)-1,1,3,3,3-pentafluoro-1-propene (Compound A); **3**, S-[2-(fluoromethoxy)-1,3,3,3-tetrafluoro-1-propenyl]glutathione; **4**, S-[2-(fluoromethoxy)-1,1,3,3,3-pentafluoropropyl]glutathione; **5**, S-[2-(fluoromethoxy)-1,3,3,3-tetrafluoro-1-propenyl]-L-cysteine; **6**, S-[2-(fluoromethoxy)-1,1,3,3,3-pentafluoropropyl]-L-cysteine; **7**, S-[2-(fluoromethoxy)-1,3,3,3-tetrafluoro-1-propenyl]-N-acetyl-L-cysteine; **8**, S-[2-(fluoromethoxy)-1,1,3,3,3-pentafluoropropyl]-N-acetyl-L-cysteine; **9**, 2-(fluoromethoxy)-3,3,3-trifluoropropanoate; GSH, glutathione; GST, glutathione transferase; GGT, γ-glutamyltransferase; DP, dipeptidases; β-lyase, cysteine conjugate β-lyase.

The controversy about the role of the β-lyase pathway in the nephrotoxicity of Compound A stems from the observation that some inhibitors of the β-lyase pathway either fail to block or even increase the nephrotoxicity of Compound A. The γ-glutamyltransferase inhibitor acivicin, which blocks the conversion of glutathione S-conjugates to cysteine S-conjugate, increases the nephrotoxicity of Compound A in rats (Kharasch et al., 1998; Martin et al., 1996). The mechanism whereby acivicin increases the nephrotoxicity of Compound A and hexachlorobutadiene has not been elucidated. (Aminooxy)acetic acid, a β-lyase inhibitor, produces inconsistent results, depending on the indices of nephrotoxicity being studied (Kharasch et al., 1998; Martin et al., 1996). The anion transport inhibitor probenecid, which inhibits the renal uptake of cysteine *S*-conjugates, blocks Compound A-induced nephrotoxicity in rats (Kharasch et al., 1998). Acivicin, (aminooxy)acetic acid, and probenecid consistently block the cytotoxicity of haloalkene-derived cysteine *S*-conjugates in isolated kidney cells or in cultured renal epithelial cells. The mechanism whereby acivicin increases the nephrotoxicity of Compound A merits further investigation.

Sevoflurane enjoys extensive clinical use: more than 65 million human subjects worldwide have been anesthetized with sevoflurane, and no cases of Compound A-associated renal injury have been reported (personal communication, Leticia Delgado Herrera, Abbott Laboratories, Abbott Park, IL). The safety of sevoflurane in man is likely attributable to the low activities of renal β-lyase in humans compared with rats: the rate of biotransformation of Compound A-derived cysteine *S*-conjugates **5** and **6** is much lower in human and nonhuman primate kidney tissue compared with rat kidney tissue (Iyer and Anders, 1996). Moreover, in contrast to many haloalkene-derived mercapturates, the Compound A-derived mercapturates **7** and **8** undergo little acylase-catalyzed hydrolysis to cysteine *S*-conjugates **5** and **6**, respectively, and show little nephrotoxicity in vivo (Uttamsingh et al., 1998). Hence, for Compound A, mercapturate formation is a detoxication pathway.

α-HALOACIDS

α-Haloacids constitute a group of chemicals that has only recently received considerable attention from the toxicological and regulatory communities. Dihaloacetates, particularly dichloroacetate (DCA) and bromochloroacetate are disinfection by-products found in drinking water consumed by an estimated 170 million individuals in the United States. DCA is hepatocarcinogenic in rats and mice; promotion of initiated cells rather than genotoxicity is thought to involved in tumor induction. In contrast to its carcinogenicity, DCA is used in the clinical management of congenital lactic acidosis and has been proposed as a cytoprotective agent in a range of pathological conditions (Stacpoole et al., 1998). 2-Chloropropionate is a neurotoxic α-haloacid that is used in chemical manufacture, and fluoroacetate is a naturally occurring α-haloacid that is highly toxic and has been used as a rodenticide.

Several studies had implicated glutathione and GSTs in the biotransformation of DCA, 2-chloropropionate, and fluoroacetate. Recent studies in our laboratory showed that the glutathione-dependent biotransformation of DCA to glyoxylate is catalyzed by glutathione transferase zeta (GSTZ1-1) (Tong et al., 1998). GSTZ1-1 is a recently described enzyme that lacks activity with substrates, *e.g.*, chlorodinitrobenzene and organic hydroperoxides, often used to assay GST activity (Board et al., 1997). Subsequent studies showed that GSTZ1-1 zeta is identical with maleylacetoacetate isomerase, which catalyzes the glutathione-dependent isomerization of maleylacetoacetate to fumarylacetoacetate, the penultimate step in tyrosine degradation (Fernández-Cañón and Peñalva, 1998).

Studies on the substrate selectivity of GSTZ1-1 showed that dihaloacetates (except difluoroacetate), 2-halopropionates (except 2-fluoropropionate), 2,2-dichloropropionate (but not 3,3-dichloropropionate), and fluoroacetate (but not fluoroacetamide or ethyl fluoroacetate) are substrates for GSTZ1-1 (Tong et al., 1998). GSTZ1-1 catalyzes the attack of glutathione on the alpha carbon of maleylacetoacetate. The observed substrate selectivity of GSTZ1-1 indicates that the enzyme requires substrates with a free carboxylate group and a good leaving group alpha to the carboxylate, indicating attack at the alpha carbon with xenobiotic substrates. Four polymorphic variants of hGSTZ1-1 that differ in their catalytic rate constants with α-haloacids as substrates have been identified (Blackburn et al., 2000).

The elimination half-life of DCA is prolonged after giving human subjects multiple doses of DCA (Curry et al., 1991). Recent studies in our laboratory demonstrated that the prolongation of the elimination half-life of DCA is attributable to a DCA-induced inactivation of GSTZ1-1. In rats given 0.3 mmol DCA/kg ip, a parallel loss of GSTZ1-1 activity and immunoreactive protein is observed (Anderson et al., 1999). Activity reaches a nadir 12 hours after giving DCA and does not return to control values for 8–9 days; these data were used to calculate that GSTZ1-1 has a half-life of 3.3 days. Structure-inactivation studies showed that only fluorine-lacking α-haloacids inactivate GSTZ1-1. In vitro studies on the DCA-induced inactivation of mouse, rat, and human GSTZ1-1 showed inactivation half-lives of 6.61, 5.44, and 22 min, respectively (Tzeng et al., 2000). As was observed in the in vivo inactivation studies, fluorine-lacking dihaloacetates produce little inactivation: the half-lives for inactivation of mouse liver GSTZ1-1 by DCA and chlorofluoroacetate amount to 6.6 and 235 min, respectively.

Studies with expressed recombinant hGSTZ1-1 were conducted to explore in more detail the mechanism of the DCA-induced inactivation of the enzyme. The four polymorphic variants of hGSTZ1-1 differ in their susceptibility to inactivation by DCA (Tzeng et al., 2000). The half-lives for the DCA-induced inactivation of hGSTZ1a-1a, hGSTZ1b-1b, hGSTZ1c-1c, and hGSTZ1d-1d are 23, 9.6, 10.1, and 9.5 min, respectively. Activity could not be restored by dialysis of the inactivated protein against 0.1 M potassium phosphate buffer (pH 7.4) overnight. To determine whether the DCA-induced inactivation of hGSTZ1-1 is associated with covalent modification of the protein, hGSTZ1c-1c was incubated with [1-^{14}C]DCA, [1-^{14}C]DCA and glutathione, [^{35}S]glutathione, and [^{35}S]glutathione and DCA. Examination of the proteins by gel electrophoresis and fluorography showed that the molecular mass of the modified protein was similar to that of the native protein and that the inactivated enzyme was labeled with ^{14}C from [1-^{14}C]DCA only in the presence of glutathione and with ^{35}S from [^{35}S]glutathione only in the presence of DCA. These data indicate that the glutathione-dependent inactivation of hGSTZ1-1 by DCA is accompanied by the binding of both the carbon skeleton of DCA and glutathione to the protein and that DCA is a mechanism-based inactivator of hGSTZ1-1.

A reaction mechanism that accounts for both the biotransformation of DCA to glyoxylate and the DCA-induced inactivation of GSTZ1-1 is shown in Figure 4. The first step in the reaction is the hGSTZ1-1 catalyzed attack of glutathione on the alpha carbon of DCA (Figure 4, 1) to give *S*-(α-chlorocarboxymethyl)glutathione (Figure 4, 2). α-Chlorosulfide 2 may lose chloride to give a carbonium-sulfonium intermediate (Figure 4, 3), which may undergo hydrolysis to give glyoxylate (Figure 4, 4) and glutathione or may react with a nucleophilic site in hGSTZ1-1 to give covalently modified enzyme (Figure 4, 5). Studies are underway to characterize the covalently modified hGSTZ1-1 by mass spectrometry.

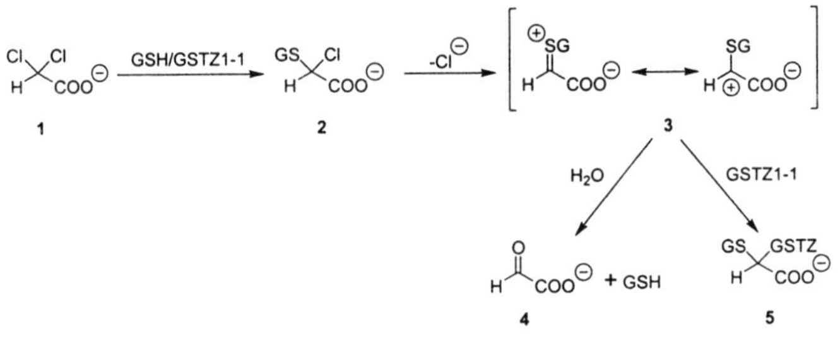

Figure 4. Glutathione transferase zeta-catalyzed biotransformation of α-haloacids. **1**, dichloroacetate; **2**, *S*-(α-chlorocarboxymethyl)glutathione; **3**, carbonium-sulfonium intermediate; **4**, glyoxylate; **5**, covalently modified and inactivated GSTZ1-1; GSH, glutathione.

Acknowledgments

Research in the author's laboratory is supported by National Institute of Environmental Health Sciences grant ES03127.

REFERENCES

Anders, M.W., and Dekant, W., 1998, Glutathione-dependent bioactivation of haloalkenes, *Ann. Rev. Pharmacol. Toxicol.* 38:501.

Anderson, W.B., Board, P.G., Gargano, B., and Anders, M.W., 1999, Inactivation of glutathione transferase zeta by dichloroacetic acid and other fluorine-lacking α-haloalkanoic acids, *Chem. Res. Toxicol.* 12:1144.

Blackburn, A.C., Tzeng, H.-F., Anders, M.W., and Board, P.G., 2000, Discovery of a functional polymorphism in human glutathione transferase zeta by expressed sequence tag database analysis, *Pharmacogenetics* 10:49.

Board, P.G., Baker, R.T., Chelvanayagam, G., and Jermiin, L.S., 1997, Zeta, a novel class of glutathione transferases in a range of species from plants to humans, *Biochem. J.* 328:929.

Casanova, M., Deyo, D.F., and Heck, H.d.A., 1992, Dichloromethane (methylene chloride): metabolism to formaldehyde and formation of DNA-protein cross-links in B6C3F1 mice and Syrian golden hamsters, *Toxicol. Appl. Pharmacol.* 114:162.

Curry, S.H., Lorenz, A., Chu, P.-I., Limacher, M., and Stacpoole, P.W., 1991, Disposition and pharmacodynamics of dichloroacetate (DCA) and oxalate following oral DCA doses, *Biopharm. Drug Dispos.* 12:375.

Fernández-Cañón, J.M., and Peñalva, M.A., 1998, Characterization of a fungal maleylacetoacetate isomerase gene and identification of its human homologue, *J. Biol. Chem.* 273:329.

Gisi, D., Leisinger, T., and Vuilleumier, S., 1999, Enzyme-mediated dichloromethane toxicity and mutagenicity of bacterial and mammalian dichloromethane-active glutathione *S*-transferases, *Arch. Toxicol.* 73:71.

Gonsowski, C.T., Laster, M.J., Eger II, E.I., Ferrell, L.D., and Kerschmann, R.L., 1994, Toxicity of compound A in rats. Effect of a 3-hour administration, *Anesthesiology* 80:556.

Gonsowski, C.T., Laster, M.J., Eger II, E.I., Ferrell, L.D., and Kerschmann, R.L., 1994, Toxicity of compound A in rats. Effect of increasing duration of administration, *Anesthesiology* 80:566.

Hashmi, M., Dechert, S., Dekant, W., and Anders, M.W., 1994, Bioactivation of [^{13}C]dichloromethane in mouse, rat, and human liver cytosol: ^{13}C Nuclear magnetic resonance spectroscopic studies, *Chem. Res. Toxicol.* 7:291.

Holme, J.A., Søderlund, E.J., Brunborg, G., Omichinski, J.G., Bekkedal, K., Trygg, B., Nelson, S.D., and Dybing, E., 1989, Different mechanisms are involved in DNA damage, bacterial mutagenicity and cytotoxicity induced by 1,2-dibromo-3-chloropropane in suspensions of rat liver cells, *Carcinogenesis* 10:49.

Iyer, R.A., and Anders, M.W., 1996, Cysteine conjugate β-lyase-dependent biotransformation of the cysteine *S*-conjugates of the sevoflurane degradation product Compound A in human, nonhuman primate, and rat renal cytosol and mitochondria, *Anesthesiology* 85:1454.

Iyer, R.A., and Anders, M.W., 1997, Cysteine conjugate β-lyase-dependent biotransformation of the cysteine *S*-conjugates of the sevoflurane degradation product 2-(fluoromethoxy)-1,1,3,3,3-pentafluoro-1-propene (Compound A), *Chem. Res. Toxicol.* 10:811.

Iyer, R.A., Frink, E.J., Jr., Ebert, T.J., and Anders, M.W., 1998, Cysteine conjugate β-lyase-dependent metabolism of Compound A (2-(fluoromethoxy)-1,1,3,3,3-pentafluoro-1-

propene) in human subjects anesthetized with sevoflurane and in rats given Compound A, *Anesthesiology* 88:611.

Jin, L., Davis, M.R., Kharasch, E.D., Doss, G.A., and Baillie, T.A., 1996, Identification in rat bile of glutathione conjugates of fluoromethyl 2,2-difluoro-1-(trifluoromethyl)vinyl ether, a nephrotoxic degradate of the anesthetic agent sevoflurane, *Chem. Res. Toxicol.* 9:555.

Kharasch, E.D., Hoffman, G.M., Thorning, D., Hankins, D.C., and Kilty, C.G., 1998, Role of the renal cysteine conjugate β-lyase pathway in inhaled compound A nephrotoxicity in rats, *Anesthesiology* 88:1624.

Kharasch, E.D., Jubert, C., Spracklin, D.K., and Hoffman, G.M., 1999, Dose-dependent metabolism of fluoromethyl-2,2-difluoro-1-(trifluoromethyl)vinyl ether (compound A), an anesthetic degradation product, to mercapturic acids and 3,3,3-trifluoro-2-(fluoromethoxy)propanoic acid in rats, *Toxicol. Appl. Pharmacol.* 160:49.

Kim, D.-H., and Guengerich, F.P., 1989, Excretion of the mercapturic acid S-[2-(N^7-guanyl)ethyl]-N- acetylcysteine in urine following administration of ethylene dibromide to rats, *Cancer Res.* 49:5843.

Livesey, J.C., Anders, M.W., Langvardt, P.W., Putzig, C.L., and Reitz, R.H., 1982, Stereochemistry of the glutathione-dependent biotransformation of vicinal-dihaloalkanes to alkenes, *Drug Metab. Dispos.* 10:201.

Martin, J.L., Laster, M.J., Kandel, L., Kerschmann, R.L., Reed, G.F., and Eger II, E.I., 1996, Metabolism of Compound A by renal cysteine-S-conjugate β-lyase is not the mechanism of Compound A-induced renal injury in the rat, *Anesth. Analg.* 82:770.

Ozawa, N., and Guengerich, F.P., 1983, Evidence for formation of an S-[2-(N^7-guanyl)ethyl]glutathione adduct in glutathione-mediated binding of the carcinogen 1,2-dibromoethane to DNA, *Proc. Natl. Acad. Sci. USA* 80:5266.

Pearson, P.G., Omichinski, J.G., Myers, T.G., Soderlund, E.J., Dybing, E., and Nelson, S.D., 1990, Metabolic activation of 1,2-dibromo-3-chloropropane to mutagenic metabolites: Detection and mechanism of formation of (Z)- and (E)-2-chloro-3-(bromomethyl)-oxirane, *Chem. Res. Toxicol.* 3:458.

Sherratt, P.J., Pulford, D.J., Harrison, D.J., Green, T., and Hayes, J.D., 1997, Evidence that human class Theta glutathione S-transferase T1-1 can catalyse the activation of dichloromethane, a liver and lung carcinogen in the mouse - Comparison of the tissue distribution of GST T1-1 with that of classes Alpha, Mu and Pi GST in human, *Biochem. J.* 326:837.

Stacpoole, P.W., Henderson, G.N., Yan, Z.M., and James, M.O., 1998, Clinical pharmacology and toxicology of dichloroacetate, *Environ. Health Perspect.* 106 (Suppl. 4):989.

Thier, R., Taylor, J.B., Pemble, S.E., Humphreys, W.G., Persmark, M., Ketterer, B., and Guengerich, F.P., 1993, Expression of mammalian glutathione S-transferase 5-5 in *Salmonella typhimurium* TA1535 leads to base-pair mutations upon exposure to dihalomethanes, *Proc. Natl. Acad. Sci. USA* 90:8576.

Tong, Z., Board, P.G., and Anders, M.W., 1998, Glutathione transferase zeta catalyzes the oxygenation of the carcinogen dichloroacetic acid to glyoxylic acid, *Biochem. J.* 331:371.

Tong, Z., Board, P.G., and Anders, M.W., 1998, Glutathione transferase Zeta-catalyzed biotransformation of dichloroacetic acid and other α-haloacids, *Chem. Res. Toxicol.* 11:1332.

Tzeng, H.-F., Blackburn, A.C., Board, P.G., and Anders, M.W., 2000, Polymorphism- and species-dependent inactivation of glutathione transferase zeta by dichloroacetate, *Chem. Res. Toxicol.* 13:231.

Uttamsingh, V., Iyer, R.A., Baggs, R.B., and Anders, M.W., 1998, Fate and toxicity of 2-(fluoromethoxy)-1,1,3,3,3-pentafluoro-1-propene (Compound A)-derived mercapturates in male, Fischer 344 rats, *Anesthesiology* 89:1174.

HUMAN EPOXIDE HYDROLASE IS THE TARGET OF GERMANDER AUTOANTIBODIES ON THE SURFACE OF HUMAN HEPATOCYTES: ENZYMATIC IMPLICATIONS.

Jacqueline Loeper,[1] Veronique De Berardinis,[1] Claude Moulis,[2] Philippe Beaune,[3] Dominique Pessayre[4] and Denis Pompon[1]

[1]UPR2167 du CNRS, F91198 Gif-sur-Yvette; Laboratoire dePharmacognosie, Université Toulouse III, F31062; INSERM Unité 490, Université Paris V, F75006; Unité 481, Hôpital Beaujon, Clichy, F92118

INTRODUCTION

Western countries observed an increased interest for the use of herbal medicines because of their supposed safety in contrast to chemical drugs. Wild germander (*Teuchrium chamaedrys L.*) was traditionally used as a folk medicine for its choleretic and antiseptic properties. In 1991, germander consumed to treat obesity, caused an epidemic of cytolytic hepatitis. Thirty cases of hepatotoxicity were first reported including cases with positive rechallenge. For these patients, an early recurrence was observed despite lack of other features of hypersensitivity (Castot and Larrey, 1992). In mice, germander toxicity required CYP3A-dependent metabolism (Loeper, *et al.*, 1994). More specifically, the metabolic activation of the furan ring of the diterpenoid teucrin A (TA) (Kouzi, *et al.*, 1994). TA-toxicity was via CYP3A-generated electrophilic metabolites that were detoxified by glutathione conjugation, depleted cellular thiols and caused apoptosis in isolated rat hepatocytes (Lekehal, *et al.*, 1996; Fau *et al.*, 1997). However, these death processes did not explain the immunological features observed in several cases of germander-induced hepatitis. To explain this immune process, patient's sera consuming germander tea in great quantity were tested by Western blot analysis. They contained autoantibodies directed against human microsomal epoxide hydrolase (hmEH), that was located both, in the endoplasmic reticulum and the plasma membrane (PM) of human hepatocyte and hmEH-expressing yeast. Germander-induced autoantibodies (GIAA) were shown to recognize hmEH on the cell surface. To implicate hmEH in the metabolic activation of TA, we used a "humanized" yeast strain expressing human P450-reductase and cytochrome b5 transformed with CYP3A4 and/or hmEH cDNAs. TA was metabolized by CYP3A4 into a metabolite which concentration diminished in presence of hmEH. Incubations of CYP3A4 and hmEH with TA inactivated hmEH in a time-dependent-manner, in agreement with formation of reactive teuchrin A-metabolite that could covalently alter and inhibit hmEH. This modified enzyme may bypass the immunologic tolerance that normally exists for hmEH.

RESULTS

hmEH is the specific target of germander-induced autoantibodies

Sera from four patients with germander-induced hepatitis (patients A, B, C and D) were screened by Western blotting. The sera recognized a single band in human liver microsomes where no protein was recognized by human control sera (Fig. 1). The GIAA recognized chromatographically-purified hmEH (Fig. 1, lane 3) and hmEH expressed in yeast microsomal and PM fractions (Fig. 1, lanes 2 and 4), while control yeast microsomes were negative (Fig. 1, lane 1). To certify their specificity, affinity-purified GIAA were used to test hepatocyte PM fractions. The purified sera recognized only one protein, hmEH, in human hepatocyte PM fractions by Western blot (Fig. 1). To confirm this recognition, GIAA immunoprecipitated hmEH but did not inhibit it (data not shown). Probably, these autoantibodies are not directed against the catalytic site of hmEH.

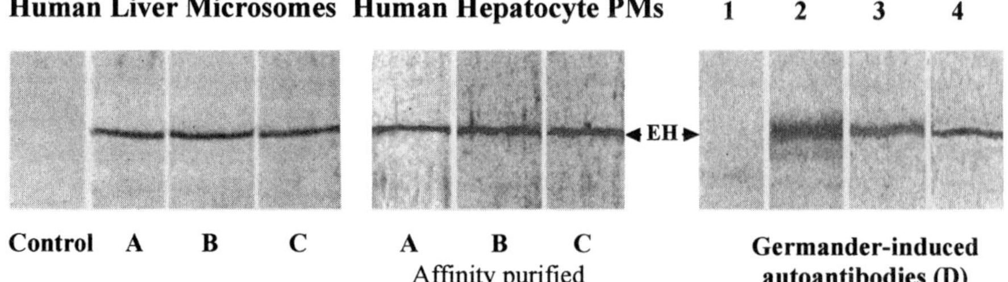

Figure 1. Western blot analysis of hmEH recognition by germander-induced autoantibodies from patients A, B, C and D. Control (lane 1) or expressing hmEH (lane 2) yeast microsomes. Transformed yeast PM (lane 4). Purified hmEH (lane 3).

hmEH is located on the external face of human hepatocyte recognized by GIAA

Electron microscopy was used to further demonstrate that hmEH was exposed on the surface of human hepatocyte and was recognized by purified GIAA. Immunolabeling was performed on uncut, unpermeabilized, fixed human hepatocytes that were then post-embedded and cut. When incubated with anti-hmEH antibodies, human hepatocyte exhibited a clear immunostaining only at the cell surface, along linear parts of PM and its microvilli (Fig. 2A). Affinity-purified GIAA gave a same positive peroxidase-labeling of PM (Fig. 2B). No PM staining was detected with normal human sera (Fig. 2C) or with preimmune rabbit sera (data not shown).

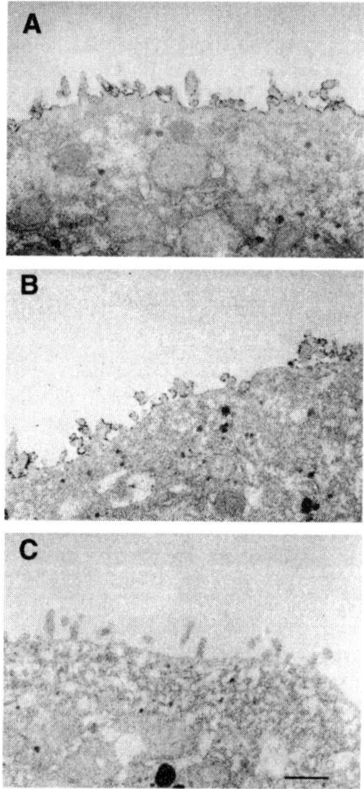

Figure 2. Location of hmEH at the surface of human hepatocytes and its recognition by purified GIAA.(A), electron micrograph of fixed human hepatocyte incubated with rabbit anti-hmEH antibodies or (B), incubated with purified GIAA or (C), incubated with human control sera. Bar = 1μm.

Metabolic activation of teucrin A

Because yeast-expressed CYP3A4 was optimally functional in the presence of both human CPR and human cytochrome b5, a "humanized" yeast strain was used in this study. In this strain, the molar ratios of CYP3A4, CPR and human cytochrome b5 (10:1:10) were close to the ratios in human liver. In the presence of TA, NADPH and yeast microsomes expressing CYP3A4 alone, the HPLC metabolite profile at 215nm showed formation of a TA-metabolite (metabolite I) (Fig 3B), that was absent in the NADPH-free control (Fig. 3A) or with microsomes from yeast transformed with vector controls (data not shown). By contrast, in the presence of hmEH metabolite I decreased, while metabolites II and III appeared (Fig. 3C). Metabolites II and III were not generated in the presence of TA with microsomes containing hmEH alone. The hmEH-mediated decrease in metabolite I could be consistent with an epoxide and the hmEH-dependent accumulation of metabolites II and III could be consistent for diol derivatives (Fig. 3C).

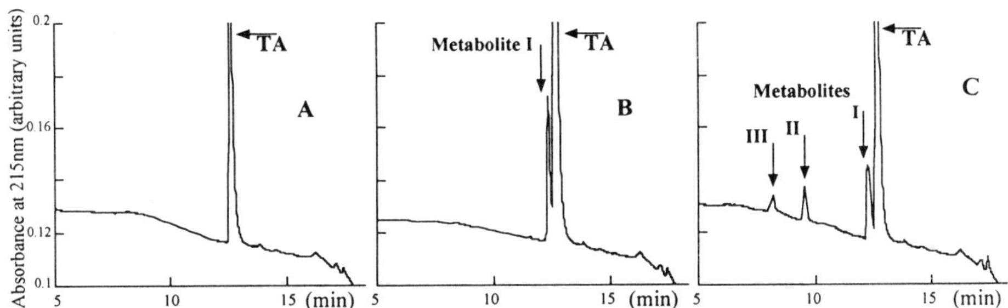

Figure 3. HPLC profiles obtained by TA activation with microsomes of the "humanized" yeast strain. A, microsomes expressing CYP3A4 and hmEH without an NADPH-regenerating system. B, microsomes expressing CYP3A4 alone. C, microsomes expressing CYP3A4 and hmEH.

Inactivation of hmEH by TA in transformed yeast microsomal and PM fractions

In order to evaluate hmEH functionality, in transformed yeast and in human hepatocytes PM its specific activities were determined from styrene oxide hydrolase activity and were 60 ± 6 nmol and 40 ± 7 nmol of styrene glycol formed/min/mg of proteins. To test hmEH inactivation, yeast microsomes or PMs containing CY3A4 and hmEH were incubated with or without TA at 37° for various times. After a 1:100 dilution, these fractions were tested for their ability to hydrolyze the B(a)P oxides formed *in situ* by CYP1A1-mediated B(a)P. In the presence of TA, microsomes showed a hmEH inactivation of 63% after 60 min of incubation (Table 1). With PMs, TA-mediated hmEH inactivation was 66 % in the same conditions (Table 1). In both, microsomes and PM, hmEH inactivation was no more observed after 45 min incubation (Table 1) implying, either inactivation of CYP3A4 by TA or by self inactivation (usually observed with CYP3A4 in yeast microsomes). This hmEH inactivation suggests that some TA reactive metabolite formed by CYP3A4 on the furan ring, alkylates and inactivates hmEH, in both fractions.

Table 1. Yeast microsomes or PM expressing CYP3A4 and hmEH were incubated with or without TA for different times. After dilution 1:100, samples were incubated for 10 min. with CYP1A1 and B(a)P. Reaction products were analyzed by HPLC. Hydrolysis of B(a)P oxides formed in situ were quantified and reported to a calibration curve of hmEH activity.

Inactivation of hmEH hydrolysis (%)[a]				
Time (min)	0	15	30	45	60
Microsomes					
With TA	5	44	52	60	63
Without TA	0	5	5	5	5
NADPH-free	5	5	6	6	6
PM					
With TA	5	35	50	64	66
Without TA	0	5	5	5	5
NADPH-free	0	5	5	5	5

[a] 100 % corresponds to 2.5 nM of microsomal hmEH or 1nM PM hmEH.

Recognition of TA-alkylated hmEH by patient's sera

In order to verify that hmEH-TA metabolite adducts generated *in vitro* included the antigen which elicited immune response in patients with germander hepatitis, yeast microsomes expressing CYP3A4 and hmEH were incubated with or without TA. Samples were then analyzed by Western blot for expression of protein antigen recognized by patient's sera. Recognition of hmEH protein-band was significantly higher with hmEH microsomes incubated with TA and an NADPH-regenerating system than either without TA or with an NADPH-free system (data not shown). This study revealed that TA-alkylated hmEH was expressed *in vitro* and was recognized by antibodies present in the patient's sera, in addition to the anti-hmEH-autoantibodies.

DISCUSSION

The present study demonstrates for the first time that a plant substance can induce autoantibody formation in some patients. Moreover, the autoantibodies are directed against hmEH, a phase II microsomal enzyme, and are not inhibitory.

To develop an immune response, the immune system must be stimulated by a non-self protein or a modified self protein (neoantigen). Covalent binding demonstrated in isolated hepatocytes incubated with TA (Lekehal *et al.*, 1996) and TA-mediated inactivation of hmEH suggest the formation of metabolite-modified neoantigen(s).

Germander furanoditerpenoids exhibit direct toxicity causing GSH and protein thiol depletion and either necrosis (Loeper *et al.*, 1994) or apoptosis (Fau *et al.*, 1997). This direct toxicity could start the immunization process, leading to the appearance of both anti-hmEH autoantibodies and also anti-hmEH-adduct antibodies. The latter possibility was observed by the higher recognition of hmEH by the human sera, with microsomes preincubated with TA than with microsomes not exposed to TA. To participate in the immune destruction of hepatocytes, autoantibodies should be able to recognize protein epitopes exposed on the outer surface of the PM. To determine whether hmEH was present on the outside of the PM. Electron microscopy with immunoperoxidase labeling showed that hmEH was indeed exposed on the outer surface of hepatocyte PM. In this respect, germander -induced hepatitis resembled tienilic acid- or dihydralazine-induced hepatitis, in which induced autoantibodies recognized their antigenic CYP targets on the outer surface of human hepatocytes (Loeper *et al.*, 1993). The presence of the antigenic molecule at the cell surface makes theoretically possible for the anti-hmEH (auto)antibodies to participate in the immune destruction of hepatocytes, through complement-induced cytotoxicity and/or antibody-dependent-cell mediated cytotoxicity. Although anti-adduct antibodies rather than autoantibodies may be involved, antibody-dependent-cell mediated cytotoxicity was demonstrated with sera containing both anti-CYP autoantibodies and anti-CYP adduct antibodies. Germander-induced hepatitis may be due to both direct toxicity and secondary immune reactions, with probably a varying contribution of these two mechanisms in different patients. An immune reaction was suspected in several patients, based on the quick onset of the relapse caused by a rechallenge, contrasting with the late onset of hepatitis during the initial exposure (Castot and Larrey, 1992). Genetic predisposition may be required to develop an immune reaction, including polymorphism in drug-metabolizing enzymes, and the polymorphism of MHC molecules.

In conclusion, for the first time, we have shown that anti-hmEH autoantibodies can develop in plant-induced hepatitis. hmEH is present and functional in the human hepatocyte PM and hmEH epitopes are exposed on the outer surface of the PM, allowing the autoantibodies to possibly participate in the immune destruction of hepatocytes. CYP3A4-mediated metabolic activation of TA into a reactive epoxide could modify and inactivate hmEH. This altered protein may bypass immune tolerance and trigger the immune response against hmEH.

REFERENCES

Castot, A and Larrey, D , 1992, Hepatitis observed during a treatment with a drug or tea containing Wild Germander. Evaluation of 26 cases reported to the Regional Centers   of Pharmacovigilance, *Gastroenterol Clin Biol* **16:**916-922

Fau, D, Lekehal, M, Farrell, G, Moreau, A, Moulis, C, Feldmann, G, Haouzi, D and Pessayre, D ,1997, Diterpenoids from germander, an herbal medicine, induce apoptosis in isolated rat hepatocytes. *Gastroenterology* **113:**1334-1346.

Loeper, J, Descatoire, V, Maurice, M, Beaune, P, Belghiti, J, Houssin, D, Ballet, F, Feldmann,G, Guengerich, FP and Pessayre, D ,1993, Cytochromes P-450 in human hepatocyte plasma membrane: recognition by several autoantibodies,   *Gastroenterology* **104:**203-216.

Loeper, J, Descatoire, V, Letteron, P, Moulis, C, Degott, C, Dansette, P, Fau, D and Pessayre, D ,1994, Hepatotoxicity of germander in mice, *Gastroenterology* **106:**464-472.

STRUCTURAL CHARACTERISATION OF THE MAIN EPICHLOROHYDRIN-GUANOSINE ADDUCTS

Jukka Mäki, Krister Karlsson, Rainer Sjöholm, and Leif Kronberg

Department of Organic Chemistry, Åbo Akademi University
Biskopsgatan 8, FIN-20500, Turku/Åbo, Finland

INTRODUCTION

Epichlorohydrin (ECH, 1-chloro-2,3-epoxypropane) is a chemical used industrially in the manufacture of *e.g.* epoxy resins, glycidyl ethers, ion exchangers, pharmaceuticals, paper, textiles and coatings. ECH has been shown to induce mutagenic and clastogenic effects in various test systems. It has been shown to be carcinogenic in animals and classified as probably carcinogenic to humans. Recently, ECH was shown to induce formation of DNA adducts in humans exposed to ECH (Plna et al., 2000). A review article concerning the toxicology of ECH has been written by Giri (1997).

In a previous report, several adducts have been proposed to be formed in the reactions of ECH with deoxyguanosine (Singh et al., 1996). However, only one adduct was adequately characterised by spectroscopic means. Therefore, we have reinvestigated the reactions, isolated the main products and characterised the adducts based on UV, ^{1}H and ^{13}C NMR spectroscopy and mass spectrometry. For most part, the results obtained in our study support the previously proposed structures for the main ECH-guanosine adducts.

EXPERIMENTAL

Guanosine was reacted with a 20-fold excess of ECH (added in small portions) in aqueous potassium phosphate buffer solution (pH 4.6) at 37 °C for four days. The course of the reaction was followed by analytical HPLC on a reversed phase (C18) column. The column was eluted with 0.01 M phosphate buffer (pH 7.1) and acetonitrile (a gradient of acetonitrile from 0 to 30 % over the course of 25 min). After the reaction had completed, the volume of the reaction mixture was reduced to half of the original by rotatory evaporation. The mixture was stored in a refrigerator overnight, during which time most of the compound abbreviated as **7-CHPGua** (Figures 1 and 2) precipitated. Compounds abbreviated as **7-CHPGuo** and **1,N^2-HP-7-CHPGuo** in Figures 1 and 2 were purified with semi-preparative HPLC on RP C18 column eluting with 0.01 M ammonium acetate and a gradient of acetonitrile as described above. The compound abbreviated as **bis-7,9-**

CHPGua (Figure 2) was obtained by reaction of pure **7-CHPGua** with an excess of ECH in phosphate buffer (pH 4.6). **Bis-7,9-CHPGua** was purified with semi-preparative HPLC eluting with 0.01 M ammonium acetate and acetonitrile gradient as above.

The structural characterisation of the compounds was carried out by UV, ^{1}H and ^{13}C NMR spectroscopy including different 2D correlation techniques (COSY, long-range COSY, HETCOR and COLOC) and mass spectrometry. The NMR-spectra were recorded in DMSO-d_6 solution and referenced internally to the signal of the solvent at $\delta = 2.50$ ppm (for ^{1}H) and $\delta = 39.50$ ppm (for ^{13}C). Due to overlapping signals in the proton spectra, as well as for the higher order couplings of the ribosyl unit and the substituents, the proton chemical shifts and coupling constants were calculated with the Perch program.[†] The elemental composition of the adducts was verified by high-resolution mass spectrometry.

RESULTS

The HPLC-analysis of the reaction mixture of ECH with guanosine in aqueous phosphate buffer (pH 4.6) showed that four major products were formed (Figure 1). In these conditions, guanosine was consumed almost quantitatively. Peaks abbreviated as **7-CHPGuo** displayed the same UV-spectra. Both compounds were isolated and analysed by ^{1}H and ^{13}C NMR spectroscopy and mass spectrometry. The spectral characteristics for both of the isolated compounds were identical. The ^{13}C NMR spectrum displayed three new signals in addition to the signals from guanosine. In the proton spectrum, several new signals in the region of $\delta = 3.5$ to 4.7 ppm were observed. These signals displayed strong couplings and were partially overlapping with the signals from the ribosyl unit. The most diagnostic signal in the proton spectrum, referring to the modification in the imidazole ring, was the signal of H-8. The signal was observed at $\delta = 9.18$ ppm, *i.e.* a remarkable downfield shift in comparison to the shift of H-8 in guanosine ($\delta \approx 8.00$ ppm). The addition of the substituent to N-7 and the consequent positive charge delocalised over the imidazole

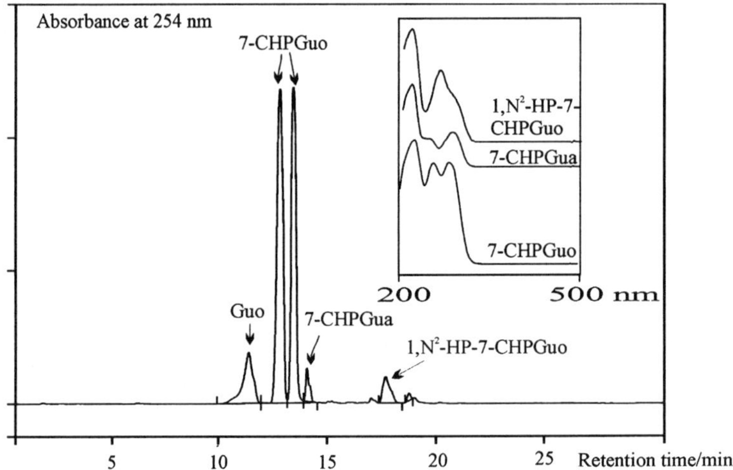

Figure 1. The HPLC chromatogram from the reaction mixture of ECH and guanosine. Reaction conditions: pH 4.6, T = 37 °C, reaction time 24 hours. Peak titled as Guo refers to the unreacted guanosine. UV-spectra (in pH 7.0) of the peaks are given in the insert.

[†] The Perch software is distributed by PERCH Project, Department of Chemistry, University of Kuopio, Kuopio, Finland.

ring was also evident from the large one-bond coupling constant of C-8 ($^1J_{C,H}$ = 222.8 Hz.). The changes in chemical shifts and coupling constants parallel the observations made in protonation studies of imidazoles, purines and ethenoadenosine (Schumacher and Günther, 1982; Sierzputowska-Gracz et al., 1986). It was concluded that the peaks **7-CHPGuo** consisted of a diastereomeric pair of 7-(3-chloro-2-hydroxypropyl)guanosine as depicted in Figure 2.

The compound abbreviated as **7-CHPGua** in Figure 1 precipitated upon storage of the original reaction mixture in a refrigerator overnight. The product was collected by filtration and reprecipitated from water. The proton NMR spectrum showed the absence of the ribosyl signals. Therefore, we account this compound as resulting from the depurination 7-**CHPGuo**. Both the NMR and mass spectrometric data are consistent with the structure 7-(3-chloro-2-hydroxypropyl)guanine (**7-CHPGua**) shown in Figure 2. Moreover, the proton NMR-data agrees with previously published data (Singh et al., 1996).

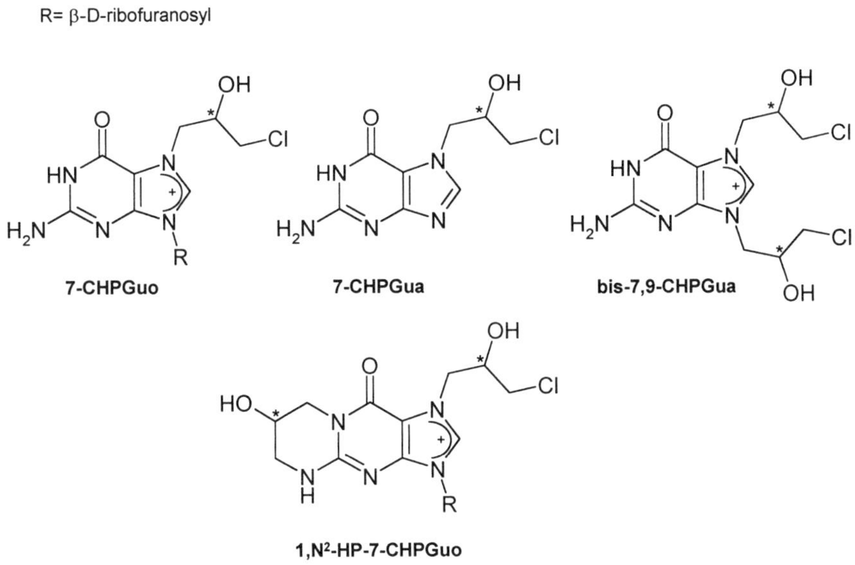

Figure 2. Structures of the main epichlorohydrin-guanosine adducts. Stereogenic centers are marked with asterisks.

The compound abbreviated as **7-CHPGua** was further modified by reaction with a 10-fold excess of ECH in an aqueous buffer solution (pH 4.6) at 50 °C. In the chromatograms from the reaction mixture, two major product peaks were observed (chromatogram not shown). Both peaks displayed similar UV-spectra and they were formed in equal amounts. The NMR-analysis of the isolated compounds showed that the spectra resembled the spectra of **7-CHPGuo**. The signal of H-8 was again observed markedly downfield (at δ = 9.18 ppm), compared with the value for the starting material **7-CHPGua** (at δ = 7.81 ppm). Also the ^{13}C chemical shifts were in accordance with the ones observed for 7-**CHPGuo**. From the isotope pattern in the mass spectrum it was evident that the compounds had two chlorine atoms. We identified these compounds as a diastereomeric pair of bis-7,9-(3-chloro-2-hydroxypropyl)guanine (**bis-7,9-CHPGua**) as depicted in Figure 2. Whether these adducts were formed also in the original reaction is unclear since the retention times in our HPLC-system for the diastereomeric pair of **bis-7,9-CHPGua** were identical with the ones for the diastereomeric pair of **7-CHPGuo**.

Finally, we isolated also the peak abbreviated as **1,N^2-HP-7-CHPGuo** in Figure 1. The NMR-analysis of the compound gave a complex mixture of signals, presumably originating from different stereoisomers of the same compound. The ^{1}H NMR spectrum clearly showed the presence of the ribosyl signals together with the signals from the (3-chloro-2-hydroxypropyl)-substituent. We have tentatively characterised this compound as 1,N^2-hydroxypropano-7-(3-chloro-2-hydroxypropyl)guanosine as depicted in Figure 2. The multiple signals can be explained by the new asymmetric carbon (in the hydroxypropano-ring) giving rise to different diastereomers. Storage of the compound in 0.01 M HCl at 50 °C gave again two identical product peaks observed in the HPLC-chromatograms. Currently, we are working with the final structural characterisation of the peak abbreviated as **1,N^2-HP-7-CHPGuo**.

REFERENCES

Giri, A.K., 1997, Genetic toxicology of epichlorohydrin: A review, *Mutation Research* 386:25.

Plna, K., Osterman-Golkar, S., Nogradi, E., and Segerbäck, D., 2000, ^{32}P-postlabelling of 7-(3-chloro-2-hydroxypropyl)guanine in white blood cells of workers occupationally exposed to epichlorohydrin, *Carcinogenesis* 21:275.

Schumacher, M., and Günther, H., 1982, ^{13}C,^{1}H Spin-spin coupling. 9. Purine, *J. Am. Chem. Soc.* 104:4167.

Sierzputowska-Gracz, H., Gopal, H. D., and Agris, P.F., 1986, Comparative structural analysis of 1-methyladenosine, 7-methylguanosine, ethenoadenosine and their protonated salts. IV: ^{1}H, ^{13}C and ^{15}N NMR studies at natural isotope abundance, *Nucleic Acids Research*, 14:7783.

Singh, U.S., Decker-Samuelian, K., and Solomon, J.J., 1996, Reaction of epichlorohydrin with 2′-deoxynucleosides: characterization of adducts, *Chemico-Biological Interactions* 99:109.

ADDUCTS OF THE CHLOROFORM METABOLITE PHOSGENE

Laura Fabrizi, Graham W. Taylor, Robert J. Edwards & Alan R. Boobis

Section on Clinical Pharmacology
Imperial College School of Medicine
W12 ONN London
UK

1. INTRODUCTION

Chlorination of water for human consumption is important for disease prevention. However, halogenated by-products, such as trihalomethanes (THM), can be formed during this process (Bunn *et al.*, 1975). THM production raises public health concerns related to potential human cancer risk (Fawell, 2000). Chloroform (trichloromethane: TCM) is a by-product that is frequently detected at relatively high concentrations (IARC, 1991), i.e. 0.1 to 300 µg/litre in finished drinking water (Uden and Miller, 1983) and up to 1 mg/litre in chlorinated swimming pool water (Lahl *et al.*, 1981).

Human exposure to TCM can occur either through ingestion or through skin absorption and inhalation. The latter routes of exposure are particularly relevant for athletes in swimming pools, but also for the general population during showers (Corley *et al.*, 2000; Kuo *et al.*, 1998). In long term studies TCM has been shown to produce tumours in the mouse liver and the rat kidney (Jorgenson *et al.*, 1985; Larson *et al.*, 1994). Further, an association has been shown between exposure to THM in drinking water and the incidence of tumours in the lower gastrointestinal tract, bladder (Morris *et al.*, 1992) and colon (Doyle *et al.*, 1997). As epidemiological studies have failed to provide conclusive evidence for these associations, an elucidation of the mode of action of TCM in carcinogenesis is needed to provide a reliable risk assessment.

Cytochrome P450 (P450) can activate TCM oxidatively to phosgene ($COCl_2$) and reductively to the dichloromethyl radical ($^\bullet CHCl_2$). Both of these reactive intermediates can bind irreversibly to cellular macromolecules (Fig. 1). Under physiological oxygenation conditions, $COCl_2$ is the major TCM metabolite formed: it is considered responsible for TCM acute toxicity (Pohl *et al.* 1984), although recent studies have shown no correlation between $COCl_2$ formation and TCM induced tumours in rodents (Vittozzi *et al.*, 2000). Most $COCl_2$ produced *in vivo* is hydrolysed to CO_2 or scavenged by glutathione (GSH) (Gemma *et al.*, 1996; Vittozzi *et al.*, 2000) and the stable product diglutathionyl dithiocarbonate (GS-CO-SG) has been isolated and characterised in chemical systems and in microsomal incubations (Pohl *et al.*, 1981). However, not only the residual $COCl_2$ but also the postulated reactive intermediate glutathionyl-chlorocarbonyl (GS-CO-Cl) can bind to cellular macromolecules. This intermediate could be produced in considerable amounts, particularly when GSH is depleted and thus it limits GS-CO-SG formation.

This study was undertaken in order to investigate the interactions of $COCl_2$ with GSH and other endogenous nucleophiles in model chemical systems.

Biological Reactive Intermediates VI, Edited by Dansette *et al.*
Kluwer Academic / Plenum Publishers, 2001

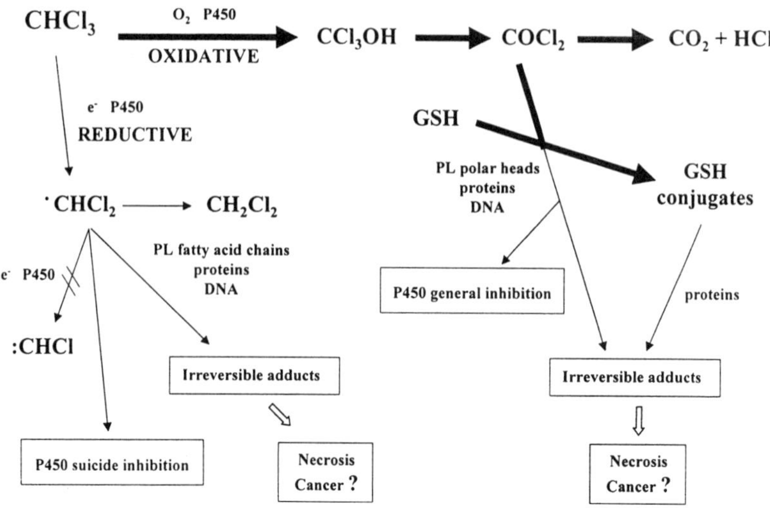

Figure 1. Chloroform metabolism: pathways and consequences.

2. METHODS

Reactions between $COCl_2$, GSH and putative biological targets were carried out for 30 minutes at room temperature in 0.4M NH_4HCO_3 buffer, pH 7.8, with shaking. $COCl_2$ was added to the mixture as a 20% solution in toluene (maximal final concentration: 60mM). The biological targets investigated were amino acids, peptides, proteins and deoxynucleotides. Reaction products and respective controls (without $COCl_2$) were analysed by positive ion mode electrospray mass spectrometry (MS) and tandem-MS on a Quattro II instrument.

3. RESULTS

3.1. Reaction of $COCl_2$ with GSH and Single Amino Acids

The reaction of $COCl_2$ in the presence of GSH and Cys produced the adduct GS-CO-Cys ($M+H^+$ at m/z 455). The formation of this adduct shows that the intermediate GS-CO-Cl can bind not only to other GSH molecules, but also to different thiol containing entities. In the same reaction mixture, peaks corresponding to the GS-CO-SG and Cys-CO-Cys adducts were also found (m/z 641 and 269, respectively). All these products were absent in the control mixtures, where $COCl_2$ was omitted. Tandem-MS confirmed the identity of the adducts: e.g. peaks corresponding to Cys (m/z 122) and GSH (m/z 308) were found as daughters of GS-CO-Cys. The products were susceptible to acetylation suggesting that the primary amines in GSH and Cys were not involved in the reaction. Under the same conditions, amino acids possessing amino-side chain groups, i.e. Lys and Arg, showed weak reactivity, forming only trace amounts of the putative (amino acid)-CO-SG adduct.

3.2. Reaction of $COCl_2$ with GSH and Peptides Containing Multiple Amino Groups

In contrast to basic amino acids, pentapeptides containing multiple amino groups, showed high reactivity with $COCl_2$, forming cyclic carbonyl adducts. For example, LKKLR ($M+H^+$, m/z 657) was transformed into a carbonyl adduct with m/z 683. LKKLG and LKGLG showed the same type of adduction. These data indicate the involvement of amino groups in adduct formation, although this requires confirmation.

3.3. Reaction of COCl₂ with GSH and Deoxyguanosine (dG)

Only the GS-CO-SG adduct was found, suggesting that $COCl_2$ does not react with dG under these conditions.

3.4. Reaction of COCl₂ with GSH and Lysozyme

Lysozyme (L_{yz}) contains eight cysteinyl residues, present as cystines in the native protein, which may be reduced to cysteines by GSH in the reaction mixture. L_{yz} was used as a model protein to investigate further GS-CO-Cl adduction to thiol groups. L_{yz} (M_r 14302) reacted with $COCl_2$ and GSH to produce both L_{yz} (CO-SG) and L_{yz} (CO-SG)$_2$ adducts, with M_r 14635 and M_r 14968, respectively (Fig. 2).

3.5. Reaction of COCl₂ with GSH and the N-terminal Peptide from Histone H2B

The N-terminal peptide (H_{pep} = PEPAKSAPAPKKGSKKAVTKAQK, M_r 2348) from human histone H2B contains multiple lysine residues, but lacks cysteinyl groups. Its reaction with $COCl_2$ and GSH produced a monocarbonyl adduct with M_r 2374 (Fig. 3).

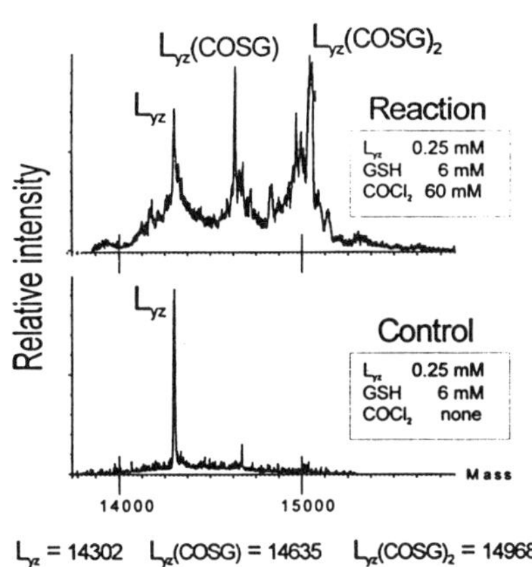

Figure 2. Transformed MS-spectrum of reaction and control mixtures with L_{yz}

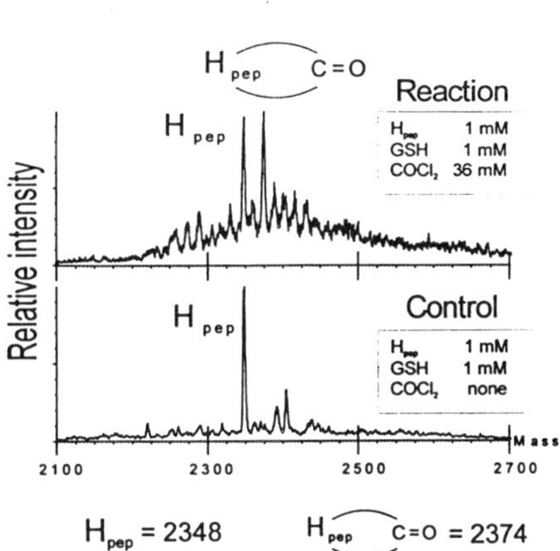

Figure 3. Transformed MS-spectrum of reaction and control mixtures with H_{pep}

4. CONCLUSIONS

Our data show that the chloroform metabolite $COCl_2$ is able to react with GSH to form a reactive intermediate (GS-CO-Cl), that binds to other GSH molecules, cysteine and thiol-containing proteins such as lysozyme. Previously, Pereira *et al.* (1984) have also shown that, in the absence of GSH, chloroform activated in microsomal incubations can produce an adduct with hemoglobin to form N-hydroxymethyl cysteine.

In the presence of peptides containing multiple amino groups, $COCl_2$ preferentially forms cyclic carbonyl products with two nucleophiles, rather than being scavenged by GSH. For example the N-terminal peptide from histone H2B, which is rich in lysine residues, covalently binds the carbonyl moiety from $COCl_2$. Chromatin assembly and gene regulation are known to be affected by the acetylation of specific lysines in the N-terminal region of histones (Grunstein, 1997). Interestingly, when Diaz Gomez and Castro (1980) treated rats with [^{14}C]CHCl$_3$ they found adduction to nuclear proteins including histones, although no DNA or RNA adducts were detected. Our data suggest that COCl$_2$ can interact with histones and this may have important biological consequences. This is currently the subject of further investigation.

REFERENCES

Bunn, W.W., Haas, B.B., Deane, E.R. and Kleopfer, R.D., 1975, Formation of trihalomethanes by chlorination of surface water. *Environ. Letters* **10**:205-213.

Corley, R.A., Gordon, S.M. and Wallace, L.A., 2000, Physiologically based pharmacokinetic modeling of the temperature-dependent dermal absorption of chloroform by humans following bath water exposure. *Toxicol. Sci.* **53**:13-23.

Diaz Gomez, M.I. and Castro, J.A., 1980, Covalent binding of chloroform metabolites to nuclear proteins – No evidence for binding to nucleic acids. *Cancer Letters* **9**:213-218.

Doyle, T.J., Zheng, W., Cerhan, J.R., Hong, C.P., Sellers, T.A., Kushi, L.H., and Folsom, A.R., 1997, The association of drinking water source and chlorination by-products with cancer incidence among postmenopausal women in Iowa: A prospective cohort study, *Am. J. Public Health* **87**:1168-1176.

Fawell, J., 2000. Risk assessment case study—chloroform and related substances, *Food and Chem Toxicol.* **38** (1 Suppl):S91-S95.

Gemma, S., Faccioli, S., Chieco, P., Sbraccia, M., Testai, E., and Vittozzi, L., 1996, *In vivo* CHCl$_3$ bioactivation, toxicokinetics, toxicity and induced compensatory cell proliferation in B6C3F1 male mice, *Toxicol. Appl. Pharmacol.* **141**:394-402.

Grunstein, M., 1997. Histone acetylation in chromatin structure and transcription, *Nature* **389**:349-352.

IARC, 1991. IARC monographs on the evaluation of carcinogenic risks to humans. Chlorinated drinking water; chlorination by-products; some other halogenated compounds; cobalt and cobalt compounds. Lyon, France: IARC.

Jorgenson, T.A., Meierhenry, E.F., Rushbrook, C.J., Bull, R.J., and Robinson, M., 1985. Carcinogenicity of chloroform in drinking water to male Osborne Mendel rats and female B6C3F1 mice, *Fund. Appl. Toxicol.* **5**:760-769.

Kuo, H.K., Chiang, T.F., Lo, I.I., Lai, J.S., Chan, C.C. and Wang, J.D., 1998, Estimates of cancer risk from chloroform exposure during showering in Taiwan. *Sci. Tot. Environ.* **218**:1-7.

Lahl, U., Batjer, K., Duszeln, J.V., Gabel, B., and Thiermann, W., 1981, Distribution and balance of volatile halogenated hydrocarbons in the water and air of covered swimming pools using chlorine for water disinfection. *Water Res.* **15**:803-814.

Larson, J.L., Wolf, D.C., and Butterworth, B.E., 1994, Induced cytolethality and regenerative cell proliferation in the livers and kidneys of male B6C3F1 mice given chloroform by gavage. *Fund. Appl. Toxicol.* **23**:537-543.

Morris, R.D., Audet, A.M., Angelillo, I.F., Chalmers, T.C. and Mosteller, F., 1992, Chlorination, chlorination by-products, and cancer: A meta-analysis. *Am. J. Public Health* **82**:955-963.

Pereira, M.A., Chang, L.W., Ferguson, J.L., and Couri, D., 1984, Binding of chloroform to the cysteine of haemoglobin. *Chem.-Biol. Interactions* **51**:115-124.

Pohl, L.R., , Branchflower, R.V., Highet, R.J., Martin, J.L., Nunn, D.S., Monks, T.J., George, J.W., and Hinson, J.A., 1981, The formation of diglutathionyl dithiocarbonate as a metabolite of chloroform, bromotrichloromethane, and carbon tetrachloride. *Drug Metab. Dispos.* **9**:334-339.

Pohl, L.R., George, J.W., and Satoh, H., 1984, Strain and sex differences in chloroform-induced nephrotoxicity. Different rates of metabolism of chloroform to phosgene by the mouse kidney. *Drug Metab. Dispos.* **12**:304-308.

Uden, P.C. and Miller, J.W., 1983, Chlorinated acids and chloral in drinking water. *J. Am. Water Work Assoc.* **75**:524-526.

Vittozzi, L., Gemma, S., Sbraccia, M., and Testai, E., 2000, Comparative characterization of CHCl$_3$ metabolism and toxicokinetics in rodent strains differently susceptible to chloroform-induced carcinogenicity. *Environ. Toxicol. Pharmacol.* **8**:103-110.

OXYGENATION OF ARACHIDONIC ACID BY CYCLOOXYGENASES GENERATES REACTIVE INTERMEDIATES THAT FORM ADDUCTS WITH PROTEINS

Olivier Boutaud[1], Junyu Li[1], Pierre Chaurand[2], Cynthia J. Brame[1], Lawrence J. Marnett[2], L. Jackson Roberts, II[1], and John A. Oates[1]

[1] Departments of Medicine and Pharmacology
[2] Department of Biochemistry
Vanderbilt University, School of Medicine, Nashville, TN 37232, USA

INTRODUCTION

Cyclooxygenase-2 (COX-2) catalyzes the oxygenation of arachidonic acid into the prostaglandin endoperoxide, PGH_2. PGH_2 is then processed by different isomerases into prostaglandins and thromboxane A_2. PGH_2 also undergoes rearrangement in aqueous solution into PGE_2, PGD_2 and into the levuglandins (LG) E_2 and D_2 [1]. LG's are highly reactive γ-ketoaldehydes that have been shown to form adducts and cross-links with proteins [2,3], primarily through their reaction with the ε-amine of lysine. The potential that these levuglandin adducts of proteins could be biologically important may be inferred from the knowledge that lipid-modification of proteins is known to influence their function and cellular localization [4-7].

In order to determine whether the expression of COX-2 in cells could lead to formation of levuglandin adducts of proteins, we previously have characterized the structures of three levuglandin adducts of lysine, the lysyl-LG Schiff base, and the pyrrole-derived lysyl-LG lactam and hydroxy lactam [8,9]. Here we demonstrate that oxygenation of arachidonic acid by either COX-1 or COX-2 leads to the formation of lysyl-LG adducts of the enzymes that are characterized as the lysyl-LG lactam and lysyl-LG Schiff base. The high degree of reactivity of the LG's is reflected in the finding that virtually all of the LG predicted to be formed from the cyclooxygenase (approximately 20% of PGH2) is adducted to the enzyme. When other proteins are placed in the solution with the cyclooxygenase, they are similarly adducted by the LG. We also established that the formation of LG adducts can lead to crosslinking between proteins and spermine. Finally, we have demonstrated that in contrast to COX-1, the autoinactivation of COX-2 was reversible over time and that it was not related to the formation of the lysyl-LG adducts.

MATERIAL AND METHODS

Incubation of [^{14}C]-arachidonic acid with COX-2

COX-1 was purified from ram seminal vesicles as previously described [10]. Mouse COX-2 was expressed in SF-9 cells (Novagen, Madison, WI) and purified as previously described [11]. Mouse COX-2 (14 μM) was preincubated at 37°C for 5 minutes in oxygenated phosphate buffer 100 mM pH 7.5, 500 μM phenol, 2 molar equivalents of hematin, then transferred to a tube containing 4.8 μCi of [^{14}C]-arachidonic acid (944 μM, final concentration), and incubated at 37°C for the time desired.

After incubation, in some experiments the enzyme was dialyzed at 4°C overnight against sodium phosphate buffer 100 mM, pH 8.0, 50 mM NaCl, 0.4 % CHAPS, 500 μM phenol. An aliquot was then chromatographed on a 4.6 x 250 mm Protein C4 VYDAC column with a linear gradient of 50% to 75% acetonitrile in water/0.1% TFA in 30 minutes at 1.0 ml/min. The radioactivity was monitored using a radioactive flow detector Flo-One beta Model CT.

MALDI-TOF mass spectrometry

Carbon embedded polyethylene (PE) membrane was used as sample support for matrix-assisted laser desorption ionization (MALDI) mass spectrometry analysis (MS) [12, 13]. The matrix was sinapinic acid 10 mg/ml in acetonitrile/H_2O 1/1, 0.1% TFA. The samples were analyzed using a Perseptive DE-STR MALDI time-of-flight mass spectrometer. The instrument was operated in the linear mode at 25 kV under optimized delayed extraction conditions. The mass spectra have been obtained by averaging 256 individual laser shots and were further processed using Data Explorer (PerSeptive Biosystem, Foster City, CA). Mass calibration was achieved using the doubly charged and molecular ions of bovine serum albumin (BSA, MW 66,430). BSA was prepared in H_2O 0.1% TFA at a final concentration of 5 pmol/μl. The centroid peak values as well as the standard deviations were determined by averaging 5 independent measurements.

Characterization of lysyl-levuglandin adducts of COX-1 and COX-2

Mouse COX-2 (14.2 μM) or ovine COX-1 (14.2 μM) was incubated with arachidonic acid (944 μM final concentration) as described above. The [$^{13}C_6$],[^{3}H]-lysyl-LG lactam adduct (14.5 ng, 22,200 cpm) and the [$^{13}C_6$],[^{3}H]-lysyl-LG reduced Schiff base adduct (20 ng, 33,200 cpm) were added as internal standards, and the reaction was stopped by adding sodium borohydride in DMF at a final concentration of 20 mM. After 30 min at room temperature, the pH of the solution was neutralized with 10X PBS, and COX was denatured at 95°C for 5 minutes. After a first digestion by pronase (0.1 mg) under argon at 37°C overnight, followed by denaturation at 95°C for 5 minutes, the peptides were further digested to single amino acids under argon by incubation for 24 h at 37°C with aminopeptidase M from porcine kidney (0.1 U). The lysyl-adducts were purified by solid phase extraction using Oasis™ cartridge. After washing it with water, the adducts were eluted with methanol.

The two adducts were then separated by C18 RP-HPLC. The radioactive fractions were pooled for the two peaks and concentrated by solid extraction using an Oasis™ SepPak column. After concentration of the eluate under nitrogen stream, the samples were analyzed by liquid chromatography / electrospray ionization-tandem mass spectrometry (LC/ESI-MS/MS). The adducts of lysine derived from protease digestion of the cyclooxygenases were chromatographed on a 2.1x15 mm XDB C8 column with a flow rate of 0.2 ml/min using an linear gradient of 10% to 90% acetonitrile in 5 mM ammonium acetate, 0.1% acetic acid. Electrospray tandem mass spectrometric analysis was carried out on a Finnigan TSQ7000. Ions were subjected to collision-induced dissociation (m/z 467.4, reduced Schiff base adduct or m/z 479.4, lactam adduct), monitoring the daughter ions at m/z 321.4 and 84.1 respectively.

Formation of arachidonic acid-derived adducts of other proteins

COX-2 was incubated with [^{14}C]-arachidonic acid in the presence of a mixture of histone fractions IIA and IIIS (Sigma) from calf thymus, or ubiquitin. After 30 min incubation, the reaction mix was denatured and loaded on a 10% acrylamide NuPAGE™ Bis-Tris gel from Novex. The proteins were separated by electrophoresis using the NuPAGE MOPS running buffer. At the end of the electrophoresis, the proteins were stained with Coomassie Blue. The radioactivity associated with the proteins was determined by autoradiography.

Participation of LGE$_2$ in covalent binding of [^{14}C]-spermine to albumin

BSA (10 µM in PBS) was incubated at 37°C in the presence of [^{14}C]-spermine (15 µM, 222,000 dpm) with or without LGE$_2$ (10 µM), or COX-2 with or without arachidonic acid. After 30 min incubation, the reaction mix was denatured and loaded on a 4-12% acrylamide NuPAGE™ Bis-Tris gel. The proteins were separated by electrophoresis using the NuPAGE MOPS running buffer. The radioactivity associated with the proteins was determined by autoradiography.

Recovery of COX-2 activity following its autoinactivation

COX-2 (260 nM) was pre-incubated in Tris HCl 100 mM, pH 8.0, 500 µM phenol, 2 molar equivalents of hematin for 5 minutes. [^{14}C]-arachidonic acid at a final concentration of 10 µM was added to inactivate the enzyme and the mixture was kept at 37°C. At different times following inactivation, an aliquot was added to cold arachidonic acid (final concentration of 10 µM) and the reaction was allowed to occur for 30 sec. At the end of the reaction, 60 µl was mixed with SDS-PAGE loading buffer for further analysis of the protein by electrophoresis. At the end of the 30 sec, the reaction was terminated by adding 2 volumes of diethyl ether/methanol/1 M citric acid (30:4:1). The products contained in the organic phase were derivatized for analysis by GC/MS or were directly analyzed by LC/ESI/MS in negative ionization.

RESULTS AND DISCUSSION

As previously has been demonstrated for COX-1 [14-16], following oxygenation of [^{14}C]-arachidonic acid by COX-2, the enzyme comigrates with the radiolabel in SDS- PAGE, and coelutes with the radiolabel on reversed-phase chromatography. The formation of arachidonate-derived adducts of COX-2 also is indicated by an increase in the mass of reacted enzyme that is observed by MALDI-TOF mass spectrometric analysis. The magnitude of adduct formation indicates that a highly reactive product is formed by the cyclooxygenases.

Characterization of the structure of the adducts was based on a hypothesis that they were derived from the levuglandins, γ-ketoaldehydes that are formed by rearrangement of PGH$_2$ [1]. In order to develop an analytical strategy for determining whether adducts of the cyclooxygenases and other proteins were derived from levuglandins, we characterized the structures of adducts of synthetic LGE$_2$ with lysine, and demonstrated the formation of the lysyl-LG Schiff base [8] as well as oxidative products of the lysyl-LG pyrrole, the lysyl-LG lactam and lysyl-LG hydroxylactam [9] (figure 1). All of these products also could be formed by incubation of PGH$_2$ with lysine, validating that this is a tangible pathway of PGH$_2$ rearrangement.

We here report evidence that a levuglandin is adducted to the PGH-synthases as a consequence of catalysis. After protease digestion of reacted COX-2 that has been treated with a reducing agent, the reduced lysyl-anhydroLG Schiff base has been isolated and characterized by co-chromatography with a [^{13}C]-labeled standard and with LC/MS/MS, based on both the molecular ion (m/z 467.4) and the characteristic daughter ion (m/z 321.4). The lysyl-LG lactam adduct also is identified, with its molecular ion (m/z 479.4) and the daughter ion of m/z = 84.1. Similar evidence indicates that lysyl-LG Schiff base and lactam adducts of COX-1 are formed following the oxygenation of arachidonic acid.

The fraction of oxygenated arachidonic acid adducted to the cyclooxygenase was assessed in two ways in the present study. HPLC of the reaction products derived from [^{14}C]-arachidonic acid demonstrated that 20% of the products of the reaction are present as adducts of the cyclooxygenase. Following oxygenation of arachidonic acid by COX-2, the shift in the average mass of the enzyme observed with MALDI-TOF is consistent with an average of 9.3 ± 0.4 molecules of levuglandin adducted to each molecule of the enzyme; with a turnover of 49.7 molecules of arachidonic acid per molecule of enzyme, the adducted levuglandins account for 18.7 % of the reaction products. Both of these measurements are considered as approximations inasmuch as they do not account for the cyclooxygenase molecules that have been crosslinked to form multimers. Thus, approximately 18.7-20% of the products of the oxygenation of arachidonic acid are found to be adducted to the enzyme, representing almost all of the levuglandins (18-22%) that would be expected to be derived from PGH$_2$[1, 8]. This is consistent with the high degree of reactivity of these potent electrophiles.

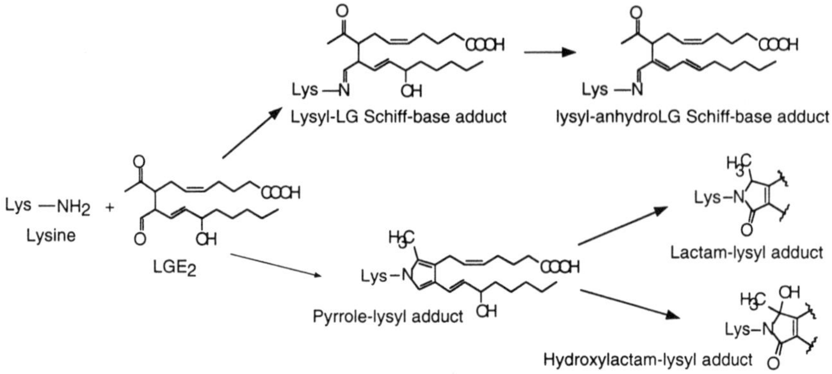

Figure 1. Lysyl-LG adducts characterized.

In an investigation of the autoinactivation of COX-1 [14], it was found that the rate of autoinactivation was disparate from the much slower formation of the adduct, suggesting that the adduct formation was not the cause of inactivation of the enzyme. It has been postulated that the inactivation process of COX-1 could be due to an active radical intermediate generated during the catalysis that leads to irreversible loss of enzyme activity [17, 18]. This study demonstrates, quite in contrast to the irreversible inactivation of COX-1 during catalysis, that the autoinactivation of COX-2 is partially reversible: after the cyclooxygenase had been almost totally inactivated 70% of the enzymatic activity of the control is recovered at 120 min. This finding characterizes yet another difference between the isoforms of the cyclooxygenase and provides additional evidence that this inactivation is not related to the extent of covalent binding of the reactive product(s) of arachidonic acid to the enzyme, inasmuch as reversal of inactivation proceeds concurrently with increasing adduct formation.

The electrophilic reactivity of sites on the levuglandin molecules leads to the formation of covalent bonds not only with the cyclooxygenases but also with other biological nucleophiles in solution with the products of the oxygenation of arachidonic acid [14]. Moreover, lysyl-LG adducts can form covalent bonds with other nucleophiles leading to intermolecular crosslinking. Such crosslinking has been demonstrated for adducts derived from synthetic levuglandin E$_2$; e.g. reaction of LGE$_2$ with albumin leads to formation of multimers of albumin [2]. The reactivity of the initial lysyl-LG adducts is clearly demonstrated by the evidence presented here that spermine will react covalently with COX-2 molecules that have been adducted by oxygenated products of arachidonic acid.

In summary, the lysyl-LG Schiff base and the pyrrole-derived lysyl-LG lactam have been characterized as adducts of the cyclooxygenases that are formed from PGH$_2$ via the levuglandin pathway. This provides a molecular basis for lipid modification of proteins and other biologic nucleophiles. Given the high degree of reactivity of the levuglandins, the biological significance of their formation will be a function of the rate of catalytic disposition of PGH$_2$ by prostanoid synthases or isomerases in different normal and pathologic cells, and of the nucleophiles that are the targets of monomolecular adduct formation or intermolecular crosslinking.

Acknowledgment

The authors gratefully acknowledge the capable technical assistance of Elizabeth Shipp, Jiemin Wang and Gwenn Sobo. This work was supported in part by NIH grants CA 68485, GM 15431, CA47479 and GM 58008-01.

References

1. R.G. Salomon, D.B. Miller, M.G. Zagorski and D.J. Coughlin, Solvent-induced fragmentation of prostaglandin endoperoxides. New aldehyde products from PGH_2 and a novel intramolecular 1,2-hydride shift during endoperoxide fragmentation in aqueous solution, *Journal of the American Chemical Society* 106:6049 (1984)
2. R.S. Iyer, S. Ghosh and R.G. Salomon, Levuglandin E2 crosslinks proteins, *Prostaglandins* 37:471 (1989)
3. K.K. Murthi, L.R. Friedman, N.L. Oleinick and R.G. Salomon, Formation of DNA-protein cross-links in mammalian cells by levuglandin E2, *Biochemistry* 32:4090 (1993)
4. E.M. Van Cott, L. Muszbek and M. Laposata, Fatty acid acylation of platelet proteins, *Prostaglandins, Leukotrienes and Essential Fatty Acids* 57:33 (1997)
5. C.T. Sigal, W. Zhou, C.A. Buser, S. McLaughlin and M.D. Resh, Amino-terminal basic residues of Src mediate membrane binding through electrostatic interaction with acidic phospholipids, *Proceedings of the National Academy of Sciences of the United States of America* 91:12253 (1994)
6. J.T. Dunphy and M.E. Linder, Signalling functions of protein palmitoylation., *Biochimica et Biophysica Acta* 1436:245 (1998)
7. C.A. Buser, C.T. Sigal, M.D. Resh and S. McLaughlin, Membrane binding of myristylated peptides corresponding to the NH2 terminus of Src, *Biochemistry* 33:13093 (1994)
8. O. Boutaud, C.J. Brame, R.G. Salomon, L.J. Roberts, 2nd and J.A. Oates, Characterization of the lysyl adducts formed from prostaglandin H2 via the levuglandin pathway, *Biochemistry* 38:9389 (1999)
9. C.J. Brame, R.G. Salomon, J.D. Morrow and L.J. Roberts, 2nd, Identification of extremely reactive gamma-ketoaldehydes (isolevuglandins) as products of the isoprostane pathway and characterization of their lysyl protein adducts, *Journal of Biological Chemistry* 274:13139 (1999)
10. L.J. Marnett, DNA adducts of alpha,beta-unsaturated aldehydes and dicarbonyl compounds, *IARC Scientific Publications (Lyon)* 151 (1994)
11. S.W. Rowlinson, B.C. Crews, C.A. Lanzo and L.J. Marnett, The binding of arachidonic acid in the cyclooxygenase active site of mouse prostaglandin endoperoxide synthase-2 (COX-2). A putative L-shaped binding conformation utilizing the top channel region, *Journal of Biological Chemistry* 274:23305 (1999)
12. T.A. Worrall, R.J. Cotter and A.S. Woods, Purification of contaminated peptides and proteins on synthetic membrane surfaces for matrix-assisted laser desorption/ionization mass spectrometry, *Analytical Chemistry* 70:750 (1998)
13. R.R. Ogorzalek Loo, C. Mitchell, T.I. Stevenson, J.A. Loo and P.C. Andrews, Diffusive transfer to membranes as an effective interface between gel electrophoresis and mass spectrometry, *International Journal of Mass Spectrometry and Ion Processes* 169/170:273 (1997)
14. R.J. Kulmacz, Attachment of substrate metabolite to prostaglandin H synthase upon reaction with arachidonic acid, *Biochemical & Biophysical Research Communications* 148:539 (1987)
15. M. Lecomte, R. Lecocq, J.E. Dumont and J.M. Boeynaems, Covalent binding of arachidonic acid metabolites to human platelet proteins. Identification of prostaglandin H synthase as one of the modified substrates, *Journal of Biological Chemistry* 265:5178 (1990)
16. A.G. Wilson, H.C. Kung, M.W. Anderson and T.E. Eling, Covalent binding of intermediates formed during the metabolism of arachidonic acid by human platelet subcellular fractions, *Prostaglandins* 18:409 (1979)
17. A. Tsai, C. Wei, H.K. Baek, R.J. Kulmacz and H.E. Van Wart, Comparison of peroxidase reaction mechanisms of prostaglandin H synthase-1 containing heme and mangano protoporphyrin IX, *Journal of Biological Chemistry* 272:8885 (1997)
18. G. Wu, C. Wei, R.J. Kulmacz, Y. Osawa and A.L. Tsai, A mechanistic study of self-inactivation of the peroxidase activity in prostaglandin H synthase-1, *Journal of Biological Chemistry* 274:9231 (1999)

USE OF ISOTOPES AND LC-MS-ESI-TOF FOR MECHANISTIC STUDIES OF TIENILIC ACID METABOLIC ACTIVATION

Maya Belghazi,[2] Pascale Jean,[1] Sonia Poli,[1] Jean-Marie Schmitter,[2] Daniel Mansuy,[1] and Patrick M. Dansette[1]

[1]Université René Descartes, CNRS UMR 8601 Paris, 45 rue des saints Pères 75270 Paris, France
[2]CNRS UMR 5472, Université de Bordeaux I, Talence, France

AIMS

Tienilic Acid (TA) is a uricosuric drug marketed in 1978 and which caused a number of rare immunoallergic hepatitis. It was withdrawn in US in 1980, in France in 1992. Early batches of tienilic acid also contained 0.1–0.5 % tienilic acid isomer. Tienilic acid isomer (TAI), has been shown to be metabolised by Cytochrome P450 into a reactive thiophene 1-oxide which either binds to proteins, or can be trapped by sulfur nucleophiles (Valadon et al. 1996). Tienilic acid is metabolized by human cytochrome P450 2C9 into 5-hydroxytienilic acid (a major metabolite representing 70% of the dose excreted in human urine) but it also forms (a) reactive metabolite(s) which binds covalently to CYP 2C9 and it is a mechanism based inhibitor of CYP 2C9. Adding glutathione to incubations decreases the covalent binding, but only to 1 mol / mol P450. However the reactive metabolite of tienilic acid is still unknown. Recently Koenigs et al. have shown using ESI-LC-MS that CYP 2C9 binds $\approx$ 2 mol of TA in absence of GSH and only one in presence of 3 mM GSH (Koenigs et al. 1999).

Two pathways of activation of the thiophene ring have been postulated: A) the arene oxide pathway and B) the thiophene 1-oxide (or thiophene S-oxide) pathway.

In the case of thiophene, benzo(b)thiophene, tienilic acid isomer and several other thiophenes, metabolism through pathway B (thiophene 1-oxide) has been clearly demonstrated. The 1-oxide has only been isolated in the case of benzothiophene.

Thus, the aim of the present study was to more carefully examine the metabolic oxidation of tienilic acid using HPLC-MS (LC-ESI-TOF mass spectroscopy) in order to determine which pathway (A or B) is involved in this oxidation. For that purpose, labelling experiment with deuterated TA, $^{18}O_2$, $^{18}H_2O$, and D_2O were also performed.

METHODS

Reconstituted recombinant cytochrome P450 2C9 RECO were purchased from PANVERA (contains cytochrome b5, CHAPS, GSH 3mM, phospholipids, N-terminal modified CYP 2C9 and NADPH cytochrome P450 reductase in Hepes buffer). Rat liver microsomes from rats pretreated with clofibrate were obtained as usual (Valadon et al 1996). Tienilic acid, Tienilic acid isomer were from Lipha-Merck. 5-D-Tienilic acid was prepared by catalytic reduction of 5-Chlorotienilic acid on Palladium (5%PdC). D_2O was obtained from Aldrich, $^{18}O_2$ (98%) and $H_2^{18}O_2$ (98%) were obtained from Eurisotop. NADPH was from Boehringer or Sigma.

A Micromass ESI-TOF spectrometer with a Linx software and a Z-spray source was coupled to a Smart system (Amersham-Pharmacia-Biotech). The column was a 2.1 mm × 100 C8 column (Amersham-Pharmacia-Biotech). The solvents were A = TFA 0.05% and B = 95% CH_3CN containing 0.05% TFA. Flow was 100 μL/min, split to 25–30 μL/min at the Z-spray source capillary.

Incubations with RECO 2C9 or Rat Microsomes CLO

100 μL sized incubations (10–50 pmol P450 Reco) or 100 pmol P450 CLO using 50 mM tris or 100 mM phosphate buffer pH 7.4 and 100 μM TA or TAI were performed for 10-30 min using NADPH to start the reaction. The reaction were stopped by 1/2 volume CH_3CN and centrifuged 5 min at 10 000g; 20 to 50 μL of supernatant were injected in the HPLC immediately after. The column was eluted with a gradient of 0% solvent B in solvent A to 100% B in 25 min at 100 μL /min.

Direct Introduction Mass Spectroscopy of Purified Metabolites

Incubations with recombinant CYP 2C9 and 2C18 expressed in yeast using 0.2 μM P450 in a final volume of 2 mL were made as above. For rat liver microsomes CLO 0.5 μM P450 was used also in a 2ml final volume. The supernatant was concentrated under vacuum and injected in two injections on a Hypersil MOS column (4,6 mm x 250 mm) and eluted with a gradient of B'= 90% CH_3CN in A'= 0.1 M ammonium acetate pH 4.6; The 5-OH-AT and AT peaks were collected, evaporated under high vacuum, taken in 100 μL MeOH and methylated with 0.5 ml etheral diazomethane. Mass spectra were obtained on a Nermag R10/10 or JEOL mass spectrometer in EI and CI mode (NH_3).

RESULTS

OXIDATION OF TA BY RECOMBINANT PURIFIED AND RECONSTITUTED CYP 2C9:
Incubation of TA 100 μM, with the 2C9 RECO system, containing purified cytochrome P450, reductase, cytochrome b_5, and 1mM GSH final were performed in presence or absence of NADPH. When 2C9 RECO alone is injected on the HPLC, all components can be detected during the elution in the following order: GSH, GSSG, cytochrome b5, heme, CHAPS, dioleylphosphatidylcholine, other phospholipids, NADPH cytochrome P450 reductase, CYP 2C9 and a poly-ethyleneoxide derivative (Emulgen 913?).
Incubations of TA in absence of NADPH show one additional peak at M_0=331 (TA+ H^+)
Incubations of TA in presence of NADPH and air show 5 new TA derived peaks: the major one is still M_0=331 (with a typical pattern of M+1 [M + H^+], M+ 23 [M+Na^+], M+ 42 [M+ H^+ + CH_3CN]. The second is 5-OH tienilic acid with M_1=347 [TA + O + H^+], then a cluster of 3 peaks at in reverse polarity order M_4=636, M_3=672, M_2=654 and finally a mass doublet of M_5= 471/ M_6= 454.

M_2 correspond to [TA + O + GSH + H^+] and M_3 = [M_2 + H_2O] and M_4 = [M_2 – H_2O].

M_5 correspond to [TA + O + dihydronicotinamide + H^+], M_6 to [M_5 - NH_3]; use of another batch of NADPH gave much less of this adduct. In addition the peaks of NADP, NADPH and GS-SG were also found. NADPH yields a [M+H^+] and a [M+H^+- NH_3]. They were used to calibrate the mass spectra to 1/100 unit.

TABLE 1. Molecular weight of TA metabolites in function of incubation or HPLC conditions

Conditions	M_0	M_1	M_2	M_3	M_4	M_5
TA + 2C9	331	347	654	672	636	471
2C9 $H_2^{18}O$	331	347	654	674 (1)	636	471
2C9 $^{18}O_2$	331	349 (1)	656 (1)	674 (1)	636	473 (1)
D_2O Eluant	333 (1)	351 (3)	664 (9)	682 (9)	644 (7)	??

( ) number of ^{18}O atoms incorporated or of protons exchanged with Deuterium during HPLC with D_2O.

These primary results, yet only qualitative, demonstrate the power of the LC-ESI-MS.

Effect of $H_2^{18}O$ and $^{18}O_2$

In order to study the activation of the thiophene ring, incubation in presence of 65% $H_2^{18}O$ were performed. M_0 331, M_1 347, M_2 654 and M_4 636 do not change. However M_3 changes to 674 and is labeled at 60–65% by $H_2^{18}O$ (90-100 % incorporation of $H_2^{18}O$).

Incubation in presence of $^{18}O_2$ 60% was performed. Because of the small size of the incubation the labelling could not be increased. However the results clearly show that M_1, M_2, M_3 and M_5 are labeled at about 55–60 % (90–100 % incorporation of one ^{18}O atom).

Use of Deuterated Tienilic Acid 5-D-TA

Incubation of deuterated TA (80% D in position 5) showed the same metabolites: M_2, M_3, M_4 and M_5/ M_6 remained labeled with deuterium in 5 position to the same extend as starting TA. On the contrary M_1 (5-0HTA) lost totally the deuterium in 5, thus not showing any apparent NIH shift. However the 4-H of 5-OHTA is known to be exchangeable.

Use of D_2O as Eluting Solvent in the HPLC

In order to count the number of exchangeable protons of each metabolite, HPLC was performed using 99.5% D_2O instead of water in the elution solvent. M_0 became 333 showing 1 exchangeable proton (330 +1 +D^+). M_1 became 350/351 showing 3 exchangeable protons (346+ 3 + D^+). Glutathione went from 308 to 315/317 showing 6 to 8 exchangeable protons (307+ 6/8 + D^+). M_2 went from 654/656/658 to an enveloppe from 661 (653 +6 + D^+) to 669 (657 +10 + D^+) centered at 664 showing from 6 to 10 exchangeable protons, most probably 9 protons. M_3 went from 672 to a cluster 680 to 688 with the same number of exchangeable protons as M_2. Finally M_4 went from 636 to a narrower cluster of 641 to 649 centered at 644 showing 7 or 8 exchangeable protons.

OXIDATION OF TA BY LIVER MICROSOMES FROM RATS PRETREATED WITH CLOFIBRATE (CLO):

Incubation of TA 100µM with rat liver microsomes CLO and NADPH without added nucleophiles showed 3 TA derived peaks: M_0, M_1 and M_5/M_6. Incubation in presence of 3mM GSH showed the same 3 peaks plus peak M_4 (636) and small amounts of M_2 and M_3.

Incubation in presence of various amount of N-acetylcysteine did not show any adduct but confirmed the formation of M_1 and M_5/M_6. Stopping the incubation with HCl 0.1 M or H_2SO_4 0.2 M demonstrated stability of peaks M_1 and M_5/M_6. Thus the adduct of dihydronicotinamide and 5-OHTA survives very acidic conditions.

Incubation in presence of morpholine at various concentrations (1-10mM) did not show any new adduct. However the M_5/ M_6 peak was still present. Changing the batch of NADPH also changed the amount of this adduct.

OXIDATION OF TA IN $^{18}O_2$ OR $H_2^{18}O$ BY MICROSOMES OF YEAST EXPRESSING CYP 2C9 OR CYP 2C18 OR BY RAT LIVER MICROSOMES (CLO)

Incubations of Tienilic acid with yeast expressed 2C9 and 2C18 and with rat liver microsomal P450 (CLO) were compared. TA and 5OHTA were isolated by HPLC and methylated with diazomethane. Then the methyl ester, methyl ether were analyzed by mass spectrometry in EI mode. The thenoyl frament ion has a mass m/z of 111 for TAMe, and 141 for the 5OMe-TAMe. For incubation in 70% $H_2^{18}O$, no incorporation of ^{18}O was found. On the contrary, incubation in presence of 85-95% $^{18}O_2$ showed a strong 143 ion demonstrating 90-100 % incorporation of one ^{18}O atom. The benzoyl fragment ion remained unchanged at 261.

DISCUSSION

Two pathways have been proposed for the metabolic oxidation of thiophene compounds.

Route A (Figure 1) involves the intermediate formation of an arene oxide which may either directly rearrange to the phenol or react with water to give a dihydrodiol. This step is often catalysed by epoxide hydrolase. GSH also reacts with the arene oxide, often with catalysis of glutathione-S-transferases. The trans-dihydrodiols and glutathione adducts are labile especially in acidic conditions and rearomatize with loss of water to phenols and aromatic glutathione adducts respectively.

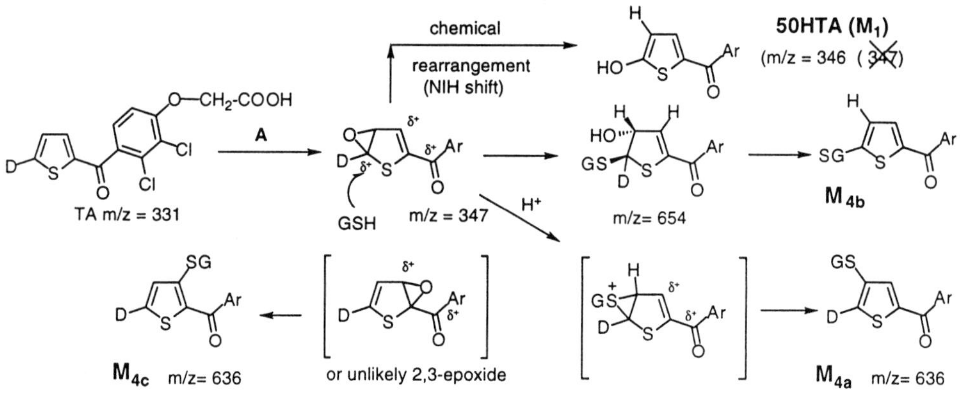

Figure 1. ROUTE A: Arene oxide path.

Route B (Figure 2) involves the intermediate formation of a thiophene S-oxide which could be followed either by attack of H_2O and rearrangement of the adduct, or by an isomerisation of the S-oxide with migration of the oxygen atom to an adjacent carbon atom.

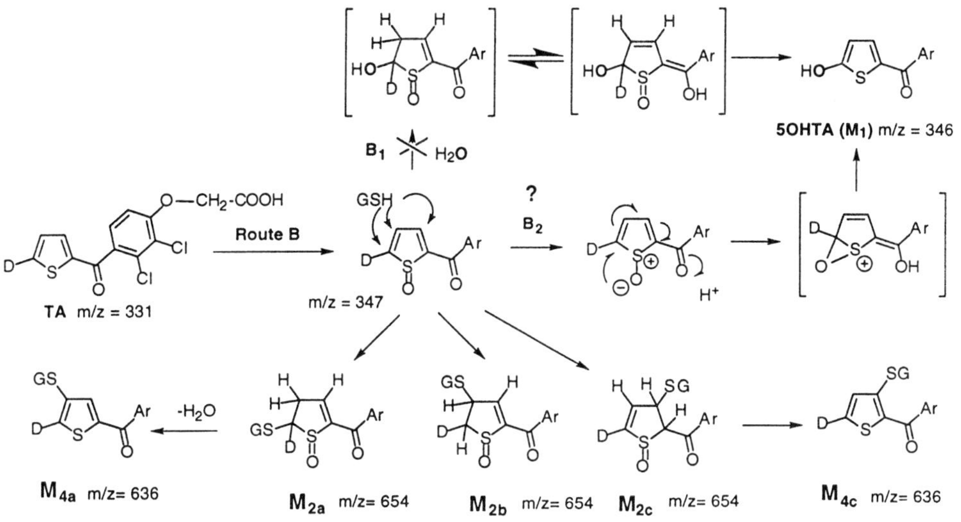

Figure 2. ROUTE B: S-oxide path.

Our results on incubation of TA with CYP 2C9 led as expected to 5OHTA, M_1, as a major metabolite. Experiments using $^{18}O_2$ and $H_2^{18}O$ clearly show that the oxygen atom introduced in TA comes from O_2 and not from H_2O. This result rules out one of the mechanisms of route B (B_1) in which H_2O adds to the thiophene S-oxide. Low kinetic isotope effects have been found (data not shown) when comparing TA and 5-deutero TA ($k_D/k_H \approx 0.8$), indicating that the breaking of the C_5-H bond of TA is not involved in the rate determining step.

Three supplementary metabolites are formed after oxidation of TA in the presence of GSH. The major one, M_2, corresponds to [TA + O + GSH], and should derive from the addition of GSH to either a thiophene S-oxide or a thiophene epoxide intermediate. Metabolite M_2 is stable after treatment of the incubate with HCl 0.1 M. Such a stability of M_2 at low pHs does not seem to be compatible with route A, that generally leeds to glutathione adducts very easily dehydrated in acidic conditions. It could thus derive from nucleophilic addition of GSH to TA S-oxide and would correspond to one of the M_2 stereoisomers shown in Fig 2. Incorporation of one oxygen atom from O_2 (experiments using $^{18}O_2$) and lack of insertion of an oxygen from $H_2^{18}O$ in the formation of M_2 are in complete agreement with such a mechanism; Metabolite M_4, that correspond to M_2 - H_2O, should derive from dehydratation of M_2 and rearomatization of the thiophene ring. The fact that M_4 formed from 5-deutero-TA keeps the deuterium atom suggests the two structures of M_4 shown in fig 2 as the most probable ones. They correspond to TA substituted at position 3 or 4 with a GS- group. However it is presently difficult to exclude an isomeric structure with GS- at position 5 which would be derived from a very complex mechanism involving migration of 5-D at another position of the thiophene ring. The third GS- containing metabolite M_3, corresponds to the addition of H_2O to M_2, in complete agreement with the results of experiments with $^{18}O_2$ and $H_2^{18}O$ (Table 1). Supplementary data are necessary for determination of the detailed structure of M_3. However a structure similar to those indicated in Fig 3, which would derive from Michael addition of H_2O to the conjugated double bond of M_{2a} (route **B**) that is substituted by two electron-withdrawing substituents, is likely. Addition of H_2O to the conjugated double bond of the glutathione-hydroxy adduct derived from route **A** appears to be much less likely, because this double bond only bears one electron-withdrawing substituent.

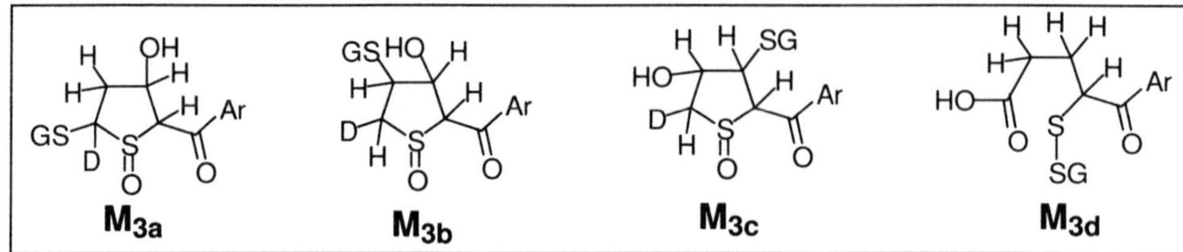

Figure 3. Possible structure for M3 (m/z = 672).

In addition the formation of the adduct M_5 is also consistent with the S-oxide path.

The aforementioned data are not sufficient to definitely discriminate between routes A and B in metabolic oxidation of TA. However they allow us to exclude route B_2 for the 5-hydroxylation of TA. Moreover, they could be in favor of the intermediate formation of a thiophene S-oxide (route B, B_1 for 5OHTA), mainly because of the stability M_2 in acidic conditions and the formation of M_3 in the incubate.

REFERENCES

Bonierbale E., Valadon P., Pons C., Desfosses B., Dansette P. M. and Mansuy D. (1999), Opposite behaviors of reactive metabolites of tienilic acid and its isomer toward liver proteins: Use of specific anti-tienilic acid-protein adduct antibodies and the possible relationship with different hepatotoxic effects of the two compounds, *Chem Res Toxicol*, <u>12</u>, 286-296. and references therein.

Jean Pascale, (1997) Mécanisme de l'inhibition des cytochromes P450 par les 2-aroylthiophenes et modèlisation moléculaire. Thèse de l' Université Paris 5.

Koenigs L. L., Peter R. M., Hunter A. P., Haining R. L., Rettie A. E., Friedberg T., Pritchard M. P., Shou M., Rushmore T. H. and Trager W. F. (1999), Electrospray ionization mass spectrometric analysis of intact cytochrome P450: Identification of tienilic acid adducts to P450 2C9, *Biochemistry*, <u>38</u>, 2312-2319.

Lopezgarcia M. P., Dansette P. M. and Mansuy D. (1994), Thiophene Derivatives as New Mechanism-Based Inhibitors of Cytochrome-P-450 - Inactivation of Yeast-Expressed Human Liver Cytochrome-P-450-2C9 by Tienilic Acid, *Biochemistry*, <u>33</u>, 166-175.

Valadon P., Dansette P. M., Girault J. P., Amar C. and Mansuy D. (1996), Thiophene sulfoxides as reactive metabolites: Formation upon microsomal oxidation of a 3-aroylthiophene and fate in the presence of nucleophiles in vitro and in vivo, *Chem Res Toxicol*, <u>9</u>, 1403-1413.

INHIBITION BY TICLOPIDINE AND ITS DERIVATIVES OF HUMAN LIVER CYTOCHROME P450. MECHANISM-BASED INACTIVATION OF CYP 2C19 BY TICLOPIDINE

Nguyêt-Thanh HA-DUONG, Sylvie DIJOLS, Anne-Christine MACHEREY, Patrick M. DANSETTE and Daniel MANSUY*

Laboratoire de Chimie et Biochimie Pharmacologiques et Toxicologiques, Université René Descartes, CNRS UMR 8601, 45 rue des Saints Pères, 75270 Paris Cedex 06, France

Ticlopidine (TCP) [1] has a wide spectrum of platelet anti-aggregating activity in man (*1*). This compound is known to inhibit the metabolism of several drugs (*2*). Recent case reports have shown that ticlopidine inhibits phenytoin clearance, which results in acute toxicity (*3-6*). Phenytoin is metabolized by CYP 2C9 and 2C19 (*7*), but ticlopidine does not affect CYP 2C9 activity (*8*) and decreased the *in vivo* activity of CYP 2C19 as measured by omeprazole metabolism (*9*), an indication of inhibition *in vivo* of CYP 2C19 activity by ticlopidine. *In vitro*, such inhibitory effects on bufuralol hydroxylation (*10*) and on S-mephenytoin 4-hydroxylation (*11*) have been shown on recombinant CYP 2C19.

This paper reports the study of the inhibitory effects of ticlopidine on several recombinant human liver cytochromes P450s expressed in yeast. A comparison of derivatives of ticlopidine, some of which have been prepared for this study, allowed us to determinate the structural factors which are important for ticlopidine recognition by the CYP 2C19 active site. The metabolism of ticlopidine by CYP 2C19 led to the identification of two metabolites, PCR37-87 and TSOD. We have also shown that TCP causes mechanism-based inactivation of CYP 2C19 and thus hypothesize that the loss of enzymatic activity of P450 2C19 is primarily due to the binding of TCP reactive intermediate to the protein.

Human liver P450 expressed in yeast were screened in order to identify which were inhibited by ticlopidine. Previous studies (*10, 11*) have found that 2C19 and 2D6 activities were inhibited by TCP equipotently. Our results confirmed the ability of TCP to inhibit these two recombinant enzymes activities expressed in yeast microsomes and indicated that in the 2C subfamily the inhibitory effect was selective to the CYP 2C19 activity. TCP acts as a

---

[1] **Abbreviations:** P450 or CYP: cytochrome P-450; TCP: ticlopidine; TSOD: Ticlopidine S-oxide dimer; OP: omeprazole; TA: Tienilic acid, TPP: 3-[2,3-dichloro-4-(2-thenoyl)phenoxy]propane-1-ol; GSH: reduced glutathione.

Biological Reactive Intermediates VI, Edited by Dansette *et al.*
Kluwer Academic / Plenum Publishers, 2001

competitive inhibitor with a K_i value of 2.9 μM. Spectral interactions studies show that TCP interacts well with purified-CYP 2C19 giving rise to a type I UV-visible difference spectrum, as expected for the formation of 2C19-inhibitor complex. The K_s value (2.8 μM) is almost equal to the K_i value found previously (Table 1), indicating that the inhibitory effects of this compound toward CYP 2C19 may result from its binding in the active site of this cytochrome.

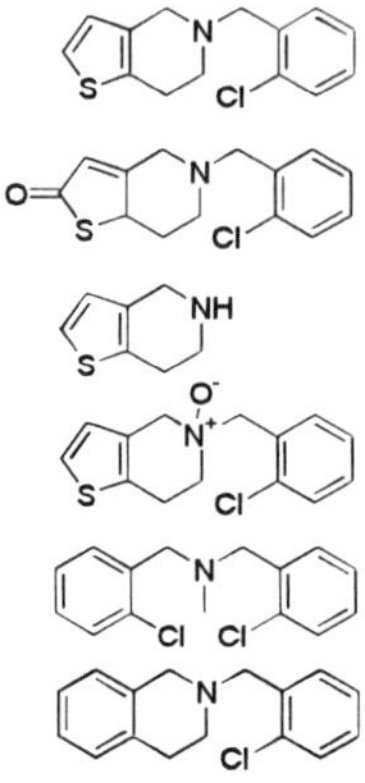

Ticlopidine

PCR37-87

PCR06-65

Ticlopidine-N-Oxyde

1

2

Figure 1. Structure of ticlopidine derivatives.

Table 1. Effects of various analogs of ticlopidine on the visible spectrum and TPP 5-hydroxylation activity of microsomes from yeast-expressing CYP 2C19.

Compounds	$\Delta A_{380-420}$ (x10^3)	K_s (μM)	IC$_{50}$ (μM)
Ticlopidine	9	2.8 ± 1.0	12
PCR37-87	13	3.4 ± 0.9	20
PCR06-65	-		>250
TCP N-Oxyde	-		>250
1	-		>250
2	-		>250

Several TCP derivatives or analogues, some of them synthetized in our laboratory, were studied to understand the origin of this particularly inhibitory effects of TCP with CYP 2C19 (Table 1). PCR37-87, PCR06-65, and ticlopidine N-Oxyde (Figure 1) were described as metabolites of ticlopidine *in vivo* in animals and man (*12*). Compounds **1** and **2** have no inhibitory effect on CYP 2C19 activity, suggesting that TCP interaction with CYP 2C19 is not due to the presence of two aromatic rings but especially to the tetrahydrothienopyridine and the chlorophenyl groups.

Incubation of TCP with CYP 2C19 in presence of NADPH led to the formation of at least 5 metabolites, the two major of which were identified as PCR37-87 and TSOD. It has been reported that many thiophene rings are metabolized into thiophene sulfoxide (*13-15*). Thiophene sulfoxides are very reactive species, which can react with nucleophiles, and participate to a Diels Alder dimerisation. Oxidation of ticlopidine by CYP 2C19, thus led to the formation of such an intermediat, the thiophene sulfoxide which gave to the metabolite TSOD by Diels Alder dimerisation. The detailed mechanism of the formation of PCR37-87, the stable tautomer of 2-hydroxy-TCP, was not elucidated. The Michaelis-Menten constants were determined for both of the metabolites, ((k_{cat} = 13 min^{-1}, K_M = 46.5 μM, and k_{cat} = 0.38 min^{-1}, K_M = 67 μM, for PCR37-87 and TSOD, respectively)

Preincubation of microsomes CYP 2C19 with TCP in the presence of NADPH resulted in a decrease in the 2-TPP-hydroxylase activity (Figure 2 D), demonstrating that the oxidative metabolism of TCP caused inactivation of P450 2C19. Similar experiments were performed with yeast-microsomes expressing CYP 2C8, 2C9 and 2C18 (Figure 2), and showed no significant decrease of their activity, indicating that P450 2C19 was selectively inactivated by TCP. To determine if the decrease in the activity of CYP 2C19 occurs via a mechanism-based inactivation process, kinetics analysis of the inactivation was performed. The pseudo firts-order kinetics for the time-dependent inactivation and saturability of inactivation with increasing TCP concentration observed (Figure 3 A) suggested such a mechanism-based inactivation of P450 2C19. The failure of GSH to decrease the inactivation rate (Figure 4 A), and the protection of the inactivation process by a competitive inhibitor, omeprazole (Figure 4 B) indicated that the inactivating event is confined to the active site of the enzyme and the inactivation was not due because the intermediate diffused out of the active site and bound elsewhere on the P450 molecule or the reductase. The kinetics parameters of the process were determinated, $t_{1/2max}$ = 3.4 min, k_{inact} = 3.2 10^{-3} s^{-1}, K_I = 87 μM, k_{inact}/K_I = 37 L. mol^{-1}. s^{-1}, partition ratio = 2.2 for TSOD and 22 for PCR37-87 (Figure 3 B).

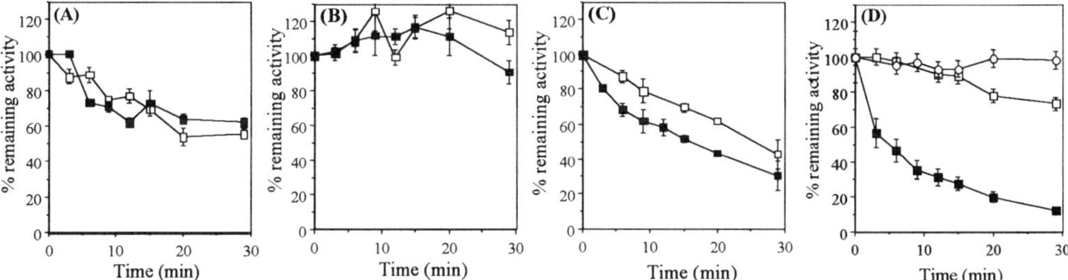

Figure 2: Time dependent of P450 2C inactivation after NADPH-dependent oxidation of ticlopidine. The residual activities are measured following different incubation times under the indicated conditions: P450 2C8 (A), 2C9 (B), 2C18 (C) and 2C19 (D) microsomes with a NADPH-generating system and in the presence (■) or absence (□) of 100 μM TCP. Control incubations (O) with CYP 2C19 (D) were performed with 100 μM TCP but without NADPH-generating system.

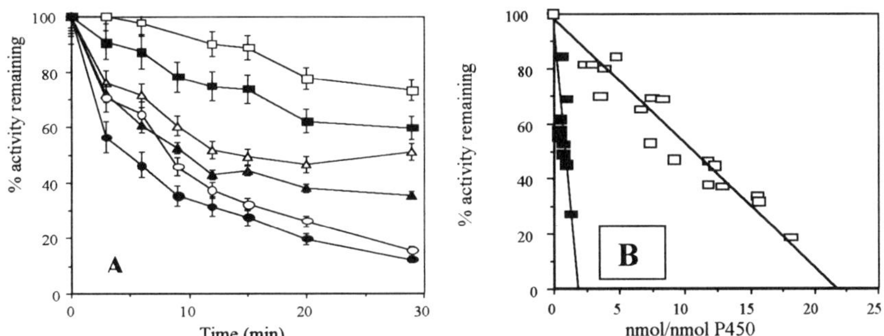

Figure 3: (A) Time- and concentration-dependent inactivation of the P450 2C19 5-TPP hydroxylation activity by ticlopidine. P450 2C19 were incubated for the indicated periods in a mixture of NADPH-generating system and 0 (□), 10 (■), 25 (ρ), 50 (□), 70 (O) or 100 μM (●) TCP. **(B) Correlation between mechanism-based P450 2C19 inactivation and efficient catalysis upon TCP oxidation**. CYP 2C19 were incubated under catalysis conditions in the presence of TCP various concentrations and for different time periods. The amount of formed metabolites ((□) PCR37-87, (■) TSOD), as well as the remaining P450 2C19 activity, was determined in parallel aliquots.

The inactivation constant (K_I) value was not far from the K_M found for TSOD, may suggest that the same reactive intermediate led to the formation of TSOD, and the inactivation of P450 2C19. We reported previously (*16*) that tienilic acid, a thiophene derivative was a mechanism-based inhibitor of CYP 2C9. A possible mechanism for suicide inactivation of P450 2C9 was proposed. The first step could be the S-oxidation of thiophene ring of TA. followed by a rearrangent of the thiophene sulfoxide into 5-OHTA. Alternatively, rather than water, a nucleophilic group of an amino acid residue of the active site could react with TA sulfoxide in a Michael-type addition reaction. Such a reaction would eventually result in covalent binding of TA to the P450 2C9 active site. We also propose for the case of inactivation of P450 2C19 by ticlopidine, that TCP, a thiophene derivative, is oxidized to the

TCP thiophene sulfoxide, which reacts on itself by Diels-Alder cycloaddition to form the metabolite TSOD, or reacts with a nucleophilic group of an amino acid residue of the active site, leading to inactivation of the enzyme. Covalent binding experiments might confirme the irreversibility of this inactivation process. In addition, using rat liver microsomes in presence of GSH, thiophene sulfoxide-GSH adducts can be trapped (data not shown)

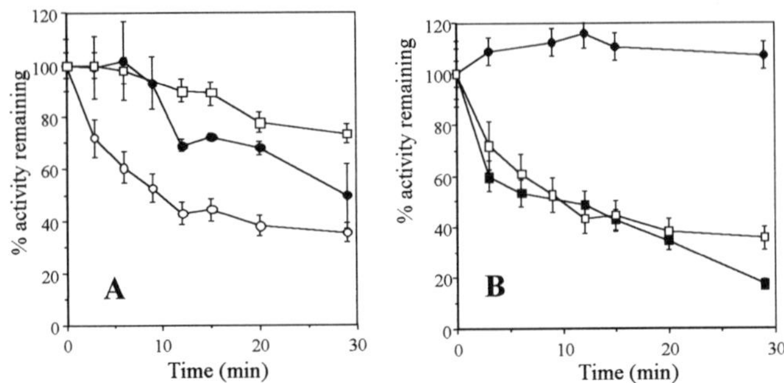

Figure 4: **(A) Protection by omeprazole of P450 2C19 inactivation by TCP**. CYP 2C19 microsomes were incubated with a NADPH-generating system, and either with 50 μM TCP (O) or with 50 μM TCP plus 100 μM OP (●). Control incubations (□) were performed in the presence of NADPH generating system without ticlopidine and in the presence of 100 μM OP. **(B) Effect of GSH on the rate of P450 2C19 inactivation by TCP.** Loss of P450 2C19 TPP 5-hydroxylase activity after NAPDH-dependent oxidation of 50 μM TCP was evaluated as a function of time in the presence (■) and absence (□) of 5mM GSH. Control incubations (●) containing the NADPH-generating system with GSH and without TCP added were also run in parallel.

In summary, this study reports that the structural properties of TCP make this molecule a good competitive inhibitor of CYP 2C19, and we demonstrated that TCP, a thiophene derivative, can act as a mechanism-based inibitor of P450 2C19.

REFERENCES

1. Defreyn, G., Bernat, A., Delebassee, D., and Maffrand, J. P. (1989) *Semin Thromb Hemost 15*, 159-66.
2. Upton, R. A. (1991) *Clin Pharmacokinet 20*, 66-80.
3. Donahue, S. R., Flockhart, D. A., Abernethy, D. R., and Ko, J. W. (1997) *Clin Pharmacol Ther 62*, 572-7.
4. Klaassen, S. L. (1998) *Ann Pharmacother 32*, 1295-8.
5. Lopez-Ariztegui, N., Ochoa, M., Sanchez-Migallon, M. J., Nevado, C., and Martin, M. (1998) *Rev Neurol 26*, 1017-8.
6. Donahue, S., Flockhart, D. A., and Abernethy, D. R. (1999) *Clin Pharmacol Ther 66*, 563-8.
7. Veronese, M. E., Doecke, C. J., Mackenzie, P. I., McManus, M. E., Miners, J. O., Rees, D. L. P., Gasser, R., Meyer, U. A., and Birkett, D. J. (1993) *Biochem. J. 289*, 533-538.
8. Gidal, B. E., Sorkness, C. A., McGill, K. A., Larson, R., and Levine, R. R. (1995) *Ther Drug Monit 17*, 33-8.
9. Tateishi, T., Kumai, T., Watanabe, M., Nakura, H., Tanaka, M., and Kobayashi, S. (1999) *Brit J Clin Pharmacol 47*, 454-457.
10. Mankowski, D. C. (1999) *Drug Metab Dispos 27*, 1024-1028.
11. Masimirembwa, C. M., Otter, C., Berg, M., Jonsson, M., Leidvik, B., Jonsson, E., Johansson, T., Backman, A., Edlund, A., and Andersson, T. B. (1999) *Drug Metab Dispos 27*, 1117-22.
12. Picard-Fraire, C. (1984) in *Ticlopidine, quo vadis ?* (verlag, B., Ed.) pp 68-75, Gordon, J. L., Basel.
13. Dansette, P. M., Thang, D. C., el Amri, H., and Mansuy, D. (1992) *Biochem Biophys Res Commun 186*, 1624-30.
14. Valadon, P., Dansette, P. M., Girault, J. P., Amar, C., and Mansuy, D. (1996) *Chem Res Toxicol 9*, 1403-13.
15. Mansuy, D., Valadon, P., Erdelmeier, I., Lopez-Garcia, P., Amar, C., Girault, J. P., and Dansette, P. M. (1991) *J Am Chem Soc 20*, 7825.
16. Lopez-Garcia, M. P., Dansette, P. M., and Mansuy, D. (1994) *Biochemistry 33*, 166-175.

MICROPEROXIDASE 8 (MP8) AS A CONVENIENT MODEL FOR HEMOPROTEINS : FORMATION AND CHARACTERISATION OF NEW IRON(II)-NITROSOALKANE COMPLEXES OF BIOLOGICAL RELEVANCE

Remy Ricoux[1], Jean-Luc Boucher[1], Daniel Mansuy[1], and Jean-Pierre Mahy[2]

[1]Laboratoire de Chimie et Biochimie Pharmacologiques et Toxicologiques, UMR 8601 CNRS, Université Paris V, 45 rue des Saints-Pères, 75270, Paris cedex 06, France
[2] Laboratoire de Chimie Bioorganique et Bioinorganique, FRE 2127 CNRS, Institut de Chimie Moléculaire d'Orsay, Université Paris-Sud XI, Bâtiment 420, 91405, Orsay cedex, France

INTRODUCTION

Microperoxidase 8 (MP8) is obtained by controlled proteolytic digestion of horse heart cytochrome c (Aron *et al.*, 1986). It consists of an iron(III)-protoporphyrin IX covalently bound through thioether links to two cysteine side chains and a histidine is axially coordinated to the heme iron, and acts as its fifth ligand. MP8 has an open active site, which leads to broad substrate specificity in the two types of catalytic activities it shows : a peroxidase-like activity (Baldwin *et al.*, 1987) and a cytochrome P450-like activity (Osman *et al.*, 1996).

Several heme proteins including hemoglobin, myoglobin (Mansuy *et al.*, 1977a), cytochrome 450 (Mansuy *et al.*, 1978), and PGH synthase (Mahy and Mansuy, 1991), have been reported to form Fe(II)-nitrosoalkane(RNO) complexes. Such complexes are not only obtained *in vitro*, either by oxidation of hydroxylamines or by reduction of nitroalkanes in the presence of a reducing agent but also *in vivo* during the oxidative metabolism of several drugs or exogenous compounds containing an amine or hydroxylamine function (Jonsson and Lindeke, 1976), or during the reduction of the corresponding nitroalkanes (Mansuy *et al.*, 1977b).The RNO ligands bind tighly to the iron (II) of the hepatic detoxifying cytochromes P-450, causing a severe inhibition of the catalytic functions of these cytochromes (Mansuy *et al.*, 1978).

$$(P)Fe^{III} \xrightarrow{\ \mathbf{RNHOH}\ } (P)Fe^{II} \leftarrow N\underset{R}{\overset{O}{\diagdown}} \xleftarrow{\ \mathbf{RNO_2}\ } (P)Fe^{III}$$

$$\text{Reductant}$$

P = hemoglobin, myoglobin, cytochrome P450,PGH synthase

This paper shows that MP8 is not only able to oxidize various aliphatic and aromatic hydroxylamines with the formation of Fe(II)-nitrosoalkane or -nitrosoarene complexes, but also that these complexes can be obtained by reduction of nitroalkanes in the presence of reducing agent. This constitutes a new activity for MP8 and further validates the use of this heme-octapeptide as a model for hemoproteins such as cytochromes P450 and peroxidases.

RESULTS AND DISCUSSION

Reaction of N-Isopropylhydroxylamine with MP8

The addition, under aerobic conditions, of 400 µM N-isopropylhydroxylamine to a 2 µM solution of MP8 in 0.1 M PBS, pH 7.4 led to the gradual replacement, with an isobestic point at 405 nm, of the soret band characteristic of iron(III)MP8 at 396 nm by a new spectrum with a soret band at 413 nm (Figure 1).

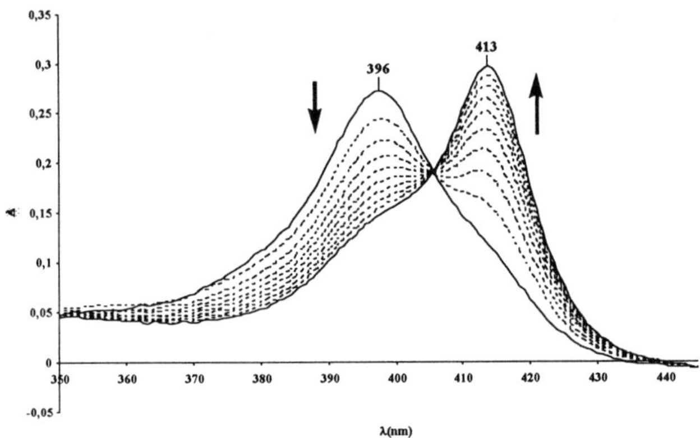

Figure 1 . Evolution of the absolute spectrum of MP8 after the addition of N-isopropylhydroxylamine. Spectra were recorded every 2 minutes after the addition of 400 µM iPrNHOH

This new spectrum could be explained by the formation of a new complex of MP8 absorbing at 413 nm. A MP8-Fe(II)-N(O)isopropyl (iPr) structure was strongly suggested for this complex by the following characteristics which were almost identical with those previously found for the nitrosoalkane complexes of other hemoproteins (Mansuy *et al.,* 1977, 1978, Mahy and Mansuy, 1991), or of iron(III)-meso-tetraphenylporphyrin (Mansuy *et al.,* 1983): (i) a great stability in the presence of an excess of sodium dithionite, (ii) an immediate regeneration of ferric MP8 upon treatment by ferricyanide, and (iii) the formation of the Fe(II)-nitrosopropane complex either by in situ oxidation of N-isopropyl-hydroxylamines or by reduction of the corresponding 2-nitropropane in the presence of dithionite . This MP8-Fe(II)-N(O)iPr structure was completely confirmed for the iPrNHOH-derived complex by the following :(i) the direct formation of this complex by coordination of 2-nitrosopropane on the iron(II) of MP8 previously reduced by sodium dithionite, (ii) its ^{1}H NMR spectrum which was characteristic of an iron(II) diamagnetic species with signals due to the protons of the nitroso-2-propane ligand almost identical to those already reported for the diamagnetic iron(II)(iPrNO)(tetraarylporphyrin) complexes (Mansuy *et al.,* 1983).

Reactions of MP8 with Various N-Substituted Hydroxylamines RNHOH

Other N-monosubstituted hydroxylamines, like N-methyl-, N-propyl-, N-(1-phenylpropyl)- (N-hydroxyamphetamine), and N-(1-p-chlorophenylpropyl)-hydroxylamine

150

reacted with MP8 to form complexes exhibiting properties similar to those of the MP8-Fe(II)-N(O)iPr complex: a fast destruction in the presence of 50 mM $Fe(CN)_6K_3$ and visible spectra with maxima of absorption around 414 and 530 nm (Table 1).
The reactivity of the various N-monosubstituted hydroxylamines RNHOH was dependent on the nature of the R substituent. The less reactive N-monosubstituted hydroxylamines were those substituted with a linear alkyl group ($R = CH_3$, $n-C_3H_7$) whereas the more reactive one was isopropylhydroxylamine bearing a branched alkyl group ($R = CH_3-CH-CH_3$). This correlated with the higher donating effect of the isopropyl substituent than that of the linear alkyl groups and could be explained by the fact that the first step of the reaction of N-monosubstituted hydroxylamines with MP8 was the reduction of MP8Fe(III) into MP8Fe(II), isopropylhydroxylamine being a better reductant than N-methyl- and N-propylhydroxylamine. Accordingly, the substitution of one of the hydrogen atoms of the isopropyl group by an electron withdrawing aryl group, like in N-(1-phenylpropyl)- and N-(1-p-chlorophenylpropyl)-hydroxylamine), led to a decrease of the initial rate of the reaction.

Table 1 :UV-visible characteristics and reactivity of MP8-Fe(II)-RNO complexes

MP8-Fe(II)-RNO R =	UV-visible λ_{max}(nm), ε(mM^{-1} cm^{-1})	Complex level [1] (%)	Stability to $Fe(CN)_6K_3$[2]
$Ph-CH_2-CH-CH_3$	415 (105), 532	63	-
$Cl-Ph-CH_2-CH-CH_3$	414 (96), 532	73	-
CH_3-	413, 531	-	-
$CH_3-CH-CH_3$	413(60), 530	67	-
$CH_3- CH_2.CH_2-$	413 (77), 530	57	-

[1] Calculated from the absorbance at 413 nm after reaction of 200 equiv of RNHOH with 2 μM MP8 Fe(III) in 0.1 M PBS Buffer pH 7.4.
[2] Fast destruction of the MP8Fe(II)-RNO complex in the presence of 50 μM $Fe(CN)_6K_3$ with regeneration of MP8-Fe(III)

By contrast, the proportion of MP8-Fe(II)-N(O)R complex formed once the equilibrium was reached, was very similar whatever the N-monosubstituted hydroxylamines RNHOH used and ranged between 57 and 73% (Table 1). This indicated that the size and hydrophobicity of the R substituent of the nitrogen atom had little effect on the binding of the RNO ligand on the iron of MP8. It could be due to a lack of control of the binding of the nitrosoalkane on the iron atom which should occur on the non-hindered face of the heme at the opposite of the octapeptide arm. This behaviour is opposite to that of hemoproteins which have a small active site, such as hemoglobin, myoglobin and catalase (Mansuy *et al.*, 1977a), which can only accomodate small RNO ligands such as CH_3NO and C_2H_5NO whereas PGHS (Mahy and Mansuy, 1991) and rat liver cytochromes(Mansuy *et al.*, 1978), which possess a wide hydrophobic active site, are able to bind bulky hydrophobic ligands such as nitrosoamphetamine but are unable to form nitroso complexes from the highly hydrophilic CH_3NHOH and C_2H_5NHOH. Finally only nitric oxide synthase was found to be able ,like MP8, to bind a wide range of RNO ligands including the smaller one CH_3NO as well as the large hydrophobic $p-Cl-Ph-CH_2-CH(NO)-CH_3$ (Renodon *et al.*, 1998).

Reactions of MP8 with Various Nitroalkanes RNO_2

Nitrosoalkane or -arene complexes of iron(II)MP8 could be prepared by reaction, in the presence of 2 mM sodium dithionite, of 2 μM MP8 in 100 mM PBS, pH 7.4, with a range of nitroalkanes or arenes (1 mM) such as nitromethane, nitroethane, 1-nitropropane, nitrobenzene, nitrohexane, and nitrocyclohexane. Their visible spectra were similar to that of the MP8-Fe(II)-N(O)iPr complex as well as their reactivity toward $Fe(CN)_6K_3$. The formation of such complexes from nitroalkanes and dithionite should be due to the concomitant reduction of MP8-Fe(III) into MP8-Fe(II) and RNO_2 into RNO by sodium dithionite (or MP8-Fe(II)) followed by the strong binding between RNO and MP8-Fe(II). The great strengh of the MP8-Fe(II)-nitrosoalkane bond explains the stability of the obtained complexes in the presence of excess dithionite.

CONCLUSION

The formation of MP8-Fe(II)-nitrosoalkane or -arene complexes, either by oxidation of N-monosubstituted hydroxylamines or by reduction of nitroalkanes in the presence of sodium dithionite, constitutes a new reaction of microperoxidase 8. This is also the first example of fully characterized iron(II)-metabolite complexes of MP8. Such complexes constitute good models for those formed not only *in vitro* but also *in vivo* during the oxidative metabolism of drugs containing an amine function such as amphetamine or macrolids (Mansuy *et al.*, 1978) and which lead to an inhibition of the catalytic functions of cytochromes P450. In addition, the aforementioned results validate the use of this mini-enzyme as a convenient model for hemoproteins of interest in toxicology and pharmacology such as cytochromes P450, peroxidases and nitric oxyde synthase.

REFERENCES

Aron, J., Baldwin, D.A., Marques, H.M., Pratt, J.M., and Adams, P.A., 1986, Hemes and hemoproteins 1: preparation and analysis of heme containing octapeptide (Microperoxidase-8) and identification of the monomeric form in aqueous solution, *J. Inorg. Biochem.* **27**, 227-243

Baldwin, D.A., Marques, H.M, and Pratt, J.M., 1987, Hemes and hemoproteins 5: Kinetics of the per-oxidatic activity of microperoxidase 8, model for the peroxidase enzymes, *J. Inorg. Biochem.* **30**, 203-217.

Mahy, J.P., and Mansuy, D., 1991, Formation of prostaglandin synthase-iron-nitrosoalkane inhibitory complexes upon in situ oxidation of N-substituted hydroxylamines, *Biochemistry* **30**, 4165-4172.

Jonsson, J., and Lindeke, B., 1976, On the formation of cytochrome P450 product complexes during the metabolism of phenylalkylamines, *Acta Pharm. Suec* **13**, 313-320.

Mansuy, D., Chottard, J.C., and Chottard, G., 1977a, Nitroalkane as Fe(II) ligands in the hemoglobin and myoglobin complexes formed from nitroalkanes in reducing Conditions., *Eur. J. Biochem.* **76**, 617-623.

Mansuy, D., Gans, P., Chottard, J.C., and Bartoli, J.F.,1977b, Nitrosoalkanes as Fe(II) ligands in the 455-nm-absorbing cytochrome P-450 complexes formed from nitroalkanes, *Eur. J. Biochem* **76**, 607-615.

Mansuy, D., Rouer, E., Bacot, C., Gans, P., Chottard, J.C., and Leroux, J.P., 1978, Interaction of aliphatic N-hydroxylamines with microsomal cytochrome P450: nature of the different derived complexes and inhibitory effects on monoxygenases activities, *Biochem. Pharmacol.* **27**, 1229-1237.

Mansuy, D., Battioni, P., Chottard, J.C., Riche, C., and Chiaroni, A., 1983, Nitrosoalkane complexes of iron-porphyrins: analogy between the bonding properties of nitrosoalkanes and dioxygen., *J. Am. Chem. Soc.* **105**, 455-463.

Osman, A.M., Koerts, J., Boersma, M.G., Boeren, S., Veeger, C., and Rietjens, I., 1996, Microperoxidase/H_2O_2-catalysed aromatic hydroxylation proceeds by a cytochrome-P450 type oxygen transfer reaction mechanism, *Eur. J. Biochem* **240**, 232-238.

Renodon, A., Boucher, J.L., Wu, C., Gachhui, R., Sari, M.A., Mansuy, D., and Stuehr D., 1998, Formation of nitic oxide synthase-iron(II) nitroalkane complexes: severe restriction of access to iron(II) site in the presence of tetrahydrobiopterin, *Biochemistry* **37**, 6367-6374.

HEMOGLOBIN ADDUCTS IN RATS CHRONICALLY EXPOSED TO ROOM-AGED CIGARETTE SIDESTREAM SMOKE AND DIESEL ENGINE EXHAUST

Regina Stabbert, Georg Schepers, Walter Stinn, and Hans-Jürgen Haussmann

INBIFO Institut für biologische Forschung,
Cologne, Germany

INTRODUCTION

Protein and DNA adducts are generally considered to be biologically effective dose markers that can indicate both carcinogen-induced cell damage and genetic susceptibility. Hemoglobin (Hb) adducts are not subject to specific enzymatic repair mechanisms like DNA modifications are; therefore, hemoglobin acts as a cumulative dosimeter capable of indicating exposure dose over the life-span of the erythrocytes, e.g., about 60 days in rats.

The objective of this chronic inhalation study was to investigate and compare the carcinogenic effects and possible related mechanistic end points of room-aged cigarette sidestream smoke (RASS) as an experimental surrogate for environmental tobacco smoke (ETS) and Diesel engine exhaust (DEE). ETS and ambient DEE are complex combustion aerosols that are widespread in the environment at similar particle concentrations and thus of similar public concern. The highest particle concentration in this study exceeds the upper limit of typical ETS and DEE exposure concentrations for humans by a factor of ~100.

As part of the study, Hb adducts considered to be indicative of the tobacco-specific N-nitrosamines (NNN/NNK) and of aromatic amines (3- and 4-aminobiphenyl/1-aminopyrene) or nitro-PAHs (3- and 4-nitrobiphenyl/1-nitropyrene) were determined.

METHODS

RASS was generated by smoking the Reference Cigarette 1R4F under standard conditions and aging the sidestream smoke for 0.5 hours in a ventilated room under controlled conditions as previously described (Haussmann et al., 1998a). DEE was generated by operating a conventional passenger car Diesel engine according to the U.S. EPA test protocol FTP72 (EPA, 1993) as previously described (Haussmann et al., 1998b). The particle mass concentrations for both aerosols were adjusted by dilution with conditioned air to 3 µg/l (low dose) and 10 µg/l(high dose).

Female and male Wistar rats were nose-only exposed to RASS and DEE as well as to filtered, conditioned air (sham-exposed group) 6 hours/day, 7 days/week, for up to 24 months followed by a 6-month postinhalation period.

The analytical methods used to determine total particulate matter (TPM), solanesol, nicotine, 3-ethenyl pyridine, tobacco-specific N-nitrosamines (TSNAs), 1-nitropyrene, elemental carbon, and particle size distribution were performed as previously described (Haussmann et al., 1998a, 1998b, 1998c). Aromatic amines were derivatized with perfluoropropionic anhydride and analyzed by CGC tandem mass spectrometry in the neutral loss scan mode.

Biological Reactive Intermediates VI, Edited by Dansette *et al.*
Kluwer Academic / Plenum Publishers, 2001

The lung soot burden in the rats after 24 months of exposure was determined by photometry of elemental carbon particles after isolation from alkaline-dissolved lung tissue (basically according to Muhle et al., 1990). The determination of the nicotine metabolites in urine collected over 24 h was performed as previously described (Rustemeier et al., 1993; Haussmann et al., 1998a).

In order to determine hemoglobin adducts, blood samples were collected 12, 18, 24, and 30 months after start of the exposure. Hb adducts indicative of aromatic amines/nitro-PAHs and TSNAs were determined by GC/MS in the negative ion chemical ionization mode according to Kutzer et al. (1997).

RESULTS AND DISCUSSION

The aerosols were characterized and possible cross-contaminations of the aerosols were excluded (Table 1). Lung soot burden and urinary nicotine metabolite data demonstrated that the rats received the correct amount of RASS and DEE during the course of this inhalation study.

In RASS-exposed rats, a dose-dependent increase of up to 2-fold was seen for 4-ABP adducts (Figure 1), which did not accumulate throughout the 24-month inhalation period; this is consistent with the 60-day life-span in rat erythrocytes. At the end of the 6-month postinhalation period, all differences in 4-ABP adduct levels between RASS and air-exposed rats disappeared. The 4-ABP adduct levels were gender-specific with higher adduct levels for female rats. 3-ABP adducts were not quantifiable (quantification limit = 0.1 ng/g Hb for male rats) in any of the blood samples. No increase over background levels was observed for 1-aminopyrene adducts or TSNA-related 4-hydroxy-1-(3-pyridyl)-1-butanone (HPB) adducts, even at the exaggerated RASS concentrations used in this study.

In DEE-exposed rats, none of the adducts investigated responded to the DEE concentrations used in this study. The lack of increased levels of 4-ABP adducts, even at exaggerated DEE exposure (at least 100-fold over human exposure conditions), is not in line with the interpretation of human data in the literature, which attributes high background levels of 4-ABP partially to DEE exposure, because nitro-PAHs should enter the same pathway as aromatic amines after reduction of a nitro group. Similarly, the lack of quantifiable 1-nitropyrene adducts (quantification limit = 0.5 ng/g Hb for male rats) under exaggerated DEE exposure questions the notion that these adducts can be used as specific biomarkers of exposure to DEE in humans (Zwirner-Baier and Neumann, 1999).

The pronounced background levels of 4-ABP Hb adducts (male rats: 0.5 ng/g Hb) and HPB Hb adducts (male rats: 29 ng/g Hb) found in fresh air-exposed rats demonstrates the lack of specificity of these adducts as biomarkers of environmental exposure. The sources of these background levels have yet to be identified.

SUMMARY

Under controlled conditions with exaggerated concentrations of environmental aerosols, the biologically effective dose markers suggested in the literature as being specific for ETS (i.e., HPB Hb adducts for TSNA exposure) and DEE (i.e., 1-aminopyrene Hb adducts for 1-nitropyrene exposure) did not respond. A slight but dose-dependent increase in 4-ABP Hb adduct levels was seen in RASS-exposed rats.

ACKNOWLEDGMENT

The authors are grateful to the staff of INBIFO—in particular to C. Biefel and G. Fleger—for their excellent technical assistance. INBIFO is a research laboratory of Philip Morris International.

Table 1   Aerosol Characterization, Nicotine Uptake, and Lung Soot Burden

Parameter		Unit of Measure	Sham	RASS		DEE		Detection Limit (DL)
				Low	High	Low	High	
particles	TPM	μg/l	<DL	3.12	10.06	3.04	9.99	0.10
RASS marker	solanesol	μg/l	<DL	37.2	129.1	<DL	<DL	1.5
	nicotine	μg/l	<DL	0.561	1.833	<DL	<DL	0.073
	3-ethenyl pyridine	μg/l	<DL	93.0	275.7	<DL	<DL	21.8
	NNN	ng/l	<DL	<DL	<DL	<DL	<DL	0.055
	NAT	ng/l	<DL	<DL	<DL	<DL	<DL	0.060
	NAB	ng/l	<DL	<DL	<DL	<DL	<DL	0.044
	NNK	ng/l	<DL	0.379	1.122	<DL	<DL	0.110
	o-toluidine	pg/l	6.6	324.4	1085.9	-	-	1
	4-ABP	pg/l	<DL	<DL	11.2	-	-	1
DEE marker	1-nitropyrene	pg/l	<DL	<DL	<DL	15.59	46.48	5.33
	elemental carbon/TPM	%	-	-	-	52.0	50.3	-
particle size	"M"MAD	μm	-	0.44	0.44	0.19	0.21	-
	GSD	-	-	1.80	1.78	1.86	1.91	-
lung soot burden[1]	male	mg/rat	<DL	<DL	<DL	8.2	28.6	0.3
	female	mg/rat	<DL	<DL	<DL	8.3	30.9	0.3
nicotine metabolites[2]	male	nmol	<DL	84.1	216.9	<DL	<DL	-[3]
	female	nmol	<DL	74.0	224.0	<DL	<DL	-[3]

[1]   retained elemental carbon after 24 months of exposure

[2]   some of 7 metabolites excreted with 24-h urine; 8 times (every third month), 6 to 8 rats per group and gender

[3]   DL: cotinine: 0.013 nmol/ml; cis-and trans-3'hydroxycotinine: 0.004 nmol/ml; 5'-hydroxycotinine: 0.020 nmol/ml; norcotinine: 0.023 nmol/ml; nornicotine: 0.005 nmol/ml; nicotine-N'oxide: 0.066 nmol/ml; volume of 24-h urine: 16.7 ± 0.3 ml (mean ± SE; N = 620; range: 2.5 to 50 ml)

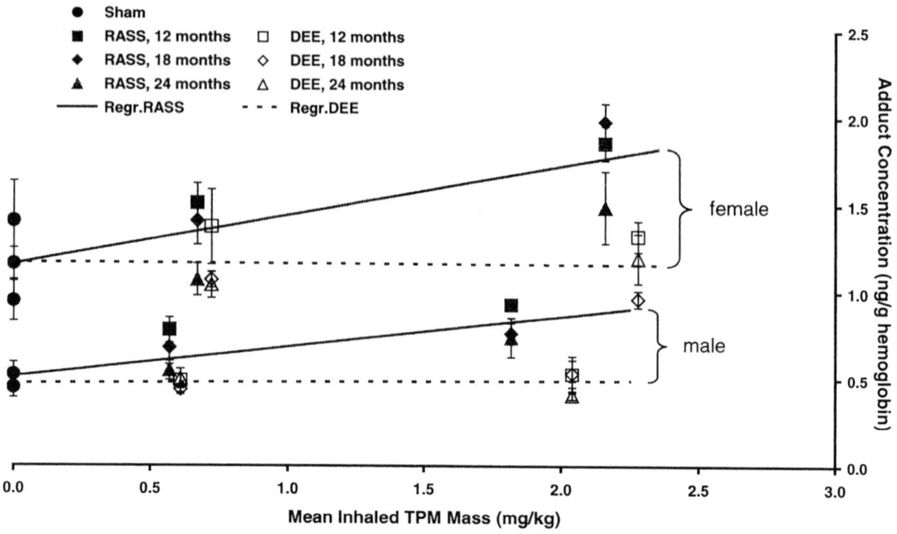

Figure 1   4-ABP Hemoglobin Adducts in RASS/DEE-Exposed Rats

REFERENCES

Environmental Protection Agency (EPA), Code of federal regulations, Federal Register, 1993

Haussmann, H.-J., Gerstenberg, B., Göcke, W., Kuhl, P., Schepers, G., Stabbert, R., Stinn, W., Teredesai, A., Tewes, F., 1998a, 12-Month inhalation study on room-aged cigarette sidestream smoke in rats, *Inhal. Toxicol.* 10: 663-697

Haussmann, H.-J., Anskeit, E., Kuhl, P., Rustemeier, K., Schepers, G., Schnell, P., Terpstra, P., The influence of diesel engine exhaust age on lung soot burden and related responses in rats, Washington: ILSI Press, 1998b, pp. 338-342

Haussmann, H.-J., Anskeit, E., Becker, D., Kuhl, P., Stinn, W., Teredesai, A., Voncken, P., Walk, R.-A., 1998c, Comparison of fresh and room-aged cigarette sidestream smoke in a subchronic inhalation study on rats, *Toxicol. Sci.* 41: 100-118

Kutzer, C., Branner, B., Zwickenpflug, W., Richter, E., 1997, ,Simultaneous solid-phase extraction and gas chromatographic-mass spectrometric determination of hemoglobin adducts from tobacco-specific nitrosamines and aromatic amines, *J. Chrom. Sci.* 35: 1-6

Muhle, H., Bellmann, B., Creutzenberg, O., 1990, Subchronic inhalation of toner in rats, *Inhal. Toxicol.* 2: 341-360

Rustemeier, K., Demetriou, D., Schepers, G., 1993, High performance liquid chromatographic determination of nicotine and its urinary metabolites via their 1,3-diethyl-2-thiobarbituric acid derivatives, *J. Chromatogr.* 613: 95-103

Zwirner-Baier, I. and Neumann, H.G., 1999, Polycyclic nitroarenes (nitro-PAHs) as biomarkers of exposure to diesel exhaust, *Mutat. Res.* 441: 135-144

DNA Microarray Reveals Increased Expression of Thioredoxin Peroxidase in Thioredoxin-1 Transfected Cells and its Functional Consequences

Bryan Husbeck, Margareta I. Berggren and Garth Powis

Arizona Cancer Center, University of Arizona, Tucson, Arizona, 85724-5024 U.S.A.

ABSTRACT

The mammalian thioredoxins are a family of small redox proteins that undergo NADPH dependent reduction by thioredoxin reductase. Reduced thioredoxins reduce oxidized cysteine groups on proteins including transcription factors to increase their binding to DNA, and is a source of reducing equivalents for enzymes such as thioredoxin peroxidase which removes H_2O_2 and alkyl peroxides. Thioredoxin-1 is over expressed in many human tumors where it is associated with aggressive tumor growth, inhibited apoptosis and decreased patient survival. Transfection of cells with thioredoxin-1 has been shown to increase cell growth and inhibit apoptosis. We have used DNA micro array to investigate the effects of thioredoxin-1 transfection on the expression of a panel of 520 redox, apoptosis and cell growth related genes in MCF-7 human breast cancer cells. One of the genes whose expression was increased as a result of thioredoxin-1 over expression was thioredoxin peroxidase-2. This increase was confirmed by Northern blotting. Transfection of mouse WEHI7.2 thymoma cells with human thioredoxin peroxidase-2 was found to protect the cells from apoptosis induced by H_2O_2 but not from apoptosis induced by dexamethasone, doxorubicin or etoposide. Thus, increased thioredoxin peroxidase-2 expression does not explain the widespread antiapoptotic effects of thioredoxin-1.

INTRODUCTION

Thioredoxins (Trxs) are a family of small redox proteins with a conserved catalytic site sequence -Trp-Cys-Gly-Pro-Cys-Lys- [for reviews see Nakamura et al (1997), Follmann and Haberlein (1995) and Powis *et al* (2000)]. There are two mammalian Trxs, Trx1 a 12 kDa, 104 amino acid protein (Deiss and Kimchi, 1991; Gasdaska *et al.*, 1994) is found predominantly in the cytoplasm and sometimes in the nucleus (Grogan *et al.*, 2000; Kawahara *et al.*, 1996) and is also secreted by cells (Ericson *et al.*, 1992), and Trx2 a 18 kDa, 166 amino acid protein with an N-terminal 60 amino acid mitochondrial translocation signal that is cleaved after import into the mitochondria (Spyrou *et al.*, 1997). Trxs undergo NADPH dependent reduction catalyzed by selenocystein (SeCys) containing flavoprotein thioredoxin reductases (TrxRs) (reviewed by Mustacich and Powis (2000)). The penultimate SeCys residue is essential for TrxRs to reduce Trx (Gasdaska *et al.*, 1999). Two two full length human TrxRs have been cloned, a 54.4 kD TrxR1 which is a predominantly cytosolic (Gasdaska *et al.*, 1995), and 56.2 kD TrxR2 which has a 33 amino acid N-terminal extension identified as a mitochondrial import sequence (Gasdaska *et al.*, 1999; Lee *et al.*, 1999; Miranda-Vizuete *et al.*, 1999). A third incomplete thioredoxin reductase sequence has been reported (Miranda-Vizueta, 1999).

The biological properties of Trx-1 are the most studied and are dependent upon the ability of the catalytic site to transfer reducing equivalents to enzymes including ribonucleotide reductase (Laurent *et al.*, 1964), vitamin K epoxide reductase (Silverman and Nandi, 1988) and thioredoxin peroxidases (Chae *et al.*, 1994; Jin *et al.*, 1997),   or to reduces key cysteine residues on transcription factors including NF-κB, the glucocorticoid receptor, AP-1 and p53 to increase their DNA binding and *trans*activating activity (Hayashi *et al.*, 1993; Peddie *et al.*, 1997; Freemerman *et al.*, 1999; Ueno *et al.*, 2000). Reduced Trx1 can also bind to enzymes to modulate their activity including apoptosis signal regulated kinase (ASK1), protein kinase C α,δ,ϵ and ζ, p40Phox and the cyteine protease inhibitor lipocalin (Saitoh *et al.*, 1998; Watson *et al.*, 1999; Nishiyama *et al.*, 1999; Redl *et al.*, 1999).

NIH 3T3 cells transfected with a nuclear localized Trx1 give foci in culture composed of cells that form tumors in mice (Mustacich *et al.*, 2000). Transfection of mouse WEHI7.2 thymoma derived cells or human MCF-7 breast cancer cells with human Trx1 causes increased growth and resistance to apoptosis (Baker *et al.*, 1997). A redox inactive mutant Trx1 acts as a dominant negative to promote apoptosis and to inhibit tumor growth (Gallegos *et al.*, 1996; Freemerman and Powis, 2000).

A variety of human primary tumors have increased levels of Trx1 compared the corresponding normal tissue including lung, colon, cervix, liver and pancreatic cancers (Gasdaska *et al.*, 1994; Berggren *et al.*, 1996; Fujii *et al.*, 1991; Nakamura *et al.*, 1992; Nakamura *et al.*, 2000). Immunohistochemical studies using paraffin embedded tissue sections have shown that Trx1 expression is increased in more than half of human primary gastric cancers. The Trx1 levels showed a highly significant positive correlation with cell proliferation measured by nuclear proliferation antigen and a highly significant negative correlation with apoptosis measured by the terminal deoxynucleotidyl transferase (Tunel) assay. In a recent immunohistochemical study of Trx1 levels in human colon cancer Trx1 protein was not increased compared to normal colonic mucosa in precancerous colon polyps but was increased 6 fold in primary colon cancers and almost 9 fold in metastatic colon tumors in adjacent lymph glands (Gallegos *et al.*, 2000). High levels of Trx1 in colon tumor was associated with decreased patient survival (Raffel, Bhattacharyya and Powis unpublished observations).

The mechanism for the growth stimulating and anti-apoptosis effects of Trx1 are unknown. In an attempt to understand more of these effects we have used DNA micro array and a panel of over 500 human redox related, cell growth and apoptosis genes to determine which genes and pathways are altered in Trx1 transfected cells. We found that one gene showing significantly increased expression was thioredoxin peroxidase-2 (TrxP2). The TrXPs belong to a conserved family of antioxidant proteins, the peroxiredoxins, that use thiol groups as reducing equivalents to scavenge oxidants (Chae *et al.*, 1994). Reduced TrxPs scavenge oxidant species such as H_2O_2 and alky peroxides, and in the process homo or heterodimerize with other family members through disulfide bonds formed between conserved cysteine residues (Chae *et al.*, 1994; Jin *et al.*, 1997). Trx reduces the oxidized TrxP to the monomeric form. Four human TrxPs have been cloned; TrxP1 (a 22 kDa monomer) also known as natural killer enhancing factor (NKEF)-B (Shau *et al.*, 1994); TrxP2 (a 22 kDa monomer) also known as NKEF-A (Pahl *et al.*, 1995), proliferation associated gene (*pag*) (Prosperi *et al.*, 1993) or heme-binding protein 23 (HBP23) (Hirotsu *et al.*, 1999); TrxP3 (a kDa) also known as human MER5 (murine erythroleukemia-associated 5)/AOP1 or SP-22 which is primarily mitochondrial (Kang *et al.*, 1998; Watabe *et al.*, 1994); and TrxP4 (a 31 kDa monomer) or human AOE372, identified as a cytosolic and secreted protein (Jin *et al.*, 1997; Matsumoto *et al.*, 1999).

Transfection of Molt-4 leukemia cells with TrxP2 has been reported to protect the cells from apoptosis (Zhang *et al.*, 1997; Shau *et al.*, 1997). This raised the question whether an increase in TrxP2 could be responsible for the anti-apoptosis effects of Trx1. To investigate this possibility we transfected mouse WEHI7.2 cells with human Trx1 and have shown that it will protect against apoptosis induced by H_2O_2 but not by other agents including dexamethasone, doxorubicin and etoposide. Thus, we conclude that the increase in TrxP2 seen following Trx1 transfection protects cells from H_2O_2 induced apoptosis but not apoptosis caused by other agents.

METHODS

Cell culture. MCF7 human breast cancer cells stably transfected with Thuman rx1 (MCF7 Trx9) or vector alone (MCF7*Neo*) (Gallegos *et al.*, 1996) were used for the DNA microarray studies. For the studies of TrxP2 transfection we used the mouse WEHI7.2 thymoma cell line because it has been extensively studied for apoptosis and we have shown that Trx1 is a potent inhibitor of apoptosis in these cells (Baker *et al.*, 1997). All cells were grown in DMEM supplemented with 10% FBS, 100 U/ml penicillin, 100 µg/ml streptomycin, and 200 µg/ml G418.

Microarray fabrication. A library of 512 DNAs (480 human clones, 24 housekeeping genes, 8 ice plant controls) were suspended in 2 x SSC at a minimum concentration of 100 µg/µl and arrayed robotically from 96-well plates onto Clontech DNA Ready Type I slides. Once dry the DNA was UV crosslinked to the slides (Stratagene Stratalinker). Fixed slides were rinsed once in 0.1% SDS for 1 min, twice in water for 1 min, and stored in a dehumidified chamber.

Probe preparation. Total RNA was first extracted from cells using TRIzol (Life Technologies). Poly A+ RNA was isolated using the Oligotex mRNA kit (Qiagen). Fluorescently labeled cDNA probes were prepared from 4.0 μg mRNA using an oligo dT primer and Superscript II reverse transcriptase (Life Technologies). Each reverse transcription reaction was performed at a final volume of 30 μl. Fluorescent nucleotides Cy3-dCTP and Cy5-dCTP (NEN) were present at a concentration of 1 mM whereas nucleotide concentrations were 5 mM for dATP, dGTP, and dTTP and 2 mM for dCTP. The reactions were carried out at 42 °C for 2 hrs. The Cy3 and Cy5 reactions were pooled and the reaction was stopped with 2.65 μl of 25 mM EDTA. 3.3 μl of 1M NaOH was added and the mixture was heated to 65 °C for 15 min to hydrolyze the RNA. Addition of 3.3 μl of 1 M HCl and 5.0 μl of 1 M Tris-HCl pH 6.8 neutralized the reaction. Unincorporated nucleotides were removed first with Microcon 30 columns and then with Qiagen nucleotide removal columns. After elution from the column the purified probe was lyophilized until just dry and stored in the dark at $-20\,°$C.

Hybridization. Microarrays were denatured in a 95 °C water bath for 3 min and fixed with ice cold ethanol. The slides were spun dry at 500 x g and placed in a hybridization chamber (TeleChem). Probes were resuspended in 6.75 μl H$_2$0, 3.75 μl 20xSSC, 3.0 μl 1% SDS, and 1.5 μl Cot-1 DNA (1 mg/ml). The probe solution was placed in boiling water bath for 3 minutes and then allowed to cool to room temperature for 1 min. The 15.0 μl hybridization solution was pipetted onto the slide and covered with a glass cover slip. The chamber was sealed and hybridization was performed at 64 °C for 18 hrs. The slides were washed once in 5 x SSC and 0 .01% SDS for 5 min, once in 6 x SSC and 0.01% SDS for 5 min, and twice in 6 x SSC for 2 min. All washing steps were performed at room temperature. Following the final wash the slides were spun dry at 500 x g.

Scanning and analysis. Hybridized slides were scanned using the GenePix 4000 microarray scanner (Axon Instruments). Raw data was uploaded into the Arizona Cancer Center microarray core facility database. The Cy3 and Cy5 channels were first normalized based on the expression of housekeeping genes and the overall hybridization to ice plant genes determined background. Those genes whose expression was greater than background were used for the calculation of expression ratios. Outliers were determined based on standard deviations.

Northern analysis. 2 μg of MCF7*Neo*, Trx9, and Serb4 poly A+ RNA or WEHI7.2 mRNA was electrophoresed in a 1.5% agarose-formaldehyde gel and transferred to a MagnaGraph nylon membrane (Osmonics Inc.). RNA was crosslinked to the membrane using the autocrosslink function on the Stratagene Stratalinker. Analysis was performed using an [γ-^{32}P]dCTP labeled fragment of TrP2 (Random Primers DNA Labeling System, Life Technologies). The 483 base pair fragment for TrxP2 was generated using TrxP2 sequence specific primers (5' TCGGCTGAATCTGAAGTC 3' and 5' GGTGCTTCTGTGCATTCT 3'). Hybridization was performed using ULTRAhyb hybridization buffer (Ambion) and all wash steps were performed according to manufacturer protocols. Blots were imaged using the MD Storm 860 phosphoimager (Molecular Dynamics) and quantified with ImageQuant software. A glyceraldehyde phosphate dehydrogenase (GAPDH) full length labeled probe was also prepared and hybridized for normalization.

Apoptosis   Apoptosis was measured using an annexin-V staining kit (Boehringer Mannheim, Indianapolis) and flow cytometry. The cells were washed and resuspended in 20 µl FITC- annexin-V/propidium iodode stain for 15 min at 22°C, diluted with 500 µl buffer and fluorescence of 10^4 cells measured using flow cytometry.

TrxP2 Transfection Full length human Trx2 was cloned from human U266 myeloma cell polyA+ mRNA by RTPCR with primers 5'GACTTGTGTTGGGACTGCTG3' and 3'GGTCTCATGGCTGCCCAC 5' using Superscript II (Gibco BRL) and inserted into the pBKCMV mammalian cloning vector (InVitrogen). WEHI7.2 cells, 10^6 in log phase growth, were transfected by electroporation with 5 µg   ofpCBKCMV/TrxP2 and cells were grown in 800 µg/ml G4118 for 3 weeks before selection in soft agarose. Thirty clones were screened for TrxP2 expression by Northern hybridization and expression was normalized to expression of GAPDH.

RESULTS

DNA Microarray showing Increased TrxP2 Expression in Trx1 Transfected Cells

DNA micro array analysis of 512 known genes with RNA from MCF-7 human breast cancer cells stably transfected with human Trx-1 revealed significantly increased expression of several genes. One of these genes was TrxP2/ NKEF-A/proliferation associated gene (*pag*). The increase in TrxP2 was confirmed by Northern hybridization (**Figure 1**).

Transfection of Cells with TrxP2. Protection Against H_2O_2 Induced Apoptosis

To further understand the role of TrxP2 in Trx1's the anti- apoptotic and growth stimulatory effects we stably transfected mouse WEHI7.2 cells with human TrxP2. A clone showing 1.7 fold increased expression of TrxP2 mRNA was chosen for the studies.

The TrxP2 transfected WEHI7.2 cells were treated for 48 hr with H_2O_2 and apoptosis was measured (**Figure 2**). Compared to wild type or empty vector transfected WEHI7.2 cells the TrxP2 transfected WEHI7.2 cells were resistant to apoptosis induced by H_2O_2.

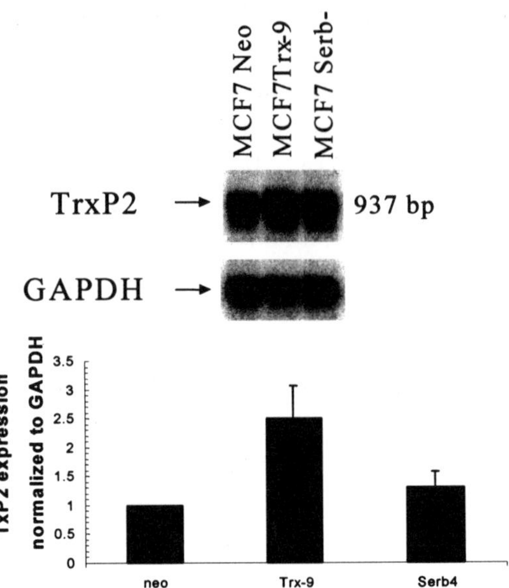

Figure 1 Northern blot showing over expression of TrxP2 in Trx1 transfected MCF-7 breast cancer cells. The upper panel shows a typical Northern hybridization of mRNA from empty vector transfected cells (MCF7*Neo*), Trx1 transfected cells (MCF7Trx9) and redox inactive Trx1 transfected cells (MCF7Serb) probed with TrxP2. The lower panel shows a histogram of data from 3 experiments.

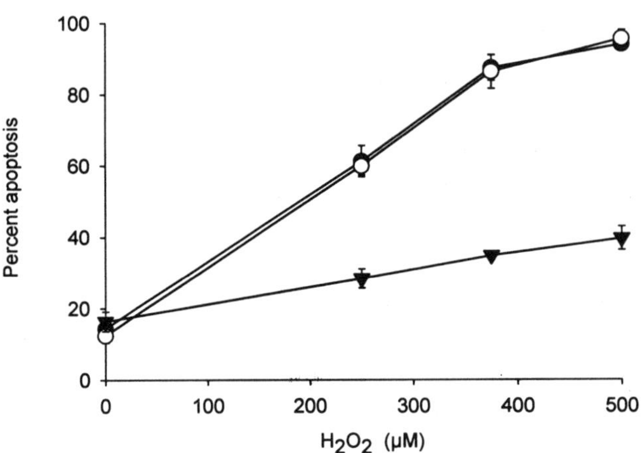

Figure 2 Effect of TrxP2 transfection on H_2O_2 induced apoptosis in WEHI7.2 cells. The cells were treated with H_2O_2 at the concentrations shown for 48 hr and apoptosis was measured by Annexin V labeling flow cytometry. (●) Wild type cells, (○) empty vector transfected cells and (▼) TrxP2 transfected cells. Each point is the mean of 3 determinations and bars are S.E.

TrxP2 Does Not Protect Cells Against Apoptosis Induced by Other Agents

The TrxP2 transfected WEHI7.2 cells did not show protection against apoptosis induced by dexamethasone, doxorubicin or etoposide. (**Figure 3**).

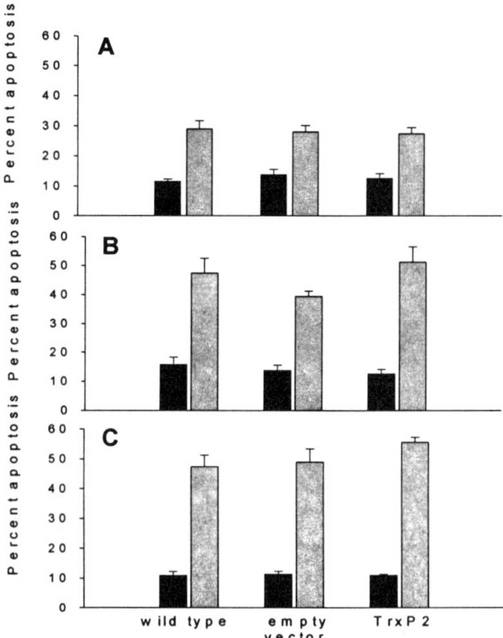

Figure 3 Effect of TrxP2 transfection on apoptosis induced by other agents. **A,** 500 nM doxorubicin; **B,** 1 µM dexamethasone; and **C,** 250 nM etoposide in wild type, empty vector transfected and TrxP2 transfected WEHI7.2 cells. The cells were treated with the agents for 24 hr and apoptosis was measured by Annexin V labeling flow cytometry. Values are the mean of 3 determinations and bars are S.E.

Effect of Selenium on H_2O_2 Induced Cell Killing

When WEHI7.2 cells were grown for 4 days in the presence of 0.1 µM sodium selenite there was protection against H_2O_2 induced cell death (**Figure 4**). The protection afforded by TrxP2 transfection was no greater and not additive with 0.1 µM sodium selenite.

DISCUSSION

TrxP2 transfected into cells has been reported to protect cells against apoptosis (Zhang *et al.*, 1997). Since we had observed an increase in TrxP2 expression by DNA micro array and confirmed by Northern analysis in Trx1 transfected cells this raised the possibility that the protection against apoptosis afforded cells by Trx1 transfection might be mediated by TrxP2. In order to address this possibility we stably transfected mouse WEHI7.2 cells with human TrxP2. The cells were protected against H_2O_2 induced apoptosis by TrxP2 transfection. The TrxPs belong to the peroxiredoxins family of antioxidant proteins that use thyl groups as reducing equivalents to scavenge oxidant species such as H_2O_2 and alky peroxides. The TrxP2 transfected cells were not, however, protected against apoptosis induced by dexamethasone, doxorubicin and etoposide. Thus, the results show that TrxP2 is not responsible for the wide spread anti apoptosis effects of Trx1 but protects against apotosis induced by H_2O_2 presumably by its H_2O_2 scavenging properties. It also appears that dexamethasone, doxorubicin and etoposide do not induce apoptosis by a mechanism involving formation of H_2O_2.

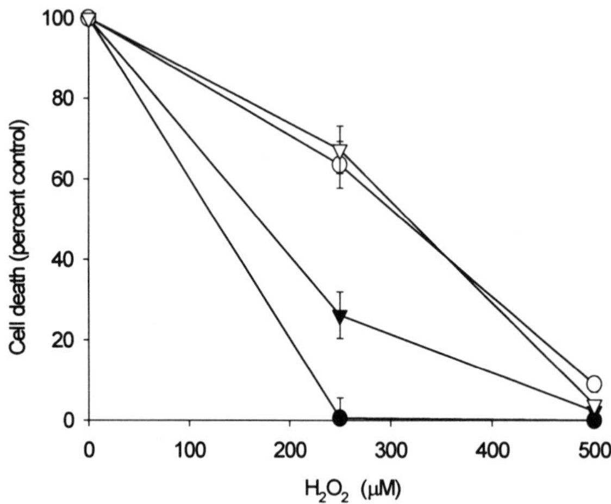

Figure 4 Effect of selenium on the protection by TrxP2 transfection against H_2O_2 induced apoptosis in WEHI7.2 cells. The cells were grown with or without 0.1 µM sodium selenite for 48 hr and then treated with H_2O_2 at the concentrations shown for 48 hr and apoptosis was measured by Annexin V labeling flow cytometry. (●) Empty vector transfected cells; (○) TrxP2 transfected cells; (▼) empty vector transfected cells with 0.1µM sodium selenite; and (▽) TrxP2 transfected cells with 0.1µM sodium selenite. Each point is the mean of 3 determinations and bars are S.E.

An interesting additional observation was that when cells were grown in the presence of physiological concentrations of selenium they were protected against H_2O_2 induced cell killing. Cell culture medium with 10% FBS has only very low levels of selenium (< 0.01 µM) compared to micro molar levels in human serum (Gallegos *et al.*, 1997) The protective effect of selenium against H_2O_2 cytotoxicity could be because the activity of glutathione peroxidase, a selenium containing enzyme, is normally very low in WEHI7.2 cells cultured without added selenium (Baker *et al.*, 1996). In the presence of sodium selenite TrxP2 transfection gave no further protection against H_2O_2 induced cell killing. TrxP2 is not a selenium containing enzyme so its activity is independent of selenium availability. Thioredoxin reductase is a selenium containing enzyme and its activity is increased in cell grown with physiological concentration of selenium (Berggren *et al.*, 1997). However, thioredoxin reductase activity does not appear to be limiting for the reduction of Trx1 in the absence of selenium . The protection by TrxP2, and perhaps other TrxPs, against H_2O_2 may only be seen when cells are grown in the absence of added selenium and where glutathione peroxidase activity is low.

164

In summary, using DNA micro array we identified TrxP2 as a gene whose expression is increased in Trx1 transfected cells. Transfection with TrxP2 protected WEHI7.2 cells against H_2O_2 induced apoptosis but did not protect against apoptosis induced by other agents. Thus, we conclude that the effects of Trx1 in protecting against apoptosis are probably not mediated by TrxP2 and that TrxP2 may only protect cells against H_2O_2 induced apoptosis when the cells are grown in the absence of selenium.

ACKNOWLEDGEMENTS

Supported by NIH grants CA48725, CA78277 and CA77204

REFERENCES

Baker, A., Payne, C.M., Briehl, M.M. and Powis, G. (1997) Thioredoxin, a gene found overexpressed in human cancer, inhibitis apoptosis *in vitro* and *in vivo*. *Cancer Res.* **57**: 5162-5167.

Baker, A.F., Briehl, M.M., Dorr, R. and Powis, G. (1996) Decreased antioxidant defense and increased oxidant stress during dexamethasone-induced apoptosis: *bcl-2* selectively prevents the loss of catalase activity. *Cell Death Differ.* **3**: 207-213.

Berggren, M., Gallegos, A., Gasdaska, J.R., Gasdaska, P.Y., Warneke, J. and Powis, G. (1996) Thioredoxin and thioredoxin reductase gene expression in human tumors and cell lines, and the effects of serum stimulation and hypoxia. *Anticancer Res.* **16**: 3459-3466.

Berggren, M., Gallegos, A., Gasdaska, J. and Powis, G. (1997) Cellular thioredoxin reductase activity is regulated by selenium. *Anticancer Res.* **17**: 3377-3380.

Chae, H.Z., Chung, S.J. and Rhee, S.G. (1994) Thioredoxin-dependent peroxide reductase from yeast. *J.Biol.Chem.* **269**: 27670-27678.

Deiss, L.P. and Kimchi, A. (1991) A genetic tool used to identify thioredoxin as a mediator of a growth inhibitory signal. *Science* **252**: 117-120.

Ericson, M.L., Hörling, J., Wendel-Hansen, V., Holmgren, A. and Rosen, A. (1992) Secretion of thioredoxin after *in vitro* activation of human B cells. *Lymphokine Cytokine Res.* **11**: 201-207.

Follmann, H. and Haberlein, I. (1995) Thioredoxins: Universal, yet specific thiol-disulfide redox cofactors. *BioFactors* **5**: 147-156.

Freemerman, A.J., Gallegos, A. and Powis, G. (1999) NF-kappaB transactivation is increased but not involved in the proliferative effects of thioredoxin overexpression in MCF-7 breast cancer cells. *Cancer Res.* **59**: 4090-4094.

Freemerman, A.J. and Powis, G. (2000) Redox-inactive thioredoxin enhances drug-induced apoptosis. *Leukemia*

Fujii, S., Nanbu, Y., Nonogaki, H., Konishi, I., Mori, T., Masutani, H. and Yodoi, J. (1991) Coexpression of adult T-cell leukemia-derived factor, a human thioredoxin homologue, and human papillomavirus DNA in neoplastic cervical squamous epithelium. *Cancer* **68**: 1583-1591.

Gallegos, A., Gasdaska, J.R., Taylor, C.W., Paine-Murrieta, G.D., Goodman, D., Gasdaska, P.Y., Berggren, M., Briehl, M.M. and Powis, G. (1996) Transfection with human thioredoxin increases cell proliferation and a dominant negative mutant thioredoxin reverses the transformed phenotype of breast cancer cells. *Cancer Res.* **56**: 5765-5770.

Gallegos, A., Berggren, M., Gasdaska, J.R. and Powis, G. (1997) Mechanisms of the regulation of thioredoxin reductase activity in cancer cells by the chemopreventive agent selenium. *Cancer Res.* **57**: 4965-4970.

Gallegos, A., Raffel, J., Bhattacharyya, A.K. and Powis, G. (2000) Increased expression of thioredoxin in human primary and metastatic colon cancer. *American Association for Cancer Research* **41**: 189

Gasdaska, J.R., Harney, J.W., Gasdaska, P.Y., Powis, G. and Berry, M.J. (1999) Regulation of human thioredoxin reductase expression and activity by 3'-untranslated region selenocysteine insertion sequence and mRNA instability elements. *J.Biol.Chem.* **274**: 25379-25385.

Gasdaska, P.Y., Oblong, J.E., Cotgreave, I.A. and Powis, G. (1994) The predicted amino acid sequence of human thioredoxin is identical to that of the autocrine growth factor human adult T-cell derived factor (ADF): Thioredoxin mRNA is elevated in some human tumors. *Biochim.Biophys.Acta* **1218**: 292-296.

Gasdaska, P.Y., Gasdaska, J.R., Cochran, S. and Powis, G. (1995) Cloning and sequencing of human thioredoxin reductase. *FEBS Lett.* **373**: 5-9.

Gasdaska, P.Y., Berggren, M.M., Berry, M.J. and Powis, G. (1999) Cloning, sequencing and functional expression of a novel human thioredoxin reductase. *FEBS Lett.* **442**: 105-111.

Grogan, T.M., Fenoglio-Prieser, C., Zeheb, R., Bellamy, W., Frutiger, Y., Vela, E., Stemmerman, G., MacDonald, J., Richter, L., Gallegos, A. and Powis, G. (2000) Thioredoxin, a puntative oncogene product, is overexpressed in gastric carcinoma and associated with increased proliferation and increased cell survival. *Human Pathol.* **31**: 475-481.

Hayashi, T., Ueno, Y. and Okamoto, T. (1993) Oxidoreductive regulation of nuclear factor kappaB: Involvement of a cellular reducing catalyst thioredoxin. *J.Biol.Chem.* **268**: 11380-11388.

Hirotsu, S., Abe, Y., Okada, K., Nagahara, N., Hori, H., Nishino, T. and Hakoshima, T. (1999) Crystal structure of a multifunctional 2-Cys peroxiredoxin heme-binding protein 23 kDa/proliferation-associated gene product. *Proc.Natl.Acad.Sci.USA* **96**: 12333-12338.

Jin, D., Zoon, H.J., Rhee, S.G. and Jeang, K. (1997) Regulatory role for a novel human thioredoxin peroxidase in NF-kappaB activation. *J.Biol.Chem.* **272**: 30952-30961.

Kang, S.W., Chae, H.Z., Seo, M.S., Kim, K., Baines, I.C. and Rhee, S.G. (1998) Mammalian peroxidoxin isoforms can reduce hydrogen peroxide generated in response to growth factors and tumor necrosis factor-α. *J.Biol.Chem.* **273**: 6297-6302.

Kawahara, N., Tanaka, T., Yokomizo, A., Nanri, H., Ono, M., Wada, M., Kohno, K., Takenaka, K., Sugimachi, K. and Kuwano, M. (1996) Enhanced coexpression of thioredixin and high mobility group protein 1 genes in human hepatocellular carcinoma and the possible association with decreased sensitivity to cisplatin. *Cancer Res.* **56**: 5330-5333.

Laurent, T.C., Moore, E.C. and Reichard, P. (1964) Enzymatic synthesis of deoxyribonucleotides VI. Isolation and characterization of thioredoxin, the hydrogen donor from *Escherichia coli* B. *J.Biol.Chem.* **239**: 3436-3444.

Lee, S., Kim, J., Kwon, K., Yoon, H.W., Levine, R.L., Ginsburg, A. and Rhee, S.G. (1999) Molecular cloning and characterization of a mitochondrial selenocysteine-containing thioredoxin reductase from rat liver. *J.Biol.Chem.* **274**: 4722-4734.

Matsumoto, A., Okado, A., Fujii, T., Fujii, J., Egashira, M., Niikawa, N. and Taniguchi, N. (1999) Cloning of the peroxiredoxin gene family in rats and characterization of the fourth member. *FEBS Lett.* **443**: 246-250.

Miranda-Vizueta, A. (1999) TrxR3, a novel human thioredoxin reductase. *GenBank Accession* **AF133519**: bp 1-2165

Miranda-Vizuete, A., Damdimopoulos, A.E., Pedrajas, J.R., Gustafsson, J.-A. and Spyrou, G. (1999) Human mitochondrial thioredoxin reductase. cDNA cloning, expression and genomic organization. *Eur.J.Biochem.* **261**: 405-411.

Mustacich, D., Coon, A. and Powis, G. (2000) Nuclear targeted thioredoxin transforms NIH 3T3 cells. *American Association for Cancer Research* **41**: 189

Mustacich, D. and Powis, G. (2000) Thioredoxin reductase. *Biochem.J.* **346**: 1-8.

Nakamura, H., Masutani, H., Tagaya, Y., Yamauchi, A., Inamoto, T., Nanbu, Y., Fujii, S., Ozawa, K. and Yodoi, J. (1992) Expression and growth-promoting effect of adult T-cell leukemia-derived factor. A human thioredoxin homologue in hepatocellular carcinoma. *Cancer* **69**: 2091-2097.

Nakamura, H., Nakamura, K. and Yodoi, J. (1997) Redox regulation of cellular activation. *Annu.Rev.Immunol.* **15**: 351-369.

Nakamura, H., Bai, J., Nishinaka, Y., Ueda, S., Sasada, T., Ohshio, G., Imamura, M., Takabayashi, A., Yamaoka, Y. and Yodoi, J. (2000) Expression of thioredoxin and glutaredoxin, redox-regulating proteins, in pancreatic cancer. *Cancer Detection and Prevention* **24**: 53-60.

Nishiyama, A., Ohno, T., Iwata, S., Matsui, M., Hirota, K., Masutani, H., Nakamura, H. and Yodoi, J. (1999) Demonstration of the interaction of thioredoxin with p40phox, a phagocyte oxidase component, using a yeast two-hybrid system. *Immunol.Lett.* **68**: 155-159.

Pahl, P., Berger, R., Hart, I., Chae, H.Z., Rhee, S.G. and Patterson, D. (1995) Localization of TDPX1, a human homologue of the yeast thioredoxin-dependent peroxide reductase gene (TPX) to chromosome 13q12. *Genomics* **26**: 602-606.

Peddie, C.M., Wolf, C.R., McLellan, L.I., Collins, A.R. and Bown, D.T. (1997) Oxidative DNA damage in CD34[+] myelodysplastic cells is associated with intracellular redox changes and elevated plasma tumor necrosis factor-α concentration. *Br.J.Haematol.* **99**: 625-630.

Powis, G., Mustacich, D. and Coon, A. (2000) The role of the redox protein thioredoxin in cell growth and cancer. *Free Rad.Biol.&Med.* (In Press)

Prosperi, M., Ferbus, D., Karczinski, I. and Goubin, G. (1993) A human cDNA corresponding to a gene overexpressed during cell proliferation encodes a product sharing homology with amoebic and bacterial proteins. *J.Biol.Chem.* **268**: 11050-11056.

Redl, B., Merschak, P., Abt, B. and Wojnar, P. (1999) Phage display reveals a novel interaction of human tear lipocalin and thioredoxin which is relevant for ligand binding. *FEBS Lett.* **460**: 182-186.

Saitoh, M., Nishitoh, H., Fujii, M., Takeda, K., Tobiume, K., Sawada, Y., Kawabata, M., Miyazono, K. and Ichijo, H. (1998) Mammalian thioredoxin is a direct inhibitor of apoptosis signal-regulating kinase (ASK) 1. *The EMBO Journal* **17**: 2596-2606.

Shau, H., Butterfield, L.H., Chiu, R. and Kim, A. (1994) Cloning and sequence analysis of candidate human nautral killer-enhancing factor genes. *Immunogenetics* **40**: 129-134.

Shau, H., Kim, A.T., Hedrick, C.C., Lusis, A.J., Tompkins, C., Finney, R., Leung, D.W. and Paglia, D.E. (1997) Endogenous natural killer enhancing factor-B increases cellular resistance to oxidative stresses. *Free Rad.Biol.&Med.* **22**: 497-507.

Silverman, R.B. and Nandi, D.L. (1988) Reduced thioredoxin: A possible physiological cofactor for vitamin K epoxide reductase. Further support for an active site disulfide. *Biochem.Biophys.Res.Commun.* **155**: 1248-1254.

Spyrou, G., Enmark, E., Miranda-Vizuete, A. and Gustafsson, J. (1997) Cloning and expression of a novel mammalian thioredoxin. *J.Biol.Chem.* **272**: 2936-2941.

Ueno, M., Masutani, H., Jun Arai, R., Yamauchi, A., Hirota, K., Sakai, T., Inamoto, T., Yamaoka, Y., Yodoi, J. and Nikaido, T. (2000) Thioredoxin-dependent redox regulation of p53-mediated p21 activation. *J.Biol.Chem.* **274**: 35809-35815.

Watabe, S., Kohno, H., Kouyama, H., Hiroi, T., Yago, N. and Nakazawa, T. (1994) Purification and characterization of a substrate protein for mitochondrial ATP-dependent protease in bove adrenal cortex. *J.Biochem.* **115**: 648-654.

Watson, J.A., Rumsby, M.G. and Wolowacz, R.G. (1999) Phage dispaly identifies thioredoxin and superoxide dismutase as novel protein kinase C-interacting proteins: thioredoxin inhibits protein kinase C-mediated phosphorylation of histone. *Biochem.J.* **343**: 301-305.

Zhang, P., Liu, B., Kang, S.W., Seo, M.S., Rhee, S.G. and Obeid, L.M. (1997) Thioredoxin peroxidase is a novel inhibitor of apoptosis with a mechanism distinct from that of Bcl-2. *J.Biol.Chem.* **272**: 30615-30618.

REACTIVE NITROGEN SPECIES AND PROTEINS: BIOLOGICAL SIGNIFICANCE AND CLINICAL RELEVANCE

Jose M. Souza, Qiping Chen, Beatrice Blanchard-Fillion, Scott A. Lorch, Caryn Hertkorn, Richard Lightfoot, Marie Weisse, Thomas Friel, Eugenia Paxinou, Marios Themistocleous, Steve Chov and Harry Ischiropoulos

Stokes Research Institute and Department of Biochemistry and Biophysics, Children's Hospital of Philadelphia and The University of Pennsylvania, Philadelphia, PA 19104

INTRODUCTION

Nitric oxide and reactive nitrogen species such as nitrogen dioxide, dinitrogen trioxide and peroxynitrite react selectively with different proteins causing covalent structural modifications that alter protein function (1-3). The predominant post-translational modification mediated by nitric oxide is the S-nitrosylation of cysteine residues whereas reactive nitrogen intermediates primarily oxidize cysteine and nitrate tyrosine residues. Glyceraldehyde-3-phosphate dehydrogenase, ryanodine receptor, p21ras, hemoglobin and caspase 3 are modified by S-nitrosylation of cysteine residues *in vivo* (4-10). The S-nitrosylation of the cysteine residues provides a selective and reversible covalent modification that regulates protein function and explains the ability of nitric oxide to regulate simultaneously different cellular pathways.
Proteins found to be modified by nitration include Mn superoxide dismutase, prostacyclin synthase, Ca^{2+}-ATPase, tyrosine hydroxylase, α-synuclein, and the plasma proteins, ceruloplasmin, transferrin, anti-chymotrypsin, α1-protease inhibitor and fibrinogen (11-18). Nitration of tyrosine residues in these proteins has been found to inhibit, have no effect or in one case enhance protein function.
The biochemical and biophysical factors for the apparent selectivity of protein modification by either nitric oxide or nitrating agents are currently being investigated. Based on the known biochemical properties of nitric oxide and nitrating agents, factors that may play a critical role in the selective biological reactivity of nitric oxide and peroxynitrite are discussed. Overall, covalent post-translational modifications by nitric oxide and other reactive nitrogen species represent a major aspect of their biological reactivity and provide a reasonable biochemical explanation for their multifaceted biological functions.

The reaction of nitric oxide with proteins

Research originating from the ability of the endothelium to induce relaxation of vascular smooth muscle cells led to the discovery of nitric oxide (19). Nitric oxide is formed by the 5-electron oxidation of one of the guanidino groups of L-arginine by nitric oxide synthases, and it is released in response to activation of endothelial cells by shear stress, ATP, acetylcholine, bradykinin and other vasodilators (19). As a small lipophilic molecule, nitric oxide can potentially diffuse through membranes to reach soluble guanylate cyclase and elevate cGMP levels in target cells (19). Alternatively, nitric oxide may react with reduced thiols to form S-nitrosothiols, which stabilize and transport nitric oxide to sites of its action such as guanylate cyclase in smooth muscle cells (1,2). As will be discussed below the formation of S-

Biological Reactive Intermediates VI, Edited by Dansette *et al.*
Kluwer Academic / Plenum Publishers, 2001

nitrosothiols may play a critical role in regulating the biological functions of nitric oxide. Other than the relaxation of smooth muscle cells, nitric oxide has been implicated in a number of important pathophysiological functions such as the regulation of apoptosis, ion channel activity and protection against oxidative stress (4-10). These diverse functions of nitric oxide appear to be derived predominantly from its ability to react with other reactive oxygen species and with proteins (1,3). The selective reactions of nitric oxide with proteins regulate protein function in a reversible manner and provide a reasonable biochemical explanation for the ability of nitric oxide to regulate different aspects of cell biology. The proteins that are selectively modified by nitric oxide fall into five major categories: 1) heme proteins, 2) proteins with zinc-thiolate clusters, 3) proteins with iron-sulfur clusters, 4) proteins containing tyrosine residues that undergo one electron oxidation to form tyrosyl radical, and 5) proteins with reduced cysteine residues.

Examples of the first category include guanylate cyclase, hemoglobin, cytochrome P450, insect salivary heme protein, and bacterial flavohemoglobin (20-25). The activation of guanylate cyclase is a well-documented signal transduction mechanism for vasodilatation (19). The insect salivary protein binds nitric oxide and delivers it locally in order to facilitate blood extraction from the host (23). The bacterial flavohemoglobins provide means for elimination of nitric oxide, where as hemoglobin was thought to provide similar function in humans (24,25). Although the binding of nitric oxide to cytochrome P450 can inhibit function (26), the implications in degradation of xenobiotics or other molecules via the P450 pathways have not been clearly established. Recently, metallothionein, cysteine-rich zinc-containing proteins, have been shown to react with nitric oxide (27). Binding of nitric oxide to metallothionein results in zinc release and represents a potential mechanism for elimination of nitric oxide that regulates the myogenic reflex in mesenteric arteries (27). The proteins modified by binding of nitric oxide to iron sulfur clusters are critical for the regulation of cellular respiration and iron metabolism. Aconitase and mitochondrial electron transport chain complexes represent iron-sulfur targets for nitric oxide (28,29). Altering the activity of aconitase and mitochondrial electron transport impacts cellular metabolism and the ability of cells to regulate iron metabolism (28,30). Ribonucleotide reductase, prostagladin H synthase-2, and prostaglandin endoperoxidase synthase are examples of proteins in which nitric oxide has been shown to react with the tyrosyl radical (31-33). Again this reactivity has been tested only under exposure of cells to nitric oxide, and it remains uncertain if nitric oxide regulates the activity of these proteins *in vivo*. In the final group glyceraldehyde-3-phosphate dehydrogenase (GAPDH), ryanodine receptor, p21ras, and caspase 3 are all modified by S-nitrosylation of functionally important reduced cysteine residues (4-10). Experimental evidence for the regulation of the activities of these proteins *in vivo* has been reported (9,10). Of great interest is the regulation of caspase-3, since this protein is one of the central executioners of apoptosis.

S-nitrosylation of protein cysteine residues

Despite the sound evidence for the formation of protein S-nitrosothiols *in vivo*, the nature of the nitrosylating agent(s) remains unclear. Several potential pathways have been proposed:

1) *The direct reaction of nitric oxide with cysteine.* This reaction is relatively slow and is not thermodynamically favored in the absence of an electron acceptor (34,35). However, since the cysteine concentration in proteins and glutathione is higher than the concentration of all other putative targets, the overall rate of this reaction may be high enough to compete with the other pathways. The electron acceptor is essential in order for the reaction to yield S-nitrosylated cysteine.
2) *Metal catalysis.* Iron dinitrosyl complexes have been detected *in vivo* and can generate S-nitrosothiol *in vitro* (36). Moreover, metaloproteins such as ceruloplasmin can generate S-nitrosylate cysteine residues (37).
3) *The direct reaction of nitric oxide with thiyl radical.* The oxidation of thiol to thiyl radical followed by coupling of the radical with nitric oxide. should be also considered. Certain cysteine residues in protein are critical targets for the oxidation and provided that nitric oxide is available the formation of S-nitrosothiol is possible.
4) *Formation of nitrogen oxides.* The formation of higher nitrogen oxides by the trimolecular reaction of nitric oxide with oxygen generates nitrosating species. However, the second order rate constant for the reaction of nitric oxide with oxygen is relatively small and hence it is unlikely to compete successfully with any of the other nitric oxide targets. However, the

reaction of nitric oxide with oxygen may take place in hydrophobic environment(s) such as cell membranes where the concentration of both reactants is higher than in the cytosol or in the extracellular environment (38). The concentration of nitric oxide and oxygen in the phospholipid bilayer may increase by as much as ten-fold and thus promote the formation of nitrogen oxides. The nitrogen oxides are reactive and hydrate easily to give nitrite and nitrate limiting their biological function to the membrane bilayer where they may be important for the formation of nitrosylated and nitrated lipid adducts (39). Another alternative is the reaction of nitric oxide with peroxynitrite to form dinitrogen trioxide (40). However, the reaction between nitric oxide and peroxynitrite is not kinetically favored since the concentration of both reactants is low and alternative targets for peroxynitrite are also present.

Nitration of tyrosine residues

A number of proteins such as Mn superoxide dismutase, prostacyclin synthase, Ca^{2+}-ATPase, tyrosine hydroxylase, and the plasma proteins, ceruloplasmin, transferrin, a1 anti-chymotrypsin, a1-protease inhibitor and fibrinogen are modified by nitration of tyrosine residues *in vivo* (11-16).
Nitration can be derived by the formation of a number of nitrating agents (3):

1) *Direct reaction of nitric oxide with tyrosyl radical.* The reaction of tyrosyl radical with nitric oxide yields an unstable nitrosotyrosine intermediate, which is converted to nitrotyrosine by a subsequent two-electron oxidation. Ribonucleotide reductase, prostagladin H synthase-2, and prostaglandin endoperoxidase synthase are examples of proteins that posses a reactive tyrosyl radical and have been shown to be nitrated *in vitro* after exposure to nitric oxide (31-33).

2) *Acidification of nitrite.* Nitrite is a stable byproduct of nitric oxide reactions and/or is present in a number of dietary ingredients and foods. At low pH the acidification of nitrite yields nitrous acid which can react slowly with proteins to 3-nitrotyrosine. This pathway maybe relevant for conditions found in the stomach or in acidic cellular compartments.

3) *The conversion of nitrite to a nitrating intermediate by peroxidases.* Myeloperoxidase, eosinophil peroxidase and other proteins with peroxidase activity have been shown to oxidize nitrite to nitrating agent capable of nitrating free tyrosine and protein tyrosine residues (41,42). The nitration efficiency can be enhanced by the presence of free tyrosine or peroxynitrite (43-45). In the case of myeloperoxidase the formation of hypochlorous acid does not appear to contribute significantly in the nitration of proteins but it specifically modifies tyrosine residues to form 3-chlorotyrosine (43,45).

4) *The reaction of proteins with peroxynitrite in the absence or presence of catalysts.* Catalysts of peroxynitrite-mediated nitration include CO_2, myeloperoxidase and other peroxidases, metal and metalloproteins, and possibly aldehydes (3).

The proximal nitrating species *in vivo* have not been completely elucidated but any of the above pathways satisfies the detection of nitrated proteins in a number of human and animal models of disease.

Specificity of cysteine and tyrosine modification by nitric oxide and nitrating agents

Both cysteine nitrosylation and tyrosine nitration appear to be specific, only few proteins and few residues in the proteins are modified by S-nitrosylation or tyrosine nitration. A number of the factors listed below attempt to provide a biochemical and biophysical rationale for this apparent selectivity. It is unlikely that the selectivity is derived by the chemical nature of the modifying species. It may be reasonable to consider that nitric oxide will modify proteins primarily by cysteine S-nitrosylation and the nitrating agent by tyrosine nitration it is unlikely that these species can select their targets. The selectivity can be derived mainly by the proximity of certain proteins to sites of nitric oxide generation, and by the biophysical properties of the target protein.

Compartmentalization and proximity to the target: Stamler et al. provided a nice discussion of and several examples indicating that the proximity of a protein to the source of nitric oxide generation is an important factor in predicting the proteins modified by cysteine S-nitrosylation (1). We have reasoned that the biological reactivity of nitrating agents is closely associated with the sites of superoxide formation and peroxidases. The superoxide dismutases maintain a low steady state level of superoxide and thus prevent reactions between superoxide and iron-sulfur proteins or nitric oxide (15). However, in local environments where the superoxide dismutase is not efficient in removing superoxide, and under conditions where higher levels of superoxide are generated, peroxynitrite will be formed and protein nitration will ensue. This may be the case with mitochondria as Poderoso and co-workers have recently reported a superoxide-dependent nitration of mitochondrial proteins (46). Similarly, the presence of peroxidases and hydrogen peroxide may be instrumental in forming nitrating agents at local sites.

Biophysical properties of the target protein. Stamler et al. are also discussed the probability of a Cys-Glu or Asp motif, which likely contains a polar amino acid (Gly, Ser, The, Cys, Tyr, Asn, Gln) in the –2 position and either acidic (Glu, Asp) or basic amino acid (Lys, Arg) in the –1 position that predicts the site of S-nitrosylation (1). Recent published work from our laboratory failed to determine a similar sequence for tyrosine nitration but determined several factors listed below that determine the sites of tyrosine nitration on different proteins. These factors include the degree of surface exposure of the tyrosine, the folding of the protein, the absence of steric hindrances, the paucity of cysteine residues, and the proximity of the tyrosine residue to negatively charged residues (47). Overall, the information appears to be is embedded in the structure and local environment afforded by specific amino acids in the protein.

We should also note that other factors such as the abundance of a protein, the abundance of cysteine and tyrosine residues and the second order rate constants for the reaction of the nitrogen species with the protein do not appear to play any significant role in the selectivity of protein modification. It is clear that proteins found to be modified by S-nitrosylation of cysteine and nitration of tyrosine residues *in vivo* are not the most abundantly expressed proteins. Moreover, S-nitrosylation of selective cysteine residues of proteins takes place in a "sea" of several mM cytosolic glutathione levels. Similarly, nitration of protein tyrosine residues takes place, even in the presence of reduce thiols and other peroxynitrite scavengers (48), indicates that simple rate constants are not the only factors determining the biological reactivity of nitric oxide and peroxynitrite.

Overall, the selective modification of cysteine and tyrosine residues may have a profound effect in the biological function of living organisms. S-nitrosylation of cysteine residues is a reversible process and thus it does not have along lasting effect on biological function but it is sufficient to induce allosteric and signal transduction events. Similarly the repair of tyrosine nitration and proteolytic degradation may prevent the long-term effects on protein function induced by nitration. The failure to repair modified proteins is likely to result in accelerated protein turnover, and other disturbances in cell function that will lead to the development of a pathogenic phenotype.

REFERENCES

1. Stamler, J.S., Toone, E.J., Lip[ton, S.A. and Sucher, N.J. (1997) (S)NO signals: Translocation, regualtion and a consensus motif. Neuron 18, 691-696
2. Simon, D.I., Mullins, M.E., Jia, L., Gaston, B., Singel, D.J., and Stamler, J.S. (1996) Polynitrosylated proteins: Characterization, bioactivity, and functional consequences. Proc. Natl. Acad. Sci. USA 93, 4736-4741
3. Ischiropoulos, H. (1998). Biological Tyrosine Nitration-A pathophysiological function of nitric oxide and reactive oxygen species. Arch. Biochem. Biophys. 356, 1-11.
4. Molina y Vedia, L., McDonald, B., Reep, B., Brune, B., Di Silvio, M., Billiar, T.R. and Lapetina, E.G. (1992) Nitric Oxide-induced S-nitrosylation of glyceraldehyde-3-phosphate dehydrogenase inhibits enzymatic activity and increases endogenous ADP-ribosylation. J Biol. Chem. 267:24929-24932.
5. Lander, H.M., Milbank, A.J., Tauras, J.M., Hajjar, D.P., Hempstead, B.L., Schwartz, G.D., Kraemer, R.T., Mirza, U.A., Chalt, B.T., Burk, S.C., and Quilliam, L.A. (1996) Redox regulation of cell signalling. Nature 381, 380-381.
6. Xu, L., Eu, J.P., Msissner, G., and Stamler, J.S. (1998) Activation of the cardiac calcium Release Channel (Ryanodine Receptor) by Poly-S-Nitrosylation. Science 279, 234-237.

7. Lipton, S.A., Choi, Y.-B., Pan, Z.-H., Lei, S.Z., Chen, H-S.V., Sucher, N.J., Loscalzo, J., Singel, D.J. and Stamler, J.S. (1993) A redox-based mechanism for the neuroprotective and neurodestructive effects of nitric oxide and related nitroso-compounds. Nature 364:626-632.
8. Gow, A. and Stamler, J. (1998) Reactions between nitric oxide and haemoglobin under physiological conditions. Nature 391, 169-173.
9. Kim YM, Talanian RV, and Billiar TR (1997) Nitric oxide inhibits apoptosis by preventing increases in caspase-3-like activity via two distinct mechanisms. J. Biol. Chem. 272, 31138-31148.
10. Mannick, J.B., Hausladen, A., Liu, L., Hess, D.T., Zeng, M., Miao, Q.X., Kane, L.S., Gow, A.J. and Stamler, J.S. (1999) Fas-induced caspase denitrosylation. Science 284, 651-654.
11. MacMillan-Crow, L.A., Crow, J.P., Kerby, J.D., Beckman, J.S. and Thompson, J.A. (1996) Nitration and inactivation of Mn superoxide dismutase in chronic rejection of human renal allografts. Proc. Natl. Acad. Sci. USA. 93, 11853-11858.
12. Zhou, M-H, Klein, T., Pasquet, J-P. and Ullrich, V. (1998) Interleukin-1β decreases prostacyclin synthase activity in rat mesanglial cells via endogenous peroxynitrite formation. Biochem. J. 336, 507-512.
13. Viner RI, Ferrington DA, Huhmer AFR, Bigelow DJ, and Schoneich C (1996) Accumulation of nitrotyrosine on the SERCA2a isoform of SR Ca-ATPase of rat skeletal muscle during aging: a peroxynitrite-mediated process? FEBS Lett. 379, 286-290.
14. Klebl, B.M., Ayoub, A.T. and Pette, D. (1998) Protein oxidation, tyrosine nitration, and inactivation of sarcoplasmic reticulum Ca2+-ATPase in low-frequency stimulated rabbit muscle. FEBS Lett. 422, 381-384.
15. Ara, J., Przedborski, S., Naini, A.B., Jackson-Lewis, V., Trifiletti, R.R., Horwitz, J. and Ischiropoulos, H. (1998) Inactivation of tyrosine hydroxylase following exposure to peroxynitrite and MPTP. Proc. Natl. Acad. Sci. USA. 95, 7659-7663.
16. Boota A, Zar H, Kim Y-M, Johnson B, Pitt B, Davies P (1996) IL-1β stimulates superoxide and delayed peroxynitrite production by pulmonary vascular smooth muscle cells. Am. J. Physiol. 271, L932-L938.
17. Gole, M.D., Souza, J.M., Choi, I., Malcolm, S., Hertkorn, C., Foust III, R.F., Finkel, B. Lanken, P.F. and Ischiropoulos, H. (2000) Plasma proteins modified by tyrosine nitration in acute respiratory distress syndrome. Am. J. Physiol., 278, L961-967.
18. Souza, J.M., Giasson, B.I., Chen, Q. Lee, V.M-Y. and Ischiropoulos, H. (2000) Dityrosine Cross-linking Promotes Formation of Stable a-synuclein Polymers: Implication of Nitrative and Oxidative Stress in the Pathogenesis of Neurodegenerative Synucleinopathies. J. Biol. Chem. 275, 18344-18349, 2000.
19. Furchgott. R.F. (1996) The discovery of EDRF and its importance in the identification of nitric oxide. JAMA 276, 1186-1188.
20. Moore, E.G. and Gibson, Q.H. (1976) Cooperativity in the dissociation of nitric oxide from hemoglobin. J Biol. Chem. 251, 2788-2794.
21. Craven, P.A. and DeRubertis, F.R. (1978) Restoration of the responsiveness of purified guanylate cyclase to nitrosoguanidine, nitric oxide, and related activators by heme and hemeproteins. J Biol. Chem. 253, 8433-8443.
22. Lancaster Jr., J.R., Langrehr, J.M., Bergonia, H.A., Murase, N., Simmons, R.L., and Hoffman, R.A. (1992) EPR detection of heme and nonheme iron-containing protein nitrosylation by nitric oxide during rejection of rat heart allograft. J. Biol. Chem. 267, 10994-10998.
23. Ribeiro, J.M.C., Hazzard, J.M.H., Nussenzveig, R.H., Champagne, D.E., and Walker, F.A. (1993) Reversible binding of nitric oxide by a salivary heme protein from a bloodsucking insect. Science 260, 539-541.
24. Gardner, P.R., Gardner, A.M., Martin, L.A., and Salzman, A.L. (1998) Nitric oxide dioxygenase: An enzymic function for flavohemoglobin. Proc. Natl. Acad. Sci. USA 95, 10378-10383.
25. Hausladen, A., Gow, A.J., and Stamler, J.S. (1998) Nitrosative stress: Metabolic pathway involving the flavohemoglobin. Proc. Natl. Acad. Sci. USA 95, 14100-14105.
26. Tsubaki, M., Hiwatashi, A., Ichikawa, Y. and Hori, H. (1987) EPR study of ferrous cytocrome P450scc-nitric oxide complexes: effects of cholesterol and its analogues. Biochemistry 26, 4527-4534.
27. Kennedy, M.C., Antholine, W.E., and Beinert, H. (1997) An EPR investigation of the products of the reaction of cytosolic and mitochondrial aconitase with nitric oxide. J. Biol. Chem. 272, 20340-20347.

28. Pearce, L.L., Gandley, R.E., Han, W., Wasserloos, K., Stitt, M., Kanai, A.J., McLaughlin, M.K., Pitt, B.R., and Levitan, E.S. (2000) Role of metallothionein in nitric oxide signaling as revealed by a green fluorescent fusion protein. Proc. Natl. Acad. Sci. 97, 477-482.
29. Cassina, A and Radi, R. (1996) Differential inhibitory action of nitric oxide and peroxynitrite on mitochondrial electron transport. Arch Biochem. Biophys. 328, 309-316.
30. Bouton, C., Raveau, M., and Drapier, J.C. (1996) Modulation of iron regulatory protein functions. J. Biol. Chem. 271, 2300-2306.
31. Gunthers, M.R., Hsi, L.C., Curtis, J.F., Giorse, J.K., Marnett. L.J., Eling, T.E., and Mason, R.P. (1997) Nitric oxide trapping of the tyrosyl radical of Prostaglandin H Synthase-2 leads to tyrosine iminoxyl radical and nitrotyrosine formation. J. Biol. Chem. 272, 17086-17090.
32. Goodwin, D.C., Gunthers, M.R., Hsi, L.C., Crews, B.C., Eling, T.E., Mason, R.P., and Marnett, L.J. (1998) Nitric oxide trapping of tyrosyl radicals generated during prostaglandin endoperoxide synthase turnover. J. Biol. Chem. 273, 8903-8909.
33. Guittet, O., Ducastel, B., Salem, J.S., Henry, Y., Rubin, H., Lemaire, G., and Lepoivre, M. (1998) Differential sensitivity of the tyrosyl radical of mouse ribonucleotide reductase to nitric oxide and peroxynitrite. J. Biol. Chem. 273, 22136-22144.
34. Gow, A., Buerk, D.G., and Ischiropoulos, H. (1997) A novel reaction mechanism for the formation of S-nitrosothiols in vivo. J. Biol. Chem. 272, 2841-2845.
35. Koppenol, W.H. (1998) The basic chemistry of nitrogen monoxide and peroxynitrite. Free Rad. Bio. Med. 25, 385-391.
36. Boese M, Mordvintcev PI, Vanin AF, Busse R, Mulsch (1995) S-Nitrosation of serum albumin by dinitrosyl-iron complex. J. Biol. Chem. 270, 29244-29249.
37. Inoue, K., Akaike, T., Miyamoto, Y., Okamoto, T., Sawa, T., Otagiri, M., Suzuki, S., Yoshimura, T., and Maeda, H. (1999) Nitrosothiol formation catalyzed by ceruloplasmin. J. Biol. Chem. 274, 27069-27075.
38. Liu, X., Miller, M.J., Johshi, M.S., Thomas, D.D. and Lancaster, Jr. J.R. (1998) Accelerated reaction of nitric oxide with oxygen within the hydrophobic interior of biological membranes. Proc. Natl. Acad. Sci. USA 95, 2175-2179.
39. O'Donnell VB, Eiserich, JP, Chumley, PH, Jablonsky, MJ, Krishna, NR, Kirk, M, Barnes, S, Darley-Usmar, VM, Freeman, BA (1998) Nitration of unsaturated fatty acids by nitric oxide-derived reactive nitrogen species peroxynitrite, nitric acid, nitrogen dioxide, and nitronium ion. Chem. Res. Toxicol. 12, 83-92
40. Wink, D.A., Cook, J.A., Kim, S.Y., Vodovotz, Y., Pacelli, R., Krishna, M.C., Russo, A., Mitchell, J.B., Jourd'heuil, D., Miles, A.M. and Grisham, M.B. (1997) Superoxide modulates the oxidation and nitrosation of thiols by nitric oxide-derived reactive intermediates. J. Biol. Chem. 272, 11147-11151.
41. Van der Vliet, A., Eiserich, J.P., Halliwell, B., and Cross, C.E. Formation of reactive nitrogen species during peroxidase-catalyzed oxidation of nitrite. J. Biol. Chem. 272:7617-7625, 1997.
42. Wu, W., Chen, Y. and Hazen, S.L. Eosinophil peroxidase nitrates protein tyrosyl residues. J. Biol. Chem. 274:25933-25944, 1999.
43. Van Dalen, C.J., Winterbourn, C.C., Senthilmohan, R. and Kettle, A.J. (2000) Nitrite as a substrate and inhibitor of myeloperoxidase. J. Biol. Chem. 275, 11638-11644.
44. Ischiropoulos, H., D. Duran, J. Nelson, A.B. Al-Mehdi. (1996) Reactions of nitric oxide and peroxynitrite with organic molecules and ferrihorseradish peroxidase: interference with the determination of hydrogen peroxide. Free Rad. Biol. Med. 20, 373-381.
45. Sampson, J.B., Ye, Y., Rosen, H. and Beckman, J.S. Myeloperoxidase and horseradish peroxidase catalyze tyrosine nitration in proteins from nitrite and hydrogen peroxide. Arch. Biochem. Biophys. 356, 207-213, 1998.
46. Schopfer F, Riobo N, Carreras MC, Alvarez B, Radi R, Boveris A, Cadenas E, Poderoso JJ. (2000) Oxidation of ubiquinol by peroxynitrite: implications for protection of mitochondria against nitrosative damage.Biochem J. 349, 35-42.
47. Souza, J.M., Daikhin, E., Yudkoff, M., Raman' C.S. and Ischiropoulos, H. Factors determining the selectivity of protein tyrosine nitration. Arch. Biochem. Biophys. 371, 169-178, 1999.
48. Alvarez, B., Ferrer-Sueta, G., Freeman, B.A., and Radi, R. (1998) Kinetics of peroxynitrite reaction with amino acids and human serum albumin. J. Biol. Chem. 274, 842-848.

BICARBONATE ENHANCES NITRATION AND OXIDATION REACTIONS IN BIOLOGICAL SYSTEMS—ROLE OF REACTIVE OXYGEN AND NITROGEN SPECIES

B. Kalyanaraman, Joy Joseph, and Hao Zhang

Biophysics Research Institute
Medical College of Wisconsin
Milwaukee, WI 53226 USA

INTRODUCTION

Bicarbonate anion (HCO_3^-) is present at high concentration (25 mM) in biological fluids. At this HCO_3^- concentration, the carbon-di-oxide (CO_2) concentration is estimated to be ca. 1.2 mM at physiological conditions. A major function of the HCO_3^-/CO_2 couple in biological systems is to regulate pH. Although HCO_3^--mediated enhancement of luminol oxidation was reported several decades ago (Hodgson and Fridovich, 1976), only recently was the role of HCO_3^- recognized in biological oxidations (Denicola *et al.*, 1996; Ischiropoulos, *et al.*, 1992; Lymar *et al.*, 1996; Singh *et al.*, 1998; Zhang *et al.*, 1997).

Peroxynitrite, a potent oxidant formed from the diffusion-controlled reaction rate between superoxide anion (O_2^{-}) and nitric oxide ($^{\bullet}NO$) (Beckman *et al.*, 1990), reacts with CO_2 to form a more selective oxidant, nitrosoperoxycarbonate ($ONOOCO_2^-$), which decomposes in part to give the carbonate anion radical (CO_3^{-}) and nitrogen-di-oxide radical ($NO_2^{\bullet}$) (Denicola *et al.*, 1996; Ischiropoulos, *et al.*, 1992; Lymar *et al.*, 1996; Zhang *et al.*, 1997). The presence of CO_2 in biological tissues, therefore, profoundly affects $ONOO^-$-induced nitration and oxidation reactions, especially those involving tyrosine and protein-bound tyrosines (Ischiropoulos, 1998).

Recently, a new perspective on the role of HCO_3^- in enhancing the peroxidase activity of copper,zinc superoxide dismutase (SOD1) was proposed (Goss *et al.*, 1999; Liochev and Fridovich, 1999). One of the proposed mechanisms of pathogenesis of amyotrophic lateral sclerosis (ALS), or Lou Gehrig's disease, involves the gain-of-function or the increased peroxidase activity of ALS SOD1 mutants (Wiedau-Pazos *et al.*, 1996). Earlier studies on oxidation reactions induced by SOD1 and ALS SOD1 mutants were performed in bicarbonate buffers (Singh *et al.*, 1998; Wiedau-Pazos *et al.*, 1996; Yim *et al.*, 1999). In the absence of HCO_3^-, the peroxidase activity of SOD1 and ALS SOD1 mutants was not detectable. This apparent dilemma was clarified when it was realized that HCO_3^-, a ubiquitous tissue and cell component, could influence the oxidative reactions via formation of CO_3^{-} radical as a reactive intermediate (Goss *et al.*, 1999; Liochev and Fridovich, 1999). The focus of this chapter is to describe how this new perspective on the role of HCO_3^- has helped us understand the peroxidative reactions of SOD1.

Biological Reactive Intermediates VI, Edited by Dansette *et al.*
Kluwer Academic / Plenum Publishers, 2001

Bicarbonate Induces Hydroxylation of Cyclic Nitrone Spin Traps, Including DMPO, in the Presence of SOD1 and H_2O_2

Previous spin-trapping studies designed to probe the peroxidase activity of SOD1 and mutants were performed in bicarbonate buffer (Singh *et al.*, 1998; Wiedau-Pazos *et al.*, 1996; Yim *et al.*, 1999). Recently it was discovered that in the absence of bicarbonate, the SOD1/H_2O_2/DMPO system did not yield DMPO-OH adduct formation (Sankarapandi and Zweier, 1999). The intensity of the DMPO-OH spectrum increased with increasing concentration of bicarbonate (Figure 1). The signal intensity increased with increasing concentrations of H_2O_2 (10-1000 μM). Addition of SOD to a freshly prepared phosphate buffer (100 mM, pH 7.4) containing DMPO, H_2O_2, and DTPA yielded very little DMPO-OH adduct (Figure 1A). This result may be attributed to the endogenous bicarbonate, as the CO_2 concentration in freshly prepared phosphate buffer has been estimated to be ca. 5-8 μM. The intensity of the DMPO-OH adduct ($a_H=a_N=15$ G) increased with increasing concentration of bicarbonate in phosphate buffer. The pH level was measured immediately before and after the addition of 25 mM bicarbonate to ensure that the bicarbonate effect was not due to pH changes. Spin-trapping experiments with [^{17}O]-labeled H_2O revealed that the ^{17}O atom was incorporated into the DMPO-OH adduct. This result lends definite proof that H_2O_2 was not responsible for DMPO-OH adduct formation came from [^{17}O]-labeled experiments. Figure 2B shows the ESR spectrum of the DMPO-OH adduct obtained by adding SOD1, H_2O_2, DMPO, and diethylenetriamine pentaacetic acid (DTPA) to phosphate buffer (prepared in [^{17}O]-labeled H_2O) and bicarbonate (25 mM)). The lines marked ● arise from ^{17}O atom incorporated into the DMPO-OH adduct from $H_2{}^{17}O$. The commercially available [^{17}O]-labeled H_2O consists of 45% [^{17}O]-H_2O and 55% [^{16}O]-H_2O. Computer simulation of the ESR spectrum of DMPO-OH (dashed line spectrum in Figure 2B) indicates 42% contribution from DMPO-^{17}OH and 58% from DMPO-^{16}OH. This result proves that hydroxylation of DMPO does not involve H_2O_2, the precursor for hydroxyl radical. Based on these data, we propose a hydroxylation mechanism involving the reaction between cyclic nitrones and $CO_3{}^{\cdot-}$ formed from oxidation of bicarbonate by SOD1/H_2O_2 (Scheme 1).

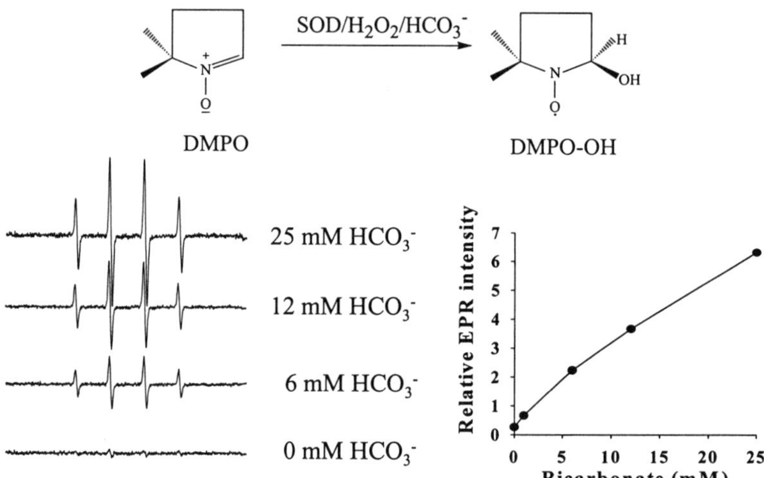

Figure 1. Bicarbonate induces hydroxylation of DMPO in the presence of SOD1 and H_2O_2. DMPO (50 mM) was incubated with SOD (1 mg/ml), H_2O_2 (1 mM), and various amount of NaHCO$_3$ in 0.1 M sodium phosphate buffer (pH 7.4) with 100 μM DTPA at room temperature. Note that DMPO-OH formation is negligible in the absence of bicarbonate.

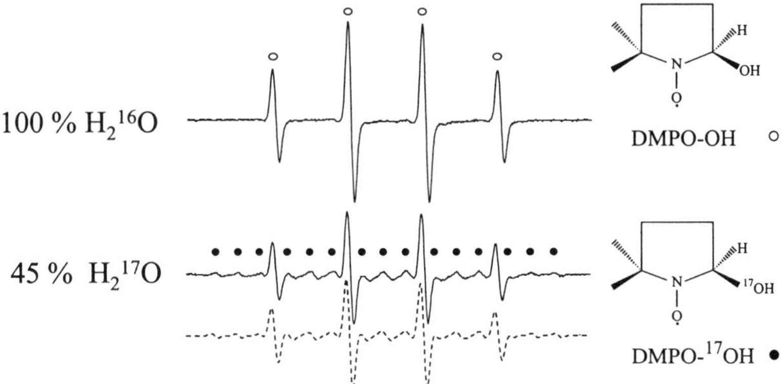

Figure 2. The effect of ^{17}O-labeled H$_2$O on SOD/H$_2$O$_2$-dependent DMPO-OH formation. DMPO (50 mM) was incubated in 0.1 M sodium phosphate buffer (pH 7.4) with 100 µM DTPA at room temperature. (A) The reaction was carried out in H$_2$^{16}O and (B) the reaction was carried out in H$_2$^{17}O (45%). Line positions from [^{16}O] (O) and [^{17}O] (●) coupling are shown. The difference between [^{16}O] and [^{17}O] atom is that the nuclear quantum number (I) equals zero for the [^{16}O] atom and 5/2 for the [^{17}O] atom. As a result, each line in the DMPO-^{17}OH adduct is split into 6 lines. Because there is 55% H$_2$^{16}O in the solution, there is still 55% contribution from DMPO-^{16}OH.

Scheme 1. Proposed mechanism of hydroxylation of DMPO. (A) Nucleophilic addition of ^{17}O-labeled H$_2$O to DMPO-carbonate radical intermediate. (B) Electron transfer mechanism involving the addition of ^{17}O-labeled H$_2$O to the DMPO cation radical. (*Reprinted with permission from: Zhang et al., J. Biol. Chem. 275, 14038-45.*)

Bicarbonate Enhances Oxidation and Nitration of Tyrosine Mediated by the Peroxidase Activity of SOD1

Tyrosine oxidation products, including dityrosine, trityrosine, pulcherosine, and isodityrosine, have been used as diagnostic marker products in radical-mediated oxidative damage in inflammatory diseases (Eiserich, *et al.*, 1998; Jacob *et al.*, 1996). Nitrotyrosine formation has been used as a diagnostic marker of peroxynitrite and other reactive nitrogen species (Eiserich, *et al.*, 1998; Ischiropoulos, 1998). We wished to determine whether bicarbonate enhances formation of these oxidation and nitration products in incubations containing SOD1, H_2O_2, tyrosine, and nitrite. Bicarbonate enhanced dityrosine formation from incubations containing SOD1 (1 mg/ml), H_2O_2 (1 mM), and tyrosine (1 mM) in a dose-dependent manner (Figure 3). In the absence of added bicarbonate, no dityrosine was detected (Figure 3C). In the presence of nitrite anion, nitrotyrosine was detected in this incubation system (Figure 4). Scheme 2 shows the postulated mechanism of oxidation and nitration of tyrosine by CO_3^- generated by the peroxidatic oxidation of bicarbonate by $SOD1/H_2O_2$.

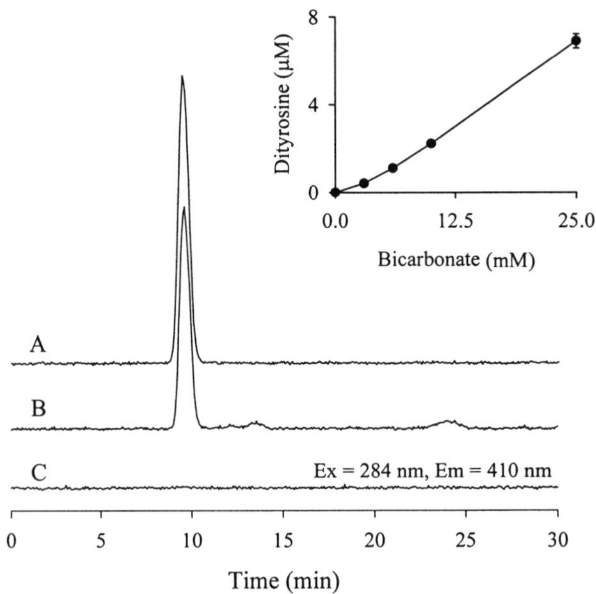

Figure 3. Bicarbonate enhances dityrosine formation during peroxidation of tyrosine by SOD. Dityrosine from SOD/H_2O_2-dependent oxidation of tyrosine was analyzed by reverse phase HPLC. (A) Typical HPLC trace showing 10 μM authentic dityrosine; retention time, 9.7 min; (B) the product of SOD/H_2O_2-dependent oxidation of tyrosine with 25 mM $NaHCO_3$; and (C) without $NaHCO_3$. Note that bicarbonate is absolutely essential for dityrosine formation. *Inset:* Dityrosine formation as the function of sodium bicarbonate concentration, tyrosine (1 mM) was incubated with SOD (1 mg/ml), H_2O_2 (1 mM), and various amounts of $NaHCO_3$ (25 mM) at 37°C for 4 hr and analyzed by HPLC (n=3 ± SD). *(Reprinted with permission from: Zhang et al., J. Biol. Chem. 275, 14038-45.)*

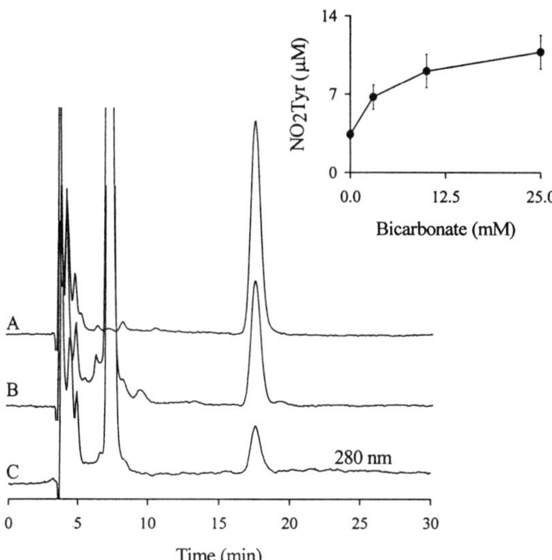

Figure 4. Bicarbonate enhances nitrotyrosine formation during oxidation of tyrosine by SOD/H_2O_2 in the presence of nitrite. Nitrotyrosine formed from SOD/H_2O_2-dependent oxidation of tyrosine was analyzed by reverse phase HPLC. (A) HPLC trace indicates 16 μM authentic nitrotyrosine, retention time, 17.5 min; (B) the product of SOD/H_2O_2-dependent oxidation of tyrosine with 25 mM $NaHCO_3$; and (C) without $NaHCO_3$ after 2 min reaction. *Inset:* Nitrotyrosine formation as the function of sodium bicarbonate concentration. Tyrosine (0.25 mM) was incubated with SOD (1 mg/ml), H_2O_2 (1 mM), and various amounts of $NaHCO_3$ at 37°C for 2 min and analyzed by HPLC (n=3±SD). (*Reprinted with permission from: Zhang et al., J. Biol. Chem. 275, 14038-45.*)

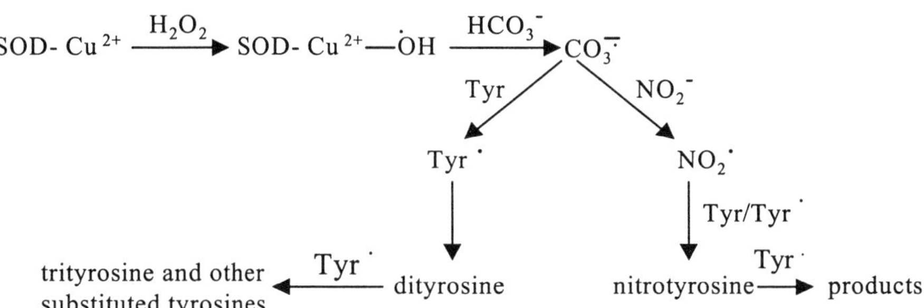

Scheme 2. Bicarbonate-mediated oxidation and nitration of tyrosine. $CO_3^{\cdot-}$ radical generated at the active site of SOD1 diffuses out of the channel and oxidizes tyrosine. $CO_3^{\cdot-}$ reacts with the nitrite anion to form $NO_2^{\cdot}$, which is also a potent nitrating agent. In the absence of bicarbonate, dityrosine formation was negligible in this system. (*Reprinted with permission from: Zhang et al., J. Biol. Chem. 275, 14038-45.*)

Oxidation of Bicarbonate by Copper-bound Hydroxyl Radical

Nearly 30 years ago, Hodgson and Fridovich (1975a, 1975b) demonstrated that the "copper-bound hydroxyl radical," $SOD-Cu^{2+}-{}^{\bullet}OH$, which is generated in the reaction between SOD1 and H_2O_2, oxidizes several anionic ligands including formate, azide, and nitrite anions. These small molecular weight anionic ligands are presumed to be oxidized by the oxidant formed at the active site of SOD1. The oxidizing potential of this putative $SOD-Cu^{2+}-{}^{\bullet}OH$ species is similar to that of the "free" hydroxyl radical, which is considerably higher than those associated with conventional peroxidases (Goss et al., 1999). Previous spin-trapping and oxidation data for the peroxidative mechanism of SOD1 and FALS SOD1 mutants have been obtained in bicarbonate buffers (Wiedau-Pazos et al., 1996; Yim et al., 1996). The one-electron oxidation potential for the HCO_3^-/carbonate radical anion ($CO_3^{\bullet-}$) couple is +1.59 V, which makes oxidation of HCO_3^- by $SOD–Cu^{2+}-{}^{\bullet}OH$ to $CO_3^{\bullet-}$ thermodynamically feasible. $CO_3^{\bullet-}$, although less reactive than the hydroxyl radical, is a more selective oxidant and can diffuse out of the active site, causing oxidation of substrates (Goss et al., 1999; Liochev and Fridovich, 1999) (Figure 5).

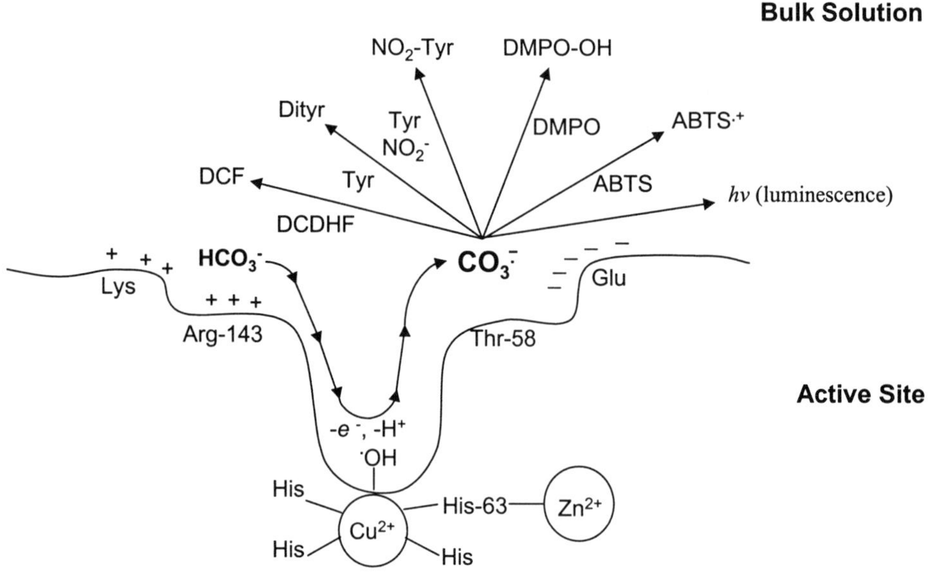

Figure 5. Oxidation of bicarbonate to the carbonate anion radical by copper-bound hydroxyl radical at the active site of SOD1. The $CO_3^{\bullet-}$ radical, a potent oxidant formed at the active site of SOD1, is able to diffuse away and oxidize several structurally different molecules in the bulk solution.

The x-ray crystal structure of SOD1 indicates that access to the active site of copper is via a narrow channel that restricts the entry of a large molecule such as ABTS or tocopherols. However, a relatively small anion such as HCO_3^- could reach the active site of SOD1 and subsequently undergo oxidation to $CO_3^{\bullet-}$, a diffusible oxidant that could leave the active site and cause oxidation of various substrates in free solution. Bicarbonate thus effectively exports oxidation from the sterically hindered active site to large molecules in bulk solution (Liochev and Fridovich, 1999). This model gives a new perspective on the peroxidative mechanism of SOD1 involving the oxidant, $CO_3^{\bullet-}$ radical.

ACKNOWLEDGEMENTS

This work was supported by grants from the ALS Association and the National Institute of Neurological Disorders and Stroke (NINDS) (NS40494) of the National Institutes of Health (NIH).

REFERENCES

Beckman, J. S., 1996, Oxidative damage and tyrosine nitration by peroxynitrite, *Chem. Res. Toxicol.* **9**:836–844.

Beckman, J. S., Beckman, T. W., Chen, J., Marshall, P. A., and Freeman, B.A., 1990, Apparent hydroxyl radical production by peroxynitrite: implications for endothelial injury from nitric oxide and superoxide, *Proc. Natl. Acad. Sci. USA* **87**:1620–1624.

Bonini, M. G., Radi, R., Ferrer-Sueta, G., Ferreira, A. M. D. C., and Augusto, O., 1999, Direct EPR detection of the carbonate radical anion produced from peroxynitrite and carbon dioxide, *J Biol. Chem.* **274**:10802–10806.

Denicola, A., Freeman, B. A., Trujillo, M., and Radi, R., 1996, Peroxynitrite reaction with carbon dioxide/bicarbonate: kinetics and influence on peroxynitrite-mediated oxidations, *Arch. Biochem. Biophys.* **333**:49–58.

Eiserich, J. P., Hristova, M., Cross, C. E., Jones, A. D., Freeman, B. A., Halliwell, B., and Van der Vliet, A., 1998, Formation of nitric oxide-derived inflammatory oxidants by myeloperoxidase and neutrophils, *Nature* **391**:393–397.

Goss, S. P. A., Singh, R. J., and Kalyanaraman, B., 1999, Bicarbonate enhances the peroxidase activity of Cu,Zn-superoxide dismutase—role of carbonate anion radical, *J. Biol. Chem.* **274**:28233–28239.

Hodgson, E. K., and Fridovich, I., 1975a, The interaction of bovine erythrocyte superoxide dismutase with hydrogen peroxide: inactivation of the enzyme, *Biochemistry* **14**:5294–5298.

Hodgson, E. K., and Fridovich, I., 1975b, The interaction of bovine erythrocyte superoxide dismutase with hydrogen peroxide: chemiluminescence and peroxidation, *Biochemistry* **14**:5299–5303.

Hodgson, E. K., and Fridovich, I., 1976, The mechanism of the activity-dependent luminescence of xanthine oxidase, *Arch. Biochem. Biophys.* **172**:202–205.

Ischiropoulos, H., 1998, Biological tyrosine nitration: a pathophysiological function of nitric oxide and reactive nitrogen species, *Arch. Biochem. Biophys.* **356**:1–11.

Ischiropoulos, H., Zu, Li., Chen, J., Tsai, M., Martin, J. C., Smith, C. D., and Beckman, J. S., 1992, Peroxynitrite-mediated tyrosine nitration catalyzed by superoxide dismutase, *Arch. Biochem. Biophys.* **298**:438–445.

Jacob, J. S., Cistola, D. P., Hsu, F. F., Muzaffar, S., Mueller, D. M., Hazen, S. L., and Heinecke, J. W., 1996, Human phagocytes employ the myeloperoxidase-hydrogen peroxide system to synthesize dityrosine, trityrosine, pulcherosine, and isodityrosine by a tyrosyl radical-dependent pathway, *J. Biol. Chem.* **271**:19950–19956.

Liochev, S. I., and Fridovich, I., 1999, On the role of bicarbonate in peroxidations catalyzed by Cu,Zn superoxide dismutase, *Free Radic. Biol. Med.* **27**:1444–1447.

Lymar, S. B., Jiang, Q., and Hurst, J. K., 1996, Mechanism of carbon dioxide-catalyzed oxidation of tyrosine by peroxynitrite, *Biochemistry* **35**:7855–7881.

Sankarapandi, S., and Zweier, J., 1999, Bicarbonate is required for the peroxidase activity of Cu,Zn-superoxide dismutase at physiological pH, *J. Biol. Chem.* **274**:1226–1232.

Singh, R. J., Goss, S. P. A., Joseph, J., and Kalyanaraman, B., 1998, Nitration of γ-tocopherol and oxidation of α-tocopherol by Cu,ZnSOD/H_2O_2/NO_2^-: role of nitrogen dioxide free radical, *Proc. Natl. Acad. Sci. USA* **95**:12912–12917.

Singh, R. J., Karoui, H., Gunther, M. R., Beckman, J. S., Mason, R. P., and Kalyanaraman, B., 1998, Reexamination of the mechanism of hydroxyl radical adducts formed from the reaction between familial amyotrophic lateral sclerosis-associated Cu,Zn superoxide dismutase mutants and H_2O_2, *Proc. Natl. Acad. Sci. USA* **95**:6675–6680.

Wiedau-Pazos, M., Goto, J. J., Rabizadeh, S., Gralla, E. B., Roe, J. A., Lee, M. K., Valentine, J. S., and Bredesen, D. E., 1996, Altered reactivity of superoxide dismutase in familial amyotrophic lateral sclerosis, *Science* **271**:515–518.

Yim, M. B., Chock, P. B., and Stadtman, E. R., 1993, Enzyme function of copper,zinc superoxide dismutase as a free radical generator, *J. Biol. Chem.* **274**:1226–1232.

Yim, M. B., Kang, J.-H., Yim, H.-S., Kwak, H.-S., Chock, P. B., and Stadtman, E. R., 1996, A gain-of-function of an amyotrophic lateral sclerosis-associated Cu,Zn-superoxide dismutase mutant: an enhancement of free radical formation due to a decrease in K_m for hydrogen peroxide, *Proc. Natl. Acad. Sci. USA* **93**:5709–5714.

Zhang, H., Squadrito, G. C., and Pryor, W. A., 1997, The mechanism of the peroxynitrite-carbon dioxide reaction proved using tyrosine, *Nitric Oxide: Biology and Chemistry* **1**:301–307.

NITRIC OXIDE AND PEROXYNITRITE IN OZONE-INDUCED LUNG INJURY

Debra L. Laskin[1], Ladan Fakhrzadeh[2] and Jeffrey D. Laskin[2]

[1,2]Environmental and Occupational Health Sciences Institute (EOHSI) and Departments of [1]Pharmacology and Toxicology, Rutgers University and [2]Environmental and Community Medicine, UMDNJ-Robert Wood Johnson Medical School, Piscataway, NJ 08854

ABSTRACT

One of the hallmarks of the inflammatory response associated with tissue injury is the accumulation of macrophages at sites of damage. These cell types release proinflammatory cytokines and cytotoxic mediators to destroy invading pathogens and initiate wound repair. However, when produced in excessive amounts, these macrophage-derived mediators may actually contribute to tissue injury. This process involves both direct damage to target tissues and amplification of the inflammatory response. One group of macrophage-derived mediators of particular interest are reactive nitrogen intermediates including nitric oxide and peroxynitrite which have been implicated in tissue injury induced by a variety of toxicants. Our laboratory has been investigating the role of reactive nitrogen intermediates in lung injury induced by oxidants such as ozone. Inhalation of ozone causes epithelial cell damage and Type II cell hyperplasia. This is associated with an accumulation of activated macrophages in the lower lungs which we have demonstrated contribute to toxicity. To analyze the role of macrophage-derived reactive nitrogen intermediates in ozone toxicity, we used transgenic mice lacking the gene for inducible nitric oxide synthase (NOSII). Treatment of wild type control animals with ozone (0.8 ppm) for 3 hr resulted in an increase in bronchoalveolar lavage (BAL) fluid protein reaching a maximum 24-48 hr after exposure. This was correlated with increased expression of NOSII protein and mRNA by alveolar macrophages and increased production of nitric oxide as well as peroxynitrite. Ozone inhalation also resulted in the appearance of nitrotyrosine residues in the lungs, an *in vivo* marker of peroxynitrite-induced damage. In contrast, in NOSII knockout mice, BAL protein was not increased demonstrating that these mice were protected from ozone-induced epithelial injury. Moreover, alveolar macrophages from the transgenic mice did not produce nitric oxide or peroxynitrite even after ozone inhalation. There was also no evidence for the formation of nitrotyrosine in lung tissue. These data indicate that ozone-induced lung injury is mediated by reactive nitrogen intermediates.

INTRODUCTION

One of the best characterized lung irritants in photochemical smog is ozone, a highly reactive oxidant. Ambient concentrations of ozone in the air during the summer months reach as high as 0.5-0.8 ppm in industrialized urban environments. A variety of human and animal models have been used to assess ozone induced lung toxicity (see Table 1 for some examples). In the lung, acute exposure to toxic concentrations of ozone can cause alveolar epithelial cell damage, airway hyperresponsiveness and pulmonary edema (reviewed in 1). Our laboratory has been interested in understanding the mechanisms by which ozone inhalation leads to pulmonary damage. It is well known that ozone toxicity is associated with an accumulation of macrophages in the lower lung. Using rodent models of lung toxicity, we have demonstrated that these cells are activated. Thus, following ozone inhalation, lung macrophages release a variety of inflammatory mediators including cytokines, reactive oxygen intermediates and reactive nitrogen intermediates (2-5). Each of these mediators has been implicated in the development of lung injury. Nitric oxide and peroxynitrite have also been reported to be involved in damage to pulmonary surfactant, pulmonary edema and emphysema (6-8).

Nitric oxide is generated in large amounts by alveolar macrophages exposed to inflammatory mediators such as interferon gamma (IFN-γ) and lipopolysaccharide (LPS), via NOSII, a calcium-independent, inducible form of nitric oxide synthase (2). NOSII activity is dependent on l-arginine and blocked by variety of inhibitors including aminoguanidine, which is relatively selective for this enzyme (9). Peroxynitrite is generated by the spontaneous reaction of nitric oxide with superoxide anion and is an even more potent cytotoxic oxidant than nitric oxide (10,11). We found that acute exposure of rats to ozone results in expression of NOSII mRNA and protein in the lung and increased production of nitric oxide by alveolar macrophages and Type II cells (2,12). In the present studies we used mice with a targeted disruption of the NOSII gene to analyze the role of reactive nitrogen intermediates in ozone toxicity. Our findings that these mice are protected from ozone-induced injury provide support for our hypothesis that reactive nitrogen intermediates contribute to tissue injury in this model.

EFFECTS OF OZONE INHALATION ON NITRIC OXIDE PRODUCTION BY WILD TYPE AND NOSII KNOCKOUT MICE

Alveolar macrophages, isolated 48 hr after exposure of mice to air or ozone (0.8 ppm for 3 hr) were cultured in the presence of LPS (100 ng/ml) and IFN-γ (100 U/ml) to induce NOSII. Cells from wild type mice were found to readily produce nitric oxide, as measured by the release of nitrite and nitrate into the culture medium. A 4-fold increase in nitric oxide production was observed in cells from ozone exposed animals when compared to air-exposed animals (9.7 $\pm$ 0.2 vs 23.9 $\pm$ 1.7 nmol nitrite/10^6 cells, $\pm$ SE, n= 3-5.). As expected, no significant nitric oxide was produced by alveolar macrophages from NOSII knockout mice. Alveolar macrophages from wild type animals were found to express NOSII as determined by western blotting. Ozone inhalation caused a marked increase in expression of this enzyme. In contrast, NOSII was not detected in macrophages from the knockout mice even after treatment of the cells with LPS and IFN-γ. Peroxynitrite is known to be an important mediator of nitric oxide-induced cellular damage (10,11,13,14). We next determined if ozone inhalation primed alveolar macrophages to produce this reactive nitrogen intermediate. Alveolar macrophages from ozone-treated, but not control animals, were found to produce significant quantities of peroxynitrite after stimulation with LPS and IFN-γ, as determined by fluorescence of 1,2,3 dihydrorhodamine (15). As observed with nitric oxide production, alveolar macrophages from NOSII knockout mice did not produce peroxynitrite and this was unaltered by ozone inhalation. Peroxynitrite has been shown to act as a nitrating agent leading to the formation of nitrotyrosine residues in proteins, a marker of peroxynitrite-mediated tissue injury (16). Following treatment of wild type mice with ozone, we found significant evidence of nitrotyrosine in isolated alveolar macrophages.

Diffuse nitrotyrosine staining was also noted in histologic sections of the lung. Alveolar macrophages stained more intensely for nitrotyrosine when compared to other cells in the tissue. Preincubation of the nitrotyrosine antibody with nitrotyrosine blocked tissue and cell staining demonstrating that the antibody was specific. No antibody staining was evident in lung sections from air exposed animals or in sections stained with non-immune IgG. Nitrotyrosine staining was also not evident in isolated alveolar macrophages or in lung sections from NOSII knockout mice from either air or ozone-exposed animals.

Table 1. Examples of Models Used to Study Acute Ozone Toxicity

Species	Level of exposure	Pulmonary Response	Reference
ferret, monkey, rat	1 ppm (8 hr)	neutrophil infiltration epithelial cell necrosis	27
guinea pig	3 ppm (1 hr)	increased epithelial permeability airway hyperreactivity nitrotyrosine staining	28
rabbit	1-2 ppm (2 hr)	increased BAL neutrophils increased chemotaxis	36,39
rat	1-2 ppm (3-6 hr)	increased BAL protein alterations in blood and tissue neutrophils, macrophage infiltration, increased TNF-α, IL-1, NOSII, fibronectin	2-5,29,30,31
mouse	0.1-2 ppm (3 hr)	increased BAL protein and neutrophils increased MIP-1, MIP-2, MCP-1	24,32,37,38
human	0.2-4 ppm (2-4 hr)	increased BAL protein and neutrophils increased IL-6, IL-8 PGE$_2$, fibronectin	33,35

We next analyzed the effects of ozone inhalation on superoxide anion production by macrophages from wild type and NOSII knockout mice. In wild type mice, inhalation of ozone resulted in a significant decrease in superoxide anion production by alveolar macrophages isolated immediately after exposure. However, these effects were transient and by 48 hr, superoxide anion release was above control levels. Interestingly, superoxide anion release by alveolar macrophages from NOSII knockout mice was significantly greater than cells from wild type animals. Ozone inhalation caused a 50% reduction in this activity, a response which was maintained for at least 48 hr post exposure.

It is clear from the above results that ozone inhalation increases   alveolar macrophage nitric oxide production and generates peroxynitrite in lung tissues of wild type mice. If nitric oxide is critical for toxicity, then NOSII knockout mice would be expected to be resistant to the toxic effects of ozone. To test this possibility, we compared ozone toxicity in wild type and NOSII knockout mice. In these studies, toxicity was assessed by measuring the accumulation of protein and leukocytes in bronchoalveolar lavage fluid (BAL) which are markers of alveolar epithelial injury. Treatment of wild type mice with ozone resulted in a time-dependent increase in the BAL fluid protein which reached a maximum 24-48 hr post exposure (Table 2 and not shown). In contrast, ozone inhalation had no effect on BAL protein in NOSII knockout mice. Similarly whereas ozone inhalation caused a significant time related increase in the number of cells recovered by bronchoalveolar lavage in control animals, no cellular accumulation was observed in NOSII knockout mice. Differential staining revealed that the majority of cells (>98%) recovered from the lung by lavage in both wild type and NOSII knockout were macrophages. Ozone inhalation had no effect on the type of cells recovered in the lung lavage in either mouse strain.

Table 2 Effects of ozone inhalation on lung damage in wild type and NOSII knockout mice.

	BAL protein (mg/ml)	Cell number x 10,000/mouse
WT		
air exposed	0.23 ± 0.2	3.5 ± 0.5
ozone exposed	0.42 ± 0.4	8.4 ± 1.4
NOSII knockout		
air exposed	0.25 ± 0.2	2.5 ± 0.6
ozone exposed	0.20 ± 0.2	3.1 ± 0.4

Female wild type (WT) B6J129SV F2 control mice or NOSII C57/BL6X129 knockout mice (8-10 weeks of age) were exposed to air or ozone (0.8 ppm, 3 hr). After 48 hr, BAL was analyzed for protein content and cell number (33). Note that NOSII knockout mice were protected from ozone induced toxicity.

DISCUSSION

A key macrophage-derived mediator of nonspecific host defense is nitric oxide (18,19). However, excessive production of this reactive nitrogen intermediate during irritant-induced inflammation can lead to tissue damage (10,11,20). Using a rat model, we previously demonstrated that ozone inhalation upregulates nitric oxide production in alveolar macrophages and type II cells (2, 12). Since nitric oxide and peroxynitrite are cytotoxic, we hypothesized that these oxidants contribute to ozone-induced lung injury. If this is the case, then animals lacking NOSII would be resistant to the toxic effects of ozone and this was examined. As observed in the rat (2), acute exposure of wild type control mice to inhaled ozone resulted in increased numbers of activated macrophages in the lungs. Of interest was the fact that these cells were primed to produce increased amounts of nitric oxide and peroxynitrite in response to inflammatory mediators. In contrast, alveolar macrophages isolated from NOSII knockout mice failed to produce nitric oxide even after ozone inhalation. As expected, this was due to an inability of these cells to produce NOSII enzyme. A significant increase in protein in BAL was also observed in wild type, but not NOSII knockout mice exposed to ozone. These findings, together with our observation of a lack of nitrotyrosine staining in histologic sections of ozone exposed NOSII knockout mice, demonstrate that these animals are resistant to ozone-induced toxicity. A similar resistance to toxicity has been observed in NOSII knockout mice treated with acetaminophen (21).

Mechanisms mediating nitric oxide toxicity are just beginning to emerge. Among its actions, nitric oxide can complex with electron rich substrates including heme and iron sulfur containing proteins (22). This can result in activation or inhibition of enzymes involved in respiration, DNA synthesis and cytotoxicity and these actions may contribute to lung tissue damage following ozone inhalation. Nitric oxide is also readily oxidized to peroxynitrite, a highly reactive free radical (10,11). Peroxynitrite can induce lipid peroxidation, oxidation of proteins and nonprotein sulfhydryls, deoxyribonucleic acid and alter DNA bases (13,14). Peroxynitrite is known to disrupt the activity of pulmonary surfactants as well as ion channel function in epithelial cells (6,23). In wild type mice, we found that alveolar macrophages produced significant quantities of peroxynitrite following ozone inhalation. This was correlated with nitrotyrosine staining *in situ* in histological sections of lung tissue. Increased superoxide anion release by alveolar macrophages was correlated with peroxynitrite production and this may account for the more prominent immunostaining of these cells in the lung when compared to other cell types.

The formation of peroxynitrite is dependent on the availability of nitric oxide and superoxide anion. In our studies, a significant increase in superoxide anion production was observed in alveolar macrophages from untreated mice lacking NOSII. The fact that these mice are protected from ozone-induced injury indicates that superoxide anion may not be a critical mediator of toxicity for this oxidant. Superoxide anion release was decreased in NOSII knockout mice 24 hr after ozone exposure. These findings are similar to those reported previously in alveolar macrophages from rats and mice immediately following ozone exposure (24). Interestingly in wild type mice, superoxide anion production increased 48 hr following ozone exposure providing further support for our model that alveolar macrophages become activated in response to ozone. The presence of excess protein and increases in the number of macrophages in BAL fluid following ozone inhalation is a hallmark of alveolar epithelial injury (1). It is generally thought that these increases are due to lysis of injured cells, transudation of proteins from the bloodstream, and/or enhanced protein secretion by cells of the respiratory tract (25). Mice lacking NOSII were protected from ozone-induced damage as shown by control levels of protein and cell number in BAL. These data firmly support our hypothesis that nitric oxide contributes to ozone toxicity.

A general pattern is emerging on the causative factors in the pathology of injury by pulmonary irritants such as ozone. Following initial tissue damage, resident and infiltrating macrophages and neutrophils release a cascade of inflammatory mediators (26). The use of transgenic mice with a targeted disruption of genes for one or more of these inflammatory mediators provide an important tool for assessing their potential role in toxicity. Our findings that NOSII knockout mice are protected from ozone-induced injury provide direct support for that model that reactive nitrogen intermediates contribute to tissue damage.

ACKNOWLEDGMENTS

Supported by NIH grants ES04738, ES06897 and the Burroughs Wellcome Fund.

REFERENCES

1. M. Lippman, Health effects of tropospheric ozone: review of recent research findings and their implications to ambient air quality standards, *J. Expos. Anal. Environ. Epidemiol.* 3: 103-129 (1993).

2. K.J. Pendino, J.D. Laskin, R.L. Shuler, C.J. Punjabi and D.L. Laskin, Enhanced production of nitric oxide by rat alveolar macrophages following inhalation of a pulmonary irritant is associated with increased expression of nitric oxide synthase, *J. Immunol.*, 151: 7196-7205 (1993).

3. K.J. Pendino, R.L. Schuler, J.D. Laskin, and D.L. Laskin, Enhanced production of interleukin-1, tumor necrosis factor-α and fibronectin by rat lung phagocytes following inhalation of a pulmonary irritant, *Am. J. Resp. Cell Molec. Biol.* 11: 279-286 (1994).

4. K.J. Pendino, T.M. Meidoff, D.E. Heck, J.D. Laskin, and D.L. Laskin, Inhibition of macrophages with gadolinum chloride abrogates ozone-induced pulmonary injury and inflammatory mediator production, *Am. J. Respir. Cell Molec. Biol.* 13: 125-132 (1995).

5. K.J. Pendino, Gardner, C.R., Shuler, R.L., Laskin, J.D., Durham, S.K., Goller, N.L., Ohnishi, S.T., Ohnishi, T., and D.L. Laskin, Inhibition of ozone-induced nitric oxide synthase expression in the lung by endotoxin, *Am. J. Respir, Cell Molec. Biol.* 14: 516-525 (1996).

6. L.Y. Haddad, H. Ischiropoulos. B.A. Holm, J.S. Beckman, J.R. Baker and S. Matalon, Mechanisms of peroxynitrite-induced injury to pulmonary surfactants, *Am. J. Physiol.* 265: L555-L564 (1993).

7. J. Beckman, P. Mehta, V. Hanks, W.H. Rowan and L. Liu, Effects of peroxynitrite on pulmonary edema and the oxidative state, *Exp Lung Res.* 26: 349-59 (2000).

8. B.M. Babior, Phagocytes and oxidative stress, *Am. J. Med.* 109: 33-44 (2000).

9. M.J.D. Griffiths, M. Messent, R.J. MacAllister and T.W. Evans, Aminoguanidine selectively inhibits inducible nitric oxide synthase, *Br. J. Pharmacol.* 110: 963-968 (1993).

10. B. Freeman, Free radical chemistry of nitric oxide. Looking at the dark side, *Chest* 105: 79S-84S (1994).

11. J. Beckman and J.P. Crow, Pathological implications of nitric oxide, superoxide and peroxynitrite formation, *Biochem. Soc. Trans.* 21: 330-334 (1993).

12. C.J. Punjabi, K.J. Pendino, J.D. Laskin and D.L. Laskin, Production of nitric oxide by rat type II pneumocytes. Increased Expression of inducible nitric oxide synthase following exposure to a pulmonary irritant, *Am. J. Resp. Cell Mol. Biol.* 11: 165-172 (1994).

13. R. Radi, J.S. Beckman, K.M. Bush and B.A. Freeman, Peroxynitrite-induced membrane lipid peroxidation: the cytotoxic potential of superoxide and nitric oxide, *Arch. Biochem. Biophys.* 288: 481-487 (1991).

14. R. Radi, J.S. Beckman, K.M. Busch and B.A. Freeman, Peroxynitrite oxidation of sulfhydryls. The cytotoxic potential of superoxide and nitric oxide, Arch. *Biochem. Biophys.* 266: 4244-4250 (1991).

15. H. Ischiropoulos, A. Gow, S.R. Thom, N.W. Kooy, J.A. Royall and J .P. Crow, Detection of reactive nitrogen species using 2,7-dichlorodihydrofluorescein and dihydrorhodamine 123, *Methods Enzymol.* 301: 367-373 (1999).

16. H. Ischiropoulos, Biological tyrosine nitration: a pathophysiological function of nitric oxide and reactive oxygen species, *Arch Biochem. Biophys.* 356: 1-11 (1998).

17. J.P. Crow and H. Ischiropoulos, Detection and quantitation of nitrotyrosine residues in proteins: in vivo marker of peroxynitrite, *Methods Enzymol.* 269: 185-94 (1996).

18. C.F. Nathan, Nitric oxide as a secretory product of mammalian cells, *FASEB J.* 6: 3051-3064 (1992).

19. C.F. Nathan and J.B. Hibbs, Role of nitric oxide synthesis in macrophage antimicrobial activity, *Curr. Opin. Immunol.* 3: 65-70 (1991).

20. C.R. Gardner and D.L. Laskin, Protective and pathologic roles of nitric oxide in tissue injury, in: *Cellular and Molecular Biology of Nitric Oxide*, J.D. Laskin, and D.L. Laskin, eds., Marcel Dekker, N.Y., 225-246 (1999).

21. C.R. Gardner, D.E. Heck, H. Chiu, J.D. Laskin, S.K. Durham, and D.L. Laskin, Decreased hepatotoxicity of acetaminophen in mice lacking inducible nitric oxide synthase, in: *Cells of the Hepatic Sinusoid, Vol. 7*, E. Wisse, D. L. Knook, R. DeZanger, and R. Fraser, eds., Kupffer Cell Foundation, Leiden, 104-105 (1999).

22. S.S. Gross and M.S. Wolin, Nitric oxide: pathological mechanisms, *Annu. Rev. Physiol.* 57: 737-769 (1995).

23. S. Matalon and H. O'Brodovich, Sodium channels in alveolar epithelial cells: molecular characterization, biophysical properties, and physiological significance, *Annu. Rev. Physiol.* 61: 627-661 (1999)

24. J.E. Ryer-Powder, M.A. Amoruso, B. Czerniecki, G. Witz, and B.D Goldstein, Inhalation of ozone produces a decrease in superoxide anion radical production in mouse alveolar macrophages, *Am. Rev. Respir. Dis.* 138: 1129-1133 (1988).

25. G.W. Hunninghake, J.E. Gadek, O. Kawanami, V.J. Ferrans, and R.G. Crystal, Inflammatory and immune processes in the human lung in health and disease: evaluation by bronchoalveolar lavage, *Am. J. Pathol.* 97: 149-206 (1979).

26. D.L. Laskin and J.D. Laskin, J.D., Phagocytes, in: *Comprehensive Toxicology, Vol. 5, Toxicology of the Immune System*, D.A. Lawrence, ed., Pergamon, N.Y., 97-112 (1997).

27. A. Sterner-Kock, M. Kock , R. Braun and D.M. Hyde, Ozone-induced epithelial injury in the ferret is similar to nonhuman primates, *Am. J. Respir. Crit. Care Med.* 162: 1152-1156 (2000).

28. J.S. Fedan, L.L. Millecchia, R.A. Johnston, A. Rengasamy, A. Hubbs, R.D. Dey, L.X. Yuan, D. Watson, W.T. Goldsmith, J.S. Reynolds, L. Orsini, J. Dortch-Carnes, D. Cutler and D.G. Frazer, Effect of ozone treatment on airway reactivity and epithelium-derived relaxing factor in guinea pigs, *J. Pharmacol. Exp. Ther.* 293: 724-734 (2000).

29. J. T. Zelikoff, G. L. Kraemer, M.C. Vogel, and R.B. Schlessinger, Immunomodulating effects of ozone on macrophage functions important for tumor surveillance and host defense, *J. Toxicol. Environ. Health*, 34: 449-467 (1991).

30. K.E. Driscoll, T.A. Vollmuth, and R.B. Schlesinger, Acute and subchronic ozone inhalation in the rabbit: response of alveolar macrophages, *J. Toxicol. Environ. Health* 21: 27-43 (1987).

31. D.J. Bassett, C. Elbon-Copp, Y. Ishii, H. Barraclough-Mitchell and H.J. Yang, Lung tissue neutrophil content as a determinant of ozone-induced injury, *J. Toxicol. Environ. Health* 60: 513-530 (2000).

32. R.E. Gordon, E. Park, D. Laskin, G.B. Schuller-Levis, Taurine protects rat bronchioles from acute ozone exposure: a freeze fracture and electron microscopic study, *Exp. Lung Res.* 24: 659-674 (1998).

33. G.B. Shuller-Levis, R.E. Gordon, E. Park, K.J. Pendino, and D.L. Laskin, Taurine protects rat bronchioles from acute ozone-induced lung inflammation and hyperplasia, *Exp. Lung Res.* 21: 877-888 (1995).

34. S.R. Kleeberger, D.J. Bassett, G.J. Jakab, and R.C. Levitt, A genetic model for evaluation of susceptibility to ozone-induced inflammation. *Am. J. Physiol.* 258: L313-320 (1990).

35. C.J. Johnston, B.R. Stripp, S.D. Reynolds, N.E. Avissar, C.K. Reed, and Finkelstein, J.N., Inflammatory and antioxidant gene expression in C57BL/6J mice after lethal and sublethal ozone exposures, *Exp. Lung Res.* 25: 81-97 (1999).

36. Q. Zhao, L.G. Simpson, K.E. Driscoll, G.D. Leikauf, Chemokine regulation of ozone-induced neutrophil and monoocyte inflammation, *Am. J. Physiol.* 274: L39-46 (1998).

37. H.Y. Cho, J.A. Hotchkiss, C.B. Bennett,and J.R. Harkema, Neutrophil-dependent and neutrophil-independent alterations in the nasal epithelium of ozone-exposed rats, *Am. J. Respir. Crit. Care Med.* 162: 629-636 (2000).

38. R.A. Jorres, O. Holz, W. Zachgo, P. Timm, S. Koschyk, B. Muller, F. Grimminger, W. Seeger, F.J. Kelly, C. Dunster, T. Frischer, G. Lubec, M. Waschewski, A. Niendorf,and H. Magnussen, The effect of repeated ozone exposures on inflammatory markers in bronchoalveolar lavage fluid and mucosal biopsies, *Am. J. Respir. Crit. Care Med.* 161: 1855-1861 (2000).

39. R.B. Devlin, W.F. McDonnell, S. Becker, M.C. Madden, M.P. McGee, R. Perez, G. Hatch, D.E. House, and H.S. Koren, Time-dependent changes of inflammatory mediators in the lungs of humans exposed to 0.4 ppm ozone for 2 hr: a comparison of mediators found in bronchoalveolar lavage fluid 1 and 18 hr after exposure, *Toxicol. Appl. Pharmacol.* 138: 176-185 (1996).

ANTIOXIDANT REACTIONS OF GREEN TEA CATECHINS AND SOY ISOFLAVONES

Daniel C. Liebler[1], Susanne Valcic[1], Arti Arora[1], Jeanne A. Burr[1], Santiago Cornejo[1], Muralee G. Nair[2] and Barbara N. Timmermann[1]

[1]Southwest Environmental Health Sciences Center
College of Pharmacy
The University of Arizona
Tucson, AZ 85721-0207

[2]Center for Food Safety and Toxicology
Michigan State University
East Lansing, MI 48824

INTRODUCTION

Green tea, a popular beverage brewed from dried leaves of the tea bush (*Camellia sinensis*), is distinguished by the presence of a group of polyphenols called flavanols or catechins (Graham, 1992). In recent years, the principal green tea catechins, i.e. (-)-epicatechin, (-)-epicatechin-3-gallate, (-)-epigallocatechin (EGC), and (-)-epigallocatechin-3-gallate (EGCG) have been recognized to be effective protectants against certain forms of cancer (Katiyar et al., 1992; Yang et al., 1993). These protective effects often have been attributed to antioxidant actions (Jovanovic et al., 1994; Terao et al., 1994; Salah et al., 1995; Kondo et al., 1999). The potential role of soy products in cancer prevention also has received much attention in recent years. Epidemiological data indicate that consumption of soybean-containing diets is associated with a lower incidence of certain human cancers in Asian compared to Caucasian populations (King et al., 1980; Locke et al., 1980). Genistein is one of the two principal isoflavones found in soy (Murphy, 1982) and, like catechins, is thought to act in large part through its ability to scavenge oxidants involved in carcinogenesis (Wei et al., 1993).

Previous structure-activity studies indicated that the presence of a galloyl ring in 3-position and a trihydroxyphenyl B-ring are most important for the antioxidant activities of catechins (Salah et al., 1995; Nanjo et al., 1996; Guo et al., 1999). Previous structure-activity studies with isoflavones suggest that hydroxyl groups at positions C-5 and C-7 on the A-ring (De Whalley et al., 1990) and C-4' of the B-ring (Chen et al., 1996; Arora et al., 1998) contribute to inhibition of lipid peroxidation. Genistein has a C-2,3 double bond in conjugation with a 4-oxo function in the C-ring, which together can participate in electron delocalization from the B-ring (Bors et al., 1990).

Despite previous structure-activity studies, the mechanisms of peroxyl radical scavenging by catechins and isoflavones remains poorly understood in mechanistic terms.

Our laboratory has studied the mechanisms of antioxidant reactions by vitamin E and carotenoids by identifying products of peroxyl radical scavenging by these compounds in biomimetic model systems. Peroxyl radicals are the principal species scavenged by chain-breaking antioxidants in biological systems. Identification of specific products of peroxyl radical scavenging reactions provides markers for subsequent studies of antioxidant action *in vivo*. Here we describe recent work on the identification of products of peroxyl radical scavenging by the green tea catechins EGCG and EGC and the soy isoflavone genistein.

PEROXYL RADICAL SCAVENGING BY GREEN TEA CATECHINS

AMVN-Initiated Peroxyl Radical Oxidations

Previous work from our laboratory and others has employed the azo initiator AMVN as a source of peroxyl radicals for studies of antioxidant chemistry. AMVN decomposes thermally to yield alkyl radicals (R$^\bullet$) which, in turn, react with molecular oxygen to generate peroxyl radicals (ROO$^\bullet$) (eqs. 1 and 2) :

$$R\text{-}N\text{=}N\text{-}R \quad \rightarrow \quad 2R^\bullet \quad + \quad N_2 \qquad (1)$$
$$R^\bullet \quad + \quad O_2 \quad \rightarrow \quad ROO^\bullet \qquad (2)$$

Advantages of using AMVN for these studies include the selectivity with which AMVN yields peroxyl radicals, the ease of controlling oxidation rates (via the AMVN concentration and temperature) and the lack of transition metals used in other oxidant generating systems, which often degrade primary oxidation products. Moreover, products of other antioxidants formed by AMVN-derived peroxyl radicals are structurally analogous to those formed by by peroxyl radicals in biological systems in vitro and in vivo (Liebler et al., 1991; Liebler et al., 1996; Ham et al., 1997).

Oxidation Products of EGC and EGCG

EGCG and EGC (**1** and **5**, respectively; Figure 1) were isolated from green tea (Valcic et al., 1996) and oxidized with peroxyl radicals generated by thermolysis of AMVN in oxygenated acetonitrile (Valcic et al., 1999). The products of catechin oxidations were purified by medium pressure liquid chromatography (MPLC) and HPLC and characterized by MS, UV-vis, CD spectroscopy, and NMR spectroscopy. Details of the characterization have been published recently (Valcic et al., 1999; Valcic et al., 2000).

Major products of EGCG oxidation included a dicarboxylic acid product (**2**), in which the B-ring underwent oxidative ring cleavage, a seven membered B-ring anhydride (**3**), and a symetric dimer, in which two EGCG molecules are joined through their oxidized B-rings (**4**). Major products of EGC oxidation included a seven membered B-ring anydride (**8**) and a symmetrical dimer (**6**), both analogous to those formed from EGCG. In addition, EGCG oxidation also yielded a second asymmetric dimer (**7**), in which two EGC molecules are joined through oxidized B-rings.

Formation of catechin oxidation products occurred with incorporation of oxygen into the products. To investigate the source of the oxygen atoms incorporated into the products, oxidations were done in acetonitrile/water mixtures containing either $H_2{}^{16}O$ or $H_2{}^{18}O$ and the products were analyzed by mass spectrometry. Incorporation of ^{18}O into products is manifested by a 2 amu increase in product molecular mass per oxygen atom incorporated and can be measured by MS. MS analyses of all of the EGCG and EGC oxidation products from incubations in $H_2{}^{18}O$ revealed no incorporation of ^{18}O label, thus indicating that the source of oxygen in these oxidation products was not the solvent, but atmospheric oxygen instead.

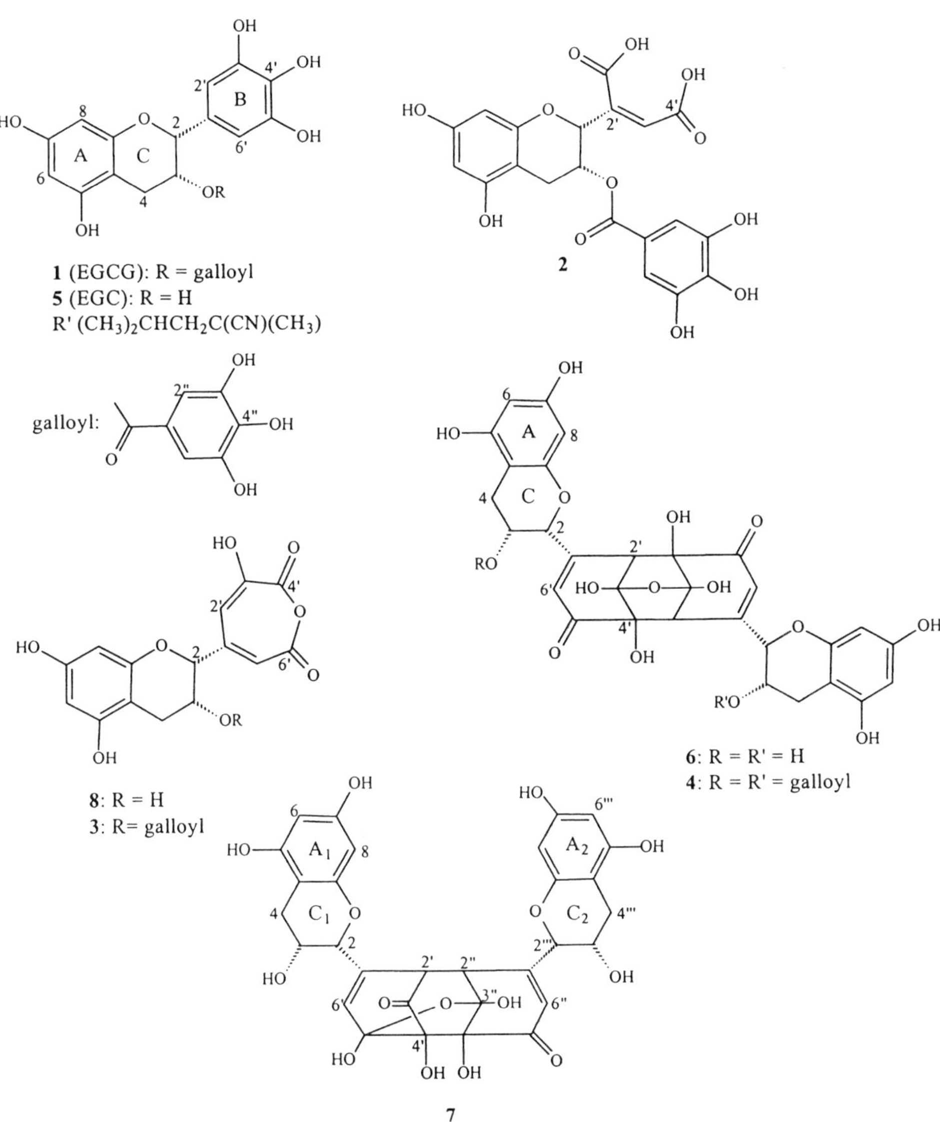

Figure 1. Structures of catechins EGCG (1), EGC (5) and their oxidation products.

Mechanisms of Catechin Antioxidant Reactions

Our previous identification of EGCG and EGC products provides unambiguous evidence that antioxidant reactions of these catechins with peroxyl radicals involve the trihydroxyphenyl B-ring, rather than the 3-galloyl moiety or the A-ring. Indeed, we have not yet detected any products resulting from oxidation of the galloyl moiety. In oxidations of EGC with peroxyl radicals, a third product was identified as an unsymmetrical dimer (7), again resulting from oxidation on the trihydroxyphenyl B-ring. We did not observe the EGCG analog of EGC product 7. This may reflect steric hindrance caused by the two galloyl rings in 4.

These studies provide new insights into the mechanisms by which catechins scavenge peroxyl radicals. First, phenoxyl radicals derived from EGCG and EGC apparently do not form stable radical addition products with AMVN-derived peroxyl radicals, as do those from simple phenolic antioxidants (Horswill et al., 1966a; Horswill et al., 1966b), α-tocopherol (Liebler et al., 1990; Yamauchi et al., 1990; Yamauchi et al., 1994) and genistein (Arora et al., 2000). Second, oxygen addition that occurs during catechin oxidation is not incorporated from water in the reaction medium, but instead most probably from atmospheric oxygen via peroxyl radicals. Although atmospheric oxygen could also be incorporated directly by oxygen addition, this appears less likely because the generation of catechin-derived peroxyl radical intermediates would result in a net peroxyl radical consumption of zero. Third, all of the oxidation products we have characterized contain oxidative modifications exclusively on the B-ring.

Based on the results of our experiments, we have proposed mechanisms for the formation of catechin oxidation products (Valcic et al., 2000). Formation of the observed products may be rationalized from competing reactions of a B-ring phenoxyl radical derived from initial one-electron oxidation of the catechins. Reaction of the phenoxyl radical with a second peroxyl radicals followed by rearrangement yields the anhydride products 3 and 8. Radical addition of the phenoxyl instead to an adjacent catechin, followed by trapping of another peroxyl radical and rearrangement yields the dimers 4, 6 and 7.

The data suggest that multiple product-forming reactions contribute to peroxyl radical scavenging by catechins. However, it appears most likely that B-ring oxidation products of EGCG and EGC would probably predominate under conditions of partial oxidation likely to occur when these compounds act as antioxidants biologically. These B-ring oxidation products are therefore the most likely potential markers for antioxidant reactions of EGCG and EGC in biological systems. Analyses of these products could thus provide a unique tool for assessing the contribution of antioxidant reactions to the disease-preventive effects of catechins.

PEROXYL RADICAL SCAVENGING BY GENISTEIN

Products of Peroxyl Radical Scavenging by Genistein

Genistein was oxidized with peroxyl radicals generated by thermolysis of AMVN in oxygenated acetonitrile. These oxidations were conducted using a 10:1 molar ratio of AMVN to genistein, which resulted in a low rate of genistein oxidation. This ensured that genistein was the major component present in the reaction mixture at all times, thereby suppressing the generation of secondary oxidation products. Analysis of the reaction mixture by reversed-phase HPLC indicated presence of residual genistein 8 and the formation of five oxidation products 9-13 (Figure 2). The products were purified by MPLC and HPLC and characterized by MS, UV-vis, CD spectroscopy, and NMR spectroscopy. Details of the characterization have been published recently (Arora et al., 2000).

A major product of genistein oxidation was identified as the B-ring hydroxylation product orobol (9). In addition to orobol, a family of related oxidation products (10-13) were identified as adducts formed between genistein and AMVN-derived radicals. In all of these products, the AMVN-derived radical underwent addition at C-4 of the isoflavone B-ring. The most prominent of these was product **12**, in which the AMVN-derived fragment is linked to the B-ring by a two oxygen bridge. Product **13**, a minor product also contained this peroxyl linkage. In two other products (**10** and **11**), the linkage to the B-ring involved a single oxygen. An unusual feature of products **10-13** is the appearance

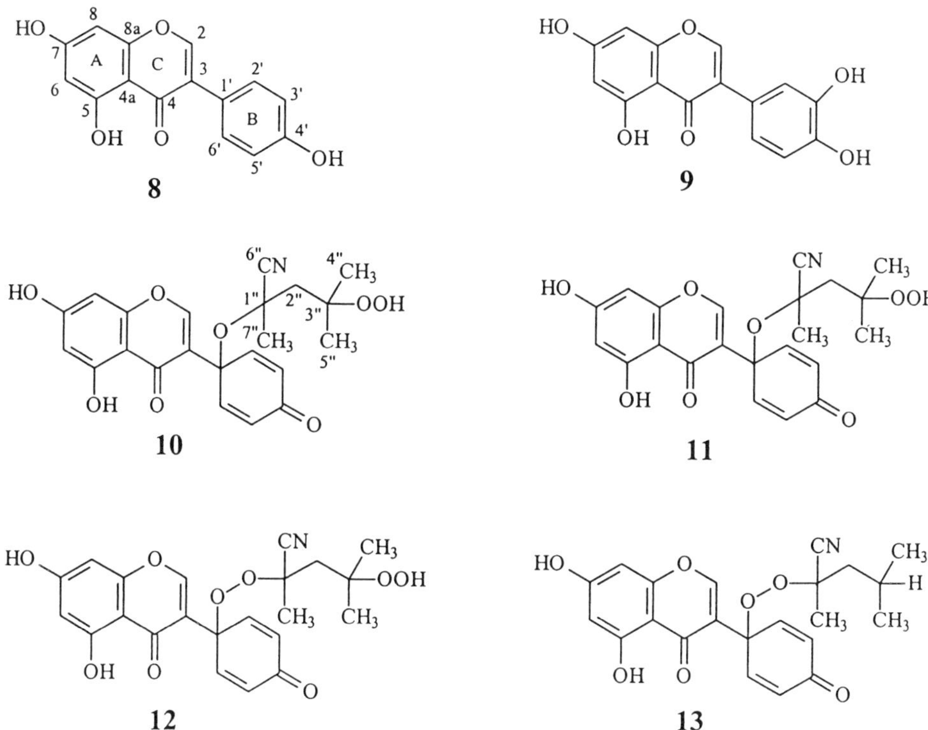

Figure 2. Structures of genistein (8) and its oxidation products.

of a hydroperoxyde group at C-3" of the AMVN-derived carbon chain. This has not been reported previously in oxidations involving AMVN and is though to reflect secondary attack of radicals on the AMVN-derived chain following the formation of the genistein-radical adducts (Arora et al., 2000).

As with the catechin oxidations, the role of solvent H_2O versus atmospheric oxygen as the source of oxygen in the products was investigated by analyzing products formed in incubations with $H_2^{18}O$. MS analysis of the products indicated that none of the genistein oxidation products incorporated ^{18}O in these oxidations. Thus, atmospheric oxygen, presumably introduced via peroxyl radicals is the most likely source of oxygen in the products.

Mechanisms of Genistein Antioxidant Reactions

The mechanisms of peroxyl radical scavenging reactions leading to the obsrved genistein oxidation products have been proposed recently (Arora et al., 2000). All of the oxidations can be rationalized on the basis of initial formation of a genistein B-ring phenoxyl radical following reaction of genistein with a peroxyl radical. Subsequent reactions of this phenoxyl may proceed according to two pathways. First, the phenoxyl radical may undergo addition of a second peroxyl radical at C-3' (i.e., *ortho* to the phenoxyl oxygen) to form an unstable adduct that rearranges to yield an elimination fragment and orobol **9**. In the second pathway, addition of a peroxyl or alkoxyl radical at C-4' (i.e., *para* to the phenoxyl oxygen) yields a stable adduct. In a secondary reaction, the radical adducts may undergo hydrogen abstraction at C-3" of the AMVN-derived carbon chain to form a carbon-centered radical that adds oxygen and hydrogen to form a hydroperoxide. This secondary oxidation has no precedent in similar studies of AMVN-initiated oxidations. In these studies, this unusual oxidation may be facilitated by non-covalent association of genistein molecules, and/or initial reaction products (such as **13**) with genistein-derived phenoxyl radicals in a juxtaposition that facilitates the secondary oxidation. Further work will be required to sort out the chemistry involved.

The characterization of these product structures provides potential markers of genistein antioxidant reactions. Sensitive assays for specific genistein oxidation products may thus be useful markers for antioxidant reactions of the isoflavone in biological systems. Currently, we are investigating whether reactions of genistein and peroxyl radicals in lipid bilayers *in vitro* and in biological membranes *in vitro* and *in vivo* yield products similar to those observed in this study in a homogeneous solution system.

CONCLUSION

The principal site of catechin and isoflavone peroxyl radical scavenging reactions is the flavonoid B-ring. What is particularly interesting about the results of these product studies with flavonoids is the variety of antioxidant chemistries displayed by the B-ring phenols. Genistein produced products that are relatively typical of well-documented phenol antioxidant reactions--B-ring hydroxylation and formation of radical addition products. In contrast, EGC and EGCG yield neither B-ring hydroxylation nor stable peroxyl radical addition products. Instead, one-electron oxidations appear to initiate radical polymerization reactions and transient radical-addition products that decompose with B-ring expansion. Thus, antioxidant chemistry of simpler phenols does not necessarily predict the course of flavonoid phenol antioxidant reactions. B-ring oxidation products observed in these model studies are logical candidates for evaluation as chemical markers for flavonoid antioxidant reactions in living systems.

ACKNOWLEDGEMENTS

This work was supported in part by NIH Grants CA75599 and ES06694.

REFERENCES

Arora, A., Nair, M. G., and Strasburg, G. M., 1998, Structure-activity relationships for antioxidant activities of a series of flavonoids in a liposomal system. *Free Radic. Biol. Med.* 24: 1355.

Arora, A., Valcic, S., Cornejo, S., Nair, M. G., Timmermann, B. N., and Liebler, D. C., 2000, Reactions of genistein with alkylperoxyl radicals. *Chem. Res. Toxicol.* 13: 638.

Bors, W., Heller, W., Michel, C., and Saran, M., 1990, Flavonoids as antioxidants: Determination of radical-scavenging activities. *Meth. Enzymol.* 186: 343.

Chen, Z. Y., Chan, P. T., Ho, K. Y., Fung, K. P., and Wang, J., 1996, Antioxidant activity of natural flavonoids is governed by number and location of their aromatic hydroxyl groups. *Chem. Phys. Lipids* 79: 157.

De Whalley, C. V., Rankin, S. M., Henlt, J. R. S., Jessup, W., and Leake, D. S., 1990, Flavonoids inhibit the oxidative modification of low-density lipoproteins by macrophages. *Biochem. Pharmacol.* 39: 1743.

Graham, H. N., 1992, Green tea composition, consumption, and polyphenol chemistry. *Prev. Med.* 21: 334.

Guo, Q., Zhao, B., Shen, S., Hou, J., Hu, J., and Xin, W., 1999, ESR study on the structure-antioxidant activity relationship of tea catechins and their epimers. *Biochim. Biophys. Acta* 1427: 13.

Ham, A. J. L., and Liebler, D. C., 1997, Antioxidant reactions of vitamin E in the perfused rat liver: product distribution and effect of dietary vitamin E supplementation. *Arch. Biochem. Biophys.* 339: 157.

Horswill, E. C., Howard, J. A., and Ingold, K. U., 1966b, The oxidation of phenols. III. The stoichiometries for the oxidation of some substituted phenols with peroxy radicals. *Can. J. Chem.* 44: 985.

Horswill, E. C., and Ingold, K. U., 1966a, The oxidation of phenols. I. The oxidation of 2,6-di-t-butyl-4-methylphenol, 2,6-di-t-butylphenol, and 2,6-dimethylphenol with peroxy radicals. *Can. J. Chem.* 44: 263.

Jovanovic, S. V., Steenken, S., Tosic, M., Marjanovic, B., and Simic, M. G., 1994, Flavonoids as antioxidants. *J. Am. Chem. Soc.* 116: 4846.

Katiyar, S. K., Agarwal, R., and Mukhtar, H., 1992, Green tea in chemoprevention of cancer. *Compr. Ther.* 18: 3.

King, H., and Locke, F. B., 1980, Cancer mortality among Chinese in the United States. *J. Natl. Cancer Inst.* 65: 1141.

Kondo, K., Kurihara, M., Miyata, N., Suzuki, T., and Toyoda, M., 1999, Scavenging mechanisms of (-)-epigallocatechin gallate and (-)-epicatechin gallate on peroxyl radicals and formation of superoxide during the inhibitory action. *Free Radic. Biol. Med.* 27: 855.

Liebler, D. C., Baker, P. F., and Kaysen, K. L., 1990, Oxidation of vitamin E: Evidence for competing autoxidation and peroxyl radical trapping reactions of the tocopheroxyl radical. *J. Am. Chem. Soc.* 112: 6995.

Liebler, D. C., Burr, J. A., Philips, L., and Ham, A. J. L., 1996, Gas chromatography-mass spectrometry analysis of vitamin E and its oxidation products. *Anal. Biochem.* 236: 27.

Liebler, D. C., Kaysen, K. L., and Burr, J. A., 1991, Peroxyl radical trapping and autoxidation reactions of alpha-tocopherol in lipid bilayers. *Chem. Res. Toxicol.* 4: 89.

Locke, F. B., and King, H., 1980, Cancer mortality risk among Japanese in the United States. *J. Natl. Cancer Inst.* 65: 1149.

Murphy, P. A., 1982, Phytoestrogen content of processed soybean products. *Food Technol.* 36: 60.

Nanjo, F., Goto, K., Seto, R., Suzuki, M., Sakai, M., and Hara, Y., 1996, Scavenging effects of tea catechins and their derivatives on 1,1-diphenylpicrylhydrazyl radical. *Free Radic. Biol. Med.* 21: 895.

Salah, N., Miller, N. J., Paganga, G., Tijburg, L., Bolwell, G. P., and Rice-Evans, C., 1995, Polyphenolic flavonols as scavengers of aqueous phase radicals and as chain-breaking antioxidants. *Arch. Biochem. Biophys.* 322: 339.

Terao, J., Piskula, M., and Yao, Q., 1994, Protective effect of epicatechin, epicatechin gallate, and quercetin on lipid peroxidation in phospholipid bilayers. *Arch. Biochem. Biophys.* 308: 278.

Valcic, S., Burr, J. A., Timmermann, B. N., and Liebler, D. C., 2000, Antioxidant chemistry of green tea catechins. New oxidation products of (-)-epigallocatechin gallate and (-)-epigallocatechin from their reactions with peroxyl radicals. *Chem. Res. Toxicol.* in press.

Valcic, S., Muders, A., Jacobsen, N. E., Liebler, D. C., and Timmermann, B. N., 1999, Antioxidant chemistry of green tea catechins. Identification of products of the reaction of (-)-epigallocatechin gallate with peroxyl radicals. *Chem. Res. Toxicol.* 12: 382.

Valcic, S., Timmermann, B. N., Alberts, D. S., Wachter, G. A., Krutzch, M., Wymer, J., and Guillen, J., 1996, Inhibitory effect of six green tea catechins and caffeine on the growth of four selected human tumor cell lines. *Anticancer Drugs* 7: 461.

Wei, H., Wei, L., Frenkel, K., Bowen, R., and Barnes, S., 1993, Inhibition of tumor promoter-induced hydrogen perxoide formation *in vitro* and *in vivo* by genistein. *Nutr. Cancer* 20: 1.

Yamauchi, R., Matsui, T., Kato, K., and Ueno, Y., 1990, Reaction products of α-tocopherol with methyl linoleate-peroxyl radicals. *Lipids* 25: 152.

Yamauchi, R., Yagi, Y., and Kato, K., 1994, Isolation and characterization of addition products of α-tocopherol with peroxyl radicals of dilinoleoylphosphatidylcholine in liposomes. *Biochim. Biophys. Acta* 1212: 43.

Yang, C. S., and Wang, Z. Y., 1993, Tea and cancer. *J. Natl. Cancer Inst.* 85: 1038.

REACTIVE OXYGEN SPECIES ANALYSIS IN GASTRITIS PATIENTS AND P53 METHYLATION ANALYSIS IN GASTRIC TUMOR CELL LINE AGS INFECTED BY HELICOBACTER PYLORI

Jochen Rudi,[1] Benedikt Bruchhausen,[1] Dirk Kuck,[1] Wolfgang Stremmel,[1] Axel von Herbay,[2] Heinrich Bauer,[3] Martin Berger,[4] Robert W. Owen,[5]

[1] Dept. of Medicine, University of Heidelberg; [2] Dept. of Pathology; [3] Dept. of Molecular Toxicology; [4] Dept. of Toxicology and Chemotherapy and [5] Division of Toxicology and Cancer Risk Factors; German Cancer Research Center, Heidelberg, Germany

Introduction

Epidemiological studies in humans and animal models support the idea that *Helicobacter pylori*, a gram-negative spiral bacterium, first described by Warren and Marshall in 1983, is one of the risk factors for the induction of gastric cancer. *H. pylori* infection is the cause of chronic gastritis, peptic ulcer disease, gastric mucosa associated lymphoid tissue lymphoma and cancer of the distal stomach (1). Virulent or less virulent *H. pylori* strains can be determined by the presence or absence of *cag* pathogenicity island (PAI) genes including the *cagA* gene and *vacA* genotypes. The expression of different genes of the *cag* PAI (total of 26 genes) seems to be essential for virulence. CagA and VacA proteins are highly antigenic (2). *H. pylori* strains expressing CagA show increased synthesis of chemokines, promote neutrophil infiltration in the gastric epithelium and stimulate synthesis of interleukin-8 (3). In the repetitive regions of the *cag* 7 gene within the cag PAI of *H. pylori*, deletions and insertions are detected representing a target for mutations (4). In addition, Covacci *et al.* (5) have suggested that new different quasi-species of *H. pylori* strains which possess or loose their cag PAI are generated during chronic infection by *H. pylori*.

The molecular mechanisms of carcinogenic effects of *H. pylori* in humans are still unknown (6). Expression of bcl2 and p53 genes in gastric lymphomas is associated with transformation from low to high grade lymphoma (7). High grade transformation of gastric mucosa associated tissue (MALT) resulted in p53 inactivation as shown by Isaacson (10). AGS gastric cancer cells show G1-S-phase inhibition by *H. pylori* infections. Nardone *et al.* could not detect differences in p53 or p21 expression in gastric tissue with respect to controls but found a reduction in p27 protein (8). Early gastric carcinomas might be independent of a *H. pylori* pathway. *H. pylori* positive and *H. pylori* negative patients showed no differences in ras, mdm2, c-erb-B-2, cyclin d1, e-cadherin and p53 expression (9). Deletions of the p16 gene were also detected after *H. pylori* infection. Human gastric mucosa biopsies were analyzed by Murakami *et al.* for p53 mutations in exons 5, 6, 7, and 8 (11). Point mutations could be detected in non-hot spot codons in exon 7 and exon 8. Mutations of p53 could not be detected in *H. pylori* negative patients. Chung *et al.* (12) performed experiments with AGS (wt p53), YCC 1 (mt p53), and YCC 3 (wt p53) cells. It could be shown that AGS cells were more invasive than YCC 1 and YCC 3 cells, but YCC 1 and YCC 3 cells were more tumorigenic. The role of apoptosis in the pathogenesi of *H. pylori* associated diseases is currently under discussion. The CD95 (APO-1/Fas) receptor and ligand system and its relevance in induction of apoptosis has been evaluated in gastric epithelial cell after *H. pylori* infection (14). The findings that a cytotoxic isolate of *H. pylori* (*H. p.* 60190; *vacA* s1a/m1, *cagA+*) but not a noncytotoxic isolate of *H. pylori* (Tx30a, *vacA* s2/m2, *cagA+*) induced CD95 upregulation in about 50% of gastric epithelial cell lines suggesting that *H. pylori* associated chronic gastritis involves apoptosis by upregulation of this system (14).

The following questions were addressed with regard to *H. pylori* infection. a) Does *H. pylori* infection increase the production of reactive oxygen species (ROS) in gastric juice, and b) is there a different methylation pattern in CpG islands of p53 in the gastric tumor cell line AGS.

Material and Methods

Culture conditions and DNA isolation

The gastric epithelial AGS cells were incubated in Ham´s culture medium F12 (500 ml). Fetal calf serum (50 ml), penicillin/streptomycin (5 ml), and glutamine (5 ml) were added. Cells were grown at 37°C. For the following isolation procedure, cells were trypsinated. DNA isolation was performed according to the Midi prep column Quiagen protocol.

H. pylori incubation procedure

AGS cells (10^7) were incubated in 12-well plates with the *H. pylori* strain 60190 (*vacA*-genotype s1a/m1, *cagA* positive) at a concentration of 10^9/ml respectively, with the supernatant of culture media of *H. pylori* (10^9/ml). The cells were incubated with 10 µl of *H. pylori* suspension or with 400ul- 500ul of the supernatant of the *H.pylori* medium for 48h-72h at 37°C.

Methylation analysis of p53 CpG islands

1 µl of Msp1-methylinsensitive- (10^4 U/ml) and 1.5 ul of HpaII –methylsensitive- (10^4 U/ml) restriction enzymes (New England Biolabs, Frankfurt) were used for DNA digestion (24h, 37°C) according to the manufacturer`s instructions. PCR amplification of a total of 14 restriction sites was performed (Perkin Elmer PCR Kit) using 0.5 ul (5U/ml) Taq DNA polymerase for each primer pair (10µmol) . MBI-and Boehringer-PCR Kits were also used. 14 CpG sites in intron 1, intron 7, exon 5, exon 6, exon 7, and exon 8 of p53 were analyzed. The separation of fragments was performed by gel-electrophoresis (2% agarose for fragments larger than 100bp, 3% agarose for fragments smaller than 100bp).

Patients and biopsy sampling

68 patients were entered into the study. During upper GI endoscopy biopsies (corpus and antrum) were taken for histology and *H. pylori* determination. Thirty-one patients were diagnosed with chronic *H. pylori* gastritis, while 37 were non-infected and had normal gastric mucosa histology. Gastric juice was taken to determine their ability to generate reactive oxygen species.

ROS determination

Generation of reactive oxygen species in gastric juice

The assays were conducted in phosphate buffer (0.1 M) containing, EDTA (500 µM), $FeCl_3.6H_2O$ (50 µM with respect to elemental iron) and salicylic acid (2 mM) suspended in millipore grade water, final pH 6.5. Gastric juice (500 µL) from each patient was freeze-dried and suspended in the assay buffer in duplicate Erlenmeyer flasks (100 ml) containing 10 ml of media and were incubated at 37°C in an orbital shaker (Grant Instuments, Cambridge, England) at 200 strokes/min. for 18 hours. After incubation the cultures (20 µl) were, following centifugation and filtration injected directly into the HPLC or else acidified with 50 µl concentrated HCl, extracted with 2 x 10 ml diethyl ether, centrifuged (5000 r.p.m. for 30 min.) and the solvent removed by aspiration. Ethereal extracts were dried over anhydrous $MgSO_4$, the solvent removed under a stream of N_2 and resuspended in 2.0 ml mobile phase. After filtration the extracts (20 µl) were analysed by HPLC.

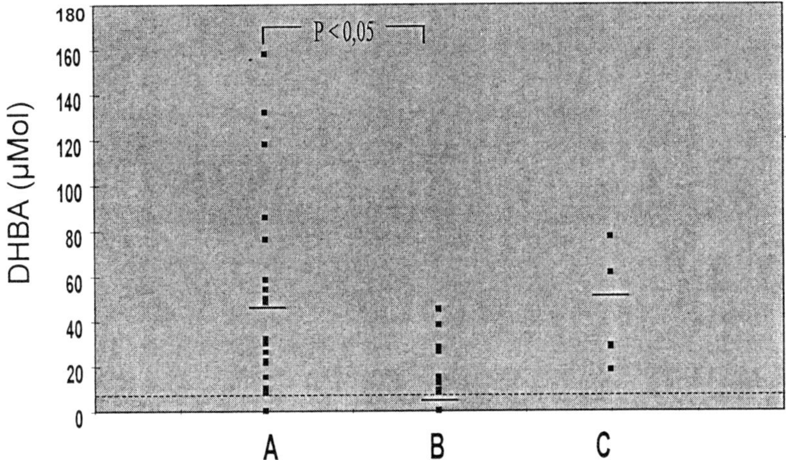

Figure 1. Dihydroxy benzoic acid in the various patient groups. (A) Gastric juice from patients with *H. pylori*-induced chronic gastritis. (B) Gastric juice from patients with normal histology (*H. pylori* negative). (C) Patients with type-C gastritis.

Analytical HPLC was conducted on a Hewlett-Packard 1090 liquid chromatograph fitted with a C-18 (Latex, Eppelheim, Germany), reverse-phase 25 cm, (5 µm) column (internal diameter, 4.0 mm). Detection of the products of reactive oxygen species attack on salicylic acid, namely 2,5-dihydroxy benzoic acid and 2,3-dihydroxy benzoic acid was as described by Owen *et al.*(15)

Results

Methylation of cytosines at CpG sites was detected in intron 1, exon 5, exon 7, and exon 8. Incubation of AGS cells with supernatant of *H. pylori* culture medium or *H. pylori* bacteria directly did not result in a higher methylation status of p53. Demethylation could only be detected at intron 1 of p53. In gastric juice samples of patients with chronic *H. pylori* gastritis or type C gastritis, 20 (64.5%) of 31 patients showed the ability to generate elevated ROS levels, but only 8 (21.6%) of 37 patients without *H. pylori* infection showed this ability (see figure 1),

Discussion

It seems to be likely that *H. pylori* infection of humans is associated with elevated levels of reactive oxygen species. Elevated methylation degrees of CpG sites of p53 were not detected, but one has to consider that we have only measured methylation of intron 1, intron 7, exon 1, exon 5, exon 6, exon 7, and exon 8 of p53. The promoter regions of p53 have to be analyzed in future experiments as well as other genes such as bcl-2. Though demethylation of p53 could only be detected in intron 1, it might be suggested that a demethylase activity of *H. pylori* is involved. There is no hint that a methylase activity of *H. pylori* is involved in the methylation of AGS cells considering the experiments performed. Future direct DNA damage measurements caused by ROS should include 8-OH-desoxyguanosine (8-OH-dG) in epithelial cells of the stomach. We are also developing a new LIF-CE (laser induced fluorescence capillary electrophoresis) based DNA adduct analysis method for 8-OH-dG.

References

1) Axon, A.T., 1999, Are all helicobacters equal?Mechanisms of gastroduodenal pathology and their clinical implications, Gut,45 Suppl.1:1-4

2) Mc Gee, D.J., May, C.A., Garner, R.M., Himpsel, J.M., and Mobley, H.L.,1999, Isolation of *Helicobacter pylori* genes that modulate urease activity, J. Bacteriol.,181:2477-84

3) Li, S.D., Kersulyte, D., Lindley, I.J., Neelam, B., Berg, D.E., and Crabtree, J.E., 1999, Multiple genes in the left half of the pathogenicity island of *Helicobacter pylori* are required for tyrosine kinase dependent transcription of interleukin-8 in gastric epithelial cells, Infect-Immun.,67:3893-9

4) Liu, G., Mc Daniel, T.K., Falkow, S., and Karlin, S., Sequence anomalies in the Cag7 gene of the *Helicobacter pylori* pathogenicity island, Proc. Natl. Acad. Sci. USA, 96:7011-6

6) 1994, Schistosomes, liver flukes, and *Helicobacter pylori*, Monogr. evaluation carcinogenic risks hum., IARC,61,Lyon

7) Dogusoy, G., Karayel, F.A., Gocener, S., and Goksel, S.,1999, Histopathological fearures and expression of Bcl-2 and p53 proteins in primary gastric lymphomas,Pathol. Oncol. Res.,5: 36-40

8) Nardone,G., Staibano, S., Rocco, A., Mezza, E., D`armiento, F.P., Insabato, L., Coppola, A., Salvatore, G., Lucariello, A., Figura, N., De Rosa, G., and Budillon, G., 1999, Effect of Helicobacter pylori infection and its eradication on cell proliferation,DNA status,and oncogene expression in patients with chronic gastritis, Gut 44: 789-99

9) Blok, P., Craanen, M.E., Offerhaus, G.J., Dekker, W., Kuipers, E.J., Meuwissen, S.G., and Tytgat, G.N., Molecular alterations in early gastric carcinomas. No apparent correlation with Helicobacter pylori status,1999, Am. J. Clin. Pathol., 111: 241-7

10) Isaacson, P.G., 1999, Gastric MALT lymphoma: from concept to cure, Ann. Oncol. :10: 637-45

11) Murakami, K., Fujioka, T., Okimoto, T., Mitsuishi, Y., Oda, T., Nishizono, A., and Nasu M.,1999, Analysis of p53 gene mutations in Helicobacter pylori-associated gastritis mucosa in endoscopic biopsy specimens,Scand. J. Gastroenterol., 34: 474-7

12) Chung, H., Rha, S., Kim, J., Noh, S., Roh, J., Min, J., and Kim, B., 1997, Different expression of biophenotypes based on p53 gene state in gastric cancer cell lines, Proc. Annu. Meet. Am. Assoc. Cancer Res., 38:A3675

14) Rudi, J., Kuck, D., Strand, S., von Herbay, A., Mariani, S.M., Krammer, P.H., Galle, P.R., and Stremmel W., 1998, Involvment of the CD95 (APO-1/Fas) receptor and ligand system in Helicobacter pylori-induced gastric epithelial apoptosis, J. Clin. Invest., 102: 1506-14

15) Owen, R. W., Spiegelhalder, B. and Bartsch H (2000). Generation of reactive oxygen species by the faecal matrix. *Gut* 46, 225-232.

BIOLOGICAL REACTIVE INTERMEDIATES THAT MEDIATE CHROMIUM (VI) TOXICITY

Jalal Pourahmad and Peter J. O'Brien

Faculty of Pharmacy, University of Toronto
Toronto, ON, M5S 2S2, Canada

ABSTRACT

1. Addition of Cr VI (dichromate) to isolated rat hepatocytes results in rapid glutathione oxidation, reactive oxygen species (ROS) formation, lipid peroxidation, decreased mitochondrial membrane potential and lysosomal membrane rupture before hepatocyte lysis occurred.
2. Cytotoxicity was prevented by "ROS" scavengers, antioxidants, and glutamine (ATP generator). Hepatocyte dichlorofluorescin oxidation (to determine ROS/Cr V formation) was inhibited by mannitol (a hydroxyl radical scavenger) or butylated hydroxyanisole and butylated hydroxytoluene (antioxidants).
3. The Cr VI reductive mechanism required for toxicity are not known. Cytotoxicity was also prevented by cytochrome P450 inhibitors, particularly CYP 2E1 inhibitors, but not inhibitors of DT diaphorase or glutathione reductase. This suggests that P450 reductase and/or reduced cytochrome P450 contributes to Cr VI reduction to Cr IV.
4. Glutathione depleted hepatocytes were resistant to Cr (VI) toxicity and much less dichlorofluorescin oxidation occurred. Reduction of dichromate by glutathione or cysteine in vitro was also accompanied by oxygen uptake and was inhibited by Mn II (a Cr IV reductant). Cr VI induced cytotoxicity and ROS formation was also inhibited by Mn II which suggests that Cr IV and Cr IV.GSH mediate "ROS" formation in isolated hepatocytes.
5. In conclusion Cr VI cytotoxicity is associated with mitochondrial / lysosomal toxicity by the biological reactive intermediates Cr IV and "ROS".

INTRODUCTION

Hexavalent chromium is recognized as an environmental and occupational carcinogen and causes DNA damage both in vivo and in vitro (Costa, 1997). Airborne exposure to chromate occurs in the chromate production industry, during leather tanning, chrome plating and welding and leads to increased incidence of lung cancer (Costa, 1997). Dermal exposure to chromate occurs in workers who handle paints, dyes, inks and detergents, and contamination of land by disposal of industrial chromate waste has increased exposure of the general public. The recognition that chromium is carcinogenic, has led to concerns about the risk associated with the leaching of chromate from the alloys used in orthopaedic implants eg. joint replacement.

Human Cr VI intoxication is also associated with hepatotoxicity, nephrotoxicity and cardiotoxicity whereas systemic Cr VI exposure of animals causes toxicity to the liver, kidneys, myocardium, testes and blood vessels. The liver is the main organ for chromium VI metabolic reduction and therefore one of the most important site for its toxicity (Stift et al., 2000), but the intracellular reductive pathway or the biological reactive chromium or oxygen intermediates responsible are not known.

Biological Reactive Intermediates VI, Edited by Dansette *et al.*
Kluwer Academic / Plenum Publishers, 2001

METHODS

Isolation of Hepatocytes

Male Sprague-Dawley rats (280-300g), fed a standard chow diet and given water ad libitum, were used in all experiments. Hepatocytes were obtained by collagenase perfusion of the liver and their viability was assessed by the trypan blue (0.2% w/v) exclusion test as described by Pourahmad and O'Brien, 2000. Approximately 85-90% of the hepatocytes excluded trypan blue. Cells were suspended at a density of 10^6 cells/ml in round bottomed flasks rotating in a water bath maintained at 37 °C in Krebs-Henseleit buffer (pH 7.4), supplemented with 12.5 mM Hepes under an atmosphere of 10% O_2 : 85% N_2: 5% CO_2.

RESULTS

Dichromate (Cr VI) induced cytotoxicity toward isolated rat hepatocytes increased with time and was dose dependent with approximately 50% cell lysis occurring in 2 hours at a concentration of 1 mM. As shown in table 1, dichromate toxicity was associated with the generation of a large amount of reactive oxygen species "ROS" and malondialdehyde (MDA). Cr VI induced cytotoxicity, MDA and "ROS" generation was prevented by the antioxidants butylated hydroxytoluene (BHT) or butylated hydroxyanisole (BHA). Hydroxyl radical scavengers, mannitol and DMSO, and the Cr IV reductant, Mn II also protected the hepatocytes.

Table 1. Preventing chromate induced hepatocyte necrosis by antioxidants, "ROS" scavengers, P450 inhibitors, pore sealing and endocytosis agents.

Addition	%Cytotoxicity 3h	"DCF" 3h	MDA 3h
Hepatocytes	20 ± 2	79 ± 7	0.41 ± 0.05
+ dichromate (1mM)	76 ± 8[a]	508 ± 6[a]	20.62 ± 2.28[a]
+ Butylated hydroxyanisole (50μM)	35 ± 5[b]	129 ± 5[b]	4.35 ± 1.15[b]
+ Butylated hydroxytoluene (50μM)	35 ± 5[b]	125 ± 7[b]	5.31 ± 1.85[b]
+ Dimethyl sulfoxide (150μM)	45 ± 4[b]	137 ± 8[b]	4.45 ± 1.14[b]
+ Mannitol (50mM)	35 ± 5[b]	109 ± 5[b]	3.35 ± 0.65[b]
+ Mn(II) (2mM)	35 ± 5[b]	129 ± 5[b]	4.35 ± 0.85[b]
+ Phenylimidazole (300μM)	37 ± 4[b]	635 ± 6[b]	4.37 ± 1.26[b]
+ Diphenyliodonium chloride (50μM)	35 ± 5[b]	629 ± 5[b]	4.35 ± 0.85[b]
+ 1,3-bis(2-chloroethyl)-1-nitrosourea (50μM)	73 ± 6	511 ± 8	20.18 ± 1.98
+ Dicumarol (30μM)	76 ± 8	505 ± 6	20.62 ± 2.28
+ Monensin (10μM)	35 ± 5[b]	129 ± 5[b]	4.35 ± 1.15[b]
+ Methylamine (30mM)	35 ± 5[b]	125 ± 7[b]	4.31 ± 1.85[b]
+ Chloroquine (100μM)	45 ± 4[b]	137 ± 8[b]	4.45 ± 1.14[b]
GSH depleted hepatocytes	26 ± 3	88 ± 8	0.73 ± 0.08
+ dichromate (1mM)	37 ± 4[b]	135 ± 4[b]	6.37 ± 1.94[b]

Hepatocytes (10^6 cells/ml) were incubated in Krebs-Henseleit buffer pH 7.4 at 37C for 3.0 hrs following the addition of potassium dichromate. Cytotoxicity was determined as the percentage of cells that take up trypan blue (Pourahmad and O'Brien, 2000).
DCF (dichlorofluorescein) formation (for "ROS"/Cr V formation) was expressed as fluorescent intensity units (Shen et al., 1996). GSH depleted hepatocytes were prepared as described by Khan and O'Brien, 1991. MDA formation was expressed as μM concentrations (Smith et al., 1982).
Values are expressed as means of three separate experiments (S.D.)
a: Significant difference in comparison with control hepatocytes ($P < 0.001$)
b: Significant difference in comparison with metal treated hepatocytes ($P < 0.001$)

Dichromate reduction is required for oxygen activation but the intracellular reductive system has not yet been identified. Table 1 shows that the DT-diaphorase inhibitor, dicumarol or the glutathione reductase inhibitor, 1,3-bis(2-chloroethyl)-1-nitrosourea (BCNU) did not show any significant effect on Cr VI induced cytotoxic alterations. However, the CYP2E1 inhibitor, phenylimidazole and the P450 reductase inhibitor, diphenyliodonium chloride (DPI) protected the hepatocyte against Cr VI induced cytotoxicity and lipid peroxidation although they both significantly increased DCF oxidation.

1 mM Cr VI oxidized 85% hepatocyte GSH in about 5 mins and the GSSG formed effluxed the cell (data not shown). Dichromate reduction by GSH or cysteine in vitro was accompanied by oxygen uptake and H_2O_2 formation and both were inhibited by Mn II (data not shown). Depletion of hepatocyte GSH beforehand prevented Cr VI induced hepatocyte toxicity, "ROS" and MDA generation (table 1).

As demonstrated in table 2, Cr VI decreased the hepatocyte mitochondrial membrane potential by 44% within 5 mins of Cr VI addition which was further decreased by 65% in 1 hr. This was prevented by addition of catalase, mannitol or dimethyl sulfoxide or the Cr IV reductant Mn II. Mitochondrial permeability transition (MPT) inhibitors, carnitine, cyclosporin and trifluoperazine also prevented the Cr VI induced decrease in mitochondrial membrane potential. No significant decrease in hepatocyte mitochondrial membrane potential was observed when 1 mM Cr VI was incubated with GSH depleted hepatocytes.

Table 2. Mitochondrial membrane potential changes during chromate induced hepatocyte injury

| Addition | Mitochondrial membrane potential decline (% initial value) | | |
Incubation Time	5 min	15 min	60 min
Hepatocytes	2 ± 1	3 ± 1	4 ± 1
+ dichromate (1mM)	44 ± 4	52 ± 5	65 ± 7[a]
+ Catalase (200 U/mL)	11 ± 1	8 ± 1	9 ± 1[b]
+ Mannitol (50mM)	12 ± 2	8 ± 1	6 ± 1[b]
+ Dimethyl sulfoxide (150μM)	11 ± 2	12 ± 2	10 ± 2[b]
+ Carnitine (2mM)	11 ± 1	10 ± 1	9 ± 1[b]
+ Cyclosporine (2μM)	12 ± 2	12 ± 1	6 ± 1[b]
+ Trifluoperazine (15μM)	11 ± 2	12 ± 2	11 ± 2[b]
+ Mn(II) (2mM)	10 ± 1[b]	9 ± 1	8 ± 1[b]
GSH depleted hepat.	2 ± 1	3 ± 1	4 ± 1
+ dichromate(1mM)	11 ± 2	8 ± 1	8 ± 2

Hepatocytes (10^6 cells/ml) were incubated in Krebs-Henseleit buffer pH 7.4 at 37°C.
Mitochondrial membrane potential was determined as the difference in mitochondrial uptake of the rhodamine 123 between control and treated cells and expressed as fluorescence intensity unit (Andersson et al., 1987). Values are expressed as means of three separate experiments (±S.D.)
a: Significant difference in comparison with control hepatocytes ($P < 0.001$)
b: Significant difference in comparison with metal treated hepatocyte ($P < 0.001$)

Table 3. Preventing chromate induced hepatocyte lysosomal membrane damage by inhibitors of oxidative stress and autophagy

| Addition | "Acridine orange" redistributed | | |
	15 min	30 min	60 min
Hepatocytes	3 ± 1	4 ± 1	4 ± 1
+ dichromate (1mM)	39 ± 4[a]	68 ± 7[a]	395 ± 8[a]
+ Catalase (200 U/mL)	3 ± 1[b]	3 ± 1[b]	35 ± 4[b]
+ SOD (100 U/mL)	6 ± 1[b]	7 ± 1[b]	71 ± 6[b]
+ Dimethyl sulfoxide (150mM)	3 ± 1[b]	4 ± 1[b]	47 ± 4[b]
+ Mannitol (50mM)	6 ± 1[b]	5 ± 1[b]	28 ± 4[b]
+ 3-Methyladenine (5mM)	5 ± 1[b]	8 ± 1[b]	40 ± 5[b]
+ Mn(II) (2mM)	6 ± 1[b]	7 ± 1[b]	38 ± 4[b]
GSH depleted hepat.	3 ± 1	4 ± 1	15 ± 3
+ dichromate (1mM)	9 ± 1[b]	9 ± 1[b]	41 ± 4[b]

Hepatocytes (10^6 cells/ml) were incubated in Krebs-Henseleit buffer pH 7.4 at 37°. Lysosomal membrane damage was determined as the alterations in intensity unit of diffuse cytosolic green fluorescence induced by acridine orange released from lysosomes (Adapted from Brunk et al., 1995 after permeabilising the hepatocytes (Bontemps and Vandenberghe, 1998)).
Values are expressed as means of three separate experiments (S.D.)
a: Significant difference in comparison with control hepatocytes ($P < 0.001$)
b: Significant difference in comparison with metal treated hepatocytes ($P < 0.001$)

When hepatocyte lysosomes were loaded with acridine orange, a massive release of acridine orange into the cytosolic fraction ensued within 60 minutes when the loaded hepatocytes were treated with Cr VI indicating leakiness of the lysosomal membranes (table 3). The Cr VI induced acridine orange release was prevented by the "ROS" scavengers dimethylsulfoxide, mannitol or catalase, superoxide dismutase or when Cr VI was incubated with GSH depleted hepatocytes. Endocytosis inhibitors including lysosomotropic agents, chloroquine and methylamine, Na^+ ionophore, monensin and the endocytosis inhibitor, 3-methyladenine also protected the hepatocytes (table 1).

DISCUSSION

Intracellular reduction appears to be critical for the expression of DNA damage and toxicity (Sugiyama et al., 1991). Enzymatic reduction by glutathione reductase (Shi and Dalal, 1989), DT-diaphorase (De Flora et al., 1985), NADPH-cytochrome c reductase, cytochrome P450 dependent systems or GSH and cysteine have been proposed (Cupo and Wetterhahn, 1985). The results presented here suggest that in intact hepatocytes, Cr VI is mostly reduced by CYP2E1 and/or P450 reductase.

Hepatocyte GSH was depleted rapidly during the metabolism of Cr VI. Of particular interest is that even though GSH is an intracellular antioxidant which prevents intracellular "ROS" formation and lipid peroxidation, depleting GSH beforehand protected the cells against Cr VI induced cytotoxicity and "ROS" or MDA formation. GSH could be required for cytotoxicity as Liu et al., 1997 have reported the formation of "ROS" arising from Cr IV-GSH formation when GSH is added to dichromate (Fig. 1). Reduction of dichromate by glutathione or cysteine in vitro was accompanied by oxygen uptake and H_2O_2 formation and both were inhibited by Mn II. Mn II has been shown to be a Cr IV reductant (Stearns and Wetterhahn, 1994)). As shown in Fig. 1, the oxygen uptake was due to Cr IV or Cr IV.GSH reacting with O_2 to form "ROS" that includes a Cr V oxo or superoxo complex (O'Brien and Woodbridge, 1997, Liu et al., 1997). Cr VI induced cytotoxicity and "ROS" formation, lipid peroxidation, mitochondrial membrane potential collapse and lysosomal membrane damage were also inhibited by Mn II which suggests that Cr IV or Cr IV.GSH mediated ROS formation and lipid peroxidation contributed to mitochondrial and lysosomal damage.

Phenylimidazole or P450 reductase inhibitors prevented Cr VI induced cytotoxicity and lipid peroxidation but not DCF oxidation. However GSH oxidation and mitochondrial toxicity occurred more rapidly than DCF oxidation or lipid peroxidation. One explanation is that DCF is oxidized by Cr V (Martin et al., 1998) or lipid peroxidation rather than by "ROS". CYP2E1 or P450 reductase may reduce Cr V to Cr IV rather than Cr III so that inhibitors of CYP2E1 or P450 reductase would increase the concentration of Cr V.

Cr VI induced a rapid decline of mitochondrial membrane potential (within 5 mins) which was prevented by catalase, dimethyl sulfoxide, mannitol or by prior depletion of hepatocyte GSH which indicates that the decline of mitochondrial membrane potential was a consequence of "ROS" formation and reductive activation of Cr VI. Collapse of the mitochondrial membrane potential results in opening of the mitochondrial permeability transition pore and causes severe mitochondrial ATP depletion. Lack of mitochondrial ATP results in intracellular acidosis and osmotic injury which leads to plasma membrane lysis (Pourahmad and O'Brien, 2000). Glutamine (a mitochondrial ATP generator) protected the cells against Cr VI induced cytotoxicity (data not shown).

Cr VI induced cytotoxicity as well as hepatocyte "ROS" formation and lipid peroxidation were prevented by the hepatocyte lysosomotropic agents methylamine (Luiken et al., 1996), chloroquine (Gaynor et al., 1998), monensin, (a Na^+ ionophore that inhibits hepatocyte endocytosis and endosomal acidification (Brunk et al., 1995)) or 3-methyladenine, an inhibitor of hepatocyte autophagy (Seglen, 1997). These results show that Cr VI overload in rat hepatocytes increases lysosomal fragility and leakiness likely as a result of lysosomal lipid peroxidation or "ROS" formation which contributes to cell death.

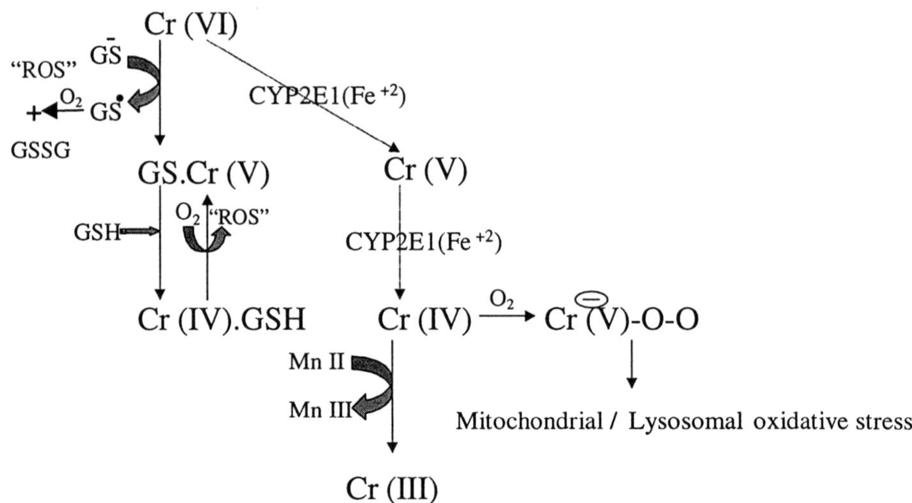

Figure 1. Proposed "ROS" mediated dichromate cytotoxic mechanism.

REFERENCES

Andersson, B.S., Aw, T.Y., Jones, D.P. (1987). Mitochondrial transmembrane potential and pH gradient during anoxia. *Am J Physiol,* **252**(4 Pt 1), C349-C355.

Bontemps, F., Van Den Berghe, G. (1998). Novel evidence for an ecto-phospholipid methyltransferase in isolated rat hepatocytes. *Biochem. J.* **330**(Pt 1), 1-4.

Brunk, U. T., Zhang, H., Roberg, K., Ollinger, K. (1995). Lethal hydrogen peroxide toxicity involves lysosomal iron-catalyzed reactions with membrane damage. *Redox Report* **1**, 267-277.

Costa, M. (1997). Toxicity and carcinogenicity of Cr(VI) in animal models and humans. *Crit. Rev. Toxicol.* **27**(5), 431-42.

Cupo D. Y. and Wetterhahn K. E. (1985). Modification of chromium (VI) induced DNA damage by glutathione and cytochromes P450 in chicken embryo hepatocytes. *Proceedings of the National Academy of Sciences of the U.S.A.* **82**, 6755-6759.

De Flora S., Morelli A., Basso C., Romano M., Serra D. and De Flora A. (1985). Prominent role of DT-diaphorase as a cellular mechanism reducing chromium (VI) and reverting its mutagenicity. *Cancer Research* **45**, 3188-3196.

Gaynor, E.C., Chen, C.Y., Emr, S.D., Graham, T.R. (1998). ARF is required for maintenance of yeast Golgi and endosome structure and function. *Mol. Biol. Cell.* **9**(3), 653-70.

Khan, S. and O'Brien, P.J. (1991). 1-bromoalkanes as new potent nontoxic glutathione depletors in isolated rat hepatocytes. *Biochem. Biophys. Res. Commun.* **179**(1), 436-41.

Liu K. J., Shi X. and Dalal N. S. (1997). Synthesis of CR(VI)-GSH, its identification and its free hydroxyl radical generation: a model compound for Cr(VI) carcinogenicity. *Biochemical and Biophysical Research Communications* **235**, 54-58.

Luiken, J.-J.; Aerts, J.-M.; Meijer, A.-J. The role of the intralysosomal pH in the control of autophagic proteolytic flux in rat hepatocytes. *Eur. J. Biochem.* **235**(3): 564-73; 1996.

Martin, B.D., Schoenhard, J.A., Sugden, K.D. (1998). Hypervalent chromium mimics reactive oxygen species as measured by the oxidant- sensitive dyes 2',7'-dichlorofluorescin and dihydrorhodamine. *Chem. Res. Toxicol.* **11**(12), 1402-10.

O'Brien, P., Woodbridge, N. (1997). Evidence for epoxidation by the chromium(VI)-glutathione system. *Polyhedron* **16**(16), 2897-2899.

Pourahmad, J., O'Brien, P.J. (2000). Contrasting role of Na(+) ions in modulating Cu(+2) or Cd(+2) induced hepatocyte toxicity. *Chem Biol Interact.* **126**(2):159-69.

Reed, D.J., Babson, J.R., Beatty, P.W., Brodie, A.E., Ellis, W.W., Potter, D.W. (1980). High-performance liquid chromatography analysis of nanomole levels of glutathione, glutathione disulfide, and related thiols and disulfides. *Anal. Biochem.* **106**(1), 55-62.

Seglen, P.O. (1997). DNA ploidy and autophagic protein degradation as determinants of hepatocellular growth and survival. *Cell Biol. Toxicol.* **13**(4-5), 301-15.

Shen, H.M., Shi, C.Y., Shen, Y., Ong, C.N. (1996). Detection of elevated reactive oxygen species level in cultured rat hepatocytes treated with aflatoxin B1. *Free Radic. Biol. Med.* **21**(2), 139-146.

Shi X. L. and Dalal N. S. (1989). Chromium (V) and hydroxyl radical formation during the glutathione reductase-catalysed reduction of chromium (VI). *Biochemical and Biophysical Research Communications* **163**, 627-634.

Smith, M.T., Thor, H., Hartizell, P., Orrenius, S. (1982). The measurement of lipid peroxidation in isolated hepatocytes. *Biochem. Pharmacol.* **31**(1), 19-26.

Stearns, D.M. and Wetterhahn, K.E. (1994). Reaction of chromium(VI) with ascorbate produces chromium(V), chromium(IV), and carbon-based radicals. *Chem. Res. Toxicol.* **7**(2), 219-30.

Stift, A., Friedl, J., Langle, F., Berlakovich, G., Steininger, R., Muhlbacher, F. (2000). Successful treatment of a patient suffering from severe acute potassium dichromate poisoning with liver transplantation. *Transplantation* **69**(11):2454-5.

Sugiyama, M., Tsuzuki, K. and Ogura, R. (1991). Effect of ascorbic acid on DNA damage, cytotoxicity, glutathione reductase, and formation of paramagnetic chromium in Chinese hamster V-79 cells treated with sodium chromate (VI). *Journal of Cell Biology* **266**, 3383-3386.

INVOLVEMENT OF PEROXYNITRITE ON THE EARLY LOSS OF P450 IN SHORT-TERM HEPATOCYTE CULTURES

Santiago Vernia, Silvia M. Sanz-González and M. Pilar López-García

Departamento de Bioquímica y Biología Molecular
Facultad de Farmacia, Universidad de Valencia, 46100 Valencia, Spain
Fax:+34-963864917; E-mail: pilar.lopez@uv.es

SUMMARY

The biological chemistry of nitric oxide (NO) in the oxygenated cellular environment is extremely complex. It involves the direct interaction of NO with specific biomolecules and the so-called indirect effects, due to secondary more potent oxidant species derived from NO which are also able to react with DNA, lipids, thiols and transition metals (Wink et al., 1996; Nathan, 1992). In addition to its regulatory role as a signalling molecule (Nathan, 1992; Moncada and Palmer, 1991) it has become evident that NO (or NO-derived species) is a critical factor involved in various toxicological mechanisms (Wink et al., 1996; Wang et al., 1998; Estevez et al., 1999; Wink et al., 1999). Some controversy exists however about the damaging vs. protective actions of NO on oxidative injury, whose biological significance in living cells and tissues remains still ill defined. Research in this laboratory (López-García, 1998; López-García and Sanz-Gonzalez, 2000) has shown that NO synthesis is significantly activated in hepatocytes from control rats following isolation by the classical collagenase-based procedure. NO overproduction appears to be due to the very early activation of liver constitutive Ca^{2+}-dependent NO synthase (cNOS). Previous results have also provided first experimental evidence for the direct involvement of endogenously generated NO as a causal factor responsible for important phenotypic changes commonly observed in short-term cultured hepatocytes, which includes the early impairment of hepatocyte mitochondrial function -i.e., transient cell energy depletion- and glucose metabolism, and the well-known quick and irreversible loss of P450 content (López et al. 1987; López-García, 1998). This study aims to further characterise the mechanisms underlying this phenomenon.

Results show that the hepatocyte isolation procedure (the commonly employed collagenase-based two step liver perfusion method) induces strong oxidative stress that lasts for at least 4 h in culture and involves both oxygen-derived (ROS) and nitrogen-derived (RNS) reactive species. On the basis of the combined use of dihydrorhodamine 123 (DHR) as a probe and L-NAME (N^G-nitro-L-arginine methyl ester) to efficiently block NO synthesis, the analysis of the amount, the time-course pattern, and the nature of the species involved support the view that peroxynitrite* (PN) is readily formed within the early culture hours. Immunodetection of protein bound 3-nitrotyrosine provides direct evidence for PN generation upon hepatocyte isolation: several nitrated protein bands -most already present after only 30 min of liver perfusion and quantitatively increasing for the first 2 hours in culture- have been identified as preferential PN protein targets in the different cellular compartments. Since the early inhibition of NO synthesis is enough to provide full maintenance of the hepatocyte initial P450 content, results support the view that PN -while not affecting cell viability and monolayer development- is the main species likely responsible for the early loss of P450 in short-term cultured hepatocytes.

EXPERIMENTAL DESIGN

Experiments were performed in very simple culture conditions, expressly avoiding supplementation with hormones and/or effectors, or the use of complex culture substrata, known to contribute to a better preservation of P450 in cultured liver cells. Hepatocytes were isolated from fed male Sprague-Dawley rats (200-250 g) by reverse liver perfusion with collagenase (Boehringer Manheim). Cells (>90% viability) were seeded on fibronectin-coated dishes in phenol red-free Ham's F12 medium plus antibiotics, 0.2% bovine serum albumin, 10^{-8}M insulin and 2% calf serum (Gibco), and cultured at 37ºC under a 5% CO_2 atmosphere (López et al., 1988). Two different culture systems were used in parallel: those in which hepatocytes were isolated in standard conditions (Control cell isolation) and those in which 1 mM L-NAME (Sigma) was present during the isolation procedure (L-NAME in cell isolation). In both conditions, L-NAME and/or DHR 123 (Sigma) were also added to the culture medium when indicated.

Nitrites (NO_2^-+NO_3^-), P450, DHR 123 oxidation to rhodamine (RH) and protein were determined by standard methods, as described (López-García, 1998; López-García and Sanz-Gonzalez, 2000). Proteins (20 mg/well) were resolved (SDS/PAGE, 4-15% gradient gels) and transferred to PVDF membranes. Immunodetection was performed using monoclonal anti-3-nitrotyrosine antibodies (Calbiochem) and chemiluminescence detection (ECL, Amersham).

RESULTS AND DISCUSSION

Short-term cultured hepatocytes isolated from control uninduced rats spontaneously produce high amounts of NO, as indicated by the accumulation of NO_2^-+NO_3^- in the medium (Fig. 1A). As previously shown (López-García and Sanz-Gonzalez, 2000), NO formation at this stage is likely the result of the very early activation of cNOS during the hepatocyte isolation procedure. Interestingly, the quick and early loss of P450 content (Fig. 1B) appears directly related to NO overproduction (López-García, 1998). While the actual stimuli for cNOS activation remains to be defined, we have focused to the molecular mechanism which relates NO formation to P450 loss in culture. DHR 123 oxidation to RH was quantified as a suitable probe to detect both ROS (through an H_2O_2-driven process, via peroxidase or metal-dependent pathways) and RNS (direct oxidation due to PN) generation. As shown (Fig. 2), oxidative stress is acutely induced from the very early stages of culture: total RH (in cells+medium) increases significantly during the first 2 h decreasing slowly thereafter.From 6 to 24 h of culture no additional RH formation is observed. Under oxidative

---

* The term peroxynitrite is used to refer to any peroxynitrite-derived intermediate (i.e., peroxynitrite anion, peroxynitrous acid and/or reactive species formed from the reaction of peroxynitrite with carbon dioxide).

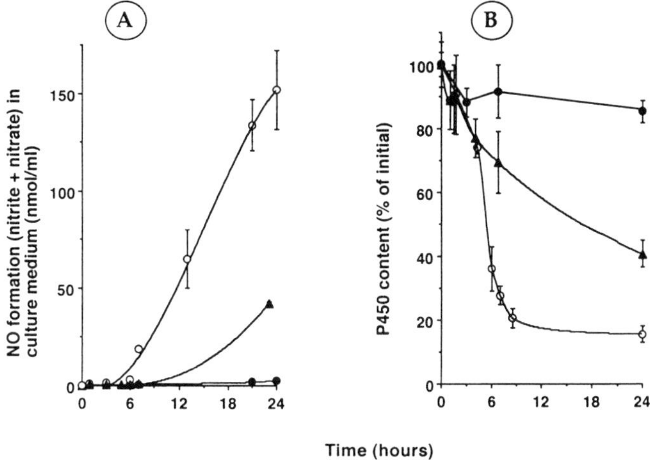

Figure 1. Evolution of NO formation (A) and P450 content (B) in short-term cultured hepatocytes.Cells were isolated in standard conditions (O) or in the presence of 1mM L-NAME in the liver perfusion medium (▲). By the time of cell plating, 1 mM L-NAME was also added (●) or not (O,▲) to the culture medium.

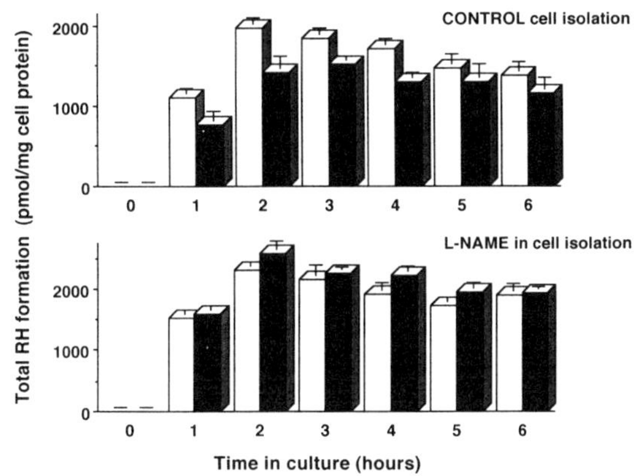

Figure 2. Time-course of total DHR 123 oxidation in CONTROL cultures and those in which hepatocytes were isolated in the presence of 1mM L-NAME. Data correspond to the amount of RH generated (cell + medium) in both conditions, as a function of time in culture in the presence (black bars) or not (clear bars) of 1 mM L-NAME.

stress conditions in which NO is also formed a role for both ROS and RNS should be considered: an indirect approach to monitor the contribution of PN to DHR 123 oxidation is simply to inhibit NO production (Ischiropoulos *et al.*, 1999). Exposure of freshly isolated hepatocytes to L-NAME concentrations able to suppress NO synthesis (Fig. 1A) in fact reduces (by about 30%) the amount of RH generated in control cells (Fig. 2). Fig. 3 shows an estimation of the total (extra+intracellular) PN formation in hepatocytes as derived from the corresponding RH cumulative curves. Data suggests that PN is significantly generated in hepatocytes at very early culture hours from the simultaneous flux of NO and superoxide. On the other hand, no PN-derived RH is detected in culture when hepatocytes are isolated in the presence of L-NAME. Also, inhibition of NO synthesis does not modify the oxidative stress depending on oxygen-derived species.

Finally, direct evidence for PN formation in cultured hepatocytes is provided by the detection of protein-bound 3-nitrotyrosine (Fig. 3), often used as a stable and specific "footprint" for PN generation in living cells. As shown, protein nitration is already significant at 0 h (after 30 min of liver perfusion) and increases for the next 2 h in culture. Three major immunoreactive bands are detected (apparent molecular masses of 40, 50 and 55 kDa) which appear to follow an identical time-course pattern.

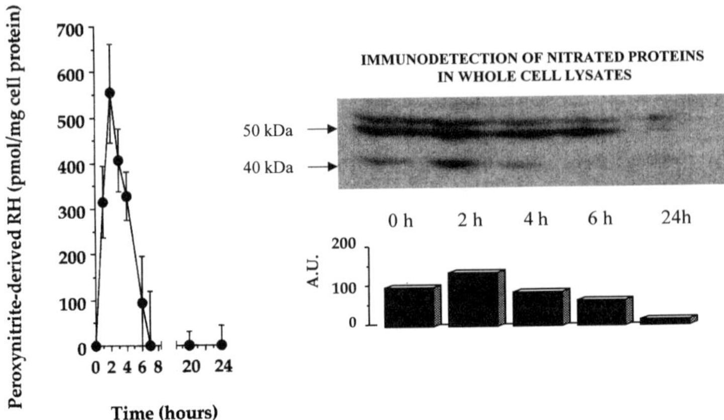

Figure 3. Evidences for PN generation in the early hours of culture in hepatocytes isolated in control conditions. PN formation is estimated as the PN-derived RH (from Fig. 2 data) and by Western blot immunodetection of nitrated proteins as a function of time in culture (See the Experimental section).

ACKNOWLEDGEMENTS

This work was supported by research grants from Generalitat Valenciana (GV-21-11896) and the Spanish MEC (SAF98-0128). S.V. is recipient of a predoctoral research grant from Generalitat Valenciana (Spain).

REFERENCES

Estevez, A.G., Spear, N., Pelluffo, H., Kamaid, A., Barbeito, L. and Beckman, J.S., Examining apoptosis in cultured cells after exposure to nitric oxide and peroxynitrite, 1999, *Methods Enzymol.* 301, 393-402.

Ischiropoulos, H., Gow, A., Thom, S.R., Kooy, N.W., Royall, J.A. and Crow, J.P., 1999, Detection of reactive nitrogen species using 2,7-dichlorodihydrofluorescein and dihydrorhodamine 123, Methods Enzymol. 301, 367-373.

López, M.P., Gómez-Lechón, M.J. and Castell, J.V., 1988, Active glycolysis and glycogenolysis in early stages of primary cultured hepatocytes. Role of AMP and fructose 2,6-bisphosphate, *In Vitro Cell. Dev. Biol.* 24, 511-517.

López-García, M.P., 1998, Endogenous nitric oxide is responsible for the early loss of P450 in cultured rat hepatocytes, *FEBS Lett.* 438, 145-149.

López-García, M.P. and Sanz-Gonzalez, S.M., 2000, Peroxynitrite generated from constitutive nitric oxide synthase mediates the early biochemical injury in short-term cultured hepatocytes, *FEBS Lett.466,* 187-191.

Moncada, S, Palmer, R.M.J. and Higgs, E.A., 1991,Nitric oxide: physiology, phatophysiology, and pharmacology, *Pharmacol. Rev.* 43, 109-142.

Nathan, C., 1992, Nitric oxide as a secretory product of mammalian cells, *FASEB J.* 6, 3051-3064.

Wang,Y.J., Ho,Y.S., Pan,M.H. and Lin,J.K., 1998, Mechanisms of cell death induced by nitric oxide and peroxynitrite in Calu-1 cells, *Environ.Toxicol.Pharmacol.* 6,35-44.

Wink, D.A., Grisham, M.B., Mitchell, J.B. and Ford, P.C., 1996, Direct and indirect effects of nitric oxide in chemical reactions relevant to biology, *Methods Enzymol.* 268, 13-31.

Wink., D.A., Vodovotz, Y., Grisham, M.B., DeGraff, W., Cook, J.C., Pacelli, R., Krishna, M. and Mitchell, J.B., 1999, *Methods Enzymol.* 301, 413-424.

LACK OF CORRELATION BETWEEN CYP2A6 GENOTYPE AND SMOKING HABITS

Thomas G. Schulz, Peter Ruhnau, and Ernst Hallier

Department of Occupational Health, Georg-August-University Göttingen, Waldweg 37, 37073 Göttingen, Germany

INTRODUCTION

Nicotine, the primary compound in tobacco that establishes and maintains tobacco dependence, is mainly metabolized to cotinine by polymorphically expressed CYP2A6. Recently it was suggested that individuals lacking fully functional CYP2A6 are significantly protected against becoming tobacco-dependent smokers (Pianezza et al., 1998).

MATERIAL AND METHODS

In a study on construction workers the consumption of cigarettes was evaluated and correlated with the genotype of CYP2A6. Two methods were used to establish the CYP2A6 genotype: namely either a two-step polymerase chain reaction (PCR) method with restriction fragment length polymorphism (RFLP) analysis (Fernandez-Salguero et al., 1995; Gullsten et al. 1997) or a recently published single-step PCR/RFLP method (Chen et al., 1999). Statistical analysis was performed using Fisher´s exact test.

Biological Reactive Intermediates VI, Edited by Dansette *et al.*
Kluwer Academic / Plenum Publishers, 2001

RESULTS

Fifty three of 365 samples were investigated in parallel experiments using both methods (Table 1). In 3 of 53 individuals, the classification as CYP2A6*3 heterozygotes by the method of Fernandez-Salguero et al. (1995) was not reproduced by the method of Chen et al. (1999). However, there was an excellent agreement in the determination of the CYP2A6*2 genotype; both methods identified the same 5 of 53 individuals.

Table 1. Comparison of two methods to determine the CYP2A6 genotype.
Method 1 (Fernandez-Salguero et al., 1995 and Gullsten et al., 1997)
Method 2 (Chen et al., 1999)

No of samples	Method 1	Method 2
43	2A6*1	2A6*1
5	2A6*1/2A6*2	2A6*1/2A6*2
3	2A6*1/2A6*3	2A6*1
2	2A6*2/2A6*3	2A6*2

The remaining samples were analysed only by the method of Chen et al. (Table 2). We did not find any individual with a CYP2A6*3 allele. All CYP2A6*2 alleles were found in heterozygotes.

Table 2. Results of genotyping of CYP2A6

	total	2A6*1	2A6*1/2A6*2	2A6*1/2A6*3
n	365	347	18	0
%		95,1	4,9	0

The comparison of CYP2A6 genotype and smoking habits was performed on the basis of CYP2A6*1 homozygotes compared with CYP2A6*2 heterozygotes. The group of smokers was stratified according to the self-reported consumption of cigarettes (Table 3). The non-smoking group was also stratified in to individuals, who never smoked, and in ex-smokers. Statistical analysis was performed in comparison with the non-smoking group or the never-smoking group. There were no differences in the frequency between any of the groups investigated.

Table 3. Smoking habits and CYP2A6 genotype

	2A6*1	in %	2A6*1/2A6*2	in %
never smoking	109	94,0	7	6,0
Ex-smoker	108	94,7	6	5,3
Non-smoker	217	94,3	13	5,7
Smoker	130	96,3	5	3,7
>9 cigarettes/day	103	95,4	5	4,6
>19 cigarettes/day	85	96,6	3	3,4
>29 cigarettes/day	20	100	0	0

CONCLUSION

Our results do not confirm the hypothesis of Pianezza et al. (1998) that individuals lacking full functional CYP2A6 are protected against becoming tobacco-dependent smokers. However, this could be due to the implementation of a new and improved method in the determination of the CYP2A6 genotype.

REFERENCES

Chen, G.F., Tang, Y.M., Green, B., Lin, D.X., Guengerich, F.P., Daly, A.K., Caporaso, N.E., Kadlubar, F.F., 1999, Low frequency of CYP2A6 gene polymorphism as revealed by a one-step polymerase chain reaction method, *Pharmacogenetics* 9:327-332.

Fernandez-Salguero, P., Hoffman, S.M.G., Cholerton, C., Mohrenweiser, H., Raunio, H., Rautio, A., Pelkonen, O., Huang, J.D., Evans, W.E., Idle, J.R., Gonzalez, F.J., 1995, A genetic polymorphism of coumarin 7-hydroxylation: sequence of the human CYP2A genes and identification of variant CYP2A6 alleles, *Am. J. Hum. Genet.* 57:651-660.

Gullsten, H., Agundez, J.A.G., Benitez, J., Laara, E., Ladero, J.M., Diaz-Rubio, M., Fernandez-Salguero, P., Gonzalez, F., Rautio, A., Pelkonen, O., Raunio, H., 1997, CYP2A6 gene polymorphism and risk of liver cancer and cirrhosis, *Pharmacogenetics* 7:247-250.

Pianezza, M.L., Sellers, E.M., Tyndale, R.F., 1998, Nicotine metabolism defect reduces smoking, *Nature* 393:750.

FREE RADICAL LIPID PEROXIDATION AND MONOOXYGENASE ACTIVITY IN EXPERIMENTAL INFLUENZA VIRUS INFECTION AFTER TREATMENT WITH RIMANTADINE

L. Tantcheva[1], E. Pavlova[5], V. Savov[5], A. Galabov[2], M. Mileva[4], A. Braykova[3]

[1]Institute of Physiology, [2]Institute of Microbiology, [3]Institute of Molecular Biology, Bulgarian Academy of Sciences, [4]Medical University-Sofia, [5]University of Sofia St. Kl. Ohridski

INTRODUCTION

Acute influenza virus infection causes accumulation of reactive oxygen species (ROS) in target organs, initiating the processes of free radical lipid peroxidation (LPO) (Peterhanse 1997, Buffinton et al. 1992, Chetverikova et al. 1996). Rimantadine (Rim), (α-methyl-1-adamantil-methylamin hydrochlorid), a universal inhibitor of type A virus, is widely used in chemotherapeutic treatment and prevention of influenza. Rim interacts with the M-2 protein of the virus, inhibits its replication and blocks the formation of some replicative components of the influenza virus (Belshe et al. 1988, Zhiravetskii 1986). At present, it is not clear how Rim can affect the processes of LPO and drug metabolizing enzyme systems (DMES) during experimental influenza virus infection.

The aim of the present work is to study the effect of Rim on the processes of LPO and drug metabolism in influenza virus infected mice.

MATERIALS AND METHODS

Animals: Albino male mice, line ICR (15 per group). Group I: healthy controls; Group II: healthy + Rim (20 mg/kg); Group III: IV infected; Group IV: IVI + Rim (20 mg/kg 24h before the inoculation). The animals were infected by intranasal inoculation with influenza virus A /Aichi/2/68 (H3N2)- (0.5 of LD_{50}).

Procedure: The mice were decapitated on the 5th day after the inoculation. Livers were perfused *in situ* with an ice-cold solution of 1.15% KCl at 4°C and homogenized in 0.1 M Na,K-phosphate buffer, pH 7.4. Homogenates were span for 20 min. at 10000 x g and supernatants were used for determination of LPO products, cyt. P 450, cyt. C reductase activity and some drug-metabolizing enzyme activities (DMES).

Egg lecitine liposomes for measurements of Rim antioxidant activity were prepared by sonication (3 x 10 sec) in an ultrasound desintegrator USD-24.

Analysis of LPO products: The concentration of the primary LPO products - conjugated dienes - was determined spectrophotometrically according to Recknagel *et al.* (1984) in total lipid extracts (Folch *et al.* 1957). The concentration of the secondary LPO products - thiobarbituric acid reactive substances (TBARS) - was determined by the procedure of Buege and Aust (1978).

The effects of Rim on TBARS production and luminol-dependent chemiluminescence (CL) in liposomes was used to evaluate its antioxidant activity.

Biological Reactive Intermediates VI, Edited by Dansette *et al.*
Kluwer Academic / Plenum Publishers, 2001

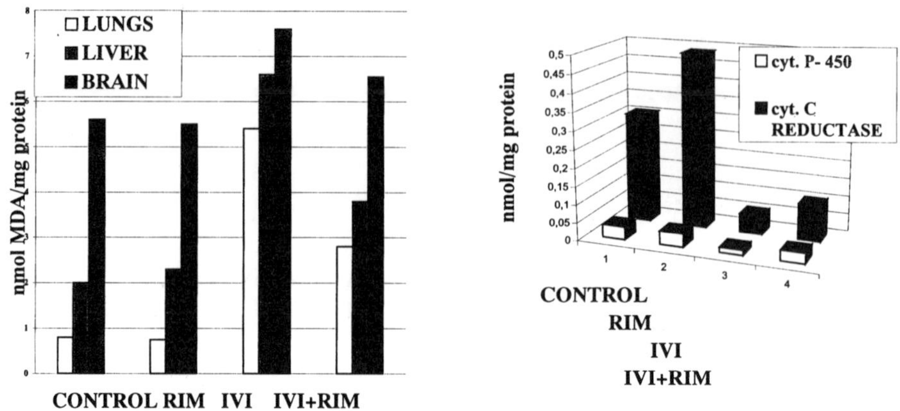

Figure 1. The effect of Rimantadin on LPO products in target organs (left panel) and on liver cyt. P-450 content and cyt. C reductase activity (right panel).

Measurements were made in nonenzymic (H_2O_2 + Fe^{2+}) and enzymic (xanthine + xanthineoxidase) induction systems, in absence or presence of ascorbic acid, SOD or catalase.

Monooxygenase enzyme activity: Aniline hydroxylase activity (AH) was determined according to Mazel (1971), N-demethylase activity with substrates ethylmorphine (EMND), amidopyrine (APND) and analgin (ANND) - according to Nash (1952), the content of hepatic cyt. P-450 - by the method of Matsubara *et al.* (1960) and the NADPH-cyt. C reductase activity - according to Roering *et al.* (1972). The enzyme activities were expressed as nmoles product . mg protein^{-1} . min^{-1} and the content of cyt. P- 450 - as nmoles . mg protein^{-1}. The protein content was determined by the Biuret method.

Statistical analysis was performed according to Student-Fisher test. The data are

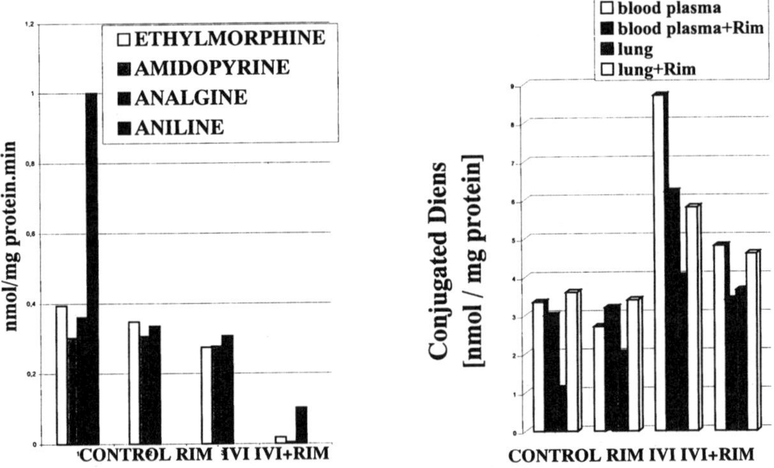

Figure 2. The effect of IVI and Rimantadin on hepatic monooxigenase activities (left panel) and on LPO products in blood plasma and lung (right panel)

presented as mean values with their SEM. The correlation coefficients were calculated according to the method of Brave and Pearson (1962).

RESULTS AND DISCUSSION

Experimental influenza virus infection is accompanied by an increase at the level of LPO products in blood plasma, liver, and lung of infected mice. Its effect is the most pronounced in blood plasma (3.5 fold) and liver (2.4 fold, Fig.1), characterizing influenza as a "free radical disease". On the 5th day of the virus inoculation the content of both conjugated dienes and TBARS is increased 2 fold. IVI reduces significantly the concentrations of cyt. P 450 (up to 50%) and monooxygenases (up to 23 - 35%) and inhibits the activity of NADPH-cyt. C-reductase (up to 50%, Figs. 1, 2) and monooxygenases (APND, EMND and AH). In fact, the whole electron-transport chain in the liver endoplasmic reticulum is inhibited by the infection. A significant correlation was observed between DMES activities and LPO products in infected animals (r = -0.986 for conjugated dienes/P-450 and r = -0.988 for conjugated dienes/cyt. C reductase). These data suggest that the reduced levels of the active cyt. P-450 form could be attributed, at least partially, to the activation of LPO processes. Rim exhibits strong preventive effect both against the increase of LPO products and the decrease of monooxygenase activities in infected animals. All parameters studied were partially restored (Fig. 1, 2). The effect of Rim on the contents of conjugated dienes was highest in the blood serum (2 fold) and in the liver (25%, Figs. 1, 2). In healthy animals Rim did not affect DMES activity. This fact allow us to exclude the possibillity for modulating effect on drug metabolism by Rim *"per se"* (but only if the pharmacokinetics and metabolism of Rim are not altered in the infected animals). The protective effect of Rim probably is not due to its antiradical activities, because in model systems *in vitro*, Rim did not exhibit any antioxidative effect (Fig. 3). Of course, we can not exclude such an effect of its metabolites formed *in vivo* (Wintermeyer et al., 1995). In our opinion, the main mechanism of the preventive effect of Rim on LPO and DMES probably is related with its antiviral activity. Some additional mechanisms can also contribute to the effect of Rim like inhibition of the processes of endocytosis, blocked interaction between virus and cell membrane; affection of the maturation of virus particles; conductivity changes of biomembranes etc. (Kharitonenkov et al. 1988, Cherny et al 1989, Zhiravetskii, 1986).

The experimental influenza virus infection-increased LPO products and the inhibitted hepatic drug-metabolizing enzyme activities are highly correlated (p < 0.001). Rimantadine treatment prevents both the activation of LPO processes and the inhibition of DMES activities in infected mice. Probably, the preventive effect of Rimantadine is related to its antiviral activity. Rimantadine has no direct antioxidative activity in model systems and had no modulating effect on DMES activity in healthy animals.

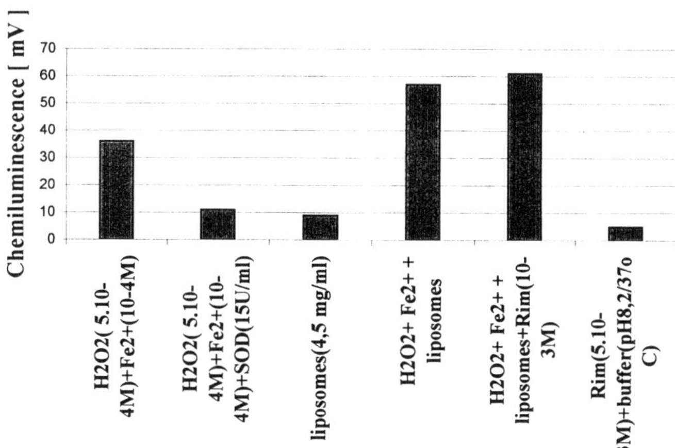

Figure 3. Luminol-dependent chemiluminescent analysis of the effect of Rimantadin on LPO processes in model membrane systems

REFERENCES

Belshe RB, Hall-Smith M, Hall CB, et. al., 1988, Genetic basis of resistance to rimantadine emergining during treatment of influenza virus infection. J. Virol., 62: 1508-1512

Buege JA, Aust SD, Microsomal lipid peroxidation. In: Methods in Enzymology. Fleisher S, Packer L (Eds). Academic Press: New York 1978, 65:302-310.

Buffinton GD, Christen S, Peterhans E, et. al. Oxidative stress in lungs of mice infected with influenza A virus. Free-Radic-Res-Commun. 1992, 16(2):99-110

Chetverikova LK, Inozemtseva LI Role of lipid peroxidation in the pathogenesis of the influenza and search for antiviral protective agents. Vestn Ross Acad Med Nauk 1996, 3:37- 40.

Folch J et al. A simple method for the isolation and purification of total lipid from animal tissues. J Biol Chem 1957, 226: 497-507

Cherny VS, Lerche D, Markin VS, The effect of rimantadine on the structure of model and biological membranes, Gen Physiol. Biophys. 1989, 8(1), 23-37

Kharitonenkov JR, El-Karadaghi S, Fedorov AN et al. Effect of remantadine and amantadine on the interaction of influenza virus proteins with model lipid membranes. Vopr. Virusol. 1988, 33(1), 22-6

Mazel P et. al., Experiments illustrating drug metabolism in vitro- determination of microsomal aniline hydroxylase. In: Fundamentals of drug metabolism and drug disposition. La Du B.N., Mandel HG, Way EL(Eds). The Willkins Co: Baltimore 1971, 546-550.

Matsubara et al. Quantitative determination of cytochrome P-450 in rat liver homogenate, Anal. Biochem. 75, 596-603, 1976

Nash TJ, The colorimetric estimation of formaldehyde by means of the Hautch reaction. Biol. Chem. 1953, 5:, 416-22.

Peterhans E, Reactive oxygen species and nitric oxide in viral diseases. Biol Trace Element Research 1997, 56: 107-115.

Recknagel R, Glende E, Spectrophotometric detection of lipid conjugated diens. Meth Enzymol 1984, 105: 331-337

Roering DL, Mascaro L Aust, S.D. Microsomal electron transport: Tetrafolium reduction by rat liver microsomal NADPH cytochrome C reductase. Arch Biochem 1972, 153: 475-9.

Zhiravetskii MM, Leont'eva MA, Vinograd JA et al. Mechanism of action of rim.hydrochl. on Sindbis virus reproduction: inhibition of virus penetration into the cell, Vopr Med Khim 1986, 32(6) 118-20

OXIDATIVE STRESS AND THE STRUCTURE/ACTIVITY RELATIONSHIPS OF ERGOPEPTIDE ALKALOIDS

Fabrice Bensaude[1], Geneviève Bouillé[2], and Marcel Delaforge[1,2]

[1]Service de Pharmacologie et d'Immunologie, Département de Recherche Médicale, Direction des Sciences du Vivant, CEA-Saclay, Bât. 136, F-91191 Gif sur Yvette, FRANCE
[2]URA 2096 CNRS « Protéines membranaires transductrices d'énergie »

INTRODUCTION

Oxidative stress is responsible for generating oxygen or non-oxygen free radicals, which are generally short-lived and highly reactive. It is well known[1] that such radicals proliferate in degenerative diseases, when their production overwhelms the body's defensive mechanisms. It has been suggested that free radicals may adversely influence the pathogenesis of Parkinson's disease (PD).

Bromocriptine (2-bromo-α-ergocriptine, BKT), an ergot alkaloid derivative, is widely used to treat PD. Yoshikawa[2] has shown that BKT has protective effects, possibly mediated by scavenging of free radicals, leading to decreased lipid peroxidation[3,4].

The present study was conducted to determine whether structure/activity relationships exist for ergopeptides and to elucidate the mechanism of defence against oxidative stress.

MATERIALS AND METHODS

Chemicals

All ergopeptides used were synthesized and kindly provided by Dr A. Jegorov (Galena Co.). Dihydroergotamine, NADP, NADPH, glucose-6-phosphate (G6P), glucose-6-phosphate dehydrogenase (G6PDH), ascorbic acid (Vitamin C, Asc.), thiobarbituric acid (TBA), trichloroacetic acid (TCA), butylated hydroxy toluene (BHT), were from Sigma. Iron (II) sulphate-7-hydrate was from Riedel de Haën Chemicals Co. Ergot alkaloid derivatives were dissolved in ethanol.

Biological Reactive Intermediates VI, Edited by Dansette *et al.*
Kluwer Academic / Plenum Publishers, 2001

Enzyme preparation

Sprague-Dawley male rats were treated intraperitoneally with dexamethasone (100 mg/kg, in corn oil, for 4 days). Rat liver microsomes were prepared as reported previously[5] and stored at – 80°C before use.

Analysis of inhibitory effects

Incubations (final volume of 1 ml) were typically carried out for 15 min at 37°C with a mixture of rat liver microsomal suspensions (equivalent to 1 µM P450), increasing amounts of ergopeptides (0-100 µM), an NADPH-generating system (NADP+ 0.5 mM, G6P 5mM and 1 U G6PDH) and 5 µL of 1 M CCl_4 in ethanol, diluted in a 0.1 M potassium phosphate pH 7.4 buffer,. $FeSO_4$/ascorbate replaced the NADPH/CCl_4 generating system. The reaction was stopped by addition of 150µL of 0.2% BHT in ethanol and addition of 2 mL of TBA test solution (15% TCA, 0.375% TBA, 0.25 M HCl in water). Samples were incubated for 15 min at 95°C and centrifuged for 10 min at 3000 rpm. The supernatant was analyzed in a Perkin Elmer-λ18 UV-visible spectrophotometer at 535nm, as described by Pré[6].

HPLC Analysis

HPLC analysis was performed as described previously[5]. Incubations were performed as described except that they were stopped by addition of 1 mL acetonitrile. Samples were centrifuged for 10 min at 3000 rpm and 100µL of the supernatant was injected.

RESULTS

Inhibitory effects of ergopeptides

Various ergot derivatives (Fig.1) are able to interact with the P450 3A active site leading to hydrophobic interaction (Type I) as shown by Peyronneau[5]. Their affinities are in the range (Ks = 0.1 to 10 µM, (Table 1). These compounds exhibit inhibitory effects toward lipid peroxidation initiated by NADPH/CCl_4 or $FeSO_4$/ascorbate by reaction of different radicals as $CCl_3°$, $OH°$ or $O_2°^-$. The values (table 1), determined by evaluation of TBA-reactive substances (TBARS) with the method described below, suggest that ergot alkaloid derivatives protect phospholipids against radical peroxidation. This protection is unaffected by the absence of or differences in the tripeptide moiety and/or in the tetracyclic structure of ergopeptides.

Moreover, the amounts of ergopeptides required, compared to the absolute values of TBA-reactive substances produced (table 2), suggest that there are two possible mechanisms explaining the inhibitory effects. In fact, ergopeptides can form either a stable complex with radicals or enter a futile cycle where the ergot-radical complex is dissociated into an unreactive form.

HPLC Analysis

Identification and characterization of metabolites of ergopeptides were described by Maurer[8]. In this study, we tried to demonstrate the presence of adducts between ergot alkaloid derivatives and free radicals produced during oxidative stress.

There were no qualitative or quantitative differences between HPLC of DHEKT observed after incubation in the presence of NADPH alone or upon addition of CCl_4. In the presence of $FeSO_4$/ascorbate, the chromatograms showed only the peak of DHEKT with no qualitative differences.

This analytical method therefore provided no proof of new compounds, even by scanning wavelengths between 210 and 380 nm with a photodiode array detector.

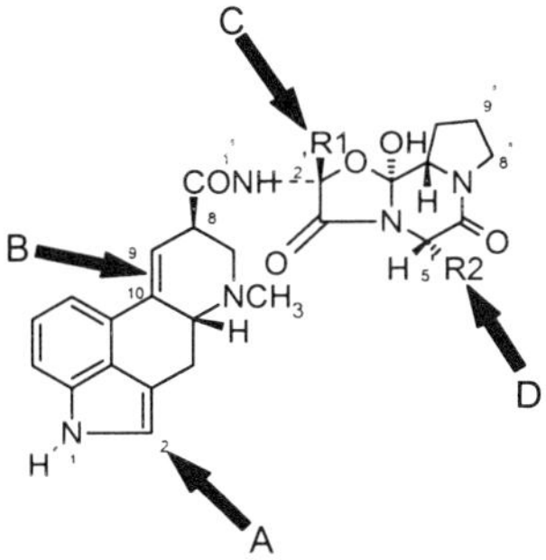

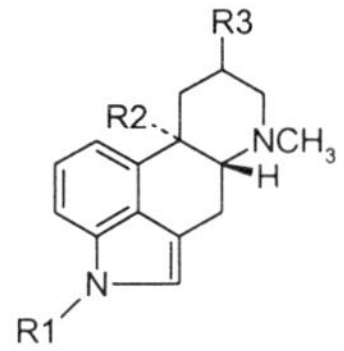

Substituents on peptide moiety

	Ergotamine	Ergocriptine
R1	CH$_3$	CH(CH$_3$)$_2$
R2	CH(CH$_3$)$_2$	CH2CH(CH$_3$)$_2$

Substitutions

A = oxidation: 2-oxo

B = saturated bound C9-C10: dihydro-

	1-MMDL	MDL	Terguride	Agroclavine
R1	CH$_3$	H	H	H
R2	OCH$_3$	OCH$_3$	H	H
R3	CH$_2$OH	CH$_2$OH	NHCON(Et)$_2$	CH$_3$
Double bond	-	-	-	C8-C9

Fig. 1. Structures of some ergot derivatives.

Table 1. Spectral interaction (A$_{max}$/P450, Ks) and inhibitory effects (IC$_{50}$) of ergopeptides with liver microsomes from dexamethasone-treated rats. Difference spectra were recorded by UV-visible spectroscopy. The reported values were calculated from the linear curve 1/ΔA = f (1/[Substrate]) for A$_{max}$/P450 and Ks, from a semilog curve % initial activity = f ([Substrate]).

Substrate	A$_{max}$/P450 (nmol^{-1})	Ks (μM)	IC$_{50}$ (μM) NADPH/CCl$_4$	FeSO$_4$/Asc
Dihydroergotamine	0.04 ± 0.02	1 ± 0.4	7	
α-ergocriptine	0.03 ± 0.01	0.4 ± 0.2	12	8
Dihydro-α-ergocriptine	0.08 ± 0.03	2.8 ± 1.8	11	25
2-oxo-α-ergocriptine	0.06 ± 0.03	1.7 ± 1.5	12	20
2-oxo-dihydro-α-ergocriptine	0.07 ± 0.01	1.2 ± 0.3	21	15
Terguride	0.035	4	60	5
Agroclavine	0.005	n.d.[1]	8	15
MMDL	0.003	n.d.[1]	>75	>75
MDL	0.003	n.d.[1]	>75	>75

[1]n.d.: not detectable

Table 2. Amounts of TBARS produced by lipid peroxidation. The amounts were determined from a standard curve plotted using 1,1,3,3 tetraethoxypropane[7].

		[ergot] (μM)	TBARS (μM)
NADPH/CCl$_4$	α-ergocriptine	0	9.6
		12	4.8
FeSO$_4$/Asc	α-ergocriptine	0	50
		8	25

Table 3. Results of HPLC chromatograms from incubations of dihydro-α-ergocriptine mesylate (DHEKT) with liver microsomes from dexamethasone-treated rats. Incubations contained 1 μM P450 from rat liver microsomes and 100 μM DHEKT and were performed at 37°C for 15 min, as described in methods. Peaks u, v, w, x, y, z were metabolites not found in blank incubations.

Incubation	t_R	17.7	19.6	23.6	24.2	25.5	26.4	27.6
	Metabolite	z	y	x	w	v	u	DHEKT
DHEKT NADPH	**% Area**	4.7	25.8	4.8	4	14.5	34	12.1
DHEKT NADPH/CCl$_4$	**% Area**	4.2	19.1	4.4	3.7	15.5	38.3	14.6

DISCUSSION

1. Ergot derivatives are potent inhibitors of lipid peroxidation induced by either CCl3°(NADPH/CCl$_4$) or OH° (FeSO$_4$/ascorbate).

2. The peptide moiety is not necessary for antioxidant activities. The peptide moiety has no effect on the nature of the modifications.

3. Similarly, the presence of conjugated double bonds and the methylation of 1-N are not absolute requisites. All these factors enhanced inhibition since the best inhibitors simultaneously possess: i) the peptide moiety, ii) conjugated aromatic + two double bonds, like α-ergocriptine.

4. Absence of detectable amounts of DHEKT-radical complexes and stoichiometric measurements suggest that a transition to a futile cycle resulted in deactivation of the radical species.

REFERENCES

1. I. N. Acworth, D. P. B. Bailey, The handbook of oxidative metabolism, *ESA, Inc.* (1996).
2. T. Yoshikawa, Y. Minamiyama, Y. Naito, and M. Kondo, Antioxidant properties of bromocriptine, a dopamine agonist, *J. Neurochem.* 62:1034 (1994).
3. M. Gomez-Vargas, S. Nishibayashi-Asanuma, M. Asanuma, Y. Kondo, E. Iwata, and N. Ogawa, Pergolide scavenges both hydroxyl and nitric oxide free radicals in vitro and inhibits lipid peroxidation in different regions of the rat brain, *Brain Res.* 790:202 (1998).
4. M. Tanaka, A. Sotomatsu, T. Yoshida, and S. Hirai, Inhibitory effects of bromocriptine on phospholipid peroxidation induced by DOPA and iron, *Neuroscience letters* 183: 116 (1995).
5. M.A. Peyronneau, M. Delaforge, R. Rivière, J.P. Renaud, and D. Mansuy, High affinity of ergopeptides for cytochromes P450 3A: importance of their peptide moiety for P450 recognition and hydroxylation of bromocriptine, *Eur. J. Biochem.* 223:947 (1994).
6. J. Pré, Revue générale: la lipoperoxidation, *Path. Biol.* 39:716 (1991).
7. K. Yagi, Simple assay for the level of total lipid peroxides in serum or plasma, *Methods in molecular biology, Free radical and antioxidant protocols*, 108:101 (1998).
8. G. Maurer, E. Scheirer, S. Delaborde, R. Nufer, and A.P. Shukla, Fate and disposition of bromocriptine in animals and man 2: Absorption, elimination and metabolism, *Eur. J. Drug. Metab. Pharmacokinet.* 8:51 (1983).

MULTIPLE OXIDATIVE STRESS PARAMETERS ARE MODULATED *IN VITRO* BY OXYGENATED POLYCYCLIC AROMATIC HYDROCARBONS IDENTIFIED IN RIVER SEDIMENTS

Luděk Bláha[1,2], Miroslav Machala[2], Jan Vondráček[2,3], and Klára Breineková[1]

[1]Department of Environmental Chemistry and Ecotoxicology, Masaryk University, Veslařská 230B, 637 00 Brno, Czech Republic

[2]Veterinary Research Institute, Hudcova 70, 621 32 Brno, Czech Republic

[3]Institute of Biophysics, Czech Academy of Sciences, 602 00 Brno, Czech Republic

INTRODUCTION

Oxidative stress has been recognized to be an important mechanism of toxic action of many compounds, such as quinones or peroxides. Increased production of reactive oxygen species (ROS) can cause significant changes in cellular redox state and may lead to: 1) adverse alterations in gene expression (Toyokuni *et al.*, 1995), or 2) oxidative damage to cellular molecules, i.e. proteins, DNA, and phospholipids (de Zwart *et al.*, 1999).

We introduced Hepa-1c1c7 cells as an *in vitro* model for studies of oxidative stress. Multiple *in vitro* parameters were followed, including both increased production of ROS and "secondary effects", i.e. changes in intracellular concentrations of reduced glutathione (GSH), and formation of 8-hydroxy-2′-deoxyguanosine (8-OH-dG). Two quinones, 2-methyl-1,4-naphthoquinone (menadione MD, a known redox-cycling compound) and 1,4-naphthoquinone (NQ, which toxicity is mostly due to arylating activity), were used as reference substances (Henry and Wallace, 1996).

Several classes of environmental pollutants have been suggested to induce oxidative stress, such as derivatives of carcinogenic polycyclic aromatic hydrocarbons, PAHs (Lehr and Jerina, 1977). Five oxygenated derivatives (oxy-PAHs), 9-fluorenone, anthracene-9-one, anthracene-9,10-dione, benz[de]anthracene-7-one, and benzo[a]anthracene-7,12-dione (Fig. 1), were originally identified in our laboratory in the river sediment extracts from the Czech Republic (Machala *et al.*, submitted). PAHs represent some of the most important environmental pollutants and our hypothesis was that oxy-PAHs, as well as the respective parent unsubstituted PAHs, can induce oxidative stress. The effects of environmentally relevant PAHs on ROS production, GSH status and the formation of oxy-DNA in vitro in Hepa-1c1c7 cells are reported in this study.

Biological Reactive Intermediates VI, Edited by Dansette *et al.*
Kluwer Academic / Plenum Publishers, 2001

MATERIALS AND METHODS

Model quinones (MD and NQ) were purchased from Fluka Chemie (Buchs, Switzerland), PAHs and their derivatives were supplied by Ehrenstorfer (Augsburg, Germany). All other chemicals were of the highest purity available. Mouse hepatoma cells Hepa-1c1c7 (ATCC CRL-2026) were cultured in Dubelcco´s MEM supplemented with 5% FBS. Prior to measurement of oxidative stress parameters, the cytotoxicity of all tested compounds was determined by MTT test (Mosmann, 1983). Dimethylsulfoxide (DMSO) was used as a negative control and its final concentration did not exceed 0.2% (v/v). Production of ROS (2×10^5 cells/mL) was measured by lucigenin-derived chemiluminescence assay (Li *et al.*, 1998). Intracellular concentrations of reduced GSH were monitored by kinetic microplate method based on conjugation of GSH with thiol-selective fluorescent probe monochlorobimane (Rice *et al.*, 1986). Oxidative damage to DNA was measured by OxyDNA Assay (Biotrin, Dublin, Ireland) using FITC-labeled antibody developed against 8-OH-dG. Fluorescence of cells was measured by flow cytometry (Becton-Dickinson FACS Calibur). Data are expressed as mean +/- standard deviation of at least two independent experiments, each performed in three replicates. Statistical calculations were performed with GraphPad Prizm (GraphPad Software, San Diego, CA, USA).

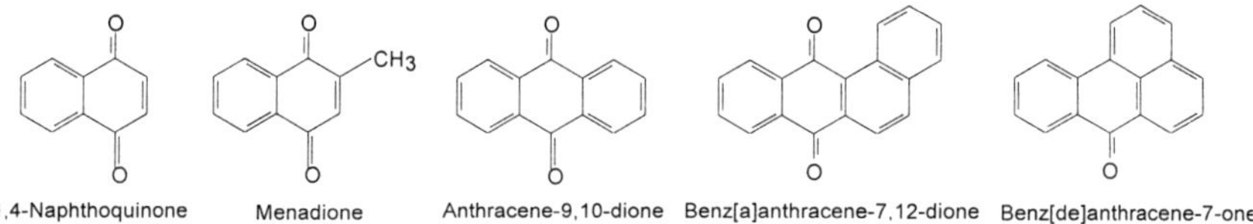

Figure 1. Structures of model quinone compounds and selected oxy-PAHs.

RESULTS AND DISCUSSION

Model system for *in vitro* studies of oxidative stress

MD (a model redox cycling quinone) and NQ (which cytotoxicity is thought to be mediated by arylating activity) were used as reference substances to evaluate an *in vitro* system of oxidative stress detection. Cytotoxicity of reference quinones was measured by the MTT assay at various exposure periods (1, 6, and 24 h) and expressed as EC_{50} values. Because cytotoxicity of quinones varied significantly (e.g., 6 h EC_{50} values were 99 and 11 µM for MD and NQ, respectively), relative "EC_{50}-derived" concentrations for each compound (i.e. $EC_{50}/2$, $EC_{50}/5$, etc.) were selected for further studies to better compare their mechanisms of toxic action. Effects of model quinones on ROS production, intracellular concentrations of GSH and formation of oxy-DNA are shown in Fig. 2. While NQ caused only a weak induction of ROS production, a more significant effect was observed after addition of MD, with maximum effect at 15 µM (i.e. $EC_{50}/10$ for 2 h exposure, Fig. 2A). Similarly, a significant concentration- and time-dependent increase in oxidative DNA damage was observed after MD, and not NQ treatment (Fig. 2B). Both MD and NQ significantly depressed GSH concentrations at similar concentrations (Fig. 2C, e.g. 50% decline after 2 h exposure was observed at 15 and 12 µM for MD and NQ, respectively). However, when these concentrations were expressed as the "EC_{50}-based" ones, there was a significant difference between both compounds (15 µM represents $EC_{50}/10$ for MD, and 12 µM represents $EC_{50}/2$ for NQ).

Our results show, that *in vitro* measurement of ROS production, as well as the less sensitive detection of oxy-DNA formation, could successfully distinguish between redox-cycling (MD) and arylating (NQ) quinone compounds. On the other hand, changes in GSH concentrations seem to be associated with both oxidative stress and with conjugations, or some other nonspecific events. Nevertheless, GSH can be considered the significant parameter related to oxidative stress, as well as to other adverse epigenetic effects, such as modulation of gene expressions (Toyokuni *et al.*, 1995), and gap junctional intercellular communication (Upham *et al.*, 1997).

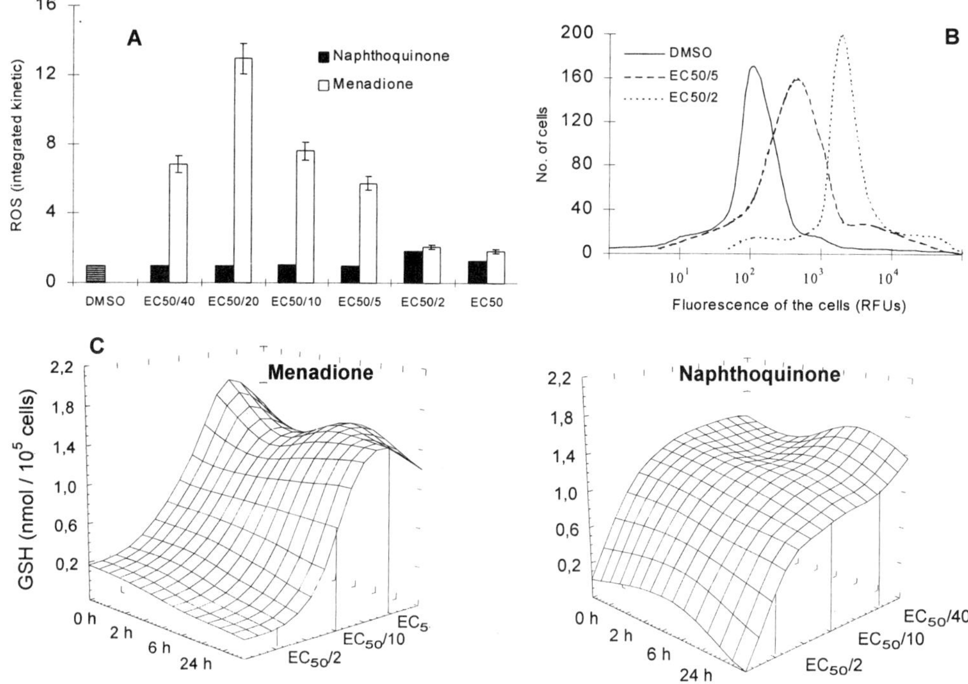

Figure 2. *In vitro* effects of reference quinones on oxidative stress parameters in Hepa-1c1c7 cells. ROS production measured as an integrated area under curve of kinetic record, (B) effect of menadione on oxy-DNA formation after 6 h exposure, (C) effects on intracellular GSH concentrations.

Modulation of oxidative stress parameters by PAHs

In vitro modulation of oxidative stress parameters of five oxy-PAHs and their parent unsubstituted PAHs (fluorene, anthracene, benz[a]anthracene) were studied using Hepa-1c1c7 cells. In MTT assay, no cytotoxicity of the tested compounds was detected using concentrations up to 100 µM concentrations and all exposure periods. Higher concentrations could not be tested correctly due to low solubility of some PAHs.

As shown in Fig. 3A, three oxy-PAHs (9-fluorenone, anthracene-9-one, and anthracene-9,10-dione) caused a significant increase in ROS production immediately after addition to cells (exposure 0 h). None of the unsubstituted PAHs induced ROS production at that time. However, measurements after prolonged exposure (6 h) showed that benzo[a]anthracene is capable to induce ROS production. Hepa-1c1c7 cells have been shown to express inducible cytochromes P450IA1, which are involved in oxidative metabolism of PAHs (Sindhu *et al.*, 1999). Thus, it seems likely that oxidative stress detected after 6 h could be induced by oxy-metabolites of benzo[a]anthracene. However, benzo[a]anthracene-7,12-dione, the tested quinone derivative of benzo[a]anthracene (originally identified in river sediments), did not appear to induce oxidative stress.

The intracellular concentrations of GSH were significantly affected by several PAHs in a time-dependent manner as shown in Fig. 3B. In accordance with ROS production, anthracene-9-one and anthracene-9,10-dione showed effects within short exposure times (up to 6 h), while benzo[a]anthracene significantly decreased GSH concentrations after prolonged incubation periods (6 h). The most cytotoxic oxy-PAH tested, benzo[de]anthracene-7-one, also induced a significant decrease in GSH concentration. Thus, similar to model quinones, GSH depletion after oxy-PAHs treatment could be affected by some nonspecific cytotoxic mechanisms. Although an increased ROS production and significant decrease in GSH concentrations were observed, none of the PAHs caused a significant oxy-DNA formation *in vitro* (data not shown). However, ROS production of PAHs was much lower than that of model quinone, MD (Fig. 2 and 3). Thus, a weak induction of ROS by PAHs, as well as possible DNA repair could result in the absence of significant oxidative damage to DNA.

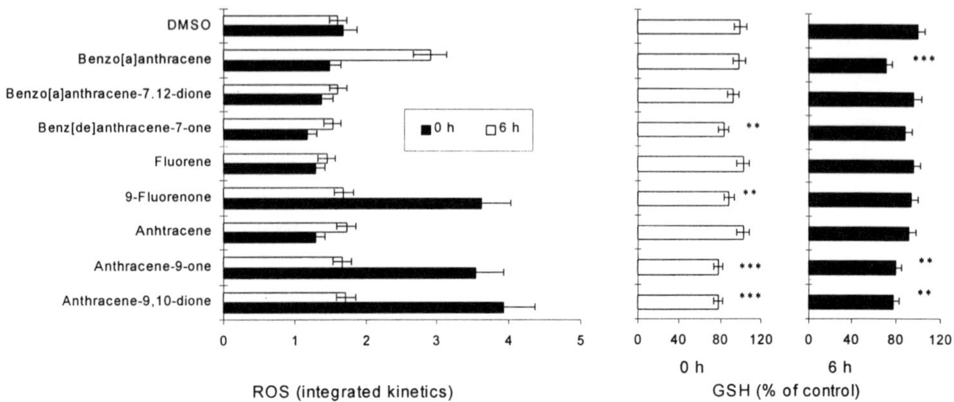

Figure 3. Effects of PAHs on oxidative stress parameters (ROS production and GSH concentrations) in Hepa-1c1c7 cells *in vitro* (**p<0.01, ***p<0.001, ANOVA with Dunnet´s post-test).

In conclusion, *in vitro* model using Hepa-1c1c7 cells can be successfully used to distinguish between redox-cycling and arylating model quinone compounds. While, detection of ROS and oxidative damage to DNA seem to be suitable oxidative stress biomarkers, a decrease in GSH reflected both oxidative stress and other epigenetic events. Studying of PAHs and their oxy-derivatives showed that besides other known effects, oxidative stress could play an important role in the mechanisms of toxic action of these environmentally relevant pollutants.

Acknowledgments
This research was supported by by Grant Agency of the Czech Republic (No. 525/98/1266) and National Agency for Agricultural Research (No. QC0194/2000).

REFERENCES

De Zwart, L. L., Meerman, J.H.N., Commandeur, J.N.M., and Vermeulen, N.P.E., 1999, Biomarkers of free radical damage - applications in experimental animals and in humans, *Free Radical Biol. Med.* **26**:202-226.

Henry, T.R., and Wallace, K.B., 1996, Differential mechanisms of cell killing by redox cycling and arylating quinones, *Arch. Toxicol.* **70**:482-489.

Lehr, R.E., and Jerina, D.M., 1977, Metabolic activation of polycyclic hydrocarbons, *Arch. Toxicol.* **39**:1-28.

Li, Y., Zhy, H., Kuppusamy, P., Roubaud, V., Zweier, J.L., and Trush, M.A., 1998, Validation of lucigenin (bis-*N*-methylacridinium) as a chemmilumigenic probe for detecting superoxide anion radicals production by enzymatic and cellular systems, *J. Biol. Chem.* **273**:2015-2023.

Machala, M., Ciganek, M., Vondráček, J., Bláha, L., and Minksová, K., Dioxin-like and estrogenic activities of oxygenated polycyclic aromatic hydrocarbons identified in extracts of river sediments, *Environ. Sci. Technol.* (submitted).

Mosmann, T., 1983, Rapid colorimetric assay for cellular growth and survival: application to proliferation and cytotoxicity assays, *J. Immunol. Meth.* **65**:55-63.

Rice, G.C., Bump., E.A., Shrieve, D.C., Lee, W., and Kovacs, M., 1986, Quantitative analysis of cellular glutathione by flow cytometry utilizing monochlorobimane: some applications to radiation and drug resistance *in vitro* and *in vivo*, *Cancer Res.* **46**: 6105-6110.

Sindhu, R.K., Rasmussen, R.E., and Kikkawa, Y., 1999, Induction of cytochrome P4501A1 by ozone-oxidized tryptophan in Hepa-1c1c7 cells, *Adv. Exp. Med. Biol.* **467**:409-18.

Toyokuni, S., Okamoto, K., Yodoi, J., and Hiai, H., 1995, Persistent oxidative stress in cancer., *FEBS Lett.* **358**:1-3.

Upham, B.L., Kang, K.-S., Cho, H.-Y., and Trosko, J.E., 1997, Hydrogen peroxide inhibits gap junctional intercellular communication in glutathione sufficient but not glutathione deficient cells. *Carcinogenesis* **18**:37-42.

1,N^6-ETHENO-2'-DEOXYADENOSINE ADDUCTS FROM *TRANS,TRANS*-2,4-DECADIENAL AND *TRANS*-2-OCTENAL

Valdemir M. Carvalho[1], Flavio Asahara[1], Paolo Di Mascio[1],
Ivan P. de Arruda Campos[2], Jean Cadet[3] and Marisa H. G. Medeiros[1]*

[1]Departamento de Bioquímica, Instituto de Química
Universidade de São Paulo, CP 26.077, CEP 05513-970, São Paulo, Brazil
[2]Instituto de Ciências Exatas e Tecnologia, Universidade Paulista, Av.
Alphaville, 3500 - Santana do Parnaíba, SP, CEP 06500-000, Brazil
[3]Département de Recherche Fondamentale sur la Matière Condensée
SCIB/LAN, CEA/Grenoble F-38054 Grenoble Cedex 9, France

INTRODUCTION

Ethenoadducts have attracted considerable interest when they were detected in the reactions between DNA and occupational carcinogens such as vinyl chloride, urethane and acrylonitrile[1]. The deleterious effects of ethenobases were further attested by studies showing their high mutagenic potential[2,3]. Interestingly, background levels of 3,N^4-etheno-2'-deoxycytidine (3,N^4-εdC) and 1,N^6-etheno-2'-deoxyadenosine (1,N^6-εdA) were detected in tissue DNA from untreated rodents and humans[4]. However, the origin of the adduct background is still open to debate. It is strongly suggested that increased oxidative stress and lipid peroxidation (LP) are implicated in this process[1]. In fact, most of LP final products are very efficient alkylating agents. Among these products, malonaldehyde and HNE, are the most extensively studied. However, only HNE was shown to form DNA ethenoadducts[5]. Due to the lack of information about other LP products, the bulk of the lesions have been attributed to HNE. Even though, other products are reported to be very cytotoxic and can also play an important role in the genotoxic effects associated with LP.

Actually, 2,4-decadienal (DDE) and 2-octenal have been reported to be highly cytotoxic[6,7]. DDE has been shown to be one of the most toxic breakdown products of lipid peroxidation to cells. This was inferred from the observed cytotoxic and lethal effects exerted by this aldehyde on different models such as human fibroblasts, endothelial and erythroleukemia cells[8-10].

In this report, we summarize the recent studies that we have performed on the identification of the etheno dAdo adducts produced by DDE and 2-octenal[11,12]. In addition, information is provided on the mechanism of their formation and reactivity towards DNA.

Biological Reactive Intermediates VI, Edited by Dansette *et al.*
Kluwer Academic / Plenum Publishers, 2001

CHARACTERIZATION OF ETHENO-2'-DEOXYGUANOSINE ADDUCTS FROM 2,4-DECADIENAL AND 2-OCTENAL

The reaction of DDE with 2'-deoxyadenosine (dA) in the presence of H_2O_2, led to the formation of five major fluorescent compounds. Besides $1,N^6$-εdA, four hydrophobic fluorescent products were identified. The characteristic UV and fluorescence spectra suggested that they could be $1,N^6$-εdA derivatives[11]. The LC-MS analyses (Figure 1) are consistent with an increase in the molecular mass of the adducts by 184, 154, 124 and 166 Da with respect to the dA moiety. This could be rationalized in terms of the formation of 1,2,3-octanetriol (adduct **I**), 1,2-heptanediol (adduct **II**), 1-hexanol (adduct **III**) and 2,3-epoxy-1-octanol (adduct **IV**) $1,N^6$-εdA derivatives, respectively. The structural assignment of the latter adducts received further support from extensive NMR measurements. Interestingly, adduct **II** is identical to the product of the HNE epoxide/dA reaction. In addition, adduct **III** was also obtained as the main product from the reaction between 2-octenal epoxide and dA.

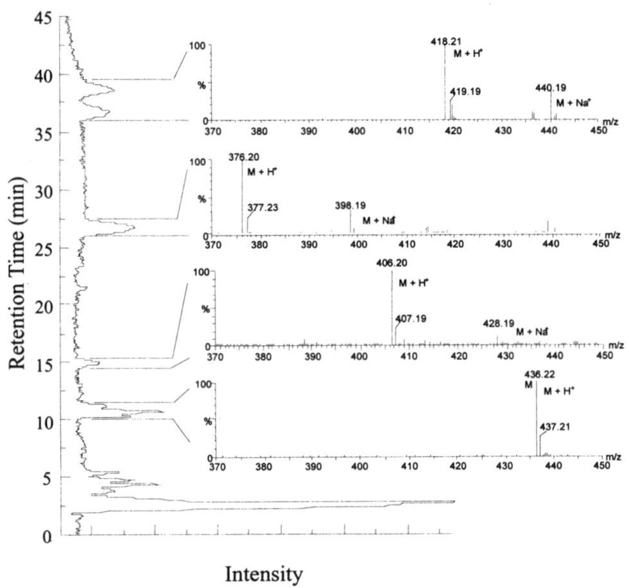

Figure 1. LC-MS Profile of the adducts formed between dAdo and DDE epoxides. Reprinted with permission from ref. 12. Copyright 2000 American Chemical Society.

The formation of the four ethenoadducts through the reaction of DDE with dA requires the oxidation of the aldehyde with the subsequent generation of a mixture of products such as epoxides, diepoxides and hydroperoxides. The proposed pathway of formation of the adducts is given in Figure 2. The double epoxidation of DDE could lead to the formation of adducts **I** and **IV**. In the first case, the 4,5-epoxy group underwent hydrolysis producing a 4,5-epoxy-2-decenal that gave rise adduct **I**. The formation of adduct **II** from this reaction suggests that 2,3-epoxy-4-hydroxynonanal is also one of the DDE oxidation products. Formation of adduct **III** may be rationalized in terms of initial production of either 2-octenal or its corresponding epoxide. A possible mechanism for 2-octenal formation upon DDE oxidation has been previously reported[13]. In the presence of peroxides, 2-octenal could form 2,3-epoxy-octanal, which, on reaction with dA, yields adduct **III**.

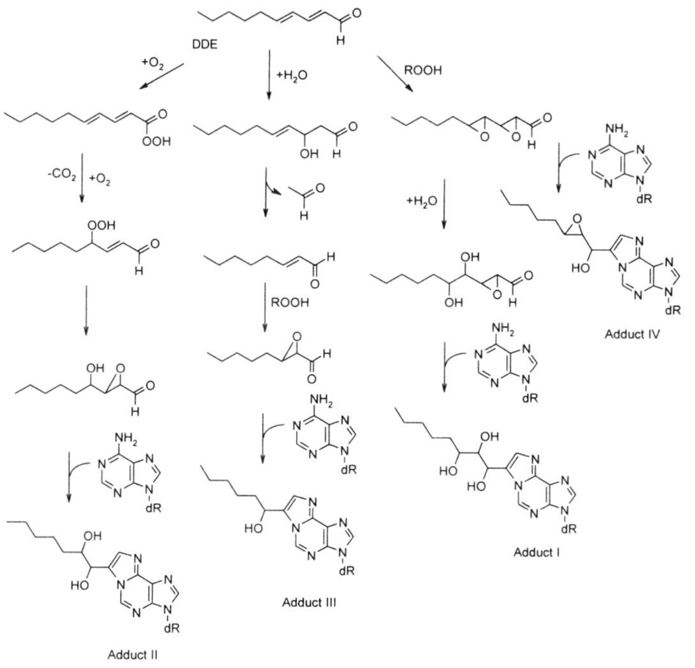

Figure 2. Mechanistic pathways for the formation of the four DDE adducts to dA. Reprinted with permission from ref. 12. Copyright 2000 American Chemical Society.

REACTIVITY OF DDE, HNE AND 2-OCTENAL TOWARDS CALF THYMUS DNA

It was shown from the characterization of the adducts that two of them are the expected products arising from the reaction of dA with two other important LP products. Adduct **II** is identical to the product of the reaction of HNE epoxide with dA[14] and adduct **III** is generated by the reaction of the corresponding 2-octenal epoxide with dA. For these reasons, attempts were made to compare the reactivity of DDE, HNE and 2-octenal towards dA within calf thymus DNA in the presence of H_2O_2 (Table 1). 2-Octenal and HNE were found to be more reactive than DDE; however the latter is a versatile alkylating agent being able to generate, after epoxidation, four different $1,N^6$-εdA derivatives in its reaction with DNA. Even though DDE generates lower levels of adducts **II** and **III** than 2-octenal and HNE, the larger variety of ethenoadducts produced by DDE reactions could account for the high cytotoxic properties of the aldehyde.

Table 1. Adducts Levels from the Reaction of DNA with Octenal HNE and DDE epoxides[1]

	Adduct level (adducts/10^3 dA)			
	I	**II**	**III**	**IV**
Control	nd[a]	nd	nd	nd
Octenal	nd	nd	114.1 ± 17.7	nd
HNE	nd	179.8 ± 2.9	nd	nd
DDE	6.5 ± 2.1	4.9 ± 0.5	26.4 ± 3.0	6.9 ± 0.4

[1]Reprinted with permission from ref. 12. Copyright 2000 American Chemical Society.

CONCLUSIONS

Several carcinogenic and mutagenic compounds were shown to form etheno adducts with DNA. The detection of background levels of etheno bases in tissues of untreated animals is suggestive of endogenous sources of reactive aldehyde intermediates, possibly from lipid peroxidation processes. Our recent data indicate that DDE after oxidation, can generate a variety of reactive intermediates and these products are able to form different $1,N^6$-εdA adducts with DNA. Indeed DDE can produce the expected adducts which are generated by HNE, 2-octenal and 4,5-dihydroxy-2-decenal, three other breakdown products of lipid peroxidation.

The increasing interest in the use of ethenoadducts as biomarkers of cancer risk associated with dietary fat intake, oxidative stress, chronic inflammatory/infectious processes, and protective dietary antioxidants should take into account the different mechanisms of adduct formation. Considering the importance of DDE as a compound present in foods and as a product of lipid peroxidation, the present results can contribute to a better understanding of the cytotoxic and genotoxic mechanisms associated with this compounds.

Acknowledgements

This work was supported by FAPESP (Brazil), CNPq (Brazil), PRONEX/FINEP (Brazil) and Comité Français d'Evaluation de la Coopération Universitaire avec le Brésil (USP-COFECUB/UC-23/96). V.M.C. and F.A. are recipients of FAPESP and CNPq fellowships, respectively.

REFERENCES

1. H. Bartsch, A. Barbin, M.-J. Marion, J. Nair, and Y. Guichard. *Drug Metab. Rev.* 26, 349–371 (1994).
2. A.M. Moriya, W. Zhang, F. Johnson, and A.P. Grollman. *Proc. Natl. Acad. Sci. USA*, 91, 11899–11903 (1994).
3. G.A. Pandya, and M. Moriya. *Biochemistry*, 35, 11487–11492 (1996).
4. J. Nair, A. Barbin, Y. Guichard, and H. Bartsch. *Carcinogenesis*, 16, 613–617 (1995).
5. F.-L. Chung, H.-J. Chen, and R.G. Nath. *Carcinogenesis* 17, 2105–2111(1996).
6. C. Nappez, S. Battu, J.L. Beneytout. *Cancer Lett.* 99, 115–119, (1996).
7. A. Benedetti, M. Comporti, R. Fulceri, and H. Esterbauer. *Biochim. Biophys. Acta.* 792, 172–181, (1984).
8. T. Kaneko, S. Honda, S. Nakano, and M. Matsuo. *Chem.-Biol.Interactions* 63, 127–137 (1987).
9. T. Kaneko, K. Kaji, and M. Matsuo. *Chem.-Biol.Interactions* 67, 295–304 (1988).
10. C. Nappez,, S. Battu, and J.L. Beneytout. *Cancer Lett.* 99, 115–119 (1996).
11. V.M. Carvalho, P. Di Mascio, I.P Arruda-Campos, T. Douki, J. Cadet, and M.H.G. Medeiros. *Chem. Res. Toxicol.* 11, 1042–1047 (1998).
12. V.M. Carvalho, F. Ashara, P. Di Mascio, I.P Arruda-Campos, J. Cadet, and M.H.G. Medeiros. *Chem. Res. Toxicol.* 13, 397–405 (2000).
13. D.B. Josephson, and R.C. Lindsay. *J. Food Sci.* 52, 1186–1190 (1987).
14. R.S. Sodum, and F.L. Chung. *Cancer Res.* 51, 137–143, (1991).

HYDROGEN PEROXIDE SUPPORTS HEPATOCYTE P450 CATALYSED XENOBIOTIC/DRUG METABOLIC ACTIVATION TO FORM CYTOTOXIC REACTIVE INTERMEDIATES

T. S. Chan, M. Moridani, A. Siraki, H. Scobie, K. Beard, M. A. Eghbal, G. Galati, and P. J. O'Brien [1]

Faculty of Pharmacy
University of Toronto
Toronto, Ontario, Canada, M5S 2S2
[1]Corresponding author. Phone: (416) 978-2716. Fax: (416) 978-8511
Email: peter.obrien@utoronto.ca

ABSTRACT

1. A H_2O_2 generating system markedly increased the cytotoxicity of catechols, hydroquinone, in isolated hepatocytes, but not in P450 inhibited hepatocytes.
2. H_2O_2 or NADPH supported microsomal catalysed GSH conjugate formation with catechols or hydroquinone. Cytochrome P450 inhibitors inhibited conjugate formation. However, superoxide dismutase inhibited NADPH, but did not affect H_2O_2 supported GSH conjugate formation. The conjugate formed with dihydrocaffeic acid was identified as a *mono*-GSH conjugate indicating that the *o*-quinone was the major metabolite formed.
3. Dopamine (a catecholamine) induced cytotoxicity was prevented by inhibitors of monoamine oxidase (MAO) or P450, but was markedly increased by hepatocyte catalase inhibition or NAD(P)H:quinone oxidoreductase inhibition. This suggests that H_2O_2 formed by the mitochondrial metabolism of monoamine oxidase then oxidised dopamine to cytotoxic *o*-quinone catalysed by P450. Dihydrocaffeic acid cytotoxicity was also increased by the monoamine oxidase substrate tyramine.
4. It is concluded that polyphenolics are oxidised by H_2O_2/P450 in hepatocytes to form quinone metabolites.

INTRODUCTION

Many xenobiotics/drugs cause tissue toxicity or initiate chemical carcinogenesis as a result of metabolic activation by cytochrome P450 to form reactive intermediates, which oxidise or alkylate essential cellular macromolecules[1]. Recently we have shown that xenobiotic drug metabolic activation by isolated rat hepatocytes can be markedly increased up to 20-fold by the addition of non-toxic concentrations of hydroperoxides[2] or the physiological hydroperoxide linoleic acid hydroperoxide[3] and was prevented by cytochrome P450 inhibitors. Organic hydroperoxides were also shown to markedly increase the cytotoxicity of 4-hydroxyanisole and its O-demethylation to the toxic benzoquinone metabolite[2]. Linoleic acid hydroperoxide also markedly increased the phenelzine cytotoxicity and its oxidative metabolism to ethylbenzene[3]. H_2O_2 was also shown to support P450 CYP1A2 catalysed N-oxidation of the heterocyclic amine 2-amine 2-amino-3-methylimidazo[4,5]quinoline to a hydroxylamine metabolite which formed a 2'-deoxyguanosine adduct[4]. However, it has not yet been shown whether extracellular or intracellular H_2O_2 generating systems can increase the oxidation of polyphenolics or catecholamines resulting in increased cytotoxicity in intact cells and if so can this be attributed to cytochrome P450 peroxidase activity[5].

MATERIALS AND METHODS

Chemicals

All chemicals were purchased from Sigma-Aldrich Chemical Co. (Oakville, ON, Canada). Collagenase (from *Clostridium histolyticum*), 4-(2-hydroxyethyl)-1-piperazine ethanesulfonic acid (HEPES), and bovine serum albumin were obtained from Boehringer-Mannheim (Montreal, QU, Canada). Deferoxamine (deferrioxamine mesylate, Desferal) was a gift from Ciba Geigy Canada Ltd (Toronto, ON, Canada).

Hepatocyte Isolation and Subcellular Organelle Preparation

The hepatocytes were prepared by collagenase perfusion of the liver and the viability of the hepatocytes was assessed by the trypan blue (0.2% w/v) exclusion test[2]. Hepatocytes were pre-incubated in Krebs-Henseleit bicarbonate buffer (pH 7.4) supplemented with 12.5mM HEPES for 30min under a carbogen atmosphere in continuously rotating 50mL round bottom flasks at 37°C[6] before the addition of chemicals[2]. Microsomal and cytosolic fractions were prepared as described[2,4].

MS Analysis of Dihydrocaffeic Acid-GSH conjugate

Rat liver microsomes (1mg/ml) were incubated for 1h at 37°C with GSH (1mM), dihydrocaffeic acid (1mM), and NADPH (1mM) or H_2O_2 generating system: glucose (1mM)/glucose oxidase (0.3 unit/ml) in a total volume of 1mL Tris buffer (0.1M, pH 7.4 containing DETAPAC 1mM). The reaction mixture was then extracted using solid phase extraction and evaporated to dryness. The mixture was re-dissolved in 1mL millipore filtered water before direct injection into the mass spectrometer (PE Sciex III, Biomolecular Mass) using a positive mode.

Statistical Analysis

One- and two-way ANOVA followed by the Scheffe's test were used for comparison amongst the multiple-treated groups and the relevant controls. Results represent the mean $\pm$ SE of 3 separate experiments.

RESULTS

As shown in Table 1, the cytotoxic effectiveness (expressed as the concentration required for 50% cell lysis in 2h) of various polyphenols were as follows: 4-nitrocatechol > quercetin > *t*-butylcatechol, hydroquinone > dihydrocaffeic, caffeic acid. However, they were 3 to 5-fold more toxic in the presence of a non-toxic concentration of the H_2O_2 generating system, glucose/glucose oxidase. Inhibiting cytochrome P450 with benzylimidazole inhibited polyphenol-induced cytotoxicity with and without enhancement by H_2O_2.

As shown in table 2, microsomes/NADPH also catalysed GSH conjugate formation by various catechols, catecholamines and hydroquinone, which was prevented by inhibiting cytochrome P450 or by the addition of superoxide dismutase (SOD). A H_2O_2 generating system could substitute for NADPH which was prevented by inhibiting cytochrome P450, but not by the addition of superoxide dismutase. The dihydrocaffeic acid conjugate formed by the H_2O_2/microsomal P450 oxidative systems was identified as the *mono*-glutathione conjugate of dihydrocaffeic acid (*m/z* [M + 1]$^+$ 488) using a mass spectrometer (PE Sciex, Biomolecular mass).

As shown in table 3, dopamine was also cytotoxic to isolated hepatocytes, which was prevented by inhibitors of cytochrome P450 or clorgyline (a MAO A inhibitor), but was markedly increased by azide (a catalase inhibitor). The MAO substrate tyramine also markedly increased dihydrocaffeic acid cytotoxicity and both dopamine and tyramine were shown to increase hepatocyte H_2O_2 formation (measured as cyanide resistant respiration).

Table 1. H_2O_2 supported P450 peroxidase/peroxygenase catalysed polyphenol xenobiotic cytotoxicity.

Xenobiotic	Approx. LC$_{50}$ (2h) mM			
	Hepatocyte	P450 inhib. hepatocyte	Hepatocyte + H_2O_2	P450 inhib. hepatocyte + H_2O_2
quercetin	0.45 ± 0.05		0.15 ± 0.01	
4-nitrocatechol	0.4 ± 0.05		0.1 ± 0.01	
t-butylcatechol	0.8 ± 0.1	> 2.0	0.2 ± 0.02	0.8 ± 0.1
dihydrocaffeic acid	7.0 ± 0.6	> 20.0	1.1 ± 0.011	11 ± 1
caffeic acid	7.0 ± 0.8	> 20.0	1.25 ± 0.09	12 ± 2
3,4-dihydrobenzoic acid	7.5 ± 0.5	> 20.0	1.5 ± 0.13	11 ± 2
hydroquinone	0.8 ± 0.1	2.0 ± 0.2	0.15 ± 0.02	1.1 ± 0.2

Reaction conditions: Hepatocytes (10^6 cells/mL) in a Krebs-Henseleit bicarbonate buffer pH 7.4 suppl. with 12.5mM HEPES under carbogen were incubated with the xenobiotic (dissolved in methanol) with or without a H_2O_2 generating system (glucose 10mM + glucose oxidase 0.6u/mL + desferoxamine 0.1mM). P450 inhib. hepatocytes were obtained by preincubating hepatocytes for 20min with 0.1mM benzylimidazole.

Table 2. H_2O_2 supported oxidation of polyphenols by rat liver microsomal P450 to form GSH conjugates.

Addition	GSH depletion			
	micros. + NADPH	P450 inhib. micro.+NADPH	micros. + NADPH+ SOD	micros. + H_2O_2 + SOD
none	2 ± 1	2 ± 1	2 ± 1	8 ± 2
hydroquinone 0.5mM	51 ± 4	18 ± 1	4 ± 2	70 ± 9
dihydrocaffeic acid 0.5mM	35 ± 3	17 ± 2	4 ± 2	49 ± 5
t-butylcatechol 0.5mM	33 ± 3	16 ± 2	4 ± 2	52 ± 6
epinephrine 0.5mM	37 ± 4	21 ± 2	5 ± 1	42 ± 5
N-acetyldopamine 1mM	48 ± 5	23 ± 2	3 ± 1	62 ± 7
dopac 1mM	43 ± 3	18 ± 2		68 ± 8
dopamine 1mM	23 ± 2	6 ± 1		31 ± 3
dopa 1mM	14 ± 1	4 ± 2		42 ± 5

Reaction conditions: rat liver microsomes (1mg/ml) in 0.1M Tris-HCl buffer were incubated with GSH (0.4mM), polyphenols (0.3-1.0mM), SOD (100u/mL) and NADPH (1mM) or H_2O_2 generating system; glucose (1mM)/glucose oxidase (0.3 unit/mL + diethylenetriaminepentaacetic acid (1mM)). The depletion of GSH was measured spectrometrically by measuring the absorbance at 412nm^6. P450 was inhibited with benzylimidazole (0.1mM).

Table 3. Endogenous H_2O_2 supported xenobiotic/biogenic amine metabolic activation mediated cytotoxicity.

Treatment	Cytotoxicity (% trypan blue)			H_2O_2 formation
Incubation time	60'	120'	180'	10'
Control hepatocytes	19 ± 2	20 ± 2	21 ± 2	2 ± 1
+ dopamine 2mM	31 ± 3	49 ± 5	68 ± 7	30 ± 1
+ clorgyline 2μM	20 ± 2	21 ± 2	22 ± 2	
+ benzylimidazole 0.1mM	21 ± 2	23 ± 2	28 ± 3	
+ dopamine 0.5mM	21 ± 2	23 ± 2	25 ± 3	12 ± 1
+ dicumarol 20μM	33 ± 3	52 ± 5	78 ± 8	
+ azide 4mM	29 ± 3	43 ± 3	67 ± 7	
+ tyramine 1mM	23 ± 2	25 ± 2	28 ± 2	14 ± 1
+ dihydrocaffeic acid 2mM	34 ± 4	58 ± 6	74 ± 8	
+ dihydrocaffeic acid 2mM	25 ± 2	28 ± 3	35 ± 3	2 ± 1

Reaction conditions: see table 1 legend. H_2O_2 formation (μM) was obtained by determining the rate of cyanide (1mM) resistant respiration in 10' with an oxygen electrode.

DISCUSSION

In the present study, we provide for the first time evidence that H_2O_2 can markedly increase the toxicity of polyphenolics towards freshly isolated rat hepatocytes and which could be prevented by P450 inhibitors. This suggests that P450 can act as an intracellular pseudoperoxidase in the metabolic activation of polyphenols. Further evidence suggesting this was that H_2O_2 and microsomal P450 catalysed the formation of diphenol:GSH conjugates, which was prevented by P450 inhibitors. A dihydrocaffeic acid-GSH conjugate (m/z $[M + 1]^+$ 488), was formed from dihydrocaffeic acid indicating that the major metabolite was an o-quinone. A similar conjugate was formed with tyrosinase. Organic hydroperoxides and cytochrome P450 also catalyse a similar peroxidase like one-electron oxidation of phenylenediamines to cation radicals[5]. However, previously H_2O_2 was found to be several orders of magnitude less effective than alkyl hydroperoxides[7], presumably because of microsomal catalase contamination. Recently, we found a K_m for H_2O_2 of 100μM for human CYP1A2 peroxygenase catalysed methoxyresorufin O-demethylation[4].

α-Methyldopa used as an antihypertensive catecholic drug, can cause chronic hepatic injury (hepatitis). Microsomes incubated with radiolabelled catechols including α-methyldopa and an NADPH-generating system resulted in irreversible binding to microsomal protein attributed to the o-quinone. Superoxide dismutase or $CO-O_2$ inhibited the binding by 60%, which indicated that cytochrome P450 generated superoxide anion that catalysed catecholamine autoxidation to form an o-quinone[8]. As shown here however superoxide dismutase inhibited microsomal catechol-GSH conjugate formation with NADPH but not H_2O_2.

The current general mechanism of P450 catalysis includes the "peroxide shunt" pathway using H_2O_2 as the oxygen source to bypass the complex NADPH system using O_2[9]. This is necessary as the ancient ancestors of today's P450 likely relied on H_2O_2 as the oxygen donor because the Earth's atmosphere originally contained little or no molecular oxygen and was relatively rich in H_2O_2 and peroxygenated organic chemicals[10]. The physiological relevance of hydroperoxide-supported cytochrome P450 peroxygenase activity is still unclear even though it was recently demonstrated in isolated hepatocytes[3]. However, a stronger case may exist for the physiological role of P450 pseudo-peroxidase activity in oxidising catechols or catecholamines to an o-quinone, as superoxide dismutase prevents oxidation by the NADPH-dependent mixed function oxidase, but does not affect oxidation by H_2O_2/P450. Furthermore, as shown here MAO substrates can increase the intracellular metabolic activation of polyphenols. Whether cytochrome P450 pseudo-peroxidase activity involves compound I and II oxidation states, analogous to peroxidases[11] remains to be determined. Chloroperoxidase exhibits P450-like activities likely because the proximal side of its heme has a cysteine axial heme ligand[12].

REFERENCES

1. F. J. Gonzalez and H. V. Gelboin. Role of human cytochrome P450 in the metabolic activation of chemical carcinogens and toxins. *Drug Metab. Rev.* 26:165 (1994).
2. M. R. Anari, S. Khan, Z. C. Liu, and P. J. O'Brien. Cytochrome P450 peroxidase/peroxygenase mediated xenobiotic metabolic activation and cytotoxicity in isolated hepatocytes. *Chem. Res. Toxicol.* 8:997 (1995).
3. M. R. Anari, S. Khan, S. D. Jatoe, and P. J. O'Brien. Cytochrome P450 dependent xenobiotic activation by physiological hydroperoxides in intact hepatocytes. *Eur. J. Drug Metab. and Pharmacokinetics* 22:305 (1997).
4. M. R. Anari, P.D. Josephy, T. Henry, and P. J. O'Brien. Hydrogen peroxide supports human and rat cytochrome P450 1A2-catalysed 2-amino-3-methylimidazo[4,5]quinoline bioactivation to mutagenic metabolites: significance of cytochrome P450 peroxygenase. *Chem. Res. Toxicol.* 10:582 (1997).
5. P. J. O'Brien. Hydroperoxides and superoxides in microsomal oxidations. *Pharmacol. Ther.* A 2:517 (1978).
6. G. L. Ellman. Tissue sulfhydryl groups. *Arch. Biochem. Biophys.* 82:70 (1959).
7. R. Penneberg, F. Scheller, K. Ruckpaul, J. Pirritz, and P. Mohr. NADPH and H_2O_2-dependent reactions of cytochrome P450$_{LM}$ compared with peroxidase catalysis. *FEBS Lett.* 96:349 (1978).
8. E. Dybing, S. D. Nelson, J. R. Mitchell, H. A. Sasame, and J. R. Gillette. Oxidation of α-methyldopa and other catechols by cytochrome P450-generated superoxide anion: possible mechanism of methyldopa hepatitis. *Molec. Pharmacol.* 12:911 (1976).
9. F. P. Guengerich, A. D. N. Vaz, E. N. Raner, S. J. Pernecky, and M. J. Coon. Evidence for a role of a perferryl-oxygen complex, FeO^{3+}, in the N-oxygenation of amines by cytochrome P450 enzymes. *Molec. Pharmacol.* 51:147 (1997).
10. V. D. Samuilov. Photosynthetic oxygen: the role of H_2O_2. A review. *Biochemistry* (Moscow). 62:451 (1997).
11. P. J. O'Brien. Peroxidases (an invited review). *Chem. Biol. Interactions* (special edition on Bioactivation).
12. M. Sundaramoorthy, J. Jenner, and T. L. Poulo. The crystal structure of chloroperoxidase: a heme-peroxidase-cytochrome P450 functional hybrid. *Structure.* 3:1367 (1995).

NAD(P)H:QUINONE OXIDOREDUCTASE (NQO1) PROTECTS ASTROGLIAL CELLS AGAINST L-DOPA TOXICITY

Benjamin Drukarch, Cornelis A.M. Jongenelen, and Freek L. van Muiswinkel

Department of Neurology
Research Institute Neurosciences Vrije Universiteit
vd. Boechorststraat 7, 1081 BT Amsterdam, The Netherlands
tel. 31-20-4448107/8103, Fax 31-20-4448100
e-mail b.drukarch.neurol@med.vu.nl

INTRODUCTION

Astroglial cells are well known to protect neurons against the toxicity of L-Dopa, a catecholaminergic drug used in the treatment of Parkinson's disease (PD) (Han *et al.*, 1996). This observation, together with the abundance of astroglial cells in the brain, has been used to explain the remarkable absence of overt neurotoxicity of L-Dopa in vivo, as opposed to its clear detrimental effect on neuronal survival in vitro (Agid, 1998). The neurotoxicity of L-Dopa in vitro is thought to be caused primarily by its autooxidative breakdown in the presence of molecular oxygen. Under these circumstances, highly unstable Dopa-quinones are formed which induce neuronal damage mainly as a result of the excessive release of reactive oxygen species (ROS), such as superoxide radicals and hydrogen peroxide, during redox-cycling of these compounds (Bindoli *et al.*, 1992). In contrast to neurons, astroglial cells possess a considerable capacity to detoxify ROS. This feature is thought to form a basis both for their relative resistance to oxidative damage and for their neuroprotective potential (Han *et al.*, 1996; Drukarch *et al.*, 1998). Astroglial ROS scavenging capacity depends to a large extent on the presence of high amounts of the thiol anti-oxidant glutathione in these cells. In fact, neuroprotection against L-Dopa toxicity has been associated with the ability of astroglial cells to increase their glutathione content upon exposure to L-Dopa (Han *et al.*, 1996).

Apart from scavenging of ROS formed as a result of redox-cycling, protection against L-Dopa-derived quinones may proceed through chemical reduction of these electrophilic species. In this context, an enzyme known as NAD(P)H:quinone oxidoreductase (EC 1.6.99.2; NQO1) is considered to be of particular interest. NQO1, a homodimeric flavoprotein formerly referred to as DT-diaphorase, is the product of the *NQO1* gene and differs from other quinone reductases in that it uses NADH or NADPH as electron donors to catalyze a two-electron reduction of quinones, including those formed during autooxidation of catecholamines like L-

Dopa. In this reaction, redox-labile quinones are turned into more stable hydroquinones which may subsequently undergo further inactivation by e.g. (enzymatic) glucuronidation or sulfation (Cadenas, 1995; Drukarch and van Muiswinkel, 2000). Thus, NQO1 is supposed to act as an anti-oxidant enzyme inter alia by averting the highly deleterious redox-cycling of quinones. In the brain, NQO1 has been shown to be expressed almost exclusively in astrocytes (Schultzberg et al., 1988, Murphy et al., 1998), suggesting a contribution of this enzyme to the detoxification of L-Dopa by these cells. Recently, we observed that, alike the glutathione content, astroglial NQO1 activity is raised following treatment with L-Dopa (van Muiswinkel et al., in press). In the experiments presented here, we investigated whether indeed NQO1 confers protection to astroglial cells against L-Dopa. For this purpose, we used the rat C6 astroglioma cell line, previously found to react in a similar manner to exposure to L-Dopa as primary cultures of rat astrocytes (van Muiswinkel et al., in press), and measured cellular survival following long-term incubation with L-Dopa in the absence or presence of dicumarol, a potent inhibitor of NQO1 catalytic activity (Chen et al., 1999).

EXPERIMENTAL METHODS

C6 astroglioma cells (American Type Culture Collection, Rockville, Maryland, USA) were seeded initially at a density of 200 cells/well in uncoated 96-well culture plates (A/S Nunc, Roskilde, Denmark), containing 100 µl of a culture medium consisting of a 1:1 mixture of DMEM and HAM's F-10 nutrient mixture supplemented with 10% fetal calf serum, glutamine (2 mM), streptomycin (100 µg/ml), and penicillin (100 IU/ml). Cultures were grown for 4 days at 37 °C in a humified atmosphere of 5% CO_2/95% air. 24 h after seeding, the culture medium was replaced by culture medium containing either solvent, L-Dopa (30 µM), dicumarol (10, 30 or 100 µM; Sigma, St. Louis, Missouri, USA) or a combination of L-Dopa with the respective concentrations of dicumarol. After incubation with drugs for 72 h, cellular survival was measured by the SRB protein assay, as described previously (van Muiswinkel et al., 1993). Statistical data analysis was performed by analysis of variance (ANOVA) followed by Bonferroni's post-hoc test to compare group means. Where appropriate, p values < 0.01 were considered significant.

RESULTS

Previously, we established that NQO1 activity in C6 astroglioma cells is highly sensitive to inhibition by dicumarol (van Muiswinkel et al., in press). Our present data show that such dicumarol-mediated blockade of NQO1 activity does not impair the survival of C6 astroglioma cells per se (Fig. 1). A similar lack of effect on C6 growth was noted upon incubation of the cultures with 30 µM L-Dopa (Fig. 1). However, treatment of the cells with a combination of L-Dopa and dicumarol, profoundly and concentration-dependently reduced cellular survival (Fig. 1).

DISCUSSION

NQO1 is regarded to play a pivotal role in the detoxication of a large array of both natural and synthetic quinones by catalyzing their reduction into less reactive hydroquinones (Cadenas, 1995). The present data illustrate, that this action confers protection to astroglial cells against L-Dopa-derived quinones, which were previously found to be released under our culture

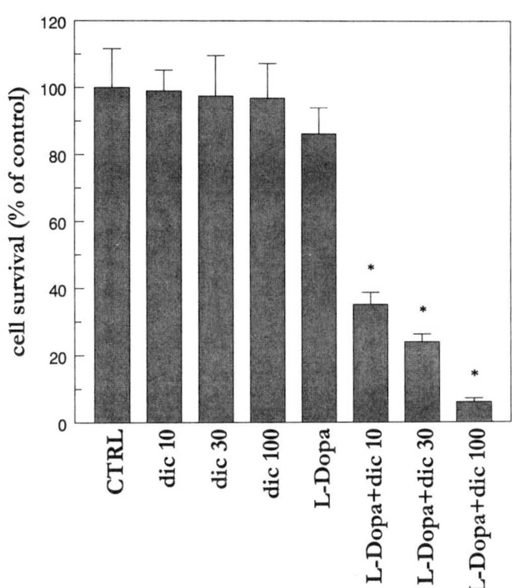

Figure 1. NQO1 protects C6 astroglioma cells against L-Dopa toxicity. Cultures were incubated for 72 h with either solvent (control; CTRL), L-Dopa (30 µM), dicumarol (dic; 10, 30 or 100 µM), or a combination of L-Dopa with dicumarol. Data represent the mean ± S.D. (n=6) and are expressed as % of cell survival under control condition. *significantly different from control and from treatment with L-Dopa alone.

conditions during prolonged incubation with L-Dopa (van Muiswinkel *et al.*, in press). Since the expression of NQO1 in the brain appears to be confined largely to the astroglial compartment (Schultzberg *et al.*, 1988; Murphy *et al.*, 1998), our data furthermore lend support to the notion that this enzyme is an important determinant of the much discussed absence of overt L-Dopa neurotoxicity as observed both in vivo as well as in in vitro preparations containing a significant number of astrocytes (Han *et al.*, 1996; Agid, 1998). Interestingly, NQO1 belongs to a group of protective proteins known collectively as phase II biotransformation enzymes, whose activity is upregulated not only by redox-cycling toxicants such as L-Dopa but also by a variety of naturally occurring anti-oxidants, including dithiolethiones and isothiocyanates (Drukarch and van Muiswinkel, 2000). Thus, NQO1 might be an ideal target for the development of neuroprotective treatment of brain diseases such as PD, in which quinone-mediated neurotoxicity is thought to be a major factor in the pathogenetic process (Drukarch and van Muiswinkel, 2000).

REFERENCES

Agid, Y., 1998, Levodopa, is toxicity a myth?, *Neurology* **50**:858-863.

Bindoli, A., Rigobello, M. P., and Deeble, D. J., 1992, Biochemical and toxicological properties of the

oxidation products of catecholamines, *Free Radic. Biol. Med.* **13**:391-405.

Cadenas, E., 1995, Antioxidant and prooxidant functions of DT-diaphorase in quinone metabolism, *Biochem.*

Pharmacol. **49**:127-140.

Chen, S., Wu, K., Zhang, D., Sherman, M., Knox, R., and Yang, C. S., 1999, Molecular characterization of binding of substrates and inhibitors to DT-diaphorase: combined approach involving site-directed mutagenesis, inhibitor-binding analysis, and computer modeling, *Mol. Pharmacol.* **56:**272-278.

Drukarch, B., and van Muiswinkel, F. L., 2000, Drug treatment of Parkinson's disease: time for phase II, *Biochem. Pharmacol.* **59:**1023-1031.

Drukarch, B., Schepens, E., Stoof, J. C., Langeveld, C. H., and van Muiswinkel, F. L., 1998, Astrocyte enhanced neuronal survival is mediated by scavenging of extracellular reactive oxygen species, *Free Radic. Biol. Med.* **25:**217-220.

Han, S. K., Mytilineou, C., and Cohen, G., 1996, L-Dopa up-regulates glutathione and protects mesencephalic cultures against oxidative stress. *J. Neurochem.* **66:**501-510.

Murphy, T. H., So, A. P., and Vincent, S. R., 1998, Histochemical detection of quinone reductase activity in situ using LY 83583 reduction and oxidation, *J. Neurochem.* **70:**2156-2164.

Schultzberg, M., Segura-Aguilar, J., and Lind, C., 1988, Distribution of DT diaphorase in the rat brain: biochemical and immunohistochemical studies, *Neuroscience* **27:**763-776.

Van Muiswinkel, F. L., Drukarch, B., Steinbusch, H. W. M., and Stoof, J. C., 1993, Chronic dopamine D2 receptor activation does not affect survival and differentiation of cultured dopaminergic neurons: morphological and neurochemical observations, *J. Neurochem.* **60:**83-92.

Van Muiswinkel, F. L., Riemers, F. M., Peters, G. J., LaFleur, M. V. M., Siegel, D., Jongenelen, C. A. M., and Drukarch, B., 2000, L-Dopa stimulates expression of the anti-oxidant enzyme NAD(P)H:quinone oxidoreductase (NQO) in cultured astroglial cells, *Free Radic. Biol. Med.* **in press.**

GSH-DEPENDENT REDOX REGULATION AND ANTIOXIDANT EN-ZYMES IN THE FORMATION OF RESISTANCE TO DOXORUBICIN IN K562 HUMAN ERYTHROLEUKEMIA CELLS

Elena Kalinina,[1] Maria Novichkova,[1] Nikolayi P.Scherbak,[2] Viktoria Solomka,[1] Anatoly N. Saprin[1]

[1]Lab.Biochemistry & Biophysics of Cancer, Institute of Chemical Physics
[2]Institute of Carcinogenesis
Kosygin Str.4, Moscow 117977, Russia
E-mail:kevsan@orc.ru

Specific changes of antioxidant enzymes during initiation and pro-motion stages are special value for the study of role and regulation of free radical processes in mechanisms of cancinogenesis for prevention and re-version of malignization. In addition, among factors controlling the bal-ance of homeostasis and formation of cellular resistance to environment free radical processes play the important role due to the including into proliferation and apoptosis which is discussed currently as one of the key events in free radical concept of chemical carcinogenesis (1).

Overexpression of a number of proteins with transporting and de-toxification functions is the typical peculiarity of cellular resistance and multidrug resistance. Nevertheless antioxidant system is non-well known as a factor involved in the formation of cellular resistance. We studied the value of antioxidant system as well as GSH dependent redox regulation for a formation of resistance of K562 human erythroleukemia cells to chemo-therapeutic agent doxorubicin (DOX).
As pro-oxidant with quinone-like structure DOX is metabolized by DT-diaphorase with the formation of semiquinone free radicals and following production of superoxide anione radicals, hydrogen peroxidase and ulti-mately hydroxyl radicals which can destroy proteins, DNA and activate lipid peroxidation. However a long exposition of cancer cells to slowly en-hancement of DOX concentration causes a rise of resistance to cytotoxic effect of the drug.

Analysis of GSH-dependent system shown that the rise of cellular resistance to DOX with 10-fold index (0,25 – 2,5 µg/ml) caused an increase of GSH maintenance upto 50% in contrast to the level of sensitive K562 cells (figure1) .

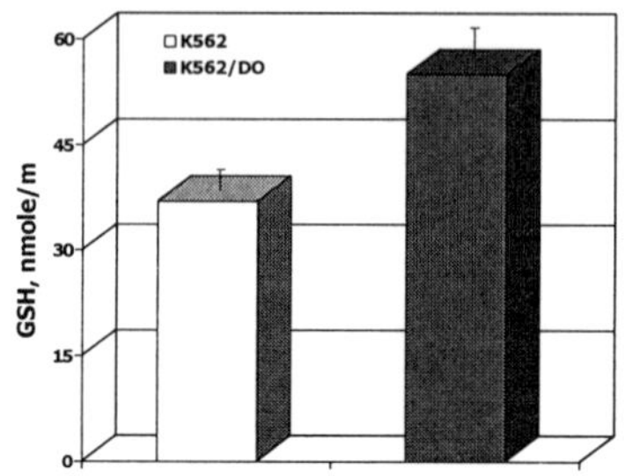

Figure 1. GSH level in sensitive and resistance to doxorubicin K562 cells.

The enhancement of activities of glutathione S-transferase and glutathione peroxidese has been found also to be a response to the increase of DOX concentration. It could be noted that activity of glutathione S-transferase towards cumene hydroperoxide was raised more rapidly then activity of glutathione peroxidase (figure 2).

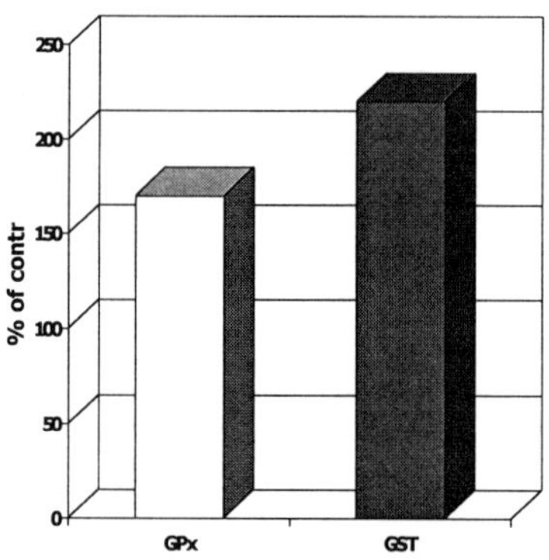

Figure 2. Enhancement of glutathione S-transferase and glutathione peroxidase activity towards cumene hydroperoxide in K562 cells with resistance to doxorubicin.

The major growth of glutathione S-transferase activity towards organic hydroperoxides confirmed by the data of Western blotting shown the significant enhancement of GSTP1-1 level (2). Moreover the raise of GSH-dependent system during the formation of cellular resistance included the elevation of glutaredoxin activity (upto 70%) and was accompanied by increasing level of thioredoxin (figure 3).

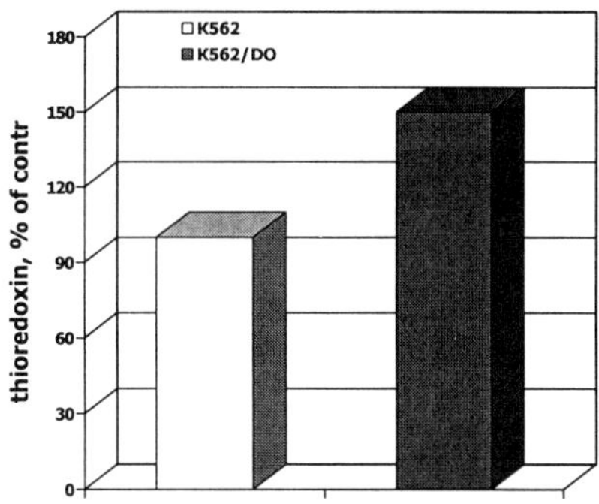

Figure 3. Relative level of thioredoxin before and after doxorubicin exposition.

This response of K562 cells to DOX cytotoxicity is compensative adaptation through activation of detoxification and antioxidant processes. The estimation of key antioxidant enzymes – superoxide dismutase and catalase shown the significant elevation of their activities: a rapid increase

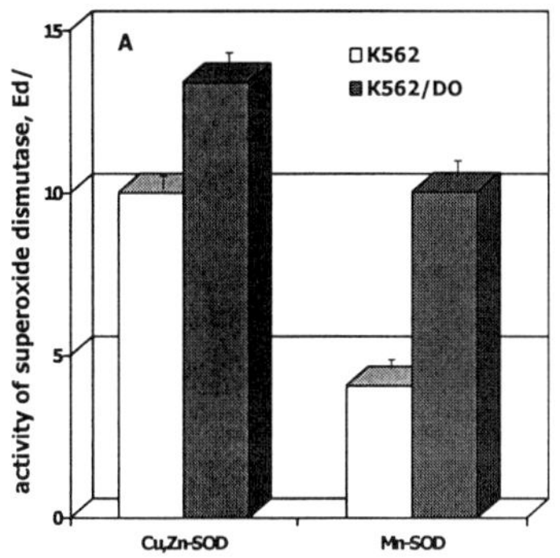
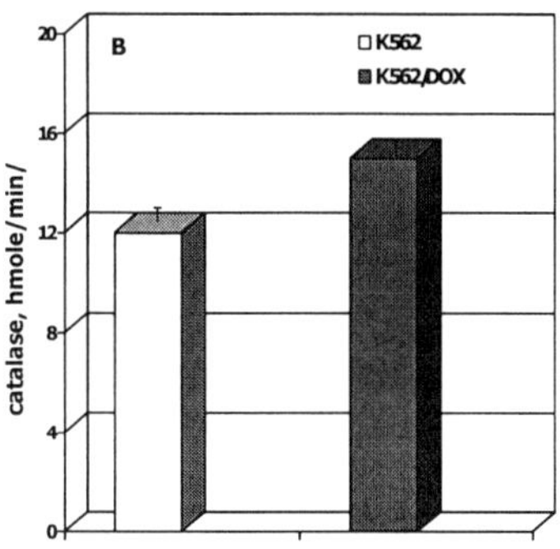

Figure 4. Increase of activities of siperoxide dismutase (A) and catalase (B) during the formation of resistance to doxorubicin in K562 cells.

of Mn-superoxide dismutase level in contrast to Cu,Zn-superoxide dismutase and more slowly increase of catalase activity (figure 4).
The raise of Mn-superoxide dismutase level is interesting as response to intensification of proliferation. In addition to slowly increase of catalase activity the rapid growth of glutathione peroxidase activity towards H_2O_2 was found.

The increase of GSH-dependent system and antioxidant enzymes level during the formation of cancer cells resistance to DOX is the part of adaptive cellular response to oxidative stress caused by chemotherapeutic drug with pro-oxidant effect. The regulation of adaptive response has been suggested to be realized through GSH-dependent control of redox status and redox-dependent signaling. One can conclude that the level of antioxidant system is the factor included in the formation of drug resistance of cancer cells as one of the key factors in the development of adaptive response.

REFERENCES

1. E.Kalinina, A.N.Saprin. Oxidative stress and its role in the mechanisms of apoptosis and development of pathological processes. *Adv.biol.chem.* 39:289 (1999).
2. E.Kalinina, M.Novichkova, N.P.Scherbak, A.Saprin. Progesterone inhibition of glutathione S-transferase P1-1 and its antiproliferative effect on human erythroleukemia K562 line cells. *Prob.oncol.* 46(1):68 (2000).

OPPOSITE EFFECTS OF OXIDATIVE STRESS ON ENDOTHELIAL CELL LINES (ECV 304 AND EAhy 926) INTERACTION WITH EXTRACELLULAR MATRIX

F. Lamari[1], M. Bernard[1], F. Braut-Boucher[2], C. Derappe[2], J. Pichon[2], M.J. Foglietti[1] and M. Aubery[2]

[1]-Laboratoire de Biochimie Générale et Glycobiologie-UFR Pharmacie ParisV.
[2]- Laboratoire de Glycobiologie et de Reconnaissance Cellulaire – UFR Saints-Pères Paris V.

INTRODUCTION

The vascular endothelium represents an important interface between the blood vessel lumen and the surrounding tissues, and regulates vasotonicity, permeability, homeostasis and angiogenesis. The adhesive interactions between endothelial cells (EC) and their surrounding extracellular matrix (ECM) regulate growth, differentiation and migration,[1]. The localization of endothelial cells at the contact with leukocytes lead to their exposure to peroxides during local inflammatory reactions. It has been reported on that many inflammatory mediators, including reactive oxygen species (ROS) and cytokines modulate EC-leukocyte interaction,[2]. In contrast little reports have focused on EC-ECM interactions. The purpose of this study was to set-up an *in-vitro* model for studying the effects of oxidative stress on EC-ECM interactions, using two immortalized cell lines derived from human umbilical vein cord (HUVEC), and classified as endothelial. ECV304 spontaneously transformed HUVEC,[3] and EA.hy926 a hybridoma of the epithelial cell line A549 and HUVEC,[4] were used to study the effects of ter-Butyl Hydroperoxide (t-BHP) on cell adhesion to collagen I, intracellular glutathione (GSH) depletion, ROS production, integrin expression and cytoskeleton organization.

MATERIAL AND METHODS

Cell Culture And Treatments

The ECV304 and EA.hy926 cell lines were cultured in RPMI 1640 and DMEM mediums respectively, supplemented with 10% fetal calf serum (FCS), 100 u/ml penicillin, 100 % µg/ml streptomycin and 2 mM L-glutamine. The ECV304 and EA.hy926 cells (2.5 x 105 cell/ml) in HBSS without phenol red (W/O PR) were incubated 30 minutes at 37°C with various concentrations of t-BHP, as a chemical source of oxidative stress. The treated

Biological Reactive Intermediates VI, Edited by Dansette *et al.*
Kluwer Academic / Plenum Publishers, 2001

cells were plated on-to collagen I-coated wells, to measure cell adhesion, ROS production or intracellular GSH. For protection assay, the cell monolayers were pretreated 18 hours with 100 µg/ml of α-D-Tocopherol before t-BHP treatment.

Cell Adhesion Assay

ECV304 or EA.hy926 cells were labeled with calcein-AM as previously decribed,[5], plated on-to 100 µg/ml collagen I-coated wells (100 µl/well) and incubated 90 minutes at 37°C for cell attachment. The fluorescence intensity was measured by a microspectrofluorimeter (Fluostar BMG), (λ_{ex} 485; λ_{em} 538 nm) before and after 3 washes with PBS to remove the non adherent cells.

ROS And GSH Measurement

Specific fluorescent probes 2',7'-Dihydro-dichlorofluoroscein diacetate (H_2DCFDA) and 5-Chloromethyl fluorescein diacetate (CMFDA) were used to measure respectively, ROS and GSH in endothelial cells,[6]. The cells in 96-wells microplates (5.10^4 cells/well) were incubated 30 min with 10 µM of H_2DCFDA or 5 µM of CMFDA at 37°C, the fluorescence intensity was measured by microspectrofluorimetry (λ_{ex} 485; λ_{em} 538 nm).

Analysis Of Cell-Surface Integrin Expression

For fluorescence-activated cell sorting analysis (FACS), cells were immunostained with the anti-α2 or anti-β1 monoclonal antibodies (1/100). After 3 washes with PBS + 0.3% BSA, the cells were incubated for 45 minutes with fluorescein isothiocianate-conjugated goat anti mouse IgG, cells were washed again and analyzed on a FACS flow cytometer (Elite-Coulter). As a control the cells were incubated with the secondary antibody alone.

Actin Filament Morphology

Untreated and t-BHP-treated cells were fixed in 4% Formaldehyde for 15 min, permeabilized in 0.2 % Triton-X100 for 5 min.and stained with 1 µM rhodamine-phalloidin for 30 min in the dark. Coverslips were mounted on to the slides with glycerol-H_2O (1:9), and viewed under a fluorescent microscope. Cells were photographed using a Kodak Ektachrom 400 film.

RESULTS

Treatment of EC-lines with 125 and 250 µM of t-BHP during 30 min induced a decrease in ECV304 attachment to collagen I by respectively, 20% and 40% compared to control, without affecting cell viability. In contrast the adhesion of EA.hy926 cells was increased by 15 and 30% after t-BHP treatment. The decrease in ECV304 adhesion induced by t-BHP was completely prevented by a pretreatment with 100 µg/ml of α -D-Tocopherol during 18 h, whereas the t-BHP-induced increase in EA.hy 926 adhesion was emphasized by 40% after α-D-Tocopherol pretreatment. In order to explain the discrepancy between the two EC-lines behavior against t-BHP, EA.hy926 cells were treated 18 h with 185 µM of ginistein before t-BHP treatment. Data showed a decrease in EA.hy926 adhesion to collagen, suggesting the involvement of a tyrosine kinase-dependant mechanism in the increased adhesion of EA.hy926 cells (Figure 1.A). To investigate if there is a difference in cell sensitivity to oxidative stress intracellular GSH and ROS production were measured in the resting cells and the t-BHP treated cells. Results showed a higher GSH content in EA.hy926 cells than in ECV304 cells, t-BHP treatment induced a decrease in GSH content in the same manner between the two cell lines; with a parallel increase in ROS production (Figure 1.B).

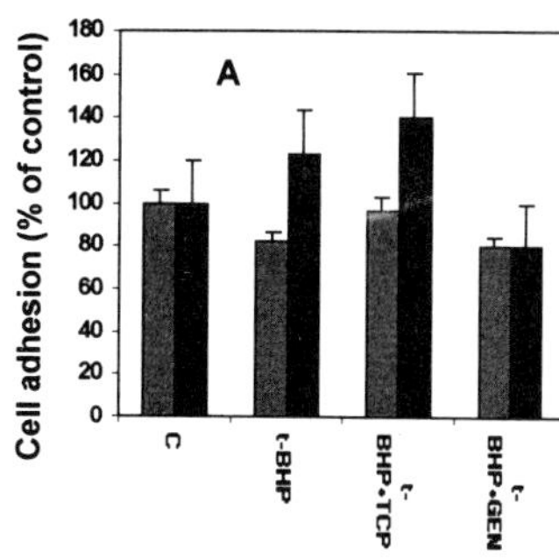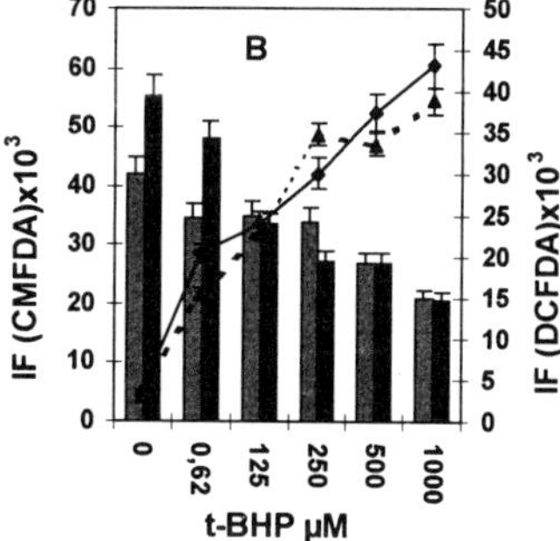

Figure 1. Effects of t-BHP on ECV304 and EA.hy926 cell-lines.
A. t-BHP effect on EC-lines adhesion to collagen I : calcein-AM labeled ECV304 (shaded bars) and EA.hy926 (dark bars) were treated (t-BHP) or not (C) with 125 µM t-BHP during 30 min and plated on-to collagen I-coated wells. Cell adhesion was measured spectrofluorimetrically after 90 min. Cells were pretreated 18 hours with 100 µg/ml a-D-Tocopherol (t-BHP+TCP) or with 185 mM genistein (t-BHP+GEN), before t-BHP treatment. Results are expressed as a mean of triplicate experiments+SD.
B. ROS and GSH measurement in EC-lines : After 30 min treatment with t-BHP, ROS production was measured using 10 µM H2DCFDA probe in ECV304 celss (---▲---) and EA.hy926 (-◆-). Intracellular GSH was measured using 5 µM CMFDA probe in ECV304 (shaded bars) and EA.hy926 (dark bars). Results are expressed as a mean of relative fluorescence intensity of triplicate experiments+SD.

Cell adhesion to collagen I was reduced by 50% when the cells were incubated with monoclonal antibodies anti-α2 and ant-β1 integrins, suggesting the involvement of $\alpha_2\beta_1$ integrin in the interaction of EC-lines with collagen I. To examine the effect of t-BHP on cell surface expression of α2 and β1 integrins, ECV304 and EA.hy926 cells were analyzed by flow cytometry. Data showed a higher β1 integrin expression on EA.hy926 cell surface than on ECV304 (Table 1). After 125 µM t-BHP treatment no change in β1 and α2 integrin expression was observed, suggesting that the t-BHP modulation of cell adhesion to collagen I is independent of $\alpha_2\beta_1$ integrin expression.

Table 1. Expression of the integrin α2 and β1 subunits on ECV304 and EA.hy926 cells. Untreated or 125µM t-BHP treated cells were incubated with the monoclonal antibodies P1E6 anti-α2 or P4C10 anti-β1 and analyzed by flow cytometry. The results are expressed as a relative fluorescence intensity (mean ± SD).

	ECV304		EA.hy926	
	untreated	t-BHP	untreated	t-BHP
α2 integrin	3.53 ± 2.17	3.39 ± 2.11	1.88 ± 0.74	1.77 ± 0.60
β1 integrin	10.3 ± 4.64	10.8 ± 4.39	15.8 ± 8.48	14.4 ± 8.28

In vitro experiments indicated that endothelial cells, when subjected to oxidative stress with t-BHP, undergo remarkable changes in morphology and in the structure of the actin cytoskeleton, often resulting in membrane blebbing. In order to observe the change in actin filaments organization, the permeabilized EC-lines were stained with rhodamine-phalloidin. Microscopic examination showed remarkable differences in the actin reorganization and membrane blebbing between the two EC-lines. In resting cells actin filaments were thiner in ECV304 than in EA.hy926, and membrane blebbing was observed in EA.hy926 cells but not in ECV304 cells. The application of 125 µM t-BHP had little effect on the actin organization of EA.hy926 cells. In contrast treatment of ECV304 cells in the same conditions resulted in stress fiber formation, and the appearance of focal adhesion sites (Figure 2).

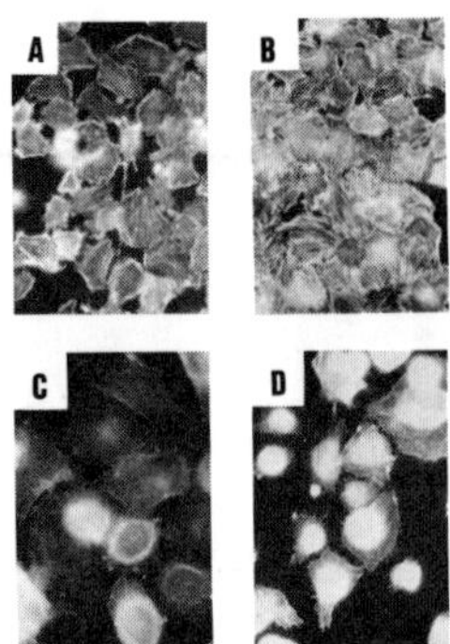

Figure 2. t-BHP effect on actin filament organization.: ECV304 (A and B) and EA.hy926 cells (C and D) plated on-to collagen I and treated (B and D) or not (A and C) 30 minutes with 125 µM t-BHP. The cells were fixed and stained with rhodamin-phalloidine. The cells were viewed under fluorescent microscope and photographed.

DISCUSSION

Immortalized EC-lines are commonly used as an *in-vitro* model for studying endothelial-leukocyte interaction. In order to set-up a useful model for endothelial cell-extracellular matrix interaction, two cell-lines classified as endothelial-derived cells were considered. Treating the cells under the same conditions with a non toxic concentration of t-BHP as an exogenous source of oxidative stress, an opposite effect was observed : the adhesion of the ECV304 to collagen I was dramatically decreased, whereas the adhesion of EA.hy926 was significantly increased. Cell protection with an antioxidant, abolished completely the effect of t-BHP on ECV304 cell adhesion. In contrast EA.hy926 treatment with the same conditions amplifies the increase in cell adhesion. This increase in EA.hy926 adhesion is abolished by a tyrosin kinase inhibitor (genistein). These data suggest the involvement of two different ways in t-BHP modulation of EC-lines interaction with ECM. The modulation of EC-ECM interaction seems to be independent of integrin expression, since no difference in $\alpha 2$ and $\beta 1$ integrin expression was observed. In contrast remarkable differences were observed between ECV304 and EA.hy926 cells when cytoskeleton organization is considered. The appearance of stress fibers in ECV304 but not in EA.hy926 cells, and the membrane blebbing observed in EA.hy926 but not ECV304 could explain the discrepancies observed in the cell behavior against oxidative stress.

REFERENCES

1. E. Ruoslahti and E. Engvall. Integrins and vascular extracellular matrix assembly. *J Clin Invest*. 99:1149 (1997).
2. J.R. Bradley, D. Johnson, and J.S. Pober. Endothelial activation by hydrogen peroxide : selective increase of intracellular adhesion molecule-1 and major histocompatibility complex class I. *Am J Pathol* 142:1598 (1993).
3. K. Takahashi, Y. Sawasaki, J. Hata, K. Mukai, and T. Goto. Spontaneous transformation and immortalization of human endothelial cells *in vitro*. *Cell Dev Biol*. 26:265 (1990).
4. C.J. Edgell, C.C. McDonald, and J.B. Graham. Permanent cell line expressing human factor VIII-related antigen established by hybridization. *Proc Natl Acad USA*. 80:3734 (1983)
5. F. Braut-Boucher, J. Pichon, P. Rat, M. Adolphe, M. Aubery, and J. Font. Anon isotopic, highly sensitive fluorimetric, cell-cell adhesion microplate assay using calcein AM-labeled lymphocytes. *J Immunol Meth*. 178:41 (1995).
6. S. Rappeneau, A. Baeza-Squiban, F. Braut-Boucher, M. Aubery, M.C. Gendron, F .Marano. Use of fluorescent probes to assess the early sulfhydryl depletion and oxidative stress induced by mechlorethamine in human bronchial epithelial cells. *Toxicol in vitro*. 13:765 (1999)

HEPATOCYTE LYSIS INDUCED BY ENVIRONMENTAL METAL TOXINS MAY INVOLVE APOPTOTIC DEATH SIGNALS INITIATED BY MITOCHONDRIAL INJURY

Jalal Pourahmad, Aleksandra Mihajlovic and Peter J. O'Brien
Faculty of Pharmacy, University of Toronto
19 Russell St., Toronto, Ont. M5S 2S2, Canada

ABSTRACT

Addition of $CdCl_2$, $HgCl_2$ or $K_2Cr_2O_7$ to isolated hepatocytes caused a rapid increase in reactive oxygen species ("ROS") formation and a decline in mitochondrial membrane potential. Later lipid peroxidation and cell lysis ensued. Cytotoxicity was prevented by "ROS" scavengers and various inhibitors of the mitochondrial permeability transition (MPT) eg. cyclosporin A, carnitine or trifluoperazine. Antioxidants prevented hepatocyte lysis induced by $CdCl_2$, $K_2Cr_2O_7$ but not $HgCl_2$. Hepatocyte lysis was also prevented by various apoptosis inhibitors eg. cycloheximide, dactinomycin and a tetrapeptide caspase 3 inhibitor which suggests that metal induced hepatocyte lysis involves apoptotic death signals initiated by MPT and "ROS".

INTRODUCTION

Human exposure to cadmium, mercury and chromium primarily occurs from the environment. Human Cd II overload states have been associated with renal tubular dysfunction , osteomalacia and anemia culminating in Itai-itai disease which was first diagnosed among the inhabitants of Toyoma, Japan (Horiguchi et al. 1994). A human Hg II overload state resulting in neurotoxicity was typified by Minamata disease, resulting from the ingestion of methyl mercury-contaminated seafood (Eto, 1997). Workers industrially exposed to Cr VI compounds developed nasal tumors,allergic dermatitis, neurotoxicity and reproductive toxicity (Miksche and Lewalter, 1997).

Parenterally administered Cd II, Hg II or Cr VI caused hepatic necrosis injury in rats (Siegers et al, 1986; Strubelt et al, 1996; Kurosaki et al, 1995). However, the molecular cell death mechanisms involved are not known. Hg II caused an increase of hepatocyte cytosolic free Ca^{+2} , collapse of mitochondrial membrane potential, ATP depletion and cell lysis (Nieminen et al, 1990). Recently we showed that Cd II caused hepatocyte "ROS" formation and a collapse of the mitochondrial membrane potential before cell lysis ensued (Pourahmad and O'Brien, 2000a). In the following the molecular cell death mechanisms of the metals Cd II, Hg II or Cr VI toward isolated rat hepatocytes have been investigated.

METHODS
Isolation of hepatocytes

Male Sprague-Dawley rats (280-300g), fed a standard chow diet and given water ad libitum, were used in all experiments. Hepatocytes were obtained by collagenase perfusion of the liver and viability was assessed by plasma membrane disruption determined by the trypan blue (0.2% w/v) exclusion test (Pourahmad and O'Brien, 2000a). Cells were suspended at a density of 10^6 cells/ml in round bottomed flasks rotating in a water bath maintained at 37 °C in Krebs-Henseleit buffer (pH 7.4), supplemented with 12.5 mM Hepes under an atmosphere of 10% O_2 : 85% N_2: 5% CO_2.

RESULTS

As shown in table 1, incubation of hepatocytes with 20 µM mercuric chloride Hg II, 20 µM cadmium chloride Cd II or 1 mM dichromate Cr VI markedly increased "ROS" formation and partly collapsed the mitochondrial membrane potential ($\Delta\psi_m$) within the first 15 minutes. Cytotoxicity (cell lysis) and lipid peroxidation (malondialdehyde (MDA) formation) ensued later. The order of effectiveness (at the metal concentration used) for inducing "ROS" was Cr VI > Hg II > Cd II whereas the order for lipid peroxidation (malondialdehyde (MDA) formation) was Cr VI > Cd II > Hg II. The "ROS" formation and $\Delta\psi_m$ collapse was delayed by the hydroxyl radical scavengers dimethylsulfoxide (DMSO) or mannitol. Cytotoxicity induced by Cd II or Cr VI was prevented by mannitol and DMSO whereas cytotoxicity induced by Hg II was more readily inhibited by mannitol than DMSO. The antioxidant diphenylphenylenediamine (DPPD) prevented lipid peroxidation and prevented Cr VI / Cd II cytotoxicity but did not significantly affect Hg II cytotoxicity, with less effect on "ROS" formation or $\Delta\psi_m$ collapse induced by these metals.

Table 1. Metal induced hepatocyte lysis results from "ROS" formation and collapse of the mitochondrial membrane potential.

Addition	"ROS"	% $\Delta\psi_m$ collapse	MDA	% Cytotoxicity	
	15 min	15 min	3 hr	1 hr	3 hr
Control Hepatocytes	78±5	3±3	0.36± 0.05	18±2	20±2
+ HgCl$_2$ (20 µM)	322±11[a]	55±5[a]	6.66± 0..07[a]	48±5[a]	82±8[a]
+ DMSO (150 µM)	164±6[b]	21±2[b]	1.89± 0.18[b]	37±4[b]	55±5[b]
+ Mannitol (30 mM)	82±8[b]	10±2[b]	0.35± 0.06[b]	33±3[b]	35±3[b]
+ DPPD (1 µM)	286±9	45±5	0.27± 0.03[b]	42±3	73±6
+ Carnitine (2 mM)	115±7[b]	11±2[b]	4.78± 0.24[b]	36±4[b]	41±5[b]
+ Cyclosporin (2 µM)	123±8[b]	12±2[b]	5.15± 0.31[b]	32±3[b]	37±4[b]
+ Trifluoperazine (15 µM)	127±9[b]	12±2[b]	4.35± 0.19[b]	31±3[b]	37±5[b]
+ CdCl$_2$ (20 µM)	225±7[a]	54±5[a]	10.89±1.06[a]	42±4[a]	79±8[a]
+ DMSO (150 µM)	127±6[b]	15±2[b]	0.77± 0.09[b]	29±3[b]	45±4[b]
+ Mannitol (30 mM)	94±6[b]	15±2[b]	0.35± 0.03[b]	19±2[b]	37±4[b]
+ DPPD (1 µM)	167±8[b]	20±2[b]	0.26± 0.03[b]	36±3[b]	58±5[b]
+ Carnitine (2 mM)	115±5[b]	10±2[b]	5.27± 0.47[b]	28±3[b]	39±5[b]
+ Cyclosporin (2 µM)	123±4[b]	8±1[b]	5.73± 0.43[b]	33±3[b]	35±4[b]
+ Trifluoperazine (15 µM)	125±7[b]	10±2[b]	3.18± 0.27[b]	31±2[b]	34±3[b]
+ K$_2$Cr$_2$O$_7$ (1 mM)	508±8[a]	52±5[a]	20.43±1.98[a]	43±4[a]	75±7[a]
+ Carnitine (2 mM)	255±6[b]	14±2[b]	10.89±1.07[b]	37±4[b]	48±4[b]
+ Cyclosporin (2 µM)	265±8[b]	12±2[b]	9.68±1.05[b]	35±3[b]	43±4[b]
+ Trifluoperazine (15 µM)	233±7[b]	11±2[b]	8.93± 0.73[b]	33±3[b]	43±5[b]
GSH Depleted Hepatocytes	79±6	3±3	0.41± 0.06	21±2	24±3
+ HgCl$_2$ (10 µM)	356±9[a]	65±7[a]	8.89± 0.63[a]	87±7[a]	100[a]
+ CdCl$_2$ (10 µM)	268±7[a]	58±6[a]	13.27±1.04[a]	50±4[a]	100[a]

"ROS" formation was expressed as fluorescent intensity units by following dichlorofluorescin oxidation (Shen et al., 1996). $\Delta\psi_m$ was determined as the difference in rhodamine 123 uptake by control and treated cells and expressed as fluorescence intensity unit (Andersson et al., 1987). GSH depleted hepatocytes were prepared as described (Khan and O'Brien, 1991). MDA formation was expressed as µM concentrations as described (Smith et al., 1982). Values are expressed as means of three separate experiments (S.D.).
a: Significant difference in comparison with control hepatocytes ($P < 0.001$)
b: Significant difference in comparison with metal treated hepatocytes ($P < 0.001$)

Table 2. Preventing metal induced hepatocyte necrosis by apoptosis inhibitory agents

Addition	% Cytotoxicity		
	1 hr	2 hr	3 hr
Control Hepatocytes	18±2	19±2	20±2
+ HgCl$_2$ (20 µM)	48±5[a]	68±6[a]	82±8[a]
+ Caspase-3 Inhibitor (1 µM)	27±3[b]	35±4[b]	47±5[b]
+ Dactinomycin (1 µg/ml)	32±3[b]	45±4[b]	55±4[b]
+ 3-Aminobenzamide (20 mM)	37±4[b]	48±5[b]	54±5[b]
+ Cycloheximide (300 µM)	29±3[b]	43±4[b]	48±5[b]
+ CdCl$_2$ (20 µM)	42±4[a]	64±6[a]	79±8[a]
+ Caspase-3 Inhibitor (1 µM)	27±3[b]	33±3[b]	47±5[b]
+ Dactinomycin (1 µg/ml)	33±3[b]	34±3[b]	45±5[b]
+ 3-Aminobenzamide (20 mM)	33±3[b]	41±4[b]	58±6[b]
+ Cycloheximide (300 µM)	31±3[b]	32±3[b]	40±3[b]
+ K$_2$Cr$_2$O$_7$ (1 mM)	43±4[a]	61±6[a]	75±7[a]
+ Caspase-3 Inhibitor (1 µM)	33±3[b]	37±4[b]	48±5[b]
+ Dactinomycin (1 µg/ml)	37±4[b]	40±4[b]	51±5[b]
+ 3-Aminobenzamide (20 mM)	35±4[b]	38±4[b]	43±4[b]
+ Cycloheximide (300 µM)	33±3[b]	35±4[b]	39±4[b]

Values are expressed as means of three separate experiments (±S.D.)
a: Significant difference in comparison with control hepatocytes ($P < 0.001$)
b: Significant difference in comparison with metal treated hepatocyte ($P < 0.001$)

As shown in table 2 however cytotoxicity induced by all three metals was delayed by caspase-3 inhibitor, the protein synthesis inhibitor cycloheximide, the RNA synthesis inhibitor dactinomycin or the poly (ADP-ribose) synthase inhibitor 3-aminobenzamide.

<u>**DISCUSSION**</u>

The order of effectiveness found for the metals investigated was Hg II > Cd II >> Cr VI for cytotoxicity. A similar order was found for "ROS" formation which is surprising as only Cr VI is a redox metal. An order of effectiveness of Hg II = Cd II >> Cr VI was found for $\Delta\psi_m$ collapse and Cd II > Hg II > Cr VI for lipid peroxidation. The sequence of events for Cd II / Cr VI was "ROS" formation / loss of mitochondrial membrane potential followed by lipid peroxidation and membrane lysis. Much of the Hg II cytotoxicity however occurred within the first hour suggesting that Hg II rapidly modified the plasma membrane / mitochondria. Furthermore preventing lipid peroxidation with antioxidants prevented Cd II and Cr VI but not Hg II induced cytotoxicity.

The generation of "ROS" at the same time as a decrease in $\Delta\psi_m$ suggests that Hg II and Cd II induced a mitochondrial permeability transition (MPT). These events were part of the cytotoxic process as "ROS" scavengers or MPT inhibitors prevented metal induced hepatocyte lysis. This could result from binding to protein thiols of the inner mitochondrial membrane resulting in the uncoupling of oxidative phosphorylation and/or inhibition of mitochondrial respiration leading to "ROS" generation. This was recently suggested as an explanation for the "ROS" generation and $\Delta\psi_m$ collapse observed when Hg II was added to human lymphocytes (Guo et al., 1998). Cd II also caused $\Delta\psi_m$ collapse in rat-1 fibroblasts (Kim et al., 2000), isolated rat hepatocytes (Pourahmad and O'Brien, 2000a) and isolated liver mitochondria at pH 6.5 (Koizumi et al., 1994). The development of a MPT resulted in a mitochondrial proton gradient loss which in turn caused intracellular acidification (Meisenholder et al., 1996) and was demonstrated for Hg II (Guo et al., 1998) and Cd II (Koizumi et al., 1994). Previously we showed that hepatocyte cytotoxicity induced by the respiratory inhibitors cyanide and antimycin resulted in cytotoxic "ROS" formation (Niknahad et al., 1995). Mitochondrial H_2O_2 formation was also induced by Hg II with rat kidney mitochondria in vitro or in vivo (Lund et al., 1993). In turn the "ROS" formed oxidatively crosslinks membrane protein thiols which in the presence of Ca^{2+} causes an MPT (Kowaltowski et al., 1998). The Cr VI induced hepatocyte "ROS" formation and $\Delta\psi_m$ collapse however may involve Cr IV redox mechanisms (Pourahmad and O'Brien, 2000b).

The MPT is defined as an abrupt permeability increase of the inner mitochondrial membrane observed after matrix Ca^{2+} accumulation and opening of a high conductance pore (that conducts solutes of molecular mass < 1500D). This collapses $\Delta\psi_m$ and uncouples respiration from ADP phosphorylation. Cyclosporin blocks the pore by binding to cyclophilin D, a pore complex component and can prevent hepatocyte necrotic cell death as well as tumor necrosis factor-α induced hepatocyte apoptosis (Lemasters et al., 1999). Carnitine or trifluoperazine also inhibit the MPT (Pastorino et al., 1993; Lemasters et al., 1999). Our finding that cyclosporin, carnitine, trifluoperazine prevented metal induced cytotoxicity and $\Delta\psi_m$ collapse further indicates that metals induce an MPT .

The onset of the MPT is believed to collapse the $\Delta\psi_m$ and release mitochondrial Ca^{2+}, an apoptosis inducing factor and cytochrome c. Cytochrome c binds to cytosolic apoptosis activating factor-1 / dATP thereby activating the apoptotic protease caspase 9, which in turn activates caspase 3. Caspase 3 activation causes poly (ADP-ribose) polymerase cleavage and internucleosomal DNA hydrolysis characteristic of apoptotic cell death (Lemasters et al., 1999). Our finding that metal induced cell lysis is prevented by caspase 3 inhibitor, even though no apoptotic morphology is apparent, suggests that apoptotic and necrotic cell death share signalling pathways. ATP is required for the highly organized sequence of events involved in the cell packaging program required for apoptosis. Apoptosis morphology would occur in vivo at metal concentrations that do not deplete ATP levels substantially (Kim et al., 2000; Habeebu et al., 1998). Further evidence suggestive of this was the delay in metal induced cytotoxicity by the protein synthesis inhibitor cycloheximide or the RNA synthesis inhibitor dactinomycin. This could explain why liver necrosis in vivo induced by various hepatotoxins is also prevented by cycloheximide (Flaks and Nicoll, 1974; Farber et al., 1971).

<u>**REFERENCES**</u>
Andersson, B.S., Aw, T.Y., Jones, D.P. (1987). Mitochondrial transmembrane potential and pH gradient during anoxia. *Am J Physiol,* **252**(4 Pt 1), C349-C355.
Eto, K. (1997) Pathology of Minamata disease. *Toxicol. Pathol.* **25**(6), 614-23.
Farber, E. (1971). Biochemical pathology. *Annu. Rev. Pharmacol.* **11**, 71-96.
Flaks, B., Nicoll, J.W. (1974). Modification of toxic liver injury in the rat. I. Effect of inhibition of protein synthesis on the action of 2-acetylaminofluorene, carbon tetrachloride, 3'-methyl-4-dimethylaminoazobenzene and diethylnitrosamine. *Chem. Biol. Interact.* **8**(3), 135-50.

Guo, T.L., Miller, M.A., Shapiro, I.M., Shenker, B.J. (1998). Mercuric chloride induces apoptosis in human T lymphocytes: evidence of mitochondrial dysfunction. *Toxicol. Appl. Pharmacol.* **153**(2), 250-7.

Habeebu, S.S., Liu, J., Klaassen, C.D. (1998). Cadmium-induced apoptosis in mouse liver. *Toxicol. Appl. Pharmacol.* **149**(2), 203-9.

Horiguchi, H., Teranishi, H., Niiya, K., Aoshima, K., Katoh, T., Sakuragawa, N., Kasuya, M. (1994). Hypoproduction of erythropoietin contributes to anemia in chronic cadmium intoxication: clinical study on Itai-itai disease in Japan. *Arch. Toxicol.* **68**(10), 632-6.

Khan, S. and O'Brien, P.J. (1991). 1-bromoalkanes as new potent nontoxic glutathione depletors in isolated rat hepatocytes. *Biochem. Biophys. Res. Commun.* **179**(1), 436-41.

Kim, M.S., Kim, B.J., Woo, H.N., Kim, K.W., Kim, K.B., Kim, I.K., Jung, Y.K. (2000). Cadmium induces caspase-mediated cell death: suppression by Bcl-2. *Toxicology* **145**(1), 27-37.

Koizumi, T., Yokota, T., Shirakura, H., Tatsumoto, H., Suzuki, K.T. (1994). Potential mechanism of cadmium-induced cytotoxicity in rat hepatocytes: inhibitory action of cadmium on mitochondrial respiratory activity. *Toxicology* **92**(1-3), 115-25.

Kowaltowski,A.J., Naia-da-Silva, E.S., Castilho, R.F., Vercesi, A.E. (1998). Ca2+-stimulated mitochondrial reactive oxygen species generation and permeability transition are inhibited by dibucaine or Mg2+. *Arch. Biochem. Biophys.* **359**(1), 77-81.

Kurasaki, K., Nakamura, T., Mukai, T., Endo, T. (1995). Unusual findings in a fatal case of poisoning with chromate compounds. *Forensic Science Intern.* **75**, 57-65.

Lemasters, J.J. (1999). Mechanisms of hepatic toxicity: necroapoptosis and the mitochondrial permeability transition. Shaped pathways to necrosis and apoptosis. *Am. J. Physiol.* **276**, G1-G6.

Lund, B.O., Miller, D.M., Woods, J.S. (1993). Studies on Hg(II)-induced H2O2 formation and oxidative stress in vivo and in vitro in rat kidney mitochondria. *Biochem. Pharmacol.* **45**(10), 2017-24.

Meisenholder, G.W., Martin, S.J., Green, D.R., Nordberg, J., Babior, B.M., Gottlieb, R.A. (1996). Events in apoptosis. Acidification is downstream of protease activation and BCL-2 protection. *J. Biol. Chem.* **271**(27), 16260-2.

Miksche, L.W., Lewalter, J. (1997). Health surveillance and biological effect monitoring for chromium-exposed workers. *Regul. Toxicol. Pharmacol.* **26**(1 Pt 2), S94-9.

Nieminen, A.L., Gores, G.T., Dawson, T.L., Herman, B., Lemaster, J.T. (1990). Toxic injury from mercuric chloride in rat hepatocytes. *J. Biol. Chem.* **265**, 2399-408

Niknahad, H., Khan, S. and O'Brien, P.J. (1995). Hepatocyte injury resulting from the inhibition of mitochondrial respiration at low oxygen concentrations involves reductive stress and oxygen activation. *Chemico-Biol. Interacns.* **98**, 27-44.

Pastorino, J.G., Snyder, J.W., Serroni, A., Hoek, J.B., Farber, J.L. (1993). Cyclosporin and carnitine prevent the anoxic death of cultured hepatocytes by inhibiting the mitochondrial permeability transition. *J. Biol. Chem.* **268**(19), 13791-8.

Pourahmad, J., O'Brien, P.J. (2000a). A comparison of hepatocyte cytotoxic mechanisms for Cu2+ and Cd2+. *Toxicology* **143**(3), 263-73.

Pourahmad, J., O'Brien, P.J. (2000b). Contrasting role of Na(+) ions in modulating Cu(+2) or Cd(+2) induced hepatocyte toxicity. *Chem. Biol. Interact.* **126**(2), 159-69.

Shen, H.M., Shi, C.Y., Shen, Y., Ong, C.N. (1996). Detection of elevated reactive oxygen species level in cultured rat hepatocytes treated with aflatoxin B1. *Free Radic. Biol. Med.* **21**(2), 139-146.

Siegers, C.P., Sharma, C.S., Younes, M. (1986) Hepatotoxicity of metals in glutathione depleted mice. *Toxicol. Lett.* **34**, 185-91

Smith, M.T., Thor, H., Hartizell, P., Orrenius, S. (1982). The measurement of lipid peroxidation in isolated hepatocytes. *Biochem. Pharmacol.* **31**(1), 19-26.

Strubelt, O., Kremer, J., Tilse, A., Keogh, J., Pentz, R., Younes, M. (1996). Comparative studies on the toxicity of mercury, cadmium, and copper toward the isolated perfused rat liver. *J. Toxicol. Environ. Health* **47**(3), 267-83.

QUANTIFICATION OF F-RING AND D-/E-RING ISOPROSTANES AND NEUROPROSTANES IN ALZHEIMER'S DISEASE

Erin E. Reich, William R. Markesbery, L. Jackson Roberts II, Larry L. Swift, Jason D. Morrow, and Thomas J. Montine

Departments of Pharmacology, Pathology, and Medicine, and the Centers for Molecular Neuroscience and Molecular Toxicology, Vanderbilt University Medical Center, Nashville, TN, and the Sanders-Brown Center on Aging, and the Departments of Pathology and Neurology, University of Kentucky, Lexington, KY

INTRODUCTION

Numerous *in vitro*, cell culture, animal, and tissue homogenate studies have implicated free radical damage in the pathogenesis of Alzheimer's disease (AD) [1, 2]. A variety of methods have been employed to assess the free radical damage; however, most of these methods suffer from either non-quantitative results, a lack of information about cellular localization, or both [3-10].

Isoprostanes (IsoPs) are exclusive products of free radical catalyzed peroxidation of arachidonic acid (C20:4ω6, AA) that are formed esterified to lipid (bound) and then are hydrolyzed (free) [11]. Unlike AA that is evenly distributed in gray and white matter, docosahexaenoic acid (C22:6ω3, DHA) is enriched in gray matter of the CNS, where it is synthesized in astrocytes and then transported and concentrated in neurons [12, 13]. Previously, we described the formation of neuroprostanes (NPs) from free radical catalyzed peroxidation of DHA via reactions analogous to IsoP generation [14, 15]. We proposed that NPs may provide more specific information on free radical damage in DHA-containing compartments, *i.e.*, neurons, and that NPs may be more sensitive markers of free radical damage because DHA is more labile to peroxidation that is AA.

IsoP and NP formation proceeds through bicyclic endoperoxide intermediates that are reduced to F-ring compounds or undergo rearrangement to D/E-ring compounds. Levels of F_2-IsoPs and F_4-NPs are increased in cerebrospinal fluid from definite AD patients compared to controls [15-18]. Additionally, tissue levels of F_2-IsoPs are increased in frontal cortex of definite AD patients compared to controls [19]. Levels of F_4-NPs, D_2/E_2-IsoPs, and D_4/E_4-NPs have not been reported in brain tissue.

In vitro, increasing the concentration of cellular reductants, such as glutathione, favors reduction of the endoperoxide intermediates resulting in greater amounts of F-ring compounds and lower amounts of D/E ring compounds [20]. Therefore, calculation of the F-ring to D/E-ring ratio supplies information on the reducing environment in which free radical-mediated damage occurred. Herein we report a comprehensive quantification of F_2-IsoP, D_2/E_2-IsoP, F_4-NPs, and D_4/E_4-NPs in 4 different brain regions from clinically and pathologically characterized definite AD patients and age matched controls.

Biological Reactive Intermediates VI, Edited by Dansette *et al.*
Kluwer Academic / Plenum Publishers, 2001

MATERIALS AND METHODS

After appropriate consent was obtained, all individuals included in this study underwent *post mortem* examination as part of a rapid autopsy program at the Alzheimer's Disease Research Center at the Sanders-Brown Center on Aging, University of Kentucky. All AD patients were diagnosed with probable AD during life and were shown by neuropathological examination to meet the criteria for definite AD [21, 22]. Controls were age- and gender-matched individuals without clinical evidence of dementia or other neurological disease. Neuropathological examination of controls showed only age-associated changes.

The following tissue sections were dissected at the time of autopsy: hippocampus at the level of the lateral geniculate nucleus, superior and medial temporal gyri (SMTG), inferior parietal lobule (IP), and cerebellar cortex. All samples were kept frozen at −80° C until used. Lipids were extracted by the method of folch. D_2/E_2-IsoPs and D_4/E_4-NPs esterified in tissue were converted to O-methyloxime derivatives in folch solution. IsoPs and NPs were hydrolyzed by chemical saponification, extracted using C-18 and silica Sep-Pak cartridges, purified by thin layer chromatography (TLC), converted to pentaflurobenzyl ester trimethylsilyl ether derivatives, and quantified by stable isotope dilution techniques employing gas chromatography (GC)/negative ion chemical ionization (NICI)/mass spectrometry (MS) using $[^2H_4]$-8-iso-$PGF_{2\alpha}$ and $[^2H_4]$-PGE_2 as internal standards as previously described [14]. The derivatized D/E-ring IsoPs or NPs co-migrate on silica TLC plates and GC and had identical masses, therefore the levels of these isomers are reported as combined values. AA and DHA concentrations were determined as previously described [23].

Statistical analyses were preformed using GraphPad Prism software (San Diego, CA). Student's t-tests was used for paired comparisons, and two-way analysis of variance (ANOVA) was used for comparison among brain regions in AD patients and control individuals.

RESULTS

Table 1 presents information on the 20 individuals included in this study. Age, gender, and PMI were not significantly different between the two groups. AD patients had characteristic average disease duration, as well as significantly lower brain weights than controls (* $p < 0.05$).

Table 1. Patient data for the 20 individuals included in this study

	n	Age (yr)	M:F	Disease Duration (yr)	Post mortem Interval (hr)	Brain Weight (g)
AD	9	78.1 ± 2.7	4:5	8.1 ± 1.1	2.6 ± 0.2	1091 ± 41*
Control	11	80.7 ± 2.5	5:6	Not Applicable	2.7 ± 0.2	1220 ± 40

AA and DHA levels were quantified in folch extracts of frozen SMTG, hippocampus, IP, and cerebellum from AD patients and controls. Overall, AA and DHA were 7.7 ± 0.4 % and 14.3 ± 0.4 %, respectively, of total fatty acids. Two-way ANOVA showed no effect of disease or brain region on AA concentrations. Two-way ANOVA for DHA tissue levels also showed no significant variance with disease, but there was significant variance with brain region ($p < 0.01$). The hippocampus had the lowest concentrations of DHA (12.0 ± 0.2 % of total fatty acids) while the IP had the highest (15.5 ± 0.4 % of total fatty acids).

Examination of IsoP and NP levels from all brain regions and all individuals revealed that the F_4-NPs were by far the most abundant, having an overall average (all individuals, all regions) level of 13.7 ± 0.8 ng/g brain tissue. The corresponding overall average level of F_2-IsoPs was 4.9 ± 0.3 ng/g brain tissue, 2.8-

fold less than F_4-NPs. Levels of D/E-ring IsoPs and NPs were the lowest, averaging 1.5 ± 0.1 and 1.4 ± 0.2 ng/g brain tissue, respectively.

Levels of F-ring plus D/E-ring compounds were determined to assess the magnitude of free radical damage to AA and DHA and then stratified by brain region and the presence of AD. Two-way ANOVA for tissue levels of NPs was significant for AD versus control ($p < 0.0001$) and for interaction between AD and brain region ($p < 0.05$), a consequence of NPs being higher in disease regions of AD brain tissue than in the cerebellum. An analogous two-way ANOVA for tissue levels of F-ring plus D/E-ring IsoPs was not significant for presence of AD or brain region ($p > 0.05$).

The ratios of F-ring to D/E-ring compounds were computed for IsoPs and NPs to assess the reducing environments in which free radical damage to AA and DHA occurred. Two-way ANOVA for the F_4- to D_4/E_4-NP ratio was highly significant for AD ($p < 0.005$), but not brain region. Brain region did not contribute to the variance in F_4- to D_4/E_4-NP ratio. Indeed, all four brain regions in AD subjects had a lower F_4- to D_4/E_4-NP ratio than controls. Neither AD nor brain region significantly contributed to the variance in IsoP ratios.

The levels and ratios of IsoP and NPs in different brain regions did not correlate with neuritic plaques or neurofibrillary tangles density in the same brain region.

DISCUSSION

We tested the hypothesis that free radical damage in gray matter from AD brain is concentrated in DHA- rather than AA-containing compartments. Our results indicate that the DHA-containing, but not AA-containing, compartments in AD cerebrum undergo significantly increased free radical damage compared to controls.

The lack of difference in total IsoP levels between AD patients and age-matched controls indicated that gray matter did not experience significantly more lipid peroxidation in AD patients compared to age-matched controls. However, neurons are only one component of gray matter. Our results with total NP levels indicated that the subset of gray matter that contains DHA did experience increased levels of free radical damage in AD patients compared to controls. In combination, these results suggest that free radical damage in AD is focused in neurons and is not evenly distributed within gray matter. Moreover, our results showed that increased free radical damage to neurons occurred in cerebral cortex and hippocampus but not cerebellum.

In addition, we quantified F- to D/E-ring ratios of IsoP and NPs, a reflection of the reducing environment in which IsoPs and NPs are formed, in different regions of AD brain. The 40 to 70% decrease in F- to D/E-ring NP ratio in AD patients with unchanged IsoP ratio indicated that neurons may have significantly diminished reducing capacity in AD. It is important to note that the F- to D/E-ring NP ratio was lower in all AD brain regions including the cerebellum. However, the levels of NPs were elevated only in brain areas affected by the disease and not in cerebellum. This comparison suggests that the lowered reducing capacity in AD neurons is not necessarily a consequence of increased free radical damage. Moreover, since cerebellum is not considered a site for AD pathological changes, our results raise the possibility that diminished reducing capacity in neurons may be a feature of patients who are vulnerable to developing AD and not an outcome of AD pathological changes.

Our results showing that abnormalities in AD brain are focused in the DHA-containing compartments are entirely consistent with numerous histochemical and immunohistochemical reports localizing increased free radical damage and diminished reducing agents in neurons. The advantage of our approach is that it allows an unbiased, robust quantification of these events.

REFERENCES

1. W.R. Markesbery and J.M. Carney. Oxidative alterations in Alzheimer's disease. *Brain Pathology*. 9:133-146 (1999).
2. M.P. Mattson. Modification of ion homeostasis by lipid peroxidation: roles in neuronal degeneration and adaptive plasticity. *Trends Neurosci*. 20:53-57 (1998).

3. J.M.C. Gutteridge and B. Halliwell. The measurement and mechanism of lipid peroxidation in biological systems. *Trends Biochem. Sci.* 15:129-135 (1990).

4. K. Moore and L.J. Roberts, II. Measurement of lipid peroxidation. *Free Radical Res.* 28:659-671 (1998).

5. K.S. Montine, S.J. Olson, V. Amarnath, W.O. Whetsell, D.G. Graham and T.J. Montine. Immunochemical detection of 4-hydroxynonenal adducts in Alzheimer's disease is associated with APOE4. *Am. J. Path.* 150:437-443 (1997).

6. K.S. Montine, E.E. Reich, S.J. Olson, W.R. Markesbery and T.J. Montine. Distribution of reducible 4-hydroxynonenal adduct immunoreactivity in Alzheimer's disease is associated with APOE genotype. *J. Neuropathol. Exp. Neurol.* 57:415-425 (1998).

7. K.S. Montine, P.J. Kim, S.J. Olsen, W.R. Markesbery and T.J. Montine. Hydroxy-2-nonenal pyrrole adducts in human neurodegenerative disease. *J. Neuropathol. Exp. Neurol.* 56:866-871 (1997).

8. L.M. Sayre, D.A. Zelasko, P.L.R. Harris, G. Perry, R.G. Salomon and M.A. Smith. Hydroxynonenal-derived advanced lipid peroxidation end products are increased in Alzheimer disease. *J. Neurochem.* 2092-2097 (1997).

9. M.A. Smith, S. Taneda, P.L. Richey, S. Miyata, S.D. Yan, D. Stern, L.M. Sayre, V.M. Monnier and G. Perry. Advanced Maillard reaction end products are associated with Alzheimer disease pathology. *Proc. Natl. Acad. Sci. USA.* 91:5710-5714 (1994).

10. M.A. Smith, L.M. Sayre, V.E. Anderson, P.L. Harris, M.F. Beal, N. Kowall and G. Perry. Cytochemical demonstration of oxidative damage in Alzheimer's disease by immunochemical enhancement of the carbonyl reaction with 2,4-dinitrophenylhydrazine. *J. Histochem. Cytochem.* 46:731-735 (1998).

11. J.D. Morrow and L.J. Roberts, II. The isoprostanes: unique bioactive products of lipid peroxidation. *Prog. Lipid Res.* 36:1-21 (1997).

12. S.A. Moore. Cerebral endothelium and astrocytes cooperate in supplying docosahexaenoic acid to neurons. *Adv. Exp. Med. Biol.* 331:229-233 (1993).

13. S.A. Moore, E. Yoder, S. Murphy, G.R. Dutton and A.A. Spector. Astrocytes, not neurons, produce docosahexaenoic acid and arachidonic acid. *J. Neurochem.* 56:518-524 (1991).

14. E.E. Reich, W.E. Zackert, C.J. Brame, Y. Chen, L J Roberts, II, D.L. Hachey, T.J. Montine and J.D. Morrow. Formation of novel D-ring and E-ring isoprostane-like compounds (D_4/E_4-neuroprostanes) *in vivo* from docosahexaenoic acid. *Biochemistry.* 39:2376-2383 (2000).

15. L.J. Roberts, II, T.J. Montine, W.R. Markesbery, A.R. Tapper, P. Hardy, S. Chemtob, W.D. Dettbarn and J.D. Morrow. Formation of isoprostane-like compounds (neuroprostanes) *in vivo* from docosahexaenoic acid. *J. Biol. Chem.* 273:13605-13612 (1998).

16. T.J. Montine, W.R. Markesbery, J.D. Morrow and L.J. Roberts, II. Cerebrospinal fluid F_2-isoprostanes are increased in Alzheimer's disease. *Ann. Neurol.* 44:410-413 (1998).

17. T.J. Montine, M.F. Beal, M.E. Cudkowicz, R.H. Brown, H. O'Donnell, R.A. Margolin, L. McFarland, A.F. Bachrach, W.E. Zackert, L.J. Roberts, II and J.D. Morrow. Increased cerebrospinal fluid F_2-isoprostane concentration in probable Alzheimer's disease. *Neurology.* 52:562-565 (1999).

18. T.J. Montine, W.R. Markesbery, W.E. Zackert, S.C. Sanchez, L.J. Roberts, II and J.D. Morrow. The magnitude of brain lipid peroxidation correlates with the extent of degeneration but not with the density of NP's or NFT's, or with APOE genotype in Alzheimer's disease patients. *Am. J. Path.* 155:863-868 (1999).

19. D. Pratico, V.M. Lee, J.Q. Trojanowski, J. Rokach and G.A. Fitzgerald. Increased F_2-isoprostanes in Alzheimer's disease: evidence for enhanced lipid peroxidation *in vivo*. *FASEB J.* 12:1777-1784 (1998).

20. J.D. Morrow, L.J. Roberts, II, V.C. Daniel, J.A. Awad, O. Mirochnitchenko, L.L. Swift and R.F. Burk. Comparison of formation of D_2/E_2-isoprostanes and F_2-isoprostanes *in vitro* and *in vivo*-effects of oxygen tension and glutathione. *Arch. Biochem. Biophys.* 353:160-171 (1998).

21. S.S. Mirra, A. Heyman, D. McKeel, S.M. Sumi, B.J. Crain, L.M. Brownlee, F.S. Vogel, J.P. Hughes, G. van Belle and L. Berg. The consortium to establish a registry for Alzheimer's disease (CERAD), Part II. Standardization of the neuropathologic assessment of Alzheimer's disease. *Neurology.* 41:479-486 (1991).

22. G. McKhann, D. Drachman, M. Folstein and Others. Clinical diagnosis of Alzheimer's disease: report of the NINCDS-ADRDA work group under the auspices of the Department of Health and Human Services Task Force on Alzheimer's disease. *Neurology.* 34:939-944 (1984).

23. T.J. Montine, K.S. Montine and L.L. Swift. Central nervous system lipoproteins in Alzheimer's disease. *Am. J. Path.* 151:1571-1575 (1997).

PREVENTIVE EFFECT OF VITAMIN E ON THE PROCESSES OF FREE RADICAL LIPID PEROXIDATION AND MONOOXYGENASE ENZYME ACTIVITY IN EXPERIMENTAL INFLUENZA VIRUS INFECTION

E. Stoeva[1], L.Tantcheva[1], M.Mileva[4], V.Savov[5], A.S.Galabov[2], A.Braykova[3]

[1]Institute of Physiology, [2]Institute of Microbiogy, [3]Institute of Molecular Biology - Bulgarian Academy of Sciences, [4]Medical University - Sofia, [5]University of Sofia "St. Kl. Ohridski", Sofia (Bulgaria)

INTRODUCTION

The processes of lipid peroxidation (LPO) in cell membranes modify effectively the lipid structure of biomembranes and the activity of drug-metabolizing enzyme systems (DMES) and specially cytochrome P-450-dependent microsomal monooxygenase (Mannering 1981). Reactive oxygen species (ROS) are generated in this system leading to chain development of LPO processes.

In some viral infections (e.g. influenza), activation of the processes of LPO has been established when the content of ROS and free radicals is increased (Peterhans, 1997, Chetverikova and Inozemtseva, 1996). The endogenous antioxidant systems are effective only at the initial stages of free radical damages. Later, generation rate of ROS and free radicals begins to increase and exceeds many-fold the capacity of endogenous utilization, hence the antioxidant protection of the body should be enhanced (Pokhilko et al. 1995, Kiselev et al. 1994). Recently, the clinicians and researchers show an enhanced interest in the protective action of antioxidants and in the possibility of their use for the prevention and treatment of the so called "free radical" diseases.

The aim of the present work is to study the effect of vitamin E on the content of LPO products and drug-metabolizing enzyme activities in liver of mice infected with influenza virus.

MATERIALS AND METHODS

Albino male mice, line ICR (14 - 16 g) received intraperitoneally (i.p.) vitamin E (vit E- alpha-tocopheryl-acetate, Sigma) (60, 120 and 240 mg/kg b.w., 5 days) and then were infected with influenza virus A/Aichi/2/68 (H3N2) (0.5 of LD_{50}), by intranasal inoculation.

On the 5th day after virus inoculation (crucial for viral infection), the animals were decapitated and the obtained 10000 x g liver supernatant was used for determination of LPO products and of some drug-metabolizing enzyme systems.

The concentration of the primary products of LPO-conjugated dienes was determined spectrophotometrically (at 232 nm) according Rechnagel and Glende (1984) after previous total lipid extraction (Folch et al. 1957). MDA-thiobarbituric acid reactive substances (MDA-TBARS) products as a secondary products of LPO processes was determined by TBA test.

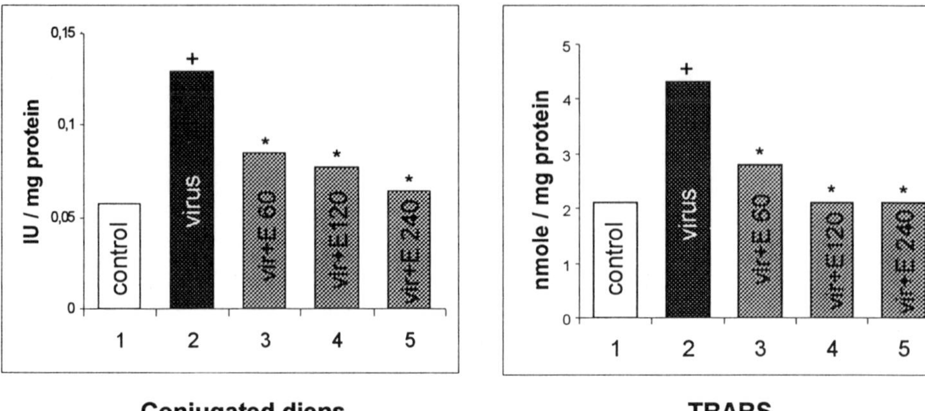

Conjugated diens   **TBARS**

Figure 1. Effect of vit E on the level of LPO products in influenza virus infected mice (5th day)

Aniline hydroxylase (AH) activity was determined according to Mazel's method (1971), N-demethylase activity with substrates ethylmorphine (EMND), amidopyrine (APND) and analgin (ANND) according Nash's method (1953), the content of hepatic cytochrome P-450 (cyt P-450) according to Matsubara et al. (1976) and NADPH-cytochrome C-reductase (CCR) according to Roering et al. (1972).

The correlation between the concentration of LPO products (nmol TBARS/mg prot) and activity of P-450-dependent microsomal monooxygenases (nmol product/mg prot/min) was calculated according to Brave and Pearson (1962).

RESULTS AND DISCUSSION

Our results confirm the important role of the products of free radical processes in the devolopment acute viral infection (Peterhans, 1997). The virus infection leads to a sharp increase (almost twice versus the controls) the primary and secondary products of LPO in the liver (Fig.1) and inhibits hepatic drug metabolism (Fig.2 and Fig.3).

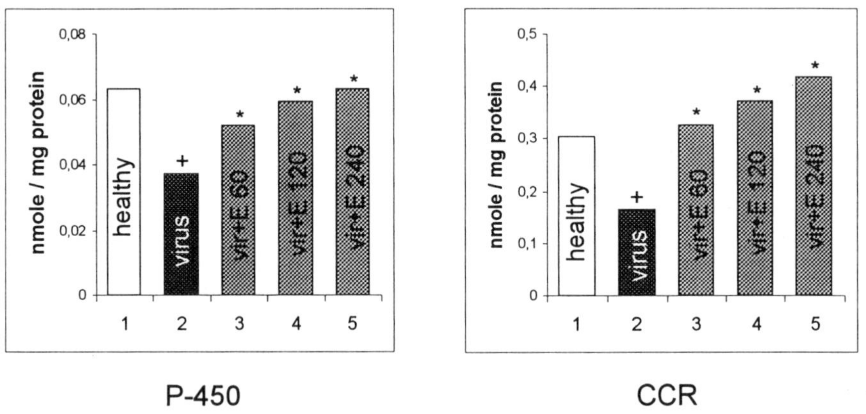

P-450   CCR

Figure 2. Effect of vit E on cyt P-450 content and CCR activity in influenza virus infected mice (5th day)

The reduced concentrations of cytochrome P-450 and CCR activity (Fig. 2) correlate inversely proportionally to the increased primary and secondary LPO products (conjugated dienes/cyt P-450 r=-0.945, and MDA-TBARS/cyt P-450 r= -0.873; conjugated dienes/CCR r= -0.671 and MDA-TBARS/CCR r= -0.796, on the 5th day). This inverse proportional dependence, with high significance (p<0.001) allows us to suggest the activation of the LPO processes during experimental influenza virus infection, can be one of the causes for DMES inhibition. At present the exact physiological role of LPO processes for DMES regulation is not well understood. In pathological condition e.g. influenza virus infection, increased LPO products can modify the lipid matrix of membranes and disorder the lipid-protein interaction. In this way the cytochrome can be transformed into its inactive form- cyt P-420 (Gorbunov et al. 1992, Eisen 1986). Farrel (1987) related this with the induction of xantinoxidase, catalizing formation of oxygen free radicals destroing cyt P-450 structure.

As result of the total inhibition of cyt P-450 chain by virus infection, the metabolism of the substrates- amidopyrine, analgin, ethylmorphine and aniline was inhibited also (Fig.3). Nevertheless the lack of data in the literature for changed drug metabolism and toxicity by influenza, we can expect some changes in pharmacological activity and toxicity of drugs, metabolized by monooxygenases and having low therapeutic index.

An protective effect of vit E during inluenza virus infection in mice was established. Vit E administration in advance reduced primary and secondary products of LPO, increased by influenza. The effect was manifested better with a dose of 240 mg/kg of vit E (Fig. 1). Our data reveal positive regenerative effect of vit E on inhibited DMES activity in infected animals. On the 5th day cyt P-450 content, CCR activity and monooxygenase activities were restored, eventhough not fully (Fig. 2 and Fig. 3). Maximal effect had vit E in dose 240 mg/kg.

Though vit E is not an agent with specific antiviral action, its antioxidant properties

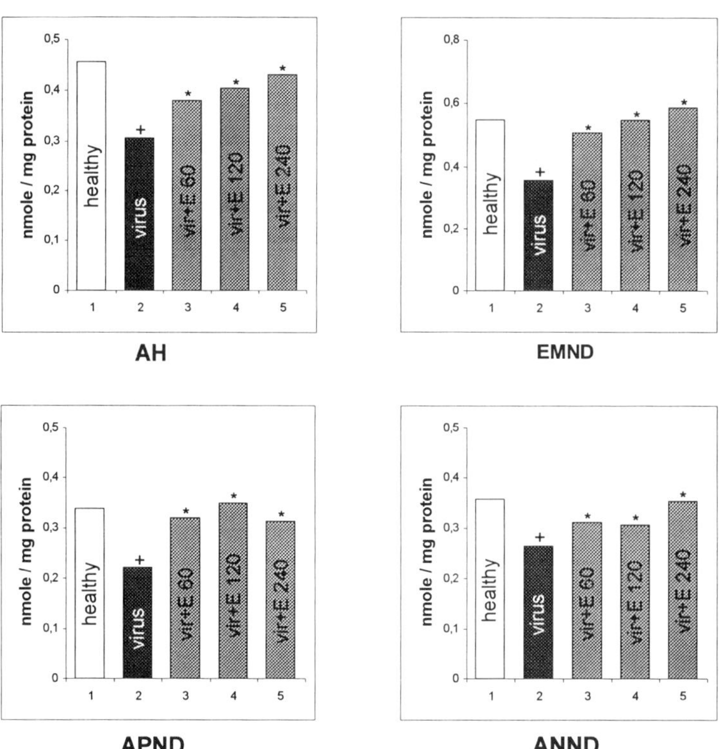

Figure 3. Effect of vit E on monooxygenase activity in influenza virus infected mice (5th day)

probably allow to moderate the course of the infection as membrane-protector. An important role of vit E in the prevention and treatment of some viral diseases can be assumed.

REFERENCES

Brave, A., Pearson, S., 1962, Statistical methods in biology, N. Bayleys. ed., New York

Chetverikova, L.K and Inozemtseva, L.I., 1996, Role of lipid peroxidation in the pathogenesis of the influenza and search for antivial protective agents, Vestn. Ross. Acad. Med. Nauk, 3: 37

Eisen, H.J., 1986, Induction of hepatic P-450 izozymes. Evidens for specific receptors, in: Cytochrome P-450 Structure, Mechanism and Biochemistry. P.R.O. de Montallano. ed., Plenum Press, New York

Farrell, G.C., 1987, Drug metabolism in extrahepatic diseases. Pharmacol. Ther., 35: 375

Folch, J., Lees, M., Shoane-Stoaley, C.H., 1957, A simple methods for the isolation and puritification of total lipid from animal tissue, J. Biol. Chem., 226: 497

Gorbunov, N.V., Volgarev, A.P., Bykova, N.O., Prozorovskaia, M.P., 1992, Microsomal hydroxilating system of the mouse liver in toxic form of influenza infection, Bull. Exp. Biol. Med. July, 114(7): 44

Kiselev, O.I, Isakov, V.F.A, Sharonov, B.P, Sukhinin, V.P., 1994, Pathogenesis of severe influenza, Vesn. Ross. Acad. Med. Nauk., 9: 32

Mannering, G.J., 1981, Hepatic cytochrome P-450-linked drug metabolizing systems, in: Concepts in Drug Metabolism. P. Jenner and B. Testa. eds., Marcel Dekker, New York

Matsubara, N., Koike, M., Touchi, A., Tochino, Y., Sugeno, K., 1976, Quantative determination of cytochrome P-450 in rat liver homogenate, Anal.Biochem. 75 (2): 596

Mazel, P., 1971, Experiments illustrating drug metabolism in vitro-determination of microsomal of aniline hydroxylase, in: Fundamentals of Drug Metabolism and Drug Disposition. La Du B.N., Mandel H.G., Way E.L. eds, The Willkins Co, Baltimore

Nash, T., 1953, The colometric estimation of formaldehyde by means of the Hautch reaction, J. Biol. Chem. 55: 416

Peterhans, E., 1997, Reactive oxygen species and nitric oxyde in viral diseases, Biol. Trace Element Research, 56: 107

Pokhillko, A.V, Kramskaia, T.A, Poliak, R.I., 1995, Stress and viral infection: dynamics of expression of influenza A virus agents during immobilization stress, Vopr-Virusol., Mar-Apr 40(2): 76

Rechnagel, R. and Glende, E., 1984, Spectrophotometric detection of lipid conjugated diens, Meth. Enzymol. 105: 331

Roering, D. L., Mascaro, L., Aust, S., 1972, Microsomal electron transport: tetrafolium reduction by rat liver microsomal NADP.H-cytochrome c-reductase. Arch. Biochem. 153: 475

LYSOSOMAL OXIDATIVE STRESS CYTOTOXICITY INDUCED BY NITROFURANTOIN REDOX CYCLING IN HEPATOCYTES

Jalal Pourahmad, Sumsullah Khan and Peter J. O'Brien

Faculty of Pharmacy, University of Toronto, Toronto, ON, M5S 2S2, Canada

ABSTRACT

1. The enzymes responsible for the reductive activation of NFT are not known. We have now shown that under aerobic conditions, inhibitors of cytochrome P450 or P450 reductase but not DT diaphorase prevented NFT induced cytotoxicity and reactive oxygen species ("ROS") formation. This suggests that NFT was reductively activated by reduced cytochrome P450 and/or P450 reductase.
2. The subcellular organelle oxidative stress effects leading to cytotoxicity are not known. Hepatocyte mitochondrial membrane potential was only slightly decreased by NFT before cytotoxicity ensued. However NFT induced lysosomal damage and hepatocyte protease activation. Endocytosis inhibitors, lysosomotropic agents or lysosomal protease inhibitors also prevented NFT induced cytotoxicity.
3. Lipid peroxidation also preceded cytotoxicity. Furthermore desferoxamine (a ferric chelator), antioxidants or ROS scavengers (catalase, mannitol, TEMPOL or dimethylsulfoxide) prevented NFT cytotoxicity.
4. It is concluded that H_2O_2 reacts with lysosomal Fe^{+2} to form "ROS" which causes lysosomal lipid peroxidation, membrane disruption, protease release and cell death.

INTRODUCTION

Nitrofurantoin (NFT) is a widely utilized urinary antimicrobial drug which has been associated with pulmonary fibrosis, neuropathy, hepatitis, and hemolytic anemia in patients with glucose-6-phosphate dehydrogenase deficiency (Rosenow, 1978) but the molecular mechanisms leading to NFT-induced cell toxicity are not understood. Previously we showed that incubation of isolated rat hepatocytes with NFT results in a nearly stoichiometric oxidation of GSH to GSSG, which rapidly effluxed the cell (Rossi et al., 1988; Silva et al., 1991). The H2O2 formed that oxidized GSH resulted from a one-electron reduction of the nitro group followed by auto-oxidation of the nitro radical anions formed (Rao et al., 1987).

METHODS

Isolation of Hepatocytes

Male Sprague-Dawley rats (280-300g), fed a standard chow diet and given water ad libitum, were used in all experiments. Hepatocytes were obtained by collagenase perfusion of the liver and their viability was assessed by the trypan blue (0.2% w/v) exclusion test as described by Pourahmad and O'Brien, 2000. Approximately 85-90% of the hepatocytes excluded trypan blue. Cells were suspended at a density of 10^6 cells/ml in round bottomed flasks rotating in a water bath maintained at 37 °C in Krebs-Henseleit buffer (pH 7.4), supplemented with 12.5 mM Hepes under a carbogen atmosphere.

RESULTS

When hepatocytes were incubated with 1mM NFT the formation of "ROS" was increased very rapidly (peak in 15 mins) in a concentration dependent fashion (table 1). Malondialdehyde (MDA) formation was much slower but preceded cell death (73% in 3 hours). The antioxidant α-tocopheryl succinate, catalase, "ROS" scavengers (tetramethyl-4-piperidinol-N-oxyl (TEMPOL), mannitol and dimethylsulfoxide (DMSO)) protected the hepatocytes against NFT induced cytotoxicity as well as MDA and "ROS" generation. On the otherhand catalase inhibitors, sodium azide and cyanamide significantly increased NFT induced cytotoxicity, "ROS" formation and MDA generation. The DT-diaphorase inhibitor, dicumarol only slightly inhibited NFT induced cell lysis. On the otherhand the CYP2E1 inhibitor phenylimidazole and P450 reductase inhibitor diphenyliodonium chloride (DPI) protected the hepatocytes against NFT. Endocytosis inhibitors including lysosomotropic agents (chloroquine and methylamine), Na^+ ionophore (monensin) and autophagy inhibitor (3-methyladenine) also protected the hepatocytes.

Table1. Preventing nitrofurantoin induced hepatocyte necrosis by inhibiting "ROS" formation and lipid peroxidation.

Addition to hepatocytes	%Cytotoxocity 3 h	"ROS" 15 min	MDA 3 h
None	19 ± 2	90 ± 5	0.48 ± 0.06
Nitrofurantoin (0.5 mM)	29 ± 3	272 ± 7^a	6.52 ± 2.28^a
Nitrofurantoin (1 mM)	73 ± 6^a	485 ± 6^a	14.62 ± 2.28^a
+ Catalase (200 U/mL)	37 ± 4^b	103 ± 6^b	3.37 ± 1.26^b
+ TEMPOL (300 μM)	37 ± 4^b	114 ± 8^b	3.68 ± 1.16^b
+ Dimethyl sulfoxide (150 mM)	37 ± 4^b	101 ± 6^b	3.47 ± 1.28^b
+ Mannitol (50 mM)	37 ± 4^b	103 ± 7^b	4.37 ± 1.16^b
+ Desferoxamine (200 μM)	37 ± 5^b	100 ± 8^b	4.37 ± 1.21^b
+ α-tocopherol succinate (100 μM)	37 ± 4^b	106 ± 5^b	3.44 ± 1.24^b
+ Sodium azide (4 mM)	83 ± 6^b	611 ± 7^b	20.58 ± 1.68^b
+ Cyanamide (100 μM)	86 ± 8^b	615 ± 8^b	20.77 ± 1.55^b
+ Phenylimidazole (300 μM)	37 ± 4^b	133 ± 6^b	4.37 ± 1.26^b
+ DPI (50 μM)	35 ± 5^b	129 ± 5^b	4.35 ± 0.85^b
+ Dicumarol (30 μM)	56 ± 4	605 ± 6	15.68 ± 1.88
+ Monensin (10 μM)	35 ± 5^b	119 ± 5^b	4.22 ± 1.12^b
+ Methylamine (30 mM)	35 ± 5^b	116 ± 7^b	4.08 ± 1.17^b
+ Chloroquine (100 μM)	45 ± 4^b	117 ± 8^b	4.15 ± 1.34^b
+ 3-Methyladenine (5 mM)	37 ± 4^b	111 ± 6^b	4.27 ± 1.20^b

Hepatocytes (10^6 cells/ml) were incubated in Krebs-Henseleit buffer pH 7.4 at 37C for 2.0 hrs following the addition of NFT. Cytotoxicity was determined as the percentage of cells that take up trypan blue (Pourahmad and O'Brien, 2000). "ROS" formation was expressed as fluorescent intensity units (Shen et al., 1996). MDA formation was expressed as μM (Smith et al., 1982). a: Significant difference in comparison with control hepatocytes, b: in comparison with NFT treated hepatocytes ($P < 0.001$). Values are expressed as means of three separate experiments (S.D.).

When hepatocyte lysosomes were preloaded with acridine orange, a release of acridine orange into the cytosolic fraction ensued within 60 minutes after treating the loaded hepatocytes with 1 mM NFT (table 2). The NFT induced acridine orange release was prevented by dimethylsulfoxide, mannitol, catalase, superoxide dismutase (SOD) or the ferric chelator desferoxamine (table 2). 3-methyladenine, phenylimidazole and DPI also inhibited NFT induced acridine orange release (Table 2).

Hepatocyte proteolysis as determined by the release of the amino acid tyrosine into the extracellular medium over 120 minutes was markedly increased when hepatocytes were incubated with NFT (Table 3). The NFT induced tyrosine release was completely prevented by the lysosomal protease inhibitors leupeptin and pepstatin, dimethylsulfoxide, mannitol, catalase or superoxide dismutase, deferoxamine. Phenylimidazole, diphenyliodonium chloride (DPI), 3-methyladenine, methylamine and chloroquine also inhibited NFT induced tyrosine release (Table 3).

Table 2. Preventing nitrofurantoin induced hepatocyte lysosomal membrane damage by inhibitors of oxidative stress

Addition to hepatocytes	"Acridine orange" redistributed		
	15 min	30 min	60 min
None	3 ± 1	4 ± 1	4 ± 1
Nitrofurantoin (1 mM)	69 ± 4[a]	118 ± 7[a]	195 ± 8[a]
+ Catalase (200 U/mL)	3 ± 1[b]	3 ± 1[b]	5 ± 1[b]
+ SOD (100 U/mL)	6 ± 1[b]	7 ± 1[b]	11 ± 2[b]
+ Dimethyl sulfoxide (150 mM)	3 ± 1[b]	4 ± 1[b]	7 ± 1[b]
+ Mannitol (50 mM)	6 ± 1[b]	5 ± 1[b]	8 ± 2[b]
+ Desferoxamine (200 μM)	3 ± 1[b]	3 ± 1[b]	3 ± 1[b]
+ Phenylimidazole (300 μM)	3 ± 1[b]	3 ± 1[b]	6 ± 1[b]
+ DPI (50 μM)	3 ± 1[b]	3 ± 1[b]	6 ± 1[b]
+ 3-Methyladenine (5 mM)	5 ± 1[b]	8 ± 1[b]	10 ± 2[b]

Hepatocytes (10^6 cells/ml) were incubated in Krebs-Henseleit buffer pH 7.4 at 37°.
Lysosomal membrane damage was determined as the alterations in intensity unit of diffuse cytosolic green fluorescence induced by acridine orange released from lysosomes (Adapted from Brunk et al., 1995 after permeabilising the hepatocytes (Bontemps and Van den Berghe, 1998)).
Values are expressed as means of three separate experiments (S.D.)
a: Significant difference in comparison with control hepatocytes ($P < 0.001$).
b: Significant difference in comparison with NFT treated hepatocytes ($P < 0.001$).

Table 3 Preventing nitrofurantoin induced hepatocyte proteolysis by inhibitors of oxidative stress or endocytosis or lysosomal protease inhibitors

Addition to hepatocytes	Hepatocyte tyrosine release (μM)		
	30 min	60 min	120 min
None	13 ± 1	15 ± 1	21 ± 2
Nitrofurantoin (1 mM)	39 ± 4[a]	48 ± 6[a]	65 ± 7[a]
+Catalase (200 u/ml)	15 ± 2[b]	17 ± 2[b]	27 ± 3[b]
+SOD (100 u/ml)	14 ± 2[b]	17 ± 2[b]	23 ± 2[b]
+Dimethyl sulfoxide (150 mM)	15 ± 2[b]	19 ± 2[b]	28 ± 3[b]
+Mannitol (50 mM)	14 ± 2[b]	16 ± 2[b]	20 ± 2[b]
+Desferoxamine (200μM)	12 ± 2[b]	14 ± 2[b]	20 ± 2[b]
+Phenylimidazole (300μM)	13 ± 2[b]	14 ± 2[b]	18 ± 2[b]
+DPI (50μM)	12 ± 2[b]	12 ± 2[b]	17 ± 2[b]
+Monensin (10μM)	7 ± 1[b]	14 ± 2[b]	15 ± 1[b]
+Methylamine (30mM)	10 ± 1[b]	15 ± 3[b]	16 ± 2[b]
+Chloroquine (100μM)	11 ± 1[b]	18 ± 5[b]	19 ± 7[b]
+3-methyladenine (30mM)	12 ± 2[b]	12 ± 2[b]	10 ± 2[b]
+Leupeptin (100μM)	12 ± 2[b]	12 ± 2[b]	17 ± 2[b]
+Pepstatin (100μM)	12 ± 2[b]	13 ± 2[b]	18 ± 2[b]

Hepatocytes (10^6 cells/ml) were incubated in Krebs-Henseleit buffer pH 7.4 at 37°.
Lysosomal induced proteolysis was determined by measuring the cellular release of tyrosine into the media (Novak et al., 1988).
Values are expressed as means of three separate experiments (S.D.)
a: Significant difference in comparison with control hepatocytes ($P < 0.001$)
b: Significant difference in comparison with NFT treated hepatocytes ($P < 0.001$)

DISCUSSION

Lipid peroxidation contributes to NFT induced cell lysis as malondialdehyde formation was markedly increased following "ROS" formation and the antioxidants α-tocopherol succinate (a plasma membrane lipid antioxidant) prevented both MDA formation and cytotoxicity.

Enzymes proposed for NFT reduction include glutathione reductase (Klee et al., 1994), DT-diaphorase (Hasspieler et al., 1997), NADPH-cytochrome c reductase and cytochrome P450 (Gram, 1997), but no cell studies have been carried out previously. As shown here treatment of hepatocytes with phenylimidazole (an inhibitor of cytochrome P450 CYP2E1) and DPI (an inhibitor of cytochrome P450 reductase) but not dicumarol prevented NFT induced cytotoxicity, lipid peroxidation and "ROS" formation which suggests that reduced CYP2E1 or P450 reductase reductively activates NFT.

NFT induced hepatocyte lysosomal disruption ensued at about 60 minutes after NFT addition. A similar release of acridine orange occurred when acridine orange loaded hepatocytes were treated with hydrogen peroxide generated by glucose/glucose oxidase (Brunk et al, 1995). Hepatocyte proteolysis markedly increased following lysosomal disruption by NFT which was inhibited by the lysosomal protease inhibitors leupeptin and pepstatin. Furthermore, NFT induced cytotoxicity was also prevented by leupeptin or pepstatin (data not shown). NFT induced cytotoxicity was also prevented by the hepatocyte lysosomotropic agents methylamine (Luiken et al., 1996), chloroquine (Graham et al., 1998) and monensin a Na^+ ionophore that inhibits hepatocyte endocytosis and endosomal acidification (Brunk et al., 1995). 3-Methyladenine, an inhibitor of hepatocyte autophagy (Seglen, 1997) prevented the NFT induced release of acridine orange and also prevented cytotoxicity. Methylamine or chloroquine or the ferric chelator desferoxamine also prevented hepatocyte cytotoxicity induced by hydrogen peroxide generated by glucose/glucose oxidase (Starke et al., 1985). We propose that free Fe^{+2} ions in the lysosomes undergo redox cycling with cysteine and form "ROS" from H_2O_2 which diffuses into the lysosomes after formation by the redox cycling of nitrofurantoin. The "ROS" formed contributes to the rupture of the membrane resulting in the intracellular release of lysosomal enzymes (Fig. 1). The H2O2 formed oxidises hepatocyte GSH stoichometrically (Silva et al., 1991).

In conclusion as shown in Fig. 1 these results suggest that NFT induced hepatocyte toxicity involves reductive activation by reduced cytochrome P450 or P450 reductase to form the nitro radical anion which undergoes futile redox cycling resulting in the formation of "ROS" which interacts with lysosomal Fe^{2+}/Cu^+ and causes lysosomal damage and protease release.

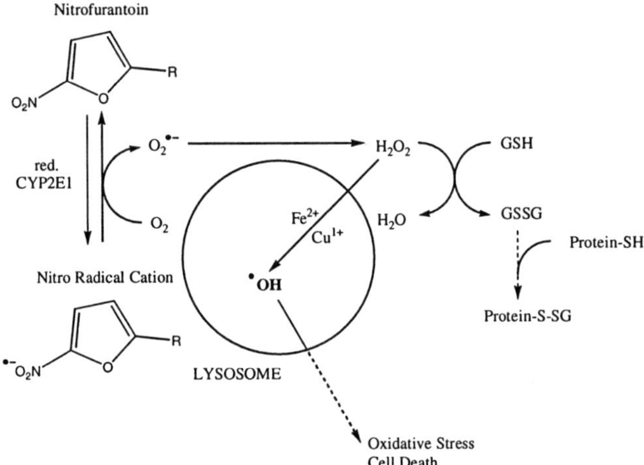

Fig. 1. Proposed lysosomal oxidative stress cell death mechanism for nitrofurantoin.

REFERENCES

Bontemps, F., Van Den Berghe, G. (1998). Novel evidence for an ecto-phospholipid methyltransferase in isolated rat hepatocytes. *Biochem. J.* **330**(Pt 1), 1-4.

Brunk, U.T., Zhang, H., Roberg, K., Ollinger, K. (1995). Lethal hydrogen peroxide toxicity involves lysosomal iron-catalyzed reactions with membrane damage. *Redox Report* **1**, 267-277.

Graham, R.M., Morgan, E.H., Baker, E. (1998). Characterisation of citrate and iron citrate uptake by cultured rat hepatocytes. *J. Hepatol.* **29**(4), 603-13.

Gram, T.E. (1997). Chemically reactive intermediates and pulmonary xenobiotic toxicity. *Pharmacol Rev.* **49**(4):297-341.

Hasspieler, B.M., Haffner, G.D., Adeli, K. (1997). Roles of DT diaphorase in the genotoxicity of nitroaromatic compounds in human and fish cell lines. *J Toxicol Environ Health.* **52**(2):137-48.

Hoener, B., Noach, A., Andrup, M., Yen, T.S. (1989). Nitrofurantoin produces oxidative stress and loss of glutathione and protein thiols in the isolated perfused rat liver. *Pharmacology* **38**(6), 363-73.

Klee, S., Nurnberger, M.C., Ungemach, F.R. (1994). The consequences of nitrofurantoin-induced oxidative stress in isolated rat hepatocytes: evaluation of pathobiochemical alterations. *Chem. Biol. Interact.* **93**(2):91-102.

Luiken, J.J., Aerts, J.M., Meijer, A.J. (1996). The role of the intralysosomal pH in the control of autophagic proteolytic flux in rat hepatocytes. *Eur. J. Biochem.* **235**(3): 564-73.

Rao, D.N., Mason, R.P. (1987). Generation of nitro radical anions of some 5-nitrofurans, 2- and 5-nitroimidazoles by norepinephrine, dopamine, and serotonin. A possible mechanism for neurotoxicity caused by nitroheterocyclic drugs. *J. Biol. Chem.* **262**(24), 11731-6.

Reed, D.J., Babson, J.R., Beatty, P.W., Brodie, A.E., Ellis, W.W., Potter, D.W. (1980). High-performance liquid chromatography analysis of nanomole levels of glutathione, glutathione disulfide, and related thiols and disulfides. *Anal Biochem.* **106**(1), 55-62.

Rosenow, E.C. 3d. (1978). Drugs that may induce pulmonary disease. *Geriatrics* **33**(1), 64-73.

Rossi, L., Silva, J.M., McGirr, L.G., O'Brien, P.J. (1988). Nitrofurantoin-mediated oxidative stress cytotoxicity in isolated rat hepatocytes. *Biochem. Pharmacol.* **37**(16), 3109-17.

Seglen, P.O. (1997). DNA ploidy and autophagic protein degradation as determinants of hepatocellular growth and survival. *Cell Biol. Toxicol.* **13**(4-5), 301-15.

Shen, H.M., Shi, C.Y., Shen, Y., Ong, C.N. (1996). Detection of elevated reactive oxygen species level in cultured rat hepatocytes treated with aflatoxin B1. *Free Radic. Biol. Med.* **21**(2), 139-146.

Silva, J.M., Khan, S., O'Brien, P.J. (1993). Molecular mechanisms of nitrofurantoin-induced hepatocyte toxicity in aerobic versus hypoxic conditions. *Arch. Biochem. Biophys.* **305**(2):362-9.

Silva, J.M., McGirr, L., O'Brien, P.J. (1991). Prevention of nitrofurantoin-induced cytotoxicity in isolated hepatocytes by fructose. *Arch. Biochem. Biophys.* **289**(2), 313-8.

Smith, M.T., Thor, H., Hartizell, P., Orrenius, S. (1982). The measurement of lipid peroxidation in isolated hepatocytes. *Biochem. Pharmacol.* **31**(1), 19-26.

Starke, P.E., Gilbertson, J.D., Farber, J.L. (1985). Lysosomal origin of the ferric iron required for cell killing by hydrogen peroxide. *Biochem. Biophys. Res. Commun.* **133**(2), 371-9.

SCAVENGING AND ANTIOXIDANT EFFECTS OF ESTROGEN DERIVATIVES IN CHOLESTEROL-FED RABBITS

Vyacheslav U. Buko,[1] Oxana Ya. Lukivskaya,[1] Yury V. Popov,[1] Elena E. Naruta,[1] Vitaly V. Sadovnichy,[1] Doris Hübler,[2] and Michael Oettel[2]

[1]Institute of Biochemistry, Grodno, 230017 Belarus
[2]Jenapharm GmbH& Co. KG., Jena, D-07745 Germany

INTRODUCTION

It is well establish that many estrogens are naturally occurring antioxidant[1]. Some $\Delta^{8,9}$-dehydro derivatives of 17α-estradiol and 17β-estradiol are effective scavengers of free oxygen radicals[2]. These estrogens are called scavestrogens because their radical scavenging effect is higher than those of the parent compounds. There are the experimental data on estrogens attenuating the progression of atherosclerosis[3]. Oxidation of low-density lipoproteins (LDL) is believed to play an important role in atherogenesis[4]. A supplementation with different antioxidants is widely used for the treatment and prevention of atherosclerosis[5]. We studied the antioxidative effect of $\Delta^{8,9}$-dehydro derivative of 17α-estradiol, 14α, 15α-methylene-8-dehydro-17α-estradiol (J861) in a comparison with 17α- and 17β-estradiols in the liver of cholesterol-fed rabbits. We also evaluated radical scavenging properties of scavestrogens and estrogens *in vitro* using spin-trapping techniques.

MATERIALS AND METHODS

Male rabbits (2.6–3.2 kg) were fed a standard diet containing stock pellet and vegetables. Rabbits were randomly divided into 8 groups of 6-7 animals. Group 1 consisted of healthy controls, which were only fed with the standard diet. The other seven groups were daily fed cholesterol (200 mg/kg) in sunflower oil (0.75 mg/kg) using intraoral intubation for 3 months. Group 2 receiving only cholesterol was the diet control group. The remaining groups were treated with 17β-estradiol, 17α-estradiol or J861(0.02 and 0.1mg/kg/day) for the last month of the study. Rabbits were treated with the tested estrogens diluted in 1 ml of sesame seed oil by oral intubation.

We also studied the *in vitro* effect of 17α-estradiol, 17β-estradiol, J861 and and the scav-estrogen, J 811 (estra-1,3,5[10],8-tetraene-3,17α-diol), on the free radical generation by isolated rabbit microsomes. Tiron (Fluka, Switzerland), which is a relatively specific scavenger of $O^{\cdot-}$[6], was used as spin-traps. The system incubated with spin trap contained 2 mg microsomal protein/ml and

Biological Reactive Intermediates VI, Edited by Dansette *et al.*
Kluwer Academic / Plenum Publishers, 2001

10^{-2} M Tiron in 0.05 M sodium phosphate buffer, pH 7.4. Free oxygen radicals were generated by applying a system containing $FeSO_4$ (5 10^{-5} M) and ascorbate (5 10^{-4} M). The tested estrogens were used in the experiment at concentrations of $5 \cdot 10^{-7}$ M, $2 \cdot 10^{-6}$ M and 10^{-5} M. All the ESR spectra were taken at room temperature (22° C) by means of PS-100 spectrometer (Belarus). The usual instrumental parameters were: the microwave power attenuation, 6 dB; the scan range, 120 G; the middle range field, 3450 G and the scan time, 60 s.

The microsomal fraction was separated by differential centrifugation at 105,000 g. The method of superoxide radical measurement is based on the reduction of nitrotetrazolium blue (NTB) to formazan by $O^{\cdot -}$[7]. The NADPH-dependent chemiluminiscence of microsomes was carried out using the chemiluminogenic probes luminol and lucigenin according to Müller-Peddinghaus and Wurl[8]. Chemiluminiscence was registered for 4 min using a BLM-100 chemiluminometer (Russia). Superoxide dismutase (SOD) activity was measured according to the method[7] based on a competition of SOD with nitroblue tetrazolium for $O^{\cdot -}$. Lipid peroxidation was measured as thiobarbitiric acid-reactive substances (TBARS) [9]. The H_2O_2 formation was determined by the ferroammonium sulfate - potassium thiocyanate method [10].

RESULTS AND DISCUSSION

Using spin trapping techniques we found that all the estrogens were effective interceptors of free oxygen radicals. The decrease in the concentration of spin trap-radical adducts demonstrated that the scavenging efficacy of the estrogens was increased in the following order: 17β-estradiol< 17 α-estradiol<J811< J861 (Fig. 1).

The feeding of cholesterol developed a prooxidant effect, increasing all the measured radical-related parameters: SOD activity and H_2O_2 production in the serum as well as NADPH-induced chemiluminiscence, enhanced by lucigenin, $O^{\cdot -}$ content and lipid peroxidation evaluated as TBARS formation in rabbit liver microsomes. (Table 1). All the investigated estrogens decreased the NADPH-induced chemiluminiscence enhanced by lucigenin, where the effects of 17α estradiol (0.1 mg/kg) and both doses of J861 did not differ from the control value.

The above dosages of 17α-estradiol and J861 also decreased $O^{\cdot -}$ content and TBARS formation in liver microsomes (Table 1). The serum SOD activity was decreased in cholesterol-fed rabbits treated with 17α-estradiol (0.1 mg/kg), whereas 17 β-estradiol (0.02 mg/kg) increased this parameter. The production of H_2O_2 in rabbit serum significantly decreased in all the estrogen-treated groups, except for the J861-treated rabbits (0.1 mg/kg). In liver microsomes this parameter was decreased in cholesterol-fed rabbits receiving 17α-estradiol (0.1 mg/kg), 17β-estradiol (0.02 mg/kg) and J861 (0.02 mg/kg), but increased under the influence of J861 at a dosage of 0.1 mg/kg. As is suggested by spin-trapping data, J861 is a more effective scavenger of free radicals in liver microsomes than the other tested estrogens. Tiron, which we used as a spin trap, preferably intercepts $O^{\cdot -}$ although some authors questioned the specificity of Tiron for $O^{\cdot -}$ [6].

It was found that chemiluminiscence attenuated by lucigenin reflects $HO^{\cdot}$ generation but can also be induced by reactions with other reactive oxygen species ($O^{\cdot -}$, singlet oxygen, hydrogen peroxide, etc.)[11]. Therefore, the decrease of lucigenin-enhanced chemiluminiscence of microsomes by the tested estrogens and, especially by, J861 is probably connected not only with an inhibition of $HO^{\cdot}$ generation, but also with a diminution of the formation of other reactive oxygen species. We believe that J861 is scavenger of both $O^{\cdot -}$ and $HO^{\cdot}$, but the definition of radical scavenging function of the scavestrogen requires further investigations. Some other free radical-related data obtained in the experiment *in vivo* also confirmed the highest free radical scavenging and antioxidative effective-ness of J861 compared with 17α-estradiol and 17β-estradiol.

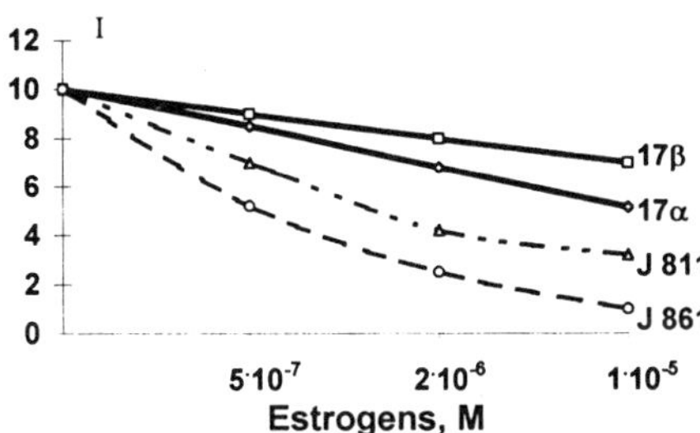

Figure 1. ESR signal intensity (I) of Tiron-radical adducts in liver microsomes preincubated with different concentrations of estrogens: 17α-estradiol; 17β-estradiol; J811; J861. The incubation mixture containing 10 mM Tyron, 0.05 mM $FeSO_4$, 0.5 mM ascorbate and 2 mg/ml microsomal protein. Estrogens at different concentrations were added and the mixtures were incubated at 24^0C for 15 min. Conditions of ESR spectroscopy see in *Materials and Methods*.

Table 1. Activity of serum superoxide dismutase (SOD) (% of inhibition/ml), production of hydrogen peroxide in serum (nmol/ml/ 1 min) and liver microsomes (nmol/mg protein/1 min), NADPH- induced chemiluminiscence enhanced by lucigenin (c.p.m./mg protein per 1 s x 10^5), superoxide anion content (nmol/mg protein) and lipid peroxidation (nmol TBARS/mg protein/min) (means $\pm$ S.E.M)

	SOD, serum	H_2O_2, serum	H_2O_2, microsomes	Chemilumini- scence	Superoxide anion	Lipid per- oxidation
Control	18.43±2.9	7.4±0.8	1.4±0.1	2.13 ± 0.33	12.6±1.22	1.4±0.16
Cholesterol	3.5±3.25*	22.3±0.6*	3.1±0.2*	16.65± 2.17*	20.0±1.39*	2.8±0.15*
+ 17α, 0.02 mg/kg	36±5.9*	17.9±0.9*@	2.7±0.4*	7.32 ± 1.77*@	18.6±1.36*	2.6±0.21*
+ 17α, 0.1 mg/kg	10.2±3.1@	14.3±0.9*@	2.0±0.3*@	3.22 ± 0.58@	13.4±1.47@	2.3±0.19*@
+ 17β, 0.02 mg/kg	49.3±7.1*@	15.4±0.6*@	2.2±0.1*@	4,80 ± 1.13 *@	18.6±1.22*	2.5±0.19*
+17β, 0.1 mg/kg	30.4±4.2*	17.9±0.6*@	3.2±0.4*	5.07 ± 0.16@	15.8±1.44	2.4±0.25*
+ J861 0.02 mg/kg	16±2.9@	10.2±0.5*@	1.7±0.1@	3.29 ± 1.32@	14.5±1.24@	2.2±0.20*@
+J 861, 0.1 mg/kg	22.4±4.6	22.9±0.6*	4.1±0.4*@	2.96 ± 0.49@	11.3±1.63@	1.7±0.18@

* - P<0.05 compared to the control group; @ - P<0.05 compared to the cholesterol-fed group.

Ever since it was shown that estrogens could attenuate the atherosclerotic lesions much interests has been focused on scavestrogens in preventing the oxidation of LDL and retarding the atherosclerotic process. Recently we found the antiatherogenic effect of J861, which was significantly higher compared to 17α-estradiol and 17β-estradiol[12]. We believe that the higher antiatherogenic effect of J861 can explain by its marked scavenging and antioxidative properties.

REFERENCES

1. A.D. Mooradian. Antioxidant properties of steroids. *J Steroid Biochem Mol Biol.* 45: 509 (1993).
2. W. Römer, M. Oettel, P. Droescher, and Schwarz S. Novel scavestrogens and their radical scavenging effects, iron-chelating and total antioxidative activities: $\Delta^{8,9}$-Dehydro derivatives of 17α -estradiol and 17β-estradiol. *Steroids.* 62: 304 (1997).
3. J.L. Hough and D.B. Zilversmit. Effect of 17 beta estradiol on aortic cholesterol content and metabolism in cholesterol-fed rabbits. *Arteriosclerosis.* 6: 57 (1986).
4. D. Steinberg, S. Parthasarathy, T.E. Carew, J.C. Khoo and J.L. Witztum. Beyond cholesterol: modification of low-density lipoprotein that increases its atherogenity. *N Engl Med J.* 320: 915 (1989).
5. D. Steinberg. Antioxidants and atherosclerosis: A current assessment. *Circulation,* 84: 1420 (1991).
6. R.C. Devlin, C.S. Lin, R.J. Perper and H. Doutherty. Evaluation of free radical scavengers in studies on lymphocyte mediated cytolysis. *Immunopharmacology.* 3:147 (1981).
7. C.A. Rice-Evans, A.T. Diplock and M.C.R. Symons. *Techniques in Free Radical Research.* Elsevier, Amsterdam- London-New York-Tokyo (1991).
8. R. Müller-Peddinhaus and M. Wurl. The amplified chemiluminiscence test to characterize antirheumatic drugs as oxygen radical scavengers. *Biochem Pharmacol.* 36: 1125 (1987).
9. J.K.A. Buege and S.D. Aust. Microsomal lipid peroxidation. *Methods Enzymol.* 52: 302 (1978).
10. A.G. Hildebrandt, I. Roots, M. Tjoe and G. Heinemeyer. Hydrogen peroxide in hepatic microsomes. *Methods Enzymol.* 52: 342 (1978).
11. D. A. Barber, N.H. Do, R.L Tackett and A.C. Capomaccia. Nonsuperoxide lucigenin enhanced chemiluminiscence from phospholipids and human saphenous vein. *Free Rad. Biol Med.* 18: 565 (1995).
12. V. Buko, A. Chirkin, O. Lukivskaya, E. Naruta , I. Chirkina, Yu. Popov , M. Oettel and D. Hübler. Regulation of atherogenesis in rabbits by oestrogens and scavostrogen J811 In: *Actuelle Fragen des Stoffwechsel,* V.Drebickas, ed. Pedagoginias Universitetas, Vilnius, (1999).

INHIBITION OF OXIDATIVE DAMAGE OF RED BLOOD CELLS AND LIVER TISSUE BY GENISTEIN-8C-GLUCOSIDE

Vyacheslav Buko,[1] Lev Zavodnik,[1] Ilya Zavodnik,[1] Elena Lapshina,[1] Alina Shkodich[1], Nikolai Laman,[2] Maria Bryszewska[3]

[1]Institute of Biochemistry, Grodno, Belarus
[2]Insitute of Experimental Botanics, Minsk, Belarus
[3]Institute of Biophysics, University of Lodz, Poland

INTRODUCTION

Genistein belongs to isoflavons, a subclass of flavonoids, the large group of polyphenolic compounds widely distributed in plants. A number of studies show a protective effect of flavonoids against cancer, toxic liver injury and cardiovascular diseases[1]. Being phytoestrogen, genistein has recently generated interest as a potential anticancer and antiatherogenic agent[2]. Several flavonoids are known as antioxidants and scavengers of free oxygen radicals[3].The antiatherogenic effect of isoflavons is associated with inhibition of myeloperoxidase in phagocytic cells producing of hypochlorite, a highly reactive chlorinated species[4].

The purpose of this work was to investigate the glycosylated derivative of genistein, genistein-8-C-glycoside (G8CG), as a prospective antioxidant and a scavenger of active chlorine. We also tried to evaluate a possible effect of G8CG on the oxidative damage of human red blood cells (RBC) caused by t-buthyl hydroperoxide (tBHP) or hypochloric acid (HClO).

MATERIALS AND METHODS

G8CG was isolated from flowers of lupine (*Lupinus luteus L.*) according to the method developed by Laman and Volynetz[5]. Isoflavon was subsequently extracted by methanol, ethylacetate, *n*-butanol and purified using a chromatographic column (3x80 cm) packed with polyamine. G8CG was eluted by ethanol-water gradient.

Blood from healthy donors was purchased from the Central Blood Bank. Blood was taken into 3% sodium citrate. After removing plasma and the leukocyte layer, erythrocytes were washed three times with cold (4° C) phosphate buffered saline (PBS: 0.15 M NaCl, 1.9 mM NaH_2PO_4, 8.1 mM Na_2HPO_4, pH 7.4). Erythrocytes were used immediately after isolation. Rat liver homogenates were prepared in 1.15% KCl (1:4; v/v). The rat liver microsomal fraction was separated by differential centrifugation at 105 000 g using a VAC-602 centrifuge (Germany).

Lipid peroxidation in RBC, liver homogenates and microsomes was induced by 2 mM t-buthylhydroperoxide (*t*BHP). Suspensions of RBC in PBS (haematocrit 10 %) were treated with different concentrations of hypochlorous acid at 22 °C for 10 min. Then the cells were washed 3 times with excess of cold PBS and resuspended in PBS (haematocrit 10 %). HOCl was added as 25 mM stock solution of NaOCl in PBS to the cell suspension. The concentration of OCl^- was determined spectrophotometrically[6] using an absorbance coefficient of $350\ M^{-1}cm^{-1}$ (292 nm) at pH 9.0.

The susceptibility of erythrocytes to oxidative damage was measured in terms of the apparent rate constant of cell haemolysis (posthaemolysis), the accumulation of thiobarbituric-reactive species (TBARS), the oxidation of intracellular oxyhaemoglobin (oxyHb) and GSH. The process of haemolysis of erythrocytes treated with oxidants was monitored by haemoglobin (Hb) release. Acetylcholinesterase (AChE) activity was measured according to Ellman et al.[7]. Cytochrome P-450 (CYP-450) content and cytochrome P-420 (CYP-420) formation in liver microsomes was determined according to Omura and Sato[8]. The content of long-living chloramines was determined using the oxidation of thionitrobenzoic acid as described earlier[6]. TBARS were determined in the acid-soluble fraction of RBC suspension[9]. GSH concentration was measured by the method of Ellman using the absorbance coefficient of $13,6\ мM^{-1}\cdot см^{-1}$ (412 нм)[10].

G8CG was added at concentrations of 0.5, 1.0, 2.0 and 5.0 mM and samples were preincubated during 30 min at 22^0C. After the preincubation with G8CG either tBHP or HOCl were added and the corresponding measurements were carried out after 20 min. Flavonoid quercetin in the experiment with tBHP was used as comparative agents at the same concentrations.

RESULTS AND DISCUSSION

Effect of G8CG on Oxidative Damage Caused by tBHP

As illustrated in Figures 1A and 1B, when 2 mM tBHP was incubated in RBC suspension, TBARS formation was dramatically increased nearly 20-fold and GSH completely disappeared. With the oxidant, 69% of Hb was oxidized to metHb. The preincubation of RBC with flavonoids, quercetin and G8CG, dose-dependently inhibited TBARS production. In this situation the effect of quercetin was especially pronounced. Neither G8CG nor quercetin developed a protective action against the Hb oxidation and GSH depletion.

The incubation of 2mM tBHP in liver homogenates led to an increase of TBARS by 30% and the decrease of GSH content by 90% (Figures 1A, 1B). Both quercetin and G8CG prevented lipid peroxidation, the effect of quercetin being especially marked. The protective effect of G8CG on the GSH depletion was dose-dependent and more pronounced compared to quercetin.

Similar effects of the oxidant and G8CG were observed in rat liver microsomes. The addition of 2 mM tBHP caused a more than 5-fold increase in the TBARS content and the amount of free protein SH-groups was lowered by 74%. The protective action of G8CG was more effective toward GSH depletion, whereas its TBARS-decreasing effect was less pronounced.

The treatment of rat liver microsomes with 150-200 μM tBHP caused a decrease of CYP-450 content with a simultaneous appearance of the inactive form, CYP-420 (Figure 1C). The pretreatment with 125 μM G8CG prevented the inactivation of CYP-450 caused by the oxidant.

RBC Damage Induced by HOCl and its Prevention by G8CG

RBC treated by HOCl underwent time-dependent lysis developing both in the presence of the oxidant (haemolysis) and after its removing (posthaemolysis). The apparent rate constant of posthaemolysis was a linear function of the oxidant concentration but the dependence of the apparent rate constant of haemolysis on the HOCl concentration was nonlinear (Figure 1D). As Figure 1D also shows, we found a dependence of the chloramines produced on the oxidant concentrations after the treatment of isolated RBC membranes with HOCl. G8CG at concentrations of 0.5-2 mM effectively inhibited HOCl-induced RBC haemolysis (Figure 1F) and at concentrations of 1-2 mM completely protected RBC against the lythic effect of HOCl. The formation of chloramines was observed at the concentrations of HOCl of 0.2 mM and higher. The correlation between the membrane-bound chloramine contents and the rate constant of cell haemolysis suggests a key role of the formed chloramines in the HOCl-induced RBC damage. The inhibition of this damage by G8CG allows us to propose that isoflavon acts as a scavenger of active chlorine forms, such as chloramines.

The pretreatment of RBC with relatively low concentrations of HOCl accompanied by an increase of AChE affinity for the substrate (the decrease of Michaelis constant, K_m) and the

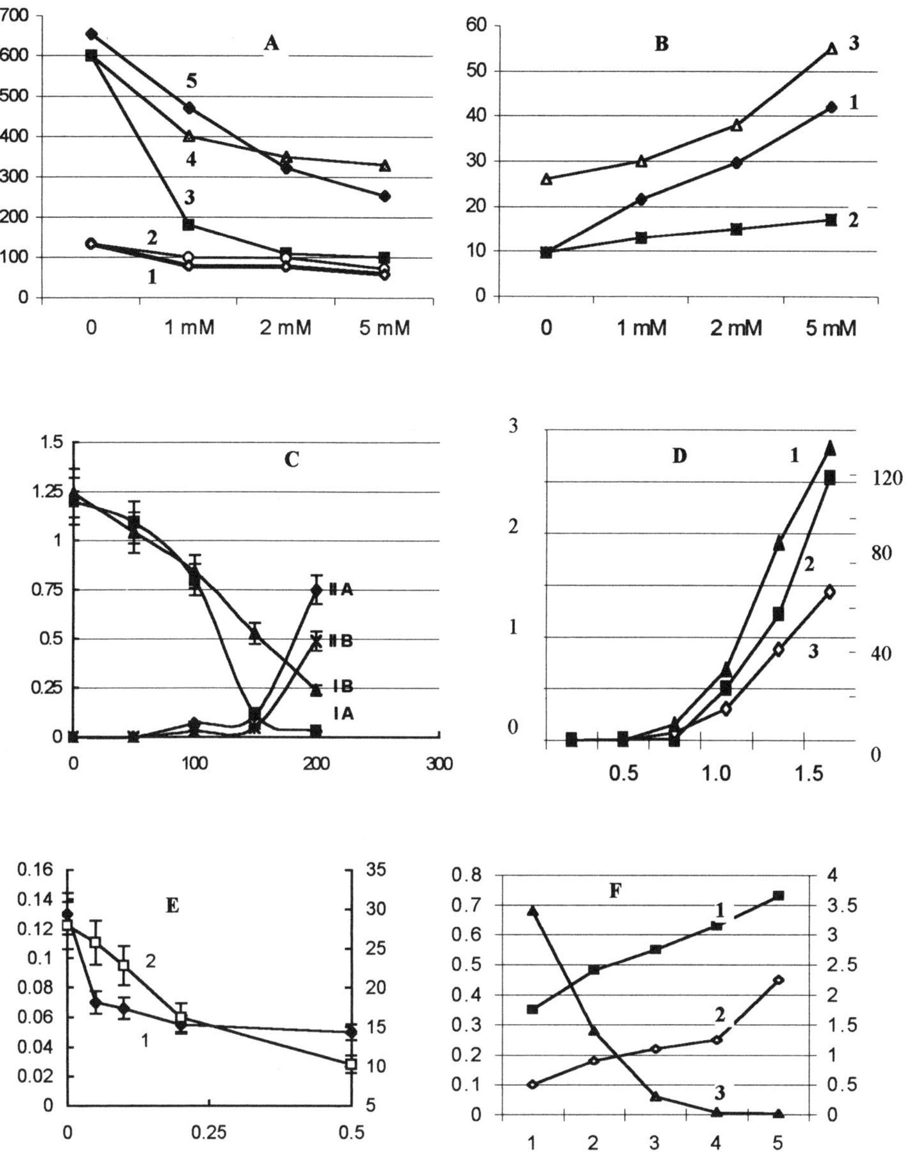

Figure 1. A. Effects of G8CG and quercetin on TBARS formation induced by tBHP at 0-5.0 mM (1-G8CG, 2- quercetin, liver homogenates; 3- quercetin; 4- G8CG, RBC; 5- G8CG, microsomes), for 1, 2, 5 - nmol/mg protein; for 3, 4 – nmol/ml packed cells. **B**. Effects of G8CG (1) and quercetin (2) on GSH content in RBC treated with tBHP at 0-5.0 mM, μmol/ml packed cells, and effect of G8CG (0-200 μmol) on microsomal protein SH-groups, μmol/mg protein (3). **C**. Inactivation of CYP-450 by tBHP at 0-200 μM (IA- CYP-450; IIA- CYP-420) and effect of G8CG (IB- cytochrome P-450; IIB- cytochrome P-420), μmol/mg protein. **D**. Chloramine concentrations (1), nmol/mg protein, the rate constants of haemolysis (2) and posthaemolysis (3), k x 10^{-3}, min^{-1}, in RBC incubated with NaOCl (0-2.0 mM). **E**. Dependence of Michaelis constant, K_m (1) and maximal reaction velocity, V_{max}, (2) of RBS membrane AChE on NaOCl concentration (0-0.5 mM). **F.** Effect of G8CG (1.0-5.0 mM) on GSH content, μmol/ml packed cells, in RBC incubated with 0.5 mM (1) and 1.0 mM (2) of NaOCl and on the rate constant of haemolysis, k x 10^{-3}, min^{-1}, (3) induced by 1 mM NaOCl.

decrease of V_{max} (Figure 1E). These data suggest a noncompetitive type of the AChE inhibition by HOCl.

The parameters of AChE enzymatic activity characterize the state of the membrane (viscosity, superficial charge, etc.). Thus, the membrane damage of preceded the RBC lysis caused by HOCl. However, we cannot exclude a direct modification of the enzyme by the oxidant.

HOCl dramatically decreased the concentration of intracellular GSH (Figure 1F). However, we did not found an effect of the oxidant on the oxyHb oxidation and lipid peroxidation evaluated as a TBARS formation. G8CG quite effectively prevented GSH depletion caused by the oxidant

In conclusion, the data obtained suggested the efficiency of G8CG to support the antioxidant defense system in different tissues (blood, liver) via its protective effect against GSH depletion. However, the inhibition of lipid peroxidation by G8CG is not so significant. We believe that the protection by G8CG against the RBC oxidative damage caused by HOCl is connected not only with the prevention of GSH depletion, but is also due to trapping of the active form of chlorine (hypochlorous acid and chloramines formed by interactions of HOCl with protein amino groups). As a result, G8CG corrects the RBC membrane damage induced by HOCl.

ACKNOWLEDGEMENTS

This work was supported by the Belarussian Foundation of Basic Research (grant B98-141).

REFERENCES

1. P.C.H. Hollman and M.B. Katan. Health effect and bioavailability of dietary flavonols. *Free Rad Res* 31:575 (1999).
2. H. Adlercreutz. Western diet and Western diseases: some hormonal and biochemical mechanisms and associations. *Scan J Clin. Lab Invest* 201: 3 (1990).
3. W Bors., W. Heller, C. Michel, and M. Saran. Flavonoids and antioxidant: determination of radical-scavanging efficiensis. *Methods in Enzymol* 180: 343 (1990).
4. H. Adlercreutz. Lignans and phytoestrogens. Possible prevention role in cancer. In: *Frontiers of Gastrointestinal Research,* Rosen P., ed. S.Karger, Basel, (1998)
5. N.A. Laman, and A.P Volynets. Flavonoids in ontogenesis of Lupinus Luteus L. *Plant Physiology* (Rus.) 21: 737 (1974).
6. M.C.M. Vissers and C.C. Winterbourn. Oxidation of intracellular glutathione after exposure of human red blood cells to hypochlorous acid. *Biochem J* 307: 57 (1995)
7. G.L Ellman,, K.D. Courtney, V.,Andres, and R.M. Featherstone. A new and rapid colorimetric determination of acetylcholinesterase activity. *Biochem Pharm* 7: 88-95 (1961).
8. T. Omura and R. Sato. The carbon monoxide binding pigment in liver microsomes. II.Solubilization, purification and properties. *J Biol Chem* 239:2379 (1964)
9. J. Stocks and T.L. Dormandy The autooxidation of human red cell lipids by hydrogen peroxide. *Br J Haemol*, 20: 95 (1971)
10. G.L. Ellman. Tissue sulfhydryl groups. *Arch Biochem Biophys* 82: 70 (1959).

EFFECTS OF SEVERAL WINE POLYPHENOLS ON LIPID PEROXIDATION AND OXYGEN ACTIVATION IN RAT LIVER MICROSOMES

Šárka Matějková and Ivan Gut

National Institute of Public Health
Šrobárova 48
10042 Praha 10, Czech Republic

INTRODUCTION

Lipid peroxidation (LP) and reactive oxygen species (ROS) are thought to play important roles in atherosclerosis, cancer, ageing and other diseases [1]. Naturally occuring polyphenols in wine, tea, vegetables, etc can protect against these diseases by scavenging lipid radicals ($R^{\bullet}$, $ROO^{\bullet}$, $RO^{\bullet}$), O_2 radicals ($O_2^{-\bullet}$, $^{\bullet}OH$), thereby preventing degradation of $NO^{\bullet}$, an important factor against atherosclerosis, by chelating transition metal ions thereby inhibiting the formation of free radicals[2], by regeneration of vitamin C and E and (vi) by influencing various enzymes. They modulate lipoprotein metabolism and inhibit carcinogenesis[1]. In the microsomes, formation of ROS and LP is provided by cytochromes P450 and b_5 and P450 reductase.

According to their structure polyphenols are classified as flavonoids (flavonols, flavanols and flavanones) and non-flavonoids (stilbenes, cinnamic and benzoic acid derivates) (Figure 1, Table 1). In our study we investigated a relation between the potency of various polyphenols to inhibit LP, $^{\bullet}OH$ radical production and the role of metal chelation in these processes.

MATERIALS AND METHODS

Chemicals

The polyphenols and most other chemicals were purest available and were obtained from Sigma-Aldrich, Czech Republic. Bovine serum albumin was from Lachema, CR and glucose-6-phosphate dehydrogenase was from Boehringer, Mannheim, Germany.

Microsomes, microsomal incubations and assays

Animal treatments and preparation of liver microsomes from untreated (CYP content 1.03 nmol/mg protein), CYP2E1-induced (1.10 nmol/mg) or CYP3A1/2-induced (2.541 nmol/ mg) rats were prepared as described before[3] and stored at $-80°C$. Protein concentration[4] and total CYP content[5] were estimated by desribed methods.

The incubation mixtures for LP, $^{\bullet}OH$ radical production and polyphenol metabolism were incubated in air in a shaking water bath for 1 hour at 37°C. They included untreated, CYP2E1- or CYP3A1/2-induced microsomes (1mg protein/ml), 0.06 M phosphate buffer (pH 7.4), NADPH-generating system[3], and various polyphenols. The production of marker of LP, malondialdehyde (MDA) was followed by estimation of TBARS (thiobarbituric acid-reactive substances as described before[6]. The $^{\bullet}OH$ radicals were quantified by photometry[7]. Polyphenol metabolism was estimated by measurement of substrate disappearance by HPLC (ECOM, CR) were carried out with Macherey-Nagel column (250x4) with Nucleosil and the mobile phase of methanol/water (65:35, v/v), flow rate 1,2ml/min and UV detection at 232nm.

Transition metal ion chelation

Metal chelation capacity of polyphenols was estimated from changes in UV-VIS polyphenol absorption spectra (220-600 nm), induced by $FeSO_4$ or $FeCl_3$, on spectrophotometer Spekord M400. Reversal of the spectral changes by EDTA confirmed chelation ability of the polyphenols. The molar ratios of polyphenols to Fe ions and EDTA, respectively, were (μM) 50:25:25, 50:50:50 or 50:100:100.

RESULTS

Effect of polyphenols on lipid peroxidation and OH radicals

The polyphenols decreased the amount of MDA in relation to their concentration. The inhibiting potency decreased in the order: quercetin > trans-resveratrol > fisetin > myricetin > morin > (+)-catechin = (-)-epicatechin > kaempferol > naringenin > p-coumaric acid > gentisic acid = caffeic acid (Table 1, Figure 2). Rutin and gallic acid were more efficient in distilled H_2O then in deionized H_2O and IC_{50} of (+)-catechin and gallic acid were 4-fold higher in CYP2E1-induced than in untreated microsomes. Quercetin and gallic acid decreased ${}^\bullet$OH production, whereas catechin and rutin did not. Trans-resveratrol did not influence ${}^\bullet$OH radical production in untreated and in CYP2E1-induced microsomes in the deionized water, but stimulated it in distilled water (Table 1, Figure 2).

Transition metal ion chelation

The aim was to evaluate the role of iron-chelation in the antioxidant effects of polyphenols. The addition of Fe^{2+} and Fe^{3+} ions increased λ_{max} (the red shift) of quercetin, in case of rutin Fe^{3+} slightly more than Fe^{2+} (Figure 3), in case of gallic acid Fe^{3+}, but not Fe^{2+}. It indicated their chelating capacity (Table 1) as well as regeneration of their spectra by subsequently added EDTA. (+)-Catechin (Figure 3) slightly chelated only Fe^{3+} and trans-resveratrol did not chelate Fe ions.

Changes in polyphenol concentration and structure during incubation with NADPH-generating system determined by HPLC

In CYP3A1/2-induced rat liver microsomes, naringenin was metabolized significantly more efficiently (450±39 pmol/min/mg protein) than in human microsomes (87.6±8.5), but the turnover rate was not different (177.3±15.6 versus 259±25 pmol/min/nmol CYP). Spotaneous trans-resveratrol conversion to cis-form was partly reversed by CYP catalysis and resveratrol was also transformed to a so far unidentified product. Therefore, in the NADPH-microsomal system, metabolism of polyphenols may influence their effects.

DISCUSSION

The order of polyphenol potency to inhibit LP we observed was basically similar to available literature. Rutin and gallic acid were more efficient in distilled water then with deionized mili-Q water indicating a possible role of transition metals in initiation. IC_{50} values of (+)-catechin and gallic acid were 4-fold higher in CYP2E1-induced than in control microsomes, possibly due to greater leakage of ROS[3].

However, we observed 20-fold lower IC_{50} for morin, 10x-30x fold lower for naringenin and 3x-4x lower for quercetin than in a similar NADPH-microsomal system[8], the AAPH assay[9] and the AA-Fe2+ assay[8,9]. The IC_{50} values reported for the Cu^{2+} - LDL system[10] were >10x lower, but relative efficiency correlated well with our data and we observed greater differences between the polyphenols.

The decrease of ${}^\bullet$OH radical production by quercetin and gallic acid corresponds to inhibition of LP, but (+)-catechin, rutin and trans-resveratrol did not quench ${}^\bullet$OH, although they decreased LP. In CYP2E1-induced microsomes incubated in distilled water, trans-resveratrol stimulated ${}^\bullet$OH production, though. This is not unusual, since hydroquinone and benzoquinone[3] and some flavonoids[11] significantly inhibited LP, but they stimulated ${}^\bullet$OH radical production. The explanation may lay in the fact that phenolic compounds, which directly inhibit LP by quenchig various intermediates, can facilitate the CYP2E1 futile cycle in which ROS are increasingly produced and leave the CYP active site. Our data confirm that quercetin[8,12,13], rutin[8,11,13], (+)-catechin[12] and gallic acid[14] chelate Fe ions, but resveratrol[15] does not. Our observation that quercetin chelates both Fe^{2+} and Fe^{3+}, rutin chelates Fe^{3+} more than Fe^{2+}, catechin only slightly chelates Fe^{3+}, and

Table 1. Comparison of antioxidant effeciency of polyphenols (mean $IC_{50} \pm SD$, (μM)).

Class	Compound	OH substitution	Inhibition of LP IC_{50} (μmol/l)	IC_{20}(μmol/l)	$^{\bullet}$OH quenching IC_{50} (μmol/l)	Iron chelation Fe^{2+}	Fe^{3+}
Flavonol	myricetin	3,5,7,3′,4′, 5′	7.5 ± 2.1				
	quercetin	3,5,7,3′,4′	1.7 ± 1.0		13.5	yes	yes
	rutin	5,7,3′,4′	3-10 ($*^1$)		> 40	yes	yes
	(quercetin rutinoside)		>> 30($*^2$)				
	fisetin	3,7,3′,4′	6.9 ± 1.0				
	kaempferol	3,5,7,4′	26 ± 15				
	morin	3,5,7,2′,4′	8.8 ± 1.3				
Flavanone	naringenin	5,7,4′	> 30	11.3 ± 4.7			
Flavanol	(+)-catechin	3,5,7,3′,4′	4.4 ($*^1$)		> 40	no	weak
			24.0 ($*^2$)				
	(-)-epicatechin	3,5,7,3′,4′	15.9 ± 1.1				
Stilbene	trans-resveratrol	3,5,4′	6.1 ± 1.8		> 40	no	no
Derivates of acids	gallic acid	3,4,5	< 5 ($*^1$)			no	yes
			> 20 ($*^2$)				
	gentisic acid	2,5	> 30	> 30			
	caffeic acid	3,4	> 30	> 30			
	p-coumaric acid	4	> 25	15.5			

($*^1$)- distilled H_2O, ($*^2$)- deionized milli-Q H_2O

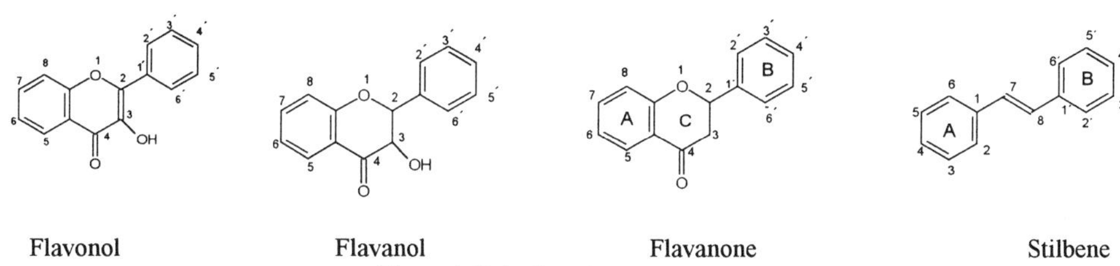

Flavonol　　　　　　Flavanol　　　　　　Flavanone　　　　　　Stilbene

Figure1. Polyphenols structures.

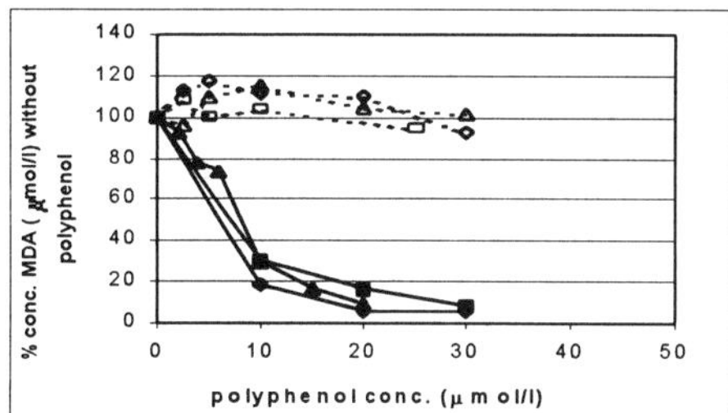

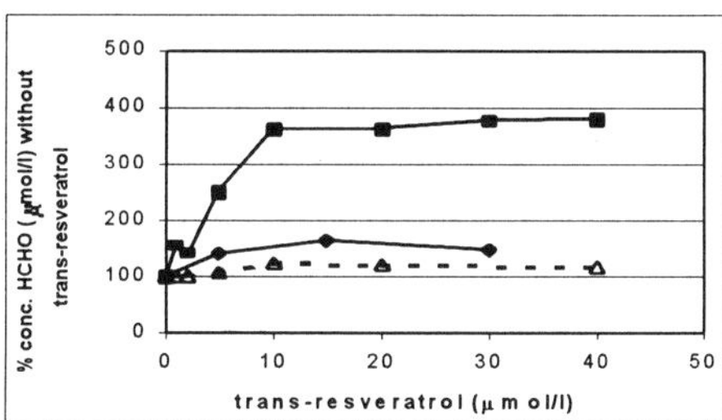

Figure 2. (left) Inhibitory effects of gentisic acid (dotted line) and fisetin (full line) on LP (three independent assays).; **(right)** Effects of trans-resveratrol on $^{\bullet}$OH quenching ($- -\Delta - -$ CYP2E1 induced microsomes, milli-Q H_2O; $- \blacklozenge -$ untreated microsomes, distilled H_2O; $- \blacksquare -$ CYP2E1 induced microsomes, distilled H_2O).

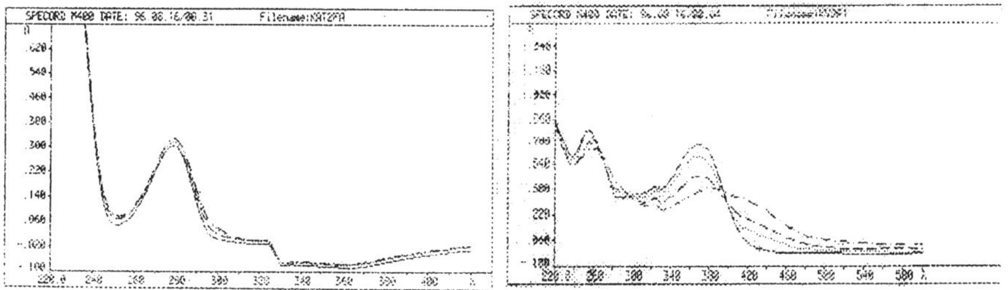

Figure 3. Absorption spectra of (+)-catechin (left) and quercetin (right), influence of Fe^{2+} addition.
—— 50 μM polyphenol; ········ 50 μM polyphenol + 25 μM Fe^{2+}; — · — 50 μM polyphenol + 50 μM Fe^{2+};
— — 50 μM polyphenol + 100 μM Fe^{2+}

gallic acid chelates only Fe^{3+} complies with the fact that Fe^{3+}-EDTA stability constant for Fe^{3+} is 10^6- fold higher than for Fe^{2+}.

It was suggested that the role of iron chelation in antioxidant effects of polyphenols is not clear[9]. Fe^{3+} chelation may not be crucial in their antioxidant effects, since we observed that Fe^{2+}, which polyphenols chelate less, was stronger initiator than Fe^{3+}. Fe^{2+} chelation may be therefore more important. The fact that quercetin chelated Fe^{2+} more than rutin and (+)-catechin did not chelate Fe^{2+} at all corresponds to their capacity to inhibit LP. The lack of Fe-chelation by the efficient antioxidant trans-resveratrol cannot be compared with the flavonoids for their different structure. Moreover, Fe-EDTA complexes (1:1) do not prevent Fe redox cycling, whereas 1:2 ratio blocks it. The Fe/polyphenol stechiometry may be therefore crucial and effective chelation probably blocks iron cycling, while it retains polyphenol antioxidant effects. In fact, Fe ions complexed with rutin or quercetin are unable to initate LP^{13}, whereas quercetin (or rutin):Fe^{2+} complexes retain their free radical scavenging activities[16] and catechol:Fe complexes scavenge $^\bullet O_2^-$, evenly more effectively than uncomplexed catechols, o-quinone and dimers, trimers, and polymers of catechols[13].

The oder of LP inhibition potency therefore complied with the reported structural requirements[9,16], which include the presence of the 2,3-double bond, the 3-hydroxyl group and the 4-oxo group (flavonols) and the 3′, 4′-dihydroxy pattern.Glycosylation of the hydroxyl group (rutin) therefore lower the ihibition potency.

REFERENCES

1. G.J. Soleas, E.P. Diamandis, D.M. Goldberg, Wine as a biological fluid : History, production and a role in disease prevention. *J. Clin. Lab. Anal.* 11:287 (1997).
2. C. Rice-Evans, Implications of the mechanisms of action of tea polyphenols as antioxidants in vitro chemoprevention in humans. *P.S.E.B.M.* 220:262 (1999).
3. I. Gut, V. Nededelcheva, P. Soucek, P. Stopka, B. Tichavska, Cytochromes P450 in benzene metabolism and involvement of their metabolites and reactive oxygen species in toxicity. *Envir. Health Persp.* 104:1211 (1996).
4. O.H. Lowry, N.J.Rosenbrough, A.J. Fass, J.L. Randall, Protein measurement with the Folin Phenol reagent. *J. Biol. Chem.*193:265 (1951).
5. T. Omura, R. Sato, The carbon monooxide-binding pigment of liver microsomes: evidence for its haemoproteine nature. *J. Biol. Chem.* 239:2370 (1964).
6. J.A.Buege, S.D. Aust, Microsomal lipid peroxidation. In: *Methods in enzymology*, Vol 52 S. Fleischer, L. Packer ed. Academic Press, New York (1978).
7. S. Khan, R. Krishnamurthy, K.P. Padnya, Generation of hydroxyl radicals during benzene toxicity. *Biochem. Pharmacol.* 39:1393 (1990).
8. S.A.B.E. van Acker, D.-J.van den Berg, M.N.L.J. Tromp, D.H. Griffionen, W.P. van Bennekom, W.J.F. Vijgh, A. Bast, Structural aspects of antioxidant activity of flavonoids. *Free Rad. Biol. Med.* 20:331 (1996).
9. S.A.B.E. van Acker, G.P. van Balen, D.-J. van den Berg, A. Bast, W.J.F. van der Vijgh, Influence of iron chelation on the antioxidant activity of flavonoids. *Biochem Pharm.*56:935 (1998).
10. J.A. Vinson, J. Jang, J. Yang, Y.A. Dabbagh, X. Liang, M.M. Serry, J. Proch, S. Cai, Vitamins and especially flavonoids in common beverages are powerful in vitro antioxidants which enrich lower density lipoproteins and increase their oxidative resistance after ex vivo spiking in human plasma. *J. Agric. Food Chem.* 47:2502 (1999).
11. Y.H. Miura, I. Tomita, T. Watanabe, T. Hirayama, S. Fukui, Active oxygens generation by flavonoids.*Biol. Pharm. Bull.* 21:93 (1998).
12. Z.S. Zhao, S.Khan, P.J. O′Brien, Catecholic iron complexes as cytoprotective superoxid scavengers against hypoxia: reoxygenation injury in isolated hepatocytes. *Biochem. Pharmacol.* 56:825 (1998).
13. I.B. Afanas′ev, A.I. Dorozhko, A.V. Brodskii, V.A. Kostyuk, A.I. Potapovitch, Chelating and free radical scavenging mechanisms of inhibitory action of rutin and quercetin in lipid peroxidation. *Biochem. Pharmacol.* 38:1763 (1989).
14. J.F. Moran, R.J. Klucas, R.J. Grayer, J. Abian, M. Becana, Complexes of iron with phenolic compounds from soybean modules and other legume tissues: prooxidant and antioxidant properties. *Free Rad. Biol. Med.* 22:861 (1997).
15. L. Belquendouz, L. Frémont, A. Linard, Resveratrol inhibits metal ion-dependent and independent peroxidation of porcine low-density lipoproteins. *Biochem. Pharmacol.* 53:1347 (1997).
16. A. Arora, M. Nair, G.M. Strasburg, Structure-activity relationship for antioxidant activities of a series of flavonoids in a liposomal system. *Free Rad. Biol.Med.* 24:1355 (1998).

CHROMOSOME DAMAGE FROM BIOLOGICAL REACTIVE INTERMEDIATES OF BENZENE AND 1,3-BUTADIENE IN LEUKEMIA

Martyn T. Smith

School of Public Health,
Division of Environmental Health Sciences,
140 Earl Warren Hall,
University of California, Berkeley, California 94720-7360
email: **martynts@uclink.berkeley.edu**

INTRODUCTION

The causes of leukemia remain largely unknown. Only ionizing radiation, benzene and various chemotherapy drugs are established causes in humans. While benzene is an established cause of acute myeloid leukemias, its ability to cause lymphocytic leukemias and the closely related lymphomas and multiple myeloma remains highly controversial (1-4). The concentration at which it significantly increases leukemia risk is also contentious. For example, it was recently claimed in a leading journal that benzene has a dose threshold of 200 p.p.m.-years for leukemia induction (4), although there is clearly information that benzene induces leukemia and other hematological changes at much lower doses than this (3, 5) and that its effects are likely to be linear down to background levels (6). Other occupational and environmental chemicals are suspected leukemogens, but are not established as such and controversy surrounds their classification as human carcinogens. One such compound is 1,3-butadiene. This compound is widely used in the rubber industry and is a product of incomplete combustion. Its' potential classification as a human carcinogen on the basis of epidemiological studies (7, 8) showing higher rates of leukemia in exposed workers is under consideration by numerous agencies (9). To throw further light on these important questions concerning the human risk posed by benzene and butadiene, we have examined the ability of benzene, butadiene and their reactive intermediates to produce the types of chromosome damage thought to be etiologically important in leukemia and lymphoma in exposed humans and in human cells *in vitro*. This paper summarizes our recent findings and discusses their impact on the risk assessments for benzene and butadiene.

Biological Reactive Intermediates VI, Edited by Dansette *et al.*
Kluwer Academic / Plenum Publishers, 2001

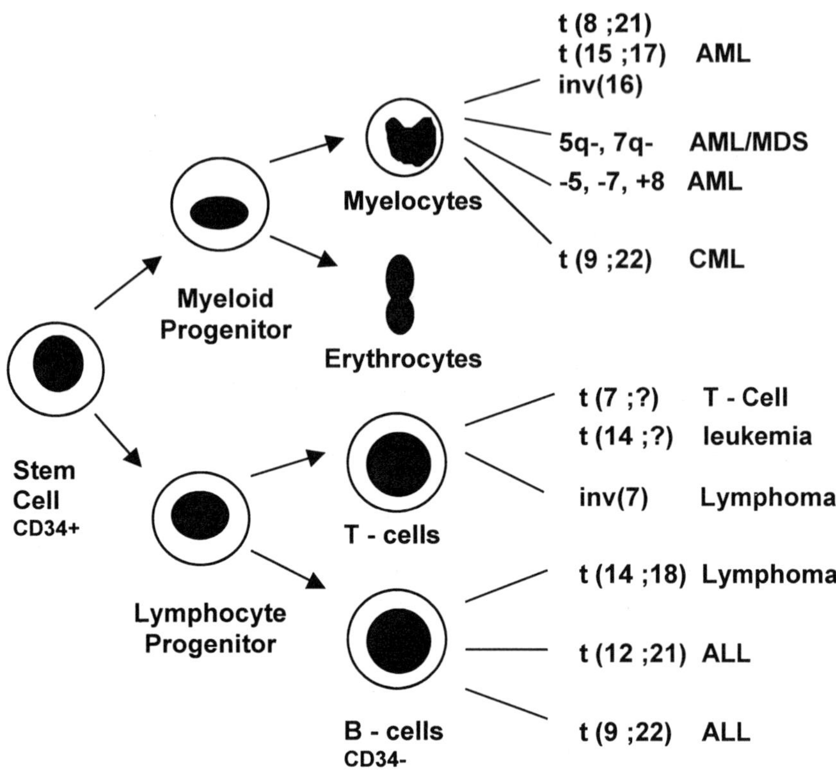

Figure 1. Clonal chromosome aberrations in the development of leukemia and lymphoma. Chromosome damage occurs in the cycling stem cells that are CD34-positive (CD34+) which progresses into leukemias of different types. AML = Acute myeloid leukemia; MDS = myelodysplastic syndromes; CML = chronic myeloid leukemia; and, ALL = acute lymphocytic leukemia

CHROMOSOME ABERRATIONS IN LEUKEMIA AND LYMPHOMA

Leukemias and lymphomas are characterized by clonal chromosomal aberrations that appear to have a central role in tumorigenesis (10, 11). In myeloid leukemia, loss of part or all of chromosomes 5 and 7 is a common early event, along with trisomy 8 and various specific translocations and inversions (Figure 1). In acute myeloid leukemia (AML), t(8;21), t(15;17) and t(11q23) are common (10, 11). Whereas in acute lymphocytic leukemia (ALL) and chronic myeloid leukemia (CML), the Philadelphia chromosome t(9;22) is relatively common. In lymphocytic leukemias and non-Hodgkins lymphomas translocations and aneuploidies are also common, with the translocation t(14;18) being associated most often with follicular lymphoma (12, 13) (Figure 1). These chromosomal changes can be detected by fluorescence *in situ* hybridization (FISH) and/or the polymerase chain reaction (PCR) (14) and may serve as biomarkers of early effect for benzene and other suspected leukemogens, including 1,3-butadiene.

SPECIFIC CHROMOSOME ABERRATIONS IN WORKERS EXPOSED TO BENZENE

We have used FISH and PCR to demonstrate elevated levels of leukemia-specific chromosome aberrations in workers exposed to high concentrations of benzene (15-18). Together with our collaborators at the Chinese Academy of Preventive Medicine in Beijing (Dr. S. Yin and G-L. Li) and at the National Cancer Institute (Drs. N. Rothman and R. Hayes) we studied a group of Chinese workers exposed to widely varying levels of benzene in Shanghai and a group of controls. Biological samples were collected from 44 healthy workers currently exposed to benzene with minimal exposure to toluene and other aromatic solvents. The same number of healthy controls without current or previous occupational exposure to benzene were enrolled from factories in the same geographic area. Controls were frequency-matched by gender and age (5 year intervals). The median benzene air level among the exposed workers was 31 ppm as an 8 hour TWA (range: 1-328 ppm). Air monitoring data were confirmed by measures of urinary benzene metabolites which showed strong, positive correlations with air benzene levels, and were substantially higher in exposed workers compared to controls (5, 19). We painted chromosomes 8 and 21 in lymphocyte metaphases from 43 workers exposed to benzene and 44 matched controls. To examine dose-response relationships the workers were divided into 2 groups at the median exposure level, a lower-exposed group ($\leq$ 31 ppm, n = 21) and a higher-exposed group (> 31 ppm, n = 22). Benzene exposure was associated with significant increases in hyperdiploidy of chromosome 8 (1.2, 1.5, 2.4 per 100 metaphases; $P_{trend} < 0.0001$) and 21 (0.9, 1.1, 1.9; $P_{trend} < 0.0001$) (15). Translocations between chromosomes 8 and 21 were increased up to 15-fold in highly exposed workers (0.01, 0.04, 0.16; $P_{trend} < 0.0001$). In one highly exposed individual these translocations were reciprocal and were detectable by reverse-transcriptase PCR (15). These data indicate a potential role for t(8;21) in benzene-induced leukemogenesis and are consistent with the hypothesis that detection of specific chromosome aberrations may be a powerful approach to identify populations at increased risk of leukemia from benzene exposure.

We also used a novel FISH procedure to determine if specific aberrations in chromosomes 1, 5 and 7 occured at an elevated rate in metaphase spreads prepared from the lymphocytes of the same benzene-exposed Chinese workers (17). We found that benzene exposure was associated with increases in the rates of monosomy 5 and 7 but not monosomy 1 ($p < 0.001$; < 0.0001; and 0.94, respectively) and with increases in trisomy and tetrasomy frequencies of all three chromosomes. Long arm deletion of chromosomes 5 and 7 was increased in a dose-dependent fashion ($p = 0.014$ and < 0.0001) up to 3.5-fold in the exposed workers. These results demonstrate that leukemia-specific changes in chromosomes 5 and 7 can be detected by FISH in the peripheral blood of otherwise healthy benzene-exposed workers. Studies are being planned to apply these methods in workers exposed to lower levels of benzene. In addition, we plan to explore the impact of inter-individual variation in genes that activate and detoxify benzene and its metabolites on these outcomes.

USING BIOMARKERS TO IMPROVE THE RISK ASSESSMENT FOR BENZENE

Toxicoepidemiology can be defined as the study of the adverse effects of chemicals using epidemiological methods. One of the goals of the above toxicoepidemiological studies is to improve the risk assessment for benzene by providing important scientific information. It is unlikely that classical epidemiological studies of cohorts of workers historically exposed to benzene will shed light on the dose response curve at exposure levels below 10 p.p.m. in air, because a very large number of workers would have to be studied (> 100,000) and the assessment of historical exposures is an inexact science in which only ranges can be estimated. Biomarker studies hold the advantage of requiring much fewer study participants and there is no need to wait for the disease to develop as a surrogate biomarker is used. The surrogate marker is assumed to be a predictor of future risk. Fortunately, for cancer and hematological malignancies especially, chromosome aberrations have been shown to be predictive of future risk (20). Our goal in using biomarkers to study a large number of workers exposed to around 1 p.p.m. benzene is therefore to establish the dose response curve for specific chromosome aberration induction in the 0.5 to 10 p.p.m. range. By inference, we suggest that this will predict the likely shape of the dose response curve for benzene induced hematological malignancies in this same low dose range.

Biomarkers can also shed light on other questions of importance to benzene risk assessment. One important question is: "Are there susceptible individuals?" Our studies in China have shown that workers with high cytochrome P4502E1 and no NQO1 activity were at a 7.8 fold higher risk of benzene hematotoxicity, a condition that predisposes to leukemia (21). This indicates that at least an 8-fold safety factor should be applied in benzene risk assessment to account for susceptible individuals. Further, it is likely that other factors will be discovered that explain inter-individual differences in susceptibility and may lead to the need to increase the size of the safety factor (14).

Another important question is: "What is the toxic metabolite(s) of benzene?" This information is needed for physiologically-based pharmacokinetic models that are commonly used in risk assessment (22). However, it is unlikely at present that biomarkers can throw much light on this issue. The exact metabolite(s) responsible for benzene's effects are unknown, but it has been suggested that a combination of at least two metabolites is involved (23). It is also possible that one metabolite may be responsible for one toxic effect of benzene and a second responsible for another toxic effect. The problem is that there is no unique pathway that can be identified as being primarily responsible for the toxic effects. There is, however, considerable evidence for the involvement of the phenolic metabolites (23). For example, we have shown that several of these metabolites, most notably hydroquinone, can produce many of the chromosomal changes found in leukemia *in vitro* in human lymphocytes (24) and most recently in human CD34-positive early progenitor / stem cells (25). Further, we demonstrated that exposure to hydroquinone produced monosomy and trisomy of chromosomes 7 and 8 to a greater extent and at lower concentrations in CD34+ than in CD34- cells (25). Particularly striking effects of hydroquinone were observed in CD34+ cells on monosomy 7 and trisomy 8, two common clonal aberrations found in myeloid leukemias, suggesting that these aneusomies produced by hydroquinone in CD34+ cells play a role in benzene-induced leukemogenesis. The reason for the greater sensitivity of CD34+ cells to hydroquinone genotoxicity is unclear at this time, but may be related to a lack of certain protective enzymes, such as NAD(P)H:quinone oxidoreductase (26, 27).

It is also possible that certain forms of DNA repair are lacking or are at lower levels in these immature progenitor cells. It has been reported, for example, that CD34+ cells have lower levels of alkylguanine DNA alkyltransferase activity (28). On the other hand, it has also been shown that CD34+ cells have higher levels of nucleotide excision repair (29). Thus, the role of DNA repair in hydroquinone sensitivity, and the DNA repair status of CD34+ cells requires further investigation.

Another issue in benzene risk assessment is which disease endpoints to use: Just AML or AML plus myelodysplastic syndromes or all leukemias or leukemias plus lymphomas (22). There is currently no consensus on this issue and uncertainty as to whether or not benzene causes non-Hodgkin's lymphoma. We have therefore examined the ability of the benzene metabolite, hydroquinone, to induce the t(14;18) translocation in human cord blood lymphocytes which possess active V(D)J recombinase. The t(14;18) translocation is frequently found in follicular lymphomas and is thought to be induced by the illegitimate action of V(D)J recombinase. Isolated lymphocyte cultures were exposed to hydroquinone for 48 h and the frequency of t(14;18) translocations determined using a highly sensitive quantitative real-time exonuclease based PCR assay. A significant increase in the frequency of t(14;18) above background levels was generated by treatment with hydroquinone. Sequencing of the PCR products revealed these to be identical or similar to breakpoint sequences previously reported for t(14:18) in cell lines and clinical samples. These findings suggest that hydroquinone has the ability to induce of t(14;18) translocations in human cells with active V(D)J recombinase and provide a mechanistic basis for the induction of non-Hodgkin's lymphoma by benzene exposure. An important question now for risk assessors is how to use this mechanistic information from *in vitro* toxicological studies.

TOXICOEPIDEMIOLOGICAL STUDIES TO IMPROVE THE RISK ASSESSMENT FOR BUTADIENE

We have also used FISH to study workers exposed to the potential leukemogen 1,3-butadiene and lymphocytes exposed *in vitro* to its metabolites. The *in vitro* studies showed that the epoxy metabolites of butadiene caused selective aneuploidy of certain chromosomes (30), with 1,2,3,4-diepoxybutane being a highly effective inducer of both aneuploidy and structural chromosome damage (30, 31). Together with Richard Hayes of the National Cancer Institute, James Swenberg's group at the University of North Carolina and Chinese investigators, including Drs. Yin and Li, we have examined if aneuploidy similar to that observed *in vitro* is also found in the blood cells of workers exposed to butadiene. We studied a group of workers at a polybutadiene rubber production facility in Yanshan, China (32). In total, 41 butadiene-polymer production workers and 38 nonexposed controls, matched for age, sex and smoking status were studied. Among the butadiene-exposed workers, the median air exposure was 2 ppm (6-hour TWA), due largely to intermittent high-level exposures. Compared to unexposed subjects, butadiene-exposed workers had greater levels of hemoglobin N-(2,3,4-trihydroxybutyl)valine (THBVal) adducts (p <.0001), and adduct levels tended to correlate, among butadiene-exposed workers, with air measures (p = .03) (32). Butadiene-exposed workers did not differ, however, from unexposed workers with respect to the frequency of aneuploidy of any of the four chromosomes 1, 7 , 8 and 12 as measured by FISH (32). Also butadiene exposure and greater THBVal levels were not associated with increases in other measures of genotoxic damage namely sister chromatid

exchanges, glycophorin A variants or lymphocyte *hprt* somatic mutation (32, 33).
Overall, this investigation in China demonstrated that exposure to butadiene, by a variety
of short-term and long-term measures, did not show specific genotoxic effects related to
that exposure.

Others studies have shown genotoxic effects in workers exposed to butadiene, but the
results have not been consistent. Among U.S. butadiene-styrene workers, dicentrics were
significantly correlated with urinary butadiene metabolites and there was evidence of
deficiencies in DNA repair by the CAT-host cell reactivation assay (34). Increased
frequency of mutations in *hprt* were also observed in U.S. butadiene workers (35) but
excesses were not found in our study (32, 33, 35) or in studies in the Czech Republic
(36). Increased frequency of chromosomal aberrations and sister chromatid exchange
were reported in the Czech Republic (36, 37), but an earlier investigation of these
subjects and others showed no excesses (38, 39). Taken together these results cast doubt
on the genotoxic potential of butadiene at low levels of occupational exposure. Our
negative results are relevant, however, only for exposures in the butadiene exposure
range studied. Also, the relatively small size of the study sample may have limited our
ability to detect modest effects. It is therefore feasible that higher levels of exposure to
butadiene may induce genotoxic effects, including aneuploidy, in humans. However, it is
also possible that humans are far less sensitive to the genotoxic effects of butadiene than
other species, as suggested by van Sittert et al. (40), because they have high liver epoxide
hydrolase activity and that our negative results reflect this. The high activity of epoxide
hydrolase in human liver affects butadiene metabolism such that proportionally very low
levels of the highly genotoxic diepoxybutane are formed from the monoepoxy metabolite
because it is rapidly detoxified to a non-genotoxic diol by the epoxide hydrolase.
Further, any diepoxybutane that is formed is metabolized by the epoxide hydrolase to
1,2-dihydroxy-3,4-epoxybutane, which is much less genotoxic that diepoxybutane. The
genotoxic and carcinogenic potential of butadiene in man therefore requires further
investigation.

ACKNOWLEDGMENTS

I wish to thank the members of my laboratory, L. Zhang, Y. Wang. W. Guo, N. Holland,
R. Turakulov and my many collaborators, most notably N. Rothman, R. Hayes, S. Yin,
G-L. Li, J. Swenberg and R. Higuchi, for their significant contributions to this work.
This work was supported by NIH grants RO1 ES 06721, P42 ES 04705 and P30 ES
01896 from the National Institute of Environmental Health Sciences, using funds
provided in part by the U.S. EPA, and funds from the National Cancer Institute.

REFERENCES

1.   Savitz, D. A., and Andrews, K. W. (1997) Review of epidemiologic evidence on
        benzene and lymphatic and hematopoietic cancers. *American Journal of
        Industrial Medicine* **31**, 287-95.
2.   Consonni, D., Pesatori, A. C., Tironi, A., Bernucci, I., Zocchetti, C., and Bertazzi,
        P. A. (1999) Mortality study in an Italian oil refinery: extension of the follow-up.
        American Journal of Industrial Medicine **35**, 287-94.
3.   Hayes, R. B., Yin, S. N., Dosemeci, M., Li, G. L., Wacholder, S., Travis, L. B.,
        Li, C. Y., Rothman, N., Hoover, R. N., and Linet, M. S. (1997) Benzene and the

dose-related incidence of hematologic neoplasms in China. Chinese Academy of Preventive Medicine--National Cancer Institute Benzene Study Group. *Journal of the National Cancer Institute* **89,** 1065-71.

4. Bergsagel, D. E., Wong, O., Bergsagel, P. L., Alexanian, R., Anderson, K., Kyle, R. A., and Raabe, G. K. (1999) Benzene and multiple myeloma: appraisal of the scientific evidence. *Blood* **94,** 1174-82.

5. Rothman, N., Li, G. L., Dosemeci, M., Bechtold, W. E., Marti, G. E., Wang, Y. Z., Linet, M., Xi, L. Q., Lu, W., Smith, M. T., Titenko-Holland, N., Zhang, L. P., Blot, W., Yin, S. N., and Hayes, R. B. (1996) Hematotoxicity among Chinese workers heavily exposed to benzene. *American Journal of Industrial Medicine* **29,** 236-46.

6. Smith, M. T. (1996) Mechanistic studies of benzene toxicity -- implications for risk assessment. *Advances in Experimental Medicine and Biology* **387,** 259-66.

7. Sathiakumar, N., Delzell, E., Hovinga, M., Macaluso, M., Julian, J. A., Larson, R., Cole, P., and Muir, D. C. (1998) Mortality from cancer and other causes of death among synthetic rubber workers. *Occupational and Environmental Medicine* **55,** 230-5.

8. Delzell, E., Sathiakumar, N., Hovinga, M., Macaluso, M., Julian, J., Larson, R., Cole, P., and Muir, D. C. (1996) A follow-up study of synthetic rubber workers. *Toxicology* **113,** 182-9.

9. IARC (1999) 1,3-Butadiene. *Iarc Monographs on the Evaluation of Carcinogenic Risks To Humans* **71 Pt 1,** 109-225.

10. Rowley, J. D. (1999) The role of chromosome translocations in leukemogenesis. *Seminars in Hematology* **36,** 59-72.

11. Rowley, J. D. (2000) Molecular genetics in acute leukemia. *Leukemia* **14,** 513-7.

12. Monni, O., Franssila, K., Joensuu, H., and Knuutila, S. (1999) BCL2 overexpression in diffuse large B-cell lymphoma. *Leukemia and Lymphoma* **34,** 45-52.

13. Stamatopoulos, K., Kosmas, C., Belessi, C., Stavroyianni, N., Kyriazopoulos, P., and Papadaki, T. (2000) Molecular insights into the immunopathogenesis of follicular lymphoma. *Immunology Today* **21,** 298-305.

14. Smith, M. T., and Zhang, L. (1998) Biomarkers of leukemia risk: benzene as a model. *Environmental Health Perspectives* **106 Suppl 4,** 937-46.

15. Smith, M. T., Zhang, L., Wang, Y., Hayes, R. B., Li, G., Wiemels, J., Dosemeci, M., Titenko-Holland, N., Xi, L., Kolachana, P., Yin, S., and Rothman, N. (1998)) Increased translocations and aneusomy in chromosomes 8 and 21 among workers exposed to benzene. *Cancer Research* **58,** 2176-81.

16. Zhang, L., Rothman, N., Wang, Y., Hayes, R. B., Bechtold, W., Venkatesh, P., Yin, S., Dosemeci, M., Li, G., Lu, W., and Smith, M. T. (1996) Interphase cytogenetics of workers exposed to benzene. *Environmental Health Perspectives* **104 Suppl 6,** 1325-9.

17. Zhang, L., Rothman, N., Wang, Y., Hayes, R. B., Li, G., Dosemeci, M., Yin, S., Kolachana, P., Titenko-Holland, N., and Smith, M. T. (1998) Increased aneusomy and long arm deletion of chromosomes 5 and 7 in the lymphocytes of Chinese workers exposed to benzene. *Carcinogenesis* **19,** 1955-61.

18. Zhang, L., Rothman, N., Wang, Y., Hayes, R. B., Yin, S., Titenko-Holland, N., Dosemeci, M., Wang, Y. Z., Kolachana, P., Lu, W., Xi, L., Li, G. L., and Smith, M. T. (1999) Benzene increases aneuploidy in the lymphocytes of exposed workers: a comparison of data obtained by fluorescence in situ hybridization in

interphase and metaphase cells. *Environmental and Molecular Mutagenesis* **34**, 260-8.

19. Rothman, N., Bechtold, W. E., Yin, S. N., Dosemeci, M., Li, G. L., Wang, Y. Z., Griffith, W. C., Smith, M. T., and Hayes, R. B. (1998) Urinary excretion of phenol, catechol, hydroquinone, and muconic acid by workers occupationally exposed to benzene. *Occupational and Environmental Medicine* **55**, 705-11.

20. Hagmar, L., Bonassi, S., Strömberg, U., Brøgger, A., Knudsen, L. E., Norppa, H., and Reuterwall, C. (1998) Chromosomal aberrations in lymphocytes predict human cancer: a report from the European Study Group on Cytogenetic Biomarkers and Health (ESCH). *Cancer Research* **58**, 4117-21.

21. Rothman, N., Smith, M. T., Hayes, R. B., Traver, R. D., Hoener, B., Campleman, S., Li, G. L., Dosemeci, M., Linet, M., Zhang, L., Xi, L., Wacholder, S., Lu, W., Meyer, K. B., Titenko-Holland, N., Stewart, J. T., Yin, S., and Ross, D. (1997) Benzene poisoning, a risk factor for hematological malignancy, is associated with the NQO1 609C-->T mutation and rapid fractional excretion of chlorzoxazone. *Cancer Research* **57**, 2839-42.

22. Smith, M. T., and Fanning, E. W. (1997) Report on the workshop entitled: "Modeling chemically induced leukemia--implications for benzene risk assessment". *Leukemia Research* **21**, 361-74.

23. Smith, M. T. (1996) The mechanism of benzene-induced leukemia: a hypothesis and speculations on the causes of leukemia. *Environmental Health Perspectives* **104 Suppl 6**, 1219-25.

24. Zhang, L., Wang, Y., Shang, N., and Smith, M. T. (1998) Benzene metabolites induce the loss and long arm deletion of chromosomes 5 and 7 in human lymphocytes. *Leukemia Research* **22**, 105-13.

25. Smith, M. T., Zhang, L., Jeng, M., Wang, Y., Guo, W., Duramad, P., Hubbard, A. E., Hofstadler, G., and Holland, N. T. (2000) Hydroquinone, a benzene metabolite, increases the level of aneusomy of chromosomes 7 and 8 in human CD34-positive blood progenitor cells. *Carcinogenesis* **21**, 1485 - 1490.

26. Moran, J. L., Siegel, D., and Ross, D. (1999) A potential mechanism underlying the increased susceptibility of individuals with a polymorphism in NAD(P)H:quinone oxidoreductase 1 (NQO1) to benzene toxicity. *Proceedings of the National Academy of Sciences of the United States of America* **96**, 8150-5.

27. Smith, M. T. (1999) Benzene, NQO1, and genetic susceptibility to cancer. *Proceedings of the National Academy of Sciences of the United States of America* **96**, 7624-6.

28. Gerson, S. L., Phillips, W., Kastan, M., Dumenco, L. L., and Donovan, C. (1996) Human CD34+ hematopoietic progenitors have low, cytokine-unresponsive O6-alkylguanine-DNA alkyltransferase and are sensitive to O6-benzylguanine plus BCNU. *Blood* **88**, 1649-55.

29. Myllyperkiö, M. H., and Vilpo, J. A. (1999) Increased DNA single-strand break joining activity in UV-irradiated CD34+ versus CD34- bone marrow cells. *Mutation Research* **425**, 169-76.

30. Xi, L., Zhang, L., Wang, Y., and Smith, M. T. (1997) Induction of chromosome-specific aneuploidy and micronuclei in human lymphocytes by metabolites of 1,3-butadiene. *Carcinogenesis* **18**, 1687-93.

31. Vlachodimitropoulos, D., Norppa, H., Autio, K., Catalán, J., Hirvonen, A., Tasa, G., Uusküla, M., Demopoulos, N. A., and Sorsa, M. (1997) GSTT1-dependent induction of centromere-negative and -positive micronuclei by 1,2:3,4-

diepoxybutane in cultured human lymphocytes [published erratum appears in Mutagenesis 1998 May;13(3):317]. *Mutagenesis* **12,** 397-403.

32. Hayes, R. B., Zhang, L., Yin, S., Swenberg, J. A., Xi, L., Wiencke, J., Bechtold, W. E., Yao, M., Rothman, N., Haas, R., O'Neill, J. P., Zhang, D., Wiemels, J., Dosemeci, M., Li, G., and Smith, M. T. (2000) Genotoxic markers among butadiene polymer workers in China. *Carcinogenesis* **21,** 55-62.

33. Hayes, R. B., Xi, L., Bechtold, W. E., Rothman, N., Yao, M., Henderson, R., Zhang, L., Smith, M. T., Zhang, D., Wiemels, J., Dosemeci, M., Yin, S., and O'Neill, J. P. (1996) hprt mutation frequency among workers exposed to 1,3-butadiene in China. *Toxicology* **113,** 100-5.

34. Hallberg, L. M., Bechtold, W. E., Grady, J., Legator, M. S., and Au, W. W. (1997) Abnormal DNA repair activities in lymphocytes of workers exposed to 1,3-butadiene. *Mutation Research* **383,** 213-21.

35. Ward, J. B., Jr., Ammenheuser, M. M., Whorton, E. B., Jr., Bechtold, W. E., Kelsey, K. T., and Legator, M. S. (1996) Biological monitoring for mutagenic effects of occupational exposure to butadiene. *Toxicology* **113,** 84-90.

36. Tates, A. D., van Dam, F. J., de Zwart, F. A., Darroudi, F., Natarajan, A. T., Rössner, P., Peterková, K., Peltonen, K., Demopoulos, N. A., Stephanou, G., Vlachodimitropoulos, D., and Srám, R. J. (1996) Biological effect monitoring in industrial workers from the Czech Republic exposed to low levels of butadiene. *Toxicology* **113,** 91-9.

37. Srám, R. J., Rössner, P., Peltonen, K., Podrazilová, K., Mra*cková, G., Demopoulos, N. A., Stephanou, G., Vlachodimitropoulos, D., Darroudi, F., and Tates, A. D. (1998) Chromosomal aberrations, sister-chromatid exchanges, cells with high frequency of SCE, micronuclei and comet assay parameters in 1, 3-butadiene-exposed workers. *Mutation Research* **419,** 145-54.

38. Adler, I. D., Cochrane, J., Osterman-Golkar, S., Skopek, T. R., Sorsa, M., and Vogel, E. (1995) 1,3-Butadiene working group report. *Mutation Research* **330,** 101-14.

39. Sorsa, M., Peltonen, K., Anderson, D., Demopoulos, N. A., Neumann, H. G., and Osterman-Golkar, S. (1996) Assessment of environmental and occupational exposures to butadiene as a model for risk estimation of petrochemical emissions. *Mutagenesis* **11,** 9-17.

40. van Sittert, N. J., Megens, H. J., Watson, W. P., and Boogaard, P. J. (2000) Biomarkers of exposure to 1,3-butadiene as a basis for cancer risk assessment. *Toxicological Sciences* **56,** 189-202.

THE ANTITUMOR AGENT ECTEINASCIDIN 743: CHARACTERIZATION OF ITS COVALENT DNA ADDUCTS AND CHEMICAL STABILITY

Laurence H. Hurley[1] and Maha Zewail-Foote[2]

[1]The University of Arizona Cancer Center, 1515 N. Campbell Ave., Tucson, Arizona 85724, USA, and [2]The University of Texas at Austin, Department of Chemistry and Biochemistry, Austin, Texas 78712, USA

1. SUMMARY

Ecteinascidin 743 (Et 743), a natural product derived from the Caribbean tunicate *Eteinascidia turbinata*, is a potent antitumor agent currently in phase II clinical trials. Et 743 binds in the minor groove of DNA, forming covalent adducts by reacting with N2 of guanine. Although DNA is considered to be the macromolecular receptor for Et 743, the precise mechanism by which Et 743 exerts its remarkable antitumor activity has not yet been elucidated. The aim of this study is to provide a rationale for the antitumor activity of Et 743 by studying its fundamental interactions with DNA at the molecular level.

First, DNA structural distortions induced by Et 743 were characterized using gel electrophoresis. Surprisingly, Et 743 bends DNA toward the major groove, a unique feature among DNA-interactive agents that occupy the minor groove. Second, in order to gain further insight into the molecular basis behind the apparent sequence selectivity of Et 743, the stability and structure of Et 743 adducts at different target sequences were determined. On the basis of this data, the overall stability of the Et 743–DNA adducts was found to be governed by the DNA target sequence, where the inability of Et 743 to form optimum bonding networks with its optimum recognition sites leads to the formation of an unstable adduct. Consequently, the reaction of Et 743 with DNA is reversible, and the rate of the reverse reaction is a function of the target and flanking sequences.

The results from this study demonstrate that Et 743 differs from other DNA alkylating agents by its effects on DNA structure and sequence-dependent chemical stability. This information provides important insight into the underlying mechanisms for its unique profile of antitumor activity.

2. INTRODUCTION

In the late 1960s, crude extracts from the colonial marine tunicate *Ecteinascidia turbinata* were found to possess antineoplastic activity *in vivo*; however, the active con-

Biological Reactive Intermediates VI, Edited by Dansette *et al.*
Kluwer Academic / Plenum Publishers, 2001

stituents of these extracts, the ecteinascidins (Ets), were not identified until 1986 (Rinehart et al., 1990). Of the Ets that have thus far been isolated, Et 743, the most abundant and most potent of this group of compounds, gained considerable attention due to its efficacy as an antitumor agent (Rinehart et al., 1990). Et 743 demonstrated LC_{50} values ranging from 1 pM to 100 pM against various NCI cell lines, including colon, CNS, melanoma, renal, and breast (Jimeno et al., 1996). Furthermore, mice with early stage MX-1 xenografts were all found to be tumor-free following treatment with Et 743 (Sakai et al., 1996). Et 743 is currently being evaluated in phase II clinical trials after showing impressive activity in phase I clinical studies with responses in breast cancer and melanoma (Bowman et al., 1998; Cvitkovic et al., 1998; Villalona-Calero et al., 1998). In phase II trials, excellent clinical responses with soft tissue sarcomas have been demonstrated (Glynn Faircloth, PharmaMar USA, private communication). The remarkable activity of Et 743, as well as its limited yield in nature (one gram per ton of tunicate), has prompted the development of a synthetic scheme (Corey et al., 1996).

Et 743 is a carbinolamine-containing antitumor antibiotic composed of three fused tetrahydroisoquinolone subunits (A–C) and is structurally related to naphthyridinomycin and the saframycin family of antibiotics (Figure 1). The main structural difference between Et

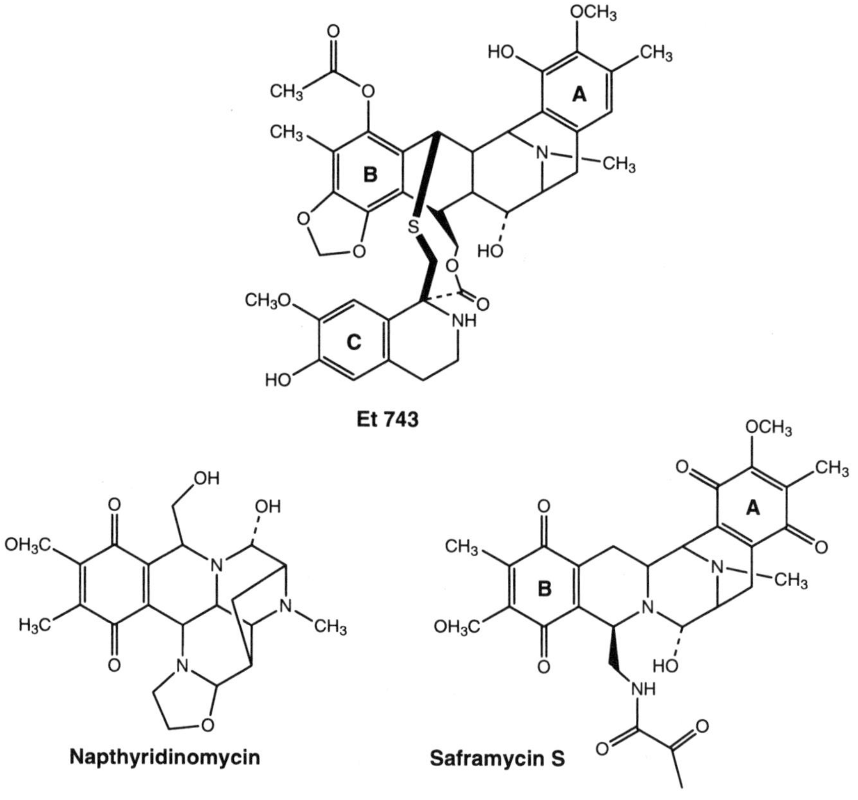

Figure 1. Structures of Et 743, naphthyridinomycin, and saframycin S, showing the A-, B-, and C-subunits. Reprinted with permission from Zewail-Foote and Hurley, 1999. Copyright 1999, American Chemical Society.

743 and saframycin is a C-subunit that is absent in saframycin S. Modification of the C-subunit effects the potency and antitumor selectivity of this class of compounds. For example, changing the C-subunit of Et 743 to a tetrahydro-β-carboline reduces the biological

activity, suggesting that the C-subunit plays an important role in the cytotoxicity (Sakai et al., 1996).

Et 743 binds in the minor groove of DNA and alkylates the N2 position of guanine. An NMR-based model of Et 743 with duplex DNA indicates that the A- and B-subunits are responsible for DNA recognition and bonding, while the C-subunit projects out of the minor groove and makes limited contacts with the DNA (Figure 2) (Moore et al., 1997). The

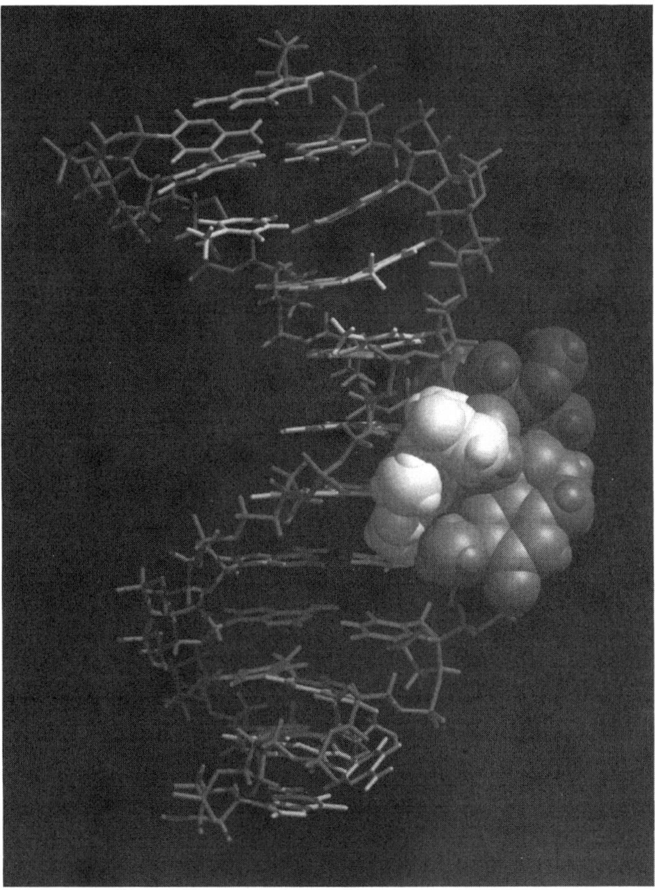

Figure 2. Molecular model of the Ecteinascidin subunits occupying different minor groove positions in DNA (5′-AGC target).

site selectivity of Et 743 is governed by a three-base-pair recognition sequence where the flanking bases immediately 5′ and 3′ to the modified guanine are involved in stabilizing the drug–DNA adduct (Pommier et al., 1996). There are a total of three hydrogen bond contacts between the A- and B-subunits of Et 743 and DNA, the most critical being the interaction of the B-subunit with the base located 3′ to the modification site. The hydrogen bond network has been proposed to direct the course of the sequence recognition by a direct readout mechanism (Seaman and Hurley, 1998). The mechanism of covalent adduct formation has been proposed to involve the reaction of Et 743 with DNA via an iminium ion intermediate caused by the intramolecular acid-catalyzed dehydration of the carbinolamine functional group (Figure 3) (Moore et al., 1998). This reaction is mechanistically different from other minor groove N2 guanine alkylators in that Et 743 contains an internal

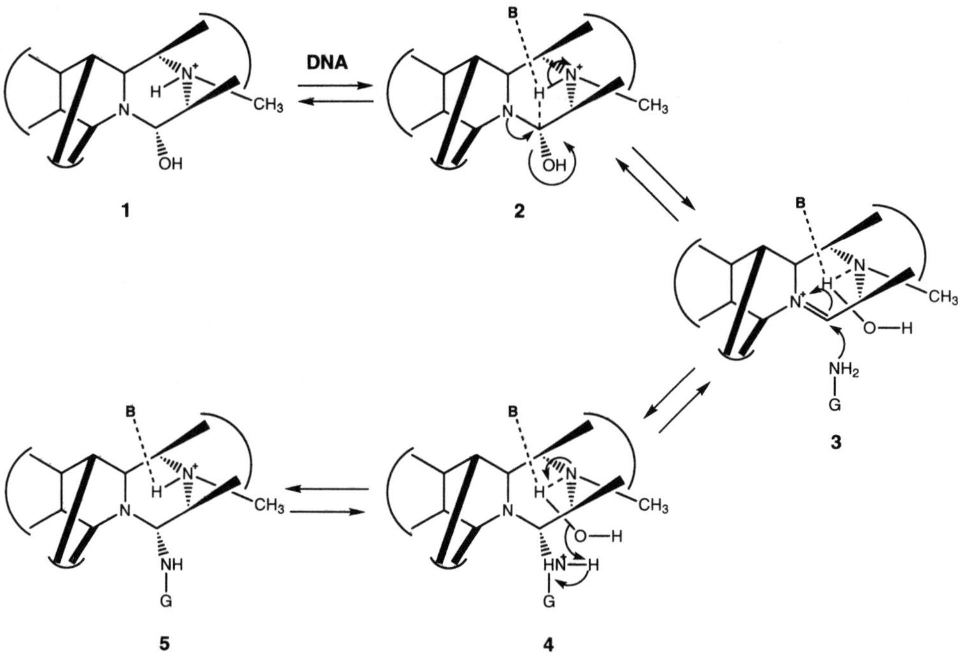

Figure 3. Reaction of Et 743 with DNA to form the Et 743–(N2-guanine)-DNA adduct. "B" is a DNA base hydrogen acceptor.

catalytic proton source in close proximity to the carbinolamine, which then catalyzes the dehydration of this moiety.

Although Et 743 is structurally similar to the saframycins, and both antibiotics react with the N2 position of guanine, Et 743 is different in that it has shown good efficacy as an antitumor agent, while the structurally related compounds have shown poor efficacy (Remers, 1988; Sakai et al., 1996). These differences in potency, despite several similarities among these drugs, suggest unique behaviors of Et 743. Deciphering these behaviors may provide the key to unlocking the secrets behind the potency of Et 743.

The mode of action of Et 743 is believed to be related to its covalent reaction with DNA through the exocyclic amino group of guanine; however, the biologically relevant targets of Et 743 have yet to be fully determined. Recent work has shown that Et 743 and related synthetic compounds induce DNA–topoisomerase I cross-linking. This may not be the primary mode of action of Et 743 since high dose levels were required to produce this effect and Et 743 was found to be equally active in topoisomerase I–deficient cell lines (Martinez et al., 1999; Takebayashi et al., 1999, 2000). Other research demonstrates that Et 743 can interfere with transcription of the MDR1 gene by inhibiting the binding of DNA binding proteins (Jin et al., 1999). In yet another study, cell lines that are deficient in nucleotide excision repair were found to be resistant to Et 743, suggesting that Et 743 could lead to repair dependent lethality (D'Incalci, 1999). Although these studies have brought to light different anecdotal behaviors, either real or implied, of Et 743, none of them have provided deterministic evidence that any of these behaviors are the fundamental mode of action responsible for its efficacy as an antitumor agent. In this article we provide additional insight into the underlying mechanisms that give rise to biological potency and clinical efficacy of Et 743.

3. STRUCTURAL EFFECTS OF ET 743 ON DNA

It is often the structural effects caused by the DNA modification of DNA-reactive drugs that give rise to the biochemical and biological consequences. Hence, alterations in DNA structure induced by Et 743 alkylation may contribute to the improved clinical efficacy. To provide further insight into the structural basis for the antitumor activity of Et 743, the effect of the covalent bonding of Et 743 on DNA structure was studied. We find that the reaction of Et 743 with DNA induces a bend in the DNA helix and that this bending directionality is toward the major groove. This is the first example of a DNA minor groove alkylator that bends DNA toward the major groove, a property that may differentiate this compound from other structurally or mechanistically similar drugs.

Gel electrophoresis is a common method used to determine alterations in DNA curvature induced by covalent adducts (Lee et al., 1991; Rink and Hopkins, 1995; Rink et al., 1996). A 21-base-pair oligonucleotide (ET21) was designed to contain only one Et 743 alkylation site (5′-AGC) by substituting inosine for guanine on the noncovalently modified strand (Figure 4).

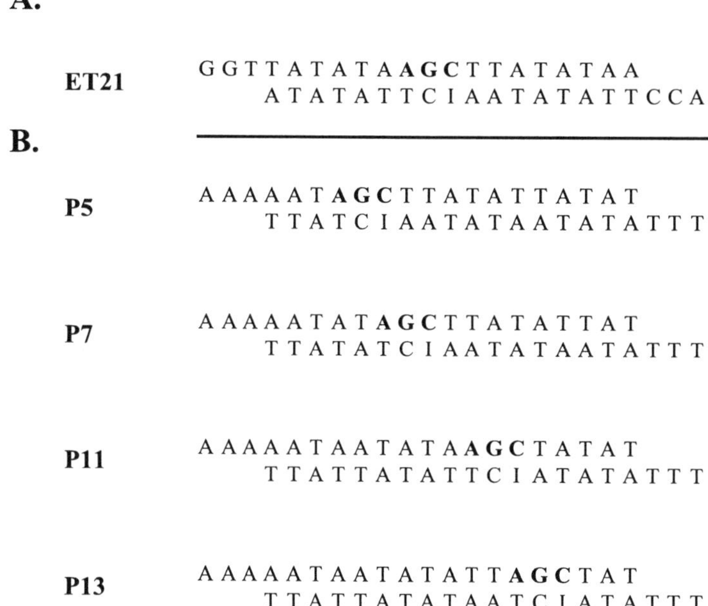

Figure 4. Sequences of the oligonucleotides used in this study. Each oligonucleotide contains one Et 743 binding site (5′-AGC): (A) 21-base-pair oligonucleotide (ET21) used to determine if Et 743 bends DNA. (B) 21-base-pair oligonucleotides used to determine the direction of Et 743–induced bending. The Et 743 modification site is positioned 5, 7, 11, or 13 base pairs away from the center of an A-tract. Reprinted with permission from Zewail-Foote and Hurley, 1999. Copyright 1999, American Chemical Society.

Nondenaturing gel analysis of the ligated 21-mers containing the Et 743–DNA adduct shows a retardation in electrophoretic mobility of the visible bands corresponding to linear DNA compared to the unmodified linear multimers (Zewail-Foote and Hurley, 1999). The ratios of the apparent length to the true length (R_L) were calculated for each ligation product (Figure 5) based on the mobility of a reference oligonucleotide. The increase in R_L with molecular weight suggests that Et 743 bends DNA. The angle of absolute curvature

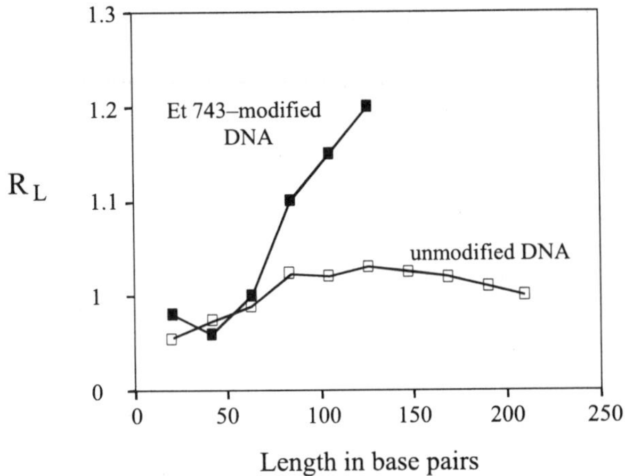

Figure 5. Plot of the relative length (R_L) as a function of total length in base pairs. The R_L values for the higher molecular weight species that exhibit double bands were calculated for the slower migration product. The experiment was repeated three times and the values were reproducible. Reprinted with permission from Zewail-Foote and Hurley, 1999. Copyright 1999, American Chemical Society.

was calculated based on the empirical relation described by Koo and Crothers (1988) and was found to be $17° \pm 3°$.

Phasing analysis was used to confirm that Et 743 induces bending and to determine the directionality of the induced bend. It has been well established that consecutive adenines (A-tracts) bend DNA toward the minor groove (Koo et al., 1986; Zahn and Blattner, 1987; Zinkel and Crothers, 1987). To determine if the Et 743–induced bending is toward the major or minor groove, the A-tract bend was used as a reference point. Oligonucleotides containing a single Et 743 alkylation site were positioned 5, 7, 11, and 13 base pairs away from the center of an A-tract. These constructs place the A-tract either "in phase" or "out of phase" with the Et 743 alkylation site. Anomalous mobility was maximum when the Et 743 alkylation site was positioned 5 base pairs away from the A-tract-induced bend (Zewail-Foote and Hurley, 1999). The R_L values were determined for each Et 743–modified oligonucleotide (Figure 6). The R_L value is at a minimum when Et 743 is

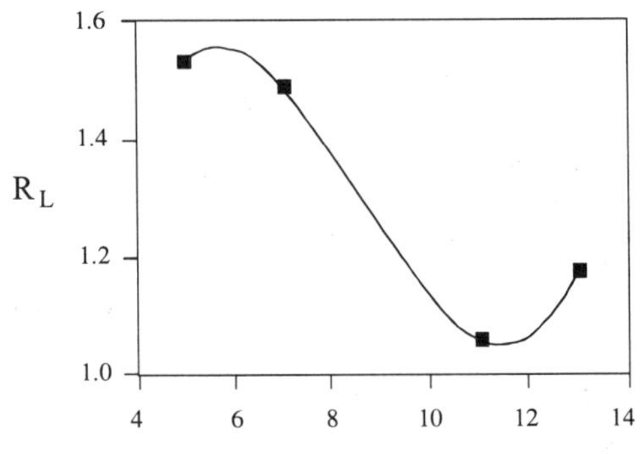

Distance between A-tract and alkylation site (base pairs)

Figure 6. Effect of distances between the center of an A-tract and the Et 743 alkylation site of R_L values. The R_L values were calculated based on the mobilities of the higher mobility bands for the 105-mer. Reprinted with permission from Zewail-Foote and Hurley, 1999. Copyright 1999, American Chemical Society.

positioned one helical turn away from the A-tract and at a maximum when Et 743 is located one-half a helical turn from the center of the A-tract. These results indicate that Et 7 43 bends the duplex DNA in the opposite direction of the A-tract. Hence, Et 743 bends DNA toward the major groove, which is a novel feature among minor groove DNA-interactive agents.

The results described in this section provide structural evidence for how Et 743 differs from the other minor groove alkylating agents and thus may provide a starting point to rationalize the improved clinical efficacy of this group of drugs. The most important finding described here is that this is the first example of a *minor groove occupancy drug that bends DNA into the major groove.*

4. ET 743–DNA ADDUCTS ARE REVERSIBLE

Drugs that covalently modify DNA, such as mitomycin C, cisplatinum compounds, and analogues of CC-1065, have been observed to be reversible under certain conditions (Borowy-Borowski et al., 1990; Gaucheron et al., 1991; Warpehoski et al., 1992).

4.1. The Rate of Release of Et 743 from Modified 5′-AGT Is Faster Than That from 5′-AGC Sequences

The first indication that the Et 743 alkylation reaction is reversible came from initial endeavors to create a 60-mer oligonucleotide containing an Et 743 site-directed adduct at either a 5′-AGT or 5′-AGC sequence. The substrates were constructed by ligating an oligonucleotide modified with Et 743 at a single guanine to flanking oligonucleotides on the 5′ and 3′ sides. Although efforts to create a site-directed adduct at a 5′-AGC sequence were successful, attempts at generating a site-directed adduct at a 5′-AGT sequence failed. After several ligation and purification steps, it was observed that the ligated 60-mer no longer contained Et 743 covalently bound at the 5′-AGT sequence. This result prompted us to compare the stability of the Et 743–DNA adduct at the 5′-AGC versus 5′-AGT sequences.

Previous reports indicate that drug modification produces a gel shift during electrophoresis (Pommier et al., 1996; Zewail-Foote and Hurley, 1999); hence, by quantifying the intensity of the retarded band, the level of Et 743 modification can be determined. This band shift assay was used to monitor the amount of modified duplex in order to determine if the covalent adduct is stable over time. An oligonucleotide containing a single modification site, either 5′-AGC or 5′-AGT, surrounded by a common sequence, was modified completely with Et 743. The Et 743–DNA adducts were then monitored at room temperature for the indicated amount of time, and the intensity of the Et 743–DNA complex was quantified (Figure 7). The level of Et 743–modified DNA decreases over time for both Et 743 target sequences. However, at the end of the 8-hour time period, the amount of the Et 743–DNA adduct at the 5′-AGC sequence was reduced by approximately 20%, whereas the amount of the Et 743–DNA adduct at the 5′-AGT sequence decreased by over 60% (compare lanes in Figure 7, A and B). This threefold difference indicates that the Et 743 adduct at the 5′-AGC sequence is more stable than the Et 743 adduct at the 5′-AGT sequence, under the conditions used here. Furthermore, the results suggest that Et 743 can reverse from DNA and that the rate of reversal depends on the base pair immediately 3′ to the site of covalent attachment of Et 743 (5′-AGC versus 5′-AGT). The observed reduction in the amount of Et 743–DNA adducts with time prompted us to look at the possibility that the released Et 743 might be able to "walk" the DNA, migrating from one sequence to another.

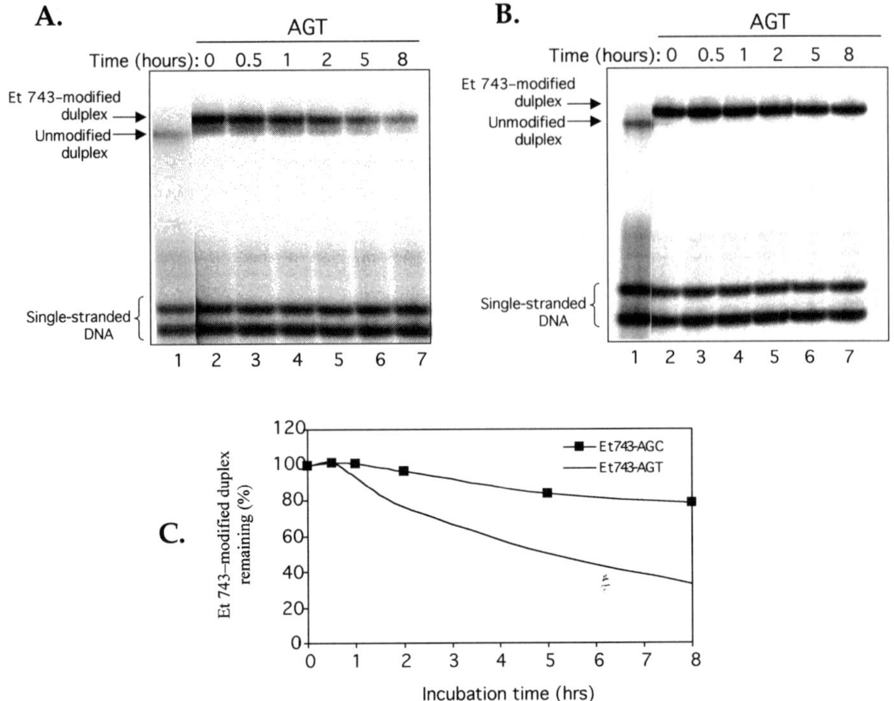

Figure 7. Time dependency of the reduction Et 743 adducts at different target sequences. An oligonucleotide containing either a single 5′-AGT (A) or 5′-AGC (B) was incubated with 20 μM Et 743. The resulting covalent adduct was incubated for the indicated amount of time at room temperature and loaded onto a native gel. (C) Plot of the amount of each drug-modified duplex, expressed as a percent, over time. The intensity of each band is normalized to the intensity at time zero, which is expressed as 100%.

4.2 Et 743 Can Reverse from Its Adduct and Bond to a Naïve Target Sequence

Once Et 743 is released from the covalent DNA adducts, it could alkylate other available target sites. To determine if Et 743 can rebond at other recognition sequences, Et 743–modified oligonucleotides (22-mer) containing the site-directed adducts at either the 5′-AGC or 5′-AGT target sequence were incubated with a different oligonucleotide (30-mer) containing either the less favored or the optimal Et 743 target sequence, respectively. Figure 8 shows that as the level of the Et 743–modified 5′-AGT adduct decreases over time, the other duplex (30-mer), containing the more favored drug target sites, becomes modified with Et 743. After 10 hours, approximately 20% of the 30-mer is Et 743 modified, suggesting that as Et 743 reverses from the less favored 5′-AGT target sequence, it can rebond at another DNA strand containing the more favored target sites. The same profile can be seen when the 5′-AGC sequence is modified, but to a much lesser extent. Here, less than 10% of the 30-mer is modified with Et 743 at 10 hours. These results support the notion that the reaction of Et 743 with duplex DNA is reversible and that in the presence of competitor DNA free drug released from the covalent DNA adducts can alkylate other available DNA sites.

This study provides evidence that the reaction of Et 743 with DNA is reversible under nondenaturing conditions. The data further suggest that the observed sequence preference for Et 743 relies on the differential rate of the reverse reaction of Et 743 from different sequences (i.e., 5′-AGC versus 5′-AGT), while the rate of alkylation appears to be independent of the DNA target sequence. We propose that the hydrogen bond network be-

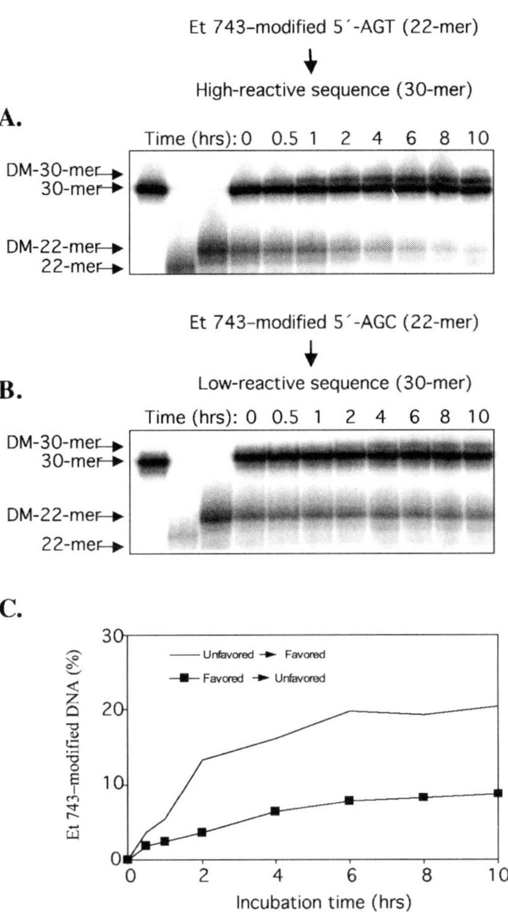

Figure 8. Translocation of covalently bound Et 743 from its adduct to another unmodified target sequence. (A) A 30-mer containing only high-reactive Et 743 target sequences was incubated for the indicated amount of time with an Et 743–modified 5′-AGT site-directed adduct (22-mer). The resulting mixture was run on a 16% native gel. Arrows point to the sites of the unmodified duplex and the Et 743–modified duplex (DM). (B) A 30-mer containing only low-reactive sequences was incubated for the indicated amount of time with an Et 743–modified 5′-AGC oligonucleotide (22-mer). (C) The amount of drug-modified duplex was quantified and plotted as a function of incubation time.

tween Et 743 and DNA governs sequence recognition, as previously reported, but that the major determinant for the sequence selectivity is the relative stability of the covalent adducts. Hence, the differences in the rate of reversibility lead to the apparent reactivity preference.

CONCLUSIONS

Et 743 bends DNA into the major groove while occupying the minor groove. It is the rate of reversibility from different sequences that gives rise to the apparent sequence selectivity, not the rate of the forward reaction.

ACKNOWLEDGMENTS

This research was supported by an Outstanding Investigator Grant from the National Institutes of Health and an American Chemical Society Fellowship to Maha Zewail-Foote. We are grateful to Dr. David M. Bishop for proofreading, editing, and preparing the manuscript and figures.

REFERENCES

Borowy-Borowski, H., Lipman, R., Chowdary, D., and Tomasz, M., 1990, Duplex oligodeoxyribonucleotides cross-linked by mitomycin C at a single site: synthesis, properties, and cross-link reversibility, *Biochemistry* **29**:2992–2999.

Bowman, A., Twelves, C., Hoekman, K., Simpson, A., Smyth, J., Vermorken, J., Höppener, F., Beijnen, J., Vega, E., Jimeno, J., and Hanauske, A.-R., 1998, Phase I clinical and pharmacokinetic (PK) study of ecteinascidin-743 (Et-743) given as a one-hour infusion every 21 days, *Ann. Oncol.* **9** (suppl. 2):119.

Corey, E. J., Gin, D. Y., and Kania, R. S., 1996, Enantioselective total synthesis of ecteinascidin 743, *J. Am. Chem. Soc.* **118**:9202–9203.

Cvitkovic, E., Mekranter, B., Taamma, A., Goldwasser, F., Beijnen, J. H., Jimeno, J., Riofrio, M., Vega, E., Misset, J. L., and Hop, P., 1998, Ecteinascidin-743 (Et-743) 24-hour continuous intravenous infusion (CI) phase I study in solid tumors (ST) patients, *Ann. Oncol.* **9** (suppl. 2):119.

D'Incalci, M., 1999, Mode of action of ecteinascidin-743 (Et-743), in *AACR–NCI–EORTC International Conference*, p. 3872, Washington, D.C.

Gaucheron, F., Malinge, J. M., Blacker, A. J., Lehn, J. M., and Leng, M., 1991, Possible catalytic activity of DNA in the reaction between the antitumor drug cis-diamminedichloroplatinum(II) and the intercalator N-methyl-2,7-diazapyrenium, *Proc. Natl. Acad. Sci. U.S.A.* **88**:3516–3519.

Jimeno, J. M., Faircloth, G., Cameron, L., Meely, K., Vega, E., Gomez, A., Sousa-Faro, J. M. F., and Rinehart, K., 1996, Progress in the acquisition of new marine-derived anticancer compounds: development of the ecteinascidin-743 (Et 743), *Drugs Future* **21**:1155–1165.

Jin, S., Gorfajn, B., Faircloth, G., and Scotto, K. W., 2000, Ecteinascidin 743, a transcription-targeted chemotherapeutic that inhibits MDR1 activation, *Proc. Natl. Acad. Sci. U.S.A.* **97**:6775–6779.

Koo, H.-S., and Crothers, D. M., 1988, Calibration of DNA curvature and a unified description of sequence-directed bending, *Proc. Natl. Acad. Sci. U.S.A.* **85**:1763–1767.

Koo, H.-S., Wu, H. M., and Crothers, D. M., 1986, DNA bending at adenine-thymine tract, *Nature* **230**:501–506.

Lee, C.-S., Sun, D., Kizu, R., and Hurley, L. H., 1991, Determination of the structural features of (+)-CC-1065 that are responsible for bending, winding, and stiffening of DNA, *Chem. Res. Toxicol.* **4**:203–213.

Martinez, E. J., Owa, T., Schreiber, S. L., and Corey, E. J., 1999, Phthalascidin, a synthetic antitumor agent with potency and mode of action comparable to ecteinascidin 743, *Proc. Natl. Acad. Sci. U.S.A.* **96**:3496–3501.

Moore, B. M., II, Seaman, F. C., and Hurley, L. H., 1997, NMR-based model of an ecteinascidin 743–DNA adduct, *J. Am. Chem. Soc.* **119**:5475–5476.

Moore, B. M., II, Seaman, F. C., Wheelhouse, R. T., and Hurley, L. H., 1998, Mechanism for the catalytic activation of ecteinascidin 743 and its subsequent alkylation of guanine N2, *J. Am. Chem. Soc.* **120**:2490–2491.

Pommier, Y., Kohlhagen, G., Bailly, C., Waring, M., Mazumder, A., and Kohn, K. W., 1996, DNA sequence- and structure-selective alkylation of guanine N2 in the DNA minor groove by ecteinascidin 743, a potent antitumor compound from the Caribbean tunicate *Ecteinascidia turbinata, Biochemistry* **35:**13303–13309.

Remers, W. A., 1988, *The Chemistry of Antitumor Antibiotics*, Volume 2, John Wiley and Sons, Inc., New York.

Rinehart, K. L., Holt, T. G., Fregeau, N. L., Stroh, J. G., Keifer, P. A., Sun, F., Li, L. H., and Martin, D. G., 1990, Ecteinascidins 729, 743, 745,7 59A, 759B, and 770: potent antitumor agents from the Caribbean tunicate *Ecteinascidia turbinata, J. Org. Chem.* **55:**4512–4515.

Rink, S. M., and Hopkins, P. B., 1995, A mechlorethamine-induced DNA interstrand cross-link bends duplex DNA, *Biochemistry* **34:**1439–1445.

Rink, S. M., Lipman, R., Alley, S. C., Hopkins, P. B., and Tomasz, M., 1996, Bending of DNA by the mitomycin C-induced, GpG intrastrand cross-link, *Chem. Res. Toxicol.* **9:**382–389.

Sakai, R., Jares-Erijman, E. A., Manzanares, I., Silva Elipe, M. V., and Rinehart, K. L., 1996, Ecteinascidins: putative biosynthetic precursors and absolute stereochemistry, *J. Am. Chem. Soc.* **118:**9017–9023.

Seaman, F. C., and Hurley, L. H., 1998, Molecular basis for the DNA sequence selectivity of ecteinascidin 736 and 743: evidence for the dominant role of direct readout via hydrogen bonding, *J. Am. Chem. Soc.* **120:**13028–13041.

Takebayashi, Y., Pourquier, P., Yoshida, A., Kohlhagen, G., and Pommier, Y., 1999, Poisoning of human DNA topoisomerase I by ecteinascidin 743, an anticancer drug that selectively alkylates DNA in the minor groove, *Proc. Natl. Acad. Sci. U.S.A.* **96:**7196–7210.

Takebayashi, Y., Goldwasser, F., Urasaki, Y., Kohlhagen, G., and Pommier, Y., 2000, Ecteinascidin 743 induces protein-linked DNA breaks in human colon carcinoma HCT116 cells and is cytotoxic independently of topoisomerase I expression, *Clinical Cancer Research*, submitted.

Villalona-Calero, M., Eckhardt, S. G., Weiss, G., Campbell, E., Hidalgo, M., Kraynak, M., Beijnen, J., Jimeno, J., Von Hoff, D., and Rowinsky, E., 1998, A phase I and pharmacokinetic study of Et-743, a novel DNA minor groove binder of marine origin, administered as a 1-hour infusion daily ×5 days, *Ann. Oncol.* **9** (suppl. 2):119.

Warpehoski, M. A., Harper, D. E., Mitchell, M. A., and Monroe, T. J., 1992, Reversibility of the covalent reaction of CC-1065 and analogues with DNA, *Biochemistry* **31:**2502–2508.

Zahn, K., and Blattner, F. R., 1987, Direct evidence for DNA bending at the lambda replication origin, *Science* **236:**416–422.

Zewail-Foote, M., and Hurley, L. H., 1999, Ecteinascidin 743: a minor groove alkylator that bends DNA toward the major groove, *J. Med. Chem.* **42:**2493–2497.

Zinkel, S. S., and Crothers, D. M., 1987, DNA bend direction by phase sensitive detection, *Nature* **328:**178–181.

DESIGN OF DNA DAMAGING AGENTS THAT HIJACK TRANSCRIPTION FACTORS AND BLOCK DNA REPAIR

John M. Essigmann,[1] Stacia M. Rink,[1,2] Hyun-Ju Park,[1] and Robert G. Croy[1]

[1]Department of Chemistry and Division of Bioengineering and Environmental Health, Massachusetts Institute of Technology, Cambridge, MA 02139 USA
[2]Department of Chemistry, Pacific Lutheran University, Tacoma, WA 98447 USA

1. INTRODUCTION

1.1. Intellectual Origins of this Project: Lessons Learned from Cisplatin

Most antitumor drugs extend life but do not fully cure cancer. Testicular cancer is one exception, as evidenced by 93% cure rates in men treated with regimens that include cisplatin (*cis*-diamminedichloroplatinum(II)) (Feuer et al., 1993). The spectacular success of cisplatin based regimens against testicular tumors and their impressive activity in delaying the progression of ovarian and other cancers (Masters et al., 1996; Loehrer and Einhorn, 1984) have stimulated much interest in the biochemical mechanism(s) of action of the drug. From the outset it has been appreciated that human cells may respond to cisplatin in unique ways that, once understood, could lead to the synthesis of novel mechanism-based drugs. It is well established that cisplatin kills cells in processes triggered by its reaction with DNA (Bruhn et al., 1990). Of possible importance to the cytotoxic mechanism of cisplatin is the observation that various cellular proteins bind to cisplatin-DNA adducts (Chu, 1994). In part through the work of our laboratory, a family of cisplatin adduct-binding proteins, most possessing a high mobility group (HMG) domain, was identified (Bruhn et al., 1992; Toney et al., 1989; Pil and Lippard, 1992; Brown et al., 1993). Significantly, these proteins display a selective affinity for the DNA adducts of therapeutically active platinum compounds (Toney et al., 1989; Pil and Lippard, 1992), suggesting that the proteins play a role potentiating cisplatin cytotoxicity.

Several models have been proposed by this laboratory (Donahue et al., 1990) to explain the possible involvement of HMG and other platinum binding proteins in the mechanism of action of cisplatin (Fig. 1). One model suggests that binding of these proteins to platinum adducts precludes access to the lesions by DNA repair enzymes (Figure 1B). Indeed, evidence has been presented both *in vitro* and *in vivo* in support of the view that HMG box proteins sensitize cells to cisplatin by shielding cisplatin adducts from repair (Brown et al., 1993; Huang et al., 1994; McA'Nulty and Lippard, 1996). A second model (Figure 1C) stems from identification of human upstream binding factor (hUBF) as the protein most strongly attracted to therapeutic platinum-DNA adducts. This transcription factor plays a key role as a regulator of ribosomal RNA synthesis, which is essential for proliferating cells. hUBF binds to a cisplatin adduct ($K_{d(app)} = 60$ pM) and to its cognate rRNA promoter sequence ($K_{d(app)} = 18$ pM) with comparable affinities, leading to the proposal that cisplatin adducts may act as molecular decoys for the transcription factor *in vivo*. Thus, by a "transcription factor hijacking" mechanism, cisplatin may deplete cells of a necessary resource for growth (Treiber et al., 1994;

Zhai et al., 1998; Jordan and Carmo-Fonseca, 1998; Vichi et al., 1997). A third model was very recently proposed based upon our observation (Mello et al., 1996) and that of Modrich (Duckett et al., 1996) that some mismatch repair proteins also bind tightly to DNA adducts formed by therapeutically useful platinum compounds. By this model, mismatch repair proteins may attempt to repair platinum adducts, but the repair reaction is aborted at mid course and results in signals (presumably nicks and gaps) that ultimately trigger programmed cell death. Alternatively, the role of mismatch repair proteins in recombination may be disrupted, which could also lead to cellular hypersensitivity to the drug (Zdraveski et al., 1999). While biochemical evidence for this new model involving mismatch repair and possibly recombination is limited, there is evidence that such a model is indeed operative in the therapeutic response to some alkylating agents (Karran and Marinus, 1982; Goldmacher et al., 1986). It is noted that the "abortive repair" model just described, in principle, could work in concert with the "repair shielding" (Figure 1B) and "hijacking" (Figure 1C) models and, in aggregate, they may contribute significantly toward explaining why cisplatin is such an effective drug for the treatment of testicular and, to a lesser extent, ovarian cancers. It is noteworthy in this regard that we have demonstrated that the mismatch repair protein hMSH2 is over expressed in testicular and ovarian tissues (Mello et al., 1996), perhaps explaining the organotropic specificity of the drug. One problem with cisplatin is its specificity for a few tumor types. It is a logical question to ask whether the body of knowledge on cisplatin could be applied to the synthesis of a selective toxin that could target other tumor types.

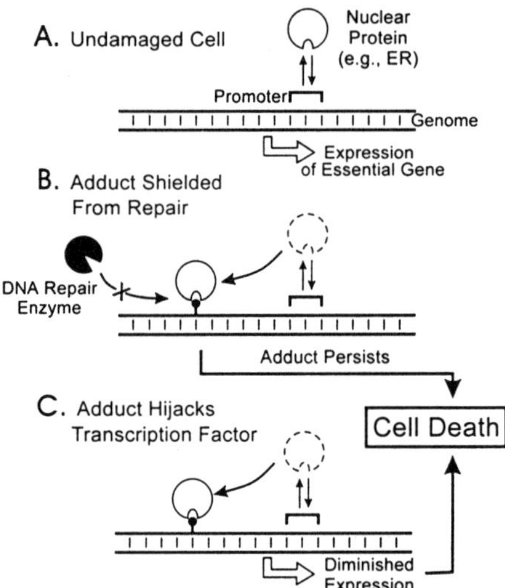

Figure 1. DNA adducts acting as decoy binding sites for nuclear proteins: possible biological consequences. (A) A genetic unit engaged in normal transcription, before DNA damage occurs (DNA damage is symbolized by the lollipop symbol). (B) A DNA lesion forms that attracts a cellular protein. The binding of the cellular protein to the adduct shields the adduct from DNA repair enzymes that would, otherwise, remove the adduct and reduce toxicity. (C) In this scenario, the protein that binds to the adduct is a transcription factor. The adduct lowers the concentration of the transcription factor at a promoter site and thereby diminishes transcription of what might be a critical gene. Figure adapted with permission from Rink et al. (1996).

1.2. Can the Lessons of Cisplatin Be Generalized to Produce Programmable Genotoxins?

The repair shielding, abortive repair and hijacking models described above are being tested in an attempt to understand the mechanism(s) by which cisplatin acts as a selective cytotoxin. While that work continues, it became of interest to determine if a selective cytotoxin could be made *de novo* – i.e., a toxin designed from first principles to work by one or more of the above mechanisms. We chose the repair shielding model as our target mechanism, and the approach taken was to make a toxic DNA adduct linked to a ligand for a protein that is over expressed in the cell type we wished to kill. Aromatic nitrogen mustards were chosen as the adduct forming DNA-interactive agents because they react slowly (they have comparatively low toxicity compared to more reactive alkyl mustards) and they form DNA adducts that are readily repaired. The mustard was linked to a molecular recognition group for a protein that is over expressed in cancer cells. The estrogen receptor (ER) was a logical choice for an over expressed cellular protein for two reasons. First, the ER is over expressed in several cancers and hence there is a practical eventual clinical target. Second, rich arrays of molecules are known that interact with the ER; these compounds were produced in the search for oral contraceptives and as antiestrogenic agents for treatment of hormone responsive cancers.

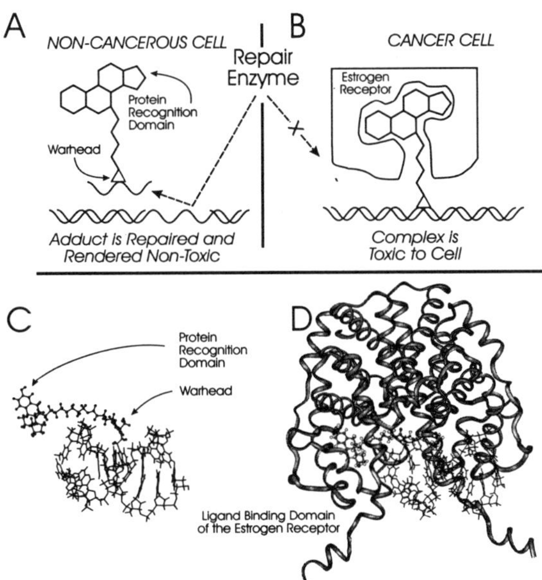

Figure 2. Exploitation of the "Repair Shielding" model. Diseased cells often have elevated levels of certain proteins; e.g., the ER (estrogen receptor) can be considered a tumor specific protein if it is over expressed. (A) A compound is prepared consisting of a chemically reactive group (triangle) that forms DNA adducts. In non-tumor cells, which lack or have low concentrations of the ER, normal repair occurs to eliminate the genotoxic insult. (B) In tumor cells expressing the ER, the ER forms a complex with the "ligand" portion of the DNA adduct (the Protein Recognition Domain). Repair enzymes have hindered access to the adduct, resulting in sluggish repair in tumor cells. (C) Energy minimized molecular model of the estradiol-mustard compound in Figure 3B attached to a DNA 8-mer duplex. Energy minimization was done before docking with the ligand binding domain of the ER (D) Molecular model of the same 8-mer in a complex with the ligand binding domain of the ER (ribbons). The DNA and adduct are embedded inside the receptor. After docking, energy minimization was performed in water.

1.3. Background on the Estrogen Receptor and its Exploitation in Conventional Cancer Therapy

The ER is a 67 KDa protein belonging to a superfamily of ligand-activated transcription factors that regulate hormone responsive genes (Evans, 1988). Regulation of the ER is partly by interaction with the natural ligand, estradiol, and additionally by phosphorylation through the Ras-MAP kinase signaling pathway (Kato et al., 1995) and association with cyclin D1 (Zwijsen et al., 1997). Immunohistochemistry has localized the ER to the cell nucleus of estrogen-responsive cells even in the absence of hormone (King and Greene, 1984; Nash et al.,

1996; Dauvois et al., 1993) (a <u>nuclear</u> location is necessary for the work described here). The ligand-binding domain of the ER has a high affinity for estradiol (K_d = 0.35 nM) as well as for a variety of structurally related small molecules that can function as agonists and/or antagonists to receptor function.

Antagonism of the ER by hormonal agents such as tamoxifen is the basis of endocrine therapy in patients with breast cancers. With breast tumors, for example, approximately half over express the ER (Ferno et al., 1990). Many of these tumors are responsive to antihormonal regimens, which put the tumor cells into growth stasis and eventually can cause cell death (Lerner and Jordan, 1990). The long-term effectiveness of therapy with tamoxifen and similar agents, however, can be complicated by the development of hormone resistant tumor phenotypes (Katzenellenbogen, 1991) and possible carcinogenesis (Jordan, 1995; Osborne et al., 1996). Moreover, the prognosis for receptor negative tumors, if not intercepted early enough by surgery, is far less favorable, and some ER positive tumors are not responsive to antiestrogens. Unfortunately there are a number of other tumors that express the ER, such as epithelial ovarian tumors, that also do not respond well to endocrine therapy. There is a need for more effective and specific agents to treat advanced ovarian cancers as well as hormone refractory breast and endometrial malignancies. In the future compounds of the type described in this review may have value for the treatment of these malignancies since they would utilize the ER to kill cells by a novel mechanism.

2. PROGRAMMABLE AGENTS DESIGNED TO KILL CELLS THAT EXPRESS THE ESTROGEN RECEPTOR

2.1. Design Principles

The salient features of the compounds being designed and their intended mechanism of action are shown in Figure 2. As indicated above, the compounds consist of a DNA reactive warhead linked to a protein recognition domain. The *warhead* functionality covalently links the molecule to DNA, forming potentially lethal damage. The *protein recognition domain* is designed to attract a protein present exclusively or predominantly in tumor cells. As shown in the right portion of Figure 2, when the repair blocking protein is present, an adduct-protein complex forms at the damaged site. The protein blocks access by the repair enzyme, preventing repair of the damaged segment of DNA. Avid binding of nuclear proteins to DNA lesions would be expected to have a negative effect on cellular welfare. Moreover, if such proteins were over expressed in tumor cells, those cells would be expected to be selectively sensitive. In normal cells in which the blocking protein is absent or expressed at lower levels (Fig. 2; left side), the adduct would be unshielded and hence susceptible to rapid repair. Thus, the lethal effects of these bifunctional toxins should be greater in tumor cells than in normal cells, resulting in a favorable therapeutic index.

2.2. Synthesis of Ligand-Warhead Composites that Form DNA Adducts that are Attractive to the Estrogen Receptor

The model system described here focuses upon the ER, which is modulated *in vivo* by endogenous estrogens (Evans, 1988). Derivatives of 2-phenylindole (2PI) interact well with the ER in competition assays with the natural ligand, estradiol (von Angerer et al., 1984). In the early phase of this work, derivatives of 2PI were linked to an aniline mustard moiety whose chemistry and ability to alkylate DNA are well understood (Lawley, 1984). A means to link the 2PI moiety to the mustard was discovered that preserves the ability of either the 2PI or the estradiol ligand to interact with the ER, even after the mustard has formed covalent lesions with DNA. The structures of 2PI-mustards prepared and tested for biological activity are shown in Figure 3 (in this Figure, m and n denote the number of methylene groups present in the linker). Figure 2C shows one mustard-ligand compound attached to a potential binding site in DNA, and Figure 2D shows the same after it was docked with the ligand binding domain of the ER. For details, see Rink et al. (Rink et al., 1996).

The relative affinities of the 2PI-mustards, both before and after reaction with DNA, for the calf uterine estrogen receptor were measured by a competitive binding assay (Korenman, 1970) with 17β-[^{3}H]estradiol. The <u>relative binding affinity (RBA)</u> is the ratio of molar concentrations of test compound to 17β-estradiol required to reduce receptor bound radioactivity by 50%, multiplied by 100. The molecular composition and length of the linker between the aromatic mustard and 2PI groups were varied to modulate the affinity of the compounds for the ER and their ability to react with DNA. 2PI-C3NC3-mustard had little, if any, ability to compete with [^{3}H] estradiol for the ER; RBA=0 (RBA=100 for estradiol). The addition of two methylene groups in 2PI-C5NC3-mustard produced a compound with an RBA of 0.6 compared to estradiol. The isomer 2PI-C6NC2-mustard, in which the secondary amine was displaced from the ER recognition domain by one additional methylene residue, showed a more than ten-fold increase in affinity for the receptor (RBA=7.1). The 2PI(OH)-C6NC2-mustard, which lacks the 4-hydroxy group on the ER recognition domain, only weakly interacted with the ER (RBA=0.1). In recent work, the 2PI ligand was replaced with a 7α-estradiol moiety, producing E2-7α-C6NC2-mustard (Fig. 3B). This compound has a very impressive RBA of 20, giving it a three fold higher affinity for the receptor than the best phenylindole derivative made thus far; this compound is 1/5th as good as estradiol at interacting with the ER.

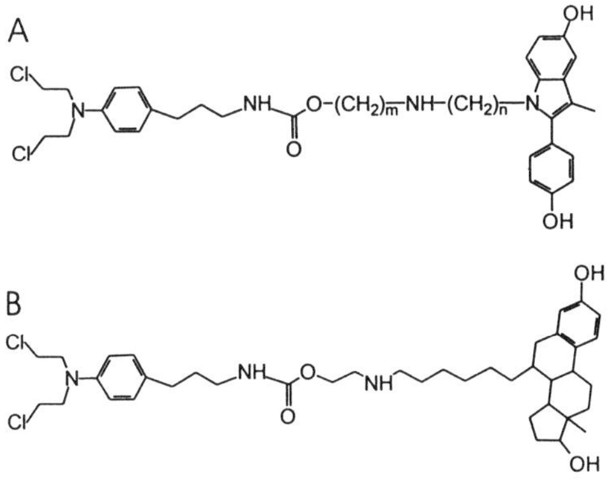

Figure 3. (A) Bifunctional toxins in which an aniline mustard is attached to a 2-phenylindole (2-PI), which acts as a ligand for the ER. In the structure shown, m and n refer to the number of methylene (-CH$_2$)- groups in the linker. In 2PI-C6NC2, for example, m=2 and n=6. (B) A bifunctional toxin in which the aniline mustard is connected by a C6NC2 linker to the 7-position of estradiol.

The observation that subtle changes in molecular architecture can have such a large effect on the RBA underscores the "programmability" of these compounds. It is important to emphasize that much has been learned about the structural features that allow good toxin-ER interactions and ultimately provide for a good selective genotoxin. To summarize briefly what has been found (Rink et al., 1996), initially a collection of molecules was made in which alkyl chains of ever increasing length separated the mustard from the 2PI (the ER ligand); a carbamate was chosen as the connecting group in the linker because of its rigidity and stability (compared to esters and amides) against degradation in biological systems (Cho et al., 1993). As the number of methylene groups was increased in the linker, the interaction with the ER *in vitro* increased, <u>but</u> the apparent ability of the compounds to get into cells and damage DNA dropped. It was suspected that the longer molecules were too hydrophobic, as partly evidenced by low solubility. Much better solubility and much better differential toxicity were obtained when a secondary amino group was incorporated into the linker. This functionality would be positively charged at physiological pH and it is believed that the charge may both bolster solubility and enhance interaction with polyanionic DNA. Finally, the placement of the positive charge appears to be critical (unpublished results); results show that placing the charge closer than five atoms from the 2PI ring is devastating to the ability to interact with the ER and to show selective toxicity.

As indicated above, several of the designed compounds showed favorable RBA's for the ER (e.g., 2PI-C6NC2). The next step was to determine whether those compounds would retain the ability to interact with the ER after the compound had been allowed to react with DNA. A synthetic 16-mer oligonucleotide was prepared and allowed to react with several structurally different mustards (Rink et al., 1996). Competition experiments with [^{3}H]-estradiol indicated that the compounds that interact most favorably with the ER in free solution form DNA adducts that attract the ER with the most impressive avidity (i.e., within an order of magnitude of the affinity of the free compound). With this optimistic result, attention was focused on one modified oligonucleotide, that modified by 2PI-C6NC2, and the relative affinities of its adducts for the ER were determined. It was shown that alkylated single strands are preferred targets over interstrand crosslinks and, by far, over unmodified DNA (Rink et al., 1996).

2.3. Toxins Programmed to Attract the ER Selectively Kill ER Positive Cells

In view of the significant affinity of the adducts for the ER, it was reasonable to expect that the binding event might be tight enough in ER positive cells to anticipate disruption of cellular homeostasis *in vivo*. The biological relevance of these associations can be estimated by a calculated estimate of the amount of ER bound to the DNA adducts in damaged cells in the presence of estradiol. This estimate was obtained by application of Equation 1 below (Long and Crothers, 1995), which describes the competitive binding of two ligands to a peptide,

$$\text{Eq. 1} \qquad \Theta = \frac{1}{2T_t}[K_t + (\frac{K_t}{K_c})C_t + P_t + T_t - \sqrt{[K_t + (\frac{K_t}{K_c})C_t + P_t + T_t]^2 - 4T_tP_t}]$$

where Θ = fractional saturation of DNA adducts by the ER, P_t = concentration of ER protein, T_t = concentration of DNA adducts, C_t = concentration of estradiol, $K_t = K_d$ of DNA adduct for the ER, and $K_c = K_d$ of estradiol for ER. To calculate Θ required estimates of the concentrations of 2PI-C6NC2-mustard adducts and estradiol in damaged cells as well as estimates of their K_d's for the ER. In cells of humans treated with alkylating drugs, DNA adduct concentrations of up to 10^{-6} M have been estimated (Reed et al., 1993). It was assumed that the intracellular concentration of estradiol is close to its K_d for the ER (0.35 nM) (Tora et al., 1989). It was not possible to obtain an exact K_d of the ER for 2PI-C6NC2-mustard DNA adducts because purified ER was unavailable and multiple adducts were present in the damaged DNA. With those constraints, the apparent dissociation constant of the monoadducted single strands for the ER was assumed based upon data (Rink et al., 1996) to be 10^3 fold greater than that of estradiol ($K_d = 0.35$ nM), implying a K_d of 350 nM for the most abundant 2PI-C6NC2-mustard adducts. Finally, the concentration of the ER in the cell nucleus was estimated to be 0.5×10^{-6} M based on its reported abundance in exponentially growing MCF-7 cells (Jakesz et al., 1984). *Using these estimated parameters one can calculate from the equation above that ~20% of the 2PI-C6NC2 adducts would be associated with an ER complex in vivo.* This calculation suggested that our compounds attracted the ER with sufficient binding energy to expect a biological result (repair shielding, hijacking, or both). It is noted, however, that the affinity of the adducts for the ER could be much improved with the application of additional synthetic chemistry.

It was next determined whether the compounds that form adducts most attractive to the ER might also selectively kill ER positive breast cancer cells, as would be predicted on the basis of either the "repair shielding" or transcription factor "hijacking" models. Clonogenic assays were performed. Preliminary experiments revealed that if the mustard moiety in 2PI-C6NC2-mustard were inactivated by hydrolysis, some toxicity to MCF-7 cells was still evident, but significant toxicity was evident only if the compound were present continuously in the growth medium (Rink et al., 1996). This result suggested that the free compound could act as an ER antagonist. Consequently, in the experiments shown, cells were exposed to the mustards for a brief period (2 h) to minimize receptor antagonism by the unreacted or hydrolyzed compound. Under these conditions the hydrolyzed 2PI-C6NC2-mustard was only slightly toxic in the clonogenic assay (~90% survival was observed), ruling out the possibility that the results observed were solely attributable to ER antagonism.

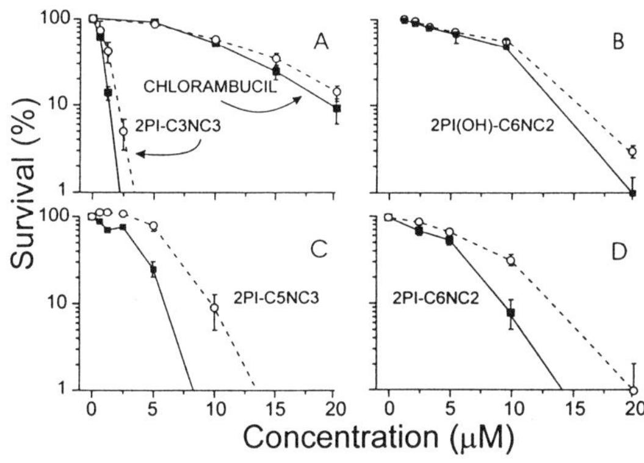

Figure 4. Survival of MCF-7 (ER positive; ■) and MDA-MB231 (ER negative; ○) cells following exposure to 2PI-mustard compounds or chlorambucil for 2 h, as determined by colony formation. (A) and (B) show the toxic effects of chlorambucil, 2PI(OH)-C6NC2 and 2PI-C3NC3, which do not interact with the ER. (C) and (D) show the toxicities of 2PI-C5NC3 and 2PI-C6NC2, which do interact with the ER. For details, see Rink et al. (1996), from which this Figure was adapted, with permission.

The data in Figure 4 indicate that the 2PI-C6NC2- and 2PI-C5NC3-mustards, which interacted well with the ER in competitive binding assays, were more toxic to the ER-expressing MCF-7 cells than to MDA-MB231 cells (panels C and D). The latter cells do not express the receptor. Compounds with minimal or no ability to interact with the ER showed roughly equal toxicities in the two cell lines (panels A and B); these three control compounds were chlorambucil (this is an aromatic nitrogen mustard that lacks a molecular recognition domain for the ER), the 2PI(OH)-C6NC2-mustard (this control compound lacks one of the two -OH groups on the "ligand" that are needed for tight binding to the ER) and 2PI-C3NC3-mustard. The results with the three control compounds made it unlikely that the differences in susceptibility of MCF-7 and MDA-MB231 cells to the 2PI-C6NC2- and 2PI-C5NC3-mustards are due to variations in detoxification of the mustards or to inherent differences in the capacity to repair DNA damage. The similar toxicities of chlorambucil, 2PI-C3NC3 and 2PI(OH)-C6NC2 in the ER+ and ER- cell lines (panels A and B) provide support for the role of the ER in the observed differential toxicity with 2PI-C6NC2 and 2PI-C5NC3 (panels C and D). The former three compounds have no or very low affinities for the ER, although they are very similar in structure to molecules that show selective toxicity in MCF-7 cells. Additional evidence for the role of the ER as a mediator of toxicity comes from a companion experiment in which the addition of estradiol at 10^{-6} M was shown to ablate the toxicity response of MCF-7 cells to the 2PI-C6NC2 mustard (data not shown). A reasonable explanation of this result is that the exogenously applied estrogen occupied the ER, making the ER less available to act as a repair shielding protein.

The toxicity differentials in panels C and D (Fig. 4) are statistically significant (Rink et al., 1996) but modest. Such differentials are absent, however, for typical alkylating drugs so it is significant that we have been able to introduce some specificity that is lacking in conventional alkylating agents. Moreover, several other ER positive and negative cell lines were obtained (the ER positive lines BT20T and T47D and the negative line HS578T) and, in these cells, the same differential toxicity in favor of killing ER positive lines has been observed. Finally, and perhaps most significantly, recent efforts have produced a compound that binds to the ER more avidly and, as expected, this compound also displays a larger toxicity differential than 2PI-C6NC2. The new compound (Fig. 3) has an aromatic mustard and C6NC2 linker but the phenylindole was replaced with estradiol. As indicated above, this compound, E2-7α-C6NC2 mustard, showed a three fold higher affinity for the ER than 2PI-C6NC2. This compound is proportionately more toxic to ER+ cells than to ER- cells (data not shown).

Several mechanisms could explain the selective toxicity of the phenylindole-mustards and the estradiol-mustard for ER positive cancer cells. A first model derives from studies in which alkylating agents and platinum derivatives were tethered to ligands for steroid receptors with the aim of using the affinity of the compounds for the receptor to facilitate delivery of the drugs to receptor positive cells (Roth et al., 1995; Lam et al., 1987; Leclercq et al., 1983). Most such molecules have an intentionally hydrolyzable internal linkage. It was anticipated that those molecules, once delivered to ER positive cells, would jettison the bulky ER ligand, liberating the alkylating group to damage DNA and kill the cell. This strategy for the selective delivery of alkylating drugs is fundamentally different from the present system in which the objective was the selective retention of the lesion within DNA of target cells.   Selective delivery is unlikely to be the mechanism by which the present molecules preferentially kill ER positive cells on the basis of preliminary data showing that the levels of interstrand crosslinks of 2PI-C6NC2-mustard in ER positive (MCF-7) and ER negative (MDA-MB231) cells are identical shortly after dosing (R. Croy, unpublished data). A second and more plausible model to explain the selective toxicity toward ER positive cells is hindered repair of mustard adducts *in vivo* brought on by the bulky ER when it is in a complex with the phenylindole portion of the DNA adducts. There are several examples of proteins that shield structurally flawed sites in DNA from repair and evidence exists that such shielding can increase cytotoxicity (Brown et al., 1993; Cheng et al., 1978; Fox et al., 1994; Devchand et al., 1993; Satoh and Lindahl, 1992). The shielding model would predict that the adducts of phenylindole-mustards (or estradiol-mustards) that interact well with the receptor after reaction with DNA would be selectively retained in ER positive cells. This result has been observed in studies using alkaline elution to monitor the kinetics of removal of interstrand crosslinks over a 48 h time period; the results obtained reveal an almost two-fold slower rate of removal of 2PI-C6NC2-mustard adducts in MCF-7 cells as compared to MDA-MB231 cells. Based on these data, we believe that the "repair shielding" model contributes to the differential toxicity observed in Figure 4.

A third model could explain the selective toxicity observed toward ER positive cells. As presented in Figure 1C, the adducts of the phenylindole- or estradiol-mustards could form decoy binding sites for the ER and thus compromise the ability of the ER to drive the transcription of necessary genes. While our data are consistent with the repair shielding model, they do not exclude "hijacking" from being operative at some level.

3. LONG RANGE THERAPEUTIC PROSPECTS

The inhibition of DNA repair <u>selectively</u> in tumor cells has not been the focus of many drug development efforts. One logical question is to ask how large a toxicity differential one could achieve between target and non-target cells? A reasonable estimate of the limiting value is at least one and more likely two orders of magnitude, which is roughly the differential in toxicity observed between DNA repair proficient and deficient cells. In the clinic, 2-5 fold differences in toxicity can make the difference between success and failure for a drug treatment regimen, so the efforts to develop DNA repair inhibitors are well spent. A second question is whether the ER is the only target to which the technology described above could be applied? In the short term one could take advantage of the known over expression of proteins for which ligands are known (e.g.,the androgen receptor is abundant in prostate cancer cells and several ligands for it are known). In the longer term, the opportunities are vast with the advent of combinatorial chemistry, which provides the means to generate 10^6-10^{13} different molecules in a population and to screen them efficiently against biological targets. Combinatorial chemistry in concert with the principles described herein could be applied to exploit, to give one example, the over abundance of mutant forms of p53 in half of all human tumors. With the application of combinatorial chemistry, one could easily envision generalizing the technology in this paper to address a multitude of different cancer, and other disease, targets.

Acknowledgments. This work was funded by NIH grants CA86061, CA77743 and ES07020.

REFERENCES

Brown, S.J., Kellett, P.J., and Lippard, S.J., 1993, Ixr1, a yeast protein that binds to platinated DNA and

confers sensitivity to cisplatin, *Science* **261**:603-605.

Bruhn, S.L., Pil, P.M., Essigmann, J.M., Housman, D.E., and Lippard, S.J., 1992, Isolation and

characterization of human cDNA clones encoding a high mobility group box protein that recognizes

structural distortions to DNA caused by binding of the anticancer agent cisplatin, *Proc. Natl. Acad.*

Sci. USA **89**:2307-2311.

Bruhn, S.L., Toney, J.H., and Lippard, S.J., 1990, Biological processing of DNA modified by platinum

compounds, *Prog. Inorg. Chem.* **38**:477-516.

Cheng, W.S., Tarone, R.E., Andrews, A.D., Whang-Peng, J.S., and Robbins, J.H., 1978, Ultraviolet light-

induced sister chromatid exchanges in xeroderma pigmentosum and in Cockayne's syndrome

lymphocyte cell lines. *Cancer Res.* **38**:1601-1609.

Cho, C.Y., Moran, E.J., Cherry, S.R., Stephans, J.C., Fodor, S.P.A., Adams, C.L., Sundaram, A., Jacobs,

J.W., and Schultz, P.G., 1993, An unnatural biopolymer, *Science* **261**:1303-1305.

Chu, G., 1994, Cellular responses to cisplatin. *J. Biol. Chem.* **269**:787-790.

Dauvois, S., White, R., and Parker, M.G., 1993, The antiestrogen ICI 182780 disrupts estrogen receptor

nucleocytoplasmic shuttling, *Journal of Cell Science* **106**, 1377-1388.

Devchand, P.R., McGhee, J.D., and van de Sande, J.H., 1993, Uracil-DNA glycosylase as a probe for protein-DNA interactions, *Nucleic Acids Res.* **21**:3437-3443.

Donahue, B.A., Augot, M., Bellon, S.F., Treiber, D.K., Toney, J.H., Lippard, S.J., and Essigmann, J.M., 1990, Characterization of a DNA damage-recognition protein from mammalian cells that binds specifically to intrastrand d(GpG) and d(ApG) DNA adducts of the anticancer drug cisplatin, *Biochemistry* **29**:5872-5880.

Duckett, D.R., Drummond, J.T., Murchie, A.I., Reardon, J.T., Sancar, A., Lilley, D.M., and Modrich, P., 1996, Human MutSalpha recognizes damaged DNA base pairs containing O6- methylguanine, O4-methylthymine, or the cisplatin-d(GpG) adduct, *Proc. Natl. Acad. Sci. USA* **93**:6443-6447.

Evans, R.M., 1988, The steroid and thyroid hormone receptor superfamily, *Science* **240**:889-895.

Ferno, M., Johansson, B.U., Olsson, N.H., Ryden, S., and Sellberg, G.. 1990, Estrogen and progesterone receptor analyses in more than 4000 human breast cancer samples, *Acta. Oncol.* **29**:129-135.

Feuer, E.J., Brown, L.M., and Kaplan, R.S., 1993, In: Miller, B.A., Ries, L.A.G., Hankey, B.F., Kosary, C.L., Harras, A., Devesa, S.S., and Edwards, B.K., editors, *SEER Cancer Statistics Review: 1973-1990*, National Cancer Institute: Bethesda, MD. p XXIV.1-XXIV.13.

Fox, M.E., Feldman, B.J., and Chu, G., 1994, A novel role for DNA photolyase: binding to DNA dmamged by drugs is associated with enhanced toxicity in *Saccharomyces cerevisiae*, *Mol. Cell. Biol.* **14**:8071-8077.

Goldmacher, V.S., Cuzick Jr., R.A., and Thilly, W.G., 1986, Isolation and partial characterization of human cell mutants differing in sensitivity to killing and mutation by methylnitrosourea and *N*-methyl-*N'*-nitro-*N*-nitrosoguanidine, *J. Biol. Chem.* **261**:12462-12471.

Huang, J.-C., Zamble, D.B., Reardon, J.T., Lippard, S.J., and Sancar, A., 1994, HMG-domain proteins specifically inhibit the repair of the major DNA adduct of the anticancer drug cisplatin by human excision nuclease. *Proc. Natl. Acad. Sci. USA* **91**:10394-10398.

Jakesz, R., Smith, C.A., Aitken, S., Huff, K., Schuette, W., Shackney, S., and Lippman, M., 1984, Influence of cell proliferation and cell cycle phase on expression of estrogen receptor in MCF-7 breast cancer cells. *Cancer Res* **44**:619-625.

Jordan, P. and Carmo-Fonseca, M., 1998, Cisplatin inhibits synthesis of ribosomal RNA in vivo, *Nucleic Acids Res.* **26**:2831-2836.

Jordan, V.C., 1995, Tamoxifen and tumorigenicity: A predictable concern, *J. Natl. Cancer Inst.* **87**(9): 623-626.

Karran, P., and Marinus, M.G., 1982, Mismatch correction at O^6-methylguanine residues in *E. coli* DNA, *Nature* **296**:868-869.

Kato, S., Endoh, H., Masuhiro, Y., Kitamoto, T., Uchiyama, S., Sasaki, H., Masushige, S., Gotoh, Y., Nishida, E., Kawashima, H., Metzger, D., and Chambon, P., 1995, Activation of the estrogen receptor through phosphorylation by mitogen-activated protein kinase, *Science* **270**: 1491-1494.

Katzenellenbogen, B.S., 1991, Antiestrogen resistance: Mechanisms by which breast cancer cells undermine the effectiveness of endocrine therapy, *J. Natl. Cancer Inst.* **83**(20):1434-1435.

King, W.J. and Greene, G.L., 1984, Monoclonal antibodies localize oestrogen receptor in the nuclei of target cells, *Nature* **307**:745-747.

Korenman, S.G., 1970, Relation Between Estrogen Inhibitory Activity and Binding to Cytosol of Rabbit and Human Uterus, *Endocrinology* **87**:1119-1123.

Lam, H.-Y.P., Ng, P.K.T., Goldenberg, G.J., and Wong, C.-M., 1987, Estrogen Receptor-Binding Affinity and Cytotoxic Activity of Three New Estrogen-Nitrosourea Conjugates in Human Breast Cancer Cell Lines In Vitro, *Cancer Treat. Rep.* **71**:901-906.

Lawley, P.D., 1984, Carcinogenesis by alkylating agents, In: Searle, C.E., editor, *Chemical Carcinogenesis, A.C.S. Monograph 182,* Washington, DC: American Chemical Society. p 325-484.

Leclercq, G., Devleeschouwer, N., and Heuson, J.C., 1983, Guide-lines in the design of new antiestrogens and cytotoxic-linked estrogens for the treatment of breast cancer, *J. Steroid Biochem. Molec. Biol.* **19**:75-85.

Lerner, L.J. and Jordan, V.C., 1990, Development of antiestrogens and their use in breast cancer: Eighth Cain Memorial Award Lecture, *Cancer Res.* **50**:4177-4189.

Loehrer, P.J. and Einhorn, L.H., 1984, Cisplatin, *Ann. Int. Med.* **100**:704-713.

Long, K.S., and Crothers, D.M., 1995, Interaction of human immunodeficiency virus type 1 tat-derived peptides with TAR RNA, *Biochemistry* **34**:8885-8895.

Masters, J.R.W.,Thomas, R., Hall, A.G., Hogarth, L., Matheson, E.C., Cattan, A.R., and Lohrer, H., 1996, Sensitivity of testis tumour cells to chemotherapeutic drugs: Role of detoxifying pathways, *Eur. J. Cancer* **32A**:1248-1253.

McA'Nulty, M.M., and Lippard, S.J., 1996, The HMG-domain protein Ixr1 blocks excision repair of cisplatin-DNA adducts in yeast, *Mutat. Res.* **362**:75-86.

Mello, J.A., Acharya, S., Fishel, R., and Essigmann, J.M., 1996, The mismatch-repair p rotein hMSH2 binds selectively to DNA adducts of the anticancer drug cisplatin, *Chemistry and Biology* **3**:579-589.

Nash, H.M., Bruner, S.D., Schaeer, O.D,. Kawate, T., Addona, T.A., Spooner, E., Lane, W.S., and Verdine, G.L., 1996, Cloning of a yeast 8-oxoguanine DNA glycosylase reveals the existence of a base-excision DNA-repair protein superfamily, *Curr. Biol.* **6**:968-980.

Osborne, M.R., Hewer, A., Hardcastle, I.R., Carmichael, P.L., and Phillips, D.H., 1996, Identification of the major tamoxifen-deoxyguanosine adduct formed in the liver DNA of rats treated with tamoxifen, *Cancer Res.* **56**:66-71.

Pil, P.M. and Lippard, S.J., 1992, Specific binding of chromosomal protein HMG1 to DNA damaged by the anticancer drug cisplatin, *Science* **256**:234-237.

Reed, E., Parker, R.J., Gill, I., Bicher, A., Dabholkar, M., Vionnet, J.A., Bostick-Bruton, F., Tarone, R., and Muggia, F.M.,1993, Platinum-DNA adduct in leukocyte DNA of a cohort of 49 patients with 24 different types of malignancies, *Cancer Res* **53**:3694-3699.

Rink, S.M., Yarema, K.J., Paige, L.A., Tadayoni, M., Solomon, M., Essigmann, J.M., and Croy, R.G., 1996, Synthesis and biological activity of DNA damaging agents that form decoy binding sites for the estrogen receptor, *Proc. Natl. Acad. Sci. USA* **93**:15063-15068.

Roth, T., Tang, W., and Eisenbrand, G., 1995, Synthesis of novel androgen-linked phosporamide mustard prodrugs and growth-inhibitory activity in human breast cancer cells, *Anticancer Drug Des.* **10**:655-666.

Satoh, M.S. and Lindahl, T., 1992, Role of poly(ADP-ribose) formation in DNA repair, *Nature* **356**:356-358.

Toney, J.H., Donahue, B.A., Kellett, P.J., Bruhn, S.L., Essigmann, J.M., and Lippard, S.J., 1989, Isolation of cDNAs encoding a human protein that binds selectively to DNA modified by the anticancer drug cis-diamminedichloroplatinum(II), *Proc. Natl. Acad. Sci. USA* **86**:8328-8332.

Tora, L,, Mullick, A., Metzger, D., Ponglikitmongkol, M., Park, I., and Chambon, P., 1989, The cloned human oestrogen receptor contains a mutation which alters its hormone binding properties, *EMBO J.* **8**:1981-1986.

Treiber, D.K., Zhai, X., Jantzen, H.-M., and Essigmann, J.M., 1994, Cisplatin-DNA adducts are molecular decoys for the ribosomal RNA transcription factor hUBF (human upstream binding factor), *Proc. Natl. Acad. Sci. USA* **91**:5672-5676.

Vichi, P., Coin, F., Renaud, J.P., Vermeulen, W., Hoeijmakers, J.H., Moras, D., and Egly, J.M., 1997, Cisplatin- and UV-damaged DNA lure the basal transcription factor TFIID/TBP, *EMBO J.* **16**:7444-7456.

von Angerer, E., Prekajac, J., and Strohmeier, J., 1984, 2-Phenylindoles. Relationship between structure, estrogen receptor affinity and mammary tumor inhibiting activity in the rat, *J. Med. Chem.* **27**:1439-1447.

Zdraveski, Z., Mello, J.A., Marinus, M.G., and Essigmann, J.M., 2000, Multiple pathways of recombination define cellular responses to cisplatin, *Chemistry & Biology* **7**:39-50.

Zhai, X., Beckmann, H., Jantzen, H.M., and Essigmann, J.M., 1998, Cisplatin-DNA adducts inhibit ribosomal RNA synthesis by hijacking the transcription factor human upstream binding factor, *Biochemistry* **37**:16307-16315.

Zwijsen, R.M.L., Wientjens, E., Klompmaker, R., van der Sman, J., Bernards, R., and Michalides, R.J.A.M., 1997, CDK-Independent activation of estrogen receptor by cyclin D1, *Cell* **88**:405-415.

THE POTENTIAL ROLE OF TOPOISOMERASE II
INHIBITION IN HYDROQUINONE-INDUCED ALTERATIONS
IN THE MATURATION OF MOUSE MYELOBLASTS

Matthew J. Hoffmann[1], David D. Kim[1], Mohammed G.K. Akbar[1],
George F. Kalf[2], and Robert Snyder[1]

[1] Department of Toxicology
Rutgers University
Piscataway, NJ 08854-8020
[2] Jefferson Medical College
Thomas Jefferson University
Philadelphia, PA 19107-6799

INTRODUCTION

The mechanism by which exposure to the industrial solvent benzene (BZ) can result in aplastic anemia or leukemia (see Snyder and Kalf, 1994 for review) has been under intensive investigation. Topoisomerase (topo) II inhibitors are used clinically as cancer chemotherapeutic agents and can also lead to secondary AML (see Felix, 1998; Leone et al., 1999 for review). Hydroquinone (HQ), a hematotoxic metabolite of BZ, has been shown to inhibit purified human topo II (Hutt and Kalf, 1996; Frantz et al., 1996). It has therefore been suggested that some of the hematotoxic effects of BZ may be mediated by the ability of some of its metabolites to inhibit topo II activity (Whysner, 2000). In the studies presented here a normal, IL-3 dependent mouse myeloblast cell line (32D.3 (G)) was used to determine whether the hematotoxic effects of BZ may result from inhibition of topo II by HQ. The 32D.3 (G) myeloblasts undergo differentiation and maturation to neutrophils upon treatment with the physiological inducer, granulocyte colony-stimulating factor (G-CSF) (Metcalf, 1985; Valtieri et al., 1987). In this system HQ treatment inhibits apoptosis, induces differentiation of myeloblasts to the myelocyte stage, and blocks maturation to the neutrophil (Hazel et al., 1995), a series of events similar to those observed in AML, which is characterized by clonal expansion of immature granulocytes (Lowenberg and Delwel, 1991). There is evidence linking topo II activity to differentiation and maturation in the HL-60, human leukemia cell line. In this system 3-nitrobenzothiazolo[3,2-a]quinolinium, an inhibitor of topo II activity, has been shown to induce incomplete myeloid differentiation (Baez and Sepulveda, 1992). As maturation progresses in HL-60 cells topo II levels steadily decrease until becoming undetectable in

Biological Reactive Intermediates VI, Edited by Dansette *et al.*
Kluwer Academic / Plenum Publishers, 2001

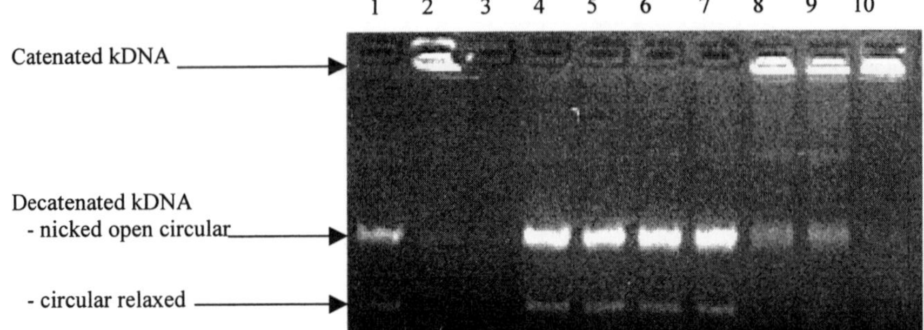

Figure 1. Agarose gel showing substrate (catenated kDNA) and product (decatenated kDNA) of topoisomerase II activity assay using nuclear lysates from 32D myeloblasts. 32D myeloblasts were treated with HQ for 30 minutes then nuclear lysates prepared and assayed for topoisomerase II activity. Gel shows decatenated kDNA marker (Lane 1), catenated kDNA marker (Lane 2), empty (Lane 3), and 0, 10, 20, 30, 40, 50, and 60 μM HQ (Lanes 4, 5, 6, 7, 8, 9, and 10, respectively).

mature granulocytes (Kaufmann et al., 1991). The objectives of our studies were first, to determine if topo II activity can be inhibited in HQ-treated 32D myeloblasts and second, to determine if etoposide, a known inhibitor of topo II, can mimic the effects of HQ on the differentiation and maturation of the 32D myeloblasts.

METHODS

Cell Culture

32D.3 (G) myeloblasts were maintained in Iscove's modified Dulbecco's medium (IMDM) containing 2 mM glutamine, 10% fetal bovine serum (FBS), and 3 U/ml recombinant murine IL-3 at 37°C, 5% CO_2. Prior to all treatments cells were washed one time in PBS and resuspended in IMDM containing 2 mM glutamine, 10% FBS, and either 0.3 U/ml IL-3 or 500 U/ml of G-CSF, at a concentration of $2.5X10^5$ cells/ml. For topo II assays cells were treated immediately with various concentrations of HQ, BQ, or etoposide for up to 24 hours, in some cases 0.003 units of peroxidase were added to the media. For differentiation experiments 4 μM HQ or 0.05 μM etoposide was added to the media 24 hours after cells were resuspended in media and cells kept in culture for 6 days.

Assay of Topoisomerase II Activity

A sample containing 1-2.5 X 10^6 cells was pelleted and washed with 1 ml of PBS. Nuclear extracts were prepared by resuspending cells in 400 μl of ice-cold buffer A [10 mM Hepes, pH 7.9, containing 10 mM KCl, 0.1 mM ethylenediaminetetraacetate (EDTA), 0.1 mM ethylene glycol-bis(β-aminoethyl ether) tetraacetate (EGTA), 1 mM dithiothreitol (DTT), and 0.5 mM phenylmethylsulfonyl fluoride (PMSF)]. Cells were incubated on ice for 15 minutes. NP-40 (25 μl of 20%) was added and cells were vortexed. The lysate was centrifuged and the nuclear pellet resuspended in 50 μl of cold buffer C (20 mM Hepes, pH 7.9, containing 0.4 M NaCl, 1 mM EDTA, 1 mM EGTA, 1 mM DTT, and 1 mM PMSF). Topo II activity was assayed using a kit from TopoGen, Inc. (Columbus, OH) which measures the release of open and/or nicked open circular kDNA from catenated kDNA.

Table 1. Inhibition of topoisomerase II activity by etoposide and benzene metabolites

Treatment	Etoposide	HQ	HQ + peroxidase	BQ	BQ + peroxidase
Inhibitory Dose[a]	50[b]	40-60	20-30	2-5	1-5

[a]Dose needed to completely inhibit topoisomerase II activity, experiments were performed three times in duplicate and the range of doses is presented.
[b]Values are expressed in μM.

Assay of Myeloblast Differentiation and Maturation

Six days after treatment cell number and viability were determined. A sample of 2.5 X 10^6 cells were fixed to microscope slides and stained with May-Grunwald and Giemsa stains. Cellular differentiation and maturation were measured by determining the morphology of 200 cells per slide using a light microscope. Data were compared using the students t-test and a $p \leq 0.01$ was considered significant.

RESULTS

Topoisomerase II Inhibition

Figure 1 shows the results of a typical topo II inhibition experiment following treatment of myeloblasts with HQ. Complete inhibition is seen only at the 60 μM HQ dose. Nuclear extracts isolated from treated and untreated 32D myeloblasts were assayed for topo II activity. Table 1 shows the doses needed to completely inhibit topo II activity. The effects of adding etoposide, HQ, and BQ alone or in combination with peroxidase were evaluated. BQ was more potent than HQ and the addition of peroxidase increased the potency of both HQ and BQ.

Differentiation and Maturation of 32D.3 (G) Myeloblasts

Table 2 shows the effects of 4 μM HQ and 0.05 μM etoposide on the differentiation and maturation of 32D myeloblasts in the presence of either IL-3 or G-CSF. In the presence of IL-3 HQ induced differentiation and minimal maturation to the neutrophil, while etoposide did not induce differentiation or maturation. However, in the presence of G-CSF both HQ and etoposide were capable of blocking terminal maturation to the neutrophil.

DISCUSSION

The ability of HQ and BQ to inhibit the activity of purified topo II has previously been shown (Hutt and Kalf, 1996; Frantz et al., 1996). The goal of our studies was to link this inhibition of topo II activity to a biological effect using the 32D myeloblast cell system. Hazel et al. (1995) demonstrated that HQ could induce differentiation in the 32D myeloblast cell system by binding to the LTD_4 receptor. It was also observed that maturation to the neutrophil was blocked in the HQ-treated myeloblasts. This block in maturation can not be readily explained by the binding of HQ to the LTD_4 receptor. The data presented in Table 2 show that while etoposide can not induce differentiation it can block maturation of myelocytes. Table 1 shows that HQ, most likely acting through BQ, and etoposide can inhibit topo II activity in 32D myeloblasts. The doses of both etoposide and HQ needed to completely inhibit topo II activity were much greater than the doses

Table 2. Differentiation and maturation of 32D cells (%)

Treatment	Differentiation	Maturation
IL-3 control	4.3 ± 0.9^a	2.4 ± 0.6
IL-3 + HQ	68.8 ± 4.9^b	13.7 ± 3.6^b
IL-3 + Etoposide	2.5 ± 1.1	0.9 ± 0.7
G-CSF control	70.6 ± 7.2	44.6 ± 2.7
G-CSF + HQ	81.6 ± 7.6	19.4 ± 3.6^c
G-CSF + Etoposide	71.2 ± 6.2	22.6 ± 3.0^c

[a]Values are expressed as mean ± standard deviation, n=3.
[b]Differs significantly from IL-3 control, $p \leq 0.01$.
[c]Differs significantly from G-CSF control, $p \leq 0.01$.

needed to block maturation. In fact, doses needed to completely inhibit topo II activity also caused a significant amount of cell death (data not shown).

These data show that an inhibitor of topo II activity, etoposide, can mimic some of the effects of the hematotoxic BZ metabolite, HQ. However, at the doses of HQ and etoposide which caused a block in maturation no inhibition of topo II activity was detected. This may in part be due to the limited sensitivity of the assay used. The data also suggest that doses of HQ and etoposide which completely inhibition of topo II activity disrupt cell function too severely to allow any differentiation or maturation to occur.

ACKNOWLEDGMENTS

This work was supported by a Rutgers Undergraduate Research Fellowship, grants from Mobil Oil Company and American Petroleum Institute, and USPHS grant ESO5022.

REFERENCES

Baez, A. and Sepulveda, J. (1992). Myeloid differentiation of HL-60 cells induced by anti-tumor drug 3-nitrobenzothiazolo[3,2-a]quinolinium. *Leuk. Res.* **16**, 363-370.

Felix, C.A. (1998). Secondary leukemia induced by topoisomerase-targeted drugs. *Biochim. Biophys.* **1400**, 233-255.

Frantz, C.E., Chen, H., and Eastmond, D.A. (1996). Inhibition of human topoisomerase II in vitro by bioactive benzene metabolites. *Environ. Health Perspect.* **104**, 1319-1323.

Hazel, B.A., O'Conner, A., Nicolescu, R., and Kalf, G.F. (1995). Benzene and its metabolite, hydroquinone, induce granulocytic differentiation in myeloblasts by interacting with cellular signaling pathways activated by granulocyte colony-stimulating factor. *Stem Cells* **13**, 295-310.

Hutt, A.M. and Kalf, G.F. (1996). Inhibition of human DNA topoisomerase II by hydroquinone and p-benzoquinone, reactive metabolites of benzene. *Environ. Health Perspect.* **104**, 1261-1269.

Kaufmann, S.H., McLaughlin, S.J., Kastan, M.B., Liu, L.F., Karp, J.E., and Burke, P.J. (1991). Topoisomerase II levels during granulocytic maturation in vitro and in vivo. *Can. Res.* **51**, 3534-3543.

Leone, G., Mele, L., Pulsoni, A., Equitani, F. and Pagano, L. (1999). The incidence of secondary leukemias. *Haematologica* **84**, 937-945.

Lowenberg, B. and Delwel, F.R. (1991). The pathology of human acute myeloid leukemia, In: Hematology, Basic Principles and Practice, Hoffman, R., Benz, E.J., Shattil, S.J., Furie, B., and Cohen, H.J. (eds.) Churchill Livingstone, New York, pp. 708-715.

Metcalf, D. (1985). Multi-CSF-dependent colony formation by cells of a murine hematopoietic cell line: specificity and action of multi-CSF. *Blood* **65**, 357-362.

Snyder, R. and Kalf, G. F. (1994). A perspective on benzene leukemogenesis. *Crit. Rev. Toxicol.* **24**, 177-209.

Valtieri, M., Tweardy, D.J., Caracciolo, D., Johnson, K., Mavilio, F., Altmann, S., Santoli, D., and Rovera, G. (1987). Cytokine-dependent granulocytic differentiation. Regulation of proliferative and differentiative responses in a murine progenitor cell line. *J. Immunol.* **138**, 3829-3835.

Whysner, J. (2000). Benzene-induced genotoxicity. *J Toxicol. Environ. Health* (in press).

DESIGN AND CHARACTERIZATION OF A NOVEL « FAMILY-SHUFFLING » TECHNOLOGY ADAPTED TO MEMBRANE ENZYME : APPLICATION TO P450s INVOLVED IN XENOBIOTIC METABOLISM

Valérie Abécassis, Denis Pompon and Gilles Truan

Centre de Génétique Moléculaire, CNRS
Avenue de la Terrasse
91198 Gif-sur-Yvette, France

INTRODUCTION

Cytochrome P450 functional diversity and their predominant role in drug and pollutant metabolism and toxicity[1] makes these enzymes particularly suitable for the design of new catalysts as well as for structure-function analysis[2]. Combinatorial molecular evolution (CME) is a powerful approach used for tuning protein functions[3;4] and for investigation of biochemical mechanisms driving substrate recognition[5] or catalysis[6]. Family-shuffling has proved to accelerate the evolution process[7]. A low content of mosaic structures was frequently reported in libraries constructed using DNase I fragmentation[8]. We designed a new strategy for family shuffling in yeast expression vectors. This procedure takes advantage of the association between *in vitro*[9] and *in vivo*[10] recombination mechanisms to build a high complexity library containing low levels of parental structures. The use of engineered yeast strains for expression of membrane proteins into an optimized redox environment[11] also allows efficient *in vivo* bioconversion. The model used is human *CYP1A1* and *CYP1A2* which share 71% identity and have distinct, while overlapping, substrate specificities.

RESULTS

Construction of yeast expression libraries by family shuffling

The system, described in figure 1, associated a redesigned PCR-based DNA shuffling step to a secondary shuffling step by *in vivo* recombination in yeast. The latter step was also used as a cloning tool. Overall, this shuffling strategy allowed direct expression and functional selection into an eucaryotic cell and did not require intermediate cloning steps into *E. coli*.

DNase I catalyzed double-strand breaks of the full expression vector were realized leading to low size DNA fragments (figure 1A). Fragments from p1A1/V60 and p1A2/V60 were mixed in equal proportion and submitted to a "progressive hybridization" PCR program to force low homology recombination. A smear including high molecular weight DNAs was formed from fragments resulting of plasmid digestion by the three DNase I concentrations. A second PCR step, involving primers located on the cDNA flanking regions, was performed and amplified bands with the expected sizes were obtained (1900 bp). Products obtained in the experiment with the lower DNase I concentration were not used for library construction because of potential contamination by undigested parental structures. The band shown in lane 5, panel C was purified and used as such to cotransform *S. cerevisiae* with linearized pYeDP60, promoting *in vivo* recombination events between PCR products and cloning into the yeast vector. The selection of transformed cells for uracil prototrophy was based only on vector recircularization events following one or multiple recombination. Typical experiments generated approximately 10,000 clones.

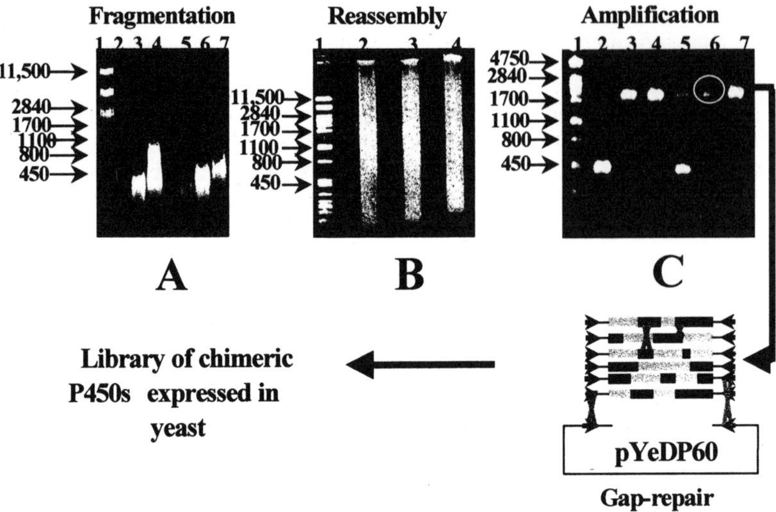

Figure 1. Principle of the library construction. Plasmidic DNA was subjected to DNase I digestion and fragments were separated on a 1% agarose gel. **A**: lane 1, DNA ladder; lanes 2, 3, 4 and 5, 6, 7 correspond to DNase I treated p1A1/V60 and p1A2/V60 respectively with decreasing concentration of DNase I. **B**: reassembly reaction. Lane 1, DNA ladder; lanes 2, 3 and 4 correspond to reassembly reactions between fragmented p1A1/V60 and p1A2/V60 mixing the reactions from lanes 2 and 5, 3 and 6, 4 and 7 respectively. **C**: amplification reaction. Lane 0, DNA ladder; lanes 1, 2 and 3 correspond to the amplification with full length plasmid pYeDP60, p1A1/V60 and p1A2/V60 respectively; lanes 4, 5 and 6 correspond to the amplification with previously reassembled DNA as a matrix (lanes B2, B3, B4 respectively).

Library characterization by DNA macro-arrays hybridization

Plasmidic DNA was prepared from the whole yeast library and used to transform *E. coli*. This step allowed segregation of individual plasmids initially present as a mixed population in yeast colonies. A matrix was built on a 384-well microtiter plate for primary structure analysis. The matrix included 378 randomly selected *E. coli* clones from the library, and the remaining wells included control plasmids (either p1A1/V60 or p1A2/V60). The six probes (table1) were chosen to match alternatively the two parental sequences in regions of low sequence similarity. Each probe was ^{32}P labeled and used for hybridization in stringent conditions.

Table 1 : 5' respective positions and sequences of the six probes used to characterize matrices.

Probes	Position
Probe 1 (24 pb)	3 (*CYP 1A2*)
Probe 2 (31 pb)	612 (*CYP 1A1*)
Probe 3 (24 pb)	683 (*CYP 1A2*)
Probe 4 (22 pb)	879 (*CYP 1A1*)
Probe 5 (21 pb)	1377 *(CYP 1A2)*
Probe 6 (24 pb)	1513 (*CYP 1A1*)

The calculated frequency of mosaic probe patterns in the library was 87 %. This value is an underestimate of the true proportion of mosaic structures as only 13 % of the sequence was probed. It can be estimated that this value likely corresponded to more than 95 % of mosaic structures. The probability of presence of each parental sequence at each of the six probed positions was calculated and found quite homogeneous (0.56 ± 0.02 for the 1A2 sequence) for all analyzed sequence segments and stands within the expected statistical error. This indicated that a very homogenous shuffling has been achieved.

To confirm DNA macro-array results, five clones were randomly selected without any functional selection criteria and five others clones were randomly picked among clones functionally selected for naphthalene hydroxylation. The ten open reading frames were fully sequenced and their structures and point mutations analyzed (table 2).

Table 2: Sequence analysis.

Average number of parental fragments composing each mosaic structure	5.4 ± 2.2
Proportion of parental segments shorter than 200 bp	60%
Smallest parental segments size observed	20 bp
Average number of mutations in functionally competent clones	8.3 ± 3.2
Average number of mutations in non-active clones	14.0 ± 4.2

Although the number of sequenced clones (10) was limited, the analysis provides interesting insight into mosaic structures confirming the high efficiency of the developed strategy. In addition, the findings deduced from the statistical analysis were confirmed by the sequence data. Analysis of the naphthalene hydroxylase activity of randomly selected clones revealed that 20 % of them were functionally competent. In all non-functional clones sequenced, at least one internal stop codon was present, thereby truncating the protein.

The association of *in vivo* shuffling to a modified *in vitro* approach led to library with higher mosaic enzyme content and featuring a high degree of complexity. Another major advantage of the developed shuffling strategy was that direct construction of expression libraries in a eucaryotic microorganism allowed functional *in vivo* selection of membrane-bound or multicomponent complexes. Transformed yeast clones derived from the shuffling step were used as such for functional screening.

CONCLUSION

We have described a particulary well-adapted method for mosaic DNA library construction when large DNA sequences with limited identity are involved. In addition, the yeast expression system is well suited for expression of membrane proteins. In addition the use of DNA macro or micro-arrays is a very efficient method to characterize the population sequence patterns during evolution cycles and to detect interesting mosaic structures.

Such technique, when applied to P450s involved in xenobiotic metabolism, is a powerful tool for substrate selectivity analysis. New informations on the functional consequences of genetic polymorphisms observed for some of these P450 are expected. This approach could contribute to improve risk assessment in cytotoxic processes or carcinogenesis.

Therefore, novel function design or existent function improvement would facilitate P450 utilization as biotechnological tools in various topics as, for example, ecology (dangerous products or pollutants detoxication) or medicine (drug synthesis, detection of reactive metabolites, prodrug activation or detoxication of toxic products).

REFERENCES

1.   Kadlubar, F.F. and G.J. Hammons, The role of cytochrome P-450 in the metabolism of chemical carcinogens. in: *Mammalian cytochrome P-450*, F.P. Guenguerich, Editor, CRC Press: Boca Raton and Florida. p. 81-130. (1987).

2.   Shao, Z. and F.H. Arnold, Engineering new functions and altering existing functions. *Curr. Opin. Struct. Biol.*6(4): p. 513-8 (1996).

3.   Giver, L. and F.H. Arnold, Combinatorial protein design by in vitro recombination. *Curr. Opin. Chem. Biol.*2(3): p. 335-8 (1998).

4.   Moore, J.C., Jin, H. M., Kuchner, O.and Arnold, F. H., Strategies for the in vitro evolution of protein function: enzyme evolution by random recombination of improved sequences. *J. Mol. Biol.* .272(3): p. 336-47 (1997).

5.   Yano, T., S. Oue, and H. Kagamiyama, Directed evolution of an aspartate aminotransferase with new substrate specificities. *Proc. Natl. Acad. Sci. USA* 95(10): p. 5511-5 (1998).

6.   Altamirano, M.M., Blackburn, J. M., Aguayo, C. and Fersht, A. R., Directed evolution of new catalytic activity using the alpha/beta-barrel scaffold . *Nature.*403(6770): p. 617-22 (2000).

7.   Crameri, A., Raillard, S. A., Bermudez, E. and Stemmer, W. P., DNA shuffling of a family of genes from diverse species accelerates directed evolution. *Nature.*391(6664): p. 288-91 (1998).

8.   Kikuchi, M., K. Ohnishi, and S. Harayama, An effective family shuffling method using single-stranded DNA. *Gene.*243(1-2): p. 133-7 (2000).

9.   Stemmer, W.P., DNA shuffling by random fragmentation and reassembly: in vitro recombination for molecular evolution. *Proc. Natl. Acad. Sci. USA.*91(22): p. 10747-51 (1994).

10.   Mezard, C., D. Pompon, and A. Nicolas, Recombination between similar but not identical DNA sequences during yeast transformation occurs within short stretches of identity. *Cell.*70(4): p. 659-70 (1992).

11.   Truan, G., Raillard, S. A., Bermudez, E. and Stemmer, W. P, Enhanced in vivo monooxygenase activities of mammalian P450s in engineered yeast cells producing high levels of NADPH-P450 reductase and human cytochrome b5. *Gene.*125(1): p. 49-55 (1993).

ONE-ELECTRON REDUCTION OF QUINONES BY THE NEURONAL NITRIC-OXIDE SYNTHASE REDUCTASE DOMAIN

M. Kitamura, H. Matsuda, S. Kimura, and T. Iyanagi

Department of Life Science, Faculty of Science, Himeji Institute of Technology
Harima Science Garden City, Hyogo 678-1297 Japan

INTRODUCTION

Quinones including antitumor quinones undergo facile reduction and oxidation. One-electron reduction of a quinone gives the semiquinone radicals while two-electron reduction gives the hydroquinone[1]. In 1969-70, Iyanagi and Yamazaki[2,3] reported the mechanism of quinone reduction by flavin enzymes, and the reduction of quinones and oxygen by flavin enzymes falls into three mechanistic categories: one-electron, two-electron and mixed-type reactions. NAD(P)H:quinone oxidoreductase(QR), also known as DT-diaphorase contains one FAD as prosthetic group, catalyzes the obligatory two-electron reduction of quinone to hydroquinone. On the other hand , microsomal NADPH-cytochrome P450 reductase (P450 reductase) and NADH-cytochome b5 reductase , and mitochondrial NADH-ubiquinone oxidoreductase, and ferredoxin: $NADP^+$ reductase catalyze typical one electron reduction of bivalent quinones. Xanthine oxidase/dehydrogenase catalyzes both reactions of one-electron and two-electron. These results have contributed greatly to the study on the formation of free-radicals in biological systems, especially in the quinone-mediated cyotoxicity .

Neuronal nitric oxide(nNOS) reductase domain, which has similar properties to those of P450 reductase[4,5], can catalyze the reduction of a several exogenous electron acceptors [7,8]. In the present study, we have examined whether recombinant neuronal Nitric-oxide synthase (nNOS) reductase domain , which contains an FAD/FMN prosthetic group pair and calmodulin (CaM)-binding site could catalyze one-electron reduction of menadione (MD), mitomycin C (MitC) and adriamycin (Adr) by a similar mechanism with that of P450 reductase, and these activities are stimulated by Ca^{2+}/CaM. Futhermore, the role of an FAD-FMN pair of the nNOS reductase domain is discussed in the context with that of the P450 reductase[2,4,6].

EXPERIMENTAL PROCEDURES

Enzymes - The rat nNOS reductase domain (amino acid residues 718-1429) was expressed in *E. coli*, strain BL21 , and was purified as described previously[9]. QR and P450 reductase were purified from rat livers . ***Methods*** - Optical spectra were measured with a Shimazu

Model MPS UV-2000 spectrophotometer. Stopped flow experiments were performed by using a Union Giken Model RA401 stopped flow spectrophotometer.

RESULTS and DISCUSSION

nNOS reductase domain can catalyze one electron reduction of quinones. NADPH oxidation by P450 reductase, which is a "one-electron transfer" enzyme[2] is greatly stimulated by the addition of MD·. The NADPH oxidation by the nNOS reductase domain is also stimulated in the presence of Ca2+/CaM and MD[10]. The NADPH-oxidase activity was very low levels in the absence of MD, even in the presence of Ca^{2+}/CaM, but MD-mediated NADPH oxidation was stimulated approximately 12-fold by the addition of Ca^{2+}/CaM. When NAD(P)H:quinone oxidoreductase (QR), which is a "two-electron transfer" enzyme[3] was added during the reaction, the rates for NADPH oxidation in the absence or the presence of Ca^{2+}/CaM were significantly decreased. The nNOS reductase domain also catalyzes one-electron reduction of Adr and Mit C. The one-elctron reduction potential of various quinones was related to their reduction raters by both the nNOS reductase domain and P450 reductase.

We next used cytochrome b_5 as a scavenger for menasemiquinone radical(MD$\cdot^-$). Cytochrome b_5 was not reduced directly by the nNOS reductase domain, but it was effectively reduced in the presence of MD. The stoichiometry of NADPH-oxidation by MD versus cytochrome b_5 reduction (as determined from $-$d[NADPH]/dt versus d[cytochrome b_5^{2+}]/dt) was correspond to 0.5 (0.5 ± 0.1, n = 5). This ratio indicates that one mole of NADPH can reduce two mole of cytochrome b_5 , and MD mediates transfer of electrons from NADPH to cytochrome b_5 as a one-electron carrier. On the other hand, QR did not stimulate effectively the reduction of cytochrome b_5 in the presence of MD, although the rate of MD reduction was the same between the nNOS reductase domain and QR. These results also confirm that the nNOS reductase domain can catalyze only one-electron reduction of MD.

We measured the reactivity of the air-stable semiquinone with MD. The air-stable semiquinone of the nNOS reductase domain did not react significantly with MD. The second order rate constant (as determined from -d[air-stable semiquinone form]/dt in the absence of Ca^{2+}/CaM) was 33.0 M^{-1} s^{-1}, but its rate constant was 28.3×10 M^{-1} s^{-1} in the presence of Ca^{2+}/CaM. The data indicate that the activity of the air-stable semiquinone with MD was increased 9-fold by the addition of Ca^{2+}/CaM. A similar rate constant, 10.5×10^2 M^{-1} s^{-1} was obtained in the reaction of MD with the air-stable semiquinone of P450 reductase.

The formation of $O_2\cdot^-$ by MQ semiquinone radical. We have studied oxidation-reduction properties of $O_2\cdot^-$ radical formed in the MD-mediated NADPH oxidation system. In the anaerobic conditions ,the cytochrome b5 was completely reduced by the nNOS reductase domain, but in the aerobic conditions the cytochrome b_5 was partially reduced by nNOS reductase domain in the presence of MD. The stedy-state level of cytochrome b_5 reduction was dramatically decreased in the presence of superoxide disumutase(SOD). The similar results were obtained in the P450 system[6]. These observations could be explaied by the following mechanism:

$$\text{reductase}$$
$$2MD + NADPH \quad \rightarrow \quad 2MD^{\cdot -} + NADP^+ + H^+ \qquad (1)$$
$$MD^{\cdot -} + Cyt\, b_5^{+3} \rightleftarrows \quad MD + Cyt\, b_5^{+2} \qquad (2)$$
$$MD^{\cdot -} + O_2 \rightleftarrows MD + O_2^{\cdot -} \qquad (3)$$
$$\text{SOD}$$
$$2O_2^{\cdot -} + 2H^+ \rightarrow H_2O_2 + O_2 \qquad (4)$$

where reactions 2 and 3 is in the equilibrium.

These data suggest that $MQ^{\cdot -}$ and $O_2^{\cdot -}$ radicals are involved as an active intermediate in this system. In the presence of SOD, the stedy-state concentration of $O_2^{\cdot -}$ was decreased by the reaction 4, and the decrease of stedy-state concentration of $MD^{\cdot -}$ is caused (reaction 3). As a consequence, the reduction level of cytochrome b5 was decreased. This level also were dependent on the concentrations of SOD. These data also suggest that reactions 2 and 3 are in the dynamic equilibrium in this system. In this system, the formation of oxy-hemoglobin from met-hemoglobin was also inhibited in the presence of SOD.

Flavin intermediates observed during oxidation of the reduced enzyme by menadione.
In the presence of Ca^{2+}/CaM, the oxidized nNOS reductase domain (FAD-FMN) was mixed with NADPH plus MD solution. The decrease in the absorbance at 457 nm paralles that at 590 nm, but the absorbance change at 504 nm was relatively constant. These results indicate that the two-electron-reduced forms of the enzyme do not accumulate measurably during the oxidation of the reduced enzyme by MD. When the air-stable semiquinone form was mixed with NADPH plus MD, the absorbance changes at 457 nm and 590 nm were increased, and the absorbance change at 504nm was also increased. The increase of absorbance at 590 nm strongly suggest the formation of another semiquinone species, probably semiquinone radical derived from FAD moiety, but such a intermediate was not observed in the P450 reductase[4]. These data strongly suggest that the air-stable semiquinone form is a predominant intermediate observed during the oxidation of the reduced enzyme by MD, but another semiquinone species is also observed during the reaction, and these semiquinone species significantly do not react with MD. These data suggest that both the two-electron reduced ($FADH_2$-FMN or FAD-$FMNH_2$) enzymes and three-electron reduced ($FADH_2$-$FMNH^{\cdot}$ or $FADH^{\cdot}$-$FMNH_2$) rapidly can donate one reducing equivalent at the catalytic cycle(Fig. 1).

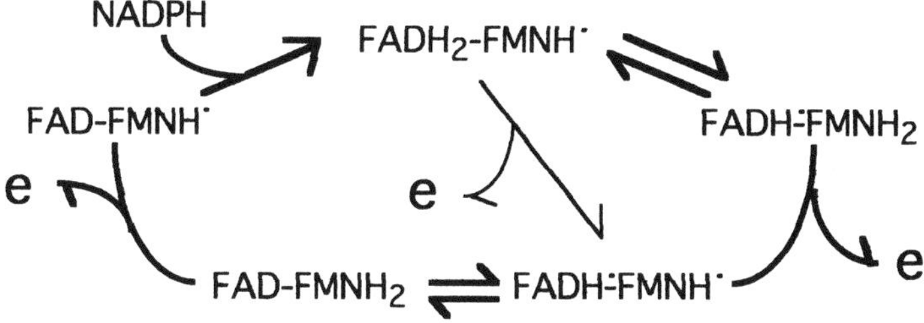

Fig. 1. Proposed mechanism of one-elctron reduction of quinones by the nNOS reductase domain(FAD-FMN). The **e** indicates one-elctron reduction of quinones.

where both the $FADH_2$ and $FNHH_2$ are the reactive species with MD. Whether or not these reactions involve all in the one-electron reduction of MD will depend upon several factors, including the presence or absence of Ca^{2+}/CaM , the reactivity of the reduced flavins with MD, the midpotential of each one-electron redox couple, and the degree of overlap of the potentials of the $FADH_2$-FADH$^\cdot$-FAD and $FMNH_2$-FMNH$^\cdot$-FMN. From the evidence presented in this study as well as from the P450 reductase[4], we now propose the role of the two flavins in one electron reduction of MD by the nNOS reductase domain. We postulate that the FAD accepts two reducing equivalent from NADPH, and both the $FADH_2$ and $FMNH_2$ an donate electrons to quinones. CaM activates intramolecular electron transfer between FAD and FMN. In the absence of Ca^{2+}/CaM, the semiquinone, FADH$^\cdot$-FMN is initially formed by one-electron oxidation by MD , but in the presence of Ca^{2+}/CaM the semiquinone, FADH$^\cdot$-FMNH$^\cdot$ is formed by the rapid equilibration between the two flavins. In the absence of Ca^{2+}/CaM , the intramolecular electron transfer from FADH$^\cdot$-FMN to FAD- FMNH$^\cdot$ could be rate limiting step in the catalytic cycle. Finally, the enzyme can function between the 1e(FAD-FMNH$^\cdot$) and 3e ($FADH_2$-FMNH$^\cdot$ $\rightleftarrows$ FAD-FMNH$^\cdot$) levels during one-electron reduction of quinones[10].

REFERENCES

1.R.L,Willson.,1990, Quinones, semiquinone free radical and one-electron transfer reactions, *Free.Rad.Res.Comms*, 8:201-217.

2.T, Iyanagi, and I,Yamazaki.,1969, One-electron reduction of quinones by microsomal flavin enzymes, *Biochim.Biophys.Acta*,172:370-381.

3.T, Iyanagi, and I,Yamazaki.,1970, Difference in the mechanism of quinone reduction by theNADH-dehydrogenase and the NAD(P)H-dehydrogenase(DT-diaphorase), Biochim. Biophys.Acta,216:282-294.

4.T.Iyanagi.,R.Makino, and F.K,Anan.,1981, Mechanism of action of hepatic NADPH-Cytochrome P450 reductase, *Biochemistry*,20:1722-1730.

5.D.S, Bredt.,P.M, Hwang., C, Glatt., C, Lowenstein., R.R, Reed., and S.H, Snyder,1991, Cloned and expressed nitric oxide synthase structural resemble cytochrome P450 reductase. *Nature*,351:714-718.

6. T.Iyanagi.,1990,On the mechanism of one-electron reduction of quinones by microsomal flavin enzymes:The kinetic analysis between cytochrome b5 and menadine, *Free.Rad. Res.Comms*, 8:259-268.

7. A.P,Garner., M.J.I, Paine., I,Rodriguz-Crespo., E.C, Chinje., P,Ortiz de Montellano., I.J, Straford., D.G, Tew., C.R, Wolf., 1999, Nitric oxide synthases catalyze the activation of redox cycling and bioreductive anticancer agents, *Cancer Res*,59:1929-1934.

8.J,Vasquez-Vivar.,P,Martasek.,N,Hogg.,B.S.S,MatersK.A,Prichard.,andB, Kalyanaraman.,1997, Endotherial nitric oxide synthase-depenent superoxide generation from adriamycin, *Biochemistry*,36:1129-11297.

9. H.Matsuda, and T.Iyanagi.,1999, Calmodulin activates intramolecular electron transfer between the two flavins of neuronal nitric oxide synthase flavin domain, *Biochim. Biophys.Acta*,1473:345-355.

10.H.Matsuda., S.Kimura, and T.Iyanagi., 2000, One-electron reduction of quinones by the neuronal nitrix-oxide synthase reductase domain, *Biochim.Biophys.Acta*, in press.

PURIFICATION, BIOCHEMICAL CHARACTERIZATION AND COMPARATIVE ENZYME KINETICS OF RECOMBINANT HUMAN CYP2D6 1 AND CYP2D6 2 VARIANTS

Aiming Yu and Robert L. Haining*

Department of Medicinal Chemistry, Univ. of Washington, Seattle, WA 98195, USA

INTRODUCTION

Several dozen human drug-metabolizing enzyme (DME) genetic polymorphisms have been characterized (Nebert et al 1997; Gonzalez et al 1994). Cytochrome P450 2D6 (CYP2D6), representing perhaps the first generally recognized polymorphic phase I DME, is involved in the metabolism of more than 50 drugs (Parkinson 1996). In fact, human CYP2D6 may be important in the processing of about 25% of drugs (Benet et al 1996) including antiarrhythmics, antihypertensives, β-blockers, opioids, antipsychotics, and tricyclic antidepressants.

The molecular basis of the CYP2D6 polymorphism has been studied intensely in recent years. Patients can be classified into four major sub-groups, commonly termed poor metabolizer (PM), intermediate metabolizer (IM), extensive metabolizer (EM) and ultrarapid metabolizer (UM) phenotypes (Daly 1995). The CYP2D6 gene, localized to chromosome 22q13.1, turns out to have numerous polymorphic alleles. These alleles are either point mutations or a combination of mutations, as well as rearrangements of genes and pseudogenes of the CYP2D6 gene cluster on the chromosome and hence, result in absent, decreased or increased activities. The CYP2D6*2 allele is present in Caucasian populations at a frequency of 0.324, compared to 0.364 for CYP2D6*1 allele (Sachse et al 1997). This gene contains three mutations ($G_{1749}C$, $C_{2938}T$, $G_{4268}C$), thus yielding the CYP2D6 2 enzyme isoform with two amino acid differences (R296C and S496T). The correlation between this allele and decreased *in vivo* capacity for dextromethorphan demethylation is well characterized in Caucasian and Gabonese populations (Sachse et al 1997; Panserat et al 1999).

In the present study we adapted the baculovirus-mediated insect cell system for the high-level expression and purification of these two major allelic variants of CYP2D6 in order to further compare their catalytic activities. The molecular weights of the isolated proteins were confirmed by liquid chromatography/electrospray mass spectrometry (LC/ES-MS). Kinetic studies (Table 1) show that this natural mutation (CYP2D6 2) causes a modest decrease in the Vm/Km for dextromethorphan O-demethylation (5-fold), codeine O-demethylation (3-fold) and fluoxetine N-demethylation (4-fold), which is

Biological Reactive Intermediates VI, Edited by Dansette *et al.*
Kluwer Academic / Plenum Publishers, 2001

consistent with the *in vivo* (Sachse et al 1997) and in vitro (Oscarson et al., 1997) data available.

MATERIALS AND METHODS

Chemicals   Dextromethorphan (DMO), dextrorphan (DOP), 3-methoxymorphinan (MEM), 3-hydroxymorphinan (HYM), fluoxetine and norfluoxetine were purchased from Research Biochemicals International (Natick, MA). Codeine, norcodeine, morphine, glycerol, sodium cholate, reduced nicotinamide adenine dinucleotide phosphate (NADPH), L-α-dilauroylphospha-tidylcholine (DLPC), dithiothreitol (DTT), phenylmethylsulfonyl fluoride (PMSF) and Octyl Sepharose CL-4B were purchased from Sigma (St. Louis, MO). DEAE Sepharose Fast-Flow (DEAE-FF) was from Pharmacia (Piscataway, NJ), and ceramic hydroxyapatite was from Bio-Rad Laboratories (Hercules, CA). Emulgen 911 was from Kao-Atlas (Tokyo, Japan). HPLC solvents and other chemicals were of the highest grade commercially available and were used as received.

Baculovirus-Mediated CYP2D6 Expression and Purification   Expression of 2D6 in T.ni suspension cultures was carried out essentially as described previously for CYP2C9 allelic variants (Haining, 1996). All purification steps were carried out at 4^0C, and all buffers were at pH 7.4. The crude insect cell pellet was homogenized in 1mL of Solubilization buffer [20%(v/v) glycerol, 1mM EDTA, 0.1mM DTT, 0.2mM PMSF and 1%(w/v) cholate in 100mM potassium phosphate buffer] per nanomole of P450 by making 5-10 passes with a Teflon homogenizer. The P450 was solublized by stirring the homogenized insect cell pellet for 30min followed by removal of insoluble material by centrifugation at 100,000g for 40min. Supernatant was loaded onto an Octyl Sepharose CL-4B column (1mL per nmol of P450) pre-equilibrated with Buffer A [20%(v/v) glycerol, 1mM EDTA, 0.1mM DTT, 0.2mM PMSF and 0.5%(w/v) cholate in 10mM potassium phosphate buffer] at 30mL/h. After the samples was loaded, the column was washed with 4 columns of Buffer A, and eluted with Buffer A plus 0.4%(v/v) Emulgen 911 (Buffer B) at 40mL/h.

The P450-containing fractions were combined and loaded onto DEAE-FF column pre-equilibrated with Buffer B without cholate (Buffer C) at 30mL/h. The column was washed with 4 columns of Buffer C, and eluted with Buffer C containing 100mM potassium phosphate. The fractions were dialyzed overnight against 200 volumes of Buffer D [20%(v/v) glycerol, 0.1mM EDTA in 3mM potassium phosphate buffer]. The dialyzed enzyme was adsorbed onto a ceramic hydroxyapatite column (1mL/10nmol) preequilibrated with Buffer D. The column was washed with Buffer D until A_{280}<0.018. The P450 was eluted with Buffer E [20%(v/v) glycerol, 0.1mM EDTA and 0.5% cholate in 350mM potassium phosphate buffer]. The fractions were dialyzed overnight against Buffer F [20%(v/v) glycerol and 0.1mM EDTA in 100mM potassium phosphate buffer]. The purified P450 sample was aliquoted and stored at -80^0C until further use.

Incubation Conditions   Incubation reactions were carried out at 37^0C in 100μM potassium phosphate, PH=7.4, containing 0.2μM CYP2D6, 0.4μM P450 reductase, 10μg DLPC, 1mM NADPH and substrate in a final volume of 300μL. Reactions were terminated by the addition of 15μL of 60% perchloric acid. The mixtures were centrifuged for 5min and the supernatant was used directly for HPLC analyses.

Instrumentation   HPLC analyses were carried out on a Hewlett Packard 1050 series instrument consisting of quaternary pump, autosampler, variable wavelength UV detector and 1046A fluorescence detector. A 150mm×4.6mm I.D. Nucleosil C18 5µm column (Sigma, USA) was used to separate the metabolites. LC/ES-MS analyses were performed on a Micromass Quattro II tandem quadrupole mass spectrometer (Micromass Ltd., Manchester, UK) coupled to an HPLC [Shimadzu LC-10AD with SPD-10AV UV-vis variable detector (Shimadzu Scientific Instruments, Inc., Colubia, MD)]. A computer running Windows NT based Micromass MassLynxNT® 3.2 software controlled the instrument.

RESULTS AND DISSCUSION

High titer baculovirus stocks, *T. ni* suspension culture, and heme addition proceeded in a manner analogous to that used for 2C9 (Haining, 1996). Cultures exhibiting greater than 50 nmol/liter equivalents of P450 were used for purification. Holoprotein yields were estimated at each stage by measuring carbon monoxide difference spectra (Estabrook, 1972). Contaminants were removed by passage through Octyl Sepharose and DEAE Sepharose. Highly purified, detergent free CYP2D6 was collected after dialysis the fractions from hydroxyapatite column (data not shown). The total yields were 52.3% and 36.4% for CYP2D6 1 and CYP2D6 2 respectively. The carbon monoxide difference spectra for the two isoforms exhibited Soret maxima at 450nm with no evidence of cytochrome P420 formation (data not shown). Meanwhile, the samples were analyzed by Western Blotting using an anti-peptide anti-2D6 antibody raised in rabbit. As expected, all CYP2D6 protein products were detected in these samples (not shown).

The isolated CYP2D6 variants were subjected to LC/ES-MS analyses. The experimentally determined molecular weight (MW) was calculated using the observed charge state distribution, or ion envelope. The experimental MWs were 55772.0 and 55731.0 for CYP2D6 1 and CYP2D6 2 respectively. Compared to the MW values of 55769.6 and 55730.6 predicted based on the amino acid sequences, the errors were only 43ppm and 7ppm for CYP2D6 1 and CYP2D6 2 respectively.

Kinetic studies of dextromethorphan O-demethylation, codeine O-demethylation and fluoxetine N-demethylation were carried out with the highly purified CYP2D6 isoforms. These experiments revealed that the CYP2D6 2 allelic variant had modestly decreased activity for all the three reactions investigated (Table 1). The apparent Km for DMO O-demethylation was nearly 9-fold higher for CYP2D6 2 compared to CYP2D6 1, and consequently the intrinsic clearance value was about 5-fold lower although an apparent corresponding increase in Vmax appears to offset this effect somewhat. Rather, the alterations in both Vmax and Km lead to 3- and 4-fold decrease in the V/K for codeine O-demethylation and fluoxetine N-demethylation respectively. The present results indicate that the decreased CYP2D6 activity observed in individuals with the CYP2D6*2 allele is caused by a CYP2D6 2 enzyme with a reduced affinity for several substrates.

In summary, the overexpression and purification of CYP2D6 allelic isoforms was achieved. The recombinant CYP2D6 2 and CYP2D6 1 isoforms were purified to high homogeneity and characterized with LC/ES-MS. These two allelic variants were also functionally studied using the substrates dextromethorphan, codeine and fluoxetine. Comparative enzyme kinetics reveal that the activity of CYP2D6 2 is decreased modestly, which provides a molecular explanation for the decreased metabolizing capacity seen in IM subjects with one or two copies of the CYP2D6*2 allele.

Table 1. Apparent Michaelis-Menten Kinetic Parameters for CYP2D6 1 and CYP2D6 2 Catalyzed Reactions

Reaction	CYP 2D6 isoform	Km (μM)	Vmax (nmol/nmol/min)	V/K	V/K Ratio[a]
Dextromethorphan O-demethylation	2D6 1	3.04±0.57	9.01±0.45	2.96	5:1
	2D6 2	26.31±0.83	14.61±0.25	0.56	
Codeine O-demethylation	2D6 1	178.1±29.5	6.24±0.29	0.035	3:1
	2D6 2	436.9±40.0	5.43±0.18	0.012	
Fluoxetine N-demethylation	2D6 1	2.84±033	0.653±0.022	0.230	4:1
	2D6 2	4.76±0.20	0.268±0.004	0.056	

[a] Ratio of CYP2D6 1 to CYP2D6 2

REFERENCES

Benet, L.Z., Kroetz, D.L., and Sheiner, L. B., 1996, Pharmacokinetics, in: *Goodman & Gilman's the Pharmacological Basis of Therapeutics.* J.G. Hardman, Gilman A. Goodman, and L.E. Limbird, eds., 9th ed., McGraw-Hill, New York.

Daly, A.K., 1995, Molecular basis of polymorphic drug metabolism, *J Mol Med*, 73:539.

Estabrook, R.W., Peterson, J., Baron, J., and Hildebrandt, A., 1972, *Methods Pharmacol* 2:303.

Gonzalez, F.J., and Idle, J.R., 1994, Evolution of the P450 gene superfamily: animal-plant "warfare," molecular drive, and human genetic differences in drug oxidation, *Trends Genet*, 6:182.

Haining, R.L., Hunter, A.P., Veronese, M.E., Trager, W.F., and Rettie, A.E., 1996, Allelic variants of human cytochrome P450 2C9: baculovirus-mediated expression, purification, structural characterization, substrate stereoselectivity, and prochial selectivities of the wild-type and I359L mutant forms, *Arch Biochem Biophys*, 333:447.

Nebert, D.W., 1997, Polymorphisms in drug-metabolizing enzymes: what is their clinical relevance and why do they exist, *Am J Hum Genet*, 60:265.

Oscarson, M., Hidestrand, M., Johansson, I., and Ingelman-Sundberg, M. 1997, A combination of mutations in the CYP2D6*17 (CYP2D6Z) allele causes alterations in enzyme function. *Mol Pharmacol* 1997 Dec;52(6):1034-40

Panserat, S., Sica, L., Gerard, N., Mathieu, H., Jacqz-Aigrain, E., and Krishnamoorthy, R., 1999, CYP2D6 polymorphism in a Gabonese population: contribution of the CYP2D6*2 and CYP2D6*17 alleles to the high prevalence of the intermediate metabolic phenotype, *Br J Clin Pharmacol*, 47:121.

Parkinson, A., 1996, An overview of current cytochrome P-450 technology for assessing the safety and efficacy of new materials, *Toxicol Pathol*, 24:45.

Sachse, C., Brockm ller, J., Baner, S., and Roots, I., 1997, Cytochrome P450 2D6 variants in a Caucasian population: allele frequencies and phenotypic consequences, *Am J Hum Genet*, 60:284.

INHIBITORY EFFECTS OF ROQUEFORTINE ON HEPATIC CYTOCHROMES P450

C. Aninat and M. Delaforge

CEA Saclay, Département de la Recherche Médicale, Service de Pharmacologie et d'Immunologie, 91191 Gif sur Yvette cedex, France

INTRODUCTION

Roquefortine, a cyclodipeptide derived from the diketopiperazine cyclo (Trp-dehydroHis), is a secondary metabolite produced by some *Penicillium* species (1,2). *P. roqueforti* is considered as one of the most important fungal contaminants of carbonated beverages, beer, wine, meats, cheese, bread. It is reported to be among the predominant species in silage.

Roquefortine has been reported to cause convulsive seizures when administrated i.p. to mice (3) and to inhibit gram-positive bacterial growth by a bacteriostatic action (4). Mechanisms responsible for its toxicity are still unknown. In comparison to cyclopeptides which are well recognized and metabolized by monooxygenase, roquefortine was investigated with regards to its interactions with cytochromes P450.

Roquefortine   cyclo (Trp-His)

Figure 1. Structure of roquefortine and cyclo (Trp-His)

Biological Reactive Intermediates VI, Edited by Dansette *et al.*
Kluwer Academic / Plenum Publishers, 2001

MATERIALS AND METHODS

Chemicals
All chemicals were commercially available except for cyclo(Trp-His), which was kindly provided by Dr. Genet.

Enzyme preparation
Male rats were treated for 3 days with various inducers and microsomes were prepared as already described (5). Yeast microsomes containing human-expressed cytochromes P450 3A4, 1A2, 2D6 and 2C8 were prepared as already described (6).

Study of substrate binding to microsomal P450s by difference visible spectroscopy
Interaction spectra were measured in a differential mode according to (5) and Ks and ΔOD_{max} were determined either by double reciprocal plot or using GraphPad Prism software.

NADPH-dependent cytochrome P450 reduction
NADPH reduction was measured at 340 nm in the presence of 0.4 mM NADPH.

HPLC Analysis
Testosterone and metabolites were detected at 254 nm by RP-HPLC 150/4.6 mm using an H_2O/CH_3CN gradient.

RESULTS

Interaction of roquefortine and analogues with P450s

The interaction of roquefortine with liver microsomes from rats pretreated with various inducers was studied by difference visible spectroscopy (Table 1). Roquefortine produced a classical type-II difference spectra with a peak around 430 nm and a trough around 410 nm. The maximal absorbance spectra were observed for microsomes from dexamethasone-treated rats. The apparent spectral dissociation constant (Ks) of roquefortine was similar for the various microsomes (around 1 μM), with the possibility of two sites for rat-DEX.
Addition of roquefortine to human P450 microsomes and to yeast expressing human P450 microsomes led to a type II difference visible spectrum. The Ks value with P4503A4 and human microsomes was always around 1 μM.

These results suggest that roquefortine interacts more strongly with P450 3A than with other P450s.

The imidazole moiety of roquefortine was involved in the binding to iron heme since interaction of cyclo(His-Phe) and cyclo(Trp-His) with microsomes led to a type II difference spectrum, whereas cyclo(Phe-Trp) and cyclo(Leu-Trp) led to a type I difference spectrum (Table 2). All these compounds exhibited a lower affinity (higher Ks) with P450 3A than roquefortine. These results indicated that the dehydrohistidine and the 1,1-dimethylallalyl moieties were directly implicated in the stronger interaction of roquefortine with the P450 active site.

Table 1. Spectral interactions of roquefortine with various rat or human P450s. Difference spectra (ΔOD_{max}) were measured as described previously (5). Apparent spectral dissociation constants, Ks, were determined from double-reciprocal plots of ΔOD versus [S]. The reported values are the mean of two to four measurements.

Microsomes	Ks (µM)	ΔOD_{max}
Rat-UT	0.48	0.053
Rat-3MC	0.64	0.077
Rat-DEX	$\mathbf{K_{s1} = 0.17}$	$\Delta OD_{max1} = 0.033$
	$K_{s2} = 2.35$	$\mathbf{\Delta OD_{max2} = 0.208}$
Rat-PB	0.62	0.039
Rat-ISO	2.24	0.095
Rat-CLO	0.48	0.069
Yeast expressing		
1A2	4.1	0.038
3A4	0.6	0.022
2D6	n.m.	0.04
2C8	n.m.	n.m.
human	0.43	0.046

Table 2. Spectral interactions of various cyclopeptides with liver microsomes from rats pretreated with dexamethasone.

	Ks (µM)	Spectrum type	ΔOD_{max}
Roquefortine	$\mathbf{K_{s1} = 0.17}$	II	$\Delta OD_{max1} = 0.033$
	$K_{s2} = 2.35$		$\mathbf{\Delta OD_{max2} = 0.208}$
Cyclo(His-Phe)	16	II	0.1
Cyclo(Trp-His)	49	II	0.09
Cyclo(Phe-Trp)	56	I	0.05
Cyclo(Leu-Trp)	69	I	0.04

Inhibition of Binding of Testosterone by Roquefortine

The binding of testosterone to P450 3A4 and rat-DEX liver microsomes gave a type I difference spectrum (peak around 390 nm and a trough around 420 nm). Roquefortine gave a type II difference spectrum, allowing competitive binding titrations of IC_{50} values. The IC_{50} of roquefortine was 20 µM for rat-DEX liver microsomes and 3 µM for P450 3A4, using 100 µM testosterone.

Inhibition of NADPH consumption by roquefortine

NADPH consumption was inhibited by roquefortine (IC_{50} around 10 µM using rat-DEX liver microsomes). The inhibition curve remained linear with time. No significant difference in NADPH consumption was observed with 50 µM roquefortine or 50 µM miconazole, a specific inhibitor of P450 3A. The same results were observed when roquefortine and testosterone were present simultaneously. Testosterone alone was able to stimulate NADPH consumption.

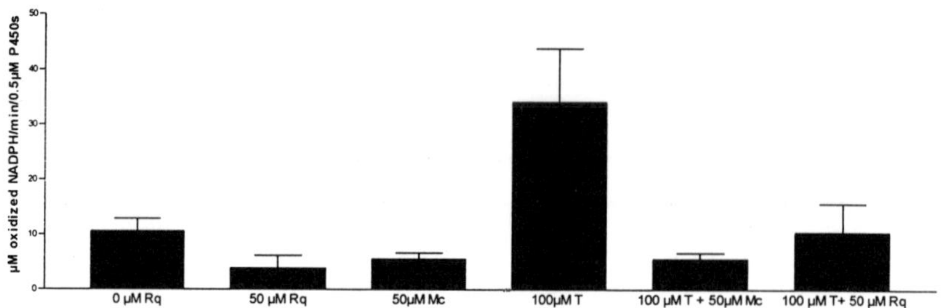

Figure 1. NADPH consumption in the presence of 50 μM substrates using microsomes from dexamethasone-treated rats

Metabolism of testosterone

Roquefortine inhibits 6β hydroxylation of testosterone formation with $IC_{50} = 10$ μM using either rat or human 3A4.

CONCLUSIONS

The various studies undertaken on roquefortine show us that is a direct inhibitor of P450s. Taking into account the concentrations at which these interaction effects are observed (lower than 1 μM), it is extremely probable that P450s are among the first targets in the case of poisoning where large quantities of roquefortine are absorbed, thus leading to hepatic impairment.

ACKNOWLEDGMENTS

This work was supported by a Ministère de l'Environnement contract N°AC009E

REFERENCES

(1) Ohmono S, Sato T, Utagawa T, Abe M. Isolation of festuclavine and three new indole alkaloids, roquefortine A, B and C from the cultures of *Penicillium roqueforti* (production of alkaloids and related substances by fungi part XII). *Agric. Chem. Soc. Japan.* 1975; 49: 615-623.

(2) Reshetilova TA, Kozlovsky AG. Synthesis and metabolism of roquefortine in Penicillium species. *J.Basic Microbiol.* 1990; 30: 109-114.

(3) Arnold DL, Scott PM, McGuire PF, Harwig J, Nera EA. Acute toxicity studies on roquefortine and PR toxin, metabolites of *Penicillium roqueforti*, in the mouse. *Fd. Cosmet. Toxicol.* 1978; 16: 369-371.

(4) Kopp-Holtwiesche B, Rehm HJ. Anitimicrobial action of roquefortine. *J. Enrion. Pathol. Toxicol. Oncol.* 1990; 10: 41-44.

(5) Peyronneau MA, Delaforge M, Rivière R, Renaud JP, Mansuy D. High affinity of ergopeptides for cytochromes P4503A. Importance of their peptide moiety for P450 recognition and hydroxylation of bromocriptine. *Eur.J.Biochem.* 1994; 223: 947-956.

(6) Peyronneau MA, Renaud JP, Truan G, Urban P, Pompon D, Mansuy D. Optimization of yeast-expressed human liver cytochrome-P450 3A4 catalytic activities by coexpressing NADPH-cytochrome-P450-reductase and cytochrome-b5. *Eur. J. Biochem.* 1992; 207: 109-116.

ASSOCIATION OF CYTOCHROMES P450 1A2 AND 2B4: ARE THE INTERACTIONS BETWEEN DIFFERENT P450 SPECIES INVOLVED IN THE CONTROL OF THE MONOOXYGENASE ACTIVITY AND COUPLING?

D. R. Davydov,[1] N. A. Petushkova,[1] E. V. Bobrovnikova,[1] T. V. Knyushko,[1] and P. Dansette[2]

[1]Institute of Biomedical Chemistry RAMS
10 Pogodinskaya, 119832, Moscow, Russia
[2]Universite Rene Descartes, CNRS UMR 8601
45 Rue des Saints Pères 75270, Paris Cedex 06, France

INTRODUCTION

The membranes of endoplasmic reticulum contain a number of co-existing isoforms of cytochrome P450. These multiple P450 species compete for the partners, namely NADPH-cytochrome P450 reductase (CPR) and cytochrome b_5, and hence have to be considered as a members of a single ensemble. Moreover, these different P450 isozimes also appear to interact with each other. Despite of numerous evidences on the oligomerization of P450s both in solution (Dean and Gray, 1982; Wendel et al., 1983; Tsuprun et al., 1986) and in the membranes (Greinert et al., 1982; Kawato et al., 1982; Schwartz et al., 1990, Alston et al., 1991), the functional significance of this phenomenon remains obscure. However, it is likely to cause several perplexing features of microsomal P450s, such as biphasic kinetics of their reduction by NADPH and dithionite (Karyakin and Davydov, 1985), multiphasic kinetics of interactions with carbon monoxide (Davydov et al., 1986) and notable allosteric behavior of several P450 isoforms (Korzekwa et al., 1998). A remarkable observation in this context we made studying the barotropic behavior of P450 2B4 (CYP2B4) (Davydov et al., 1992,1995). We have found that only about 65-70% of the ferrous carbonyl complex of this oligomeric protein in solution is exposed to pressure-induced P450→P420 inactivation. The same non-uniform barotropic behavior was also observed for the oligomers of ferric CYP2B4, where only about 30-35% of the hemoprotein participates in the substrate binding and related spin transitions, being, at the same time, insensitive to the pressure-induced inactivation (Davydov et al., 1995). As these irregularities disappear at the P450 monomerization in the presence of detergent, we suggest them to reflect some peculiarities of the oligomer architecture resulting in inequality of the subunits in conformation and/or orientation. This peculiarity might be related to the asymmetry of the dimeric crystallization unit of the heme-containing domain of the cytochrome P450BM-3 (BMP), where two constituting BMP molecules have significantly different conformations, one with a more open substrate- and water-access to the heme moiety than the other (Ravichandran et al., 1993). Although P450BM-3 is believed to be monomeric in solution, the organization of its dimeric crystallization unit is likely to be related to the architecture of the oligomers of eukaryotic P450s. We suppose that this apparent inequality of the P450 subunits may serve in the control of the activity, degree of coupling and production of reactive oxygen species (ROS) in microsomal monooxygenase (MMO). This apparent mechanism may function through the substrate-modulated formation of the mixed oligomers of several P450 species, where the isozymes lacking the substrate are hidden from the interactions with their red/ox partners.

The present study of the mutual effects of CYP1A2 and CYP2B4 on their activity and interactions with CPR was designed to probe this hypothesis.

Biological Reactive Intermediates VI, Edited by Dansette *et al.*
Kluwer Academic / Plenum Publishers, 2001

MATERIALS AND METHODS

Electrophoretically homogeneous cytochromes P450 1A2, P450 2B4 and NADPH-cytochrome P450 reductase (CPR) were purified from rabbit liver by published procedures (Imai and Sato, 1974, Alterman, e.a., 1990, Kanaeva, et al., 1992). The procedure used to introduce the 7-ethylamino-3-(4'-maleimidilphenyl)-4-methylcoumarin maleimide (CPM) probe into CPR was detaily described earlier (Davydov et al., 2000). Essentially the same technique was employed to attach the fluorescent maleimide probes, either CPM or N-(1-pyrenyl)maleimide (PM), to cytochromes P450. The detailed procedure of this modification will be described elsewhere. Labeling of P450 by fluorescein isothiocyanate (FITC) was performed as described by Bernhardt et al. (1983). To monitor protein-protein interactions we employed the fluorescence energy transfer (FRET) in PM/CPM, CPM/FITC or CPM/heme donor/acceptor pairs (Davydov et al., 2000). O-Dealkylation of 7-ethoxy-resorufin (EROD) and 7-pentoxyresorufin (PROD) was measured at 30°C by direct fluorescent assay (Perrin et al., 1990). Analysis of the spectra were done using principal component analysis (PCA) technique (Davydov et al., 2000). All experiments were done in 0.1M Na-Hepes buffer (pH 7.4) containing 1mM DTE, 1 mM EDTA.

RESULTS AND DISCUSSION

Process of Subunit Exchange in Homo- and Hetero-Oligomers of CYP1A2 and CYP2B4. We monitored the process of subunit exchange in P450 oligomers in solution using FRET technique (Erijman and Weber, 1991) between cytochrome P450 molecules labeled with two different fluorescent maleimide probes. In these experiments we heve used either CPM/FITC or CPM/PM donor/acceptor pairs. In the absence of detergent the equilibrium of oligomerization was totally shifted towards oligomers and subunit exchange was very slow (*fig. 1a*). Introductdion of low concentration (0.025%.-0.05%) of Emulgen-913 into the system importantly facilitates the process of subunit exchange (*fig. 1b*). In these conditions the monomers and oligomers of P450 eisxst in the dynamic equilibrium although the constant of dissociation of oligomers remain to be rather low (K_d=0.2 μM). The solution of P450 1A2 and P450 2B4 hemoproteins, taken in micromolar concentrations in the presence of detergent (0.05% Emulgen-913), was chosen for further studies of the interactions of these hemoproteins.

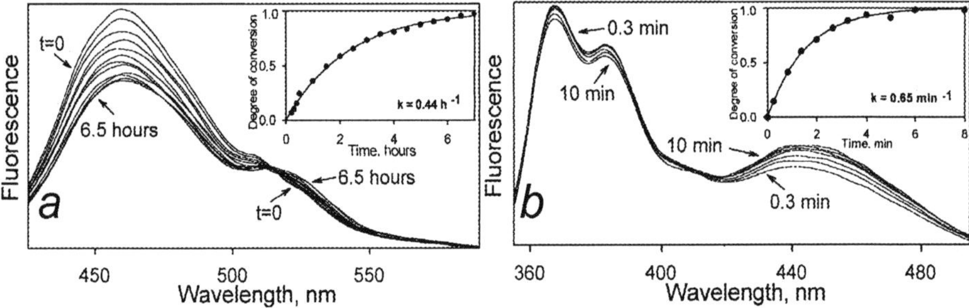

Fig.1. Fluorescence energy transfer upon the process of subunit exchange in P450 oligomers. *(a)* Series of spectra (excitation at 385 nm) recorded in the intervals of 10 - 20 minutes after mixing of CYP2B4-CPM with CYP2B4-FITC (1 μM of each) in the absence of detergent. Inset shows the kinetics of the process and the results of the fitting by the first reaction equation, which gives the rate constant of the decay of oligomers (k_{off}). *(b)* Series of spectra (excitation at 325 nm) recorded in 30 sec. intervals after mixing of CYP1A2-PM with CYP2B4-CPM (1 μM of each) in the presence of 0.05% Emulgen-913. Inset shows the kinetic curve of the process.

Mutual Effects of CYP1A2 and CYP2B4 on Their Interactions with NADPH-cytochrome P450 Reductase. Interactions of CPM-labeled CPR with CYP1A2 and CYP2B4 were monitored by a decrease in the intensity of fluorescence due to energy trabsfer from CPM to the heme chromophore (Davydov et al., 2000). Taken separately, CYP1A2 and CYP2B4 exhibited similar values of the dissociation constant (K_d) of their complexes with CPR (0.044 ± 0.02 μM and 0.037 ± 0.01 μM, correspondingly). However, when the reductase was titrated by the CYP1A2+CYP2B4 mixtures in the presence of 7-ethoxyresorufin (ER), a highly specific substrate of CYP1A2, the dependence of the apparent K_d of the P450-CPR complex on the molar ratio of the P450 species is given by an asymmetric bell-shaped curve (Fig. 3). The maximal value of K_d of 0.5 ± 0.2 μM is reached at the CYP1A2 : CYP2B4 molar ratio of 1:5 – 1:3. These results strongly supports a hypothesys that the

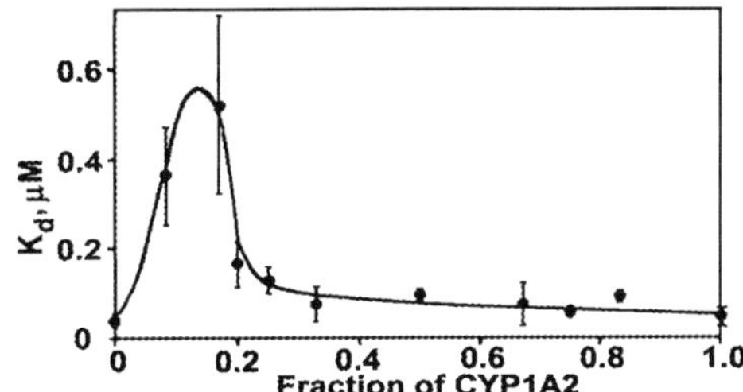

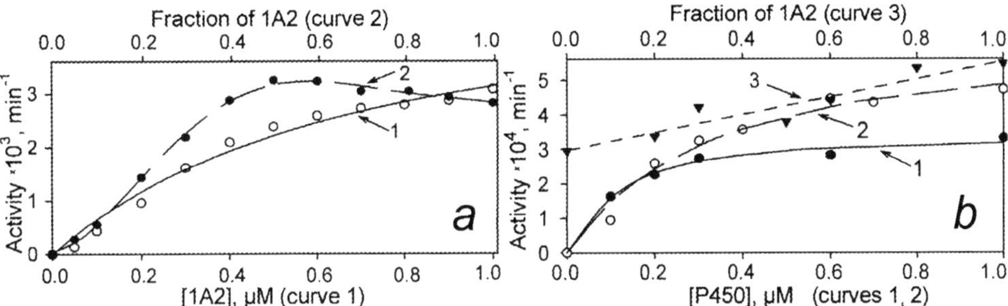

Fig. 2. Effect of the content of CYP1A2 in its mixture with CYP2B4 on the effective K_d of their complex with CPR in the presence of 7-etoxyresorufin (..mM). K_d was determined by titration of CPM-labeled CPR (0.02 µM) by 1A2, 2B4 or their mixture. Conditions: 0.1M Hepes pH 7.4, 0.05% Emulgen-913, 10µM 7-ethoxyresorufin. Each point represents an average of 4 - 5 repetitive experiments. The confidence intervals shown for each point are cqlculated for P≤0.05.

formation of the heterooligomers of CYP2B4 and CYP1A2 hemoproteins in the presence of ER partially hides the P450 2B4 enzyme, which is inactive in the oxidation of this substrate, from the interactions with the flavoprotein partner.

Effect of the CYP1A2-CYP2B4 Interactions on Their Activity. To probe the reciprocal effect of these enzymes on their functional interactions with CPR we have studied the catalytic activities of the mixtures of these two hemoprtoteins taken at various ratios but at the constant total concentration (1 µM). CYP1A2 was highly active in the reaction of the oxidative deethylation of ER (EROD), while the activity of CYP2B4 in this reaction was negligibly low. If EROD activity of CYP1A2 is not affected by its interaction with CYP2B4, the apparent specific activity of CYP1A2+CYP2B4 mixture has to be proportional to the molar content of CYP1A2. However, the plot of the EROD activity versus CYP1A2 content importantly deviates from linearity and exhibits an apparent activation of the CYP1A2 enzyme in the mixture with CYP2B4 (*fig. 3a, curve 2*). Therefore, in the presence of CYP1A2 and ER, CYP2B4 appears to be at least partially excluded from the interaction with the flavoprotein.

Fig. 3. Effect of the CYP1A2-CYP2B4 interactions on their association with CPR monitored by catalytic activity of P450. *(a)* Dealkylation of etoxyresorufin (CYP1A2-specific). Curve 1 shows the titration of 0.02 µM CPR by 1A2 fitted by the equation of binary association (K_d=0.3µM); curve 2 was obtained at the constant total concentration of the cytochromes (1 µM), but variable CYP1A2:CYP2B molar ratio, at 0.02µM CPR. *(b)* Dealkylation of pentoxyresorufin. Curves 1and 2 show the titration of 0.02 µM CPR by CYP1A2 (K_d=0.06 µM, curve 1) and CYP2B4 (K_d=0.25µM, curve 2); curve 3 was obtained at the constant total concentration of the cytochromes (1 µM), but variable CYP1A2:CYP2B molar ratio. at 0.02uM CPR. Other conditions were as those specified for *Fig. 2*.

At the same time, when ER was replaced by pentoxyresorufin (PR), which is metabolized by both enzymes, the plot of the activity versus the content of CYP1A2 was linear (*fig. 2b, curve 3*). Thus, in this case when both enzymes were capable to bind and methabolize the substrate, P450-P450 interactians has apparently no effect on their activity and interactions with the reductase.

These results appear to be in good agreement with the above data on the reciprocal effects of P450 1A2 and P450 2B4 on their interactions with reductase. Taken together, these data support our initial hypothesis on the substrate-modulated distribution of P450 isophorms between "open" and "closed" positions in the oligomer. The occupancy of the "open" subunit location is promoted by the substrate binding. In the presence of ER, which is a specific substrate for 1A2, occupancy of the "open" locations is preferred for the substrate-bound 1A2, whereas in the presence of PR both "closed" and "open" locations appear to be randomly distributed between 1A2 and 2B4 hemoproteins. This mechanism might by of high importance for the regulation of the degree of coupling of MMO and production of ROS in this system.

ACKNOWLEDGEMENTS

This research was supported in part by a Russian Foundation of Basic Research (RFBR) Grant 97-4-49132 to D.R.D. and INTAS Research Collaborative Grant 96-1343 to D.R.D. and. G.H.B.H.

REFERENCES

Alston, K., Robinson, R.C., Park, S.S., Gelboin, H.V., and Friedman, F.K., 1991, Interactions among cytochromes P-450 in the endoplasmic reticulum. Detection of chemically cross-linked complexes with monoclonal antibodies, *J. Biol. Chem.* **266:**, 735-739.

Alterman, MA., and Dowgii A.I.. , 1990, A simple and rapid method for the purification of cytochrome P-450 (form LM4), *Biomed. Chromatogr.* **4:** 221-222.

Bernhardt R., Ngoc Dao N.T., Stiel H., Schwarze W., Friedrich J., Janig G.R., and Ruckpaul K., 1983, Modification of cytochrome P-450 with fluorescein isothiocyanate, *Biochim Biophys Acta* , **745:** 2140-2148

Davydov, R.M., Khanina, O.Yu., Iagofarov, S., Uvarov ,V. Yu., and Archakov, A.I., 1986, Effect of lipids and substrates on the kinetics of binding of ferrocytochrome P-450 to CO. *Biokhimia* **51:** 125-129

Davydov, D.R., Knyshko, T.V., and Hui Bon Hoa, G., 1992, High pressure induced inactivation of ferrous cytochrome P-450 LM2 (IIB4) CO-complex: Evidence for the presence of two conformers in the oligomer, *Biochem. Biophys. Res. Commun.* **188:** 216 - 221.

Davydov, D.R., Deprez, E., Hui Bon Hoa, G., Knyushko, T.V., Kuznetsova, G.P., Koen, Y.M., and Archakov, A.I., 1995, High-Pressure-Induced Transitions in Microsomal Cytochrome P450 2B4 in Solution - Evidence for Conformational Inhomogeneity in the Oligomers, *Arch. Biochem. Biophys.* **320:** 330-344.

Davydov D.R., Kariakin A.A., Petushkova N.A., and Peterson J.A., 2000, Association of cytochromes P450 with their reductases: opposite sign of the electrostatic interactions in P450BM-3 as compared with the microsomal 2B4 system. *Biochemistry*, **39:** 6489-6497

Erijman L. and Weber G., 1993, Use of sensitized fluorescence for thestudy of the exchange of subunits in protein aggregates, *Photochemistry and Photobiology*, .**57:** 411-415.

Imai Y, and Sato, R, 1974, A gel electrophoretically homogenious preparation of cytochrome P-450 from liver microsomes of phenobarbital pretreated rabbits, *Biochem. Biophys. Res. Commun*, **60:** 8-14.

Kanaeva, I.P., Dedinskii, I.R., Skotselyas, E.D., Krainev A.G., Guleva, I.V., Sevryukova, I.F., Koen, Y.M., Kuznetsova, G.P., Bachmanova, G.I., and Archakov, A.I., 1992, Comparative study of monomeric reconstituted and membrane microsomal monooxygenase systems of the rabbit liver. I. Properties of NADPH-cytochrome P450 reductase and cytochrome P450 LM2 (2B4) monomers, *Arch Biochem Biophys*, **298:** 395-402

Kariakin A.V. Davydov D.R., 1988, Kinetics of the electron transfer reactions in the monooxygenase system. *Vestnik Akad. Med. Nauk SSSR.* **1988**(1): 53-62.

Korzekwa K.R., Krishnamachary N., Shou M., Ogai A., Parise R.A., Rettie A.E., Gonzalez F.J., and Tracy T.S., 1998, Evaluation of atypical cytochrome P450 kinetics with two-substrate models: evidence that multiple substrates can simultaneously bind to cytochrome P450 active sites, *Biochemistry*, **37:** 4137-4147.

Perrin, R., Minn, A., Ghersi-Egea, J.F., Grassiot, M.C., and Siest, G., 1990, Distribution of cytochrome P450 activities towards alkoxyresorufin derivatives in rat brain regions, subcellular fractions and isolated cerebral microvessels, *Biochem Pharmacol*, **40:** 2145-2151.

Ravichandran, K.G., Boddupalli, S.S., Hasermann, C.A., Peterson, J.A., and Deisenhofer, J., 1993, Crystal structure of hemoprotein domain of P450BM-3, a prototype for microsomal P450's, *Science* **261:**731-736.

Schwarz, D., Pirrwitz, J., Meyer, H.W., Coon, M.J., and Ruckpaul K., 1990, Membrane topology of microsomal cytochrome P-450: saturation transfer EPR and freeze-fracture electron microscopy studies, *Biochem. Biophys. Res. Commun.* **171:** 175-181

Tsuprun, V.L., Myasoedova, K.N., Berndt, P., Sograf, O.N., Orlova, E.V., Chernyak, V.Ya., Archakov, A.I., and Skulachev, V.P., 1986, Quaternary structure of the liver microsomal cytochrome P-450, *FEBS Lett.* **205:** 35-40.

Wendel, I., Behlke, J., and Janig, G.R. , 1983, Hydrodynamic studies on the association of cytochrome P-450, *Biomed. Biochim. Acta* **42:** 633-640.

INACTIVATION OF POLYMORPHIC VARIANTS OF HUMAN GLUTATHIONE TRANSFERASE ZETA (hGSTZ1-1) BY MALEYLACETONE AND FUMARYL-ACETONE

Hoffman B. M. Lantum,[1] Philip G. Board,[2] and M.W. Anders[1]

[1]Department of Pharmacology and Physiology
University of Rochester
Rochester, NY 14642, USA
[2]Division of Molecular Medicine
John Curtin School of Medical Research
Canberra, ACT 2601, Australia

INTRODUCTION

Glutathione transferase zeta (GSTZ1-1),[1] which is identical to maleylacetoacetate isomerase (EC 5.2.1.2),[2] is the penultimate enzyme in the tyrosine degradation pathway. GSTZ1-1 catalyzes the glutathione (GSH)-dependent *cis-trans* isomerization of maleylacetoacetate (MAA) to fumarylacetoacetate (FAA). Maleylacetone (MA), a decarboxylation product of MAA, is a substrate for the *Vibrio 01* isomerase and is converted to fumarylacetone (FA).[3] No other endogenous role for GSTZ1-1 is known, but GSTZ1-1 also catalyzes the GSH-dependent biotransformation of a range of α-haloacids, such as dichloroacetic acid (DCA) and chlorofluoroacetic acid (CFA).[4]

DCA and other fluorine-lacking α-haloalkanoic acids inactivate hGSTZ1-1.[5] Rats given DCA excrete MA.[6] With the rat ortholog of hGSTZ1-1, the isomerization of MAA to FAA was linear at low substrate concentrations but deviated from linearity at high substrate concentrations, which was attributed to substrate inhibition of the enzyme.[7] Wong and Seltzer also observed that ^{14}C-labeled MA and FA became covalently bound to the *Vibrio 01* isomerase via a non-Schiff base mechanism.[8] MAA and MA are substrates for rat and *Vibrio 01* GSTZ1-1, but their inhibitory or inactivating effects have not been elucidated. Four polymorphic variants of hGSTZ1-1 (1a-1a, 1b-1b, 1c-1c, and 1d-1d) have been described, which have different activities with DCA and CFA as substrates as well as differences in susceptibility to DCA-induced inactivation.[9,10]

These observations indicated that MA and FA may be both substrates and inhibitors or inactivators of hGSTZ1-1. The objective of this study was to determine the effects of MA and FA on the biotransformation of CFA by hGSTZ1-1 polymorphic variants.

MA and FA, which were a gift from Dr. Peter Dedon, MIT, were synthesized by the method of Fowler and Seltzer,[11] and their properties were confirmed by NMR- and UV-spectral analysis.

Recombinant N-terminal His-tagged hGSTZ1-1 polymorphic variants were expressed in *E. coli* M15[pREP4] cells and purified with nickel affinity columns.[10] Activity of hGSTZ1-1 variants was determined spectrophotometrically by measuring the rate of conversion of CFA to glyoxylate in the presence of GSH in 0.1 M phosphate buffer (pH 7.4) at 37 °C.[4] Except where otherwise mentioned, 2 µg of purified protein, 0.5 mM GSH, and 0.01–1 mM CFA was used.

In the inactivation studies, hGSTZ1-1 variants were incubated with MA and FA (0–50 µM) for various times (0–30 min) in 0.1 M phosphate buffer (pH 7.4), and activity was determined with 2 µg of protein, 1 mM CFA, and 1 mM GSH. To determine the reversibility of inhibition, 100 µg/ml of each variant was incubated with 50 µM MA, FA, and N-ethylmaleimide (NEM) in 0.1 M phosphate buffer (pH 7.4) at 37 °C. After 30 min, 20 µl of the solution (2 µg of protein) was added to 1 ml of phosphate buffer, and the activity was measured.

The kinetic data were fitted to the Michaelis-Menten equation with EnzFitter software (BioSoft®). K_{ic} and K_{iu} of MA and FA for the different variants were obtained from Dixon plots ($1/V$ vs. [I]) and Cornish-Bowden plots ([S]/V vs. [I]), respectively.

RESULTS

All polymorphic variants of hGSTZ1-1 catalyzed the biotransformation of CFA to glyoxylate (Table 1). The activities were different for the four variants: the activity of hGSTZ1a-1a was 3-fold higher than the activity of the other variants. Variant 1c-1c had the lowest activity, which was not significantly different from the activities of 1b-1b and 1d-1d.

The pH-dependence of the reaction was also determined in 0.1 M phosphate solution buffered at pHs ranging from 6–9. Variants 1a-1a, 1b-1b, and 1c-1c had maximal activities at pH 8.5, and variant 1d-1d had a pH optimum of 7.4. The activities were thus pH-dependent and variants 1a-1a, 1b-1b, and 1c-1c had pH optima that were similar to those of other glutathione transferases.

Table 1. Effects of MA and FA on the kinetics of the biotransformation of CFA by hGSTZ1-1 polymorphic variants.

	0 µM MA				25 µM MA				50 µM MA			
	1a-1a	1b-1b	1c-1c	1d-1d	1a-1a	1b-1b	1c-1c	1d-1d	1a-1a	1b-1b	1c-1c	1d-1d
V_{max}	5450 ± 287	2180 ± 50	2105 ± 40	2477 ± 49	6190 ± 175	2030 ± 13	1639 ± 37	2410 ± 37	5083 ± 212	1828 ± 39	1518 ± 9	2241 ± 23
K_s	1023 ± 71	145 ± 9	167 ± 8	161 ± 8	1386 ± 57	168 ± 3	165 ± 10	153 ± 6	1388 ± 85	182 ± 9	209 ± 9	149 ± 4

	0 µM FA				25 µM FA				50 µM FA			
	1a-1a	1b-1b	1c-1c	1d-1d	1a-1a	1b-1b	1c-1c	1d-1d	1a-1a	1b-1b	1c-1c	1d-1d
V_{max}	6435 ± 458	2514 ± 54	1918 ± 47	2539 ± 16	4778 ± 165	1946 ± 22	1531 ± 36	2305 ± 6	4299 ± 25	1688 ± 36	1277 ± 4	2438 ± 7
K_s	1294 ± 140	159 ± 9	174 ± 11	158 ± 3	1179 ± 62	157 ± 5	176 ± 9	168 ± 1	1241 ± 77	170 ± 10	175 ± 1	190 ± 1

MA and FA inhibited all hGSTZ1-1 polymorphic variants but by different mechanisms. MA was a competitive inhibitor of variant 1a-1a: the K_{ic} was 124 µM and the K_{iu} was infinite. MA was a mixed inhibitor of variants 1b-1b and 1c-1c. MA was an uncompetitive inhibitor of variant 1d-1d: the K_{ic} was infinite, whereas the K_{iu} was 425 μM. Variant 1c-1c had the lowest K_{ic} (71 μM) and the lowest K_{iu} (119 μM).

FA was a mixed inhibitor of all variants: the K_{ic} and K_{iu} were finite for all variants. The K_{ic} and K_{iu} for variants 1a-1a, 1b-1b, and 1c-1c were similar with an average of about 100 μM and 104 μM, respectively, indicating that FA was a noncompetitive inhibitor. Variant 1d-1d showed a 2-fold higher K_{ic} and an 8-fold higher K_{iu} for FA than the other variants.

The time- and concentration-dependence of the inhibitory effects of MA and FA were examined with variant 1c-1c, which is the most common hGSTZ1-1 variant.[9] The specific activity of 1c-1c was markedly reduced within 30 sec of mixing 50 μM MA or FA with 2 or 4μg of the enzyme in the absence of GSH and CFA. These preliminary studies indicated that MA and FA inactivated hGSTZ1-1 in the absence of GSH. The magnitude of the inactivation increased with longer incubation times but was constant beyond 30 min of incubation.

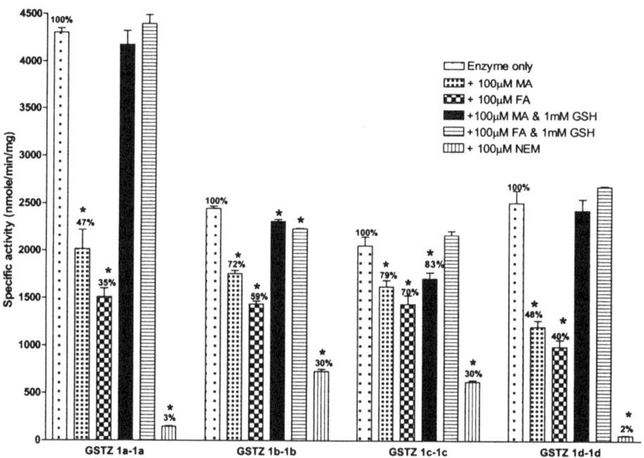

Figure 1. Inactivation of hGSTZ1-1 polymorphic variants by MA, FA, and NEM Protein (100 μg/ml) was incubated with 100 μM of the inhibitor at 37 °C for 30 min and was then diluted 50-fold to determine activity. Percentages are relative to "enzyme-only" values for the corresponding variant. One-way ANOVA non-parametric and Dunnett's multiple comparison tests were done for each group ($*p < 0.05$).

Time-dependent inactivating effects were also observed when CFA was present in the MA- or FA-enzyme incubation mixture, indicating that CFA did not block the inactivation caused by MA and FA. When 50 μM MA or FA and 1 mM GSH were incubated for various times before adding 2 μg of enzyme, activity was reduced by only 10% and was not time-dependent. This indicated that GSH formed adducts with MA and FA and that the adducts were not responsible for the inactivation. Furthermore, mixing MA or FA and CFA before adding the enzyme and GSH did not inhibit the enzyme, indicating that the inactivation was not associated with an adduct of CFA and MA or FA.

The concentration-dependent effects of MA and FA on the activity of variant 1c-1c were determined and compared with the inhibitory effects of N-ethylmaleimide (NEM) and succinylacetone (SA), the saturated analogue of MA and FA. MA, FA, and NEM caused a concentration-dependent decrease of the activity of variant 1c-1c with CFA as substrate, but with different potencies (NEM > FA > MA). SA did not inhibit hGSTZ1c-1c. Dilution of the reaction mixture did not reverse the observed inhibition, indicating that MA, FA, and NEM inactivated the enzyme. Incubation of the enzyme with 1 mM GSH for 5 min before addition of 100 μM MA or FA completely blocked the inactivation of 1a-1a and 1d-1d and partially for 1b-1b and 1c-1c (Figure 1).

We tested the ability of MA and FA to react with GSH and L-cysteine at pH 7.4 in 0.1 M phosphate buffer. Addition of 10 mM GSH or L-cysteine to 100 µM MA or FA caused a slow decrease in absorbance of MA (λ_{max}^{MA} = 312 nm) and a rapid decrease in absorbance of FA (λ_{max}^{FA} = 345 nm) indicating that both MA and FA reacted nonenzymatically with GSH and L-cysteine.

DISCUSSION

These experiments show that MA is a substrate as well as an inactivator of hGSTZ1-1 variants. MA, FA, and NEM may form covalent adducts with the thiol moiety of a cysteine residue of hGSTZ1-1 and, thereby, inactivate the enzyme. Wong and Seltzer also observed labeling of the enzyme by [^{14}C]MA and [^{14}C]FA.[8]

Our findings indicate that the inactivation of hGSTZ1-1 in vivo by DCA may be exacerbated by the bioaccumulation of MA and FA. GSH, however, protects hGSTZ1-1 from inactivation by MA and FA, indicating that in the presence of constitutive GSH concentrations, little inactivation may be expected. The lowest K_{ic} was observed with 1c-1c, which is the most common variant,[9] indicating that hGSTZ1c-1c has the highest activity for MAA compared with other variants.

ACKNOWLEDGEMENTS

This work was partly supported by the Wilmot Cancer Research Fellowship Program (HBML), Cancer Center, University of Rochester, and NIEHS grant ES03127.

REFERENCES

1. P.G. Board, R.T. Baker, G. Chelvanayagam and L.S. Jermiin, Zeta, a novel class of glutathione transferases in a range of species from plants to humans. *Biochem J* 328:929 (1997).
2. J.M. Fernandez-Canon and M.A. Penalva, Characterization of a fungal maleylacetoacetate isomerase gene and identification of its human homologue. *J Biol Chem* 273:329 (1998).
3. S. Seltzer, Purification and properties of maleylacetone cis-trans isomerase from *Vibrio 01*. *J Biol Chem* 248:215 (1973).
4. Z. Tong, P.G. Board and M.W. Anders, Glutathione transferase zeta-catalyzed biotransformation of dichloroacetic acid and other α-haloacids. *Chem Res Toxicol* 11:1332 (1998).
5. W.B. Anderson, P.G. Board, B. Gargano and M.W. Anders, Inactivation of glutathione transferase zeta by dichloroacetic acid and other fluorine-lacking alpha-haloalkanoic acids. *Chem Res Toxicol* 12:1144 (1999).
6. R. Cornett, M.O. James, G.N. Henderson, J. Cheung, A.L. Shroads and P.W. Stacpoole, Inhibition of glutathione S-transferase zeta and tyrosine metabolism by dichloroacetate: a potential unifying mechanism for its altered biotransformation and toxicity. *Biochem Biophys Res Commun* 262:752 (1999).
7. W.E. Knox and S.W. Edwards, Maleylacetoacetate isomerase. *Methods Enzymol* 2:295 (1955).
8. G. Wong and S. Seltzer, Maleylacetone cis-trans isomerase reaction via a non-Schiff base mechanism. *Brookhaven National Laboratory Report 18123*: 17 (1972).
9. A. Blackburn, H. Tzeng, M. Anders and P. Board, Discovery of a functional polymorphism in human glutathione transferase zeta by expressed sequence tag database analysis. *Pharmacogenetics* 10:49 (2000).
10. H.F. Tzeng, A.C. Blackburn, P.G. Board and M.W. Anders, Polymorphism- and species-dependent inactivation of glutathione transferase zeta by dichloroacetic acid. *Chem Res Toxicol* 13:231 (2000).
11. J. Fowler and S. Seltzer, The synthesis of model compounds for maleylacetoacetic acid: maleylacetone. *J Org Chem* 35:3529 (1970).

STRUCTURE-ACTIVITY RELATIONSHIPS OF CYCLOTETRAPEPTIDES: INTERACTION OF TENTOXIN DERIVATIVES WITH THREE MEMBRANE PROTEINS

Nicolas Loiseau,[1] Marcel Delaforge,[1] Claire Minoletti,[2] François André,[2] Alexia Garrigues,[2] Stéphane Orlowski,[2] and Jean-Marie Gomis[2]

[1] Département de Recherche Médicale
[2] Département de Biologie Cellulaire et Moléculaire
Direction des Sciences du Vivant, CEA-Saclay
F-91191 Gif sur Yvette Cedex, France.

INTRODUCTION

Tentoxin[1] (TTX) **1** is a natural cyclic tetrapeptide (cyclo(L-NMeAla-L-Leu-NMeΔ(Z)Phe-Gly)) produced by the phytopathogenic fungus *Alternaria alternata*. This cyclopeptide is a selective weedkiller which causes the chlorosis of some higher plants[2]. A possible mechanism involves the inhibition of photophosphorylation in the chloroplasts. *In vitro* and at low concentrations (10^{-8}-10^{-7} M), TTX inhibits the F1 moiety of H^+ ATP-synthase, but at higher concentrations it stimulates the ATPase activity [3,4]. TTX is the only natural effector known to inhibit the catalytic part of chloroplast ATP-synthase, and is non-competitive for nucleotide binding. In order to understand this mechanism we developed a large number of TTX analogues, either by total synthesis and chemical modification, or by biotransformation. This report describes the study of two close analogues: dihydro-tentoxin (DH-TTX) **2**, a metabolite also isolated from *Alternaria alternata* and presumably precursor of TTX during its biosynthesis, and iso-tentoxin (Iso-TTX) **3**, a photochemical derivative of TTX.

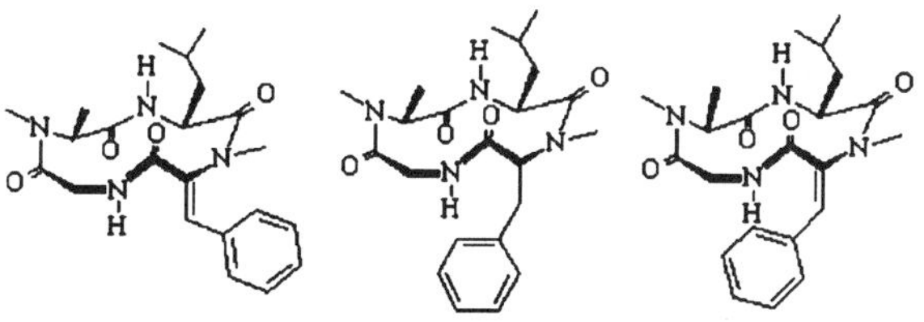

Scheme 1. Tentoxin analogue formulas.

Biological Reactive Intermediates VI, Edited by Dansette *et al.*
Kluwer Academic / Plenum Publishers, 2001

Recently we discovered that TTX also interacts with two mammalian membrane proteins: cytochrome P450-3A[5] and P-glycoprotein (P-gp). The cytochromes P450-3A are hepatic proteins involved in the metabolism of a large number of drugs. P-gp is responsible for the transport of exogenic molecules out of the cell. These two proteins have numerous substrates in common[6]. Here we report our main conclusions concerning the structural factors involved in the recognition of various substrates by these enzymes.

MATERIALS & METHODS

Natural tentoxin and dihydro-tentoxin were purified from *Alternaria alternata*. Isotentoxin was prepared by photoisomerization of natural tentoxin upon ultraviolet irradiation at 254 nm, according to the procedure described previously[7]. Cytochrome P450 affinities were determined using UV-visible modification upon access of the substrate to the active site[8]. CF1 ATPase and Pgp activities were determined according to published methods.[9,10]

RESULTS AND DISCUSSION

TTX, Iso-TTX and DH-TTX were favoured in this study for the following reasons. In addition to their affinities for the three enzymatic systems, these molecules have very similar structures (Scheme 1). DH-TTX is the biological precursor of TTX, hydrogenated on the dehydrophenylalanyl moiety. Iso-TTX is obtained by the double bond isomerization of the dehydro-residue. TTX, Iso-TTX and DH-TTX differ only by the aromatic nature of their third amino acid, respectively NMeΔZPhe, NMeΔEPhe and NMePhe. The three-dimensional structures[11,12] of these substrates determined by ^{1}H 2D-NMR at 500 MHz (COSY, TOCSY, NOESY and ROESY) indicate that TTX exists in aqueous solution as three conformers A, B and C, whereas Iso-TTX and DH-TTX present two conformers A and B (Figure 1). All these conformers exhibit a general structure of the cycle in the BOAT UP form. The transformation of the A to B conformer is realized by rotation of the peptide bond NMe-Ala-Leu, and the passage from the B to C conformer is realized by rotation of the peptide bond between the third amino acid and the glycine.

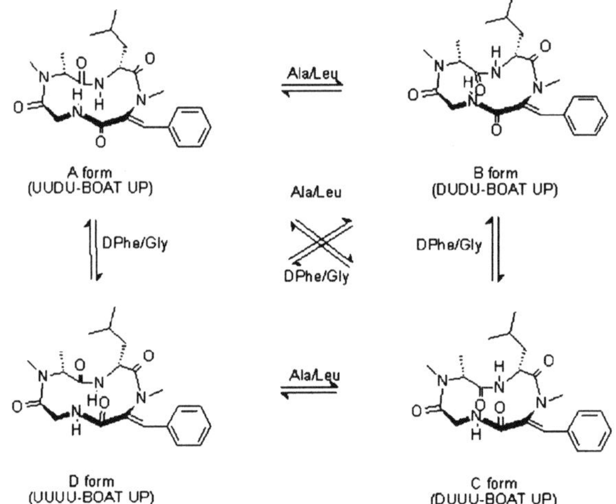

Figure 1. Schematic representation of the four tentoxin forms in chemical exchange in aqueous solution. The notation U (D) indicates that the carbonyl is directed above (below) the average plane of the cycle.

We first determined the affinity of these substrates for P450-3A, CF1-ATPase and P-gp. Using male rat liver microsomes, treated with dexamethasone to induce P450-3A isozymes, we measured substrate access to the active site close to the heme by UV/visible differential spectrophotometry. Dissociation constants were found to be 20 and 80 µM for TTX and Iso-TTX, respectively (Table 1, Figure 2), indicating greater affinity for the first molecule. Similarly, the spectral intensity decreased from 35 to 10 mAU from TTX to Iso-TTX, indicating that less P450 was involved in the binding with Iso-TTX than with TTX. Similar results were obtained using human P450 3A4 expressed in yeast. We used the purified part CF1-ε of chloroplast F0-F1 ATPase to measure the ATP hydrolysis, with various concentrations of ligand. With P-glycoprotein, we monitored NADH consumption, measured by UV (Figure 2), according to the substrate concentration. This allowed us to determine the affinity and the influence of the substrate on this protein. As for P450, the measured affinities for CF1-ε and P-gp were lower for Iso-TTX and DH-TTX than for TTX (Table 1, Figure 2).

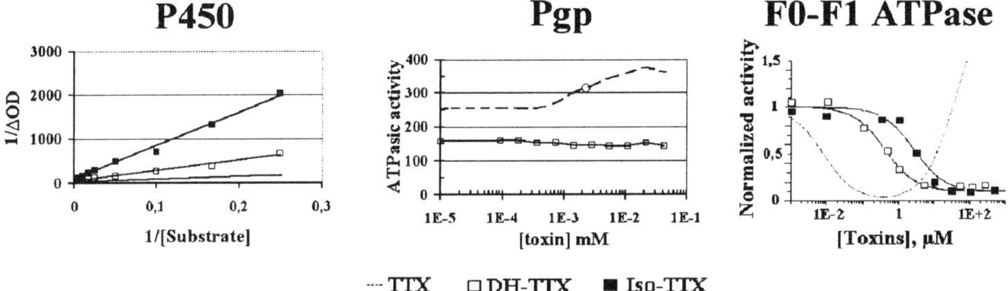

Figure 2. Effects of tentoxin analogues on enzymatic systems.

Table 1. Affinities (in µM) of tentoxin analogues for various enzymatic systems. (n.d.: not determined).

Substrates	P450 3A	ATPase CF1-ε	Pgp
TTX	23	0.008	3
DHTTX	50	0.5	n.d.
IsoTTX	85	3	>50

Noteworthily, the affinities of these cyclopeptides for the three enzymes follow the same order: TTX > DH-TTX > Iso-TTX. The nature of the aromatic residue controls the molecular recognition by these enzymes. The more the aromatic nucleus is directed towards the outside of the molecule, the better the recognition. This can be explained by the presence of $\pi-\pi^*$ or hydrophobic interactions between the drug and the protein. We thus calculated the hydrophobic potential surfaces of each molecule using the MOLCAD module of SYBYL molecular modelling software (Figure 3).

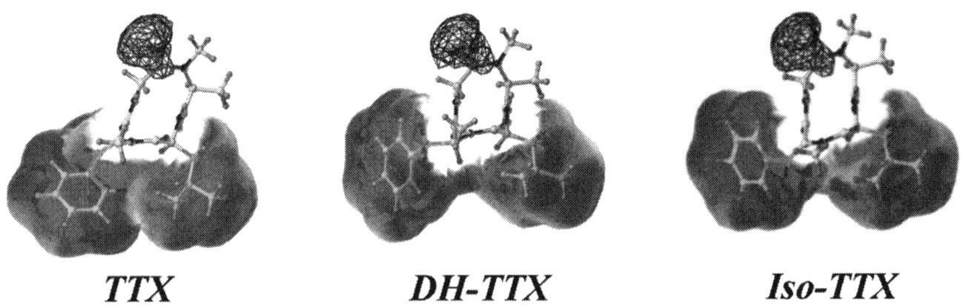

Figure 3. Common recognition points of tentoxin analogues. The hydrophobic area (in grey) is presumably responsible for the recognition by the three proteins. The hatched area is presumed to be involved in the modulation of ATPase activities of F0-F1 ATPase and Pgp.

The three analogues present a marked hydrophobic potential around the aromatic nucleus. This hydrophobic area is likely to be a common point of recognition in the active sites. This confirms the role of the cyclopeptide aromatic residue in the affinity. The difference in affinity comes from the relative position with respect to the aromatic nucleus compared to the cyclopeptide. The analysis of the tentoxin metabolites produced by P450-3A and the molecular docking studies also suggest that this hydrophobic zone is responsible for the substrate recognition by P450-3A.

F0-F1 ATPase and P-gp show a modulation of their ATPase activity. Using other tentoxin analogues for F0-F1 ATPase and other substrates for P-gp, we concluded that ATP hydrolysis regulation cannot be explained only by the interaction of this hydrophobic zone. It may be possible that a second site interacts with the substrate via hydrogen bonds for example. If we study the capacity of each molecule to form hydrogen bonds, we see that the glycine carbonyl is an ideal acceptor site because it affords good accessibility to a solvent molecule or an active site residue. Thus we can make the assumption that the glycine carbonyl of tentoxin analogues is involved in the modulation of the ATPase activity both for F0-F1 ATPase and P-gp.

REFERENCES

(1)   N.D. Fulton, K. Bollenbacher, and G.E. Templeton, A metabolite from *Alternaria tenuis* that inhibits chlorophyll production, *Phytopathology*, **55:** 49 (1965).

(2)   R.D. Durbin, and T.F. Uchytil, A survey of plant insensitivity to tentoxin, *Phytopathology*, **67**: 602 (1977).

(3)   I. Dahse, S. Pezennec, G. Girault, G. Berger, F. André, and B. Liebermann, The interaction of tentoxin with CF1 and CF1-ε isolated from spinach chloroplast, *J. Plant Physiol.* **143**: 615 (1994).

(4)   J.A. Steele, T.F. Uchytil, R.D. Durbin, P. Bathnagar, and D.H. Rich, Chloroplast coupling factor 1: a species-specific receptor for tentoxin, *Proc. Natl. Acad. Sci.* USA, **73**: 2245 (1976).

(5)   M. Delaforge, F. André, M. Jaouen, H. Dolgos, H. Benech, J.M. Gomis, J.P. Noël, F. Cavelier, J. Verducci, J.L. Aubagnac, and B. Liebermann, Metabolism of tentoxin by hepatic cytochrome P-450 3A isozymes, *Eur. J. Biochem.*, **250**: 150 (1997).

(6)   C. Wandel, R.B. Kim, S. Kajiji, F.P. Guengerich, G.R. Wilkinson, and A.J.J. Wood, P-glycoprotein and cytochrome P-450 3A inhibition: dissociation of inhibitory potencies, *Cancer research* **59**: 3944 (1999).

(7)   B. Liebermann, R. Ellinger, and E. Pinet, Isotentoxin, a conversion product of the phytotoxin tentoxin, *Phytochemistry (Oxf.)* **42**: 1537 (1996).

(8)   J.B. Schenkman, S.G. Sligar, and D.L. Cinti, Substrate interaction with cytochrome P450, *Pharmacol. Ther.* **42**: 43 (1981).

(9)   G. Berger, G. Girault, and J.M. Galmiche, Application of HPLC to the study of the chloroplast ATPase Mg^{2+} dependent mechanism, *J. Liquid. Chrom.* **13**: 4067 (1990).

(10) M. Garrigos, J.Jr. Belehradek, L.M. Mir, and S. Orlowski, Absence of cooperativity for MgATP and verapamil effects on the ATPase activity of P-glycoprotein containing membrane vesicles, *Biochem. Biophys. Res. Commun.* **196**: 1034 (1993).

(11) E. Pinet, Etude par résonance magnétique nucléaire de la tentoxine et d'analogues tétrapeptidiques cycliques pour l'élaboration d'une plate-forme structurale d'étude de l'ATP synthase chloroplastique, Doctoral thesis, Université PARIS VI (1996).

(12) J. Santolini, Etude de la relation structure-fonction d'une phytotoxine naturelle, la tentoxine, et de ses dérivés synthétiques. Interaction avec l'ATP-synthase chloroplastique, Doctoral thesis, Université PARIS VI (1999).

TRIAZOLAM SUBSTRATE INHIBITION

Evidence of Competition for Heme-Bound Reactive Oxygen within the CYP3A4 Active Site

Michael L. Schrag and Larry C. Wienkers

Drug Metabolism Research
Pharmacia Corporation, Kalamazoo, MI 49007

INTRODUCTION

The active site of CYP3A4 is generally considered to be spacious as evidenced by its ability to oxidize large polycyclic aromatic hydrocarbons and macrolides such as cyclosporin. In deference to the large CYP3A4 active site, it has been hypothesized that two substrates may physically occupy the active site simultaneously (Shou et al., 1994). A number of studies have provided evidence which supports this hypothesis and a kinetic model has been established which rationalizes many of the anomalous kinetics associated with CYP3A4 catalytic activity (Korzekwa et al., 1998).

Recently it was reported that the biotransformation of triazolam by CYP3A4 was characterized by "substrate inhibition" (Perloff et al., 2000). Moreover, midazolam, a close structural analog of triazolam, has also been shown to display the same kinetic phenomenon (Gorski et al., 1994; Kronbach et al., 1989; Perloff et al., 2000). The current study explores the hypothesis that "substrate inhibition" is a consequence of multiple binding of triazolam within the active site of CYP3A4. We find evidence for a model in which triazolam binds concurrently in two specific orientations (leading to separate oxidative products) which compete for activated oxygen.

EXPERIMENTAL RESULTS

In vitro microsomal oxidation of triazolam generated both 1'-hydroxytriazolam (1'OHTz) and 4-hydroxytriazolam (4OHTz) metabolites. During preliminary kinetic studies it was observed that the substrate-velocity curve for 1'OHTz formation was non-hyperbolic at higher substrate concentrations. Figure 1 shows that the rate of 1'OHTz formation was found to decline at elevated substrate concentrations, a characteristic which is commonly referred to as 'substrate inhibition' (Segal, 1975).

Additional kinetic experiments were conducted with the objective of identifying the mechanism of behind 1'OHTz substrate inhibition. One possibility was that product inhibition could account for the decrease in 1'OHTz formation at higher concentrations. Consequently, incubations were supplemented with non-radio-labeled authentic metabolite standards prior to the initiation of the reaction. The addition of both metabolites up to a concentration of 30 μM had no significant effect on $[^{14}C]1'OHTz$ formation. Thus, indicating that product inhibition was not a likely explanation for 1'OHTz substrate inhibition.

Biological Reactive Intermediates VI, Edited by Dansette *et al.*
Kluwer Academic / Plenum Publishers, 2001

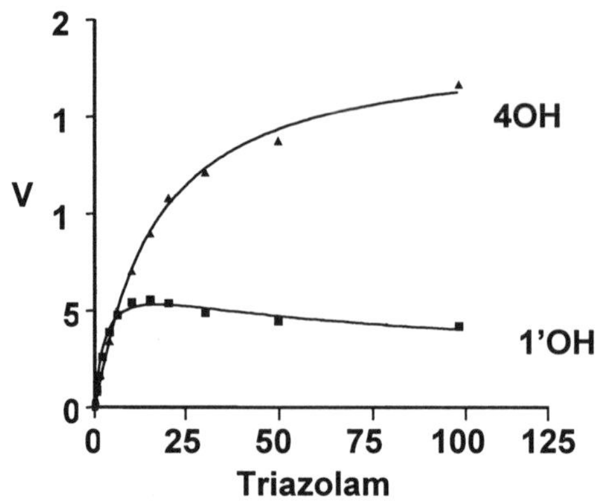

Figure 1. Substrate velocity curves for triazolam metabolites.

The potential for CYP3A4 to bind multiple substrates offered a second possible explanation for triazolam substrate inhibition. This was tested by competitive inhibition experiments where it was hypothesized that if triazolam bound to separate active site locations (each producing one of the two triazolam metabolites) then an inhibitor could potentially demonstrate product selective inhibition. Selectivity could result from an inhibitor that may also have a propensity to bind in one of the proposed triazolam binding regions. In a subsequent inhibition screen it was observed that steroids, such as testosterone and progesterone, selectively inhibited the formation of 1'OHTz over 4OHTz. Hence competitive inhibition experiments provided evidence that triazolam bound in at least two spatially distinct locations within the CYP3A4 active site.

In this light, we hypothesized that 'substrate inhibition' might be the result of competition for reactive oxygen between the two separate triazolam binding orientations within a single CYP3A4 active site, Figure 2. Such a hypothesis was deemed reasonable in light of isotope effect experiments which have shown that the substrate rotational and transnational movement within a P450 active site can be faster than the rate of the substrate oxidation (Jones et al., 1986; Korzekwa et al., 1989).

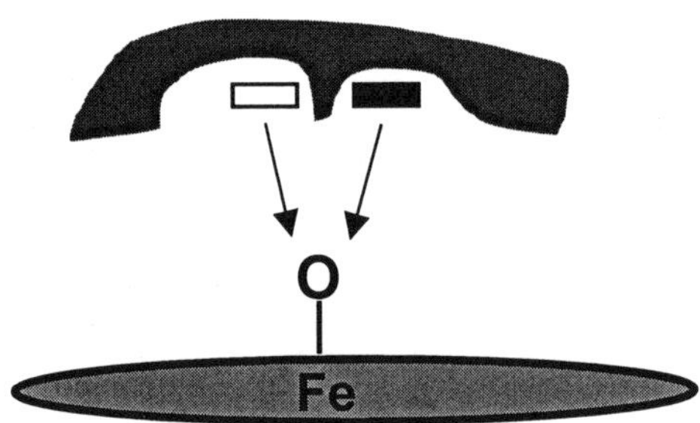

Figure 2. Hypothetical scheme depicting multiple triazolam molecules in a the CYP3A4 active site.

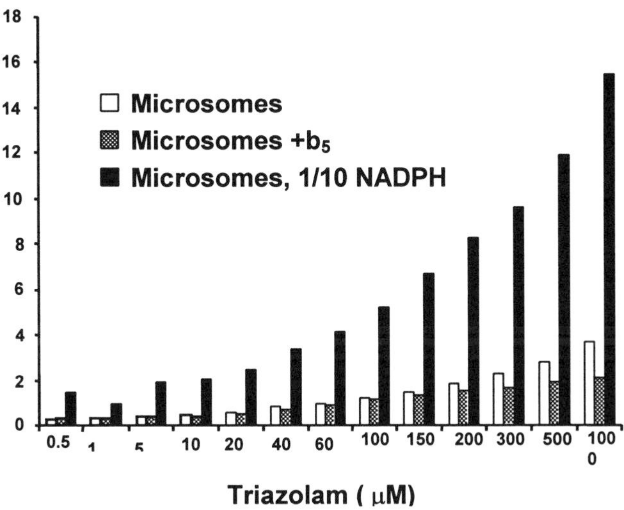

Figure 3. Effect of cytochrome b_5 and reduced NADPH concentration on the 4OHTz/1'OHTz ratio.

Thus, it appeared likely that triazolam might occupy the active site in a single (ES) or double (ESS) binding mode and that the relatively faster time scale of binding would allow for equilibration between two (or more) binding locations within the CYP3A4 active site. Moreover, these distinct orientations could potentially compete for reactive oxygen.

Supplementation of incubations with cytochrome b_5 provided an indirect means to test this idea. It was anticipated that the addition of cytochrome b_5 would increase the efficiency of CYP3A4 reactive oxygen production and therefore enhance the availability of heme bound oxygen for either binding orientation. Previous studies with CYP3A4 have demonstrated that cytochrome b_5 can decrease the uncoupling of the catalytic cycle and increase the partition of the iron-oxygen complex towards substrate oxidation (Perret and Pompon, 1998). In Figure 3, the ratio of 4OHTz/1'OHTz is plotted as a function of substrate concentration and it can be observed that, compared to control, when exogenous cytochrome b_5 is added to human liver microsomes the ratio decreases at higher substrate concentrations. Indeed, when this same data (b_5 supplemented) is re-plotted as a substrate-velocity curve, 1'OHTz substrate inhibition is no longer apparent. One potential interpretation is that at low substrate concentrations, 1'OHTz production effectively competes with 4OHTz because the 4OHTz binding orientation has not yet saturated (apparent K_m = 171 μM). However, as occupation of the 4OHTz site increases this in turn leads to increased formation of 4OHTz with increased competition between the two sites for heme-bound reactive oxygen. The result is that the 4OHTz/1'OHTz ratio increases with higher substrate concentrations. At the highest concentrations, where most of the enzyme is driven towards the doubly occupied enzyme (ESS), formation of 1'OHTz is suppressed due to competition. The result is apparent 'substrate inhibition'. The addition of cytochrome b_5 increases the availability of heme-bound reactive oxygen and therefore decreases the observed substrate inhibition as visualized by a change in the ratio of 4OHTz/1'OHTz at saturating substrate concentrations.

Since the interaction CYP3A4 with cytochrome b_5 may also involve an enzyme conformational change, this could complicate the interpretations proposed above. In a separate experiment, triazolam substrate-velocity data was obtained in the presence of 100 μM NADPH, 1/10 the amount of cofactor typically used in an in vitro incubation. In this instance, it was rationalized that by using a reduced concentration of NADPH, the presentation of heme-bound oxygen to either triazolam orientation would become limited and hence the competition between binding orientations would be increased. The data in

Figure 3 confirmed this prediction, as reduced NADPH concentration greatly enhanced substrate   competition and markedly increased the concentration dependent ratio of 4OHTz /1'OHTz   over the entire substrate range. The substantial ratio increase supported the conclusion that electronic factors were predominant and not protein conformational change.

In conclusion, the enzymatic properties of CYP3A4 are complex and at present, knowledge of active site dynamics is limited. However, based upon data presented, it is likely that one factor involved in CYP3A4 active site dynamics does stem in part from competition between two triazolam binding orientations for heme-bound reactive oxygen within a single CYP3A4 active site.

REFERNCES

Gorski, J.C., Hall, S.D., Jones, D.R., VandenBranded, M. and Wrighten, S.A., 1994, Regioselective biotransformation of midazolam by members of the human cytochrome P450 3A (CYP3A) subfamily. *Biochem Pharm* **47**: 643.

Jones, J.P., Korzekwa, K.R., Rettie, A.E. and Trager, W.F., 1986, Isotopically sensitive branching   and its effect on the observed intermolecular isotope effects in cytochrome P-450 catalyzed reactions: A new method for the estimation of intrinsic isotope effects. *J Am Chem Soc* **108**:7074.

Korzekwa, K.R., Trager, W.F. and Gillette, J.R., 1989, Theory for the observed isotope effects from enzymatic systems that form multiple products via branched reaction pathways: cytochrome P450. *Biochemistry* **28**:9012.

Korzekwa, K.R., Krishnamachary, N., Shou, M., Ogai, A., Parise, R.A., Rettie, A.E., Gonzalez, F.J. and Tracy, T.S., 1998, Evaluation of atypical cytochrome P450 kinetics with two-substrate models: Evidence that multiple substrates can simultaneously bind to cytochrome P450 active sites. *Biochemistry* **37**:4137.

Kronbach, T., Mathys, D., Umeno, M., Gonzalez, F.J. and Meyer, U.A., 1989, Oxidation of midazolam and triazolam by human liver cytochrome P450IIIA4. *Mol Pharm* **36**:89.

Perret, A. and Pompon, D., 1998, Electron shuttle between membrane-bound cytochrome P450 3A4 and b$_5$ rules uncoupling mechanisms. *Biochemistry* **37**:11412.

Segal, I.H. (1975) *Enzyme kinetics* pp 384, John Wiley & Sons, New York.

Shou, M., Grogan, J., Mancewics, J.A., Krausz, K.W., Gonzales, .F.J, Gelboin, H.V. and Korzekwa, K.R., 1994) Activation of CYP3A4: Evidence for the simultaneous binding of two substrates in a cytochrome P450 active site. *Biochemistry* **33**:6450.

EFFECT OF THE MICROSOMAL SYSTEM ON QUINONE REDOX CYCLING, OXYGEN ACTIVATION, AND LIPID PEROXIDATION

Pavel Souček[*] and Ivan Gut[*]

[*] National Institute of Public Health, Center of Occupational Diseases, Šrobárova 48, 100 42 Praha 10, Czech Republic, email: psoucek@szu.cz

INTRODUCTION

Our previous results indicated that benzoquinone caused cytochrome P450 destruction[1-3]. Metabolites of benzene - catechol, hydroquinone, and paticularly its oxidized form, benzoquinone, exert toxic effects on biomacromolecules, eg by covalent binding to proteins, formation of DNA adducts[4], and destruction of cytochrome P450 (CYP)[1,3]. Benzene is an established human and animal carcinogen[5] acting through an epigenetic mechanism. The main goal of this study was to provide a more detailed insight into toxicology of benzene. Interconversions among hydroquinone (HQ), semiquinone radical (SQ), and benzoquinone (BQ) in both aqueous solutions and in the presence of the microsomal system were followed up. The production of hydroxyl radicals and the initiation of lipid peroxidation was investigated and the role of iron and antioxidant ascorbate was assessed. Microsomes from rats pretreated with CYP2E1 inducer were used in all experiments because this CYP isoform has been suggested to produce enhanced amounts of reactive oxygen species and increased the rate of lipid peroxidation *in vitro*[6]. The results presented here bring an additional evidence for the recently proposed novel mechanism of toxicity induced by metabolites of benzene[3].

MATERIALS AND METHODS

HQ or BQ (for concentrations used see figures) were incubated in aqueous solution (50 mM potassium phosphate, pH 7.4) or in the presence of microsomes with chemically induced CYP2E1 (1 mg of protein/ml) with or without 1000 µM NADPH (for usual NADPH-generating system see reference 1)in shaking water bath at 37 °C.

After various periods of time, the concentration of the following agents was measured:
- hydroquinone and benzoquinone by spectrophotometry[7]
- semiquinone radicals by electron spin resonance (ESR; see 1 - 3 and references cited therein)
- malondialdehyde, marker of lipid peroxidation, by colorimetric method of Buege and Aust[8]
- OH radicals by colorimetric method of Khan *et al.*[9]

Biological Reactive Intermediates VI, Edited by Dansette *et al.*
Kluwer Academic / Plenum Publishers, 2001

In some experiments, ascorbate (100, 500, 1000, 3000, 5000 μM), EDTA (5, 10, 30, 50, 100 μM), or FeCl₃ (5, 25, 100, 500 μM) were added to the incubation mixture of microsomes with HQ/BQ and/or NADPH. After 60 min. incubation, the production of OH radicals and lipid peroxidation was measured. Chelex-100 was used for removal of transition metals from all solutions but not microsomes in experiments designed to study the influence of FeCl₃.

RESULTS AND DISCUSSION

In buffered aqueous solution, HQ was slowly oxidized to BQ via a SQR. This conversion was slowed down by the addition of NADPH and stopped by the complete system, i.e. microsomes and NADPH (Figure 1). BQ was reduced to SQR and HQ at a significantly higher rate and this conversion was stimulated by NADPH and more effectively by microsomes plus NADPH while SQR was quenched there (Figure 2).

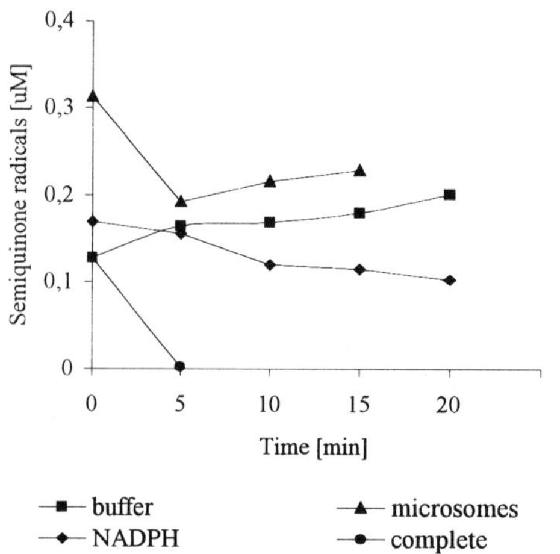

Figure 1. SQR formation in solution of 1000 μM HQ - influence of NADPH and microsomes.

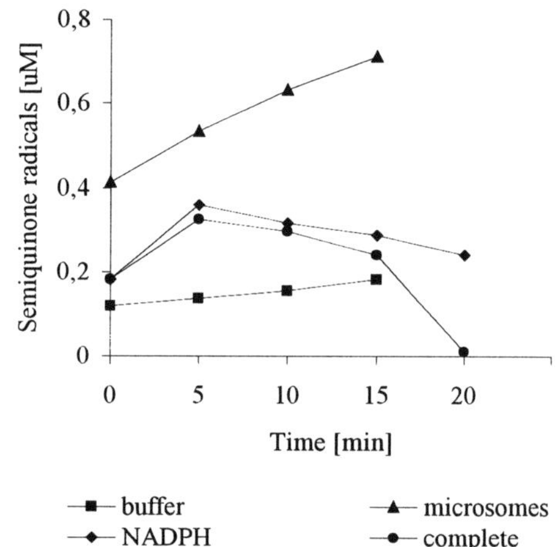

Figure 2. SQR formation in solution of 1000 μM BQ - influence of NADPH and microsomes.

Our data indicate the participation of microsomal NADPH-dependent enzymes in reduction of BQ. The role of various microsomal bioreductive enzymes in quinone toxicity was implicated, e.g. microsomal NADPH-P450 reductase, NADH-cytochrome b₅ reductase, and other flavoenzymes acting as single-electron donors. Also the influence of a mostly cytosolic two-electron donor NAD(P)H-quinone oxidoreductase cannot be excluded. These data together with the previously observed rapid destruction of CYP in microsomal samples, where BQ reduction was slow in the absence of NADPH[1] suggested that BQ was the main CYP destructive agent.

In microsomes with NADPH, both HQ and BQ stimulated the formation of OH radicals (Figure 3) but inhibited peroxidation of lipids (Figure 4). This finding suggested that processes involving production of OH radicals and lipid peroxidation were independent. In microsomes incubated with HQ or BQ in the absence of NADPH, similar level of CYP destruction as in the presence of NADPH was observed[1,3]. Thus, it seems evident that BQ does damage CYP neither by the initiation of lipid peroxidation nor by enhanced production of OH radicals but the main mechanism of destruction is covalent binding to the protein and destruction of CYP heme we detected before[1].

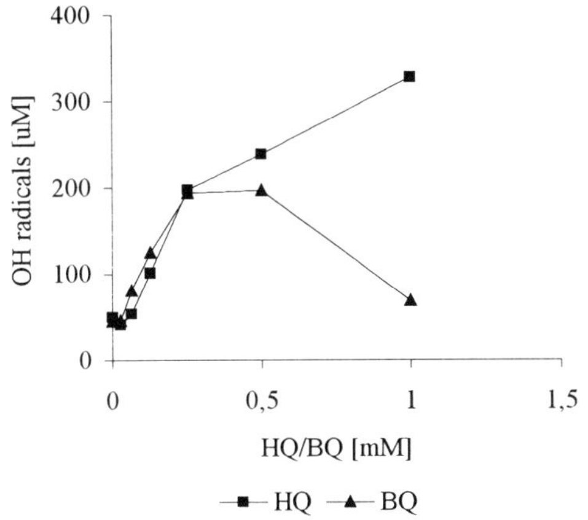
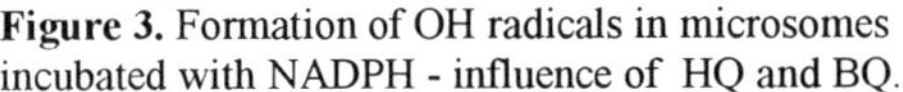

Figure 3. Formation of OH radicals in microsomes incubated with NADPH - influence of HQ and BQ.

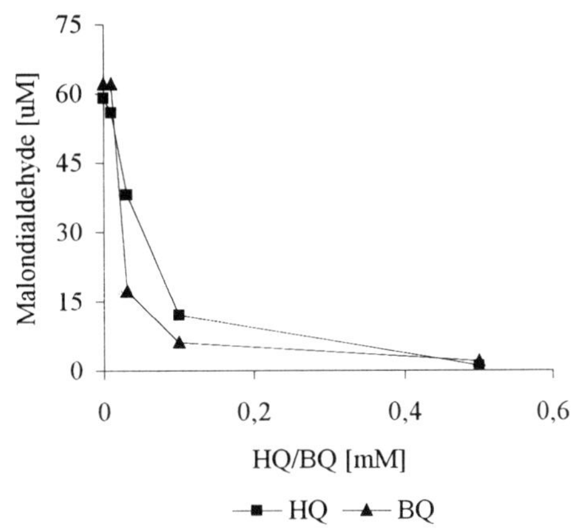

Figure 4. Peroxidation of lipids in microsomes incubated with NADPH - influence of HQ and BQ.

In an attempt to analyze the observed synergic effect of HQ/BQ and NADPH on the production of OH radicals in microsomes we have employed antioxidant ascorbate. We have reported previously that 1 mM ascorbate caused significant CYP destruction in microsomes while 5 mM ascorbate did not[1]. Ascorbate at 0.5 - 5 mM concentration produced significant generation of OH radicals in microsomes. Neither HQ nor BQ did change this ascorbate effect. On the contrary, 0.1 - 1.0 mM ascorbate stimulated peroxidation of lipids in microsomes whereas presence of HQ or BQ completely inhibited this deleterious effect of ascorbate (Figure 5). These results suggest that lipid peroxidation could be responsible for the reported CYP damage by ascorbate[1]. On the other hand, the production of OH radicals probably does not contribute to CYP destruction because a low extent of CYP destruction was reported when ascorbate and HQ or BQ were incubated together with microsomes[1].

We have also used EDTA as chelator in order to investigate the influence of traces of transition metals, especially iron which, play a significant role in OH radicals formation and initiation of lipid peroxidation[10]. EDTA also offered protection of CYP enzymes in experiments

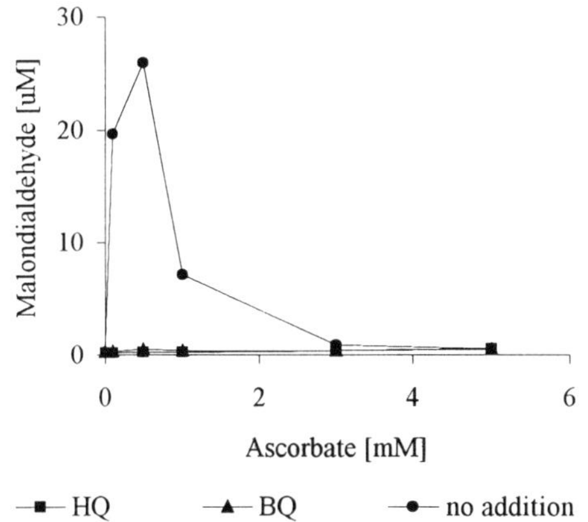

Figure 5. Lipid peroxidation in microsomes incubated with 500 µM HQ or BQ without NADPH - influence of ascorbate.

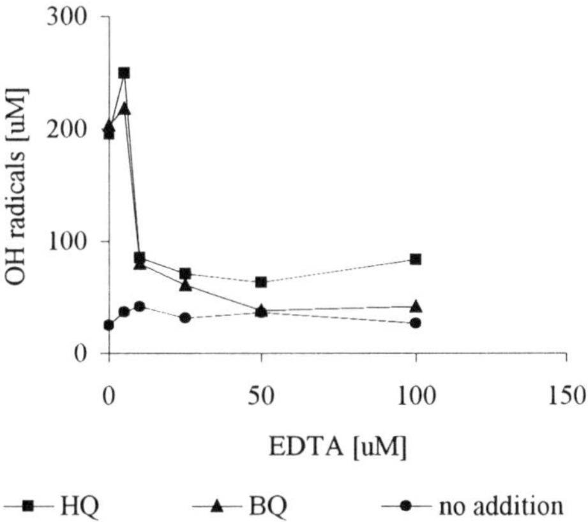

Figure 6. Formation of OH radicals in microsomes incubated with 250 µM HQ or BQ without NADPH - influence of EDTA.

described previously[1]. EDTA inhibited lipid peroxidation and formation of OH radicals in microsomes incubated with NADPH and HQ/BQ (Figure 6). Addition of Fe^{3+} to metal-free solutions stimulated production of OH radicals (Figure 7). Our data on the absence of production of OH radicals in microsomes incubated with the metal-free buffer solutions of BQ and NADPH, and production of OH radicals proportional to $FeCl_3$ up to 500 µM concentration, supported the view that iron played an important role there. In contrast, lipid peroxidation was not influenced by iron in these conditions.

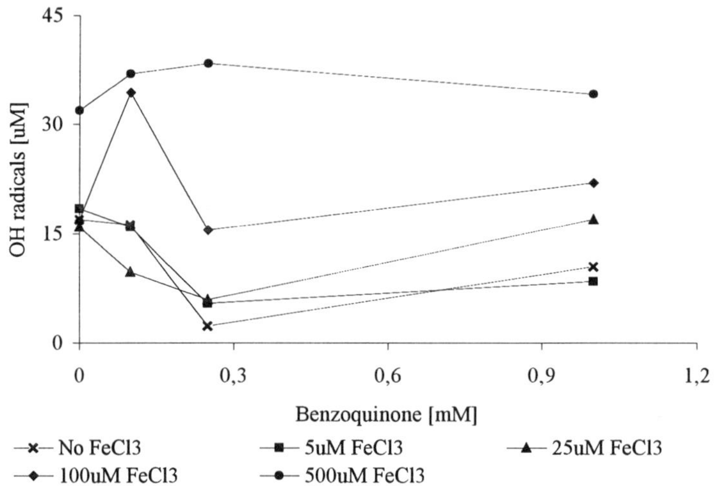

Figure 7. Formation of OH radicals in aqueous solution of benzoquinone - influence of $FeCl_3$.

It can be concluded that interconversions between HQ and BQ are influenced by NADPH and more effectively by the complete microsomal system exerting mostly a reducing action. Both HQ and BQ are very efficient inhibitors of lipid peroxidation most probably by binding of free or enzymatically bound transition metals. On the other hand, when a sufficient amount of NADPH is present, overproduction of OH radicals may be observed due to the presence of especially reduced form, i.e. HQ. Thus, the CYP destruction observed before[1-3] was caused mainly by direct attack of BQ to CYP heme and protein.

Acknowledgements

The work at this project was supported by grant IGA, No.: 6095-3.

References

1. P. Souček, B. Filipcová, and I. Gut, *Biochem. Pharmacol.*, 47: 2233-42, (1994).
2. I.Gut, V.Nedelcheva, P.Souček, P.Stopka, and B.Tichavská, *Env. Health Perspect.*, 104, Suppl. 6:1211-8, (1996)
3. P. Souček, *Chem-Biol. Interact.*, 121: 223-36, (1999).
4. D. Ross, *Eur. J. Hematol.*, 57 (suppl): 111-8 (1996).
5. *IARC Monographs on the Evaluation of Carcinogenic Risk to Humans.* Vol 1-42, Suppl. 7, Lyon, France, (1987)
6. G. Ekström, and M. Ingelman-Sundberg, *Biochem. Pharmacol.*, 38: 313-9 (1989).
7. R.D. Irons, D.A. Neptun, and R.W. Pfeifer, *J. Reticuloendothel. Soc.*, 30: 359-72 (1981).
8. J.A. Buege and S.D. Aust, *Methods Enzymol.*, 52: 302-10, (1978).
9. R. Khan, K.P. Krishnamurthy, and K.P. Padnya, *Biochem. Pharmacol.*, 39: 1393-5 (1990).
10. Y.H. Miura, I. Tomita, T. Hirayama, and S. Fukui, *Biol. Pharm. Bull.*, 21: 93-6 (1998).

REACTION OF NUCLEIC ACIDS WITH TRIFORMYLMETHANE; A NOVEL DNA-MODIFYING AGENT

Koissi Niangoran,[*] Neuvonen Kari and Lönnberg Harri

Department of Chemistry, University of Turku, FIN-20014 Turku, Finland

INTRODUCTION

A large body of compelling evidence either confirms or implicates various environmental factors in the development of wide range of malignancies. Among the key factors are chemical and physical carcinogens. It is generally accepted that DNA adducts of chemical carcinogens play a role in carcinogenesis; however, the relationship between DNA adducts and carcinogenesis is not fully understood.

Triformylmethane (TFM, **1**)[a] is the simplest tricarbonyl compound. It has a pK_a around 2.3 in water at room temperature. It may be classified as a derivative of MDA bearing a formyl group, which makes it rather stable. Surprisingly, for a compound first prepared nearly four decades ago (Arnold and Zemlicka, 1960), there are only few references in the literature dealing with nucleic acid chemistry. Neuvonen *et al.* (Neuvonen *et al.*, 1996) first showed that TFM formed adducts with nucleic acid bases, adenine and cytosine. Recently, we have described an efficient synthesis route of fluorescent compounds by the condensation of TFM with guanosine in separated accounts (Koissi *et al.*, 1999).

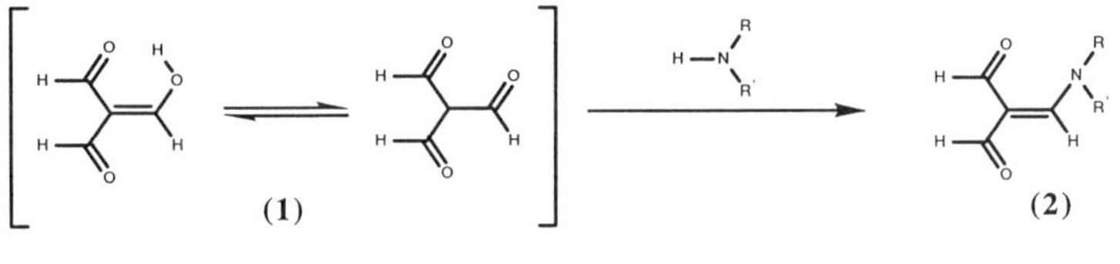

Scheme 1: General reaction of TFM with primary amine

In light of our considerable expansion of TFM research and our effort to understand its chemical and physical properties, we have focused on its mechanism of reaction with

---

[*] e-mail: niko, Tel: (358) 2-3336786; Fax: (358) 2-3336700

[a] Abbreviations: TFM: Triformylmethane, MDA: Malonaldehyde, Malondialdehyde

Biological Reactive Intermediates VI, Edited by Dansette *et al.*
Kluwer Academic / Plenum Publishers, 2001

nucleic acids. In this paper, we present some data on the primary condensation products from the reaction of adenosine and guanosine with TFM. Characterization of the adducts were based on ^{1}H and ^{13}C NMR spectroscopy, UV absorbance, and mass spectrometry data.

RESULTS AND DISCUSSION

The course of the reaction of TFM with primary amines depends both on the character of the amine and the reaction conditions. Essential for a facile formation of compounds described hereby, is an equimolar ratio of both reactants, or only a slight excess of one of them. TFM exhibits keto-enol tautomerism. Indeed, in these reactions, TFM behaves more as a hydroxyvinyl dicarboxaldehyde than as a tricarbonyl molecule. The adduct formation is probably due to the electron attractive effect of the two formyl groups at the 1-position of the diformyl-2-hydroxyethylene as a tautomeric form of (**1**) (see scheme 1). In principle, it can undergo reactions known to be typical for either aldehydes or enols and its synthetic potential is much based on this fact, but has not been yet satisfactorily evaluated. The expected product from such condensation with primary amine is an aminomethylene derivative (**2**), which could cyclized. To our knowledge, no studies of DNA adduct formation specific to TFM have been published so far.

The mechanism of DNA adduct formation by substituted malondialdehyde derivative, with nucleic acid bases affords an additional arm, which could ring closure to yield a propeno like bridge (Moschel and Leonard, 1976). At adequate conditions, it has been possible to generate in good yield some adducts. Cyclic and / or acyclic adducts from guanosine and adenosine reactions have been structurally characterized. A monoadduct derived from TFM at adenine residue has been observed. We were able to identify an open-chain structure like as previously reported with MDA (Marnett, 1999). Its isolation has not yet been possible, but it was easily assigned as (**3**) (scheme 2)[‡].

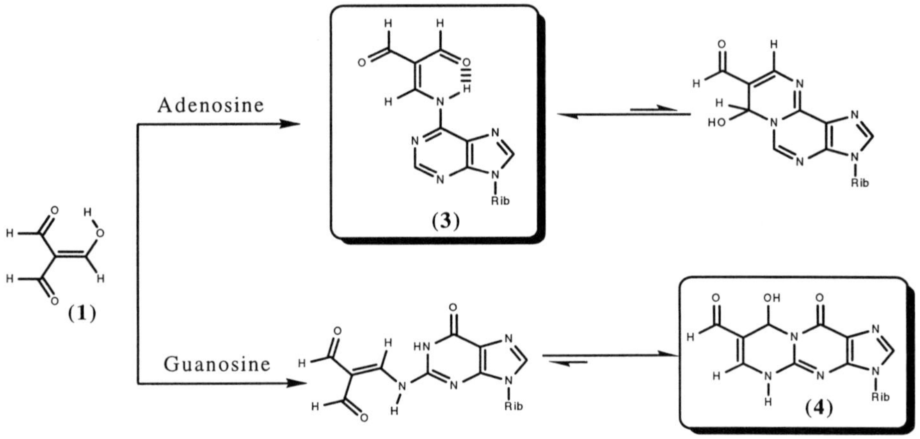

Scheme 2: Reactions of adenosine and guanosine with TFM

[‡]**TFM-Adenosine adduct (3):** ^{1}H NMR (500 MHz, DMSO): δ (ppm); 12.33 (NH); 9.95 (CHO); 9.68 (CHO); 9.13 (HC); 8.88 (H-2); 8.76 (H-8); 6.05 (H-1´); 4.62 (H-2´); 4.22 (H-3´); 4.01 (H-4´); 3.70- 3.62 (H-5´, 5´´)
TFM-Guanosine adduct (4): ^{13}C NMR (500 MHz, DMSO): δ (ppm); 61.30 (C-5'), 66.78 (C-8), 70.37 (C-3'), 73.98 (C-2'), 85.52 (C-4'), 86.63 (C-1'), 114.97 (C-7), 119.11 (C-9a), 137.98 (C-2), 143.06 (C-6), 145.62 (C-4a), 148.05 (C-3a), 154.99 (C-9), 187.33 (CHO)

By contrast to the reaction with Guanosine, the open-chain product with adenosine could be identified as a major compound in nearly 95 % yield, whereas the cyclized one was not stable enough to be isolated in pure form. Dehydration during cyclization of the adducts yields a schiff base intermediate which undergoes covalent hydration, producing an equilibrium mixture of diastereomers. In our reaction condition, the covalent hydrate phenomenon is essential and strongly marked. (**4**), which is rather stable compound, has been assigned in equilibrium with its dehydrated analog (**5**) (Koissi *et al.*). The literature cited below clearly supports the DNA adducts structures of the reaction products formed from the investigated dicarbonyl and DNA bases. The chemical properties, especially the stability of the cyclic compounds, and their close structural relationship to the previously reported adducts is a strong data (Mao *et al.* 1999; Nechev *et al.* 2000).

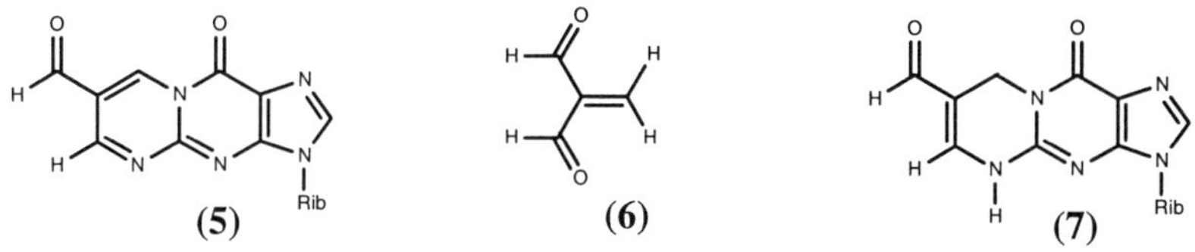

Figure 1: Chemical structures of some exocyclic guanosine adducts

Currently, the mechanism of decomposition of TFM is not well known, and it is therefore difficult to decide whether TFM really can generate certain intermediates. Nevertheless, structural data collected on some compounds such as (**7**), which derived from secondary reaction of TFM and guanosine, suggested the decomposition of TFM into MDA, which is a well-known, toxic and mutagenic metabolite. Although no extensive studies on the presence of MDA have been undertaken so far, it is most likely that the mechanism involved such intermediate. While the mechanism of formation of that compound remains obscure, it is worth noting that a plausible mechanism has been tentatively formulated (Koissi *et al.*). It is established that, the self-condensation of TFM generates loss of formic acid. Therefore, an aldol condensation type reaction between MDA and formaldehyde and subsequent dehydration would initially give (**6**) (Gómez-Sánchez *et al.* 1993). A Michael type addition of N-1 of guanosine, followed by intramolecular cyclization through N^2 would yield (**7**) (figure 1). Most compounds are isolated in equilibrium with their tautomeric parent. Some of the products exhibit remarkable fluorescence properties which, could be of great importance.

CONCLUSIONS

The elucidated structures and the information on the stability of the modified bases should be the basis for development of methods to detect and quantify TFM-derived DNA adducts in biological systems. The apparent evidence for the significant reactivity could help in a better understanding of its probable carcinogenicity and the evaluation of its toxicological aspects. Furthermore, these adducts allow derivatization through their reactive aldehydic function.

REFERENCES

Arnold, Z. and Zemlicka, J., 1960, Synthetische reaktionen von dimethylformamid VII, darstellung von triformylmethan. *Collect. Czech. Chem. Commun., 25: 1318-1322*

Gómez-Sánchez, A., Hermosín, I., Lassaletta, J-M. and Maya, I., 1993, Cleavage and oligomerization of malondialdehyde. *Tetrahedron, 49: 1237-1250*

Koissi, N., Neuvonen, K., Munter, T., Kronberg, L. and Lönnberg, H., 1999, A novel fluorescent base modification by reaction of guanosine with triformylmethane. *Collection Symposium Series. Chemistry of nucleic acid components. 2: 228-229*

Koissi, N., Neuvonen, K., Munter, T., Kronberg, L. and Lönnberg, H., Reaction of malonaldehyde derivative, triformylmethane with guanosine. *Eur. J. Org. Chem.,* submitted

Mao, H., Schnetz-Boutaud, N.C., Weisenseel, J.P., Marnett, L.J. and Stone, M.P. 1999, Duplex DNA catalyzes the chemical rearrangement of a malondialdehyde deoxyguanosine adduct. *Proc. Natl. Acad. Sci.U.S.A, 96: 6615* and references therein

Marnett.L.J., 1999. Chemistry and biology of DNA damage by malondialdehyde. In Singer, B. and Bartsch, H., eds. *Exocyclic DNA adducts in mutagenesis and carcinogenesis* (IARC Sci. Publ. No 150) Lyon, IARC, pp. 17-27

Moschel, R.C. and Leonard, N.J., 1976, Fluorescent modification of guanine. Reaction with substituted malondialdehydes. *J. Org. Chem. 41: 294-300*

Nechev, L.V., Harris, C.M., and Harris, T.M., 2000, Synthesis of nucleosides and oligonucleotides containing adducts of acrolein and vinyl chloride. *Chem. Res. Toxicol., 13: 421-429*

Neuvonen, K., Zewi, C. and Lönnberg, H., 1996, Condensation of triformylmethane with heteroaromatic amines, including nucleic acid bases. A novel example of ring-chain tautomerism. *Acta Chem. Scand. 50: 1137-1142*

ARE BLOOD-BRAIN INTERFACES EFFICIENT IN PROTECTING THE BRAIN FROM REACTIVE MOLECULES ?

Jean-François Ghersi-Egea, Nathalie Strazielle, Audrey Murat, Jonathan Edwards, and Marie-Françoise Belin

INSERM U 433
Faculté de Médecine R.T.H. Laennec
Rue G. Paradin
69008 Lyon, France

INTRODUCTION

It has been known from more than a century that the central nervous system is isolated from the rest of the body by a barrier between the blood and the brain. The brain is a complex organ, formed by numerous anatomically defined structures, which requires the maintenance of a finely tuned homeostasis to fulfill its functions. Besides the blood circulation, the brain possesses a circulatory system of its own : the cerebrospinal fluid, mainly synthesized by the choroid plexuses, circulates within the ventricular system of the brain, then in the subarachnoid space, before being reabsorbed in the venous blood, or draining in the lymphatic system. These special anatomical features, together with the homeostatic requirements of the brain, explain that not one, but several structures are responsible for the barrier between the blood and the brain, and that these barriers should be viewed more as active interfaces regulating the exchanges between the brain and the periphery. One function of these structures collectively referred to as the blood-brain interfaces (BBI) is to achieve brain protection against active, potentially deleterious endo- and xenobiotics.

MORPHO-FUNCTIONAL CHARACTERISTICS OF THE BLOOD-BRAIN INTERFACES

The interface between the blood and the brain parenchyma is located at the endothelial cells of brain capillaries and microvessels. This BBI is the most important in term of surface area and is usually referred to as the blood-brain barrier (BBB). The main interface between the blood and the CSF (blood-CSF barrier or BCSFB) is located at the choroid plexuses, whose main function is the secretion of most of the CSF which results from the active transport of inorganic ions and a subsequent movement of water through the choroidal epithelium. The arachnoid membrane (constituting the intermediate layer of the meninges) forms the interface between the downstream subarachnoid CSF and the venous blood. Exchange processes between the ventricular CSF and the adjacent parenchyma occur through a specialized epithelium, the ependyma, that cover the ventricular walls. The exchange between the subarachnoid CSF and most adjacent brain parenchymal structures is more limited, due to a thick multilayered network from glial origin called the glia limitans, covered externally by a pial membrane lying on a basal lamina. By contrast to the situation observed at the BBB or the BCSFB, the ependymal cells and the cells forming the glia limitans are not sealed by tight junctions. Thus the CSF compartment is part of the CNS, although this does not mean that the CSF composition is a true reflection of the extracellular fluid composition of the cerebral parenchyma (Ghersi-Egea et al., 1996). Figure 1 summarizes the different interfaces present in the brain which can influence the cerebral biodisposition of drugs and toxic compounds.

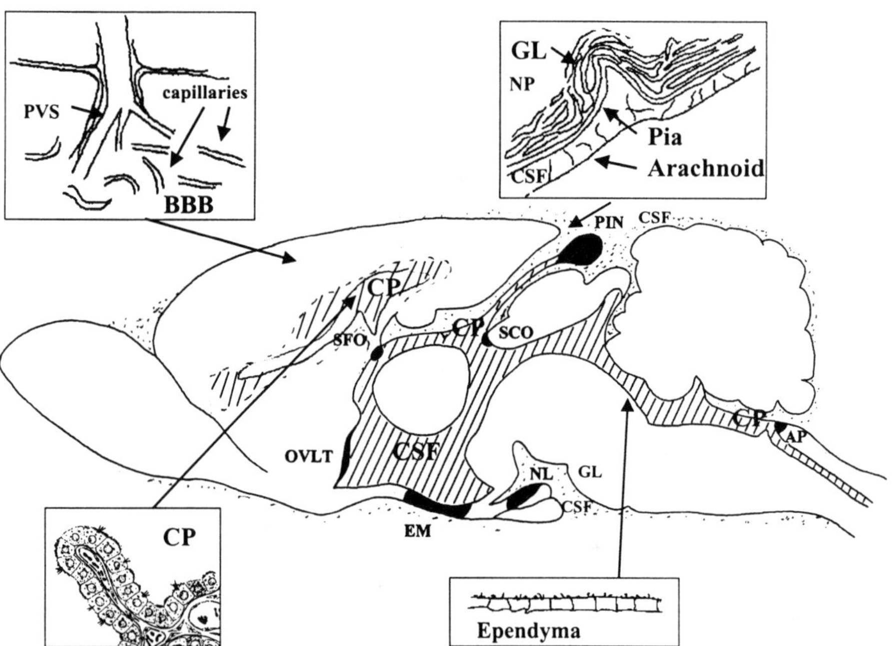

Figure 1 : Schematic representation of the rat brain showing the different interfaces between the blood, the brain parenchyma and the cerebrospinal fluid.
Hatched and dotted area represent ventricular cerebrospinal fluid (CSF) and subarachnoid CSF, respectively; CP indicates where the three types of choroid plexuses approximately locate; dark areas represent circumventricular organs : area postrema (AP), pineal gland (PIN), subcommisural organ (SCO), subfornical organ (SFO), organum vasculosum laminae terminalis (OVLT), median eminence (EM), neural lobe of the hypophysis (NL); other abbreviations : PVS : perivascular spaces; NP : neuropil, GL: glia limitans.

The Blood-Brain Barrier

The endothelial cells of the cerebral microvessels lie on a basal lamina which also engulfs pericytes, and are almost completely surrounded by astrocytic processes. The intercellular junctions that join these endothelial cells are tight (of the zonula occludens type), and of high electric resistance, and therefore prevent the paracellular passage of solutes into the brain. In addition, cerebral endothelial pinocytosis is strongly limited. Thus, brain penetration across the BBB occurs only along the transcellular pathway by passive diffusion unless specific mechanisms are involved. This is the case for hydrophilic nutrients that cannot dissolve in the lipid membranes, such as glucose or amino acids which enter the brain by crossing both luminal and abluminal membranes via specific transporters. Uptake of larger molecules such as transferrin or low density lipoproteins occurs via receptor-mediated transcytosis. Most drugs or toxic compounds will enter the brain parenchyma only as a function of their liposolubility and size. In addition, efflux mechanisms that prevent or limit the entry of xenobiotics in the brain also exist at the BBB (reviewed in Strazielle and Ghersi-Egea, 2000b). P-glycoprotein, a multidrug resistance associated protein physiologically expressed in cerebral capillaries is responsible for the efflux of a large range of structurally different compounds from within the endothelial cells back into the blood. As a result, the net brain influx of hydrophobic drugs such as vinca alcaloid antineoplastic drugs, ivermectin, cyclosporin A, digoxin or loperamide to name a few, is very low (Schinkel *et al.*, 1996). In addition, other, yet unidentified, transport protein (s) seem (s) to be involved in the efflux at the BBB of more hydrophilic compounds such as the organic anions cefodizime and valproic acid, azidothymidine or baclofen (Susuki *et al.*, 1997). Such a typical BBB exist in all but 7 stuctures involved in neuroendocrine regulation, collectively referred to as the circumventricular organs (Gross and Weindl, 1987; Fig 1). These structures are therefore particularly sensitive to toxic insults.

The Blood-Cerebrospinal Fluid Barrier

Three types of choroid plexuses are found, differing in their localization (lateral, third or fourth ventricle) and their morphology (Strazielle and Ghersi-Egea, 2000a). However, they are all composed of a choroidal epithelium protruding into the ventricle, and delimiting a conjunctive stroma that contain sinusoidal, fenestrated capillaries. In the choroid plexus, the effective barrier between the blood and the brain compartment (CSF) is localized at the choroidal epithelium whose cells are joined by tight junctions. In addition to these barrier properties and its role in CSF secretion from blood, the choroidal epithelium has numerous other functions such as nutrient and electrolyte transport, clearance of toxic neurotransmitter metabolites, secretory activities related to developmental and other hormonal regulation, and neuro-immune mediation (Strazielle and Ghersi-Egea, 2000a).

As at the BBB, the rate of entry of a xenobiotic into the CSF through the tight choroidal epiuthelium is mainly governed by its physico-chemical parameters (mainly the size and lipid solubility). However, the tightness of the interepithelial junctions is not as complete as for the BBB endothelial cells, and a low rate of paracellular influx of drug and toxicant may occur. With respect to efflux mechanisms, a rapid and specific efflux from CSF has been evidenced at the choroid plexus site for different compounds such as benzylpenicillin, quinolones, anionic pesticides and the cationic compound cimetidine (reviewed in Susuki *et al.*, 1997). Several organic anion transport proteins, of both the oatp and OAT families have being demonstrated at the apical membrane of the choroidal epithelium, and are likely to be involved in the efflux of these compounds (Ghersi-Egea and Strazielle, 2000). Choroidal drug metabolism and metabolite disposition may also modulate the cerebral biodisposition of drugs and reactive species and will be reviewed below.

DETOXIFICATION MECHANISM AT THE BLOOD-BRAIN INTERFACES

Different parameters factors will influence the cerebral bioavailability or the neurotoxicity of a drug or a reactive and toxic compound. These factors include not only the presence of tight junctions and efflux transporters which limit the entry of the xenobiotics in the CNS, as previously mentionned, but also plasma protein and red blood cell binding, and local cerebral blood-flow (Fenstermacher et al., 1993). In addition, the brain has, to some extent, an enzymatic capacity to detoxify reactive compounds, glutathione-dependent biochemical pathways being largely involved in these processes (Monks et al., 1999). A more pronounced localization of several cerebral drug metabolizing and antioxidant enzymes at the different BBI suggests another facet for the neuroprotective functions of these interfaces (reviewed in Strazielle and Ghersi-Egea, 2000b). Significant activities and/or expression of several phase I (i.e. functionalization) and phase II (i.e. conjugation) enzymes, have been recorded in cerebral microvessels, arachnoid, and choroid plexuses (Ghersi-Egea et al., 1994, Strazielle and Ghersi-Egea, 1999). Thus, as previously demonstrated for endogenous neurotransmitters, the BBI may form an enzymatic barrier between the blood and the brain for different active or reactive xenobiotics also. The most striking observation has been the very high level of enzymes such as epoxide hydrolase, UDP-glucuronosyl transferase (UGT), glutathione-S-transferase (GST), in the murine choroid plexus. Indeed these activities reached hepatic levels, thus pointing out the choroid plexus as the main site of enzymatic detoxification in the brain. The activity of glutathione peroxidase, an antioxidant enzyme which inactivates hydrogen peroxide, and different other organic peroxides, is also much higher in the choroid plexus than in the brain parenchyma. Of importance in clinical investigation, preliminary observations have shown that human choroid plexus shares at least some of these enzymatic specificities (unpublished results). The fate of formed conjugates has also been investigated, and found to occur mainly by cellular extrusion at the basolateral, i.e. blood-facing membrane by a transporter that belongs to the multidrug resistance protein family (Strazielle and Ghersi-Egea, 1999). Accordingly, at least one member of this family, MRP-1, has been shown to be largely expressed at the choroid plexus epithelium and to localize at the basolateral membrane. (Rao et al., 1999, Nishino et al., 1999).

Finally, drug metabolism/detoxifying enzymes have been shown to be associated with the ependyma, and, among parenchymal cells, with glial rather than neuronal cells (reviewed in Monks et al., 1999, Strazielle and Ghersi-Egea, 2000b). Thus, although the paracellular pathway between the CSF and the brain parenchyma is not impeded by tight junctions, both the ependyma and the glia limitans may also be active players in the overall neuroprotection mechanisms.

IN VITRO MODELS TO STUDY DRUG METABOLISM AND TRANSPORT IN THE CENTRAL NERVOUS SYSTEM.

The functional relevance of the transport and metabolic processes at the BBI can be studied in vivo. The design of informative in vivo experiments is however difficult, and does not allow to easily distinguish between the effective role of the different interfaces. In particular, the study of blood-CSF exchange at the CP is difficult because the CSF is a circulating fluid which follows complex ventricular, velae and subarachnoid pathways of circulation, and undergoes a partial mixing from one compartment to another, as well as exchanges with the surrounding cerebral parenchyma (Ghersi-Egea et al., 1996). The cells forming the BBI can be grown on permeable filters, thus delineating two compartments, one representing the central side, the other the peripheral (blood) side. Such bicameral in vitro cellular models should be useful to study the mechanisms of drug and toxin transport/efflux and metabolism in the CNS. Several models of the BBB have been developed in the past years. With the exception of some studies related to p-glycoprotein functions, few data, however, have been generated regarding the characterization of drug transporters and detoxification processes at the BBB. In view of the central importance of the choroid plexuses in cerebral drug metabolism and detoxification, such cellular model of the choroidal epithelium was called for. We developed a bicameral in vitro model which reproduces the main features of the choroidal epithelium in vivo, namely its specific cell morphology and polarity, efficient tight junctions, specific transport systems, and a choroidal phenotype visualized by the high expression of transthyretin and drug metabolizing enzymes. It was used to demonstrate that the choroidal epithelium can act as an complete metabolic blood-to-CSF barrier towards some neurotoxic xenobiotics, by glucurono- or glutathione conjugation reactions coupled to MRP-dependent efflux process at the basolateral membrane of the choroid plexus epithelium (Strazielle and Ghersi-Egea, 1999). Evidence for an active efflux process of anionic toxic pesticides from CSF to blood was also obtained in vitro. Finally this model allowed to demonstrate that metabolic activation processes can also occur at the BBI (unpublished results), thus pointing out a risk for a challenge of the integrity of the neuroprotective barriers following BBI targeted xenobiotic bioactivation.

Acknowledgments

This work was supported by grants from the Institut National de la Santé et de la Recherche Médicale (PRISME 98-01), and ARSEP. N.S. is a recipient of ARSEP.

REFERENCES

Fenstermacher, J.D., Bereczki, D., Wei, L., Chen, J.L., Hans, F.J., Acuff, V., and Patlak, C.S., 1993, The pharmacology of the blood-brain barrier and cerebral extracellular fluid, in *CNS Barriers and Modern CSF Diagnostics* (K. Felgenhauer, M. Holzgraefe, and H.W. Prange, eds.), VCH Verlags, Weinheim, New York, pp.31-40.

Ghersi-Egea, J.F., Leininger-Muller, B., Suleman, G., Siest, G. and Minn, A., 1994,. Localization of drug-metabolizing enzyme activities to blood-brain interfaces and circumventricular organs, *J. Neurochem.* **62**:1089-1096.

Ghersi-Egea, J.F., Finnegan, W., Chen, J.L., and Fenstermacher, J.D., 1996, Rapid distribution of intraventricularly administered sucrose into cerebrospinal fluid cisterns via subarachnoid velae in rat, *Neuroscience*, **75**:1271-88.

Ghersi-Egea, J.F. and Strazielle, N., 2000, Brain drug delivery, drug metabolism and multidrug resistance at the choroid plexus. Microscopy Research and Technique, in press.

Gross, P.M. and Weindl, A., 1987, Peering through the windows of the brain, *J. Cereb. Blood Flow Metabol.*, **7**:663-672.

Monks, T.J., Ghersi-Egea, J.F., Philbert, M., Cooper, A.J., and Lock, E.A., 1999, Symposium overview: the role of glutathione in neuroprotection and neurotoxicity, *Toxicol. Sci.*, **51**:161-77.

Nishino, J., Suzuki, H., Sugiyama, D., Kitazawa, T., Ito, K., Hanano, M., and Sugiyama, Y., 1999, Transepithelial transport of organic anions across the choroid plexus: possible involvement of organic anion transporter and multidrug resistance-associated protein, *J. Pharmacol. Exp. Ther.*, **290**:289-94.

Rao, V.V., Dahlheimer, J.L., Bardgett, M.E., Snyder, A.Z., Finch, R.A., Sartorelli, A.C., and Piwnica-Worms, D., 1999, Choroid plexus epithelial expression of MDR1 P glycoprotein and multidrug resistance-associated protein contribute to the blood- cerebrospinal-fluid drug-permeability barrier, *Proc. Natl. Acad. Sci. U S A*, **96**:3900-5.

Schinkel, A.H., Wagenaar, E., Mol, C.A., and van Deemter, L., 1996, P-glycoprotein in the blood-brain barrier of mice influences the brain penetration and pharmacological activity of many drugs, *J. Clin. Invest.*, **97**:2517-24.

Strazielle, N., and Ghersi-Egea, J.F., 1999, Demonstration of a coupled metabolism-efflux process at the choroid plexus as a mechanism of brain protection toward xenobiotics, J. Neurosci., **19**:6275-89.

Strazielle, N., and Ghersi-Egea, J.F., 2000a, Choroid plexus in the central nervous system biology and physiopathology, J. Neuropathol. Exp. Neurol., **59**:561-574.

Strazielle, N., and Ghersi-Egea, J.F., 2000b, Implication of blood-brain interfaces in cerebral drug metabolism and drug metabolite disposition, in: *Molecular drug metabolism and toxicology*, chap. 12 (G. Williams, and O.I. Aruoma, eds.), OICA International, Saint Lucia, London, in press.

Suzuki, H., Terasaki, T., and Sugiyama, Y., 1997, Role of efflux transport across the blood-brain and blood-cerebrospinal fluid barrier on the disposition of xenobiotics in the central nervous system, *Adv. Drug Deliv. Rev.*, **25**:257-285.

THE ROLES OF P-GLYCOPROTEIN AND MRP1 IN THE BLOOD-BRAIN AND BLOOD-CEREBROSPINAL FLUID BARRIERS

Alfred H. Schinkel

Division of Experimental Therapy
The Netherlands Cancer Institute
Amsterdam, The Netherlands

INTRODUCTION

The brain is one of the most critical and sensitive organs in vertebrates. It is therefore understandably well shielded from various disturbing influences, including the occurrence of potential toxins that may be present in the bloodstream. It has long been recognized that there is a selective barrier between the bloodstream and the brain tissue. This barrier, called the blood-brain barrier, allows efficient uptake of nutrients in order to meet the extensive metabolic needs of the brain, but at the same time withholds many compounds from entering brain tissue, whereas these same compounds can readily penetrate most other tissues in the body (for reviews see Pardridge, 1998). The specialized structure of the endothelial cells in the blood capillaries of the brain is largely responsible for these unique barrier properties. Recent work has provided compelling evidence that at least one active efflux drug transporter in the blood-brain barrier, P-glycoprotein (P-gp), makes an important contribution to the special barrier properties.

Compounds may also penetrate the brain from the cerebrospinal fluid (CSF) which effectively surrounds the brain and spinal cord. Not surprisingly, there is also a blood-CSF barrier that limits penetration of compounds from the bloodstream into the CSF, and thus indirectly restricts access of these compounds to the brain. It was recently found that in the blood-CSF barrier there also occurs at least one active drug transporter that contributes to the barrier properties. However, in this case the responsible protein is the so-called multidrug resistance-associated protein, MRP1.

This review aims to summarize the recent work that has unequivocally established the roles of P-gp and MRP1 in blood-brain and blood-CSF barriers, respectively. It should be kept in mind that there may be additional active transporters that also contribute to the respective barriers, but their role has not been sorted out as thoroughly yet.

Biological Reactive Intermediates VI, Edited by Dansette *et al.*
Kluwer Academic / Plenum Publishers, 2001

ROLE OF P-GLYCOPROTEIN IN THE BLOOD-BRAIN BARRIER

P-glycoprotein Properties

P-glycoprotein was originally discovered by its ability to confer multidrug resistance to tumor cells. This protein is localized in the plasma membrane and consists of an internally duplicated structure, with two intracellular ATP-binding domains and two sets of six putative transmembrane segments. Based on its structure it belongs to the so-called ATP-binding cassette (ABC) family of membrane transporters. Extensive biochemical analysis (for review see e.g. Gottesman and Pastan, 1993) has demonstrated that P-glycoprotein can actively extrude its substrates from cells, even against steep concentration gradients, using ATP hydrolysis as its energy source. Many structurally different cytotoxic anticancer drugs can be exported from cells by P-glycoprotein, thus explaining why P-glycoprotein confers multidrug resistance: decreased drug concentration at intracellular drug target sites results in tolerance for higher external drug concentrations. Transported anticancer dugs include Vinca alkaloids (vincristine, vinblastine), epipodophyllotoxins (etoposide, teniposide), anthracyclines (doxorubicin, daunorubicin), taxanes (paclitaxel, docetaxel), and many others.

One of the most intriguing aspects of P-glycoprotein function is the huge structural diversity of substrates that can be transported. Efficiently transported substrates range in size from 200 to more than 1800 Da and are usually weakly basic or uncharged, but also at times positively or negatively charged, or zwitterionic. The only common denominator found in all verified substrates is that they are amphipathic in nature (i.e. having spatially separated hydrophobic and hydrophilic domains), and thus usually able to partition into membranes. The mechanistic background of recognition of such a diversity of structures is not well understood, but it means that a large collection of clinically important drugs and toxins is affected by P-glycoprotein activity.

In view of the possible contribution of P-glycoprotein to drug resistance in cancer, many compounds have been developed that can efficiently inhibit P-glycoprotein activity. Some of these, such as PSC 833 and GF120918, can be given in high dosages to patients, and they are currently tested in various clinical trials to modulate P-glycoprotein activity *in vivo* (see e.g. Sikic, 1997).

P-glycoprotein Tissue Distribution

The tissue distribution of P-glycoprotein indicates that it could have an important role in protection of the organism against a range of xenotoxins that it might encounter in nature. High levels of P-glycoprotein were found in the apical membrane of the luminal epithelial cells of large and small intestine, and in the bile canalicular membrane of hepatocytes (Thiebaut et al., 1987). At these locations, P-glycoprotein could prevent penetration of substrates present in the intestinal lumen from entering the bloodstream, and it could mediate active excretion of substrates present in the blood into the bile and into the intestinal lumen.

Cordon-Cardo et al. (1989) found that P-glycoprotein is also present at high levels in the small blood capillaries of the brain and testis, i.e. at the location of the blood-brain and blood-testis barriers. Although there is some discussion about the exact location of P-glycoprotein in these blood capillaries, most available evidence indicates that it is primarily located in the luminal membrane of the endothelial cells (for overviews see Pardridge, 1998 and Schinkel, 1999). This location would be strategically optimal to restrict P-glycoprotein substrates present in the blood from entering the surrounding tissue. Finally, high levels of

P-glycoprotein were found in the luminal membrane of placental trophoblasts at the materno-fetal barrier, suggesting that it might prevent penetration of substrates from the maternal blood into the fetal circulation (Sugawara et al., 1988; Lankas et al., 1998; Smit et al., 1999).

Blood-Brain Barrier Effects Observed in P-glycoprotein Knockout Mice

The generation and characterization of mice deficient for the drug-transporting P-glycoproteins clearly demonstrated that P-glycoprotein could indeed have a very important role in protecting the organism from toxins and drugs (reviewed in Schinkel, 1997 and 1999). Whereas the P-glycoprotein knockout mice appeared physiologically overall normal, they were far more sensitive to exposure to a range of drugs and toxins. In fact, all of the putative functions described for P-glycoprotein in the previous section could be directly demonstrated in the knockout mice with various drugs and toxins. However, the most pronounced effects were observed for penetration of drugs and toxins across the blood-brain barrier. A range of drugs and toxins displayed at least 10-fold higher accumulation in the brain of P-glycoprotein knockout mice as compared to wild-type mice. Examples are the neurotoxic pesticide and anthelmintic ivermectin, the cardiac glycoside digoxin, the anticancer drug vinblastine, the immunosuppressive agent cyclosporin A, the non-sedative opioid asimadoline, a range of HIV protease inhibitor drugs (Kim et al., 1998), and the antidiarrheal opioid loperamide. In several cases, differences in brain accumulation between wild-type and knockout mice reached up to 50- or 100-fold. Given the promiscuity of P-glycoprotein there will be many more clinically important drugs that are substantially affected by P-glycoprotein in the blood-brain barrier.

Structural Basis for the Pronounced P-glycoprotein Effects in the Blood-Brain Barrier

Two factors contribute to the large effect of P-glycoprotein on restricted brain penetration of many substrate drugs. One factor is the architecture of the endothelial cells at the blood-brain barrier. These cells cover the capillary wall completely, and they are very tightly linked to each other by tight junctions. Thus, only very small molecules can diffuse between the endothelial cells to reach the brain compartment from the bloodstream. All larger molecules, including most P-glycoprotein substrates, have to pass through the endothelial cell body. For many nutrient molecules and some drugs there are specific carrier systems that allow passage through the endothelial cell membrane (Pardridge, 1998), but most P-glycoprotein substrates are in principle sufficiently hydrophobic to diffuse across the endothelial cell membranes at an appreciable rate, and this is what is observed in the P-glycoprotein knockout mice. However, when P-glycoprotein is present in the luminal membrane of the endothelial cells, it very effectively pumps all penetrating substrate molecules back to the blood. Thus, it is the combination of the continuous lipophilic barrier formed by the endothelial cells, and the active back-transport by P-glycoprotein that makes for the highly efficient barrier function of P-glycoprotein in the blood-brain barrier.

Clinical Implications of Blood-Brain Barrier P-glycoprotein Function

For many drugs, the increased brain penetration in P-glycoprotein-deficient mice resulted in qualitatively different toxicities that would prohibit the normal clinical use of these drugs. For instance, the anti-emetic drug domperidone turned out to have a strong neuroleptic effect in the knockout mice, whereas the antidiarrheal loperamide gave rise to strong opiate-like behavioral abnormalities, indicating clear CNS effects of this opioid drug (Schinkel, 1999). The very large increase in ivermectin toxicity (about 100-fold) resulted in nearly 100% lethality of knockout mice exposed to a normally very safe routine spray treatment for mite infestation. Thus, the presence of blood-brain barrier P-glycoprotein function is an important positive determinant for the current clinical use of many drugs.

On the other hand, in some cases it may be desirable that a drug displays enhanced brain penetration, for instance when treating small brain metastases of tumors that are positioned behind the blood-brain barrier with chemotherapy, or when treating pathogens for which the brain is potentially a pharmacological sanctuary, such as HIV. In this respect, it is interesting to note that others and we have demonstrated that, using effective P-glycoprotein blockers such as PSC 833 or GF120918, it is possible to completely or nearly completely abrogate blood-brain barrier P-glycoprotein function in animal models (reviewed in Schinkel, 1999). It remains to be established whether this principle can also be applied safely in humans.

ROLE OF MRP1 IN THE BLOOD-CEREBROSPINAL FLUID BARRIER

Properties of MRP1

In 1992, six years after the cloning of the P-glycoprotein gene, another multidrug resistance gene encoding an ABC transporter protein was cloned. This protein, called multidrug resistance-associated protein or multidrug resistance protein (MRP1), displays several interesting similarities and differences with the P-glycoprotein multidrug transporter (for recent reviews see Hipfner et al., 1999 and König et al., 1999). Like P-glycoprotein, MRP1 is primarily located in the plasma membrane, and it confers multidrug resistance to tumor cells by actively pumping a range of anticancer out of the cell, using ATP hydrolysis as an energy source. Although the core structure of MRP1 appears to be similar to that of P-glycoprotein (i.e. an internally duplicated structure of two ATP-binding sites and two times six putative transmembrane segments), it contains an additional N-terminal domain with five putative transmembrane segments. Another important difference is that MRP1 needs a cellular co-factor, the ubiquitous negatively charged tripeptide glutathione (GSH) in order to pump out hydrophobic anticancer drugs such as etoposide, vincristine and doxorubicin. Further analysis demonstrated that MRP1 primarily transports amphipathic organic anions, including GSH-, glucuronide-, and sulfate-conjugates of many drugs, toxins and endogenous compounds, including the inflammatory mediator leukotriene C_4.

Tissue Distribution of MRP1

Another striking difference between MRP1 and P-glycoprotein is the subcellular localization in most polarized epithelia: whereas P-glycoprotein consistently routes to the apical membrane of polarized cells, MRP1 is primarily found in the basolateral membrane. This implies that in epithelial cells, where both P-glycoprotein and Mrp1 are present, the two proteins would transport their substrates in opposite directions. In human tissues, MRP1 appears to be more ubiquitously distributed than P-glycoprotein, with expression found in e.g. many epithelia, lung bronchioles, alveolar macrophages, heart, skeletal and smooth muscle, adrenal cortex, epidermis and salivary gland ducts (Flens et al., 1994). This generalized tissue distribution made it difficult to speculate on normal functions of MRP1.

Relevant for this review, MRP1, or its rat and mouse homologues Mrp1, could be readily detected in the epithelium of the choroid plexus, whereas Mrp1 was not detectable in brain capillaries of rat or mouse. As expected, MRP1 appeared to localize primarily to the basolateral side of the choroid plexus epithelium (Rao et al., 1999; Nishino et al., 1999; Wijnholds et al., 2000). The choroid plexus epithelium is the major site for production of cerebrospinal fluid (CSF). At the same time, it constitutes the blood-CSF barrier, since the blood capillaries feeding the choroid plexus are fenestrated and highly permeable, lacking the endothelial cell structure characteristic for blood-tissue barriers (see e.g. reviews in Pardridge, 1998).

Characterization of Mrp1 Knockout Mice

Generation and characterization of Mrp1 knockout mice demonstrated that, like P-gp, this protein is not essential for normal viability and fertility of mice (Wijnholds et al., 1997; Lorico et al., 1997). However, these mice were somewhat more sensitive to etoposide toxicity (about 1.5- to 2-fold), and they displayed a partial deficiency in a leukotriene C_4-mediated inflammatory response. The latter phenotype suggests a role for Mrp1 as the main leukotriene C_4 exporter in leukotriene-synthesizing cells. Detailed analysis of the etoposide toxicity indicated that Mrp1 contributes to the protection of the oropharyngeal mucosal layer and the seminiferous tubules of the testis (Wijnholds et al., 1998). As Mrp1 is abundant in the basal layers of the oropharyngeal mucosa, and in the basal membrane of the Sertoli cells that completely surround the testicular tubules, it appears very likely that normally Mrp1 restricts access of etoposide to these tissues.

Presence and Functional Activity of Mrp1 in the Blood-Cerebrospinal Fluid Barrier

Several lines of evidence indicate that Mrp1 in the choroid plexus can functionally contribute to the blood-CSF barrier. For a proper understanding of Mrp1 function it is useful to have a closer look at the choroid plexus structure (Segal, 1998). Choroid plexuses are present in the two lateral ventricles and the third and fourth ventricles of the brain. The cell layer facing the CSF consists of modified ependymal cells that have many characteristics of transporting epithelia. The apical (CSF) side of the cells is covered with microvilli, and the cells are closely joined by tight junctions, strongly restricting the flow of molecules from the basolateral (blood) side to the apical side. The choroid plexuses have their own blood supply and extend into the lateral ventricles as leaflike structures with a highly convoluted surface. As the blood vessels in the choroid plexus do not have blood-tissue barrier characteristics, the blood-CSF barrier is effectively formed by the ependymal epithelial-like cell layer of the choroid plexus.

Two recent studies demonstrated that MRP1 is expressed in the epithelial cells of the choroid plexus of man, mouse and rat, and provided evidence that MRP1 might contribute to transport of substrates from the CSF to the blood side of the choroid plexus. Rao et al. (1999) demonstrated that an intravenously injected radioactive marker, ^{99m}Tc-sestamibi, a substrate for both P-glycoprotein and MRP1, accumulated to a high extent in the choroid plexus, but not in the surrounding CSF or brain tissue. As ^{99m}Tc-sestamibi is by itself highly membrane permeable, this suggested a functional barrier between blood and CSF and between blood and brain tissue. Subsequent experiments with a microdialysis probe inserted into the lateral ventricle of a rat confirmed that ^{99m}Tc-sestamibi hardly enters the CSF. Interestingly, when P-glycoprotein in the blood-brain barrier was blocked with the P-glycoprotein inhibitor GF120918, the CSF demonstrated a delayed (20-40 min) large increase in drug concentration. This was interpreted as a consequence of the blocking of blood-brain barrier P-glycoprotein, resulting in diffusion of ^{99m}Tc-sestamibi into the brain interstitial fluid, and from there via the brain ependymal cell layer (which lacks tight junctions) into the ventricular CSF. Furthermore, polarized transport experiments with cultured rat choroid plexus epithelial cells suggested some contribution of basolaterally localized Mrp1 to ^{99m}Tc-sestamibi translocation. Immunohistochemistry and Western blotting confirmed high expression of Mrp1 in choroid plexus of man, rat and mouse. The data also indicated that there is some P-glycoprotein in the choroid plexus epithelium, but since this would transport its substrates towards the CSF, this would rather counteract than support a blood-CSF barrier function.

Experiments by Nishino et al. (1999) showed that isolated rat choroid plexus could readily take up the negatively charged 17β-estradiol 17β-D-glucuronide ($E_2$17βG), a shared substrate for the organic anion-transporting polypeptide (oatp) cellular uptake system, and Mrp1. Uptake was most likely mediated by oatp. When $E_2$17βG was administered directly into the lateral ventricle, it was very rapidly cleared from the CSF, in contrast to the control marker inulin. When the same experiment was repeated in the presence of probenecid, a known inhibitor of Mrp1 (but also of other organic anion transporters), $E_2$17βG clearance decreased to the same level as that of inulin. Since these authors also found a high level of Mrp1 expression in the choroid plexus, they concluded that basolaterally localized Mrp1 might be responsible for the high $E_2$17βG clearance from CSF.

Direct Evidence for Mrp1 Contribution to the Blood-CSF Barrier

The most compelling and direct evidence for a contribution of choroid plexus Mrp1 to the blood-CSF barrier came from an elegant experiment carried out by De Lange and Wijnholds from the groups of Breimer and Borst (Wijnholds et al., 2000). These authors made use of compound P-glycoprotein and Mrp1 knockout mice, obtained by inbreeding of the separate knockout strains. These so-called "triple knockout" mice lacking both P-glycoproteins and Mrp1 were compared to mice lacking both murine drug-transporting P-glycoproteins ("double knockout" mice). The principal advantage of this approach is that it avoids confounding effects that might arise from P-glycoprotein activity in the endothelial blood-brain barrier, or in the choroid plexus epithelium. This was necessary since the probe drug of choice, etoposide, is both a P-glycoprotein and an Mrp1 substrate.

To assess Mrp1 activity in the blood-CSF barrier, etoposide was given intravenously to double and triple knockout mice. At the same time, CSF was collected over a 1 hour period by slow perfusion of the lateral ventricle, and analysis of etoposide concentration in the perfusate indicated the extent to which etoposide penetrated the blood-CSF barrier. Absence of Mrp1 in the triple knockout mice resulted in a 10-fold higher penetration of etoposide in the CSF, whereas the plasma and brain concentration in both mouse strains at 1 hour was not significantly different. These results indicate that Mrp1 in the basolateral membrane of the choroid plexus epithelium restricts penetration of etoposide across the blood-CSF barrier.

It is interesting that the brain concentrations of etoposide were not different at the 1 hour time point between the two mouse strains. This can reflect that under these circumstances the CSF concentration is not an important determinant for the brain concentration, for instance because entry across the blood-brain barrier lacking P-glycoprotein is already quite high. It may also indicate a delay in equilibration between CSF and brain interstitial fluid, as previously suggested by the ^{99m}Tc-sestamibi data (Rao et al., 1999). In any case, the data do demonstrate that the CSF concentration of a drug does not necessarily accurately reflect the brain concentration of the drug, in spite of the connection between the two compartments.

PERSPECTIVE

The studies reviewed in this paper demonstrate that at least two active drug transporters, P-glycoprotein and MRP1, can significantly restrict the penetration of drugs into the brain and the CSF, respectively. They also suggest that the behavior of both compartments (brain and CSF) with regard to drug penetration can be qualitatively different, since different transporters, with in part very different substrate specificities contribute to their respective blood-tissue barriers. The data also indicate that, despite the connection between the CSF and the brain interstitial fluid via the "leaky" brain ependymal cell layer, drug equilibration between the two compartments can be significantly delayed. One obvious implication from these findings is that CSF drug concentration, which is at times used to gauge the CNS penetration of drugs, may not always accurately reflect the true brain penetration - even if used only in a qualitative manner.

It should further be kept in mind that there may be additional active drug transporters in blood-brain and blood-CSF barrier that affect penetration of drugs in both compartments. The possible contribution of these transporters still needs to be sorted out.

However that may be, knowledge of P-glycoprotein and MRP1 function in blood-brain and blood-CSF barriers may perhaps in the future be used to advantage in drug therapy. Increased brain penetration of P-glycoprotein drug substrates by using P-glycoprotein inhibitors was already demonstrated and discussed above. Whether increased CSF penetration of some drugs by pharmacological MRP1 inhibitors is feasible remains to be determined. So far, it has not been easy to identify efficient Mrp1 inhibitors that can be applied at sufficiently high dosages to inhibit MRP1 activity in vivo. Nevertheless, the good viability of P-glycoprotein, Mrp1, and also the combined triple knockout mice (Wijnholds et al., 2000) gives hope that in principle simultaneous inhibition of both P-glycoprotein and MRP1 might be acceptable also in humans. Any type of drug therapy for which the blood-brain and blood-CSF barriers form substantial hurdles, and where the best drugs are good P-glycoprotein and MRP1 substrates is a potential candidate for simultaneous inhibition. Examples that come to mind are for instance brain tumors or viral (HIV) and bacterial infections of brain or CSF. One of our tasks will be to establish for what forms of brain/CSF drug therapy this approach would be sufficiently promising to initiate trials in patients.

ACKNOWLEDGEMENTS

I thank Hans Jonker, Hans Smit, Maarten Huisman and John Allen for critical review of the manuscript. Much of the work described in this paper was supported by grants from the Dutch Cancer Society.

REFERENCES

Cordon-Cardo, C., O'Brien, J. P., Casals, D., Rittman-Grauer, L., Biedler, J.L., Melamed, M.R., and Bertino, J.R., 1989, Multidrug-resistance gene (P-glycoprotein) is expressed by endothelial cells at blood-brain barrier sites, *Proc. Natl. Acad. Sci. U.S.A.* **86**:695-698.

Flens, M.J., Zaman, G.J.R., van der Valk, P., Izquierdo, M.A., Schroeijers, A.B., Scheffer, G.L., van der Groep, P., de Haas, M., Meijer, C.J.L.M., and Scheper, R.J., 1994, Tissue distribution of the multidrug resistance protein, *Am. J. Pathol.* **148**:1237-1247.

Gottesman, M.M., and Pastan, I., 1993, Biochemistry of multidrug resistance mediated by the multidrug transporter, *Annu. Rev. Biochem.* **62**:385-427.

Hipfner, D.R., Deeley, R.G., and Cole, S.P.C., 1999, Structural, mechanistic and clinical aspects of MRP1, *Biochim. Biophys. Acta* **1461**:359-376.

Kim, R.B., Fromm, M.F., Wandel, C., Leake, B., Wood, A.J.J., Roden, D.M., and Wilkinson, G.R., 1998, The drug transporter P-glycoprotein limits oral absorption and brain entry of HIV-1 protease inhibitors, *J. Clin. Invest.* **101**:289-294.

König, J., Nies, A. T., Cui, Y., Leier, I., and Keppler, D., 1999, Conjugate export pumps of the multidrug resistance protein (MRP) family: localization, substrate specificity, and MRP2-mediated drug resistance, *Biochim. Biophys. Acta* **1461**:377-394.

Lankas, G.R., Wise, L.D., Cartwright, M.E., Pippert, T., and Umbenhauer, D.R., 1998, Placental P-glycoprotein deficiency enhances susceptibility to chemically induced birth defects in mice, *Reprod. Toxicol.* **12**:457-463.

Lorico, A., Rappa, G., Finch, R.A., Yang, D., Flavell, R.A., and Sartorelli, A.C., 1997, Disruption of the murine MRP (multidrug resistance protein) gene leads to increased sensitivity to etoposide (VP-16) and increased levels of glutathione, *Cancer Res.* **57**:5238-5242.

Nishino, J.-I., Suzuki, H., Sugiyama, D., Kitazawa, T., Ito, K., Hanano, M., and Sugiyama, Y., 1999, Transepithelial transport of organic anions across the choroid plexus: possible involvement of organic anion transporter and multidrug resistance-associated protein, *J. Pharmacol. Exp. Ther.* **290**:289-294.

Pardridge, W.M. (ed.), 1998, *Introduction to the Blood-Brain Barrier. Methodology, Biology and Pathology*, Cambridge University Press, Cambridge, U.K.

Rao, V.V., Dahlheimer, J.L., Bardgett, M.E., Snyder, A.Z., Finch, R.A., Sartorelli, A.C., and Piwnica-Worms, D., 1999, Choroid plexus epithelial expression of MDR1 P-glycoprotein and multidrug resistance-associated protein contribute to the blood-cerebrospinal-fluid drug-permeability barrier, *Proc. Natl. Acad. Sci. U.S.A.* **96**:3900-3905.

Schinkel, A.H., 1997, The physiological function of the drug-transporting P-glycoproteins, *Semin. Cancer Biol.* **8**:161-170.

Schinkel, A.H., 1999, P-glycoprotein, a gatekeeper in the blood-brain barrier, *Adv. Drug Deliv. Rev.* **36**:179-194.

Segal, M., 1998, The blood-CSF barrier and the choroid plexus, in: *Introduction to the Blood-Brain Barrier. Methodology, Biology and Pathology*, Pardridge W.M., ed., Cambridge University Press, Cambridge, U.K.

Sikic,B.I., 1997, Pharmacologic approaches to reversing multidrug resistance, *Semin. Hematol.* **34**:40-47.

Smit, J.W., Huisman, M.T., van Tellingen, O., Wiltshire, H.R., and Schinkel, A.H., 1999, Absence or pharmacological blocking of placental P-glycoprotein profoundly increases fetal drug exposure, *J. Clin. Invest.* **104**:1441-1447.

Sugawara, I., Kataoka, I., Morishita, Y., Hamada, H., Tsuruo, T., Itoyama, S., and Mori, S., 1988, Tissue distribution of P-glycoprotein encoded by a multidrug-resistant gene as revealed by a monoclonal antibody, MRK 16, *Cancer Res.* **48**:1926-1929.

Thiebaut, F., Tsuruo, T., Hamada, H., Gottesman, M.M., Pastan I., and Willingham, M.C., 1987, Cellular localization of the multidrug resistance gene product in normal human tissues, *Proc. Natl. Acad. Sci. U.S.A.* **84**:7735-7738.

Wijnholds, J., Evers, R., van Leusden, M.R., Mol, C.A.A.M., Zaman, G.J.R., Mayer, U., Beijnen, J.H., van der Valk, M., Krimpenfort, P., and Borst, P., 1997, Increased sensitivity to anticancer drugs and decreased inflammatory response in mice lacking the multidrug resistance-associated protein, *Nature Med.* **3**:1275-1279.

Wijnholds, J., Scheffer, G.L., van der Valk, M., van der Valk, P., Beijnen, J.H., Scheper, R.J., and Borst, P., 1998, Multidrug resistance protein 1 protects the oropharyngeal mucosal layer and the testicular tubes against drug-induced damage, *J. Exp. Med.* **188**:797-808.

Wijnholds, J., de Lange, E.C.M., Scheffer, G.L., van den Berg, D.-J., Mol, C.A.A.M., van der Valk, M., Schinkel, A.H., Scheper, R.J., Breimer, D.D., and Borst, P., 2000, Multidrug resistance protein 1 protects the choroid plexus epithelium and contributes to the blood-cerebrospinal fluid barrier, *J. Clin. Invest.* **105**:279-285.

ARE DOPAMINE, NOREPINEPHRINE, AND SEROTONIN PRECURSORS
OF BIOLOGICALLY REACTIVE INTERMEDIATES INVOLVED IN
THE PATHOGENESIS OF NEURODEGENERATIVE BRAIN DISORDERS?

Glenn Dryhurst

Department of Chemistry and Biochemistry
University of Oklahoma, Norman, OK 73019
U.S.A.

Introduction

During the past two or three decades an immense amount of research has been carried
out to understand the fundamental molecular mechanisms that underlie the neurotoxicity
evoked by 1-methyl-4-phenyl-1,2,3,6-tetrahydropyridine (MPTP), methamphetamine (MA)
and related amphetamine drugs of abuse. The neurotoxicity of MPTP is of particular interest
because it not only mimics the symptoms and major pathobiochemical changes that occur in
Parkinson's disease (PD) but also arguably provides the best model of PD in animals
(Gerlach et al., 1991; Gerlach and Riederer, 1996; Royland and Langston, 1998). Interest in
the neurotoxicity evoked by MA derives not only from the fact that this compound is a widely
abused psychoactive drug but in animals it also mimics many of the major pathobiochemical
changes that occur in PD (Gerlach and Riederer, 1996) although it is a less selective
neurotoxin than MPTP. Transient cerebral ischemia, or ischemia-reperfusion (I-R), can also
lead to neurodegeneration although this is even less selective than MA. Nevertheless, there
are a rather striking number of similar factors that appear to be key steps in a complex
cascade of processes that compromise the neurotoxic mechanisms evoked by MPTP, MA and
I-R. In this communication similarities associated with the neurotoxicity evoked by MPTP,
MA and I-R will be briefly reviewed. Subsequently, these will be integrated into a working
hypothesis for the underlying neurotoxic mechanisms in which one or more of the
neurotransmitters dopamine (DA), norepinephrine (NE) and 5-hydroxytryptamine (5-HT;
serotonin) are proposed to be the precursors of biologically reactive intermediates in the
pathological processes.

MPTP

MPTP is one of the most selective neurotoxins which causes the degeneration of
nigrostriatal DA neurons (Gerlach et al., 1991; Gerlach and Riederer, 1996). The selectivity
of MPTP can be traced to its oxidation by monoamine oxidase-B (MAO-B) in glia to its
active metabolite 1-methyl-4-phenylpyridinium (MPP$^+$) (Ransom et al., 1987) which is
translocated into the cytoplasm of dopaminergic neurons by the plasma DA transporter (DAT)
(Javitch et al., 1985). MPP$^+$ is then concentrated by an energy-dependent process into
mitochondria and reversibly (Cooper and Schapira, 1997) inhibits complex I (Ramsey et al.,
1986) and α-ketoglutarate dehydrogenase (α-KGDH) (Mizuno et al., 1987) causing rapid
ATP depletion (Chan et al., 1991). Microdialysis experiments demonstrate that perfusion of
neurotoxic concentrations MPP$^+$ evokes a massive virtually instantaneous release of DA
(Rollema et al., 1986; Giovanni et al., 1994). Although the mechanisms by which MPP$^+$
evokes rapid DA release are not fully understood they probably include depolarization of the

Biological Reactive Intermediates VI, Edited by Dansette *et al.*
Kluwer Academic / Plenum Publishers, 2001

neuronal membrane (Chiueh and Huang, 1991) and possibly reversal of the plasma membrane DAT (Matsubara et al., 1996). When MPP[+] perfusions are discontinued extracellular concentrations of DA begin to decline back towards basal levels without corresponding increases of extracellular concentrations of its metabolites 3,4-dihydroxyphenylacetic acid (DOPAC) or homovanillic acid (HVA) suggesting that increasing ATP production initiates repolarization of the neuronal membrane and resultant DAT-mediated reuptake of DA (Cao et al., 1990). During MPP[+] perfusions into the rat striatum extracellular levels of L-glutamate (Glu) remain at or close to basal levels and increase only when MPP[+] perfusions are discontinued (Carboni et al., 1990; Matarredona et al., 1997).

The selective degeneration of nigrostriatal DA neurons by MPTP/MPP[+] is accompanied by a significant fall of glutathione (GSH) levels (Ferraro et al., 1986; Yong et al., 1986; Sriram et al., 1997) but without corresponding increases of glutathione disulfide (GSSG) (Oishi et al., 1993).

Methamphetamine

MA is a somewhat less selective neurotoxin evoking the degeneration of nigrostriatal DA neurons (Gibb et al., 1994; Sonsalla et al., 1996), serotonergic terminals in a number of brain regions (Ricaurte et al., 1980; Axt and Molliver, 1991) and a population of glutamatergic cell bodies in the somatosensory cortex (Pu et al., 1996). MA-induced neurotoxicity does not appear to be caused directly by the drug or its known metabolites (Gibb et al., 1994, 1997). A number of lines of evidence implicate perturbations of neuronal energy metabolism and rapid ATP depletion with MA-induced neurotoxicity (Chan et al., 1994; Albers et al., 1996; Callahan et al., 1998). This energy impairment probably results from a combination of metabolic perturbations, exchange of MA with DA such that the DAT operates in a highly activated energy-consuming state, disruption of ion gradients (Huether et al., 1997) and, particularly, hyperthermia (Madl and Allen, 1995). Among the immediate central effects of neurotoxic doses of MA is a massive release of DA and 5-HT (O'Dell et al., 1991; Gibb et al., 1994) by mechanisms that are believed to include reversal of the DAT (Raiteri et al., 1979; Lew et al., 1998) and 5-HT transporter (5-HTT) (Heikkila et al., 1975; Berger et al., 1992) and possibly neuronal depolarization (Della Donna and Sonsalla, 1994). Similar to MPTP/MPP[+], MA also causes a decrease in striatal GSH levels (Moszczynska et al., 1998) and a delayed increase of extracellular Glu which appears to be an important factor in the underlying neurotoxic mechanism (Nash and Yamamoto, 1992; Stephens and Yamamoto, 1994).

Ischemia-Reperfusion

I-R can lead to the degeneration of striatal DA terminals, serotonergic and noradrenergic terminals in many brain regions (Weinberger et al., 1983) and other selectively vulnerable cells (probably primarily glutamatergic) particularly in certain regions of the hippocampus and cortex (Globus et al., 1992; Sims and Zaiden, 1995). During ischemia, brain ATP levels rapidly decrease (Sims and Zaidan, 1995) evoking a massive release of DA (Globus et al., 1988), NE (Globus et al., 1989) and 5-HT (Globus et al., 1992). Upon reperfusion (reoxygenation) ATP begins to rise (Pulsinelli and Duffy, 1983; Sims and Zaiden, 1995) and reuptake of DA, NE and 5-HT is initiated presumably owing the repolarization of neuronal membranes and reinstatement of the activity of the transporters for these neurotransmitters. Ischemia also leads to a marked decrease of GSH levels without increased GSSG that persists into the reperfusion phase (Cooper et al., 1980; Rehncrona et al., 1980; Slivka and Cohen, 1993). Furthermore, ischemia evokes increased extracellular concentrations of Glu (Benveniste et al., 1984) some of which, together with elevated extracelluar L-cysteine (CySH) (Slivka and Cohen, 1993; Puka-Sundvall et al., 1997), appears to arise from the degradation of GSH (Yang et al., 1995).

GSH in the nervous system

As noted above, one common change evoked by the neurotoxins MPTP/MPP[+], MA and I-R is a reduction of GSH levels in vulnerable regions of the brain but without corresponding increases of GSSG. Before discussing the implications of this effect it is worthwhile briefly reviewing GSH in the nervous system. In the brain much evidence suggests that neuronal GSH is dependent on the export of the tripeptide from glia (Raps et al., 1989; Sagara et al.,

1993a, 1993b; Wang and Cynadar, 2000). However, neurons are unable to import GSH. Thus, extracellular GSH, released from glia, must be degraded first by γ-glutamyl transpeptidase (γ-GT) which transfers the glutamyl residue of GSH to acceptor amino acids/peptides (transpeptidation) or water (hydrolysis) giving the respective γ-glutamyl amino acid/peptide or Glu, respectively, and cysteinylglycine (CysGly) (Tate and Meister, 1985). CysGly can then by hydrolyzed by dipeptidases (DP) to CySH and L-glycine (Gly). CySH and CysGly are then translocated into neurons to provide intraneuronal CySH, the rate limiting substrate for GSH biosynthesis (Sagara et al., 1993b; Dringen et al., 1999; Wang and Cynader, 2000). A recent study suggests that reduction of extracellular cystine by GSH released from glia might non-enzymatically provide CySH which is translocated into neurons for GSH biosynthesis (Wang and Cynader, 2000). Maintenance of intracellular GSH is energy dependent. Thus energy impairment and depolarization of the neuronal and mitochondrial membranes evokes release of both cytoplasmic and, probably, mitochondrial GSH (Mithöfer et al., 1992; Zängerle et al., 1992; Reed and Savage, 1995).

One explanation for the fall of GSH caused by MPTP, MA and I-R is that they deplete ATP needed for its synthesis (Seelig and Meister, 1985). However, shortly after MPTP administration to mice mitochondrial complex I activity in the striatum is transiently inhibited and then recovers (Sriram et al., 1997), possibly reflecting the reversible inhibition of this respiratory complex by MPP$^+$ (Cooper and Schapira, 1997), whereas GSH decrements persist (Ferraro et al., 1986; Yong et al., 1986, Oishi et al., 1993; Thomas et al., 2000). Subsequently, complex I inhibition again develops in both the striatum and substantia nigra pars compacta (SN_c) (Sriram et al., 1997). Similarly, following almost complete ATP depletion during ischemia brain energy returns to near normal levels during early reperfusion (Pulsinelli and Duffy, 1983; Sims and Zaidan, 1995) yet GSH decrements remain (Cooper et al., 1980; Rehncrona et al., 1980). Subsequently, clear indications of altered mitochondrial respiration develop which include decreased state 3, cytochrome c oxidase (COX, complex IV) and pyruvate dehydrogenase complex (PDHC) activities (Sims, 1991; Zaidan et al., 1998; Wagner et al., 1990) which approximately coincide with the appearance of neuronal damage. MA administration also leads to decreased COX activity in several brain regions without altered succinate dehydrogenase activity (Prince et al., 1998; Burrows et al., 2000) although effects on other mitochondrial enzymes have not been studied. The mitochondrial enzyme defects evoked by I-R clearly cannot be attributed to an exogenous toxicant and therefore raise the possibility that the fall of GSH without increased GSSG may be related to these defects. Similarly, the second delayed decrease of complex I activity following MPTP administration together with the fall of GSH suggests that these effects may be related.

Other similarities between the central effects evoked by MPTP/MPP$^+$, MA, and I-R

Under conditions of impaired neuronal ATP production evoked by MPP$^+$, MA or ischemia, extracellular excitatory amino acids (EAAs) such as Glu, especially when elevated, can activate N-methyl-D-aspartate (NMDA) and α-amino-3-hydroxy-5-methyl-4-isoxazolepropionic acid (AMPA) receptors causing an influx of Ca^{2+} (Novelli et al., 1988; Greene and Greenamyre, 1995; Cebers et al., 1998) which in turn mediates neuronal superoxide (O_2-·) generation (Lafon-Cazal et al., 1993; Sensi et al., 1999) and neuronal nitric oxide synthase (nNOS) activation with resultant nitric oxide (NO·) and thence peroxynitrite (ONOO$^-$) formation (Schulz et al., 1995). The resistance of transgenic mice that overexpress cytoplasmic or mitochondrial superoxide dismutase (SOD) to the neurotoxic effects of MPTP/MPP$^+$ (Przedborski et al., 1992; Klivenyi et al., 1998), MA (Cadet et al., 1994; Hirata et al., 1995) and I-R (Yang et al., 1994) all implicate O_2-· as a key participant in the neurotoxic mechanism. Increased 3-nitrotyrosine (3-NT) immunoreactivity in striatal and SN_c neurons of animals administered MPTP which is blocked by nNOS inhibitors (Ara et al., 1998; Ferrante et al., 1999) and the ability of such inhibitors to attenuate dopaminergic lesions (Schulz et al., 1997), an effect also observed in nNOS knockout mice (Matthews et al., 1997), and the fact that MPTP-induced protein tyrosine nitration is blocked in Cu/Zn SOD transgenic mice (Ara et al., 1998) implicate ONOO$^-$ as another key participant in the neurotoxic mechanism. MA also induces NO· and ONOO$^-$ generation and agents that destroy ONOO$^-$ or inhibit nNOS protect against its DA and 5-HT neurotoxicity (Ali et al., 1998; Imman et al., 1999). Similarly, transient cerebral ischemia activates nNOS leading to increased NO· and ONOO$^-$ production (Forman et al., 1998) and 3-NT immunoreactivity (Eliasson et al., 1999). Conversely, increased SOD activity (Keller et al., 1998) and nNOS inhibitors block 3-NT formation and are neuroprotective (Hirabayashi et al., 1999).

Despite early conflicting results (see Royland and Langston, 1998), more recent evidence suggests that NMDA receptor antagonists protect dopaminergic SN_c cells against MPTP/MPP$^+$-induced damage (Loschmann et al., 1994; see also Zeevalk et al., 2000) whereas striatal terminals are protected by AMPA receptor antagonists (Merino et al., 1999). NMDA receptor antagonists attenuate the DA and 5-HT neurotoxicity of MA (Johnson et al., 1989; Sonsalla et al., 1989). Similarly, NMDA and/or AMPA receptor antagonists attenuate neuronal damage caused by I-R (Gill et al., 1988; Kato et al., 1990; Lee et al., 1999).

MPTP/MPP$^+$ (Chiueh et al., 1994; Smith and Bennett, 1997; Thomas et al., 2000), MA (Giovanni et al., 1995; Yamamoto and Zhu, 1998; Kita et al., 1999) and I-R (Lancelot et al., 1995; Yang et al., 1996; Ste-Marie et al., 2000) all induce extracellular hydroxyl radical (HO·) generation. There are a number of mechanisms that could mediate such HO· generation. For example, O_2^{-}· (Flint et al., 1993; Yoshida et al., 1995), NO· and ONOO$^-$ (Dorrepaal et al., 1997; Gardner et al., 1997; Keyer and Imlay, 1997) release Fe^{2+} from iron-containing proteins and the neurotoxicity of MPTP (Lan and Jiang, 1997; Santiago et al., 1997), MA (Yamamoto and Zhu, 1998) and I-R (Patt et al., 1990; Palmer et al., 1994) involves low molecular weight iron mobilization. Thus, NMDA/AMPA receptor activation by extracellular EAAs under conditions of impaired neuronal energy metabolism with resultant O_2^{-}·, NO· and ONOO$^-$ generation and release of Fe^{2+} in the presence of ascorbate, implicated in the neurotoxcity of MPTP (Matarredona et al., 1997; Revuelta et al., 1997) and MA (Matsuda et al., 1987), and H_2O_2 provide conditions ideal for HO· generation (Gutteridge, 1996). Furthermore, decomposition of ONOO$^-$ can form HO· (Crow et al., 1994). Indeed, nNOS inhibitors dramatically attenuate extracellular HO· formation in the rat striatum in response to MPP$^+$ without affecting the release of DA (Smith et al., 1994; Rose et al., 1999). The latter observation is important because it implies that autoxidation or MAO-mediated metabolism of released DA is probably not a major source of extracellular reactive oxygen species (ROS).

DA appears to be a key participant in the dopaminergic neurotoxicity of MPTP/MPP$^+$, MA and I-R because its synthesis and reuptake inhibitors and DA depletion are all neuroprotective (Clemens and Phebus, 1988; Martin et al., 1991; Marek et al., 1990). Furthermore, mice lacking the DAT are resistant to MPTP-induced DA neurotoxicity (Bezard et al., 1999) whereas vesicular monoamine transporter 2 (VMAT2) knockout mice are more susceptible to this damage (Gainetdinov et al., 1998). One interpretation of these observations is that the DAT potentiates the neurotoxicity of MPTP by accumulating intraneuronal MPP$^+$ whereas the VMAT2 is protective by sequestering MPP$^+$ from its primary target of action in mitochondria (Edwards, 1993). However, DAT knockout mice are also resistant to the DA neurotoxicity of MA (Fumagalli et al., 1998) and VMAT2 knockout mice have increased susceptibility (Fumagalli et al., 1999). Furthermore, DA released by MPP$^+$, MA or ischemia may subsequently potentiate its reuptake by interaction with the DA2 receptor (Meiergerd et al., 1993). Indeed, DA2 receptor antagonists attenuate MA-induced DA neurotoxicity (Sonsalla et al., 1986). Taken together, these lines of evidence strongly implicate the DAT-mediated reuptake of released DA and increased cytoplasmic levels of this neurotransmitter as a key step in the dopaminergic neurotoxicity evoked by MPTP, MA and I-R. The 5-HT uptake inhibitor fluoxetine similarly blocks MA-induced serotonergic neurotoxicity but not its dopaminergic neurotoxicity (Ricaurte et al., 1983). Fluoxetine administration does not block hyperthermia caused by MA (Fleckenstein et al., 1997), which also appears to play an important role in the neurotoxic effects of this drug (Lew et al., 1998). Thus, the release and/or subsequent reuptake of 5-HT appears to represent a key step in MA-induced serotonergic neurotoxicity.

In summary, several processes appear to represent key steps that contribute to the neurotoxic effects evoked by MPTP/MPP$^+$, MA and I-R. These include: 1) an initial but transient impairment of energy metabolism and depolarization of the plasma membrane of affected neurons that, together with other mechanisms, evokes a massive release of one or more biogenic amine neurotransmitters; 2) as the energy impairment subsides repolarization of the neuronal membrane initiates reuptake of the released neurotransmitters; 3) a fall of GSH without corresponding increases of GSSG that appears to persist for long periods of time; 4) elevated neuronal O_2^{-}·, NO· and ONOO$^-$ generation implying NMDA and/or AMPA receptor activation under conditions of neuronal energy impairment; 5) extracellular (perhaps also intraneuronal) HO· generation; and, 6) a key role for the DAT, 5-HTT and possibly the NE plasma transporter (NET) in the dopaminergic, serotonergic and noradrenergic neurotoxicity, respectively, evoked by one or more of these brain insults.

Hypothetical Neurotoxic Mechanism

The preceding lines of evidence can be integrated into a hypothetical, and experimentally testable, mechanism for the dopaminergic neurotoxicity evoked by MPTP/MPP$^+$, MA and I-R, the serotonergic neurotoxicity evoked by MA and I-R and the noradrenergic neurotoxicity evoked by I-R in which one or more of the biogenic amine neurotransmitters play a central role. This hypothetical mechanism is schematically illustrated in Fig. 1A-C. Figure 1A is a representation of a monoaminergic neuron and glial cell prior to the neurotoxic insult. Under such circumstances most mitochondrial ATP is utilized by ion pumps such as Na$^+$/K$^+$ATPase and Ca^{2+}ATPase, the high activity of which are necessary to maintain the normal neuronal membrane potential so that it is polarized (Greene and Greenamyre, 1996). When the neuronal membrane is polarized the Mg^{2+} blockade of the NMDA receptor channel (Nowak et al., 1984) prevents extracellular EAAs such as Glu from transducing a signal. The initial step in the hypothetical neurotoxic mechanism evoked by MPTP, MA and I-R is a large but transient impairment of neuronal energy metabolism. Regardless of cause, the energy impairment must be sufficient to depolarize the neuronal and possibly also the mitochondrial membrane, as does MPP$^+$ (Wu et al., 1990), evoking a massive release of DA (MPP$^+$, MA, ischemia), 5-HT (MA and ischemia) or NE (ischemia). Of equal importance, however is that the energy impairment and neuronal depolarization (together perhaps with other mechanisms) should also evoke a massive release of cytoplasmic and mitochondrial GSH (Fig. 1B). It will be demonstrated subsequently that intraneuronal GSH would be expected to scavenge and detoxify putative intraneuronal toxicants and, hence, the depolarization-mediated release of GSH represents an essential step in the hypothesized neurotoxic mechanism.

When neurons are depolarized, the voltage-dependent Mg^{2+} block of NMDA receptors is relieved and they can be activated by extracellular EAAs. A sufficiently large neuronal energy impairment permits much lower extracellular concentrations of EAAs to activate NMDA receptors (Novelli et al., 1988; Zeevalk and Nicklas, 1991) and AMPA receptors (Greene and Greenamyre, 1995) thus mediating an influx of Ca^{2+} that triggers neuronal O$_2$-·, NO· and thence ONOO$^-$ generation. Accordingly, during the profound neuronal energy impairment evoked by MPP$^+$, MA or ischemia, NMDA/AMPA receptor activation by basal extracellular levels of EAAs is proposed to generate O$_2$-· in excess of the scavenging capacity of SOD that reacts with NO· to form ONOO$^-$ (Fig. 1B). ONOO$^-$ readily crosses the plasma membrane and its subsequent decomposition would provide one source of extracellular HO·. Similarly, O$_2$-·, which can also cross the neuronal membrane (Lafon-Cazal et al., 1993), and ONOO$^-$ can also release-Fe^{2+} from iron-containing proteins that in the presence of extracellular ascorbate or other strong reductants catalytically decomposes H$_2$O$_2$ to provide yet another source of extracellular HO·. Being the major HO· scavenger in the brain (Cooper, 1998), extracellular GSH (basal plus that released from neurons affected by the energy impairment) should be immediately oxidized (Fig. 1B). Similarly, even more easily oxidized CySH should be oxidized by HO·. The resultant depletion of extracellular GSH and CySH should, therefore, trigger release of GSH from glia as a mechanism to protect neighboring neurons against direct damage by highly cytotoxic HO· (Cooper, 1998) and to replenish extracellular CySH.

As the energy impairment begins to subside (Fig. 1C), increasing ATP production should *initiate* repolarization of the neuronal membrane and reinstatement of the Mg^{2+} block of NMDA receptors which would act to decrease Ca^{2+} influx through these and AMPA receptors attenuating O$_2$-·, NO·, ONOO$^-$ and thus HO· generation. More importantly, membrane repolarization should initiate reuptake of released DA, 5-HT or NE which are all readily oxidized by O$_2$-· and ONOO$^-$ (Spencer et al., 1998; Wrona and Dryhurst, 1998; Kerry and Rice-Evans, 1999). Thus, by scavenging intraneuronal O$_2$-· and ONOO$^-$ returning DA, 5-HT or NE should completely block HO· formation. Consequently the HO·-mediated oxidation of extracellular GSH and CySH should cease. However, in order to replenish intraneuronal GSH, released and oxidized during the period of maximal energy impairment necessitates its continued release from glia and degradation by γ-GT/DP to provide the CySH (and CysGly) needed for intraneuronal GSH biosynthesis. In the event that γ-GT functions at least in part as a hydrolase enzyme during this period then the degradation of released glial GSH by γ-GT/DP would form not only CySH (and CysGly) and Gly but also Glu. This raises the possibility that the elevation of extracellular Glu that occurs when perfusions of neurotoxic concentrations of MPP$^+$ are discontinued (Carboni et al., 1990) or the delayed elevation of Glu following MA administration (Nash and Yamamoto, 1992) and evoked by I-R (Ikeda et al., 1989; Yang et al., 1995) may result from

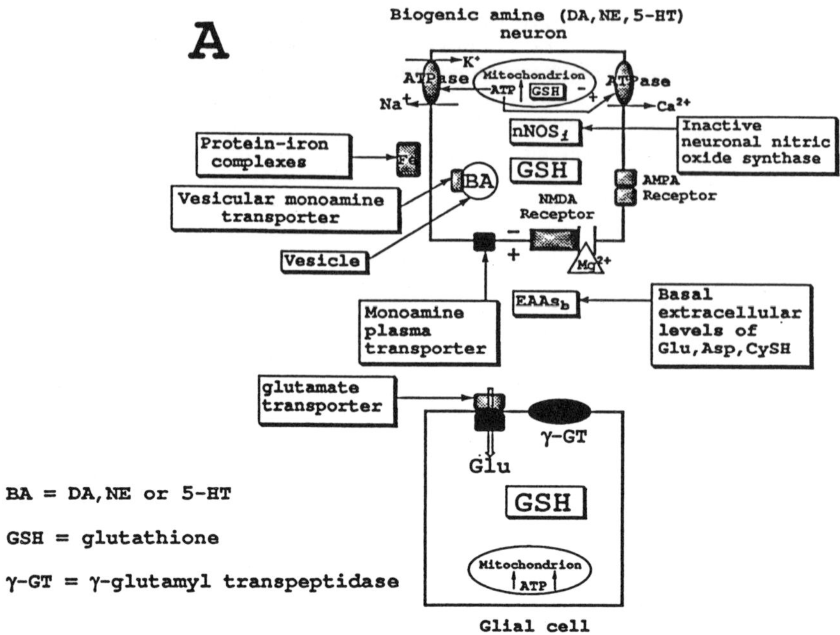

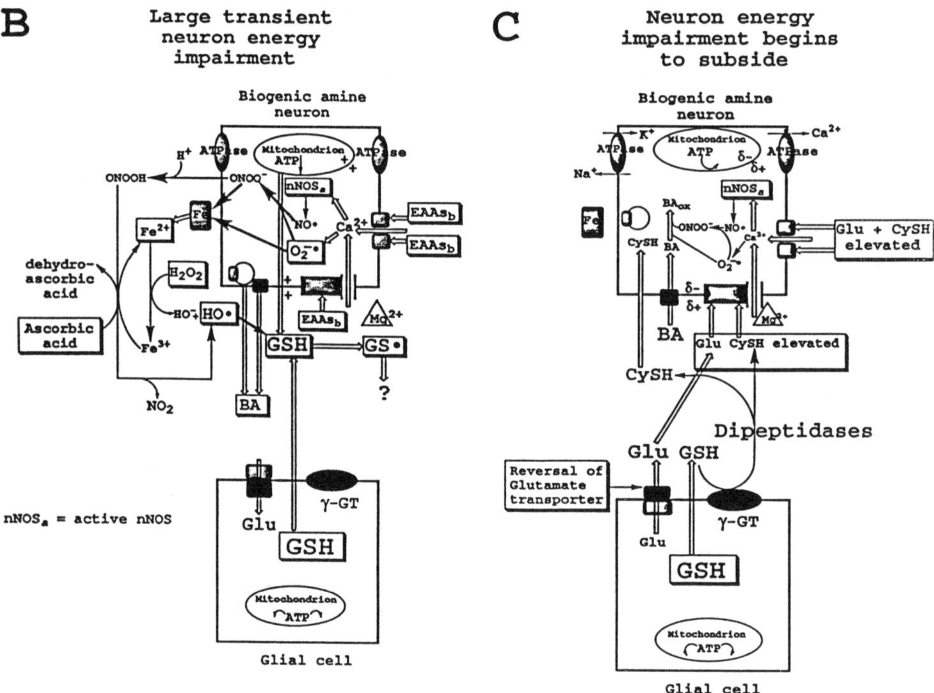

Figure 1. (A) Schematic representation of a biogenic amine (DA, NE, 5-HT) neuron and glial cell; (B) hypothesized effects evoked during a large but transient neuronal energy impairment; and (C) hypothesized effects that occur as the neuronal energy impairment begins to subside.

the γ-GT-mediated hydrolysis of released glial GSH as the neuronal energy impairment subsides. The delayed rise of extracellular Glu levels evoked by MPP$^+$, MA and I-R and activation of NMDA and/or AMPA receptors with resultant intraneuronal generation of O$_2$-·, NO· and ONOO$^-$ and oxidation of DA, 5-HT or NE as they return via their transporters to the cytoplasm of their parent neurons may, therefore, lead to formation of endogenous toxic metabolites that are ultimately responsible for or contribute to neuronal death.

Experimental Support for the Hypothesis Neurotoxic Mechanism

Preliminary experiments designed to explore the hypothesis outlined in Fig. 1 have employed microdialysis to monitor extracellular levels of GSH and CySH in the rat striatum and SN$_c$ before, during, and after perfusions of neurotoxic concentrations of MPP$^+$ (Han et al., 1999). As previously discussed, during such MPP$^+$ perfusions, depolarization of dopaminergic neurons evokes an almost instantaneous release of DA (Rollema et al., 1986) and extracellular HO· generation (Rose et al., 1999) and would be expected to mediate release of intraneuronal GSH (Zängerle et al., 1992). However, during MPP$^+$ perfusion into the rat striatum or SN$_c$ extracellular GSH and CySH remain at basal levels (Fig. 2) (Han et

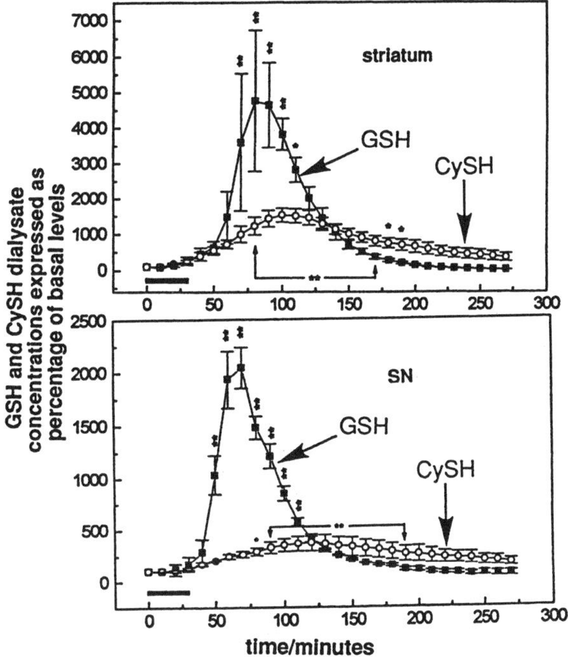

Figure 2. Time-dependent effects of a 30 min perfusion (horizontal black bar) of MPP$^+$ (2.5 mM) into the rat striatum and substantia nigra (SN) on microdialysate concentrations of GSH and CySH. Data are mean ± SEM (vertical bars) percentages of basal GSH and CySH levels. (Reprinted from Han *et al.*, 1999 with permission from the *Journal of Neurochemistry*).

al., 1999), both estimated to be approximately 2 μM (Lada and Kennedy, 1997). This may indicate that during the MPP$^+$-induced dopaminergic energy impairment released neuronal GSH is oxidized by HO· and perhaps also O$_2$-· (Wefers and Sies, 1983) and ONOO$^-$ (Quijano et al., 1997) that, in turn, triggers release of GSH from surrounding glia in amounts sufficient to scavenge most ROS and maintain normal basal extracellular levels of the tripeptide. The fact that extracellular CySH also remains at basal levels during MPP$^+$ perfusion may be the result of its protection from HO·-mediated oxidation by released GSH, replenishment by γ-GT/DP-mediated degradation of released GSH, or as a result of cystine reduction by released GSH. These interpretations appear to be supported by the results of experiments in which Fe^{2+} was perfused into the rat striatum as a method to directly generate extracellular HO· (Xie et al., 1995). Thus, during Fe^{2+} (HO·) perfusion extracellular GSH and CySH remain at basal levels (Han et al., 1999). This implies that during Fe^{2+} perfusion, GSH is released from a brain compartment, probably glia, in quantities sufficient to scavenge most HO· and maintain basal extracellular concentrations of both GSH and CySH. The initial step in the HO·-mediated oxidation of extracellular GSH and CySH during Fe^{2+} or MPP^+ perfusion would be expected to form the thiyl radicals GS· and Cys·, respectively. However, it remains to be determined whether GS· and Cys· subsequently dimerize or undergo more extensive oxidation chemistry.

When MPP^+ perfusions are discontinued and the dopaminergic neuronal energy impairment begins to subside, DA reuptake appears to be initiated based upon decreasing extracellular concentrations of this neurotransmitter without increased concentrations of its metabolites DOPAC and HVA. The hypothesis suggests that as DA, 5-HT or NE return via their plasma membrane transporters to the cytoplasm of their parent neurons they are oxidized by and hence scavenge intraneuronal $O_2^{-\cdot}$ and $ONOO^-$ and therefore block extracellular HO· formation and oxidation of extracellular GSH and CySH. Indeed, microdialysis experiments demonstrate that as high extracellular levels of MPP^+-induced released DA begin to decline (because of its reuptake) extracellular levels of GSH begin to massively increase (Fig. 2). However, after reaching peak concentrations extracellular GSH rather rapidly declines back to basal levels. After discontinuing MPP^+ perfusions, extracellular CySH also begins to increase but more slowly and peaks later than GSH before slowly declining towards basal levels (Fig. 2). These observations suggest that as the MPP^+-induced dopaminergic neuronal energy impairment subsides, elevated extracellular CySH either derives from the γ-GT/DP-mediated degradation of GSH released from glia or that released GSH reduces cystine (Wang and Cynander, 2000). That the former mechanism is the principal source of elevated extracellular CySH appears to be supported by the fact that inhibition of γ-GT with acivicin (Stole et al., 1994) prior to MPP^+ perfusion blocks the elevation of extracellular CySH when MPP^+ perfusions are discontinued but potentiates the increase of extracellular GSH (Han et al., 1999) (Fig. 3). Furthermore, γ-GT inhibition blocks the rather rapid

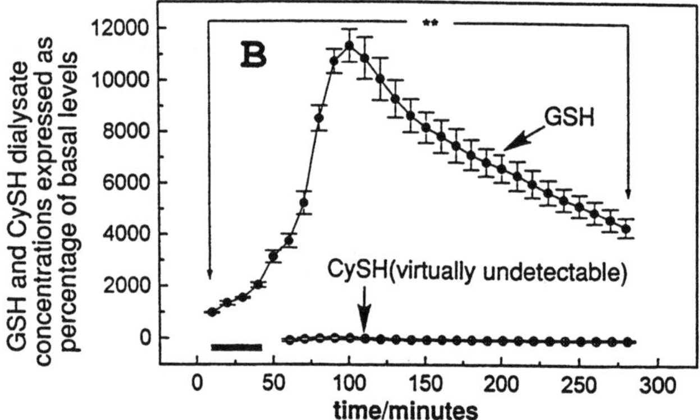

Figure 3. Effects of a 30 min coperfusion of MPP^+ (2.5 mM) and acivicin (1.0 mM) into rat striatum, following prior perfusion of acivicin for 3 h, on microdialysate levels of GSH and CySH. The horizontal black bar shows the time during which MPP^+ entered the brain. Data are mean ± SEM (bars) percentages of normal basal GSH and CySH levels (n ≥ 4). $*p < 0.05$, $**p < 0.01$. (Reprinted from Han *et al.*, 1999 with permission from the *Journal of neurochemistry*.)

decline of extracellular GSH which remain very significantly above basal levels for many hours. In view of the fact that increased extracellular concentrations of CySH that occur when a MPP^+-induced DA neuron energy impairment subsides result from the γ-GT/DP-mediated degradation of released glial GSH raises the possibility that the delayed increase of extracellular Glu (Carboni et al., 1990) might also be derived from released GSH. Preliminary experiments in this laboratory, however, indicate that inhibition of γ-GT by acivicin fails to block the delayed elevation of extracellular Glu when perfusions of MPP^+

into the rat striatum are discontinued. Similarly, γ-GT inhibition potentiates ischemia-induced release of GSH but does not affect elevation of extracellular Glu (Li et al., 1999). Thus, the delayed release of Glu evoked by MPP$^+$ (Carboni et al., 1990) and MA (Nash and Yamamoto, 1992) appears to be from another source such as energy-impaired glia (Zeevalk et al., 1998) and/or glutamatergic neurons (Storey et al., 1992; Srivastava et al., 1993). Nevertheless, Glu released in response to cerebral ischemia rat appears, in part, to be derived from GSH (Yang et al., 1995).

During the period of recovering but still reduced neuronal ATP production, elevated extracellular levels not only of Glu but also CySH, another excitotoxin (Pullan et al., 1987; Olney et al., 1990), would be expected to continue NMDA and AMPA receptor activation with consequent O_2-·, NO· and ONOO$^-$ generation which, we suggest, oxidizes DA, 5-HT or NE as they are returned to their parent neurons via their plasma membrane transporters.

Possible consequences of intraneuronal oxidation of biogenic amine neurotransmitters by O_2-· and ONOO$^-$.

Intraneuronal O_2-·/ONOO$^-$-mediated oxidation of DA by the mechanism just described would generate initially DA-o-quinone (DAQ) (Scheme 1). In vitro, DAQ impairs mitochondrial respiration by decreasing state 3 and dramatically increasing state 4, which together should significantly decrease ATP production (Berman and Hastings, 1999). DAQ also causes mitochondrial swelling and opening of the permeability transition pore (Berman and Hastings, 1999). DAQ inactivates tyrosine hydroxylase (TH) (Xu et al., 1998), the rate limiting enzyme for DA biosynthesis. Interestingly, MA administration induces inhibition of TH (Hotchkiss and Gibb, 1980). Nitration of TH in response to MPTP administration to mice has been reported to account for inhibition of this enzyme (Ara et al., 1998) although, in vitro, nitration of tyrosine residues of TH by ONOO$^-$ has no influence on its activity (Kuhn et al., 1999) suggesting that DAQ is responsible for this effect. The effects of DAQ appear to result from its covalent addition to active site protein cysteine sulfhydryl residues of mitochondrial enzyme complexes and TH. Indeed, neurotoxic doses of MA lead to significantly increased levels of DA covalently bound to cysteinyl residues of striatal proteins (LaVoie and Hastings, 1999). Furthermore, in vitro, the ability of DAQ to impair mitochondrial respiration and inactivate TH is blocked by excess GSH. This is apparently because GSH rapidly reacts with DAQ to form glutathionyl conjugates of DA.

The data presented in Fig. 2 indicates that when MPP$^+$ perfusions into rat striatum or SN$_c$ are discontinued the total release of GSH is significantly greater than that of extracellular CySH. There are several possible explanations for this effect such as the facile autoxidation of CySH to cystine or only partial DP-mediated hydrolysis of CysGly formed by the γ-GT-mediated degradation of released GSH. We have suggested that it may reflect partial translocation of extracellular CySH into dopaminergic neurons, normally the first step necessary to replenish intraneuronal GSH released during the MPP$^+$-induced neuronal energy impairment. This, in turn, would imply that as the neuronal energy impairment subsides, NMDA/AMPA receptor activation by elevated extracellular levels of Glu and CySH would mediate intraneuronal oxidation of DA by O_2-·/ONOO$^-$ in the presence of translocated CySH. In vitro studies indicate that under such circumstances DAQ would be scavenged by CySH (or DA· by CyS·) to form initially 5-S-cysteinyldopamine (5-S-CyS-DA, Scheme 1) (Shen and Dryhurst, 1996). However, 5-S-CyS-DA is appreciably more easily oxidized than DA to give o-quinone **1**, the precursor of the dihydrobenzothiazine DHBT-1 and benzothiazines BT-1 and BT-2 (Scheme 1) (Shen and Dryhurst, 1996). When incubated in vitro with intact rat brain mitochondria for 5 min, DHBT-1 causes a dose-dependent inhibition of complex I respiration with an IC$_{50}$ of approximately 0.8 mM (Fig. 4) but is without effect on succinate-supported (complex II) respiration (Li and Dryhurst, 1997). When incubated with freeze-thawed rat brain mitochondria (mitochondrial membranes) DHBT-1 causes a slower, time-dependent irreversible inhibition of NADH-coenzyme Q_1 (CoQ$_1$) reductase (Fig. 5). The time-dependence of the irreversible inhibition of NADH-CoQ$_1$ reductase by DHBT-1 appears to be connected to the oxidation of this putative intraneuronal metabolite by an as yet unknown constituent of the inner mitochondrial membrane (Li et al., 1998). In this reaction DHBT-1 is first oxidized to o-quinone imine **2** which rapidly tautomerizes and decarboxylates to BT-1 and BT-2, respectively (Scheme 2). However, BT-1 and BT-2 also evoke a time-dependent irreversible inhibition of NADH-CoQ$_1$ reductase when incubated with mitochondrial membranes. This effect is related to the mitochondrial membrane-catalyzed oxidation of BT-1 and BT-2. However, when this reaction is monitored by HPLC analysis, low molecular weight oxidative metabolites of BT-1

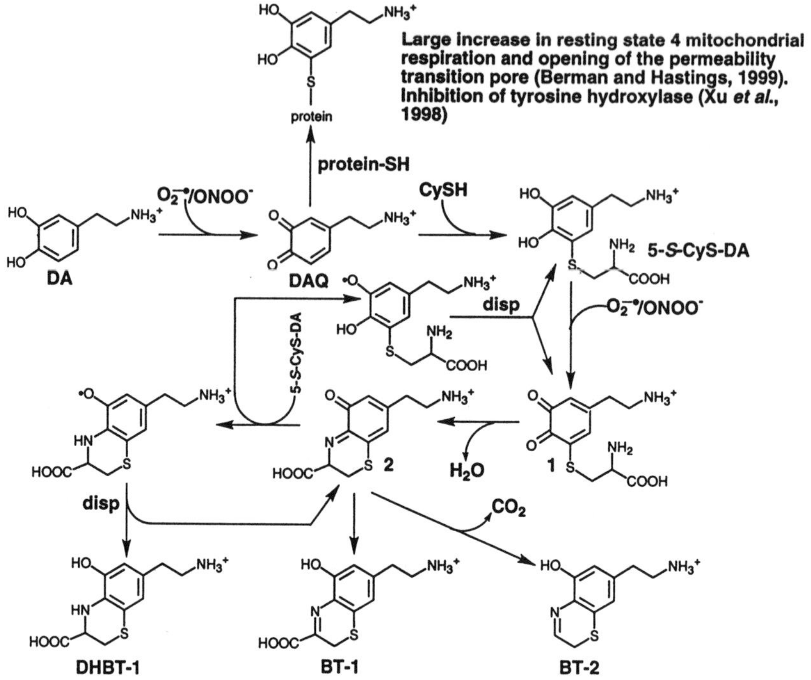

Scheme 1

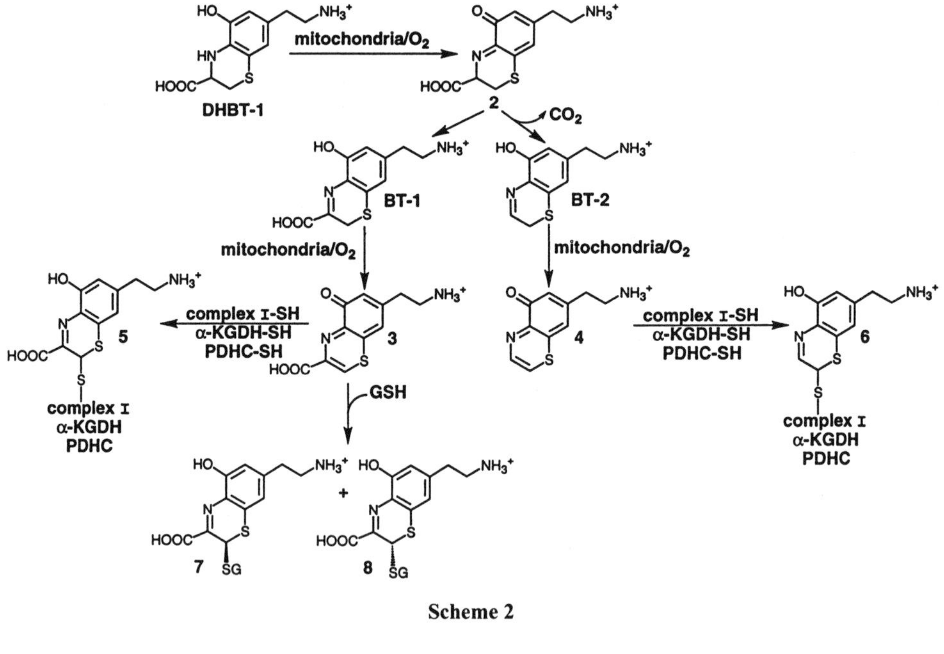

Scheme 2

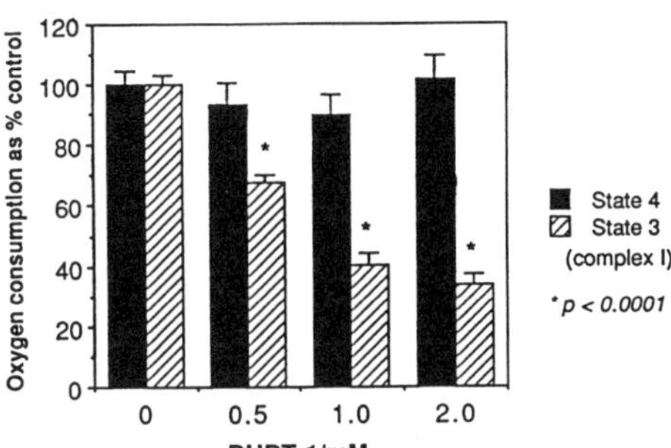

Figure 4. Effects of DHBT-1 on state 4 and state 3 (complex I) oxygen consumption (respiration). DHBT-1 was incubated for 5 min at 27°C with intact rat brain mitochondria prior to addition of malate (2.5 mM) and pyruvate (2.5 mM). State 4 oxygen consumption was then measured (2 min), and then adenosine 5'-diphosphate was added to stimulate state 3 (complex I) respiration.

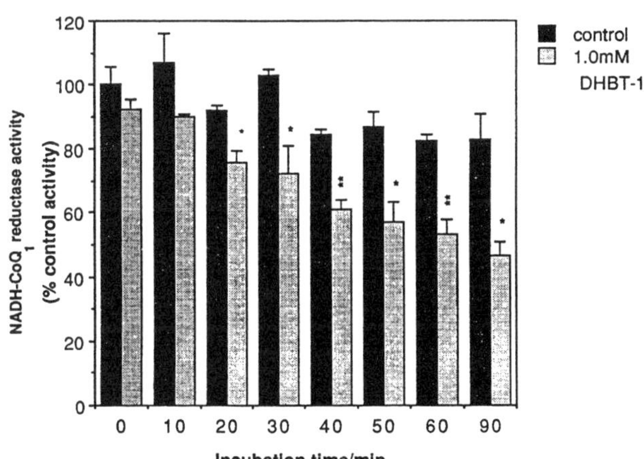

Figure 5. Time-dependent inhibition of NADH-CoQ$_1$ reductase by DHBT-1 (1 mM) when incubated with freeze-thawed rat brain mitochondrial (mt membranes) in 20 mM potassium phosphate buffer (pH 8.0).

and BT-2 are not observed suggesting that their proximate oxidation products may bind covalently to mitochondrial membrane proteins. Support for this suggestion derives from the fact that > equimolar concentrations of GSH completely block the irreversible inhibition of NADH-CoQ$_1$ reductase by DHBT-1 and BT-1/BT-2 although without affecting the rate of their mitochondrial membrane-catalyzed oxidations. Furthermore, following incubation of DHBT-1 or BT-1/BT-2 with mitochondrial membranes in the presence of GSH new metabolites are formed which include **7** and **8** (Scheme 2), epimeric 2-*S*-glutathionyl conjugates of BT-1. Thus, it appears that a constituent of the mitochondrial membrane catalyzes the oxidation of BT-1 and BT-2 to the highly electrophilic *o*-quinone imine intermediates **3** and **4** which may covalently bind to active site cysteinyl sulfhydryl residues of complex I, forming adducts conceptualized as **5** and **6**, respectively in Scheme 2, thus evoking irreversible inhibition of NADH-CoQ$_1$ reductase.

DHBT-1 and BT-1/BT-2 also cause a time-dependent inhibition of α-KGDH when incubated with rat brain mitochondrial membranes but have no effect on COX activity (Shen et al., 2000). Similarly, DHBT-1 and BT-1/BT-2 inhibit PDHC (unpublished results). Again, the time dependence of α-KGDH and PDHC inhibition by DHBT-1 is dependent on its oxidation catalyzed by mitochondrial membranes to BT-1 and BT-2 that are further oxidized to **3** and **4** which bind to active site cysteinyl residues of these enzyme complexes (Scheme 2). The inhibition of α-KGDH and PDHC by DHBT-1 and BT-1/BT-2 is blocked by GSH with resultant formation of metabolites such as **7** and **8** (Scheme 2).

The massive release of GSH and its γ-GT/DP-mediated degradation to extracellular CySH which may react with intraneuronal DAQ forming 5-*S*-CyS-DA, DHBT-1, BT-1 and BT-2 provides a plausible, although presently unproven, rationale for the fall of GSH without increased GSSG evoked by MPTP/MPP$^+$, MA and I-R. However, as the dopaminergic energy impairment evoked by MPP$^+$, MA and ischemia subsides NMDA/AMPA receptor activation by elevated extracellular levels of Glu and CySH, the latter derived from degradation of released GSH by γ-GT/DP, with resultant intraneuronal O$_2$-·, NO· and ONOO$^-$ generation provides several possible mechanisms for neuronal damage. These include direct damage to key proteins by O$_2$-·, ONOO$^-$ and HO·, covalent modification of proteins by DAQ and by electrophilic intermediates formed by oxidation of DHBT-1, BT-1 and BT-2. A role for the latter putative metabolites depends upon translocation of CySH into dopaminergic neurons coincident with intraneuronal oxidation of DA. Intraneuronal GSH would be neuroprotective against each of these toxins by scavenging reactive oxygen and nitrogen species, DAQ and electrophilic intermediates **3** and **4** formed by oxidation of DHBT-1, BT-1 and BT-2 (Scheme 2). This points to the central importance in the hypothesized neurotoxic mechanism of a massive release of neuronal GSH during the dopaminergic energy impairment evoked by MPP$^+$, MA and ischemia. Indeed, elevation of brain GSH or other thiols attenuates the dopaminergic neurotoxicity of MPTP (Oishi et al., 1993; Weiner et al., 1988), MA (Steranka and Rhina, 1987) and cerebral ischemia (Yamamoto et al., 1993). Conversely, depletion of brain GSH potentiates the dopaminergic neurotoxicity of MPTP/MPP$^+$ (Wullner et al., 1996; McNaught and Jenner, 1999).

In vitro, concentrations of DAQ, DHBT-1, and BT-1/BT-2 required to evoke statistically significant impairment of mitochondrial respiration or to inhibit respiratory and other enzyme complexes range from approximately 20 μM - 1 mM. Thus, an important question is whether such putative neurotoxicants derived from DA might reach such intraneuronal concentrations in vivo. Generally, microdialysis-based estimates of basal extracellular concentrations of DA in the rat striatum, for example, fall in the single-digit nanomolar range (Olson and Justice, 1993). However, recent evidence suggests that basal extracellular DA levels are probably in the micromolar range (Yang et al., 2000). Indeed, the instantaneous concentration of DA released by exocytosis has been estimated to be approximately 70 μM (Hochstetler et al., 2000). These basal extracellular DA levels appear to be reasonable in view of its very high concentration (> 0.8 M) in individual vesicles (Hochstetler et al., 2000). Furthermore, concentrations of DA within striatal terminals and cell bodies in the SN$_c$ have been estimated to be approximately 50 mM and more than 1 mM, respectively (Andén et al., 1966). Thus, the massive release of DA evoked by MPP$^+$, MA and ischemia and its subsequent reuptake as the energy impairment subsides might well lead to cytoplasmic concentrations of DAQ and/or electrophilic intermediates formed by oxidation of DHBT-1, BT-1 and BT-2 sufficient to evoke irreversible damage to mitochondrial and other essential proteins.

There is presently no information bearing on the interactions of NE-*o*-quinone (NEQ) with mitochondria or other noradrenergic enzymes. However, oxidation of NE in the presence of CySH generates 5-S-cysteinylnorepinephrine and thence DHBTs and BTs in a reaction analogous to that presented in Scheme 1 (Shen and Dryhurst, 1996; Xin et al., 2000). Furthermore, these DHBTs and BTs evoke irreversible inhibition of mitochondrial complex I (NADH-CoQ$_1$ reductase), α-KGDH and PDHC by mechanisms virtually identical to those discussed in connection with DHBT-1, BT-1 and BT-2 (Scheme 2) (Xin et al., 2000). Thus, these putative DHBT and BT metabolites, together with NEQ, might contribute to the degeneration of noradrenergic neurons evoked by I-R.

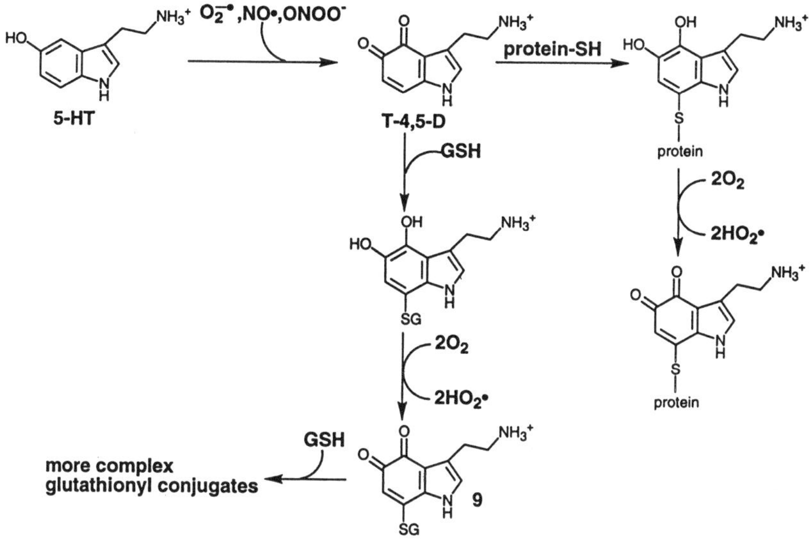

Scheme 3

The 5-HTT-mediated reuptake of 5-HT as the neuronal energy impairment evoked by MA or ischemia subsides and its oxidation by O$_2$-· (Wrona and Dryhurst, 1998), NO· or ONOO$^-$ (unpublished observations) would form tryptamine-4,5-dione (T-4,5-D) as a major reaction product (Scheme 3). When incubated with intact rat brain mitochondria, T-4,5-D uncouples respiration and inhibits malate + pyruvate-supported state 3 respiration (Jiang et al., 1999). Experiments with rat brain mitochondrial membranes confirm that T-4,5-D irreversibly inhibits NADH-CoQ$_1$ reductase and COX apparently by covalently modifying active site cysteine sulfhydryl residues of these enzyme complexes (Scheme 3). GSH blocks the inhibition of these respiratory enzyme complexes by scavenging T-4,5-D to form, initially, 7-S-glutathionyl-tryptamine-4,5-dione (**9**, Scheme 3) and then, more slowly, more complex glutathionyl conjugates (Wong et al., 1993). Micromolar concentrations of T-4,5-D also rapidly and irreversibly inactivate tryptophan hydroxylase (TPH), the rate-limiting enzyme for 5-HT biosynthesis, also as a result of its covalent modification of active site cysteinyl residues (unpublished results). The activity of TPH exposed to T-4,5-D cannot be restored by anaerobic reduction with dithiothreitol (DTT) and Fe^{2+} indicating that the inactivation is irreversible. Glutathionyl conjugate **9** also inactivates TPH, but the activity of this enzyme can be largely restored by anaerobic reduction with DTT and Fe^{2+}. These results are interesting because decreased activity of TPH evoked by a single non-neurotoxic dose of MA can be restored in vitro by incubation with DTT/Fe^{2+} (Stone et al., 1989). However, at longer times following multiple doses of MA which evoke serotonergic neurotoxicity, TPH activity cannot be restored by anaerobic reduction. Thus, it is conceivable that a single dose of MA is insufficient to evoke massive depletion of GSH from the cytoplasm of serotonergic neurons. Thus, as the energy impairment subsides and 5-HT returns to its parent neurons it may be oxidized to T-4,5-D which reacts with remaining GSH forming **9** which reversibly inhibits TPH. Multiple neurotoxic doses of MA, however, might severely deplete intraneuronal GSH. Thus, as the energy impairment subsides, intraneuronal oxidation of 5-HT might generate T-4,5-D which, in the absence of cytoplasmic GSH, irreversibly inhibits TPH. Other studies with T-4,5-D suggest it is a neurotoxin (Crino et al., 1989), mediates release of 5-HT (Chen et al., 1989), possibly by covalent modification of cysteinyl residues of guanine nucleotide-binding regulatory proteins (Fishman et al., 1991). As discussed in connection with the dopaminergic and noradrenergic neurotoxic mechanism, the preceding results again suggest that a very important step in serotonergic neurotoxicity is the release of intraneuronal GSH during the neuronal energy impairment.

Possible relevance of the hypothesized neurotoxic mechanism to neurodegenerative brain disorders

Many factors implicated in the dopaminergic neurotoxicity of MPTP/MPP+, MA and I-R also appear to be involved in the selective degeneration of nigrostriatal DA neurons in PD (Hornykiewicz and Kish, 1980). The pathological processes in PD are believed to occur in neuromelanin-pigmented dopaminergic cells in the SN_C and include decreased activity of mitochondrial complex I (Schapira et al., 1990) and possibly α-KGDH (Mizuno et al., 1994). A rise of SOD (Saggu et al., 1989) and 3-NT formation (Good et al., 1998) in the parkinsonian SN_C are consistent with elevated intraneuronal O_2-· generation, nNOS activation with resultant NO· and $ONOO^-$ generation, effects consistent with NMDA receptor activation under conditions of impaired DA neuronal energy metabolism. Increased accumulation of iron by neuromelanin (Good et al., 1992) is consistent with O_2-·/NO·/$ONOO^-$-mediated transient mobilization of Fe^{2+} from iron-containing proteins. Evidence for oxidative stress (Jenner and Olanow, 1996) implies Fe^{2+}- and $ONOO^-$-mediated HO· formation in the parkinsonian SN_C. Another characteristic change in PD is a massive fall of nigral GSH levels without a corresponding increase of GSSG (Sian et al., 1994a) that occurs early in the pathogenesis of the disorder (Dexter et al., 1994). Increased activity of γ-GT (Sian et al., 1994b) also implicates altered metabolism of GSH as a key step in the neuropathological mechanism.

Based on the preceding lines of evidence together with studies of the dopaminergic neurotoxicity evoked by MPTP/MPP+, MA and I-R permits at least a plausible sequence of events that might account for the degeneration of pigmented dopaminergic SN_C cells in PD. Thus, the decline of neuronal energy metabolism with aging (Beal, 1995), periodic exposure to environmental toxins that interfere with mitochondrial respiration (Cooper and Schapira, 1997; Corrigan et al., 2000), particularly in view of age-based (Le Couteur and McClean, 1998) and genetically-determined (Wood, 1997) impairments of xenobiotic metabolism, superimposed on systemic complex I (Swerdlow et al., 1996) and α-KGDH defects (Kobayashi et al., 1998) might represent a combination of factors that evoke a profound transient dopaminergic energy impairment in the PD brain. This energy impairment could then evoke release of DA and GSH from pigmented SN_C cells and trigger the hypothetical mechanism conceptualized in Fig. 1. Subsequent DAT-mediated reuptake of DA as the energy impairment begins to subside and NMDA receptor activation by elevated extracellular levels of Glu and CySH (formed by γ-GT/DP-mediated degradation of released glial GSH) would then mediate intraneuronal O_2-·, NO· and $ONOO^-$ generation that oxidizes DA to DAQ. A possible link between the loss of nigral GSH, increased γ-GT activity and increased neuronal O_2-·, NO· and $ONOO^-$ generation is provided by reports that 5-*S*-CyS-DA, normally a minor metabolite of DA, becomes a major metabolite in the parkinsonian SN_C as shown by a large increase in the 5-*S*-CyS-DA/DA concentration ratio (Fornstedt et al., 1989; Spencer et al., 1998). This rise of the 5-*S*-CyS-DA/DA ratio suggests not only an accelerated rate of DA oxidation to DAQ but also increased availability of CySH, normally present in very low concentrations in the brain, in the parkinsonian SN_C. Further oxidation of 5-*S*-CyS-DA would be expected to form DHBT-1 (Scheme 1) which might account for the anatomically selective decrease of mitochondrial complex I activity in the parkinsonian SN_C.

Similarly, DAQ is also a mitochondrial toxicant (Berman and Hastings, 1999) and formation of this biologically reactive intermediate is consistent with the presence of DA-modified proteins in PD patients (Rowe et al., 1998).

The neurodegeneration in Alzheimer's disease (AD) affects multiple neurotransmitter systems but only in remarkably anatomically selective regions of the brain. These include the association areas of the cortex, hippocampus and amygdala, regions that are richly innervated with serotonergic and noradrenergic terminals which degenerate (Hardy et al., 1985). Individuals at high risk to develop AD typically show impairments of cerebral metabolic rate even before the appearance of clinical disease (Blass et al., 2000). This impairment of energy metabolism might, therefore, trigger the neurotoxic mechanism proposed in Fig. 1 with resultant intraneuronal formation of T-4,5-D, NEQ or DHBT/BT metabolites that contribute to the defects in PDHC, α-KGDH and COX that occur in the AD brain (Blass et al., 2000). Indeed, abnormal oxidized forms of 5-HT have been detected in the cerebrospinal fluid of AD patients (Volicer et al., 1985).

Acknowledgments

The work from this laboratory reported in this communication was supported by NIH grants GM32367 and NS29886. Additional support was provided by the Vice President for Research at the University of Oklahoma.

References

Albers, D.S., Zeevalk, G.D., and Sonsalla, P.K., 1996, Damage to dopaminergic nerve terminals in mice by combined treatment of intrastriatal malonate with systemic methampehtamine or MPTP. *Brain Res.* **718**: 217-220.

Ali, S.F., and Itzhak, Y., 1998, Effects of 7-nitroindazole, an nNOS inhibitor, on methamphetamine-induced dopaminergic and serotonergic neurotoxicity in mice. *Ann. N.Y. Acad. Sci.* **844**: 122-130.

Andén, N.-E., Fuxe, B., Hamberger, B., and Hökfelt, T., 1966, A quantitative study on the nigro-neostriatal dopamine neuron system in the rat. *Acta Physiol. Scand.* **67**: 306-312.

Ara, J., Przedborski, S., Naini, A.B., Jackson-Lewis, V., Trifiletti, R.R., Horwitz, J., and Ischiropoulos, H., 1998, Inactivation of tyrosine hydroxylase by nitration following exposure to peroxynitrite and 1-methyl-4-phenyl-1,2,3,6-tetrahydropyridine (MPTP). *Proc. Nat. Acad. Sci. U.S.A.* **95**: 7659-7663.

Axt, K.J., and Molliver, M.E., 1991, Immunocytochemical evidence for methamphetamine-induced serotonergic axon loss in the rat brain. *Synapse* **9**: 302-313.

Beal, M.F., 1995, *Mitochondrial Dysfunction and Oxidative Damage in Neurodegenerative Diseases,* R.G. Landes Co., Austin, TX.

Benveniste, H., Drejer, J., Schousboe, A., and Diemer, N., 1984, Elevation of extracellular concentrations of glutamate and aspartate in rat hippocampus during transient cerebral ischemia monitored by intracerebral microdialysis. *J. Neurochem.* **43**: 1369-1374.

Berger, U.V., Gu, X.F., and Azmitia, E.C., 1992, The substituted amphetamines 3,4-methylenedioxymethamphetamine, methamphetamine, *p*-chloroamphetamine and fenfluramine induce 5-hydroxytrytpamine release via a common mechanism blocked by fluoxetine and cocaine. *Eur. J. Pharmacol.* **215**: 153-160.

Berman, S.B., and Hastings, T.G., 1999, Dopamine oxidation alters mitochondrial respiration and induces permeability transition in brain mitochondria: implications for Parkinson's disease. *J. Neurochem.* **73**: 1127-1137.

Bezard, E., Gross, C.E., Fournier, M.C., Dovero, S., Bloch, B., and Jaber, M., 1999, Absence of MPTP-induced neuronal death in mice lacking the dopamine transporter. *Exp. Neurol.* **155**: 268-273.

Blass, J.P., Sheu, R.K., and Gibson, G.E., 2000, Inherent abnormalities in energy metabolism in Alzheimer disease. Interaction with cerebrovascular compromise. *Ann. N.Y. Acad. Sci.* **903**: 204-221.

Burrows, K.B., Gudelsky, G., Yamamoto, B.K., 2000, Rapid and transient inhibition of mitochondrial function following methamphetamine or 3,4-methylenedioxymethamphetamine administration. *Eur. J. Pharmacol.* **398**: 11-18.

Cadet, J.L., Sheng, P., Ali, S., Rothman, R., Carlson, E., and Epstein, C., 1994, Attenuation of methamphetamine-induced neurotoxicity in copper/zinc superoxide dismustase transgenic mice. *J. Neurochem.* **62**: 380-383.

Callahan, B., Yuan, J., Stover, G., Hatzidimitriou, G., and Ricaurte, G., 1998, Effects of 2-deoxy-D-glucose on methamphetamine-induced dopamine and serotonin neurotoxicity. *J. Neurochem.* **70**: 190-197.

Cao, C.J., Eldefrawi, A.T., and Eldefrawi, M.E., 1990, ATP-regulated neuronal catecholamine uptake: a new mechanism. *Life Sci.* **47**: 655-667.

Carboni, S., Melis, F., Pani, L., Hadjiconstantinou, M., and Rossetti, Z.L., 1990, The non-competitive NMDA receptor antagonist MK-801 prevents the massive release of glutamate and aspartate from rat striatum induced by 1-methyl-4-phenylpyridinium (MPP$^+$). *Neurosci. lett.* **117**, 129-133.

Cebers, G., Cebere, A., and Liljequist, S., 1998, Metabolic inhibition potentiates AMPA-induced Ca^{2+} fluxes and neurotoxicity in rat cerebellar granule cells. *Brain Res.* **779**: 194-204.

Chan, P., Delanney, L.E., Irwin, I., Langston, J.W., and Di Monte, D., 1991, Rapid ATP loss caused by MPTP in mouse brain. *J. Neurochem.* **57**: 348-351.

Chan, P., Di Monte, D.A., Luo, J.-J., Delanney, L.E., Irwin, I., and Langston, J.W., 1994, Rapid ATP loss caused by methamphetamine in the mouse striatum: relationship between energy impairment and dopaminergic neurotoxicity. *J. Neurochem.* **62**: 2484-2487.

Chiueh, C.C., and Huang, S.J., 1991, MPP$^+$ enhances potassium evoked striatal dopamine release through an Ω-conotoxin-insensitive, tetrodotoxin- and nimodipine-sensitive calcium dependent mechanism. *Ann. N.Y. Acad. Sci.* **635**: 393-396.

Chiueh, C.C., Wu, R.-M., Mohanakumar, K.P., Sternberger, L.M., Krishna, G., Obata, T., and Murphy, D.L., 1994, In vivo generation of hydroxyl radicals and MPTP-induced dopaminergic toxicity in the basal ganglia. *Ann. N.Y. Acad. Sci.* **738**, 25-36.

Clemens, J.A., and Phebus, L.A., 1988, Dopamine depletion protects striatal neurons from ischemia-induced cell death. *Life Sci.* **42**: 707-713.

Cooper, A.J.L., 1998, Role of astrocytes in maintaining cerebral glutathione homeostasis and in protecting the brain against xenobiotics and oxidative stress, in *Glutathione in the Nervous System* (C.A. Shaw, Ed.), Taylor and Francis, Washington, D.C., pp. 91-115.

Cooper, J.M., and Schapira, A.H.V., 1997, Mitochondrial dysfunction in neurodegeneration. *J. Bioenerg. Biomembr.* **29**: 175-183.

Cooper, A.J.L., Pulsinelli, W.A., and Duffy, T.E., 1980, Glutathione and ascorbate during ischemia and postischemic reperfusion in rat brain. *J. Neurochem.* **35**, 1242-1245.

Chen, J.-C., Crino, P.B., To, A.C.S., and Volicer, L., 1989, Increased serotonin efflux by a partially oxidized serotonin: tryptamine-4,5-dione. *J. Pharmacol. Exp. Therap.* **250**: 141-148.

Corrigan, F.M., Wienberg, C.L., Shore, R.F., Daniel, S.E., and Mann, D., 2000, Organochlorine insecticides in substantia nigra in Parkinson's disease. *J. Toxicol. Environ. Health* **59**: 229-234.

Crino, P.B., Vogt, B.A., Chen, J.-C., and Volicer, L., 1989, Neurotoxic effects of partially oxidized serotonin: tryptamine-4,5-dione. *Brain Res.* **504**: 247-257.

Crow, J.P., Spruell, C., Chen, J., Gunn, C., Ischiropoulos, H., Tsai, M., Smith, C.D., Radi, R., Koppenol, W.H., and Beckman, J.S., 1994, On the pH-dependent yield of hydroxyl radical products from peroxynitrite. *Free Rad. Biol. Med.* **16**: 331-338.

Della Donne, K.T., and Sonsalla, P.K., 1994, Protection against methamphetamine-induced neurotoxicity to neostriatal dopamine neurons by adenosine receptor activation. *J. Pharmacol. Exp. Therap.* **271**: 1320-1326.

Dexter, D.T., Sian, J., Rose, H., Hindmarsh, J.-G., Mann, V.M., Cooper, J.M., Wells, F.R., Daniel, S.E., Lees, A.J., Schapira, A.H.V., Jenner, P., and Marsden, C.D., 1994, Indices of oxidative stress and mitochondrial function in individuals with incidental Lewy body disease. *Ann. Neurol.* **35**: 38-44.

Dorrepaal, C.A., van Bel, F., Moison, R.M., Shadid, M., van de Bor, M., Steedijk, P., and Berger, H.M., 1997, Oxidative stress during post-hypoxic-ischemia reperfusion in the newborn lamb: the effect of nitric oxide synthesis inhibition. *Pediatr. Res.* **41**: 321-326.

Dringen, R., Pfeiffer, B., and Hamprecht, B., 1999, Synthesis of the antioxidant glutathione in neurons: supply by astrocytes of CsyGly as precursor for neuronal glutathione. *J. Neurosci.* **19**: 562-569.

Edwards, R.H., 1993, Neural degeneration and the transport of neurotransmitters. *Ann. Neurol.* **34**: 638-645.

Eliasson, M.J., Huang, Z., Ferrante, R.J., Sasamata, M., Molliver, S.E., Snyder, S.H., and Moskowitz, M.A., 1999, Neuronal nitric oxide synthase activation and peroxynitrite formation in ischemic stroke linked to neuronal damage. *J. Neurosci.* **19**: 5910-5918.

Ferrante, R.J., Hantraye, P., Brouillet, E., and Beal, M.F., 1999, Increased nitrotyrosine immunoreactivity in substantia nigra neurons in MPTP treated baboons is blocked by inhibition of neuronal nitric oxide synthase. *Brain Res.* **823**: 177-182.

Ferraro, T.N., Golden, G.T., DeMattei, M., Hare, T.A., and Fariello, R.G., 1986, Effect of 1-methyl-4-phenyl-1,2,3,6-tetrahydropyridine (MPTP) on levels of glutathione in the extrapyramidal system of the mouse. *Neuropharmacology* **25**: 1071-1074.

Fishman, J.B., Rubins, J.B., Chen, J.-C., Dickey, B.F., and Volicer, L., 1991, Modification of brain guanine nucleotide-binding regulatory proteins by tryptamine-4,5-dione, a neurotoxic derivative of serotonin. *J. Neurochem.* **56**, 1851-1854.

Fleckenstein, A.E., Beyeler, M.L., Jackson, J.C., Wilkins, D.G., Gibb, J.W., and Hanson, G.R., 1997, Methamphetamine-induced decrease in tryptophan hydroxylase activity: role of 5-hydroxytryptaminergic transporters. *Eur. J. Pharmacol.* **324**, 179-186.

Flint, D.H., Tuminello, J.F., and Emptage, M.H., 1993, The inactivation of Fe-S cluster containing hydrolyases by superoxide. *J. Biol. Chem.* **268**, 22369-22376.

Forman, L.J., Liu, P., Nagele, R.G., Yin, K., and Wong, P.Y., 1998, Augmentation of nitric oxide, superoxide, and peroxynitrite production during cerebral ischemia and reperfusion in the rat. *Neurochem. Res.* **23**: 141-148.

Fornstedt, B., Brun, A., Rosengren, E., and Carlsson, A., 1989, The apparent autoxidation rate of catechols in dopamine-rich regions of human brain increases with the degree of depigmentation of substantia nigra. *J. Neural Transm. [P-D Dementia Sect.]* **1**: 279-295.

Fumagalli, F., Gainetdinov, R.R., Valenzano, K.J., and Caron, M.G., 1998, Role of dopamine transporter in methamphetamine-induced neurotoxicity: evidence from mice lacking the transporter. *J. Neurosci.* **18**: 4861-4869.

Fumagalli, F., Gainetdinov, R.R., Wang, Y.M., Valenzano, K.J., Miller, G.W., and Caron, M.G., 1999, Increased methamphetamine neurotoxicity in heterozygous vesicular monoamine transporter 2 knockout mice. *J. Neurosci.* **19**: 2424-2431.

Gainetdinov, R.R., Fumagalli, F., Wang, Y.M., Jones, S.R., Levey, A.I., Miller, G.W., and Caron, M.G., 1998, Increased MPTP neurotoxicity in vesicular monoamine transporter 2 heterozygote knockout mice. *J. Neurochem.* **70**: 1973-1978.

Gardner, P.R., Constantino, G., Szabo, C., and Salzman, A.L., 1997, Nitric oxide sensitivity of aconitases. *J. Biol. Chem.* **272**: 25071-25076.

Gerlach, M., and Riederer, P., 1996, Animal models of Parkinson's disease: an empirical comparison with the phenomenology of the disease in man. *J.Neural Transm.* **103**, 987-1041.

Gerlach, M., Riederer, P., Przuntek, H., and Youdim, M.B.H., 1991, MPTP mechanisms of neurotoxicity and their implications for Parkinson's disease. *Eur. J. Pharmacol.* **208**: 273-286.

Gibb, J.W., Hanson, G.R., and Johnson, M., 1994, Neurochemical mechanisms of toxicity, in: *Amphetamine and Its Analogs, Psychopharmacology, Toxicology and Abuse* (A.K. Cho, and D.S. Segal, Eds.), Academic Press, San Diego, pp. 269-289.

Gibb, J.W., Johnson, M., Elayan, I., Lim, H.K., Matsuda, L., and Hanson, G.R., 1997, Neurotoxicity of amphetamines and their metabolites, in *Pharmacokinetics, Metabolism, and Pharmaceutics of Drugs of Abuse* (R.S. Rapaka, N. Chiang, and B.R. Martin, Eds.), *NIDA Research Monograph 173*, pp. 128-145.

Gill, R., Foster, A.C., and Woodruff, G.N., 1988, MK-801 is neuroprotective in gerbils when administered during the post-ischemic period. *Neuroscience* **25**: 847-855.

Giovanni, A., Sonsalla, P,K., and Heikkila, R.E., 1994, Studies on species sensitivity to the dopaminergic neurotoxin 1-methyl-4-phenyl-1,2,3,6-tetrahydropyridine. Part 2. Central administration of 1-methyl-4-phenylpyridinium. *J. Pharmacol. Exp. Therap.* **270**: 1008-1014.

Giovanni, A., Liang, L.P., Hastings, T.G., and Zigmond, M.J., 1995, Estimating hydroxyl radical content in rat brain using systemic and intraventricular salicylate: impact of methamphetamine. *J. Neurochem.* **64**: 1819-1825.

Globus, M.Y.-T., Busto, R., Dietrich, W.D., Martinez, E., Valdés, I., and Ginsberg, M.D., 1988, Instraischemic extracellular release of dopamine and glutamate is associated with striatal vulnerability to ischemia. *Neurosci. Lett.* **91**: 36-40.

Globus, M.Y.-T., Busto, R., Dietrich, W.D., Martinez, E., Valdés, I., and Ginsberg, M.D., 1989, Direct evidence for acute and massive norepinephrine release in the hippocampus during transient ischemia. *J. Cereb. Blood Flow Metab.* **9**: 892-896.

Globus, M.Y.-T., Wester, P., Busto, R., and Dietrich, W.D., 1992, Ischemia-induced extracellular release of serotonin plays a role in CA1 neuronal death in rats. *Stroke* **23**: 1595-1601.

Good, P.F., Olanow, C.W., and Perl, D.P., 1992, Neuromelanin-containing neurons of the substantia nigra accumulate iron and aluminum in Parkinson's disease: a LAMMA study. *Brain Res.* **593**: 343-346.

Good, P.F., Hsu, A., Werner, P., Perl, D.P., and Olanow, C.W., 1998, Protein nitration in Parkinson's disease. *J. Neuropath. Exp. Neurol.* **57**: 338-342.

Greene, J.G., and Greenamyre, J.T., 1995, Exacerbation of NMDA, AMPA, and L-glutamate excitotoxicity by the succinate dehydrogenase inhibitor malonate. *J. Neurochem.* **64**: 2332-2338.

Greene, J.G., and Greenamyre, J.T., 1996, Bioenergetics and excitotoxicity: the weak excitotoxic hypothesis, in *Neurodegeneration and Neuroprotection in Parkinson's Disease* (C.W. Olanow, P. Jenner and M. Youdim, Eds.,), Academic Press, New York, pp. 125-142.

Gutteridge, J.M.C., 1996, Hydroxyl radical, iron, oxidative stress, and neurodegeneration. *Ann. N.Y. Acad. Sci.* **738**: 201-213.

Han, J., Cheng, F.-C., Yang, Z., and Dryhurst, G., 1999, Inhibitors of mitochondrial respiration, iron (II), and hydroxyl radical evoke release and extracellular hydrolysis of glutathione in rat striatum and substantia nigra: potential implications to Parkinson's disease. *J. Neurochem.* **73**: 1683-1695.

Hardy, J., Adolfsunn, R., Alafuzoff, L., Bucht, G., Marcusson, J., Nyberg, P., Perdahl, E., Wester, P., and Winblad, B., 1985, Transmitter defects in Alzheimer's disease. *Neurochem. Int.* **7**: 545-563.

Heikkila, R.E., Orlansky, H., and Cohen, G., 1975, Studies on the distinction between uptake inhibition and release of (^{3}H) dopamine in rat brain tissue slices. *Biochem. Pharmacol.* **24**: 847-852.

Hirabayashi, H., Takizawa, S., Fukuyama, N., Nakazawa, H., and Shinohara, Y., 1999, 7-Nitroindazole attenuates nitrotyrosine formation in the early phase of cerebral ischemia-reperfusion in mice. *Neurosci. Lett.* **268**, 111-113.

Hirata, H., Landenheim, B., Rothman, R.B., Epstein, C., and Cadet, J.L., 1995, Methamphetamine-induced serotonin neurotoxicity is mediated by superoxide radicals. *Brain Res.* **677**: 345-347.

Hochstetler, S.E., Puopolo, M., Gustincich, S., Raviola, E., and Wightman, R.M., 2000, Real-time amperometric measurements of zeptomole quantities of dopamine released from neurons. *Anal. Chem.* **72**: 489-496.

Hornykiewicz, O., and Kish, S.J., 1980, Biochemical pathophysiology of Parkinson's disease. *Adv. Neurol.* **45**: 19-34.

Hotchkiss, A., and Gibb, J.W., 1980, Long-term effects of multiple doses of methamphetamine on tryptophan hydroxylase and tyrosine hydroxylase in rat brain. *J. Pharmacol. Exp. Therap.* **214**, 257-263.

Huether, G., Zhou, D., and Rüther, E., 1997, Causes and consequences of the loss of serotonergic presynapses elicited by the consumption of 3,4-methylenedioxymethamphetamine (MDMA, "ecstasy") and its congeners. *J. Neural Transm.* **104**: 771-794.

Ikeda, M., Nakazawa, T., Abe, K., Takeru, K., and Yamatsu, K., 1989, Extracellular accumulation of glutamate in the hippocampus induced by ischemia is not calcium dependent-in vitro and in vivo evidence. *Neurosci. Lett.* **96**, 202-206.

Imman, S.Z., Crow, J.P., Newport, G.D., Islam, F., Slikker, W., and Ali, S.F., 1999, Methamphetamine generates peroxynitrite and produces dopaminergic neurotoxicity in mice: protective effects of peroxynitrite decomposition catalyst. *Brain Res.* **837**: 15-21.

Javitch, J.A., D'Amato, R.J., Strittmatter, S.M., and Snyder, S.H., 1985, Parkinsonism-inducing neurotoxin, *N*-methyl-4-phenyl-1,2,3,6-tetrahydropyridine: uptake of the metabolite *N*-methyl-4-phenylpyridine by dopamine neurons explains selective toxicity. *Proc. Nat. Acad. Sci. U.S.A.* **82**: 2173-2177.

Jiang, X.-R., Wrona, M.Z., and Dryhurst, G. 1999, Tryptamine-4,5-dione, a putative endotoxic metabolite of the superoxide-mediated oxidation of serotonin, is a mitochondrial toxin: possible implications in neurodegenerative brain disorders. *Chem. Res. Toxicol.* **12**: 429-436.

Johnson, M., Hanson, G.R., and Gibb, J.W., 1989, Effect of MK-801 on the decrease in tryptophan hydroxylase induced by methamphetamine and its methylenedioxy analog. *Eur. J. Pharmacol.* **165**, 315-318.

Kato, H., Araki, T., and Kogure, K., 1990, Role of excitotoxic mechanism in the development of neuronal damage following repeated brief cerebral ischemia in the gerbil: protective effects of MK-801 and pentobarbital. *Brain Res.* **516**: 175-179.

Keller, J.N., Kindy, M.S., Holtsberg, F.W., St. Clair, D.K., Yen, H.C., Germeyer, A., Steiner, S.M., Bruce-Keller, A.J., Hutchins, J.B., and Mattson, M.P., 1998, Mitochondrial manganese superoxide dismutase prevents neural apoptosis and reduces ischemic brain injury: suppression of peroxynitrite production, lipid peroxidation and mitochondrial dysfuction. *J. Neurosci.* **18**: 687-697.

Kerry, N., and Rice-Evans, C., 1999, Inhibition of peroxynitrite-mediated oxidation of dopamine by flavonoid and phenolic antioxidants and their structural relationships. *J. Neurochem.* **73**, 247-253.

Keyer, K., and Imlay, J.A., 1997, Inactivation of dehydratase [4Fe-4S] clusters and disruption of iron homeostasis upon cell exposure to peroxynitrite. *J. Biol. Chem.* **272**: 27652-27659.

Kita, T., Takahashi, M., Kubo, K., Wagner, G.C., and Nakashima, T., 1999, Hydroxyl radical formation following methamphetamine administration to rats. *Pharmacol. Toxicol.* **85**: 133-137.

Klivenyi, P, St. Clair, D., Wermer, M., Yen, H.C., Oberley, T., Yang, L., and Beal, M.F., 1998, Manganese superoxide dismutase overexpression attenuates MPTP toxicity. *Neurobiol. Dis.* 5: 253-258.

Kobayashi, T., Matsumme, H., Matuda, S., and Mizuno, Y., 1998, Association between the gene encoding the E2 subunit of the α-ketoglutarate dehydrogenase complex and Parkinson's disease. *Ann. Neurol.* 43, 120-123.

Kuhn, D.M., Aretha, C.W., Geddes, T.J., 1999, Peroxynitrite inactivation of tyrosine hydroxylase: mediation by sulfhydryl oxidation, not tyrosine nitration. *J. Neurosci.* 19: 10289-10294.

Lada, M.W., and Kennedy, R.T., 1997, In vivo monitoring of glutathione and cysteine in rat caudate nucleus using microdialysis on-line with capillary zone electrophoresis-laser induced fluorescence detection. *J. Neurosci. Meth.* 72: 153-159.

Lafon-Cazal, M., Pletri, S., Culcasi, M., and Bockaert, J., 1993, NMDA-dependent superoxide production and neurotoxicity. *Nature* 364: 535-537.

Lan, J., and Jiang, D.H., 1997, Desferrioxamine and vitamin E protect against iron and MPTP-induced neurodegeneration in mice. *J. Neural Transm.* 104: 469-481.

Lancelot, E., Callebert, J., Revaud, M.L., Boulu, R.G., and Plotkine, M., 1995, Detection of hydroxyl radicals in rat striatum during transient focal cerebral ischemia: possible implications in tissue damage. *Neurosci. Lett.* 197: 85-88.

LaVoie, M.J., and Hastings, T.G., 1999, Dopamine quinone formation and protein modification associated with the striatal neurotoxicity of methamphetamine: evidence against a role for extracellular dopamine. *J. Neurosci.* 19, 1484-1491.

Le Couteur, J.M., and McClean, A.J., 1998, The ageing liver: drug clearance and an oxygen diffusion barrier hypothesis. *Clin. Pharmacokinetics* 34: 359-373.

Lee, J.-M., Zipfel, G.J., and Choi, D.W., 1999, The changing landscape of ischaemic brain injury mechanisms. *Nature* 399: A7-A14.

Lew, R., Malberg, J.E., Ricaurte, G.A., and Seiden, L.S., 1998, Evidence for and mechanism of action of neurotoxicity of amphetamine related compounds, in *Highly Selective Neurotoxins: Basic and Clinical Applications* (R.M., Kostrzewa, Ed.), Humana Press, Totowa, N.J., pp. 235-268.

Li, H., and Dryhurst, G., 1997, Irreversible inhibition of mitochondrial complex I by 7-(2-aminoethyl)-3,4-dihydro-5-hydroxy-2H-1,4-benzothiazine-3-carboxylic acid (DHBT-1): a putative nigral endotoxin of relevance to Parkinson's disease. *J. Neurochem.* 69: 1530-1541.

Li, H., Shen, X.-M., and Dryhurst, G., 1998, Brain mitochondria catalyze the oxidation of 7-(2-aminoethyl)-3,4-dihydro-5-hydroxy-2H-1,4-benzothiazine-3-carboxylic acid to intermediates that irreversibly inhibit complex I and scavenge glutathione: potential relevance to the pathogenesis of Parkinson's disease. *J. Neurochem.* 71: 2049-2062.

Li, X., Wallin, C., Weber, S.G., and Sandberg, M., 1999, Net efflux of cysteine, glutathione and related metabolites form rat hippocampal slices during oxygen/glucose deprivation: dependence on γ-glutamyl transpeptidase. *Brain Res.* 815: 81-88.

Loschmann, P.A., Lange, K.W., Wachtel, H., and Turski, L., 1994, MPTP-induced degeneration: interference with glutamatergic toxicity. *J. Neural Transm.* 43: 133-143.

Madl, J.E., and Allen, D.L., 1995, Hyperthermia depletes adenosine triphosphate and decreases glutamate uptake in rat hippocampal slices. *Neuroscience* 69: 395-405.

Marek, G., Vosmer, G., and Seiden, L.S., 1990, Dopamine uptake inhibitors block long-term neurotoxic effects of methamphetamine upon dopaminergic neurons. *Brain Res.* 513: 274-279.

Martin, F.R., Sanchez-Ramos, J., and Rosenthal, M., 1991, Selective and non-selective effects of MPTP on oxygen consumption in rat striatal and hippocampal slices. *J. Neurochem.* 57, 1340-1346.

Matarredona, E.R., Santiago, M., Machado, A., and Cano, J., 1997, Lack of involvement of glutamate-induced excitotoxicity in MPP$^+$ toxicity in striatal dopamine terminals: possible involvement of ascorbate. *Br. J. Pharmacol.* 121: 1038-1044.

Matsubara, K., Idzu, T., Kobayashi, Y., Gonda, T., Okunishi, H., and Kimura, K., 1996, Differences in dopamine efflux induced by MPP$^+$ and β-carbolinium in the striatum of conscious rats. *Eur. J. Pharmacol.* 315: 145-151.

Matsuda, L.A., Schmidt, C.J., Gibb, J.W., and Hanson, G.R., 1987, Ascorbic acid-deficient condition alters central effects of methamphetamine. *Brain Res.* 400: 176-180.

Matthews, R.J., Beal, M.F., Fallon, J., Fedorchak, K., Huang, P.L., Fishman, M.C., and Hyman, B.T., 1997, MPP$^+$-induced substantia nigra degeneration is attenuated in nNOS knockout mice. *Neurobiol. Dis.* 4: 114-121.

McNaught, K.S., and Jenner, P., 1999, Altered glial function causes neuronal death and increased neuronal susceptibility to 1-methyl-4-phenylpyridinium- and 6-hydroxydopamine-induced toxicity in astrocytic/ventral mesencephalic co-cultures. *J. Neurochem.* 73: 2469-2476.

Meiergerd, S.M., Patterson, T.A., and Schenk, J.O., 1993, D2 receptors may modulate the function of the striatal transporter for dopamine: kinetic evidence from studies in vitro and in vivo. *J. Neurochem.* **61**: 764-767.

Merino, M., Vizuete, M.L., Cano, L., and Machado, A., 1999, The non-NMDA glutamate receptor antagonists 5-cyano-7-nitroquinoxaline-2,3-dione and 2,3-dihydroxy-6-nitrosulfamoyl-benzo(f)quinoxaline, but not NMDA antagonists, block the intrastriatal neurotoxic effect of MPP$^+$. *J. Neurochem.* **73**: 750-757.

Mithöfer, K., Sandy, M.S., Smith, M.T., and Di Monte, D., 1992, Mitochondrial poisons cause depletion of reduced glutathione in isolated hepatocytes. *Arch. Biochem. Biophys.* **295**: 132-136.

Mizuno, Y., Saitoh, T., and Sone, N., 1987, Inhibition of mitochondrial α-ketoglutarate dehydrogenase by 1-methyl-4-phenylpyridinium ion. *Biochem. Biophys. Res. Commun.* **143**: 971-976.

Mizuno, Y., Matuda, S., Yoshino, H., Mori, H., Hattori, N., and Ikebe, S.-J., 1994, An immunohistochemical study on α-ketoglutarate dehydrogenase complex in Parkinson's disease. *Ann. Neurol.* **35**: 204-210.

Moszczynska, A., Turenne, S., and Kish, S.J., 1998, Rat striatal levels of the antioxidant glutathione are decreased following binge administration of methamphetamine. *Neurosci. Lett.* **355**: 49-52.

Nash, J.F., and Yamamoto, B.K., 1992, Methamphetamine neurotoxicity and striatal glutamate release: comparison to 3,4-methylenedioxymethamphetamine. *Brain Res.* **581**: 237-243.

Novelli, A., Reilly, J.A., Lysko, P.G., and Henneberry, R.C., 1988, Glutamate becomes neurotoxic when intracellular energy levels are reduced. *Brain Res.* **451**: 205-212.

Nowak, L., Bregestovski, P., Ascher, P., Herbet, A., and Prochiantz, A., 1984, Magnesium gates glutamate-activated channels in mouse central neurons. *Nature* **307**: 462-465.

O'Dell, S.J., Weihmuller, F.B., and Marshall, J.F., 1991, Multiple methamphetamine injections induce marked increases in extracellular dopamine which correlate with subsequent neurotoxicity. *Brain Res.* **564**: 256-260.

Oishi, T., Hasegawa, E., and Murai, Y., 1993, Sulfhydryl drugs reduce neurotoxicity of 1-methyl-4-phenyl-1,2,3,6-tetrahydropyridine (MPTP) in the mouse. *J. Neural Transm. [P-D Dement. Sect.]*, **6**: 45-52.

Olney, J.W., Zorumski, C., Price, M.T., and Labruyere, J., 1990, L-Cysteine, a bicarbonate-sensitive excitotoxin. *Science* **248**: 596-599.

Olsen, R.J., and Justice, J.B., 1993, Quantitative microdialysis under transient conditions. *Anal. Chem.* **65**: 1017-1022.

Palmer, C., Roberts, R.L., and Bero, C., 1994, Deferoxamine posttreatment reduces brain injury in neonatal rats. *Stroke* **25**: 1039-1045.

Patt, A., Horesh, I.R., Berger, E.M., Harken, A.H., and Repine, J.E., 1990, Iron depletion or chelation reduces ischemia/reperfusion-induced edema in gerbil brains. *J. Pediatr. Surg.* **25**: 224-228.

Prince, J.A., Yassin, M.S., and Oreland, L., 1998, Normalization of cytochrome c oxidase activity in the rat brain by neuroleptics after chronic treatment with PCP or methamphetamine. *Neuropharmacology* **36**: 1665-1678.

Przedborski, S., Kostic, V., Jackson-Lewis, V., Naini, A.B., Simonettá, S., Fah, S., Carlson, E., Epstein, C.J., and Cadet, J.L., 1992, Transgenic mice with increased Cu/Zn superoxide dismutase activity are resistant to N-methyl-4-phenyl-1,2,3,6-tetrahydropyridine-induced neurotoxicity. *J. Neurosci.* **12**: 1658-1667.

Pu, C., Broening, H.W., and Vorhees, C.V., 1996, Effect of methamphetamine on glutamate-positive neurons in the adult and developing rat somatosensory cortex. *Synapse* **23**: 328-334.

Puka-Sundvall, M., Sandberg, M., and Hagberg, H., 1997, Brain injury after hypoxia-ischemia in newborn rats: relationship to extracellular levels of excitatory amino acids and cysteine. *Brain Res.* **750**: 325-328.

Pullan, L.M., Olney, J.W., Price, M.T., Compton, R.P., Hood, W.F., Michel, J., and Monahan, J.B., 1987, Excitatory amino acid receptor potency and subclass specificity of sulfur-containing amino acids. *J. Neurochem.* **49**: 1301-1307.

Pulsinelli, W.A., and Duffy, T.E., 1983, Regional energy balance in the rat brain after transient forebrain ischemia. *J. Neurochem.* **40**: 1500-1503.

Quijano, C., Alvarez, B., Gatti, R.M., Augusto, O., and Radi, R., 1997, Pathways of peroxynitrite oxidation of thiol groups. *Biochem. J.* **322 (Part 1)**: 167-173.

Raiteri, M., Cerrito, F., Cervoni, A.M., and Levi, G., 1979, Dopamine can be released by two mechanisms differentially affected by the dopamine transporter inhibitor nomifensine. *J. Pharmacol. Exp. Therap.* **208**: 195-202.

Ramsay, R.R., Dadgar, J., Trevor, A., and Singer, T.P., 1986, Energy-driven uptake of N-methyl-4-phenylpyridine by brain mitochondria mediates the neurotoxicity of MPTP. *Life Sci.* **39**: 581-588.

Ransom, B.R., Kunis, D.M., Irwin, I., and Langston, J.W., 1987, Astrocytes convert the parkinsonian inducing neurotoxin, MPTP, to its active metabolite, MPP$^+$. *Neurosci. Lett.* **75**: 323-328.

Raps, S.P., Lai, J.C.K., Hertz, L., and Cooper, A.J.L., 1989, Glutathione is present in high concentrations in cultured astrocytes but not cultured neurons. *Brain Res.* **493**: 398-401.

Reed, D.J., and Savage, M.K., 1995, Influence of metabolic inhibitors on mitochondrial permeability transition and glutathione status. *Biochim. Biophys. Acta* **1271**: 43-45.

Rehncrona, S., Folbergrová, J., Smith, D.S., and Siesjö, B., 1980, Pronounced incomplete cerebral ischemia and subsequent recirculation on cortical concentrations of oxidized and reduced glutathione in the rat. *J. Neurochem.* **34**: 477-486.

Revuelta, M., Romero-Ramos, M., Venero, J.L., Millan, F., Machado, A., and Cano, J., 1997, Less-induced MPP$^+$ neurotoxicity on striatal slices from guinea pigs fed with a vitamin C lacking diet. *Neuroscience* **77**: 167-174.

Ricaurte, G.A., Schuster, C.R., and Seiden, L.S., 1980, Long-term effects of repeated methamphetamine administration on dopaminergic and serotonergic neurons in the rat brain: a regional study. *Brain Res.* **193**: 153-160.

Ricaurte, G.A., Fuller, R.W., Perry, K.W., Seiden, L.S., and Schuster, C.R., 1983, Fluoxetine increases long-lasting neostriatal dopamine depletion after administration of d-methamphetamine and d-amphetamine. *Neuropharmacology* **22**: 1165-1169.

Rollema, H., Damsma, G., Horn, A.S., De Vries, J.B., and Westerink, B.H.C., 1986, Brain dialysis in conscious rats reveals an instantaneous massive release of striatal dopamine in response to MPP$^+$. *Eur. J. Pharmacol.* **126**: 345-346.

Rose, S., Hindmarsh, J.G., and Jenner, P., 1999, Neuronal nitric oxide synthase inhibition reduces MPP$^+$-evoked hydroxyl radical formation but not dopamine efflux in rat striatum. *J. Neural Transm.* **106**, 477-486.

Rowe, D.B., Le, W., Smith, R.G., and Appel, S.H., 1998, Antibodies from patients with Parkinson's disease react with protein modified by dopamine oxidation. *J. Neurosci. Res.* **53**: 551-558.

Royland, J.E., and Langston, J.W., 1998, MPTP: A dopaminergic neurotoxin, in *Highly Selective Neurotoxins: Basic and Clinical Applications* (R.M. Kostrzewa, Ed.), Humana Press, Totowa, N.J., pp. 141-194.

Sagara, J., Miura, K., and Bannai, S., 1993a, Cystine uptake and glutathione level in fetal brain cells in primary culture and suspension. *J. Neurochem.* **61**: 1667-1671.

Sagara, J., Miura, K., and Bannai, S., 1993b, Maintenance of neuronal glutathione by glial cells. *J. Neurochem.* **61**: 1672-1676.

Saggu, H., Cooksey, J., Dexter, D., Wells, F.R., Lees, A.J., Jenner, P., and Marsden, C.D., 1989, A selective increase in particulate superoxide dismutase activity in parkinsonian substantia nigra. *J. Neurochem.* **53**: 692-697.

Santiago, M., Matarredona, E.R., Granero, L., Cano, J., and Machado, A., 1997, Neuroprotective effects of the iron chelator desferrioxamine against MPP$^+$ toxicity on striatal dopaminergic terminals. *J. Neurochem.* **68**: 732-738.

Schapira, A.H.V., Mann, V.M., Cooper, J.N., Dexter, D., Daniel, S.E., Jenner, P., Clark, J.B., and Marsden, C.D., 1990, Anatomic and disease specificity of NADH-CoQ$_1$ reductase (complex I) deficiency in Parkinson's disease. *J. Neurochem.* **55**: 2142-2145.

Schulz, J.B., Matthews, R.T., and Beal, M.F., 1995, Role of nitric oxide in neurodegenerative diseases. *Curr. Opin. Neurol.* **8**: 480-486.

Schulz, J.B., Matthews, R.T., Klockgether, T., Dichgans, J., and Beal, M.F., 1997, The role of mitochondrial dysfunction and neuronal nitric oxide in animal models of neurodegenerative diseases. *Mol. Cell. Biochem.* **174**: 193-197.

Seelig, G.F., and Meister, A., 1985, Glutathione biosynthesis; γ-glutamylcysteine synthetase from rat kidney. *Meth. Enzymol.* **113**: 379-390.

Sensi, S.L., Yin, H.Z., Carriedo, S.G., Rao, S.S., and Weiss, J.H., 1999, Preferential Zn^{2+} influx through Ca^{2+} permeable AMPA/kainate channels triggers prolonged mitochondrial superoxide production. *Proc. Nat. Acad. Sci. U.S.A.*, **96**: 2414-2419.

Shen, X.-M., and Dryhurst, G., 1996a, Further insights into the influence of L-cysteine on the oxidation chemistry of dopamine: reaction pathways of potential relevance to Parkinson's disease. *Chem. Res. Toxicol.* **9**: 751-763.

Shen, X.-M., and Dryhurst, G., 1996b, Oxidation chemistry of (-)-norepinephrine in the presence of L-cysteine. *J. Med. Chem.* **39**, 2018-2029.

Shen, X.-M., Li, H., and Dryhurst, G., 2000, Oxidative metabolites of 5-*S*-cysteinyldopamine inhibit the α-ketoglutarate dehydrogenase complex: possible relevance to the pathogenesis of Parkinson's disease. *J. Neural Transm.* **00**: 00-00.

Sian, J.,Dexter, D., Lees, A., Daniel, S., Agid, Y., Javoy-Agid, F., Jenner, P., and Marsden, C.D., 1994a, Alterations in glutathione levels in Parkinson's disease and other neurodegenerative disorders affecting the basal ganglia. *Ann. Neurol.* **36**: 348-355.

Sian, J., Dexter, D.T., Lees, A.J., Daniel, S., Jenner, P., and Marsden, C.D., 1994b, Glutathione-related enzymes in the brain in Parkinson's disease. *Ann. Neurol.* **36**: 356-361.

Sims, N.R., 1991, Selective impairment of respiration in mitochondria isolated from brain subregions following transient forebrain ischemia in the rat. *J. Neurochem.* **56**: 1836-1844.

Sims, N.R., and Zaidan, E., 1995, Biochemical changes associated with selective neuronal death following short-term cerbral ischaemia. *Int. J. Biochem. Cell Biol.* **27**: 531-550.

Slivka, A., and Cohen, G., 1993, Brain ischemia markedly elevates levels of the neurotoxic amino acid, cysteine. *Brain Res.* **608**: 33-37.

Smith, T.S., and Bennett, J.P., 1997, Mitochondrial toxins in models of neurodegenerative disease. I. In vivo brain hydroxyl radical production during systemic MPTP treatment or following microdialysis infusion of methylpyridinium or azide ions. *Brain Res.* **765**: 183-188.

Smith, T.S., Swerdlow, R.H., Parker, W.D., and Bennett, J.P., 1994, Reduction of MPP$^+$-induced hydroxyl radical formation and nigrostriatal MPTP toxicity by inhibiting nitric oxide synthase. *Neuroreport* **5**: 2598-2600.

Sonsalla, P.K., Gibb, J.W., and Hanson, G.R., 1986, Roles of D$_1$ and D$_2$ dopamine receptor subtypes in mediating the methamphetamine-induced changes in monoamine systems. *J. Pharmacol. Exp. Therap.* **238**: 932-937.

Sonsalla, P.K., Nicklas, W.J., and Heikkila, R.E., 1989, Roles for excitatory amino acids in methamphetamine-induced nigrostriatal dopaminergic neurotoxicity. *Science* **243**: 398-400.

Sonsalla, P.K., Jochnowitz, N.D., Zeevalk, G.D., Oostveen, J.A., and Hall, E.D., 1996, Treatment of mice with methamphetamine produces cell loss in the substantia nigra. *Brain Res.* **738**, 172-175.

Spencer, J.P.E., Jenner, P., Daniel, S.E., Lees, A.J., Marsden, C.D., and Halliwell, B., 1998, Conjugates of catecholamines with cysteine and GSH in Parkinson's disease. Possible mechanisms of formation involving reactive oxygen species. *J. Neurochem.* **71**: 2112-2122.

Sriram, K., Pai, K.S., Boyd, M.R., and Ravindranath, V., 1997, Evidence for generation of oxidative stress in brain by MPTP: in vitro and in vivo studies in mice. *Brain Res.* **749**: 44-52.

Ste-Marie, L., Vachon, P., Vachon, L., Bemeur, C., Guertin, M.C., and Montgomery, J., 2000, Hydroxyl radical production in the cortex and striatum in a rat model of focal cerebral ischemia. *Can. J. Neurol.* **27**, 152-159.

Stephens, S.E., and Yamamoto, B.K., 1994, Methamphetamine-induced neurotoxicity: roles for glutamate and dopamine efflux. *Synapse* **17**: 203-209.

Steranka, L.R., and Rhina, A.W., 1987, Effect of cysteine on the persistent depletion of brain monoamines by amphetamine, *p*-chloroamphetamine and MPTP. *Eur. J. Pharmacol.* **133**: 191-197.

Stole, E., Smith, T.K., Manning, J.M., and Meister, A., 1994, Interaction of γ-glutamyl transpeptidase with acivicin. *J. Biol. Chem.* **269**: 21435-21439.

Stone, D.M., Johnson, M.. Hanson, G.R., and Gibb, J.W., 1999, Acute inactivation of tryptophan hydroxylase by amphetamine analogs involves oxidation of sulfhydryl sites. *Eur. J. Pharmacol.* **172**: 93-97.

Srivastava, R., Brouillet, E., Beal, M.F., Storey, E., and Hyman, B.T., 1993, Blockade of 1-methyl-4-phenylpyridinium (MPP+) nigral toxicity in the rat by prior decortication or MK-801 treatment: a stereological estimate of neuronal loss. *Neurobiol. Aging* **14**: 295-301.

Storey, E., Hyman, B.T., Jenkins, B., Brouillet, E., Miller, J.M., Rosen, B.R., and Beal, M.F., 1992, 1-Methyl-4-phenylpyridinium produces excitotoxic lesions in rat striatum as a result of impairment of oxidative metabolism. *J. Neurochem.* **58**: 1975-1978.

Swerdlow, R.H., Parks, J.H., Miller, S.W., Tuttle, J.B., Trimmer, P.A., Sheehan, J.P., Bennett, J.P., Davis, R.E., and Parker, W.D., 1996, Origin and functional consequences of the complex I defect in Parkinson's disease. *Ann. Neurol.* **40**: 663-671.

Tate, S.S., and Meister, A., 1985, γ-Glutamyl transpeptidase from kidney. *Meth. Enzymol.* **113**: 400-437.

Thomas, B., Muralikrishnan, D., and Mohanakumar, K.P., 2000, In vivo hydroxyl radical generation in the striatum following systemic administration of 1-methyl-4-phenyl-1,2,3,6-tetrahydropyridine in mice. *Brain Res.* **852**: 221-224.

Volicer, L., Langlais, P.J., Matson, W.R., Mark, K.A., and Gamache, P.H., 1985, Serotoninergic system in dementia of the Alzheimer type: abnormal forms of 5-hydroxytryptophan and serotonin in cerebrospinal fluid. *Arch. Neurol.* **42**: 1158-1161.

Wagner, K.R., Kleinholz, M., and Myers, R.E., 1990, Delayed onset of neurologic deterioration following anoxia/ischemia coincides with appearance of impaired brain mitochondrial respiration and decreased cytochrome oxidase activity. *J. Cereb. Blood Flow Metab.* **10**: 417-423.

Wang, X.F., and Cynader, M.X., 2000, Astrocytes provide cysteine to neurons by releasing glutathione. *J. Neurochem.* **74**: 1434-1442.

Wefers, H., and Sies, H., 1983, Oxidation of glutathione by the superoxide radical to the disulfide and sulfonate yielding singlet oxygen. *Eur. J. Biochem.* **137**: 29-36.

Weinberger, J., Cohen, G., and Nieves-Rosa, J., 1983, Nerve terminal damage in cerebral ischemia: greater susceptibility of catecholamine nerve terminals relative to serotonergic nerve terminals. *Stroke* **14**: 986-989.

Weiner, H.L., Hashim, A., Lajtha, A., and Sershen, H., 1988, (-)-2-Oxo-4-thiazolidine carboxylic acid attenuates 1-methyl-4-phenyl-1,2,3,6-tetrahydropyridine induced neurotoxicity. *Res. Commun. Subst. Abuse* **9**: 53-68.

Wong, K.-S., Goyal, R.N., Wrona, M.Z., Blank, C.L., and Dryhurst, G., 1993, 7-*S*-Glutathionyl-tryptamine-4,5-dione: a possible aberrant metabolite of serotonin. *Biochem. Pharmacol.* **46**: 1637-1652.

Wood, N., 1997, Genes and parkinsonism. *J. Neurol. Neurosurg. Psychiatry* **62**: 305-309.

Wrona, M.Z., and Dryhurst, G., 1998, Oxidation of serotonin by superoxide radical: implications to neurodegenerative brain disorders. *Chem. Res. Toxicol.* **11**: 639-650.

Wu, E.Y., Smith, M.T., Bellomo, G., and Di Monte, D., 1990, Relationship between mitochondrial transmembrane potential, ATP concentration, and cytotoxicity in isolated rat hepatocytes. *Arch. Biochem. Biophys.* **282**, 358-362.

Wullner, U., Loschmann, P.A., Schulz, J.B., Schmid, A., Dringen, R., Eblen, F., Turski, L., and Klockgether, T., 1996, Glutathione depletion potentiates MPTP and MPP$^+$ toxicity in nigral dopaminergic neurones. *Neuroreport* **7**: 921-923.

Xie, C.X., St. Pyrek, J., Porter, W.H., and Yokel, R.A., 1995, Hydroxyl radical generation in rat brain is initiated by iron not aluminum, as determined by microdialysis with salicylate trapping and GS-MS analysis. *Neurotoxicology* **16**: 489-496.

Xin, W., Shen, X.-M., Li, H., and Dryhurst, G., 2000, Oxidative metabolites of 5-*S*-cysteinylnorepinephrine are irreversible inhibitors of mitochondrial complex I and the α-ketoglutarate dehydrogenase and pyruvate dehydrogenase complexes: possible implications for neurodegenerative brain disorders. *Chem. Res. Toxicol.* **00**, 00-00.

Xu, Y.M., Stokes, A.H., Roskoski, R., and Vrana, K.E., 1998, Dopamine, in the presence of tyrosinase, covalently modifies and inactivates tyrosine hydroxylase. *J. Neurosci.* **54**: 691-697.

Yamamoto, M., Sakamoto, N., Iwai, A., Yatsugi, S., Hidaka, K., Noguchi, K., and Yuasa, T., 1993, Protective actions of YM737, a new glutathione analog, against cerebral ischemia in rats. *Res. Commun. Chem. Pathol. Pharmacol.* **81**: 221-232.

Yamamoto, B., and Zhu, W., 1998, The effects of methamphetamine on the production of free radicals and oxidative stress. *J. Pharmacol. Exp. Therap.* **287**: 107-114.

Yang, G., Chan, P.H., Chen, J., Carlson, E., Chen, S.F., Weinstein, P., Epstein, C.J., and Kamii, H., 1994, Human copper-zinc superoxide dismutase transgenic mice are highly resistant to reperfusion injury after focal cerebral ischemia. *Stroke* **25**: 165-170.

Yang, C.S., Lin, N.N., Liu, L., Tsai, P.J., and Kuo, J.S., 1995, Lowered brain glutathione by diethylmaleate decreased the glutamate release by cerebral ischemia in the anesthetized rat. *Brain Res.* **698**, 237-240.

Yang, C.S., Lin, N.N., Tsai, P.J., Liu, L., and Kuo, J.S., 1996, In vivo evidence of hydroxyl radical formation induced by elevation of extracellular glutamate after cerebral ischemia in the cortex of anesthetized rats. *Free Rad. Biol. Med.* **20**, 245-250.

Yang, H., Peters, J.L., Allen, C., Chern, S.-S., Coalson, R.D., and Michael, A.C., 2000, A theoretical description of microdialysis with mass transport coupled to chemical events. *Anal. Chem.* **72**: 2042-2049.

Yong, V.W., Perry, T.L., and Krisman, A.A., 1986, Depletion of glutathione in brainstem of mice caused by *N*-methyl-4-phenyl-1,2,3,6-tetrahydropyridine is prevented by antioxidant pretreatment. *Neurosci. Lett.* **63**: 56-60.

Yoshida, T., Tanaka, M., Somomatsu, A., and Hirai, S., 1995, Activated microglia cause superoxide-mediated release of iron from ferritin. *Neurosci. Lett.* **190**: 21-24.

Zaidan, E., Sheu, K.F., and Sims, N.R., 1998, The pyruvate dehydrogenase complex is partially inactivated during early recirculation following short-term forebrain ischemia in rats. *J. Neurochem.* **70**: 233-241.

Zängerle, L., Cuenod, M., Winterhalter, K.H., and Do, K.Q., 1992, Screening of thiol compounds: depolarization-induced release of glutathione and cysteine from rat brain slices. *J. Neurochem.* **59**: 181-189.

Zeevalk, G.D., and Nicklas, W.J., 1991, Mechanisms underlying initiation of excitotoxicity associated with metabolic inhibition. *J. Pharmacol. Exp. Therap.* **257**: 870-878.

Zeevalk, G.D., Davis, N., Hyndman, A.G., and Nicklase, W.J., 1998, Origins of the extracellular glutamate released during total metabolic blockade of the immature retina. *J. Neurochem.* **71**: 2373-2381.

Zeevalk, G.D., Manzino, L., and Sonsalla, P.K., 2000, NMDA receptors modulate dopamine loss due to energy impairment in the substantia nigra but not striatum. *Exp. Neurol.* **161**: 638-646.

SEROTONERGIC NEUROTOXICITY OF METHYLENEDIOXYAMPHETAMINE AND METHYLENEDIOXYMETAMPHETAMINE

Terrence J. Monks, Fengju Bai, R. Timothy Miller, and Serrine S. Lau

Center for Molecular and Cellular Toxicology
Division of Pharmacology and Toxicology
College of Pharmacy
University of Texas at Austin
Austin, Texas 78712-1074

INTRODUCTION

3,4-($\pm$)-Methylenedioxyamphetamine (MDA) and 3,4-($\pm$)-methylenedioxymeth-amphetamine (MDMA, "Ecstasy") are ring-substituted amphetamine derivatives that have stimulant and hallucinogenic properties (1, 2). MDA and MDMA are popular recreational drugs and their abuse is increasing in both the United States (3) and Europe (4). In recent years their clandestine manufacture and appearance on the street have made them popular drugs of abuse (5,6) for their ability to induce "a state of sensory amplification and enhancement without appreciable sympathomimetic stimulation" (7) and have been reported as useful adjuncts to psychotherapy (8). After misuse, chronic paranoid psychosis has been reported, which is persistent and resistant to treatment with haloperidol (9). In experimental animals, including primates, toxicity is also manifest as a selective serotonergic neurotoxicity. The actions of MDA and MDMA are biphasic, initially causing an acute release of 5-hydroxytryptamine (5-HT) (10) followed by prolonged depletion of 5-HT and 5-hydroxyindoleacetic acid (5-HIAA), inhibition of tryptophan hydroxylase (TPH) (11,12), and structural damage to 5-HT terminal and preterminal axons in various regions of the central nervous system (11,13). The immediate 5-HT release caused by these compounds can be blocked *in vitro* by 5-HT uptake inhibitors (14). The long term neurotoxicity can also be blocked *in vivo* by 5-HT uptake inhibitors (15) and by 5-HT receptor antagonists, but is potentiated by L-dopa (16). The predominant adverse consequences of MDMA and MDA abuse in humans include convulsions, hyperthermia, rhabomyolysis, and acute liver and renal failure (17).

The neurotoxic component(s) of MDA and MDMA is(are) not known, but roles for endogenous dopamine (18) and for the 5-HT$_2$ receptor (19,20) have been proposed. The neurotoxic effects of MDA and MDMA are dependent on the route and frequency of drug administration (21). Direct injection of MDA or MDMA into the brain does not reproduce the acute or long-term effects observed after peripheral administration, suggesting an important role for systemic metabolism in the development of toxicity (22-24). In support of this view, pretreatment of rats with SKF-525A, an inhibitor of cytochrome P450 (P450), attenuates MDMA-mediated depletions in 5-HT, whereas pretreatment with phenobarbital enhances 5-HT depletion (25). Moreover, the inability of MDMA to inhibit TPH activity *in vitro*, supports a requirement for metabolic activation (23). However several major metabolites of MDA and MDMA either fail to reproduce the serotonergic neurotoxicity or fail to exhibit specificity for the serotonergic system (26,27). Thus, administration of α-

Biological Reactive Intermediates VI, Edited by Dansette *et al.*
Kluwer Academic / Plenum Publishers, 2001

methyldopamine (α-MeDA) or 3-O-methyl-α-MeDA, major metabolites of MDA and MDMA, into brain also fails to produce long-term serotonergic neurotoxicity (26). Although direct central injection of 2,4,5-trihydroxyamphetamine or 2,4,5-trihydroxy-methamphetamine, putative *in vivo* metabolites of MDA and MDMA, are toxic to the serotonergic neurotransmitter system, they also target the dopaminergic system (27-29) and thus do not exhibit the selectivity of the parent amphetamines. In addition, mechanisms by which these metabolites gain access to the brain have not been determined.

MDA is metabolized to α-MeDA (30,31) and MDMA to N-methyl-α-MeDA and α-MeDA (32,33) reactions catalyzed by CYP2B, CYP2D, and CYP3A (34). Both of these catechols can undergo oxidation to the corresponding *ortho*-quinones, followed by the reductive addition of glutathione (GSH) to form GSH conjugates (35,36) (Figure 1). Conjugation of polyphenols with GSH frequently results in either preservation of, or enhancement of biological (re)activity (37). For example, quinone-thioethers retain the ability to redox cycle and produce reactive oxygen species (38) and to arylate tissue macromolecules (39). In addition, quinone-thioethers have been shown to be substrates for, or inhibitors of enzymes which utilize GSH and/or quinones as substrates (35). Therefore, because (i) neither MDA or MDMA produce neurotoxicity when injected directly into brain, (ii) icv administration of some major metabolites of MDA fail to reproduce the neurotoxicity, (iii) α-MeDA is a metabolite of both MDA and MDMA, (iv) α-MeDA is readily oxidized to the corresponding quinone which can undergo conjugation with GSH and, (v) quinone-thioethers exhibit a variety of toxicological activities, we investigated the potential role of thioether metabolites of α-MeDA in the neurotoxicity of MDA and MDMA.

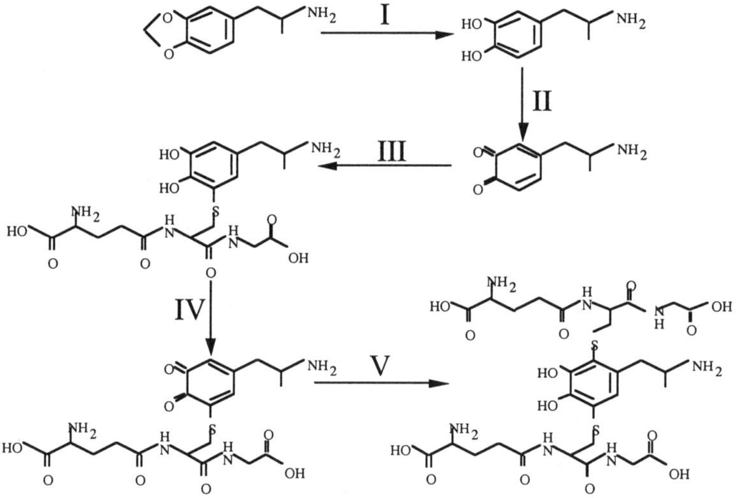

Figure 1.
Metabolism of MDA to potentially neurotoxic thioether metabolites of α-MeDA. MDA is (**I**) demethylenated by CYP to the catechol, α-methyldopamine, which (**II**) is readily oxidized to the corresponding *ortho*-quinone. The reductive addition of GSH to *ortho*-quinone (**III**) can occur either spontaneously, or in a reaction catalyzed by GST. A second oxidation/GSH conjugation cycle (**IV** and **V**) gives rise to 2,5-*bis*-(glutathion-S-yl)-α-MeDA. A similar series of reactions is envisioned for MDMA.

RESULTS AND DISCUSSION

Behavioral Effects of α-MeDA And Its Thioether Metabolites

We initially examined the overt behavioral changes induced in rats following direct intracerebroventricular (icv) administration of α-MeDA (2.4 and 3.0 μmol), 5-(glutathion-S-yl)-α-MeDA (720 nmol), 5-(N-acetylcystein-S-yl)-α-MeDA (7 and 100 nmol), and 2,5-

bis-(glutathion-S-yl)-α-MeDA (475 nmol). Differing behavioral effects were observed between treatment groups. Behaviors appeared rapidly (1-2 min) following a single icv administration of all compounds, and peaked in intensity during the first 30 min. α-MeDA-treated animals became docile. All of the animals treated with α-MeDA-thioether conjugates became hyperactive, aggressive, displayed forepaw treading and Straub tails (40), behaviors consistent with those seen after administration of established 5-HT-releasers such as MDA and MDMA. Following multiple i.c.v. administration of 5-(glutathion-S-yl)-α-MeDA and 5-(N-acetylcystein-S-yl)-α-MeDA most animals also displayed splayed hind limbs (41). Some animals began circling away from the side of the icv injection of 5-(N-acetylcystein-S-yl)-α-MeDA. With each successive dose, behaviors became less apparent. Following multiple icv administrations of 2,5-*bis*-(glutathion-S-yl)-α-MeDA, animals became hyperactive and displayed Straub tails (beginning 1-2 minutes post-dose). Five to ten minutes post-dose the animals exhibited behaviors similar to "wet dog shakes" and approximately 30 minutes post-dose the animals were salivating profusely (41).

The finding that α-MeDA does not produce the same behavioral profile as it's thioether conjugates (40) suggests that the behavioral changes induced by the conjugates are not simply a consequence of the catecholamine function. Although the cytoprotective effects of GSH are well established, GSH and related enzymes participate in the protection of neurons from a variety of stresses. Additional roles for GSH in brain function are consistent with associations between changes in GSH metabolism and the development of certain neurodegenerative processes of the brain, including ischemia and Parkinson's disease (42). Thus, GSH (and GSSG) appear to play important functional roles in the CNS. The unique structure of polyphenolic-GSH conjugates such as 5-(glutathion-*S*-yl)-α-MeDA permits toxicological and pharmacological activity as a consequence of either the polyphenolic (catecholamine) or GSH moeity. Toxicological sequelae may result from the electrophilic and redox properties of the quinone, whereas neuropharmacological changes may result from the either the catecholamine or GSH moiety.

Effects of 5-(Glutathion-*S*-yl)-α-MeDA, 5-(N-Acetylcystein-S-yl)-α-MeDA, and 2,5-*bis*-(Glutathion-S-yl)-α-MeDA on Rat Brain Neurotransmitter Levels Following Icv Injection

In addition to the behavioral changes described above, a single icv injection of 5-(glutathion-*S*-yl)-α-MeDA also caused short-term alterations in the dopaminergic, serotonergic and noradrenergic systems (40). An increase in dopamine synthesis has been implicated as a prerequisite for the long-term depletion of brain 5-HT following MDMA administration. In support of this view, pharmacological interventions which either inhibit dopamine synthesis or the reuptake of synaptic dopamine, block the long-term depletion of brain 5-HT by MDMA (18). The tyrosine hydroxylase inhibitor, α-methyl-*p*-tyrosine also significantly attenuated MDMA mediated striatal dopamine release, and blocked the long-term depletion of striatal 5-HT, but inhibition of tryptophan hydroxylase with *p*-chlorophenylalanine failed to protect against the long-term neurotoxic effects of MDMA (50). Thus, acute depletion of dopamine, but not of 5-HT, is protective against MDMA neurotoxicity, and supports the hypothesis that dopamine plays a major role in the serotonergic toxicity of MDMA. However, although 5-(glutathion-*S*-yl)-α-MeDA reproduced some of the effects of MDA on the dopaminergic system, and was capable of causing acute increases in 5-HT turnover, icv administrartion did not cause long-term serotonergic toxicity.

5-(Glutathion-S-yl)-α-MeDA crosses the blood brain barrier (40) and is rapidly metabolized within all regions of the CNS to 5-(cystein-S-yl)-α-MeDA and 5-(N-acetyl-L-cystein-S-yl)-α-MeDA (Figure 2, [51]), the latter being the final redox active metabolite with an apparent ability to persist in brain tissue. Regional differences in the distribution of γ-glutamyl transpeptidase (γ-GT) correlated with the formation of 5-(cystein-*S*-yl)-α-MeDA (51). Remarkably, 5-(N-acetyl-L-cystein-S-yl)-α-MeDA was able to reproduce the overt behaviors of MDA at a dose of only 7 nmol (0.03% dose of MDA) (41). Based on these observations, and the fact that a single icv administration of 5-(glutathion-*S*-yl)-α-MeDA did not produce neurotoxicity, we examined a multiple dose regimen with the α-

MeDA-thioether conjugates, with the expectation that such a protocol might result in the accumulation of metabolites to toxic concentrations. However, no long-term serotonergic

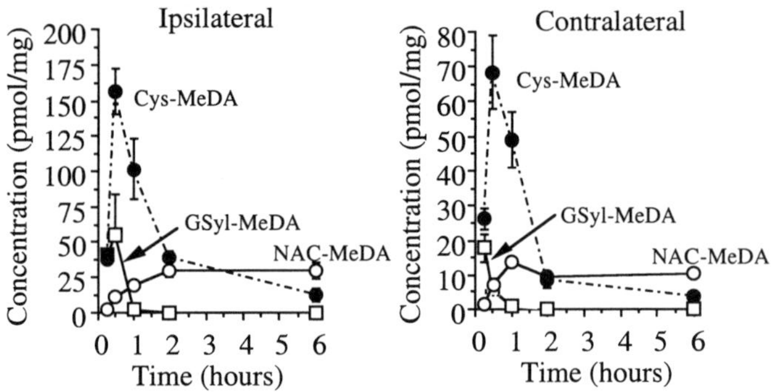

Figure 2: Metabolism of 5-(glutathion-*S*-yl)-α-MeDA (720 nmol) in rat striatum ipsilateral and contralateral to the site of icv injection (From Miller *et al.*, 1995).

neurotoxicity was observed with either 5-(glutathion-*S*-yl)-α-MeDA or 5-(N-acetylcystein-*S*-yl)-α-MeDA (41) implying that the acute behavioral changes can be dissociated from the long-term neurotoxicity. Since 5-(glutathion-S-yl)-α-MeDA is able to reoxidize and form 2,5-*bis*-(glutathion-S-yl)-α-MeDA, and since the potency of polyphenolic-GSH conjugates increases with the degree of GSH substitution (37) we also examined the effects of 2,5-*bis*-(glutathion-S-yl)-α-MeDA on brain 5-HT concentrations. In contrast to 5-(glutathion-*S*-yl)-α-MeDA and 5-(N-acetylcystein-*S*-yl)-α-MeDA, 2,5-*bis*-(glutathion-S-yl)-α-MeDA (4 X 475 nmol) produced serotonergic neurotoxicity (41). The relative sensitivity of the striatum, hippocampus and cortex to 2,5-*bis*-(glutathion-S-yl)-α-MeDA was the same as that observed for MDA, although the absolute effects were greater with MDA (41). The greater sensitivity of the hippocampal serotonergic system to 2,5-*bis*-(glutathion-S-yl)-α-MeDA compared to the cortex may be a consequence of the proximity of the lateral ventricles to the hippocampus, thus permitting greater access to 2,5-*bis*-(glutathion-S-yl)-α-MeDA. Consistent with this view, concentrations of 5-(glutathion-S-yl)-α-MeDA, at the earliest time-point (15 min) measured following icv injection, were highest in the hippocampus > striatum > cortex (44).

5-(Glutathion-S-yl)-α-MeDA is rapidly cleared from the brain (44) and it is therefore likely that the neurotoxicity of 2,5-*bis*-(glutathion-S-yl)-α-MeDA is mediated by downstream metabolites such as 2,5-*bis*-(cystein-S-yl)-α-MeDA and 2,5-*bis*-(N-acetylcystein-S-yl)-α-MeDA. The toxicity of these metabolites can be regulated by intramolecular cyclization reactions that occur subsequent to oxidation (45,46) therefore it may be important that there are regional differences in the distribution of cysteine conjugate *N*-acetyl transferase and *N*-acetylcysteine conjugate deacetylase in the brain (44). Cyclization of 2,5-*bis*-(cystein-S-yl)-α-MeDA may occur in one of two ways. Following oxidation, the side chain (alanine-derived) amino group can cyclize to give the 5,6-dihydroxyindole or the cysteinyl amino group can condense with the quinone carbonyl to give a benzothiazolyl-like compound. Only the latter reaction removes the reactive quinone function, since the dihydroxyindole can undergo further oxidation. Because the cysteinyl amino groups are blocked in 2,5-*bis*-(N-acetylcystein-S-yl)-α-MeDA it can no longer undergo cyclization following oxidation, and this metabolite will maintain redox activity. The ratio of *N*-acetylation to *N*-deacetylation in the hippocampus is more than double that in the striatum and 5-(N-acetylcystein-S-yl)-α-MeDA appears to persist in the brain after icv administration of 5-(glutathion-S-yl)-α-MeDA (51). Experiments on the metabolism and distribution of 2,5-*bis*-(glutathion-S-yl)-α-MeDA following icv administration are required to address these questions.

The inability of 5-(glutathion-*S*-yl)-α-MeDA and 5-(N-acetylcystein-*S*-yl)-α-MeDA to reproduce MDA-mediated serotonergic neurotoxicity following icv administration may be a consequence of the complex pharmacokinetics following icv administration, and the inherent reactivity of their metabolites which may limit their ability to reach 5-HT nerve terminal sites. However, both of 5-(glutathion-S-yl)-α-MeDA and 5-(N-acetylcystein-S-yl)-α-MeDA produce a similar overt behavioral response to that seen following peripheral administration of MDA (40,41) indicating that these metabolites share some properties in common with the parent amphetamine. In addition, preliminary experiments indicate that icv administration of either 5-(glutathion-S-yl)-α-MeDA or 5-(N-acetylcystein-S-yl)-α-MeDA induces the activation of microglia in striatum, cortex and hippocampus[2], evidence that these metabolites do cause neuronal damage at these 5-HT nerve terminal enriched sites.

Effects of 5-(Glutathion-S-yl)-α-MeDA, and 5-(N-acetylcystein-S-yl)-α-MeDA, and 2,5-*bis*-(glutathion-S-yl)-α-MeDA on Rat Brain Neurotransmitter Levels Following Direct Intrastriatal or Intracortical Administration

Although 5-(glutathion-S-yl)-α-MeDA and 5-(N-acetylcystein-S-yl)-α-MeDA produce neurobehavioral changes similar to those seen with MDA and MDMA, and acute changes in brain 5-HT and dopamine concentrations, as noted above, neither conjugate causes long-term decreases in 5-HT concentrations (40,41). Because of the inherent reactivity of polyphenolic-thioethers, icv injection may result in concentrations in target areas insufficient to produce toxicity. We therefore determined the effects of 2,5-*bis*-(glutathion-S-yl)-_-MeDA (4 X 150 nmol; 4 X 300 nmol), 5-(glutathion-S-yl)-_-MeDA (4 X 200 nmol; 4 X 400 nmol), and 5-(N-acetylcystein-S-yl)-_-MeDA (4 X 7 nmol; 4 X 20 nmol), on monoaminergic neurotransmitter concentrations following their direct injection into the striatum, cortex, and hippocampus (47). Direct intrastriatal or intracortical administration of the _-MeDA thioether conjugates caused significant decreases in striatal and cortical 5-HT concentrations (7 days following the last injection). Interestingly, intrastriatal injection of 5-(glutathion-S-yl)-_-MeDA or 2,5-bis-(glutathion-S-yl)-_-MeDA, but not 5-(N-acetyl-cystein-S-yl)-_-methyldopamine, also caused decreases in 5-HT concentrations in the ipsilateral cortex. The same pattern of changes was seen when the conjugates were injected into the cortex (47). The effects of the thioether conjugates of α-MeDA were confined to 5-HT nerve terminal fields, since no significant changes in monoamine neurotransmitter levels were detected in brain regions enriched in 5-HT cell bodies (midbrain/diencephalon/ telencephalon and pons/medulla). In addition, the effects of the conjugates were selective to the serotonergic system, as no significant changes were seen in dopamine or norepinephrine concentrations (47). The results indicate that thioether conjugates of α-MeDA are selective serotonergic neurotoxicants. Nonetheless, a role for these conjugates in the toxicity observed following systemic administration of MDA and MDMA remains to be demonstrated (see below).

The relative neurotoxic potency of the thioether metabolites of α-MeDA is 5-(N-acetylcystein-S-yl)-α-MeDA >> 2,5-*bis*-(glutathion-S-yl)-α-MeDA > 5-(glutathion-S-yl)-α-MeDA, which is in accordance with their ability to produce an acute "serotonin behavioral syndrome" (40,41). The greater potency of the mercapturic acid may be due to its relative persistence in brain (44) and its ability to maintain redox activity by limiting intramolecular cyclization (45). A study using CuZn-superoxide dismutase (CuZn-SOD) transgenic mice showed that homozygous SOD-transgenic mice that carry two copies of the human CuZn-SOD gene are resistant to the depletion of dopamine and DOPAC following MDMA administration (48), indicating an important role for ROS in the biochemical effects of MDMA. The ability of α-MeDA thioethers to redox-cycle and generate ROS provides a basis for their biological reactivity, and their relative potency is likely determined by their ability to generate ROS and their persistence in the tissue.

Brain Uptake of 5-(Glutathion-S-yl)-α-MeDA

The catechol-GSH metabolites of MDA and MDMA must be capable of crossing the BBB and the BCSFB in order to produce toxicity. Brain microvascular endothelial cells maintain tight junctions (see above) and possess a high density of mitochondria, which supply the high energy requirements for the transport of water-soluble substances through the endothelial barrier *via* specific transporters. The directionality of ion transport across the BBB is achieved by a polarized distribution of ion-channels on the endothelial cell surfaces. Therefore, endothelial cells are polarized in a manner similar to other transport interfaces, such as renal epithelia, with the preferential localization of specific transport systems and receptors either on the luminal or the antiluminal side of vessel walls (49). For amino acids, at least three different carrier systems have been identified within the BBB (50) and a variety of neurotoxicants are transported into brain across the BBB *via* these amino acid carriers. The cysteine conjugate of dichloroacetylene, a potent nephrotoxicant (51) and neurotoxicant (52) is also transported across the BBB by the Na^+-independent system L-transporter for neutral amino acids, while uptake of the corresponding GSH conjugate is mediated by an as yet unknown carrier system (53). The saturable, carrier-mediated transport of GSH across the blood-brain barrier has also been reported (54). Thus, transport systems that can facilitate the uptake of GSH and cysteine conjugates of α-MeDA into the brain have been described. Because γ-GT is present in high concentrations in the brain, particularly on endothelial cells that form the blood-brain barrier (55) and because there appears to be a transporter capable of transferring GSH conjugates from the circulation into the brain (54), systemic formation of 5-(glutathion-*S*-yl)-α-MeDA followed by uptake into, and metabolism by the brain may provide a mechanism to explain the role of metabolism in MDA- and MDMA-mediated neurotoxicity. The BUI for 5-(glutathion-*S*-yl)-[³H]-α-MeDA was 7.35 ± 0.50% (40), which compares with the value of 8.3 ± 3.2% (56) and 10.9 ± 0.2% to 12.2 ± 1.6% for GSH (54). GSH (1mM) decreased the BUI for 5-(glutathion-*S*-yl)-[³H]-α-MeDA by 45% (40), suggesting that GSH and 5-(glutathion-*S*-yl)-[³H]-α-MeDA may share the same carrier. Pretreatment of animals with acivicin (AT-125, 18 mg/kg) 20 min prior to injection of 5-(glutathion-*S*-yl)-[³H]-α-MeDA caused a substantial (~6-fold) increase in the BUI. Inhibition of γ-GT at the blood brain barrier may decrease the metabolism of 5-(glutathion-*S*-yl)-[³H]-α-MeDA and increase the pool available for transport by the putative GSH carrier. This is an important finding because we have previously reported that structurally related polyphenolic-GSH conjugates decrease the activity of γ-GT (37). Thus, prolonged exposure of the blood brain-barrier to 5-(glutathion-*S*-yl)-α-MeDA and other thioether metabolites, may result in decreased γ-GT activity with a subsequent increase the uptake of the conjugates into brain (Figure 3).

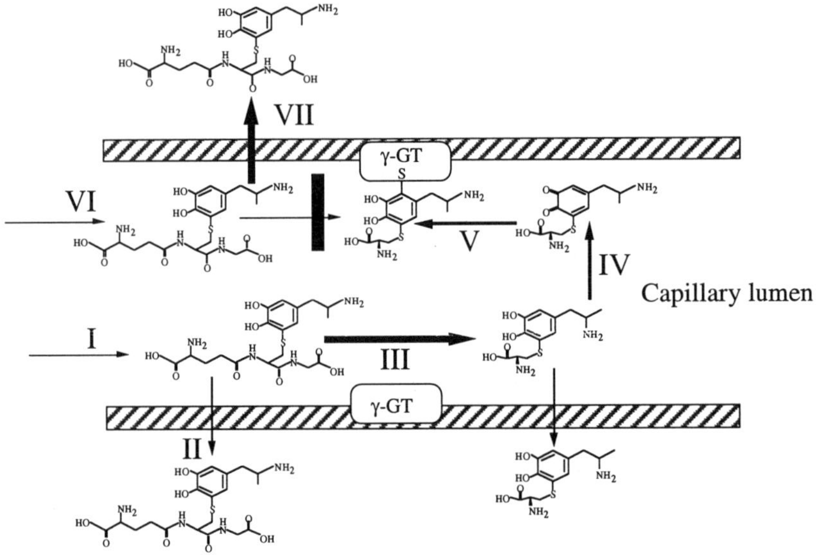

Figure 3

We therefore reasoned that inhibition of γ-GT at the blood-brain barrier should potentiate MDA-mediated neurotoxicity. Consistent with this hypothesis, pretreatment of animals with acivicin, which caused an ~60% decrease in brain capillary endothelial cell γ-GT activity, potentiated MDA-mediated decreases in brain 5-HT and 5-HIAA concentrations (Bai *et al.*, unpublished data). These data implicate a role for metabolites that are substrates for γ-GT in the neurotoxicity that occurs subsequent to the systemic administration of MDA. The mechanism(s) by which the thioether metabolites of α-MeDA produce selective serotonergic neurotoxicity are unclear, but are the subject of ongoing studies.

Doses of the α-MeDA thioethers required to produce serotonergic neurotoxicity are much lower (0.03 - 1.72%) than doses of MDA producing a similar degree of toxicity, even accounting for differences in the route of administration. Moreover the doses used in this study likely fall within the range of α-MeDA thioethers present in the brain following MDA (93 μmol/kg, sc) administration. Thus, about 1.6% of a dose of MDA (23 μmol; sc) was excreted in bile as 5-(glutathion-S-yl)-α-MeDA, within 5 h[4]. This translates ~50% of the dose that caused both neurobehavioral and neurochemical changes (40). Because of the reactivity of polyphenolic-GSH conjugates, quantitation of their biliary and urinary excretion represents a minimum estimate of *in vivo* formation. Assuming a minimum 1.6% conversion of MDA 5-(glutathion-S-yl)-α-MeDA[4] and a minimum brain uptake index of ~7.5% (40) a minimum of 28 nmol of 5-(glutathion-S-yl)-α-MeDA may gain access to the brain. Because only 7 nmol of 5-(N-acetyl-L-cystein-S-yl)-α-MeDA (icv) produces behaviors identical to those produced by MDA[4], the metabolism data fall within the range necessary to support a potential role for these metabolites in MDA-mediated neurotoxicity. If one also considers (i) the multiple dosing regimens typical in animal models of MDA and MDMA neurotoxicity (ii) metabolism of MDA to polyphenolic-GSH conjugates > 1.6% (iii) and the potential inhibition of γ-GT by 5-(glutathion-S-yl)-α-MeDA (and its metabolites) will increase the BUI for 5-(glutathion-S-yl)-α-MeDA, then some combination of these factors will increase the likelihood for these metabolites to cause adverse effects in the brain following peripheral MDA administration.

Potential Factors Contributing to Inter-Individual Susceptibility to MDA and MDMA - Mediated Neurotoxicity

Potential differences in the activity of the enzymes that catalyze the demethylenation of MDA and MDMA are likely to play an important role in predisposing individuals to the adverse effects of these drugs. Although neither the mechanism nor the enzymes involved in the oxidation of the resulting catechols, or their thioether conjugates, are known, variability in this reaction will also contribute to individual susceptibility to neurotoxicity. Although conjugation of the corresponding *ortho*-quinone to GSH occurs non-enzymatically, it may also be catalyzed by GSH S-transferase. While it is not known which isoforms are involved in catalyzing this reaction, perhaps the μ class isoforms may be particularly important, since GSH S-transferase M2-2 participates in the very specific conjugation of the dopamine metabolite aminochrome to GSH (57,58). Variations in the transport of the conjugates into the brain will also be an important determinant of susceptibility, and we know very little about human variability in γ-GT at the blood-brain barrier, and virtually nothing about the activity of the intact GSH transporter at the blood-brain barrier. The ratio of N-acetylation and N-deacetylation of the thioether metabolites is also very important, because maintaining the conjugate in the N-acetylated form (i.e. as the mercapturic acid) may result in persistence of the metabolite in brain, and retention of the redox activity of the conjugate. Variability in DA and 5HT$_2$ receptor number and activity may be an important determinant of the response to MDA and MDMA, because DA receptors (18) and 5HT$_2$ receptors (19,20) may play important roles in the development of the serotonergic neurotoxicity. Finally, variability in antioxidant defenses may also predispose to the neurotoxic response. A variety of different factors may therefore predispose certain individuals within the population to the potential adverse effects of these amphetamine analogues.

Acknowledgements

The authors acknowledge the support of NIH grant DA 10832. RTM was supported by an award from the NIEHS (T32 ES 07247).

REFERENCES

1.   Thiessen, P. N. and Cook, D.A. (1973) The properties of 3,4-methylene-dioxyamphetamine (MDA). I. A review of the literature. *Clin. Toxicol.* **6**, 45-52.

2.   Kovar, K.A. (1998) Chemistry and pharmacology of hallucinogens, entactogens and stimulants. *Pharmacopsychiatry* **31(Suppl 2)**, 69-72.

3.   Cuomo, M., Dyment, P. and Gammino, V. (1994) Increasing use of ecstasy (MDMA) and other hallucinogens on a college campus. *J. Am. Coll. Health* **42**, 271.

4.   Johnson, L.D., O'Malley, P.M., and Bachman, J.G. (1997) National survey results on drug use from the Monitoring the Future study, 1975-1995. Vol. 2. College students and young adults. Rockville, MD. National Institutes on Drug Abuse.

5.   Baum, R. M. (1985) New variety of street drugs poses growing problem. *Chem. Eng. News* **63**, 7-16.

6.   Peroutka, S..J. (1987) Incidence of recreational use of 3,4-methylenedioxymethamphetamine (MDMA, "ECSTACY") on an undergraduate campus. *New Eng. J. Med.* **317**, 1542-1543.

7.   Shulgin, A. T. (1981) Hallucinogens. In: *Burger's Medicinal Chemistry*. (Wolff, M. E. Ed.) 4th Ed., part III, John Wiley and Sons, New York.

8.   Greer, G. and Strassman, R. J. (1985) Information on "ecstacy". *Am. J. Psychiat.* **142**, 1391.

9.   Winstock, A.R. (1991) Chronic paranoid psychosis after misuse of MDMA. *British Med. J.* **302**, 1150-1151.

10.   Fuller, R. W. (1976) Pharmacology of p-chloroamphetamine and analogs. *Psychopharmacol. Bull.* **12**, 55-57.

11.   Ricaurte, G. A., Bryan, G., Strauss, L., Seiden, L., and Schuster, C. (1985) Hallucino-genic amphetamine selectively destroys brain serotonin nerve terminals. *Science.***229**, 986-988.

12.   Ricaurte, G. A., Martello, A. L., Katz, J. L. and Martello, M. B. (1992) Lasting effects of 3,4-methylenedioxymethamphetamine (MDMA) on central serotonergic neurons in nonhuman primates: Neurochemical observations. *J. Pharmacol. Exp. Ther.* **261**, 616-622.

13.   Axt, K.J., Mullen, C.A. and Molliver, M.E. (1992) Cytopathologic features indicitive of 5-hydroxytryptamine axon degeneration are observed in rat brain after administration of D- and L-methylenedioxyamphetamine. In: *Neurotoxins and Neurodegenerative Disease, Ann. NY Acad. Sci.*, **648**, 245-247.

14.   Berger, U.V., Gu, X.F. and Azmitia, E.C. (1992) The substituted amphetamines 3,4-methylenedioxymethamphetamine, methamphetamine, p-chloroamphetamine and fenfluramine induce 5-hydroxytryptamine release *via* a common mechanism blocked by fluoxetine and cocaine. *Eur. J. Pharmacol.* **215,** 153-160.

15.   Hashimoto, K., Maeda, H. and Goromaru, T. (1992) Effects of benzylpiperazine derivatives on the neurotoxicity of 3,4-methylenedioxymethamphetamine in rat brain. *Brain Research* **590**, 341-344.

16.   Schmidt, C.J., Black, C.K. and Taylor, V.L. (1991) L-Dopa potentiation of the serotonergic deficits due to a single administration of 3,4-methylenedioxymethamphetamine, *p*-chloroamphetamine or methamphetamine to rats. *Eur. J. Pharmacol.* **203**, 41-49.

17.   Henry, J.A., Jeffreys K.J., and Dawling, S. (1992) Toxicity and deaths from 3,4-methylenedioxymethamphetamine. *Lancet* **340**: 384-387.

18.   Stone, D. M., Johnson, M., Hanson, G. R., and Gibb, J. W. Role of endogenous dopamine in the central serotonergic deficits induced by 3,4-methylenedioxymethamphetamine. *J. Pharmacol. Exp. Ther.* 247 (1988) 79-87.

19.   Schmidt, C. J., Black, C. K., Abbate, G. M., and Taylor, V. L. Methylenedioxymethamphetamine-induced hyperthermia and neurotoxicity are independently mediated by 5-HT$_2$ receptors. *Brain Res.* 529 (1990) 85-90.

20.   Schmidt, C. J., Taylor, V. L., Abbate, G. M., and Nieduzak, T. R. 5-HT$_2$ antagonists stereoselectively prevent the neurotoxicity of 3,4-methylenedioxymethamphetamine by blocking the acute stimulation of dopamine synthesis: Reversal by L-DOPA. *J. Pharmacol. Exp. Ther.* 256 (1991) 230-235.

21.   Ricuarte, G.A., Delanney, L.E., Irwin, I. and Langston, J. W. (1988) Toxic effects of MDMA on central serotonergic neurons in primate: importance of route and frequency of drug administration. *Brain Research* **446**, 165-168.

22. Molliver, M.E., O'Hearn, E., Battaglia, G., and DeSouza, E.B. (1986) Direct intracerebral administration of MDA and MDMA does not produce serotonin neurotoxicity. *Soc. Neurosci. Abstr.* **12**, 1234.

23. Schmidt, C.J. and Taylor, V.L. (1988) Direct central effects of acute methylenedioxymethamphetamine on serotonergic neurons. *Eur. J. Pharmacol.* **156**, 121-131.

24. Paris, J.M., Cunningham, K.A. (1992) Lack of serotonin neurotoxicity after intraraphe microinjection of (+)-3,4-methylenedioxymethamphetamine (MDMA). *Brain. Res. Bull.* **28**, 115-119.

25. Gollamudi, R., Ali, S.F., Lipe, G., Newport, G., Webb, P., Lopez, M., Leakey, J.E., Kolta, M., and Slikker, W. Jr. (1989) Influence of inducers and inhibitors on the metabolism *in vitro* and neurochemical effects *in vivo* of MDMA. *Neurotoxicology* **10**, 455-466.

26. McCann, U.A., and Ricaurte, G.A. (1991) Major metabolites of 3,4-methylene-dioxyamphetamine (MDA) do not mediate its toxic effects on brain serotonin neurons. *Brain Research* **545**, 279-282.

27. Zhao, Z., Castagnoli, N. Jr., Ricaurte, G.A., Steele, T., and Martello, M. (1992) Synthesis and neurotoxicological evaluation of putative metabolites of theserotonergic neurotoxin 2-(methylamine)-1-[3,4-methylenedioxyl)]propane [(methylenedioxy)methamphetamine]. *Chem. Res. Toxicol.* **5**, 89-94.

28. Johnson, M., Elayan, I., Hanson, G. R., Foltz, R. L., Gibb, J. W., and Lim, H. K. Effects of 3,4-dihydroxymethamphetamine and 2,4,5-trihydroxymethamphetamine, two metabolites of 3,4-methylenedioxymethamphetamine, on central serotonergic and dopaminergic systems. *J. Pharmacol. Exp. Ther.* 261 (1992) 447-453.

29. Elayan, I., Gibb, J. W., Hanson, G. R., Foltz, R. L., Lim, H. K., and Johnson, M. Long-term alteration in the central in the central monoaminergic systems of the rat by 2,4,5-trihydroxyamphetamine, but not by 2-hydroxy-4,5-methylenedioxymeth-amphetamine or 2-hydroxy-4,5-methylenedioxyamphetamine. *Eur. J. Pharmacol.* 221 (1992) 281-288.

30. Marquardt, G. M., DiStefano, V., and Ling, L. L. Metabolism of β-3,4-methylene-dioxyamphetamine in the rat. *Biochem. Pharmacol.* 27 (1978) 1503-1505.

31. Midha, K. K., Hubbard, J. W., Bailey, K., and Cooper, J. K. α-Methyldopamine, a key intermediate in the metabolic disposition of 3,4-methylenedioxyamphetamine *in vivo* in dog and monkey. *Drug Metab. Dispos.* 6 (1978) 623-630.

32. Lim, H. K. and Foltz, R. L. *In vivo* and *in vitro* metabolism of 3,4-(methylene-dioxy)methamphetamine in the rat: Identification of metabolites using an ion trap detector. *Chem. Res. Toxicol.* 1 (1988) 370-378.

33. Kumagai, Y., Wickham, K.A., Schmitz, D.A., and Cho, A.K. Metabolism of methylenedioxyphenyl compounds by rabbit liver preparations. Participation of different cytochrome P450 isozymes in the demethyleneation reaction. *Biochem. Pharmacol.* 42 (1991) 1061-1067.

34. Kumagai, Y., Schmitz, D.A. and Cho, A.K. (1992) Cytochrome P450 isozymes responsible for the metabolic activation of methylenedioxymethamphetamine (MDMA) in rat. *FASEB J.* **6**, A2567.

35. Hiramatsu, M., Kumagai, Y., Unger, S. E., and Cho, A. K. Metabolism of methylenedioxymethamphetamine: Formation of dihydroxymethamphetamine and a quinone identified as its glutathione adduct. *J. Pharmacol. Exp. Ther.* 254 (1990) 521-527.

36. Patel, N., Kumagai, Y., Unger, S. E., Fukuto, J. M., and Cho, A. K. Transformation of dopamine and α-methyldopamine by NG108-15 cells: Formation of thiol adducts. *Chem. Res. Toxicol.* 4 (1991) 421-426.

37. Monks, T.J., and Lau, S.S. (1997) Biological reactivity of polyphenolic-glutathione conjugates. *Chem. Res. Toxicol.* **10**, 1296-1313.

38. Wefers, H. and Sies, H. (1983) Hepatic low-level chemiluminesence during redox cycling of menadione and the menadione-glutathione conjugate: Relation to glutathione and NAD(P)H: quinone reductase (DT diaphorase) activity. *Arch. Biochem. Biophys.* **224**, 568-578.

39. Monks, T. J., Highet, R. J. and Lau, S. S. (1988) 2-Bromo-(diglutathion-S-yl)hydro-quinone nephrotoxicity: Physiological, biochemical and electrochemical determinants. *Molec. Pharmacol.* **34**, 492-500.

40. Miller, R.T., Lau, S.S.and Monks, T.J. (1996) Effect of 5-(Glutathion-S-yl)-α-methyldopamine on dopamine, serotonin and norepinepherine concentrations following intracerebroventricular administration to male Sprague-Dawley rats. *Chem.Res.Toxicol.* **9**, 457-465

41. Miller, R.T., Lau, S.S. and Monks, T.J. (1997) 2,5-*bis*-(Glutathion-S-yl)-α-methyl-dopamine, a putative metabolite of ($\pm$)-3,4-methylenedioxyamphetamine, decreases brain serotonin concentrations. *Eur. J. Pharmacol.* **323**, 173-180.

42. Monks, T.J., Ghersi-Egea, J-F., Philbert, M.A., Cooper, A.J.L., and Lock, E.A The role of glutathione in neuroprotection and neurotoxicity. *Toxicol. Sci.* **51**, 161-177,1999.

43. Brodkin, J., Malyala, A. and Nash, J. F. (1993) Effect of acute monoamine depletion on 3,4-methylenedioxymethamphetamine-induced neurotoxicity. *Pharmacol. Biochem. Behav.* **45**(3), 647-653.

44. Miller, R. T., Lau, S. S., and Monks, T. J. Metabolism of 5-(glutathion-S-yl)-α-methyldopamine following intracerebroventricular administration to male Sprague-Dawley rats. *Chem. Res. Toxicol.* 8 (1995) 634-641.

45. Monks, T.J., Highet, R.J. and Lau, S.S. Oxidative cyclization, 1,4-benzothiazine formation and dimerization of 2-bromo-3-(glutathion-*S*-yl)hydroquinone. *Molec. Pharmacol.*, **38**: 121-127, 1990.

46. Monks, T. J., Lo, H. H. and Lau, S. S. (1994) Oxidation and acetylation as determinants of 2-bromocystein-*S*-ylhydroquinone-mediated nephrotoxicity. *Chem. Res. Toxicol.* **7**, 495-502.

47. Bai, F., Lau, S.S., and Monks, T.J. (1999) Glutathione and N-acetylcysteine conjugates of α-methyldopamine produce serotonergic neurotoxicity. Possible role in methylenedioxyamphetamine-mediated neurotoxicity. *Chem. Res. Toxicol.*, **12**, 1150-1157.

48. Cadet, J.L., Ladenheim, B., Hirata, H., Rothman, R.B., Ali, S., Carlson, E., Epstein, C., and Moran, T.H. (1995) Superoxide radicals mediate the biochemical effects of methylenedioxymethamphetamine (MDMA): evidence from using CuZn-superoxide dismutase transgenic mice. *Synapse* **2**, 169-176.

49. Schlosshauer, B. (1993). The blood-brain barrier: morphology, molecules, and neurothelin. *Bioessays*, **15**, 341-346.

50. Oldendorf, W.H. and Szabo, J. (1976) Amino acid assignment to one of three blood-brain barrier amino acid carriers. *Am. J. Physiol.* **230**, 94-98.

51. Jackson, M. A., Lyon, J. P. and Siegel, J. (1971) Morphological changes in kidneys of rats exposed to dichloroacetylene-ether. *Toxicol. Appl. Pharmacol.* **18**, 175-184.

52. Reichert, D., Liebaldt, G., and Henschler, D. (1976) Neurotoxic effects of dichloroacetylene. *Arch. Toxicol.* **37**, 23-38.

53. Patel, N. J., Fullone, J., and Anders, M. W. (1993) Brain uptake of *S*-(1,2-dichlorovinyl)-glutathione and *S*-(1,2-dichlorovinyl)-L-cysteine. *Mol. Brain Res.* **17**, 53-58.

54. Kannan, R., Kuhlenkamp, J.F., Jeandidier, E., Trinh, H., Ookhtens, M., and Kaplowitz, N. (1990) Evidence for carrier-mediated transport of glutathione across the blood-brain barrier in the rat. *J. Clin. Invest.*, **85**, 2009-2013.

55. Wolff, J. E. A., Belloni-Olivi, L., Bressler, J. P. and Goldstein, G. W. (1992) γ-Glutamyl transpeptidase activity in brain microvessels exhibits regional heterogeneity. *J. Neurochem.* **58**, 909-915.

56. Cornford E.M., Braun, L.D., Crane, P.D., and Oldendorf, W.H. (1978) Blood-brain barier restriction of of peptides and the low uptake of enkephalins. *Endocrinology*, **103**, 1297-1303.

57. Baez, S., Segura-Aguilar, J., Widersten, M., Johansson A.S., and Mannervik, B. (1997) Glutathione transferases catalyse the detoxication of oxidized metabolites (o-quinones) of catecholamines and may serve as an antioxidant system preventing degenerative cellular processes. *Biochem. J.* **324**, 25-28.

58. Segura-Aguilar, J., Baez, S., Widersten, M., Welch, C.J., and Mannervik, B. (1997) Human class Mu glutathione transferases, in particular isoenzyme M2-2 catalyze detoxication of the dopamine metabolite aminochrome. *J. Biol. Chem.* **272**, 5727-5731.

CASPASE CASCADES IN CHEMICALLY-INDUCED APOPTOSIS

Shawn B. Bratton and Gerald M. Cohen

MRC Toxicology Unit, Hodgkin Building, University of Leicester
P O Box 138, Lancaster Road, Leicester, LE1 9HN UK

INTRODUCTION

Caspases play a central role in the execution phase of apoptosis and are responsible for many of the morphological features normally associated with this form of cell death. Toxicants appear to activate caspases primarily through perturbation of mitochondria and the subsequent formation of an Apaf-1/caspase-9 apoptosome complex. In this model, release of cytochrome c (cyt. c) from mitochondria is required to initiate assembly of the apoptosome. Release of cyt. c is promoted by pro-apoptotic Bcl-2 family members, such as Bax, Bak and Bid, and inhibited by anti-apoptotic Bcl-2 proteins, including Bcl-2 and Bcl-x_L. Toxicants may also, in some cases, upregulate death receptors and their cognate ligands leading to autocrine/paracrine-related apoptosis through assembly of Death Inducing Signaling Complexes (DISCs) and activation of caspase-8. Each of these basic caspase cascades will be discussed in detail.

Biological Reactive Intermediates VI, Edited by Dansette *et al.*
Kluwer Academic / Plenum Publishers, 2001

APOPTOSIS: PHYSIOLOGICAL AND TOXICOLOGICAL SIGNIFICANCE

Apoptosis is an evolutionarily conserved and energy-dependent form of cell death controlled by a very specific set of genes[1]. Unlike necrosis, which is generally considered "accidental", apoptosis is a "programmed" form of cell death with defined morphological and biochemical characteristics. Apoptosis acts in concert with mitosis to regulate cell number[2,3]. Apoptosis is important in the maintenance of the immune system and has been implicated in many different diseases including cancer, autoimmune diseases, AIDS, certain neurodegenerative disorders and ischemic injury[4].

Toxic chemicals also induce apoptosis in a variety tissues[5]. In fact, the majority of cancer chemotherapeutic agents (including antimetabolites, nitrogen mustards, vinca alkaloids and topoisomerase inhibitors) administered to humans for the treatment of disease induce apoptosis[6]. In addition, a number of industrial chemicals and environmental contaminants (including benzene and its relevant metabolites, various quinones, halogenated alkenes and heavy metals) also appear to induce apoptosis[7,8,9,10]. Importantly, toxicants often target particular subcellular organelles, including the nucleus, mitochondria and endoplasmic reticulum (ER), and yet each induces apoptosis in a morphologically indistinguishable manner. This suggest that diverse "death signals" must converge at some point upon one or more common pathways for the actual execution of apoptosis.

BIOCHEMICAL AND MORPHOLOGICAL CHANGES
ASSOCIATED WITH APOPTOSIS

A cell undergoing apoptosis exhibits characteristic morphological and biochemical changes that are distinct from necrosis[11,12]. Many of these changes result from the activation of caspases (see below) and their proteolytic cleavage of cellular proteins. Prominent nuclear changes include chromatin condensation and internucleosomal DNA cleavage. The latter appears to arise from caspase-mediated cleavage of a nuclease inhibitor, ICAD (inhibitor of caspase-activated DNase)/DFF45 so releasing an endonuclease, CAD (caspase-activated DNase)/DFF40[13,14]. Chromatin condensation may be due to the action of one or more factors depending on the cell type and the apoptotic stimulus and include apoptosis inducing factor (AIF), CAD and acinus[15,16]. Changes at the membrane surface, such as externalization of phosphatidylserine from the inner leaflet of the plasma membrane, result in recognition and phagocytosis of the apoptotic cell, so ensuring that an inflammatory response does not occur[17].

CASPASES: EXECUTIONERS OF APOPTOSIS

The execution phase of apoptosis generally involves the activation of caspases, a class of cysteine proteases which were first implicated in cell death following the discovery that the *Caenorhabditis elegans* death gene, *ced-3*, was related to the mammalian enzyme, interleukin-1β-converting enzyme (ICE/caspase-1)[18]. Eleven human and three murine caspases have now been identified. Each recognizes a tetrapeptide motif within a substrate and is characterized by almost absolute specificity for aspartic acid in the P_1 position. Caspases are synthesized as zymogens or proenzymes (~30-50 kDa) containing an N-terminal prodomain together with one large (~20 kDa) and one small (~10 kDa) subunit; however, the crystal structures of caspases-1 and -3 suggest the active enzymes are heterotetramers, composed of two small and two large subunits. Importantly, specific aspartic acid cleavage sites exist between the prodomain of a caspase and each of its subunits, allowing for the activation of one caspase by another[19,20]. Caspases can be subdivided based on a number of criteria including phylogenetic analysis, substrate specificity and the length of their prodomains. "Initiator" caspases, including caspases-8 and -9, contain long prodomains, which facilitate their interaction with specific adapter proteins. Such interactions bring initiator caspases in close proximity to one another and promote the activation of one zymogen by another[21,22]. Initiator caspases are responsible for either directly or indirectly activating various "effector caspases", including caspases-3, -6, and -7, which contain short prodomains. Effector caspases cleave a number of structural and regulatory proteins (e.g. DFF45/ICAD, PARP, lamins, fodrin, gelsolin and PKCδ) and are directly responsible for dismantling the cell during apoptosis and thus, many of the morphological features described above[20].

THE APOPTOSOME: STRESS-INDUCED ACTIVATION OF CASPASES

In *Caenorhabditis elegans*, at least three specific genes (*ced-3*, *ced-4* and *ced-9*) regulate developmental cell death[23]. CED-4 is an adaptor protein which undergoes oligomerization and subsequently recruits and facilitates activation of the death protease, CED-3[24]. In contrast, CED-9 is an anti-apoptotic protein that inhibits cell death by binding to CED-4 and inhibiting activation of CED-3. This complex of proteins has been coined the "apoptosome". In mammalian cells, a number of cellular signals induce perturbations in the mitochondria which results in release of pro-apoptotic molecules, including AIF and more importantly cyt. *c*, from the intermembrane space into the cytoplasm[16,25]. Cyt. *c* interacts with apoptotic protease activating factor (Apaf-1), dATP or ATP, and procaspase-9 to form a similar "apoptosome" complex[26]. Apaf-1 was the first identified mammalian homologue of CED-4. It is an ~130 kDa protein with an N-terminal CARD (caspase recruitment domain), followed by a region homologous to CED-4 and a C-terminal domain containing multiple WD-40 repeats (WDR)[27]. In a mechanism not clearly understood, cyt. *c* and dATP/ATP act as cofactors and stimulate Apaf-1 self-oligomerization. Procaspase-9 binds in a 1:1 ratio to the N-terminal CARD of Apaf-1, which initiates clustering of the zymogens and trans-catalytic activation. Once activated, caspase-9 can activate effector caspases-3 and -7[22,26,27]. The importance of the Apaf-1→caspase-9→caspase-3 pathway has been verified in studies using Apaf-1[-/-], caspase-9[-/-] and caspase-3[-/-] mice. In each case, cells derived from these animals are resistant to chemically-induced apoptosis[28].

The apoptosome has been reconstituted using purified recombinant proteins [29,30]. In these studies, Apaf-1 oligomerized into an ~1.4 MDa complex in the presence of dATP and cyt. *c* and subsequently recruited and processed caspase-9. Caspase-9 was then released from the complex where it apparently activated caspase-3. In contrast, studies in our laboratory indicate that in native lysates activated with dATP, Apaf-1 oligomerizes into both ~1.4 MDa and ~700 kDa complexes which recruit and process caspase-9. However, only the latter complex activates caspases-3 and -7, and initial activation of these effector caspases takes place within the apoptosome complex[31,32]. Crystallographic studies indicate that Apaf-1 strongly binds caspase-9 through CARD-CARD interactions[33], and indeed, caspase-9 must remain bound to Apaf-1, even after its processing, in order to maintain catalytic activity[34]. Thus, Apaf-1 likely induces a conformational change in caspase-9 that is necessary for its catalytic activity. The multiple WDRs in Apaf-1 are thought to form β-propeller-like structures, which mediate protein-protein interactions similar to G-proteins. Therefore, the WDRs may provide a "docking" region for effector caspases. Indeed, caspases-3 and -7 associate with the apoptosome, and truncated Apaf-1 proteins that lack WDRs process caspase-9 normally, but not caspase-3[35]. Therefore, we currently envision the apoptosome as a multi-caspase activating complex. Apaf-1 probably requires both dATP/ATP and cyt. *c* in order to assume a conformation that promotes appropriate self-oligomerization. Once the core complex is formed, the CARD domains of Apaf-1 recruit and bind very tightly to caspase-9, the catalytic subunit in the enzyme complex, and the WDR domains of Apaf-1 recruit caspases-3 and -7. Caspase-9 then processes these effector caspases, initiating their release from the complex (Figure 1).

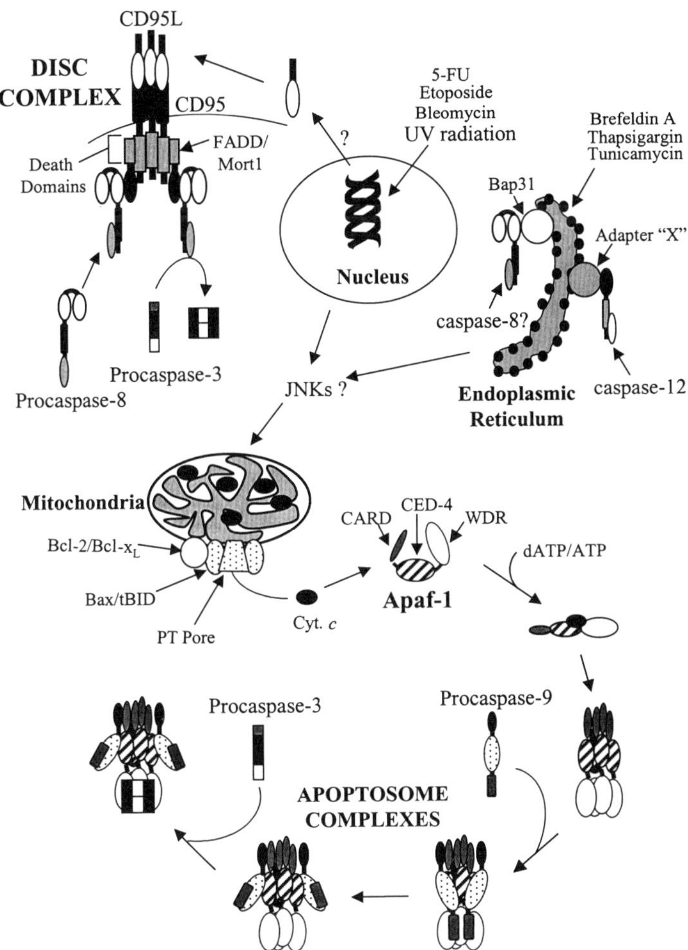

Figure 1. Stress-induced caspase cascades. Mitochondrial stress induces formation of Apaf-1 apoptosome complexes with caspase-9 as the apical caspase. Certain chemicals may upregulate death receptors and/or ligands and initiate receptor-mediated caspase cascades. Trimerization of CD95 is induced by CD95L resulting in formation of a DISC and activation of caspases-8 and -3. Additional caspase cascades may also be initiated on the outer membrane of the ER following ER stress.

MITOCHONDRIA: CONVERGENCE POINT FOR TOXICANT-INDUCED APOPTOSIS?

Because the release of cyt. *c* is such a critical step in the formation of apoptosome complexes, the cell must tightly regulate its release from mitochondria. Toxicologists have recognized for some time that disruption of normal mitochondrial function (*e.g.* inhibition of the mitochondrial respiratory chain) can lead to a substantial increase in intracellular reactive oxygen species (ROS), as well as a loss in ATP, and that oxidative stress and energy depletion often precede necrosis. However, as already noted, release of cyt. *c* from the intermembrane space during cellular stress can lead to apoptosis through the activation of caspases. Unfortunately, the mechanism(s) that mediate cyt. *c* release from mitochondria are contentious and not well understood. In many apoptotic models, a loss in the mitochondrial inner transmembrane potential ($\Delta\psi$m) occurs when the mitochondrial permeability transition (PT) pore is activated[36]. This pore is formed from a complex of adenine nucleotide translocator (ANT), voltage-dependent anion channel (VDAC; *also* termed porin) and cyclophilin-D, at contact sites between the inner and outer mitochondrial membranes[37].

ANT and VDAC act in concert to form the PT pore, which is relatively nonselective and allows for the passage of small molecules or proteins (<1.5 kDa). When the pore opens, there is a loss in the normal proton gradient (hypopolarization) across the inner membrane, and the respiratory chain is uncoupled. During this process an influx of fluid into the matrix also occurs, which results in swelling and rupturing of the outer membrane, because the surface area of the outer membrane is less than that of the convoluted, cristae-containing inner membrane. Disruption of the outer membrane then results in the release of proteins from the intermembrane space into the cytosol[37]. Indeed, inhibitors of the PT pore, including cyclosporin A and bongkrekic acid, which bind and inhibit cyclophilin D and ANT respectively, inhibit cell death in some models of apoptosis[38]. Nevertheless, a number of reports indicate that cyt. *c* release occurs before a loss in the $\Delta\psi$m, and that in fact, caspases induce opening of the PT pore, suggesting that a change in the $\Delta\psi$m is a relatively late apoptotic event[39,40]. However, a rapid opening and closing of the PT pore at its low conductance state could maintain the $\Delta\psi$m, while still allowing for a loss in outer membrane integrity and release of cyt. *c*[41]. This is an attractive explanation, since ATP production would also be maintained, and apoptosis is an energy-dependent process. Alternatively, it is still possible that an as yet undiscovered specific cyt. *c* transporter could exist in the outer mitochondrial membrane.

Bcl-2 family proteins can either inhibit (Bcl-2, Bcl-x$_L$) or induce (Bax, Bak) apoptosis *via* their effects on mitochondria[25,42]. All these proteins can dimerize to form ion selective channels when added to synthetic membranes (particularly at low pH) and share pore-forming domains that are similar in structure to those observed in diphtheria toxin and colicins[43,44,45]. Several of the Bcl-2 family proteins are anchored in the outer membranes of mitochondria by their hydrophobic C-termini and are oriented towards the cytosol[46]. Bcl-2 and Bcl-x$_L$ inhibit activation of the PT pore, apparently through their ability to inhibit opening of VDAC, whereas Bax and Bak promote opening of this channel[47,48]. Interestingly, a recent report suggests that BH3-only pro-apoptotic Bcl-2 family members, such as Bid and Bik, which do not bind to VDAC, induce cyt. *c* release in a calcium-independent, cyclosporin-insensitive, and respiration-independent manner[49]. Thus, the loss in mitochondrial membrane potential may depend upon the specific involvement of particular pro-apoptotic Bcl-2 family members. Surprisingly, despite its location in the outer mitochondrial membrane, Bax has also been reported to directly interact with ANT to form atractyloside-responsive channels in a Bcl-2 inhibitable manner, and these Bax/ANT channels appear to trigger the mitochondrial permeability transition[50]. Bax also forms pH and voltage-dependent channels, which require the presence of a functional F$_o$F$_1$-ATPase (the proton pump located in the inner membrane of mitochondria)[51,52], whereas Bcl-2 and Bcl-x$_L$ modulate the pH of the intermembrane space by controlling the efflux of protons[53]. In addition, Bcl-2 and Bcl-x$_L$ inhibit release of calcium from the mitochondrial matrix (and ER) and enhance calcium buffering capacity[54,55]. Thus, formation of any number of channels in the outer and/or inner membrane of mitochondria may account for the release of cyt. *c*.

As most toxicants induce cyt. *c* release and apoptosis in a Bcl-2/Bcl-x$_L$-inhibitable manner, it appears that the mitochondria represents a convergence point for most forms of toxicant-induced apoptosis. Of course, the key question is, how do toxicants with vastly different mechanisms of action induce cyt. *c* release? Certainly, one possibility is that they may alter the balance between anti- and pro-apoptotic Bcl-2 proteins. However, another intriguing possibility involves activation of the ubiquitously expressed c-Jun NH$_2$-terminal kinases (JNKs; also known as stress-induced kinases or SAPKs). JNKs are activated in response to numerous apoptotic stimuli including DNA damage, reactive oxygen species and glutathione/thiol depletion. A recent study indicates that *jnk1$^{-/-}$ jnk2$^{-/-}$* cells are highly resistant to a number of chemicals/chemotherapeutic agents[56]. Moreover, JNK appears to redistribute from the cytosol to the mitochondria during apoptosis[57]. Therefore, the possibility exists that JNK may modulate the release of cyt. *c via* phosphorylation of a specific mitochondrial protein, such as Bcl-x$_L$.

Does this mean that all toxicants will induce apoptosis directly or indirectly by altering mitochondrial function and releasing cyt. *c*? The answer to this question is probably no. It now appears that the ER may also possess one or more caspase activating complexes. Bap31 is an ER protein that binds to Bcl-x$_L$ and appears to mediate activation of caspase-8[58]. In addition, recent studies show that caspase-12 is localized to the ER and activated by ER stressors, including accumulation of excess proteins in the ER and disruption of microsomal calcium homeostasis. The presence of caspase-12 in the ER appears physiologically important, given that caspase-12$^{-/-}$ cells are more resistant to chemicals which specifically induce ER stress, such as brefeldin A, tunicamycin and thapsigargin. Moreover, caspase-12, which is specifically expressed in renal proximal tubular cells but not in glomerular cells, is important in tunicamycin-induced renal toxicity[59]. Thus, activation of caspase-12 appears to be important in ER stress-induced apoptosis, but the adapter protein(s) required for its activation are currently unknown.

THE DISC: UPREGULATION OF DEATH LIGANDS AND RECEPTORS IN TOXICANT-INDUCED APOPTOSIS

In addition to the apoptosome-mediated pathway already described, toxicants may in some cases also activate caspases and induce apoptosis through upregulation of various death receptor pathways. Within the plasma membrane of many cells are death receptors, which when triggered by their corresponding death ligands, initiate rapid activation of caspases and the induction of apoptosis. Death receptors are members of the tumor necrosis factor (TNF) receptor superfamily. They possess both cysteine-rich extracellular domains and an intracellular cytoplasmic sequence known as the death domain (DD). The best-known death receptors are CD95/Fas/Apo1, TNFR1, TNFR2, DR3/Wsl-1/Tramp, DR4/TRAIL-R1 (TNF related apoptosis inducing ligand receptor 1), DR5/TRAIL-R2/ TRICK2/Killer and DR6. Triggering of death receptors with their cognate ligands or agonistic antibodies results in receptor trimerization and recruitment of adapter proteins[60,61,62]. For example, CD95 ligand (CD95L) interacts with and induces trimerization of CD95 receptors, resulting in clustering of the receptor's cytosolic DD and recruitment of the adapter molecule, FADD (Fas-associated death domain; also termed Mort1). FADD contains a C-terminal DD which enables it to bind trimerized CD95 receptors through DD-DD interactions, as well as an N-terminal DED (death effector domain) which can associate with similar DEDs located in the prodomain of caspase-8. This complex of proteins is referred to as the DISC (death-inducing signaling complex), and it is proposed that as more procaspase-8 molecules are recruited to this complex, they begin to cluster and undergo trans-catalysis to generate active caspase-8[21,63].

Several studies have suggested that many anticancer agents, including doxorubicin, etoposide, methotrexate, cisplatin and bleomycin, may induce apoptosis by inducing synthesis of CD95L which binds to CD95 and activates the death receptor pathway. The increase in CD95L mRNA may be a result of drug-induced activation of the transcription factors NF-κB and AP-1. In some studies increases in CD95L protein have also been observed. However, the relevance of these findings to drug-induced cytotoxicity has been questioned partly because of the high concentration of drugs required[6]. Thus, some toxicants may induce apoptosis in an autocrine/paracrine fashion by upregulating expression of the death receptor and/or ligand, however, this still remains to be proven in the majority of examples. One possible exception is 5-fluorouracil (5-FU), whose active metabolite inhibits thymidylate synthase and causes thymine depletion. This results in a p53-dependent increase in CD95L expression and ultimately in toxicity. Thus in the case of 5-FU, there appears to be much stronger evidence supporting the involvement of CD95 signalling[6,64].

However, even in circumstances where toxicant induced apoptosis involves death receptors, the involvement of the apoptosome will likely still play a role in the activation of caspases and the induction of apoptosis. Toxicant-induced death signals could stimulate the release of cyt. c from the mitochondria as already described, or if death receptor-mediated pathways become involved, activation of the pro-apoptotic Bcl-2 protein Bid could become important. Caspase-8-mediated cleavage of Bid during receptor-induced apoptosis produces a truncated protein that translocates from the cytosol to the mitochondria where it potently stimulates release of cyt. c[65]. In effect, Bid may couple death receptor stimulation to the formation of an apoptosome and activation of caspase-9. Thus, a caspase amplification loop can be established, whereby both caspases-8 and -9 can contribute to the activation of effector caspases-3, -6, and -7, and ensure destruction of the cell. However, the true physiological importance of Bcl-2, Bcl-x_L and Bid in CD95L-induced apoptosis is controversial, because in some studies cyt. c is not released during CD95-induced apoptosis and Bcl-2/Bcl-x_L over-expression provides little protection[66]. This discrepancy led to the proposal that two types of cells exist—those in which death ligands induce apoptosis without mitochondrial involvement (Type I) and those that require cyt. c release in order to induce apoptosis (Type II)[67]. However, even this proposal remains highly controversial[68].

CONCLUDING REMARKS

There is an increasing awareness that many toxicants induce cell death through apoptotic mechanisms. In this chapter, we have discussed in detail a few of the primary apoptotic pathways. The release of cyt. *c* from mitochondria following cellular stress and the subsequent formation of Apaf-1/caspase-9 apoptosome complexes are clearly important processes in most chemically-induced models of cell death. However, it should be emphasized that additional caspase-cascades may exist within subcellular compartments, such as the ER, and these pathways may be selectively activated by particular toxicants. Moreover, the specific "death signals" induced by toxicants, and how these signals converge upon the mitochondria, are largely unknown and will undoubtedly remain the focus of future research.

REFERENCES

1. Steller, H., Mechanisms and genes of cellular suicide, *Science* 267:1445 (1995).

2. Jacobson, M. D., Weil, M., Raff, M. C., Programmed cell death in animal development, *Cell* 88:347 (1997).

3. Kerr, J. F., Wyllie, A. H., Currie, A. R., Apoptosis: a basic biological phenomenon with wide-ranging implications in tissue kinetics, *Br. J. Cancer* 26:239 (1972).

4. Thompson, C. B., Apoptosis in the pathogenesis and treatment of disease, *Science* 267:1456 (1995).

5. Searle, J., Lawson, T. A., Abbott, P. J., Harmon, B., Kerr, J. F., An electron-microscope study of the mode of cell death induced by cancer-chemotherapeutic agents in populations of proliferating normal and neoplastic cells, *J. Pathol.* 116:129 (1975).

6. Kaufmann, S. H., Earnshaw, W. C., Induction of apoptosis by cancer chemotherapy, *Exp. Cell Res.* 256:42 (2000).

7. Bratton, S. B., Lau, S. S., Monks, T. J., The putative benzene metabolite 2,3,5-tris-(glutathion-S-yl)hydroquinone depletes glutathione, stimulates sphingomyelin turnover, and induces apoptosis in HL-60 cells, *Chem. Res. Toxicol.* in press (2000).

8. Chrestensen, C. A., Starke, D. W., Mieyal, J. J., Acute cadmium exposure inactivates thioltransferase (glutaredoxin), inhibits intracellular reduction of protein-glutathionyl mixed disulfides, and initiates apoptosis, *J Biol Chem* 18:in press (2000).

9. Moran, J. L., Siegel, D., Sun, X. M., Ross, D., Induction of apoptosis by benzene metabolites in HL60 and CD34+ human bone marrow progenitor cells, *Mol. Pharmacol.* 50:610 (1996).

10. Zhan, Y., van de Water, B., Wang, Y., Stevens, J. L., The roles of caspase-3 and bcl-2 in chemically-induced apoptosis but not necrosis of renal epithelial cells, *Oncogene* 18:6505 (1999).

11. Arends, M. J., Wyllie, A. H., Apoptosis: mechanisms and roles in pathology, *Int Rev Exp Pathol* 32:223 (1991).

12. Raffray, M., Cohen, G. M., Apoptosis and necrosis in toxicology: a continuum or distinct modes of cell death?, *Pharmacol Ther.* 75:153 (1997).

13. Enari, M., Sakahira, H., Yokoyama, H., Okawa, K., Iwamatsu, A., et al., A caspase-activated DNase that degrades DNA during apoptosis, and its inhibitor ICAD, *Nature* 391:43 (1998).

14. Liu, X., Zou, H., Slaughter, C., Wang, X., DFF, a heterodimeric protein that functions downstream of caspase-3 to trigger DNA fragmentation during apoptosis, *Cell* 89:175 (1997).

15. Sahara, S., Aoto, M., Eguchi, Y., Imamoto, N., Yoneda, Y., et al., Acinus is a caspase-3-activated protein required for apoptotic chromatin condensation, *Nature* 401:168 (1999).

16. Susin, S. A., Lorenzo, H. K., Zamzami, N., Marzo, I., Snow, B. E., et al., Molecular characterization of mitochondrial apoptosis-inducing factor, *Nature* 397:441 (1999).

17. Fadok, V. A., Bratton, D. L., Rose, D. M., Pearson, A., Ezekewitz, R. A., et al., A receptor for phosphatidylserine-specific clearance of apoptotic cells, *Nature* 405:85 (2000).

18. Yuan, J., Shaham, S., Ledoux, S., Ellis, H. M., Horvitz, H. R., The C. elegans cell death gene ced-3 encodes a protein similar to mammalian interleukin-1 beta-converting enzyme, *Cell* 75:641 (1993).

19. Cohen, G. M., Caspases: the executioners of apoptosis, *Biochem. J.* 326:1 (1997).

20. Nicholson, D. W., Caspase structure, proteolytic substrates, and function during apoptotic cell death, *Cell Death Differ.* 6:1028 (1999).

21. Muzio, M., Stockwell, B. R., Stennicke, H. R., Salvesen, G. S., Dixit, V. M., An induced proximity model for caspase-8 activation, *J. Biol. Chem.* 273:2926 (1998).

22. Srinivasula, S. M., Ahmad, M., Fernandes-Alnemri, T., Alnemri, E. S., Autoactivation of procaspase-9 by Apaf-1-mediated oligomerization, *Mol. Cell* 1:949 (1998).

23. Hengartner, M. O., Horvitz, H. R., Programmed cell death in Caenorhabditis elegans, *Curr. Opin. Genet. Dev.* 4:581 (1994).

24. Yang, X., Chang, H. Y., Baltimore, D., Essential role of CED-4 oligomerization in CED-3 activation and apoptosis, *Science* 281:1355 (1998).

25. Green, D. R., Reed, J. C., Mitochondria and apoptosis, *Science* 281:1309 (1998).

26. Li, P., Nijhawan, D., Budihardjo, I., Srinivasula, S. M., Ahmad, M., et al., Cytochrome c and dATP-dependent formation of Apaf-1/caspase-9 complex initiates an apoptotic protease cascade, *Cell* 91:479 (1997).

27. Zou, H., Henzel, W. J., Liu, X., Lutschg, A., Wang, X., Apaf-1, a human protein homologous to C. elegans CED-4, participates in cytochrome c-dependent activation of caspase-3, *Cell* 90:405 (1997).

28. Yeh, W. C., Hakem, R., Woo, M., Mak, T. W., Gene targeting in the analysis of mammalian apoptosis and TNF receptor superfamily signaling, *Immunol. Rev.* 169:283 (1999).

29. Saleh, A., Srinivasula, S. M., Acharya, S., Fishel, R., Alnemri, E. S., Cytochrome c and dATP-mediated oligomerization of Apaf-1 is a prerequisite for procaspase-9 activation, *J. Biol. Chem.* 274:17941 (1999).

30. Zou, H., Li, Y., Liu, X., Wang, X., An APAF-1.cytochrome c multimeric complex is a functional apoptosome that activates procaspase-9, *J. Biol. Chem.* 274:11549 (1999).

31. Cain, K., Brown, D. G., Langlais, C., Cohen, G. M., Caspase activation involves the formation of the aposome, a large (approximately 700 kDa) caspase-activating complex, *J. Biol. Chem.* 274:22686 (1999).

32. Cain, K., Bratton, S. B., Langlais, C., Walker, G., Brown, D. G., et al., Apaf-1 Oligomerizes into Biologically Active ~700-kDa and Inactive ~1.4- MDa Apoptosome Complexes, *J. Biol. Chem.* 275:6067 (2000).

33. Qin, H., Srinivasula, S. M., Wu, G., Fernandes-Alnemri, T., Alnemri, E. S., et al., Structural basis of procaspase-9 recruitment by the apoptotic protease-activating factor 1, *Nature* 399:549 (1999).

34. Rodriguez, J., Lazebnik, Y., Caspase-9 and APAF-1 form an active holoenzyme, *Genes Dev.* 13:3179 (1999).

35. Hu, Y., Benedict, M. A., Ding, L., Nunez, G., Role of cytochrome c and dATP/ATP hydrolysis in Apaf-1-mediated caspase- 9 activation and apoptosis, *EMBO J.* 18:3586 (1999).

36. Marzo, I., Brenner, C., Zamzami, N., Susin, S. A., Beutner, G., et al., The permeability transition pore complex: a target for apoptosis regulation by caspases and bcl-2-related proteins, *J. Exp. Med.* 187:1261 (1998).

37. Crompton, M., The mitochondrial permeability transition pore and its role in cell death, *Biochem J* 341:233 (1999).

38. Zamzami, N., Marchetti, P., Castedo, M., Hirsch, T., Susin, S. A., et al., Inhibitors of permeability transition interfere with the disruption of the mitochondrial transmembrane potential during apoptosis, *FEBS Lett* 384:53 (1996).

39. Bossy-Wetzel, E., Newmeyer, D. D., Green, D. R., Mitochondrial cytochrome c release in apoptosis occurs upstream of DEVD-specific caspase activation and independently of mitochondrial transmembrane depolarization, *Embo J* 17:37 (1998).

40. Marzo, I., Susin, S. A., Petit, P. X., Ravagnan, L., Brenner, C., et al., Caspases disrupt mitochondrial membrane barrier function, *FEBS Lett.* 427:198 (1998).

41. Ichas, F., Mazat, J. P., From calcium signaling to cell death: two conformations for the mitochondrial permeability transition pore. Switching from low- to high-conductance state, *Biochim Biophys Acta* 1366:33 (1998).

42. Adams, J. M., Cory, S., The Bcl-2 protein family: arbiters of cell survival, *Science* 281:1322 (1998).

43. Minn, A. J., Velez, P., Schendel, S. L., Liang, H., Muchmore, S. W., et al., Bcl-xL forms an ion channel in synthetic lipid membranes, *Nature* 385:353 (1997).

44. Muchmore, S. W., Sattler, M., Liang, H., Meadows, R. P., Harlan, J. E., et al., X-ray and NMR structure of human Bcl-xL, an inhibitor of programmed cell death, *Nature* 381:335 (1996).

45. Schendel, S. L., Montal, M., Reed, J. C., Bcl-2 family proteins as ion-channels, *Cell Death Differ.* 5:372 (1998).

46. Krajewski, S., Tanaka, S., Takayama, S., Schibler, M. J., Fenton, W., et al., Investigation of the subcellular distribution of the bcl-2 oncoprotein: residence in the nuclear envelope, endoplasmic reticulum, and outer mitochondrial membranes, *Cancer Res* 53:4701 (1993).

47. Shimizu, S., Narita, M., Tsujimoto, Y., Bcl-2 family proteins regulate the release of apoptogenic cytochrome c by the mitochondrial channel VDAC, *Nature* 399:483 (1999).

48. Shimizu, S., Konishi, A., Kodama, T., Tsujimoto, Y., BH4 domain of antiapoptotic Bcl-2 family members closes voltage-dependent anion channel and inhibits apoptotic mitochondrial changes and cell death, *Proc. Natl. Acad. Sci. U S A* 97:3100 (2000).

49. Shimizu, S., Tsujimoto, Y., Proapoptotic BH3-only Bcl-2 family members induce cytochrome c release, but not mitochondrial membrane potential loss, and do not directly modulate voltage-dependent anion channel activity, *Proc. Natl. Acad. Sci. U S A* 97:577 (2000).

50. Marzo, I., Brenner, C., Zamzami, N., Jurgensmeier, J. M., Susin, S. A., et al., Bax and adenine nucleotide translocator cooperate in the mitochondrial control of apoptosis, *Science* 281:2027 (1998).

51. Antonsson, B., Conti, F., Ciavatta, A., Montessuit, S., Lewis, S., et al., Inhibition of Bax channel-forming activity by Bcl-2, *Science* 277:370 (1997).

52. Matsuyama, S., Xu, Q., Velours, J., Reed, J. C., The Mitochondrial F0F1-ATPase proton pump is required for function of the proapoptotic protein Bax in yeast and mammalian cells, *Mol. Cell* 3:327 (1998).

53. Shimizu, S., Eguchi, Y., Kamiike, W., Funahashi, Y., Mignon, A., et al., Bcl-2 prevents apoptotic mitochondrial dysfunction by regulating proton flux, *Proc. Natl. Acad. Sci. U S A* 95:1455 (1998).

54. Baffy, G., Miyashita, T., Williamson, J. R., Reed, J. C., Apoptosis induced by withdrawal of interleukin-3 (IL-3) from an IL-3-dependent hematopoietic cell line is associated with repartitioning of intracellular calcium and is blocked by enforced Bcl-2 oncoprotein production, *J Biol Chem* 268:6511 (1993).

55. Lam, M., Bhat, M. B., Nunez, G., Ma, J., Distelhorst, C. W., Regulation of Bcl-xl channel activity by calcium, *J. Biol. Chem.* 273:17307 (1998).

56. Tournier, C., Hess, P., Yang, D. D., Xu, J., Turner, T. K., et al., Requirement of JNK for stress-induced activation of the cytochrome c-mediated death pathway, *Science* 288:870 (2000).

57. Kharbanda, S., Saxena, S., Yoshida, K., Pandey, P., Kaneki, M., et al., Translocation of SAPK/JNK to mitochondria and interaction with Bcl-x(L) in response to DNA damage, *J Biol Chem* 275:322 (2000).

58. Ng, F. W., Shore, G. C., Bcl-XL cooperatively associates with the Bap31 complex in the endoplasmic reticulum, dependent on procaspase-8 and Ced-4 adaptor, *J. Biol. Chem.* 273:3140 (1998).

59. Nakagawa, T., Zhu, H., Morishima, N., Li, E., Xu, J., et al., Caspase-12 mediates endoplasmic-reticulum-specific apoptosis and cytotoxicity by amyloid-β, *Nature* 403:98 (2000).

60. Ashkenazi, A., Dixit, V. M., Death receptors: signaling and modulation, *Science* 281:1305 (1998).

61. Bratton, S. B., MacFarlane, M., Cain, K., Cohen, G. M., Protein complexes activate distinct caspase cascades in death receptor and stress-induced apoptosis, *Exp. Cell Res.* 256:27 (2000).

62. Wallach, D., Varfolomeev, E. E., Malinin, N. L., Goltsev, Y. V., Kovalenko, A. V., et al., Tumor necrosis factor receptor and Fas signaling mechanisms, *Annu. Rev. Immunol.* 17:331 (1999).

63. Medema, J. P., Scaffidi, C., Kischkel, F. C., Shevchenko, A., Mann, M., et al., FLICE is activated by association with the CD95 death-inducing signaling complex (DISC), *EMBO J.* 16:2794 (1997).

64. Houghton, J. A., Harwood, F. G., Tillman, D. M., Thymineless death in colon carcinoma cells is mediated via fas signaling, *Proc. Natl. Acad. Sci. U S A* 94:8144 (1997).

65. Li, H., Zhu, H., Xu, C. J., Yuan, J., Cleavage of BID by caspase 8 mediates the mitochondrial damage in the Fas pathway of apoptosis, *Cell* 94:491 (1998).

66. Strasser, A., Harris, A. W., Huang, D. C., Krammer, P. H., Cory, S., Bcl-2 and Fas/APO-1 regulate distinct pathways to lymphocyte apoptosis, *EMBO J.* 14:6136 (1995).

67. Scaffidi, C., Fulda, S., Srinivasan, A., Friesen, C., Li, F., et al., Two CD95 (APO-1/Fas) signaling pathways, *EMBO J.* 17:1675 (1998).

68. Huang, D. C., Hahne, M., Schroeter, M., Frei, K., Fontana, A., et al., Activation of Fas by FasL induces apoptosis by a mechanism that cannot be blocked by Bcl-2 or Bcl-x(L), *Proc. Natl. Acad. Sci. U S* 96:14871 (1999)..

SPERMATOGENESIS BY SISYPHUS: PROLIFERATING STEM GERM CELLS FAIL TO REPOPULATE THE TESTIS AFTER 'IRREVERSIBLE' INJURY

Kim Boekelheide and Heidi A. Schoenfeld

Department of Pathology and Laboratory Medicine
Brown University
Providence, RI 02912 USA

ABSTRACT

2,5-Hexanedione is the toxic metabolite resulting from oxidation of the commonly used solvents n-hexane and methyl n-butyl ketone. Exposure to 2,5-hexanedione or its precursors results in a slowly progressive peripheral polyneuropathy and testicular injury. The chemical basis of the injury involves reaction of 2,5-hexanedione with protein amines, such as the ε-amine of lysine, to form pyrroles which further react to form protein-protein crosslinks. The target cell of injury in the testis is the supportive cell in the seminiferous epithelium, the Sertoli cell. A major function of the Sertoli cell is to nurture the dependent germ cell population by secreting seminiferous tubule fluid. 2,5-Hexanedione-induced crosslinking of the microtubule subunit protein, tubulin, leads to altered Sertoli cell microtubule-dependent transport and deficient formation of seminiferous tubule fluid, compromising germ cell viability.

In an established model of testicular injury, rats are exposed to 1% 2,5-hexanedione in the drinking water for a period of 3 – 5 weeks. Three weeks after initiating exposure, decreased seminiferous tubule fluid secretion initiates a wave of germ apoptosis which peaks during the 5th week. The germ cell content of the injured testis continues to decline after cessation of the exposure, reaching a nadir during the 12th week. From this time onward, the testis is severely atrophic with less than 1% of seminiferous tubules in a testicular cross section containing germ cells more advanced than spermatogonia.

Interestingly, this persistent state of post-injury 'irreversible' atrophy in the rat is characterized by the presence of a proliferating stem germ cell population which produces differentiating spermatogonia which then die by apoptosis. Serial cross sections of bromodeoxyuridine-labeled testis were analyzed to determine the kinetics of stem germ cell proliferation. Approximately 40% of stem cells (identified as single cells in the seminiferous epithelium) were actively proliferating with a cell cycle time of 8-14 days. Analysis of the total germ cell population present and modeling using the known cell cycle times of differentiating spermatogonia indicated a block in differentiation at the level of type A_3/A_4 spermatogonia. Quantitation of the frequency of apoptosis indicated that all of the germ cells died prematurely by this mechanism.

Biological Reactive Intermediates VI, Edited by Dansette *et al.*
Kluwer Academic / Plenum Publishers, 2001

Leuprolide is a gonadotropin-releasing hormone agonist which produces a profound suppression of testosterone levels with chronic administration. When delivered as a series of 3 depot injections 24 days apart, leuprolide resulted in a partial reversal of the 2,5-hexanedione-induced persistent atrophy. The reinitiation of spermatogenesis follows a lowering of the intratesticular testosterone concentration, indicating that intratesticular testosterone is at least partially responsible for the persistent atrophy. The efficacy of leuprolide-induced reversal of the persistent atrophy decreases with time after injury, suggesting that atrophic seminiferous tubules are initially capable of recovery and then enter a state of irreversible injury. Injection of ethane dimethane sulfonate at the beginning of leuprolide treatment eliminated Leydig cells during therapy and ablated the recovery of spermatogenesis, indicating that a Leydig cell-associated paracrine factor is required to restart spermatogenesis.

The rat, therefore, has multiple states of testicular germ cell proliferation: normal spermatogenesis and at least two forms of persistent atrophy (leuprolide reversible and leuprolide non-reversible). Partial reversal of the persistent atrophy can be achieved by lowering intratesticular testosterone. Ongoing experiments are designed to address the role of the Leydig cell in post-injury recovery, and to further characterize the molecular events contributing to the different states of persistent atrophy.

OVERVIEW

The human testis is a well known target organ for injury resulting from exposure to both environmental and chemotherapeutic agents. A feared outcome of exposure is the development of irreversible testicular injury characterized by long lasting azoospermia and infertility. Although irreversible testicular injury was originally assumed to result from a total depletion of germ cells (Meistrich, 1986), it is now apparent that germ cells may be present in the persistently atrophic testis (Boekelheide and Hall, 1991), raising the possibility of therapeutic intervention to restart spermatogenesis. Rats mimic this human condition, entering a state of irreversible testicular atrophy after toxicant exposure even though their testes contain germ cells. The recognition that apoptosis is the usual path of germ cell death and that hormonal manipulations can reestablish spermatogenesis after irreversible testicular injury are two important new insights guiding the ongoing research effort in this area.

BACKGROUND

2,5-Hexanedione exposure

2,5-Hexanedione is the toxic γ-diketone metabolite of n-hexane and methyl n-butyl ketone. Exposure of rats to 2,5-hexanedione produces a slowly progressive syndrome of selective nervous system and testicular dysfunction. The temporal sequence of biochemical and morphological testicular alterations has been studied in detail in young adult rats exposed for 5 weeks to 1% 2,5-hexanedione in the drinking water. The earliest alteration detected, two weeks after initiating exposure, is enhanced polymerization of purified rat testis tubulin, which is predominantly of Sertoli cell origin (Boekelheide, 1988b). By 3 weeks after initiating exposure, seminiferous tubule fluid formation is decreased (Richburg et al., 1994), possibly due to decreased microtubule-dependent transport in the Sertoli cell (Redenbach et al., 1994). By 4 weeks after initiating exposure, Sertoli cell vacuolization (Chapin et al., 1983) and a profound decrease in seminiferous tubule fluid formation is followed by a rapid and progressive apoptosis of germ cells (Boekelheide, 1988a;

Blanchard et al., 1996)). After exposure and return of the rats to normal drinking water, the testis continues to lose germ cells and, by 12 weeks after initiating exposure, enters a persistent state of irreversible testicular injury (Boekelheide and Hall, 1991).

Rescue

Recently, suppression of the pituitary-gonadal hormonal axis by various treatments, including gonadotropin-releasing hormone antagonists or agonists, antiandrogens, or systemic androgens and estrogens, has been successful in reestablishing spermatogenesis in the persistent testicular atrophy produced by cyclophosphamide, procarbazine, and x-irradiation (Kangasniemi et al., 1995a; Kangasniemi et al., 1995b; Meistrich et al., 1995; Meistrich, 1998). The hormonal effect common to these treatments is a reduction of the intratesticular testosterone concentration, although it is not understood why high levels of intratesticular testosterone are detrimental to spermatogenesis. An interesting feature of this hormonal manipulation is that the intervention is effective whether performed at the time of toxicant injury or much later after atrophy is established (Kurdoglu et al., 1994; Meistrich, 1998).

Apoptosis

Apoptosis, also called programmed cell death, is a form of actively induced cell death which occurs during both normal physiological processes and after toxicant exposure. Classically, apoptosis has been identified morphologically by the characteristic condensation of the cytoplasm, margination of chromatin to the nuclear membrane, and fragmentation of the cell into apoptotic bodies (Hale et al., 1996). The oligonucleosomal DNA cleavage pattern characteristic of apoptosis is seen as a DNA ladder by DNA electrophoresis or as positive *in situ* staining when using the terminal deoxynucleotide transferase-mediated dUTP nick end labeling (TUNEL) technique (Gavrieli et al., 1992). Apoptosis has been divided into two phases (Steller, 1995): 1) a *signaling phase* in which cells are initiated to die by various signals, and 2) an *execution phase* in which cells rapidly execute a death program. Many different types of signals—growth factor withdrawal, cell cycle perturbations, or DNA damage, to name a few—can initiate apoptosis. Apparently, the cell uses various sensing mechanisms to interpret and integrate the different death-inducing signals and then funnels the initiation message to a common cysteine protease-dependent execution phase pathway.

In the adult rat, physiological apoptosis of type A_2, A_3, and A_4 spermatogonia reduces the number of germ cells produced to 25% of that expected if all A_1 spermatogonial progeny were to survive (Allan et al., 1975; Clermont and Hermo, 1975; Huckins, 1978). A likely explanation for this germ cell culling is that Sertoli cells can support only a limited number of germ cells; when this supportive capacity is exceeded, then germ cells are eliminated by apoptosis (Huckins, 1978). Various types of testicular injuries—including heat exposure (Allan et al., 1975), Sertoli cell toxicants such as 2,5-hexanedione (Blanchard et al., 1996) and mono-(2-ethylhexyl) phthalate (Richburg and Boekelheide, 1996), and germ cell toxicants like x-irradiation (Hasegawa et al., 1997)—all result in germ cell apoptosis. Together, these observations indicate that the seminiferous epithelium responds to most adverse environmental conditions by eliminating germ cells through programmed cell death.

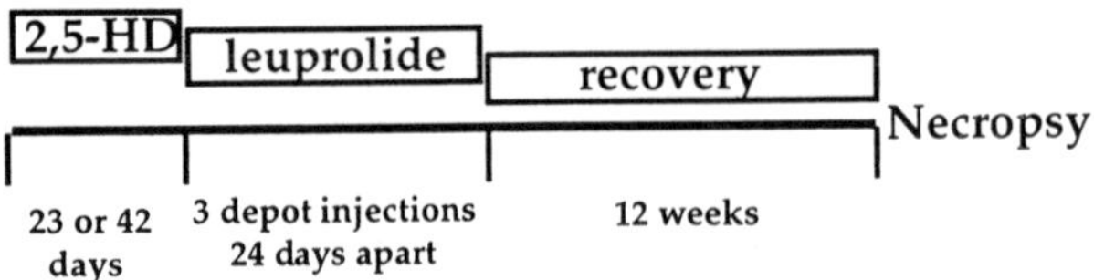

Figure 1. Rats were exposed to 1% 2,5-hexanedione (2,5-HD) in the drinking water for 23 or 42 days followed by 3 depot injections of leuprolide 24 days apart. After a post-treatment recovery period of 12 weeks, the rats were killed and evaluated for testis weight and germ cell repopulation of the seminiferous epithelium (Blanchard et al., 1998).

RESULTS

Stem and committed germ cell kinetics in atrophic testes

The goal of these experiments was to determine, quantitatively, the number, type and kinetic behavior of those spermatogonia which populate the irreversibly injured testis. These data address fundamental questions, such as: Is the stem cell mass normal? Are stem cells dividing and, if so, are they dividing at a normal rate? Where is the block to germ cell differentiation? By what mechanism are germ cells dying?

A series of papers (Allard et al., 1995; Allard and Boekelheide, 1996; Blanchard et al., 1996) characterized the germ cell population during and after 2,5-hexanedione exposure. During acute injury, the germ cells died en masse by apoptosis, as determined by morphology, DNA laddering, and TUNEL staining(Blanchard et al., 1996). After this phase of massive germ cell loss, the 2,5-hexanedione-exposed testes entered a persistent state of irreversible atrophy in which shrunken seminiferous tubules were populated by Sertoli cells and a few germ cells. By using serial sections and quantitative analysis, we showed that the stem cell population was reduced in the irreversibly injured testis (~0.38 stem cells/100 Sertoli cells compared to the normal ~2.00 stem cells/100 Sertoli cells). Modeling predicted that the differentiation of germ cells in the irreversibly injured testis was blocked at the level of A_3/A_4 spermatogonia, and also that all of the germ cells were dying by apoptosis (Allard and Boekelheide, 1996). Interestingly, labeling studies indicated that the stem cells were dividing approximately once per cycle in the atrophic epithelium with a growth fraction of 0.42 (Allard et al., 1995). Exposure of young rats to mono-(2-ethylhexyl)phthalate, a fast acting Sertoli cell toxicant, suggested that apoptosis of germ cells is the usual response to toxic injury in the testis (Richburg and Boekelheide, 1996).

Leuprolide rescues 2,5-hexanedione-induced irreversible testicular injury

In the last several years, a variety of hormonal manipulations which lower intratesticular testosterone concentrations have been used to reestablish spermatogenesis in atrophic rat testes injured by x-irradiation or exposure to radio-mimetic chemotherapeutic compounds (Kangasniemi et al., 1995a; Kangasniemi et al., 1995b; Meistrich et al., 1995; Meistrich, 1998). Taking another approach to rescue of the atrophic testis, we showed that the size of germ cell clones significantly increased after infusion of stem cell factor into testes injured by 2,5-hexanedione exposure (Allard et al., 1996). Based on these past results, we decided to examine the response of the atrophic 2,5-hexanedione-exposed testis to leuprolide injection. Leuprolide is a gonadotropin-releasing hormone agonist which, after an initial upregulation of the hypothalamic-pituitary-gonadal axis, produces a sustained suppression of testosterone secretion. The question asked with these experiments was: Can the irreversible 2,5-hexanedione-induced testicular atrophy be rescued by lowering intratesticular testosterone concentrations?

The experimental paradigm for these experiments is illustrated in Figure 1, and the results have been published (Blanchard et al., 1998). Fischer rats were exposed to 1% 2,5-hexanedione in the drinking water for 23 or 42 days followed by a series of three subcutaneous injections 23 - 24 days apart of leuprolide (Lupron depot, kindly provided by TAP Pharmaceuticals Inc.; 1.5 mg/rat in 0.5 ml diluent). Twelve weeks after the last leuprolide injection, the rats were killed. A comparison group of rats received 2,5-hexanedione alone. Leuprolide treatment significantly increased ($p < 0.05$) testis weight (1.02 ± 0.05 *versus* 0.62 ± 0.03) and the percentage of recovered seminiferous tubules (~90% *versus* 0.1%) in rats exposed to 2,5-hexanedione for 23 days. In rats exposed to 2,5-hexanedione for 42 days, no increase in testis weight accompanied leuprolide rescue, although the percentage of recovered seminiferous tubules (25% versus 0%) improved significantly.

Ongoing Experiments

Preliminary data from our laboratory indicates that the ability of leuprolide to reverse 2,5-hexanedione-induced testicular atrophy decreases with increasing time after the onset of toxicant exposure and germ cell loss. While ~90% of seminiferous tubules recover if leuprolide therapy is begun immediately after the onset of injury, the recovery is reduced to between 10 and 20% if therapy is delayed until 12 weeks after the onset of toxicant exposure. These findings are consistent with reports of reduced recovery at increased time points after hormonal treatment following radiation (Meistrich, 1998).

This decrease in recovery potential with time after injury suggests that two or more states of testicular atrophy exist, one in which the injury is potentially reversible followed by an irreversible state. Interestingly, there are no morphologically apparent differences in germ cell content early compared to late after toxicant-induced testicular atrophy (Boekelheide and Hall, 1991). Time course studies are in progress to characterize these changes in the state of reversibility, and to identify factors that contribute to these two states.

While the mechanisms mediating the initial injury to seminiferous tubules by 2,5-hexanedione have been well characterized, the factors contributing to the irreversible nature of the injury are poorly understood. Intratesticular testosterone levels are elevated 2.5 fold in the atrophic testis, 30 weeks after 2,5-hexanedione treatment. The observation of increased intratesticular testosterone in the atrophic testis, coupled with a reversal of testicular atrophy by testosterone-suppressing leuprolide therapy, suggests that the stimulation of spermatogenesis is mediated by a suppression of intratesticular testosterone levels. Based on these observations, we hypothesized that ablation of testosterone-producing Leydig cells might also stimulate spermatogenesis. Twelve weeks after the initiation of 2,5-hexanedione treatment, rats were administered either: 1) a single injection of the Leydig cell toxicant ethane dimethane sulfonate, or 2) a single injection of ethane dimethane sulfonate and 3 injections of leuprolide. Treatment with ethane dimethane sulfonate alone produced a transient ablation of Leydig cells yet failed to stimulate spermatogenesis. When ethane dimethane sulfonate was administered simultaneously with leuprolide, Leydig cell ablation and intratesticular testosterone suppression were maintained for the duration of leuprolide therapy. However, the combination of ethane dimethane sulfonate and leuprolide failed to stimulate a reversal of testicular atrophy. These results indicate that the repopulation of atrophic tubules by leuprolide therapy cannot be explained by a suppression of intratesticular testosterone levels alone, and that paracrine-acting factors, produced by Leydig cells, also mediate the repopulation. Both gonadotropin-releasing hormone and testosterone receptors are present on the Leydig cell, and the effects of leuprolide on Leydig cell release of paracrine acting factors may be mediated through either receptor. A role for testosterone-dependent Leydig cell signaling is plausible, as it is generally accepted that testosterone receptors are present in Leydig cells in addition to Sertoli and peritubular myoid cells, but not germ cells. However, relatively little is known regarding the signaling events downstream of testosterone binding in the testis, and how these signaling events modulate differentiation or apoptosis in germ cells. Ongoing studies are aimed at isolating Leydig cell factors which contribute to the repopulation of the atrophic testis.

DISCUSSION

We know from these initial studies (Blanchard et al., 1998) of leuprolide-induced reversal of testicular atrophy that the total dose of 2,5-hexanedione exposure is a critical determinant of recovery potential. Another important consideration is the timing of leuprolide treatment relative to toxicant exposure; apparently, recovery from irreversible injury occurs even if hormonal suppression is delayed for months after the injury (Meistrich, 1998), although the extent of recovery declines with time.

Apoptosis plays a critical role in germ cell loss during both acute and irreversible toxicant-induced testicular injury. For both 2,5-hexanedione exposure of adult rats and mono-(2-ethyhexyl)phthalate exposure of young rats, all of the germ cells die by "programmed cell death" during the acute phase of injury. Since both 2,5-hexanedione and mono-(2-ethyhexyl)phthalate are Sertoli cell toxicants, our original view was that germ cell loss was a passive consequence of Sertoli cell dysfunction, such as the loss of growth factor support. The demonstration that apoptosis occurred synchronously in the germ cell population following exposure to a Sertoli cell toxicant raised the alternative possibility that germ cell loss was actively signaled by the Sertoli cell. Our recent demonstration that the Fas system is upregulated by these toxicants (Lee et al., 1997; Lee et al., 1999) supports the view that death factors are actively involved in causing germ cell apoptosis after toxicant exposure.

The ability of leuprolide treatment to reestablish spermatogenesis in the irreversibly injured 2,5-hexanedione-exposed testis is an exciting observation. The finding of decreased germ cell apoptosis and reversal of testicular atrophy following suppression of testosterone by a gonadotropin-releasing hormone agonist runs contrary to the known requirement for testosterone as a maintenance factor for spermatogenesis and progression of round spermatids into elongate spermatids. It has been suggested that testosterone may be a dual acting factor whose effects are dictated in a cell specific fashion. Thus, testosterone may stimulate proliferation and differentiation in advanced germ cell types while inducing apoptosis in type A spermatogonia. Leuprolide treatment provides a means to identify the underlying elements which modulate spermatogenesis after injury, since the outcome of toxicant exposure can be dynamically altered and the system can be coaxed into recovery. A careful analysis of the balance between testicular growth and death factors and their influence on germ cell survival during leuprolide-induced rescue from irreversible testicular injury will enhance understanding of the paracrine processes which regulate germ cell output, maintain testicular homeostasis, and cause toxicant-induced irreversible testicular injury.

ACKNOWLEDGEMENTS

This publication was made possible by grant number R01 ES05033 from the National Institute of Environmental Health Sciences.

REFERENCES

Allan, D.J., Harmon, B.V., and Kerr, J.F.R., 1987, Cell death in spermatogenesis, in: *Perspectives on Mammalian Cell Death*, C.S. Potten, ed., Oxford University Press, Oxford, pp. 229-258.

Allard, E.K., and Boekelheide, K., 1996, Fate of germ cells in 2,5-hexanedione-induced testicular injury. II. Atrophy persists due to a reduced stem cell mass and ongoing apoptosis, *Toxicol. Appl. Pharmacol.* 137:149-156.

Allard, E.K., Hall, S.J., and Boekelheide, K., 1995, Stem cell kinetics in rat testis after irreversible injury induced by 2,5-hexanedione, *Biol. Reprod.* 53:186-192.

Allard, E.K., Blanchard, K.T., and Boekelheide, K., 1996, Exogenous stem cell factor (SCF) compensates for altered endogenous SCF expression in 2,5-hexanedione-induced testicular atrophy, *Biol. Reprod.* 55:185-193.

Blanchard, K.T., Allard, E.K., and Boekelheide, K., 1996, Fate of germ cells in 2,5-hexanedione-induced testicular injury. I. Apoptosis is the mechanism of germ cell loss, *Toxicol. Appl. Pharmacol.* 137:141-148.

Blanchard, K.T., Lee, J.-W., and Boekelheide, K., 1998, Leuprolide, a gonadotropin-releasing hormone agonist, reestablishes spermatogenesis after 2,5-hexanedione-induced "irreversible" testicular injury in the rat resulting in normalized stem cell factor expression, *Endocrinology* 139:236-244.

Boekelheide, K., 1988a, Rat testis during 2,5-hexanedione intoxication and recovery. I. Dose response and the reversibility of germ cell loss, *Toxicol. Appl. Pharmacol.* 92:18-27.

Boekelheide, K., 1988b, Rat testis during 2,5-hexanedione intoxication and recovery. II. Dynamics of pyrrole reactivity, tubulin content, and microtubule assembly, *Toxicol. Appl. Pharmacol.* 92:28-33.

Boekelheide, K., and Hall, S.J., 1991, 2,5-Hexanedione exposure in the rat results in long-term testicular atrophy despite the presence of residual spermatogonia, *J. Androl.* 12:18-26.

Chapin, R.E., Morgan, K.T., and Bus, J.S., 1983, The morphogenesis of testicular degeneration induced in rats by orally administered 2,5-hexanedione, *Exp. Mol. Pathol.* 38:149-169.

Clermont, Y., and Hermo, L., 1975, Spermatogonial stem cells in the albino rat, *Am. J. Anat.* 142:159-176.

Gavrieli, Y., Sherman, Y., and Ben-Sasson, S.A., 1992, Identification of programmed cell death in situ via specific labeling of nuclear DNA fragmentation, *J. Cell Biol.* 119:493-501.

Hale, A.J., Smith, C.A., Sutherland, L.C., Stoneman, V.E., Longthorne, V.L., Culhane, A.C., and Williams, G.T., 1996, Apoptosis: molecular regulation of cell death, *Eur. J. Biochem.* 236:1-26.

Hasegawa, M., Wilson, G., Russell, L.D., and Meistrich, M.L., 1997, Radiation-induced cell death in the mouse testis: relationship to apoptosis, *Rad. Res.* 147:457-467.

Huckins, C., 1978, The morphology and kinetics of spermatogonial degeneration in normal adult rats: an analysis using a simplified classification of the germinal epithelium, *Anat. Record* 190:905-926.

Kangasniemi, M., Wilson, G., Parchuri, N., Huhtaniemi, I., and Meistrich, M.L., 1995a, Rapid protection of rat spermatogenic stem cells against procarbazine by treatment with a GnRH antagonist (Nal-Glu) and an antiandrogen (flutamide), *Endocrinology* 136:2881-2888.

Kangasniemi, M., Wilson, G., Huhtaniemi, I., and Meistrich, M.L., 1995b, Protection against procarbazine-induced testicular damage by GnRH-agonist and antiandrogen treatment in the rat, *Endocrinology* 136:3677-3680.

Kurdoglu, B., Wilson, G., Ye, W.-S., Parchuri, N., and Meistrich, M.L., 1994, Protection from radiation-induced damage to spermatogenesis by hormone treatment, *Radiat. Res.* 139:97-102.

Lee, J.-W., Richburg, J.H., Younkin, S.C., and Boekelheide, K., 1997. The Fas system is a key regulator of germ cell apoptosis in the testis, *Endocrinology* 138:2081-2088.

Lee, J.-W., Richburg, J.H., Shipp, E.B., Meistrich, M.L., and Boekelheide, K., 1999, The Fas system, a regulator of testicular germ cell apoptosis, is differentially up-regulated in Sertoli cell versus germ cell injury of the testis, *Endocrinology* 140:852-858.

Meistrich, M.L., 1986, Critical components of testicular function and sensitivity to disruption, *Biol. Reprod.* 34:17-28.

Meistrich, M.L., 1998, Hormonal stimulation of the recovery of spermatogenesis following chemo- or radiotherapy, *APMIS* 106:37-46.

Meistrich, M.L., Parchuri, N., Wilson, G., Kurdoglu, B., and Kangasniemi, M., 1995, Hormonal protection from cyclophosphamide-induced inactivation of rat stem spermatogonia, *J. Androl.* 16:334-341.

Redenbach, D.M., Richburg, J.H., and Boekelheide, K., 1994, Microtubules with altered assembly kinetics have a decreased rate of kinesin-based transport, *Cell Motil. Cytoskel.* 27:79-87.

Richburg, J.H., and Boekelheide, K., 1996, Mono-(2-ethylhexyl) phthalate rapidly alters both Sertoli cell vimentin filaments and germ cell apoptosis in young rat testes, *Toxicol. Appl. Pharmacol.* 137:42-50.

Richburg, J.R., Redenbach, D.M., and Boekelheide, K., 1994, Seminiferous tubule fluid secretion is a Sertoli cell microtubule-dependent process inhibited by 2,5-hexanedione exposure, *Toxicol. Appl. Pharmacol.* 128:302-309.

Steller, H., 1995, Mechanisms and genes of cellular suicide, *Science* 267:1445-1450.

THE INTERACTION OF 1,4-BENZOQUINONE, A BIOREACTIVE INTERMEDIATE OF BENZENE, WITH THREE PROTEINS ESSENTIAL FOR DIFFERENTIATION/ MATURATION OF THE MOUSE MYELOID STEM CELL

George F. Kalf,[1] Betsy A. Hazel,[1*] Matthew J. Hoffmann,[2] David D. Kim,[2] and Robert Snyder[2]

[1]Department of Biochemistry and Molecular Pharmacology
Jefferson Medical College of Thomas Jefferson University
Philadelphia, PA
[2]Department of Pharmacology and Toxicology
Rutgers University
Piscataway, NJ

INTRODUCTION

Exposure of humans to benzene, a Class I carcinogen, causes acute myeloid leukemia (see Snyder and Kalf, 1994 for review). There is no animal model with which to study benzene-induced leukemia. Acute administration of benzene or hydroquinone (HQ), a major metabolite found in the bone marrow, to mice causes a severe depression of bone marrow cellularity except for the cells of the granulocytic lineage which not only survive (Niculescu and Kalf, 1995), but undergo differentiation and increase in number. The granulocytic progenitor cells may survive because HQ can substitute for granulocyte colony-stimulating factor (G-CSF), the requisite cytokine for granulopoiesis and induce differentiation in the myeloid stem cell (Hazel et al., 1995; Hazel, Baum and Kalf, 1996; Hazel and Kalf, 1996). The ability to alter cytokine-dependent growth and differentiation in hematopoietic progenitor cells appears to be a property of agents that have a potential for causing secondary leukemia in humans (Irons & Stillman, 1993).

Because of the association between benzene exposure and an increased incidence of acute myeloid leukemia (Snyder & Kalf, 1994), the objective of the studies reported here was to determine whether HQ could affect differentiation/maturation of the restricted myeloid stem cell, the myeloblast, in such a way as to induce changes typically seen in the phenotype of leukemia cells in culture.

As our model system we used the IL-3-dependent mouse myeloblastic cell line, 32D clone 3(G) derived by Greenberger et al. (1983) from normal bone marrow of C3H/HeJ mice. It has many features that permit a determination of whether HQ can affect myeloid differentiation by interacting with a cytokine-induced signal transduction system.

---

* Present address: School of Health Professions, The University of Arizona, Tucson, AZ.

Biological Reactive Intermediates VI, Edited by Dansette *et al.*
Kluwer Academic / Plenum Publishers, 2001

The cell line is: 1) myeloblastic and has a normal karyotype; 2) nonleukemic as indicated by its inability to cause tumors in nude mice or to generate cytokine-independent clones; 3) IL-3-dependent for survival and proliferation (Metcalf, 1985); and 4) is induced to undergo terminal granulocytic differentiation in the presence of G-CSF (Valtieri et al., 1987).

METHODS

Cell Culture

32D.3(G) Myeloblasts were maintained in Iscove's Modified Dulbecco's Medium (IMDM) in the presence of 10 % fetal calf serum (FBS), 2 mM L-glutamine, and 3 U/ml recombinant murine IL-3. In experiments to induce differentiation, a concentration of IL-3 (0.13 U/ml) was used to sustain survival and a level of proliferation that did not overly compete with the induction of differentiation by the inducer (Hazel et al., 1995; Hazel & Kalf, 1996). Cells were cultured at 37°C, 5% CO_2 with biweekly replacement of medium and adjustment of cell concentration to $2.5X10^5$ cells/mL for optimal growth. Cell number was determined with a hemocytometer and cell viability was assessed by trypan blue exclusion. Cultures were demonstrated to be free of Mycoplasma contamination by periodic testing of the culture supernatant with a radiolabeled Mycoplasma cDNA probe.

Assessment of Differentiation

Morphological assessment of granulocytic (neutrophil) differentiation was performed on Cytospin prepared May-Grunwald/Giemsa-stained cell monolayers. The slides were examined under oil at 100X. Total differentiation (Figure 1) was defined as the combined percentages of promyelocytes, myelocytes, metamyelocytes and banded/segmented, or terminally differentiated cells, and was determined by averaging the percentages of total differentiation out of 200 cells on each of triplicate slides. Terminal differentiation was assessed by counting only segmented granulocytes. Maturation was determined by assessing the number of myelocytes that matured to segmented granulocytes.

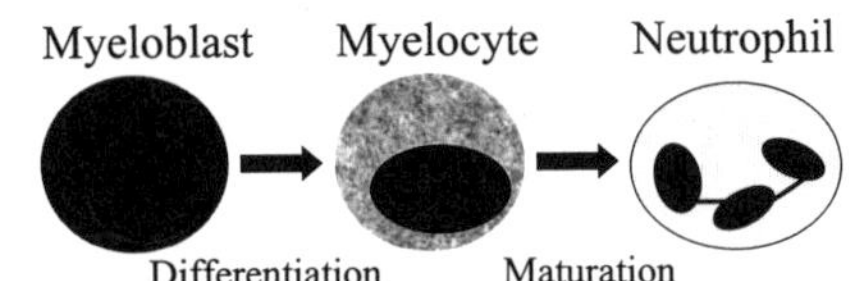

Figure 1. Differentiation and maturation of myeloblasts.

The induction of terminal granulocytic differentiation was verified by demonstrating that G-CSF or HQ caused the induction of various functional properties of granulocytes (Figure 2) such as superoxide production (measured as nitroblue tetrazolium (NBT) reduction), the expression of a specific enzyme (i.e., chloroacetate esterase), and the appearance of ganulocytic-specific cell surface antigens. No monocytes were induced. IL-3, while sustaining survival and proliferation of the myeloblasts, did not induce differentiation.

Assessment of Apoptosis

The presence of apoptosis was assessed by ascertaining the acquisition of micronuclei, surface blebbing and apoptotic bodies. These changes occurred too rapidly to use as a quantitative morphologic marker of apoptosis. Internucleosomal fragmentation manifested by a DNA ladder upon gel electrophoresis was also determined. DNA was extracted from

2×10^5 cells with DNAzol, purified by 2 ethanol precipitations, solubilized in NaOH, and quantified spectrophotometrically. The presence of DNA ladders was ascertained in a 1.5 % agarose/EtBr gel. DNA ladders were visualized in comparison with a 1Kb standard DNA ladder by exposure of the EtBr-intercalated DNA to ultraviolet light.

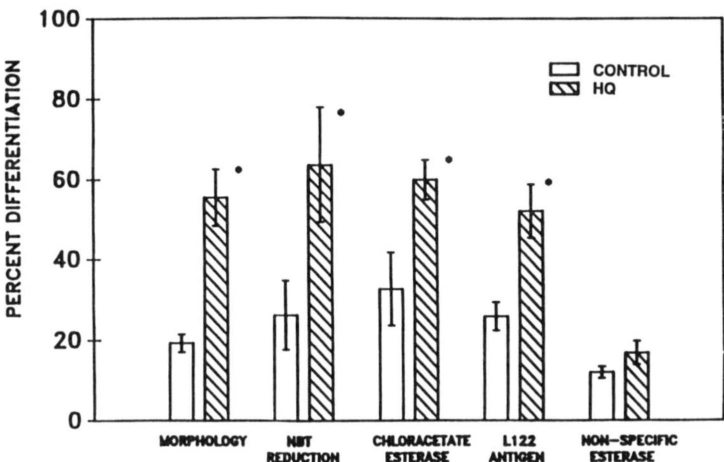

Figure 2. HQ-induced granulocytic differentiation in myeloblasts. Cells (2.5×10^5 mL) were incubated with 2 μM HQ in PBS/2 mM glucose (PBS-A) for 30 min at 37°C, 5 % CO_2. The cells were collected, suspended in IMDM with 10 % FBS/0.13U/ mL IL-3 and incubated for 7 d. The cells were tested for granulocytic differentiation by measuring the characteristics presented in the figure The value represents the mean ± SD of three experiments. Reprinted from Hazel et al., (1995) with permission of AlphaMed Press.

Assay of Protease Activity of Caspace-3

Cloned caspace-3 was incubated in IL-1b converting enzyme assay buffer with DEVDAMC peptide substrate with or without 2.5 μM benzoquinone (BQ) in 100 μL of reaction mixture (Femandez-Alnemri et al., 1995). The release of AMC (7-amino-4-methyl-coumarin) was measured by spectrofluorometry.

Treatment of Myeloblasts with HQ, G-CSF or Leukotriene D₄ (LTD₄)

32D Myeloblasts (2.5×10^5/mL) were cultured in IMDM/10% FBS/2mM L-glutamine (incubation medium) supplemented with 3 U/mL rMuIL-3 and/or inducing agent. For HQ-induced differentiation, the cells were pretreated with 2μM HQ in PBS-A for 30 min at 37°C, collected, washed with PBS, and placed into culture. In some experiments the HQ was added to the culture medium. In that case twice the concentration of HQ was required to cause a comparable result due to the propensity of HQ to bind to proteins in the medium. rHuG-CSF, LTD₄, or the LTD₄ receptor antagonist, MK-571, was added to the culture medium at time zero in amounts indicated in the figure legends. Conditions for LTD₄ treatment were altered in experiments requiring concomitant addition of LTD₄ and HQ, in which case the cells were pretreated with HQ in PBS with no effect on the differentiation observed. After incubation for 5 to 7 days, a sample of each culture was removed for cell counting, determination of viability and slide preparation. The concentrations of inducers used were not cytotoxic and the viability of the cells after treatment was > than 98%.

Statistical Analysis

Data between groups were analyzed using Student's t-test. A p $\leq$ 0.01 was considered significant. Each experiment was repeated at least two times with similar or identical results. In figures where the individual data points do not show error bars, it is because the variation was too low to be indicated by the plot scale used.

RESULTS AND DISCUSSION

HQ-Induced Differentiation of Myeloblasts

HQ (1-2 μM) was shown to induce total granulocytic differentiation in myeloblasts. (Table 1, Figure 2). As previously reported, the putative inducing agent is the bioreactive intermediate, BQ, derived from the myeloperoxidase-mediated oxidation of HQ in the myeloblast (Hazel & Kalf, 1996).

G-CSF, induces myeloblast differentiation by the up-regulation of the 5-lipoxygenase (LPO) pathway for the production of LTD_4, the intracellular mediator of G-CSF transmembrane signaling (Miller, Weiner & Ziboh, 1986; Snyder & Desforges, 1986; Ziboh et al., 1986). With the use of highly specific LPO inhibitors we determined that BQ did not up-regulate the LPO pathway for the production of LTD_4 in the myeloblast (Hazel & Kalf, 1996). Because of the ability of BQ to interact covalently with proteins containing sulfhydryl groups, it was anticipated that BQ might directly activate the LTD_4 receptor thus obviating the requirement for G-CSF or LTD_4.

Concomitant addition of the specific ligand-binding site receptor antagonist, MK-571 with HQ or G-CSF showed inhibition of differentiation induced by both HQ and G-CSF (Table 2).

Table 1. Induction of granulocytic differentiation in myeloblasts by HQ

System	Morphology	NBT reduction
	Percent of cells counted	
Control, IL-3 only	7.5 ± 3.2	4.3 ± 0.6
+ G-CSF	30.8 ± 2.5	36.7 ± 1.7
+ HQ	67.8 ± 3.3	66.8 ± 1.4

Cells (2.5×10^5/mL) were pretreated with 2 μM HQ in PBS-A or PBS-A only for 30 min at 37° C. The cells were harvested, suspended in IMDM containing 3 U/mL rMuIL-3 and 10 % FBS. rHuG-CSF (0.15ng/mL was added to the PBS only cells. After 3 days the medium was changed, the cells diluted to the original number and the incubation continued for 4 additional days and granulocytic differentiation assessed. The values represent the mean SD of the results of triplicate incubations. Data are presented as the percentage of cells showing differentiation out of a total of 200 cells.

As can be seen in Figure 3, inhibition of HQ or LTD_4–induced differentiation by the LTD_4 receptor antagonist, MK-571, was concentration dependent. Higher concentrations of antagonist were required to inhibit HQ-induced differentiation, suggesting that HQ binds more tightly to the ligand binding site on the receptor than does LTD_4.

System	Morphology	NBT Reduction
	Percent of cells counted	
Control	7.5 ± 2.6	12.5 ± 3.6
+ G-CSF	70.8 ± 3.5	48.2 ± 16.6
+ G-CSF + R antagonist	32.8 ± 6.8	26.8 ± 6.2
+ HQ	81.5 ± 1.3	48.6 ± 5.7
+ HQ + R antagonist	24.1 ± 1.8	15.2 ± 2.5

Cells (2.5 × 10^5/mL) were pretreated with a final concentration of 2 μM HQ or treated with rHuG-CSF (0.15ng/mL) as described in Table 1. The LTD$_4$ receptor antagonist, MK 571 was added to each appropriate sample at a 1 μM final concentration. Morphogical analysis was performed as described in the legend for Table 1.

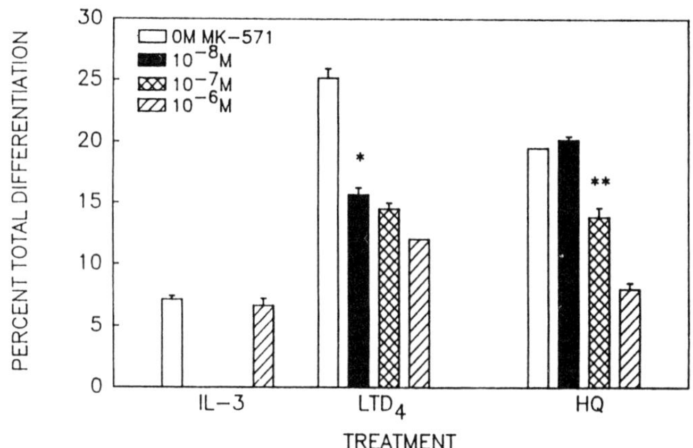

Figure 3. Effect of an LTD$_4$ receptor antagonist (MK-571) on granulocytic differentiation induced in myeloblasts by LTD$_4$ or HQ. Myeloblasts, incubated for 15 min in the presence or absence of various concentrations of MK-571, were treated with 4 μM LTD$_4$ or 2 μM HQ and incubated for an additional 30 min. The cells were collected, suspended in incubation medium and allowed to undergo differentiation for 5 d. Cytospin preparations were made, stained with May Grunwald/Giemsa stain and analyzed morphologically for the percentage of total differentiation by determining the number of differentiated cells (promyelocytes and higher differentiated forms) in a population of 200 cells on each of triplicate slides. Where error bars are not seen on individual data points, the SDs were too close to plot. * Significance at p ≤ 0.001 when compared to results obtained with LTD$_4$. ** Significance at p≤0.001 when compared with HQ treatment alone. Reprinted from Hazel and Kalf (1996) with permission of Humana Press, Inc.

HQ, compared to LTD$_4$ showed a qualitatively similar concentration-dependent induction of myeloblast differentiation, (Figure 4) allowing for comparison of the nature of the differentiation induced by the binding of each of these ligands to the LTD$_4$ receptor (Figure 3).

G-CSF, LTD$_4$ and HQ each induced approximately 97% total differentiation, whereas HQ induced significantly less terminal differentiation than did either G-CSF or LTD$_4$ (Figure 5).

The fact that both HQ and LTD$_4$ putatively work through the same receptor, but cause significantly different levels of terminal differentiation led us to morphologically analyze/compare the kinetics of stage-specific granulocytic differentiation induced by these agents over a 6-day period (Figure 6). Differentiation induced by LTD$_4$ showed predominately segmented granulocytes at Days 2 through 6, whereas, differentiation induced by HQ was arrested at the myelocyte stage, and showed few mature granulocytes.

Receptor antagonist effectively competed with LTD$_4$ for the receptor as measured by total differentiation induced (data not presented), however, only the highest level of antagonist could compete with HQ for the receptor suggesting that BQ may be binding covalently to and constitutively activating the receptor.

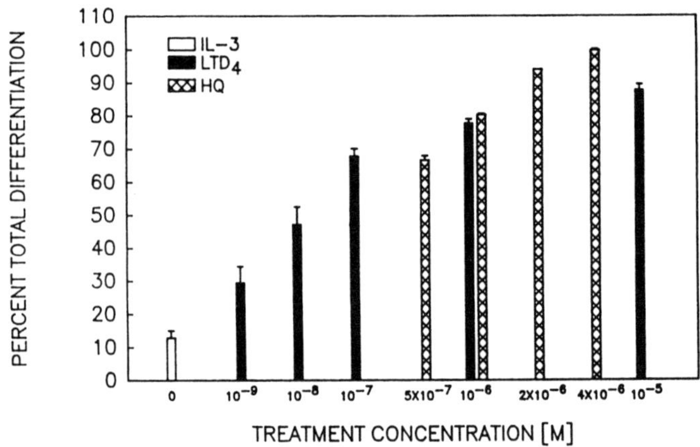

Figure 4. Concentration-dependent induction of granulocytic differentiation in myeloblasts induced by LTD$_4$ or HQ. Myeloblasts (2.5 × 10^5 mL) were pretreated with PBS-A or PBS-A and HQ as indicated in the figure for 30 min at 37° C. The cells were collected and suspended in FBS/ 2 mM L-glutamine/3U/mL IL-3. The cells that received PBS-A only were treated with 4 µM LTD$_4$. All cells were incubated for 4 d, harvested, and analyzed for differentiation as described in the legend for Figure 2. Reprinted from Hazel and Kalf (1996) with permission of Humana Press, Inc.

HQ Inhibition of Apoptosis

When HQ, at a concentration that induces granulocytic differentiation, is added to a culture of myeloblasts growing in the presence of IL-3, a 2- to 3-fold increase in the number of myeloblasts is seen. This increase does not result from an ability of HQ to synergize with IL-3 to increase its proliferative signal (data not presented). The relevant question then becomes whether the increase in the number of myelocytes can be attributed only to the intrinsic proliferative capacity of the myelocyte or, in addition, to an ability of HQ to prevent or delay apoptosis. As can be seen in Figure 7, HQ, at concentrations that induce differentiation in myeloblasts, inhibits myelocyte apoptosis induced by IL-3 withdrawal.

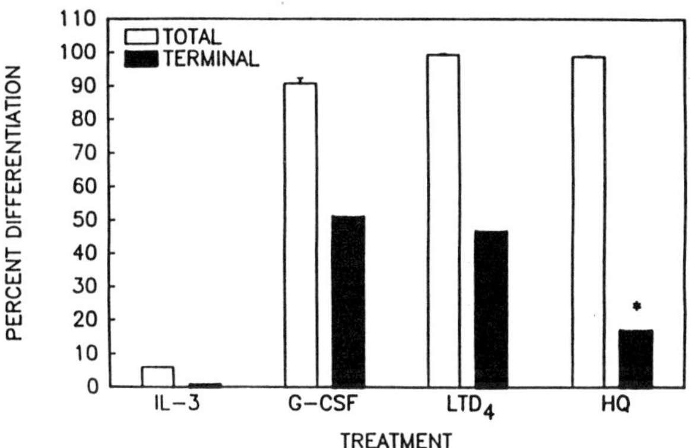

Figure 5. Effects of inducing agents on terminal granulocytic differentiation in myeloblasts. Myeloblasts were treated with PBS-A only or with PBS-A and 2 μM HQ for 30 min at 37° C. The cells were collected, suspended in incubation medium. Cells pretreated with PBS-A only received 3 U/mL IL-3, 500 U/mL G-CSF, or 4 μM LTD_4 Cells were incubated for 6 d then treated as described in the legend for Figure 2. The percentage of terminal differentiation was calculated using the ratio of terminally differentiated granulocytes (band and segmented forms) to the total number of differentiated cells counted. * Indicates significance at the p≤ 0.001 when compared to the G-CSF and LTD_4 results. Reprinted from Hazel and Kalf (1996) with permission of Humana Press, Inc.

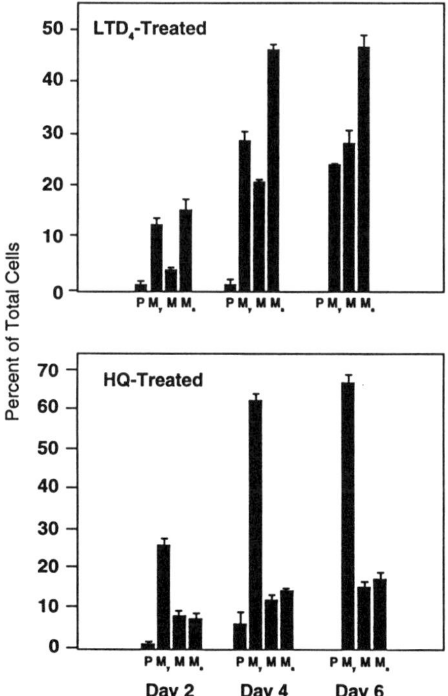

Figure 6. Comparison of stage-specific differentiation of myeloblasts induced by LTD_4 or HQ. Myeloblasts (2.5 × 10^5 /mL) were treated, morphologically analyzed, and the results calculated as described in the legend for Figure 5. Samples of each culture were collected every 2 d for the determination of stage-specific granulocytic differentiation. P, promyelocyte; M_y, myelocyte; M, metamyelocyte; M_a, mature band and segmented forms.

BQ has been shown to inhibit the activity of IL-1β converting enzyme (ICE), a member of a family of caspace proteases involved in the final steps in the regulation of apoptosis (Niculescu et al., 1995). In the myeloid lineage, BQ (2µM) was shown to inhibit recombinant caspace-3, a cysteine protease that plays a major role in programmed cell death in myeloid progenitors, by 50 per cent in a kinetic assay involving a specific fluorescent peptide containing the active site sequence of the caspace family of proteases (Figure 8). These results suggest that HQ, via conversion to BQ, inhibits apoptosis in a proliferating population of myeloblasts and early myeloid progenitors.

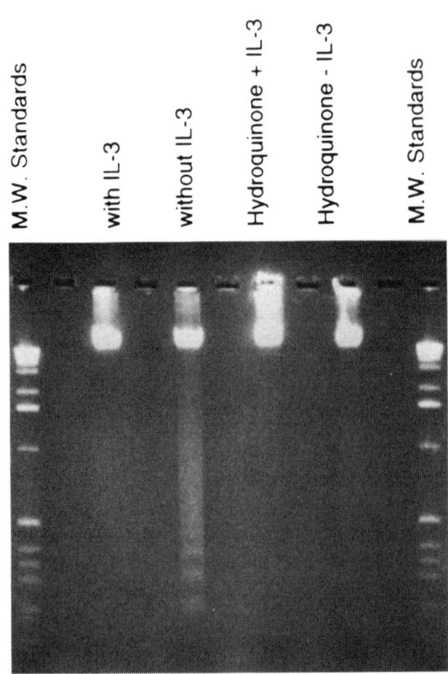

Figure 7. Inhibition of DNA fragmentation in IL-3 deprived myeloblasts by HQ. Myeloblasts (5 × 10⁵/mL were pretreated with 3 µM HQ/PBS-A or PBS-A alone for 1 h at 37°C after which they were suspended in IMDM/10% FBS/2mM glutamine in the presence or absence of 3U/mL IL-3 and incubated for 18 h. Cells were collected (1-2 × 10⁷), and DNA was extracted and solubilized in NaOH as described under Methods. DNA (1µg) was loaded per lane in a 1.5 % agarose gel containing 1 mg/mL EtBr and electrophoresis was carried out in TAE buffer for 1 h, 40 min at 100v. Reproduced from Hazel, Baum and Kalf (1996) with permission of AlphaMed Press.

Inhibition of Maturation of Myelocytes by HQ

HQ, when present concomitantly with LTD$_4$, appears to have the ability to inhibit maturation of myelocytes such that the differentiation profile seen in the presence of LTD$_4$ shifts from one of terminal granulocytes to proliferating myelocytes. (Figure 9). In the bone marrow, HQ may out-compete LTD$_4$ for the LTD$_4$ receptor and thus may induce myeloblasts to differentiate into an expanding clone of myelocytes that is incapable of both programmed cell death and maturation to mature granulocytes. However, the ability of HQ to block the maturation of myelocytes induced by G-CSF cannot be explained by the interaction of HQ with the LTD$_4$ receptor.

Topoisomerase II (topo II) has been reported to be involved in differentiation and maturation of leukemia cells (Baez and Supulveda, 1992; Constantinou et al., 1992) Additionally, as maturation of HL-60 promyelocytic cells progresses topo II levels steadily decrease and are undetectable in mature granulocytes (Kaufmann et al., 1991).

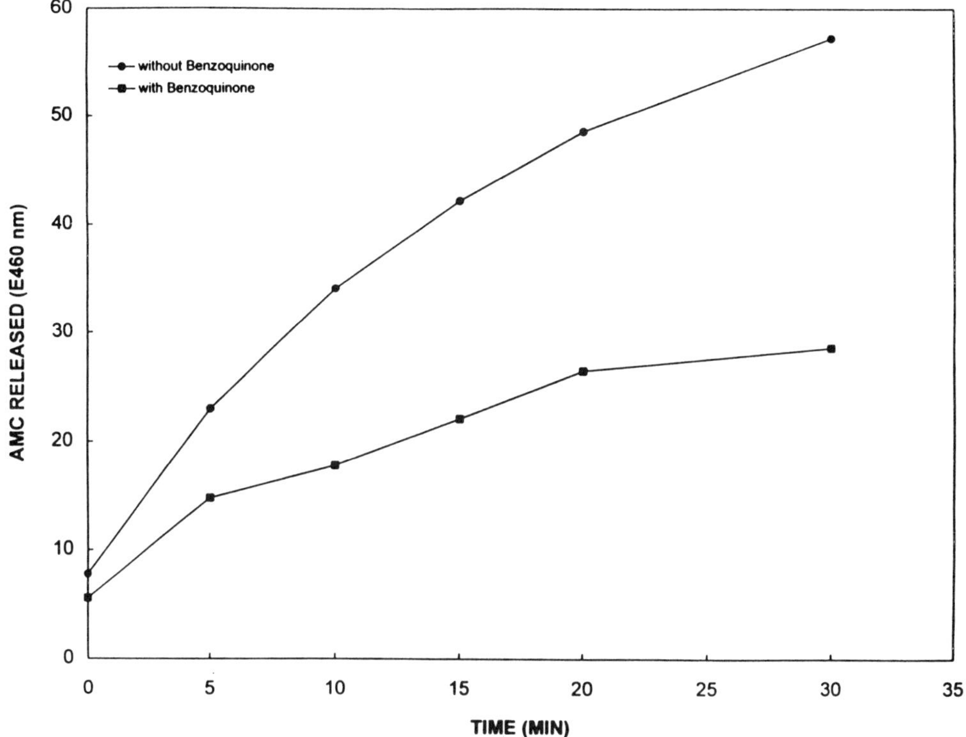

Figure 8. Inhibition of the apoptotic protease caspace-3 by BQ. DEVD-AMC peptide (50 μM) was incubated with cloned caspace-3 (1 μg) in buffer with or without 2.5 μM BQ. Protease activity was measured by following the release of AMC (7-amino-4-methylcoumarin) over a 30 min period spectrofluorometrically. Control buffer gave a value of 3.9 in both cases. Reprinted from Hazel, Baum and Kalf (1996) with permission from AlphMed Press.

Both HQ and BQ have been shown to inhibit topo II activity in vitro (Hutt and Kalf, 1996; Franz, Chen and Eastman, 1996). We compared the ability of HQ and etoposide, a topo II inhibitor, both induce differentiation in myeloblasts and prevent maturation of myelocytes in comparison with etoposide, a topo II inhibitor. As can be seen in Table 3, HQ on its own and with G-CSF induced differentiation in myeloblasts and prevented myelocytes from maturing, whereas the topo II inhibitor did not induce differentiation in myeloblasts, but prevented maturation of myelocytes.

In summary, BQ, the highly bioreactive metabolite of benzene, appears to interact in the myeloid stem cell with: 1) the LTD$_4$ receptor protein to induce the myeloblast to differentiate into proliferating myelocytes; 2) caspace-3, a protease essential for physiological cell death to allow the myelocytes to clone out; and 3) topo II, an enzyme known to play a role in gene expression, to prevent maturation of myelocytes to mature granulocytes. Thus HQ causes a series of events similar to those observed in AML which is characterized by clonal expansion of immature myeloid progenitor cells (Lowenberg and Delwel, 1991) and confers upon the myelocyte some of the phenotypic features of a leukemia cell in culture.

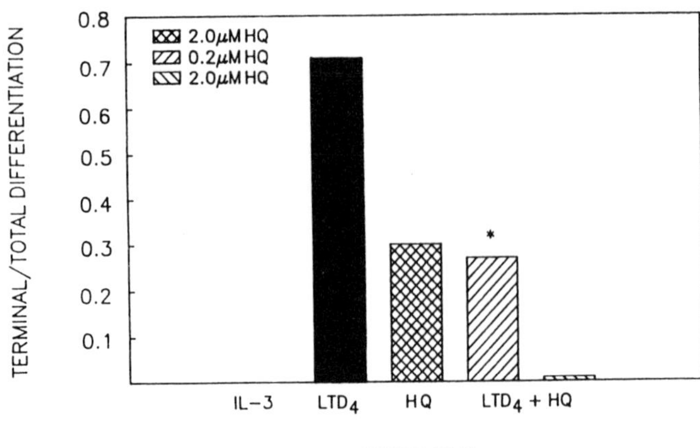

Figure 9. Concentration-dependent inhibition of LTD$_4$-induced granulocytic differentiation in myeloblasts by HQ. Myeloblast (2.5 × 10^5/mL) were treated with PBS-A only, PBS-A and LTD$_4$ (4 μM), HQ (2 μM), or LTD$_4$ (4 μM) concomitantly with either 0.2 or 2 μM HQ for 30 min at 37° C. The cells were collected, suspended in incubation medium as described in the legend for Figure 3. Cytospin preparations were made and triplicate slides were morphologically analyzed for differentiation as described in the legend for Figure 2. In each case SD values used in the determination of ratios were ±5 %. *Indicates significance a p $\leq$ 0.001 when compared with the results of treatment with LTD$_4$ alone. Reprinted from Hazel and Kalf with permission of Humana Press, Inc.

Table 3. A Comparison of the Inhibition of Maturation of Myelocytes by HQ and the Topo II Inhibitor, Etoposide

System	Differentiation	Maturation
	Percent	
Control (IL-3)	8.6 ± 2.0	1.8 ± 0.7
+ G-CSF	89.9 ± 6.8	61.1 ± 7.9
+ HQ	68.8 ± 4.9	13.7 ± 3.6
+ G-CSF + HQ	83.2 ± 4.2	12.8 ± 2.8
+ G-CSF + ETO	9.8 ± 4.8	9.2 ± 1.1

Cells 2.5 × 10^5/mL were incubated with 4 μM HQ or 0.05 μM etoposide (ETO) for 6 d. rHuG-CSF was added at 500 U/mL Cells were fixed, stained and morphologically analyzed as described in Methods.

REFERENCES

Baez, A., and Sepulveda, J., 1992, Myeloid differentiation of HL-60 cells induced by the anti-tumor drug 3-nitrobenzothiazolo[3,2-a]quinolinium, *Leuk. Res.*16: 363-370.

Constantinou, A., Grdina, D., Kiguchi, K., and Huberman, E., 1992, The effect of topoiomerase inhibitors on the expression of differentiation markers and cell cycle progression of human K-562 leukemia cells, *Exp. Cell Res.* 203:100-106.

Femandez-Alnemri, T., Litwack, G., and Alnemri, E., 1995, Mch-2, a new member of the apoptotic Ced-3/ICE cysteine protease gene family, Cancer *Res.* 55:2737-2742.

Frantz, C. E., Chen, H., and Eastmond, D. A, 1996, Inhibition of human topoisomerase II in vitro by bioactive benzene, *Environ. Health. Perspect.* Suppl.6, 104: 1319-1322.

Greenberger, J.S., Sakakeeny, M.A., Humphries, R.K., Eaves, C.J., Eckner, R.J., 1983, Demonstration of factor-dependent multipotential (erythroid, neutrophil, basophil) hematopoietic cell lines, *Proc. Natl. Acad. Sci. USA,* 80:2931-2935.

Hazel, B. A., O'Connor, A., Niculescu, R., and Kalf, G.F., 1995, Benzene and its metabolite, hydroquinone induce granulocytic differentiation in myeloblasts by interacting with cellular signaling pathways activated by granulocyte-stimulating factor, *Stem Cells.* 13:295-310.

Hazel, B. A., and Kalf, G.F., 1996, Induction of granulocytic differentiation in myeloblasts by hydroquinone, a metabolite of benzene, involves the leukotriene D_4 receptor. *Recept. Signal Transduct.* 6:1-12.

Hazel, B.A., Baum, C., and Kalf, G.F., 1996, Hydroquinone, a metabolite of benzene inhibits apoptosis in myeloblasts, *Stem Cells.* 14:730-742.

Hutt, A.M., and Kalf, G.F., 1996, Inhibition of human topoisomerase II by hydroquinone and benzoquinone, reactive intermediates of benzene. *Environ. Health Perspect..* Suppl, 6,104: 1265-1269.

Irons, R.D., and Stillman, W.S., 1993, Cell proliferation and differentiation in chemical leukemogenesis, Stem *Cells.* 11:235-242.

Kaufmann, S.H., McLaughlin, S.J. Kastan, M.B., Liu, L.F., Karp, J.E., and Burke, P.J., 1991, Topoisomerase II levels during granulocytic maturation in vivo and in vitro, *Cancer Res.* 51:3534-3543.

Lowenberg, B., and Delwel, F.R., 1991, The Pathology of human acute myeloid leukemia, In: *Hematology, Basic Principles and Practice,* Hoffman, R., Benz, E.J., Shattil, S.J., Furie, B., and Cohen, H.J. (eds.) Churchill Livingstone, New York, pp 708-715.

Metcalf, D., 1985, Multi-CSF-dependent colony formation by cells of a murine hematopoietic cell line specificity and action of multi-CSF, Blood. 65:357-362.

Miller, A.M., Weiner, R.S., and Ziboh, V.A., 1986, Evidence for the roles of leukotienes C_4 and D_4 as essential intermediates in CSF-stimulated human myeloid colony formation, *Exp. Hematol.* 14:760-765.

Niculescu, R., Bradford, H.N., Colman, R., and Kalf, G.F., 1995, Inhibition of the conversion of pre-interleukins-lα and -1β to mature cytokines by p-benzoquinone, a metabolite of benzene, *Chem. Biol. Interact.* 98:211-222.

Niculescu, R., and Kalf, G.F., 1995, A morphological analysis of the short-term effects of benzene on the development of the hematological cells in the bone marrow of mice and the effects of interleukin-1 on the process, *Arch. Toxicol.* 69:141-148.

Snyder, D.S., and Deforges, J.F., 1986, 5-Lipoxygenase metabolites of arachidonic acid modulate hematopoiesis, *Blood.* 67:1675-1679.

Snyder, R., and Kalf, G.F., 1994, A perspective on benzene leukemogenesis,. *Crit. Rev. Toxicol* 24:177-209.

Valtieri, M., Tweardy, D.J., Caracciolo, D., Johnson, K., Mavilio, F., Altmann, S., Santoli, D., Rovera, G., 1987, Cytokine-dependent granulocytic differentiation. Regulation of proliferative and differentiative responses in a murine progenitor cell line,. *J. Immunol.* 138:3829-3835.

Ziboh, V. A., Kobb, S.M., and McTiernan, 1990, Regulation of HL-60 cell differentiation by lipoxygenase pathway metabolites in vitro, *Cancer Res.* 50:7257-7260.

HEMATOPOIETIC STEM AND PROGENITOR CELLS AS TARGETS FOR BIOLOGICAL REACTIVE INTERMEDIATES

Richard D. Irons,[1] David W. Pyatt,[2] Sherilyn A. Gross,[3] and Wayne S. Stillman[4]

Molecular Toxicology and Environmental Health Sciences Program
Department of Pharmaceutical Sciences, School of Pharmacy[1,2,4]
Department of Pathology, School of Medicine[1]
Experimental Hematology Laboratory, School of Medicine[3]
Comprehensive Cancer Center[1]
University of Colorado Health Sciences Center
Denver, Colorado USA

INTRODUCTION

Over the past two decades it has become increasingly apparent that xenobiotic agents can directly influence transcriptional regulation of gene expression without producing detectable changes in the structure of DNA. The response of a cell to its environment is a dynamic process in which differentiation-driven paradigms for gene expression can be modulated by both endogenous and exogenous extracellular agents. Typical outcomes include: altered cell function, changes in proliferation and/or regulation of cell growth or cell death. Altered gene expression is often the result of a physiologically programmed interaction between the cell and a regulatory hormone or cytokine. Alternatively, altered gene expression can result from an interaction of the cell with a xenobiotic molecule. In many cases the consequences of these apparently disparate events, as well as their biochemistry, do not appear to be remarkably different. Our laboratory has been investigating the role of altered transcriptional regulation in the response of hematopoietic and immune cells to xenobiotic agents as one approach to understanding molecular mechanisms of toxicity and also with a view toward exploring whether altered patterns of transcriptional regulation can be used as a tool for predicting the target organ toxicity of xenobiotic agents. Results of these studies suggest that the effects of biological reactive intermediates on transcriptional regulation of gene expression are cell-type specific and that the response of individual target cells is at least in part determined by differentiation-dependent patterns of signal transduction and gene expression.

Biological Reactive Intermediates VI, Edited by Dansette *et al.*
Kluwer Academic / Plenum Publishers, 2001

CELL-SPECIFIC PATTERNS OF DIMETHYLDITHIOCARBAMATE-MEDIATED TOXICITY

Dithiocarbamates are Potent Immunosuppressive Agents

Dithiocarbamates (DTC) are a class of thiono-sulfur containing compounds that exhibit complex chemical and biological properties. Dithiocarbamates are potent but selective inhibitors of a variety of enzymes including: cytochrome P450 IIE1 (Guengerich et al., 1991), cholesterol 7a-hydroxylase (Zon et al., 1991), superoxide dismutase (Heikkila et al., 1978), prostaglandin synthase (Marnett et al., 1977) and aldehyde dehydrogensase (Hald and Jacobsen 1948), among others. Dithiocarbamates are also potent metal chelators and ionophores (Fredga 1950). Although DTC have been generally regarded as anti-oxidants, they can produce both pro-oxidant and anti-oxidant effects, demonstrating the potential to simultaneously inhibit hydroxyl radical formation and to oxidize protein thiols in biological systems (Fredga 1950). The literature is conflicted over the effects of DTC on immune response, these compounds variously having been reported to be "immunorestorative," "immunostimulatory," or "immunosuppressive," depending on the study (Reviewed in Irons & Pyatt (Irons and Pyatt 1998)). The initial suggestion that DTC might be "immunorestorative" apparently was based on reports indicating that administration of DTC ameliorated the immunosuppressive effects of cyclophosphamide in mice. Unfortunately, these studies did not control for the inhibition of enzymatic activation of cyclophosphamide by DTC which may account for these observations. Alternatively, DTC have been reported to produce immunosuppression in mice following both oral and dermal exposure (Padgett et al., 1992; Pruett et al., 1992). Several reports have characterized the ability of pyrolinedithiocarbamate (PDTC) to alter transcriptional regulation in T lymphocytes through the inhibition of activation of the transcription factor, nuclear factor kappa B (NF-kB) (Schreck et al., 1992; Ziegler-Heitbrock et al., 1993). In general, DTC have been shown to be potent inhibitors of NF-kB in a variety of cell types (Brewton et al., 1989; Pyatt et al., 1998; Brennan and O'Neill 1996). Cyclosporin A (CsA) is an immunosuppressive agent and well known inhibitor of NF-κB and nuclear factor of activated T lymphocytes (NFAT) in T lymphocytes. Inhibition of these two transcription factors in T lymphocytes is accompanied by a suppression of T lymphocyte activation and suppression of cell-mediated immune response. Since previous studies have indicated that PDTC is a potent inhibitor of NF-kB in T lymphocytes (Schreck et al., 1992; Ziegler-Heitbrock et al., 1993) we decided to examine the effects of dimethyldithiocarbamate (DMDTC) and diethyldithiocarbamate (DEDTC) on transcriptional regulation and cytokine production in human CD4+ T lymphocytes (Pyatt et al., 1998).

DTC Inhibit Activation of NF-kB, NFAT and Cell Activation in Human T Lymphocytes

Activation of NF-kB is a requirement for activation of CD4+ T lymphocytes, and is known to occur following exposure of T cells to a variety of stimuli including: antigen stimulation, anti-CD3 administration, ligand interaction with CD28, exposure to calcium ionophore, PHA, lectins, oxygen species, TNF-α and virus (Bagasra et al., 1992; Osborn et al., 1989; Thelen and Wirthmueller 1994; Oakes et al., 1994; Antoni et al., 1994). Activation of CD4+ T lymphocytes is normally accompanied by a rapid down-regulation of CD4+ surface expression, increased expression of IL-2 receptor, and production of IL-2 in responding T cells. DMDTC inhibits NF-kB and NFAT in human CD4+ lymphocytes (Figure 1)(Pyatt et al., 1998). This alteration in transcriptional regulation, which is identical to that observed for CsA, is also accompanied by a decrease in production of IL-2 and inhibition of IL-2R expression. Particularly noteworthy is the potency of DMDTC inhibition of lymphocyte activation, occurring over a concentration range of 0.1-0.5 uM.

However, down-regulation of CD4+ expression, which is a dynamic event associated with early signalling in helper T lymphocyte activation, occurs in the presence of DMDTC, suggesting that DTC inhibition of T lymphocyte activation is not associated with general cytotoxicity.

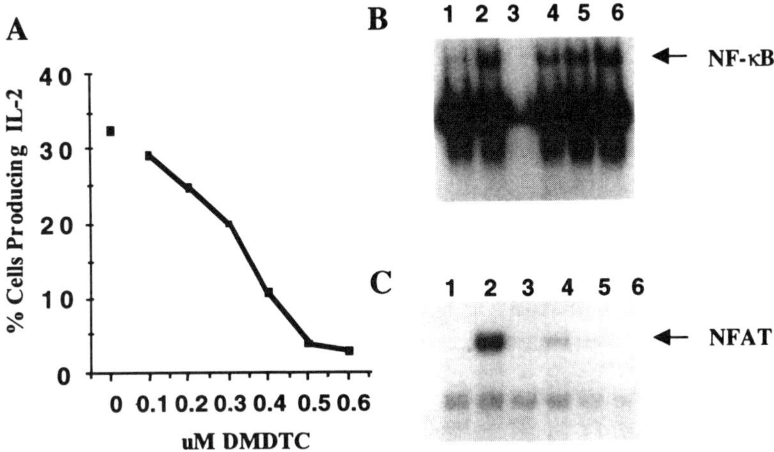

Figure 1. Comparison of the effects of DMDTC on IL-2 production and activation of NF-kB and NFAT in primary CD4+ human lymphocytes. A) Effect of DMDTC on IL-2 production; B) on activation of NF-kB; and C) NFAT (Pyatt et al., 1998).

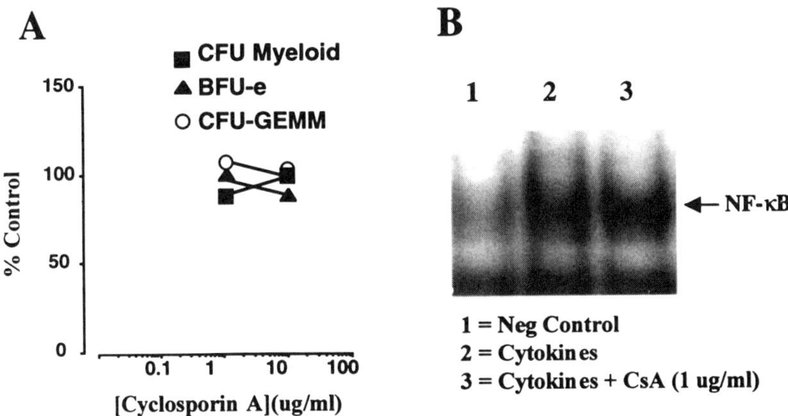

Figure 2. Effect of Cyclosporin A on (A) Clonogenic response and activation of NF-kB in cytokine-stimulated CD34+ human bone marrow cells. (Pyatt et al., 1998).

DTC are Myelotoxic in Human Bone Marrow Cells

Controversy also exists regarding the myelotoxicity of DTC. Some authors have reported DTC to be myeloprotective (East et al., 1992; Schmalbach and Borch 1989a; Schmalbach and Borch 1989b), while others have found them to be myelotoxic (Thompson et al., 1982; Shenep et al., 1994; Corke 1984; Tsvetkova et al., 1992). Additionally, bone marrow and hematopoietic toxicity have also been reported as a side effect of many thiono-sulfur-containing compounds (Neal and Halpert 1982; Tsvetkova et al., 1992), and suppression of neutrophil, monocyte and platelet counts have been reported following DEDTC treatment of HIV-infected subjects (Shenep et al., 1994). We have previously demonstrated an essential role for NF-kB in human CD34+ bone marrow cell survival (Pyatt et al., 1999). Therefore, we compared the effects of CsA and a variety of DTC on transcription factor activation and function in human CD34+ bone marrow cells. Cyclosporin A is often considered a "prototype" inhibitor of NF-κB. However, in sharp contrast to its effects on T lymphocytes, CsA has no effect on either activation of NF-kB or clonogenic response in bone marrow progenitor cells (Figure 2).

DTC also do not inhibit NF-kB in CD34+ bone marrow cells. However, both DMDTC and DEDTC are potent inhibitors of clonogenic response (Figure 3). DTC stimulate activation of AP.1 in cytokine-treated bone marrow cells (Figure 4). It is known that various pro-oxidant stimuli can activate AP.1 through the up-regulation of *fos* (Aragonés et al., 1996). This is consistent with our observations that indicate c-*fos* is the predominant AP-1 member expressed. The role of AP.1 activation in DTC myelotoxicity is not completely understood, but activation of this dimeric transcription factor suggests that DTC are capable of inducing a pro-oxidant signal in bone marrow cells.

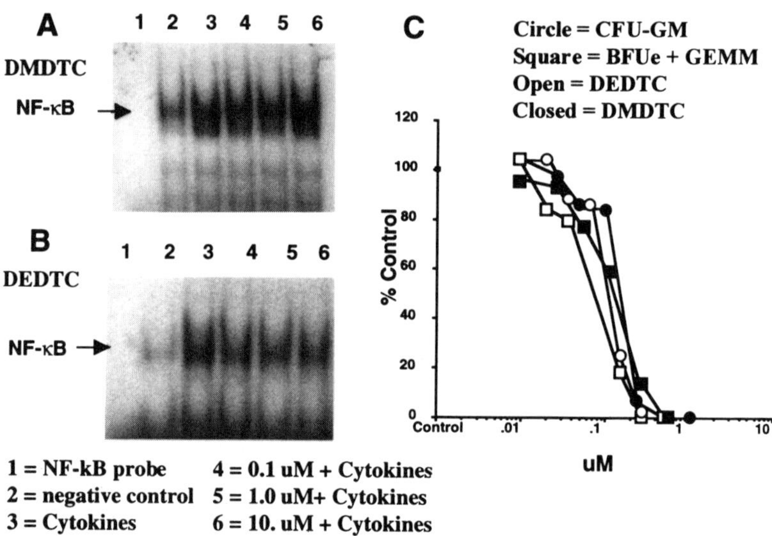

Figure 3. Comparison of the effects of (A) DMDTC and (B) DEDTC on activation of NF-kB and (C) Clonogenic response in cytokine-stimulated human CD34+ bone marrow cells.

DTC Myelotoxicity Appears to be Mediated via Copper

Nobel et al have previously demonstrated that the addition of copper potentiates the toxicity of DTC in thymocytes via a pro-oxidant mechanism (Nobel et al., 1995). Therefore, we examined the effects of copper on DTC induced suppression of cytokine stimulated colony formation. The addition of 1 uM cupric sulfate greatly potentiates the toxicity of DTC in myeloid progenitor cells (Figure 5A) and to a lesser extent in erythroid progenitor cells (data not shown), while the same concentration of copper alone had no observable effect. Further experiments were conducted to investigate the effects of copper chelation on DTC toxicity. The addition of 10 uM of 2,9-dimethyl-4, 7-diphenyl-1, 10-

phenanthroline disulfonic acid (BCPS), a copper specific chelator, resulted in complete protection against the toxicity of DTC (Figure 5B). These findings confirm previous studies that copper plays a critical role in DTC toxicity but also reveal that resulting alterations in the pattern of transcriptional regulation are cell-type specific. These results illustrate the lineage-specific nature of transcriptional regulation and demonstrate that similar, if not identical, mechanisms of toxicity may result in different patterns of altered transcription in different cell populations.

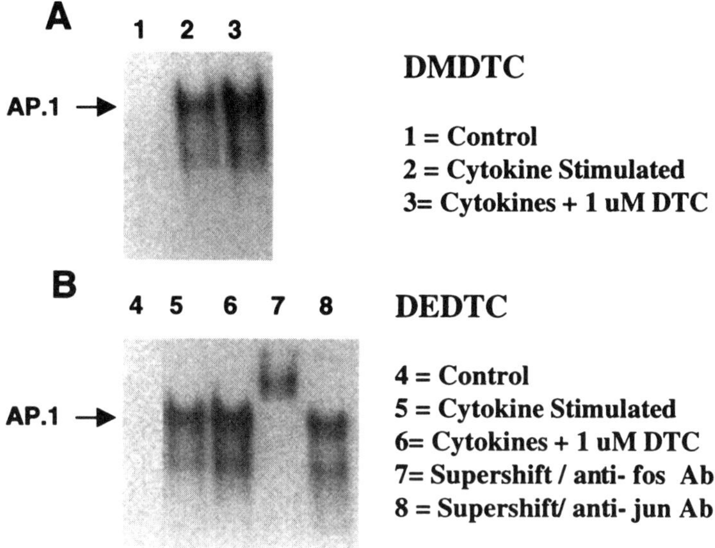

Figure 4. Effects of DMDTC (A) and DEDTC (B) on AP.1 activation in cytokine-stimulated human CD34+ bone marrow cells.

MECHANISMS OF HYDROQUNINONE-INDUCED ALTERATIONS IN GENE-EXPRESSION AND DIFFERENTIATION IN HUMAN CD34+ BONE MARROW CELLS

Clonal Expansion of Myeloid Progenitor Cells Correlates with PU.1-DNA Binding

Over the past twenty years it has become apparent that exogenous agents can directly influence regulation of gene expression without producing any detectable changes in the structure of DNA, and a number of well documented carcinogenic mechanisms lead to changes in gene expression, cell proliferation and subsequent development of neoplasia in the absence of detectable changes in DNA (Schüller 1991). A number of studies have demonstrated increased myeloid differentiation at the expense of erythroid differentiation in animals exposed to benzene (Cronkite et al., 1989; Dempster and Snyder 1990). This observation is associated with an increase in the number of granulocyte/macrophage progenitor cells (CFU-GM)(Cronkite et al., 1989). We have previously demonstrated that exposure of murine or human bone marrow cells to hydroquinone (HQ), a major metabolite of benzene, results in a selective increase in clonogenic response to the hematopoietic cytokine, granulocyte/macrophage-colony stimulating factor (GM-CSF) (Irons et al., 1992; Irons and Stillman 1996) (Figure 6), The ability to act synergistically with GM-CSF to increase clonogenic response of hematopoietic progenitor cells (HPC) is a characteristic shared by a diverse group of leukemogenic agents (Irons and Stillman 1993). The promotion of early myeloid differentiation in HPC increases cell turnover in a cell population that is at risk for replication-dependent error and subsequent events in transformation. The population of HPC cells exhibiting increased clonal expansion in response to HQ is known to be a subpopulation of CD34+ cells in human bone marrow (Irons and Stillman 1996). In order to further characterize this subpopulation we isolated a variety of different CD34+ bone marrow cell phenotypes using multiparameter flow cytometry and high resolution cell sorting, these subpopulations of HPC were then exposed to HQ and assayed for the frequency of CFU-GM forming cells. Results reveal that GM-CSF synergistic response in the presence of HQ is restricted to a population of HPC that are CD34+,CD19-,CD15+. Treatment with HQ also results in an increase in CD15 antigen density on these cells. This phenotype coincides with a multilineage myeloid progenitor cell capable of giving rise to granulocyte, erythroid, myeloid and megakaryocytic lineages (CFU-GEMM) (Ball 1995).

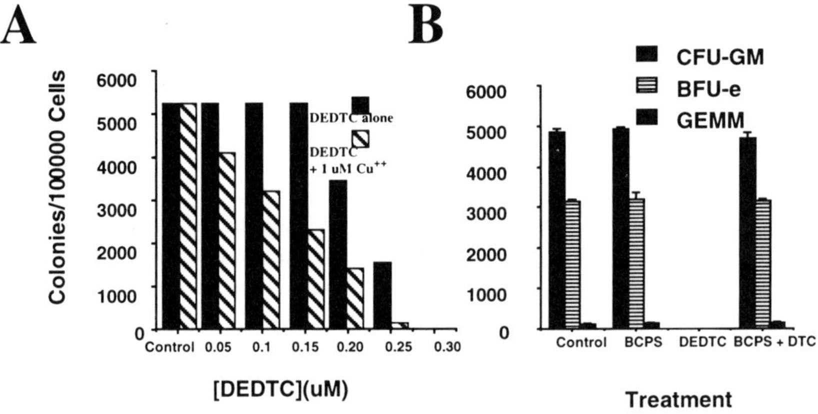

Figure 5. Contrasting effects of copper and copper chelation on DEDTC inhibition of clonogenic response in human CD34+ bone marrow cells. (A) Dose-response of DEDTC on CFU-GM with and without the addition of 1uM cupric sulfate. (B) Effect of BCPS on DEDTC inhibition of clonogenic response.

PU.1 is a member of the Ets family of transcription factors (Klemsz et al., 1990), and is known to be expressed in early lymphoid and myeloid populations (Voso et al., 1994; Fisher and Scott 1998). PU.1 is regarded as a master regulator of myeloid differentiation and is upregulated in myeloid differentiation and downregulated during erythroid differentiation (Voso et al., 1994). This pattern is inverted for the transcription factor, GATA1, which is upregulated in erythroid differentiation and downregulated in myeloid differentiation (Sposi et al., 1992).

We have been investigating the respective roles of PU.1 and GATA 1 in HQ-stimulated myeloid differentiation. Results of these studies reveal that HQ results in a dose-dependent activation of PU.1 in human CD34+ bone marrow cells. PU.1 is also required for CD18 surface expression and alterations in PU.1-DNA binding may lead to alterations in CD18 expression, surface distribution or function (Rosmarin et al., 1995). Consistent with a role for PU.1 in myeloid differentiation, alterations in PU.1 DNA binding following treatment of CD34+ bone marrow cells with HQ are accompanied by changes in the distribution and/or binding affinity of CD18 and a decrease in GATA 1 expression. Postranslational modification of PU.1, primarily associated with its phosphorylation state, is known to be required for PU.1-DNA binding and myeloid differentiation (Celada et al., 1996; Zhang et al., 1996; Rosmarin et al., 1995). Additional studies in our laboratory have identified post-translational changes in the phosphorylation state of PU.1, suggesting that HQ-induced changes in PU.1-DNA binding are most likely the result of altered phosphorylation of PU.1.

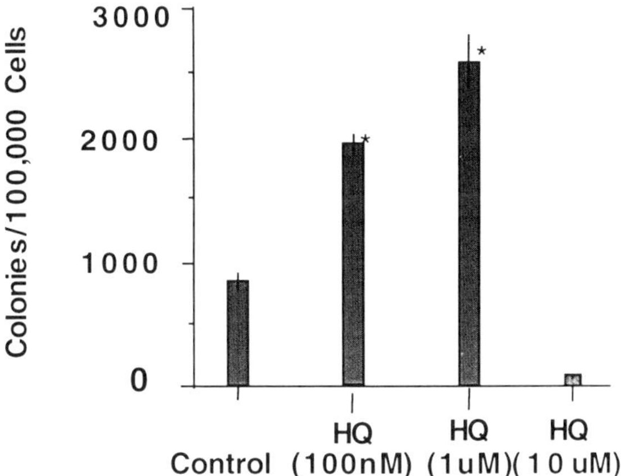

Figure 6. Synergistic effect of HQ on colony formation in human CD34+ bone marrow cells following stimulation with GM-CSF (Irons and Stillman 1996).

PERSPECTIVE

These studies illustrate that the regulation of transcription factor expression and activation in human cells is cell-type specific and largely governed by master paradigms of gene expression important in cell differentiation. As a result, agents that apparently act through similar, if not identical, mechanisms of toxicity can produce different patterns of altered transcriptional regulation in different cells. Therefore, the utility of using altered patterns of transcriptional regulation as a predictor of the effects of biological reactive intermediates appears to be relevant only in the context of target-organ specific toxicity. Alternatively, the systematic analysis of the regulation of transcription in primary human cells provides a powerful tool for characterizing cell-specific mechanisms of altered gene expression in target organ toxicity.

ACKNOWLEDGMENTS

This project was supported by Grant ES06258 from the National Institute of Environmental Health Sciences, National Institute of Health, as well as a grant from the American Chemical Council and the support of the University of Colorado Cancer Center Flow Cytometry Core (Grant 2 P30 CA 46934-09) and the Clinical Investigation Core.

REFERENCES

Antoni BA, Rabson AB, Kinter A, Bodkin M, Poli G. 1994. NF-kB-dependent and -independent pathways of HIV activation in a chronically infected T cell line. *Virology* 202:684-94.

Aragonés J, López-Rodríquez C, Corbí A, Del Arco PG, López-Cabrera M, De Landázuri MO, Redondo JM. 1996. Dithiocarbamates trigger differentiation and induction of CD11c gene through AP-1 in the myeloid lineage. *J.Biol.Chem.* 271:10924-31.

Bagasra O, Wright SD, Seshamma T, Oakes JW, Pomerantz RJ. 1992. CD14 is involved in control of human immunodeficiency virus type 1 expression in latently infected cells by lipopolysaccharide. *Proc.Natl.Acad.Sci.USA* 89:6285-9.

Ball ED. 1995. Introduction: Workshop summary of the CD15 monoclonal antibody panel from the fifth international workshop on leukocyte antigens. *European Journal of Morphology* 33:95-100.

Brennan P, O'Neill LAJ. 1996. 2-mercaptoethanol restores the ability of nuclear factor kappaB (NFkappaB) to bind DNA in nuclear extracts from interleukin 1-treated cells incubated with pyrollidine dithiocarbamate (PDTC) - Evidence for oxidation of glutathione in the mechanism of inhibition of NFkappaB by PDTC. *Biochem.J.* 320:975-81.

Brewton GW, Hersh EM, Rios A, Mansell PWA, Hollinger B, Reuben JM. 1989. AIDS RESEARCH COMMUNICATIONS: A pilot study of diethyldithiocarbamate in patients with acquired immune dificiency syndrome (AIDS) and the AIDS-related comples. *Life Sci* 45:2509-20.

Celada A, Borras FE, Soler C, Lloberas J, Klemsz M, Van Beveren C, McKercher S, Maki RA. 1996. The transcription factor PU.1 is involved in macrophage proliferation. *J.Exp.Med.* 184:61-9.

Corke CF. 1984. The influence of diethyl-dithiocarbamate ('Imuthiol") on mononuclear cells in vitro. *Int.J.Immunopharmacol.* 6:245-7.

Cronkite EP, Drew RT, Inoue T, Hirabayashi Y, Bullis JE. 1989. Hematotoxicity and carcinogenicity of inhaled benzene. *Environ.Health Perspect.* 82:97-108.

Dempster AM, Snyder CA. 1990. Short Term Benzene Exposure Provides a Growth Advantage for Granulopoietic Progenitor Cells Over Erythroid Progenitor Cells. *Arch.Toxicol.* 64:539-44.

East CJ, Abboud CN, Borch RF. 1992. Diethyldithiocarbamate induction of cytokine release in human long-term bone marrow cultures. *Blood* 80:1172-7.

Fisher RC, Scott EW. 1998. Role of PU.1 in hematopoiesis. *Stem Cells* 16:25-37.

Fredga A. 1950. Some observations on tetraethylthiuram disulphide. *Recueil* 69:416-20.

Guengerich FP, Kim D-H, Iwasaki M. 1991. Role of human cytochrome P-450 IIE1 in the oxidation of many low molecular weight cancer suspects. *Chem.Res.Toxicol.* 4:168-79.

Hald J, Jacobsen E. 1948. The formation of acetaldehyde in the organism after ingestion of Antabuse (tetraethylthiuramdisulphide) and alcohol. *Acta Pharmacol.* 4:305-10.

Heikkila RE, Cabbat FS, Cohen G. 1978. Inactivation of superoxide dismutase by several thiocarbamic acid derivatives. *Experientia* 34:1553-4.

Irons RD, Pyatt DW. 1998. Dithiocarbamates as potential confounders in butadiene epidemiology. *Carcinogenesis* 19:539-42.

Irons RD, Stillman WS. 1993. Cell proliferation and differentiation in chemical leukemogenesis. *Stem Cells* 11:235-42.

Irons RD, Stillman WS. 1996. Impact of benzene metabolites on differentiation of bone marrow progenitor cells. *Environ.Health Perspect.* 104 Suppl. 6:1247-50.

Irons RD, Stillman WS, Colagiovanni DB, Henry VA. 1992. Synergistic action of the benzene metabolite hydroquinone on myelopoietic stimulating activity of granulocyte/macrophage colony-stimulating factor *in vitro*. *Proc.Natl.Acad.Sci.USA* 89:3691-5.

Klemsz MJ, McKercher SR, Celada A, Van Beveren C, Maki RA. 1990. The macrophage and B cell-specific transcription factor PU.1 is related to the *ets* oncogene. *Cell* 61:113-24.

Marnett LJ, Reed GA, Johnson JT. 1977. Prostagland in synthetase dependent benzo(A)pyrene oxidation: Products of the oxidation and inhibition of their formation by antioxidants. *Biochem.Biophys.Res.Commun.* 79:569-76.

Neal RA, Halpert J. 1982. Toxicology of thiono-sulfur compounds. *Annu.Rev.Pharmacol.Toxicol.* 22:321-39.

Nobel CSI, Kimland M, Lind B, Orrenius S, Slater AFG. 1995. Dithiocarbamates induce apoptosis in thymocytes by raising the intracellular level of redox-active copper. *J.Biol.Chem.* 270:26202-8.

Oakes JW, Bagasra O, Duan L, Pomerantz RJ. 1994. Association of alterations in NF-kB moieties with HIV type 1 proviral latency in certain monocytic cells. AIDS Res.Hum.Retroviruses 10:1213-9.

Osborn L, Kunkel S, Nabel GJ. 1989. Tumor necrosis factor a and interleukin 1 stimulate the human immunodeficiency virus enhancer by activation of the nuclear factor kB. *Proc.Natl.Acad.Sci.USA* 86:2336-40.

Padgett EL, Barnes DB, Pruett SB. 1992. Disparate effects of representative dithiocarbamates on selected immunological parameters in vivo and cell survival in vitro in female B6C3F1 mice. *J.Toxicol.Environ.Health* 37:559-71.

Pruett SB, Barnes DB, Han YC, Munson AE. 1992. Immunotoxicological characteristics of sodium methyldithiocarbamate. *Fundam.Appl.Toxicol.* 18:40-7.

Pyatt DW, Gruntmeir J, Stillman WS, Irons RD. 1998. Dimethyldithiocarbamate inhibits in vitro activation of primary human CD4$^+$ T lymphocytes. *Toxicology* 128:83-90.

Pyatt DW, Stillman WS, Yang YZ, Gross S, Zheng JH, Irons RD. 1999. An essential role for NF-kappaB in human CD34$^+$ bone marrow cell survival. Blood 93:3302-8.

Rosmarin AG, Caprio D, Levy R, Simkevich C. 1995. CD18 (b$_2$ leukocyte integrin) promoter requires PU.1 transcription factor for myeloid activity. *Proc.Natl.Acad.Sci.USA* 92:801-5.

Schmalbach TK, Borch RF. 1989a. Diethyldithiocarbamate modulation of murine bone marrow toxicity induced by cis-diammine (cyclobutanedicarboxylato) platinum (II). *Cancer Res.* 49:6629-33.

Schmalbach TK, Borch RF. 1989b. Myeloprotective effect of diethyldithiocarbamate treatment following

1,3-Bis(2-chloroethyl)-1-nitrosourea, adriamycin, or mitomycin C in mice. *Cancer Res.* 49:2574-7.

Schreck R, Meier B, Mannel DN, Droge W, Baeuerle PA. 1992. Dithiocarbamates as potent inhibitors of nuclear factor kB activation in intact cells. *J.Exp.Med.* 175:1181-94.

Schüller HM. 1991. The signal transduction model of carcinogenesis. *Biochem.Pharmacol.* 42:1511-23.

Shenep JL, Hughes WT, Flynn PM, Roberson PK, Behm FG, Fullen GH, Kovnar SG, Guito KP, Brodkey TO. 1994. Decreased counts of blood neutrophils, monocytes, and platelets in human immunodeficiency virus-infected children and young adults treated with diethyldithiocarbamate. *Antimicrob.Agents Chemother.* 38:1644-6.

Sposi NM, Zon LI, Carè A, Valtieri M, Testa U, Gabbianelli M, Mariani G, Bottero L, Mather C, Orkin SH et al.,. 1992. Cell cycle-dependent initiation and lineage-dependent abrogation of GATA-1 expression in pure differentiating hematopoietic progenitors. *Proc.Natl.Acad.Sci.USA* 89:6353-7.

Thelen M, Wirthmueller U. 1994. Phospholipases and protein kinases during phagocyte activation. *Curr.Opin.Immunol.* 6:106-12.

Thompson CC, Tacke RB, Woolley LH, Bartley MH. 1982. Purpuric oral and cutaneous lesions in a case of drug-induced thrombocytopenia. *JADA* 105:465-7.

Tsvetkova T, Andonova S, Zvetkova E, Blagoeva S. 1992. Aspects of granulocyte function in workers professionally exposed to pesticides. *Int.Arch.Occup.Environ.Health* 64:275-9.

Voso MT, Burn TC, Wulf G, Lim B, Leone G, Tenen DG. 1994. Inhibition of hematopoiesis by competitive binding of transcription factor PU.1. *Proc.Natl.Acad.Sci.USA* 91:7932-6.

Zhang D-E, Hohaus S, Voso MT, Chen H-M, Smith LT, Hetherington CJ, Tenen DG. 1996. Function of PU.1 (Spi-1), C/EBP, and AML1 in early myelopoiesis: Regulation of multiple myeloid CSF receptor promoters. *Curr.Top.Microbiol.Immunol.* 211:137-47.

Ziegler-Heitbrock HWL, Sternsdorf T, Liese J, Belohradsky B, Weber C, Wedel A, Schreck R, Bauerle P, Strobel M. 1993. Pyrrolidine dithiocarbamate inhibits NF-kB mobilization and TNF production in human monocytes. *J.Immunol.* 151:6986-93.

Zon LI, Youssoufian H, Mather C, Lodish HF, Orkin SH. 1991. Activation of the erythropoietin receptor promoter by transcription factor GATA-1. *Proc.Natl.Acad.Sci.USA* 88:10638-41.

METABOLISM OF 4-(METHYLNITROSAMINO)-1-(3-PYRIDYL)-1-BUTANONE
(NNK) IN A/J MOUSE LUNG AND EFFECT OF CIGARETTE SMOKE EXPOSURE
ON *IN VIVO* METABOLISM TO BIOLOGICAL REACTIVE INTERMEDIATES

A.R. Tricker[1], B.G. Brown[2], D.J. Doolittle[2], and E. Richter[3]

[1] FTR, CH-2003 Neuchâtel, Switzerland,
[2] RJ Reynolds Tobacco Company, Winston-Salem, NC, USA, and
[3] Walther Straub Institute of Pharmacology and Toxicology, Nussbaumstr.
26, D-80336 Munich, Germany

INTRODUCTION

It has been proposed that the tobacco-specific *N*-nitrosamine 4-(methyl-nitrosamino)-1-(3-pyridyl)-1-butanone (NNK) may be involved in the causation of human lung cancer (Hecht and Hoffmann, 1988), but direct evidence for the involvement of NNK in human lung cancer is lacking (Hecht and Tricker, 1999). In the A/J mouse lung tumor model, a single intraperitoneal (i.p.) injection of 10 μmole [2.07 mg] NNK/mouse results in 7-12 lung tumors per mouse after 16 weeks (Hecht et al., 1989). The initial events in NNK-induced A/J mouse lung tumorigenesis are believed to be metabolism of NNK to biological reactive intermediates (BRI) with the potential to react (via methylation of DNA) to produce O^6-methylguanine (O^6MeG), GC→AT transitional mispairing, and subsequent activation of the K-*ras* proto-oncogene (Ronai et al., 1993). α-Hydroxylation of the methylene carbon atoms adjacent to the *N*-nitroso group in NNK and NNAL (Figure 1) yields unstable BRI which spontaneously decompose to methanediazohydroxide with the potential to react with DNA to produce 7-methylguanine (7-MeG), O^4-methylthymidine (O^4MeT) and O^6MeG adducts. α-Hydroxylation of the NNK methyl group yields a BRI with the potential to pyridyloxobutylate DNA, while α-methyl hydroxylation of NNAL is not known to result in DNA adduct formation. Other metabolic transformations represent detoxification pathways for NNK and NNAL.

Inhalation of mainstream cigarette smoke (Finch et al., 1996) and sidestream cigarette smoke (Witschi et al., 1995) fail to induce lung tumors in A/J mice. Furthermore, exposure of A/J mice to mainstream cigarette smoke does not promote NNK-induced tumorigenesis (Finch et al., 1996). However, a tumorigenic response is observed after whole-body inhalation exposure of A/J mice to high doses of 11% mainstream cigarette smoke and 89% sidestream cigarette smoke, used as an experimental surrogate for environmental tobacco smoke (ETS) (Witschi et al., 1997a, 1997b). Although it is possible to completely prevent development of lung tumors in A/J exposed to tobacco smoke, agents otherwise shown to prevent NNK-induced lung tumor formation are ineffective against tobacco smoke (Witschi et al., 2000).

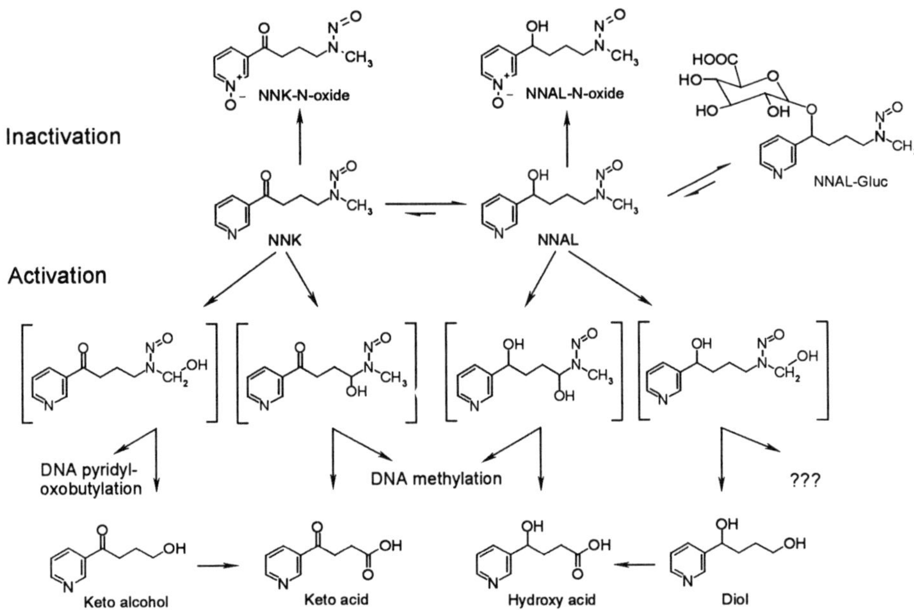

Figure 1. NNK metabolism pathways based on studies in laboratory animals.

IN VITRO METABOLISM OF NNK IN A/J MOUSE AND HUMAN LUNG

In vitro metabolism of 0.003-250 µM [5-^{3}H]NNK was determined under identical experimental conditions in A/J mouse (n=6) and human (n=7) lung tissues in dynamic organ culture (Richter et al., 2000). A/J mouse lung showed a high capacity for metabolic activation of NNK to BRI via α-hydroxylation to keto acid and keto alcohol by low *Km* (high affinity) pathways (Table 1). In contrast, human lung tissues extensively metabolize NNK by reduction to NNAL in the absence of significant metabolism by either α-hydroxylation to BRI or pyridine-*N*-oxidation.

Table 1. Michaelis-Menton kinetics of NNK metabolism by lung tissues[1]

Metabolite	A/J mouse lung			Human lung		
	K_m	V_{max}	Cl_{int}	K_m	V_{max}	Cl_{int}
Keto acid	0.6±0.3	27±7	49	1.8±2.1	21±8.1	12
Keto alcohol	0.6±0.3	60±18	105	7.1±10	48±58	7
Hydroxy acid	-			-		
Diol	1.6±0.9	32±12	20	1.6±1.3	18±1.3	11
NNK-*N*-oxide	0.6±0.3	68±16	124	0.9±1.7	11±10	12
NNAL-*N*-oxide	4.2±15	31±79	7	No saturation		
NNAL	93±35	1289±198	14	3.5±5.4	611±731	175

[1]Calculated in the concentration range 0.003-2.8 µM (NNAL in A/J mouse lung: 0.003-250 µM) K_m, µM; V_{max}, fmol/min/mg protein; Cl_{int}, intrinsic metabolic clearance (V_{max}/K_m)

EFFECT OF CIGARETTE SMOKE ON *IN VIVO* NNK METABOLISM IN A/J MICE

The effect of tobacco smoke exposure on NNK metabolism was studied in female A/J mice (4-5 weeks old), randomly assigned on body weight to either a control group which was sham-exposed by 2 h nose-only inhalation to HEPA-filtered air, or an experimental group which was exposed by 2 h nose-only inhalation to mainstream cigarette smoke from the Kentucky 1R4F reference research cigarette (Brown et al., 1999). Mainstream smoke was generated under Federal Trade Commission smoking conditions of a 2 s puff of 35 ml volume, and a frequency of one puff per min. The mainstream smoke exposure concentration was 600 mg total suspended particulate (TSP)/m^3 with a relative humidity of 40-70% and a temperature of 18-26°C. Immediately following 2 h exposure to either HEPA-filtered air or mainstream cigarette smoke, animals were given a single i.p. dose of 7.5 μmole [5-^{3}H]NNK in physiological saline, and 24 h urine collected in metabolic cages. NNK metabolite profiles were determined by HPLC with solid-phase radioactivity monitoring (Richter and Tricker, 1994).

Total urinary excretion of NNK was not different between sham-exposed (67.4±2.8%) and smoke-exposed animals (64.5±2.8%). However, cigarette smoke exposure significantly reduced metabolism to BRI, evident from the reduced excretion of α-hydroxylation metabolites hydroxy acid and keto acid to 85% (p=0.003) and 58% (p<0.0001) of NNK-treated control mice, respectively (Table 2).

Table 2. Effect of cigarette smoke exposure on NNK excretion in A/J mice

Metabolite	Percentage urinary excretion (mean±SD) in 24 h urine		p^1
	Control mice (n=12)	Smoke-exposed mice (n=12)	
Keto acid	16.4±1.8	9.4±1.9	<0.0001
Keto alcohol	1.1±0.8	1.4±0.7	NS
Hydroxy acid	15.5±1.8	13.2±1.5	0.003
NNK-*N*-oxide	0.8±0.6	0.05±0	NS
NNAL-*N*-oxide	9.3±2.0	8.8±1.2	NS
NNAL	33.1±3.6	41.3±4.3	<0.0001
NNAL-Gluc	20.6±2.6	21.0±2.1	NS
Σ α-hydroxylation^2	33.0±3.6	24.1±1.6	<0.0001

[1]Significance (two-sided *t*-test), [2]Sum of hydroxy acid, keto acid, and keto alcohol.

DISCUSSION

In the A/J mouse, inhalation exposure to mainstream cigarette smoke (Finch et al., 1996) and sidestream cigarette smoke (Witschi et al., 1995) fails to induce a tumorigenic response in the lung. In contrast to this, a tumorigenic response is observed under certain experimental exposure protocols in which A/J mice are exposed whole-body to 11% mainstream cigarette smoke and 89% sidestream cigarette smoke, used as an experimental surrogate for ETS (Witschi et al., 1997a, 1997b, 2000). However, the relevance of the high experimental exposure conditions to the ETS surrogate (78, 87 or 137 mg TSP/m^3, 6 h/day, 5 day/week for 5 months, followed by a 4 month recovery period in air) are unclear; this, given that personal exposure concentrations among nonsmokers exposed to particulate matter associated with ETS has been estimated to range from 18 to 64 μg/m^3 (Guerin et al., 1992).

The relevance of NNK to the induction of lung tumors in cigarette smoke-exposed A/J mice has also been questioned (Witschi et al., 1997b, 2000). Mechanistic studies show that exposure of A/J mice to mainstream cigarette smoke (0, 400, 600, or 800 mg TSP/m^3) does not result in detectable levels of O^{6}MeG in either lung or liver (Brown et al., 1999). Moreover, co-exposure of A/J mice to mainstream cigarette smoke (0, 400, 600, or 800 mg TSP/m^3) and a single i.p. administration of NNK (0, 3.75, or 7.5 µmole/mouse) results in a significant dose-dependent reduction in NNK-induced lung and liver O^{6}MeG (Brown et al., 1999). The current studies confirm that NNK is extensively metabolized *in vitro* by the A/J mouse lung to BRI with the potential to methylate DNA. However, co-exposure to tobacco smoke significantly inhibits the *in vivo* metabolic activation of NNK by α-hydroxylation to BRI, a critical step in the induction of lung tumorigenesis in the A/J mouse.

While this study provides no explanation for the observed increase in lung tumorigenesis in A/J mice following exposure to high concentrations of tobacco smoke under certain experimental conditions (Witschi et al., 1997a, 1997b, 2000), the current results, and those from a previous study (Brown et al., 1999), do not support the hypothesis that NNK contributes to the induction of lung tumorigenesis in cigarette smoke-exposed A/J mice. Furthermore, the observed differences in NNK metabolism in A/J mouse and human lung, in particular the lack of significant activation of NNK to BRI in the human lung, suggest that NNK is an unlikely causal agent for human lung cancer.

REFERENCES

Brown, B.G., Chang, C.-J.G., Ayres, P.H., Lee, C.K. and Doolittle, D.J., 1999, The effect of cotinine or cigarette smoke co-administration on the formation of O^6-methylguanine adducts in the lung and liver of A/J mice treated with 4-(methylnitrosamino)-1-(3-pyridyl)-1-butanone (NNK). *Toxicol. Sci.*, 47:33.

Finch, G.L., Nikula, K.J., Belinsky, S.A., Barr, E.B., Stoner, G.D. and Lechner, J.F., 1996, Failure of cigarette smoke to induce or promote lung cancer in the A/J mouse. *Cancer Lett.*, 99:161.

Guerin, M.R., Jenkins, R.A. and Tomkins, B.A., 1992, *The Chemistry of Environmental Tobacco Smoke: Composition and Measurement.* Lewis, Boca Raton, FL.

Hecht, S.S., and Hoffmann, D., 1988, Tobacco-specific nitrosamines, an important group of carcinogens in tobacco and tobacco smoke. *Carcinogenesis*, 9:875.

Hecht, S.S. and Tricker, A.R., 1999, Nitrosamines derived from nicotine and other tobacco alkaloids. In Gorrod, J.W. and Jacob, III, P. (eds.), *Analytical Determination of Nicotine and Related Compounds and their Metabolites.* Elsevier, Amsterdam, pp. 421-488.

Hecht, S.S., Morse, M.A., Amin, S., Stoner, G.D., Jordan, K.G., Choi, C.I. and Chung, F.-L., 1989, Rapid single-dose model for lung tumor induction in A/J mice by 4-(methylnitrosamino)-1-(3-pyridyl)-1-butanone and the effect of diet. *Carcinogenesis*, 10:1901.

Richter, E. and Tricker, A.R., 1994, Nicotine inhibits the metabolic activation of the tobacco-specific nitrosamine 4-(methylnitrosamino)-1-(3-pyridyl)-1-butanone in rats. *Carcinogenesis*, 15:1061.

Richter, E., Friesenegger, S., Engl, J. and Tricker, A.R., 2000, Use of precision-cut tissue slices in organ culture to study metabolism of 4-(methylnitrosamino)-1-(3-pyridyl)-1-butanone (NNK) and 4-(methylnitrosamino)-1-(3-pyridyl)-1-butanol (NNAL) by hamster lung, liver and kidney. *Toxicology*, 114:83.

Ronai, Z.A., Gradia, S., Peterson, L.A. and Hecht, S.S., 1993, G to A transitions and G-T transversions in codon 12 of the Ki-ras oncogene isolated from mouse lung tumors induced by 4-(methyl-nitrosamino)-1-(3-pyridyl)-1-butanone (NNK) and related DNA methylating and pyridyloxo-butylating agents. *Carcinogenesis*, 14:2419.

Witschi, H.P., Oreffo, V.I.C. and Pinkerton, K.E., 1995, Six-month exposure of strain A/J mice to cigarette sidestream smoke: Cell kinetics and lung tumor data. *Fundam. Appl. Toxicol.*, 26:32.

Witschi, H., Espiritu, I., Peake, J.L., Wu, K., Maronpot, R.R. and Pinkerton, K.E., 1997a, The carcinogenicity of environmental tobacco smoke. *Carcinogenesis*, 18:575.

Witschi, H., Espiritu, I., Maronpot, R.R., Pinkerton, K.E. and Jones, A.D., 1997b, The carcinogenic potential of the gas phase of environmental tobacco smoke. *Carcinogenesis*, 18:2035.

Witschi, H., Uyeminami, D., Moran, D., and Espiritu, I., 2000, Chemoprevention of tobacco-smoke lung carcinogenesis in mice after cessation of smoke exposure. *Carcinogenesis*, 21:977.

PHARMACOKINETICS OF BENZENE FOLLOWING AN ORAL OR INTRADERMAL DOSE IN FVB AND Tg.AC MICE

Matthew J. Hoffmann[1], Patrick J. Sinko[2], Robert J. Meeker[3], and Robert Snyder[1]

[1] Department of Toxicology
[2] Department of Pharmaceutics
Rutgers University
Piscataway, NJ 08855-8020
[3] UMDNJ
Robert Wood Johnson Medical School
Piscataway, NJ 08855-8020

INTRODUCTION

Chronic occupational exposure to benzene (BZ) is associated with an increased risk for the development of acute myelogenous leukemia (AML) (see Snyder and Kalf, 1994 for review). Attempts to reproduce this BZ-induced granulocytic leukemia in laboratory animals has by and large been unsuccessful. The recently developed Tg.AC transgenic mouse model appears to be the first reliable animal model for the development of BZ-induced granulocytic leukemia (French et al., 2000). The Tg.AC mouse was created by insertion of a v-Ha-ras construct into the genome of the FVB mouse (Leder et al., 1990). Tg.AC mice develop a granulocytic leukemia in a dose dependent manner following 26 weeks of dermal BZ application, but not after 26 weeks of oral BZ exposure at doses up to 200 mg/kg (French et al., 2000; Tennant et al., 1999). The leukemia was not observed in the parental FVB strain after 26 weeks of exposure by either route (French et al., 2000; Tennant et al., 1999).

There is a clear association between the metabolism of BZ and its hematotoxic effects (Andrews et al., 1977; Sammett et al., 1979). Species and strains of animals which produce more metabolites of BZ, especially polyhydroxylated and ring open metabolites, are more susceptible to its toxicity (see Ross, 2000 for review). The present studies were carried out to compare the pharmacokinetics and metabolism of BZ in both FVB and Tg.AC mice following either an oral or intradermal BZ exposure. It was necessary to use intradermal dosing as opposed to dermal BZ application to ensure consistent and accurate dosing and to limit the amount of radiolabeled BZ lost due to evaporation. Comparisons between the FVB and Tg.AC mouse strains should help determine if insertion of the v-Ha-ras construct

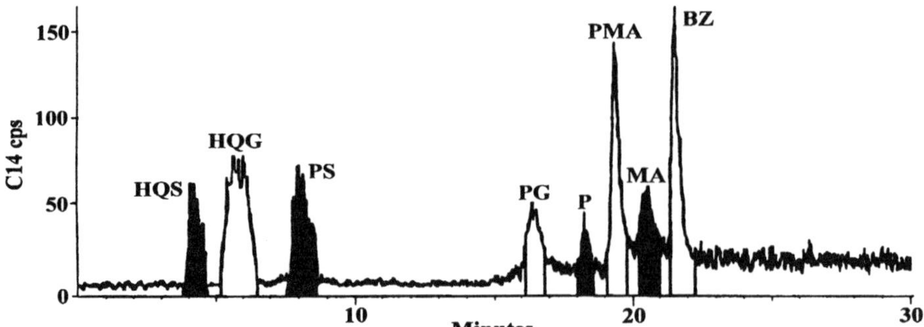

Figure 1. An HPLC chromatogram of radioactivity extracted from a blood sample taken from an FVB mouse 60 minutes following oral benzene exposure. Samples were spiked with benzene metabolite standards, which were detected by UV absorbance (chromatogram not shown). Radioactivity was measured as ^{14}C counts per second (cps). [Hydroquinone sulfate (HQS), hydroquinone glucuronide (HQG), phenyl sulfate (PS), phenyl glucuronide (PG), phenol (P), phenylmercapturic acid (PMA), muconic acid (MA), and benzene (BZ)]

led to any alterations in the pharmacokinetics of BZ. Differences between the two routes of exposure may provide insight into why the leukemia was observed following dermal BZ application but not after oral exposure.

METHODS

Animals for Pharmacokinetic Studies

FVB or Tg.AC mice were given a single dose of 220 mg/kg of ^{14}C]BZ in propylene glycol either via oral gavage or intradermal injection. Specific activity of BZ in the dosing solution was always 17.73 mCi/mmol. The target site for intradermal dosing was between the dermis and the epidermis. Ten, 30, 60, 90, 120, or 240 minutes after dosing animals were anesthetized, blood, liver, spleen, lung, kidney, stomach, large and small intestine, epididymal fat pads, and the skin around the dosing site (for intradermally dosed animals) were removed and weighed. Organs were placed into an extraction solution (20 mM ascorbic acid with 50% methanol and 1% Zinc Sulfate) and homogenized. Femurs were removed, cleaned of connective tissue, the ends removed and marrow flushed out with 1 ml of extraction solution.

HPLC Analysis

All organ samples were kept in extraction solution at -20°C for 24 hours and centrifuged for 20 minutes at 15,000 X g, to precipitate proteins. Samples were diluted at least 1:4 with HPLC grade water containing 5 μg each of nonradioactive BZ metabolite standards. Samples were injected onto a 4.6 X 150 mm 5 μm Microsorb C18 analytical column (Rainin), fitted with a 5 μm C18 analytical guard column; UV absorbance and radioactivity were measured. Metabolite peaks were separated using two HPLC solutions, 90% water acidified to pH 2.1 with formic acid/10% methanol (solution A) and 10% acidified water/90% methanol (solution B). Separation was achieved by using a program of 10 minutes of 100% solution A, then 10 minutes of a linear gradient to 100% solution B, and then 10 minutes of 100% solution B. Peaks were identified using UV standards, enzymatic digestion, and/or comparison to previously reported data. Figure 1 shows a typical chromatogram and the radiolabeled peaks which could be separated and identified.

Table 1. Absorption rate constants for benzene

Organ	FVB Oral	FVB Dermal	Tg.AC Oral	Tg.AC Dermal
Blood	19.9 ± 9.22[a]	6.59 ± 3.51	13.8 ± 14.0	4.25 ± 2.30
Bone Marrow	23.7 ± 11.5	3.92 ± 0.98	15.6 ± 11.6	3.60 ± 1.09
Fat	3.68 ± 1.14	3.73 ± 0.69	5.28 ± 3.23	2.52 ± 0.86
Kidney	7.51 ± 0.64	2.31 ± 0.33	6.44 ± 2.22	2.70 ± 1.02
Liver	16.4 ± 12.0	3.40 ± 0.067	13.9 ± 9.00	2.32 ± 0.019
Lung	3.81 ± 1.45	2.86 ± 2.62	8.88 ± 8.16	2.26 ± 0.47

[a] values are expressed as mean $\pm$ standard deviation in hours^{-1}.

Pharmacokinetic and Statistical Analysis

The area under the curve (AUC) from 0 to 240 minutes was calculated for BZ and each of the metabolites in each organ. Elimination (K_{el}) and absorption (K_a) rate constants were calculated for BZ in each of the organs. Data from each individual mouse was randomly assigned to one of three sets for each treatment group and the AUC and rate constants determined for each set of data and are expressed as mean $\pm$ standard deviation.

RESULTS

All of the parameters measured revealed few differences between the strains following BZ administration. In addition to unmetabolized BZ, phenol (P), P sulfate, P glucuronide, hydroquinone sulfate (HQS), hydroquinone glucuronide (HQG), muconic acid (MA), and phenylmercapturic acid were detected in most organs regardless of the strain or route of exposure. Little difference was observed in the K_{el} values for each of the different groups (data not shown). Table 1 shows the K_a values for BZ, the lower values following intradermal dosing than after oral dosing indicate slower BZ absorption in the intradermally treated mice. This is true for nearly every organ evaluated in both strains.

Determining the exposure of each treatment group to the potentially hematotoxic metabolites was done by comparing AUC values of metabolites which are formed directly from hematotoxic intermediates. The main focus was on the sulfate and glucuronide conjugates of hydroquinone (HQ) and MA, derived from trans,trans muconaldehyde. The most consistent difference observed was higher AUC values for HQS, shown in Figure 2, in the intradermally treated mice than in the orally treated mice. This was seen in both strains of mice and in nearly every organ examined. A similar result was seen with HQ glucuronide (data not shown). The levels of MA detected in the intradermally treated animals were generally lower than those in the orally treated mice (data not shown).

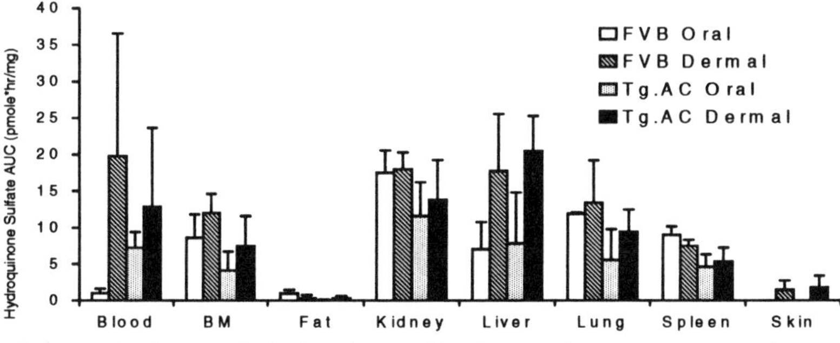

Figure 2. Area under the curve for hydroquinone sulfate in the various organs, expressed as pmoles of hydroquinone sulfate*hour/mg of wet organ weight.

DISCUSSION

Metabolism is required for the development of BZ-induced hematotoxicity (Andrews et al., 1977; Sammett et al., 1979). Specifically, it is believed that the polyhydroxylated and the ring opened metabolites of BZ are involved in its hematotoxicity. Using the newly developed Tg.AC mouse model, French et al. (2000) have been able to induce a dose dependent granulocytic leukemia following 26 weeks of dermal BZ application. Since this leukemia is not observed in orally treated Tg.AC mice or in FVB mice treated by either route there may be differences in the pharmacokinetics of BZ between the two routes or between the two strains which can account for the differences in toxicity.

In the present studies there were few differences observed in the metabolism of BZ between the two strains. Two differences were observed between the routes of exposure, first, the rate of BZ absorption was slower following intradermal dosing than after oral gavage, and second, AUC values for conjugates of HQ, and theoretically HQ itself, were greater in the intradermally treated mice as compared to orally treated mice. The slower rate of BZ absorption following intradermal dosing may be responsible for the increase in HQ formation. The difference in BZ absorption rates is effectively a difference in the exposure rate, with oral dosing leading to a shorter, higher level exposure than intradermal dosing. Using inhalation as the route of exposure it has been shown that shorter, higher level exposures lead to the formation of less HQ conjugates than longer, lower level exposures (Sabourin et al., 1989). Since HQ has been shown to be hematotoxic (see Ross, 2000 for review) an increase in the amount of HQ formed may very well be linked to the differences in hematotoxicity observed between the two routes of exposure.

ACKNOWLEDGMENTS

This research was partially supported by a fellowship from the Pharmaceutical Research and Manufacturers of America Foundation. The Tg.AC mice used in these studies were generously donated by the NIEHS.

REFERENCES

Andrews, L.S., Lee, E.W., Witmer, C.M., Kocsis, J.J. and Snyder, R., 1977. Effects of toluene on the metabolism, disposition, and hematopoietic toxicity of [3]HBenzene. *Biochem. Pharmacol.* **26**, 293-300.

French, J.E., Spalding, J.W., Hansen, L.A., Seeley, J., Trempus, C., Tice, R.R., Furedi-Machachek, M., Mahler, J., and Tennant, R.W. (2000). Topical application of benzene to Tg.AC (v-Ha-ras) transgenic mice induces skin tumors and leukemia. *Carcinogenesis* (in press).

Leder, A., Kuo, A., Cardiff, R.D., Sinn, E., and Leder, P. (1990). v-Ha-ras transgene abrogates the initiation step in mouse skin tumorigenesis: Effects of phorbol esters and retinoic acid. *Proc. Natl. Acad. Sci. USA* **87**, 9178-9182.

Ross, D. (2000). The role of metabolism and specific metabolites in benzene-induced toxicity; evidence and issues. *J Toxicol. Environ. Health* (in press).

Sabourin, P.J., Bechtold, W.E., Griffith, W.C., Birnbaum, L.S., Lucier, G., and Henderson, R.F. (1989). Effect of exposure concentration, exposure rate and route of administration on metabolism of benzene by F344 rats and B6C3F1 mice. *Toxicol. Appl. Pharmacol.* **99**, 421-444.

Sammett, D., Lee, E.W., Kocsis, J.J., and Snyder, R. (1979). Partial hepatectomy reduces both metabolism and toxicity of benzene. *J. Toxicol. Environ. Health.* **5**, 785-792.

Snyder, R. and Kalf, G.F. (1994). A perspective on benzene leukemogenesis. *Crit. Rev. Toxicol.* **24**, 177-209.

Tennant, R.W., Stasiewicz, S., French, J.E., and Spalding, J.W. (1999). Genetically altered mouse models for identifying carcinogens. In: <u>The use of Short- and Medium-term Tests for Carcinogenic Hazard Evaluation.</u> IARC, Lyon, pp. 123-150.

METABOLISM OF THE FOOD MUTAGEN 2-AMINO-3,8-DIMETHYLIMIDAZO[4,5-*f*]QUINOXALINE IN HUMAN HEPATOCYTES

Sophie Langouët[1], Dieter H. Welti[2], Nathalie Kerriguy[1], Laurent B. Fay[2],
F. Peter Guengerich[3], André Guillouzo[1], and Robert J. Turesky[2]

[1]INSERM U456, Faculté de Pharmacie, 35043 Rennes, France
[2]Nestlé Research Center, Nestec Ltd., 1000 Lausanne 26, Switzerland
[3]Dept. of Biochemistry and Center in Molecular Toxicology,
 Vanderbilt University School of Medicine, Nashville, TN 37232

INTRODUCTION

2-Amino-3,8-dimethylimidazo[4,5-*f*]quinoxaline (MeIQx) is a heterocyclic aromatic amine formed during cooking of fish and meat (1). It is mutagenic and carcinogenic in different animal species and is suspected to play a role in the etiology of human cancer (2). Studies performed in animals and with purified human enzymes demonstrated that MeIQx is metabolized by cytochrome P450 (CYP) 1A2 in the liver and CYP1A1 and 1B1 in extrahepatic tissues (3). However the metabolic pathways of MeIQx remain incompletely elucidated in humans.

As the liver is the primary target organ for MeIQx-induced tumors in the rat (4), the aim of the present study was to identify biotransformation pathways of MeIQx in human liver parenchymal cells. Indeed, human hepatocytes in primary culture represent a suitable system to investigate metabolic pathways of chemicals since co-factors are present at physiological concentrations and *in vitro*/*in vivo* correlations have been well established (5). Using this approach, we were able to elucidate the involvement of CYP1A2 activity in both the activation and detoxication of this food mutagen.

MATERIALS AND METHODS

Cell Isolation and Culture

Human liver samples were obtained from four patients undergoing liver resection for primary or secondary hepatomas, in agreement with French laws and fulfilled the requirements of the local Ethics committee. Hepatocytes were isolated by a two-step collagenase perfusion procedure as previously described (6).

Human liver parenchymal cells were seeded at a density of 10^6 viable cells/35 cm^2 dish in 2 ml of a Williams' medium, supplemented with 0.2% bovine insulin, 3.2% bovine serum albumin, 1% glutamine, 0.1% penicillin/streptomycin, 0.2% gentamycin and 10 % fetal calf serum (all w/v). This medium, supplemented with 7×10^{-5} M hydrocortisone hemisuccinate but lacking FCS, was renewed daily. [2-^{14}C]MeIQx (10 mCi/mmole, dissolved in dimethylsulfoxide) was added 36-48 h after cell seeding during various times (2 h to 48 h) at concentrations of 1, 10, or 50 μM without medium changes. The incubations were terminated by collecting the culture medium. Cells were rinsed three times with PBS; both media and cells were immediately stored at −80 °C until analysis.

Purification and Analysis of Metabolites

Cell extracts were added to 3 volumes of chilled CH_3CN, followed by removal of the precipitated material by centrifugation. After evaporation to dryness, they were analyzed by a C-18 reverse phase HPLC with UV and radioactivity detection. Metabolites were eluted at a flow rate of 1 mL/min using a gradient of 50 mM $NH_4CH_3CO_2$ buffer (pH 5.0) and 11% (v/v) CH_3OH for 10 min. This was followed by a linear gradient which reached 20% (v/v) CH_3OH at 40 min, followed by an increase to 100% CH_3OH at 51 min and held at 100% CH_3OH for an additional 5 min. Recovery of radioactivity from hepatocyte extracts and spiking experiments with ^{14}C-labelled reference metabolites of MeIQx was ≥85%.

CYP1A2 Activity Measurement

Ethoxyresorufin O-deethylase (EROD) associated with CYP1A1/2 activity was measured directly in cultured hepatocytes essentially as described by Burke and Mayer (7). The reaction rates were linear with time and proportional to protein concentration. Cellular protein content was estimated by the Bradford procedure (8). Values are means ± SD of triplicate measurements.

RESULTS AND DISCUSSION

Identification of Phase II Metabolites

We were able to identify six metabolites of MeIQx formed in human hepatocytes. The reactive, carcinogenic metabolite, 2-hydroxyamino-3,8-dimethylimidazo[4,5-f]quinoxaline (HNOH-MeIQx) was found to be transformed into the N^2-glucuronide conjugate, N^2-(β-1-glucosiduronyl)-N-hydroxy-2-amino-3,8-dimethylimidazo[4,5-f]quinoxaline (HON-MeIQx-N^2-Gl) (scheme 1). Phase II conjugates N^2-(3,8-dimethylimidazo[4,5-f]quinoxalin-2-yl)sulfamic acid (MeIQx-N^2-SO$_3^-$) was the most important metabolite formed at 10 μM MeIQx substrate concentration (Table1). The N^2-(β-1-glucosiduronyl)-2-amino-3,8-dimethylimidazo[4,5-f]quinoxaline (MeIQx-N^2-Gl) was also identified in culture media at all concentrations of MeIQx used.

Characterization of Oxidative Products

Interestingly, we were able to identify for the first time the 7-oxo derivatives of MeIQx and N-desmethyl-MeIQx, 2-amino-3,8-dimethyl-6-hydroimidazo[4,5-f]quinoxalin-7-one (7-oxo-MeIQx), and 2-amino-8-methyl-6-hydroimidazo[4,5-f]quinoxalin-7-one (N-desmethyl-7-oxo-MeIQx), previously thought to be formed exclusively by the intestinal flora.

A novel metabolite was characterized as 2-amino-3,8-dimethylimidazo[4,5-f]quinoxaline-8-carboxylic acid (MeIQx-8-COOH) (scheme 1), and it was the predominant metabolite formed in hepatocytes exposed to MeIQx at levels approaching human exposure (data not shown) (9). This metabolite is a detoxication product, which also appears to be the principal oxidation product of MeIQx excreted in human urine (10).

Table 1. Distribution of MeIQx metabolite formation in human hepatocytes. Three different cell populations (HL-1 to HL-3) were treated during 24 hr with 10 μM [2-^{14}C]-MeIQx. The metabolites were analyzed as described in the material and methods section, and the data were expressed as the percentage of the initial radioactivity of ^{14}C-MeIQx used for cell treatment. CYP1A2 activity was expressed as pmole/min/mg protein

Hepatocyte Preparation	MeIQx-8-COOH	MeIQx-N^2-Gl	HON-MeIQx-N^2-Gl	MeIQx-N^2-SO$_3^-$	N-desmethyl-7-oxo-MeIQx	7-oxo-MeIQx	MeIQx	CYP1A2 activity
HL-1	9	9	8	23	3	9	14	2.5 ± 0.5
HL-2	3	2	8	19	0	7	27	3.2 ± 0.3
HL-3	4	3	9	22	4	9	13	2.3 ± 1.0

Influence of CYP1A2 Activity on MeIQx Metabolism

The percentage of total metabolism correlated with CYP1A2 activity as illustrated in Table1. MeIQx-8-COOH formation is catalyzed by CYP1A2, while 7-oxo-MeIQx is formed by an non CYP-dependent reaction.

In summary, our study has elucidated the principal metabolic pathways of MeIQx in human liver; it is a prerequisite for assessing the role of this heterocyclic aromatic amine in human carcinogenesis.

Scheme 1: Major metabolic pathways of MeIQx in primary human hepatocytes

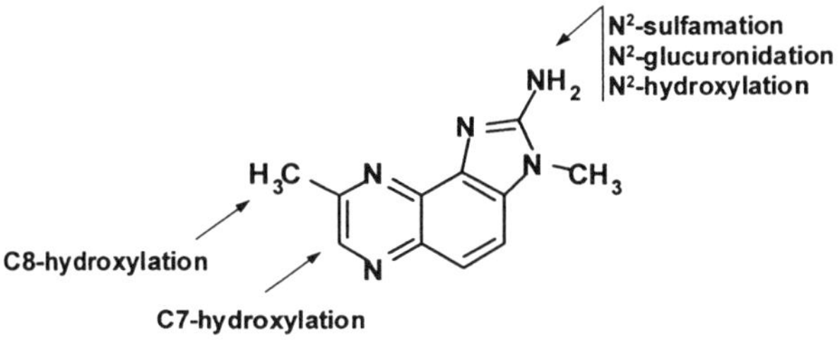

Scheme 1. Major metabolic pathways of MeIQx in primary human hepatocytes.

ACKNOWLEDGEMENTS

This work was supported in part by the Institut National de la Santé et de la Recherche Médicale, the Association pour la Recherche contre le Cancer and United States Public Health Service Grants R35 CA44353 and P30 ES00267. S. Langouët was a recipient of a fellowship from the Association pour la recherche contre le cancer and N. Kerriguy was supported by the EEC contract EUROCYP, BM-CT96-0254.

REFERENCES

1. Wakabayashi, K., Ushiyama, H., Takahashi, M., Nukaya, H., Kim, S. B., Hirose, M., Ochiai, M., Sugimura, T., and Nagao, M. Exposure to heterocyclic amines, Environ Health Perspect. *99:* 129-34, 1993.
2. Muscat, J. E. and Wynder, E. L. The consumption of well-done red meat and the risk of colorectal cancer, Am J Public Health. *84:* 856-8, 1994.
3. Turesky, R. J., Constable, A., Richoz, J., Varga, N., Markovic, J., Martin, M. V., and Guengerich, F. P. Activation of heterocyclic aromatic amines by rat and human liver microsomes and by purified rat and human cytochrome P450 1A2, Chem Res Toxicol. *11:* 925-36, 1998.

4.   Kato, T., Ohgaki, H., Hasegawa, H., Sato, S., Takayama, S., and Sugimura, T. Carcinogenicity in rats of a mutagenic compound, 2-amino-3,8- dimethylimidazo[4,5-*f*]quinoxaline, Carcinogenesis. *9:* 71-3, 1988.

5.   Guillouzo, A., Morel, F., Langouet, S., Maheo, K., and Rissel, M. Use of hepatocyte cultures for the study of hepatotoxic compounds, J Hepatol. *26:* 73-80, 1997.

6.   Guguen-Guillouzo, C., Campion, J. P., Brissot, P., Glaise, D., Launois, B., Bourel, M., and Guillouzo, A. High yield preparation of isolated human adult hepatocytes by enzymatic perfusion of the liver, Cell Biol Int Rep. *6:* 625-8, 1982.

7.   Burke, M. D. and Mayer, R. T. Differential effects of phenobarbitone and 3-methylcholanthrene induction on the hepatic microsomal metabolism and cytochrome P-450- binding of phenoxazone and a homologous series of its n-alkyl ethers (alkoxyresorufins), Chem Biol Interact. *45:* 243-58, 1983.

8.   Bradford, M. M. A rapid and sensitive method for the quantitation of microgram quantities of protein utilizing the principle of protein-dye binding, Anal Biochem. *72:* 248-54, 1976.

9.   Langouet, S., Welti, D. H., Kerriguy, N., Fay, L. B., Markovic, J., Guengerich, F. P., Guillouzo, A., and Turesky, R. J. Metabolism of 2-amino-3,8-dimethylimidazo[4,5-*f*]quinoxaline in human hepatocytes: 2-amino-3,8-dimethylimidazo[4,5-*f*]quinoxaline-8-carboxylic acid is a major detoxication pathway catalysed by cytochrome P450 1A2, Chem Res Toxicol. *14*: 211–221, 2001.

10.   Turesky, R. J., Garner, R. C., Welti, D. H., Richoz, J., Leveson, S. H., Dingley, K. H., Turteltaub, K. W., and Fay, L. B. Metabolism of the food-borne mutagen 2-amino-3,8-dimethylimidazo[4,5-*f*]quinoxaline in humans, Chem Res Toxicol. *11:* 217-25, 1998.

PREVENTIVE EFFECT OF SELENIUM ON T-2 TOXIN MEMBRANE TOXICITY

Seyed Ali Keshavarz[1], Abbas Memarbashi[1], Mehdi Balali[1]

[1]Dept. of Nutrition and Biochemistry,
School of Public Health and
Institute of Public Health Research,
Tehran University of Medical Sciences
and Health Services ,
P.O.Box 6446-14155 , Tehran, I.R.IRAN

ABSTRACT

T-2 toxin, one of the major toxic trichothecene mycotoxines, has been shown to cause effects such as inhibition of protein synthesis and impairement of mitochondrial function.The use of T-2 toxin as chemical warfare in south east Asia and Iran has been reported . It has been suggested that T-2 toxin may mediate its toxic effect via the cell membrane, but mechanism of action is poorly understood . In cytotoxicity studies, erythrocytes are an excellent model system. In the present study different doses of sodium selenite were injected into male albino mice for 6 days every 48h. Blood samples were taken from experimental and control groups (normal saline). The red cells were counted in isotonic phosphate buffer containing different doses of T-2 toxin. The mixture was incubated at $37^{\circ C}$ for 4h. The results indicate that selenium is able to prevent erythrocyte membrane damage induced by T-2 toxin. The protective effect of selenium may be due to its membrane stabilizing properties, although inhibition of lipid peroxidation is likely, too.

INTRODUCTION

Antioxidant nutrients, e.g., selenium, play several roles in physiology. Over the last ten years, great progress in identification of the role of these nutrients has been achieved [1,2.]
The results of much reserch indicate the protective role of antioxidant nutrients against the damages and pathologic states caused by harmful agents, specially free radicals[3]. Free radicals are produced constantly and their effects nutralizied by physiologic mechanisms in the body. Antioxidant nutrients, as shown by Fernandez-bunares[4],are the main component of such mechanisms. According to Kawamura[5] , the only harmful agent groups is mycotoxins, which contaminate foods. In this study,T-2 toxin(a cytotoxin agent), the most important fungal toxin of fusarium species, was studied [5,6]. Chanarin[7] believes

that it releases free radicals . For the first time in this study. Probable preventive effect of selenium in cellular toxicity (hemolysis) was examined. It is hoped that the results of this study can clarify some ambiguities about the role of seleniumin this connection.

MATERIALS AND METHODS

This was an experimental study in which the in vitro effect of selenium in the prevention of hemolysis due to T-2 toxin was examined. The study was conducted at the Pharmacology Department of the Medical Faculty of Mashhad University of Medical Sciences. Nine groups of male albino mice (30-35 g),5 in each group, were used. Experimental group 1 received, intraperitonealy, 2 mg/kg body weight selenium (as sodium selenite salt), 1 hour and 25 hours before the study. Experimental groups 2,3 and 4 received 2, 0.5 and 0.125 (equivalent to Se requirement) mg/kg body weight selenium, respectively, every 48h for 6 days. The control groups received normal saline. Blood samples were drawn and the erythrocytes were washed 4 times with phosphate buffer (PH=7.2) and counted by a sysmex hematologic apparatus. Finally, duplicate suspensions were prepared with 10×10^6 RBC/300µl phosphate buffer. Different doses of T-2 toxin were added and the samples were incubated at 37°C for 4h. After incubation, the samples were centrifuged, the supernatants aspirated, and hemoglobin concentration was determined by the Crosby W.H. method[7]. Data were analysed using the two-way analysis of variance and SPSS.

RESULTS

In all the experimental groups there was a strong positive correlation between hemolysis and toxin concentration (r=0.96 to 1.00; p<0.01);the hemolysis rate was lower as compared to the controls. Results of the control group with low, medium and high concentrations of the toxin also indicate a positive correlation between hemolysis and toxin concentration(r=+1). Treatment with selenium caused significant decreases in the hemolysis rate (p<0.01).

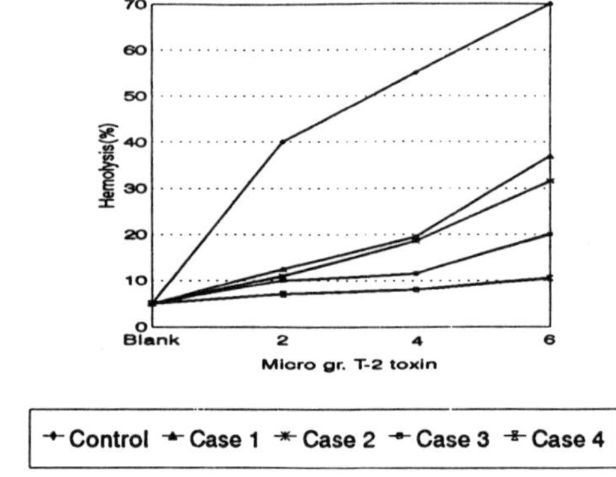

Figure1. Comparison of hemolysis rate in experimental groups 1,2,3,4 and the control group.

At the begining of study the weight of all of the animals had about 3035 g. After one week of acetimalization (and 12 hour darking period) daily weighting at 7 o'clock was done (Fig.2).

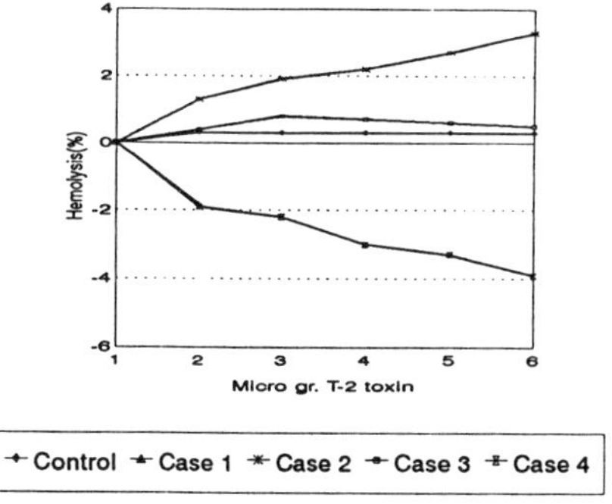

Figure2. Weight variation of experimental groups 1,2,3,4 and control group.

DISCUSSION

Because of the low concentration of glutathione peroxidase in the membrane and its inability to prevent lipid peroxidation [8,9,10], the decrease in hemolysis observed in this study could not be related to its increased activity or concentration. But the role of phospholipidhydroperoxide glutathione peroxidase (GSH-PXII) is more likely. Fuji[9] and Mcleod[10] reported that when rats were fed for six days a diet containing 5 ppm selenium, no significant increase in glutathione peroxidase was seen.

Yang[8] and Wo[11] reported that the effects of selenium on cell membrane (in vitro) are due to: 1- Increased fluidity of the membran's fat; 2- Increased stability of the membran's spectrin oligomers; 3- Reattachment of spectrin to the inner surface of the membrane's outer layer; and 4- Increased activity of Na^+/K^+ ATPase enzyme.

Raparot[12] , Reglinski[13] and Shiviro[14] believe that spectrin is a type of membrane's outer layer protein which has an important role in the stability of cellular structure and firmness. Due to complete washing of the red cells in our study, selenium was not present in this medium, and the results are probably solely due to direct effect of selenium on the structure and composition of red cells, especially their membrane. On the other hand, some scientists[14] reported that T-2 toxin toxicity is caused by attachment to the cellular surface receptors and by means of cell membrane. So it is probable that the selenium's effect is due to the competitive inhibition of the cellular surface receptors. In this study, the hematologic indices of mice before intervention were measured and we concluded that all of the indices were normal. High selenium infusion caused decreased weight and appetite, but infusion of amounts of selenium equal to daily requirement (0.125mg/kg) caused increased weight, so it may be that these mice were selenium-deficient or that low selenium infusion has an appetizing action.

CONCLUSION

The results indicate the important in vitro role of selenium in the prevention of hemolysis due to T-2 toxin. Because of the considerable decrease in hemolysis in experimental group 1 at 25 hours, there is probably a nonenzymatic effect of selenium on the cell stability; this is confirmed by other studies[8,10] It seems that the selenium's role in the prevention in experimental group 2,3,4 hemolysis is mediated by decreased lipid peroxidation, the most important effect of T-2 toxin. Beacause of glutathione peroxidase's inability to decrease lipid peroxidation, two reasons can be assumed for decreased hemolysis :

1- A considerable increase in the phospholipid hydroperoxide glutatione peroxidase enzyme in the membrane after selenium infusion ;

2- Production of an unidentified selenoenzyme at the membrane. Considering the results of Wo and Yang[8] about the effect of selenium on red cells membrane, the following reasons should be considered too :

- Increased fluidity of the membrane's fat;
- Increased stability of the membrane's spectrin oligomers;
- Reattachment of spectrin to the inner surface of the membrane's outer layer; and
- Increased activity of the Na^+/K^+ ATPase enzyme.

The final conclusion of this study is that selenium can prevent cell injuries caused by T-2 toxin and also by other harmful agents with a similar mechanism.

REFERENCES

1- Tamura, T. and Stadtman, TC.,1998, Selenium, in: *Advanced Nutrition Micronutrients*.First ed. (I.Wolinsky, ed),CRC Press, USA, pp:203-207.

2- Hallberg, L. Sandstrom, B.,2000, Selenium, in: *Human nutrition and dietetics*. 10th ed, (J.S.Garrow, W.P.T. James, A. Ralph, eds.), Churchil Livingstone, Edinburg, pp:200-202.

3- Pence, B.C.,1991, Dietary selenium and antioxidant status: Toxic effects of 1,2-Dimethylhydrazine in rats, *J. Nutr.* **121**: 138-144.

4- Fernandez-bunares, F., Dolz, C., Mingorance, M.A., Cabre, E. and Gassual, M.A.,1990, Low serum selenium concentration in a healthy population resident in Cataluya: A preliminary report. *Europ J. Clin. Nutr.* **44**:225-229.

5- Kawamura L. and Osamu,1990, Survey of T-2 toxin in cereals by an indirect Elisa. *Food Agric. Immunol***2**:173-180 (abs).

6- Babich, H. and Borenfreud, E.,1991, Cytotoxicity of T-2 toxin and its metabolites determined with the neutral red cell viability assay. *Appl. Environ. Microbiol.* **57**:2101-2103.

7- Chanarin, D.,1991, The blood count, its quality control and related methods. in: *Laboratory Hematology* , (I.Chanarin,ed), pp:26-27.

8- Yang, F.Y. and Wo, W.H.,1987, Role fo Se in stabilization of human erythrocyte membrane skeleton. *Biochem. Inter.* **15**:175-188.

9- Fujii, S., Dale, G.L. and Beutler, E.,1984, Glutathione-dependent protection against oxidative damage of the human red cell membrane. *Blood.***63**:1096-1101.

10- Mcleod, R.,1997, Protection conferred by selenium-deficiency against aflatoxin B1 in the rat is associated with hepatic... . *Cancer Research,* **57**:4257-4266.

11- Fujii, S. and Kaneco, T.,1992, Selenium, Glotathione Peroxidase and oxidative hemolysis in sickle cell disease. *Acta Haematologica,* **87**:105-106.

12- Rapaport, S.I., Ronney, H. and Taetle, R.,1990, Blood, in: *Physiological basis of medical practise,* (J.B.West, ed). pp:371-375.

13- Reglinski, J., Hoey, S., Smith, W.E. and Sturrock, R.D.,1998, Cellular response to oxidative stress at sulfhydryl group receptor site on the erythrocyte membrane. *J. Biol. Chem.* **263**:12360-12366.

14- Shiviro, Y. and Shaklai, N.,1987, Glutathione as a scarvenger of hemin:A mechanism of preventing red cell membrane damage. *Biochem. Pharmacol.***36**:3801-3807.

MODULATION INFLUENCE OF p-CHLOROMERCURIBENZOATE ON PLASMA MEMBRANE Na^+-Ca^{2+} EXCHANGER OF THE SECRETORY CELLS OF *CHIRONOMUS* LARVAE SALIVARY GLAND

Nataly V. Fedirko, Myron Yu. Klevets, and Volodymyr V. Manko

Ivan Franko National University of Lviv, 4, Grushevsky Str., 79005, Lviv, Ukraine

INTRODUCTION

Investigations the role of SH-groups of Ca^{2+}-transport systems molecules in secretory cells functioning have special significance because of: (i) Ca^{2+} cations have extraordinary importance in secretory process; (ii) SH-groups have importance to maintain of lipid bilaer structure [5,11]; (iii) one of the most sensitive links, obviously, for all membrane Ca^{2+}-transporting systems are SH-groups of aminoacid`s residues since its: * have high reactive ability [9]; ** determine fourth structure of protein molecule [9], besides its are form part of regulatory domains of these transport systems and especially Na^+-Ca^{2+}-exchanger molecule dogs cardiomyocites [1]. It has also known that SH-groups are form part of large intracellular loop f of Na^+-Ca^{2+} exchanger molecule [10]. Availability of SH-groups in the Na^+-Ca^{2+} exchanger molecule of secretory cell membrane in *Chironomus* larvae salivary gland we have supposed on the base of changes in Na^+-Ca^{2+} exchange current at alkalisation of external solution [6,8] and influence of Cd^{2+}, Ni^{2+} and La^{3+} [3]. Therefore the target of this study was to investigate the influence of SH-groups specific blocator – p-chloromercuribenzoate (PCMB) on the Na^+-Ca^{2+} exchange in the secretory cell membrane of investigated glands.

MATERIALS AND METHODS

Salivary gland of *Chironomus* larvae belong to group of exocrine glands with the low cell quantity. Physiological incubation medium contains (mM): NaCl - 136,9; KCl - 5,36; $CaCl_2$ - 1,76; $MgCl_2$ - 0,49; Na_2HPO_4 - 0,35; KH_2PO_4 - 0,44; glucose - 5,55; pH=7,2. Measurements of Ca^{2+} content in gland tissue carrying out using arsenazo III, total protein concentration in incubate as an indicator of secretion - by Lowry method. Intensity of dyeing of arsenazo III and Ca^{2+} complex was measured fotocolorymetrycally at wave length 650 nm. The changes of Ca^{2+} content in gland tissue directly testifies about the modulation of Ca^{2+}-transport systems functioning. Since the secretion level is determined by increasing in cytoplasmic Ca^{2+} level so additional information about its changes could be obtained by analyse of changes in total protein content in the glands incubation medium. Investigations of PCMB influence on Na^+-Ca^{2+}-exchange was carried out on the base of analyse of

Biological Reactive Intermediates VI, Edited by Dansette *et al.*
Kluwer Academic / Plenum Publishers, 2001

changes in salivary gland's cells Ca^{2+} content and total protein secretion owing to adding of specific thiol reagent to physiological (136 Na^+) and hyposodium (35 Na^+) incubation mediums since it makes possible to study inhibitor influence on forward (Ca^{2+}-efflux) and reverse (Ca^{2+}-influx) modes of Na^+-Ca^{2+}-exchanger functioning.

RESULTS AND CONCLUSIONS

According to the first experimental protocol PCMB was added to the incubation medium with physiological Na^+ concentration It has been shown increasing on cells Ca^{2+} content because of PCMB adding: at 1.0; 2.5; 5.0 and 10.0 µM on 6.5; 64.1; 71.8 and decrease on 17.95 % comparing with control (without inhibitor) which accompanied by slight arising of protein secretion level: by 6.5; 7.5; 10,7 % accordingly and at 10 µM PCMB - protein content in incubation medium has no difference to control. We suppose, low concentrations PCMB suppress Na^+-Ca^{2+} exchange or/and plasma membrane Ca^{2+}-pump functioning. Effects of 10 µM inhibitor connected with distinctive diminishing of Ca^{2+} influx into the secretory cells because of endoplasmic reticulum Ca^{2+}-pump suppression.

Then we have analysed changes in glands Ca^{2+} accumulation and secretion in hyposodium medium (Fig.1, curves 1 and 2), since at these conditions the Na^+-Ca^{2+}-exchanger transports Ca^{2+} into the cells, operating in reverse mode. Addition of 1 µM PCMB to this medium caused increase in Ca^{2+} content on 27.4% and decrease of secretion on 15.8% but 2.5; 5 and 10 µM PCMB decreased Ca^{2+} content on 39.3, 49.4 and 48.8% and secretion – on 9.0; 23.3; 18.48% accordingly. Increasing of Ca^{2+} content in glands cells at 1 µM PCMB could be also evoked by Ca^{2+}-pump suppression or/and stimulation of Na^+-Ca^{2+}-exchanger (Ca^{2+}-influx), that is clearly confirmed secretion level enhancement.

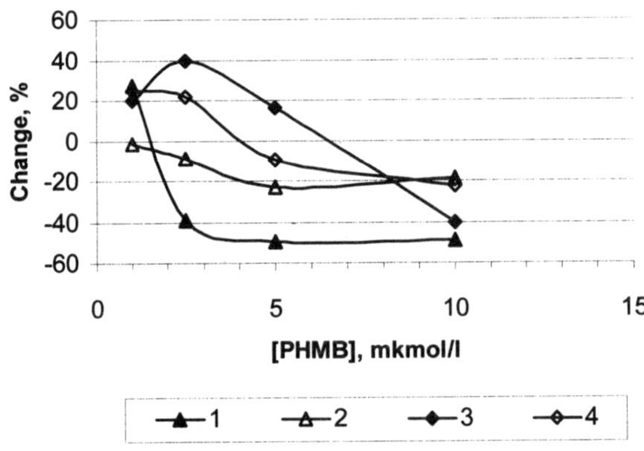

Figure 1. PHMB evoked changes of cells Ca^{2+} content (curves 1 and 3) and protein content in incubation hyposodium medium (curves 2 and 4) of secretory cells in Chironomus plumosus salivary gland. 1 and 2 – without eosin Y; 3 and 4 – on the background of eosin Y.

Testing which Ca^{2+}-transporting system is much more effected by low concentration PCMB to the hyposodium medium with PCMB specific inhibitor of Ca^{2+}-pump - eosin Y (5 µM) was added. At these conditions 1 µM PCMB evoked increase in Ca^{2+} content and secretion (Fig.1., curves 3 and 4): on 20.2 and 25.1%, 2.5 µM – on 39.4 and 22.0%, 5 µM – 16.2% and secretion decreased on 9.8%, at 10 µM – Ca^{2+} content and secretion decreased on 40.4 and 22.3%. We suppose that diminishing of the Ca^{2+} content in the investigated glands by PCMB in 2,5 µM and higher concentrations in control (35 Na^+ without eosin Y; Fig.1, curves 1 and 2) appear to be evoked by suppression endoplasmic reticulum Ca^{2+}-

468

pump. Its inhibition caused unfavourable conditions for gland cells Ca^{2+} accumulation in hyposodium medium even plasma membrane Ca^{2+}-pump is completely suppressed. Since as it has well known Ca^{2+}-pump functioning providing with ATP-ase activity it's molecule and SH groups which consisting of all enzyme systems have specific high sensitivity to mercaptidoforming thiol reagents [2]. Recent studies have confirmed the former by establishing that depression of Ca^{2+}-Mg^{2+}-ATP-ase by mercurial reagents [7,12,16] and after that it stimulation by SH-groups protectors [13] evoked by suppression of SH-groups consisting of ATP-ase active centres through which are carrying out enzyme activation and modulation by different substances [13].

Therefore indicated above findings clearly confirmed that 1 and 2.5 μM PCMB effects was evoked by suppression of Ca^{2+}-pump but partly (remain effect that took place on the background of eosin Y) - Na^+-Ca^{2+}-exchange (Ca^{2+}-influx mode) stimulation. The former was confirmed by enhancement of total protein secretion level (at the conditions: 35 Na^+ + eosin Y) at indicated above PCMB concentrations. Additionally it has been confirmed by our previous data that Cd^{2+} (0.01-0.5 mM) which is well known characterised by high affinity to SH-groups of aminoacids radicals [16], has evoked increasing of inward Na^+-Ca^{2+}-exchange current [6,8] and stimulated by hyposodium solution Ca^{2+} accumulation as well as protein secretion [3] in the secretory cells of *Chironomus plumosus* salivary gland. We have made a prediction that Cd^{2+} cations after the binding to the exchanger and transporting inside of the cell could interact with SH-groups in regulatory centre of the exchanger and because of that changes of exchanger molecule conformation and, obviously, "unmasking" of hided SH-groups [14] happened that resulted in changes in exchanger activation centre pK and velocity of its molecule moving [15]. Since it has shown stimulation of Na^+-Ca^{2+}-exchanger functioning by low concentration PCMB we suggest that activity of Na^+-Ca^{2+}-exchanger depends on slight number of SH-groups [2]. Moreover, for the other objects was established that Hg^{2+} and mercurial reagents effectively stimulate Ca^{2+} influx into the cells through the potentialdependent Ca^{2+}-channels [16]. Preliminary was shown that through the exocrine secretory cells membrane of *Chironomus plumosus* larvae salivary gland carry out steady-state Ca^{2+} entry for maintain of basal secretion by cells. This Ca^{2+} influx provide by population of open Ca^{2+}-channels, sensitive to blocators of potential-dependent Ca^{2+}-conduction of the plasmatic membrane [4]. So, hypothetically we could predict that effects of higher concentrations PCMB evoked by potentialdependent Ca^{2+} entrance but this question is discussed.

Essential decreasing of Ca^{2+} accumulation in cells at all experimental protocols detailed above testifies on decreasing of Ca^{2+} enter into the cells at the high PCMB concentrations. The former are evoked by suppression of endoplasmic reticulum Ca^{2+}-pump and also Na^+-Ca^{2+}-exchanger although in physiological conditions effects of endoplasmic reticulum Ca^{2+}-pump suppression exceed but in hyposodium − inhibition of Ca^{2+}-influx through the Na^+-Ca^{2+}-exchanger because of this effects remains after Ca^{2+}-pump suppression using eosin Y. Moreover, revealed in all experiments changes of secretory activity as a role correlate with changes in cells Ca^{2+} content but were less expressed that testifies on slight fluctuations of cytosolic free Ca^{2+} comparing with them total cells content and thereby confirming participating intracellular Ca^{2+} compartmentalising structures in PCMB evoked changes in Ca^{2+} homeostasis. Finally, PCMB changes of Ca^{2+}-transporting systems functioning of exocrine secretory cells membrane. SH-groups are extremely important for the regulation of functional activity of Ca^{2+}-tansporting systems and this way to maintain of physiological Ca^{2+} oscillations needed to normal level of secretory activity.

ACKNOWLEDGEMENT

This work was supported by West-Ukrainian BioMedical Research Center.

REFERENCES

1. Beauge, L., DiPolo, R., 1992, Chemical, *Biophys. J., 61: A388.*
2. Dixon, M., Webb, E.C., 1958, *Enzymes*, Longmans, Gree and Co. London-New York-Toronto.
3. Fedirko, N., Klevets, M., Manko, V., 1998, Modulation, *Pathophys., 5, Suppl. 1:134.*
4. Fedirko, N.V., M.Yu.Klevets, 1999, Evidence, *Physiol. J., 45 (4):84.*
5. Ibarra, C., Ripoche, P., Bourguet, J., 1989, Effect, *Membr. Biol., 110(2):115.*
6. Klevets, M.Yu., V.V. Manko, N. V. Fedirko, 1996, Investigation, *Neurophys., 28(4/5):193.*
7. Lee, C., Okabe, E., 1995, Hydroxyl, *J. Pharmac., 67(1):21.*
8. Manko, V.V., 1998, Influence, *Ukr. Biochem. J., 70(6):3.*
9. Monte, D., Bellomo, G., Thor, H., Nicotera, P., Orrenius, S., 1984, Menadione-induced, *Arch. Bioch. Bioph., 235(2):343.*
10. Nicoll, D.A., Longoni, S., Philipson, K.D., 1990, Molecular, *Science, 250:78.*
11. Rakowska, M.., Jasirska, R., Lenard, J., Komarnska, I., Makowski, R., Dygas, A., Pikula, S., 1997, Membrane, *Mol.Cell.Biochem.,168(1-2):163.*
12. Senchuk, V.V., Pikulev, A.T., Dashkevich, I.N., 1990, Study, *Biokhimia, 55(9):1648.*
13. Takahashi, H., Yamaguechi, H., 1999, Role, *J.Cell Biochem., 74(4):663*
14. Torchinskij, Yu.M., 1977, *Sulfur In Proteins,* Nauka, Moskow.
15. Valenzuela, Bender, 1971, Ref. Lenindger, A., 1974, *Biochemistry,*Mir, Moskow.
16. Viarengo, A., Nicotera, P., 1991, Possible…, *Comp. Biochem. Physiol.C., 100(1-2):81.*

TOBACCO TOXICOLOGY REVISITED

Hanspeter Witschi

Institute of Toxicology and Environmental Health and
Department of Molecular Biosciences, School of Veterinary Medicine
University of California
Davis CA 95616

INTRODUCTION

Worldwide, the diseases caused by inhalation of tobacco smoke represent one of the major health problems of our times. Of particular concern are lung cancer, chronic obstructive pulmonary disease and cardiovascular disease. In the Western World aggressive campaigning and education against tobacco use by both private interest groups and government agencies have slowed down, if not decreased, the incidence of lung cancer, the most lethal disease. Nevertheless, there is some concern that teen smoking might reverse this trend (Wingo et al., 1999). In many other countries tobacco smoking continues to be widespread and it is anticipated that in certain regions of the world, particularly in Asia, the toll on public health eventually will amount to millions of people annually (Liu et al., 1998; Pisani et al., 1999).

Tobacco toxicology has received attention ever since the hazards of smoking were recognized. It also became a source of major frustration that it proved to be exceedingly difficult, if not impossible to produce lung cancer in laboratory animals exposed to tobacco smoke. In 1986, IARC reviewed the then available evidence. Tobacco smoke induced consistently preneoplastic lesions as well as benign and malignant tumors in the larynx of hamsters (but not in the deep lung). Experiments with rats and mice showed in general that tobacco smoke had only weak carcinogenic activity, if any (IARC, 1986). More recently, a mouse and a rat study were conducted by exposing animals to comparatively high concentrations of cigarette smoke. The mouse study was negative (Finch et al., 1996). Prevalence of lung cancers in rats did not exceed 10% (Finch et al., 1995). Coggins concluded that animal models of tobacco smoke failed to reflect epidemiological findings (Coggins, 1998).

The realization that it is not only active smoking that causes untoward health effects, but that inhalation of the smoke produced by active smokers may be harmful to bystanders (National Cancer Institute, 1999) spawned a new round of research. The potential health hazards caused by environmental tobacco smoke (ETS) were explored (Witschi et al., 1997c). In a series of experiments we examined whether exposure of strain A mice to ETS would produce an increase in lung tumor multiplicity and also in lung tumor incidence. As

Biological Reactive Intermediates VI, Edited by Dansette *et al.*
Kluwer Academic / Plenum Publishers, 2001

shown in Table 1, in seven independently conducted experiments we found a significant increase in lung tumor multiplicity in all seven, and a significant increase in lung tumor incidence in five out of the seven (Witschi et al., 1997a, 1997b, 1998, 1999, 2000). Mouse lung tumors share many common features with human peripheral lung adenocarcinoma (Malkinson, 1992, 1998) a lung tumor that is on the increase (Thun et al., 1997). A few observations made in the course of these studies provided a few clues that some assumptions in tobacco smoke carcinogenicity might have to be reexamined.

Table 1. Summary of Lung Tumor Data

Exposure (mg TSP/m^3)	Tumor Multiplicity[a]		Tumor Incidence[b]		Reference[c]
	Filtered air	Smoke	Filtered air	Smoke	
87	0.5 ± 0.2 (24)	1.4 ± 0.2 (24)[d]	38%	83% [e]	Witschi et al., 1997a
79	0.5 ± 0.1 (24)	1.3 ± 0.3 (26)[d]	42%	58%	Witschi et al., 1997b
83	0.9 ± 0.2 (29)	1.3 ± 0.2 (33)[d]	69%	73%	Witschi et al., 1998
132	0.6 ± 0.1 (30)	2.1 ± 0.3 (38)[d]	50%	86% [e]	Witschi et al., 1999
137	0.9 ± 0.2 (30)	2.8 ± 0.2 (38)[d]	60%	100% [e]	Witschi et al., 2000
137	1.0 ± 0.1 (54)	2.4 ± 0.3 (28)[d]	65%	89% [e]	Witschi et al., 2000
134	1.2 ± 0.2 (25)	2.3 ± 0.3 (26)[d]	60%	88% [e]	(unpublished data)

a) Mean ± SE, number of animals in parenthesis.
b) No. of tumor bearing animals as percentage of all animals at risk.
c) Data reproduced with permission from Oxford University Press.
d) Significantly different (p < 0.05) compared to air controls by Welch's alternate test.
e) Significantly different (p < 0.05) compared to air controls by Fisher's exact test.

THE ROLE OF POLYCYCLIC AROMATIC HYDROCARBONS

The first experimental evidence that implicated tobacco smoke or at least some of its constituents as a carcinogen came from skin painting studies with tobacco smoke condensate ("tar"). Benzo(a)pyrene (B(a)P), a potent animal carcinogen was assumed to be the most active ingredient. By inference then it was assumed to play a decisive role in human tobacco smoke carcinogenesis. B(a)P has been extensively studied. When it was shown in 1996 that B(a)P preferentially formed adducts at P53 mutational hotspots in HeLa and human bronchial epithelial cells (Denissenko et al., 1996), an accompanying editorial called this discovery "the smoking gun". An unequivocal causal link seemed to have been established between tobacco smoke and human lung cancer.

While there is no question that B(a)P is present in tobacco smoke and thus may readily produce biomarkers of exposure, there are, nevertheless, some doubts on how important it is as a carcinogen in tobacco smoke. In 1961, Druckrey stated: "But on the account of the very low concentration of only 1 mg B(a)P per kg of smoke condensate, there remains some doubt as to its decisive importance". Wynder, who had pioneered skin painting studies with tobacco smoke condensate, estimated that B(a)P might account at most for 3% of the carcinogenic activity. Nevertheless, it was the focus on the tar phase and its carcinogens, the most prominent being B(a)P, that led to the development and aggressive marketing of low-tar filter cigarettes, in the hope that such cigarettes would decrease lung cancer risk. The premise appears to have been wrong. In 1996 an official statement of the American Thoracic Society read: "Epidemiologic studies have shown that brands of cigarettes that contain less tar and nicotine only marginally reduce the risk of lung cancer mortality. Similarly, little difference has been found for lifelong filter versus nonfilter smokers and for persistent smokers who switch from nonfilter to filter cigarettes."

Actually, there was experimental evidence available as early as 1970 that could have cast some doubt on the importance of the tar phase in tobacco smoke carcinogenesis. Mice exposed to the gas phase of tobacco smoke, i.e. to filtered tobacco smoke from which all particulate material had been removed, developed as many lung tumors (both benign and malignant) as did animals exposed to unfiltered tobacco smoke (Leuchtenberger and Leuchtenberger, 1970; Leuchtenberger and Leuchtenberger, 1974). Similar observations were made by Harris et al. (1974), and Witschi et al., (1997b), fully confirming the original observation.

Analysis of the chamber atmosphere showed that filtered ETS contained 0.04 $\mu g/m^3$ of B(a)P whereas unfiltered smoke contained almost 50 times as much (1.7 $\mu g/m^3$). If B(a)P was indeed one of, if not the most important carcinogens, then a 50 times reduction in chamber concentration should have reduced substantially lung tumor burden. An additional comparison involved calculating the amount of B(a)P inhaled by the mice during the exposure period and compare this number to the amount of B(a)P required to produce one lung tumor in the strain A mouse (Stoner and Shimkin, 1985). The ratio of inhaled dose to required dose was 0.0002 for filtered smoke and 0.009 for full smoke. These observations suggest that B(a)P might not be the most decisive carcinogen present in cigarette smoke and that the gas phase might contain some rather potent, but not yet identified or recognized as such carcinogens (Witschi et al., 1997b).

TOBACCO SMOKE SPECIFIC NITROSAMINES

These days, tobacco specific nitrosamines are considered to be as important as polycyclic aromatic hydrocarbons. There is an excellent rationale for this view. While the introduction of low-nicotine, low-tar cigarettes had not a major impact on over-all lung cancer incidence, there occurred and continues to occur a gradual shift in lung cancer types. A decrease in squamous cell carcinomas and an increase in adenocarcinomas, first observed in 1961 (Herman and Crittenden, 1961) has been attributed to the "changing cigarette" (Wynder and Hoffmann, 1994). Nicotine regulates smoking patterns (Djordjevic et al.,1997). With the advent of low-nicotine filter cigarettes, addicted people found it increasingly difficult and frustrating to satisfy their nicotine cravings. Consequently, cigarettes were smoked to a shorter butt and the smoke was more deeply inhaled and retained in the peripheral lung. Potent carcinogens residing in the peripheral lung include tobacco specific nitrosamines, particularly NNK 4-(methylnitrosamino)-1-(3pyridyl)-1-butanone). Their chemistry, metabolism and interaction with key cellular constituents of tobacco smoke specific nitrosamines, have been extensively explored (Hecht and Hoffmann, 1988; Hecht, 1999). Nitrosamines are known to produce adenocarcinomas in experimental animals.

Lung cancer would cease to be a major health problem should people no longer smoke. However, it is recognized that this is, for several reasons, including political ones, an unrealistic goal to achieve in the present society and socio-economic structures. Chemoprevention might benefit to smokers who are unable to quit, to smokers who have quit, and where their risk to develop lung and other cancers could be further reduced and perhaps people exposed to ETS at home or in the workplace. Chemoprevention of lung and other cancers is considered to be a major goal (Hong and Sporn, 1997, 1999). There has been extensive work in several laboratories to develop efficient chemopreventive agents. The preferred model to study their impact on lung cancer are strain A mice treated with a carcinogen (Stoner et al., 1993, 1997). Agents that have been examined for their effect against lung carcinogenesis include tea phenols, isothiocyanates, organoselenium compounds, naturally occurring agents such as chalcones, inositols, glucocorticoid hormones, N-acetylcysteine, non-steroidal anti-inflammatory drugs and others. In the majority of these experiments, NNK was used as the pulmonary carcinogen, mostly because it was thought to be a representative of tobacco smoke.

In our laboratory, we have exposed strain A mice for 5 months to ETS, followed by a 4 month recovery period in air. Chemopreventive agents were added to the diet, which the animals could freely access during the entire experimental period. Agents and dosage regimens were selected when they had been found to be effective in the laboratories of others to be effective against lung tumor induction in the strain A mouse. In each experiment, we included a positive control group in which we duplicated the protocol that had been successfully used by others.

The findings of our studies can be summarized as follows. Green tea, N-acetylcysteine or acetylsalicylic acid had no or only very little effects at all on animals treated with NNK or with urethane. The agents were also ineffective against full tobacco smoke. On the other hand, D-limonene, phenethyl isothiocyanate (PEITC) or p-XSC (1,4-phenylenebis (methylene) selenoisocyanate) were highly effective in the positive control experiments, reducing lung tumor multiplicities to 20-60% of animals treated with NNK and fed control diets. However, they proved ineffective against full tobacco smoke and so was a mixture of benzyl isothiocyanate (BITC) and PEITC. There was a slight reduction in tumor multiplicities (to about 85% of controls), but the difference was never statistically significant (Witschi et al., 1998, 1999, 2000). Also ineffective against tobacco smoke-induced lung tumorigenesis was beta carotene in the diet, fed at concentrations of 0.05%, 0.5% and 5% (unpublished observations).

It could be argued that lung tumor multiplicities following exposure to tobacco smoke are not high enough (from 1.3 to 2.8 tumors/lung) to enable us to detect comparatively small reductions produced by the chemopreventive agents. While this may be true, it also should be pointed out that yet another chemopreventive regimen we tried, exposure in the diet to a mixture of myoinositol and dexamethasone, gave highly significant results in reducing lung tumor multiplicities in smoke exposed animals to values that were found in controls that never had been exposed to smoke (Witschi et al., 1999, 2000).

These experiments seem to raise several questions about the importance of NNK as a tobacco smoke carcinogen. The question is not whether NNK is a potent animal lung carcinogen; it undoubtedly is. It is also present in tobacco smoke and evidence of its presence and its metabolites can readily be found in the tissues and fluids of animals exposed to tobacco smoke and, more importantly, in people who smoke. Several chemopreventive agents, such as PEITC, are highly specific in counteracting metabolism and DNA adduct formation of NNK. Their efficiency against NNK-induced mouse lung tumors is very high and yet they have no or at best only a marginal effect in the same animal model against full tobacco smoke. The question needs to be considered: "if a given chemopreventive agent inhibits tumor formation by agent X, but has no or only a marginal effect against a mixture containing agent X, then how important is agent X to the over-all carcinogenic activity of the mixture?" Our data suggest that NNK (and p-XSC) might have a small effect, although statistical significance was not achieved. Assuming, however, that these agents reduce lung tumor multiplicity by 15%, what advantage would it be to use them if we have 2 chemopreventive agents which are highly effective and reduce lung tumor multiplicities to unexposed control values? Myoinositol is a naturally occurring agent that already has been given to people in very high dosages without any observable untoward effects. Dexamethasone is a drug. Recent experiments by Wattenberg suggest that intratracheal instillation is highly effective against NNK-induced mouse lung tumors at dosage levels that approach the ones used in the chromic treatment of asthma (Wattenberg et al., 2000).

THE JANUS HEAD OF THE COMPLEX MIXTURE OF TOBACCO SMOKE

In one of the first experiments with ETS, we injected strain A mice with a single dose of urethane or of methylcholanthrene and then exposed them to tobacco smoke. In animals that were killed after a 5 months exposure to ETS, significantly fewer lung tumors were found than in animals kept in air. However, if the animals treated with a carcinogen were allowed to recover from the smoke exposure for an additional 4 months in air, tumor counts became comparable to control values. The experiment concluded that some as yet unidentified constituents in ETS were capable of suppressing or preventing the growth of chemically induced lung tumors in mice (Witschi et al., 1997a).

We then reasoned that some tobacco smoke constituents might suppress the growth of tumors induced by carcinogens present in the tobacco smoke. In animals kept for 5 months in ETS and killed, lung tumor multiplicity was not higher than it was in animals kept in air. However, if the animals were killed after an additional recovery period in air, lung tumor multiplicities and incidences were significantly higher in the ETS exposed animals than in the air controls. On the other hand, in animals kept continuously for 9 months in ETS, lung tumor multiplicities were not significantly higher than in air controls, although lung tumor incidences were (Witschi, unpublished observations). This suggested that tobacco smoke would slow down the growth of lung tumors. More recently, de Flora et al. have made a similar observation: in 2 studies they found that exposure of strain A mice followed for 6 or 5 months to ETS and then given a 4 month recovery period in air, results in a significant increase in lung tumor multiplicity, whereas exposure for 9 months to smoke has no such effect (deFlora S., personal communication).

Of course, these data should be no means imply that tobacco smoke might have a favorable influence on lung tumor development, although an unexpected "beneficial" effect of cigarette smoking has been reported in at least 2 epidemiological studies (Axelson and Sundell, 1978; Weiss, 1980). The observation has some importance for the conduct of inhalation studies with tobacco smoke. A recovery period following smoke exposure appears to be necessary to allow the growth of tumors and hence to reveal the full carcinogenic potential of tobacco smoke. A study in which A mice were not allowed to recover for more than 5 weeks after cessation of smoke exposure showed negative results (Finch et al., 1996). Practically all carcinogenesis studies with rats and mice conducted prior to 1986 and showing mostly negative results were done under lifetime exposure conditions, i.e. without recovery period.

The events underlying this observation need to be clarified. It has been described that nicotine inhibits apoptosis in several normal and transformed cell lines induced by selected agents such as UV light, tumor necrosis factor or calcium ionophore. It was concluded that nicotine might have tumor promoting properties (Maneckjee and Minna, 1994;Wright et al., 1993). This seems unlikely in view of our observations that tumors do not grow as long as the animals are exposed to an atmosphere containing nicotine. It is possible that cytotoxic constituents in tobacco smoke are capable to offset at least partially cell growth. Compounds capable of having such an effect might be acrolein, formaldehyde and other gases. While being exposed to ETS, animals fail to gain weight at the same rate as do the controls. Decreased weight gain is known to reduce lung tumor burden in mice (Larsen and Heston, 1945), although usually much more severe food restriction is needed than seen in our experiments to create a measurable decrease in lung tumor multiplicity (Pashko and Schwartz, 1996). Stress caused by chronic inhalation of high concentrations of ETS also might adversely affect the development of lung (Schwartz et al., 1986).

The observation that lung tumors apparently begin to grow once the animals have been removed from tobacco smoke might have some implications for chemoprevention. We have shown that administration of a mixture of myoinositol and dexamethasone in the diet effectively prevents lung tumor development in strain A mice even if the regimen is started once the animals have been removed from the smoke atmosphere (Witschi et al., 2000). It has been found repeatedly, in epidemiologic studies, that active smokers who quit smoking actually have an increased risk to develop lung cancer compared with current smokers (Hammond, 1966; Postmus, 1998; Wynder and Stellman, 1977). A logical explanation for this observation is that people quit smoking because they do not feel well and already might have acquired the conditions that would lead to lung cancer development. Active chemoprevention might be indicated in smokers who have quit but do not yet warrant treatment for diagnosed cancer of the lung. Chemoprevention in this particular group might further reduce the risk.

CONCLUSIONS

Many of our currently held assumptions and beliefs on the toxicity of tobacco smoke, particularly mechanisms of carcinogenesis and its prevention, come from studies with selected individual constituents of tobacco smoke, often examined in cultured cell and other in vitro systems. In 1997, the Presidential/Congressional Commission on Risk Assessment recommended that complex mixtures should be tested as such. Demonstrably, tobacco smoke is the most widespread and the most lethal complex mixtures to which humans are voluntarily or involuntarily exposed. In revisiting tobacco smoke carcinogenesis in an animal model, we have made several observations that might call for a reevaluation of some previous assumptions and beliefs.

ACKNOWLEDGMENTS

I wish to thank Imelda Espiritu, Dale Uyeminami, Kent E. Pinkerton, and Robert R. Maronpot for their help in performing these studies. This publication was made possible by grants numbers ES07908, ES07499, and ES05707 from the National Institute of Environmental Health Sciences (NIEHS). Its contents are solely the responsibility of the authors and do not necessarily represent the official views of the NIEHS, NIH.

REFERENCES

American Thoracic Society, 1996, Cigarette smoking and health. *Am. J. Respir. Crit. Care Med.* 153:861-865.

Axelson, O., and Sundell, L., 1978, Mining, lung cancer and smoking, *Scand J Work.Environ Health*, 4:46-52.

Coggins, C.R.E., 1998, A review of the chronic inhalation studies with mainstream cigarette smoke in rats and mice, *Toxicologic Pathology*, 26:307-314.

Denissenko, M.F., Pao, A., Tang, M., and Pfeifer, G.P., 1996, Preferential formation of benzo[a]pyrene adducts at lung cancer mutational hotspots in P53, *Science*, 274:430-432.

Djordjevic, M.V., Hoffmann, D., and Hoffmann, I., 1997, Nicotine regulates smoking pattern, *Preventive Medicine*, 26:435-440.

Druckrey, H., 1961, Experimental investigations on the possible carcinogenic effects of tobacco smoking, *Acta Med Scand*, Suppl 369, 24-41.

Finch, G.L., Nikula, K.J., Barr, E.B., Bechtold, W.E., Chen, B.T., Griffith, W.C., Hobbs, C.H., Hoover, M.D., and Mauderly, J.L., 1995, Lung tumor synergism between 239PuO2 and cigarette smoke inhaled by Fisher 344 rats, The Toxicologist 15, 47.

Finch, G.L., Nikula, K.J., Belinsky, S.A., Barr, E.B., Stoner, G.D., and Lechner, J.F., 1996, Failure of cigarette smoke to induce or promote lung cancer in the A/J mouse, *Cancer Letters*, 99:161-167.

Hammond, E.C., 1966, Smoking in relation to the death rates of one million men and women, *Natl. Cancer Inst Monogr.* 19:127-204.

Harris, R.J., Negroni, G., Ludgate, S., Pick, S.R., Chesterman, F.C., and Maidment, B.J., 1974, The incidence of lung tumors in c57bl mice expsoed to cigarette smoke: air mixtures for prolonged periods, Int. J. Cancer 14:130-136.

Hecht, S.S., 1999, Tobacco smoke carcinogens and lung cancer, J. Natl. Cancer. Inst. 91:1194-1210.

Hecht, S. S. and Hoffmann, D., 1988, Tobacco-specific nitrosamines, an important group of carcinogens in tobacco and tobacco smoke, *Carcinogenesis*, 9:875-884.

Herman, D.L. and Crittenden, M., 1961, Distribution of primary lung carcinomas in relation to time as determined by histochemical techniques, *J Natl Cancer Inst.* 27:1227-1271.

Hong, W.K. and Sporn, M.B., 1997, Recent advances in chemoprevention of cancer, *Science*, 278:1073-1077.

Hong, W. and Sporn M.B., 1999, Prevention of cancer in the next millenium: Report of the chemoprevention working group to the American Association for Cancer Research, *Cancer Res*, 59:4743-4758.

IARC (International Agency for Research on Cancer) 1986, Biological Data Relevant to the Evaluation of Carcinogenic Risk to Humans. 1. Carcinogenicity Studies in Animals, in *IARC Monographs on the Evaluation of the Carcinogenic Risk of Chemicals to Humans. Tobacco Smoking. Volume 38*, WHO IARC, ed., IARC, Lyon, 127-139.

Larsen, C.D. and Heston, W.E., 1945, Effects of cysteine and caloric restriction on the incidence of spontaneous pulmonary tumors in strain A mice, *J Natl Cancer Inst.* 6:31-40.

Leuchtenberger, C. and Leuchtenberger, R., 1970, Effects of chronic inhalation of whole fresh cigarette smoke and of its gas phase on pulmonary tumorigenesis in Snell's mice, in *Morphology of Experimental Respiratory Carcinogenesis. Proceedings of a Biology Division, Oak Ridge National Laboratory, conference held in Gatlinburg, Tennessee, May 13-16, 1970.* P. Nettesheim, M.G. Jr. Hanna, and J.W. Jr. Deatherage, eds. US Atomic Energy Commission, Division of Technical Information, Washington DC, 329-346.

Leuchtenberger, C. and Leuchtenberger, R., 1974, Differential response of Snell's and C57 black mice to chronic inhalation of cigarette smoke. Pulmonary carcinogenesis and vascular alterations in lung and heart, *Oncology*, 29:122-138.

Liu, B.Q., Peto, R., Chen, Z.M., Boreham, J., Wu, Y.P., Li, J.Y., Campbell, T.C., and Chen, J.S., 1998, Emerging tobacco hazards in China: 1. Retrospective proportional mortality study of one million deaths *BMJ*, 317:1411-1422.

Malkinson, A.M., 1992, Primary lung tumors in mice: an experimentally maniputable model of human adenocarcinoma, *Cancer Res*, 52:2670s-2676s.

Malkinson, A.M., 1998, Molecular comparison of human and mouse pulmonary adenocarcinomas, *Experimental Lung Research*, 24:541-555.

Maneckjee, R. and Minna, J.D., 1994, Opioids induce while nicotine suppresses apoptosis in human lung cancer cells, *Cell Growth Differ.* 5:1033-1040.

National Cancer Institute 1999, *Health Effects of Exposure to Environmental Tobacco Smoke: The Report of the California Environmental Protection Agency* Smoking and Tobacco Control Monograph No. 10. U.S. Department of Health and Human Services, National Institutes of Health, National Cancer Institute, NIH Pub. No. 99-4645, Bethesda, MD.

Pashko, L.L. and Schwartz, A.G., 1996. Inhibition of 7,12-dimethylbenz(a)anthracene-induced lung tumorigenesis in A/J mice by food restriction is reversed by adrenalectomy, *Carcinogenesis*, 17:209-212.

Pisani, P., Parkin, D.M., Bray, F., and Ferlay, J., 1999, Estimates of the worldwide mortality from 25 cancers in 1990, *Int.J Cancer*, 83:18-29.

Postmus, P.E., 1998, Epidemiology of Lung Cancer, in *Fishman's Pulmonary Diseases and Disorders*, Third edn, A.P. Fishman et al., eds. McGraw - Hill, New York, 1707-1717.

Schwartz, A.G., Pashko, L.L., and Whitcomb, J.M., 1986, Inhibition of tumor development by dehydroepiandros-terone and related steroids, *Toxicologic Pathology*, 14:357-362.

Stoner, G.D. and Shimkin, M.B., 1985, Lung tumors in strain A mice as a bioassay for carcinogenicity, in *Handbook of Carcinogen Testing*, H. Milman and E.K. Weisburger, eds. Noyes Publications, Park Ridge NJ, 179-214.

Stoner, G.D., Adam-Rodwell, G., and Morse, M.A., 1993, Lung tumors in strain A mice: application for studies in cancer chemoprevention, *J Cell Biochem Suppl*, 17F, 95-103.

Stoner, G.D., Morse, M.A., and Kelloff, G.J., 1997, Perspectives in cancer chemoprevention, *Environ.Health Perspect.* 105 Suppl 4, 945-954.

The Presidential/Congressional Commission on Risk Assessment and Risk Management 1997, *Risk Assessment and Risk Management in Regulatory Decision Making.*

Thun, M.J., Lally, C.A., Flannery, J.T., Calle, E.E., Flanders, W.D., and Heath, C.W., Jr. 1997, Cigarette smoking and changes in the histopathology of lung cancer. *J. Natl. Cancer Inst.* 89:1580-1586.

Wattenberg, L.W., Wiedmann, T.S., Estensen, R.D., Zimmerman, C.L., Galbraith, A.R., Steele, V.E., and Kelloff, G.J., 2000, Chemoprevention of pulmonary carcinogenesis by brief exposures to aerosolized budesonide or beclomethasone dipropionate and by the combination of aerosolized budesonide and dietary myo-inositol, *Carcinogenesis*, 21:179-182.

Weiss, W., 1980, The cigarette factor in lung cancer due to chloromethyl ethers, *J Occup Med.* 22:527-529.

Wingo, P.A., Ries, L.A., Giovino, G.A., Miller, D.S., Rosenberg, H.M., Shopland, D.R., Thun, M.J., and Edwards, B.K., 1999, Annual report to the nation on the status of cancer, 1973-1996, with a special section on lung cancer and tobacco smoking. *J. Natl. Cancer Inst.* 91:675-690.

Witschi, H.P., Espiritu, I., Peake, J.L., Wu, K., Maronpot, R.R., and Pinkerton, K.E., 1997a, The carcinogenicity of environmental tobacco smoke, *Carcinogenesis*, 18, No. 3, 575-586.

Witschi, H.P., Espiritu, I., Maronpot, R.R., Pinkerton, K.E., and Jones, A.D., 1997b, The carcinogenic potential of the gas phase of environmental tobacco smoke, *Carcinogenesis*, 18:2035-2042.

Witschi, H.P., Joad, J.P., and Pinkerton, K.E., 1997c, ETS Toxicity, *Ann. Rev. Pharmacol. Toxicol.* 37:29-51.

Witschi, H., Espiritu, I., Yu, M., and Willits, N.H. 1998, The effects of phenethyl isothiocyanate, N-acetylcysteine and green tea on tobacco smoke-induced lung tumors in strain A/J mice, *Carcinogenesis*, 19:1789-1794.

Witschi, H., Espiritu, I., and Uyeminami, D., 1999, Chemoprevention of tobacco smoke-induced lung tumors in A/J strain mice with dietary myo-inositol and dexamethasone, *Carcinogenesis*, 20:1375-1378.

Witschi, H., Uyeminami, D., Moran, D., and Espiritu, I., 2000, Chemoprevention of tobacco-smoke lung carcinogenesis in mice after cessation of smoke exposure, *Carcinogenesis*, 21: 977-982.

Wright, S.C., Zhong, J., Zheng, H., and Larrick, J.W., 1993, Nicotine inhibition of apoptosis suggests a role in tumor promotion, *FASEB J.* 7: 1045-1051.

Wynder, E.L., 1961, Laboratory contributions to the tobacco-cancer problem, *Acta Med Scand.* Suppl 369, 63-95.

Wynder, E.L. and Hoffmann, D., 1994, Smoking and lung cancer: scientific challenges and opportunities, *Cancer Res.* 54:5284-5295.

Wynder, E.L. and Stellman, S.D., 1977, Comparative epidemiology of tobacco-related cancers, *Cancer Research*, 37, No.12, 4608-4622.

FUNCTIONAL GENOMICS OF OXIDANT-INDUCED LUNG INJURY

George D. Leikauf, Susan A. McDowell, Cindy J. Bachurski, Bruce J. Aronow, Kelly Gammon, Scott C. Wesselkamper, William Hardie, Jonathan S. Wiest, John E. Leikauf, Thomas R. Korfhagen, and Daniel R. Prows

Departments of Environmental Health, Pediatrics, & Medicine, University of Cincinnati, & Divisions of Pulmonary Biology, Information Services, & Pulmonary Medicine, Children's Hospital Medical Center, Cincinnati, Ohio

INTRODUCTION

Our underlying hypothesis is that gene-environmental interactions control individual susceptibility to oxidant-induced acute lung injury. Although genetic pre-disposition is likely to be important in many lung diseases, this presentation focuses on acute lung injury, which is characterized by compromised gas exchange following macrophage activation, surfactant disruption, and epithelial perturbation (Lewis and Jobe, 1993). Activated macrophages release cytokines, and reactive oxygen and nitrogen species that disrupt endothelial function. Together these events bring about the key clinical manifestations of this syndrome, pulmonary edema, cellular infiltration, and airway collapse (Pittet et al., 1997). We reasoned that because mortality varies between individuals classified with equivalent severity, persons could vary innately in their ability to withstand oxidative injury during periods of severe stress.

Traditionally, clinician scientists have investigated and cataloged numerous environmental causes of lung diseases such as silica-induced pulmonary fibrosis, nickel-induced chronic obstructive pulmonary disease (COPD), or asbestos-induced mesothelioma. The scientific basis for this approach was to identify hazards, understand mechanisms, and evaluate dose-response relationships so that human exposure could be limited and disease could be prevented. Although individual susceptibility or resistance (e.g., "healthy worker" effect) has been suspected to influence these diseases, few scientists would evoke genetic causal factors, and until recently, avoided systematic investigation of genetic risk factors in acute diseases.

Similarly, geneticists have taken a limited approach to lung disease. Environmental factors typically are viewed among these scientists as confounders to assessment of disease causality and mechanisms of heredity. For example, initial studies of inheritance in lung biology have focused mainly on diseases with strong genetic risk factors (e.g., alpha-1 antitrypsin deficiency in COPD). The clinical motivation for this approach was that predictions of outcomes among offspring are more reliable in diseases with a single, strong genetic component [i.e. those that follow simple Mendelian inheritance patterns produced by single (or a few) allelic variant(s)]. Such predictions of heritability are useful in genetic counseling to inform parents and children of their individual risks. This approach has been used for respiratory diseases like cystic fibrosis, an autosomal recessive disease determined

Biological Reactive Intermediates VI, Edited by Dansette *et al.*
Kluwer Academic / Plenum Publishers, 2001

by a mutation (codon deletion) in the cystic fibrosis transmembrane regulator (CFTR) (Widdicombe and Wine, 1991; Pilewski and Frizzell, 1999). Although cystic fibrosis is one of the more common genetic conditions (1 in 2500 live births among Caucasians), the number of total cases is low compared to other respiratory diseases. Complex respiratory diseases such as acute lung injury, asthma, and COPD, affect large portions of the population. These diseases, however, are likely to be controlled by a few or several genes making prediction of inheritance difficult (Lander and Schork, 1994).

Because most of lung disease phenotypes have a strong set of environmental risk factors, we sometimes failed to consider the genetic risk factors. Nonetheless, gene-environmental interactions are critical to lung disease because each phenotype is primarily expressed when induced by an exposure (i.e. penetrance dependent on environmental exposures). The relationships of individual exposure with individual risk are therefore very complex. An excellent example is cigarette smoking, a major cause of respiratory disease. Yet, not everyone who smokes cigarettes (even in dosages >2 packs/day) develops COPD, cardiovascular disease, or tobacco-related cancers (Burrows et al., 1987). This is remarkable in that the amount of particulate matter depositing in the lung from 40 cigarettes exceeds 150 mg/day (assuming 40% deposition of approximately 10 mg/cigarette). This is 15 times greater than highest possible occupational exposure allowed for the least toxic, "nuisance" dust (respirable particles not otherwise classified). (Estimated to be 10 mg/day from the threshold limit value of 3 mg/m^3 x 8 h/d x NCRP light work daily ventilation of 1.3 m^3/h classified and assuming 30% tracheobronchial and pulmonary deposition). These observations support the contention that dose alone does not determine who will develop lung diseases induced by environmental agents and strongly suggest that genetic factors can influence susceptibility.

RESULTS AND DISCUSSION

Acute lung injury can be initiated by a diverse array of precipitating factors. To investigate the mechanisms afoot in this disease process, several animal models of acute lung injury have been developed, including induction by lipopolysaccharide, hyperoxia, embolism, and oleic acid (Levy et al., 1995). To begin to examine genetic determinants of acute lung injury, we compared the response between inbred mouse strains during continuous high concentrations of ozone (Prows et al., 1997), ultrafine polytetrafluoroethylene (PTFE) (Hardie et al., 1999), and submicrometer NiSO$_4$ (Wesselkamper et al., 2000).

Our strategy involves new approaches that include linkage analysis in model species followed by a homology search in humans. The purpose of linkage analysis is to delineate gene(s) by demonstrating co-segregation of a phenotype with polymorphic DNA markers distributed throughout the genome. Mice can be very useful to environmental toxicologists interested in dissecting genetic determinants of complex traits, like oxidant-induced lung injury. Recently, selected mouse strains have been used to investigate linkage to oxidant-induced mortality (Prows et al., 1997, Prows et al., 1999) and inflammation (Kleeberger et al., 1997).

To identify loci that may control resistance to acute lung injury induced by oxidative stress, quantitative trait locus (QTL) analysis was initially performed with the backcross mice generated from sensitive [A/J: (A)] and resistant [C57BL/6J: (B6)] mouse strains following exposure to ozone. The genomic DNA was isolated from the liver and polymerase chain reaction (PCR) performed for simple sequence length polymorphisms (SSLPs). Primers were chosen based on polymorphisms between the A and B6 strains and purchased from Research Genetics (Huntsville, AL). Using 65 SSLPs distributed at 20-30 cM intervals across the mouse genome, QTL analysis derived six regions of interest on chromosomes 1, 3, 7, 11, 12 and 13. These chromosomes were then analyzed with an additional 50 SSLPs to obtain a denser map, and the results were then analyzed by MAPMAKER/QTL (V1.1b) (Lander et al., 1987). Significant linkage was found on chromosome 11 (proposed as _A_cute _l_ung _i_njury-1, *Ali1*) and 13 (*Ali2*) with lod (log of odds) scores of 4.1 (located at *D11Mit263*) and 3.3 (*D13Mit59*), respectively. The first value is above the level of significance of 3.3 for backcross analysis, as recommended by Lander and Kruglyak (1995), and explains 13% of the genetic variance. (Note a lod score of 3.3 is essentially a p-value of 0.0005). The second QTL, *Ali2,* also had a lod score (3.3) significant for linkage and explained 10% of the genetic variance. In addition, this analysis determined suggestive linkages (with lod scores 1.7-2.3) on chromosomes 7 and 12.

To further assess the genetic determinants of this response, we used Recombinant Inbred (RI) mice. RI mice are derived by inbreeding the F_2 progeny from two different strains of mice for at least 20 generations to fix the initial segregation of each allele. The resultant RI strains contain a unique combination of genetic material derived from the progenitor strains. The probability that each RI strain has alleles of one progenitor strain at two loci is related to the distance between the loci in the original material inherited by the strain. This provides a tool for genetic mapping. The AXB and BXA RI strains are particularly useful because the progenitors differ in susceptibility to over 30 infectious or chronic diseases (Marshall et al., 1992; Sampson et al., 1998) and 31 live RI strains (and DNA of 41 strains) generated from these crosses are commercially available from Jackson Laboratory. In addition, strain distribution patterns (SDPs) of over 700 SSLPs and other loci have been genotyped for this RI set (Mu et al., 1993; Higgins et al., 1997; Naggert et al., 1997; Prows and Leikauf, 1998; Sampson et al., 1998). The mean survival times values of the A and B6 progenitor strains were separated by seven standard deviations. If resistance was controlled by a single dominant gene, then each RI strain should demonstrate a discrete sensitive or resistant phenotype. However, the RI strains demonstrated a continuous distribution, consistent with susceptibility to acute lung injury from oxidative damage being controlled by more than one gene, or modified by at least one other gene. Moreover, at least one RI strain had a survival time lower than the A progenitor strain. This also suggests that this is an oligo- or polygenic trait. The SDP for survival time in ozone was compared with all other SDPs available in the AXB, BXA RI set using MapManager (Manly and Elliott, 1991). This analysis yielded another genetic region with significant linkage (lod score 3.5) to survival time on mouse chromosome 17, which we proposed as *Ali3.*

We subsequently performed another QTL analysis on the F_2 generation of mice derived from the B6 and A strains (Prows et al., 1999). These data support that acute lung injury induced by severe oxidative stress is a complex oligogenetic trait in the mouse. The primary finding strengthened our previous finding that a major QTL on mouse chromosome 11 contains a gene (or genes) involved in sensitivity to acute lung injury. This QTL had a more narrow range and a maximum lod score of 6.8. In addition, *Ali1* maps to a region on mouse chromosome 11 that is highly conserved [having major areas of synteny with human chromosome 17q21-q22 (Copeland et al., 1993; Dietrich et al. 1996; Nusbaum et al., 1999)]. This region contains several candidate genes, including nerve growth factor receptor, retinoic acid receptor-α, and granulocyte-colony stimulating factor. This QTL also contains several small inducible cytokines including chemokine genes similar to those induced in the mouse lung by ozone exposure (Driscoll et al., 1993; Zhao et al., 1998).

Together these studies have identified *Ali1* as a region on chromosome 11 containing a major gene controlling ozone-induced acute lung injury. Evidence supportive of modifier genes on chromosome 17 (*Ali3*) was obtained by the RI and the F_2 analyses. Additional QTLs on chromosome 5 or 6 may play modifying roles in the overall phenotype, whereas the QTL on chromosome 13 (*Ali2*) was found only in the backcross analysis. The QTL interval of *Ali1* on chromosome 11 can be further narrowed using congenic strains and a positional candidate gene approach.

More recently our studies have focused on $NiSO_4$ because it is an occupational contaminant, a component of cigarette smoke, and ambient particulate matter (Fishbein, 1976; Nriagu, 1980; Leikauf et al., 1995), and because it is a potent respiratory irritant (Camner and Johansson, 1992; Benson et al., 1995). In our study, 0.2 μm MMAD $NiSO_4$ exposure modeled acute lung injury in mice, producing perivascular distension, alveolar epithelial damage, hemorrhage, neutrophil infiltration, and pulmonary edema (Wesselkamper et al. 2000).

To compare inbred mouse strains, eight different strains were exposed continuously to $NiSO_4$, PTFE, or O_3, and survival time recorded. Regardless of the irritant, the strain phenotype pattern was similar with the A mouse strain being the most sensitive, the C3H/He (C3) intermediate, and the C57BL/6 (B6) resistant to lung injury. The phenotype of the A and B6 offspring ($B6AF_1$) resembled the resistant B6 parental strain with strains exhibiting sensitivity in the following order ($A>C3>B6=B6AF_1$). Pulmonary pathology was comparable for the A and B6 mice indicating that each strain succumbed to severe edema (acute lung injury). The $NiSO_4$ concentrations studied were below the current occupational limit for soluble nickel $100\mu gNi/m^3$ with continuous exposure to 15 $\mu gNi/m^3$ producing 20% mortality (14 day study) in the sensitive A strain. The inter-strain sensitivity to protein ($B6>C3>A$) or polymorphonuclear leukocytes (PMNs) ($A\geq B6>C3$) recovered in lung lavage differed from that of survival time. Thus, these phenotypes were discordant, suggesting that they are not causally linked and are controlled by arrays of independent genes. As was noted for acute lung injury, offspring from a cross of the respective sensitive and resistant strain ($B6C3F_1$) exhibited phenotypes (lavage protein or PMNs) resembling the resistant parental strain. The agreement of the strain phenotype pattern with these irritants suggests that each agent may be initiating acute lung injury by a common mechanism, possibly through genes controlling macrophage activation and oxidative stress. In addition, the response of offspring suggests that sensitivity is inherited as a recessive trait.

Subsequently, we characterized gene expression in this model by analyzing 8734 sequence-verified murine cDNAs using microarray (McDowell et al. 2000). In this study, mRNA levels were assessed in B6 mouse lung following exposure to particulate $NiSO_4$ for 0 (control), 3, 8, 24, 48 or 96 hours. Lung polyadenylated mRNA was isolated, reverse transcribed and fluorescent labeled. Samples from exposed mice (Cy5 labeled) were competitively hybridized against samples from unexposed, control mice (Cy3 labeled), to microarrays containing murine cDNAs. To assess the temporal pattern of gene expression, genes were clustered according to similarities in expression with time. To evaluate the pathophysiology of changes in gene expression, genes were categorized according to function. Changes in the expression of 5 selected genes were also analyzed by conventional (Northern blot and S1 nuclease protection) assays. Clustering of co-regulated genes (displaying similar temporal expression patterns) revealed the increased expression of relatively few genes including those associated with macrophage activation, oxidative stress, anti-proteolytic function, and extracellular matrix repair followed by delayed decrease in surfactant proteins.

To further evaluate the genetics of $NiSO_4$–induced acute lung injury, we performed a genome wide analysis with 307 mice generated from the backcross of resistant B6xA F_1 with susceptible A strain (Prows and Leikauf, submitted). Unlike ozone, significant linkage was limited to a single region on chromosome 6 (proposed as *Aliq4*). Suggestive linkages were identified on chromosomes 1, 8, and 12. Mice containing a genotype of either all A or all B6 alleles in each of four regions (combined with two additional possible modifying loci, chromosome 9 and 16) had phenotypes that were as different from each other in survival time as that noted when comparing the A and B6 parental strains. This suggests that the interplay of genes in these six chromosomal regions, together, are sufficient to explain this phenotype. Interestingly, the region on chromosome 6 (*Aliq4*) with significant linkage to acute lung injury exposed to NiSO$_4$ had previously been identified as a region with significant linkage in a F$_2$ population of mice exposed to ozone (Prows et al., 1999). Several interesting candidate genes exist in the QTL on chromosome 6 including aquaporin 1, surfactant protein B, and transforming growth factor-alpha (TGF-α).

A polypeptide member of a protein family that includes the epidermal growth factor (EGF), TGF-α is a ligand for the EGF receptor and shares many similar biological properties with EGF (Derynck, 1988). The precise physiologic role of TGF-α in lung injury is not completely understood but TGF-α and EGF both appear to play a role in initiating and sustaining tissue repair. Experiments in animals and humans demonstrate that TGF-α and EGF treatment enhance healing of a variety of wounds. Topical administration of either recombinant TGF-α or EGF in pigs and humans accelerated the rate of epidermal regeneration in skin burns (Brown et al., 1989; Schultz et al., 1991), increased the tensile strength of injured rabbit corneas (Woost et al., 1992), and accelerated healing in patients with corneal defects (Daniele et al., 1979). Parental pretreatment with TGF-α or EGF decreased gastric mucosal damage induced by strong irritants including absolute ethanol, acidified aspirin or stress in rodents (Konturek et al.,1992; Tarnawski et al., 1992, Romano et al. 1994). Complementary studies using TGF-α knockout and induced overexpressing mice found increased and decreased, respectively, dextran-sulfate-induced colitis compared with wild type mice (Egger et al., 1997; Egger et al. 1998).

Administration of TGF-α in models of lung injury has had moderate success. For example, Kheradmand et al. (1994) using denuded rat type II epithelial cells in vitro demonstrated that TGF-α increased and that a monoclonal antibody to TGF-α in the presence of serum decreased the rate of "wound closure". To study the effects of TGF-α on lung injury in vivo, we developed transgenic mice overexpressing human TGF-α in the distal lung under control of the surfactant protein-C (SP-C) promoter (Korfhagen et al. 1994; Hardie et al. 1997). Mice from four TGF-α transgenic lines were exposed to ultrafine PTFE particles that induces rapid lung injury (Oberdörster et al., 1995) and is fatal within 8 hours of administration in wild type mice (Hardie et al., 1999). TGF-α transgenic mice demonstrated increased survival, reduced inflammation and pulmonary edema compared with wild type mice, indicating that hTGF-α expression in the lung could protect against acute lung injury.

To further investigate whether TGF-α protects against acute lung injury, four transgenic mouse lines with varying amount of expression of hTGF-α directed to the lung were exposed to NiSO$_4$. Acute lung injury was characterized by assessing pulmonary histology, bronchoalveolar lavage (BAL) fluid cell differential and protein, lung wet to dry weight ratios, and lung homogenate cytokine levels. Length of survival of four separate TGF-α transgenic mouse lines was significantly longer than that of nontransgenic control mice, and survival correlated with the levels of TGF-α expression in the lung. The transgenic line expressing the highest level of TGF-α (line 28) and non-transgenic controls were compared for differences in lung histology, BAL protein and cell differential, lung wet to dry weight ratios and lung homogenate pro-inflammatory cytokines at 24, 48 and 72 hrs after initiating nickel exposure. Within 72 hrs of nickel exposure, line 28 mice demonstrated reduced histological changes of lung inflammation and edema, decreased BAL protein and neutrophils, reduced lung wet to dry weight ratios and reduced production of interleukin-1β, interleukin-6 and macrophage inflammatory protein-2 levels compared with non-transgenic controls. In this transgenic mouse model, TGF-α protects against nickel-induced acute lung injury, at least in part, by attenuating the inflammatory response.

SUMMARY AND CONCLUSIONS

In summary, acute lung injury is a severe (>40% mortality) respiratory disease associated with numerous precipitating factors. Despite extensive research since its initial description over 30 years ago, questions remain about the basic pathophysiological mechanisms and their relationship to therapeutic strategies. Histopathology reveals surfactant disruption, epithelial perturbation and sepsis, either as initiating factors or as secondary complications, which in turn increase the expression of cytokines that sequester and activate inflammatory cells, most notably, neutrophils. Concomitant release of reactive oxygen and nitrogen species subsequently modulates endothelial function. Together these events orchestrate the principal clinical manifestations of the syndrome, pulmonary edema and atelectasis. To better understand the gene-environmental interactions controlling this complex process, we examined the relative sensitivity of inbred mouse strains to acute lung injury induced by ozone, ultrafine PTFE, or fine particulate $NiSO_4$ (0.2 μm MMAD, 15-150 $\mu g/m^3$). Measuring survival time, protein and neutrophils in bronchoalveolar lavage, lung wet: dry weight, and histology, we found that these responses varied between inbred mouse strains, and susceptibility is heritable. To assess the molecular progression of $NiSO_4$-induced acute lung injury, temporal relationships of 8734 genes and expressed sequence tags were assessed by cDNA microarray analysis. Clustering of co-regulated genes (displaying similar temporal expression patterns) revealed the altered expression of relatively few genes. Enhanced expression occurred mainly in genes associated with oxidative stress, anti-proteolytic function, and repair of the extracellular matrix. Concomitantly, surfactant proteins and Clara cell secretory protein mRNA expression decreased. Genome wide analysis of 307 mice generated from the backcross of resistant B6xA F_1 with susceptible A strain identified significant linkage to a region on chromosome 6 (proposed as *Aliq4*) and suggestive linkages on chromosomes 1, 8, and 12. Combining of these QTLs with two additional possible modifying loci (chromosome 9 and 16) accounted for the difference in survival time noted in the A and B6 parental strains. Combining these findings with those of the microarray analysis has enabled prioritization of candidate genes. These candidates, in turn, can be directed to the lung epithelium in transgenic mice or abated in inducible and constitutive gene-targeted mice. Initial results are encouraging and suggest that several of these mice vary in their susceptibility to oxidant-induced lung injury. Thus, these combined approaches have led to new insights into functional genomics of lung injury and diseases.

ACKNOWLEDGMENTS

The authors thank Dr. Alvaro Puga, Ph.D. and Dr. Pratim Biswas for helpful advice and technical assistance. This study was supported by NIH HL65612, HL65213, ES06096, and ES10562, and by the Health Effects Institute. The Health Effects Institute is an organization jointly funded by the US Environmental Protection Agency, Assistance Agreement X-812059, and automotive manufacturers. The contents of this article do not necessarily reflect the views of the Health Effects Institute or the policies of the US Environmental Protection Agency or automotive manufacturers. Scott Wesselkemper was a recipient of a University of Cincinnati (Ohio) Graduate assistantship and STARA fellowship, and this work is in partial fulfillment of graduate degree requirements at the University of Cincinnati. Susan McDowell received a University of Cincinnati Graduate Assistantship, Ryan Fellowship, and USEPA STAR Award, and this work is in partial fulfillment of graduate degree requirements at the University of Cincinnati. For complete listing of expression changes of cDNAs contained in the manuscript, address correspondence to: George Leikauf, Ph.D. Professor of Environmental Health, Director, Division of Molecular Toxicology College of Medicine, University of Cincinnati, P.O. Box 670056, Cincinnati, OH 45267-0056, United States of America, Telephone (513) 558-0039, FAX (513) 558-4397, email leikaugd@uc.edu .

REFERENCES

Benson, J. M., I. Y. Chang, Y. S. Cheng, F. F. Hahn, C. H. Kennedy, E. B. Barr, K. R. Maples and M. B. Snipes. 1995. Particle clearance and histopathology in lungs of F344/N rats and B6C3F1 mice inhaling nickel oxide or nickel sulfate. *Fundam. Appl. Toxicol.* 28: 232-44.

Brown, G.L., L.B. Nanney, J. Griffen, A.B. Cramer, J.M. Yancey, L.J. Curtsinger 3d, L. Holtzin, G.S. Schultz, M.J. Jurkiewicz, and J.B. Lynch. 1989. Enhancement of wound healing by topical treatment with epidermal growth factor. *New Engl. J. Med.* 321: 76-9.

Burrows, B., J.W. Bloom, G.A.Traver, and M.G. Cline. 1987. The time course and prognosis of different forms of chronic airways obstruction in a sample from the general population. *New Eng. J. Med.* 317: 1309-1314.

Camner, P. and A. Johansson. 1992. Reaction of alveolar macrophages to inhaled metal aerosols. *Environ. Health Perspect.* 97: 185-8

Copeland, N.G., N.A. Jenkins, D.J. Gilbert, J.T. Eppig, L.J. Maltais, J.C. Miller, W.F. Dietrich, A. Weaver, S.E. Lincoln, R G. Steen, et al. 1993. Genetic linkage map of the mouse: current applications and future prospects. *Science* 262: 57-66.

Daniele, S., L. Frati, C.Fiore, and G. Santoni. 1979. The effect of the epidermal growth factor (EGF) on the corneal epithelium in humans. *Albrecht Von Graefes Arch Klin Exp Ophthalmol.* 210: 159-65.

Derynck, R.1988. Transforming growth factor-alpha. *Cell* 54: 593-5.

Dietrich W.F., J. Miller, R. Steen, M.A. Merchant, D. Damron-Boles, Z. Husain, R. Dredge, M.J. Daly, K.A. Ingalls, J.T. O'Connor, et al. 1996. A comprehensive genetic map of the mouse genome. *Nature* 380: 149-52.

Driscoll, K.E, L.G. Simpson, J. Carter, D. Hassenbein, and G.D. Leikauf. 1993. Ozone inhalation stimulates expression of the neutrophil chemotactic peptide, MIP-2. *Toxicol. Appl. Pharmacol.* 119: 306-309.

Egger, B., F. Procaccino, J. Lakshmanan, M. Reinshagen, P. Hoffmann, A. Patel, W. Reuben, S. Gnanakkan, L. Liu, L. Barajas L, and V.E. Eysselein. 1997. Mice lacking transforming growth factor alpha have an increased susceptibility to dextran sulfate-induced colitis. *Gastroenterology.* 113: 825-32.

Egger, B., H.V. Carey, F. Procaccino,N-N. Chai, E.P. Sandgren, J. Lakshmanan, V.S. Buslon, S.W. French, M.W. Büchler, and V.E. Eysselein. 1998. Reduced susceptibility of mice overexpressing transforming growth factor α to dextran sodium sulphate induced colitis. *Gut* 43: 64-70.

Fishbein, L. 1976. Environmental metallic carcinogens: an overview of exposure levels. *J. Toxicol. Environ. Health.* 2: 77-109

Hardie, W.D., M.D. Bruno, K.M. Huelsman, H.S. Iwamoto, P.E. Carrigan, G.D. Leikauf, J.A. Whitsett, and T.R. Korfhagen. 1997. Postnatal lung function and morphology in transgenic mice expressing transforming growth factor-α (TGF-α). *Am. J Path.* 151: 1075- 1083.

Hardie, W.D., D.R. Prows, G.D. Leikauf, and T.R. Korfhagen T.R. 1999. Attenuation of acute lung injury in transgenic mice expressing human transforming growth factor-α. *Am. J. Physiol.: Lung Cell. Mol. Physiol.* 277: L1045-L1050.

Hardie, W.D., D.R. Prows, A. Piljan-Gentle, S.C. Wesselkamper, G.D. Leikauf, and T.R. Korfhagen. Submitted. Prevention of nickel-induced lung injury in transgenic mice overexpressing human transforming growth factor-α.

Higgins, D.C., and B. Paigen. 1997.An additional 150 SSLP markers typed for the AXB and BXA recombinant inbred mouse strains. *Mamm. Genome* 8:846-9.

Kheradmand, F., H.G. Folkesson, L. Shum, R. Derynk, R. Pytela, M.A. Matthay. 1994. Transforming growth factor-alpha enhances alveolar epithelial cell repair in a new in vitro model. *Am. J. Physiol.: Lung Cell. Mol. Biol.* 267: L728-38.

Kleeberger, S.R., R.C. Levitt, L.-Y. Zhang, M. Longphre, J. Harkema, A. Jedlicka, S.M. Eleff, D. DiSilvestre, and K.J. Holroyd. 1997. Linkage analysis of susceptibility to ozone-induced lung inflammation in inbred mice. *Nat. Genet.* 17: 475-478.

Konturek, S.J., T. Brzozowski, J. Majka, A. Dembinski, A. Slomiany, and B.L. Slomiany. 1992. Transforming growth factor alpha and epidermal growth factor in protection and healing of gastric mucosal injury. *Scand. J. Gastroenterol.* 27: 649-55

Korfhagen T.R., R.J. Swantz, S.E. Wert, J.M. McCarty, C.B. Kerlakian, S.W. Glasser, and J.A. Whitsett. 1994. Respiratory epithelial cell expression of human transforming growth factor-alpha induces lung fibrosis in transgenic mice. *J. Clin. Invest.* 93:1691-9

Lander, E.S., P. Green, J. Abrahamson, A. Barlow, M.J. Daly, S.E. Lincoln, and L. Newburg. 1987. MAPMAKER: an interactive computer package for constructing primary genetic linkage maps of experimental and natural populations. *Genomics* 1:174-81.

Lander, E.S., and N.J.Schork. 1994. Genetic dissection of complex traits. *Science* 265: 2037-2048.

Lander, E., and L. Kruglyak. 1995. Genetic dissection of complex traits: guidelines for interpreting and reporting linkage results. *Nat. Genet.* 11:241-7.

Leikauf, G. D., S. Kline, R. E. Albert, C. S. Baxter, D. I. Bernstein, and C. R. Buncher. 1995. Evaluation of a possible association of urban air toxics and asthma. *Environ. Health Perspect.* 103 Suppl 6: 253-71.

Levy, P.C., M.J. Utell, J.Z. Sickel, and M.J. Apostolakos. 1995. The acute respiratory distress syndrome: current trends in pathogenesis and management. *Compr. Ther.* 21: 438-44.

Lewis, J.F., and A. Jobe. 1993. Surfactant and the adult respiratory distress syndrome. *Am. Rev. Respir. Dis.* 147: 218-33.

Manly, K.F., and R.W. Elliott. 1991. RI Manager, a microcomputer program for analysis of data from recombinant inbred strains. *Mamm. Genome* 1:123-6.

Marshall, J.D., J.L. Mu, Y.C. Cheah, M.N. Nesbitt, W.N. Frankel, and B. Paigen. 1992. The AXB and BXA set of recombinant inbred mouse strains. *Mamm. Genome* 3:669-80.

McDowell, S.A., K. Gammon, C.J. Bachurski, J.S. Wiest, J.E. Leikauf, D.R. Prows, and G.D. Leikauf. 2000. Differential gene expression in the initiation and progression of nickel-induced acute lung injury. *Am. J. Respir. Cell Mol. Biol.* (In Press)

Mu, J.L., J.K. Nagger, P.M. Nishina, Y.C. Cheah, and B. Paigen. 1993. Strain distribution pattern in AXB and BXA recombinant inbred strains for loci on murine chromosomes 10, 13, 17, and 18. *Mamm. Genome* 4:148-52.

Naggert, J.K., K.L. Svenson, L. Lin, Y. Cheah, P.M. Nishina, J. Mu, T.R. Devereux, M. You, and B. Paigen B. 1997. An additional 136 SSLP markers typed for the AXB and BXA recombinant inbred mouse strains. *Mamm. Genome* 8:209-11.

Nriagu, J. O. 1980. Nickel in the Environment. John Wiley and Sons, New York.

Nusbaum, C., D.K. Slonim, K.L. Harris, B.W. Birren, R.G. Steen, L.D. Stein LD, J. Miller W.F. Dietrich, R. Nahf, V. Wang, O. Merport, A.B. Castle, Z. Husain, G. Farino, D. Gray, M.O. Anderson, R. Devine, L.T. Horton Jr., W.Ye, X. Wu, V. Kouyoumjian, I.S. Zemsteva, Y. Wu , A.J. Collymore, D.F. Courtney et al. 1999. A YAC-based physical map of the mouse genome. *Nat. Genet.* 22:388-93.

Oberdörster, G., R.M. Gelein, J. Ferin, and B. Weiss. 1995. Association of particulate air pollution and acute mortality; involvement of ultrafine particles? *Inhalation Toxicol.* 7: 111-124.

Pilewski J.M. and R.A. Frizzell. 1999. Role of CFTR in airway disease. *Physiol. Rev.* 79:S215-55.

Pittet, J.F., R.C. Mackersie, T.R. Martin, and M.A. Matthay. 1997. Biological markers of acute lung injury: prognostic and pathogenetic significance. *Am J Respir Crit Care Med.* 155: 1187-205

Prows, D.R., H.G. Shertzer, M.J. Daly, C.L. Sidman, and G.D. Leikauf. 1997. Genetic analysis of ozone-induced acute lung injury in sensitive and resistant strains of mice. *Nat. Genet.* 17: 471-4.

Prows D.R., and G.D. Leikauf. 1998. Genotypes for 66 microsatellite markers in the AXB and BXA recombinant inbred mouse strains. *Mamm. Genome* 9:160-1.

Prows, D.R., M.J. Daly, H.G. Shertzer, and G.D. Leikauf. 1999. Ozone-induced acute lung injury: Genetic analysis of F_2 mice generated from A/J and C57BL/6J strains. *Am. J. Physiol.: Lung Cell. Mol. Physiol.* 277: L372-L380.

Prows. D.R., and G.D. Leikauf. Submitted: Quantitative trait locus analysis of nickel-induced acute lung injury in mice. *Am. J. Respir. Cell Mol. Biol.*

Romano, M., E.R. Kraus, C.R. Boland CR, and R.J. Coffey. 1994. Comparison between transforming growth factor alpha and epidermal growth factor in the protection of rat gastric mucosa against drug-induced injury. *Ital. J. Gastroenterol.* 26: 223-8.

Sampson, S.B., D.C. Higgins, R.W. Elliot, B.A. Taylor, K.K. Lueders, R.A. Koza, and B. Paigen. 1998. An edited linkage map for the AXB and BXA recombinant inbred strains of mice. *Mamm. Genome* 9: 688-694.

Schultz, G., D.S. Rotatori, and W. Clark. 1991. EGF and TGF-alpha in wound healing and repair. *J. Cell. Biochem.* 45: 346-52.

Tarnawski, A., S.Y. Lu, J. Stachura, and I.J. Sarfeh. 1992. Adaptation of gastric mucosa to chronic alcohol administration is associated with increased mucosal expression of growth factors and their receptor. *Scand. J. Gastroenterol. Suppl.* 1992;193:59-63.

Wesselkamper, S.C., D.R. Prows, P. Biswas, K. Willeke, E. Bingham, and G.D. Leikauf. 2000. Genetic susceptibility to irritant-induced acute lung injury in mice. *Am. J. Physiol.: Lung Cell. Mol. Biol.* 279: L575-L582.

Widdicombe, J.H., and J.J. Wine. 1991. The basic defect in cystic fibrosis. *Trends Biochem Sci.* 16:474-477.

Woost, P.G., M.M. Jumblatt, R.A. Eiferman, and G.S. Schultz. 1992. Growth factors and corneal endothelial cells: I. Stimulation of bovine corneal endothelial cell DNA synthesis by defined growth factors. *Cornea* 11: 1-10.

Zhao, Q., L.G. Simpson, K.E. Driscoll, and G.D. Leikauf. 1998. Chemokine regulation of ozone-induced neutrophil and monocyte inflammation. *Am. J. Physiol. Lung Cell. Mol. Physiol.* 274: L39-46.

MITOCHONDRIAL-DERIVED OXIDANTS AND QUARTZ ACTIVATION OF CHEMOKINE GENE EXPRESSION

Kevin E. Driscoll,[1] Brian W. Howard,[1] Janet M. Carter,[1] Yvonne M.W. Janssen,[2] Brooke T. Mossman,[2] and Robert J. Isfort[1]

[1]Procter & Gamble Pharmaceuticals
Health Care Research Center
Mason, Ohio
[2]Department of Pathology
University of Vermont
Burlington, Vermont

Key words: chemokines, quartz, nuclear factor k B, mitochondria, oxidative stress, macrophage inflammatory protein

ABSTRACT

Macrophage inflammatory protein 2 (MIP-2) is a chemotactic cytokine which mediates neutrophil recruitment in the lung and other tissues. Pneumotoxic particles such as quartz increase MIP-2 expression in rat lung and rat alveolar type II epithelial cells. Deletion mutant analysis of the rat MIP-2 promoter demonstrated quartz-induction depended on a single NFκB consensus binding site. Quartz activation of NFκB and MIP-2 gene expression in RLE-6TN cells was inhibited by anti-oxidants suggesting the responses were dependent on oxidative stress. Consistent with anti-oxidant effects, quartz was demonstrated to increase RLE-6TN cell production of hydrogen peroxide. Rotenone treatment of RLE-6TN cells attenuated hydrogen peroxide production, NFκB activation and MIP-2 gene expression induced by quartz indicating that mitochondria-derived oxidants were contributing to these responses. Collectively, these findings indicate that quartz and crocidolite induction of MIP-2 gene expression in rat alveolar type II cells results from stimulation of an intracellular signaling pathway involving increased generation of hydrogen peroxide by mitochondria and subsequent activation of NFκB.

INTRODUCTION

Macrophage inflammatory protein 2 (MIP-2) is a CXC chemokine which is selectively chemotactic for neutrophils; has been shown to stimulate an oxidative burst in neutrophils; and act as a mitogen for alveolar epithelial cells (Driscoll et al., 1995; Frevert et al., 1995). MIP-2 is a key mediator of neutrophilic inflammation in rodent lungs. Respiratory tract inflammation in response to a variety of conditions including bacterial

Biological Reactive Intermediates VI, Edited by Dansette *et al.*
Kluwer Academic / Plenum Publishers, 2001

infection, endotoxin exposure, inhalation of the oxidizing air pollutant ozone or exposure to pneumotoxic particles (e.g., quartz, crocidolite asbestos) is associated with marked increases in lung MIP-2 gene expression (Driscoll et al., 1993a,b; Xing et al., 1994). Also, pretreatment of rats with neutralizing antibody to MIP-2 has been shown to markedly inhibit the neutrophilic inflammatory response in the lung to quartz exposure or bacterial infection (Greenberger et al., 1996; Driscoll et al., 1996).

Lung epithelial cells appear to be an important source of MIP-2 in response to pneumotoxic particles such as quartz. The expression of MIP-2 is increased in rat bronchiolar and alveolar type II epithelial cells within 2 hours of quartz exposure (Driscoll et al., 1996). Direct, *in vitro* treatment of rat alveolar type II cells with quartz, activates MIP-2 gene transcription and protein production (Driscoll et al., 1996). Recent studies have shown that the oxidant responsive transcription factor NFkB is critical for the regulation of MIP-2 gene expression (Widmer et al., 1993). In the present studies we characterized the role of oxidative stress in quartz-induced activation of NFkB and MIP-2 gene expression in rat alveolar epithelial cells. Our results provide evidence that oxidants resulting from exposure to quartz contribute to nuclear translocation of NFκB and activation of MIP-2 gene transcription in rat alveolar type II epithelial cells. In addition, results suggest a role for mitochondria-derived oxidants in particle-induced NFκB activation and MIP-2 gene expression.

METHODS

Cell culture and exposure to test particles. The rat alveolar type II cell line RLE-6TN (Driscoll et al., 1995) was used for all cell culture experiments. These cells retain characteristics of alveolar type II cells including chemokine expression in response to a variety of stimuli (Driscoll et al., 1996). RLE-6TN cells were grown in RLuE media (Biological Research Facility and Faculty, Ijamsville, MD). The test particle examined was quartz, (median diameter (± GSD) = 0.9 ± 1.8 μm; surface area = 4.5 m^2/g). Recombinant murine TNFα (Genzyme, Cambridge, MA) and bacterial lipopolysaccharide (LPS; E. coli 055:B5, Sigma, St. Louis, MO) were used in some experiments as positive controls for RLE-6TN activation. Cells were exposed to particles for 6 hr at concentrations ranging from 1-60 μg/cm^2. The range of particle exposure concentrations used did not produce cytotoxicity determined as the release of lactate dehydrogenase into the culture media. In some experiments cells were incubated with 0.1-2% dimethyl sulfoxide (DMSO, Sigma); 0.04-1% ethanol; 2-50 mM tetramethylthiourea (TMTU, Sigma); 6-60 μM pyrrolidine dithiocarbamate (PDTC, Sigma); and 2.5 and 25μM rotenone (Sigma) two hr prior to adding quartz.

Electrophoretic Mobility Shift Assay (EMSA). EMSA was used to characterize nuclear protein binding activity for specific DNA sequences. Briefly, RLE-6TN cells were rinsed twice with cold phosphate buffered saline (pH=7.2) and the nuclear extracts collected as described by Staal et al. 1990. Nuclear protein extracts (4 μg) were incubated for 20 minutes in DNA-binding buffer (40 mM Hepes buffer, 4% Ficoll 400, 200 ng polyI:polyC per μl, 1mM MgCl2, 0.1 mM dithiothreitol and 0.175 pmol) containing ^{32}P-end-labeled doublestranded oligonucleotide with a consensus NFκB site (Promega, Madison, WI). Samples were loaded onto a 4% polyacrylamide gel and electrophoresed in 0.25X Tris borate-EDTA buffer for 2 hours at 120 V. Gels were dried and visualized by exposure to Kodak X-Omatfilm.

RT-PCR analysis of lung MIP-2 mRNA: MIP-2 mRNA transcript levels were assessed by polymerase chain reaction (PCR) amplification of the MIP-2 cDNA as described by Driscoll et al 1993a. Briefly, RNA was extracted from the left lung lobe of 2 cell cultures/treatment. Glyceraldehyde-3-phosphate dehydrogenase (GAPDH) mRNA was evaluated as an internal control. The primers were designed from the published sequences for MIP-2 (Driscoll et al., 1995a) and GAPDH (Fort et al., 1985), were as follows:

```
MIP-2: 5' - GGCACATCAGGTACGATCCAG - 3';5' - ACCCTGCCAAGGGTTGACTTC - 3';

GAPDH: 5' - CAGGATGCATTGCTGACAATC - 3'; 5' - GGTCGGTGTGAACGGATTTG - 3'
```

Measurement of hydrogen peroxide. Hydrogen peroxide was determined as described by Pick and Mizel, 1981 based on the horseradish peroxidase-dependent conversion of phenol red into a compound with absorbance at 600 nm. Briefly, RLE-6TN cells were suspended in Earl's Balanced Salt Solution(EBSS)-phenol red free (Gibco/BRL, Gaithersburg, MD) and seeded into 96 well tissue culture plates (Corning, Corning, NY) at 1×10^5 cells/well and allowed to adhere overnight. Quartz particles (2-20 μg/cm^2) and 100 μl of phenol red solution (140mM NaCl, 10mM KH_2PO_4, 5.5mM dextrose, 0.56mM Phenol red)(Sigma, St. Louis) containing 6U/ml type VI horseradish peroxidase (Sigma) was added to cell cultures and incubated for 6 hr followed by addition of 10μl 1N NaOH. Absorbance at 600nm was determined using an EL311 Biotek Microplate Reader and hydrogen peroxide concentration calculated from a standard curve.

Statistical Analysis: Where appropriate, data were analyzed by ANOVA with group differences determined using the Newman-Keuls test (Zar, 1984).

RESULTS

Nuclear NFκB in RLE-6TN cells exposed to particles. Previous studies have demonstrated that quartz activates MIP-2 gene transcription in rat alveolar type II cells while titanium dioxide does not. Using EMSA we characterized the ability of these particles to increase RLE-6TN nuclear proteins that bind to oligomers containing an NFκB consensus DNA binding sequence. As show in Figure 1A, exposure to quartz increased NFκB-like binding activity while similar doses of titanium dioxide did not. TNFα and LPS increased nuclear binding activity for the NFκB consensus binding sequence. Increases in RLE-6TN nuclear NFκB binding activity could be detected within one hour of quartz exposure, with a peak response at 6 and 12 hr and regression of the response by 24 hr after exposure (figure 1B).

Anti-oxidants and nuclear translocation of NFκB and MIP-2 gene expression. To investigate the possible role of oxidants in quartz-induced activation of NFκB and MIP-2 gene expression, RLE-6TN cells were exposed to particles in the presence of several chemically diverse oxygen radical scavengers. As shown in Figure 2A, all the anti-oxidants examined attenuated nuclear NFκB binding activity. Consistent with a role for NFκB in the regulation of quartz-induced MIP-2 gene transcription, treatment of RLE-6TN cells with the anti-oxidants attenuated quartz-induced increases in MIP-2 mRNA (Figure 2B).

Quartz exposure and production of hydrogen peroxide. Experiments were conducted to determine if quartz treatment increased production of hydrogen peroxide by RLE-6TN. As shown in Figure 3 treatment of RLE-6TN cells with quartz resulted in a dose-related increase in hydrogen peroxide production.

<u>Mitochondria-derived oxidants and quartz activation of NFκB and MIP-2 expression.</u>
Experiments were conducted using mitochondrial respiratory chain inhibitors to determine
if the hydrogen peroxide produced by RLE-6TN cells was derived from mitochondria. The
complex I inhibitor rotenone attenuated quartz-induced increases in nuclear NFκB binding
activity and MIP-2 gene expression.

A. Effect of quartz and TiO2 on nuclear NF B binding activity

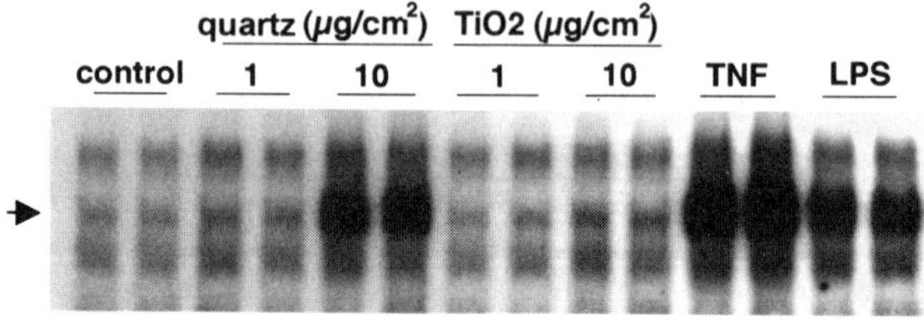

B. Time-course of quartz-induced NFkB nuclear binding activity

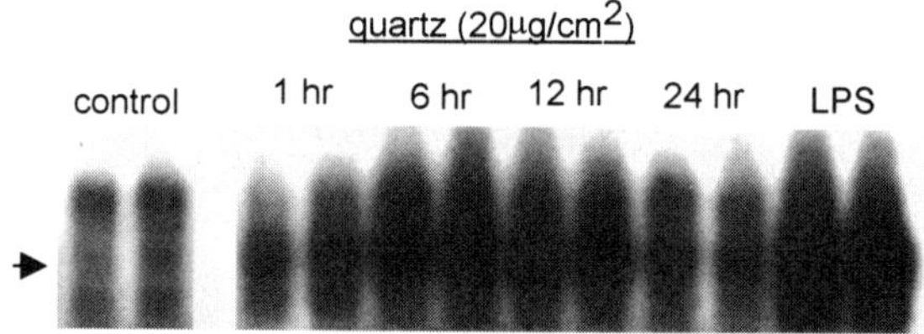

Figure 1. Effect of particles on nuclear NFκB binding activity in RLE-6TN cells. A. EMSA for NFκB
binding activity in nuclear protein extracts from cells exposed to quartz and titanium dioxide. B. Time-course
of increases in nuclear NFκB binding activity in RLE-6TN cells exposed to 20 µg/cm² quartz particles.

DISCUSSION

The rat chemokine MIP-2 exhibits structural and functional homology with the mouse MIP-
2 and the human gro chemokines. MIP-2 plays a key role regulating respiratory tract
inflammation in response to a variety of agents including endotoxin, ozone, quartz and
crocidolite asbestos (Driscoll et al., 1993a, b; Xing et al., 1994; Greenberger et al., 1996;
Zhao et al., 1998). Rat MIP-2 can be produced by immune cells such as macrophages and
neutrophils and nonimmune cells including lung fibroblasts and epithelial cells. *In vivo*,
airway and alveolar epithelial cells appear to be key sources of MIP-2 in the lung after
exposure to pneumotoxic particles. Recently, we reported that rat alveolar type II epithelial
cells respond *in vitro* to pneumotoxic particles such as quartz and crocidolite with increased
MIP-2 gene transcription but are not similarly activated by relatively innocuous particles
(Driscoll et al., 1993b). Since rat MIP-2 is a key factor in host defense and because
epithelial cells appear to discriminate between some toxic and nontoxic particles regarding
MIP-2 expression, we investigated the regulation of rat MIP-2 gene expression in alveolar
type II epithelial cells exposed to particles.

Gel retardation assays demonstrated that quartz increases nuclear NFκB binding activity in RLE-6TN cells. NFκB is a dimeric protein complex comprised of members of the Rel protein family (Schreck et al., 1992; Siebelist et al., 1994). Rel family members contain a Rel homology region which includes DNA binding and dimerization domains and a nuclear localization signal. Dimers of several Rel family proteins have been described which can bind to the same decameric DNA sequence called the κB motif, however, NFκB is a specific heterodimer of the Rel proteins p50 and p65. We detected increases in nuclear NFκB binding activity by 1 hr after quartz exposure of RLE-6TN cells, this response exhibited an apparent peak at 6 to 12 hr and regressed by 24 hr after exposure. The decrease in nuclear NFκB binding activity at 24 hr occurred even though the quartz particles remained on the cells suggesting a down-regulation of the NFκB response. In this respect, the promoter of the IκBα protein has been shown to contain several active NFκB binding sites and expression of IκBα mRNA is upregulated by NFκB (De Martin et al., 1993). Thus, it is possible that induction of IκBα expression and subsequent binding of IκBα to NFκB serves to decrease nuclear NFκB levels even with continued quartz exposure. This temporal pattern of quartz activation of NFkB in rat lung epithelial cells is consistent with our previous observations *in vivo* which demonstrated increase expression of the MIP-2 gene in rat lung epithelial cells 2 hr after quartz exposure with less expression present at 24 hr (Driscoll et al., 1993b).

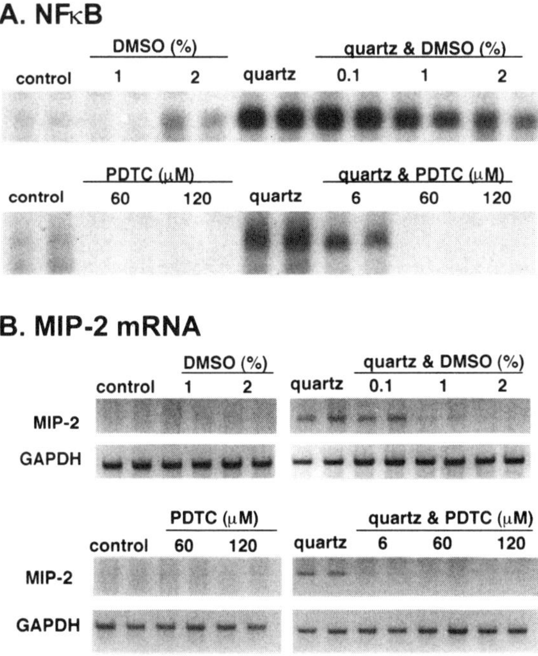

Figure 2. Effect of antioxidants on quartz-induced nuclear NFκB binding activity and MIP-2 gene expression. RLE-6TN cells were treated with the anti-oxidants dimethylsulfoxide (DMSO), ethanol (EtOH), tetramethylthiourea (TMTU), or pyrrolidine dithiocarbamate (PDTC) and exposed to quartz (20 μg/cm^2 × 6 hr); cells exposed to the antioxidants without quartz or cells without exposure (control) were used as controls. A. EMSA of nuclear protein extracts to determined NFκB binding activity. B. RT-PCR characterizing MIP-2 and GAPDH mRNA expression.

Several studies using pro-oxidant or antioxidant molecules provide evidence that reactive oxygen species can contribute to the activation of NFκB (Schreck et al., 1992). In this respect, quartz has been associated with the production of an oxidative stress (Kamp et al., 1992) which suggests a mechanism whereby these mineral particles could activate epithelial NFκB. We observed that quartz particles induced an oxidant response of RLE-6TN cells demonstrated by increases in hydrogen peroxide. The finding that a panel of chemically dissimilar antioxidants molecules (DMSO, ethanol, TMTU, PDTC) can each inhibit quartz-induced increases in nuclear NFκB binding activity as well as MIP-2 gene expression, strongly indicates the oxidant response stimulate by quartz contributes to their activation of NFκB and MIP-2 gene transcription in RLE-6TN cells.

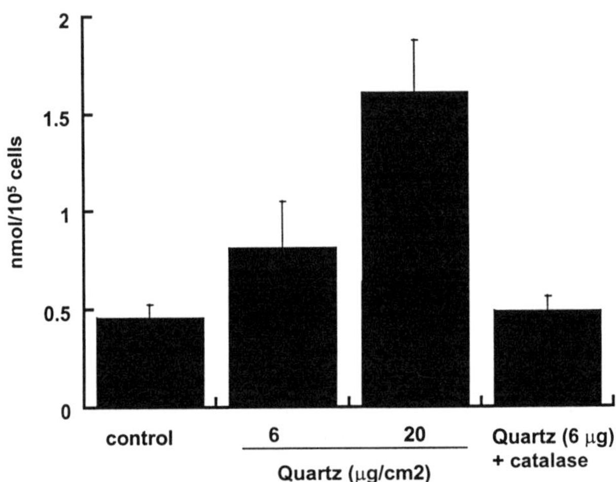

Figure 3. Production of hydrogen by quartz exposed RLE-6TN cells. RLE-6TN cells were given either no exposure (control) or exposed to quartz (6-20 $\mu g/cm^2$ × 6hr) and release of hydrogen peroxide determined. Catalase was added to some cultures exposed to the high dose of quartz to demonstrate specificity of the hydrogen peroxide. Results represent the X+/- SD of 3-4 cultures. (*) denotes a statistically significant difference from the control cultures; $p<0.05$.

There are several potential sources of reactive oxygen within cells, however, our findings provide evidence that a key source of hydrogen peroxide in quartz exposed RLE-6TN cells is the mitochondria. Rotenone is a selective inhibitor of the mitochondrial respiratory chain, blocking the transfer of electrons from complex I to complex III (Forman and Boveris, 1982; Turrens et al., 1985). The finding that rotenone markedly inhibits quartz-induced increases in epithelial cell hydrogen peroxide production and also attenuates nuclear translocation of NFκB and MIP-2 gene expression, suggest mitochondria-derived oxidants contribute to a signaling pathway underlying particle activation of rat alveolar type II epithelial cells. In this respect, Schulze-Ostoff and colleagues (Schulze-Ostoff et al., 1992) reported that rotenone inhibits TNFα-induced nuclear translocation of NFκB in L929 fibroblasts and that mitochondria deficient fibroblast lines were hyporesponsive to TNFα with respect to NFκB activation. Thus, it is possible that quartz and TNFα share, at some level, a common a signaling pathway involving oxidant generation by mitochondria leading to the activation of NFκB and transcription of NFκB regulated genes.

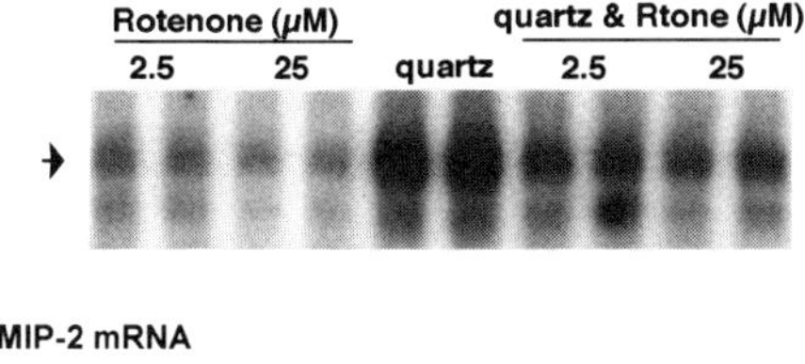

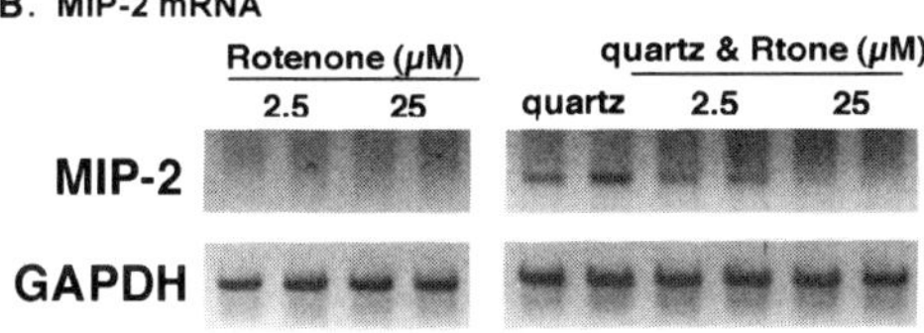

Figure 4. Rotenone inhibits quartz-induced activation of RLE-6TN cells. RLE-6TN cells were pretreated with the mitochondrial respiratory chain complex I inhibitor rotenone at the indicated concentrations and then exposed to quartz (20 µg/cm^2 × 6hr). Nuclear NFκB binding activity and MIP-2 mRNA expression were determined. A. EMSA for nuclear NFκB binding activity. B. RT-PCR analysis of MIP-2 and GAPDH mRNA expression.

In summary we have demonstrated that quartz exposure of RLE-6TN cells stimulates intracellular oxidant production and this oxidative stress contributes to the activation of NFκB and MIP-2 gene transcription in these cells. The increase in hydrogen peroxide production by quartz exposed epithelial cells appears to be derived, at least in part, from mitochondria. Inhibitors of the mitochondrial respiratory chain increase or decrease hydrogen peroxide production, NFκB activation and MIP-2 gene expression in RLE-6TN cells in a manner which implicates complex III of the mitochondrial respiratory chain as a key source of the oxidative stress. Collectively, these findings indicate that quartz exposure of RLE-6TN cells activates an intracellular signaling pathway(s) which results in the increased generation of hydrogen peroxide by mitochondria and contributes to the activation of NFκB and transcription of the gene for the pro-inflammatory chemokine, MIP-2.

REFERENCES

De Martin, R., B. Vanhove, Q. Cheng, E. Hofer, V. Csizmadia, H. Winkler, and F.H. Bach, 1993. Cytokine-inducible expression in endothelial cells of an IkB-a-like gene is regulated by NFkB. EMBO J. 12: 2773-2779.

Driscoll, K.E., D.G. Hassenbein, J.M. Carter, J. Poynter, T.N. Asquith, R.A. Grant, J. Whitten, M.P. Purdon, and R. Takigiku. 1993a. Macrophage inflammatory proteins 1 and 2: expression by rat alveolar macrophages, fibroblasts, and epithelial cells and in rat lung after mineral dust exposure. Am. Rev. Resp. Cell Mol. Biol., 8:311-318.

Driscoll, K.E., D.G. Hassenbein, J.M. Carter, J. Poynter, T.N. Asquith, R.A. Grant, J. Whitten, M.P. Purdon, and R. Takigiku. 1993b. Macrophage inflammatory proteins 1 and 2: expression by rat alveolar macrophages, fibroblasts, and epithelial cells and in rat lung after mineral dust exposure. Am. Rev. Resp. Cell Mol. Biol., 8:311-318.

Driscoll, K.E., D.G. Hassenbein, B.W. Howard, R.J. Isfort, D. Cody, M.H. Tindal, and J.M. Carter, 1995a. Cloning, expression, and functional characterization of rat macrophage inflammatory protein 2: a neutrophil chemoattractant and epithelial cell mitogen. J. Leuk. Biol. 58:359-364.

Driscoll, K.E., P.T. Iype, H.L. Kumari, J.M. Carter, M.J. Aardema, L.L Crosby, M.H. Chestnut, and R.A. LeBoeuf. 1995b. Establishment of immortalized alveolar type ii cell lines from adult rat lungs. *In Vitro* Cell Develop. Biol. 31:516-527.

Driscoll, K.E., B.W. Howard, J.M. Carter, T.A. Asquith, P. Detilleux, C. Johnston, S.L. Kunkel, D. Paugh, and R.J. Isfort, 1996. Chemokine expression by rat lung epithelial cells: effects of *in vitro* and *in vivo* mineral dust exposure. Am. J. Pathol. 149:1627-1637, 1996.

Forman, H.J. and A Boveris. Superoxide radical and hydrogen peroxide in mitochondria. In: Free Radicals in Biology, New York, Academic Press, 1982, pp 65-90.

Fort, P., L. Marty, L. Piechaczyk, S.E. Sabrouty, C. Dani, P. Jeanteur, P., and J.M. Blanchard. 1985. Various rat adult tissues express only one major mRNA species from the glyceraldehyde-3-phosphate dehydrogenase multigenic family. Nuclei Acid Res. 13:1431-42.

Frevert, C.W., A. Farone, H. Danaee, J.D. Paulauskis and L. Kobzik. 1995. Functional characterization of rat chemokine macrophage inflammatory protein 2. Inflammation 19:133-142.

Greenberger, M.J.R., R.M. Strieter, S.L. Kunkel, J.M. Danforth, L.L. Laichalk, D.C. McGillicuddy, and T.J. Standiford. 1996. Neutralization of macrophage inflammatory protein 2 attenuates neutrophil recruitment and bacterial clearance in murine Kelsiella pneumonia. J. Infect. Dis. 173:159-165.

Kamp DW, P. Graceffa, W.A. Pryor and S.A. Weitzman. 1992. The role of free radicals in asbestos-induced diseases. Free Radical Biol Med 12:293-315.

Pick, E, and D. Mizel. 1981. Rapid assays for the measurement of superoxide and hydrogen peroxide production by macrophages in culture using an automatic enzyme immunoassay reader. J. Immunol. Meth. 46:211-226.

Siebenlist, U., G. Franzoso, and K. Brown. 1994. Structure, regulation and function of NF-kB. Annu. Rev. Cell Biol. 10: 705-16.

Schreck, R., K. Albermann, and P.A. Baeuerle. 1992. Nuclear factor kB: an oxidant stress-responsive transcription factor of eukaryotoc cells. Free Rad. Res. Comm. 17:221-237.

Schulze-Osthoff, K., A.C. Bakker, B. Vanhaesebroeck, R. Beyaert, W.A. Jacob, and W. Fiers. 1992. Cytotoxic activity of tumor necrosis factor is mediatd by early damage of mitochondrial functions. J. Biol. Chem. 267: 5317-5323.

Staal, F.J.T., M. Roederer, L.A. Hersenberg, and L.A. Herzenberg, 1990. Intracellular thiols regulate activation of nuclear factor NF-kB and transcription of human immunodeficiency virus. Proc. Natl. Acad. Sci. 87: 9943-9947.

Turrens, J.F., A. Alexandre, and A.L. Lehninger. 1985. Ubisemiquinone is the electron donor for superoxide formation by complex III of heart mitochondria. Arch. Biochem. Biophys. 237:408-411.

Widmer, U., K.R. Manoque, A. Cerami, and B. Sherry, 1993. Genomic Cloning of Promoter Analysis of Macrophage Inflammatory Protein (MIP)-2, MIP-1α, and MIP-1β, Members of the Chemokine Superfamily of Proinflammatory Cytokines J. Immunol. 150:4996-5012.

Xing, Z., M. Jordana, H. Kirpalani, K.E. Driscoll, T.J. Schall, and J. Gauldie. 1994. Cytokine expression by neutrophils and macrophages *in vivo*: endotoxin induces tumor necrosis factor , macrophage inflammatory protein-2, interleukin 1 and interleukin-6, but not RANTES or transforming growth factor b, mRNA expression in acute lung inflammation. Am. J. Mol. Cell Biol. 10:148-153.

Zar, J.H. Biostatistical Analysis. 2nd ed. Englewood Cliffs; Prentice-Hall; 1984.

Zhao, Q., L. Simpson, K.E. Driscoll, and G.D. Leikauf. 1998. Chemokine regulation of ozone-induced neutrophil and monocyte inflammation. Am. J. Physiol. Lung Cell Mol Biol. 274:L39-L46.

QUINOIDS AS REACTIVE INTERMEDIATES IN ESTROGEN CARCINOGENESIS

Judy L. Bolton and Minsun Chang

Department of Medicinal Chemistry and Pharmacognosy (M/C 781)
College of Pharmacy
University of Illinois at Chicago
833 S. Wood St.
Chicago, Illinois
60612-7231

INTRODUCTION

The risk/benefits of traditional estrogen replacement therapy. Recent data indicate that only 30% of postmenopausal women in the United States are currently receiving estrogen replacement therapy (Harris et al., 1990). This is in spite of the proven health benefits of estrogen replacement therapy that has been shown not only to improve the quality of life for women but also has extended the life span for women. The benefits of estrogen replacement therapy include the relief of menopausal symptoms, a substantial reduction in the risk of cardiovascular disease and osteoporosis (Grodstein et al., 1997) and a possible correlation between the reduced risk of stroke (Paganinihill, 1995) and Alzheimer's disease (Wickelgren, 1997; Henderson, 1997; Haskell, 1997). However, many women fear traditional estrogen replacement therapy because of the clear association between excessive exposure to estrogens and the development of cancer in several tissues including breast, endometrium, liver, and kidney (Henderson et al., 1988; Liehr, 1990; Bolton et al., 1998). As a result, the consequences of estrogen deficiency versus the potential risks of estrogen replacement therapy, especially the ability of estrogens to initiate and/or promote the carcinogenic process, remains a highly controversial issue.

Most of the epidemiology studies on estrogen replacement therapy and cancer risk have been conducted with Premarin® (Wyeth-Ayerest) which remains the estrogen replacement treatment of choice and one of the most widely prescribed drugs in North America. Although Premarin® was approved by the Food and Drug Administration in the 1940s, very little is known about the metabolism and potential toxic metabolites that could be produced from the various equine estrogens which make up Premarin® (Purdy et al., 1982; Li et al, 1985; Sarabia et al., 1997; Zhang et al., 1999). It is known that treating hamsters for 9 months with either estrone, equilin + equilenin, or sulfatase-treated Premarin®, resulted in 100% kidney tumor incidences and abundant tumor foci (Li et al., 1995). Further, in a recent clinical trial of 596 postmenopausal women, a significant increase in endometrial hyperplasia was found in those women receiving a daily dose of 0.625 mg of Premarin® (Judd et al., 1996).

The molecular mechanism(s) involved in the carcinogenic action of estrogens still remains both controversial and elusive (Service, 1998). There are at least three different potential mechanisms of estrogen carcinogenesis which have been proposed. These include hormonal feedback inhibition, oxidative damage to DNA, and/or alkylation of DNA by electrophilic metabolites of estrogens (Yager and Liehr, 1996; Li et al., 1996; Zhu and Conney, 1998; Roy and Liehr, 1999). The first mechanism involves a break in the pituitary-gonadal hormone feedback loop, leading to a failure of the estrogens to suppress the pituitary production of gonadotropins. The excessive production of gonadotropins and their constant stimulation of the target organ may result in neoplasia. However, this cannot be the only mechanism because the hormonal potencies of various estrogens <u>do not</u> correlate with the incidence of tumor induction (Liehr, 1983; Li and Li, 1984). Rather, metabolic activation of estrogens to either redox active and/or electrophilic metabolites probably plays an etiologic role in tumor formation. For example, it is known that both estrogen and 17β-estradiol are readily oxidized to 2- and 4-hydroxylated catechols as shown in Figure 1. Once formed, these catechols can be oxidized by virtually any oxidative enzyme and/or metal ion giving *o*-quinones (Yager and Liehr, 1996). *o*-Quinones are potent redox active compounds as well as alkylating agents and formation of these reactive metabolites *in vivo* likely plays a role in estrogen carcinogenesis (Bolton et al., 1998; Bolton et al., 2000).

RESULTS AND DISCUSSION

Metabolism of estrogens to o-quinones. Interestingly, increasing unsaturation in the B ring of equine estrogens leads to a change in metabolism from predominately 2-hydroxylation for the endogenous estrogens estrone and 17β-estradiol to mainly 4-hydroxylation for equilin and exclusively 4-hydroxylation for equilenin (Figure 2). This could be problematic since 2-hydroxylation of endogenous estrogens is regarded as a benign metabolic pathway whereas 4-hydroxylation could lead to carcinogenic metabolites (Liehr et al., 1986; Li and Li, 1987). In fact, based on our progress to date, it is our <u>strong belief</u> that metabolism of equilenin or equilin (and their 17β-hydroxylated metabolites) to 4-hydroxyequilenin and 4,17β-dihydroxyequilenin represents a major carcinogenic pathway for equine estrogens.

In contrast to endogenous catechol estrogens, the 4-hydroxylated equine catechol estrogens (4-hydroxyequilenin and 4-hydroxyequilin) both autoxidize to *o*-quinones without the need for enzymatic or metal ion catalysis (Figure 2, Shen et al., 1997; Zhang et al., 1999). The *o*-quinone formed from 4-hydroxyequilenin is much more stable ($t_{1/2}$ = 2.3 h, Shen et al., 1997) than the endogenous catechol estrogen *o*-quinones. It appears that the adjacent aromatic ring stabilizes 4-hydroxyequilenin-*o*-quinone through extended π-conjugation. In support of this it has been shown that the catechol metabolite of benzo[a]pyrene rapidly undergoes air oxidation to yield a very stable *o*-quinone, benzo[a]pyrene-7,8-dione (Smithgall et al., 1988; Penning, 1993).

The 4-hydroxyequilin-*o*-quinone readily isomerizes to 4-hydroxyequilenin-*o*-quinone (Figure 2). As a result, most of the biological effects caused by catechol metabolites of equilin are likely due to 4-hydroxyequilenin-*o*-quinone formation (Zhang et al., 1999). Finally, although 2-hydroxylation does occur with equilin producing 2-hyroxyequilin which will isomerize to 2-hydroxyequilenin, the latter catechol does not autoxidize to an *o*-quinone at any appreciable rate (Zhang and Bolton, 1999). This suggests that similar to

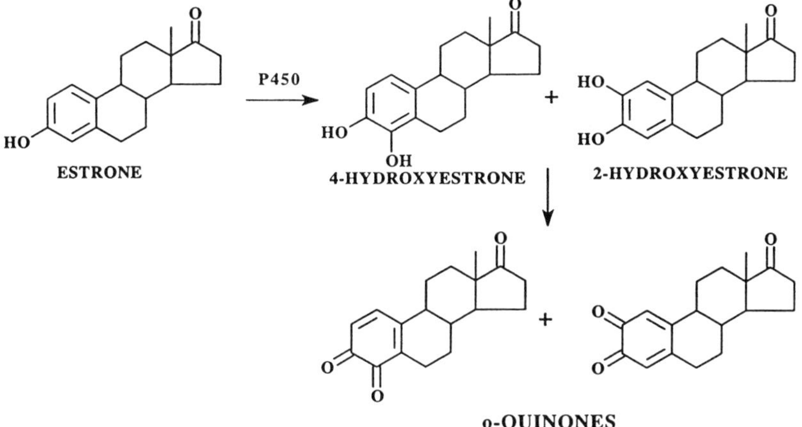

Figure 1. Phase I metabolism of endogenous estrogens.

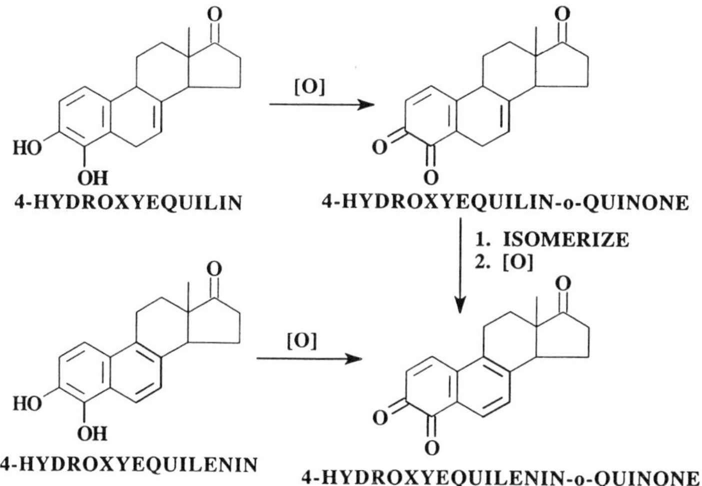

Figure 2. Primary phase I metabolic pathway for equine estrogens.

what has been observed with endogenous catechol estrogens, 2-hydroxylation is likely a benign metabolic pathway for equilin.

Consequences of estrogen quinoid formation. *(i) Protein alkylation.* Chemical or enzymatic activation of estrone or 17β-estradiol and their catechol metabolites leads to protein alkylation which may be responsible for the toxic effects of estrogen *o*-quinones (Nelson et al., 1976; Freyberger, A. and Degen, 1989). For the equine estrogens, we have found that both 4-hydroxyequilenin and 4-hydroxyequilin are potent irreversible inactivators of glutathione S-transferase (Chang et al., 1998). Preliminary LC-MS analysis of the intact enzyme showed that 4-hydroxyequilenin covalently modifies one amino acid residue which is likely to be Cys47 or Cys101 (Figure 3). Site directed mutagenesis is in progress to identify which cysteine residue is modified by 4-hydroxyequilenin. Finally, alkylation of DNA by electrophilic estrogen *o*-quinones may also play a role in initiation of the carcinogenic process (see below).

(ii) Formation of reactive oxygen species. *o*-Quinones are also potent redox active compounds (Monks et al., 1992, Bolton et al., 2000). They can undergo redox cycling with the semiquinone radical generating superoxide radicals mediated through cytochrome P450/P450 reductase (as shown in Figure 4 for 4-hydroxyequilenin). The reaction of superoxide anion radicals with hydrogen peroxide, formed by the enzymatic or spontaneous dismutation of superoxide anion radical, in the presence of trace amounts of iron or other transition metals gives hydroxyl radicals. The hydroxyl radicals are powerful oxidizing agents that may be responsible for damage to essential macromolecules. For example, oxidation of cysteine residues in proteins leads to disulfide bond formation which can dramatically alter structure and function. Hydroxyl radicals can also catalyze oxidation of lipids generating lipid hydroperoxides and elevated levels in lipid hydroperoxides have been detected in the kidneys of hamsters dosed with estrogens (Wang and Liehr, 1994; Wang and Liehr, 1995). Lipid peroxide-derived malondialdehyde DNA adducts were also elevated and these endogenous DNA adducts could contribute to estrogen carcinogenesis (Wang and Liehr, 1994). Finally, biomarkers for oxidative damage to DNA include the formation of the mutagenic lesion, 8-oxo-2'-deoxyguanosine (8-oxo-dG) (Shigenaga and Ames, 1991).

Catechol estrogen-*o*-quinone-induced DNA oxidation. The excessive production of reactive oxygen species in breast cancer tissue has been linked to metastasis of tumors in women with breast cancer (Floyd, 1990; Malins et al., 1996). The source of reactive oxygen species has been suggested to be the result of redox cycling between the *o*-quinones and their semiquinone radicals generating superoxide, hydrogen peroxide, and ultimately reactive hydroxyl radicals which cause oxidative cleavage of the phosphate-sugar backbone as well as oxidation of the purine/pyrimidine residues of DNA (Han and Liehr, 1994). In support of this mechanism, various free radical toxicities have been reported in hamsters treated with 17β-estradiol including DNA single strand breaks (Han and Liehr, 1994, Nutter et al., 1991), 8-oxo-dG formation (Han and Liehr, 1994), and chromosomal abnormalities (Li et al., 1993, Banerjee et al., 1994). Recently, we have shown that 4-hydroxyequilenin is also capable of causing DNA single strand breaks and oxidative damage to DNA bases (Chen et al., 1998). Treating λ phage DNA with 4-hydroxyequilenin resulted in extensive single strand breaks that were concentration and time dependent. By including scavengers of reactive oxygen species in the incubations, DNA could be completely protected from 4-hydroxyequilenin-mediated damage. In contrast, NADH and $CuCl_2$ enhanced the ability of 4-hydroxyequilenin to cause DNA single strand breaks presumably due to redox cycling between 4-hydroxyequilenin and the semiquinone radical generating hydrogen peroxide and ultimately copper peroxide

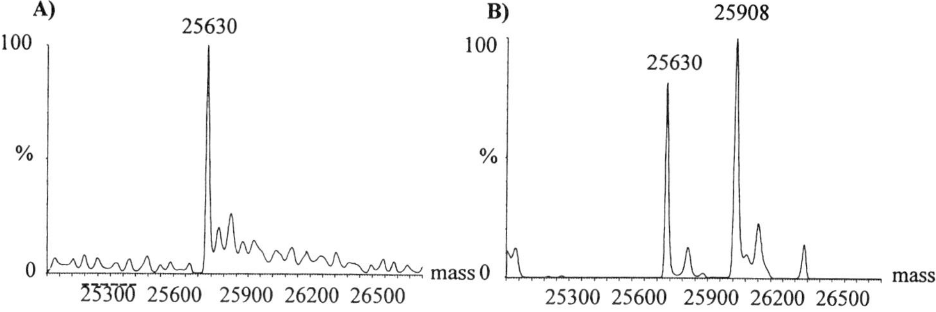

Figure 3. (A) Electrospray ionization maximum entropy transformed spectrum of wild type hexahistidine tagged human glutathione S-transferase P1-1. (B) Electrospray ionization maximum entropy transformed spectrum of wild type hexahistidine tagged human glutathione S-transferase P1-1 incubated with 1 molar equivalent of 4-hydroxyequilenin. The molecular mass addition relative to control protein of 278 Da corresponds to addition of 4-hydroxyequilenin o-quinone.

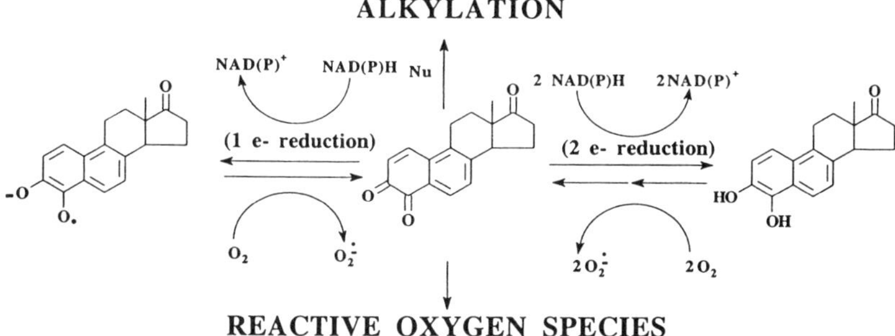

Figure 4. Potential toxic mechanism(s) of 4-hydroxyequilenin quinoids.

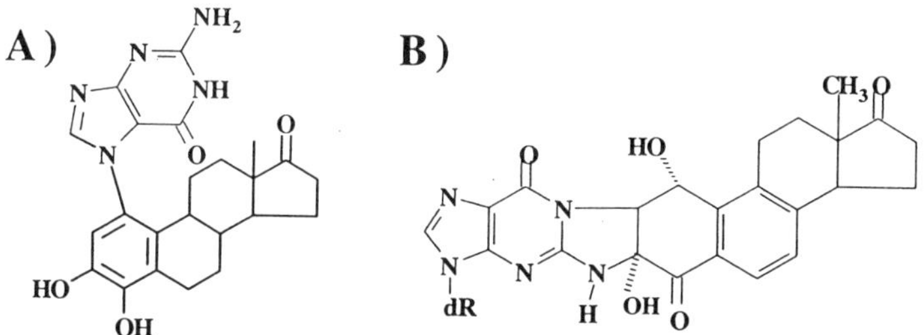

Figure 5. DNA adducts formed by estrogen quinones. A) 4-hydroxyestrone N-7 guanine adduct. B) Cyclic deoxyguanosine adducts formed from 4-hydroxyequilenin quinoids.

complexes. It was confirmed that 4-hydroxyequilenin could oxidize DNA bases because hydrolysis of 4-hydroxyequilenin treated calf thymus DNA and HPLC separation with electrospray-MS detection revealed oxidized deoxynucleosides including 8-oxo-dG and 8-oxo-dA. 4-Hydroxyequilenin also caused a dose-dependent increase in the mutagenic lesion 8-oxo-dG in breast cancer cells as determined by LC-MS-MS (Chen et al., 2000). In support of this, previous reports have shown that incubations with 4-hydroxyequilenin-*o*-quinone, DNA, and hamster liver microsomes also enhanced 8-oxo-dG formation (Han and Liehr, 1995). Using the single cell gel electrophoresis assay (comet assay) to measure DNA damage, we found that 4-hydroxyequilenin causes concentration-dependent DNA single strand cleavage in breast cancer cell lines and this effect could be enhanced by NADH or diethyl maleate (Chen et al., 2000). These and other data are evidence for a mechanism of estrogen-induced tumor initiation/promotion by redox cycling of estrogen metabolites generating reactive oxygen species which damage DNA.

DNA adducts formed by catechol estrogen-*o*-quinones. The formation of covalent DNA adducts *in vivo* is usually regarded as the initiation event in the carcinogenic process. With estrogens, DNA adducts have been detected *in vivo* using ^{32}P-postlabeling methods in susceptible target organs such as the Syrian hamster kidney (Liehr et al., 1986; Liehr et al., 1987). Model studies with catechol estrogen *o*-quinones and deoxynucleosides or bases showed that different types of adducts are obtained depending on the reaction conditions and the reactivity of the *o*-quinone. For example, when 4-hydroxyestrone-*o*-quinone was reacted with adenine in organic solvent under reductive conditions, an adduct was isolated consistent with coupling between the C-1 position of 4-hydroxyestrone and the C-8 position of adenine (Abul-Hajj et al., 1995). In contrast, incubating 4-hydroxyestrone-*o*-quinone with dA in an acidic environment resulted in no detectable adduct formation (Stack et al., 1996). Reaction with dG gave an adduct that had lost the ribose moiety resulting from reaction of the N^7 position of dG with the 4-hydroxyestrone-*o*-quinone at the C-1 position (Figure 5). N^7 adducts are extremely unstable and readily depurinate leading to mutations (Loeb and Preston, 1986). DNA adducts have also been isolated from rats treated with 4-hydroxyestrone-*o*-quinone. This *o*-quinone was directly injected into the rat mammary gland, the mammary tissue was subjected to Soxhlet extraction, and the extracts were analyzed by HPLC to determine the extent of depurinating adducts formed *in vivo* (Cavalieri et al., 1997). The adduct detected was 4-hydroxyestrone-N^7-guanine, identical to that obtained from the model studies with 4-hydroxyestrone-*o*-quinone and dG. DNA also was isolated from the mammary tissue, hydrolyzed to deoxynucleosides, and analyzed by HPLC. No stable adducts were detected under these conditions which suggests that only apurinic sites would be formed from reaction of 4-hydroxyestrone-*o*-quinone with DNA *in vivo*.

The quinoids of the equilenin metabolite 4-hydroxyequilenin reacted with 2'-deoxynucleosides generating very unusual cyclic adducts (Figure 5, Shen et al., 1997; Shen et al., 1998; Bolton et al., 1998). Since 4-hydroxyequilin is converted to 4-hydroxyequilenin and 4-hydroxyequilenin-*o*-quinone, the same adducts were observed during incubations with 4-hydroxyequilin and deoxynucleosides or DNA (Zhang et al., 1999). Deoxyguanosine (dG), dA, or dC all gave four isomers but no product was observed for thymidine under similar physiological conditions. With DNA, significant apurinic sites were produced as 4-hydroxyequilenin-adenine adducts were detected in the ethanol wash prior to hydrolysis. When the DNA was hydrolyzed to deoxynucleosides and analyzed by electrospray mass spectrometry, only a single isomer of 4-hydroxyequilenin-dG and 4-hydroxyequilenin-dC was observed. These data suggest that several different types of DNA lesions could be expected from 4-hydroxyequilenin including apurinic sites and bulky stable adducts, in addition to the reported oxidized damage to DNA (Han and

Liehr, 1995; Chen et al., 1998; Chen et al., 2000) caused by 4-hydroxyequilenin. If similar adducts are formed *in vivo*, which are not repaired efficiently, mutations could result leading to initiation of the carcinogenic process in the endometrium or the breast.

Role of the estrogen receptor in estrogen carcinogenesis. It is well known that the estrogen receptor plays a major role in the mechanism of estrogen-induced carcinogenesis (Li et al., 1996; Henderson and Feigelson, 2000). The theory is that excessive binding to the estrogen receptor leads to an increase in cell proliferation in hormone-sensitive target tissues such as the breast and endometrium. In these rapidly dividing cells, the chances for mutations to occur increases dramatically leading to initiation/promotion of the carcinogenic process. Recently, we have hypothesized that the estrogen receptor could also play a role in catechol estrogen/*o*-quinone-induced carcinogenesis. It is quite possible that the catechol estrogen and/or *o*-quinone binds to the estrogen receptor, which carries it directly to estrogen sensitive genes, where DNA damage occurs resulting in mutations.

We have preliminary data that this mechanism may play a role in catechol estrogen-induced toxicity, DNA damage, and apoptosis (Chen et al., 2000). Recently, we have obtained two breast cancer cell lines either without the estrogen receptor (MDA-MB-231) or the same cell line stably transfected with ERα (S30 cells). Using the comet assay to measure DNA damage, we found that 4-hydroxyequilenin causes concentration dependent DNA single strand cleavage in both cell lines and this effect could be enhanced by agents which catalyze redox cycling (NADH) or deplete cellular GSH (diethyl maleate). In addition, the ER^+ cell line (S30) was considerably more sensitive to induction of DNA damage by 4-hydroxyequilenin compared to the ER^- cells (MDA-MB-231). 4-Hydroxyequilenin also caused a concentration dependent increase in the mutagenic lesion 8-oxo-dG in the S30 cells as determined by LC-MS-MS. Cell morphology assays showed that 4-hydroxyequilenin induces apoptosis in these cell lines. As observed with the toxicity assay and the comet assay, the ER^+ cells were more sensitive to induction of apoptosis by 4-hydroxyequilenin as compared to MDA-MB-231 cells. Finally, the endogenous catechol estrogen metabolite 4-hydroxyestrone was considerably less effective at inducing DNA damage and apoptosis in breast cancer cell lines as compared to 4-hydroxyequilenin. Our data suggest that the cytotoxic effects of 4-hydroxyequilenin could be related to its ability to induce DNA damage and apoptosis in hormone sensitive cells *in vivo,* and these effects may be potentiated by the estrogen receptor.

Finally, it should be noted that a new estrogen receptor has recently been discovered called ERβ (Gustafsson, 1999). The DNA-binding domains of ERα and ERβ show a high degree of homology whereas the ligand-binding domain only shows only 59% homology. This means it should be possible to develop ERα and ERβ specific ligands. In addition to differences in binding domains, the expression of these receptors is also tissue specific. As a result, it appears that ERα is present in much higher amounts in the breast and endometrium whereas ERβ maybe much more important in the cardiovascular and central nervous systems. At present it is not known what the relative affinity of the catechol estrogens for each receptor and this will be a major focus of future work.

CONCLUSIONS

The roles of quinones in mediating the adverse effects of estrogens have not been investigated in detail. It is possible for these electrophilic/redox active quinones to cause damage within cells by a variety of different pathways. Oxidative enzymes, metal ions, and in some cases molecular oxygen can catalyze quinoid formation, so alkylation of cellular nucleophiles (GSH, proteins, DNA) by these species may occur to a significant extent in many tissues. In addition, the formation of reactive oxygen species especially through redox cycling between the quinones and semiquinone radicals, could contribute to the adverse properties of the parent compounds. Redox cycling can cause lipid peroxidation, consumption of reducing equivalents, oxidation of DNA, and DNA strand breaks. DNA binding occurs, but the sites of alkylation and relationships to cytotoxicity are dependent on the chemical structure of the quinones and the cellular environment in which they are formed. Finally, ERα and/or ERβ may play a role in potentiating the deleterious effects of catechol estrogens and/or *o*-quinones. Given the direct link between excessive exposure to estrogens, metabolism of estrogens, and increased risk of breast cancer, it is crucial that factors which affect the formation, reactivity, and cellular targets of estrogen quinoids be thoroughly explored.

ACKNOWLEDGMENT

This research was supported by NIH grant CA73638.

REFERENCES

Abul-Hajj, Y. J., Tabakovic, K. and Tabakovic, I., 1995, An estrogen-nucleic acid adduct. Electroreductive intermolecular coupling of 3,4-estrone-o-quinone and adenine, *J. Am. Chem. Soc.* **117:** 6144-6145.

Banerjee, S. K., Banerjee, S., Li, S. A. and Li, J. J., 1994, Induction of chromosome aberrations in Syrian hamster renal cortical cells by various estrogens, *Mutat. Res.* **311**: 191-197.

Bolton, J. L., Pisha, E., Zhang, F. and Qiu, S., 1998, Role of quinoids in estrogen carcinogenesis, *Chem. Res. Toxicol.* **11:** 1113-1127.

Bolton, J. L., Trush, M. A., Penning, T. M., Dryhurst, G. and Monks, T. J., 2000, Role of quinones in toxicology, *Chem. Res. Toxicol.* **13:** 135-160.

Cavalieri, E. L., Stack, D. E., Devanesan, P. D., Todorovic, R., Dwivedy, I., Higginbotham, S., Johansson, S. L., Patil, K. D., Gross, M. L., Gooden, J. K., Ramanathan, R., Cerny, R. L. and Rogan, E. G., 1997, Molecular origin of cancer: Catechol estrogen-3,4,-quinones as endogenous tumor initiators, *Proc. Natl. Acad. Sci. USA* **94**: 10937-10942.

Chang, M., Zhang, F., Shen, L., Pauss, N., Alam, I., van Breemen, R. B., Blond, Y. S., and Bolton, J. L., 1998, Inhibition of glutathione S-transferase activity by the quinoid metabolites of equine estrogens, *Chem. Res. Toxicol.* **11:** 758-765.

Chen, Y., Shen, L., Zhang, F., Lau, S. S., van Breemen, R. B., Nikolic, D. and Bolton, J. L., 1998, The equine estrogen metabolite 4-hydroxyequilenin causes DNA single strand breaks and oxidation of DNA bases *in vitro*, *Chem. Res. Toxicol.* **11**: 1105-1111.

Chen, Y., Liu, X., Pisha, E., Constantinou, A. I., Hua, Y., Shen, L., van Breemen, R. B., Elguindi, E. C., Blond, S. Y., Zhang, F. and Bolton, J. L., 2000, A metabolite of equine estrogens, 4-hydroxyequilenin, induces DNA damage and apoptosis in breast cancer cell lines, *Chem. Res. Toxicol.* **13:** 342-350.

Fishman, J., 1983, Aromatic hydroxylation of estrogens, *Ann. Rev. Physiol.* **45:** 61-72.

Floyd, R. A., 1990, The role of 8-hydroxyguanine in carcinogenesis, *Carcinogenesis* **11**: 1447-1450.

Freyberger, A. and Degen, G. H., 1989, Covalent Binding to Proteins of Reactive Intermediates Resulting from Prostaglandin H Synthase-Catalyzed Oxidation of Stilbene and Steroid Estrogens, *J. Biochem. Toxicology* **4**: 95-103.

Grodstein, F., Stampfer, M. J., Colditz, G. A., Willett, W. C., Manson, J. E., Joffe, M.,

Rosner, B., Fuchs, C., Hankinson, S. E., Hunter, D. J., Hennekens, C. H. and Speizer, F. E., 1997, Postmenopausal hormone therapy and mortality, *New Eng. J. Med.* **336:** 1769-1775.

Gustafsson, J. A., 1999, Estrogen receptor beta--a new dimension in estrogen mechanism of action, *J. Endocrinol.* **163:** 379-383.

Han, X. and Liehr, J. G., 1994, 8-Hydroxylation of guanine bases in kidney and liver DNA of hamsters treated with estradiol. Role of free radicals in estrogen-induced carcinogenesis, *Cancer Res.* **54**: 5515-5517.

Han, X. and Liehr, J. G., 1995, Microsome-mediated 8-hydroxylation of guanine bases of DNA by steroid estrogens: correlation of DNA damage by free radicals with metabolic activation to quinones, *Carcinogenesis* **16**: 2571-2574.

Harris, R. B., Laws, A., Reddy, F. M., King, A. and Haskell, W. L., 1990, Are women using postmenopausal estrogens? A community survey, *Am. J. Public Health* **80:** 1266-1268.

Haskell, S. G., Richardson, E. D. and Horwitz, R. I., 1997, The effect of estrogen replacement therapy on cognitive function in women: A critical review of the literature, *J. Clin. Epidemiology* **50:** 1249-1264.

Henderson, B. E., Ross, R. and Bernstein, L., 1988, Estrogens as a cause of human cancer: The Richard and Hinda Rosenthal Foundation Award Lecture, *Cancer Res.* **48:** 246-253.

Henderson, B. E. and Feigelson, H. S., 2000, Hormonal carcinogenesis, *Carcinogenesis* **21:** 427-433.

Judd, H. L., Mebane-Sims, I., Legault, C., Wasilauskas, C., Johnson, S., Merino, M., Barrett-Connor, B. and Trabal, J., 1996, Effects of hormone replacement therapy on endometrial histology in postmenopausal women, *JAMA* **275:** 370-375.

Li, J. J. and Li, S. A., 1984, Estrogen-induced tumorigenesis in hamsters: Roles for hormonal and carcinogenic activities, *Arch. Toxicol.* **55:** 110-118.

Li, S. A., Klicka, J. K. and Li, J. J., 1985, Estrogen 2- and 4-hydroxylase activity, catechol estrogen formation, and implications for estrogen carcinogenesis in the hamster kidney, *Cancer Res.* **45:** 181-185.

Li, J. J. and Li, S. A., 1987, Estrogen carcinogenesis in Syrian hamster tissues: role of metabolism, *Fed. Proc.* **46:** 1858-1863.

Li, J. J., Gonzalez, A., Banerjee, S., Banerjee, S. K. and Li, S. A., 1993, Estrogen carcinogenesis in the hamster kidney: role of cytotoxicity and cell proliferation, *Environ. Health Perspect.* **5**: 259-264.

Li, J. J., Li, S. A., Oberley, T. D. and Parsons, J. A., 1995, Carcinogenic activities of various steroidal and nonsteroidal estrogens in the hamster kidney: relation to hormonal activity and cell proliferation, *Cancer Res.* **55:** 4347-4351.

Li, J. J., Li, S. A., Gustafsson, J. A., Nandi, S. and Sekely, L. I. (1996) Hormonal Carcinogenesis II. Springer-Verlag, New York.

Liehr, J. G., 1983, 2-Fluoroestradiol. Separation of Estrogenicity from Carcinogenicity, *Mol. Pharm.* **23:** 278-281.

Liehr, J. G., Fang, W. R., Sirbasku, D. A. and Ari-Ulubelen, A., 1986, Carcinogenicity of catechol estrogens in Syrian hamsters, *J. Steroid Biochem.* **24:** 353-356.

Liehr, J. G., Avitts, T. A., Randerath, E. and Randerath, K., 1986, Estrogen-induced endogenous DNA adduction: Possible mechanism of hormonal cancer, *Proc. Natl. Acad. Sci. USA* **83:** 5301-5305.

Liehr, J. G., Hall, E. R., Avitts, T. A., Randerath, E. and Randerath, K., 1987, Localization of estrogen-induced DNA adducts and cytochrome P450 activity at the site of renal carcinogenesis in the hamster kidney., *Cancer Res.* **47:** 2156-2159.

Liehr, J. G., 1990, Genotoxic effects of estrogens., *Mutation Res.* **238:** 269-276.

Loeb, L. A. and Preston, B. D., 1986, Mutagenesis by apurinic/apyrimidinic sites, *Ann. Rev. Genet.* **20:** 201-203.

Maclusky, N. J., Naftolin, F., Krey, L. C. and Franks, S., 1981, The catechol estrogens, *J. Steroid Biochem.* **15**: 111-124.

Malins, D. C., Polissar, N. L. and Gunselman, S. J., 1996, Progression of human breast cancers to the metastatic state is linked to hydroxyl radical-induced DNA damage, *Proc. Natl. Acad. Sci. USA* **93**: 2557-2563.

Monks, T. J., Hanzlik, R. P., Cohen, G. M., Ross, D. and Graham, D. G., 1992, Contemporary issues in toxicology: Quinone chemistry and toxicity, *Toxicol. Appl. Pharmacol.* **112**: 2-16.

Nelson, S. D., Mitchell, J. R., Dybing, E. and Sasame, H. A., 1976, Cytochrome P450-mediated oxidation of 2-hydroxyestrogens to reactive intermediates, *Biochem. Biophys. Res. Commun.* **70**: 1157-1165.

Nutter, L. M., Ngo, E. O. and Abul-Hajj, Y. J., 1991, Characterization of DNA damage induced by 3,4-estrone-o-quinone in human cells, *J. Biol. Chem.* **266**: 16380-16386.

Paganinihill, A., 1995, Estrogen replacement therapy and stroke, *Progess Cardiovascular Diseases* **38**: 223-242.

Penning, T. M., 1993, Dihydrodiol dehydrogenase and its role in polycyclic aromatic hydrocarbon metabolism, *Chem.-Biol. Interact.* **89**: 1-34.

Purdy, R. H., Moore, P. H., Williams, M. C., Goldzheher, H. W. and Paul, S. M., 1982, Relative rates of 2- and 4-hydroxyestrogen synthesis are dependent on both substrate and tissue, *FEBS Lett.* **138**: 40-44.

Roy, D. and Liehr, J. G., 1999, Estrogen, DNA damage and mutations, *Mutat. Res.* **424**: 107-115.

Sarabia, S. F., Zhu, B. T., Kurosawa, T., Tohma, M. and Liehr, J. G., 1997, Mechanism of cytochrome P450-catalyzed aromatic hydroxylation of estrogens, *Chem. Res. Toxicol.* **10**: 767-771.

Service, R. F., 1998, New role for estrogen in cancer?, *Science* **279**: 1631-1633.

Shen, L., Pisha, E., Huang, Z., Pezzuto, J. M., Krol, E., Alam, Z., van Breemen, R. B. and Bolton, J. L., 1997, Bioreductive activation of catechol estrogen-ortho-quinones: Aromatization of the B ring in 4-hydroxyequilenin markedly alters quinoid formation and reactivity, *Carcinogenesis* **18**: 1093-1101.

Shen, L., Qiu, S., van Breemen, R. B., Zhang, F., Chen, Y. and Bolton, J. L., 1997, Reaction of the Premarin® metabolite 4-hydroxyequilenin semiquinone radical with 2'-deoxyguanosine: Formation of unusual cyclic adducts, *J. Am. Chem. Soc.* **119**: 11126-11127.

Shen, L., Qiu, S., Chen, Y., Zhang, F., van Breemen, R. B., Nikolic, D. and Bolton, J. L., 1998, Alkylation of 2'-deoxynucleosides and DNA by the Premarin® metabolite 4-hydroxyequilenin semiquinone radical, *Chem. Res. Toxicol.* **11**: 94-101.

Shigenaga, M. K. and Ames, B. N., 1991, Assays for 8-hydroxy-2'-deoxyguanosine: A biomarker of in vivo oxidative DNA damage, *Free Radical Biol. Med.* **10**: 211-216.

Smithgall, T. E., Harvey, R. G. and Penning, T. M., 1988, Spectroscopic identification of ortho-quinones as the products of polycyclic aromatic trans-dihydrodiol oxidation catalyzed by dihydrodiol dehydrogenase, *J. Biol. Chem.* **263**: 1814-1820.

Stack, D. E., Byun, J., Gross, M. L., Rogan, E. G. and Cavalieri, E. L., 1996, Molecular characteristics of catechol estrogen quinones in reactions with deoxyribonucleosides, *Chem. Res. Toxicol.* **9**: 851-859.

Wang, M. and Liehr, J. G., 1994, Identification of fatty acid hydroperoxide cofactors in the cytochrome P450-mediated oxidation of estrogens to quinone metabolites. Role and balance of lipid peroxides during estrogen-induced carcinogenesis, *J. Biol. Chem.* **269**: 284-291.

Wang, M. Y. and Liehr, J. G., 1995, Induction by estrogens of lipid peroxidation and lipid peroxide-derived malonaldehyde-DNA adducts in male Syrian hamsters: role of lipid peroxidation in estrogen-induced kidney carcinogenesis, *Carcinogenesis* **16**: 1941-1945.

Wickelgren, I., 1997, Estrogen: A new weapon against Alzheimer's, *Science* **276**: 676-677.

Yager, J. D. and Liehr, J. G., 1996, Molecular mechanisms of estrogen carcinogenesis, *Annu. Rev. Pharmacol. Toxicol.* **36**: 203-232.

Zhang, F. and Bolton, J. L., 1999, Synthesis of the equine estrogen metabolites 2-hydroxyequilin and 2-hydroxyequilenin, *Chem. Res. Toxicol.* **12**: 200-203.

Zhang, F., Chen, Y., Pisha, E., Shen, L., Xiong, Y., van Breemen, R. B. and Bolton, J. L., 1999, The major metabolite of equilin, 4-hydroxyequilin autoxidizes to an o-quinone which isomerizes to the potent cytotoxin 4-hydroxyequilenin-o-quinone, *Chem. Res. Toxicol.* **12**: 204-213.

Zhu, B. T. and Conney, A. H., 1998, Functional role of estrogen metabolism in target cells: review and perspectives, *Carcinogenesis* **19**: 1-27.

ASSESSING UNDERLYING MECHANISMS OF QUINOID-INDUCED HEMATO-POIETIC CELL TOXICITY

Michael A. Trush, Hong Zhu and Yunbo Li
Department of Environmental Health Sciences
Johns Hopkins University School of Hygiene and Public Health
Baltimore, MD 21205

INTRODUCTION

Within the bone marrow are a number of cell populations that can serve as potential targets of xenobiotics and/or their metabolites (Trush et al., 1996). These include the hematopoietic and lymphopoietic stem cells, committed progenitors, mature functional blood cells and the cells that comprise the bone marrow stromal microenvironment. The altered function of different populations probably results in the manifestation of different toxicities. Two environmental chemicals that have been linked with toxicity to the bone marrow are benzene and benzo(a)pyrene (BP). There is continued concern about human exposure to benzene arising from a variety of sources including certain occupational settings, hazardous waste sites, automobiles and cigarette smoking. Exposure of humans to benzene is associated with the development of aplastic anemia and leukemia. Polycyclic aromatic hydrocarbons (PAHs), such as BP, have been shown to be myelotoxic in animals (Legraverend et. al., 1983). For example DBA/2 mice, with the Ah^d/Ah^d genotype, develop an aplastic anemia-like condition and leukemia in response to orally administered benzo(a)pyrene. In this regard, diet is considered a major route of exposure to PAHs in humans (Buckley and Lioy, 1992). Whether dietary PAHs pose a risk to human bone marrow in poorly investigated from either an experimental or an epidemiologic perspective. The bone marrow toxicity of both benzene and BP has been linked to their biotransformation. Interestingly, both compounds are metabolized to quinone derivatives: benzene to hydroquinone; BP to 1,6-, 3,6-, and 6,12-quinone. Because of the importance of the bone marrow stromal microenvironment, we have studied the toxicity of hydroquinone and BP-derived quinones to stromal cells obtained from DBA/2 mice (Twerdok et al., 1992; Zhu et al., 1995). The observations obtained from these studies are summarized in Table 1. Based on these observations we propose that the conversion of hydroquinone to benzoquinone and the conversion of BP-quinones to a semiquinone redox cycling radical occur in different cellular compartments. The purpose of this study was to further examine the bioactivation of hydroquinone in ML-1 cells, a human myeloid leukemia cell line and the bioactivation of BP-derived quinones in DBA/2 bone marrow stromal cells.

Biological Reactive Intermediates VI, Edited by Dansette *et al.*
Kluwer Academic / Plenum Publishers, 2001

Table 1 Comparison of the Toxicity of Hydroquinone (HQ) and BP-1,6-Quinone to Bone Marrow Stromal Cells

	HQ	BP-1,6-Quinone
Selectivity Toxic to Stromal Macrophages	Yes	No
Induction of GSH or Quinone Reductase Decreases Toxicity	Yes	No
Depletion of GSH or Inhibition of Quinone Reductase Increases Toxicity	Yes	No

METHODS

Cellular glutathione (GSH) was determined fluorometrically as described by Zhu et al., 1995b. To modulate CuZn SOD activity undifferentiated ML-1 cells were transfected with either the sense or the antisense (5'-CACACGGCCTTCGTCGCCATAACTCGCTAG-3') phosphorothioate oligodeoxynucleotide of CuZnSOD. Cultures were analyzed for CuZnSOD activity and then subjected to various concentrations of HQ to determine if antisense treatment modulates the bioeffects of HQ. Cells were also treated with cyanide to inhibit CuZnSOD, or azide to inhibit MPO. Cellular ATP content was measured using the luciferase-luciferin procedure (Zhu et al., 1995), and cellular oxygen consumption was monitored in a Clark-type electrode (Li et al., 1996). Reactive oxygen species generation was assessed by lucigenin- and luminol-derived chemiluminescence (CL) as described by Li et al., (1999a,b).

RESULTS AND DISCUSSION

Our previous studies with mouse stromal cells indicated that mitochondria morphology and function were being altered by the addition of BP-derived quinones (Zhu et al., 1995). This was consistent with the data presented in Table 1 in that modulation of cytoplasmic GSH or quinone reductase did not affect BP-1,6-quinone-induced cytotoxicity. Upon addition of 10μM BP-1,6-quinone to stromal cells a stimulation of cyanide-resistant oxygen consumption was observed, which was inhibitable by rotenone. This observation indicates that BP-1,6-quinone was undergoing redox cycling in the mitochondria generating reactive oxygen species (ROS). To test this possibility we used lucigen-derived CL to assess intramitochondrial superoxide generation and luminol-derived CL to assess extramitochondrial hydrogen peroxide as previously described by Li et al., (1999a, b). As shown in Figure 1 the addition of 1μM BP-1,6-quinone to intact stromal cells resulted in a significant increase in both superoxide and hydrogen peroxide generation which could be modulated by the mitochondrial electron transport chain inhibitors rotenone, antimycin A and myxothiazol. Similar BP-1,6-quinone-induced stimulation of ROS generation could be observed with mitochondria isolated from stromal cells or intact undifferentiated ML-1 cells, a human myeloblastic cell line.

In contrast to BP-1,6-quinone, the addition of up to 1mM hydroquinone to cells did not result in an increase in either lucigenin- or luminol-derived CL. This observation suggests that the mechanism of toxicity of hydroquinone to cells is probably different than that elicited by BP-1,6-quinone. A likely possibility is the conversion of hydroquinone to the highly protein reactive electrophile benzoquinone. This is consistent with the data presented in Table 1 indicating that modulation of cytoplasmic GSH and quinone reductase affects hydroquinone-

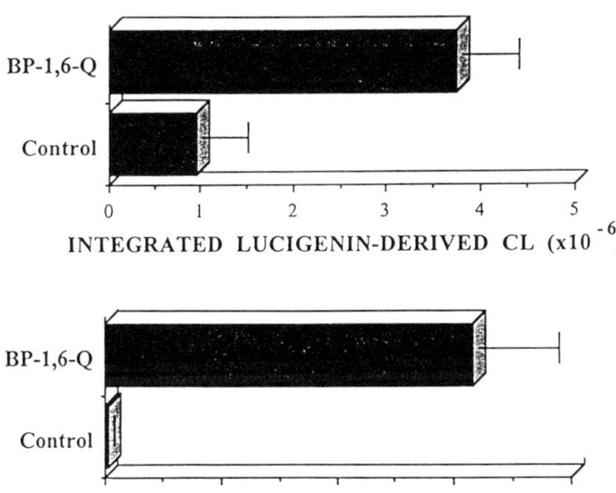

Figure 1. Assessment of ROS generation by 10^6 stromal cells in the absence or presence of 1µM BP-1-6-quinone.

induced toxicity to stromal cells. Likewise, modulation of GSH affects the toxicity of hydroquinone to undifferentiated ML-1 cells (Li et al., 1994; Trush et al., 1996). In addition, hydroquinone cytotoxity to both stromal cells and ML-1 cells is preceeded by a decrease in cytoplasmic GSH (Li and Trush, 1992; Li et al., 1994). This suggests that the bioactivation of hydroquinone may be cytoplasmic. Previously, we have demonstrated that CuZn SOD could facilitate this process (Li et al., 1996a,b). To evaluate this possibility we used undifferentiated ML-1 cells and assessed hydroquinone-induced GSH depletion and toxicity. As indicated by the data presented in Table 2 these effects could be reduced by the addition of 100µM KCN, a CuZnSOD inhibitor, but not by 100µM azide, an MPO inhibitor. KCN can be used in this experiment since undifferentiated ML-1cells lack mitochondrial respiration. A preliminary experiment showed that ML-1 cells transfected with an CuZnSOD antisense oligonucleotide exhibited less hydroquinone-induced GSH depletion and cytotoxicity than control cells.

Table 2. Effect of KCN or Azide on HQ-Induced GSH Depletion or Cytotoxicity in ML-1 Cells.

Treatment	GSH Content %[1]	Viability %[2]
None	25	100
50µM HQ	3	1
50µM HQ+.lmM KCN	15	40
50µM HQ+.lmM Azide	3	1

1 GSH content nmoles/10^7 cells determined at 3 hrs.
2 Cellular viability determined at 48 hrs.

Based on these comparative studies, it appears that hydroquinone and BP-1,6-quinone are bioactivated in hematopoietic cells in different compartments, cytoplasmic versus mitochondrial respectively. As such, molecular targets for these two quinoids are likely to differ. Upon differentiation of ML-1 cells to macrophages there is an increase in mitochondrial respiration as well as significant increases in cytoplasmic GSH and quinone reductase (Trush et al, 1996). In this regard, we have observed that BP-1,6-quinone is more cytotoxic to differentiated ML-1 cells as compared to undifferentiated cells, whereas hydroquinone is more cytotoxic to undifferentiated ML-1 cells.

ACKNOWLEDGEMENTS

We gratefully acknowledge financial support from ES 03760, ES 03818 and ES 08078.

REFERENCES

Buckley, T.J., and Lioy, P.J., 1992, An examination of the time course from human dietary   exposure to polycyclic aromatic hydrocarbons to urinary elimination of 1-hydroxypyrene. *Brit. J. Ind. Med.,* 49:113.

Legraverend, D., Harrison, D.E., Ruscetti, F.W., and Nebert, D.W., 1983, Bone marrow toxicity induced by oral benzo(a)pyrene: protection resides at the level of the intestine and liver. *Toxicol. Appl. Pharmacol.,* 70:390.

Li, Y., and Trush, M.A., 1992, Alteration of cellular glutathione as a factor in hydroquinone-induced cytotoxicity to primary cultured bone marrow stromal cells from DBA/2 mice. *In Vitro Toxicology: A Journal of Molecular and Cellular Toxicology,* 5:59.

Li, Y., Lafuente, A., and Trush, M.A., 1994, Characterization of quinone reductase, glutathione and glutathione S-transferase in human myeloid cell lines: induction by 1,2-dithiole-3-thione and effects on hydroquinone-induced cytotoxicity. *Life Sciences,* 54:901.

Li, Y., Kuppusamy, P., Zweier, J.L., and Trush, M.A., 1996, Role of CuZn-superoxide dismutase (Cu,ZnSOD) in xenobiotic activation: I. Chemical reactions involved in the CuZnSOD-accelerated oxidation of the benzene metabolite, 1,4-hydroquinone. *Mol. Pharmacol.,* 49:404.

Li, Y., Kuppusamy, P., Zweier, J.L., and Trush, M.A., 1996, Role of CuZn-superoxide dismutase (CuZnSOD) in xenobiotic activation: II. biological effects resulting from the CuZnSOD-accelerated oxidation of the benzene metabolite, 1,4-hydroquinone. *Mol. Pharmacol.,* 49:412.

Li, Y., Zhu, H., and Trush, M.A., 1999a, Detection of mitochondria-derived reactive oxygen species production by the chemilumigenic proges lucigenin and luminol. *Biochem. Biophys. Acta,* 1428:1.

Li, Y., Stansbury, K.H., Zhu, H., and Trush, M.A., 1999b, Biochemical characterization of lucigenin (bis-N-methylacridinium) as a unique chemiluminescent probe for detecting intramitchondrial superoxide anion radical production. *Biochem. Biophys. Res. Commun.,* 262:80.

Twerdok, L.E., Rembish, S.J., and Trush, M.A., 1992, Induction of quinone reductase and glutathione in bone marrow stromal cells by 1,2-dithiole-3-thione: effect on hydroquinone-induced cytotoxicity. *Toxicol. and Appl. Pharmachol.,* 112:273.

Trush, M.A., Twerdok, L.E., Rembish, S.J., Zhu, H., and Li, Y., 1996, Analysis of target cell susceptibility as a basis for the development of a chemoprotective strategy against benzene-induced hematotoxicities. *Env. Health Perspectives,* 104 (Suppl 6):1227.

Zhu, H., Li, Y., and Trush, M.A., 1995a, Characterization of benzo(a) pyrene quinone-induced toxicity to primary cultured bone marrow stromal cells from DBA/2 mice: potential role of mitochondrial dysfunction. *Toxicol. Appl. Pharmachol.,* 130:108.

Zhu, H., Li, Y., and Trush, M.A., 1995b, Differences in xenobiotic detoxifying activites between bone marrow stromal cells from mice and rats: implications for benzene-induced hematoxicity. *J. Toxicol. Env. Health,* 46:183.

STRUCTURE OF THE MALONDIALDEHYDE DEOXYGUANOSINE ADDUCT M_1G WHEN PLACED OPPOSITE A TWO-BASE DELETION IN THE $(CpG)_3$ FRAMESHIFT HOTSPOT OF THE *SALMONELLA TYPHIMURIUM hisD3052* GENE

Nathalie C. Schnetz-Boutaud[1], Samir Saleh[2], Lawrence J. Marnett[3] and Michael P. Stone[1]

Departments of Chemistry[1], Pharmacology[2], and Biochemistry[3], Center of Molecular Toxicology, and the Vanderbilt-Ingram Cancer Center, Vanderbilt University, Nashville, TN 37235

ABSTRACT

Malondialdehyde (MDA) is a toxic and mutagenic metabolite produced by lipid peroxidation, and prostaglandin biosynthesis. MDA induces frameshift mutations in tester strains of *Salmonella typhimurium*. It reacts with DNA, and at physiological pH the major adduct is a pyrimidopurinone formed by reaction with guanine: M_1G [3-(2'-deoxy-β-D-erythro-pentofuranosyl)pyrimido[1,2-α]-purin-10(3H)-one]. When site-specifically incorporated into a duplex oligodeoxynucleotide containing a frameshift-prone $(CG)_3$ repeat derived from the Salmonella *typhimurium hisd3052* gene, spontaneous opening of M_1G to the N^2-(3-oxo-1-propenyl)-dG species occurred. In this work d(ATCGCMCGGCATG), $(M=M_1G)$ was annealed to d(CATGCCGCGAT) to model the putative strand slippage intermediate which would precede a two base deletion in the $(CG)_3$ iterated repeat. ^1H NMR studies indicate that in contrast to the duplex DNA structure, M_1G remains intact. A single bulge conformation exists. M_1G and its 3'-neighbor cytosine are unpaired. The M_1G is intrahelical and stacked, whereas the unpaired cytosine is poorly stacked and appears to be extrahelical.

INTRODUCTION

Malondialdehyde (MDA) is a toxic and mutagenic metabolite produced by lipid peroxidation and prostaglandin biosynthesis [for a review, see (*1*)]. MDA exists in solution primarily as β-hydroxyacrolein. It reacts with DNA and at physiological pH the major adduct is a pyrimidopurinone formed by reaction with guanine: M_1G [3-(2'-deoxy-β-D-erythro-pentofuranosyl)pyrimido[1,2-α]-purin-10(3H)-one] (*2*). Alternatively, M_1G arises as a consequence of DNA oxidative damage, resulting in the formation of base propenals that can transfer their oxopropenyl group to deoxyguanosine (*3*).

Biological Reactive Intermediates VI, Edited by Dansette *et al.*
Kluwer Academic / Plenum Publishers, 2001

Our interest in the structure of M_1G embedded into the $(CpG)_3$ frameshift hotspot of the *Salmonella typhimurium hisD3052* gene arose because MDA, a small alkylating agent, induced frameshift mutations (*4*). This was unexpected since frameshifts are commonly associated with intercalating agents. Frameshift mutations are strongly DNA sequence-dependent. These mutations are often associated with iterated bases, palindromes, and tandem repeats. The *hisD3052* mutation, located within the histidinol dehydrogenase gene from *Salmonella typhimurium*, resulted from deletion of a cytosine by ICR-191 (*5,6*). A CG deletion in the iterated sequence $(CG)_4$ represents one of the most common reversion events (*7*).

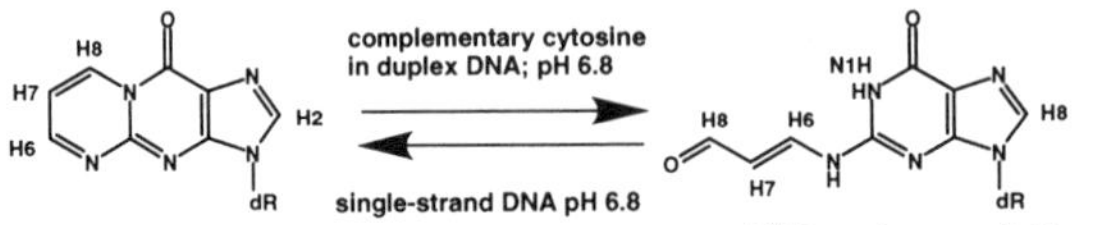

M_1G spontaneously converted to its N^2-(3-oxo-1-propenyl)-dG derivative when inserted into duplex DNA opposite cytosine in the complementary strand. This ring-opening was reversible, such that upon denaturation of the DNA duplex, M_1G was regenerated (*8*). A refined structure was obtained for N^2-(3-oxo-1-propenyl)-dG in d(ATCGCXCGGCATG)•d(CATGCCGCGCGAT), $X = N^2$-(3-oxo-1-propenyl)-dG, containing the *hisD3052* gene d(CpG)₃ frameshift hotspot. The modified guanine maintained stacking interactions with neighboring bases. It was not Watson-Crick hydrogen bonded. The cytosine complementary to N^2-[3-oxo-1-propenyl]-dG was pushed toward the major groove but maintained partial stacking with its neighboring bases. The modified guanine remained in the *anti* conformation, while the 3-oxo-1-propenyl moiety was positioned in the minor groove of the duplex.

$$5\text{'}-A^{-1}T^{-2}C^1\ G^2\ C^3\ M^4\ C^5\ G^6\ G^7\ C^8\ A^9\ T^{10}G^{11}-3\text{'}$$
$$3\text{'}-T^{22}A^{21}G^{20}C^{19}G^{18}\qquad C^{17}G^{16}G^{15}T^{14}A^{13}C^{12}-5\text{'}$$

The M_1G-2BD Oligodeoxynucleotide Derived From the $(CG)_3$ Iterated Repeat Hotspot of the *Salmonella typhimurium hisD3052* Gene and Numbering of the Nucleotides. $M = M_1G$. The unmodified oligodeoxynucleotides were synthesized and purified by anion exchange chromatography (Midland Certified Reagent Co., Midland, TX). M_1G was incorporated into oligodeoxynucletoides as described (*12,13*).

The present work utilizes high-field ^{1}H NMR to examine the consequences of introducing a dinucleotide deletion opposite M_1G located in the $(CpG)_3$ frameshift hotspot of the Salmonella *typhimurium hisD3052* gene. We constructed a M_1G-modified oligodeoxynucleotide (designated M_1G-2BD) to model the putative strand slippage intermediate which would precede a two base base-pair deletion in the *hisD3052* sequence. ^{1}H NMR studies indicate that a single bulge conformation exists, consistent with M_1G and its 3'-neighbor cytosine being unpaired. The M_1G is intrahelical and stacked, whereas the unpaired cytosine is poorly stacked and appear to be extrahelical.

RESULTS AND DISCUSSION

The M_1G-2BD oligodeoxynucleotide yielded well resolved ^{1}H NMR spectra, indicative of a single conformation in solution. Duplex DNA exhibits a characteristic pattern of sequential NOEs between adjacent base pairs. Figure 1 shows these NOEs for the M_1G-2BD oligodeoxynucleotide. For the modified strand these were interrupted at the C^3 to M^4 step, where the cross-peak between C^3 H1' and M^4 H2 (the guanine imidazole proton of M_1G) was missing. The cross-peak between C^5 H1' and G^6 H8 was weak, indicating a greater-than-normal distance between these two protons. The C^5 H1' to C^5 H6 and C^5 H6 to M^4 H1' cross-peaks were also weaker than expected. For the complementary strand the sequential NOEs were interrupted between C^{17} H1' and G^{18} H8 indicating also a greater-than-normal distance between these two protons. Collectively, these observations suggested that the 2-nucleotide bulge is localized at the M^4C^5 position in the M_1G-2BD duplex. The observation an NOE of normal intensity between M^4 H2 and M^4 H1' indicates that M_1G adopts an *anti* conformation with respect to the glycosidic bond.

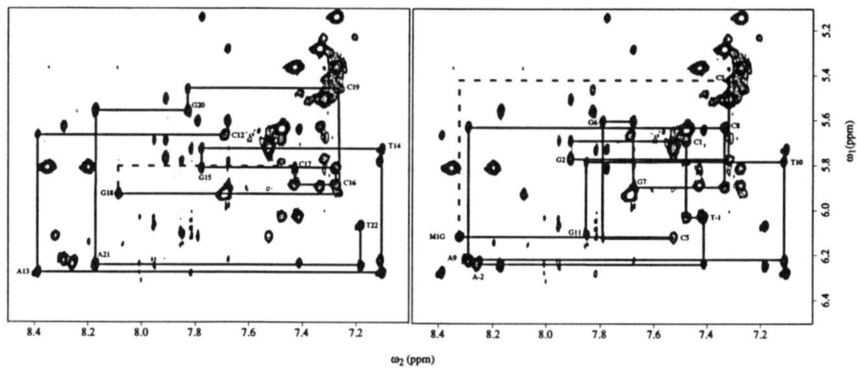

Figure 1. A 600.13 MHz NOESY spectrum in D_2O (mixing time=200 ms) for M_1G-2BD showing sequential NOEs from aromatic protons to deoxyribose H1' protons. **Left**: The unmodified strand. **Right**: The modified strand. The dotted lines represent NOE connectivities normally observed in duplex DNA but not observed in the M_1G-2BD sample. The missing NOEs between C^3 and M^4 in the modified strand and between C^{17} and G^{18} in the complementary strand suggest that the 2-nucleotide bulge is localized at the M^4C^5 position in the M_1G-2BD duplex. The duplex concentration was 2 mM. The buffer was 10 mM NaH_2PO_4, 0.1 M NaCl, 50 µM Na_2EDTA (pH 7.0). The temperature was 20 °C. Chemical shifts were referenced to water at 4.81 ppm.

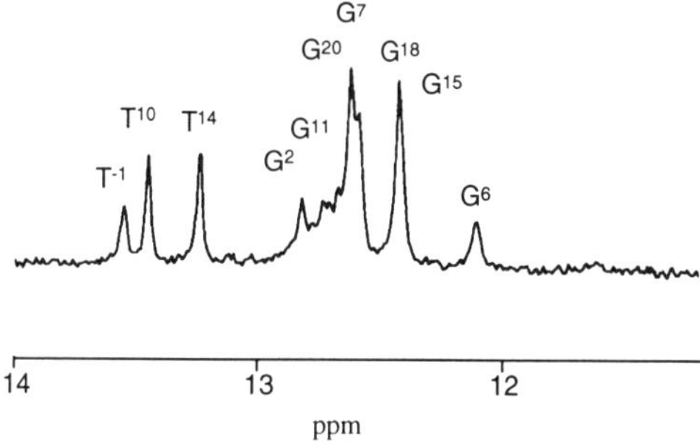

Figure 2. ^1H NMR spectrum for M_1G-2BD in 9:1 $H_2O:D_2O$ at 5 °C. The buffer was 10 mM NaH_2PO_4, 0.1 M NaCl, 50 µM Na_2EDTA (pH 7.0). The chemical shifts are referenced to water at 4.97 ppm.

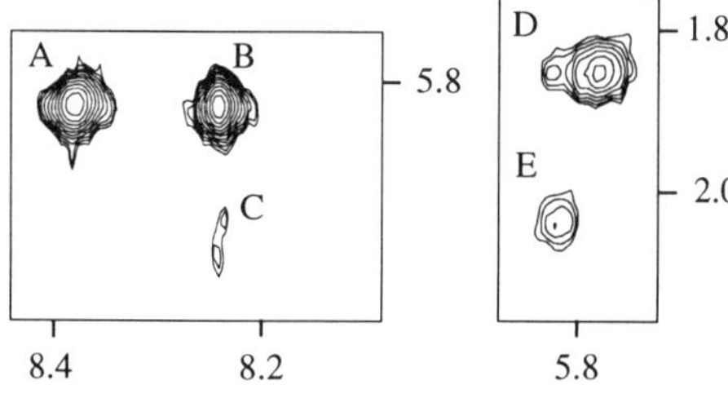

Figure 3. Assignment of the M_1G H6, H7, and H8 resonances establishes that M_1G is stable in the M_1G-2BD oligodeoxynucleotide at pH=7. NOE cross-peaks between M_1G and DNA : (A) M^4 H8 to M^4 H7, (B) M^4 H6 to M^4 H7, (C) M^4 H6 to G^{18} H1', (D) M^4 H7 to C^{17} H2', (E) M^4H7 to C^{17} H2".

This conclusion was corroborated by inspection of data for the imino protons involved in Watson-Crick base pairing (Figure 2). Watson-Crick base pairing between guanine N1 and cytosine N3 or between adenine N1 and thymine N3 slows the exchange of these protons with water and allows them to be observed in the far-downfield region of the NMR spectrum. The M_1G base is prevented from participating in Watson-Crick hydrogen bonding due to the presence of the $1,N^2$ exocyclic ring; no imino resonance was observed for M_1G. The N1 imino resonances of nucleotides G^2, G^6, and G^{18} were identified by analysis of NOE data. The observation of these resonances supports the conclusion that the position of the bulge is localized at M^4C^5. The identification of additional imino resonances from T^{-1}, G^7, T^{10}, G^{11}, T^{14}, G^{15}, and G^{20} indicated formation of a stable duplex on both sides of the unpaired bases.

The exocyclic protons H6, H7, and H8 of M_1G were identified from COSY- and NOESY-type spectra as the characteristic aromatic spin system (Figure 3). M_1G H8 was observed at 8.4 ppm, M_1G H7 was observed at 5.8 ppm, and M_1G H6 was observed at 8.2 ppm. There was no evidence for quantitative conversion of M_1G to its ring-opened N^2-(3-oxo-1-propenyl)-dG derivative. Thus, M_1G was stable in the M_1G-2BD oligodeoxynucleotide at pH=7, in contrast to its behavior in the fully complementary duplex (*9*). Significant increases in shielding were observed for each of the M_1G protons, as compared to the chemical shifts in the single-stranded oligodeoxynucleotide. M^4 H8 shifted upfield by ~0.9 ppm, H7 shifted upfield by ~1.75 ppm, and H6 shifted upfield by ~0.6 ppm. Theseupfield chemical shifts were attributed to ring-current effects and led to the conclusion that the M_1G moiety must be inserted into the DNA duplex.

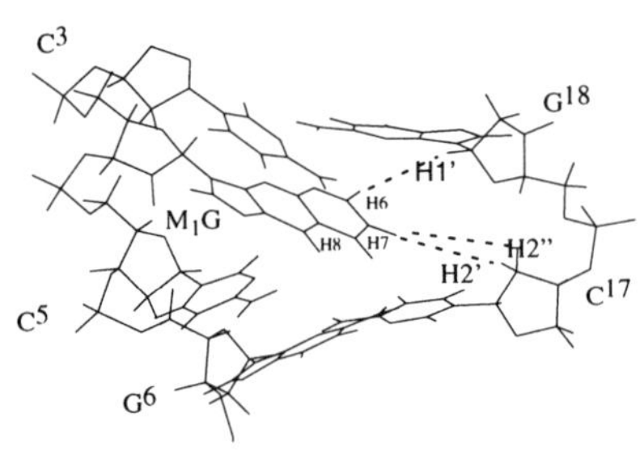

Figure 4. Representation of the 2-base bulge in M_1G-2BD. The dashed lines indicate observed NOEs. Structural refinement is in progress.

The observation of cross-strand NOEs between M_1G protons and the DNA confirmed that M_1G was inserted into the duplex. These were observed from M^4 H7 to C^{17} H2' (1.85 ppm) and H2" (2.03 ppm). A weak cross-strand NOE was observed from M^4 H6 to G^{20} H1' (5.93 ppm). The NOEs between M_1G and DNA also helped to establish the orientation of M_1G at the location of the 2-base bulge in M_1G-2BD. In Figure 4, the dashed lines indicate the observed NOEs at the location of the 2-base bulge. The NMR data suggest that base pair $C^3 \bullet G^{18}$, located above M_1G in the model, remains Watson-Crick hydrogen-bonded. Likewise, base pair $G^6 \bullet C^{17}$, located below the 2-base bulge, also remains hydrogen

is not well-defined at the present level of modeling, but this base appears to be extruded towards the major groove.

In conclusion, NMR studies reveal the formation of a stable structure in which M_1G is inserted opposite a two-base deletion in the complementary strand. The unpaired 5'-neighbor cytosine adopts a poorly stacked or extrahelical conformation. This dinucleotide bulge containing M_1G results in a localized structural perturbation. The structure is similar to that of PdG in the same sequence (*10,11*). In contrast to the fully complementary sequence, where M_1G was spontaneously ring-opened by cytosine in the opposite strand (*9*), M_1G was ring-closed in this two-base deletion sequence.

REFERENCES

1. Marnett, L. J. (1999) in "Exocyclic DNA Adducts in Mutagenesis and Carcinogenesis", *IARC Sci. Publ.* **150**, Singer, B., and Bartsch, H., Eds., pp. 17-27, IARC Publications, Lyon, France
2. Reddy, G. R. and Marnett, L. J. (1996) *Chem. Res. Toxicol.* **9**, 12-15.
3. Dedon, P. C., Plastaras, J. P., Rouzer, C. A., and Marnett, L. J. (1998) *Proc. Natl. Acad. Sci. U S A* **95**, 11113-11116.
4. O'Hara, S. M. and Marnett, L. J. (1991) *Mutat. Res.* **247**, 45-56.
5. Oeschger, N. S. and Hartman, P. E. (1970) *J. Bacteriol.* **101**, 490-504.
6. Hartman, P. E., Ames, B. N., Roth, J. R., Barnes, W. M., and Levin, D. E. (1986) *Environ. Mutagen.* **8**, 631-641.
7. Isono, K. and Yourno, J. (1974) *Proc. Natl. Acad. Sci.USA* **71**, 1612-1617.
8. Mao, H., Schnetz-Boutaud, N. C., Weisenseel, J. P., Marnett, L. J., and Stone, M. P. (1999) *Proc. Natl. Acad. Sci. U S A* **96**, 6615-6620.
9. Mao, H., Reddy, G. R., Marnett, L. J., and Stone, M. P. (1999) *Biochemistry* **38**, 13491-13501.
10. Moe, J. G., Reddy, G. R., Marnett, L. J., and Stone, M. P. (1994) *Chem. Res. Toxicol.* **7**, 319-328.
11. Weisenseel, J. P., Moe, J. G., Reddy, G. R., Marnett, L. J., and Stone, M. P. (1995) *Biochemistry* **34** , 50-64.
12. Reddy, G. R. and Marnett, L. J. (1995) *J. Am. Chem. Soc.* **117**, 5007-5008.
13. Schnetz-Boutaud, N. C., Mao, H., Stone, M. P., and Marnett, L. J. (2000) *Chem. Res. Toxicol.* **13**, 90-95.

A SENSITIVE PROCEDURE FOR THE DETERMINATION OF PROTEIN BOUND 3,4-DIHYDROXYPHENYL-ALANINE AS A MARKER FOR POSTTRANSLATIONAL PROTEIN HYDROXYLATION IN HUMAN FRONTAL CORTEX, LIVER, AND RED BLOOD CELLS

Roland Harth[2], Manfred Gerlach[2], Peter Riederer[2], and Mario E. Götz[1,2]

1) Department of Pharmacology and Toxicology, D-97078 Würzburg, Germany
2) Department of Psychiatry, D-97080 Würzburg, Germany

INTRODUCTION

Oxidative stress has been implicated as an initiating or at least aggravating factor in neurodegenerative diseases, atherosclerosis, stroke, and inflammatory diseases. Since reactive oxygen species (ROS) are short lived reactive species, the involvement of ROS in biological tissues has to be determined by the evaluation of ROS-modified marker molecules. In principle, ROS react unspecifically with nearly any kind of biomolecule depending on the site of their production. One of these relatively stable markers of oxidative stress is supposed to be protein-bound 3,4-dihydroxyphenylalanine (PB-DOPA; Gieseg et al., 1993). Current fluorimetric assays for the detection of PB-DOPA (Armstrong and Dean, 1995) turned out to be less sensitive than the electrochemical detection of PB-DOPA. The former studies mainly used model proteins to investigate the biochemistry of *in vitro* oxidation of tyrosine induced by the iron / EDTA / ascorbate system in proteins. We now tried to further improve the sensitivity of the available methods in order to be able to determine the basal levels of PB-DOPA in various human tissues with high oxygen turnover rates (frontal cortex, liver and red blood cells) from healthy controls. In addition, we determined tyrosine levels in order to be able to establish a tissue specific ratio of PB-DOPA to tyrosine in analogy to the well known ratio of 8-hydroxy-2'deoxyguanosine / 2'-deoxyguanosine for the determination of oxidative stress related DNA damage. This should allow the comparison of oxidative stress related protein damage in different organs and could be of value if ROS-mediated actions following the exposure to toxicants that cause free radical production shall be investigated.

EXPERIMENTAL SECTION
Protein isolation from tissue

The tissue was homogenized with o-phosphoric acid containing a chelator for transition metals by pulsed ultrasound. The homogenate was centrifuged to pellet the hydrophobic proteins. The supernatant containing the hydrophilic proteins was transferred into molecular weight filters cut off 3 kDa. Peptides, free amino acids and salts were removed by centrifugation. After this step the cleaned fraction of hydrophilic proteins was dried by lyophilisation.

The protein pellet consisting of membrane proteins and lipids was washed with water and extracted with methanol/chloroform to remove the lipids. The residual membrane proteins were dried in a gentle stream of nitrogen.

Protein isolation from red blood cells

Instead of sonification, the cells were lysed with a buffer containing NH_4Cl, $KHCO_3$, EDTA, and processed according to the method described above.

Hydrolysis of proteins

The method was based on the hydolysis procedure described by Gieseg et al., 1993. This method works in the gas phase and was performed in a glas exicator. The hydrolysis reagent consisted of 6 M hydrochloric acid containing phenol and thioglycolic acid as antioxidants. After hydrolysis the dry samples were diluted with a defined volume of hydrochloric acid in order to prepare for injection into the HPLC system without any further purification step. The recovery of DOPA and tyrosine during the hydrolysis procedure was determined to be nearly 100%.

HPLC-Separation

The separation of DOPA and tyrosine was performed on a C-18 reversed phase column. The mobile phase consisted of pentane sulfonic acid as ion-pairing reagent, disodium EDTA, and diethylamine dissolved in water. The pH was adjusted to 2,2 with o-phosphoric acid. Tyrosine and DOPA were detected simultaneously. Tyrosine was detected with a fluorescence detector whereas DOPA was subsequently electrochemically detected.

RESULTS

Following pilot experiments (not shown) aimed at optimizing the procedures for protein isolation and hydrolysis, as well as for electrochemical detection of DOPA we were able to detect PB-DOPA at concentrations as low as 0.1 ng from pure BSA and from various organs. In our experiments levels of PB-DOPA in pure BSA did never exceed 10 ng / mg. This corresponds to one molecule of PB-DOPA per 7000 tyrosine or 285 molecules of BSA giving a ratio of PB-DOPA / tyrosine of approximately $1.8*E-4$ in BSA. When BSA was incubated with increasing concentrations of ferrous sulfate (10-100 µM) prior to lyophilisation levels of PB-DOPA increased dose dependently and significantly. Incubation of BSA with 100 µM ferrous sulfate for 180 min increased the PB-DOPA / tyrosine ratio 3 times. This result emphasizes that iron may critically interact with protein tyrosine in vitro giving rise to artifactual DOPA formation if care is not taken to remove soluble iron from tissue samples. Thus, we homogenized tissue samples in the presence of an excess of the metal chelator (DTPA) and washed all our protein pellets twice with bidistilled water in order to remove most of the non-tightly bound transition metals from protein samples.

Comparing the content of PB-DOPA and tyrosine of frontal cortex, liver, and red blood cells, reveals significant differences in background hydroxylation levels of tyrosine in these tissues (Table I). In liver and frontal cortex ratios of PB-DOPA / tyrosine are quite similar. However, red blood cells contain in the fraction of the hydrophobic proteins about half the amount of PB-DOPA per tyrosine as compared to liver and frontal cortex from humans. All tissue values in the fraction of the hydrophobic proteins are about 6 times higher than pure BSA. In the hydrophilic protein fraction levels of PB-DOPA / tyrosine ratios are similar to those found in BSA without stimulation by iron.

DISCUSSION

The method mentioned may be very useful in determining protein hydroxylation in tissues vulnerable to oxidation such as red blood cells, brain tissue and liver if care is taken to prevent artifactual in vitro oxidation. Thus, prepurification of proteins prior to hydrolysis by removing lipids and salts is recommended. Hydrolysis should be performed under reductive conditions in the absence of oxygen in hydrochloric acid vapour. The establishment of PB-DOPA / tyrosine tissue ratios should make this method valuable to compare oxidative stress to proteins occurring in different tissues and species because it is independent on the amount and type of protein analysed whilst only dependent on the overall amount of tyrosine in the tissue. This should facilitate the evaluation of the relevance of protein hydroxylation in a given disease or experimental model in comparison to adequate controls.

Table I: Contents of DOPA and tyrosine as well as DOPA / tyrosine ratios in native frontal cortex, liver, and red blood cells from humans.

	ng / mg BRAIN	ng / mg LIVER	ng / 10e9 red BLOOD CELLS
DOPA in membranes	0.52 ± 0.12	0.63 ± 0.06	0.29 ± 0.05
DOPA in cytosols	0.36 ± 0.01	0.41 ± 0.10	260,72 ± 17.39
Total DOPA	0.88 ± 0.12	1.04 ± 0.11	261.01 ± 17.43
Tyrosine in membranes	355 ± 36	513 ± 62	431 ± 63
Tyrosine in cytosols	1636 ± 155	1941 ± 212	786,954 ± 52,846
Total tyrosine	1992 ± 141	2455 ± 267	787,385 ± 52,808
	ratio*10e-4	ratio*10e-4	ratio*10e-4
DOPA/tyrosine in membranes	13.1 ± 2.4	12.1 ± 2.3	6.2 ± 0.7
DOPA/tyrosine in cytosols	2.0 ± 0.2	2.1 ± 0.7	3.1 ± 0.4
Total DOPA/tyrosine	4.1 ± 0.7	4.1 ± 0.9	3.1 ± 0.4

Data are expressed as means ± SEM of four independent determinations processed according to the procedures described in the experimental section.

REFERENCES

Armstrong, S.G., and Dean, R.T., 1995, A sensitive fluorometric assay for protein-bound DOPA and related products of radical-mediated protein oxidation, *Redox Report 1: 291-298.*

Gieseg, S.P., Simpson, J.A., Charlton, T.S., Duncan, M.W., and Dean, R.T., 1993, Protein-bound 3,4-dihydroxyphenylalanine is a major reductant formed during hydroxyl radical damage to proteins, *Biochemistry 32: 4780-4786.*

REACTIVE METABOLITE SCREEN FOR REDUCING CANDIDATE ATTRITION IN DRUG DISCOVERY

Weichao G. Chen, Chenghong Zhang, Michael J. Avery and Hassan G. Fouda
Drug Metabolism Development
Pfizer Central Research
Groton, CT 06340

INTRODUCTION

Toxicity of drug candidates accounts for a significant portion of attrition during exploratory development. One common mode of toxicity is the formation of electrophilic reactive metabolites, which manifest their toxicity by covalent binding to nucleophilic groups present in vital cellular proteins and nucleic acids (1-2). Although not all toxicological manifestations are attributable to reactive metabolites, a vast body of literature suggests that inadequate detoxification of chemically reactive metabolites formed as a result of drug bioactivation is a pathogenic mechanism for tissue necrosis (3-4), carcinogenicity (5), teratogenicity (6) and immune-mediated toxicity (7).

A potential high throughput method of identifying candidates with the potential to produce reactive metabolites was presented. It exploits one of the natural mechanisms for eliminating reactive intermediates: conjugation with glutathione (8). The proposed analytical method utilizes tandem mass spectrometry. During collision induced dissociation, all glutathione adducts undergo the neutral loss of the pyroglutamic acid moiety (129 Da). The diagnostic neutral loss of 129 Da allows the specific detection of any glutathione conjugates that have been generated via metabolic activation followed by conjugation (9). In the current study, we applied this HTS reactive metabolite screen to 20 commercially available therapeutic agents with known toxicological profiles.

EXPERIMENTAL PROCEDURES

Human liver microsomal (HL-mix-11) incubation mixture containing 500 μM substrate, 1 mM glutathione (GSH), 1 μM P450 and 100 mM potassium phosphate buffer (pH 7.4) was pre-incubated for 3 min at 37°C. The reaction was initiated by the addition of an NADPH-generating system (0.54 mM NADP$^+$, 10 mM MgCl$_2$, 6.2 mM DL-isocitric acid and 0.5 U/ml isocitric dehydrogenase). The final incubation volume was 1 mL. Samples without NADPH or substrates are used as negative controls. After 30 min incubation at 37°C, the incubation mixture was centrifuged at 3,500 rpm for 10 min. The supernatant was prepared by automated 96-well solid phase extraction (10). This included removing proteins, washing 3 times with 100 µl of water and eluting with 100 µl of acetonitrile. Following solvent evaporation, the residues were dissolved in 100 µl of starting mobile phase.

Chromatographic separations utilized a HP 1100 quaternary HPLC pump after injection by a CTC PAL autosampler. Aliquots (20 µl) of prepared samples were injected onto a 2x30 C18 column packed with 3µm particles. The analyses were performed using a mobile phase flow rate of 0.2 ml/min and a fast gradient of from 5/95 acetonitrile/10 mM ammonium acetate to 80/20 over a 5 min period after an initial hold of 1 min. The HPLC column eluant was introduced into the TurboIonspray source of a SCIEX API 3000 triple quadrupole mass spectrometer. The mass spectrometer was operated in the neutral loss mode, scanning over a range of m/z 320 to 800 in about 2.4 sec.

The current method was validated by application to 20 commercially available therapeutic agents representing diverse chemical structures and toxicity profiles. Half of the compounds (ampiclillin, verapamil, sulfamethazine, ibuprofen, citric acid, caffeine, levodopa, isoniazid, quinidine and aspirin) are known to posses desirable safety profiles. The other half (acetaminophen, clozapine, carbamazepine, indomethacin, 4-hydroxyanisole, m-cresol, p-cresol, 4-isopropylphenol, valproic acid and phenytoin) are known to produce reactive metabolites.

RESULTS AND DISCUSSION

A typical total ion chromatogram of LC/MS/MS neutral loss scan is shown in Figure 1. Each of the 10 compounds that posses desirable safety profile produced negative responses (data not shown). This suggests that the method is not likely to produce false positive responses.

Eight out of the 10 compounds that are known to generate reactive metabolites produced positive responses (Fig. 2). These positive responses cover a broad range of reactive metabolites, including quinone imine (acetaminophen and indomethacin), nitrenium (clozapine), epoxide (carbamazepine), quinone (4-hydroxyanisole) and quinone methide (m-cresol, p-cresol and 4-isproylphenol). The formation of the reactive metabolite from indomethacin requires several sequential oxidation steps (*11*). These results suggest that the current method detects most reactive metabolites including those resulting from multiple oxidation steps.

The two compounds that are known to generate reactive metabolites but did not produce positive responses in the current assay are valproic acid and phenytoin. However, it is well documented that the formation of valproic acid reactive metabolite (2,4-diene-VPA) requires not only microsomal P450-oxidation to form 4-ene VPA but also β-oxidation catalyzed by a mitochondrial Coenzyme A dependent process to form 2,4-diene-VPA (*12*). For phenytoin, the reactive metabolite is a free radical instead of an epoxide (*13*). GSH can reduce the free radical and reduce covalent binding of phenytoin, but GSH can not form a stable adduct with it. Clearly the current screen detects only those reactive metabolites formed by bioactivation via microsomal P450 systems and specifically those forming stable adducts with glutathione.

CONCLUSION

The method described here is reliable, rapid, simple and amenable to high throughput automation. The method will detect most toxic compounds whose toxicity is mediated by reactive metabolites generated by bioactivation with drug metabolizing oxidative P450

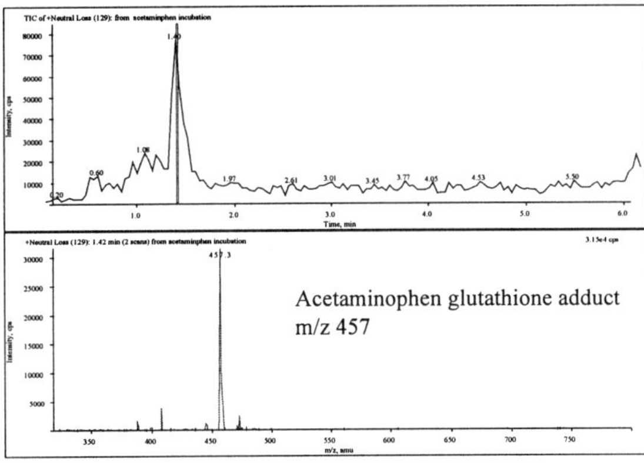

Figure 1. Total ion chromatogram from LC/MS/MS analysis (neutral loss of 129) of human liver microsomal incubation containing acetaminophen.

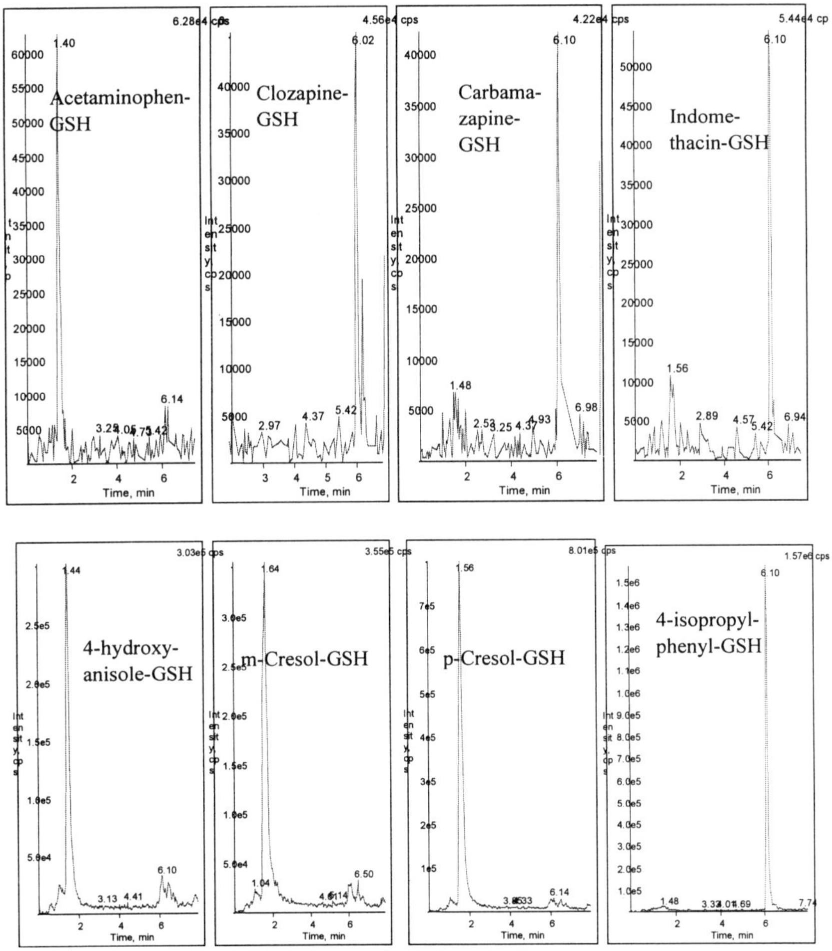

Figure 2. Total ion chromatograms from LC/MS/MS analysis (neutral loss of 129) of human liver microsomal incubations containing acetaminophen, clozapine, carbamazapine, indomethacin, 4-hydroxyanisole, m-cresol, p-cresol and 4-isopropylphenyl, respectively.

systems. It will not detect reactive metabolites that does not form stable adducts with glutathione (such as free radicals) or those formed by non-microsomal enzymes.

A positive response in the above would be a warning of potential toxicology implications and not necessarily a rationale to discard a compound unless another equally desirable prototype without potential toxicity liability is available. If the formation of potential reactive metabolites is a minor metabolic pathway, a mere positive response in the above screen may not be a serious concern. The screen described here promises to provide valuable insight at the lead seeking or candidate selection stages. Its potential for reducing candidate attrition rates should be further evaluated.

REFERENCES

1. Nelson, S. D. (1982) Metabolic activation and drug toxicity. J. Med. Chem. 25, 753-65.
2. Jollow, D. J., Kocsis, J., Snyder, R., Vainio H. (1977): Biological Reactive Intermediates. New York, Plenum Press.
3. Pirmohamed, M., Madden, S. and Park, B. K. (1996) Idiosyncratic drug reactions. Clin. Pharmacokinet. 31, 215-230.
4. Prescott L. F. (1983) Reactive metabolites as a cause of hepatotoxicity. Int. J. Clin. Pharm. 3, 437-441.
5. Guengerich, F. P. (1992) Metabolic activation of carcinogens. Pharmacol. Ther. 54, 17-61.
6. Juchau, M. R., Lee, Q. P., Fantel, A. G. (1992) Xenobiotic biotransformation/ bioactivation in organogenesis-stage conceptal issues: implications for embryotoxicity and teratogenesis. Drug Metab. Rev. 24, 195-238.
7. Hess, D. A. and Rieder, M. J. (1997) The role of reactive drug metabolites in immune-mediated adverse drug reactions. Ann. Pharmacother. 31, 1378-1387.
8. Chasseaud, L. F. (1977): In Arias, I. M., Jakoby, W. B. (eds), Glutathione: Metabolism and Function, p. 77. New York, Raven Press.
9. Baillie, T. A., Davis, M. R. (1993) Mass spectrometry in the analysis of glutathione conjugates. Biol. Mass Spectrom. 22, 319-325.
10. Janiszewski, J, Schneider, R. P.,.Hoffmaster, K., Swyden, M., Wells, D., Fouda, H. G., Automated Sample Preparation Using Membrane Microtiter Extraction for Bioanalytical Mass Spectrometry. Rapid Comm. Mass Spectrom. 11,9 (1997).
11. Ju, C.; Uetrecht, J. P. (1998) Oxidation of a metabolite of indomethacin (desmethyldeschlorobenzoylindomethacin) to reactive intermediates by activated neutrophils, hypochlorous acid, and the myeloperoxidase system. Drug Metab. Dispos. 26(7), 676-680.
12. Kassahun, Kelem; Hu, Pei; Grillo, Mark P.; Davis, Margaret R.; Jin, Lixia; Baillie, Thomas A. (1994) Metabolic activation of unsaturated derivatives of valproic acid. Identification of novel glutathione adducts formed through coenzyme A-dependent and -independent processes. Chem.-Biol. Interact. 90(3), 253-75.
13. Mays, Dennis C.; Pawluk, Lew J.; Apseloff, Glen; Davis, W. Bruce; She, Zhi-Wu; Sagone, Arthur L.; Gerber, Nicholas. Metabolism of phenytoin and covalent binding of reactive intermediates in activated human neutrophils. Biochem. Pharmacol. (1995), 50(3), 367-80.

METABOLISM OF *R,S* ENANTIOMERS OF 3-PHENYLAMINO-1,2-PROPANEDIOL, A COMPOUND ASSOCIATED WITH THE TOXIC OIL SYNDROME, IN C57BL/6- AND A/J-STRAIN MICE

Margarita G. Ladona, Jordi Bujons, Angel Messeguer, Coral Ampurdanés, and Anna Morató

Consejo Superior de Investigaciones Científicas (CSIC)
C/ Jordi Girona Salgado, 18-26
Barcelona 08034, Spain
Email: mglqob@cid.csic.es

INTRODUCTION

The Toxic Oil Syndrome (TOS) appeared in Spain in spring 1981 as a massive food-borne intoxication in the central and northwestern areas of the country. A huge population of over 20,000 people was afflicted and over 400 deaths occurred[1]. The clinical features of TOS resembled an allergic-toxic syndrome in the acute phase and an autoimmune condition in the chronic phase[1,2]. Although some patients recovered after a long period of time (several months), many continue to suffer severe sequel or mild symptoms.

Epidemiological studies provided conclusive evidence that the epidemic resulted from the ingestion of adulterated rapeseed oil[1,2]. Rapeseed oil denatured with 2% aniline was imported for industrial use and illegally refined and delivered for human consumption. Aniline derivatives such as fatty acid anilides and fatty acid esters of 3-phenylamino-1,2-propanediol (PAP) were identified in toxic oil batches and strongly implicated in TOS[1,4-5]. Fatty acid anilides are good markers of toxic oils; however, they could not have been the toxic agent since they were present in oils from the Catalonia circuit distribution with no reported toxicity in consumers. In contrast, careful epidemiological investigations combined with chemical analyses have clearly established that PAP esters were formed during the refining process and were only detected in oil batches coming from one refinery; thus the epidemic is believed to have emerged from a single source in that refinery[6]. In particular, levels of 1,2-di-oleyl ester of 3-phenylamino-1,2-propanediol (OOPAP) identified in toxic oil batches correlate strongly with patient incidence in corresponding households. Furthermore, OOPAP was undetected in oils from the Catalonia circuit. Thus, OOPAP is currently considered to be the toxic candidate[7,8]. To date it is assumed that TOS patients ingested considerable amounts of anilides and PAP esters in their edible toxic oils; these chemical species are the only aniline derivatives identified so far in oil batches.

PAP and its mono-oleyl-ester were toxic to mice when administered intraperitoneally[9], and evidence of OOPAP intestinal hydrolysis yielding PAP and its mono-oleyl-ester has been reported in rats[10]. Moreover, the A/J mouse strain intraperitoneally treated with

Biological Reactive Intermediates VI, Edited by Dansette *et al.*
Kluwer Academic / Plenum Publishers, 2001

oleanilide developed an immunological pattern similar to that reported in TOS patients[11,12] Thus, mouse strains may be useful candidates for developing a TOS animal model.

Therefore, in order to obtain an animal model for in-depth analysis of TOS pathogenic mechanisms, full comprehension of metabolism and clearance of the putative substances is required.

We previously studied the in vivo clearance of *rac*-PAP in these mouse strains[13]. The present investigation studies the metabolism and clearance of the PAP enantiomers. (**Chart 1**).

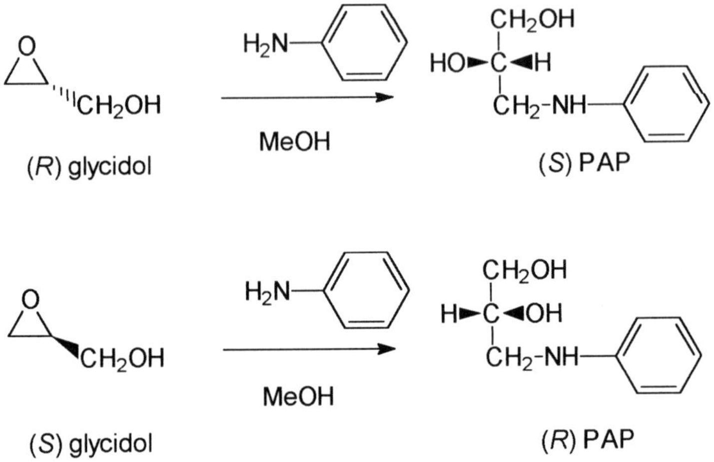

Chart 1. Synthesis of PAP enantiomers.

MATERIALS AND METHODS

A/J and C57BL/6 mice (9-10 weeks old) were obtained from Charles River. Three animals of each strain were placed in metabolic cages for 3 days of acclimatization on a 12 h light-dark schedule with food and water ad libitum. Induction treatment with β-naphthoflavone (β-NF) prior to administration of PAP was performed as described[13]. A single total dose of either *(R)- (S)* -PAP (250 mg/kg; average animal weight in each group: 22-24 g) was intraperitoneally given. The dose was spiked with 4 µCi of [U-^{14}C]-*rac*-PAP to monitor metabolic species. Urine and feces were collected over a 24-hour period and animals euthanized. Organs (lung, liver, kidney, heart, muscle and spleen) were collected in cold PBS (pH 7.4) and stored at -80 °C until analysis. [U-^{14}C]-PAP was prepared from [U-^{14}C]aniline hydrochloride to 5.7 mCi/mmol specific activity. PAP-enantiomers were independently synthesized by reaction of aniline with (R)-and (S)-glycidol according to procedures used for the preparation of the racemate[14]. The compounds were purified by reversed-phase preparative HPLC.

Tissue and feces were homogenized with cold PBS (pH 7.4). Urine or tissue aliquots were mixed with scintillation cocktail and counted. Total ^{14}C-dpm in urine, feces and organs was calculated to estimate the total radioactivity clearance.

HPLC analyses were conducted on a Waters system equipped with a DAD detector coupled on-line to a Berthold LB 507 radioactivity monitor system. C-18 Kromasil 100, 5 µm particle size, 250 × 4.6 mm columns were used with 10 mM TEAA pH 6.8 (solvent A) and acetonitrile (solvent B) as mobile phases[13]. Urine samples were used as reported previously[13]. For the separation of polar metabolites from PAP, urine samples were buffered with 100 mM sodium phosphate (pH 8.0) and extracted with a mixture of 9:1 dichloromethane-isopropyl alcohol containing 1% ammonia.

GC/MS analyses were performed on collected peak samples, lyophilized and treated with BSTFA prior to injection into the GC/MS system. Some samples were treated with HCl-MeOH and either injected directly or further treated with BSTFA before injection into the GC/MS system[13]. The EI-MS-identified metabolites were confirmed using corresponding synthesized standards. Detailed analytical procedures were reported recently[13].

Table 1. ^{14}C-PAP clearance in urine and feces

| | S- PAP | | R- PAP | |
| | β-NF treated mice[a] | | β-NF treated mice[a] | |
	A/J	C57BL/6	A/J	C57BL/6
^{14}C-PAP dose[b]	3.1×10^{7}	3.1×10^{7}	2.5×10^{7}	2.4×10^{7}
URINE (mL)	2.00	2.00	1.23	1.80
% of dose	59 %	52 %	54 %	63 %
FECES (g)	3.60	3.10	4.06	3.58
% of dose	0.45 %	0.54 %	1.8 %	0.8 %

[a]Three mice in each group. [b]^{14}C-PAP total dose (dpm) administered to each animal group.

RESULTS AND CONCLUSIONS

The radioactivity clearance observed in the mouse strains treated with each PAP-enantiomer is depicted in Table 1. As can be seen, urine is the major elimination route over a 24-hour period, whereas neither feces nor organ tissues contributed significantly to radioactivity clearance. Total radioactivity content in tissues was less than 1% of the administered dose. No differences in urine elimination were observed between the enantiomeric doses.

Urine analyses of methanol-diluted samples revealed similar chromatographic profiles in both strains in each PAP-enantiomer dose (Figure1). The major metabolite observed at 15.5 minute, with both enantiomers of PAP, was identified as 2-hydroxy-3-phenylaminopropanoic acid previously reported as the major metabolite in *rac*-PAP dosage[13]. Less abundant chemical species were present in the 3-7 and 16-18 minute chromatographic regions, principally in the *R*-PAP dosage-experiment in the A/J mouse strain. For comparison purposes, urine samples were adjusted to similar radioactivity content in each strain and each enantiomer to be injected into the chromatograph (approx. 4,000 dpm per injection). Untransformed *R*-PAP (18.5 min peak, A/J mouse) was detected and identified by GC/MS analysis.

All metabolites remained in the aqueous phase of CH_2Cl_2-extracted urine at pH 8.0 whereas PAP partitioned towards the organic phase. Phenolic species derived from PAP and its acid metabolite were identified by GC/MS analysis of collected peak fractions which yielded EI-MS spectra identical to their synthetic standards, as we previously reported for *rac*-PAP administration (Figure 2)[13]. The GC/MS analysis of the 3-7 and 16-18 minute peaks collected by HPLC were tentatively assigned to oxidized species at the alkyl chain or at the aromatic ring. These proposed structures reflecting further oxidation of the alkyl chain and iminoquinones are underway to be confirmed by independent synthesis.

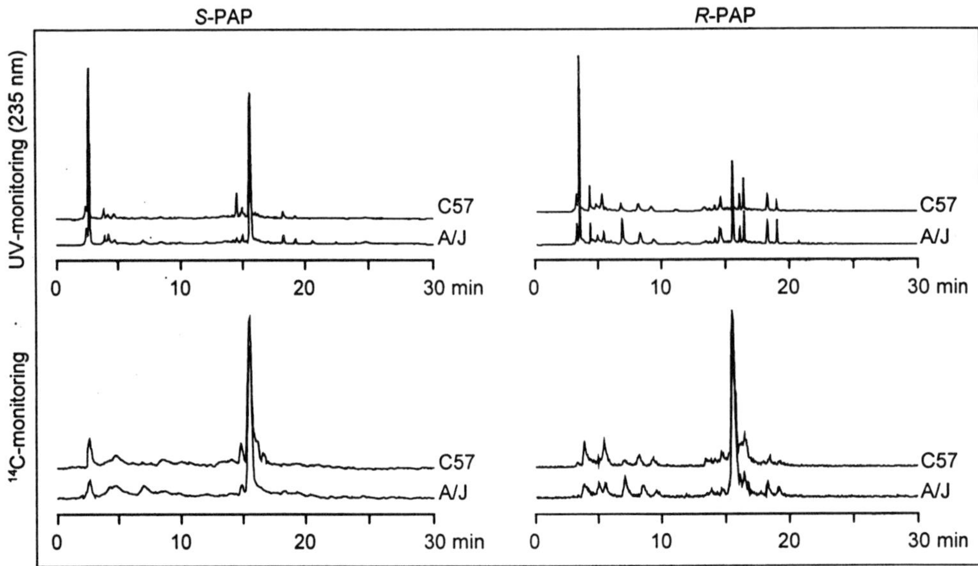

Figure 1. Typical HPLC chromatograms of methanol-diluted urine samples from S- and R-[^{14}C]PAP-treated β-NF-induced A/J and C57BL/6 mice.

Finally, it is worth noting that PAP acid metabolite was also detected in fecal samples of A/J *R*-PAP- treated mice. Preliminar in vitro experiments in liver homogenates show formation of PAP acid metabolite under incubation conditions for alcohol dehydrogenases and cytochrome P450 oxidation. Formation of these metabolites by human cDNA-expressed P450 isoenzymes is also under study.

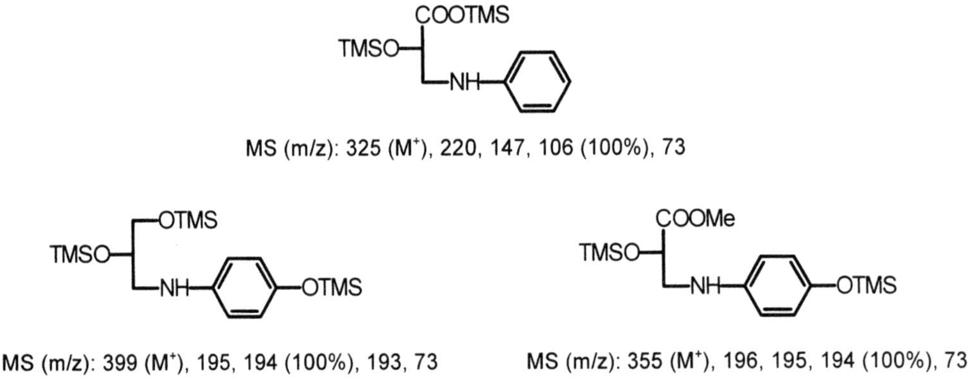

MS (m/z): 325 (M$^+$), 220, 147, 106 (100%), 73

MS (m/z): 399 (M$^+$), 195, 194 (100%), 193, 73

MS (m/z): 355 (M$^+$), 196, 195, 194 (100%), 73

Figure 2. Structures and EI-MS spectra of identified PAP-metabolites.

In summary: PAP, a very polar substance, is highly metabolized in mice and excreted principally in urine in the form of the 2-hydroxy-3-phenylaminopropanoic acid of each enantiomer. Thus, the major route of PAP elimination in these strains is alkyl chain oxidation. In particular, *S*-PAP is eliminated principally in the form of that metabolite, whereas *R*-PAP enantiomer showed further oxidized species at the aromatic ring and alkyl chain, yielding potential decarboxylated compounds and iminoquinones. All these metabolites may have toxicologic implications. On the other hand, OOPAP intestinal hydrolysis in favour of one PAP enantiomer might be expected since lipases show chiral hydrolysis (unpublished data, manuscript in preparation). In this respect, enantiomeric distribution and metabolic differences should be taken into account in the toxicokinetics of these compounds and their potential association with Toxic Oil Syndrome symptoms.

ACKNOWLEDGEMENTS

This work was supported by Grant FIS-97/1317-SAT of the Fondo de Investigaciones Sanitarias from the Spanish Ministry of Health.

REFERENCES

1. P. Gradjean, and S. Tarkowski eds. Toxic Oil Syndrome: mass food poisoning in Spain. Report of a WHO Meeting: Madrid 21-25 March 1983. World Health Organization Regional office for Europe, Copenhagen (1984).
2. J.M. Tabuenca. Toxic allergic syndrome caused by ingestion of rapeseed oil denatured with aniline. *Lancet* 2, 567-568 (1981).
3. J. Nadal and S. Tarkowski. Toxic Oil Syndrome: current knowledge and future perspectives. WHO Regional Publications, European series no 42, Copenhagen (1992).
4. A. Vázquez-Roncero, C. Janer del Valle, R. Maestro-Duran and E. Graciane-Constante. New aniline derivatives in cooking oils associated with the toxic oil syndrome. *Lancet* ii, 1024-1025 (1983).
5. J.T. Bernert, E.M. Kilbourne, J.R. Akins, M. Posada de la Paz, N.K. Meredith, I. Abaitua-Borda and S. Wages. Compositional analysis of oil samples implicated in the Spanish toxic oil syndrome. *J. Food Sci.* 52, 1562-1569 (1987).
6. M. Posada de la Paz, R.M. Philen, I. Abaitua Borda, J.T. Bernert, J.C. Bada-Gancedo, P.J. DuClos, E.M. Kilbourne. Toxic Oil Syndrome: traceback of the toxic oil and evidence for a point source epidemic. *Food. Chem. Toxicol.* 34, 251-257 (1996).
7. H.H. Schurz, R.H. Hill, M. Posada de la Paz, R.M. Philen, I. Abaitua-Borda, S.L. Bailey and L.L. Needham. Products of aniline and triglycerides in oil samples associated with the Toxic Oil Syndrome. *Chem. Res. Toxicol.* 9, 1001-1006 (1996).
8. M. Posada de la Paz, R.M. Philen, H.H. Schurz, R.H. Hill, O. Gimenez-Ribota, A. Gómez de la Cámara, E.M. Kilbourne, I. Abaitua-Borda. Epidemiologic evidence for a new class of compounds associated with Toxic Oil Syndrome. *Epidemiology* 10, 130-134 (1999).
9. A.Vázquez-Roncero, R. Maestro-Duran, and V. Ruiz-Gutierrez. New aniline derivatives in oils related to the Toxic Syndrome. Toxicity in mice of 3-phenylamino-1,2- propanediol and its fatty acid mono- and diesters. *Grasas y Aceites* 35, 330-331 (1984).
10. V. Ruiz-Gutierrez and R. Maestro-Duran. Lymphatic absorption of 3-phenyalmino-1,2-propanediol and its esters. *Exp. Toxic. Pathol.* 44, 29-33 (1992).
11. C. Berking, M.V. Hobbs, R. Chatelain, M. Meurer and S.A. Bell. Strain-dependent cytokine profile and susceptibility to oleic acid anilide in a murine model of the Toxic Oil Syndrome. *Toxicol. Appl. Pharmacol.* 148, 222-228 (1998).
12. V. del Pozo, B. de Andrés, S. Gallardo, B. Cardaba, E. de Arruda-Chaves, I. Cortegano, A. Jurado, P. Palomino, H. Oliva, B. Aguilera, M. Posada and C. Lahoz. Cytokine mRNA expression in lung tissue from Toxic Oil Syndrome patients: a Th2 immunological mechanism. *Toxicology* 118, 61-70 (1997).
13. M.G. Ladona, J. Bujons, A. Messeguer, C. Ampurdanés, A. Morató and J. Corbella. Biotransformation and clearance of 3-(phenylamino)propane-1,2-diol, a compound present in samples related to Toxic Oil Syndrome in C57BL/6 and A/J mice. *Chem. Res. Toxicol.* 12, 1127-1137 (1999).
14. M. Ferrer, M. Galceran, F. Sanchez-Baeza, J. Casas and A. Messeguer. Synthesis of aniline derivatives with potential toxicological implications to the Spanish Toxic Oil Syndrome. *Liebigs Ann. Chem.* 507-511 (1993).

CYTOCHROME P450-CATALYSED IRREVERSIBLE BINDING EXAMINED IN PRECISION-CUT ADRENAL SLICE CULTURE

Örjan Lindhe, Lizette Granberg and Ingvar Brandt

Dept of Environmental Toxicology, EBC, Uppsala University
Norbyvägen 18A, S-752 36 Uppsala, SWEDEN
E-mail Orjan.Lindhe@ebc.uu.se

INTRODUCTION

In order to examine cell-specific localisation of cytochrome P450 (CYP) catalysed binding and toxicity in human and wild animal tissues, we required a suitable tissue culture system. Precision-cut tissue slice culture was introduced in the early 1980s (*1*). It provides a test system where cell-specific effects can be studied *in vitro* during several days (*2-4*).

The adrenal cortex is highly susceptible to chemical insult due to its high lipid content, resulting in uptake and retention of lipophilic xenobiotics in the cells. The adrenal cortex has a high content of CYPs and is highly vascularised (*5*).

The persistent adrenocorticolytic DDT metabolite 3-methylsulphonyl-2,2′-bis(4-chlorophenyl)-1,1′-dichloroethene (MeSO$_2$-DDE) was originally identified in Baltic grey seal, a population suffering from adrenocortical hyperplasia (*6*). In human tissue, a number of methylsulfones of DDT and PCB have been found. Among these, MeSO$_2$-DDE is a predominating metabolite in human adipose tissue (*7*). Another compound of interest is 7,12-dimethylbenz(a)anthracene (DMBA), a well known adrenocorticolytic substance that induces necrosis and bleedings in zona fasciculata and reticularis in rats (*8, 9*).

In this study we have developed and applied precision-cut adrenal slice culture to examine CYP catalysed irreversible binding of MeSO$_2$-DDE and DMBA in mouse and rat tissue.

MATERIAL AND METHODS

Chemicals

^{3}H-7,12-Dimethylbenz(a)anthracene (^{3}H-DMBA, 52.0–74.0 Ci/mmol) was purchased from Amersham Life Science (Amersham, England). 3-methylsulphonyl-2,2′-bis(4-chloro-^{14}C-phenyl)-1,1′-dichloroethene (MeSO$_2$-^{14}C-DDE; 13.4 mCi/mmol) (*10*) and 3,3′,4,4′,5-pentachlorobiphenyl (PCB 126) were prepared by Dr. Åke Bergman, Stockholm University, Sweden. 1-ethynylpyrene was a kind gift from Dr. William Alworth, Tulane University, New Orleans, USA. Other chemicals were obtained from commercial sources.

Biological Reactive Intermediates VI, Edited by Dansette *et al.*
Kluwer Academic / Plenum Publishers, 2001

Preparation and Incubation of Tissue Slices

Mouse and rat adrenals were excised and placed in ice-cold PBS-buffer. Immediately before slicing the adrenals were embedded in 3% agarose. Precision-cut slices (200 μm) were prepared in a Krumdieck tissue slicer (Alabama Research and Development, USA) in ice-cold PBS buffer. Following sectioning, the slices were put on titanium inserts in a 6-well plate and incubated in FDMEM medium (2.5 ml) at 38°C (5% CO_2). The medium was supplemented with ^{3}H-DMBA (0.4 μM, 74 μCi) or MeSO$_2$-^{14}C-DDE (7.5 μM, 0.25 μCi), DMSO (0.1%), FBS (2%), gentamycin (0.1%), L-glutamine (2 mM), mercaptoethanol (50 μM), penicillin (100 U/ml) and streptomycin (100 g/ml). Slices were incubated for 4 h with ^{3}H-DMBA or 24 h with MeSO$_2$-^{14}C-DDE. Some slices were pre-incubated 30 min with CYP inhibitors before addition of MeSO$_2$-^{14}C-DDE or ^{3}H-DMBA. Ellipticine, α-naphthflavone or 1-ethynylpyrene were used to inhibit CYP1A1/1B1 and metyrapone to inhibit CYP11B1.

Hormone Analysis

Corticosterone content in the medium was measured with HPLC using UV detection (241 nm). Medium (1 ml) was removed and steroid hormones were extracted twice with chloroform:methanol (2:1, 1.5 ml). The combined chloroform phases were evaporated to dryness, redissolved in acetonitrile (50%) and injected into the HPLC system (Lichrosorb RP 18 column, 20 cm, 5 μm particle size). The steroid products were separated using a linear gradient of 40–80% acetonitrile (1 ml/min) and mixed with 40% methanol over 25 min. The amounts of steroid were expressed as nmol/slice. The detection level of corticosterone was 5 pmol/ml medium.

Microautoradiography

Following incubation, the slices were fixed in buffered formaldehyde (4%, pH 7) for 24 h and dehydrated in 70% ethanol (7 d) – 95% ethanol (1 d) – 99.5% ethanol (3 h). This extraction procedure was considered to remove unbound substance, leaving the irreversibly bound fraction of radioactivity in the tissue. The extracted slices were then embedded in methacrylate and sectioned (2 μm) in a rotation microtome. The glass slides were dipped in liquid film emulsion (NTB-2:H_2O, 1:1, Kodak) and after 6–60 weeks of exposure at 4°C, the autoradiograms were developed and the sections stained with toluidine blue.

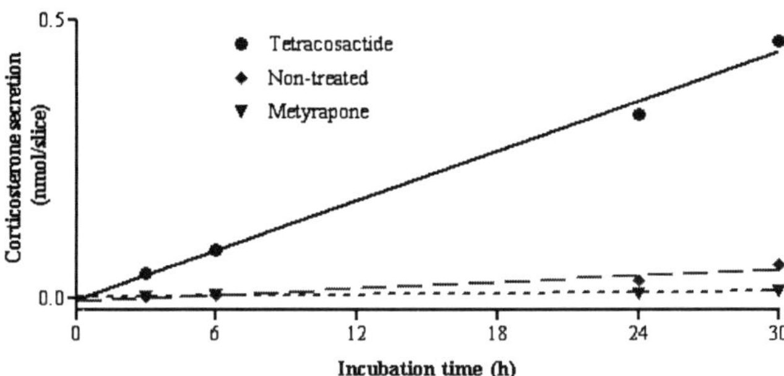

Figure 1 Corticosterone secretion to the medium by four adrenal slices originating from the same mice. Curves represents secretion from slices exposed to tetracosactide (11 nM), metyrapone (50 μM) and non-treated slices. Tetracosactide induced secretion 8-fold while metyrapone reduced secretion 4-fold.

Corticosterone secretion from adrenal slices could readily be measured in the culture medium. The rate of secretion was constant for more than 30 h. Corticosterone secretion was induced 8-fold by exposure to the ACTH-analogue tetracosactide (11 nM). Addition of the CYP11B1 inhibitor metyrapone (50 µM) reduced corticosterone secretion 4-fold (Fig. 1). Other steroids, such as aldosterone, 11-deoxycorticosterone and pregnenolone could also be measured in the medium.

Irreversible binding of $MeSO_2$-^{14}C-DDE was confined exclusively to the zona fasciculata in mouse adrenal slices (Fig. 2 A). Rat adrenal slices showed a very weak binding in these cells. As determined by phosphorautoradiography, the CYP11B1 inhibitor metyrapone reduced irreversible $MeSO_2$-DDE binding by 70%. We were also able to observe mitochondrial degeneration (electron microscopy) as well as inhibition of corticosterone secretion from mouse adrenal slices. These results are in accordance with previous observations made in vivo (*11, 12*).

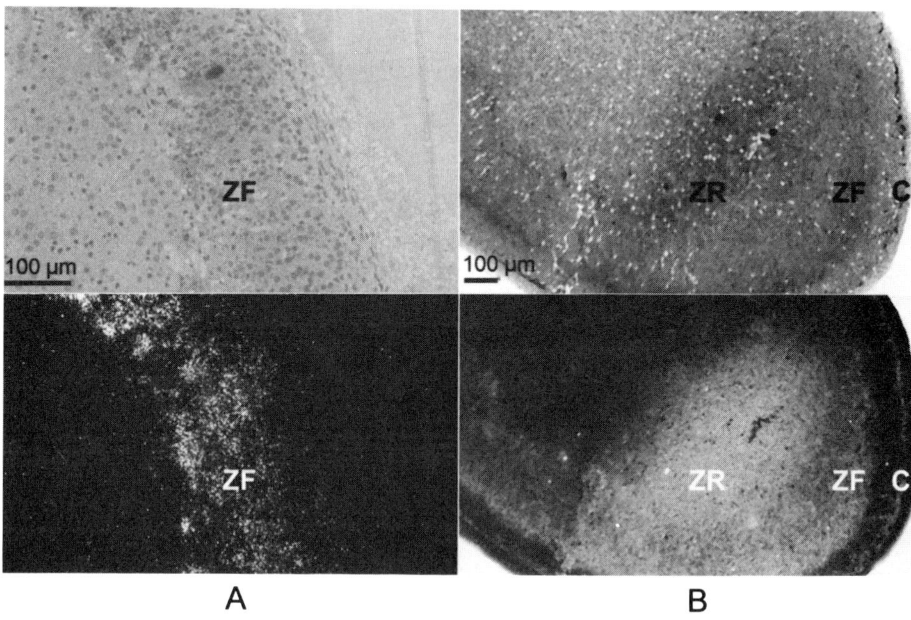

Figure 2 A) Irreversible binding of $MeSO_2$-^{14}C-DDE in a mouse adrenal slice. Binding was confined exclusively to zona fasciculata (ZF). **B)** Irreversible ^3H-DMBA binding in an adrenal slice from a rat pre-treated in vivo with PCB 126. Binding was localised to zona reticularis (ZR), zona fasciculata and the endothelium of the capsule (C). Top row represents bright-field images and bottom row dark-field images.

Irreversible binding of ^3H-DMBA was confined to zona fasciculata (preferentially to the outer part bordering zona glomerulosa) and zona reticularis in rat adrenal slices (Fig. 2 B). No pronounced binding was observed in the mouse adrenal cortex. PCB 126 treatment in vivo resulted in a roughly twofold induction of binding in rat adrenal cortex. No such induction was observed in mouse adrenal slices. The cortical binding of DMBA in the rat was partially inhibited by ellipticine, α-naphthflavone, 1-ethynylpyrene and metyrapone.

Huggins and Morii (*8*) reported that DMBA causes adrenal apoplexy and necrosis in zona fasciculata and zona reticularis in rats, while zona glomerulosa and medulla remained intact. The first sign of injury to the adrenal was found to be a selective destruction of hormone secreting cells. This was later confirmed by electron microscopy, where the first pathologic changes were seen in the mitochondria of zona fasciculata and zona reticularis cells (*13*). Our results show that the sites of metabolite binding correlates with the reported sites of toxicity.

Interestingly a selective DMBA binding was observed in endothelial cells in the capsule and zona glomerulosa in both rat and mouse adrenal slices exposed to PCB 126 in vivo (Fig. 2 B). Unlike the partial inhibition in the cortex, endothelial binding was completely blocked by ellipticine, α-naphthflavone or 1-ethynylpyrene but remained unaffected by metyrapone. This indicates involvement of CYP1A1/1B1 and excludes CYP11B1 as activating enzymes in these endothelial cells.

We conclude that precision-cut tissue slice culture is a useful test system to examine cell-specific irreversible binding of adrenocorticolytic environmental pollutants. The advantage of the system is the maintained cellular organisation combined with the easy culture method and control of exposure. Steroid secretion can be modulated and used as a marker for viability and an endpoint for toxicity. Human and wild animal tissue can be examined in cultured slices.

REFERENCES

1. Krumdieck CL, Santos JE, Ho K-J. A new instrument for the rapid preparation of tissue slices. Anal Biochem:118-123(1980).
2. Brendel K, Fisher RL, Krumdieck CL, Gandolfi AJ. Precision-cut rat liver slices in dynamic organ culture for structure-toxicity studies. In: Methods in Toxicology, vol 1A:Academic Press Inc., 1993;222-230.
3. Gandolfi AJ, Brendal K, Fisher RL, Michaud JP. Use of tissue slices in chemical mixture toxicology and interspecies investigations. Toxicology 105:285-290(1995).
4. Parrish AR, Gandolfi AJ, Brendel K. Precision-cut tissue slices: Applications in pharmacology and toxicology. Life Sci 57:1897-1901(1995).
5. Vainio H. Role of hepatic metabolism. In: Concepts in Drug Metabolism (Testa B, Jenner P, eds). New York:Dekker, 1980;251-84.
6. Bergman A, Olsson M. Pathology of Baltic grey seal and ringed seal females with special reference to adrenocortical hyperplasia: Is environmental pollution the cause of a widely distrubuted disease syndrome? Finn Game Res:47-62(1985).
7. Weistrand C, Norén K. Methylsulfonyl metabolites of PCBs and DDE in human tissue. Environ Health Perspect 105:644-649(1997).
8. Huggins, Morii. Selective adrenal necrosis and apoplexy induced by 7,12-dimethylbenz(a)antracene. J Exp Med 114:741(1961).
9. Huggins CB, Sugiyama T. Production and prevention of two distinctive kinds of destruction of adrenal cortex. Nature 206:1310-4(1965).
10. Bergman Å, Wachtmeister CA. Synthesis of methanesulfonyl derivates of 2,2-bis(4-chlorophenyl)-1,1-dichloroethylene (p,p′-DDE) present in seal from the Baltic. Acta Chem Scand:90-91(1977).
11. Lund BO, Bergman Å, Brandt I. Metabolic activation and toxicity of a DDT-metabolite, 3-methylsulphonyl-DDE, in the adrenal zona fasciculata in mice. Chem Biol Interact 65:25-40(1988).
12. Jönsson C-J, Rodriguez Martinez H, Lund BO, Bergman Å, Brandt I. Adrenocortical toxicity of 3-methylsulfonyl-DDE in mice. II. Mitochondrial changes following ecologically relevant doses. Fundam Appl Toxicol 16:365-74(1991).
13. Belloni AS, Mazzocchi G, Robba C, Gambino AM, Nussdorfer GG. An ultrastructural, morphometric and autoradiographic study of the effects of 7,12-dimethylbenzanthracene on the rat adrenal cortex. Virchows Arch B Cell Pathol 26:195-214(1978).

ELEVATION OF GLUTATHIONE LEVELS BY COFFEE COMPONENTS AND ITS POTENTIAL MECHANISMS

Gerlinde Scharf, Sonja Prustomersky, and Wolfgang W. Huber
Institute of Cancer Research
Borschkegasse 8a
A - 1090 Vienna
Austria

BACKGROUND & INTRODUCTION

The tripeptide glutathione (L-γ-glutamyl-L-cysteinylglycine) is found ubiquitous in microorganisms, plants and animals. In mammalian cells, where the tripeptide fulfils numerous functions, concentrations range from 0.5 to 10 mM (Meister & Tate, 1976; Meister, 1984; Redegeld et al., 1990). Glutahione is involved, for example, in the synthesis of proteins and DNA, in the regulation of enzyme activity, in the transport and reservoir of amino acids. A very important function of glutathione is the protection of cells, for instance as an antioxidant or as a co-factor in the conjugation of xenobiotics (Meister & Anderson, 1983; Redegeld et al., 1990).

A disturbed glutathione status which may be a consequence of cytotoxic events usually occurs as a lowering of the concentration of the reduced form of glutathione (= GSH) and, in case of oxidative stress, can be accompanied by a simultaneous increase of its oxidized form (= GSSG) (Redegeld et al., 1990).

The synthesis of glutathione takes place within the cell, mainly in the liver, and to a lesser extent in kidney, brain, small intestine, lens, muscle and erythrocytes. The enzymes involved in the reactions shown below are γ-glutamylcysteine synthetase (1) and glutahione synthetase (2) (Meister & Tate, 1976).

$$\text{L-Glutamate} + \text{L-Cysteine} + \text{ATP} \leftrightarrow \text{L-γ-Glutamyl-L-cysteine} + \text{ADP} + \text{Pi} \qquad (1)$$

$$\text{L-γ-Glutamyl-L-cysteine} + \text{Glycine} + \text{ATP} \leftrightarrow \text{Glutathione} \qquad (2)$$

γ-Glutamylcysteine synthetase (= GCS, 1) is the rate limiting enzyme. It is not very specific and can bind various amino acids, L-cysteine and L-α-aminobutyrate displaying the greatest affinity. The enzyme consists of two subunits. The heavy subunit is responsible for its catalytic activity and the light subunit for the regulation of the non-allosteric feedback inhibition by glutathione itself. GCS can either be inhibited or induced by several xenobiotic substances which usually results in changes of glutathione level. An inhibition has been observed, for example, with buthionine-, prothionine-, and methionine sulfoximines (Richman et al., 1973; Griffith, 1979; Griffith 1982), cystamine (Lebo & Kredlich, 1978), S-sulfocysteine, and S- sulfohomocysteine (Moore et al., 1987).

Induction of GCS may on the one hand occur as an adaptive response to chemically induced oxidative stress as shown with quinones (Shi et al., 1994, Liu et al., 1996; Tian et al., 1997), hydrogen peroxides (Ochi, 1995), and methyl mercury (Woods et al., 1992). Importantly, a simultaneous GSSG and GCS elevation by hydrogen peroxides was observed in hamster V79 cells by Ochi (1995). On the other hand, butylated hydroxyanisole (Eaton & Hamel, 1994) and nitric oxide (Moellering et al., 1998) were found to induce GCS without any evidence of concomitant oxidative stress.

Glutathione synthetase catalyzes the conversion of γ-glutamylcysteine to glutathione. The enzyme is highly specific for glycine and the COOH-terminal of the γ-glutamyl amino acid substrate. In contrast to GCS glutathione synthetase consists of two subunits of identical molecular weights (Oppenheimer et al., 1979; Meister, A., 1985) and, again in contrast to GCS, was not affected by butylated hydroxyanisole (Eaton and Hamel,1994) .

Huber et al. (1997) showed that a 1:1 mixture of the coffee ingredients Kahweol and Cafestol (K/C) causes a significant reduction of DNA adducts that were formed by the cooked food mutagen PhIP (2-amino-1-methyl-6-phenylimidazo [4,5-*b*]pyridine) in several organs of the rat. This chemoprotective effect of K/C, which has been found with aflatoxine B1 instead of PhIP (Cavin et al.,1998), may be attributed to several enzymatic K/C-related changes such as the induction of glutrathione-S-transferases (GSTs). Since GSTs use GSH as a cofactor we checked for effects of K/C on GSH levels and metabolism, primarily GCS, in order to evaluate a possible role of these mechanisms in the chemoprotective potential of K/C against mutagenic and carcinogenic damage. Furthermore, we investigated whether oxidative stress had any significance in the effects of K/C.

MATERIALS AND METHODS

K/C was a kind gift of Nestle Research Center, Lausanne, Switzerland.
Male F 344 rats (6-8 per group) were exposed for 10 days to K/C in the diet at 4 different doses (free alcohols, equimolar to 0.2, 0.1, 0.04, 0.02% of palmitates). After sacrifice of the animals by decapitation organs were quickly excised, weighed and frozen at −70°C until final processing. Cytosols (Turesky et al., 1991) and homogenates were prepared for the determination of GCS and GSH/GSSG/TBARS, respectively, according to the methods mentioned below.

GSH (Boyne & Ellmann, 1972), GSSG (Sacchetta et al., 1986) and TBARS (Lieners et al., 1989) were all measured by spectrophotometry GCS was measured by HPLC according to Nardi & Cipollaro, 1990. GCS and TBARS were related to cytosol or homogenate protein (Bradford, 1976).

RESULTS & DISCUSSION

We observed a dose-dependent increase of GSH in the liver, being significant even at the lowest investigated dose of 0.02% K/C. In the 0.2% K/C group the increase in GSH was almost 2-fold. GSH levels in the colon, a target organ of PhIP which is important in human carcinogenesis, showed a dose-dependent increase as well, although the effect was significant at the highest dose only (Fig. 1 and 2).

Simultaneously with this elevation of GSH, there was an increase in GCS in both liver and colon (Fig. 1 and 2). Thus, an induction of the rate limiting enzyme of GSH synthesis may serve as a good explanation for the observed increase in GSH.
Oxidative stress in the liver was measured with two indicators, i. e. TBARS and the oxidized form of glutathione (GSSG). While K/C produced no significant increase in TBARS, there was an elevation of the GSSG level exclusively at the highest dose (Fig. 3). However as the most accurate indicator of oxidative stress one should consider the ratio of the oxidized and reduced forms respectively (GSSG/GSH) rather than the GSSG levels alone and a clear tendency towards a reduction of this ratio is recognized after treatment with K/C (Table 1). Thus neither the data with TBARS nor those with GSSG gave any indication of oxidative stress under K/C-treatment.

In summary, our results suggest that induction of GCS is an important mechanism by which K/C enhances GSH, however, without the involvement of oxidative stress.

ACKNOWLEDGEMENTS

The authors thank Robert Turesky, Nestle Research Center, Lausanne, Switzerland.

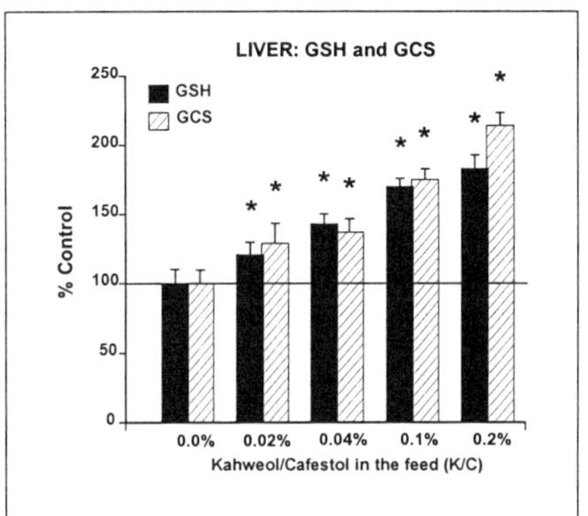

Figure 1: GSH and GCS in rat liver after 10 days of K/C-treatment $* p = <0.05$

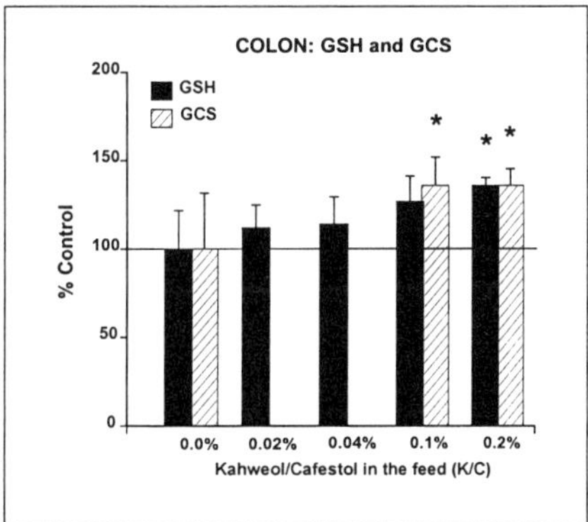

Figure 2: GSH and GCS in rat colon after 10 days of K/C-treatment. Because only a significant increase of GSH in the 0.2% K/C group has been observed, GCS was only measured in the 0.1% and 0.2% K/C group; $* = p<0.05$.

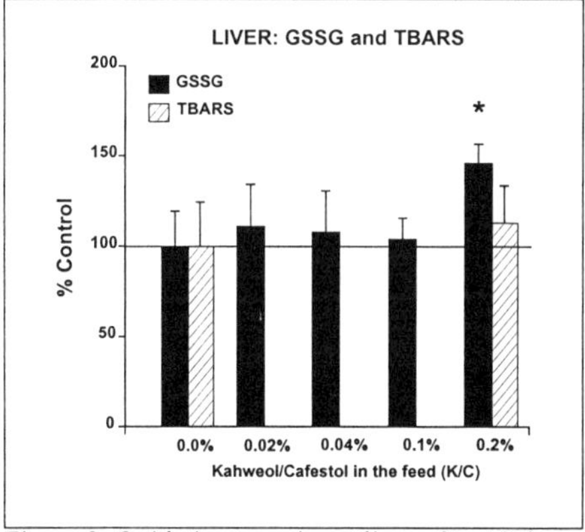

Figure 3: Oxidative stree in rat liver after 10 days of K/C-treatment. $* = p<0.05$.

Table 1: GSSG/GSH-ratio; $* p = <0.05$

K/C dose (in % of the diet)	GSSG/GSH
0.00 (Control)	1.00
0.02	0.93
0.04	0.78
0.10	0.61*
0.20	0.81

REFERENCES

Boyne, A. F., and Ellmann, G. L., 1972, A methodology for analysis of tissue sulfhydryl components, *Analytical Biochemistry* 46(2): 639.

Bradford, M. M., 1976, A rapid and sensitive method for the quantification of microgram quantities of protein utilizing the principle of protein-dye binding, *Analytical Biochemistry* 72: 248.

Cavin, C., Holzhäuser, D., Constable, A., Huggett, A. C., and Schilter, B., 1998, The coffee-specific diterpenes cafestol and kahweol protect against aflatoxin B1-induced genotoxicity through a dual mechanism, *Carcinogenesis* 19(8): 1369.

Eaton, D. L. and Hamel, D. M., 1994, Increase in γ-glutamylcysteine synthetase activity as a mechanism for butylated hydroxyanisole-mediated elevation of hepatic glutathione, *Toxicology and Applied Pharmacology* 126: 145.

Griffith, O. W., Anderson, M. E., and Meister, A., 1979, Inhibition of glutathione biosynthesis by prothionine sulfoximine, a selective inhibitor of γ-glutamylcysteine synthetase, *The Journal of Biological Chemistry* 254(4): 1205.

Griffith, O. W., 1982, Mechanism of action, metabolism, and toxicity of buthionine sulfoximine and its higher homologs, potent inhibitors of glutathione, *The Journal of Biological Chemistry* 257(22): 13704.

Huber, W. , McDaniel, L. P., Kaderlik, K. R., Teitel, C. H., Lang, N. P., and Kadlubar, F. F. ,1997, Chemoprotection against the formation of colon DANN adducts from the food-borne carcinogen 2-amino-1-methyl-6-phenylimidazo[4,5-b]pyridine in the rat, *Mutation Research* 376: 115.

Huang, C.-S., Moore, W. R., and Meister, A., 1988, On the active site thiol of γ-glutamylcysteine synthetase: Relationships to catalysis, inhibition, and regulation, *Processings of the National Academy of. Science.* 85: 2464.

Lebo, R. V. and Kredlich, N. M., 1978, Inactivation of human gamma-glutamylcysteine synthetase by cystamine. Demonstration and quantification of enzyme-ligand complexes, *Journal of Biological Chemistry* 253: 2615.

Lieners, C., Redl, H., Molnar, H., Furst, W., Hallstrom, S., and Schlag, G., 1989, Lipidperoxidation in a canine model of hypovolemic-traumatic shock, *Progressions in Clinical and Biological Research* 308: 345-50.

Liu, R.-M., Hu, H., Robison, T. W., and Forman, H. J., 1996, Increased γ-glutamylcysteine synthetase and γ-glutamyl transpeptidase activities enhance resitance of rat lung epithelial L2 cells to quinone toxicity, *American Journal of Respiratory Cell Molecular Biology* 14: 192.

Meister, A. and Tate, S. S., 1976, Glutathione and related γ-glutamyl compounds: Biosynthesis and utilization, *Annual Review of Biochemistry* 45: 559.

Meister, A. and Anderson, M. E., 1983, Glutathione, *Annual Reviews in Biochemistry* 52: 711.

Meister, A., 1984, New aspects of glutathione biochemistry and tranport: selective alteration of glutathione metabolism, *Federation Proceedings* 43(15): 3031.

Meister, A, 1985, Glutathione synthetase from rat kidney, *Methods in Enzymology* 113: 393.

Moellering, D., McAndrew, J., Patel, R. P. , Cornwell, T., Lincoln, T., Cao, K., Messina, J. L., Forman, H. J., Jo, H., and Darley-Usmar, V. M., 1998, Nitric oxide-dependent induction of glutathione synthesis through increased expression of γ-glutamylcysteine synthetase, *Archives of Biochemistry and Biophysics* 358(1): 74.

Moore, W., Wiener, H. L., and Meister, A., 1987, Inactivation of gamma-glutamylcysteine synthetase, but not of glutamine synthetase, by S-sulfocysteine and S-sulfohomocysteine, *The Journal of Biological Chemistry* 262(35): 16771.

Nardi, G. and Cipollaro, M, 1990, Assay of γ-Glutamylcysteine synthetase and glutathione synthetase in erythrocytes by high-performance liquid chromatography in fluorimetric detection, *Journal of Chromatography* 530: 122.

Oppenheimer, L., Wellner, V. P., Griffith, O. W., and Meister, A., 1979, Glutathione synthetase, *The Journal of Biological Chemistry* 254(12):5184.

Ochi, T. (1995) Hydrogen peroxide increases the activity of γ-Glutamylcysteine synthetase in cultured chinese hamster V79 cells, *Archives of Toxicology* 70: 96.

Redegeld, F. A. M., Koster, A. S., van Bennekom, W. P., 1990, Determination of tissue Glutathione. in: *Glutathione: Metabolism and Physiological Functions.* J. Vina ed., CRC Press, Boca Raton.

Richman, P. G., Orlowski, M., and Meister, A., 1973, Inhibition of gamma-glutamylcysteine synthetase by L-methionine-S-sulfoximine, *The Journal of Biological Chemistry* 248: 6684.

Richman, P. G. and Meister, A., 1975, Regulation of γ-glutamylcysteine synthetase by nonallosteric feedback inhibition by glutathione, *The Journal of Biological Chemistry* 250(49): 1422.

Romero, F. J. and Galaris, D., 1990, Compartmentation of cellular glutathione in mitochondrial and cytosoloic pools. in: *Glutathione: Metabolism and Physiological Functions.* J. Vina ed., CRC Press, Boca Raton.

Sacchetta, P., Di Cola, D., and Federici, G., Alkaline hydrolysis of N-ethylmaleimide allows a rapid assay of glutathione disulfide in biological samples, *Analytical Biochemistry* 154(1):205.

Shi, M. M., Kugelman, A., Iwamoto, T., Tian, L., and Forman, H. J., 1994, Quinone-induced oxidative stress elevates glutathione and induces γ-glutamylcysteine synthetase activity in rat lung epithelial L2 cells, *The Journal of Biological Chemistry* 269(42): 26512.

Tian, L., Shi, M. M., and Forman, H. J., 1997, Increased transcription of the regulatory subunit of γ-glutamylcysteine synthetase in rat lung epithelial L2 cells exposed to oxidative stress of glutathione depletion, *Archives of Biochemistry and Biophysics* 342(1): 126.

Turesky, R. J., Lang, N. P., Butler, M. A., Teitel, C. H., and Kadlubar, F. F., Metabolic activation of carcinogenic heterocyclic aromatic amines by human liver and colon, *Carcinogenesis* 12(10): 1839.

Woods, J. S., Davis, H. A., and Baer, R. P., 1992, Enhancement of γ-glutamylcysteine synthetase mRNA in rat kidney by methyl mercury. *Archives of Biochemistry and Biophysics* 296(1): 350

INHIBITION OF CARCINOGENESIS AND TOXICITY BY DIETARY CONSTITUENTS

Chung S. Yang

Laboratory for Cancer Research
College of Pharmacy
Rutgers, The State University of New Jersey
Piscataway, New Jersey 08854-8020

INTRODUCTION

Diet is known to play a key role in affecting carcinogenesis and toxicity. Although some dietary chemicals have been shown to be toxic or carcinogenic, many dietary constituents have been shown to inhibit toxicity and carcinogenesis in animal models. In this article, results from our lab will be used to illustrate the mechanisms by which dietary constituents inhibit toxicity and carcinogenesis. The possible application of these mechanisms in the reduction of human risk for toxicity and cancer and the limitations of this approach will be discussed.

Most chemicals require metabolic activation for conversion to reactive metabolites before exerting their toxic or carcinogenic effects. In theory, the toxicity of a chemical is reduced if the formation of the reactive intermediates, usually catalyzed by phase I enzymes, is inhibited or the elimination of the chemical and its reactive intermediates, by phase II enzymes, is enhanced. Dietary chemicals may also inhibit toxicity and carcinogenesis by affecting the toxic or carcinogenic events beyond the biotransformation of xenobiotics.

INHIBITION OF CYP2E1 BY ORGANOSULFUR COMPOUNDS

Garlic (*Allium salivum*) is rich in organosulfur compounds, primarily in the form of alliin (*S*-allylcysteine sulfoxide) which is derived from the amino acid cysteine.

Biological Reactive Intermediates VI, Edited by Dansette *et al.*
Kluwer Academic / Plenum Publishers, 2001

Crushing the garlic causes the release of the enzymes allinase which catalyze the formation allicin and other *S*-allyl containing compounds such as diallyl disulfide, diallyl sulfide (DAS), and methyl allyl sulfide (Block, 1985). These compounds are enriched in garlic oil preparations and give the strong garlic odor. For example, DAS is used as a food-flavoring agent.

Treatment of rats and mice with DAS caused a time-dependent inhibition of CYP2E1 activity (Brady, et al., 1991). At a dose of 1.75 mmol/kg (i.g.) to the rat, the CYP2E1 activity was reduced to 50 and 17% at 5 and 15 h, respectively. Additional studies indicated that DAS is oxidized to diallyl sulfoxide (DASO) and diallyl sulfone (DASO$_2$) in the microsomes by CYP enzymes and the flavin-dependent monooxygenase. All three compounds are competitive inhibitors of CYP2E1. In addition, DASO$_2$ is a suicide inhibitor of CYP2E1. These concepts are illustrated in Figure 1. In theory, these compounds should be able to block the formation of reactive metabolites from CYP2E1 substrates and the associated toxicity. Indeed, the inhibition of carbon tetrachloride and *N*-nitrosodimethylamine induced toxicity by DAS has been demonstrated (Brady, et al., 1991).

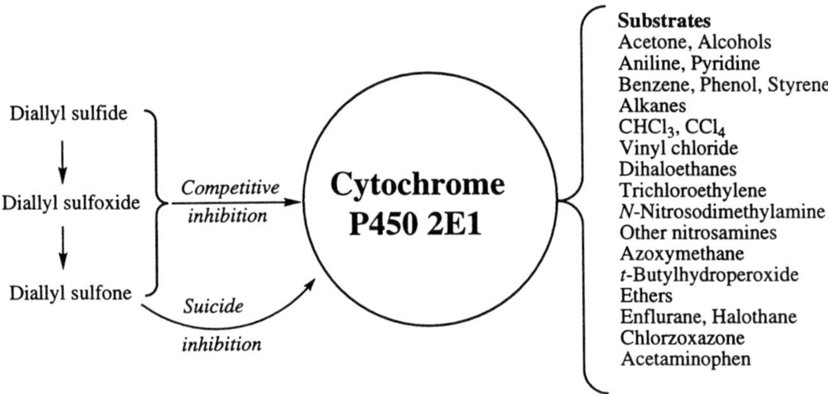

Fig. 1. Inhibition of CYP2E1-catalyzed reactions by DAS and its metabolites. The figure illustrates that DAS is oxidized to DASO and DASO$_2$. All three compounds are competitive inhibitors of CYP2E1, and DASO$_2$ is also a suicide inhibitor. Thus, they inhibit the metabolism of the CYP2E1 substrates listed.

EFFECTS OF GARLIC RELATED ORGANOSULFUR COMPOUNDS ON THE METABOLISM AND TOXICITY OF ACETAMINOPHEN

Acetaminophen (APAP), the leading analgesic used in the United States, is usually efficiently eliminated after conjugation with glucuronic acid and sulfate; only a few percent is oxidized by CYP2E1 and other enzymes (Patten, et al., 1993) to the reactive *N*-acetyl-*p*-benzoquinone imine intermediate which is eliminated by conjugating with glutathione. The drug is rather safe under normal conditions. Toxicity is produced when cellular glutathione is exhausted by the production of large quantities of quinone imine due to an overdose of APAP or elevated levels of CYP2E1, for example in alcoholics. Although APAP is also activated by CYP3A, CYP2E1 is the major activating enzyme in alcoholic humans and fasting mice and rats. We studied the inhibition of APAP toxicity by organosulfur compounds in a mouse model (Wang, et al., 1996, Lin, et al., 1996, Hu, et al., 1996). When given orally before or immediately after an APAP dose, DAS and $DASO_2$ effectively inhibited APAP induced hepatotoxicity, as judged by hepatopathology and serum lactic dehydrogenase levels. The effect was weaker when DAS or $DASO_2$ was given one or two hours after APAP, possibly because the glutathione was already depleted and tissue damage already initiated. Rats appeared to have a large reserve of glutathione than mice; the inhibitory effect of DAS can be demonstrated when given 3 hours after APAP administration (our unpublished results). Various organosulfur compounds were examined for their abilities to inhibit APAP toxicity in mice (Wang, et al., 1996). Diallyl disulfide, methyl allyl sulfide, and *S*-allyl cysteine, were less effective than DAS. The results agree with the concept that the *S*-allyl structure is needed for the activity.

Fresh garlic homogenate, when given before or immediately after a dose of APAP, also protected the mice from APAP hepatotoxicity in a dose-dependent manner. A complete protection was observed at a dose of 5 g fresh garlic per kg body weight in mice. Cooked garlic homogenate did not produce such a protective effect (Wang, et al., 1996). The idea that these organosulfur compounds and fresh garlic homogenates inhibited APAP toxicity by blocking the oxidative activation pathways is supported by the decreased levels of APAP-glutathione and APAP-cysteine, which are derived from *N*-acetyl-*p*-benzoquinone imine, in the mice.

The above results demonstrate the potential use of $DASO_2$ or DAS for the protection against APAP overdose toxicity. It can be used as an antidote, in combination with the presently used antidote, *N*-acetyl cysteine, to be given to patients as soon as an APAP overdose is discovered. $DASO_2$ prevents the further production of toxic metabolites from APAP while *N*-acetyl cysteine generates glutathione, preventing the further development of toxicity. $DASO_2$ can also be incorporated into the formulation of APAP tablets. When overdose of the APAP tablets occurs, the individual has also taken a high dose of CYP2E1 inhibitor, blocking the activation of APAP to its toxic metabolites, and thus preventing toxicity. $DASO_2$ is also a weak inhibitor of CYP3A and can inhibit the activation of APAP by this enzyme.

INHIBITION OF APAP METABOLISM AND TOXICITY BY PHENETHYL ISOTHIOCYANATE (PEITC) AND WATERCRESS

PEITC and other isothiocyanates, exist as glucosinolates in cruciferous and other vegetables. When the vegetables are crushed, the myrosinase catalyze the release of PEITC and other isothiocyanate. PEITC has been found to be a competitive and suicide inhibitor of CYP2E1. When PEITC (19-150 mg/kg) was given to mice (i.g.) 1h before or immediately prior to a toxic dose of APAP, the APAP-induced hepatotoxicity was significantly decreased or completely prevented (Li, et al., 1997). PEITC treatment is believed to inhibit the formation of *N*-acetyl-*p*-benzoquinone imine because it prevented the depletion of hepatic glutathione levels and decreased the formation APAP-glutathione, APAP-cysteine, and APAP-*N*-acetylcysteine in mice. PEITC and cruciferous vegetables also induced phase II enzymes in rodents (Guo, et al., 1992).

CAN BENZENE TOXICITY BE PREVENTED BY VEGETABLES WHICH CONTAIN CYP2E1 INHIBITORS?

The above studies with APAP suggest that DAS, $DASO_2$, and PEITC can also be used to block the toxicity of other compounds that are activated by CYP2E1. One of the possible applications is in the prevention of benzene toxicity. Benzene-containing glue is still being used in shoe-making and other industries in third world countries such as

China. Benzene toxicity among these workers is well documented, and the problem is expected to exist for many years. Whereas improving the working environment is the best solution, use of chemoprotective agents would benefit these workers before harmful benzene exposure can be eliminated. A possible approach is to advise the workers to take more allium and cruciferous vegetables. This goal may be accomplished by eating a combination of different vegetables rather than consuming high doses of a particular vegetable. These vegetables need to be consumed in forms that DAS, PEITC, and other CYP2E1 inhibitors are bioavailable, possibly by eating their uncooked or prepared in a special way to produce and preserved these compounds. In a pilot study, the effect can be assessed by measuring the ratio between benzene and its metabolites. These organosulfur compounds also induce CYP2B1 and affect CYP1A2 and CYP3A in rodents (Brady, et al., 1991, Wang, et al., 1996, Lin, et al., 1996, Li, et al., 1997, Ishizaki, et al., 1990). At modest doses, these compounds or the vegetables are not expected to affect the normal functions of key CYP enzymes in humans. The induction of NADP(H)-quinone reductase and phase II enzymes by allium and cruciferous vegetables (Guo, et al., 1992) may provide an additional protective effect.

INHIBITION OF LUNG TUMORIGENESIS BY DAS, DASO$_2$, AND PEITC

Administration of DAS (200 mg/kg) or DASO$_2$ (100 mg/kg) prior to NNK treatment significantly decreased lung tumor incidence (by 50-62%) and lung tumor multiplicity (by > 90%) in mice (Hong, et al., 1992, Hong, et al., 1994). The protective effect of DAS and DASO$_2$ against NNK-induced lung tumorigenesis appears to be due to the inhibition of NNK activation. *In vivo* and *in vitro* studies demonstrated that DAS decreased the metabolic activation of NNK in mouse lung and liver microsomes (Hong, et al., 1992). This decrease in NNK activation may be due to the inhibitory effect of DAS on CYP enzymes. In the mouse lung, CYP 2A and CYP 2B are involved in the activation of NNK (Smith, et al., 1990, Smith, et al., 1993). It is possible that DAS or its metabolites are inhibiting one or both of these enzymes. The inhibition of lung tumorigenesis by fresh garlic homogenates (given before the NNK dose) has also been demonstrated (Hong, et al., 1994).

INHIBITION OF CARCINOGENESIS BY TEA IN ANIMAL MODELS

Using an NNK-induced lung tumorigenesis model, the inhibitory effects of tea was demonstrated when green and black tea preparations were administered orally (through drinking fluid) to the A/J mouse during either the initiation period (starting two weeks before the NNK dose, for three weeks) or the post-initiation period (starting one day or one week after the NNK dose until the termination of the experiment) (Wang, et al., 1992, Yang, et al., 1997). At 16 weeks after a dose of NNK, almost all the mice developed adenomas. The progression of adenoma to adenocarcinoma was inhibited when black tea was administered orally to mice during weeks 16 to 52 of the experiment (Yang, et al., 1997). Black tea and green tea extracts have also been shown to inhibit spontaneously developed lung tumors and rhabdomyosarcomas in A/J mice (Landau, et al., 1998). The inhibition of the growth and progression of skin tumors has also been clearly demonstrated (Katiyar and Mukhtar, 1996, Huang, et al., 1997, Lu, et al., 1997, Conney, et al., 1999). These and other experiments demonstrate that tea constituents posses broad inhibitory activities against carcinogenesis. The activities against tumor growth and progression are especially worth considering for future applications in human cancer prevention.

MECHANISMS OF ANTI-CARCINOGENIC ACTIONS OF TEA

The possible mechanisms of the inhibitory action of tea against tumorigenesis have been discussed in several reviews (Katiyar and Mukhtar, 1996, Yang and Wang, 1993, Yang, et al., 2000). In terms of the inhibition of tumor growth and progression, tea polyphenols may inhibit key signal transduction pathways which lead to the inhibition of cell proliferation, transformation, and angiogenesis as well as enhanced apoptosis. These and other possible mechanisms are summarized in Figure 2. The inhibition of transformation of the JB6 cells by epigallocatechin-3-gallate (EGCG), a major polyphenolic compound in green tea, and theaflavins (characteristic constituents in black tea) showed LC_{50} of ~5 µM for both compounds (Dong, et al., 1997). The inhibition of transformation is closely related to the inhibition of the AP-1 activity. In the H-ras transformed 30.7 b Ras 12 cells, the AP-1 activity was inhibited by many tea related polyphenolic compounds (Chung, et al., 1999). Structure-activity relationship studies suggest that the trihydroxy structure on the B-ring and the gallate ester structure significantly contribute to the AP-1 inhibitory activity. For example, EGCG has the highest and EC the lowest activities in our study. The inhibitory activity is attributed to the inhibition of the MAP-kinase activities; for example, the phosphorylation of c-jun by JNK and the phosphorylation of Erk by MEK. These inhibitory activities may or may not be related to the antioxidative properties of tea polyphenols. In the lung adenomas developed at 16 weeks after the NNK dose, the immunohistochemical staining intensity of the phospho-c-jun and-Erk in the green tea treated group was significantly lower than the positive control group (Yang *et al.* unpublished results).

Possible Mechanisms

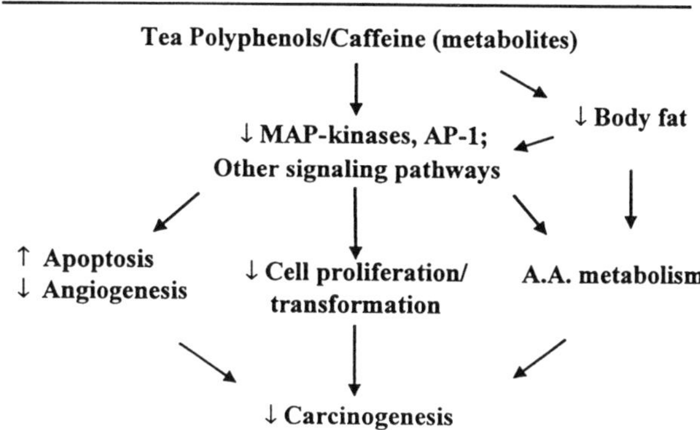

Fig. 2. Possible mechanisms for the inhibition of carcinogenesis by tea. Tea polyphenols are proposed to inhibit carcinogenesis by modulating key signal transduction pathways, which lead to the inhibition of cell proliferation and transformation or the induction of apoptosis. The body fat-lowering effect of tea constituents (mainly due to caffeine) may also contribute to the inhibition of carcinogenesis. Inhibition of arachidonic acid metabolism is also proposed to be a key mechanism involved.

Inhibition of cell proliferation and enhancement of apoptosis by tea polyphenols have been observed in many cell lines (reviewed in (Yang and Chung, 1999)). In our studies with the human lung cancer H661 cells and the SV-40-immortalized (33 BES) and H-ras transformed (21BES) human bronchial epithelial cells, the EGCG-induced apoptosis was associated with the production of H_2O_2, and cell apoptosis was prevented by the inclusion of catalase in the culture medium (Yang, et al., 1998, Yang, et al., 2000). The results suggest that the "pro-oxidative" property, rather than the "antioxidative" property, of EGCG induces apoptosis. It is not clear whether such a mechanism is applicable *in vivo*.

In our NNK-induced mouse lung carcinogenesis model, however, enhanced apoptosis was observed in the lung adenomas of mice who had received green tea in drinking fluid for 16 weeks. The apoptosis was scored based on the TUNEL assay and the appearance of nuclear debris in morphological analysis. In this experiment, tea treatment also inhibited angiogenesis in the adenomas, based on the number of new blood vessals (immunohistochemistry using antibodies against von Willebrand factor) and vascular endothelial growth factor (VEGF) expression. The anti-angiogenesis activity of EGCG has been suggested by Cao *et al.* based on studies in cultured cells and the mouse retina (Cao and Cao, 1999). Our demonstration of such an activity in a carcinogenesis model will set the stage for many important studies.

In our studies on the inhibition of spontaneously generated tumors in A/J mice by tea (Landau, et al., 1998), treatment of mice with 1% green tea and 1 and 2% black tea infusion for 52 weeks significantly decreased the body weights, and in particular body fat. These mice consumed about the same amount of food as the control mice. The lowering of body weight is probably due to the lower absorption of fat and other nutrients as well as enhanced energy utilization (mainly due to caffeine consumption) of the tea treatment groups. The concept that the body fat-lowering effect may be closely associated with the inhibition of tumorigenesis has been developed by Dr. Allan Conney based on careful observation in his studies on the inhibition of UV-induced skin tumorigenesis by tea (personal communications). The situation may be similar to the experiments with caloric restriction and the entailed mechanisms are rather elusive. One possibility is that the lower body fat represents a lower supply of arachidonic acid and its precursors, and this may result in a lower level of pro-carcinogenic metabolites formed from arachidonic acid.

Inhibition of arachidonic acid metabolism by tea constituents may, by itself, contribute to the inhibition of carcinogenesis, in a similar manner to that proposed for aspirin and other anti-inflammatory agents. In a recent study, we have demonstrated that ingestion of green tea (equivalent to 1-3 cups) by human volunteers significantly decreased the levels of prostaglandin E_2 in rectal biopsy samples taken 4 h and 8 h after tea consumption (August, et al., 1999). This is probably due to the inhibition of the phospholipase A_2 activities by tea polyphenols. Preliminary experiments indicate that the release of arachidonic acid from cell membrane of human colon cancer H-29 cells was inhibited by EGCG and epicatechin-3-gallate with estimated LC_{50} values ~10 µM. These tea constituents also inhibited the cyclooxygenase and lipoxygenase activities of human colonic tissues and colon cancer samples as well as in cell lines, but the LC_{50} values were higher, ~ 50 µM.

CONCLUDING REMARKS

Dietary constituents may inhibit toxicity and carcinogenesis by a variety of mechanisms. These mechanisms as well as their possible applications and limitations are discussed as follows:

1)   Inhibition of the activation of toxic chemicals - Strong inhibitors will effectively block the formation of reactive intermediates and thus prevent chemical

toxicity. It is important to know the identity of the toxic chemical that the specified population is exposed to, and then specific non-toxic inhibitors can be used. General inhibitors of CYP enzymes are not useful, because they may affect the normal physiological function of CYP enzymes such as in sterol metabolism. Even if the inhibitors are of dietary origin, toxicity at high doses is still a major concern. A combination of different agents at low or moderate doses may be a more effective approach, but the possible interactions (expected additive effects) among these agents need to be demonstrated. The possible use of dietary inhibitors of CYP2E1 to prevent benzene toxicity in humans is a challenging opportunity.

2) Induction of phase II enzymes - In theory, the induction of phase II enzymes will facilitate the elimination of reactive intermediates or their precursors. Caution should be applied, however, in the application of this concept to reduce human risks for toxicity and cancer. The induction of phase II enzyme is probably an adaptive response to potentially harmful agents. For example, some of the phase II enzymes are electrophiles and some are pro-oxidants; at moderate doses, the effects may be beneficial; but toxic effects may be produced at high doses. In addition, some of the phase II enzymes participate in the activation of certain chemicals.

3) Reduction of oxidative stress - Reactive oxygen and nitrogen species are known to play a major role in chemical toxicity and carcinogenesis. In principle, antioxidants which can quench these species or prevent their formation are expected to reduced the risk of chemical toxicity and carcinogenesis. However, this concept may have been over-used in the interpretation of many existing results and in the prediction of the protective effects of dietary antioxidants. Many of the "antioxidants" also have other activities. In addition, many of the dietary chemicals, such as polyphenolic compounds, may have poor systemic bioavailability or may not reach the specified tissue sites to quench the specific reactive species.

4) Inhibition of arachidonic acid metabolism - Metabolites of arachidonic acid (and linoleic acid) results in the production of many pro-inflammatory or pro-carcinogenic metabolites including prostaglandins and reactive oxygen species. Dietary inhibitors of cyclooxygenase and lipooxygenase are potentially beneficial. The beneficial effects of ω-3 fatty acid have been attributed to such a mechanism. The potential use of phospholipase A_2 inhibitors remains to be explored.

5) Modulation of signal transduction pathways which lead to inhibition of cell proliferation, transformation, and angiogenesis as well as to induction of apoptosis - With the rapid increase in our understanding of the signal transduction pathways, more results on the modulation of signal transduction by dietary chemicals will become available. In relating results obtained in culture cells to *in vivo* situation, comparison of the effective concentration of a particular agent in cell line systems with those achievable in animal and human tissues is of great importance.

Additional research is needed to elucidate the mechanisms by which dietary constituents inhibit toxicity and carcinogenesis. In future work, it is important to study the effects of these chemicals at concentrations approximate to the levels achievable through the diet as well as their possible additive effect and interactions. For assessing the applicability of this concept for the prevention of toxicity and carcinogenesis in humans, more epidemiological and interventions studies are needed.

ACKNOWLEDGEMENTS

This work was supported by NIH grants ES03938 and CA56673. I thank Dr. Janelle M. Landau, Ms. Dorothy Wong, and Ms. Hnin P. Phyu for their assistance in the preparation of this manuscript.

REFERENCES

1. Block, E. The chemistry of garlic and onions. *Sci. Am.*, 252, 114-119 (1985).
2. Brady, J.F., Wang, M.-H., Hong, J.-Y., Xiao, F., Li, Y., Yoo, J.-S.H., Ning, S.M., Fukuto, J.M., Gapac, J.M. and Yang, C.S. Modulation of rat hepatic microsomal monooxygenase activities and cytotoxicity by diallyl sulfide. *Toxicol. Appl. Pharmacol.*, 108, 342-354 (1991).
3. Patten, C.J., Thomas, P.E., Guy, R.L., Lee, M., Gonzalez, F.J., Guengerich, F.P. and Yang, C.S. Cytochrome P450 enzymes involved in acetaminophen activation by rat and human liver microsomes and their kinetics. *Chem. Res. Toxicol.*, 6, 511-518 (1993).
4. Wang, E.-J., Li, Y., Lin, M., Chen, L., Stein, A.P., Reuhl, K.R. and Yang, C.S. Protective effects of garlic and related organosulfur compounds on acetaminophen-induced hepatotoxicity in mice. *Toxicol. Appl. Pharmacol.*, 136, 146-154 (1996).
5. Lin, M.C., Wang, E.-J., Patten, C., Lee, M.-J., Xiao, F., Reuhl, K. and Yang, C.S. Protective effect of diallyl sulfone against acetaminophen induced hepatotoxicity in rats and mice. *J. Biochem. Toxicol.*, 11, 11-19 (1996).
6. Hu, J.J., Yoo, J.-S.H., Lin, M., Wang, E.-J. and Yang, C.S. Protective effects of diallyl sulfide on acetaminophen-induced toxicities. *Food Chem.Toxicol.*, 34, 963-969 (1996).
7. Li, Y., Wang, E.-J., Chen, L., Stein, A.P., Reuhl, K.R. and Yang, C.S. Effects of phenethyl isothiocyanate on acetaminophen metabolism and hepatotoxicity in mice. *Toxicol. Appl. Pharmacol.*, 144, 306-314 (1997).
8. Guo, Z., Smith, T.J., Wang, E., Sadrieh, N., Ma, Q., Thomas, P.E. and Yang, C.S. Effects of phenethyl isothiocyanate, a carcinogenesis inhibitor, on xenobiotic-metabolizing enzymes and nitrosamine metabolism in rats. *Carcinogenesis*, 13, 2205-2210 (1992).
9. Ishizaki, H., Brady, J.F., Ning, S.M. and Yang, C.S. Effect of phenethyl isothiocyanate on microsomal *N*-nitrosodimethylamine (NDMA) metabolism and other monooxygenase activities. *Xenobiotica*, 20, 255-264 (1990).
10. Hong, J.-Y., Wang, Z.-Y., Smith, T., Zhou, S., Shi, S. and Yang, C.S. Inhibitory effects of diallyl sulfide on metabolism and tumorigenicity of the tobacco-specific carcinogen 4-(methylnitrosamino)-1-(3-pyridyl)-1-butanone (NNK) in A/J mouse lung. *Carcinogenesis (Lond.)*, 13, 901-904 (1992).
11. Hong, J.-Y., Lin, M.C., Wang, E.-J., Yang, C. and S. Inhibition of chemical toxicity and carcinogenesis by diallyl sulfide and diallyl sulfone, in *Food Phytochemicals for Cancer Prevention I*, Huang, M.-T., Osawa, T., Ho, C.-T. and Rosen, R.T. (eds.) ACS Symposium, Washington, D.C. (1994) .
12. Smith, T.J., Guo, Z.-Y., Thomas, P.E., Chung, F.-L., Morse, M.A., Eklind, K. and Yang, C.S. Metabolism of 4-(methylnitrosamino)-1-(3-pyridyl)-1-butanone in mouse lung microsomes and its inhibition by isothiocyanates. *Cancer Res.*, 50, 6817-6822 (1990).
13. Smith, T.J., Guo, Z., Li, C., Ning, S.M., Thomas, P.E. and Yang, C.S. Mechanisms of inhibition of 4-(methylnitrosamino)-1-(3-pyridyl)-1-butanone

(NNK) bioactivation in mouse by dietary phenethyl isothiocyanate. *Cancer Res.*, 53, 3276-3282 (1993).

14. Wang, Z.Y., Hong, J.-Y., Huang, M.-T., Reuhl, K., Conney, A.H. and Yang, C.S. Inhibition of *N*-nitrosodiethylamine and 4-(methylnitrosamino)-1-(3-pridyl)-1-butanone-induced tumorigenesis in A/J mice by green tea and black tea. *Cancer Res.*, 52, 1943-1947 (1992).

15. Yang, G.-Y., Liu, Z., Seril, D.N., Liao, J., Ding, W., Kim, S., Bondoc, F. and Yang, C.S. Black tea constituents, theaflavins, inhibit 4-(methylnitrosamino)-1-(3-pyridyl)-1 butanone (NNK)-induced lung tumorigenesis in A/J mice. *Carcinogenesis*, 18, 2361-2365 (1997).

16. Yang, G.-Y., Wang, Z.-Y., Kim, S., Liao, J., Seril, D., Chen, X., Smith, T.J. and Yang, C.S. Characterization of early pulmonary hyperproliferation, tumor progression and their inhibition by black tea in a 4-(methylnitrosamino)-1-(3-pyridyl)-1 butanone (NNK)-induced lung tumorigenesis model with A/J mice. *Cancer Res.*, 57, 1889-1894 (1997).

17. Landau, J.M., Wang, Z.-Y., Yang, G.-Y., Ding, W. and Yang, C.S. Inhibition of spontaneous formation of lung tumors and rhabdomyosarcomas in A/J mice by black and green tea. *Carcinogenesis*, 19, 501-507 (1998).

18. Katiyar, S.K. and Mukhtar, H. Tea in chemoprevention of cancer: epidemiologic and experimental studies (review). *Intl. J. Oncol.*, 8, 221-238 (1996).

19. Huang, M.-T., Xie, J.-G., Wang, Z.-Y., Ho, C.-T., Lou, Y.-R., Wang, C.-X., Hard, G.C. and Conney, A.H. Effects of tea, decaffeinated tea, and caffeine on UVB light induced complete carcinogenesis in SKH-1 mice: demonstration of caffeine as a biologically important constituent of tea. *Cancer Res.*, 57, 2623-2629 (1997).

20. Lu, Y.-P., Lou, Y.-R., Xie, J.-G., Yen, P., Huang, M.-T. and Conney, A.H. Inhibitory effect of black tea on the growth of established skin tumors in mice: effects on tumor size, apoptosis, mitosis, and bromodeoxyuridine incorporation into DNA. *Carcinogenesis*, 18, 2163-2169 (1997).

21. Conney, A.H., Lu, Y.-P., Lou, Y.-R., Xie, J.-G. and Huang, M.-T. Inhibitory effect of green and black tea on tumor growth. *Proc. Soc. Exptl. Biol. Med.*, 220, 229-233 (1999).

22. Yang, C.S. and Wang, Z.-Y. Tea and cancer: a review. *J. Natl. Cancer Inst.*, 58, 1038-1049 (1993).

23. Yang, C.S., Chung, J.Y., Yang, G.-Y., Chhabra, S.K. and Lee, M.-J. Tea and tea polyphenols in cancer prevention. *J. Nutr.*, 130, 472S-478S (2000).

24. Dong, Z., Ma, W.-Y., Huang, C. and Yang, C.S. Inhibition of tumor promoter-induced AP-1 activation and cell transformation by tea polyphenols, (-)-epigallocatechin gallate and theaflavins. *Cancer Res.*, 57, 4414-4419 (1997).

25. Chung, J.Y., Huang, C., Meng, X., Dong, Z. and Yang, C.S. Inhibition of activator protein 1 activity and cell growth by purified green tea and black tea polyphenols in H-*ras*-transformed cells: structure-activity relationship and mechanisms involved. *Cancer Res.*, 20, 1810-1807 (1999).

26. Yang, C.S. and Chung, J.Y. Growth inhibition of human cancer cell lines by tea polyphenols. *Curr. Pract. Med.*, 2, 163-166 (1999).

27. Yang, G.-Y., Liao, J., Kim, K., Yurkow, E.J. and Yang, C.S. Inhibition of growth and induction of apoptosis in human cancer cell lines by tea polyphenols. *Carcinogenesis*, 19, 611-616 (1998).

28. Yang, G.-Y., Liao, J., Li, C., Chung, J., Yurkow, E.J., Ho, C.-T. and Yang, C.S. Effect of black and green tea polyphenols on c-jun phosphorylation and H_2O_2 production in transformed and non-transformed human bronchial cell lines:

Possible mechanisms of cell growth inhibition and apoptosis induction. *Carcinogenesis*, (in press) (2000).

29. Cao, Y. and Cao, R. Angiogenesis inhibited by drinking tea. *Nature*, 398, 381 (1999).

30. August, D.A., Landau, J., Caputo, D., Hong, J., Lee, M.-J. and Yang, C.S. Ingestion of green tea rapidly decreases prostaglandin E_2 levels in rectal mucosa in humans. *Cancer Epidemiol. Biomark. Prev.*, 8, 709-713 (1999).

MODELLING THE RESPONSES TO BIOLOGICAL REACTIVE INTERMEDIATES

Establishing the Borderlines of Risk

P. J. van Bladeren, J. J. P. Bogaards, N. H. P. Cnubben, and E. M. Hissink

TNO Nutrition and Food Research, PO Box 360
Utrechtseweg 48, 3700 AJ, Zeist, The Netherlands
Email: bladeren@voeding.tno.nl

INTRODUCTION

Biotransformation enzymes modify xenobiotic compounds into more hydrophilic products (metabolites) which can be efficiently excreted from the body. Phase 1 enzymes, of which the cytochromes P450 are the most important, perform functionalization reactions creating a reactive centre in the molecule through oxidation or reduction. Phase 2 enzymes, such as glutathione S-transferases, UDP-glucuronyltransferases, N-acetyltransferases and sulphotransferases couple an endogenous molecule to a functional moiety in the molecule or metabolite from the phase 1 reaction, generally leading to more hydrophilic metabolites that are easily excreted with urine or bile.

An important point that will be stressed in the remainder of this presentation is the fact that all xenobiotic-metabolizing enzyme families consist of multiple isoenzymes with different expression patterns in different individuals, different organs and different species. Xenobiotic-metabolizing enzymes in principle are meant to detoxify and excrete foreign compounds. In some cases use has been made of these enzymes to activate prodrugs to the biologically active therapeutic agent, but in many cases the toxic potential of xenobiotics is enhanced through the metabolic transformation to reactive intermediates. This process is called bioactivation and occurs quite regularly. Thus, biotransformation in the form of detoxication or activation is involved in the toxicity of most compounds to which man is exposed and the balance between detoxification and bioactivation eventually determines drug efficacy or the expression of adverse effects upon xenobiotic exposure.

INTERINDIVIDUAL DIFFERENCES IN BIOTRANSFORMATION

For most of the xenobiotic-metabolizing enzymes a considerable interindividual variation in expression exists. Variability in the level of expression of these enzymes is firstly due to the existence of genetic polymorphisms.

Secondly, enzyme induction and inhibition by endogenous or exogenous factors is the other important cause of interindividual variation.

The regulation of biotransformation enzymes is very complex, involving various mechanisms such as altered transcription of specific genes through the interaction of a xenobiotic with a regulatory region of genes (e.g. Ah receptor, ARE, XRE), degradation of mRNA, translation processes and protein degradation/stabilization. Compounds can bind to the active site of the enzymes serving as a substrate or (non-) competitive inhibitor, or may act by a so-called suicide mechanism, where a reactive metabolite modifies the protein.

Numerous factors play a role in the induction and/or inhibition of xenobiotic-metabolizing enzymes like gender, age, smoking, nutrition, environment, medication, disease state and coexistent exposures (drug-drug interactions). Some of the cytochromes P450 are inducible by compounds like phenobarbital (CYP2B), PAHs in cigarette smoke and coffee (CYP1A), rifampicin (CYP3A4) and ethanol (CYP2E1) (Barry and Feely, 1990). Examples of non-nutritive dietary compounds known to affect xenobiotic-metabolizing enzymes are for instance indole-3-carbinol isolated from Brussels sprouts (CYP1A1/1A2), and diallyl sulphide, a component of garlic oil (P450 2B1). (van Iersel et al 1999)

The nutritional status (the content of protein, lipid and carbohydrate or fasting and dietary restriction) of an organism is also known to influence the metabolism significantly (Yang et al., 1992). A well-known example of a food-drug interaction is the presence of flavonoids (naringenin, quercetin and kaempherol) in grapefruit juice, which have been shown to influence the pharmacokinetic behaviour of therapeutic agents like nifedipine and midazolam through interaction with the P450 system (Fuhr and Kummert, 1995). Compounds that induce GSTs are extremely diverse and include for instance PAH, azodyes, phenolic food antioxidants, flavonoids and thiocarbamates (Hayes and Pulford, 1995). Eugenol, a compound present in essential oils of cinnamon, basil and nutmeg, or the diuretic drug ethacrynic acid are able to inhibit some isoenzymes of the GST family. (van Iersel et al 1999)

The physiologically based pharmacokinetic, or PBPK, models are computer-generated and can describe biological systems - physiology and anatomy (e.g. Andersen et al. 1987, Gerlowski and Jain 1983). For instance, in the testing of animals, real organ volumes and blood flows can be used to describe a rat or mouse, and also metabolism of testing animals and humans can be described. The mathematical models are used to predict the kinetic behaviour of compounds and their metabolites in any particular species. Since inter-individual differences in metabolism can be described using these models, the inter-individual differences that exist in the formation of toxic metabolites, and thus the differences in susceptibility to toxicity, can in principle be predicted as well.

A strategy was investigated for this prediction by the use of human *in vitro* metabolic parameters, obtained through the combined use of individual enzyme preparations and a human liver microsome bank.

This strategy consisted of the following of steps: (1) estimation of enzyme kinetic parameters K_m and V_{max} for the drug metabolizing enzyme involved with the usage of individual human isoenzymes, e.g. of microsomes containing individual P450 enzymes. (2) scaling of the enzyme levels present in these systems to the levels present in a set of human liver microsomal or cytosolic preparations by the use of enzymatic activities towards enzyme-selective model compounds. In this way, the metabolism of a particular compound is scaled-up to the levels of individual enzymes present in these liver samples. (3) Incorporation into a PBPK model. (4) Comparison of predicted blood plasma levels to those found in human volunteers or literature data. The scaling procedure described above can in principle be validated by measuring the enzyme activities in the same set of human liver preparations, and comparing these activities with the rates calculated from the single enzyme Km and Vmax and the scaling factors. (Hissink, et al. 1997)

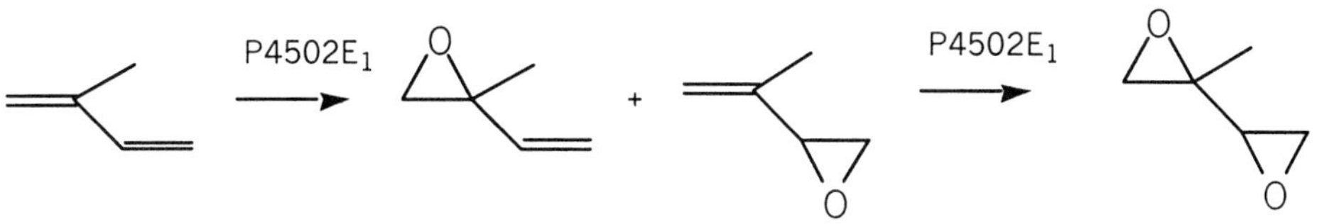

P4502E$_1$ + P4502E$_1$

Figure 1

ISOPRENE

Isoprene is one of several related compounds, such as 1,3-butadiene and vinylcyclohexene, used in the rubber industry. Isoprene is also produced endogenously in rats and mice, an emission product of many plant species and the major endogenous hydrocarbon in human breath. Isoprene is oxidized by cytochrome P450 into 3,4-epoxy-3-methyl-butene (epox-I) and to a lesser extent into 3,4-epoxy-2-methyl-1-butene (epox-II) (DelMonte *et al.*, 1985; Longo *et al.*, 1985; Bogaards *et al.*, 1996). Subsequently, by a second oxidation step, the diepoxide of isoprene can be formed, which has been shown to be highly mutagenic in the Ames test (Gervasi and Longo, 1990. CYP2E1 showed the highest rates of formation of the isoprene monoepoxides 3,4-epoxy-3-methyl-1-butene and 3,4-epoxy-2-methyl-1-butene, followed by CYP2B6. (Bogaards *et al.*, 1996) CYP2E1 was the only enzyme showing detectable formation of the diepoxide of isoprene, 2-methyl-1,2;3,4-diepoxybutane. Both isoprene monoepoxides were oxidized by CYP2E1 to the diepoxide at similar enzymatic rates.

Detoxication of the epoxides can occur through epoxide hydrolase-catalyzed hydrolysis as well as through GST-catalyzed conjugation with glutathione. To investigate species differences with regard to the role of epoxide hydrolase in the metabolism of isoprene monoepoxides, the epoxidation of isoprene by human liver microsomes was compared to that of mouse and rat liver microsomes (Bogaards *et al.*, 1996). The amounts of monoepoxides formed as a balance between epoxidation and hydrolysis, was measured in incubations with and without the epoxide hydrolase inhibitor cyclohexene oxide. Inhibition of epoxide hydrolase resulted in similar rates of monoepoxide formation in mouse, rat and man. Without inhibitor, however, the total amount of monoepoxides present at the end of the incubation period was twice as high for mouse liver microsomes than for rat and even 15 times as high as for human liver microsomes. Thus, differences in epoxide hydrolase activity between species may be of crucial importance for the toxicity of isoprene in the various species.

In addition to hydrolysis, isoprene epoxides may also be conjugated with glutathione. All GSTs showed high Km values towards the conjugation of both isoprene mono-epoxides (Km values >1000 μM). The μ and θ class glutathione S-transferases were much more efficient than the π and χ class glutathione S-transferases, both for rat and man. Rat glutathione S-transferases were more efficient than human glutathione S-transferases: rat GST T1-1 showed about 2.1 to 6.5-fold higher activities than human GST T1-1 for the conjugation of both epox-I and epox-II, while rat GST M1-1 and GST M2-2 showed about 5.2 to 14-fold higher activities than human GST M1a-1a (Bogaards etcl. 1999). Because the μ- and θ-class glutathione S-transferases are only expressed in about 50% and 40-90% of the human population, respectively, this may have significant consequences for the detoxification of isoprene monoepoxides in individuals who lack these enzymes.

It is clear that not the activation but the detoxication steps seem to determine interspecies differences in the putative toxic effects of isoprene, and presumably also interindividual difference in man. To establish the relative role of epoxide hydrolase and the

GST's in the inactivation of isoprene monoepoxide, PBPK modelling seems the only answer and work is currently in progress on this problem.

ETHYLENE DIBROMIDE

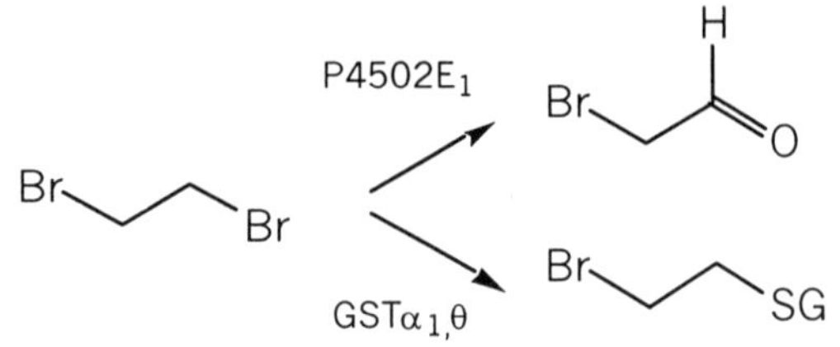

P4502E$_1$

GSTα$_{1,θ}$

Figure 2

The use of a PBPK model together with human in vitro metabolic parameters to assess individual risk, may be illustrated with the model compound ethylene dibromide (EDB)(Ploemen et al., 1997). EDB is a widely used "anti-knock" additive in gasolines and fumigant, and has been found to be carcinogenic in animal studies. The compound is metabolized by two routes: a conjugative route catalyzed by GSTs and an oxidative route catalyzed by cytochrome P450. The glutathione conjugate formed by the GST mediated conjugation of EDB, is a sulfur mustard analog that reacts via a reactive episulfonium ion. This episulfonium ion has been identified as the main DNA-reacting species, and is believed to be the ultimate carcinogen (Koga et al., 1986). Thus, dealing with the risk assessment of EDB, a central issue is the significance of the conjugative pathway.

In the case of EDB, several GST isoenzymes have been shown to be active towards EDB, of which the (polymorphic) theta class enzymes show the highest activity, for rat as well as man. However, due to the higher expression of the alpha class enzymes in the liver, the total contribution to EDB metabolism of these enzymes is larger. Of the cytochromes P450, CYP2E1 is by far the most important enzyme involved in EDB metabolism, displaying saturable kinetics. Large interindividual differences of CYP2E1 activity towards EDB exist (Wormhoudt et al., 1996).

One of the benefits of PBPK modelling is that individual metabolic pathways can be easily incorporated. Thus, once the principal enzymes involved in metabolism of a compound have been identified, the rate of metabolism can be scaled to the whole liver, based on the expression level of each enzyme. This liver metabolism (or metabolism in other organs if relevant), can then be integrated in a PBPK model. Applying this strategy for EDB, the relative and overall contribution of critical metabolic pathways in vivo in man, can be predicted. Moreover, interindividual differences in CYP2E1 and GST activity can be modeled to determine "extreme" cases.

It was shown that of 21 cases, the individual with the highest GST/CYP ratio, showed an approximately 1.5 fold increase of the amount of conjugative metabolites formed, compared to the average of these 21 individuals. Furthermore, it was shown that the saturation of the CYP route occurred faster in the rat than in man, resulting in relatively more glutathione conjugates per gram liver in the rat. However, even at low EDB concentrations, the GST route was shown to remain significantly active in man. (Hissink et al, in press)

TRICHLOROETHYLENE

Trichloroethylene has been in extensive use as an industrial solvent since the 1920s. Exposure to TRI, of the population in general and as an occupational hazard in particular, is of concern because of potential carcinogenic and toxic risks. Glutathione (GSH) S-transferase and cytochrome P450 are biotransformation enzymes which play a crucial role in the detoxification and toxification of TRI. Inter-individual differences in the expression and activity of these enzymes, caused either by environmental (e.g. induction by alcohol) or genetic (i.e. genetic polymorphism) factors, therefore have a significant bearing on health risks related to exposure to tri.

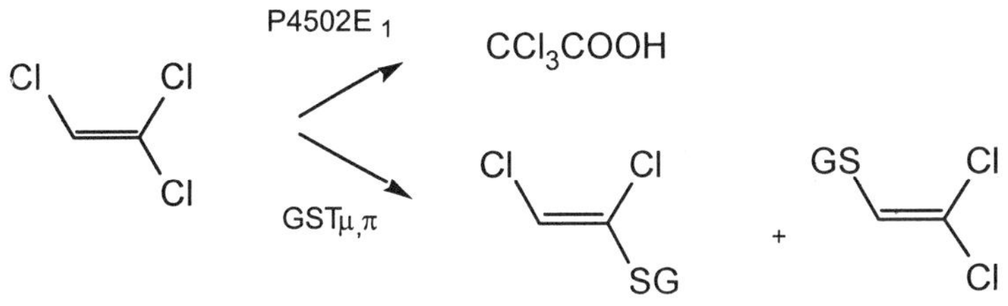

P4502E$_1$

CCl_3COOH

GSTμ,π

Figure 3

Distinct differences exist in the organ toxicity of TRI in rats and mice: kidney toxicity and carcinogenicity have been demonstrated to occur in rats but not in mice, whereas TRI-induced lung and liver tumour formation seems to be specific to mice. TRI is conjugated with glutathione (GSH, catalysed by glutathione S-transferase, GST) or oxidised by cytochrome P450. The kidney toxicities specific to the male rat appear to result from the GSH-conjugation route, which is the formation of the toxic metabolites 1,2- and 2,2-dichlorovinylcysteine (DCVC). The toxicities specific to mice are probably not relevant for humans since the mechanisms involved are not, or are only rarely, present in man (Goeptar et al. 1997). The PBPK modelling approach is able to explain the inter-species differences in sensitivity to the compound since formation of the toxic metabolites can be predicted per species. It is also capable of examining the inter-individual variation in metabolism and "the range of humans at risk".
Firstly the metabolic activities of rat and human glutathione S-transferases (GSTs) involved in GSH conjugation and cytochrome P450 enzymes involved in the oxidative metabolism of TRI was determined, Mu class an Pi class are the main GST's involved (GSTM 1-1 1 5.2 pmol/mg/hr and GSTP 1-1 2.1 pmol/mg/hr)

The only P 450 enzym involved for both rat and man is CYP2E 1

Conjugation and oxidation rates were subsequently scaled to the whole liver in order to incorporate these values in the PBPK model which predicted the fate of TRI at 24 hours after exposure to the MAC (maximal acceptable concentration) value of 35 ppm for an eight-hour period. These predictions are shown in the table. As already mentioned, the GSH route is responsible for the formation of the toxic metabolite. For man, this GSH conjugation is primarily catalysed by the μ class GST (M1-1). Since a wide variation exists in expression of this GST class - and even 50% of the human population do not have this enzyme - a variation in formation of GSH conjugate is to be expected. This variation is also shown in the table.

Table 1. Fate of TRI at 24 hours after exposure to the MAC value (35 ppm) for 8 hours, for rat and man

	Rat	Human
% metabolised of total	84	63
% metabolised by P450 (of total)	99.999	99.9998-100
% metabolised by GST (of total metabolised)	0.00057	μ neg.: 0.000007
		μ pos.: 0.000021- 0.00021
% exhaled	15	28
% in fat	0.0006	9

The extent of formation of the GSH conjugate by humans with GST M1-1 was about 20% of that formed by rats; in humans without GST M1-1, it was only 1% of that of rats. Based on body weight, these figures will be even lower since a rat inhales much more TRI per kg BW than humans due to their higher ventilation rate. Thus, a risk assessment performed in rats will overestimate the risk expected for humans after exposure to TRI. (E. Hissink, P.J. van Bladeren, N.P.E. Vermeulen, in preparation)

CONCLUSIONS

Because of the heterogeneity of the human population, it is to be expected that there will be a broad range of observed susceptibilities to the biological effects of exposure to chemicals or drugs. Insight into the sources and magnitude of variability in susceptibility in the human population is a central issue for making quantitative estimates of risk.

Non-cancer risk assessments often address the population variability by dividing the experimentally determined acceptable exposure level by an uncertainty factor of 10 to protect sensitive individuals; cancer risk assessments typically do not address this issue quantitatively. PBPK-modelling provides the capability to quantitatively describe the potential impact of metabolic and kinetic factors on the variability of individual risk. In particular, PBPK models can be used to determine the impact of differences in key metabolism enzymes, whether due to genetic polymorphisms, or to environmentally induced variations within the general population. In each case, the PBPK model provides a quantitative structure for determining the effect of these various factors on the relationship between the external (environmental) exposure and the internal (biologically effective) target tissue exposure. Thus, the range of target tissue concentrations can be predicted.

Similarly, the occurrence of interspecies differences can be addressed. By firstly determining the key steps in the formation and detoxication of the active metabolite and the potential alternative pathways, and secondly comparing the properties (both quantitative and qualitative) of the (iso)enzymes involved in these steps between species, a knowledge-based risk extrapolation can be performed.

REFERENCES

Andersen, M.E., H.J.Clewell, M.L. Gargas, F.A. Smith and R.H. Reitz, 1987, Physiologically-based pharmacokinetics and the risk assessment process for methylene chloride., Toxicol. Appl. Pharmacol, 87, 185-205

Barry, M. and J. Feely, Enzyme induction and inhibition, 1990, Pharmac. Ther. 48, 71-94

Bogaards J.J.P., Venekamp J.C., and van Bladeren P.J. (1996). The biotransformation of isoprene and the two isoprene monoepoxides by human cytochrome P450 enzymes, compared to mouse and rat liver microsomes. *Chemico-Biological Interactions* **102**, 169-182.

J.J.P. Bogaards, J.C. Venekamp, F.G.C. Salmon, P.J. van Bladeren, (1999) Conjugation of isoprene monoepoxides with glutathione, catalyzed by χ, μ, π and θ-class glutathione S-transferases of rat and man. Chem. Biol. Interact 117, 1-14

Del Monte M., Citti L. and Gervasi P.G. (1985). Isoprene metabolism by liver microsomal mono-oxygenases. *Xenobiotica* **15**, 591-597.

Fuhr, U. and A.L. Kummert, 1995, The fate of naringing in humans: A key to grapefruit juice-drug interactions? Clin. Pharmacol Ther. 58, 365-373

Goeptar, A.R.. Commandeur, J.M.M Van Ommen, B. Van Bladeren, P.J. Vermeulen M.P.E. (1995) Metabolism and kinetics of trichloroethylene in relation tot toxicity and carcinogenic Chem. Res. Tox. 8, 3-21

Gerlowski, L.E. and R.J. Jain, 1983, Physiologically-based pharmacokinetic modeling: principles and applications., J. Pharm. Sci., 72, 1103-

Gervasi P.G. and Longo V. (1990). Metabolism and Mutagenicity of Isoprene. *Environ. Health Perspect.* **86**, 85-87.

Gonzalez, F.J. and U.A. Meyer U.A., 1991, Molecular genetics of the debrisoquin-sparteine polymorphism, Clinical Pharmacology and Therapeutics, 50, 233-238

Hayes, J.D. and D.J. Pulford, 1995, The glutathione S-transferase supergene family: regulation of GST and the contribution of the isoenzymes to cancer chemoprotection and drug resistance, Critical reviews in biochemistry and molecular biology, 30, 445-600

Hissink, A.M., B. van Ommen, J. Krüse and P.J. van Bladeren, 1997, A physiologically-based pharmacokinetic (Pb-PK) model for 1,2-dichlorobenzene linked to two possible parameters of toxicity., Toxicol. Appl. Pharmacol., 145, 301-310

Hissink, A.M., Wormhoudt, L.W., Sheraratt, P.J., Hayes, J.D., Commandeur, J.N.M., Vermeulen, N.P.E., Van Bladeren, P.J., A physiologically-based pharmacokinetic (PB-PK) model for ethylene dibromide; relevance of extrahepatic metabolism. Food Chem in Tox in press.

Houston, J.B., 1994, Relevance of in vitro kinetic parameters to in vivo metabolism of xenobiotics., Toxicology in vitro, 8, 507-512

Van Iersel, M.L.P.S. Verhagen, H. Van Bladeren, P.J. (1999)
The role of biotransformation in dietory (anti) carcinogenesis Mut. Res. 443, 259 - 270 ·

Koga, N., P.B. Inskeep, T.M. Harris and F.P.Guengerich, 1986, The major DNA adduct formed from 1,2-dibromoethane.,Biochemistry 25,2191-2198

Longo V., Citti L. and Gervasi P.G. (1985). Hepatic microsomal metabolism of isoprene in various rodents. *Toxicol. Lett.* **29**, 33-37

Meyer, D.J., B. Coles, S.E. Pemble, K.S. Gilmore, G.M. Fraser and B.Ketterer, 1991, Theta, a new class of glutathione transferase purified from rat and man. Biochem. J., 274, 409-414

Nelson H.H., Wiencke J.K., Christiani D.C., Cheng T.J., Zuo Z.F., Schwartz B.S., Lee B.K., Spitz M.R., Wang M, Xu X., and Kelsey K.T. (1995). Ethnic differences in the prevalence of the homozygous deleted genotype of glutathione S-transferase theta. *Carcinogenesis* **16**, 1243-1245.

Parkinson, A., 1996, An overview of current cytochrome P450 technology for assessing the safety and efficay of new materials, Toxicologic pathology, 24, 45-57

Ploemen, J.H.T.M., L.W. Wormhoudt, G.R.M.M. Haenen, M.J. Oudshoorn, J.N.M. Commandeur, N.P.E. Vermeulen, I. De Waziers, P.H. Beaune, T. Watanabe and P.J. Van Bladeren, 1997, The use of in vitro metabolic parameters to explore the risk assesment of hazardous compounds:the case of ethylene dibromide. Toxicol.Appl.Pharmacol., 143, 56-69

Smith G., L.A. Stanley, E. Sim, R.C. Strange and C.R. Wolf, 1995, Metabolic polymorphisms and cancer susceptibility, Cancer Surveys 25, 27-65

Wormhoudt, L.W., J.H.T.M. Ploemen, I. De Waziers, J.H.M. Commandeur, P.H. Beaune, P.J. Van Bladeren and N.P.E. Vermeulen, 1996, Interindividual variability in the oxidation of 1,2-dibromoethane: use of hetereologously expressed human cytochrome P450 and human liver microsomes. Chem. Biol. Interact., 101,175-192

Yang C.S., J.F. Brady and J.Y. Hong, 1992, Dietary effects on cytochrome P450, xenobiotic metabolism and toxicity, The FASEB Journal, 6, 737-744

Pemble S., Schroeder K.R., Spencer S.R., Meyer D.J., Hallier E., Bolt H.M., Ketterer B. And Taylor J.B. (1994). Human glutathione S-transferase Theta (GSTT1): cDNA cloning and the characterization of a genetic polymorphism. *Biochem J.* **300**, 271-276.

Nelson H.H., Wiencke J.K., Christiani D.C., Cheng T.J., Zuo Z.F., Schwartz B.S., Lee B.K., Spitz M.R., Wang M, Xu X., and Kelsey K.T. (1995). Ethnic differences in the prevalence of the homozygous deleted genotype of glutathione S-transferase theta. *Carcinogenesis* **16**, 1243-1245.

INTERINDIVIDUAL DIFFERENCES IN RESPONSE TO CHEMOPROTECTION AGAINST AFLATOXIN-INDUCED HEPATOCARCINOGENESIS: IMPLICATIONS FOR HUMAN BIOTRANSFORMATION ENZYME POLYMORPHISMS

David L. Eaton, Theo K. Bammler and Edward J. Kelly

Center for Ecogenetics and Environmental Health
Department of Environmental Health
University of Washington
Box 354695
Seattle, WA 98195

INTRODUCTION

Aflatoxin B_1 as an Animal and Human Carcinogen

Aflatoxin B_1 (AFB) is a difuranocoumarin fungal metabolite produced by certain strains of *Aspergillus flavus* and related species that commonly contaminate corn, peanuts and other commodities (Newberne and Butler, 1969; Fig 1).

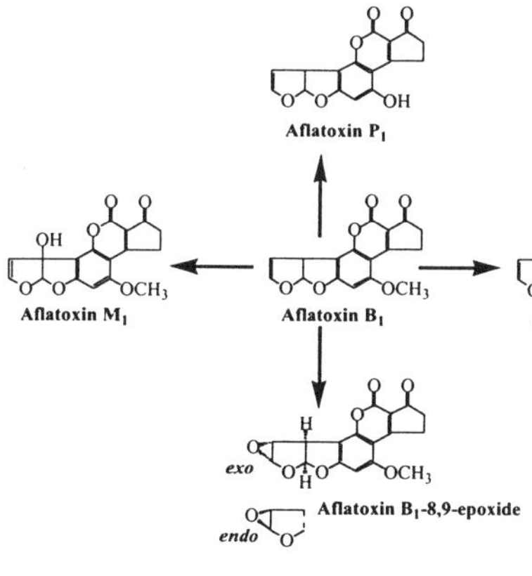

Figure 1. Chemical Structures of Aflatoxin B_1 (AFB), and its primary oxidative metabolites, Aflatoxin-8,9-epoxide (AFBO), Aflatoxin Q_1 (AFQ), Aflatoxin M_1 (AFM) and Aflatoxin P_1 (AFP).

Biological Reactive Intermediates VI, Edited by Dansette *et al.*
Kluwer Academic / Plenum Publishers, 2001

In addition to being a potent hepatotoxin, AFB is among the most potent hepatocarcinogens ever identified. In laboratory rats, daily doses as low as 15 ppb induce a high incidence of liver cancer (for reviews see Eaton and Gallagher, 1994; Massey et al., 1995). The growth of the mold on food products is highly dependent on storage conditions; high temperature and humidity favor production of the mold and toxin in improperly stored grains and nuts. Although generally not considered a significant public health risk in developed countries because of adequate monitoring and regulation, epidemiological studies have definitively identified AFB as a human liver carcinogen (Hall and Wild, 1994). In some regions of China, liver cancer is a leading cause of cancer-related deaths, and AFB is thought to contribute significantly to this burden. A strong interaction between AFB exposure and hepatitis B virus (HBV) has been demonstrated. Qian et al (1994) found that dietary exposure to AFB increased liver cancer risk by about 3-fold in the absence of HBV, HBV antigen seropositivity alone increased liver cancer risk by about 7-fold, but the two risk factors together increased risk by 60-fold.

Because there is strong evidence that AFB is a potent human carcinogen, and because there are large populations unavoidably exposed to it via the diet, AFB serves as a model carcinogen to explore avenues of chemointervention strategies to reduce human cancer burden. Most chemointervention studies to date have focused on efforts to modify biotransformation of AFB to less toxic metabolites.

BIOTRANSFORMATION OF AFLATOXIN AS A DETERMINANT OF SPECIES AND INDIVIDUAL SENSITIVITY

AFB requires metabolic activation to a reactive epoxide in order to be carcinogenic. Activation can occur through several pathways, including cytochrome P450 (CYP) oxidation and lipoxygenase mediated co-oxidation (Fig. 2) (Roy and Kulkarni, 1997). However, CYP oxidation is the most important for activation of AFB to the ultimate carcinogenic form, AFB-8,9-epoxide (AFBO). CYP-mediated activation of AFB can result in two different stereo-configurations of the epoxide, AFB-8,9-*endo*-epoxide (*endo*-AFBO) and AFB-8,9-*exo*-epoxide (*exo*-AFBO) (Raney et al., 1992a, b). Harris and colleagues (Iyer et al., 1994) demonstrated that only the *exo*-AFBO has high affinity for DNA adduction. Previous studies have demonstrated that activated AFB binds extensively and selectively to the N7-position of guanine on DNA, and it is this adduction that is believed to be the initiating step in hepatocarcinogenesis. Thus, one chemointervention strategy would be to reduce the CYP-mediated formation of *exo*-AFBO. To do this, it is critical to know which of the numerous P450s expressed in human liver is responsible for activation of AFB to *exo*-AFBO.

Role of Specific Cytochrome P450s in AFB Activation and Detoxification

We and others have previously demonstrated that several human hepatic CYP enzymes have the capability to form AFBO (Ramsdell et al., 1991; Gallagher et al, 1994, 1996; Raney et al., 1992c). Among the major forms of CYP expressed in human liver, both CYP1A2 and CYP3A4 are capable of activating AFB to AFBO. In mouse, CYP2A5 has been shown to have substantial AFB activating capacity (Pelkonen et al., 1994), but cDNA expressed human CYP2A6 has very little AFBO activity. Human CYP2B and 2C forms also have some activity toward AFB, although the specific activities are quite low compared to CYP1A2 and 3A4 (Bammler et al., 1999; Bammler and Eaton unpublished data). In addition to forming the epoxide, both CYP1A2 and 3A4 form other AFB oxidation products. CYP1A2 forms the 9-alpha hydroxylation product known as AFM (Fig. 1), whereas CYP3A4 forms the 3-hydroxylation product, AFQ (Fig. 1).

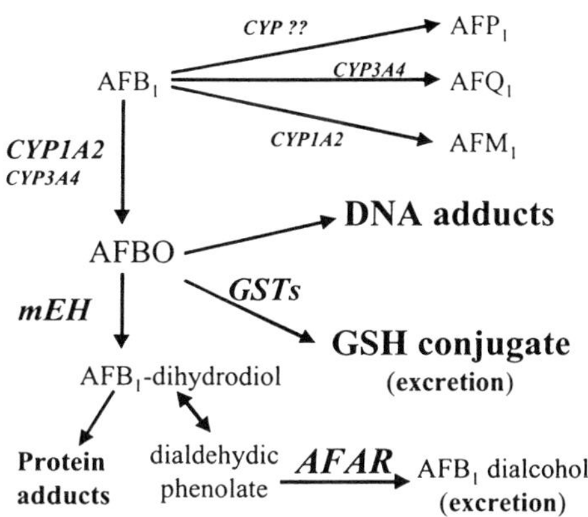

Figure 2. Biotransformation pathways for activation and detoxification of aflatoxin B₁.

A third monohydroxylated metabolite of AFB, called AFP, results from O-demethylation of AFB, but the specific enzyme(s) responsible for this have yet to be identified. As AFM, AFQ and AFP are all considered to be substantially less carcinogenic than AFB, these are generally considered detoxification pathways. However, AFM is considerably more carcinogenic than either AFQ or AFP, and is only about 10-fold less potent a carcinogen than AFB (Neal et al., 1998).

Kinetic studies have demonstrated that human CYP1A2 forms both AFBO and AFM, in an approximate ratio of 3:1 (Eaton et al., 1995; Gallagher et al., 1996). However, it forms approximately equal amounts of *endo* and *exo*-AFBO such that about one third of the CYP1A2-mediated oxidation of AFB yields the carcinogenic form. Human CYP1A2-mediated formation of both AFM and AFBO follows Michaelis-Menten kinetics, with an apparent Km of approximately 40 µM for both products (Eaton et al., 1995; Gallagher et al., 1996).

Human CYP3A4 forms both AFQ and AFBO, but with a ratio of detoxification (AFQ) to activation (AFBO) of approximately 10:1. In contrast to that formed by CYP1A2, CYP3A4 produces almost exclusively *exo*-AFBO. However, CYP3A4-mediated oxidation of AFB follows Hill, rather than Michaelis-Menten kinetics (Gallagher et al., 1996). This second order reaction kinetics is similar to kinetics for enzymes that have allosteric activation, i.e. the apparent affinity of the enzyme for the substrate (AFB) varies with substrate concentration. At low concentrations (e.g., less than 1 µM *in vitro*)

CYP3A4 has very low apparent activity toward activation of AFB. Thus, at the concentrations of AFB that could result in human liver following ingestion of even highly AFB-contaminated diets, CYP1A2 is likely to be the single most important CYP enzyme in human liver for the activation of AFB to *exo*-AFBO. One caveat to this conclusion might be the presence of natural planar aromatic hydrocarbons or flavonoids that could serve to activate CY3A4. For example, alpha-napthoflavone has been demonstrated to fully activate CYP3A4, thus increasing its activity toward AFB (Gallagher et al., 1994; Raney et al., 1992c).

These factors are important when considering the potential effects of chemointervention strategies. From the *in vitro* work noted above, a logical target for chemoprevention of AFB would be inhibitors of CYP1A2. At the same time, one would want to consider whether chemopreventive compounds, particularly flavones and flavonoids found in the diet, might paradoxically enhance AFBO formation through activation of CYP3A4.

Role of Specific Glutathione S-transferases in AFBO Detoxification

Once formed, AFBO can be subject to detoxification via conjugation with glutathione (Fig. 2). Athough *exo*-AFBO is highly reactive, with an apparent half-life in an aqueous environment of less than 1 second (Johnson et al., 1996), it does not readily react non-enzymatically with glutathione. Rather, the formation of the AFB-SG conjugate relies upon the presence of certain forms of glutathione S-transferase with catalytic activity toward AFBO. Remarkable species differences in GST activity toward AFBO exist. Soon after the discovery of AFB as a potent rat hepatocarcinogen, it was recognized that mice were highly resistant to this effect. Indeed, early feeding studies of doses of AFB + AFG as high as 10,000 ppb produced no evidence of liver tumors in mice (Wogan, 1973). Studies in our laboratory demonstrated that the resistance of mice to the hepatocarcinogenic effects of AFB was due largely, if not exclusively, to the constitutive expression of one form of alpha class GST. This GST, initially designated as mYcYc but now identified as mGSTA3-3, has been cloned and sequenced (Hayes et al., 1992; Buetler and Eaton, 1992; Buetler et al., 1992), and appears to be constitutively expressed in most if not all murine strains (Borroz et al., 1991). The cDNA sequence for *mGSTA3* is 86% homologous to the primary constitutively expressed alpha class GST in rats, *rGSTA3*. Although the two homodimeric enzymes that result from expression of these rat and mouse cDNAs have similar activity toward the universal GST substrate, CDNB, they differ by nearly 1,000-fold in their specific activities toward *exo*-AFBO (Table 1). However, a second inducible rat liver alpha class GST with 91% cDNA and protein sequence homology to *mGSTA3*, termed *rGSTA5*, has substantial AFBO-conjugating activity (Hayes et al., 1992, 1994; Table 1). Although the *in vitro* activity of rGSTA5-5 toward microsomally-generated AFBO is several hundred times higher than that of rGSTA3-3, the activity is only approximately 20% of that for mGSTA3-3. Based on these activity comparisons and sequence alignments, 15 amino acid residues were identified as potentially being important determinants of AFBO activity. Rat-mouse chimeric enzyme constructs, site-directed mutagenesis and molecular modeling studies subsequently identified 6 residues that seemed to account for most of the selective activity toward AFBO seen in the mGSTA3-3 protein (Van Ness et al., 1994, 1998) (Table 1).

The identification of the specific forms of GST responsible for AFBO detoxification has important implications in the design and interpretation of human chemointervention studies. Because certain rodent GSTs are highly inducible, one viable strategy for chemointervention for AFB hepatocarcinogenicity would be to induce human GSTs with catalytic detoxification activity toward *exo*-AFBO. However, early studies with human liver cytosol failed to identify any significant human hepatic GST activity toward microsomally generated AFBO (Ketterer et al., 1983; Moss and Neal 1985; Slone et al. 1995). Subsequent studies with human alpha class GST protein expressed from cDNAs proved that human GSTA1-1 and A2-2 proteins had very little, if any, activity toward AFBO, even though the enzymes were highly active toward a variety of other substrates (Eaton and Bammler, 1999; Table 1). When compared with mGSTA3-3, human GSTA1-1 was at least 1000 times less active toward microsomally generated AFBO, although the activity toward CDNB was approximately 6 times greater than mGSTA3-3. As alpha class GSTs are the major GST isoforms constitutively expressed in human liver (van Ommen at al. 1990; Rowe et al., 1997), these results were consistent with previous observations that human liver cytosolic fractions had little if any GST activity toward AFBO.

Table 1. Activities of various glutathione S-transferases toward AFBO (SE)

	AFBO Activity[1]	CDNB Activity[2]	Ref.
Mouse GSTA3-3	265.0 (11.11)	10.0 (0.26)	Van Ness et al., 1998
Rat GSTA3-3	BDL [3]	18.4 (1.04)	Van Ness et al., 1998
Rat GSTA5-5	57.0 (2.32)	9.8 (0.79)	Van Ness et al., 1998
Human GSTA1-1	BDL [3]	57.1 (4.60)	Van Ness et al., 1998
Human GSTM1-1	0.08 (0.006)	109.5 (8.7)	Bammler and Eaton, unpublished data
Human GSTM2-2	0.82 (0.035)	294.9 (9.5)	Bammler and Eaton, unpublished data

[1] Activity measured by generating excess AFBO with a mouse microsomal generating system. Units are expressed as nmol AFB-SG formed per minute per mg GST protein.
[2] Activitity toward 1-chloro-2,4-dinitrobenzene. Units are expressed as µmol per minute per mg protein.
[3] BDL, below detection limit (~ 1 pmol/min/mg)

Although alpha class GSTs are the predominant GST proteins in human liver, other GSTs are also expressed. *hGSTM1* is polymorphic in the human population, such that approximately 50% of Caucasians are homozygous for a gene deletion of *hGSTM1* (Seidegard et al., 1988). However, for those individuals that possess one or two copies of the gene, it is constitutively expressed in relatively small amounts in human liver (van Ommen et al., 1990, Rowe et al., 1997). Guengerich and colleagues (Raney et al., 1992b) found that hGSTM1a-1a protein had measurable activity toward AFBO, although the majority of the activity was toward the *endo* form. More recent studies from our laboratory have found that human GSTM2-2 protein has substantially higher GST activity toward AFBO than GSTM1-1 (Table 1). Relatively little is known about the pattern and extent of expression of *hGSTM2*, although there is some evidence that it is expressed in human liver. The significance, or lack thereof, of human mu class GSTs as important mediators of AFB detoxification remains uncertain. Because mu class GSTs are inducible in rodent (Buetler et al., 1995) and marmoset monkey (*Callithrix jacchus*) liver, it is possible that induction of certain mu class GSTs could be an effective avenue for chemoprevention of AFB-induced liver cancer in AFB exposed populations.

Role of Microsomal Epoxide Hydrolase in AFBO Detoxification

Another potential mechanism for detoxification of AFBO is hydrolysis of the epoxide to the corresponding dihydrodiol by microsomal epoxide hydrolase (mEH). Previous studies in rodents have provided mixed results regarding the role of mEH in hydrolysis of AFBO (Chi'ih et al., 1983a,b; Wilson et al., 1997; Guengerich et al., 1996; Johnson et al., 1997). The epoxide is highly unstable in an aqueous environment and will rapidly and spontaneously react with water, and thus the hydrolysis reaction clearly does not require an enzyme to proceed. However, an epidemiological study by McGlynn et al. (1995) reported a significant association between risk of hepatocellular carcinoma and occurrence of a polymorphic variant of mEH. There is some evidence to suggested that the expressed protein from the variant mEH allele is less stable than protein from the common allele, resulting in lower steady-state protein levels and thus less hepatic activity (Hassett et al., 1994).

Because it is possible that enzymatic hydrolysis could be important in the lipid microenvironment, where the reactive epoxide is formed in the intact cell, we evaluated the role of mEH in AFB genotoxicity by heterologous expression in the eukaryotic organism *S. cerevisiae*. By co-expressing human mEH with the enzymatic pathway of bioactivation by CYP1A enzymes we directly examined mEH function without membrane reconstitution because both CYP1A and mEH enzymes co-localize in yeast endoplasmic reticulum (Eugster and Sengstag, 1993).   To ensure that manipulations of the yeast while inserting the mEH plasmid did not alter AFB activation or mutagenesis responsiveness, we created a site-directed mutant of mEH with no catalytic activity by replacing aspartic acid 226 with glycine (Arand et al, 1999). The parental yeast strain used for heterologous expression is diploid and auxotrophic for the amino acid tryptophan due to mutations in the *trp5* alleles. If these two alleles undergo mitotic recombination, gene conversion may occur, allowing yeast to grow in the absence of tryptophan, and previous studies have demonstrated that activated AFB is a potent inducer of mitotic recombination in yeast expressing human CYP1A activity (Sengstag and Wurgler, 1994). Thus, the mutagenic responsiveness of these various recombinant yeast to AFB exposure was determined in three ways: 1) by measuring Trp- recombination events in response to AFB; 2) by measuring AFB-DNA adducts using ^{3}H-AFB; and 3) by measuring the mutagenic response of Salmonella bacteria to AFB using microsomal fractions of the various yeast strains.

CYP and mEH Activities in Recombinant Yeast

Yeast that did not contain human CYP1A1 or 1A2 had little capacity to activate AFB as measured directly by activity assays of yeast microsomes for AFBO, AFM or AFQ formation (Table 2).   Yeast that expressed human CYP1A1 had high AFM-forming activity, but formed very little AFBO. Less than 1% of total AFB oxidative activity of yeast-expressed human CYP1A1 was to the mutagenic epoxide. Consistent with previous results from our laboratory (Gallagher et al., 1996), human CYP1A2 expressed in these yeast had relatively high activity toward AFB, forming approximately equivalent amounts of AFM and AFBO (Table 2). The wild-type yeast had no measurable microsomal epoxide hydrolase activity toward cis-stilbene oxide (CSO). However, yeast expressing the recombinant human mEH had large amounts of mEH activity (Table 2). The Asp226Gly mutant of mEH (ΔmEH) lacked measurable CSO activity, even though Western blot analysis demonstrated that the quantity of mEH protein expressed in these yeast was similar to that in the yeast that expressed the functional human mEH cDNA.

While it is difficult to directly compare cytochrome P450 content and enzymatic activity of CYP1A2 heterologously expressed in yeast with that observed in human liver, the yeast P450 content, in terms of pmol/mg microsomal protein is within the range observed for human liver microsomes (Guengerich, 1995). The specific activity towards cis-stilbene oxide (CSO) was similar for all mEH-expressing strains with the exception of CYP1A2/ΔmEH that lacked any mEH activity (Table 2). Taking these two values (CYP1A2 and mEH) together would suggest that the ability of these recombinant yeast microsomes to activate and hydrolyze AFB may be within the range seen in human liver microsomes.

Table 2. CYP and mEH activities of various recombinant yeast strains

	hCYP1A1	hCYP1A1 + mEH	hCYP1A2	hCYP1A2 + mEH	hCYP1A2 + ΔmEH
AFM[1]	10.99 ± 1.06	10.60 ± 0.83	3.53 ± 0.36	3.75 ± 0.39	3.94 ± 0.30
AFBO[2]	0.05 ± 0.01	0.07 ± 0.03	3.30 ± 0.14	3.46 ± 0.17	3.86 ± 0.12
CSO hydrolysis[3]	< 0.2	4.7 ± 0.3	<0.2	5.3 ± 0.9	< 0.2
P450 content [4]	36.6 ± 4.3	34.8 ± 3.7	13.9 ± 0.8	12.8 ± 1.2	13.2 ± 3.0

1. Aflatoxin M1 formation, in units of nmol AFM per minute per nmol CYP
2. Aflatoxin 8,9-epoxide (both *endo* and *exo*) formation, in units of nmol AFBO per minute per nmol CYP
3. Microsomal epoxide hydrolase activity toward cis-stilbene oxide, in units of nmol diol formed per min mg microsomal protein (Gill et al., 1983).
4. P450 content is expressed in units of pmol cytochrome P50 per mg microsomal   protein, determined by CO difference spectra.

<u>Mutagenic Responsiveness of Recombinant Yeast to AFB</u>

To determine if mEH expression could offer any protection against AFB genotoxicity, we analyzed formation of AFB-DNA adducts in exponentially growing yeast exposed to [³H] AFB (Kelly et al, 1999). Expression of CYP1A1 did not promote adduct formation over background, consistent with our previous observations that CYP1A1 preferentially forms AFM and only small amounts of AFBO. Expression of CYP1A2 resulted in a significant increase in AFB-DNA adduct levels over background with a dose-dependent increase in adduct levels (Fig. 3). When cells expressing both CYP1A2 and mEH were exposed to [³H]-AFB, they too formed DNA adducts above background levels but, at 1.25 µM AFB, the levels of adducts were significantly lowered relative to CYP1A2 alone.

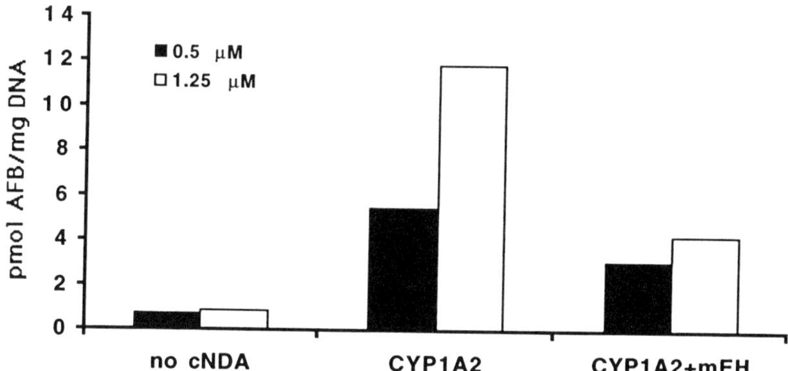

Fig. 3. AFB-DNA adduct formation in recombinant yeast.

When cells expressing either CYP1A1 and mEH or CYP1A2 and mEH were exposed to AFB, they too exhibited gene conversion at the *trp5* locus in a dose dependent manner, but at significantly lower levels as compared to strains expressing CYP1A1 or CYP1A2 alone (Fig 4). Furthermore, the inactive mEH mutant did not confer any protection against AFB genotoxicity as measured by *trp5* gene conversion, suggesting that the presence of the enzymatically active protein, rather than some unrelated event associated with non-functional mEH protein expression, is responsible for the reduction in mitotic recombination. When microsomes were isolated from the indicated strains and used as an activating system in the Ames assay with tester strain TA98, both CYP1A1 and CYP1A2 containing microsomes produced a dose-dependent increase in frame-shift mutations as scored by recovery of histidine revertants (data not shown). Similar to what was observed with *trp5* gene conversion, co-expression of mEH with either CYP1A1 or CYP1A2 resulted in significant protection against AFB mutagenicity, and the catalytically inactive mutant of mEH conferred no protection. For both measures of genotoxicity, mEH was able to afford a greater degree of protection in cells expressing CYP1A2 as compared to cells expressing CYP1A1.

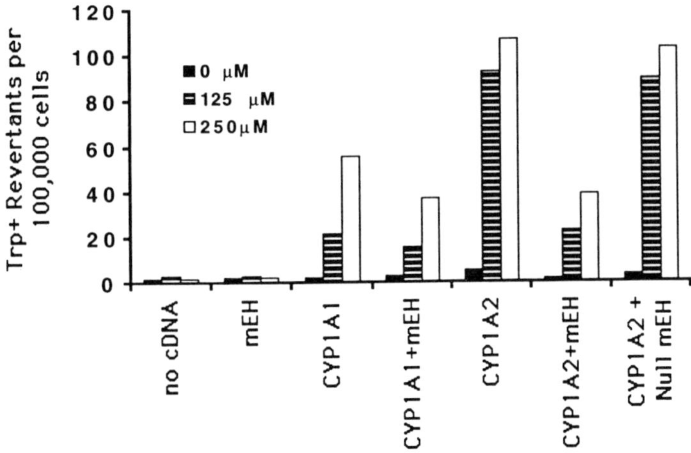

Figure 4. AFB-induced Trp reversion in recombinant yeast

Thus, it appears that mEH could provide some protection against the genotoxic actions of AFB in liver cells that do not have an effective GST conjugating system. Oesch et al. (2000) provided evidence at this conference that human mEH may tightly bind to some mEH substrates, but do not 'turn over' (release the hydrolyzed product) the substrate efficiently. Because the molar quantity of mEH protein in human microsomes, and by inference our yeast microsomes expressing the human mEH cDNA, is quite high, it is possible that the binding and effective sequestration of the reactive epoxide could be responsible for most or perhaps all of the apparent protective effects of mEH. However, for this to be the case, the Asp226Gly mutant of human mEH would have to lack substrate binding affinity as well as catalytic activity. Guengerich and colleagues (Johnson et al., 1996, 1997) have suggested that the kinetics of AFB-diol formation from AFBO in the presence of human mEH is such that enzymatic 'turn over' of the *exo*-AFBO would be unlikely to contribute significantly to detoxification of AFBO in vivo. However, their model did not consider the possibility suggested by Oesch et al (2000) that human mEH could effectively bind and sequester AFBO, thereby protecting DNA from adduction. Whether the protection from AFB-induced DNA damage by human mEH observed in our study is due to catalytic turn-over of *exo*-AFB or to sequestration remains uncertain and will require detailed kinetic studies of both the wild-type and Asp226Gly mutant to differentiate between these two possible mechanisms.

Because of the relatively high GST-conjugating activity toward AFBO in mice, and to a lesser extent rats, it is likely that mEH is of little, if any, importance in rodents. However, because the cytosolic GST activity toward AFBO in human liver is at least 1,000-fold lower than that seen in rodents, the modest protection offered by human mEH against AFB genotoxicity, regardless of mechanism, may be of toxicological significance, as suggested by the molecular epidemiology study of McGlynn et al. (1995). It should be noted that a recent study by Wild et al (2000) failed to find any difference in AFB-albumin adducts between individuals with the mEH variant versus those who were homozygous for the common mEH allele. However, because albumin adducts are formed via Schiff base formation between free amino groups of lysine residues and the phenolate ion that spontaneously forms from the AFB-dihydrodiol (Sabbioni et al. 1987), a protective effect of mEH against AFBO-DNA adducts would not necessarily be reflected in changes in AFB-dihydrodiol-lysine residues in albumin.

Role of Aflatoxin B_1-aldehyde Reductase in AFB Metabolism

As noted previously, both the *endo* and *exo* stereoisomers of AFBO can hydrolyze spontaneously or via mEH to form aflatoxin-8,9-dihydrodiol. At physiological pH, the dihydrodiol undergoes opening of the furan rings to yield a dialdehydic phenolate ion which forms Schiff bases with primary amine groups in proteins (Sabbioni and Wild, 1991). Judah et al. (1993) were the first to describe an inducible aldehyde reductase in rat liver that catalyzes the conversion of the dialdehydic phenolate to AFB-dialcohol, hence, this enzyme was named AFB-aldehyde reductase (AFB-AR). In contrast to the dialdehydic phenolate ion, AFB-dialcohol does not form Schiff bases with primary amine groups in proteins and, therefore, is likley to attenuate cytotoxicity (Hayes et al., 1993). More recently, two distinct cDNAs encoding human aldo-keto reductases have been isolated and shown to catalyze the formation of AFB-dialcohol (Ireland et al., 1998; Knight et al., 1999). Based on kinetic parameters, Knight and co-workers proposed that human AFB-AR has the potential to catalyze the reduction of AFB-dihydrodiol to AFB-dialcohol in the context of various competing pathways of AFB metabolism *in vivo* (Knight et al., 1999).

CHEMOPREVENTION STRATEGIES BASED ON ALTERATIONS IN AFB ACTIVATION AND/OR DETOXIFICATION

Based on the discussion above, it is evident that chemicals that modify CYP, GST and/or mEH activity may have an important impact on the dose-response relationship for AFB exposure and liver cancer. Numerous previous studies have identified both synthetic and naturally occurring dietary constituents that may prove to be effective in chemointervention against AFB hepatocarcinogenicity.

Dietary treatment with the antioxidant ethoxyquin (Cabral and Neal, 1983; Kensler et al., 1986) and the synthetic dithiolthione oltipraz (Kensler et al., 1987), protect rats from AFB-induced hepatocarcinogenesis by inducing the alpha class rGSTA5 gene (also known as Yc2). The corresponding gene product also detoxifies *exo*-AFBO (Hayes et al., 1994). In addition, treatment of mice with glutathione depleting agents dramatically potentiates AFB genotoxicity (Monroe and Eaton, 1988). Both rodent species activate AFB, with the mouse generating the *exo*-AFBO even more efficiently (Monroe and Eaton, 1987). Together, these data strongly suggest that GST mediated aflatoxin B_1-glutathione conjugating (AFB-SG) activity is a major determinant of species susceptibility to the carcinogenic effects of AFB.

It is not known whether chemoprevention strategies based on GST induction that are effective in rodents are also effective in primates. Although there is some evidence that diets high in cruciferous vegetables can induce alpha class GSTs in humans (Nijhoff et al., 1995; Bogaards et al., 1994), we and others have failed to detect significant AFB-SG activity in human liver cytosol (Slone et al., 1995; Moss and Neal, 1985; Moss et al., 1985). Furthermore, purified recombinant hGSTA1-1 has no detectable activity toward AFBO (Buetler et al.,1996), although human mu class GSTs do have some measurable activity toward AFBO (Raney et al., 1992b) (Table 1).

The question arises whether chemointervention affords protection against AFB-induced hepatocarcinogenesis in human subjects that are exposed to high levels of AFB in their diet, as is the case in large parts of Africa and China. The Qidong region, in the Republic of China, is such an area, and was recently chosen for a Phase IIa chemointervention trial, in which 234 adults participated (Kensler et al., 1998; Wang et al., 1999). The FDA approved antischistosomal drug oltipraz (4-methyl-5-[*N*-2-pyrazynil]-1,2-dithiole-3-thione), which has been shown to protect rats from the carcinogenic effects of AFB (Kensler et al.,1987), was used in that study. While this trial is crucial in assessing the efficacy of this drug *in vivo* in humans, it was limited to measuring AFB-albumin adducts, urinary AFM and AFB-mercapturic acid and, for obvious reasons, is not able to measure effects directly in the target organ (liver).

Recently we evaluated whether dietary oltipraz and/or ethoxyquin could modulate hepatic AFB metabolism *in vivo* in a way that was protective against its genotoxic and potentially carcinogenic effects in a non-human primate model (Bammler et al., 2000). To identify a non-human primate species with a hepatic aflatoxin B_1 (AFB) metabolism most similar to human, we analyzed hepatic microsomal and cytosolic fractions prepared from untreated adult male macaques (*Macaca nemestrina*), male marmosets (*Callithrix jacchus*) and humans (*Homo sapiens*). While an *in vitro* comparison of marmosets, macaques and humans showed similar oxidative metabolic profiles, their capacity to conjugate AFBO with glutathione was markedly different (Bammler et al., 2000). Both humans and marmosets lacked any constitutive AFB-SG activity, whereas the macaques expressed measurable constitutive GST activity toward AFBO.

Oltipraz when given in the diet at a dose of 18 mg/kg/day for 12 days produced a significant reduction in AFB-DNA adduct formation in marmoset liver, compared to a control group of animals (Bammler et al., 2000). In two of the four marmoset monkeys treated with oltipraz and two of the three ethoxyquin treated animals, hepatic cytosolic AFB-SG activity was identified. AFB-GST activity was identified in only one of 7 non-treated marmoset monkey livers, suggesting that oltipraz and ethoxyquin were capable of inducing a GST with AFBO activity in some, but not all animals, of this species of non-human primates. Subsequently, we cloned and characterized a (our data suggest that there might be an additional mu class GST with AFB-SG activity in marmoset liver) marmoset hepatic GST with AFBO activity, and identified it as a mu class GST with high homology to human GSTM2 (Bammler et al., 1998).

In that same study, we demonstrated that olitpraz was an effective inhibitor of marmoset liver CYP-mediated activation of AFB to AFBO, presumably through inhibition of marmoset CYP1A2-like enzymes. Langouet and co-workers (1995) demonstrated that OPZ not only induces certain GST isoforms in human primary hepatocytes, but also inhibits both CYPs 1A2 (Ki=10 μM) and 3A4 (Ki=80 μM), the two major enzymes activating AFB in humans (Raney et al., 1992c; Gallagher et al., 1994).

It is informative to compare our results in non-human primates to the phase IIa chemointervention trial conducted by Kensler and colleagues (Kensler et al., 1998, Wang et al., 1999) in an aflatoxin-exposed Chinese population. The participants of that trial received either a placebo, a daily dose of 125 mg oltipraz, or a once weekly dose of 500 mg oltipraz. Assuming an average body weight of 70 kg, these doses translate into a daily dose of 1.8 mg/kg or a weekly dose of 7.1 mg/kg, compared to the daily dose of 18 mg/kg used in our study. In addition, the human subjects received oltipraz for 8 weeks, whereas the marmosets received this compound for a total of 12 days. Wang *et al.* (1999) provided evidence that 500 mg of oltipraz administered once weekly to Chinese subjects decreased urinary AFM significantly. Consistent with this report, we found that OPZ inhibited the formation of AFBO by marmoset hepatic microsomes at concentrations as low as 10 μM (Bammler et al., 2000). Furthermore, based on the report by Gupta *et al.* (1995), we estimated that the transient OPZ concentration *in vivo* in the plasma of the marmosets treated with this compound was approximately 50 μM. Based on our *in vitro* studies, such a concentration would provide substantial inhibition of CYP-mediated AFB oxidation, if the hepatic concentration equaled or exceeded that of plasma. Thus, the decrease in DNA adducts found in the animals treated with OPZ may, at least in part, be rationally explained by inhibition of CYP-mediated AFBO formation. Because the treatment produced both inhibition of activation and induction of detoxification pathways, it is difficult to know for certain the relative contributions of each toward the lowering of AFB-DNA adducts seen *in vivo* in oltipraz-treated marmosets. Regression analysis of the data suggested that alterations of both pathways contributed to the apparent decrease in AFB-DNA adducts seen in the treated animals (Bammler et al., 2000).

SUMMARY

It is now evident that most, if not all, of the remarkable species differences in susceptibility to AFB hepatocarcinogenesis is due in large part, if not exclusively, to differences in biotransformation. Certainly the relative rate of oxidative formation of the proximate carcinogen, AFB-8,9-*exo*-epoxide, is an important determinant of species and interindividual differences in susceptibility to AFB. However, mice produce relatively large amounts of *exo*-AFBO, yet are highly resistant to AFB-hepatocarcinogenesis because they express a particular form of GST with remarkably high catalytic activity toward the *exo*-epoxide of AFB. Rats, which are highly susceptible to AFB hepatocarcinogenesis, can be made resistant through dietary induction of an orthologous form of GST that is normally expressed in only very small amounts. Based on these findings in laboratory animal models, there is great interest in identifying chemicals and/or specific dietary constituents that could offer protection against AFB-hepatocarcinogenesis to humans. Current experimental strategies have focused on the antiparasitic drug, oltipraz, which induces protection in rats and has also shown some promise in humans. The mechanism of protection in rats appears to be via induction of an alpha class GST with high catalytic activity toward AFBO (rGSTA5-5), yet human alpha class GST proteins that are constitutively expressed in the liver (hGSTA1 and hGSTA2) have little, if any activity toward AFBO. Rather, it appears that mu class GSTs may be responsible for the very low, but potentially significant, detoxification activity toward AFBO. Oltipraz and certain dietary constituents may induce mu class GSTs in human liver, and this could afford some protection against the genotoxic effects of AFBO. However, it also appears that oltipraz, and perhaps certain dietary constituents, act as competitive inhibitors of human CYP1A2. As CYP1A2 appears to mediate most of the activation of AFB to *exo*-AFBO in human liver at low dietary concentrations of AFB encountered in the human diet, much of the putative protective effects of oltipraz could be mediated via inhibition of CYP1A2 rather than induction of GSTs. There is now evidence that human microsomal epoxide hydrolase (mEH) could play a role in protecting human DNA from the genotoxic effects of AFB, although the importance of this detoxification pathway, relative to mu class GSTs, remains to be elucidated. Oltipraz is an effective inducer of mEH in rats (Lamb Franklin, 2000), and thus induction of this pathway in humans could also potentially contribute to the protective effects of this drug toward AFB genotoxicity. Because the dihydrodiol of AFB may contribute indirectly to the carcinogenic effects of AFB via protein adduction and subsequent hepatotoxicity, the recently characterized human aflatoxin aldehyde reductase (AFAR) may also offer some protection against AFB-induced carcinogenicity in humans. Current and future dietary and/or chemointervention strategies aimed at reducing the carcinogenic effects of AFB in humans should consider all of the possible mechanistic approaches for modifying AFB-induced genotoxicity.

ACKNOWLEDGEMENTS

This work was supported in part by R01 ES-05780 and NIEHS Center grant ES-07033. The authors would like to thank Mr. Dennis Slone for his excellent technical assistance.

REFERENCES

Arand, M., Müller, F., Mecky, A., Hinz, W., Urban, P., Pompon, Kellner, R. and Oesch, F., 1999, Catalytic triad of microsomal epoxide hydrolase: replacement of Glu404 with Asp leads to a strongly increased turnover rate, *Biochem. J.* 337:37-43.

Bammler, T.K., Slone, D.H., Board, P.G., Listowsky, I., Patskovsky, Y.V. and Eaton, D.L., 1998, The role of mu-class glutathione S-transferases in the detoxification of aflatoxin B1-8,9-epoxide, *The Toxicologist.* 42 (1-S):910

Bammler, T.K., Slone, D.H. and Eaton, D.L., 2000, Effects of dietary oltipraz and ethoxyquin on aflatoxin B$_1$ biotransformation in non-human primates, *Toxicol. Sci.* 54:30-41

Bammler, T.K., Thompson, S.J., Gallagher, E.P., Sengstag, C., Slone, D.H., Haining, R.L., Rettie., A.E. and Eaton, D.L., 1999, Aflatoxin B1 oxidation by human cytochromes P450 CYP1A1, 2A6, 2C8, 2C9, 2C18 and 2C19, *The Toxicologist.* 48(1-S):1054.

Bogaards, J.J., Verhagen, H., Willems, M.I., van-Poppel, G. and van-Bladeren, P.J., 1994, Consumption of brussels sprouts results in elevated alpha-class glutathione S-transferase levels in human blood plasma, *Carcinogenesis.* 15:1073-1075

Borroz, K.I., Ramsdell, H.S. and Eaton, D.L., 1991, Mouse strain differences in glutathione S-transferase activity and aflatoxin B1 biotransformation, *Toxicol. Lett.* 58:97-105.

Buetler, T.M. and Eaton, D.L., 1992, Complementary DNA cloning, messenger RNA expression, and induction of alpha-class glutathione S-transferases in mouse tissues, *Cancer Res.* 52:314-318.

Buetler, T.M., Bammler, T.K., Hayes, J.D. and Eaton, D.L., 1996, Oltipraz-mediated changes in aflatoxin B1 biotransformation n rat liver: implications for human chemointervention, *Cancer Res.* 56, 2306-2313.

Buetler, T.M, Gallagher, E.P, Wang, C-H, Stahl, D.L, Hayes, J.D. and Eaton, D.L., 1995, Induction of phase I and phase II drug metabolizing enzyme mRNA, protein and activity by BHA, ethoxyquin and oltipraz, *Toxicol. Appl. Pharmacol.* 135:45-57.

Buetler, T.M., Slone, D. and Eaton, D.L., 1992, Comparison of the aflatoxin B1-8,9-epoxide conjugating activities of two bacterially expressed alpha class glutathione S-transferase isozymes from mouse and rat, *Biochem. Biophys. Res. Commun.* 188:597-603.

Cabral, J.R. and Neal, G.E., 1983, The inhibitory effects of ethoxyquin on the carcinogenic action of aflatoxin B1 in rats, *Cancer. Lett.* 19:125-132.

Ch'ih, J.J., Lin, T. and Devlin, T.M., 1983a, Activation and deactivation of aflatoxin B1 in isolated rat hepatocytes, *Biochem. Biophys. Res. Commun.* 110:668-674.

Ch'ih, J.J., Lin, T. and Devlin, T.M., 1983b, Effect of inhibitors of microsomal enzymes on aflatoxin B1-induced cytotoxicity and inhibition of RNA synthesis in isolated rat hepatocytes, *Biochem. Biophys. Res. Commun.* 115:15-21.

Eaton, D.L and Bammler, T.K., 1999, Concise review of the glutathione S-transferases and their significance to toxicology, *Toxicol. Sci.*, 49:156-164.

Eaton, D.L. and Gallagher, E.P., 1994, Mechanisms of aflatoxin carcinogenesis. *Annu. Rev. Pharmacol. Toxicol.* 34:135-172.

Eaton, D.L, Gallagher, E.P, Bammler, T.K. and Kunze K.L., 1995, Role of Cytochrome P450 1A2 in chemical carcinogenesis: Implications for human variability in expression and enzyme activity, *Pharmacogenetics.*5:259-274.

Eugster, H.P. and Sengstag, C., 1993, *Saccharomyces cerevisiae*: An alternative source for human microsomal liver enzymes and its use in drug interaction studies, *Toxicology.* *82*: 61-73.

Gallagher, E.P., Kunze, K.L., Stapleton, P.L. and Eaton, D.L., 1996, The kinetics of aflatoxin B-1 oxidation by human cDNA-expressed and human liver microsomal cytochromes P450 1A2 and 3A4, *Toxicol. Appl. Pharmacol.* 141:595-606.

Gallagher, E.P., Wienkers, L.C., Stapleton, P.L., Kunze, K.L. and Eaton, D.L., 1994, Role of human microsomal and human complementary DNA-expressed cytochromes P4501A2 and P4503A4 in the bioactivation of aflatoxin B1, *Cancer Res.* 54:101-108.

Gill, S.S., Ota, K. and Hammock, B.D., 1983, Radiometric assays for mammalian epoxide hydrolases and glutathione *S*-transferase. *Anal. Biochem.* 131: 273-282.

Guengerich, F.P., 1995, Human cytochrome P450 enzymes, in: Cytochrome P450: Structure, Mechanism and Biochemistry, P.R. Ortiz de Montellano, ed., Plenum, New York.

Guengerich, F.P., Johnson, W.W., Ueng, Y.F., Yamazaki, H. and Shimada, T., 1996, Involvement of cytochrome P450, glutathione *S*-transferase, and epoxide hydrolase in the metabolism of aflatoxin B1 and relevance to risk of human liver cancer. *Environ. Health Perspect.*, 104(Suppl 3):557-562.

Gupta, E., Olopade, O.I., Ratain, M.J., Mick, R., Baker, T.M., Berezin, F.K., Benson, A.B. and Dolan, M.E., 1995, Pharmacokinetics and Pharmacodynamics of oltipraz as a chemopreventive agent, *Clin. Cancer Res.* 1:1133-1138.

Hall, A.J and Wild, C.P., 1994, Epidemiology of aflatoxin-related disease, in: *The Toxicology of Aflatoxins: Human Health, veterinary and Agricultural Significance*, D.L. Eaton and J.D. Groopman, eds., Academic Press, Inc., San Diego.

Hassett, C., Aicher, L., Sidhu, J.S. and Omiecinski, C.J., 1994, Human microsomal epoxide hydrolase: genetic polymorphism and functional expression in vitro of amino acid variants. *Hum. Molec. Genetics.* 3:421-428.

Hayes, J.D., Judah, D.J. and Neal, G.E., 1993, Resistance to aflatoxin B1 is associated with the expression of a novel aldo-keto reductase which has catalytic activity towards a cytotoxic aldehyde-containing metabolite of the toxin, *Cancer. Res.* 53:3887-3894.

Hayes, J.D., Judah, D.J., McLellan, L.I. and Neal, G.E., 1991, Contribution of the glutathione S-transferases to the mechanisms of resistance to aflatoxin B1, *Pharmacol. Ther.* 50:443-472.

Hayes, J.D., Judah, D.J., Neal, G.E. and Nguyen, T., 1992, Molecular cloning and heterologous expression of a cDNA encoding a mouse glutathione S-transferase Yc subunit possessing high catalytic activity for aflatoxin B1-8,9-epoxide, *Biochem. J.* 285:173-180.

Hayes, J.D., Nguyen, T., Judah, D.J., Petersson, D.G. and Neal, G.E., 1994, Cloning of cDNAs from fetal rat liver encoding glutathione S-transferase Yc polypeptides. The Yc2 subunit is expressed in adult rat liver resistant to the hepatocarcinogen aflatoxin B1, *J. Biol. Chem.* 269:20707-20717.

Ireland, L.S., Harrison, D.J., Neal, G.E., Hayes, J.D., 1998, Molecular cloning, expression and catalytic activity of a human AKR7 member of the aldo-keto reductase superfamily: evidence that the major 2-carboxybenzaldehyde reductase from human liver is a homologue of rat aflatoxin B-1 aldehyde reductase, *Bichem. J.* 332:21-34.

Iyer, R.S., Coles, B.F., Raney, K.D., Thier R., Guengerich, F.P and Harris T.M., 1994, DNA adduction by the potent carcinogen aflatoxin B1: Mechanistic studies, *J. Am. Chem. Soc.* 116:1603-1609.

Johnson, W.W., Harris, T.M. and Guengerich, F.P., 1996, Kinetics and mechanism of hydrolysis of aflatoxin B-1 exo-8,9-epoxide and rearrangement of the dihydrodiol, *J. Am. Chem. Soc.* 118:8213-8220.

Johnson, W.W., Yamazaki, H., Shimada, T., Ueng, Y.F. and Guengerich, F.P., 1997, Aflatoxin B-1 8,9-epoxide hydrolysis in the presence of rat and human epoxide hydrolase, *Chem. Res. Toxicol.* 10:672-676.

Judah, D.J, Hayes, J.D, Yang, J.C, Lian, L.Y, Roberts, G.C, Farmer, P.B, Lamb, J.H, Neal, G.E., 1993, A novel aldehyde reductase with activity towards a metabolite of aflatoxin B1 is expressed in rat liverduring carcinogenesis and following the administration of an anti-oxidant, *Biochem. J.* 292:13-18.

Kelly, E.J., Sengstag, C. and Eaton, D.L., 1999, Expression of human microsomal epoxide hydrolase protects against aflatoxin B1-induced genotoxicity in yeast co-expressing human CYP1A enzymes, *The Toxicologist.* 48 (1-S):315.

Kensler, T.W., Egner, P.A., Davidson, N.E., Roebuck, B.D., Pikul, A. and Groopman, J.D., 1986, Modulation of aflatoxin metabolism, aflatoxin-N7-guanine formation, and hepatic tumorigenesis in rats fed ethoxyquin: role of induction of glutathione S-transferases, *Cancer. Res.* 46:3924-3931.

Kensler, T.W., Egner, P.A., Dolan, P.M., Groopman, J.D. and Roebuck, B.D., 1987, Mechanism of protection against aflatoxin tumorigenicity in rats fed 5-(2-pyrazinyl)-4-methyl-1,2-dithiol-3-thione (oltipraz) and related 1,2-dithiol-3-thiones and 1,2-dithiol-3-ones, *Cancer. Res.* 47:4271-4277.

Kensler, T.W., He, X., Otieno, M., Egner, P.A., Jacobson, L.P., Chen, B.B., Wang, J.S., Zhu, Y.R., Zhang, B.C., Wang, J.B., Wu, Y., Zhang, Q.N., Qian, G.S., Kuang, S. Y., Fang, X., Li, Y. F., Yu, L.Y., Prochaska, H.J., Davidson, N.E., Gordon, G.B., Gorman, M.B., Zarba, A., Enger, C., Munoz, A., Helzlsouer, K.J. Groopman, J. D., 1998, Oltipraz chemoprevention trial in Qidong, People's Republic of China: Modulation of serum aflatoxin albumin adduct biomarkers, *Cancer Epidemiol. Biomarkers Prev.* 7:127-134.

Ketterer, B., Coles, B, Meyer, D. and Garner, C., 1983, Detoxication of aflatoxin B1 by glutathione transferase, *Proc. Am. Assoc. Cancer Res.* 24:64.

Knight, L. P., Primiano, T., Groopman, J.D., Kensler, T. W. and Sutter, T. R. (1999). cDNA cloning, expression and activity of a second human aflatoxin B1-metabolizing member of the aldo-keto reductase superfamily,AKR7A3. *Carcinogenesis* 20,1215-1223.

Lamb J.G. and Franklin M.R., 2000, events in the induction of rat hepatic UDP-glucuronosyltransferases, glutathione S-transferase, and microsomal epoxide hydrolase by 1,7-phenanthroline: comparison with oltipraz, tert-butyl-4-hydroxyanisole, and tert-butylhydroquinone, *Drug Metab Dispos.* 28, 1018-1023.

Langouet, S., Coles, B., Morel, F., Becquemont, L., Beaune, P., Guengerich, F.P., Ketterer, B. and Guillouzo, A., 1995, Inhibition of CYP1A2 and CYP3A4 by oltipraz results in reduction of aflatoxin B1 metabolism in human hepatocytes in primary culture, *Cancer Res.* 55:5574-5579.

Massey, T.E., Stewart, A.K., Daniels, J.M. and Liu, L., 1995, Biochemical and molecular aspects of mammalian susceptibility to aflatoxin B1 carcinogenicity, *Proc. Soc. Exp. Biol. Med.* 208:213-217.

McGlynn, K.A., Rosvold, E.A., Lustbader, E.D., Hu, Y., Clapper, M.L., Zhou, T., Wild, C.P., Xia, X.L., Baffoe-Bonnie, A., Ofori-Adjei, D., Chen, G-C., London, W.T., Shen, F-M. and Buetow, K.H., 1995, Susceptibility to hepatocellular carcinoma is associated with genetic variation in the enzymatic detoxification of aflatoxin B1, *Proc. Natl. Acad. Sci. U.S. A.* 92:2384-2387.

Monroe, D.H. and Eaton, D.L., 1987, Comparative effects of butylated hydroxyanisole on hepatic in vivo DNA binding and *in vitro* biotransformation of aflatoxin B1 in the rat and mouse, *Toxicol. Appl. Pharmacol.* 90:401-409.

Monroe, D.H. and Eaton, D.L., 1988, Effects of modulation of hepatic glutathione on biotransformation and covalent binding of aflatoxin B1 to DNA in the mouse, *Toxicol. Appl. Pharmacol.* 94:118-127.

Moss, E.J. and Neal, G.E., 1985, The metabolism of aflatoxin B1 by human liver, *Biochem. Pharmacol.* 34:3193-3197.

Moss, E.J., Neal, G.E. and Judah, D.J., 1985, The mercapturic acid pathway metabolites of a glutathione conjugate of aflatoxin B1, *Chem. Biol. Interact.* 55:139-155.

Neal, G.E, Eaton, D.L, Judah, D.J and Verma, A., 1998, Metabolism and toxicity of aflatoxins M_1 and B_1 in human-derived in vitro systems, *Toxicol. Appl. Pharmacol.* 151:152-158.

Newberne, P.M. and Butler, W.H., 1969, Acute and chronic effects of aflatoxin on the liver of domestic and laboratory animals: a review, *Cancer Res.* 29:236-250.

Nijhoff, W.A., Grubben, M.J., Nagengast, F.M., Jansen, J.B., Verhagen, H., van-Poppel, G. and Peters, W.H., 1995, Effects of consumption of Brussels sprouts on intestinal and lymphocytic glutathione S-transferases in humans, *Carcinogenesis.* 16:2125-2128.

Oesch, F, Herrero, M.E., Lohmann, M, Hengstler, J.G. and Arand, M., 2000, Sequestration of biologgical reactive intermediates by trapping as covalent enzyme-intermediate complex, in: *Chemical and Biological Mechanisms in Susceptibility to and Prevention of Environmental* Diseases, 6th International Symposium on Biological Reactive Intermediates, Proceedings, p.L47, Paris.

Pelkonen, P, Lang, M.A, Wild, C.P, Negishi, M, Juvonen, R.O., 1994, Activation of aflatoxin B1 by mouse CYP2A enzymes and cytotoxicity in recombinant yeast cells, *Eur. J. Pharmacol.* 292:67-73.

Qian, G.S., Ross, R.K., Yu, M.C., Yuan, J.M., Gao, Y.T., Henderson, B.E., Wogan, G.N. and Groopman, J.D., 1994, A follow-up study of urinary markers of aflatoxin exposure and liver cancer risk in Shanghai, People's Republic of China, *Cancer Epidemiol. Biomarkers Prev.* 3:3-10.

Ramsdell, H.S., Parkinson, A., Eddy, A.C. and Eaton, D.L., 1991, Bioactivation of aflatoxin B1 by human liver microsomes: role of cytochrome P450 IIIA enzymes, *Toxicol. Appl. Pharmacol.* 108:436-447.

Raney, K.D., Coles, B., Guengerich, F.P. and Harris, T.M., 1992a, The endo-8,9-epoxide of aflatoxin B1: a new metabolite, *Chem. Res. Toxicol.* 5:333-335.

Raney, K.D., Meyer, D.J., Ketterer, B., Harris, T.M. and Guengerich, F.P., 1992b, Glutathione conjugation of aflatoxin B1 exo- and endo-epoxides by rat and human glutathione S-transferases, *Chem. Res. Toxicol.* 5:470-478.

Raney, K.D., Shimada, T., Kim, D.H., Groopman, J.D., Harris, T.M. and Guengerich, F.P., 1992c, Oxidation of aflatoxins and sterigmatocystin by human liver microsomes: significance of aflatoxin Q1 as a detoxication product of aflatoxin B1, *Chem. Res. Toxicol.* 5:202-210.

Rowe, J.D, Nieves, E, Listowsky, I., 1997, Subunit diversity and tissue distribution of human glutathione S-transferases: interpretations based on electrospray ionization-MS and peptide sequence-specific antisera, *Biochem J.* 325:481-486.

Roy, S.K, Kulkarni, A.P., 1997, Aflatoxin B1 epoxidation catalysed by partially purified human liver lipoxygenase, *Xenobiotica.* 27:231-241.

Sabbioni, G., Skipper, P.L., Buchi, G. and Tannenbaum, S.R., 1987, Isolation and characterization of the major serum albumin adduct formed by aflatoxin B1 in vivo in rats, *Carcinogenesis.* 8:819-824.

Sabbioni, G. and Wild, C.P., 1991, Identification of an aflatoxin G1-serum albumin adduct and its relevance to the measurement of human exposure to aflatoxins, *Carcinogenesis*, 12:97-103.

Seidegard, J., Vorachek., W.R., Pero, R.W., Pearson., W.R., 1988, Hereditary differences in the expression of the human glutathione transferase active on trans-stilbene oxide are due to a gene deletion, *Proc. Natl. Acad. Sci. U S A.* 85: 7293-7297.

Sengstag, C. and Würgler, F.E., 1994, DNA recombination induced by aflatoxin B1 activated by cytochrome P450 1A enzymes, *Molec. Carcinogenesis.*11: 227-2351.

Slone, D.H., Gallagher, E.P., Ramsdell, H.S., Rettie, A.E., Stapleton, P.L., Berlad, L.G. and Eaton, D.L., 1995, Human variability in hepatic glutathione S-transferase-mediated conjugation of aflatoxin B1-epoxide and other substrates. *Pharmacogenetics.* 5:224-233.

Van Ness, K.P., Buetler, T.M. and Eaton, D.L., 1994, Enzymatic characteristic of chimeric mYc/rYC1 glutathione S-transferases, *Cancer Res.* 54:4573-4575.

Van Ness, K.P., McHugh, T.M., Bammler, T.K. and Eaton, D.L., 1998, Identification of amino acid residues essential for high aflatoxin B_1-8,9-epoxide conjugation activity in Alpha class glutathione S-transferases through site-directed mutagenesis, *Toxicol. Appl. Pharmacol.* 152:166-174.

van Ommen, B, Bogaards J.J., Peters, W.H., Blaauboer, B. and van Bladeren, P. J., 1990, Quantification of human hepatic glutathione S-transferases., *Biochem. J.* 269:609-613.

Wang, J-S., Shen, X.,. He, X., Zhu, Y-R., Zhang, B-C., Wang, J-B., Qian, G-S., Kuang, S-Y, Zarba, A., Egner, P. A., Jacobson, L. P., Munoz, A., Helzlsouer, K. J., Groopman, J. D. and Kensler T.W., 1999, Protective alterations in phase 1 and 2 metabolism of aflatoxin B-1 by oltipraz in residents of Qidong, People's Republic of China, *J. Natl. Cancer Inst.* 91:347-354.

Wild, C.P., Yin, F., Turner, P.C., Chemin, I., Chapot, B., Mendy, M., Whittle, H., Kirk., G.D., Hall., A.J., 2000, Environmental and genetic determinants of aflatoxin-albumin adducts in the Gambia, *Int. J. Cancer.* 86:1-7.

Wilson, A.S., Williams, D.P., Davis, C.D., Tingle, M.D. and Park, B.K., 1997, Bioactivation and inactivation of aflatoxin B1 by human, mouse and rat liver preparations: effect on SCE in human mononuclear leucocytes, *Mutat. Res., 373*:257-264.

Wogan, G. N. 1973, Aflatoxin carcinogenesis, *Methods Cancer Res.* 7, 309-344.

SEQUESTRATION OF BIOLOGICAL REACTIVE INTERMEDIATES BY TRAPPING AS COVALENT ENZYME-INTERMEDIATE COMPLEX

Franz Oesch, Maria Elena Herrero, Matthias Lohmann, Jan Georg Hengstler, and Michael Arand

Institute of Toxicology, University of Mainz, Obere Zahlbacher Str. 67, D-55131 Mainz, Germany

Abstract

One important class of biological reactive intermediates arising in the course of human xenobiotic metabolism are arene and alkene oxides. The major safeguard against the potential genotoxic effects of these compounds is the microsomal epoxide hydrolase (mEH). This enzyme has a broad substrate specificity but - on the first sight - seems to be inadequately suited for this protection task due to its low turnover number with most of its substrates. The recent progress in the understanding of the mechanism of enzymatic epoxide hydrolysis has shed new light on this apparent dilemma: Epoxide hydrolases convert their substrates *via* the intermediate formation of a covalent enzyme-substrate complex, and it has been shown that the formation of the intermediate proceeds by orders of magnitudes faster than the subsequent hydrolysis, i.e. the formation of the terminal product. Thus, the enzyme acts like a molecular sponge by binding and inactivating the dangerous metabolite very fast while the subsequent product release is considerably slower, and quantification of the latter heavily underestimates the speed of detoxification. Usually, the slow enzyme regeneration does not pose a problem, since the mEH is highly abundant in human liver, the organ with the highest capacity to metabolically generate epoxides. Computer simulation provides evidence that the high amount of mEH enzyme is crucial for the control of the steady-state level of a substrate epoxide and can keep it extremely low. Once the mEH is titrated out under conditions of extraordinarily high epoxide concentration, the epoxide steady-state level steeply rises, leading to a sudden burst of the genotoxic effect. This prediction of the computer simulation is in perfect agreement with our experimental work. V79 Chinese Hamster cells that we have genetically engineered to express human mEH at about the same level as that observed in human liver are well protected from any measurable genotoxic effect of the model compound styrene oxide (STO) up to an apparent threshold level of 100 µM in the cell culture medium. In V79 cells that do not express mEH, STO triggers the formation of DNA strand breaks in a dose-dependent manner with no apparent threshold. Above 100 µM, the genotoxic effect of STO in the mEH-expressing cell line parallels the one in the parental cell line.

Biological Reactive Intermediates VI, Edited by Dansette *et al.*
Kluwer Academic / Plenum Publishers, 2001

Epoxides - genotoxic intermediates in the metabolism of xenobiotics

Lipophilic compounds that enter the body have to be transformed to water soluble metabolites in order to allow their excretion via urine or bile. The innumerable xenobiotics that are present in the environment have led to the development of a large network of xenobiotic metabolizing enzymes during the evolution of complex organisms (reviewed in (Oesch and Arand, 1999)). A general strategy to increase water solubility is to enhance the chemical reactivity of the target compound by chemical functionalization and subsequent conjugation of the resulting intermediate with a strongly hydrophilic endogenous chemical building block. Functionalization in the so-called phase I of xenobiotic metabolism is usually afforded by oxidation, reduction or hydrolysis of the mother compound.

One important class of phase I metabolites that arise from oxidation of arenes or alkenes comprises the epoxides. Their structure, a strained three-membered ring system with an electron-withdrawing oxygen as the hetero atom, confers an electrophilic reactivity to these compounds, the strength of which is modulated by the substitution pattern of the ring. In general, asymmetric substitution enhances the reactivity of the resulting compound. Since electrophilic reactive compounds can chemically modify DNA, reactive epoxides posses an intrinsic genotoxic potential that is modified by their substitution pattern.

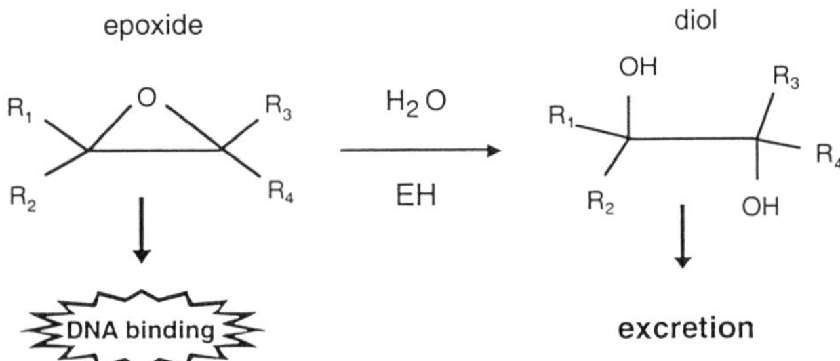

Fig.1 Epoxide hydrolases (EH) detoxify epoxides.

Two enzyme systems have evolved that protect higher organisms against the genotoxic effects of epoxides. The specialized family of epoxide hydrolases converts these compounds to the usually less harmful diols (Armstrong, 1999; Hammock et al., 1997; Oesch, 1973) (Fig. 1). Glutathione S-transferases, the more general safeguard against electrophilic agents, are also capable of conjugating epoxides, among other compounds, with the tripeptide glutathione, which also usually leads to inactivation (Hayes and Pulford, 1995). In general, glutathione conjugation becomes important either at high epoxide concentrations - or in case that the given epoxide is no substrate for epoxide hydrolases.

In contrast to many other xenobiotic metabolizing enzymes, epoxide hydrolases do not occur as a large family of isoenzymes. The majority of xenobiotic epoxides is hydrolyzed by a single EH, the endoplasmic reticulum resident microsomal epoxide hydrolase (mEH) (Oesch, 1973; Oesch and Bentley, 1976). There is one additional EH in human proficient for xenobiotic epoxide hydrolysis, the soluble epoxide hydrolase (sEH) that, in contrast to mEH, can take *trans*-substituted epoxides and is mainly responsible for the metabolism of fatty acid epoxides (Guenthner et al., 1981; Moghaddam et al., 1997; Ota and Hammock, 1980).

Here is a dilemma: on the one hand, the low complexity of the EH family demands a broad substrate specificity for structurally diverse compounds while on the other hand, a high affinity to these substrates is mandatory to efficiently remove these potentially genotoxic agents from the circulation already at low concentrations. This task for nature resembles the attempt to construct a quadrate circle since it is to be expected that optimization towards one of these leads almost inevitably impairs the performance of the enzyme with respect to the other requirement. Nevertheless, nature has found a successful way to fulfil both of the above requirements, the strategy of which is the subject of this paper.

The enzymatic mechanism of epoxide hydrolysis

Recently, detailed insights in the enzymatic mechanism of epoxide hydrolysis have been obtained by kinetic (Armstrong, 1999; Lacourciere and Armstrong, 1993; Rink and Janssen, 1998; Tzeng et al., 1996) and structural analysis (Arand et al., 1999; Arand et al., 1996; Argiriadi et al., 1999; Laughlin et al., 1998; Nardini et al., 1999; Tzeng et al., 1998; Zou et al., 2000) of the enzyme: Analysis of these data reveal the unique strategy that has evolved to fulfil the above demands coincidently.

On the basis of their primary sequence, EHs have been grouped into the structural family of α/β hydrolase fold enzymes, leading to the previously unexpected recognition that EH-catalyzed epoxide hydrolysis proceeds via formation of an enzyme-substrate ester intermediate (Fig. 2) (Arand et al., 1994; Lacourciere and Armstrong, 1994; Pries et al., 1994).

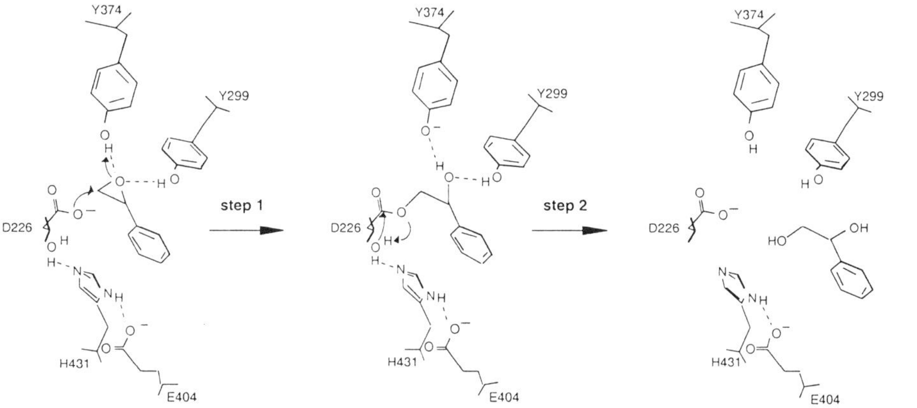

Fig. 2 Enzymatic mechanism of epoxide hydrolysis. The present example depicts the catalytic center of mammalian microsomal epoxide hydrolase. For further details see text.

The important elements of the catalytic center of epoxide hydrolases are (i) the catalytic triad that is common to all α/β hydrolase fold enzymes, and (ii) a pair of tyrosine residues as revealed by X-ray analysis of epoxide hydrolase structures. Initially, the two tyrosines catch the compound by hydrogen bonding to the ring oxygen. In the first enzymatic reaction step, the catalytic nucleophile of the triad, invariably an aspartic acid residue, attacks a carbon atom of the oxirane ring. At the same time, the bond of this carbon atom to the ring oxygen is released and the latter is saturated by proton transfer from one of the tyrosines, in the sense of a classic push-pull mechanism. Thus, an ester intermediate between enzyme and substrate is formed. This covalent intermediate is hydrolysed in the second step of the enzymatic reaction by a water activated via proton abstraction by the second member of the triad, always a histidine. This histidine is positioned and activated through hydrogen bonding to the third member of the catalytic triad, an acidic residue. This is a glutamic acid in animal microsomal epoxide hydrolases and an aspartic acid in all other epoxide hydrolases investigated so far. This hydrolytic step liberates the product of the enzymatic reaction and regenerates the active enzyme.

Kinetics of enzymatic epoxide hydrolysis

An important step towards the understanding of the enzymatic mechanism of epoxide hydrolysis was the observation by Armstrong and co-workers, that the formation of the ester intermediate proceeds by about three orders of magnitudes faster than the subsequent hydrolysis, during the turnover of glycidyl-4-nitrobenzoate (Tzeng et al., 1996). The authors used the change in the intrinsic fluorescence of the purified epoxide hydrolase that occurred on covalent binding of the substrate to directly measure the rate constants of the two reaction steps. We have provided indirect evidence that the same holds true for two other, structurally unrelated substrates of mEH, namely styrene-7,8-oxide and 9,10-epoxystearic acid, by comparing the kinetic constants obtained with the wild type enzyme and a $Glu_{404}Asp$ mutant of rat mEH (Fig. 3, (Arand et al., 1999)). What we observed is that in both cases, V_{max}

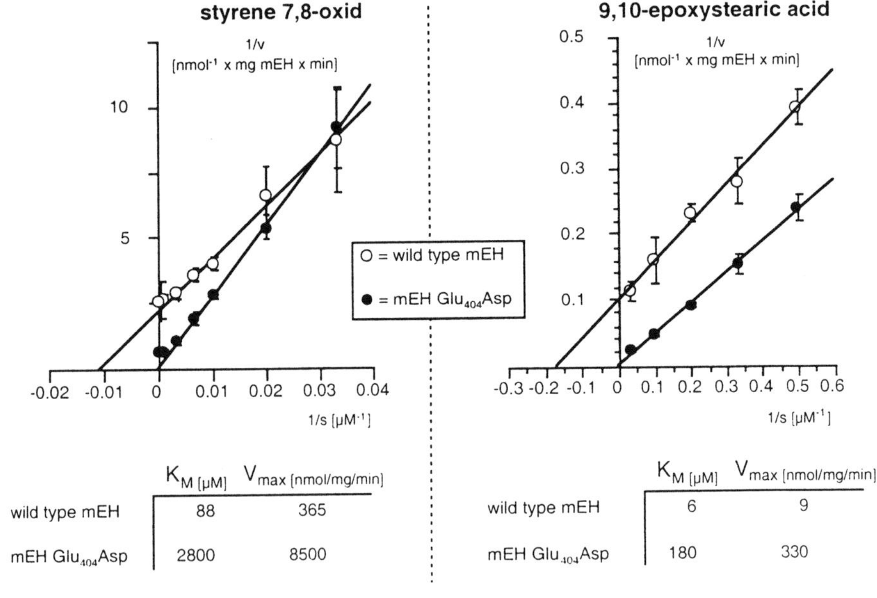

	K_{M} [µM]	V_{max} [nmol/mg/min]
wild type mEH	88	365
mEH Glu$_{404}$Asp	2800	8500

	K_{M} [µM]	V_{max} [nmol/mg/min]
wild type mEH	6	9
mEH Glu$_{404}$Asp	180	330

Fig. 3 Comparative kinetics of wild type and mutant (E404D) microsomal epoxide hydrolase

increased by a factor of about 30-fold, paralleled by a similar increase in K_M. Since Glu$_{404}$ is the member of the catalytic triad supporting the histidine in water activation we can expect that it is mainly, if not exclusively, decisive in the second step of the enzymatic reaction. Its exchange towards aspartic acid should therefore selectively affect the rate constant of step 2 of the enzymatic reaction. For an enzymatic reaction

$$E + S \xrightleftharpoons{K_D} ES \xrightarrow{k_1} E\cdot S \xrightarrow{k_2} E + P$$

where ES is the Michaelis-Menten complex and E•S is the ester intermediate, the relationship between K_M, K_D, and the rate constants k_1 and k_2 for the two separate enzymatic reactions is

$$K_M = K_D \cdot \frac{k_2}{k_1 + k_2}$$

Consequently, the ratio of k_1 to k_2 dictates the effect of modulation of k_2 on V_{max} and K_M. If we look at the three principal options for this ratio

(1) $\quad k_1 \ll k_2 \quad => \quad \dfrac{k_2}{k_1 + k_2} \approx 1 \quad => \quad K_M = K_D; V_{max} \sim k_1$

(2) $\quad k_1 \gg k_2 \quad => \quad \dfrac{k_2}{k_1 + k_2} \approx \dfrac{k_2}{k_1} \quad => \quad K_M \sim k_2; V_{max} \sim k_2$

(3) $\quad k_1 = k_2 \quad => \quad \dfrac{k_2}{k_1 + k_2} = \dfrac{1}{2} \quad => \quad K_M = \dfrac{1}{2} \cdot K_D; V_{max} \sim k_1, k_2$

it becomes evident that the observed increase in K_M when speeding up V_{max} is exclusively compatible with scenario 2, as only here a correlation between K_M and V_{max} is expected.

These deductions imply an accumulation of the ester intermediate during the enzymatic hydrolysis of any of the above substrates by mEH. Using [^{14}C]-labelled 9,10-epoxystearic acid, we were, indeed, able to demonstrate this accumulation (Fig. 4, (Müller et al., 1997)) of the covalent intermediate.

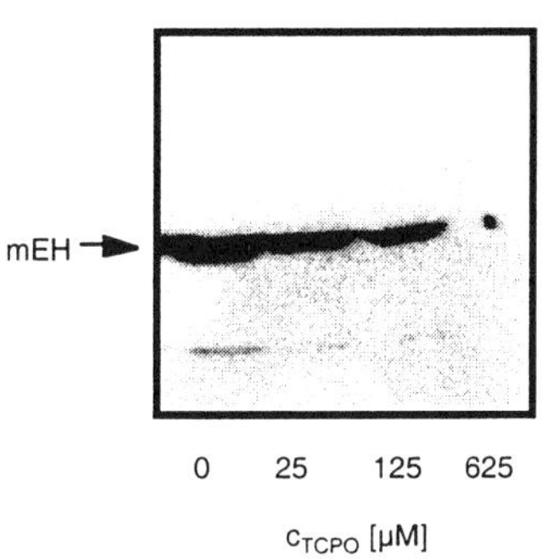

Fig. 4 Visualization of the covalent enzyme-substrate intermediate. After short incubation of rat liver microsomes with [^{14}C]-labelled 9,10-epoxystearic acid, proteins were precipitated and resolved on an SDS polyacrylamide gel. The subsequent autoradiography of the dried gel reveals a strongly labelled band at the migrational speed of mEH. The fact that the labelling intensity was significantly reduced by preincubation of the microsomes with the competitive mEH inhibitor 3,3,3-trichloropropene oxide (TCPO) provides further evidence that the target of the label was, indeed, mEH.

Implications of the enzymatic mechanism of epoxide hydrolysis

The above outlined enzymatic mechanism is a very efficient solution for the problem to combine broad substrate specificity with high affinity. From the structural work on EHs we see three major determinants of substrate selectivity: (i) the tyrosine pair that searches a hydrogen bonding partner; (ii) the shape of the substrate binding cavity that brings in some constraints as towards the permissive geometry of the substrate (e.g. no *trans*-substituted epoxides seem to fit in the active site of mEH); and (iii) the catalytic nucleophile that looks for an electrophilic reaction partner in the vicinity of the hydrogen bonded substrate moiety. It is important to note that, apparently, not the perfect fit of the substrate into the catalytic pocket dominates the determination of the substrate specificity. As long as the substrate is not hindered in entering the active site, the chemical nature and reactivity of the compound, i.e. being an epoxide, is decisive. Thus, the substrate specificity within the chemical class of epoxides can be unusually broad with a single enzyme. The surprisingly high affinity towards a broad range of structurally different compounds of mammalian mEH is afforded by a kind of trick. The important step for substrate detoxification is the formation of the ester intermediate, since this already neutralizes the chemical reactivity of the substrate. Consequently, the enzymatic reaction has been optimized for a rapid first step, which leads to k_1 being orders of magnitudes higher than k_2, as discussed above. Since under these conditions the relationship between K_M and K_D simplifies to

$$K_M = K_D \cdot \frac{k_2}{k_1}$$

the ratio of the apparent affinity K_M to the real K_D of the enzyme/substrate combination is equal to the ratio of k_2 to k_1, and therefore K_M is magnitudes smaller than K_D, mimicking a higher affinity. The relatively slow enzyme regeneration caused by the slow k_2 has to be compensated by a high enzyme concentration, as observed, for instance, with mEH in human liver.

The consequences of the enzymatic mechanism for the efficacy of substrate detoxification can be visualized by computer simulation (Fig. 5). As a first lesson to learn, we see that the rate of product formation does not even approximately reflect the detoxification speed of the enzyme (Fig. 5A), because the product formation rate is determined by k_2 (small) while the detoxification rate is dependent on k_1 (large). As a result, the AUC of the substrate, i.e. the toxic load of the organism, is largely overestimated if product formation is taken as the base for the evaluation instead of substrate disappearance. In the present example, the styrene-7,8-oxide, we observe a 3-fold underestimation. This factor may even be substantially larger in the case of high affinity substrates, such as the epoxides derived from polycyclic aromatic hydrocarbons. A second, particularly impressive observation is how strongly the covalent binding in the first reaction step influences the kinetics. Fig. 5B compares the AUC resulting from the established two-step mechanism of enzymatic epoxide hydrolysis with that resulting from a hypothetical one-step reaction involving direct hydrolysis under otherwise identical kinetic conditions. The third, yet not mechanism-dependent observation is that, if under steady state conditions the substrate formation rate reaches approximately 70% of the product formation rate, the steady state concentration of the substrate starts to rise exponentially with a further increase in the substrate formation rate (Fig. 5C), which may represent the

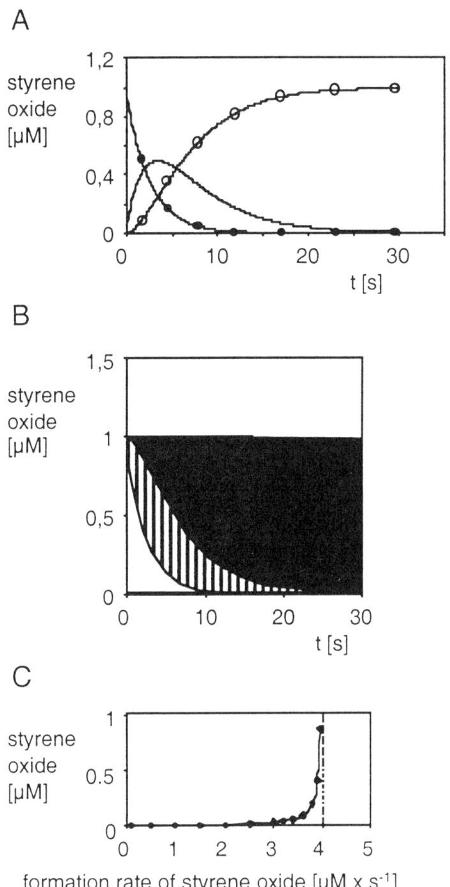

Fig. 5 Computer simulation of styrene oxide hydrolysis by human microsomal epoxide hydrolase. A. The concentrations of free substrate (closed circles), the covalent ester intermediate between enzyme and substrate (no symbols), and the terminal reaction product (diol; open circles) are computed over time after addition of 1 µM styrene oxide into a single compartment. Rapid clearance is observed due to the essentially irreversible first step of the enzymatic mechanism, despite of the relatively low affinity of the enzyme to the substrate. In **B**, the area under the concentration-time curve (AUC) of styrene oxide is plotted as obtained from different simulation conditions. The white area represents the styrene oxide burden calculated assuming the above conditions. The hatched area plus the white and the black area gives the AUC for styrene oxide that would result from a one step mechanism using the same equilibrium constant and hydrolysis rate as for the two step simulation. It is three orders of magnitude above the AUC for the two-step mechanism and thus heavily exceeds the scale. The hatched area indicates the amount of overestimation of styrene oxide burden with a two-step mechanism if calculated from the rate of product formation instead of substrate disappearance, which would result in a 3-fold underestimation of the detoxification efficacy of the EH, in the present case. **C** shows the dependence of steady state concentration of styrene oxide on the rate of styrene oxide formation under the present conditions.

Once the formation rate significantly rises above 2 µM x s⁻¹, corresponding to 50% of maximum hydrolysis rate in the system, the styrene oxide steady state concentration rises steeply. Above the maximum capacity, no EH-controlled equilibrium is possible. Computer simulation was performed using Microsoft Excel on a Macintosh G3 desktop computer. The respective spreadsheet containing the computations is available from the authors via email (arand@mail.uni-mainz.de). Results obtained with this approach resembled those produced with the software package Kinsim (Barshop et al., 1983), yet offer somewhat more flexibility, as long as the kinetics are kept simple. In order to match the experimental data available for the human mEH (Jenkins Sumner and Fennell, 1994; Lu et al., 1979; Mendrala et al., 1993), mEH enzyme concentration, equilibrium constant K_s, rate constant k_1 for the ester formation and the rate constant k_2 for ester hydrolysis were set to 20 µM, 10 mM, 200 x s⁻¹, and 0.2 x s⁻¹, respectively. A slow back reaction from the ester to the epoxide that has recently been postulated by others on the basis of indirect evidence (Rink and Janssen, 1998; Tzeng et al., 1996) is neglected in the present scenario because (i) it is, in our view, energetically highly unfavoured and (ii) it complicates computation without qualitatively changing the outcome of the simulation.

basis for a practical threshold in the susceptibility of an epoxide hydrolase-protected organism towards epoxide genotoxicity. An experimental proof for this result of the computer simulation is displayed in Fig. 6, that shows exactly the expected threshold response of V79 Chinese hamster cells towards styrene-7,8-oxide-induced DNA damage, but only if these cells express mEH.

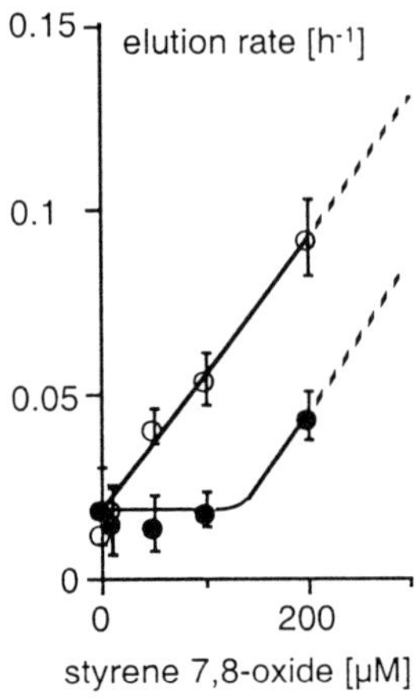

Fig. 6 Expression of human mEH protects V79 Chinese Hamster cells from styrene oxide-induced DNA damage and introduces a threshold. Parental mEH-deficient V79 Chinese hamster cells (open circles) and a stable V79-derived cell line expressing human mEH (closed circles) (Herrero et al., 1997) were treated with increasing concentrations of styrene 7,8-oxide in culture and the effect of the DNA damaging agent was assessed by alkaline filter elution of the DNA as described earlier (Hengstler et al., 1992). The onset of the genotoxic effect caused by styrene oxide in the recombinant cell line only at concentrations above 100 µM clearly demonstrates that mEH expression introduces a threshold for styrene oxide genotoxicity in V79 cells. Data points represent the average of three independent determinations. Error bars indicate the respective standard deviations. At a styrene oxide concentration of 500 µM, a strong cytotoxic effect on the parental V79 cells and a moderate toxic effect on the recombinant cells was observed.

In conclusion, the recent insights in the enzymatic mechanism of epoxide hydrolysis reveal a fascinating catalytic strategy and very well explain the high efficacy of this low turnover number enzyme.

References

Arand, M.,Grant, D. F.,Beetham, J. K.,Friedberg, T.,Oesch, F., and Hammock, B. D., 1994, Sequence similarity of mammalian epoxide hydrolases to the bacterial haloalkane dehalogenase and other related proteins. Implication for the potential catalytic mechanism of enzymatic epoxide hydrolysis, *FEBS Lett.* **338**: 251-256.

Arand, M.,Müller, F.,Mecky, A.,Hinz, W.,Urban, P.,Pompon, D.,Kellner, R., and Oesch, F., 1999, Catalytic triad of microsomal epoxide hydrolase: replacement of Glu404 with Asp leads to a strongly increased turnover rate, *Biochem. J.* **337**: 37-43.

Arand, M.,Wagner, H., and Oesch, F., 1996, Asp[333], Asp[495], and His[523] form the catalytic triad of rat soluble epoxide hydrolase, *J. Biol. Chem.* **271**: 4223-4229.

Argiriadi, M. A.,Morisseau, C.,Hammock, B. D., and Christianson, D. W., 1999, Detoxification of environmental mutagens and carcinogens: structure, mechanism, and evolution of liver epoxide hydrolase, *Proc Natl Acad Sci U S A* **96**: 10637-42.

Armstrong, R. N., 1999, Kinetic and chemical mechanism of epoxide hydrolase, *Drug Metabolism Reviews* **31**: 71-86.

Barshop, B. A.,Wrenn, R. F., and Frieden, C., 1983, Analysis of numerical methods for computer simulation of kinetic processes: development of KINSIM--a flexible, portable system, *Anal. Biochem.* **130:** 134-145.

Guenthner, T.,Hammock, B. D.,Vogel, U., and Oesch, F., 1981, Cytosolic and microsomal epoxide hydrolase are immunologically distinguishable from each other in the rat and mouse, *J. Biol. Chem.* **256:** 3163-3166.

Hammock, B. D.,Storms, D. H., and Grant, D. F., 1997, Epoxide hydrolases, in *Biotransformation* (F. P. Guengerich, ed.), Pergamon Press, Oxford, pp. 283-305.

Hayes, J. D., and Pulford, D. J., 1995, The glutathione S-transferase supergene family: regulation of GST and the contribution of the isoenzymes to cancer chemoprotection and drug resistance, *Crit Rev Biochem Mol Biol* **30:** 445-600.

Hengstler, J. G.,Fuchs, J., and Oesch, F., 1992, DNA strand breaks and DNA cross-links in peripheral mononuclear blood cells of ovarian cancer patients during chemotherapy with cyclophosphamide/carboplatin, *Cancer Res* **52:** 5622-5626.

Herrero, M. E.,Arand, M.,Hengstler, J. G., and Oesch, F., 1997, Recombinant expression of human microsomal epoxide hydrolase protects V79 Chinese hamster cells from styrene oxide - but not from ethylene oxide-induced DNA strand breaks, *Environ. Mol. Mutagen.* **30:** 429-439.

Jenkins Sumner, S., and Fennell, T. R., 1994, Review on the metabolic fate of styrene, *Crit. Rev. Toxicol.* **24:** S11-S33.

Lacourciere, G. M., and Armstrong, R. N., 1993, The catalytic mechanism of microsomal epoxide hydrolase involves an ester intermediate, *J. Am. Chem. Soc.* **115:** 10466-10467.

Lacourciere, G. M., and Armstrong, R. N., 1994, Microsomal and soluble epoxide hydrolases are members of the same family of C-X bond hydrolase enzymes, *Chem. Res. Toxicol.* **7:** 121-124.

Laughlin, L. T.,Tzeng, H.-F.,Lin, S., and Armstrong, R. N., 1998, Mechanism of microsomal epoxide hydrolase. Semifunctional site-specific mutants affecting the alkylation half-reaction, *Biochemistry* **37:** 2897-2904.

Lu, A. Y.,Thomas, P. E.,Ryan, D.,Jerina, D. M., and Levin, W., 1979, Purification of human liver microsomal epoxide hydrase. Differences in the properties of the human and rat enzymes, *J Biol Chem* **254:** 5878-5881.

Mendrala, A. L.,Langvardt, P. W.,Nitschke, K. D.,Quast, J. F., and Nolan, R. J., 1993, In vitro kinetics of styrene and styrene oxide metabolism in rat, mouse, and human, *Arch Toxicol* **67:** 18-27.

Moghaddam, M. F.,Grant, D. F.,Cheek, J. M.,Greene, J. F.,Williamson, K. C., and Hammock, B. D., 1997, Bioactivation of leukotoxins to their toxic diols by epoxide hydrolase, *Nature Medicine* **3:** 562-566.

Müller, F.,Arand, M.,Frank, H.,Seidel, A.,Hinz, W.,Winkler, L.,Hänel, K.,Blee, E.,Beetham, J. K.,Hammock, B. D., and Oesch, F., 1997, Visualization of a covalent intermediate between microsomal epoxide hydrolase, but not cholesterol epoxide hydrolase, and their substrates, *Eur. J. Biochem.* **245:** 490-496.

Nardini, M.,Ridder, I. S.,Rozeboom, H. J.,Kalk, K. H.,Rink, R.,Janssen, D. B., and Dijkstra, B. W., 1999, The X-ray structure of epoxide hydrolase from *Agrobacterium radiobacter* AD1: an enzyme to detoxify harmful epoxides, *J. Biol. Chem.* **274:** 14579-14586.

Oesch, F., 1973, Mammalian epoxide hydrases: Inducible enzymes catalysing the inactivation of carcinogenic and cytotoxic metabolites derived from aromatic and olefinic compounds, *Xenobiotica* **3:** 305-340.

Oesch, F., and Arand, M., 1999, Xenobiotic metabolism, in *Toxicology* (H. Marquardt,S. Schäfer,D. McLellan, and C. Welsch, eds.), Academic Press, Inc., San Diego, pp. 83-110.

Oesch, F., and Bentley, P., 1976, Antibodies against homogeneous epoxide hydratase provide evidence for a single enzyme hydrating styrene oxide and benz(a)pyrene 4,5-oxide, *Nature* **259:** 53-55.

Ota, K., and Hammock, B. D., 1980, Cytosolic and microsomal epoxide hydrolases: Differential properties in mammalian liver, *Science* **207:** 1479-1481.

Pries, F.,Kingma, J.,Pentenga, M.,van Pouderoyen, G.,Jeronimus-Stratingh, C. M.,Bruins, A. P., and Janssen, D. B., 1994, Site-directed mutagenesis and oxygen isotope incorporation studies of the nucleophilic aspartate of haloalkane dehalogenase, *Biochemistry* **33:** 1242-1247.

Rink, R., and Janssen, D. B., 1998, Kinetic mechanism of the enantioselective conversion of styrene oxide by epoxide hydrolase from *Agrobacterium radiobacter* AD1, *Biochemistry* **37:** 18119-18127.

Tzeng, H.-F.,Laughlin, L. T., and Armstrong, R. N., 1998, Semifunctional site-specific mutants affecting the hydrolytic half-reaction of microsomal epoxide hydrolase, *Biochemistry* **37:** 2905-2911.

Tzeng, H.-F.,Laughlin, L. T.,Lin, S., and Armstrong, R. N., 1996, The catalytic mechanism of microsomal epoxide hydrolase involves reversible formation and rate-limiting hydrolysis of the alkyl-enzyme intermediate, *J. Am. Chem. Soc.* **118:** 9436-9437.

Zou, J.,Hallberg, B. M.,Bergfors, T.,Oesch, F.,Arand, M.,Mowbray, S. L., and Jones, T. A., 2000, Structure of Aspergillus niger epoxide hydrolase at 1.8 A resolution: implications for the structure and function of the mammalian microsomal class of epoxide hydrolases, *Structure Fold. Des.* **8:** 111-122.

CHEMOPROTECTION AND INTERINDIVIDUAL DIFFERENCES IN RESPONSE TO BIOLOGICAL REACTIVE INTERMEDIATES

R. Thier

Institute of Occupational Physiology at the University of Dortmund
Ardeystr. 67
D-44139 Dortmund, Germany

INTRODUCTION

In humans and experimental animals, carcinogenesis is a complex process in which normal cell growth is modified. Carcinogenesis is divided into three main stages: initiation, promotion and progression. Chemopreventive interaction in carcinogenesis offers two major strategies. The first strategy will inhibit or at least slow down carcinogenesis by blocking its progress. This might occurr at all stages of carcinogenesis. This strategy includes the scavenging of bioreactive intermediates (BRIs), induction or inhibition of enzymes of the metabolism of xenobiotics which create or detoxify the BRIs, of enzymes of DNA-repair or of other enzymes. A second strategy aims at reversing the process of tumour formation either by redifferentiation of transformed cells or the elimination of precarcinogenic clones. Both strategies cover one or several steps of carcinogenesis and approach the subject matter in more general terms without consideration of individual susceptibility to particular cancers (Kelloff et al., 1999).

Molecular Epidemiologic Studies on Individual Susceptibility

Recently, many epidemiological studies have been conducted to investigate differences in individual susceptibility. One of the aims in such studies was always to identify genetic biomarkers of individual susceptibility such as polymorphic metabolic enzymes (PMEs) e.g. cytochrome P450 CYP2D6, *N*-acetyltransferase NAT2 or glutathione transferase (GST) GSTM1. PMEs can lead to several different phenotypes with altered metabolic kinetics.

Biological Reactive Intermediates VI, Edited by Dansette *et al.*
Kluwer Academic / Plenum Publishers, 2001

Faster endogenous formation or slower detoxification of BRIs might increase the risk for mutagenesis, a substantial event in carcinogenesis.

When investigating a single genetic biomarker in molecular epidemiologic studies the investigators find in most studies an increased risk by a factor in the range of about 2 for the more susceptible group. Investigations considering combinations of several genetic bio-markers can identify subgroups being at considerably higher risk. The study by Hirvonen et al., (2000) investigating the breast cancer risk of Finnish women (n=965) presents a very impressive example of such a study. The GSTM1 null genotype increased the risk of of postmenopausal breast cancer 2.2fold and the mutation of at least one GSTM3 allele increased the risk of premenopausal advanced breast cancer 2.4fold. The combined genotypes appeared to have a much higher impact regarding the susceptibility to breast cancer. In contrast, substantially increased risks of advanced breast cancer were seen for premenopausal women lacking the GSTM1 gene and carrying the GSTP1*AA genotype together with at least one GSTM3*B allele (OR: 8.3; 95% CI: 1.4-49.3) or with the GSTT1 null genotype (OR: 10.5; 95% CI: 1.7-65.5). The most remarkable risk of advanced breast cancer was observed for premenopausal women concurrently carrying the GSTM3*B allele, the GSTT1 null genotype, and the GSTP1*AA genotype (OR: 25.8; 95% CI: 2.5-263).

The limitations of such molecular epidemiologic studies include usually the number of patients and the lack of information regarding exposure to chemicals either by environment, occupation or life style. In some cases there might just be a coincidence between a particular subgroup and the disease under investigation. It is safe to assume that exogenous exposure varies considerably from individual to individual depending on life style and occupation. This explains why it is highly unlikely to find a single chemopreventive agent that will protect an entire population from carcinogenesis.

This is different in defined populations as the ones you willl find at workplaces. Generally, the exposure is well known and monitored - even though it might be a complex mixture of chemicals. Individuals are closely observed during the time of occupation and very often also afterwards. Thus, a more specific chemoprotection approach as opposed to the more general chemoprevention approach is conceivable and also - for the time being - more likely to succeed.

ACRYLONITRILE

Exposure to toxic chemicals at the workplace is in general and for many compounds specifically regulated and, therefore, severe intoxications occurr only accidently. Acrylonitrile (ACN) is a compound where accidental intoxication can be fatal (Buchter and Peter 1984; Steffens et al., 1998). ACN is used as a monomer for the production of fibres and resins in huge quantities. While it is carcinogenic in animal studies epidemiologic data on carcinogenicity to humans have been evaluated to be inadequate by IARC (1991). Great concern regarding accidental peak exposure towards ACN exists because of its high acute toxicity.

Metabolism of Acrylonitrile

ACN is metabolised via two pathways (Fennell et al., 1991; Kedderis et al., 1993). The GSH conjugation leading to several products of the mercapturic acid pathway that are excreted in the urine is considered to be detoxifying. Oxidation of ACN catalyzed by CYP2E1 leads to the formation of cyanoethylene oxide. The epoxide is hydrolyzed by the microsomal epoxide hydrolase or conjugated with glutathione (GSH) by GST. The GSH conjugation results in two different isomers the 2-*S*-glutathionyl-1-cyanoethanol and the 1-*S*-glutathionyl-1-cyanoethanol. The latter will be metabolized and excreted along the mercapturic acid pathway. The 2-*S*-glutathionyl-1-cyanoethanol conjugate releases cyanide (CN⁻). Its GSH conjugate moiety is excreted via the mercapturic acid pathway. The cyanide is converted to thiocyanate (SCN) by rhodanese (Figure 1).

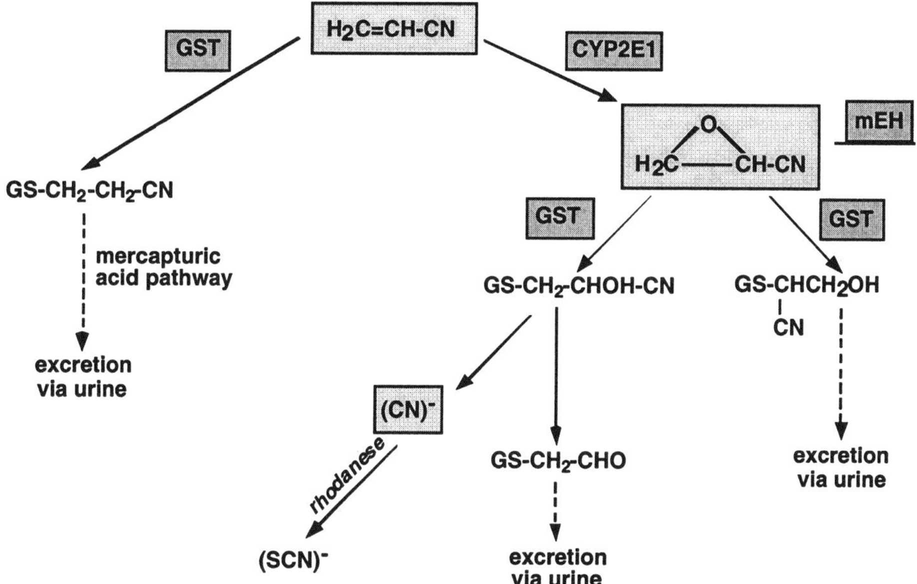

Figure 1 Metabolic pathways of acrylonitrile (according to Fennell et al., 1991 and Kedderis et al., 1993, modified)

ACN itself reacts with cellular proteins (Peter and Bolt 1981; Benz et al., 1997a). The epoxide has been shown to form DNA and protein adducts and, thus, is held responsible for the carcinogenicity of ACN in rats (Peter et al., 1983; Recio and Skopek 1988). The cyanide is of immediate interest because of its acute toxicity.

Chemoprotection Approach Against Acute ACN Intoxication

Chemoprotection regarding the acute toxicity of the cyanide and the carcinogenic effect might be achieved by prevention of glutathione depletion via supplementation with *N*-acetylcysteine (NAC). Protection against cyanide toxicity can also be gained by methemoglobin formation with dimethylaminophenol (DMAP) or sodium thiosulfate ($Na_2S_2O_3$). These different candidate antidotes were tested for their effectiveness. Male rats (groups of 12) were exposed to 4800 ppm ACN for 30 min. Each of the three groups recieved either no antidote, cyanide antidote ($Na_2S_2O_3$) or NAC as chemoprotective agent. All animals recieving NAC while none of the animals in the groups recieving $Na_2S_2O_3$ or no antidote survived the ACN exposure. From these results it was concluded that acute ACN toxicity resulted from its intrinsic biological reactivity rather than from cyanide (Appel et al., 1981). Consequently, in Germany high intravenous doses of NAC were recommended for treatment of accidental intoxications in ACN workers (Buchter et al., 1984). In contrast, in the United States suggested antidotal measures focus solely on cyanide as metabolite (MEDITEXT (R) Medical Management 1996) which is considered to be unfortunate (Benz et al., 1997).

Intoxication Cases

Two cases of acute ACN overexposure or intoxication occurred within a group of males workers (n=59) having industrial contact with acrylonitrile. In the entire group of workers the exposure to acrylonitrile was well below current official exposure standards, as demonstrated by hemoglobin adduct monitoring (Thier et al., 1999). ACN and its metabolite cyanide were analyzed using gas chromatographic procedures. ACN in blood was quantified using GC-NPD, and cyanide in plasma was assayed using GC-MSD after derivatisation with pentafluorobenzyl bromide. Both methods have been validated by the Deutsche Forschungsgemeinschaft (DFG 1983, 1988).

The first acute intoxication to ACN occured through skin contact. After 2 hours the male subject, patient 1, complained about headache, nausea, fatigue and generally impaired wellbeing and received treatment at the industrial plant's Medical Department. The blood ACN level was 824 µg/liter. A very high cyanide blood level (4300 µg/liter) was observed at the same time. The patient recieved sequential antidote treatment. In accordance with current recommendations in Germany (Buchter et al., 1984) NAC (Fluimucil® Antidot, 150 mg/kg bodyweight) was administered twice, 80 and 55 min after intoxication, supplemented by sodium thiosulfate (10 ml of a 10% solution), 90 and 95 min after intoxication, in view of the high cyanide levels. After having received the antidote treatment the general health improved considerably, and specific subjective symptoms were no longer complained by the subject.

In connection with the case described above, a similar intoxication was supposed to have happened in a collegue who was also transferred to the Medical Department of the industrial plant. His initial ACN blood level turned out even higher (1250 µg/liter blood) when compared to that of patient 1, but no clinical symptoms were recorded. Specifically, there was no reporting of subjective symptoms of illness, as was in patient 1. Because of the absence of any symptoms the further recording of biological monitoring parameters was discontinued (Steffens et al., 1998).

Both cases reported here showed similar initial blood levels of ACN. However, only patient 1 developed severe clinical signs of intoxication. This patient suffered three such ACN intoxications, so far, all of them displaying the same pattern of symptoms and blood levels of ACN and cyanide. The symptoms ceased upon antidote treatment with NAC and sodium thiosulfate. The initial cyanide blood levels were, based on animal experimental data (Yamamoto et al., 1979), very close to the acutely lethal range and the clinical symptoms reported in case 1 are entirely compatible with those of cyanide. This demonstrates the practical importance of human individual differences in susceptibility (Lewalter and Neumann 1998). Uncommon high cyanide levels were observed in about five out of 30 workers (Lewalter, *personal communication*). In general, the present observations point to important metabolic species differences of ACN metabolism and toxicity between humans and rodents and to the likelyhood of significant human individual differences.

From the metabolic scheme (Figure 1) several hypotheses can be deduced. ACN is directly detoxified by a polymorphic GST. Because of their deletion polymorphism, resulting in different distinct phenotypes hGSTM1 and hGSTT1 appeared to be the most obvious candidate genes but hGSTP1 was also considered.

Alternatively, the cyanothylene epoxide might be detoxified by a particular polymorphic GST. The formation of the glutathionyl-1-cyanoethanol isomer might be preferred by this GST over to the formation of the 2-glutathionyl-1-cyanoethanol. Other enzymes could be involved as well e.g. the epoxide hydrolase. In addition, the elimination of the cyanide via rhodenase needs to be considered.

Interindividual Differences in ACN Protein Adduct Levels

In a study on 59 workers exposed to ACN we investigated several genetic biomarkers that might explain these individual differences. The smoking behaviour, globin adduct level of ACN (cyanoethylvaline, CEV) and the genotype of hGSTM1, hGSTT1 were determined of each individual. Additionally, the genotypes of GSTP1 were determined according to Harries et al., (1997) and Harris et al., (1998).

The biological adduct monitoring procedures were conducted in accordance with the methods and procedures recommended by the Deutsche Forschungsgemeinschaft (DFG 1996). After isolation of erythrocytes and globin the levels of N-terminal CEV were determined using the method described in detail by Lewalter (1996). Protein adduct analyses resulted in levels between 2 and 231 μg CEV/l blood (mean $\pm$ S.D. 59.8 $\pm$ 51.5 μg) The wide ranges supports the hypothesis regarding an interindividual difference in the detoxification capacity of AN. There were no significant correlations between the smoking behaviour, the hGSTM1 or hGSTT1 genotype and globin adduct levels though (Thier et al., 1999).

The genotyping of GSTP1 revealed twenty-five workers with at least one mutated allele at the hGSTP1 $A_{1404}G$ locus, 5 were homozygotes. The CEV adduct levels correlated significantly ($p<0.05$; Jonckheere-Terpstra-Test) with the presence of the mutation (w/w<w/m<m/m). All subjects showing a mutated GSTP1 $C_{2294}T$ locus were also carrying at least one $A_{1404}G$ mutation; no homozygous mutant was found at the $C_{2294}T$ locus. As

expected from these results the hemoglobin CEV adduct levels were also significantly higher in the GSTP1 $C_{2294}T$ heterozygotes.

Since CEV is formed only by the reaction of ACN with hemoglobin the first metabolic step appears to be important for the detoxification of ACN. Thus, GSTP1 does contribute to the ACN detoxication in humans and individual differences in susceptibility regarding its acute intoxication might be resulting from the GSTP1 polymorphism.

On the basis of the case reports which are now available, as well as on increased theoretical background knowledge, a combination of NAC with sodium thiosulfate seems an appropriate measure for antidote therapy of acute ACN intoxications.

METHYL BROMIDE

One question still remains: Is NAC a suitable candidate for a general chemoprotection approach?

This might not be true for acute intoxication with methyl bromide (MB). MB is carcinogenic in rats and mice due to its high reactivity towards DNA (Gansewendt et al., 1991) While MB is slightly neurotoxic a metabolite (methanethiol) formed by the GSTT1 depending pathway is severely neurotoxic.

Two workers acutely exposed to methyl bromide with inadequate respiratory protective devices suffered intoxication. Seven weeks after the accident, blood samples were drawn from both patients. GSTT1 activity in erythrocytes and measurement of binding products of methyl bromide with blood proteins were determined. Duration and intensity of exposure were identical for both patients as they worked together with the same protective devices and with similar physical effort. However, one patient had very severe poisoning, whereas the other developed only mild neurotoxic symptoms. The GSTT1 activity, of the first patient was normal whereas it was undetectable in the erythrocytes of the second patient, who consequently had higher concentrations of S-methylcysteine adduct in albumin (149 *versus* 91 nmol/g protein) and in globin (77 *versus* 30 nmol/g protein). These results suggest a mechanism for methyl bromide neurotoxicity which could be related to the transformation of methylglutathione into toxic metabolites such as methanethiol and formaldehyde. If such metabolites are the ultimate toxic species, NAC treatment could have a toxifying rather than a detoxifying effect (Garnier et al., 1996).

CONCLUSION

These examples demonstrate very clearly the complexicity of chemoprotection strategies. A successful chemoprotection approach designed for a specific purpose - e.g. the prevention of fatal acrylonitrile intoxication by administration of NAC and sodium thiocyanate - might prove detrimental in a different situation. Therefore, it must be concluded that more general chemoprevention approaches are more likely to fail. Thus, successful chemoprotection strategies depend on the exposure scenario *and* on the population at risk.

REFERENCES

Appel, K.E., Peter, H., and Bolt, H.M., 1981, Effect of potential antidotes on the acute toxicity of acrylonitrile, *Int. Arch. Occup. Environ. Hlth* 49:157.

Benz, F.W., Nerland, D.E., Corbett, D., and Li, J., 1997, Biological markers of acute acrylonitrile intoxication as a function of dose and time, *Fundam. Appl. Toxicol.* 36:141.

Benz, F.W., Nerland, D.E., Li, J., and Corbett, D., 1997a, Dose dependence of covalent binding of acrylonitrile to tissue protein and globin in rats, *Fundam. Appl. Toxicol.* 36:149.

Buchter, A. and Peter, H. 1984, Clinical toxicology of acrylonitrile. *G. Ital. Med. Lav.* 6:83.

Buchter, A., Peter, H., and Bolt, H.M., 1984, N-Acetyl-Cystein als Antidot bei akzidenteller Acrylnitril-Intoxikation, *Int. Arch. Occup. Environ. Hlth* 53:311.

DFG, 1983, Head-Space-Technik (Dampfraumanalyse), in: Analytische Methoden zur Prüfung gesundheitsschädlicher Arbeitsstoffe. Analysen im biologischen Material, Band 2, 7. Lieferung, pp. D1-D2, VCH, Weinheim.

DFG, 1988, Cyanid, in: Analytische Methoden zur Prüfung gesundheitsschädlicher Arbeitsstoffe, Analysen im biologischen Material, Band 2, 9. Lieferung, pp. D1-D3, VCH, Weinheim.

DFG, 1996, *N*-2-Cyanoethyl-Valin, *N*-2-Hydroxyethyl-Valin, *N*-Methyl-Valin (zum Nachweis einer Belastung/Beanspruchung durch Acrylnitril, Ethylenoxid sowie methylierende Substanzen, in: Analytische Methoden zur Prüfung gesundheitsschädlicher Arbeitsstoffe. Analysen im biologischen Material, Band 2, 12. Lieferung, pp. D1-D9, VCH, Weinheim.

Fennell, T.R., Kedderis, G.L., and Sumner, S.C.J., 1991, Urinary metabolites of [1,2,3-^{13}C]acrylonitrile in rats and mice detected by ^{13}C nuclear magnetic resonance spectroscopy, *Chem. Res. Toxicol.* 4:678.

Gansewendt, B., Foest, U., Xu, D., Hallier, E., Bolt, H.M., and Peter, H., 1991, Formation of DNA adducts in F-344 rats after oral administration or inhalation of [^{14}C]methyl bromide, *Food Chem. Toxicol.* 29:557.

Garnier, R., Rambourg-Schepens, M.O., Muller, A., and Hallier, E., 1996, Glutathione transferase activity and formation of macromolecular adducts in two cases of acute methyl bromide poisoning, *Occup. Environ. Med.* 53:211.

Harries, L.W., Stubbins, M.J., Forman, D., Howard, G.C.W., and Wolf, C.R., 1997, Identification of genetic polymorphisms at the glutathione S-transferase Pi locus and association with susceptibility to bladder, testicular and prostate cancer, *Carcinogenesis* 18:641.

Harris, M.J., Coggan, M., Langton, L., Wilson, S.R., and Board, P.G., 1998, Polymorphism of the Pi class glutathione S-transferase in normal populations and cancer patients, *Pharmacogenetics* 8:27.

Hirvonen, A., Bouchardy, C., Mitrunen, K., Kataja, V., Eskelinen, M., Kosma, V.M., Saarikoski, S.T., Jourenkova, N., Anttila, S., Dayer, P., Uusitupa, M., and Benhamou, S., 2000, Polymorphic GSTs and cancer predispotiotion, In: GST 2000, International Conference on GlutathioneTransferases, Uppsala, Sweden, Abstract Book: L10.

IARC, 1999, Acrylonitrile, in: IARC Monographs on the Evaluation of Carcinogenic Risks to humans, IARC, Lyon.

Kedderis G.L., Sumner S.C.J., Held S.D., Batra R., Turner Jr. M.J., Roberts A.E., and Fennell T.R., 1993, Dose-dependent urinary excretion of acrylonitrile metabolites by rats and mice, *Toxicol. Appl. Pharmacol.* 120:288.

Kelloff, G.J., Sigman, C.C., and Greenwal, P., 1999, Cancer chemoprevention: progress and promise, *Eur. J. Cancer* 35:2031.

Lewalter J., 1996, N-Alkylvaline levels in globin as a new type of biomarker in risk assessment of alkylating agents, *Int. Arch. Occup. Environ. Hlth* 68:519.

Lewalter J. and Neumann H.G., 1998, Biologische Arbeitsstoff-Toleranzwerte (Biomonitoring). Teil XII. Die Bedeutung der individuellen Empfindlichkeit beim Biomonitoring, *Arbeitsmed. Sozialmed. Umwelmed.* 33:352.

MEDITEXT (R) Medical Management, 1996, Acrylonitrile, in: *System,* A.H. Hall and B.H. Rumack, eds. MICROMEDEX, Inc.:Englewood, CO, USA

Peter H. and Bolt H.M., 1981, Irreversible protein binding of acrylonitrile, *Xenobiotica* 11:51.

Peter H., Schwarz M., Mathiasch B., Appel K.E., Bolt H.M., 1983, A note on synthesis and reactivity towards DNA of glycidonitrile, the epoxide of acrylonitrile, *Carcinogenesis* 4:235.

Peter H. and Bolt, H.M., 1984, Experimental pharmacokinetics and toxicology of acrylonitrile, *G. Ital. Med. Lav.* 6:77.

Recio L. and Skopek T.R., 1988, Mutagenicity of acrylonitrile and its metabolite 2-cyanoethylene oxide in human lymphoblasts in vitro, *Mut. Res.* 206:297.

Steffens, W., Sibbing, D., Kiesselbach, N., and Lewalter, J., 1998, Behandlung von Patienten mit Vergiftungen durch aliphatische oder olefinische Nitrile, *Arbeitsmed. Sozialmed. Umweltmed.* 33:11.

Thier R., Lewalter J., Kempkes M., Selinski S., Brüning T., and Bolt H.M., 1999, Haemoglobin adducts of acrylonitrile and ethylene oxide in acrylonitrile workers, dependent on polymorphisms of the glutathione transferases GSTT1 and GSTM1, *Arch. Toxicol.* 73:197.

Yamamoto K., Yamamoto Y., and Kuwahara C., 1979, A blood cyanide distribution study in the rabbits intoxicated by oral route and by inhalation, *Z. Rechtsmed.* 83:313.

OXIDATIVE STRESS, GLUCOSE-6-PHOSPHATE DEHYDROGENASE AND THE RED CELL

David J. Jollow and David C. McMillan

Department of Cell and Molecular Pharmacology
Medical University of South Carolina
Charleston, SC 29425

INTRODUCTION

The red blood cell is highly specialized for the transport of oxygen to tissues and the removal of carbon dioxide. Lacking a nucleus, mitochondria and a ribosomal protein synthetic mechanism, and hence the capacity for "self-renewal", the mammalian red cell has a fixed life span of 50–60 days in rats and about 120 days in humans. However, the apparent structural simplicity is misleading; the red cell is well designed for its purpose, consisting of a highly viscous hemoglobin "liquid crystal" surrounded by a skeletal protein meshwork underlying, and attached to, a lipid bilayer. During its lifetime, the human red cell will travel approximately 250 km and undergo continual transition between the biconcave disc of the large blood vessels and the sheer-stress elongated forms of the microcirculation. Restoration of biconcave discoidosity is an incompletely understood ATP-requiring active process, dependent in turn on the presence of a fully functional glycolytic pathway. Additional ATP production is required to support the various pumps essential to maintain the electrolyte milieu of the red cell.

Although about 1% of the human red cell mass (equal to about 3×10^9 cells/kg) is removed from the circulation per day, the process by which this occurs is not well defined. It is known that the activity of the metabolic machinery declines with the age of the cell. However, in even the oldest cells, the activity of the glycolytic pathway and other key enzymes are still in excess of demand, and hence loss of enzymatic activity *per se* does not appear to be the direct cause of removal from the circulation. Under normal conditions, the major site of removal of senescent cells appears to be the sinusoidal vasculature of the spleen and to involve uptake and degradation by splenic macrophages. The process by which old cells are recognized and sequestered by macrophages is the subject of much current research.

Biological Reactive Intermediates VI, Edited by Dansette *et al.*
Kluwer Academic / Plenum Publishers, 2001

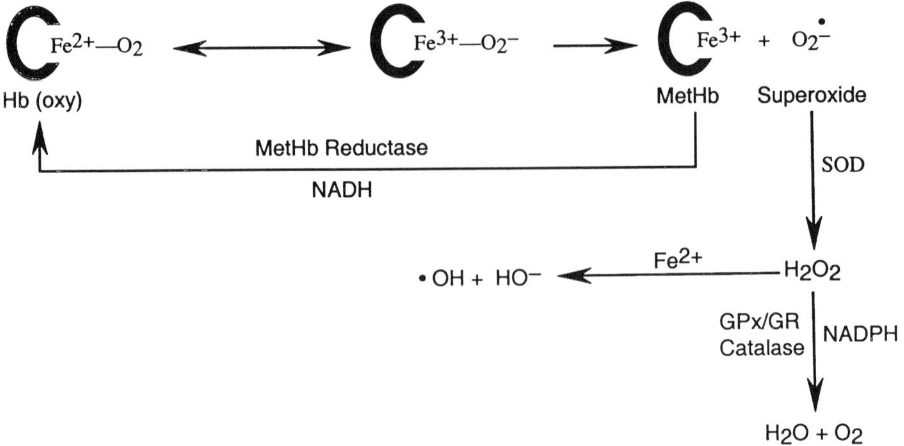

Figure 1. Normal production of active oxygen species within the red cell as a consequence of ferric iron-superoxide anion dissociation from oxyhemoglobin. Hb, oxyhemoglobin; MetHb, methemoglobin; SOD, superoxide dismutase; GPx/GR, glutathione peroxidase/glutathione reductase.

Free Radical Production and Removal in the Normal Red Cell

Much current thinking about senescence in red cells is directed to the role of free radical oxidative reactions in the development of irreversible damage to the cellular architecture, and how such internal changes might lead to an external signal for macrophage sequestration. Of note, the hemoglobin content of the red cell is ca. 5 mM (34 g/dl) which corresponds to a heme-iron content of about 20 mM. Thus, the oxygen-rich environment is also rich in the metal ion that catalyzes active oxygen formation. Indeed, when the ferrous iron of deoxyhemoglobin complexes with molecular oxygen, the oxygen can acquire one of the unpaired electrons of the outer (d) shell of the heme iron such that the complex may be regarded as an equilibrium mixture of $Fe(II)-O_2$ and $Fe(III)-O_2^-$ (Figure 1). Shift to the R (relaxed) conformational state is considered to promote dissociation of the ferric-superoxide anion to yield methemoglobin and superoxide anion radical (for review, see Mansouri and Lurie, 1993). Since between 0.1% and 0.5% of human hemoglobin is converted to methemoglobin every hour (Chiu *et al.*, 1985), this dissociation represents a relatively slow but constant formation of active oxygen within the red cell.

The red cell is well equipped to deal with this constant low-level production of active oxygen because of the activities of superoxide dismutase, catalase, glutathione peroxidase and glutathione reductase. Of importance, both catalase and glutathione peroxidase/reductase have an obligatory requirement for NADPH. Catalase requires NADPH to reactivate complex II (Kirkman and Gaetani, 1984) and glutathione reductase utilizes NADPH as co-substrate.

Glucose-6-Phosphate Dehydrogenase

The sole source of NADPH in the red cell is the hexose monophosphate shunt. This pathway normally consumes only a few percent of cellular glucose-6-phosphate, but under conditions of oxidative stress when the demand for NADPH is increased, the activity of this pathway may increase by up to 20-fold (Pandolfi *et al.*, 1996). The activity of the pathway is regulated by the first enzyme in the sequence, glucose-6-phosphate dehydrogenase (G6PD; EC 1.1.1.49), which in turn is controlled by the NADP+/NADPH ratio. While the precise mechanism underlying this control is not yet completely understood, there appears to be two separate components. Human G6PD consists of 514 amino acids and has binding sites for NADP+ and glucose-6-phosphate (G6P) (Camardella *et al.*, 1995). The monomeric form is catalytically inactive and must be converted to a dimer or tetramer; this conversion requires the presence of NADP+. Thus in the presence of high NADP+, monomer to dimer conversion would promote G6PD and hence shunt activity. A more important control appears to lie in a potent product inhibition by NADPH. The K_i for NADPH inhibition is further dependent on the substrate (G6P) concentration such that the efficiency of inhibition is greatest when the concentration of G6P in the cell is low (Yoshida and Lin, 1973). Collectively, these controls have been estimated to limit the activity of G6PD in the normal cell to less than 1% of the V_{max} value of the enzyme (Yoshida and Lin, 1973).

G6PD deficiency is considered to be the most common human enzymopathology, and is widespread in areas that have a history of endemic malaria. The deficiency is X-linked and hence most marked in males. Recent studies have confirmed that the African variant of G6PD deficiency is associated with significant reduction in susceptibility to severe malaria in both male homozygotes and female heterozygotes (Ruwende *et al.*, 1995). Complete absence of G6PD activity has not been observed, suggesting that it is a housekeeping gene essential for growth and development. G6PD deficiency has been classified by the World Health Organization into three groups: Class I deficiency is rare and is the most severe, with only about 0.1% residual activity in red cells. This class is not associated with malarial areas. Class I patients typically present with chronic non-spherocytic hemolytic anemia. Class II deficiencies have 10% or less residual activity in their red cells and comprise the bulk of human G6PD deficiencies. The Mediterranean and African variants have about 1% and 10% of normal activity, respectively, and show no gross pathology under unstressed conditions. Class II variants are well known to show a profound hemolytic response to a variety of drugs and environmental chemicals as well as to infection and other stressful episodes. Class III variants have 10-60% residual activity in their red cells and are not associated with any specific pathology.

The Mediterranean and African variants are genetically and biochemically well characterized. Both variants are point mutations that do not significantly alter catalytic activity but do decrease the life span of the enzyme in the mature red cell. Both mutations result in amino acid changes in the interface between dimer subunits, and are considered to lead to decreased NADP+ binding, dimer stability and perhaps inappropriate folding (Mason, 1996). Overall, these variants, and the red cells that contain them, have markedly reduced life spans. Mediterranean variant red cells older than 20 days, and African variant red cells older than 55 days, are essentially devoid of the capacity to generate NADPH and are highly susceptible to oxidative stress and the normal aging process.

Mechanisms of Red Cell Injury and Removal

As noted above, oxyhemoglobin is subject to auto-oxidation, generating methemoglobin and superoxide. It has long been known that a variety of chemicals, both naturally-occurring and man-made, can exacerbate the normal low-level oxidative stress in the red cell and induce premature removal of the cells from the circulation. Extensive studies in the 1950s and 1960s (for review, see Beutler, 1969) led to the appreciation that these chemicals caused the transient appearance of methemoglobin and Heinz bodies in circulating red cells, as well as lipid peroxidation, the loss of erythrocytic GSH, and the appearance of glutathione disulfide. Of importance, G6PD-deficient red cells showed markedly enhanced susceptibility to the hemotoxicity of these chemicals. Collectively, these observations led to the formulation of the concept of oxidative stress and the recognition of the key role played by G6PD/NADPH in resisting oxidant damage.

The cellular and molecular mechanism(s) of capture and erythrophagocytosis of senescent erythrocytes are still not known in detail, and a great deal of controversy currently exists regarding what the signal for senescence might be. Extensive studies have been published in support of three competing viewpoints (Bratosin *et al.*, 1998): 1), unmasking of a cryptic antigenic site on the external cell surface, either by proteolysis or by spatial remodeling of membrane proteins induced by cytoskeletal protein changes, leading to opsonization by circulating autologous IgG; 2), unmasking of carbohydrate epitopes by desialylation of glycoproteins, such as glycophorin A, which is recognized by a ß-galectin on the macrophage membrane; and 3), exposure of phosphatidylserine in the outer leaflet of the red cell membrane, leading to unmasking of a lipid epitope.

A number of studies (Morrison *et al.*, 1983; Waugh *et al.*, 1987; Turrini *et al.*, 1991) suggest a key role for the binding of hemoglobin (or hemichrome) to cytoskeletal protein in the mechanism underlying red cell senescence. Morrison *et al.* (1983) detected hemichromes associated with the membranes of red cells just prior to senescence. Low and colleagues (Turrini *et al.*, 1991) showed that hemoglobin binding to the membrane of red cells stimulated the deposition of autologous antibodies and complement, and that these cells were ingested by cultured monocytes. These investigators have raised the hypothesis that formation of protein aggregates of hemoglobin and skeletal proteins (i.e., band 3 protein) on the inside of the cell membrane generates a senescent antigenic site on the exterior surface of the cell. When a critical number of antigenic sites are generated and bound by circulating autologous antibodies, it is thought that the cells are bound by complement (Turrini *et al.*, 1991), which is recognized by immunoglobulin and complement receptors on splenic macrophages, and triggers phagocytosis of the senescent cell (Clark, 1988).

During the last 30 years, a variety of studies have focused on the relative importance of lipid peroxidation *vs.* protein oxidation in the hemolytic process, and the relationship between the initial "hit" and the subsequent removal of the cell from the circulation. The

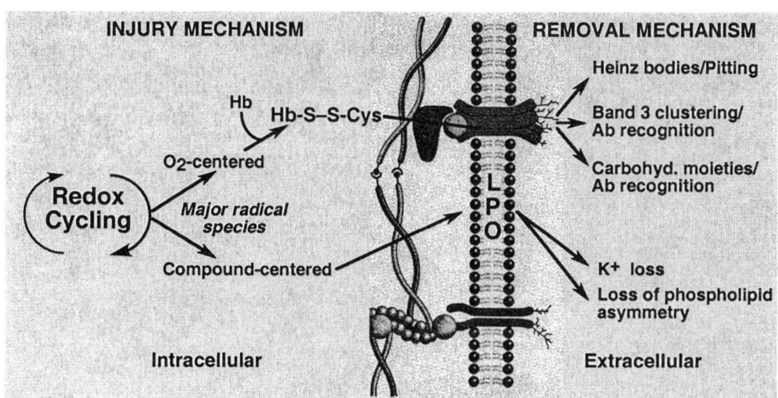

Figure 2. Conceptual relationship between injury occurring within the red cell and external cell surface changes that initiate the process of removal of damaged cells by the spleen. Hb, hemoglobin; Hb-S–S-Cys, disulfide-linked hemoglobin-skeletal protein adduct.

conceptual relationships are illustrated in Figure 2. Phenylhydrazine has been used extensively for this purpose as it is directly hemotoxic and hence its *in vivo* hemolytic activity can be compared with *in vitro* cellular effects. Incubation of red cells with phenylhydrazine is well known to result in the formation of phenyldiazine, which is thought to generate oxygen- and sulfur-centered free radicals, as well as compound-centered free radical species such as the phenyl radical (Goldberg and Stern, 1977; Shetlar and Hill, 1985; Maples *et al.*, 1988). Induction of lipid peroxidation, loss of spectrin and the formation of high molecular weight protein aggregates (>240 kDa) after phenylhydrazine are well documented (Goldberg and Stern, 1977; Arduini and Stern, 1985; Arduini *et al.*, 1989; Ferrali *et al.*, 1992). Of importance, Comporti and colleagues have shown the release of "redox-active" iron by phenylhydrazine and other direct hemolytic agents (Ferrali *et al.*, 1992), and thus provide the basis for a Haber-Weiss-type formation of hydroxyl radical. Thus, phenylhydrazine causes the production of both oxy/thiyl- and compound-centered free radicals, and promotes both lipid and protein oxidation.

To examine the role of lipid peroxidation, Goldberg and Stern (1977) incubated human red cells with phenylhydrazine in the presence of potassium ferricyanide in order to generate extracellular phenyldiazine and derived radical species. They reported that a rapid development of lipid peroxidation products preceding frank lysis and concluded that the active hemolytic agent was a phenyldiazine-derived free radical. Similar frank lysis has been reported by Sato et al. (1999), who incubated human red cells with a membrane-located lipophilic radical initiator and a hydrophilic initiator. These workers concluded that lipid peroxidation alone did not account for the hemotoxicity and that oxidative modification of band 3 protein was also essential. However, the relevance of these observations to *in vivo* hemotoxicity is unclear because the fate of the injured red cell in the intact animal appears to be splenic sequestration rather than frank intravascular lysis. Further, Goldstein *et al.* (1980) observed no enhancement of phenylhydrazine-induced acute-phase decrease in hematocrit

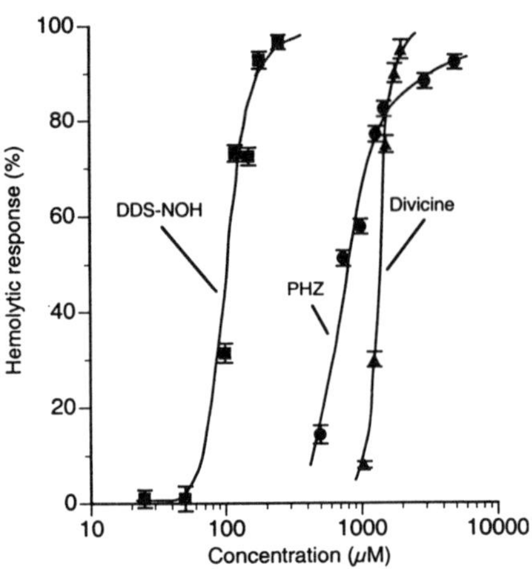

Figure 3. Concentration dependence of the hemolytic activity of DDS-NOH (■), divicine (▲) and phenyl-hydrazine (●). Data points are means ± SD (n=4).

in rats that were deficient in vitamin E, and concluded that in the intact animal, lipid peroxidation was of relatively minor consequence.

RESULTS AND DISCUSSION

Comparison of Dapsone Hydroxylamine, Divicine and Phenylhydrazine Hemotoxicity

We have recently revisited this long-standing problem by comparing the effects of an arylhydroxylamine, dapsone hydroxylamine (DDS-NOH) with those of divicine, an active principle derived from fava beans, and phenylhydrazine, the classical hemotoxicant. In these studies, the hemolytic activity was assessed by incubating ^{51}Cr-tagged rat red cells with various concentrations of these direct-acting hemotoxins for 2 hrs, followed by washing and re-administration of the tagged cells to isologous rats. Serial blood samples were obtained and the rate of loss of tagged cells from the circulation was determined (Harrison and Jollow, 1986). This allowed us to generate concentration response curves (Fig. 3) and hence select concentrations of equal hemotoxicity for the *in vitro* comparison studies. Of importance, the radiolabel appeared almost exclusively in the spleen with relatively little uptake into the liver (Table 1). Since cell fragments are taken up by the RES in general and the Kupffer cells of the liver in particular (Jandl *et al.*, 1956; Harris and Kellermeyer, 1970), the high spleen/liver ratio indicates removal of *intact* damaged cells by splenic sequestration and little or no intravascular frank lysis.

Lipid peroxidation studies (McMillan *et al.*, 1998, and unpublished observations) demonstrated that while phenylhydrazine and, to a lesser extent divicine, induced lipid peroxidation, equi-hemotoxic concentrations of DDS-NOH were without effect. Similarly, incubation with phenylhydrazine resulted in spherocytic echinocytes that contained prominent Heinz bodies, whereas equi-hemotoxic concentrations of DDS-NOH caused non-

600

Table 1. Fate of ^{51}Cr in rats given ^{51}Cr-labeled red cells previously treated *in vitro* with hemolytic agents.

| | % of Administered Dose of [51]Cr | | |
Treatment	Spleen	Liver	Spleen/Liver
DDS-NOH (0.15 mM)	44.0 ± 2.2	10.4 ± 0.8	4.2
Divicine (1.4 mM)	37.0 ± 9.4	5.8 ± 0.7	6.4
Phenylhydrazine (1 mM)	20.6 ± 3.9	1.5 ± 1.4	13.7
Phenylhydrazine (5 mM)	73.3 ± 8.8	5.0 ± 0.1	14.7

[51]Cr-Labeled red cells were incubated with the test compounds at the concentrations indicated for 2 hr in the presence of isotonic phosphate-buffered saline supplemented with glucose. The cells were then washed, resuspended and administered to isologous rats. Animals were sacrificed 2 days after treatment and the radioactivity in the spleen, liver and blood was determined. Data points are means ±SD (n=4).

spherocytic echinocyte formation without evidence of Heinz bodies (Grossman *et al.*, 1992). Since all three treatments resulted in similar rapid splenic uptake on re-administration to isologous rats, we conclude that neither Heinz body formation nor lipid peroxidative damage *per se*, are essential prerequisites for hemolytic response.

Protein oxidation was assessed in two ways. Previous studies with DDS-NOH had shown that exposure of rat red blood cells to hemotoxic concentrations of DDS-NOH caused the formation of hemoglobin adducts with the skeletal protein meshwork (Grossman *et al.*, 1992; McMillan *et al.*, 1995). Adduct formation was reversed by incubating the pre-exposed "damaged" cells with disulfide reducing agents such as cysteamine, indicating that the hemoglobin monomers were linked to the skeletal proteins via disulfide bonds. In addition, hemoglobin monomers were linked via disulfide bonds to form dimers, trimers, and tetramers. The effect of equi-hemotoxic concentrations of DDS-NOH, divicine and phenylhydrazine on the electrophoretic pattern of the skeletal proteins of membrane ghosts prepared from treated red cells are shown in Figure 4. While all three treatments show evidence of adduct formation, the effect is most pronounced with DDS-NOH and least with phenylhydrazine.

The second approach to measurement of the extent of protein oxidation is based on the observation that the disappearance of GSH in DDS-NOH treated red cells is quantitatively accompanied by the appearance of mixed disulfides between glutathione and soluble cellular protein (Jensen *et al.*, 1986). Glutathione-protein mixed disulfide formation parallels hemolytic damage and thus may be used as a quantitative measure of oxidant stress in the red cell (McMillan *et al.*, 1995). Comparison of the efficacy of equi-hemolytic concentrations of DDS-NOH, divicine and phenylhydrazine for mixed disulfide formation indicated that both DDS-NOH and divicine caused extensive and about equal (equivalent to the binding of about 2 mM glutathione) protein oxidation. In marked contrast, mixed disulfide formation after phenylhydrazine was minimal and an order of magnitude less than that seen with the other hemotoxicants.

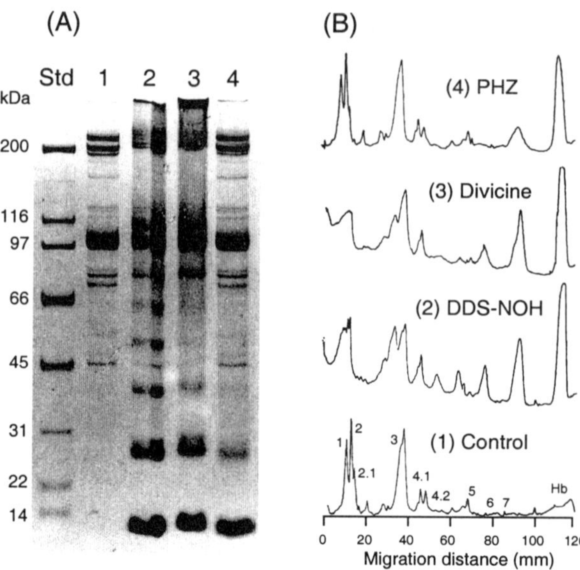

Figure 4. Effect of hemolytic agents on the SDS-PAGE pattern of rat erythrocyte membrane skeletal proteins. Rat red cells were incubated for 2 hr at 37°C in the presence of the vehicle (lane 1), 175 μM DDS-NOH (lane 2), 1.8 mM divicine (lane 3) and 2.5 mM phenylhydrazine (lane 4). (A) Coomassie-blue-stained gel; (B) densitometric scan of the stained gel.

These data are consistent with the possibility that lipid peroxidation, either alone or in combination with protein oxidation, could be crucial for phenylhydrazine hemotoxicity. To examine this possibility further, advantage was taken of the fact that cysteamine can "reverse" the protein oxidation caused by DDS-NOH by removing disulfide-linked hemoglobin monomers from skeletal protein. Provided that this is done at a relatively early time of incubation (i.e., within 30 min), the hemolytic response is substantially reduced and the cells may be regarded as having been "rescued". On the other hand, cysteamine cannot "reverse" any lipid peroxidation that has taken place. Thus if the hemolytic sequence is initiated by lipid peroxidation, cysteamine "rescue" should be ineffectual.

As shown in Table 2, red cells treated with DDS-NOH for 30 min followed by washing and exposure to cysteamine (5 mM) showed significant reduction in their hemolytic response as compared with similarly treated DDS-NOH cells that were washed and incubated without cysteamine. Parallel studies performed with phenylhydrazine indicated that after exposure for 30 min, the cells could not be rescued by cysteamine. These data argue strongly that protein oxidation plays little role in the phenylhydrazine-initiated hemolytic sequence.

SUMMARY AND CONCLUSIONS

As discussed above, the process by which normal senescent red cells are selected for removal from the circulation is the subject of much ongoing research and is not yet well

Table 2. Effect of cysteamine on the hemolytic activity of DDS-NOH and phenylhydrazine.

Treatment	^{51}Cr-T_{50} (days)	% Reduction in ^{51}Cr-T_{50}
Control	7.23 ± 0.37	—
DDS-NOH (0.18 mM)	2.83 ± 0.24	60.8 ± 3.3
DDS-NOH + cysteamine	4.38 ± 0.11	42.2 ± 5.9
Phenylhydrazine (1.5 mM)	1.47 ± 0.01	79.8 ± 0.1
Phenylhydrazine + cysteamine	1.78 ± 0.01	75.4 ± 1.5

^{51}Cr-Labeled red cells were incubated with the indicated concentrations of the test compounds for 30 min at 37°C in glucose-buffered saline. The cells were then incubated with 5 mM cysteamine for 1 hr, washed and administered to isologous rats. ^{51}Cr-T_{50} and percent reduction in ^{51}Cr-T_{50} were determined as described previously (Harrison and Jollow, 1986). Values are means ±SD (n=4).

understood. This in turn creates a problem for studies on the enhanced removal that occurs in xenobiotic-induced hemolytic states; specifically, whether the enhanced removal should be considered as an increase in rate of the normal sequestration mechanism or as an unrelated process, in part or in whole. This difficulty bears directly on the interpretation of much of the mechanistic hemolytic literature. Because of its dual *in vivo* and *in vitro* hemolytic capability, and because of its capacity to induce frank lysis in the incubation mixture, phenylhydrazine has been used extensively as a model compound for mechanistic studies. These data have contributed heavily to our current concepts of how chemicals induce damage in the red cell.

The comparison studies presented above cast doubt on the relevance of many of these phenylhydrazine studies for the *in vivo* hemolytic response. Phenylhydrazine, like divicine and DDS-NOH, shows an overwhelming predominance of uptake into the spleen, as distinct from removal by the RES system in general, as evidenced by relatively low liver uptake. This suggests strongly that damaged cells are removed intact by the spleen and do not lyse or fragment in the general circulation, at least to any significant extent.

The studies with DDS-NOH indicate that neither Heinz body formation nor lipid peroxidation *per se* are essential steps in the process by which damaged red cells are removed from the circulation in the rat. It is not yet clear whether this lack of obligatory involvement of Heinz bodies and lipid peroxidation is peculiar to the arylhydroxylamine-induced hemolytic state or whether it will prove to be of general applicability.

On the other hand, cysteamine failed to reverse the hemolytic damage caused by phenylhydrazine. Since cysteamine "rescued" DDS-NOH treated cells under the same experimental conditions, this observation raises the possibility that protein-thiol oxidation *per se* is also not an obligatory step in the sequence of events leading to premature sequestration.

Clearly, the ratio of lipid to protein oxidation is markedly different in these three examples of hemotoxic compounds. DDS-NOH showed high protein oxidation with no discernible lipid oxidation, divicine showed both high protein and high lipid oxidation, and phenylhydrazine showed high lipid and low protein oxidation. While the significance of these markedly different patterns of injury is far from clear, it seems reasonable to conclude that there is more than one way by which chemicals damage the red cell. It is intriguing that these apparently different chemical insults within the red cell result in a common "message" on the outside of the cell, such that the cell appears as "prematurely" aged. Although the pattern of injury inside the cell may be significantly different, the process by which the three hemotoxic compounds enhance uptake by splenic macrophages may remain the same. That is, there may be a variety of insults sustained within the red cell that lead by different pathways to similar "recognition-specific" changes on the external surface of the red cell. Clearly, comparison of the effects of the three hemotoxic compounds will shed light on both the hemolytic process and on normal red cell sequestration mechanisms.

Acknowledgments

The authors gratefully acknowledge Jennifer M. Schulte for her assistance in the preparation of this manuscript. This work was supported by NIH grants HL30038 and DK47423.

REFERENCES

Arduini, A. and Stern, A., 1985, Spectrin degradation in intact red blood cells by phenylhydrazine, *Biochem. Pharmacol.* **34**:4283-4289.

Arduini, A., Storto, S., Belfiglio, M., Scurti, R., Mancinelli, G. and Federici, G., 1989, Mechanism of spectrin degradation induced by phenylhydrazine in intact human erythrocytes, *Biochim. Biophys. Acta* **979**:1-6.

Beutler, E., 1969, Drug-induced hemolytic anemia, *Pharmacol. Rev.* **21**:73-103.

Bratosin, D., Mazurier, J., Tissier, J. P., Estaquier, J., Huart, J. J., Ameisen, J. C., Aminoff, D. and Montreuil, J., 1998, Cellular and molecular mechanisms of senescent erythrocyte phagocytosis by macrophages. A review, *Biochimie* **80**:173-195.

Camardella, L., Damonte, G., Carrotore, V., Benarri, U., Tonetti, M. and Moneti, G., 1995, Glucose-6-phosphate dehydrogenase from human erythrocytes: Identification of N-acetylalanine at the N-terminus of the mature protein, *Biochem. Biophys. Res. Commun.* **207**:331-338.

Chiu, D., Kuypers, F. and Lubin, B., 1985, Lipid peroxidation in human red cells, *Semin. Hematol.* **26**:257-276.

Clark, M. R., 1988, Senescence of red blood cells: progress and problems, *Physiol. Rev.* **68**:503-554.

Ferrali, M., Signorini, C., Ciccoli, L. and Comporti, M., 1992, Iron release and membrane damage in erythrocytes exposed to oxidizing agents, phenylhydrazine, divicine and isouramil, *Biochem. J.* **285**:295-301.

Goldberg, B. and Stern, A., 1977, The mechanism of oxidative hemolysis produced by phenylhydrazine, *Mol. Pharmacol.* **13**:832-839.

Goldstein, B. D., Rozen, M. G. and Kunis, R. L., 1980, Role of red cell membrane lipid peroxidation in hemolysis due to phenylhydrazine, *Biochem. Pharmacol.* **29**:1355-1359.

Grossman, S. J., Simson, J. and Jollow, D. J., 1992, Dapsone-induced hemolytic anemia: Effect of N-hydroxy dapsone on the sulfhydryl status and membrane proteins of rat erythrocytes, *Toxicol. Appl. Pharmacol.* **117**:208-217.

Harris, J. W. and Kellermeyer, R. W., 1970, *The Red Cell. Production, Metabolism, Destruction: Normal and Abnormal*, Harvard University Press, Cambridge, MA.

Harrison, J. H., Jr. and Jollow, D. J., 1986, Role of aniline metabolites in aniline-induced hemolytic anemia, *J. Pharmacol. Exp. Ther.* **238**:1045-1054.

Jandl, J. H., Greenberg, M. S., Yonemoto, R. H. and Castle, W. B., 1956, Clinical determination of the sites of red cell sequestration in hemolytic anemias, *J. Clin. Invest.* **35**:842-867.

Jensen, C. B., Grossman, S. J. and Jollow, D. J., 1986, Improved method for determination of cellular thiols, disulfides and protein mixed disulfides using HPLC with electrochemical detection, *Adv. Exp. Med. Biol.* **197**:407-413.

Kirkman, H. N. and Gaetani, G. F., 1984, Catalase: a tetrameric enzyme with four tightly bound NADPH molecules, *Proc. Natl. Acad. Sci. USA* **81**:4343-4347.

Mansouri, A. and Lurie, A. A., 1993, Concise review: Methemoglobinemia, *Am. J. Hematol.* **42**:7-12.

Maples, K. R., Jordan, S. J. and Mason, R. P., 1988, In vivo rat hemoglobin thiyl free radical formation following administration of phenylhydrazine and hydrazine-based drugs, *Drug Metab. Disp.* **16**:799-803.

Mason, P., 1996, New insights into G6PD deficiency, *Br. J. Haematol.* **94**:581-591.

McMillan, D. C., Jensen, C. B. and Jollow, D. J., 1998, Role of lipid peroxidation in dapsone-induced hemolytic anemia, *J. Pharmacol. Exp. Ther.* **287**:868-876.

McMillan, D. C., Simson, J. V., Budinsky, R. A. and Jollow, D. J., 1995, Dapsone-induced hemolytic anemia: effect of dapsone hydroxylamine on sulfhydryl status, membrane skeletal proteins and morphology of human and rat erythrocytes, *J. Pharmacol. Exp. Ther.* **274**:540-547.

Morrison, M., Jackson, C. W., Mueller, T. J., Huang, T., Dockter, M. E., Walker, W. S., Singer, J. A. and Edwards, H. H., 1983, Does cell density correlate with red cell age?, *Biomed. Biochim. Acta.* **42**:S107-S111.

Pandolfi, P. P., Sonati, F., Rivi, R., Mason, P., Grosveld, F. and Luzzatto, L., 1996, Target disruption of the housekeeping gene encoding glucose-6-phosphate dehydrogenase (G6PD): G6PD is dispensable for pentose synthesis but essential for defense against oxidative stress, *Embo J.* **14**:5209-5215.

Ruwende, C., Khoo, S. C., Snow, R. W., Yates, S. N., Kwiatkowski, D., Gupta, S., Warn, P., Allsop, C. E., Gilbert, S. C., Peschu, N., Newbold, C. I., Greenwood, B. M., Marsh, K. and Hill, A. V., 1995, Natural selection of hemi- and heterozygotes for G6PD deficiency in Africa by resistance to severe malaria, *Nature* **376**:246-249.

Sato, Y., Sato, K. and Suzuki, Y., 1999, Mechanism of free radical-induced hemolysis of human erythrocytes: Comparison of calculated rate constants for hemolysis with experimental rate constants, *Arch. Biochem. Biophys.* **366**:61-69.

Shetlar, M. D. and Hill, H. A., 1985, Reactions of hemoglobin with phenylhydrazine: A review of selected aspects, *Environ. Health Perspect.* **64**:265-281.

Turrini, F., Arese, P., Yuan, J. and Low, P. S., 1991, Clustering of integral membrane proteins of the human erythrocyte membrane stimulates autologous IgG binding, complement deposition, and phagocytosis, *J. Biol. Chem.* **266**:23611-23617.

Waugh, S. M., Walder, J. A. and Low, P. S., 1987, Partial characterization of the copolymerization reaction of erythrocyte membrane band 3 with hemichromes, *Biochemistry* **26**:1777-1783.

Yoshida, A. and Lin, M., 1973, Regulation of glucose-6-phosphate dehydrogenase activity in red blood cells from hemolytic and nonhemolytic variant subjects, *Blood* **41**:877-891.

HEPATOCARCINOGENESIS IN THE CONTEXT OF STRAIN DIFFERENCES IN ENERGY METABOLISM BETWEEN INBRED STRAINS OF MICE (C57BL/6J AND C3H/He)

Monika Chabicovsky[1], Katrin Staniek[2], Walter Rossmanith[1], Wilfried Bursch[1], Hans Nohl[2] and Rolf Schulte-Hermann[1]

[1]Institute of Cancer Research, University of Vienna, Vienna
[2]Institute of Pharmacology and Toxicology, Veterinary University of Vienna, Vienna

1. INTRODUCTION

Liver neoplasia in mice is one of the most frequent tumor target tissue endpoints observed in 2-year carcinogenicity studies of the US National Toxicology Program (NTP). B6C3F1 mice are the hybrids of male C3H/He, which are highly sensitive (hepatomas in 72-91% at 14 months), and C57Bl/6J females, which exhibit a low susceptibility to spontaneous and chemically induced hepatocarcinogenesis (Heston, 1963; Drinkwater and Bennett, 1991; Dragani *et al.*, 1995; Wastl *et al.*, 1998). Thus, the validity of mouse liver tumor endpoint in assessing the potential hazards of chemical exposure to humans is a controversial issue, since tumor susceptibility varies even within inbred strains of mice. Therefore our studies aimed to elucidate strain specifities that may modulate tumor development to improve the assessment for the relevance of mouse liver tumor induction in toxicological testing of chemicals.

A possible explanation for the high spontaneous incidence of liver tumors in C3H/He mice are previously described strain differences in thyroid hormone metabolism. It has been shown that these mouse strains differ in their capacity to deiodinate iododioxine and iodothyronines. Hepatic deiodination is ~20-fold lower in the C3H/HeJ than in the C57Bl/6J mouse strain, caused by a decreased level of dio1 (5'deiodinase) mRNA. This different quantity of hepatic dio1 mRNA is due to a 21 base pair insert in the dio1 C3H gene located in the 5'-flanking region (Maia *et al.*, 1995a), and causes a 2-3-fold decrease in DIO1 expression in C3H/HeJ mice (Maia *et al.*, 1995b), paralleled by a twofold higher serum free L-thyroxine (T4) level than in C57Bl/6J mice, while free 3,3',5-triiodo-L-thyronine (T3) levels are similar (Berry *et al.*, 1993; Maia *et al.*, 1995b).

In the context of strain specific susceptibility to tumor induction this difference in thyroid hormone metabolism is of special interest since the influence of thyroid hormones on growth and tumor promotion in the liver has been described in cell culture as well as in animal experiments (Leffert and Alexander, 1976; Short *et al.*, 1980; Lemaire *et al.*, 1981; Mishkin *et al.*, 1981; Dombrowski *et al.*, 2000).

2. STRAIN DIFFERENCES IN PHYSIOLOGICAL PARAMETERS

Under physiological conditions a stimulating effect of thyroid hormones on hepatocyte proliferation is regulated within the network of energy metabolism, and increases in oxygen consumption and heat production by thyroid hormones are consistent properties of many mammalian tissues including the liver (Harper and Brand, 1993; Curcio *et al.*, 1999). To find out whether strain specific differences in hepatocarcinogenesis and thyroid hormone metabolism are accompanied by changes in energy metabolism in C57Bl/6J and C3H/He mice, food consumption, body temperature, mitochondrial respiration parameters and hepatic UCP2 expression were compared. The cellular targets for this regulation are mitochondria, which consume more than 90% of cellular oxygen, thereby conserving energy in form of ATP. The tightly coupled process of oxygen consumption (respiration) and energy fixation (ATP synthesis) can become inefficient due to the presence of uncoupling proteins (UCP) in the inner mitochondrial membrane. UCP2 belongs to the family of proton transporters and is widely expressed in numerous tissues and cell types, including the liver (Fleury *et al.*, 1997).

2.1. Food Consumption, Body Weight and Body Temperature

An aspect, severely influenced by thyroid hormones is food consumption. Recording of relative food consumption revealed a significant ($p < 0.001$) strain difference in mice aged 7 - 40 weeks: C3H/He mice ate nearly 30% more food than C57Bl/6J, independent on the age of the mice, whereas body weight was approximately the same in both strains. Further, body temperature was elevated in C3H/He compared with C57Bl/6J mice ($p < 0.05$) (Table 1).

Table 1. Body weight, food consumption and body temperature

	C57Bl/6J			C3H/He		
	n	mean	SD	n	mean	SD
[1] Body weight	48	29.47	2.75	47	29.90	2.99
[2] Relative food consumption	48	14.13*	2.34	47	19.02	2.45
[3] Body temperature	39	37.21**	0.57	37	38.83	0.78

Male C57Bl/6J and C3H/He mice were purchased from the Forschungsinstitut für Versuchstierzucht und -haltung (Himberg, Austria), housed individually under standardized conditions (Macrolon cages, controlled temperature and humidity, natural 12 hours light/dark rhythm, food (Altromin 1321N, Germany) and tap water ad libitum). Body weight and food consumption were recorded at regular intervals (i.e. 2 to 7 times per week) at the beginning of the light period. Mice were sacrificed between 9 and 10 a.m. by cervical dislocation under isoflurane anesthesia (Forane®, ABBOTT Laboratories Ltd., Great Britain), nitrous oxide and oxygen. Body temperature was measured at the beginning of the light period (8:00 a.m.), with a semiconductor sensor (thermocouple type T [Cu-CuNi]), accuracy: ± 0.3°C, resolution: 0.1°C). The sensor was inserted 17 mm into the rectum for about ~30 seconds. Values given in Table 1 demonstrate strain differences in mice 17-20 weeks of age (n: number of animals, SD: standard deviation). [1] Mean of body weight in [g], calculated from a mean of 3-7 measurements per animal. [2] Mean of relative food consumption [g/100g body weight per day], calculated from a mean of 3-7 measurements per animal. [3] Mean of two measurements of body temperature [°C] within one week.

2.2. Mitochondrial Uncoupling

2.2.1. Oxygen Consumption

It is known from the literature (Horst *et al.*, 1989) that pretreatment of isolated livers from hypothyreotic rats with T3 or T4 induces higher oxygen consumption rates. The question arises whether strain specific differences in endogenous thyroid hormone levels could change bioenergetic activities of mouse liver mitochondria. For this purpose the respiratory control parameter determined as the ratio of oxygen consumption rate during active respiration (ATP synthesis) and resting respiration (without ATP production) was compared in both strains. This parameter was significantly (p<0.001) lower in liver mitochondria from C3H/He mice compared with C57Bl/6J mice using glutamate/malate or succinate as substrates. This was predominantly due to increased O₂ consumption during resting respiration, state 4 (without ATP synthesis). ATP/oxygen ratios were also reduced in mitochondria from C3H/He strain (Figure 1).

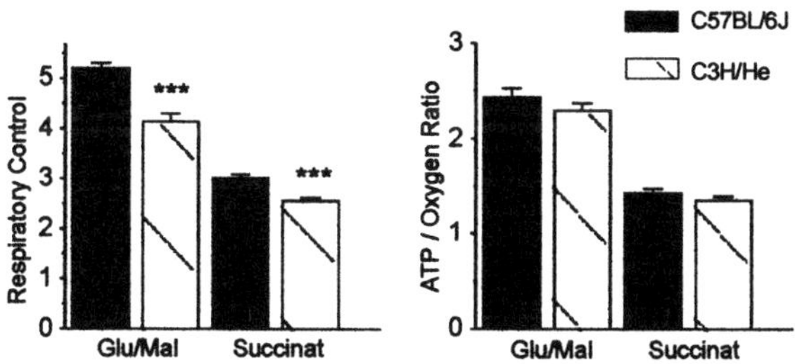

Figure 1. Respiratory control parameter of isolated liver mitochondria from C57Bl/6J and C3H/He mice, 14 weeks of age. The livers were immediately excised and plunged into ice-cold isolation buffer containing 0.25 M sucrose, 20 mM triethanolamine, 1 mM EDTA-Na2, pH 7.4. For details of mitochondrial preparation see (Staniek, 1999). Final concentrations used in the assay: 0.5 mg/ml BSA, 4 mM inorganic phosphate, 5 mM glutamate / 5 mM malate, 10 mM succinate plus 2 µg/ml rotenone, 210 µM ADP, 0.8-1.3 mg/ml mitochondrial protein.

2.2.2. Hepatic UCP2 Expression

UCP2 expression in the liver has been linked to energy expenditure and body weight regulation by increasing thermogenesis and reducing fat stores. Therefore we compared UCP2 expression in the liver of both strains. RNA was isolated from the same livers, as used for respiratory experiments, and strain differences were compared using RNase protection assay. Densitometric analysis revealed a threefold higher expression of UCP2 in the liver of C3H/He mice (Figure 2).

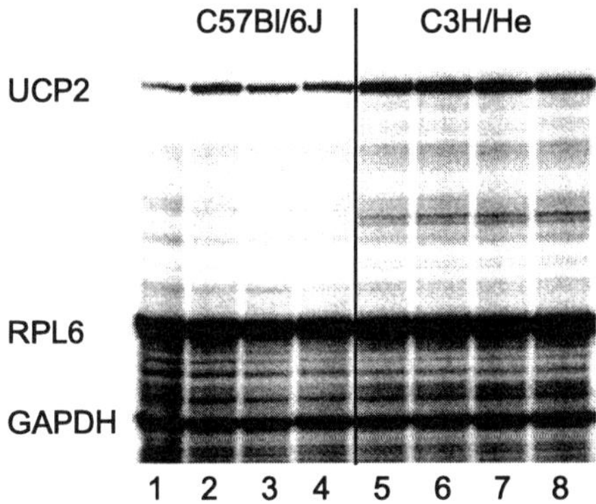

Figure 2. Strain difference in UCP2 RNA between C57Bl/6J and C3H/He mice. Lane 1-4 represent values from C57Bl/6 mice, lane 5-8 are values from C3H/He mice. GAPDH (glycerol aldehyde-3-phosphate-dehydrogenase) and RPL6 (ribosomal protein L-6) were used to assess equal RNA loading of the lanes.

Summarizing data, the food consumption was found to be much higher in C3H/He mice, although body weight did not differ from that observed in C57Bl/6J. Instead we found mitochondrial uncoupling, hepatic UCP2 expression and body temperature elevated in C3H/He mice. These observations suggest that increased energy uptake (food) is consumed for heat production instead of ATP synthesis due to inefficient coupling in mitochondrial respiration, leading thereby to elevated body temperature.

3. MITOCHONDRIAL UNCOUPLING AND HEPATOCARCINOGENESIS

For understanding the carcinogenic mechanism of thyroid hormones mitochondrial respiration may be important. ATP synthesis is coupled to oxygen consumption by the electrochemical proton gradient established across the mitochondrial inner membrane. A leakage of protons across this membrane can occur under different physiological and pathophysiological conditions (Brand *et al.*, 1988). Thyroid hormones, for example, may stimulate O_2 consumption (Horst *et al.*, 1989) and increase proton leak (Porter *et al.*, 1999). Further, UCP2 may uncouple oxidative phosphorylation (Fleury *et al.*, 1997). Mitochondrial uncoupling may influence carcinogenesis by at least two mechanisms.

(1) Mitochondrial respiration is a predominant site of production of reactive oxygen species, and may play a role in the development of hepatic lesions by damaging cellular constituents (Chance *et al.*, 1979). Therefore liver sections were compared to look for strain specifities in the modulation of a biomarker of oxidative damage to DNA (4-HNE). Preliminary data from immunohistochemical staining did not establish a definite strain difference in the liver of mice (data not shown).

(2) ATP depletion, due to the dissipation of energy as heat during the process of uncoupling, is connected to cell viability and cancer. It has been described that an excess of T4 leads to dissipation of ATP. Severe hyperthyroidism is correlated with ATP deficiency due to uncoupling of oxidative phosphorylation (Popovici *et al.*, 1980). More recently UCP2 induction was found to reduce the efficiency of ATP synthesis, enhancing the vulnerability of hepatocytes to necrosis (Rashid *et al.*, 1999). Chavin and colleagues described cells which are highly expressing UCP2 to be more vulnerable to ATP depletion and necrosis (1999). These finding are important for the understanding of a connection between mitochondrial uncoupling and hepatocarcinogenesis, thus, further examinations have to be done to study the meaning of the marked strain difference in mitochondrial uncoupling in the context of ATP depletion.

LITERATURE

Berry, M. J., Grieco, D., Taylor, B. A., Maia, A. L., Kieffer, J. D., Beamer, W., Glover, E., Poland, A., and Larsen, P. R., 1993, Physiological and genetic analyses of inbred mouse strains with a type I iodothyronine 5' deiodinase deficiency, *J. Clin. Invest.* **92:**1517-1528.

Brand, M. D., Hafner, R. P., and Brown, G. C., 1988, Control of respiration in non-phosphorylating mitochondria is shared between the proton leak and the respiratory chain, *Biochem. J.* **255:**535-539.

Chance, B., Sies, H., and Boveris, A., 1979, Hydroperoxide metabolism in mammalian organs, *Physiol. Rev.* **59:**527-605.

Chavin, K. D., Yang, S., Lin, H. Z., Chatham, J., Chacko, V. P., Hoek, J. B., Walajtys Rode, E., Rashid, A., Chen, C. H., Huang, C. C., Wu, T. C., Lane, M. D., and Diehl, A. M., 1999, Obesity induces expression of uncoupling protein-2 in hepatocytes and promotes liver ATP depletion, *J Biol. Chem.* **274:**5692-5700.

Curcio, C., Lopes, A. M., Ribeiro, M. O., Francoso, O. A., Jr., Carvalho, S. D., Lima, F. B., Bicudo, J. E., and Bianco, A. C., 1999, Development of compensatory thermogenesis in response to overfeeding in hypothyroid rats, *Endocrinology* **140:**3438-3443.

Dombrowski, F., Klotz, L., Hacker, H. J., Li, Y., Klingmuller, D., Brix, K., Herzog, V., and Bannasch, P., 2000, Hyperproliferative hepatocellular alterations after intraportal transplantation of thyroid follicles, Am *J. Pathol.* **156:**99-113.

Dragani, T. A., Manenti, G., Gariboldi, M., De Gregorio, L., and Pierotti, M. A., 1995, Genetics of liver tumor susceptibility in mice, *Toxicol. Lett.* **82-83:**613-619.

Drinkwater, N. R., and Bennett, L. M., 1991, Genetic control of carcinogenesis in experimental animals, *Prog. Exp. Tumor. Res.* **33:**1-20.

Fleury, C., Neverova, M., Collins, S., Raimbault, S., Champigny, O., Levi Meyrueis, C., Bouillaud, F., Seldin, M. F., Surwit, R. S., Ricquier, D., and Warden, C. H., 1997, Uncoupling protein-2: a novel gene linked to obesity and hyperinsulinemia, *Nat. Genet.* **15:**269-272.

Harper, M. E., and Brand, M. D., 1993, The quantitative contributions of mitochondrial proton leak and ATP turnover reactions to the changed respiration rates of hepatocytes from rats of different thyroid status, *J. Biol. Chem.* **268:**14850-14860.

Heston, W. E., 1963, Genetics of neoplasia, in: *Methodology in mammalian genetics* (W. J. Burdette, ed.), Holden-Day, San Francisco, pp. 247-268.

Horst, C., Rokos, H., and Seitz, H. J., 1989, Rapid stimulation of hepatic oxygen consumption by 3,5-di-iodo-L-thyronine, *Biochem. J.* **261:**945-950.

Leffert, H. L., and Alexander, N. M., 1976, Thyroid hormone metabolism during liver regeneration in rats, *Endocrinology* **98:**1241-1247.

Lemaire, M., Baeyens, W., de Saint Georges, L., and Baugnet Mahieu, L., 1981, Thyroid influence on the growth of hepatoma HW-165 in Wistar rats, *Biomedicine* **34:**133-139.

Maia, A. L., Berry, M. J., Sabbag, R., Harney, J. W., and Larsen, P. R., 1995a, Structural and functional differences in the dio1 gene in mice with inherited type 1 deiodinase deficiency, *Mol. Endocrinol.* **9:**969-980.

Maia, A. L., Kieffer, J. D., Harney, J. W., and Larsen, P. R., 1995b, Effect of 3,5,3'-Triiodothyronine (T3) administration on dio1 gene expression and T3 metabolism in normal and type 1 deiodinase-deficient mice, *Endocrinology* **136:**4842-4849.

Mishkin, S. Y., Pollack, R., Yalovsky, M. A., Morris, H. P., and Mishkin, S., 1981, Inhibition of local and metastatic hepatoma growth and prolongation of survival after induction of hypothyroidism, *Cancer Res.* **41:**3040-3045.

Popovici, D., Mihai, N., and Urbanavicius, V., 1980, Abnormalities of oxidative phosphorylation due to excess of deficiency of thyroid hormones, *Endocrinologie* **18:**143-147.

Porter, R. K., Joyce, O. J., Farmer, M. K., Heneghan, R., Tipton, K. F., Andrews, J. F., McBennett, S. M., Lund, M. D., Jensen, C. H., and Melia, H. P., 1999, Indirect measurement of mitochondrial proton leak and its application, *Int. J. Obes. Relat. Metab. Disord.* **23 Suppl. 6:**12-18.

Rashid, A., Wu, T. C., Huang, C. C., Chen, C. H., Lin, H. Z., Yang, S. Q., Lee, F. Y., Diehl, A. M., Gong, D. W., He, Y., and Reitman, M. L., 1999, Mitochondrial proteins that regulate apoptosis and necrosis are induced in mouse fatty liver. Genomic organization and regulation by dietary fat of the uncoupling protein 3 and 2 genes, *Hepatology* **29:**1131-1138.

Short, J., Wedmore, R., Kibert, L., and Zemel, R., 1980, Triidothyronine: on its role as a specific hepatomitogen, *Cytobios.* **28:**165-177.

Staniek, K., and Nohl, H., 1999, H(2)O(2) detection from intact mitochondria as a measure for one-electron reduction of dioxygen requires a non-invasive assay system, *Biochim. Biophys. Acta* **1413:**70-80.

Wastl, U. M., Rossmanith, W., Lang, M. A., Camus Randon, A. M., Grasl Kraupp, B., Bursch, W., and Schulte Hermann, R., 1998, Expression of cytochrome P450 2A5 in preneoplastic and neoplastic mouse liver lesions, *Mol. Carcinog.* **22:**229-234.

CHARACTERIZATION OF *hOGG1* PROMOTER STRUCTURE, EXPRESSION DURING CELL CYCLE AND OVEREXPRESSION IN MAMMALIAN CELLS

Andreia Dhénaut,[1][*] Stephan Hollenbach,[2] Inge Eckert,[2] Bernd Epe,[2] Serge Boiteux,[1] and Juan Pablo Radicella[1]

[1]CEA, UMR 217 CEA/CNRS
Département de Radiobiologie et Radiopathologie
BP 6, 92265 Fontenay-aux-Roses, France
[2]Institute of Pharmacy
University of Mainz, Germany

Oxygen radicals are produced in all cells either by the normal cellular metabolism or by the exposure to external mutagens. The reactive oxygen species (ROS) generated can induce DNA damage. Among the principal lesions found in DNA due to ROS is an oxidized form of guanine, 8-oxo-7,8-dihydroguanine (8-oxoG). The biological relevance of this lesion has been unveiled by the study of *Escherichia coli* and *Saccharomyces cerevisiae* genes involved in the neutralization of the mutagenic effects of 8-oxoG (Cabrera et al., 1988; Nghiem et al., 1988; Radicella et al., 1988; van der Kemp et al., 1996). These genes, *fpg* and *mutY* for *E. coli* and *OGG1* for yeast, code for DNA glycosylases. Inactivation of any of those genes leads to a spontaneous mutator phenotype characterized by the exclusive increase of GC->TA transversions. In yeast, the *OGG1* gene encodes a DNA glycosylase/AP lyase that excises 8-oxoG from DNA. A similar activity has been established in mammals by the recent cloning of the human and rodent *OGG1* (reviewed in Boiteux and Radicella (2000)). Consistent with the yeast findings, knock-out mice for *OGG1* accumulate abnormal levels of 8-oxoG in their genome and show a moderately elevated spontaneous mutation rate in non proliferative tissues (Klungland et al., 1999). In human cells, the *OGG1* gene is located on chromosome 3p25 and encodes two forms of hOgg1 protein resulting from an alternative splicing of a single RNA. The α–*hOgg1* protein has a nuclear localization whereas the β–*hOgg1* is targeted to the mitochondrion (Nishioka et al., 1999).

[*] Corresponding author. Tel: 33 1 46 54 98 57; Fax: 33 1 46 54 88 59; Email: silvacos@dsvidf.cea.fr

Biological Reactive Intermediates VI, Edited by Dansette *et al.*
Kluwer Academic / Plenum Publishers, 2001

RESULTS

Cloning and Sequence Analysis of *hOGG1*

We have isolated the genomic sequences of *OGG1* from a human genomic library in phage λ. The genomic fragment, spanning approximately 10.5 kb, covers the hole transcribed region for the nuclear form of the protein as well as a substancial part of the 5' flanking region. The exon/intron structure found agrees with the one recently reported by Ishida et al. (1999) and is very similar in number of exons as well as in sizes to the mouse homolog. The start of transcription was determined by primer extension. This allowed the definition of the promoter sequences of the gene. Two CpG islands and an Alu repeat were identified within the promoter and the 5' sequence of the transcribed region. The lack of TATA or CAAT boxes suggests that *OGG1* is a house keeping gene (Dhénaut et al., 2000, Mutation Research, in press).

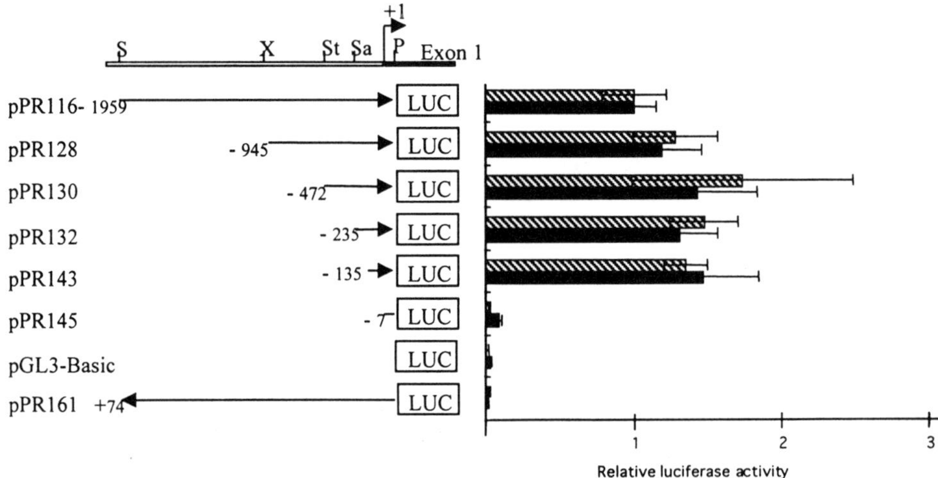

Figure 1. Functional analysis of the *hOGG1* promoter. HeLa (black bars) or H1299 (striped bars) cells were transiently transfected with the *hOGG1* promoter-luciferase fusion constructs represented on the left of the figure together with plasmid pRL-CMV expressing the Renilla luciferase from a constitutive promoter as a transfection efficiency reporter. All constructs used the initiation codon provided by the reporter vector (pGL3-Basic) and the results were normalized for transfection efficiency and the size of the plasmids. pPR161 harbors the same promoter sequences present in pPR116 but in the opposite orientation. The promoter activity is displayed as the ratio between the firefly / Renilla luciferases, normalized with respect to the construct including the 2 kb 5' of the *OGG1* gene : pPR116. Each data point corresponds to the average of five or more independent transfection experiments +/- one standard deviation (taken from Dhénaut et al., 2000 ,Mut. Res., *in press*). S = *SacI*, X = *XhoI*, St = *StuI*, Sa = *SacII*, P = *PpuMI*.

Expression of *hOGG1*

In order to evaluate the transcriptional activity of the 5' region of the *hOGG1* gene, the firefly luciferase (LUC) gene fused to the putative promoter sequences was used as a reporter after transient expression in Hela and H1299 cells. Within the 2.1 kb fragment of upstream sequences analyzed, the most 3' 135 bp have full promoter activity in both cell lines (Figure 1) (Dhénaut et al., 2000).

The human and mouse *OGG1* genes are ubiquitously expressed but the level of expression varies from tissue to tissue. However, *hOGG1's* expression, measured as the transcription from the promoter or as the enzymatic activity in cultured human fibroblast cell lines, does not vary during the cell cycle (not shown). All these results are consistent with *hOGG1* being a house keeping gene.

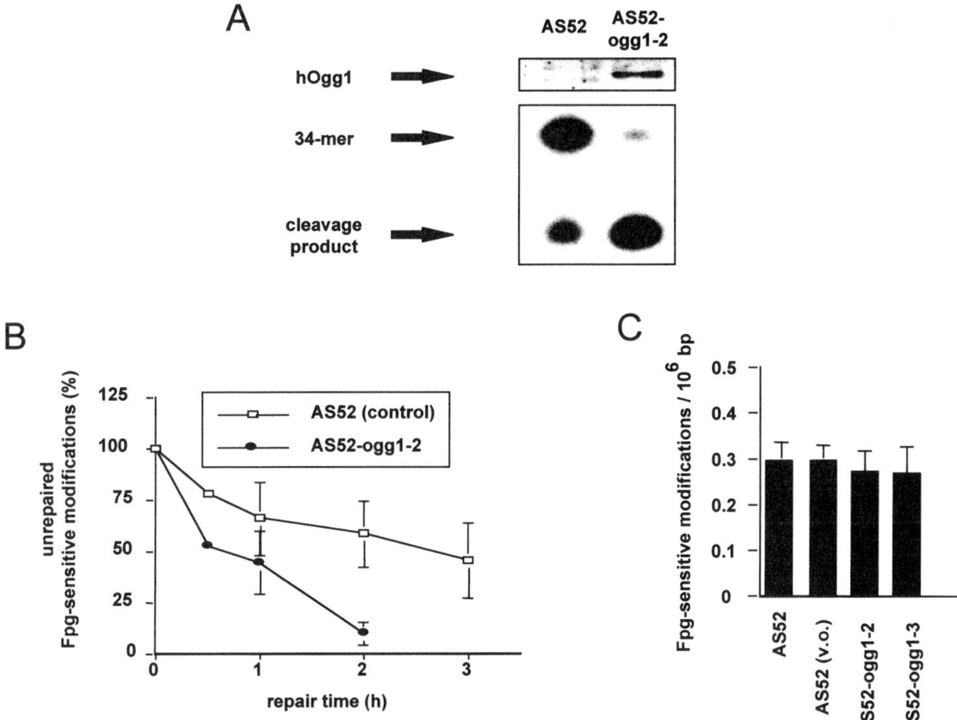

Figure 2. (A) Expression of *hOGG1* in transfected (clone AS52-ogg1-2) and parental AS52 chinese hamster ovary cells detected in total protein extracts by (i) western blotting and (ii) cleavage of a ^{32}P-labelled oligonucleotide containing a single 8-oxoG/C pair, separation by gel electrophoresis and subsequent autoradiography. **(B)** Repair of Fpg-sensitive DNA modifications induced by the photosensitizer RO19-8022 plus light in transfected and parental AS52 cell lines. **(C)** Steady state levels of DNA modifications recognized by Fpg protein in parental and *OGG1*-transfected AS52 cells (taken from Hollenbach et al. (1999)).

In order to study the *in vivo* activity of *hOGG1*, Chinese hamster ovary cell lines were stably transfected to overexpress the nuclear form of the protein. This induced a repair rate of 8-oxoG residues due to damaging agents up to 3-fold more rapid than in the parental cell lines. However, in cells not exposed to damaging agents the steady-state (background) levels of DNA base modifications sensitive to Fpg protein, which include 8-oxoG, were not reduced by the overexpression of hOgg1 protein (Hollenbach et al., 1999) (Figure2).

CONCLUSIONS

The results summarized here show that *hOGG1* has no TATA nor CAAT boxes, has a CpG Island spanning the promoter sequence, is expressed in all cell types and is not regulated during cell cycle. All these characteristics are common to house keeping genes.The ubiquitous expression of the gene is consistent with its antimutator activity by repairing 8-oxoG induced by ROS. Overexpression of the human gene has shown that the Ogg1 protein alone may not be rate limiting in the repair of 8-oxoG in normal growth conditions but it accelerates the repair of substrate modifications induced by exogenous oxidants.

REFERENCES

Boiteux, S., and Radicella, J. P. (2000). The human *OGG1* gene : structure, functions, and its implication in the process of carcinogenesis. Arch Biochem Biophys *377*, 1-8.

Cabrera, M., Nghiem, Y., and Miller, J. H. (1988). mutM, a second mutator locus in *Escherichia coli* that generates G.C--->T.A transversions. J Bacteriol *170*, 5405-7.

Dhénaut, A., Boiteux, S., and Radicella, J. P. (2000). Characterisation of the hOGG1 promoter and its expression during the cell cycle, Mutation Research *461(2)*, 109–18.

Hollenbach, S., Dhénaut, A., Eckert, I., Radicella, J. P., and Epe, B. (1999). Overexpression of Ogg1 in mammalian cells : effects on induced and spontaneous oxidative DNA damage and mutagenesis. Carcinogenesis *20*, 1863-8.

Ishida, T., Hippo, Y., Nakahori, Y., Matsushita, I., Kodama, T., Nishimura, S., and Aburatani, H. (1999). Structure and chromosome location of human *OGG1*. Cytogenet Cell Genet *85*, 232-6.

Klungland, A., Rosewell, I., Hollenbach, S., Larsen, E., Daly, G., Epe, B., Seeberg, E., Lindahl, T., and Barnes, D. E. (1999). Accumulation of premutagenic DNA lesions in mice defective in removal of oxidative base damage. Proc Natl Acad Sci U S A *96*, 13300-5.

Nghiem, Y., Cabrera, M., Cupples, C. G., and Miller, J. H. (1988). The mutY gene: a mutator locus in Escherichia coli that generates G.C--->T.A transversions. Proc Natl Acad Sci U S A *85*, 2709-13.

Nishioka, K., Ohtsubo, T., Oda, H., Fujiwara, T., Kang, D., Sugimachi, K., and Nakabeppu, Y. (1999). Expression and differential intracellular localization of two major forms of human 8-oxoguanine DNA glycosylase encoded by alternatively spliced *OGG1* mRNAs. Mol Biol Cell *10*, 1637-52.

Radicella, J. P., Clark, E. A., and Fox, M. S. (1988). Some mismatch repair activities in *Escherichia coli*. Proc Natl Acad Sci U S A *85*, 9674-8.

van der Kemp, P. A., Thomas, D., Barbey, R., de Oliveira, R., and Boiteux, S. (1996). Cloning and expression in *Escherichia coli* of the OGG1 gene of *Saccharomyces cerevisiae*, which codes for a DNA glycosylase that excises 7,8-dihydro-8-oxoguanine and 2,6-diamino-4-hydroxy-5-N-methylformamidopyrimidine. Proc Natl Acad Sci U S A *93*, 5197-202.

hOGG1 GENE ALTERATIONS IN HUMAN CLEAR CELL CARCINOMAS OF THE KIDNEY : EFFECT OF SINGLE MUTATIONS IN *hOGG1* GENE ON SUBSTRATE SPECIFICITY OF THE hOgg1 PROTEIN

M. Audebert[*], C. Levalois, S. Chevillard, M. Dizdaroglu[1], S. Boiteux and J. P. Radicella.

CEA, UMR 217 CEA/CNRS, Département de Radiobiologie et Radiopathologie, BP 6, 92265 Fontenay-aux-Roses, France.
[1]- Chemical Science and Technology Laboratory, National Institute and Technology, Gaithersburg, MD 20899-8311, USA.
* Corresponding author. Tel: 33 1 46 54 98 57; Fax: 33 1 46 54 88 59;
Email: audebert@dsvidf.cea.fr

INTRODUCTION

Damage to DNA by oxygen free radicals is postulated to cause mutations that are associated to the initiation and / or the progression of human cancers (Breimer, 1990). Oxidative damage-induced mutations can activate oncogenes or inactivate tumor suppressor genes altering the cell growth control (Fearon, 1997). An oxidatively damaged guanine, 7,8-dihydro-8-oxoguanine (8-OxoG) is abundantly produced in DNA as a consequence of the cellular oxidative metabolism, exposure to ionizing radiation or chemical carcinogens. The presence of 8-OxoG in DNA has been shown to be mutagenic since it preferentially pairs to adenine during *in vitro* DNA synthesis (Shibutani et al., 1991) and generates a mutator phenotype characterized by a high frequency of G:C to T:A transversions (reviewed in Boiteux et al. (Boiteux and Radicella, 2000). In mammalian cells, the *OGG1* gene codes for an 8-OxoG DNA glycosylase/AP lyase that has the capacity to excise this oxidized guanine from DNA. Mice lacking a functional Ogg1 protein accumulate abnormal levels of 8-OxoG in their genomes and display a moderately elevated spontaneous mutation rate in nonproliferative tissues. Since the inactivation of the *OGG1* gene in mammalian cells causes a mutator phenotype, it can be expected that cells lacking the Ogg1 activity could have enhanced probability to undergo cancer transformation (Loeb, 1991). The validation of this hypothesis requires the identification of human tumors where both alleles of the *OGG1* are non functional.

Biological Reactive Intermediates VI, Edited by Dansette *et al.*
Kluwer Academic / Plenum Publishers, 2001

The analysis of the sequence changes in p53 suppressor gene showed a bias in favor of GC to TA transversions in lung and kidney cancers (Hollstein et al., 1996). This type of mutations would be expected in cells incapable of eliminating 8-OxoG from their DNA and therefore likely to be deficient in the *OGG1* gene. Furthermore, human chromosome 3p cytogenetic abnormalities and loss of heterozygosity (LOH) have been observed at high frequency in sporadic forms of renal cell carcinoma (RCC), and human *OGG1* gene has been located on chromosome 3p25.

Here we analyze a series of 99 clear cell kidney tumors, and matched normal tissues, for mutations in *OGG1* as well as for LOH in the region of chromosome 3p harboring this gene. Different variant alleles of the Ogg1 protein found in tumors were expressed and purified from *E. coli* and analyzed for their enzymatic activities and substrate specificities.

RESULTS

Tumor analysis

We analysed 99 RCC by DGGE method for the five first exons of the *OGG1* gene and by sequencing PCR amplified fragments of the genomic DNA for the two last ones.

Five different polymorphisms were detected in this series. Two of them, at codons 220 and 323, are silent base pair substitutions not described previously.

Two types of mutations were found in the *OGG1* mRNA from these tumors. Not all can be found in normal tissue. We find 6 cases of deletion or insertion mutants (Table 1). Analysis of the chromosomal sequences failed to show alterations in the genomic DNA. This is consistent with the fact that mutant RNAs result from abnormal splicing of the primary transcript.

Four single base changes were also found. All of them are missense mutations and affect codons 12, 46, 169 and 232 respectively. Two show LOH (R36 and R48), one is non-informative (R37) and the remaining one (R16) is heterozygous in the tumoral sample (Table 1).

Table 1. Alterations in *OGG1* from 99 kidney tumors (from article submitted to cancer research). [a] The alteration found is consistent with an aberrant splicing. [b] Somatic mutations.

Case no.	Exon	Codon	cDNA alteration	Amino acid alteration	LOH
R44[a]	3/4		+GGTGAGTA	frame shift	Yes
R57[a]	5		Δ1056-1085	Δ 250-259	Yes
R69[a]	3/4		+GGTGAGTA	frame shift	Yes
R71[a]	5		Δ exon 5	Δ 250-300	Yes
R94[a]	5		Δ exon 5	Δ 250-300	Yes
R106[a]	5		Δ exon 5	Δ 250-300	Yes
R16[b]	4	232	1002 T to C	Ser to Thr	No
R36[b]	1	12	343 G to A	Gly to Glu	Yes
R37[b]	1/2	46	445 G to A	Arg to Gln	NI[c]
R48[b]	3	169	814 G to A	Arg to Gln	Yes

Study of mutant proteins

After site directed mutagenesis of a bacterial vector allowing the expression of the human *OGG1* open reading frame, we have initiated the analysis of the enzymatic properties of the hOgg1 purified proteins carrying the mutations found in different human tumors (Audebert et al., 2000).

We investigated the substrate specificity of the wild type hOgg1-Ser[326] (wt-hOgg1) and its two mutants with Arg-46→Gln (α-hOgg1-Gln[46]) and Arg-154→His (hOgg1-His[154]) found in human kidney tumor and a gastric cancer cell, respectively.

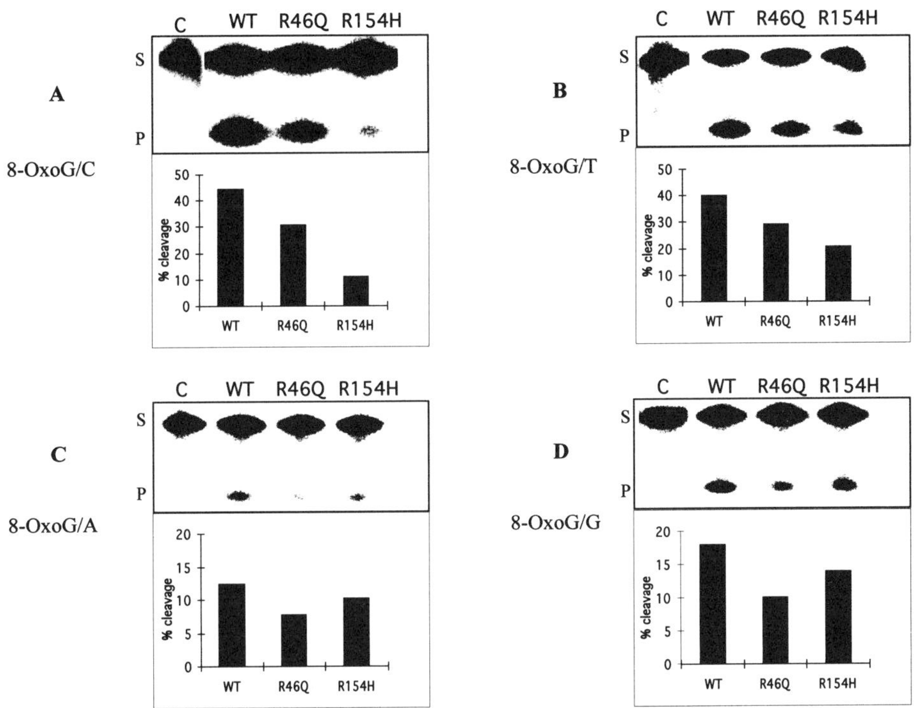

Figure 1. Effect of the base opposite 8-OxoG on the cleavage activity of the Ogg1 proteins. Duplex 34-mers carrying each of the four DNA bases opposite 8-OxoG were incubated for 15 minutes at 37°C in the presence of 10 ng (A) or 20 ng (B, C and D) of each of the proteins. Products of the reaction were resolved on denaturing 20% PAGE (upper panels). S: substrate, P: product. The gels were quantified and the results were ploted (bottom pannels). Note the different scales for pannels A and B, and Pannels C and D. R = Arg, Q = Gln, H = His.

We analyzed the ability of wt-hOgg1 and the two mutants found in human tumors to excise modified bases from irradiated DNA. Using GC/IDMS, 17 modified bases can be identified and quantified in DNA γ-irradiated under N_2O. Of these modified bases, wt-hOgg1 protein and its mutants efficiently excised FapyGua and 8-OxoG. No other modified base was excised significantly under the conditions used in this work. The specificity constants (k_{cat}/K_M) of the excision of FapyGua and 8-OxoG by wt-hOgg1 were significantly greater than those by the mutant enzymes. A significant difference was also noted between the k_{cat}/K_M values of the excision of 8-OxoG by hOgg1-Gln[46] and hOgg1-His[154].

The specific activities for each of the 8-OxoG pairs show that hOgg1-Gln[46] consistently exhibited a diminished activity on all the substrates when compared to wt-hOgg1 (Figure 1). However, there was a more significant loss of the specificity for hOgg1-His[154]. There is a 4-fold reduction of the activity of the mutant form with respect to the wild type when the substrate has an 8-OxoG/Cyt pair (A), whether the activity of the mutant protein on an 8-OxoG/Thy substrate was only 2-fold lower (B). More strikingly, when the base opposite the lesion was a purine, there was no difference in the efficiency of repair between wt-hOgg1 and hOgg1-His[154] (C and D). These results are in agreement with the role proposed for amino acid 154 in the recognition of the base opposite the 8-OxoG.

CONCLUSIONS

The development of a mutator phenotype in a cell has been proposed to be an important factor in the initiation and / or the progression of the carcinogenic process. The antimutator function of the Ogg1 protein together with the localization of the gene to chromosome 3p25 renders *OGG1* a good candidate as a cancer predisposition gene for kidney tissue which have an unusually high rate of oxygen metabolism.

Our analysis of 99 cases of renal cancers has identified four somatic mutations in *OGG1*, along with several polymorphisms. It is important to note that similar studies for other types of cancer failed to identify any mutation for *OGG1* in primary tumors.

Therefore, the presence of the Arg-154→His mutant form in a cell would have a double mutator effect by reducing the repair rate of 8-OxoG in the context of an 8-OxoG/Cyt pair without affecting the repair rate of 8-OxoG/Ade, which would then lead to G/C→T/A transversion and could contribute to the tumorigenesis process.

REFERENCES

Audebert M., Radicella J. P., and Dizdaroglu M. (2000). *Nucleic Acids Res.* (in press).

Boiteux S., and Radicella J. P. (2000). The human OGG1 gene: structure, functions, and its implication in the process of carcinogenesis. Arch *Biochem Biophys. 377:* 1-8.

Breimer, L. H. (1990). Molecular mechanisms of oxygen radical carcinogenesis and mutagenesis: the role of DNA base damage. Mol Carcinog. *3:* 188-97.

Fearon, E. R. (1997). Human cancer syndromes: clues to the origin and nature of cancer, Science. *278:* 1043-50.

Hollstein, M., Shomer, B., Greenblatt, M., Soussi, T., Hovig, E., Montesano, R., and Harris, C. C. (1996). Somatic point mutations in the p53 gene of human tumors and cell lines: updated compilation, Nucleic Acids Res. *24:* 141-6.

Loeb, L. A. (1991). Mutator phenotype may be required for multistage carcinogenesis. Cancer Res. *51:* 3075-3079.

Shibutani, S., Takeshita, M., and Grollman, A. P. (1991). Insertion of specific bases during DNA synthesis past the oxidation- damaged base 8-oxodG. Nature. *349:* 431-4.

PROINFLAMMATORY CYTOKINES mRNA EXPRESSION IN DEPENDENCE OF SUPPRESSIVE EPITOPE OF RETROVIRAL TRANSMEMBRANE p15E PEPTIDE ACTIVATION AT MULTIPLE SCLEROSIS PATIENTS

Irina A. Goldina,* Marina N. Tuzova,* Alexander A. Smagin,** Vitaly V. Morozov,**Michel S. Lubarsky,** Konstantin V. Gaidul,* Vladimir A. Kozlov*

* Institute of Clinical Immunology, Novosibirsk, Russia
**Institute of Clinical and Experimental Lymphology, Novosibirsk, Russia
Yadrintcevskaya 14, Novosibirsk, Russia 630091

INTRODUCTION

Multiple sclerosis (MS) is a chronic inflammatory demyelinating disease of the central nervous system with the autoimmune component. It is characterised by discrete, multifocus lesions of myelin degradation. All phenotypes of lymphocytes and activated macrophages are represented in the inflammatory infiltrates.
Myelin basic protein reactive T cells, macrophages and microglia produce a wide range of factors, wich may influence oligodendrocyte function during the demyelinating process. The complex of proinflammatory cytokines is, probably, the major factor of demyelination (1,3,7).
Different bacterial and viral agents may play a role of triggering factors, predecessing the autoimmune reactions with myelin antigens (9, 15). Not virus, per se, is a key factor in the disease development, but the immune reaction on its persistence (13). In this context, we pay attention on endogenous retroviruses,due to retroviral genes products play the certain role in development of some immunopathilogical conditions. Genomes of vertebrates carry a variety of endogenous retrovirus sequences (ERV), as normal constituents in their genome (6). Most often endogenous retrovirus sequences are defective, some of them contains stop codons in the reading frams, or truncated, preventing their expression as viral particles (14). At the same time, in some conditions, endogenous retroviruses may exert their activity. The observations, that endogenous retroviruses are involved in the regulation of immune response have focussed attention on these genetic elements as factors, potentially involved in the development of autoimmune diseases (4,5). In particular, some interrelations have

Biological Reactive Intermediates VI, Edited by Dansette *et al.*
Kluwer Academic / Plenum Publishers, 2001

been found between rheumatoid arthritis and ERV – 3 (12). A characteristic feature of infections with retroviruses is the incorporation of retroviral DNA into the genome of the infected cells. On the other hand, it is possible that retroviruses have been involved from normal genetic elements.

Several endogenous retroviruses are normally transcribed in peripheral blood mononuclear cells and in brain tissue – HRES – 1, HERV – K, ERV3, ERV3 – Kruppel (10). Others mediate immunosupression through the production of immunosuppressive retrovirus – related proteins, like p15E (2, 16).

The aim of our study was to investigate the level of expression of mRNA for the suppressive epitope of retroviral transmembrane p15E peptide in peripheral blood mononuclear cells of patients with relapsing – remitting form of multiple sclerosis; to estimate the level of expression of proinflammatory cytokines in infected and non – infected patients; to evaluate the results of the treatment by the profile of proinflammatory cytokine expression in peripheral blood mononuclear cells from infected and non – infected patients.

MATERIALS AND METHODS

Samples of peripheral blood were taken from 44 multiple sclerosis patients 23 – 67 years old, 28 of them were women, 16 – men. In all cases it was clinically definite multiple sclerosis, the relapsing – remitting form. The disease duration of these patients ranged from 6 months to 18 years. All of this patients suffer from different degree of neurological disorders.

 Mononuclear cells were obtained from the samples of peripheral blood by Ficoll – Paque gradient density centrifugation (pharmacia, Uppsala, Sweeden). Cytoplasmic RNAs were isolated by the method, described by (8). The predetermined amounts of RNA were treated with reverse transcriptase (Medigen Lab, Russia), in amounts, corresponding to 20U per 1 μg of RNA, in mixture of 10 μl, containing 1 mM each dNTP (Medigen Lab., Russia), 5mM magnesium chloride (Sibensim, Russia), 50 mM potassium chloride (Sibensim, Russia), 5U RNAse inhibitor (Boehringer Mannheim, Biochemica), oligonucleotides mixture (Vector – Best, Russia) and 10mM Tris – HCl, pH 8,3 (Medigen Lab, Russia) (11). After the incubation at 42°C for an 1 h, the samples were boiled for 5 min and then placed on ice. Then the samples were exposed to amplification in reaction mixture, containing 0,4 μM of upstream as well as downstream primers (Vector – Best, Russia), 0,625 U of Taq DNA polymerase (Medigen Lab., Russia), 0,2 mM of each dNTP (Medigen Lab., Russia), 2 mM magnesium chloride (Sibensim, Russia), 50 mM potassium chloride (Sibensim, Russia), 10mM Tris – HCl, ph 8,3 (Medigen, Russia). For the amplification a temperature cycling device from DNA – Technology Lab., Russia, was used. Amplifications were carried out for 35 cycles each consisting of 1 min at 94° C, 1 min at 58° C, and 1 min at 72° C. After the amplification the amounts of 10 μl of the amplified samples were subjected to agarose gel electrophoresis. Visualisation of the amplification products have been done with ethidium bromide.

RESULTS

After the amplification with the β-actin primers and subsequent electrophoretic examination, all samples from peripheral blood mononuclear cells produced the fragment with the expected size. These fragments were registered as bands with the similar intensity of staining in the agarose gel. 20 samples of peripheral mononuclear cells from 44 were amplifiable with the primers. All of them produced a fragment of the expected size (692 bp). This group of patients were immunosuppressed. The quantity of CD3 lymphocytes were 729 ± 121 cells in 1 μl (N 1000 – 2000 cells/μl), CD4 lymphocytes – 375 ± 98 cells/μl (N 600 – 1000 cells/μl), CD8 lymphocytes – 274 ±87 cells/μl (N 300 – 700 cells/μl), B – lymphocytes – 74 ±19 cells/μl (N 100 – 500 cells/μl), the phagocytic activity was reduced. The group of patients without the activation of the endogenous retrovirus genome characterised with higher quantity of T and B lymphocytes, but it is not exceed the normal level. In infected group of patients the level of expression of proinflammatory cytokines was the same: IL – 1 at 16 patients (80%), IL – 6 at 14 patients (70%), TNF - α at 12 patients (60%). Non – infected group of patients demonstrate the lower level of proinflammatory cytokines expression: IL – 1 at 5 patients (20,9%), IL – 6 at 7 patients (29,0%), TNF - α at 5 patients (20,9%). After the treatment the expression of mRNA for the suppressive epitope of retroviral transmembrane p15E peptide observed at16 of patients (36,4%). At patients, expressing the mRNA for the suppressive epitope of retroviral transmembrane p15E peptide, the level of proinflammatory cytokines mRNA expression was: IL – 1 at12 patients (75%), IL – 6 at 11 patients (69%), TNF - α at 15 patients (94%). At non – infected patients the level of mRNA for proinflammatory cytokines expression was significantly lower: IL – 1 at 4 patients (14,2%), IL – 6 at 5 patients (17,9%), TNF - α at 3 patients (10,8%).

At the group of patients, expressed the mRNA for the immunosuppressive epitope of retroviral transmembrane p15E peptide notwithstanding to the treatment, the quantity of lymphocytes remains low: CD3 903±117 cells/μl, CD4 650±114 cells/μl, CD8 290±101 cells/μl, B lymphocytes 138±66 cells/μl. At the same time, the non – infected group of patients demonstrates the increasing of the quantity of lymphocytes to normal level. The increasing of the quantity of lymphocytes and decreasing of the level of mRNA for proinflammatory cytokines expression correlates with the reduction of neurological disorders: the defeats of pyramidal tract, coordination disorders, sensitivity violations, pelvium organs defeats, optical nerve disorders, decreasing of the size of demyelinating lesions.

DISCUSSION

At the present study we have investigated the level of expression of mRNA for the suppressive epitope of retroviral transmembrane p15E peptide and compared the expression of mRNA for proinflammatory cytokines at relapsing – remitting multiple sclerosis patients infected and non-infected with endogenous retrovirus, using RNA – PCR followed by agarose gel electrophoresis and ethidium bromide staining of the products of amplification. Our results indicate that the mRNA for the suppressive epitope of retroviral transmembrane p15E peptide expresses in perpheral blood mononuclear cells at 45% of relapsing – remitting multiple sclerosis patients with high level of proinflammatory cytokines mRNA expression. Due to the important role of the proinflammatory cytokines in myelin damage this data indicates that this group of patients possesses more hard course of the disease. Moreover, we achieved evidence that the expression of mRNA for the suppressive epitope of retroviral transmembrane p15E peptide prevented the decreasing of the mRNA expression for the proinflammatory cytokines in peripheral blood mononuclear cells after the treatment. This observations permit to elaborate new approaches to the therapy of relapsing – remitting forms of multiple sclerosis, direct to the inactivation of endogenous retroviruses and subsequent decreasing of the level of proinflammatory cytokines.

REFERENCES

1. Brosnan C. F., Raine C. S., 1996, Mechanisms of immune injury in multiple sclerosis. Brain Pathol. 6 (3): 243 – 257.
2. Gaidul K. V., Chernukhin I. V., Khaldoianidi S. K., Svinarchuk F. P., e.a. 1995, Functional activity of T-, B-lymphocytes and macrophages during suppression of expression of the env gene from an endogenous retrovirus genome. Mol. Biol., Moscow, 29(3):612 – 618.
3. Hermans G., Stinissen P., Hauben L., Van den Berg – Loonen E., e.a. 1997. Cytokine profile of myelin basic protein – reactive T cells in multiple sclerpsis and healthy individuals. Ann. Neurol. 42: 18 – 27.
4. Krieg A. M., Steinberg A.D. 1990, Retroviruses and autoimmunity. J. Autoimmunity 3: 137 – 166.
5. Kreig A.M., Steinberg A.D. 1992, Endogenous retroviruses: potential etiologic agents in autoimmunity. FASEB J. 6:2537 – 2544.
6. Larsson E., Kato N., Cohen M. 1989, Human endogenous proviruses. Current topics in microbiology and immunology, Volume 148. Berlin: Springer – Verlag, 115 – 132.
7. Ledeen R.W., Chakraborty G. 1998, Neurochemical Researh, Volume 23 (3): 277 – 289.
8. Maniatis T., Fritsch E. R., Sambrook J. 1982, Molecular cloning: a laboratory manual. Cold Spring Harbor Laboratory Press, New York.
9. Nielsen l., Larsen A. M., Munk M., Vestergaard B. F. 1997, Human herpesvirus-6 immunoglobulin G antibodies in patients with multiple sclerosis. Acta Neurol. Scand. Suppl. 169: 76 – 78.
10. Rasmussen H.B., Clausen J. 1997, Possible involvement of endogenous retroviruses in the development of autoimmune disorders, especially multiple sclerosis. Acta Neurol. Scand. Suppl. 169: 32 – 37.
11. Rasmussen H. B., Geny C., Deforges L., Perron H., e. a.1997, Expression of endogenous retroviruses in blood mononucleal cells and brain tissue from multiple sclerosis patients. Acta Neurol. Scand., Suppl. 168:38 – 44.
12. Rubin L.A., Siminovitch K.A., Shi M.N., Cohen M. 1991, A novel retroviral gene assotiation with rheumatoid arthritis. Arthritis Rheum. Suppl. 34:60.
13. Simon J., Neubert W. J. 1996, The pathogenesis of multiple sclerosis: reconsideration of the role of viral agents and defence mechanisms. Med. Hypotheses 46(6): 537 - 543.
14. Stoye J., Coffin J. 1985, Endogenous retroviruses. In: WeissR., Teich N., Varmus H., Coffin J.,ed. RNA tumor viruses, Volume 2. Cold Spring Harbor
15. Talbot P. J., Paquette J.S., Ciurli C. e. a. 1996, Myelin basic protein and human coronavirus 229E cross – reactive T cells in multiple sclerosis. Ann. Neurol. 39 (2): 233 – 240.
16. Turbeville M. A., Rhodes J. C., Hyams D. M., e.a. 1996, Expression of a putative immunosuppressive protein in human tumors and tissues. Pathobiology 64(5): 233 – 238.

DETERMINATION OF NUCLEOTIDE EXCISION REPAIR CAPACITY OF LIVER CELLS *IN VIVO* AND *IN VITRO* BY A CELL-FREE ASSAY

Lydie Sparfel,[1] Sophie Langouët,[1] Alain Fautrel,[1] Bernard Salles,[2] and André Guillouzo[1]

[1]INSERM U456, Détoxication et Réparation Tissulaire, Faculté des Sciences Pharmaceutiques et Biologiques, Université de Rennes I, 35043 Rennes, France
[2]IPBS UMR 5089 CNRS, Institut de Pharmacologie et de Biologie structurale, 31077 Toulouse, France

INTRODUCTION

Exposure to physical and chemical agents may cause severe forms of DNA damage. To counteract mutagenic and lethal effects of DNA damage, organisms have developed diverse enzymatic repair processes. Among these, nucleotide excision repair (NER) represents the main pathway by which mammalian cells remove bulky DNA lesions induced by ultraviolet light, cis-dichlorodiammine platinum (Cis-Pt) and various chemical carcinogens (Sancar *et al.*, 1994 ; Friedberg *et al.*, 1995).

The liver is a chief organ involved in the metabolism of carcinogens ; however its DNA repair capacity is poorly documented. The present study was designed to explore NER activity of hepatic tissues and liver cell lines from rat or human origin, by using a cell-free assay in which a plasmid containing cisplatinum DNA adducts, known to be predominantly repaired by NER (Friedberg *et al.*, 1995), served as a substrate for repair enzymes contained in hepatic protein extracts. NER activity was demonstrated in all protein extracts prepared from either rat and human livers or human hepatoma cell lines.

MATERIALS AND METHODS

Protein extraction

Male Wistar rat livers, peritumoral human hepatic liver fragments resected for secondary tumors and two human hepatoma cell lines : HepG2, a hepatoblastoma cell line (Knowles *et al.*, 1980) and B16A2, a hepatocarcinoma cell line (Glaise *et al.*, 1998) were used.

Biological Reactive Intermediates VI, Edited by Dansette *et al.*
Kluwer Academic / Plenum Publishers, 2001

Briefly, frozen hepatic samples and cell pellets (corresponding to 10^8 cells) were pulverized with a pestle and placed in a hypotonic lysis buffer containing 10 mM Tris-HCl (pH 8.0), 1 mM EDTA and 5 mM dithiothreitol (DTT) according to Manley *et al.* (1983) with minor modifications (Coudore *et al.*, 1997). After a first precipitation of cell lysates by 10% ammonium sulfate and centrifugation at 45 000 rpm for 3 hours, cellular proteins of the supernatants were precipitated by addition of 0.45 g/ml ammonium sulfate and collected by centrifugation. The resulting pellet was then dialyzed for 12 hours at 4°C against a buffer containing 25 mM Hepes-KOH (pH 7.8), 0.1 M potassium glutamate, 2 mM EDTA, 2 mM DTT and 17% glycerol. Extracts were immediately frozen and stored at –80°C. Protein concentrations were determined using the Bradford's procedure (Bradford *et al.*, 1976).

In vitro repair synthesis assay

The damaged pBluescript KS$^+$ plasmid (2958 bp) obtained by a mole to mole reaction with Cis-Pt as previously described (Hansson *et al.*, 1989) and the derivative untreated plasmid pHM (3740 bp) were incubated in a 50 µl reaction mixture (Wood *et al.*, 1988) containing 200 µg protein extracts prepared from either liver tissues or human hepatoma cells. Optimal reaction conditions were obtained by adding 150 mM and 100 mM potassium glutamate to the reaction mixtures containing tissue and cell extracts respectively. After 3 hours at 30°C with 2 µCi of [α-^{32}P]dCTP (Amersham, Buck, UK), plasmids were extracted, then linearized with 5U EcoRV (Promega, Madison, USA) and electrophoresed overnight on a 1% agarose gel containing 0.03% ethidium bromide. The extents of incorporation of radiolabeled deoxynucleotide quantified by densitometry of the autoradiography were normalized to the amounts of recovered DNA and the results were expressed as the mean of arbitrary units +/- S.D.

RESULTS AND DISCUSSION

Most chemical carcinogens are activated in the liver. While metabolism activities of this organ are well documented, little is known about its DNA repair capacity. Only two studies have been previously performed to estimate NER activity in different rat tissues, including liver, by an *in vitro* cell-free assay initially developed with Hela cells (Wood *et al.*, 1988). The liver was found to exhibit more NER activity than the other organs studied, especially lung and spleen (Jones *et al.*, 1994 ; Coudore *et al.*, 1997). The main difficulty in setting up the NER assay for the liver is related to the presence of non-specific nuclease activity in the extracts, thereby requiring the use of well defined experimental conditions. Extensive preliminary studies led us to add different potassium glutamate concentrations to the reaction mixture, respectively 150 mM for tissues and 100 mM for cells. Using these experimental conditions, we obtained reproducible intra- and inter-assay results with all protein extracts investigated, prepared from either liver fragments or hepatoma cells. As evidenced by the greater incorporation of radiolabeled deoxynucleotide into the damaged plasmid compared to the control corresponding, an active NER exists in both rat and human livers (figure 1A and 1B). In both tissues, we measured specific NER activity *versus* non-specific one ; the activity ratio obtained should reach at least 1.5 to demonstrate an efficient NER. The values of activity ratios for human and rat samples were closed, being equal to 2.5 and 3.1 respectively.

By using the same approach, we analyzed NER activity in two human hepatoma cell lines. In order to obtain optimal reaction conditions, HepG2 cells were harvested 4 days after seeding at subconfluency, whereas B16A2 cells were collected after 21 days of confluency, the time required for obtaining the highest levels of the liver-specific functions expressed in this specific cell clone (Glaise *et al.*, 1998). As shown in figure 1C,

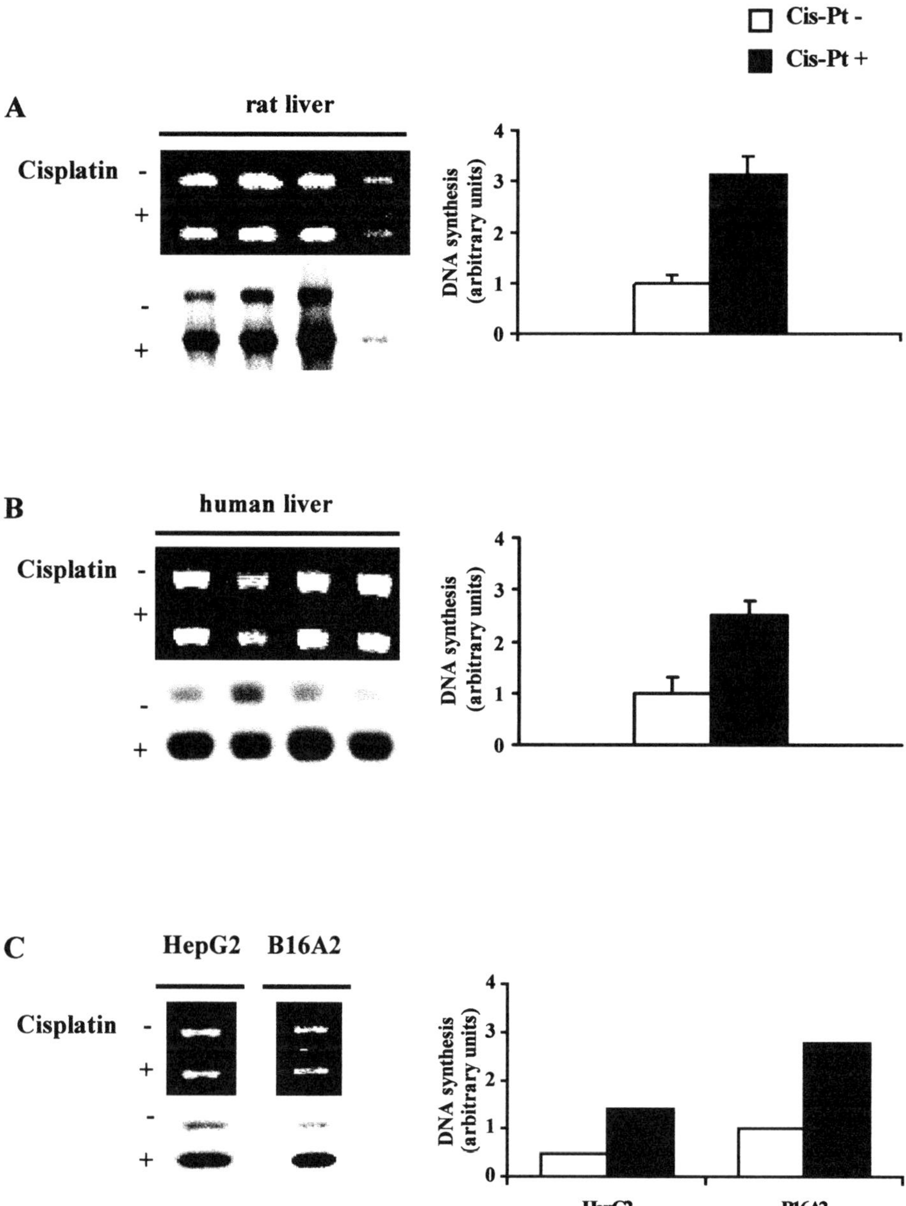

Figure 1 : NER activity in protein extracts from rat livers (A), human livers (B) and HepG2 and B16A2 human hepatoma cell lines (C).

Left : Photograph of the ethidium bromide stained agarose gel and autoradiograph of the dried gel.

Right : Quantification of DNA synthesis into cisplatin damaged plasmid and undamaged plasmid. The extent of incorporation was normalized to the amount of recovered DNA. The results were expressed as densitometry arbitrary units and are the mean +/- S.D. of four different preparations (A and B) or the mean of duplicates (C).

an active NER of platinum DNA adducts was found in both cells. However, while activity ratios were similar for both cell lines (about 3), specific platinum repair activity of B16A2 extracts was 1.8 fold higher than that of HepG2 ones.

Taken altogether, our results indicate that NER activity can be easily measured in liver cells both *in vivo* and *in vitro* using a cell-free assay. This approach was recently used to demonstrate the absence of influence of the chemopreventive agent oltipraz on NER activity in the liver (Sparfel *et al.*, submitted).

ACKNOWLEDGMENTS

We thank O. Musso for providing human hepatic biopsies. This work was supported by the Institut National de la Santé et de la Recherche Médicale. L. Sparfel was a recipient fellowship from the Ministère de la Recherche et de l'Enseignement Supérieur and S. Langouët from the Association pour la Recherche sur le Cancer.

REFERENCES

Bradford, M.M.A., 1976, Rapid and sensitive method for the quantification of microgram quantities for protein utilizing the principle of protein dye binding, *Anal. Biochem.* 72 : 248-254.

Coudoré, F., Calsou, P. and Salles, B., 1997, DNA repair activity in protein extracts from rat tissues, *FEBS Lett.* 414 : 581-584.

Friedberg, E.C., Walker, G.C. and Siede, W., 1995, DNA repair and mutagenesis, *Washington DC : ASM Press.*

Glaise, D., Ilyin, G.P., Loyer, P., Cariou, S., Bilodeau, M., Lucas, J., Puisieux, A., Ozturk, M. and Guguen-Guillouzo, C., 1998, Cell cycle gene regulation in reversibly differentiated new human hepatoma cell lines, *Cell Growth Differ.* 9 : 165-176.

Hansson, J. and Wood, R.D., 1989, Repair synthesis by human cell extracts in DNA damaged by *cis*- ans *trans*-diamminedichloroplatinum, *Nucleic Acids Res.* 17 : 8073-8091.

Jones, S.L. and Harnett, P.R., 1994, Heterogeneous repair of platinum-DNA adducts by protein extracts from mammalian tissues, *Biochem. Pharmacol.* 48 : 1662-1665.

Knowles, B.B., Horve, C.C. and Aden, D.P., 1980, Human hepatocellular carcinoma lines secrete the major plasma proteins and hepatitis B surface antigen, *Science.* 209 : 497-499.

Manley, J.L., Fire, A., Samuels, M. and Sharp, P.A., 1983, *In vitro* transcription : whole-cell extract, *Methods Enzymol.* 101 : 568-582.

Sancar, A., 1994, Mechanisms of DNA excision repair, *Science.* 266 : 1954-1956.

Sparfel, L., Langouët, S., Salles, B. and Guillouzo, A., 2000, Oltipraz does not increase the nucleotide excision repair capacity in the liver, submitted.

Wood, R.D., Robins, P. and Lindahl, T., 1988, Complementation of the *Xeroderma pigmentosum* DNA repair in cell-free extracts, *Cell.* 53 : 97-106.

DIETARY INDUCTION OF PHASE II ENZYMES: A PROMISING STRATEGY FOR PROTECTION AGAINST DNA-REACTIVE INTERMEDIATES IN MAN ?

Hans Steinkellner[1], Sylvie Rabot[2], Fekadu Kassie[1] and Siegfried Knasmüller[1]

[1]Institute of Cancer Research, University of Vienna, Austria
[2]Institute de la Recherché Agronomique, Jouy-en-Josas, France

INTRODUCTION

Over the last decades, strong efforts have been made to identify chemoprotective agents in *in vitro* models with mammalian cells and in laboratory rodents and numerous molecular studies have been carried out to elucidate the mechanisms of prevention of DNA-damage and tumour formation by dietary constituents. Among the most important protective mechanisms identified are inhibition of activating- and induction of deactivating enzymes. Since humans differ substantially in their drug metabolism from laboratory rodents, it is unclear if animal data on chemoprotective agents can be extrapolated to man. The present article gives a brief overview on the impact of dietary factors on the activity and induction of important phase II enzymes in humans.

1. Glutathione *S*-transferases (GSTs)

Glutathione S-transferases are enzymes that play a major role in the detoxification of electrophilic molecules by catalysing the conjugation with glutathione which prevents binding to DNA and leads to enhanced excretion of electrophiles (Hayes and Pulford,1995). More than twenty years ago, Wattenberg and his coworkers showed that induction of GST in laboratory rodents by dietary constituents is associated with a decrease in chemically induced tumours (Wattenberg et al., 1979; Wattenberg, 1983). Based on these findings it was hypothesised that GST induction by dietary constituents might be protective in humans as well. Only a few human intervention studies have been carried out so far to clarify if GST can be induced by nutrients. All these studies are hampered by the low number of participants and the design as matched control studies (for review see Steinkellner et al., 2000).

Recently we have carried out seven crossover intervention studies to study if cruciferous vegetables and coffee can induce GST. 300 g cooked Brassica vegetables or 1l coffee were found in any of the trials. In all participants the GST-μ and GST-π genotypes were determined (Bell et al., 1992; Helzlsour et al., 1998) and no correlations were found between gender, genotypes and GST induction.

Table 1. Induction of GSTs in human intervention studies

Diet	GST-activity[1]	π-GST[2]	α-GST[2]
Brussels sprouts "Cyrus"	↑*	↑↑*	↔
Red cabbage "Roxy"	↑↑*	↑↑↑*	↔
Broccoli "Montop"	↔	↔	↔
White cabbage "Kilor"	↔	↔	↔
Red cabbage "Reliant"	↑	↑↑↑*	↔
Coffee "Brazil"	↑↑*	↑↑↑*	↔

[1]GST activities in plasma were determined according to Habig et al.1974. [2]GST-α and GST-π in plasma were determined with immunoadsorbant assays (Biotrin, Dublin). ↑=induction; ↑↑=induction >50%; ↑↑↑=induction >100 %; ↔ = no induction. * -Indicates statistically significant differences in plasma samples before and after diets in a group of 10 subjects (Wilcoxon test, $p< 0.05$ was considered as statistically significant).

Induction of GSTs could also have adverse effects since they are involved in the activation of dihalogenated carbonhydrates. Besides very rare occupational exposures, these compounds do not play a role in human health.

It has been reported recently that a point mutation in the gene encoding for GST-π which leads to decreased GST activity is strongly associated with cancer in the oral cavity, bladder, lung, testicles, larynx and breast (Taningher et al.1999). This can be taken as an indication that GST induction has indeed a protective effect towards cancer in humans.

2. Uridine-di-phospho-glucuronic acid transferases (UDPGTs)

UDPGTs are microsomal phase II enzymes that conjugate nucleophilic metabolites such as hydroxylated amines with glucuronic acid and thus prevent formation of DNA-reactive metabolites. Some findings showed that these enzymes can be induced in laboratory rodents. For example it was seen that UDPGT activity could be stimulated by the food additive butylated hydroxyanisole (BHA) which appears to be the mechanism for the inhibition of diethylnitrosamine induced gamma-glutamyltranspeptidase-positive foci in rat liver (Sato et al., 1984). UDPGT was also shown to be inducible in laboratory rodents by drugs like phenobarbital or 3-methylcholanthrene (Bock, 1991). We recently carried out an animal study where we found DNA-protective effects of garden cress juice towards the carcinogenic heterocyclic aromatic amine IQ. This protective effect was associated with an induction of UDPGT (Kassie et al., to be submitted).
To our knowledge no direct evidence on induction of UDPGT in man by dietary factors is available and to date evidence of UDPGT induction in humans is only reported from paracetamol (Bock et al., 1987). Further evidence for inducibility comes from observations in human derived hepatoma cells (Doostdar et al., 1988). In a human intervention study we found a pronounced decrease in urinary mutagenicity in *Salmonella typhymurium* YG 1024 after consumption of fried beef when red cabbage was given prior to the hamburger meal (Fig. 1).

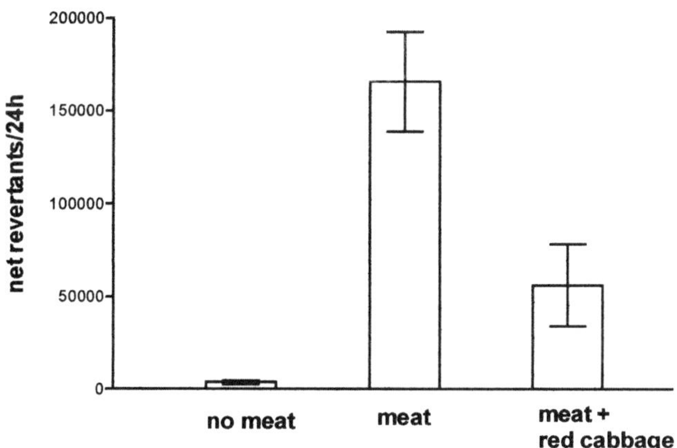

Figure 1: Mutagenicity of XAD2-extracts (Yamasaki and Ames, 1977) of human urine in *S. typhimurium* YG 1024. Bars represent numbers of his+-revertants (means and standard deviations from 3 plates) induced by XAD-2 extracts of 24h urine collected from one individual who for two weeks consumed a diet devoid of fried meat (no meat), after ingestion of 300 g fried beef (meat) and after consumption of 300g red cabbage for 5 consecutive days prior to the meat meal (meat+red cabbage).

A similar human intervention study was carried out by another group with fried chicken and broccoli. Antimutagenic effects were not monitored in this investigation but chemical anlalyses showed a strong increase in glucuronidation products in urine after vegetable consumption (personal communication, Knize, Livermore, U.S.A.). It is likely that the antimutagenic effects seen in our study can be attributed to enhanced glucuronidation of the heterocyclic aromatic amines contained in the meat meals as well.

3. Other enzymes

The role of **N-acetyltransferases (NATs)** in cancer development is ambiguous since they catalyse not only the detoxification but also the formation of carcinogenic metabolites. Rapid and slow N-acetylation in bladder and colon cancer aetiology has been extensively reviewed by Taningher et al., (1999). To our knowledge, no studies on induction of N-acetyltransferases by dietary factors in humans are available. **Quinone reductases (QRs)** play an important role in the inactivation of DNA reactive intermediates (e.g. by reducing semichinones) and can be induced in humans by oltipraz (Kensler 1997) and upon enhanced consumption of broccoli and coffee (Sreerama et al.,1995). **Epoxide hydrolases (EHs)** were enhanced in laboratory animals and humans after phenobarbital and phenytoin administration but no data is available on their induction by dietary constituents (Seidegard and Ekström 1997). No reports on inducibility of **sulfotransferases (SULTs)** which play a role in the metabolism of phenolics and steroid hormones and are also involved in the activation of carcinogens such as heterocyclic aromatic amines are available at present.

OUTLOOK

Overall evidence is accumulating that induction of protective enzymes by dietary constituents is a promising strategy to protect man against DNA-damage and its consequences. A number of animal studies show that induction of specific phase II by vegetable diets and dietary constituents indeed reduce formation of chemically induced tumours. Our findings give clear-cut evidence that GSTs, which are probably the most important detoxyfying enzymes, can be induced in humans by vegetable diets. Our experimental models can also be used for the detection of highly protective vegetable species and cultivars, assessment of optimal food processing procedures and can also help to elucidate if recently developed rapid *in vitro* screening methods (Gerhäuser et al., 1997; Zhang and Talalay, 1994) can be extrapolated to humans.

ACKNOWLEDGEMENTS

This work was funded by an European Union research grant („Effects of Food Borne Glucosinolates"-EFGLU-FAIR CT 973029).

REFERENCES

Bell, D. A., Thompson, C. L., Taylor J., Miller, C. R., Perera, Hsieh, F. L. L., and Lucier, G. W., 1992 Genetic monitoring of human polymorphic cancer susceptibility genes by polymerase chain reaction: application to glutathione transferase µ, *Env Health Perspect.* 98: 113.

Bock, K. W., Wiltfang, J., Blume, R. , Ullrich, D., and Bircher, J., 1987, Paracetamol as a test drug to determine glucuronide formation in man. Effects of inducers and of smoking, *Eur J Clin Pharmacol.* 31: 677.

Bock, K. W., 1991, Roles of UDP-glucuronosyltransferases in chemical carcinogenesis, *Crit Rev Biochem Molec Biol.* 26 (2): 129.

Doostdar, H., Duthie S. J., Gray, A. G., Burke M. D., 1988, The influence of culture medium composition on drug metabolizing enzyme activities of the human liver-derived Hep-G2 hepatoma cell line, *FEBS Lett.* 241:15.

Gerhäuser, C., You, M., Liu, Jinfang, Moriarty, R., M., Hawthorne, M., Mehta, R. G., Moon, R. C., and Pezzuto J. M., 1997, Cancer chemopreventive potential of sulforamate, an novel analogue of sulforaphane that induces phase 2 drug-metabolizing enzymes, *Cancer Res.* 57:272.

Habig, W. H., Pabst, M. J., and Jakoby, W. B., 1974, Glutathione *S*-transferases. The first enzymatic step in mercapturic acid formation, *J Biol Chem.* 249: 7130.

Hayes, J. D., and Pulford, D. J., 1995, The glutathione S-transferase supergene family: Regulation of GST and the contribution of the isoenzymes to cancer chemoprotection and drug resistance, *Crit Rev in Biochem. and Molec Biol.* 30 (6): 445.

Helzlsouer, K.J., Selmin, O., Huang, H-Y., Strickland, P. T., Hoffman, S., Alberg, A.J., Watson, G.W., Comstock, P., and Bell, D., 1998 Association between glutathione -*S*-transferase, M1, P1, and T1 genetic polymorphisms and development of breast cancer, *J Natl Cancer Inst*, 90, 7: 512.

Kensler, T. W., 1997, Chemoprevention of carcinogen detoxication enzymes, *Env Health Perspect.* 105, Suppl 4: 965.

Seidegard, J, and Ekström, G, 1997, The role of human glutathione transferases and epoxide hydrolases in the metabolism of xenobiotics, *Environ Health Persp.* 105, Suppl 4: 791.

Sato, K., Kitahare, A., Yin, Z., Waragi, F., Nishimura, K., Hatayama, I., Ebina, T., Yamazaki, T., Tsuda, H., and Ito, N., 1984, Induction by butylated hydroxianisole of specific molecular forms of glutathione *S*-transferase and UDP-glucuronosyl-transferase and inhibition of development of gamma-glutamyl transpeptidase-positive foci in rat liver, *Carcinogenesis*, 5:473.

Sreerama, L., Hedge, M. W., and Sladek, N. E., 1995 Identification of a class 3 aldehyde dehydrogenase in human saliva and increased levels of this enzyme, glutathione *S*-transferases, and DT-diaphorase in the saliva of subjects who continually ingest large quantities of coffee or broccoli, *Clin Cancer Res.* 1: 1153.

Steinkellner, H., Hietsch, G., Sreerama, L., Haidinger, G., Gsur, A., Kundi, M. and Knasmüller, S., 2000, Induction of glutathione *S*-transferases in humans by vegetable diets. in: *Food and Cancer Prevention IV: Chemical and Biological Aspects.* K.W. Waldron, I.T. Johnson and G. R. Fenwick. ed., The Royal Society of Chemistry, Cambridge.

Taningher, M., Malacarne, D., Izzotti, A., Ugolini, D., and Parodi, S., 1999 Drug metabolism polymorphisms as modulators of cancer susceptibility, *Mutat Res.* 436: 227.

Wattenberg, L. W., 1983 Inhibition of neoplasia by minor dietary constituents, *Cancer Res.* 43: 2448.

Yamasaki, E., and Ames, B. N., 1977, Concentration of mutagens from urine by adsorption with the non-polar resin XAD-2: cigarette smokers have mutagenic urine, *Proc Natl Acad Sc..*74: 3555.

Zhang, Y., and Talalay, P., 1994, Anticarcinogenic activities of organic isothiocyanates: Chemistry and mechanisms, *Cancer Res.* 54: 1976.

CONCURRENT FLAVIN-CONTAINING MONOOXYGENASE DOWN REGULATION AND CYTOCHROME P450 INDUCTION BY DIETARY INDOLES IN THE RAT: IMPLICATION FOR DRUG-DRUG INTERACTIONS

David E. Williams[1], Sirinmas Katchamart[1], Shelley A. Larsen-Su[1], David M. Stresser[2], Shangara S. Dehal[2] and David Kupfer[2]

[1]Department of Environmental and Molecular Toxicology and the Linus Pauling Institute, Oregon State University, Corvallis, OR 97331
[2]Department of Pharmacology and Molecular Toxicology, University of Massachusetts Medical Center, Worcester, MA 01655

INTRODUCTION

Indole-3-carbinol (I3C), a major component of cruciferous vegetables[1], is effective as a chemopreventive agent in a number of animal models[2-6] and is being considered for chemoprevention of breast cancer in women[7]. *In vivo*, in the acidic environment of the stomach, I3C undergoes a series of condensation reactions, resulting in the production of an oligomeric mixture of dimers, trimers and tetramers[8]. A major acid condensation product is 3,3'-diindolylmethane (DIM). DIM is an Aryl Hydrocarbon Receptor (AHR) agonist and induces cytochrome P450 (CYP) 1A1 and 1A2. In addition, DIM functions as an anti-estrogen in MCF-7 cells[9].

Previous work by our laboratory had documented the ability of dietary I3C to markedly inhibit the expression and activity of hepatic and intestinal flavin-containing monooxygenase (FMO) form 1 while inducing CYP1A1 expression over 20-fold[10]. A number of drugs and other xenobiotics are substrates for both monooxygenase systems; however, the products are most often distinct. Specific to the present work, tertiary amines are N-oxygenated by FMO to the tertiary amine N-oxide, whereas, CYP activity results in N-dealkylation.

The hypothesis under test was that, for rats administered I3C or DIM, the *in vitro* hepatic microsomal metabolism of tertiary amines that are substrates for both FMO and CYP should exhibit a marked shift in metabolic profile. The substrates chosen for testing this hypothesis were N,N'-dimethylaniline (DMA), (S)-nicotine (NIC) and tamoxifen (TAM).

METHODS

Male (4 weeks old) Fischer 344 rats were fed powdered AIN76A diets containing 0, 1000 or 2500 ppm I3C or DIM *ad lib* for 4 weeks. After euthanasia, liver microsomes were prepared and FMO1 and CYP1A1 quantified by western blotting. ^{14}C-DMA N-oxygenation

Biological Reactive Intermediates VI, Edited by Dansette *et al.*
Kluwer Academic / Plenum Publishers, 2001

(FMO) and N-dealkylation (CYP) were assayed by HPLC with radioisotope detection as described previously[11]. Metabolites of ^{3}H-(S)-NIC were also resolved and quantified by HPLC with radioisotope detection[12]. ^{3}H-TAM metabolites were separated by TLC and quantified by radioscanning[13].

RESULTS

Consistent with our previous findings, dietary administration of I3C to rats markedly induced liver microsomal CYP1A1 with concurrent down-regulation of FMO1 (Fig. 1A). At the highest dose (2500 ppm), FMO1 levels were decreased by 90%. The degree of CYP1A1 induction could not be quantified as no constitutive expression was observed. The induction appeared to be at maximum levels (210 pmol/mg) at the lowest dose (1000 ppm) of I3C tested.

DIM is more potent in repression of FMO1 compared to I3C, but less potent as a CYP1A1 inducer (Fig. 1B). FMO1 protein levels in liver microsomes from rats fed 1000 or 2500 ppm DIM were reduced by 84 and 97%, respectively. The induction of CYP1A1 at 1000 ppm DIM was relatively modest but reached maximum levels at the 2500 ppm dose level.

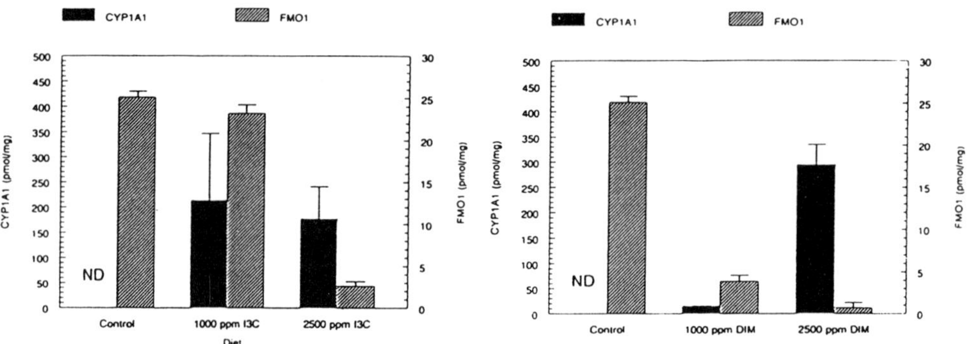

Figure 1. Immunoquantification of Liver Microsomal FMO1 and CYP1A1 from Rats Fed I3C (left panel) or DIM (right panel)

Consistent with the induction of CYP and repression of FMO, the metabolic profile of DMA was markedly shifted in rats fed I3C or DIM from an approximately equal contribution by FMO and CYP to predominantly the CYP-mediated N-dealkylation pathway (Fig. 2). A similar effect was observed with (S)-NIC as the substrate. The FMO-mediated N'-oxygenation of (S)-NIC was essentially eliminated at the highest doses of I3C and at both doses of DIM, whereas, the CYP-mediated pathways were unaltered (data not shown).

TAM N-oxygenation by rat liver microsomes was significantly inhibited in rats fed the I3C or DIM diets (Fig. 3). The ratio of FMO/CYP-mediated TAM metabolism decreased from near unity in animals fed control diet to 0.25-0.35 when the rats consumed the indole-containing diets.

DISCUSSION

The dietary indoles, I3C and DIM, phytochemicals found in cruciferous vegetables, modulate the expression of FMO and CYP monooxygenases in the liver, with the former being markedly repressed and the latter induced. CYP1A1 is the isoenzyme induced to the greatest degree, but more modest induction of CYP1A2, CYP2B1/2 and CYP3A1/2 is also observed[14]. To date, we have only characterized effects on FMO1 in rats. Primates differ from many other mammals in that FMO1 is not the major isoform expressed in adult liver. Sometime after birth, fetal liver FMO1 expression is repressed and FMO3 becomes the major isoform expressed[15].

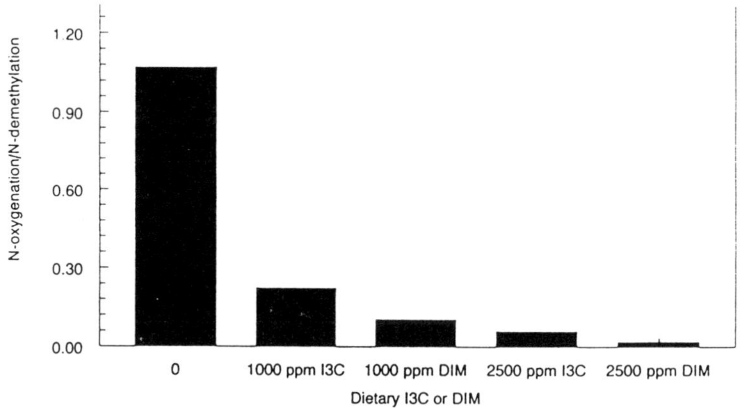

Figure 2. Alterations in FMO/CYP-Mediated DMA Metabolism in Liver Microsomes from Rats Fed I3C or DIM

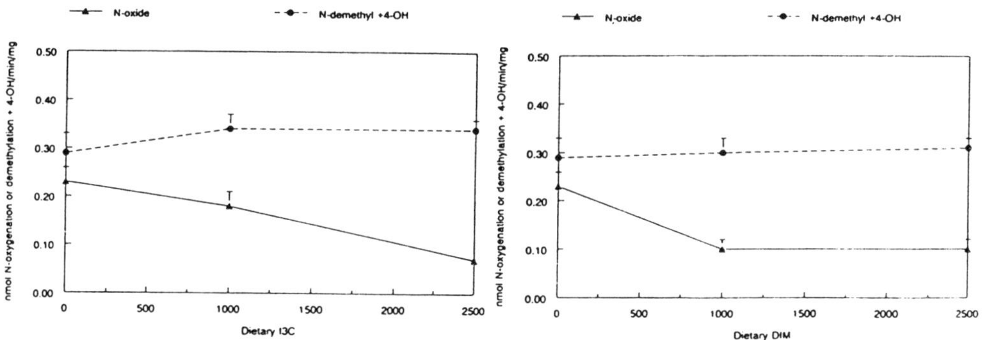

Figure 3. FMO and CYP-Mediated Metabolism of TAM by Liver Microsomes from Rats Fed I3C (left panel) or DIM (right panel)

A number of drugs are substrates for both monooxygenases[16,17]. FMO- and CYP-mediated metabolism is distinct, producing metabolites with markedly different physiological, pharmacological and toxicological properties. Given the dramatic shift in expression of FMO and CYP in rat liver following exposure to I3C or DIM in the diet, we predicted a marked alteration in the metabolic profile of tertiary amines and this was found to be the case especially with DMA and NIC, for which N-oxygenation by FMO was almost eliminated. N-oxygenation of TAM, on the other hand, was only inhibited to a maximum of 55-70%, suggesting other enzymes contribute to TAM N-oxide production.

This phenomenon could represent a potential adverse drug-drug interaction in human liver drug metabolism if FMO3 responds to these indoles in a similar fashion. However, even if FMO3 expression is not inhibited, I3C and DIM have been shown to directly inhibit human

FMO3 catalytic activity with a K_i in the low µM range[18]. For these reasons, we believe that the potential exists for adverse drug-drug interactions to occur in individuals taking certain therapeutic or recreational drugs if they are also supplementing with high doses of these indoles.

ACKNOWLEDGEMENTS

We thank Dr. Dan Ziegler for purified pig FMO1 and antibody, Dr. Mark Shigenaga for radiolabeled (S)-nicotine, BioResponse L.L.C. for DIM and Dr. Mei-Fei Yueh for her assistance with animal care. This work was supported by USPHS grants HL38650 and ES00834.

REFERENCES

1. R. McDanell, A.E. McLean, A.B. Hanley, R.K. Heaney and G.R. Fenwick. Chemical and biological properties of indole glucosinolates (glucobrassicin): a review. *Fd. Chem. Toxicol.* 26:59 (1988).
2. G.S. Bailey, R.H. Dashwood, A.T. Fong, D.E. Williams, R.A. Scanlan and J.D. Hendricks. Modulation of mycotoxin and nitrosamine carcinogenesis by indole-3-carbinol: quantitative analysis of inhibition versus promotion, in *Relevance to Human Cancer of N-Nitroso Compounds*, I.K. O'Neill, J. Chen and H. Bartsch, ed., IARC, Lyon (1991).
3. C.J Grubbs, V.E. Steele, T. Casebolt, M.M. Juliana, I. Eto, L.M. Whitaker, K.H. Dragnev, G.J. Kelloff and R.L. Lubet. Chemoprevention of chemically-induced mammary carcinogenesis by indole-3-carbinol. *Anticancer Res.* 15:709 (1995)
4. T. Kojima, T. Tanaka and H. Mori. Chemoprevention of spontaneous endometrial cancer in female Donryu rats by dietary indole-3-carbinol. *Cancer Res.* 54:1446 (1994).
5. M.A. Morse, S.A. LaGreca, S.G. Amin and F.L. Chung. Effects of indole-3-carbinol on lung tumorigenesis and DNA methylation induced by 4-(methylnitrosamino)-1-(3-pyridyl)-1-butanone (NNK) and on the metabolism and disposition of NNK in A/J mice. *Cancer Res.* 50:2613 (1990).
6. D. Guo, H.A.J. Schut, C.D. Davis, E.G. Snyderwine, G.S. Bailey and R.H. Dashwood. Protection by chlorophyllin and indole-3-carbinol against 2-amino-1-methyl-6-phenylimidazo[4,5-b]pyridine (PhIP)-induced DNA adducts and colonic aberrant crypts in the F344 rat. *Carcinogenesis* 16:2931 (1995).
7. G.Y. Wong, L. Bradlow, D. Sepkovic, S. Mehl, J. Mailman and M.P. Osborne. Dose-ranging study of indole-3-carbinol for breast cancer prevention. *J. Cell. Biochem. Suppl.* 28-29:111 (1997).
8. L.F. Bjeldanes, J.-Y. Kim, K.R. Grose, J.C. Bartholmoew and C.A. Bradfield. Aromatic hydrocarbon responsiveness-receptor agonists generated from indole-3-carbinol *in vitro* and *in vivo*: comparison with 2,3,7,8-tetrachlorodibenzo-p-dioxin. *Proc. Natl. Acad. Sci (USA)* 88:9543 (1991).
9. I. Chen, A. McDougal, F. Wang and S. Safe. Aryl hydrocarbon receptor-mediated antiestrogenic and antitumorigenic activity of diindolylmethane. *Carcinogenesis* 19:1631 (1998).
10. S. Larsen-Su and D.E. Williams. Dietary indole-3-carbinol inhibits FMO activity and the expression of flavin-containing monooxygenase form 1 in rat liver and intestine. *Drug Metabol. Dispos.* 24:927 (1996).
11. M.-Y. Lee, J.E. Clark and D.E. Williams. Induction of flavin-containing monooxygenase (FMO B) in rabbit lung and kidney by sex steroids and glucocorticoids. *Arch. Biochem. Biophys.* 302:332 (1993).
12. D.E. Williams, M.K. Shigenaga and N. Castagnoli, Jr. The role of cytochromes P-450 and flavin-containing monooxygenase in the metabolism of (S)-nicotine by rabbit lung. *Drug Metabol. Dispos.* 18:418 (1990).
13. S.S. Dehal and D. Kupfer. CYP2D6 catalyzes tamoxifen 4-hydroxylation in human liver. *Cancer Res.* 57:3402 (1997).
14. D.M. Stresser, G.S. Bailey and D.E. Williams. Indole-3-carbinol and beta-naphthoflavone induction of aflatoxin B_1 metabolism and cytochrome P450 associated with bioactivation and detoxification of aflatoxin B_1 in the rat. *Drug Metabol. Dispos.* 22:383 (1994).
15. J.R. Cashman. Structural and catalytic properties of the mammalian flavin-containing monooxygenase. *Chem. Res. Toxicol.* 8:165 (1995).
16. D.M. Ziegler. Functional groups bearing nitrogen, in *Metabolic Basis of Detoxication*, W.B. Jakoby and J. Caldwell, ed., Academic Press, NY (1982).
17. D.M. Ziegler. S-Oxygenases, I: chemistry and biochemistry, in *Sulfur-Containing Drugs and Related Organic Compounds*, L.A. Damani, ed., Ellis Horwood, Chichester (1989).
18. J.R. Cashman, Y. Xiong, H. Verhagen, P.J. van Bladeren, S. Larsen-Su and D.E. Williams. *In vitro* and *in vivo* inhibition of human FMO3 in the presence of dietary indoles. *Biochem. Pharmacol.* 58:1047 (1999).

REACTIVE INTERMEDIATES IN BIOLOGICAL SYSTEMS: WHAT HAVE WE LEARNED AND WHERE ARE WE GOING?

F. Peter Guengerich, Hongliang Cai, William W. Johnson, and Asit Parikh

Department of Biochemistry and Center in Molecular Toxicology
Vanderbilt University School of Medicine
Nashville, TN 37232

INTRODUCTION

During the past 25 years and the six meetings in this series, there have been many advances in the study of reactive intermediates and their relevance to health. Many issues of risk assessment remain controversial. Nevertheless, the science has advanced considerably and has enabled real progress in many difficult areas. Two areas of major advancement in the field involve the availability of new technology and approaches—analytical chemistry and recombinant DNA technology (Guengerich, 2000). Since 1975 we have seen extensive development of HPLC, and both NMR and mass spectrometry are orders of magnitude more sensitive. Recombinant DNA technology has not only greatly facilitated studies of the involved enzymology but has also made possible the development of model cellular systems to deal with questions that could not have been addressed before.

This chapter will deal with advances in this laboratory in the past 5-year period with three chemicals: AFB_1,[1] TCE, and heterocyclic amines (esp. MeIQ). These accounts are provided to demonstrate the application of kinetics, mass spectrometry, and genomics to problems with these and other chemicals.

AFB_1

AFB_1 is a well-known carcinogen with important consequences for human health (Busby and Wogan, 1984). AFB_1 requires bioactivation for biological activities and provides an interesting paradigm for the balance of activation and detoxication. Relevant reactions are shown in Figure 1.

Biological Reactive Intermediates VI, Edited by Dansette *et al.*
Kluwer Academic / Plenum Publishers, 2001

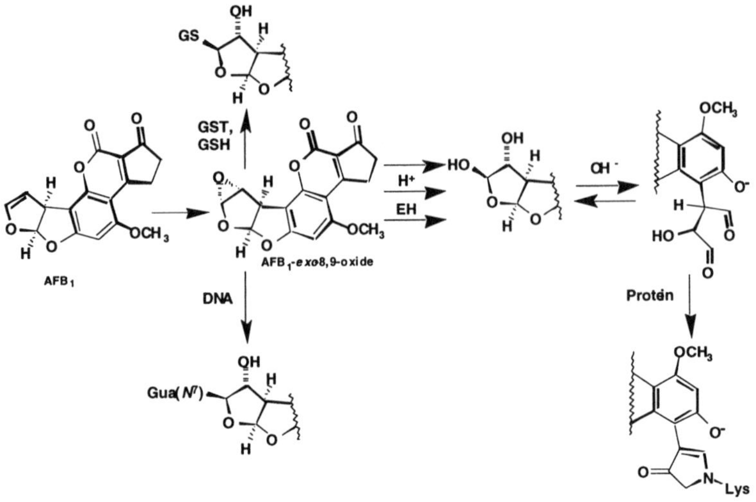

Figure 1. Reactions involved in the metabolism of AFB$_1$.

At the previous meeting in this series (1995) we presented information about the involvement of the enzymes P450 3A4 and GST M1-1 in this reaction, along with the stereoselective modification of DNA by the *exo* and *endo* isomers (Guengerich *et al.*, 1996). Since 1995 we have evaluated the kinetics of most of the reactions shown in Figure 1 and developed a better understanding of the most relevant enzymes in the process.

The oxidation of AFB$_1$ to the *exo* epoxide is catalyzed by several human P450 enzymes, particularly P450 3A4 (k_{cat} = 0.9 min^{-1}, K_m = S$_{50}$ = 27 μM) and P450 1A2 (k_{cat} = 0.15 min^{-1}, K_m = 10 μM) (Ueng *et al.*, 1995; Ueng *et al.*, 1997). These rates must be considered in the context of P450 concentrations in the small intestine and liver, the existence of other detoxication products (i.e., AFQ$_1$ for P450 3A4, AFM$_1$ and *endo* epoxide for P450 1A2), and the potential influence of flavonoids on these reactions (Raney *et al.*, 1992; Ueng *et al.*, 1997).

The non-enzymatic rate of hydrolysis of AFB$_1$ *exo*-8,9-oxide is 0.6 s^{-1} at 23 °C and neutral pH (i.e., $t_{1/2}$ = 1 s) (Johnson *et al.*, 1996). The acid-catalyzed rate is 2 × 10^3 M^{-1}s^{-1}. Rat epoxide hydrolase provided some catalysis of hydrolysis (k_{cat} = 0.17 s^{-1}, K_m = 0.08 μM) but human epoxide hydrolase did not (Johnson *et al.*, 1997). The difficulty of catalysis of hydrolysis of the *exo* epoxide is seen when the rate of non-enzymatic hydrolysis [0.6 s^{-1} (Johnson *et al.*, 1996)] is compared with the rate-limiting step (hydrolysis of enzyme-substrate ester intermediate) in the fastest reported reactions of the enzyme (1 s^{-1}) (Tzeng *et al.*, 1996).

A critical reaction in detoxication involves the conjugation of the *exo* epoxide with GSH. This reaction was not observed directly but k_{cat} and K_d for conjugation by different human GSTs could be estimated by fitting of yield results, as a function of GST concentration (Johnson *et al.*, 1997). The results support a major role of GST M1-1 in humans, consonant with previous studies on polymorphisms in human hepatocytes (Langouët *et al.*, 1995).

The covalent reaction of AFB$_1$ *exo*-8,9-oxide with DNA occurs almost exclusively at the N7 atom of guanine (Essigmann *et al.*, 1977; Iyer *et al.*, 1994) and is probably critical to mutagenesis (Bailey *et al.*, 1996). In our kinetic studies we found that DNA catalyzes the hydrolysis of the epoxide (Johnson and Guengerich, 1997). This effect is postulated to be related to a proton field that surrounds the polyanionic surface of helical DNA (Lamm *et al.*, 1996; Lamm and Pack, 1990). DNA not only facilitates the hydrolysis of the epoxide but also intercolates the epoxide between bases to facilitate the reaction with the guanine (Gopalakrishnan *et al.*, 1989). This reaction has been observed directly (Johnson and

Guengerich, 1997). Thus, the interaction of DNA with epoxide can be described by the parameters $k_{\text{hydrolysis}} = 0.6$ s^{-1} (the supplement to hydrolysis provided by the DNA surface is 0.2 s^{-1} resulting in a total of 0.8 s^{-1}), $K_d = 0.43$ mg ml^{-1} or 1.4 mM monomer equivalents, and $k_{\text{adduction}} = 42$ s^{-1} (Johnson and Guengerich, 1997). These results explain why AFB$_1$ *exo* epoxide can be bound so effectively with DNA despite having a $t_{1/2}$ of only 1 s in water (Johnson and Guengerich, 1997; Johnson *et al.*, 1996).

We have already described the equilibrium of the AFB$_1$ dihydrodiol and the ring-opened dialdehyde product, as well as the kinetics of the individual steps (Figure 1) (Johnson *et al.*, 1996). The reactions are much more complex than depicted in Figure 1, with multiple steps in both directions. An effective k = 2.3×10^3 M^{-1}s^{-1} dominates the base-catalyzed ring opening (Johnson *et al.*, 1996). A pKa of 8.2 can be estimated, arguing that the ring-opened form should not be predominant under physiological conditions. There are two reactions that have not yet been well-characterized. One is the reaction of the dialdehyde with proteins (Sabbioni *et al.*, 1987). Presently the contribution of a direct reaction of the epoxide with protein cannot be ruled out. The other reaction of interest is the reduction of the dialdehyde to a diol, catalyzed by an inducible cytosolic dehydrogenase (Judah *et al.*, 1993), which is thought to play a role in protection from toxicity. The kinetics are complicated by the equilibrium of the dialdehyde.

TCE

TCE is an important industrial solvent and one of the most ubiquitous contaminants of waste dumps (Brüning and Bolt, 2000; Huff, 1971; Kimbrough *et al.*, 1985; Bruckner *et al.*, 1989). The compound is known to be a liver and kidney carcinogen when administered to rodents at high concentrations (Brüning and Bolt, 2000). Several pathways of bioactivation are of interest. The GST-catalyzed activation is very slow and its contribution to human health is controversial (Brüning and Bolt, 2000; Goeptar *et al.*, 1995). Our interest has been directed towards TCE oxide, the major reactive product of oxidative metabolism.

In 1982 we reported that the major oxidation product of TCE, chloral, was produced by P450s in a reaction that did not involve the epoxide as an intermediate (Miller and Guengerich, 1982). TCE oxide was proposed to be derived from a separate collapse of a P450-TCE intermediate. However, events related to how covalent modification of biological nucleophiles occurs remained uncharacterized (Miller and Guengerich, 1983).

TCE oxide was synthesized and its degradation was characterized by tracking the incorporation of ^{18}O and ^{2}H labels from H$_2$O by mass spectrometry. On-line HPLC/MS proved most effective in analyzing some of the products that were difficult to derivatize, e.g. formic acid. The mechanism shown in Figure 2 was deduced (Cai and Guengerich, 1999a). Key support for this pathway included the incorporation of two oxygen atoms from H$_2$O into HCO$_2$H and none into CO and the incorporation of one H$_2$O oxygen atom into Cl$_2$CHCO$_2$H and glyoxylic acid. ^{18}O and ^{2}H labeling were also used to deduce the mechanism of the reaction of TCE Oxide with Lys to generate the amide products (Figure 3) (Cai and Guengerich, 1999a).

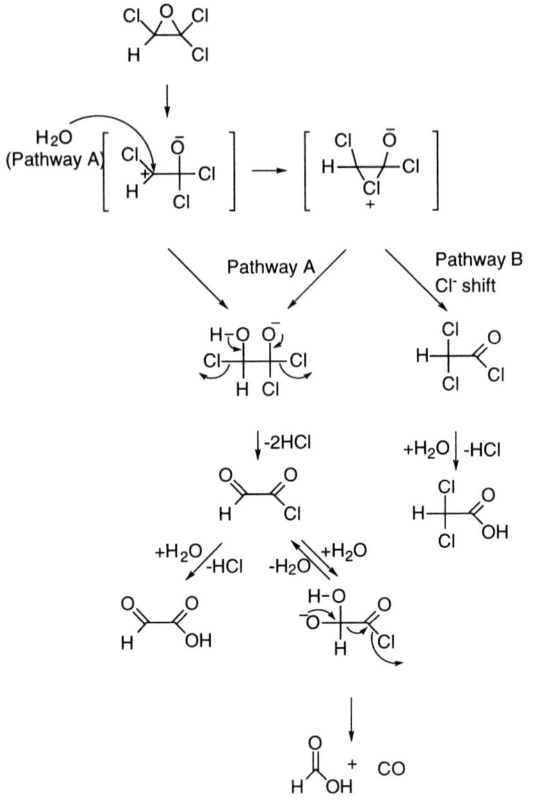

Figure 2. Hydrolysis of TCE oxide (Cai and Guengerich, 1999a).

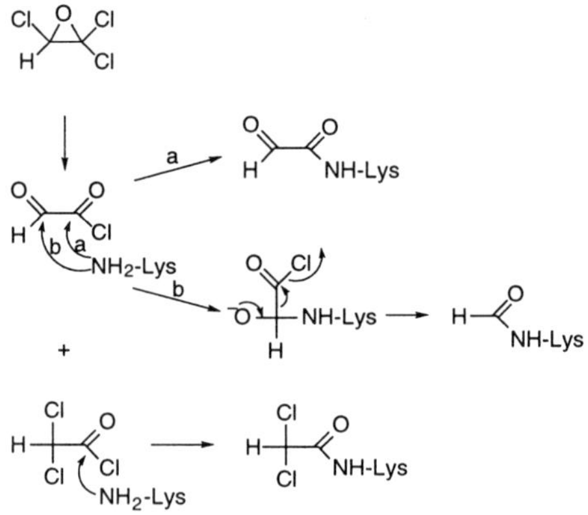

Figure 3. Reaction of TCE oxide with Lys (Cai and Guengerich, 1999a).

The work is of interest in that direct attack of a nucleophile on TCE oxide can yield only unstable geminally-substituted products, and the solvolysis of the epoxide, to form reactive acyl chlorides, precedes reaction with nucleophiles (Cai and Guengerich, 1999a). Further evidence for this view is provided by the effect of GSH in blocking protein binding, which requires the free thiol (Cai and Guengerich, 1999b). We also found that GSH does not affect the $t_{1/2}$ of TCE oxide ($t_{1/2}$ = 12 s in neutral solution at 23 °C), arguing against "direct" conjugation (Cai and Guengerich, 1999b).

HPLC/MS (electrospray) was also used to measure the Lys adducts found in proteins, following enzymatic digestion (Figure 4).

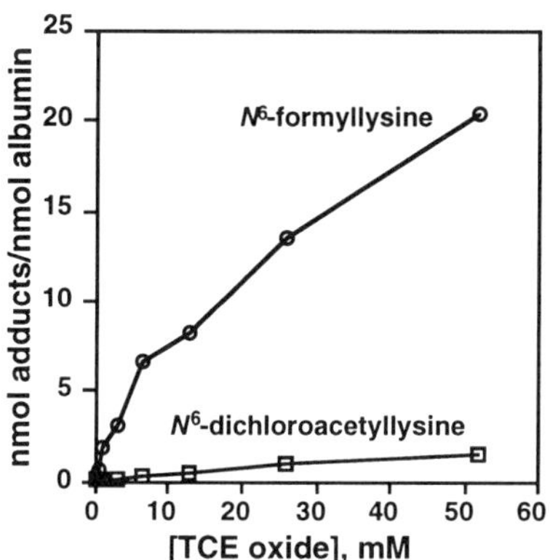

Figure 4. Lys adducts found in BSA following reaction with TCE oxide (Cai and Guengerich, 1999b).

N^6-Glyoxyl Lys was not found in any proteins examined (Cai and Guengerich, 1999b). The ratio of N^6-formyl Lys to N^6-dichloroacetyl Lys varied among different proteins but the former was always the major Lys adduct (Cai and Guengerich, 1999b).

Preliminary work with serine and tyrosine O-acyl esters demonstrated their lability. We utilized electrospray MS to estimate the fraction of non-Lys adducts. Reactions of small proteins (e.g., insulin) with TCE oxide and immediate analysis allowed direct observation of formyl and dichloroacetyl adducts (Figure 5).

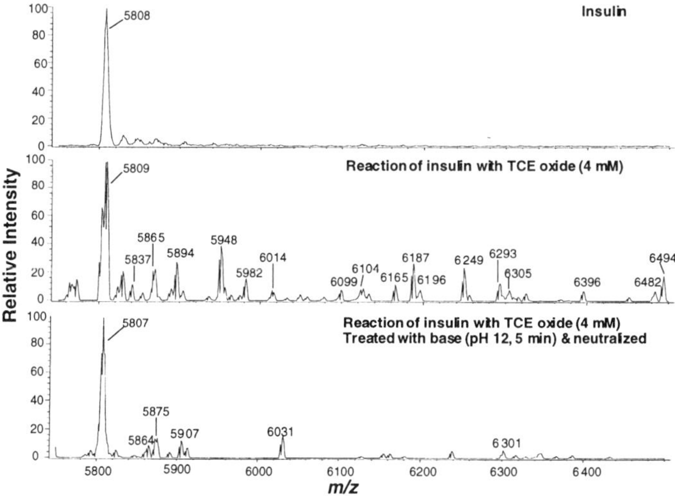

Figure 5. Electrospray mass spectra of insulin. (A) Insulin. (B) Insulin treated with 4 mM TCE oxide. (C) Insulin treated with 4 mM TCE oxide and then pH 12 buffer (for 5 min).

Treatment of the sample before analysis with mild base (pH 12) removed 75-80% of the adducts, indicating that this fraction represents unstable ester adducts. When insulin or ACTH (residues 1-24) was modified with TCE oxide and analyzed by MS as a function of time, a $t_{1/2}$ of ~ 1 h was estimated for the release of putative ester thioesters (at pH 7, 23 °C). Aldolase, which contains a critical active site Lys (Rutter *et al.*, 1961), was treated with sufficient TCE oxide to inhibit 50% of activity; the loss of activity persisted (Figure 6A).

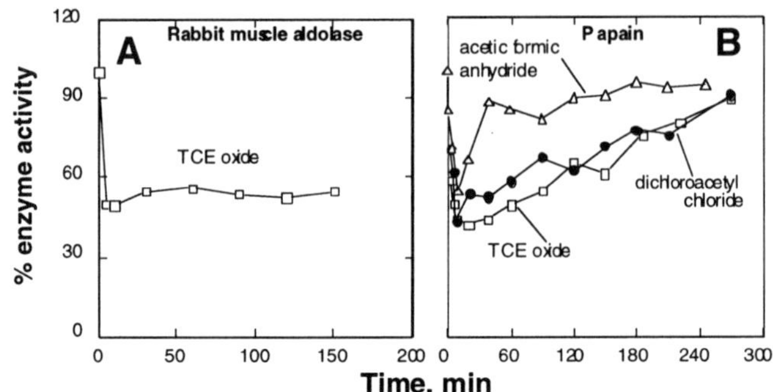

Figure 6. Catalytic activity of two model enzymes as a function of time following treatment with TCE oxide. (A) Rabbit muscle aldolase. (B) Papain.

Papain contains a critical cysteine residue (Slama *et al.*, 1984). When this protease was treated either TCE oxide or other acylating agents (dichloroacetyl chloride, acetic formic anhydride), there was a recovery with a $t_{1/2}$ of ~ 1 h (Figure 6B), consistent with the loss of unstable adducts from other proteins seen with MS.

Varying reports on the genotoxicity of TCE have appeared (Bruckner *et al.*, 1989). Treatment of the four deoxyribonucleosides with TCE oxide and subsequent HPLC/UV analysis indicated only an adduct with deoxyguanosine. MS indicated the addition of a formyl group (+28), and the similarity of the UV spectrum with that of dGuo suggests that the site of modification is the N2 atom (or possibly a ribose hydroxyl). We treated a synthetic oligonucleotide (8-mer, containing 2 residues each of A, C, G, and T) with TCE oxide; direct MS analysis showed the adduction of two formyl groups (+56). The adducts were unstable, and 2/3 of the labeling was lost over 2.5 h, suggesting a $t_{1/2}$ of ~ 30 min in the oligomer (results not shown).

Collectively, these and other results obtained with TCE show the role of acyl chlorides derived from hydrolysis of TCE oxide as electrophiles, the reaction with Lys to yield stable adducts, and the major fraction of esters, thioesters, and possibly other unstable amino acid and nucleoside adducts that are lost rapidly. Other work showed that (rat) P450 2B1 was more effective in generating TCE oxide than human P450 2E1 or other human P450s (Cai and Guengerich, 1999b). A general conclusion is that at least some of the rodent results overestimate the risk of TCE. Another interesting aspect is the rather transient nature of much of the damage.

HETEROCYCLIC AMINES AND GENOMICS

Today there is considerable interest in the areas of pharmacogenomics and toxicogenomics. In the area of drug metabolism, dramatic differences have been observed in the pharmacological effects of drugs due to polymorphisms in P450s and other enzymes involved in drug metabolism (Meyer, 1994; Evans and Relling, 1999). Although many allelic variants are known with many of the enzymes (e.g. see <http://www.imm.ki.se/cypalleles>) the fraction of genetic polymorphisms that actually result in alterations of function is relatively low, probably about 10%. Thus, traditional strategies are deficient in that most of the effort put into sequence analysis is wasted (Figure 7A,B).

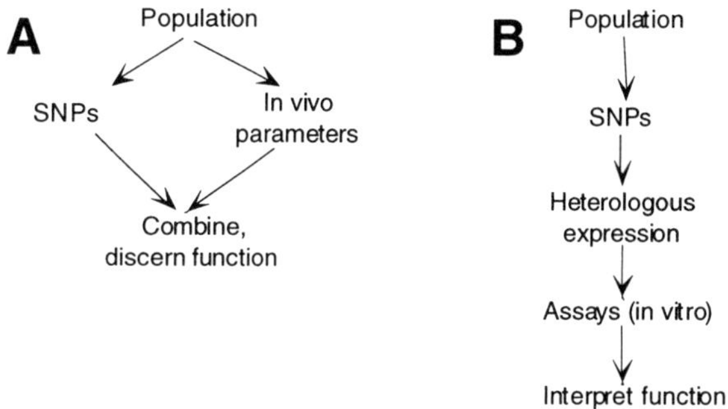

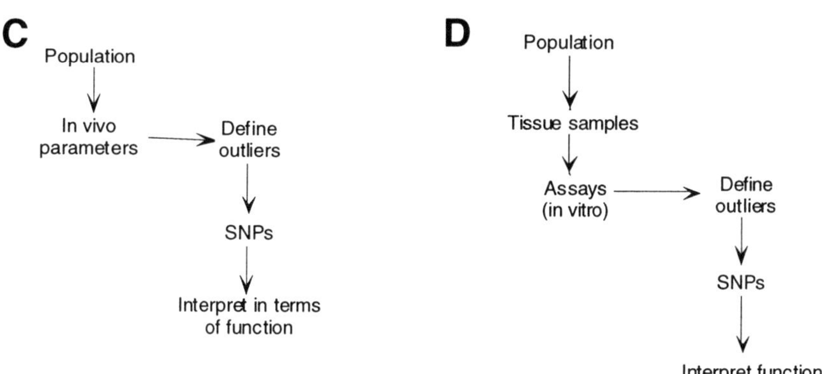

Figure 7. "Traditional" strategies for identification of functional polymorphisms in enzymes: (A) *in vivo*, (B) *in vitro*. Approaches working "backwards" from outlines to sequence analysis. (C) *in vivo*. (D) *in vitro*.

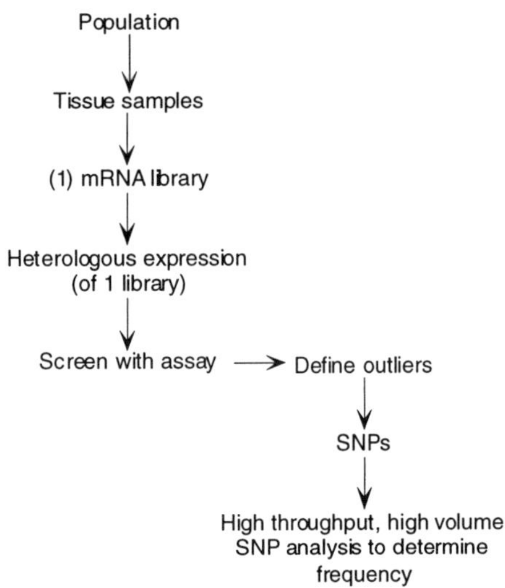

Figure 8. A strategy for *in vitro* identification of functional polymorphisms in coding regions of genes.

Moreover, when SNPs are in protein coding regions there is the added complication that function A may change but function B may not (e.g., the oxidation of individual substrates may be affected differently).

There are some ways to work "backwards" in SNP analysis, in order to reduce the fraction of wasteful sequence and expression analysis (Figure 7C,D). The problem with the *in vivo* approach (Figure 7C) is that it is not prospective and cannot be used to predict problems. Major problems with the *in vitro* version (Figure 7D) are the need to process and analyze large numbers of tissues, the lack of sensitivity with heterozygotes (also for Figures 7A,C), and the influence of any tissue deterioration. We propose an alternate approach in Figure 8.

As in the case of Figure 7D, this approach also begins with a large collection of tissue samples but utilizes an mRNA library made from the pool of samples. Expression of the library in a simple vector (e.g. bacteria) allows analysis of individual alleles (not heterozygotes) and screening for function prior to sequence analysis.

We have begun to utilize this approach, coupling it with previous work on large-scale screening done in this laboratory (Parikh *et al.*, 1999). An application of this work is seen in consideration with heterocyclic amines. These compounds are highly mutagenic, cause cancer in rodents and primates, and are consumed by many people in their diets (Sugimura, 1986). Knowledge that these compounds are activated primarily by P450 1A2 (N-hydroxylation) and subsequent acetylation has been available for a number of years (Kadlubar and Hammons, 1987). We demonstrated that human P450 1A2 in the N-hydroxylation of MeIQx and PhIP (Turesky *et al.*, 1998), and such differences may have implications in extrapolation of risk assessment data to humans. The difference points out the significance of direct examination of enzyme variants for activity with each substrate of interest.

An example of how an approach can be implemented comes from our other studies on the (collaborative) development of genotoxicity test systems and random mutagenesis of P450s (Josephy *et al.*, 1998; Parikh*et al.*, 1999). *Escherichia coli* DJ4309 (Figure 9) contains plasmids for the expression of P450 1A2, NADPH-P450 reductase, and (bacterial) N-acetyltransferase plus an F′episomal target for *lac* mutagenesis.

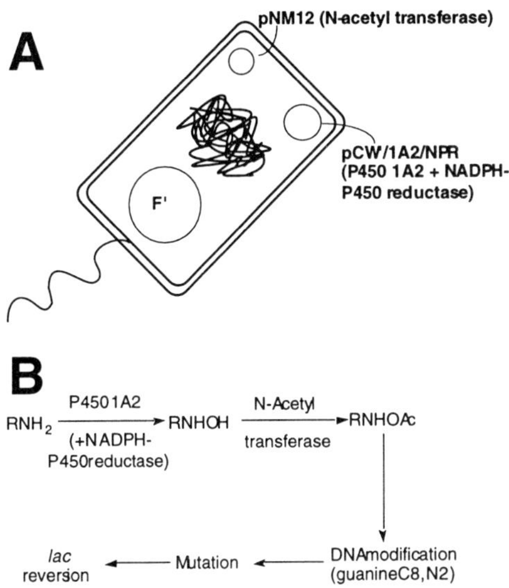

Figure 9. (A) E. coli DJ 4309 (Josephy *et al.*, 1998). (B) Basis of response to heterocyclic amines in DJ 4309.

We have utilized this assay in analysis of libraries of human P450 1A2 mutants (Parikh *et al.*, 1999), but the same system can be used with other P450s (Nakamura *et al.*, 1999) and variants of a human P450, NADPH-P450 reductase, or N-acetyltransferase. The F′ episome may also be changed from a –2 frameshift reporter to any of the basepair changes (Cupples *et al.*, 1990). Further, variants of other enzymes (e.g. GST) could be used to study other chemicals that are converted to mutagens (e.g. dihaloalkanes) (Thier *et al.*, 1993).

An example of the use of such a screening system is shown in the work presented in Figure 10 (Parikh *et al.*, 1999). The mutants were picked on the basis of either high or low P450 1A2 activity towards MeIQ in the basic screen described in Figure 9. Membranes were isolated from *E. coli* preparations, in which all of the P450 1A2 mutants were expressed, and normalized on the basis of P450 concentration. As indicated, the catalytic efficiency of these single residue mutants varied three orders of magnitude, with some reproducibly more active than wild type (Figure 10).

Further the effect of the substitution on catalytic activity depended upon which reaction was being monitored.

The work presented here shows some of the approaches that can be applied to issues of functional genomics in the area of chemical activation and detoxication. What is clear is that larger scale, faster, and more efficient approaches will be needed to deal with questions in the area of genomics and particularly toxicogenomics. Success in this area will come with the introduction of more technology and also intelligent new approaches to the collection and analysis of large collections of data.

ACKNOWLEDGMENTS

This work was supported in part by United States Public Health Service grants R35 CA44353, P30 ES00267, F32 ES05663, and T32 GM07347.

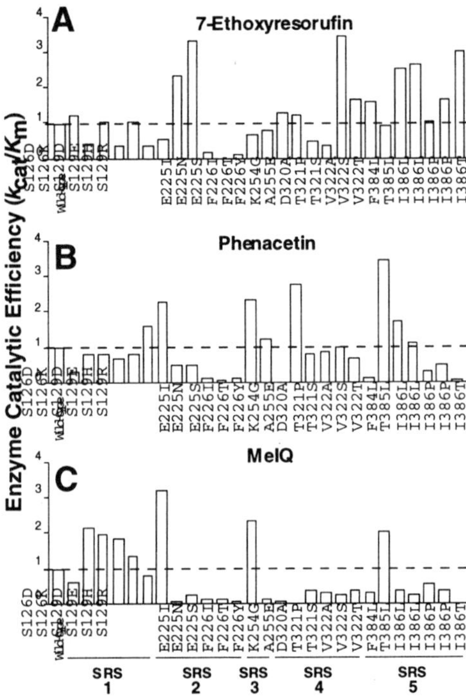

Figure 10. Comparison of k_{cat}/K_m for wild-type human P450 1A2 and 27 P450 1A2 mutants (selected from an experimental single-codon random library) with regard to marker reactions (Parikh *et al.*, 1999). "SRS" numbers indicate the putative substrate recognition sequence regions

[1]The abbreviations used are: AF, aflatoxin; TCE, trichloroethylene; MeIQ, 2-amino-3,5-dimethylimidazo[4,5-*f*]quinoline; GSH, glutathione; GST, GSH transferase; HPLC, high performance liquid chromatography; MS, mass spectrometry; Lys, lysine; MeIQx, 2-amino-3,8-dimethylimidazo[4,5-*f*]quinoxaline; PhIP, 2-amino-1-methyl-6-phenylimidazo[4,5-*b*]pyridine.

REFERENCES

Bailey, E.A., Iyer, R.S., Stone, M.P., Harris, T.M., and Essigmann, J.M., 1996, Mutational properties of the primary aflatoxin B1-DNA adduct, *Proc. Natl. Acad. Sci. USA* **93**:1535-1539.

Bruckner, J.V., Davis, B.D., and Blancato, J.N., 1989, Metabolism, toxicity, and carcinogenicity of trichloroethylene, *Crit. Rev. Toxicol.* **20**:31-50.

Brüning, T. and Bolt, H.M., 2000, Renal toxicity and carcinogenicity of trichloroethylene: key results, mechanisms, and controversies, *Crit. Rev. Toxicol.* **30**:253-285.

Busby, W.F. and Wogan, G.N., 1984, Aflatoxins, in: *Chemical Carcinogens* (Searle, C.E., ed.) Amer. Chem. Soc., Washington, D.C., pp. 945-1136.

Cai, H. and Guengerich, F.P., 1999a, Mechanism of aqueous decomposition of trichloroethylene oxide, *J. Am. Chem. Soc.* **121**:11656-11663.

Cai, H. and Guengerich, F.P., 1999b, Acylation of protein lysines by trichloroethylene oxide, *Chem. Res. Toxicol.* **13**:327-335.

Cupples, C.G., Cabrera, M., Cruz, C., and Miller, J.H., 1990, A set of *lacZ* mutations in *Escherichia coli* that allow rapid detection of specific frameshift mutations, *Genetics* **125**:275-280.

Essigmann, J.M., Croy, R.G., Nadzan, A.M., Busby, W.F.,Jr., Reinhold, V.N., Büchi, G., and Wogan, G.N., 1977, Structural identification of the major DNA adduct formed by aflatoxin B1 *in vitro*, *Proc. Natl. Acad. Sci. USA* **74**:1870-1874.

Evans, W.E. and Relling, M.V., 1999, Pharmacogenomics: translating function genomics into rational therapeutics, *Science* **286**:487-491.

Goeptar, A.R., Commandeur, J.N.M., van Ommen, B., van Bladeren, P.J., and Vermeulen, N.P.E., 1995, Metabolism and kinetics of trichloroethylene in relation to toxicity and carcinogenicity. Relevance to the mercapturic acid pathway, *Chem. Res. Toxicol.* **8**:3-21.

Gopalakrishnan, S., Byrd, S., Stone, M.P., and Harris, T.M., 1989, Carcinogen-nucleic acid interactions: equilibrium binding studies of aflatoxin B1 with the oligodeoxynucleotide d(ATGCAT)$_2$ and with plasmid pBR322 support intercalative association with the B-DNA helix, *Biochemistry* **28**:726-734.

Guengerich, F.P., Ueng, Y-F., Kim, B-R., Langouët, S., Coles, B., Iyer, R.S., Thier, R., Harris, T.M., Shimada, T., Yamazaki, H., Ketterer, B., and Guillouzo, A., 1996, Activation of chemicals by cytochrome P450 enzymes: regio- and stereoselective oxidation of aflatoxin B$_1$, *Adv. Exp. Biol. Med.* **387**:7-15.

Guengerich, F.P., 2000, Metabolism of chemical carcinogens, *Carcinogenesis* **21**:345-351.

Huff, J.E., 1971, New evidence on the old problems of trichloroethylene, *Industr. Med.* **40**:25-33.

Iyer, R.S., Voehler, M.W., and Harris, T.M., 1994, Adenine adduct of aflatoxin B$_1$ epoxide, *J. Am. Chem. Soc.* **116**:8863-8869.

Johnson, W.W. and Guengerich, F.P., 1997, Reaction of aflatoxin B$_1$ *exo*-8,9-epoxide with DNA: kinetic analysis of covalent binding and DNA-induced hydrolysis, *Proc. Natl. Acad. Sci. USA* **94**:6121-6125.

Johnson, W.W., Harris, T.M., and Guengerich, F.P., 1996, Kinetics and mechanism of hydrolysis of aflatoxin B$_1$ *exo*-8,9-oxide and rearrangement of the dihydrodiol, *J. Am. Chem. Soc.* **118**:8213-8220.

Johnson, W.W., Ueng, Y-F., Mannervik, B., Widersten, M., Hayes, J.D., Sherratt, P.J., Ketterer, B., and Guengerich, F.P., 1997, Conjugation of highly reactive aflatoxin B$_1$ 8,9-*exo*-epoxide catalyzed by rat and human glutathione transferases: estimation of kinetic parameters, *Biochemistry* **36**:3056-3060.

Johnson, W.W., Ueng, Y-F., Yamazaki, H., Shimada, T., and Guengerich, F.P., 1997, Role of microsomal epoxide hydrolase in the hydrolysis of aflatoxin B$_1$ 8,9-epoxide, *Chem. Res. Toxicol.* **10**:672-676.

Josephy, P.D., Evans, D.H., Parikh, A., and Guengerich, F.P., 1998, Metabolic activation of aromatic amine mutagens by simultaneous expression of human cytochrome P450 1A2, NADPH-cytochrome P450 reductase, and *N*-acetyltransferase in *Escherichia coli*, *Chem. Res. Toxicol.* **11**:70-74.

Judah, D.J., Hayes, J.D., Yang, J.-C., Lian, L.-Y., Roberts, G.C.K., Farmer, P.B., Lamb, J.H., and Neal, G.E., 1993, A novel aldehyde reductase with activity towards a metabolite of aflatoxin B$_1$ is expressed in rat liver during carcinogenesis and following the administration of an anti-oxidant, *Biochem. J.* **292**:13-18.

Kadlubar, F.F. and Hammons, G.J., 1987, The role of cytochrome P-450 in the metabolism of chemical carcinogens, in: *Mammalian Cytochromes P-450, Vol. 2* (Guengerich, F.P., ed.) CRC Press, Boca Raton, FL, pp. 81-130.

Kimbrough, R.D., Mitchell, F.L., and Houk, V.N., 1985, Trichloroethylene: an update, *J. Toxicol. Environ. Health* **15**:369-383.

Lamm, G. and Pack, G.R., 1990, Acidic domains around nucleic acids, *Proc. Natl. Acad. Sci. USA* **87**:9033-9030.

Lamm, G., Wong, L., and Pack, G.R., 1996, DNA-mediated acid catalysis: calculations of the rates of DNA-catalyzed hydrolyses of diol epoxides, *J. Am. Chem. Soc.* **118**:3325-3331.

Langouët, S., Coles, B., Morel, F., Becquemont, L., Beaune, P.H., Guengerich, F.P., Ketterer, B., and Guillouzo, A., 1995, Inhibition of CYP1A2 and CYP3A4 by oltipraz results in reduction of aflatoxin B$_1$ metabolism in human hepatocytes in primary culture, *Cancer Res.* **55**:5574-5579.

Meyer, U.A., 1994, Pharmacogenetics: the slow, the rapid, and the ultrarapid, *Proc. Natl. Acad. Sci. USA* **91**:1983-1984.

Miller, R.E. and Guengerich, F.P., 1982, Oxidation of trichloroethylene by liver microsomal cytochrome P-450: evidence for chlorine migration in a transition state not involving trichloroethylene oxide, *Biochemistry* **21**:1090-1097.

Miller, R.E. and Guengerich, F.P., 1983, Metabolism of trichloroethylene in isolated hepatocytes, microsomes, and reconstituted enzyme systems containing cytochrome P-450, *Cancer Res.* **43**:1145-1152.

Nakamura, K., Yoshihara, S., Shimada, T. and Guengerich, F.P., 1999, Selection and characterization of human cytochrome P450 1B1 mutants by random mutagenesis, in: *Abstracts, 9th North Amer. Meetg. Int. Soc. Study Xenobiotics/Amer. Chem. Soc.-Div. Chem. Toxicol.* (24-28 October) Nashville, p. 62.

Parikh, A., Josephy, P.D., and Guengerich, F.P., 1999, Selection and characterization of human cytochrome P450 1A2 mutants with altered catalytic properties, *Biochemistry* **38**:5283-5289.

Raney, K.D., Shimada, T., Kim, D-H., Groopman, J.D., Harris, T.M., and Guengerich, F.P., 1992, Oxidation of aflatoxins and sterigmatocystin by human liver microsomes: significance of aflatoxin Q$_1$ as a detoxication product of aflatoxin B$_1$, *Chem. Res. Toxicol.* **5**:202-210.

Rutter, W.J., Richards, O.C., and Woodfin, B.M., 1961, Comparative studies of liver and muscle aldolase. I. Effect of carboxypeptidase on catalytic activity, *J. Biol. Chem.* **236**:3193-3197.

Sabbioni, G., Skipper, P.L., Büchi, G., and Tannenbaum, S.R., 1987, Isolation and characterization of the major serum albumin adduct formed by aflatoxin B_1 *in vivo* in rats, *Carcinogenesis* **8**:819-824.

Slama, J.T., Radziejewski, C., Oruganti, S.R., and Kaiser, E.T., 1984, Semisynthetic enzymes: Characterization of isomeric flavopapains with widely different catalytic efficiencies, *J. Am. Chem. Soc.* **106**:6778-6785.

Sugimura, T., 1986, Studies on environmental chemical carcinogenesis in Japan, *Science* **233**:312-318.

Thier, R., Pemble, S.E., Taylor, J.B., Humphreys, W.G., Persmark, M., Ketterer, B., and Guengerich, F.P., 1993, Expression of mammalian glutathione *S*-transferase 5-5 in *Salmonella typhimurium* TA1535 leads to base-pair mutations upon exposure to dihalomethanes, *Proc. Natl. Acad. Sci. USA* **90**:8576-8580.

Toxicological Profile for Trichloroethylene, Oak Ridge:Oak Ridge National Laboratory, 1988.

Turesky, R.J., Constable, A., Richoz, J., Vargu, N., Markovic, J., Martin, M.V., and Guengerich, F.P.,1998, Differences in activation of heterocyclic aromatic amines by rat and human liver microsomes and by rat and human cytochromes P450 1A2, *Chem. Res. Toxicol.* **11**:925-936.

Tzeng, H-F., Laughlin, L.T., Lin, S., and Armstrong, R.N., 1996, The catalytic mechanism of microsomal epoxide hydrolase involves reversible formation and rate-limiting hydrolysis of the alkyl-enzyme intermediate, *J. Am. Chem. Soc.* **118**:9436-9437.

Ueng, Y-F., Kuwabara, T., Chun, Y-J., and Guengerich, F.P., 1997, Cooperativity in oxidations catalyzed by cytochrome P450 3A4, *Biochemistry* **36**:370-381.

Ueng, Y-F., Shimada, T., Yamazaki, H., and Guengerich, F.P., 1995, Oxidation of aflatoxin B_1 by bacterial recombinant human cytochrome P450 enzymes, *Chem. Res. Toxicol.* **8**:218-225.

THE ROLE OF TOXICOLOGICAL SCIENCE IN RISK ASSESSMENT AND RISK MANAGEMENT

Bernard D. Goldstein

Environmental and Occupational Health Sciences Institute (EOHSI)
170 Frelinghuysen Road
Piscataway, NJ 08854

INTRODUCTION

Toxicological science plays a central role in protecting public health and the environment against environmental risk. It does so directly through providing tools to prevent the development and release of toxic chemicals, and indirectly through its impact on risk assessment. Yet it appears that the influence of toxicological science on the management of risks, appears to be decreasing. This is harmful to the ability of society to make decisions that protect public health and the environment. In part this represents the failure of risk assessment to impact on risk management decisions; in part this is due to the dilution of the impact of toxicological science on risk assessment and on primary prevention, the most effective approach to protecting public health and the environment.

The apparent diminution of the role of toxicology in risk management is occurring while the science of toxicology is moving rapidly ahead. The Sixth International Symposium on Biological Reactive Intermediates provided many examples of a dynamic field. There are exciting new findings, there are exciting new methods, and most importantly, there are exciting new concepts.

The two fields of toxicological science and of risk assessment are intimately related, with risk assessment largely resting on core toxicological principles. The field of risk assessment is also moving rapidly ahead. The number of scientists involved in the field are growing. There are now many international versions of the Society of Risk Analysis, the initial learned society in the field. The harmonization of risk assessment procedures has increasingly been a topic of international attention, and the use of risk assessment to approach such problematic issues as the safety assessment of genetically modified organisms is also being debated.

In many ways risk assessment has become the primary pathway by which toxicological science is able to justify its societal importance beyond that of being a valued handmaiden to pharmacology and its classic role in forensic science. In this paper I explore concerns about the broader environmental health role of toxicological science resulting from threats to the value of risk assessment as a risk management tool, and resulting from the dilution of the role of toxicology in risk assessment and in primary prevention, the most effective form of risk management.

THE RELATION BETWEEN TOXICOLOGICAL SCIENCE AND RISK ASSESSMENT

A long term goal of toxicological science is to impact positively on decisions that lead to the protection of public health and the environment from chemical and physical agents. Risk assessment, of course, has the same goal as its primary objective, although usually over a shorter time frame.

The four steps in risk assessment (Table 1) are closely related to two of the "laws" of toxicology. Dose response assessment is the basis of toxicology, firmly grounded on the "dose makes the poison". Hazard identification reflects the toxicological principle of the specificity of the effects of chemicals related both to inherent chemical structure and to the biological niches which are central to structure activity relationships. Exposure assessment aims at capturing information about dose, and risk characterization in essence is the process of describing this risk (NAS, 1983).

Table 1. Four components of risk assessment

Hazard Identification
Dose Response Evaluation
Exposure Assessment
Risk Characterization

The impact of toxicological science on environmental health extends beyond that of risk assessment and has been far more effective. Risk assessment in essence is a form of secondary prevention (Goldstein, 1996). Classically, risk assessment is used to determine the extent of an existing problem. It is far less effective in providing the information needed to perform primary prevention, to prevent a problem before it occurs. In contrast, toxicological science has been very effective in the primary prevention of adverse environmental health consequences. The basic science understanding that led to the Ames test and to other safety assessment techniques has been the foundation for the early exclusion of potentially dangerous chemicals from production and from introduction into society. Unfortunately, it is not possible to quantify the number of lives that have been saved, or fish kills that did not occur, because toxicological science led to safer and cleaner chemicals being preferentially developed. As toxicologists we need to do a much better job of explaining the value of what we do in primary prevention or we will be locked into a risk assessment paradigm that is best suited to help clean up after it is already dirty.

DIRECT THREATS TO THE FUTURE OF TOXICOLOGICAL SCIENCE

Toxicological science is performed both *in vitro* and *in vivo*. Our classic approach is to use an *in vitro* system as a model to design tests for toxicity *in vivo*, to use an animal model to predict human toxicity and, where ethically appropriate, to obtain data in humans exposed to the agent of concern. To be most effective, the process is iterative, with *in vivo* findings being validated and amplified by *in vitro* studies, and vice versa. Unfortunately, there are forces that would like to ban all *in vivo* research, both animal and human. And these forces are growing in strength.

The campaign against animal research is long-standing. So-called animal rights activists have already been effective in substantially driving up the costs of performing research, costs that are borne directly and indirectly by all of society. Direct costs vary from the need for increased security measures to protect against animal rights extremists, to the time and paper burden of the many documents and committees required to conform to regulations. Indirect costs are even more substantial, but harder to measure. These include the loss of benefits that would have accrued from the research that has not been done because funds and time have been diverted from research to defense against animal rights activists.

We must, of course, continue to support the ethical treatment of animals in research, wherever possible reducing pain and developing valid methods to reduce and replace the extent to which safety assessment for chemicals depends upon animal testing. But of importance is that we must pay much more attention to the opponents of animal research. Perhaps it is the many concessions that we have made, or been forced to make, over a period of decades that has made us overly confident about the longer term viability of toxicological research in laboratory animals. In fact, we are under an increasingly sophisticated attack, and in my judgement are losing. We are losing in large part because we are not taking the opposition seriously enough, because we have not made sufficient effort to not only respond to the scurrilous charges of routine mistreatment of laboratory animals, but also to proactively reach out to those who are most concerned, by telling what to them would be our best story about the value of animal research. Our best story may well be the value of toxicological science to animal pets.

A major thrust of animal rights organizations is to convince pet owners to actively oppose animal research and to contribute money to animal rights organizations. There are many millions of pet owners and they are genuinely concerned about the welfare of their pets. We need to do a much better job of telling pet owners about the value of toxicological science to their pets. Not only has laboratory science provided major advances of value to human well-being, it has also been of signal importance to the health and longevity of our animal pets. We also must educate pet owners about therapeutic drug development. The inevitable corollary to insisting that new veterinary drugs not be first tested on live animals is that the next time a new drug is developed that might alleviate their own pet's suffering, their pet may be the first living animal to receive the new drug, with all of the risk that this implies.

Toxicological science performed in humans is also under attack. Until recently, standard interpretation of the ethics of human research permitted a narrow range of toxicological research in humans, research which has had great importance to understanding potential health effects of exogenous agents in humans. This research, and the value it provides to protecting public health, is now under attack both on ethical and political grounds. In our own program we have had research turned down on "ethical" grounds that would have exposed individuals to the levels of benzene otherwise occurring while refueling an automobile. The leukemia risk would be about 1/1000 of that due to a chest x-ray, which is usually allowable if appropriate for the protocol. At this BRI meeting Dr. van Bladeren discussed the importance of obtaining human data for PBPK models to understand the source and magnitude of variability and susceptibility in the human population.

It is ironic that at a time when our science is providing many new concepts relevant to understanding human toxicology, and biological markers are being developed which can be readily applied to small amounts of ethically obtained human tissues, we are faced with a substantial attack on human research.

Another problem leading to concern about the support of toxicological sciences by governmental agencies relates to the not infrequent overreliance on epidemiology. There are a number of examples, at least in the United States, in which relatively weak human epidemiology data drive out consideration of toxicology. IARC has attempted to bring mechanistic toxicology into its deliberations but it is still too early to know the success of this approach (IARC, 1991).

Two excellent examples of the value of toxicology in comparison to epidemiology were demonstrated at BRI VI. Dr. Hanspeter Witschi clearly demonstrated a lack of chemoprevention by beta carotene in an animal lung cancer model. These studies would best have been done before the decision to launch three major multi-million dollar epidemiology studies–two of which needed to be abruptly discontinued because the beta carotene groups had a statistically significant <u>increase</u> in lung cancer and the third showed no effect (Omaye et al., 1997).

Another example from BR VI is the intriguing finding of Dr. Martyn Smith who in both an *in vitro* model and in Chinese workers exposed to benzene found evidence of chromosomal abnormalities that are relatively specific to human Non-Hodgkin's Lymphoma (NHL). The importance of Dr. Smith's findings, if confirmed, relate to a recently published meta-analysis of over 300,000 petroleum workers which is claimed to show that benzene does not cause NHL (Wong and Raabe, 2000). In contradiction to the basic tenets of toxicology, this epidemiology study completely ignores exposure issues. As just one example, there is inadequate benzene exposure in the cohorts assembled for the meta analysis to even show an increase in acute myelogenous leukemia (AML), a cancer known to be caused by benzene. If the exposure levels for this population are too low to show an increase in AML, it is an unsuitable population in which to ask the question of whether benzene causes NHL. There is a similar recent putative negative meta analysis of multiple myeloma in petroleum refinery workers (Bergsagel et al, 1999). Both of these published meta analyses are notable for their lack of significant consideration of the toxicological evidence supporting a causal relation of lymphatic tumors to benzene (Goldstein and Shalat, 2000; Goldstein, 1990).

The problem we face is not just defending the importance to risk assessment of toxicological science. Risk assessment itself is under fire from a number of directions. Some critics would replace risk assessment by approaches that would be far less dependent upon understanding the toxicity of environmental chemicals. In the United States we have replaced a risk-based, toxicology intensive approach to regulating most air pollutants with one that is based on using maximal available control technology, irrespective of risk. There has also been a major drive worldwide, but particularly in Europe, to use the Precautionary Principle as a prescription for control without more than a minimal understanding of a potential threat.

One problem with making governmental regulatory decisions with only minimal data is the reluctance to ever change this decision, even if shown to be erroneous. An example from the United States of a decision taken without adequate data for which there has been enormous governmental and industry resistance to change is that of the use of oxygenated fuels, particularly methyl tert butyl ether (MTBE). Exposure of about 100 million Americans to MTBE, at levels of 11-15% by volume in gasoline, as well as to other oxyfuels, was begun even before the results of a routine safety assessment were completely evaluated. The controversial finding of cancer at high doses, coupled with acute non-specific complaints in a subset of individuals, led to more than a dozen official governmental reviews of the issue in less than a decade. In essence, all asked for the same needed research. Yet almost none of this research was funded, reflecting the preference of the regulatory agency, the U.S. Environmental Protection Agency, to defend the status quo. A clear lesson from this ongoing oxyfuel problem is that for compounds for which there will be extensive human exposure we need a thorough toxicological understanding rather than just routine safety assessment (Goldstein and Erdal, 2000).

By use of the term toxicological science I intend to clearly distinguish basic laboratory science, as presented at the BRI VI meeting, from routine toxicological safety assessment. This is an important distinction. Toxicological safety assessment is a very valuable but limited means to develop information about the potential rules of chemicals. Better safety assessment techniques inevitably depend on better toxicological science. Yet there are enormous expenditures on routine safety assessment in comparison to the relatively minuscule expenditures on toxicological science. For example, the international chemical industry is now funding a very welcome major expansion in the number of chemicals in commerce for which there will be routine safety assessment data. This effort, responding to environmental groups, OECD, USEPA and others, is unfortunately unaccompanied by any similar effort to enhance the tools of safety assessment through improving toxicological science.

CONCLUSION

Toxicological science has never been stronger. But our ability to build upon this strength, and upon our superb track record of contributions to environmental health, is in peril. We need to recognize these threats, to treat the sources of these threats with respect, and to join together with risk assessors and risk managers to respond.

ACKNOWLEDGMENTS

I acknowledge with deep thanks my associates at EOHSI including Robert Snyder, Michael Gallo and Debra and Jeffrey Laskin, and I am grateful for the expert technical assistance of Betty Davis and Janet Huang.

REFERENCES

Bergasagel, D.E., Wong, O., Bergsagel, P.L., et al., 1999. Benzene and multiple myeloma: appraisal of the scientific evidence. *Blood.* 94:1174-1162.

Goldstein, B.D., 1990. Is exposure to benzene a cause of human multiple myeloma? *Ann NY Acad Sci.* 609:225-230.

Goldstein, B.D., 1996. Editor: R.W. Hahn. *Risk Assessment as an Indicator for Decision Making, Risks, Costs, and Lives Saved: Getting Better Results from Regulation.* Oxford University Press, New York, NY, 907-914.

Goldstein, B.D., and Erdal, S. MTBE: A Policy Review. Accepted for publication in *Ann. Rev. of Energy and the Environment.* February, 2000.

Goldstein, B.D., and Shalat, S.L, 2000. The causal relation between benzene exposure and multiple myeloma. *Blood.* 95(4):1512-4. Letter.

IARC, 1991. *A Consensus Report of an IARC Monographs Working Group on the Use of Mechanims of Carcinogenesis in Risk Identification* (IARC Intern. Tech. Rep. No. 91/002), Lyon.

National Academy of Science, 1983. *Risk Assessment in the Federal Government: Managing the Process.* National Academy Press, Washington, D.C.

Omaye, S.T., Krinsky, N.I., Kagan, V.E., Mayne, S.T., Liebler, D.C., Bidlack, W.R., 1997. ß-Carotene: Friend or Foe? *Fundamental and Applied Toxicology.* 40:163-174.

Wong, O., and Raabe, G.K., 2000. Non-Hodgkin's lymphoma and exposure to benzene in a multinational cohort of more than 308,000 petroleum workers, 1937 to 1996. *J Occup Environ Med.* 42(5):554-68.

COVALENT BINDING OF CHEMICAL RESIDUES: HEALTH IMPACT

Anthony Y.H. Lu

Laboratory for Cancer Research, Department of Chemical Biology, College of Pharmacy, Rutgers, The State University of NJ, Piscataway, N.J. 08854

INTRODUCTION

Risk assessment of covalent binding of reactive intermediates to proteins in humans is dependent on whether the exposure of these chemicals is direct or indirect. It is well established that direct exposure of drugs and environmental chemicals capable of covalent binding to proteins in laboratory and food-producing animals and humans could result in potential risk. The degree of toxicological concern is dependent on the dose, chemical nature of the reactive intermediates, protein targets and other factors. Humans could also be exposed to food containing covalently bound chemical residues, mostly meat from drug-treated food-producing animals, and agriculture products from pesticide-treated crops. Health impact from this type of indirect exposure (i.e. protein bound residues rather than the free parent compound) is more difficult to evaluate in humans.

In this paper, I shall focus on the evaluation of toxicological potential of covalently bound chemical residues in food-producing animals treated with veterinary medicines. Based on the degree of exposure, the chemical stability of the protein adduct in general, and the lack of regeneration of reactive species from protein bound drug residues in most of the studies, one could conclude that the toxicological potential of covalently bound drug residues is extremely low, if any.

BOUND DRUG RESIDUES: ISSUES AND CONCERNS

Veterinary medicines are routinely used for the prevention and treatment of diseases in food-producing animals. Some veterinary drugs developed prior to 1980 were genotoxic. When used in food-producing animals, these drugs can be activated to reactive intermediates capable of binding to macromolecules. Unlike the parent drug and other metabolites, covalently bound drug residues are not readily extracted from tissues by organic solvents and thus are called bound residues (1,2). Since meat containing bound residues is consumed by humans, questions are raised about the health impact of these covalently bound residues by regulatory agencies (1,3). The critical safety issue is whether bound residues can be digested in meat-consuming humans to amino acid, small peptide or nucleotide adducts which are bioavailable and can exert toxic effects. Such concern is based on the fact that some cysteine

Biological Reactive Intermediates VI, Edited by Dansette *et al.*
Kluwer Academic / Plenum Publishers, 2001

and glutathione adducts, mostly halogenated hydrocarbons, are toxic when studied in *in vitro* systems and in the intact animal (4,5).

Since the reactive intermediates generated from genotoxic drug (representing the worst case of the bound residue problem) are immobilized on protein molecules, many *in vivo* (short term toxicity and long term carcinogenicity studies) and in vitro (genotoxicity, hepatotoxicity and nephrotoxicity) methods can not be used for evaluating the toxicological potential of bound residues. Hydrolysis of the bound residue containing tissues could, in principle, release the bound adducts. However, hydrolyzed tissue samples are generally not suitable for these tests due to low concentrations of released metabolites and high concentrations of other substances that interfere with the assays. If the amino acid adduct structure is known, then the synthesized adduct can be directly tested for toxicity in various test systems.

TOXICITY EVALUATION OF BOUND RESIDUES

The toxicological potential of bound drug residues depends on the bioavailability of these residues and whether reactive intermediates can be regenerated from bound residues to exert toxic effects (6). If bound residues are not absorbed, then the toxicological potential of these residues would be very low. In this type of study, food-producing animals are treated with radiolabeled drug. After removing parent drug and non-protein bound metabolites, tissues containing labeled bound residues are fed to rats to determine the radioactivity in bile, urine and feces. The bioavailability of several radiolabeled bound residue studies in rats following oral administration is generally very low (6), less than 10%. These results indicate that the exposure of bound drug residues to humans should also be low.

Another approach used to evaluate the toxicological potential of bound residues is to combine bioavailability and the measurement of regeneration of active species from the residues. For example, aflatoxin B_1 is one of the most potent environmental carcinogens. Reactive intermediates derived from aflatoxin B_1 can covalently bind to proteins and nucleic acids. When rat liver tissues containing protein-bound ^{14}C-labeled aflatoxin B_1 are fed to rats, no significant radioactivity can be detected in the DNA isolated from the liver (7). These results demonstrate that reactive species are not generated from the aflatoxin protein adducts to bind to new cellular targets. Thus, despite the fact that aflatoxin B_1 is a potent carcinogen, bound residues derived from this compound possess virtually no toxicological potential.

If the drug DNA or protein adduct structures are known, then *in vitro* studies can be carried out to determine if reactive intermediates can be regenerated from the bound residues. For example, ethylene dibromide is mutagenic and carcinogenic and its glutathione conjugate binds to DNA, *via* the reactive intermediate of episulfonium ion, to form the N^7-guanylethyl glutathione adduct (8). To determine if the episulfonium ion could be generated from this adduct, Inskeep et al. (9) performed *in vitro* studies under a variety of conditions and found no detectable DNA adduct formation when N^7-guanylethyl glutathione was incubated with DNA. These results indicate that the putative episulfonium ion could not be regenerated once the adduct is formed.

Reversibly bound drug residue, although rare, has been reported. In the presence of nitrite and acidic conditions in the stomach, diazonium cation can be formed from sulfamethazine (10), a widely used veterinary drug. Part of the sulfamethazine diazonium ion can covalently bind to proteins, whereas the rest is converted to desaminosulfamethazine. *In vitro* and *in vivo* studies showed that a small amount of sulfamethazine can be regenerated from the protein bound residues. In this case, when tissues containing sulfamethazine-bound residues are consumed by humans, the possibility exists that humans could be exposed to small amounts of diazonium cation generated from sulfamethazine released from the bound residues.

RONIDAZOLE BOUND RESIDUES: A MECHANISTIC APPROACH

In vitro studies designed to investigate the mechanism of metabolic activation, reactive intermediate formation and covalent binding of drugs to protein or DNA represent a sensible approach to generate adducts for structural identification and subsequent toxicity evaluation, provided that the established mechanism also operates in drug targeted animals *in vivo*. This approach was used to study the ronidazole bound residue formation and toxicological evaluation of the adducts (6,11,12).

Ronidazole (Fig. 1), a substituted 5-nitroimidazole, is an effective veterinary drug used for the treatment of swine dysentery. In swine, ronidazole is extensively metabolized.

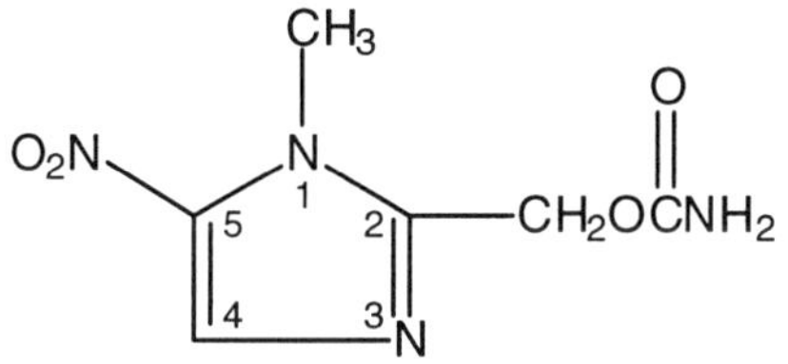

Figure 1. Structure of ronidazole

After several days, a large portion of the total radioactivity in liver and muscle is protein-bound and no parent drug can be detected. Like many of the nitro-containing compounds, ronidazole is mutagenic in Ames' test and carcinogenic in rodents. Therefore, meat containing bound residues from ronidazole-treated swine may not be safe for human consumption. To answer this question, *in vitro* mechanism studies were conducted to define the chemical and biochemical events leading to the covalent binding of ronidazole to proteins. In addition, structure-activity relationship and model compounds were used to define the mutagenic potential (an easily measured parameter) of ronidazole bound residues.

In a series of studies in rat liver microsomes (13-18), it was established that covalent binding of ronidazole to proteins involves 4-electron reduction of the nitro group, the loss of carbamic acid and the proton at C-4 position, and the attachment to cysteine *via* the 2-methylene (major) and/or the C-4 position. These results suggest a hydroxylamine as the reactive intermediate, consistent with an amino group as the final product of the reduction of ronidazole and the interaction of the reduced molecule with a sulfhydryl group. The protein bound residue should have the following characteristics: (1) the nitro group has been reduced to the amine; (2) imidazole ring remains intact; and (3) binding to cysteine of protein occurs predominantly at the 2-methylene position. A critical step in this mechanistic approach is the demonstration that ronidazole covalent adducts generated from *in vitro* studies are identical to those obtained *in vivo* in rats and in swine, the target animal, based on the results of radiolabeled studies and hydrolysis product patterns (19).

One approach that can be used to evaluate the mutagenic potential of a bound residue is to determine the structural requirements for the mutagenicity of the parent drug and then to determine whether such structural features still remain in the bound residue (20). While ronidazole is highly mutagenic, removal of the carbamoyl group at the 2-methylene position decreases the mutagenic activity by a factor of 10. Substitution at the C-4 position reduces the mutagenic activity by another factor of 10. The reduction of the nitro group to the amine totally eliminates the mutagenic activity of 5-nitroimidazoles. Therefore, the presence of a nitro group (most important), a carbamoyl group, and an unsubstituted C-4 position are essential for the full mutagenic activity of ronidazole. Interestingly, mechanism studies have shown that all three of these sites have been modified during ronidazole activation and subsequently covalent binding. Since structural features essential for the mutagenic activity of ronidazole no longer exist in the protein-bound product, the ronidazole-protein adduct should not be mutagenic.

During ronidazole activation, 95% of the reactive intermediate formed is converted to water-soluble breakdown products that are not mutagenic while the other 5% is protein-bound (6,11,20). The following observations support the conclusion that ronidazole protein adducts are not mutagenic: (1) the structural features essential for the mutagenic activity no longer exist in the protein adduct; (2) the ronidazole cysteine adduct is not mutagenic nor can it be activated to mutagenic product; and (3) enzymatic hydrolysis of ronidazole-bound proteins fails to generate any mutagenic products. Thus, despite the fact that ronidazole is genotoxic, bound residues generated from ronidazole are not.

CONCLUSION

From the available data, a strategy can be developed to evaluate the toxicological potential of protein bound drug residues. This strategy consists of three elements:
1. Bioavailability study to measure the potential exposure of bound residues to humans.
2. *In vitro* studies to determine if reactive intermediates can be regenerated from bound residues under various conditions.
3. *In vitro* studies to measure the intrinsic reactivity and toxicity of the amino acid adducts based on the structure information and the mechanism of adduct formation.

Overall, the toxicological potential of protein bound residues derived from toxic drugs should be very low or zero. Furthermore, no more genotoxic drugs are being developed for veterinary uses at the present time under current regulatory policies.

ACKNOWLEDGEMENT

I would like to express my sincere thanks to Ms. Florence Florek for the preparation of this manuscript, and to Drs. Gerald Miwa and Peter Wislocki for their important contributions to the ronidazole bound residue studies.

REFERENCES

1. N.E. Weber, Persistent residues: Interface with regulatory decisions, *J. Environ. Path. Toxicol.*, 3:35 (1980).
2. H.W. Dorough, Classification of radioactive pesticide residues in food-producing animals, *J. Environ. Path. Toxicol.*, 3:11 (1980).
3. T.M. Farber, Problems in the safety evaluation of tissue residues, *J. Environ. Path. Toxicol.*, 3:73 (1980).
4. M.W. Anders, L. Lash, W. Dekant, A. Elfarra, and D.R. Dohn, Biosynthesis and biotransformation of glutathione S-conjugates to toxic metabolites, *CRC Crit. Rev. Toxicol.*, 18:311 (1988).
5. W. Dekant, S. Vamvakas, and M.W. Anders, Bioactivation of nephrotoxic haloalkenes by glutathione conjugation: formation of toxic and mutagenic intermediates by cysteine conjugate β-lyase, *Drug Metab. Rev.*, 20:43 (1989).

6.   A.Y.H. Lu, G.T. Miwa, and P.G. Wislocki, Toxicological significance of covalently bound drug residues, *Rev. Biochem. Toxicol.*, 9:1 (1988).

7.   W. Jaggi, W.K. Lutz, J. Luthy, U. Zweifel and C. Schlatter, *In vivo* covalent binding of aflatoxin metabolites isolated from animal tissue to rat liver DNA, *Fd. Cosmet. Toxicol.*, 18:257 (1980).

8.   N. Ozawa and F.P. Guengerich, Evidence for formation of an S-[2-(N^7-guanyl)ethyl]glutathione adduct in glutathione-mediated binding of the carcinogen 1,2-dibromoethane to DNA, *Proc. Natl. Acad. Sci. U.S.A.*, 80:5266 (1983).

9.   P.B. Inskeep, N. Koga, J.L. Cmarik, and F.P. Guengerich, Covalent binding of 1,2-dihaloalkanes to DNA and stability of the major DNA adduct, S-[2-(N^7-guanyl)ethyl]glutathione, *Cancer Res.*, 46:2839 (1986).

10. G.G. Paulson, The influence of nitrite on sulfonamide drug metabolism in animals, *Drug Metab. Rev.*, 18:137 (1987).

11. P.G. Wislocki and A.Y.H. Lu, Formation and biological evaluation of ronidazole bound residues, *Drug Metab. Rev.*, 22:649 (1990).

12. A.Y.H. Lu, S.H.L. Chiu, and P.G. Wislocki, Development of a unified approach to evaluate the toxicological potential of bound residues, *Drug Metab. Rev.*, 22:891 (1990).

13. S.B. West, P.G. Wislocki, K.M. Fiorentini, R. Alvaro, F.J. Wolf and A.Y.H. Lu, Drug residue formation from ronidazole, a 5-nitroimidazole. I. Characterization of *in vitro* protein alkylation, *Chem. -Biol. Interact.*, 41:265 (1982).

14. S.B. West, P.G. Wislocki, F.J. Wolf and A.Y.H. Lu, Drug residue formation from ronidazole, a 5-nitroimidazole. II. Involvement of microsomal NADPH-cytochrome P-450 reductase in protein alyklation *in vitro*, *Chem. -Biol. Interact.*, 41:281 (1982).

15. G.T. Miwa, S.B. West, J.S. Walsh, F.J. Wolf, and A.Y.H. Lu, Drug residue formation from ronidazole, a 5-nitroimidazole. III. Studies on the mechanism of protein alkylation *in vitro*, *Chem. -Biol. Interact.*, 41:297 (1982).

16. P.G. Wislocki, E.S. Bagan, W.J.A. VandenHeuvel, R.W. Walker, R.F. Alvaro, B.H. Arison, A.Y.H. Lu, and F.J. Wolf, Drug residue formation from ronidazole, a 5-nitroimidazole V. Cysteine adducts formed upon reduction of ronidazole by dithionite or rat liver enzyems in the presence of cysteine, *Chem. -Biol. Interact.*, 49:13 (1984).

17. G.L. Kedderis, L.S. Argenbright, and G.T. Miwa, Covalent interaction of 5-nitroimidazole with DNA and protein in vitro: Mechanism of reductive activation, *Chem. Res. Toxicol.*, 2: 146 (1989).

18. R.F. Alvaro, P.G. Wislocki, G.T. Miwa, and A.Y.H. Lu, Drug residue formation from ronidazole, a 5-nitroimidazole. VIII. Identification of the 2-methylene position as a site of protein alkylation. *Chem.-Biol. Interact.*, 82:21 (1992).

19. G.T. Miwa, R.F. Alvaro, J.S. Walsh, R. Wang, and A.Y.H. Lu, Drug residue formation from ronidazole, a 5-nitroimidzole. VII. Comparison of protein-bound products formed *in vitro* and *in vivo*, *Chem. -Biol. Interact.*, 50: 189 (1984).

20. P.G. Wislocki, E.S. Bagan, M.M. cook, M.D. Bradley, F.J. Wolf and A.Y.H. Lu, Drug residue formation from ronidazole, a 5-nitroimidazole. VI. Lack of mutagenic activity of reduced metabolites and derivatives of ronidazole, *Chem. -Biol. Interact.*, 49: 27 (1984).

IDENTIFICATION OF HEPATIC PROTEIN TARGETS OF THE REACTIVE METABOLITES OF THE NON-HEPATOTOXIC REGIOISOMER OF ACETAMINOPHEN, 3'-HYDROXYACETANILIDE, IN THE MOUSE *IN VIVO* USING TWO-DIMENSIONAL GEL ELECTROPHORESIS AND MASS SPECTROMETRY

Yongchang Qiu[1], Leslie Z. Benet[2,3], and A. L. Burlingame[1,3]

Departments of [1]Pharmaceutical Chemistry
and [2]Biopharmaceutical Sciences,
and [3]the Liver Center
University of California
San Francisco, CA 94143-0446

INTRODUCTION

The hepatotoxicity observed following an overdose of acetaminophen is due to the excessive production of the reactive metabolite N-acetyl-*p*-benzoquinone imine (NAPQI) by cytochrome P450[1-6]. At therapeutic doses NAPQI is efficiently trapped by conjugation with glutathione; however, after a hepatotoxic dose, both cytosolic and mitochondrial glutathione pools in liver cells are depleted and NAPQI covalently binds to numerous liver proteins[7] before centralobular liver cell necrosis develops via an unknown mechanism. Because NAPQI is a powerful electrophile as well as an oxidant, both protein covalent binding and oxidative stress hypotheses have been invoked to address the biochemical processes responsible for the hepatotoxicity of acetaminophen.

The covalent binding theory postulates that the toxicity is initiated by the covalent binding of electrophilic metabolites to liver proteins critical in maintaining homeostatsis[8]. Support lies in the observation of a close correlation between the extent of protein covalent binding and the severity of liver cell necrosis measured by both autoradiographic[8-12] and immunochemical methods[13,14].

The second hypothesis suggests that oxidative damage to essential macromolecules resulting from intracellular oxidant formation is responsible for the initiation of the hepatotoxicity[15-18]. Since both GSSG formation[19] and lipid peroxidation[15] temporally occur long after the maximum in protein covalent binding, and because inhibition or stimulation of lipid peroxidation within the time frame of maximum covalent binding did

Biological Reactive Intermediates VI, Edited by Dansette *et al.*
Kluwer Academic / Plenum Publishers, 2001

not influence hepatic damage[20-22], oxidative stress is often considered a consequence rather than a cause of damage[23].

The covalent binding hypothesis as a general mechanism for drug-induced liver toxicity gains further credence from studies of liver cytotoxicity induced by other chemical agents, including bromobenzene[24], furosemide[25], isoniazid[26,27], iproniazid[28], and cocaine[29,30] where similar correlations between protein covalent binding and hepatic damage have been observed. However, there are examples of related compounds which extensively form covalent protein adducts but do not cause hepatotoxicity. These include hydroquinone and 3'-hydroxyacetanilide[31-33], 2'-hydroxyacetanilide[34,35], methoxychlor[36], isaxonine[37], and a number of substituted bromobenzenes[38-40], including p-bromophenol[41] and o-bromophenol[42]. Among them, 3'-hydroxyacetanilide (AMAP) has attracted most attention due to its chemical similarity to acetaminophen.

As a regioisomer of acetaminophen (4'-hydroxyacetanilide), 3'-hydroxyacetanilide also has analgesic and antipyretic activities[34,35]. Even though it cannot be directly oxidized to a quinone imine like acetaminophen because of the meta-position of the hydroxy substitution[43], many secondary metabolites, such as 2-acetamido-p-benzoquinone, 4-acetamido-o-benzoquinone and N-acetyl-3-methoxy-p-benzoquinone imine, are also reactive quinone derivatives, which like NAPQI avidly react with protein nucleophiles to form covalent adducts[44]. Thus, it is surprising that AMAP is not hepatotoxic because total protein covalent binding levels are comparable following equimolar doses of AMAP and APAP. Even when the maximum AMAP metabolite binding was nearly twice that observed for a hepatotoxic dose of APAP (after administration of its highest nonlethal dose), no hepatic damage from AMAP ensued[45,46]. Hence, this obvious dissociation between total xenobiotic covalent protein binding *per se* and hepatotoxicity provides an important opportunity to dissect binding that is of toxicological significance from that which is adventitious.

Previously, investigators compared the binding pattern of proteins arylated by both AMAP and APAP reactive metabolites and found many similarities as well as significant differences[47,48]. However, these studies failed to reveal the identities of the target proteins.

Recently our laboratory has undertaken studies aimed at developing a detailed understanding of the identity of liver target proteins *in vivo*, the structural nature of covalent drug binding and its impact on protein function, and the elucidation of those factors of relevance in the induction of hepatotoxicity. Our initial work involved establishing the identity of the complete suite of mouse protein targets observed following administration of a hepatotoxic dose of acetaminophen *in vivo*[49]. This was achieved using a comprehensive new strategy based on 2D-PAGE separation of liver homogenates, recent developments in mass spectrometry for protein identification, and gene/protein and EST database interrogation. Using a similar strategy, we now report the comparative identification of hepatic protein targets for reactive metabolites of AMAP. This study provides the first comprehensive insight into putative deleterious acetaminophen-protein covalent modifications, and reveals that the hepatic protein targets of primary toxicological significance reside in the mitochondria.

EXPERIMENTAL PROCEDURES

Dosing Protocol

B6C3F1 mice were obtained from Simonsen (Gilroy, CA). Prior to the administration of acetaminophen, mice were given phenobarbital as a 0.1% solution in their drinking water for five days and food was withheld for 15 hours prior to dosing. Both radiolabeled and non-labeled AMAP or APAP were given ip as warm aqueous solutions (30 mg/ml) at doses of 600 mg/kg. The mice were killed by cervical dislocation two hours after dosing.

Covalent Binding

Covalent binding to proteins in liver homogenate was determined at 2.32 and 2.19 nmol/mg proteins for APAP and AMAP, respectively, as described previously[45]. Briefly, exhaustive washing was performed with cold methanol:diethylether (3:1, v/v) to remove reversibly bound drug until the radioactivity in the supernatant became two times less than the background level. The remaining protein pellet was dissolved in 0.25M potassium hydroxide solution at 80°C for 1 hr. The clear solution was then subjected to scintillation counting.

Materials, animals, sample preparation, binding, and gel separation and mass spectrometry have been described in detail in our previous APAP study[49].

RESULTS

Two-dimensional Gel Electrophoresis and Autoradiography

Samples from mice treated individually with APAP or AMAP were prepared under the same conditions and run as a single gel batch to facilitate the comparison of APAP and AMAP binding to hepatic proteins. Gels prepared this way were virtually identical, thus readily permitting inter-gel spot correlation. Examples of Coomassie blue stained 2D gels obtained from whole liver homogenates of phenobarbital-induced B6C3F1 mice treated with 600 mg/kg ^{14}C-labeled 3'-hydroxyacetanilide and acetaminophen are shown in Figs. 1a and 1c together with the corresponding film exposed counterparts in Figs. 1b and 1d, respectively. The radioactive gel for mouse treated with 600 mg/kg ^{14}C-labeled acetaminophen was exposed to film for only two weeks(Fig. 1d). It is virtually identical to the analogous film obtained from mice treated with 375 mg/kg ^{14}C-labeled acetaminophen[49]. Twenty-eight major ^{14}C-containing spots were revealed.

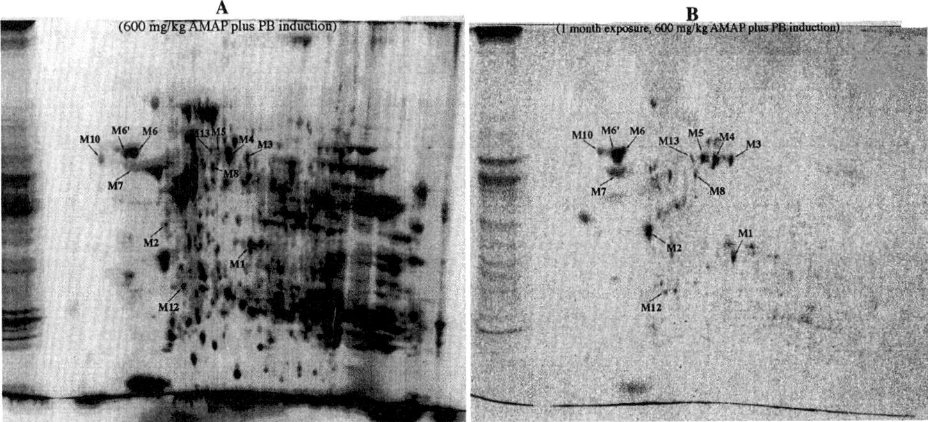

Figure 1. A. Coomassie blue R-250-stained 2D preparative gels of B6C3F1 mouse phenobarbital induced liver proteins after treatment with radio-labeled 3'-hydroxyacetanilide (600 mg/kg) via ip injection, numbered as in **B** and Table 1. **B.** Autoradiogram of the 2D preparative gel. About 2.4 mg protein in 200 ul whole liver homogenate was loaded in the gel. Protein samples were focused (x-axis, cathode on the right) and then separated by SDS-PAGE (y-axis, dye front at the bottom).

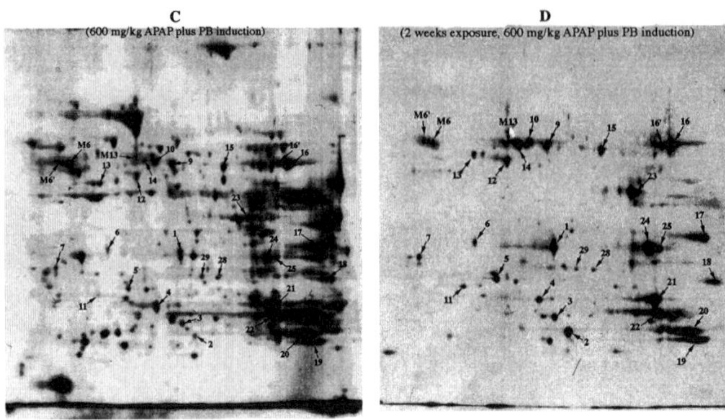

Figure 1. C. Coomassie blue R-250-stained 2D preparative gels of B6C3F1 mouse phenobarbital induced liver proteins after treatment with radio-labeled acetaminophen (600 mg/kg) via ip injection, numbered as in **D** and Table 1. **D.** Autoradiogram of the 2D preparative gel. About 2.4 mg protein in 200 ul whole liver homogenate was loaded in the gel. Protein samples were focused (x-axis, cathode on the right) and then separated by SDS-PAGE (y-axis, dye front at the bottom).

However, the comparable radioactive gel for mouse treated with 600 mg/kg ^{14}C-labeled 3'-hydroxyacetanilide had to be exposed for one month to obtain a radiogram (Fig. 1b) having a similar level of film spot intensity (compared to Fig. 1d), despite the fact that the total protein covalent binding by reactive metabolites of 3'-hydroxyacetanilide (2.19 nmol/mg protein) was approximately equal [reactive metabolites of acetaminophen (2.32 nmol/mg protein)]. As can be seen from inspection of Figs. 1b and 1d, twelve major AMAP binding spots are present and are all located within the acidic (left) side of the 2D gel. For comparison, in our previous studies many major APAP binding proteins are also found in the basic range (right side, see Figs. 1a and 1c). Judging from the 2D gel coordinates, about eight radioactive protein spots are observed in common, which was confirmed after identification of the proteins in these gel spots (see Table 1). Spot-to-spot coincidence permitted in-gel digestion to be performed on analogous gels obtained from mouse liver homogenate after administration of the same amount of non-radiolabelled drugs.

Identification of Target Proteins for AMAP and Their Comparison to Those Observed for APAP

AMAP target proteins were identified using the same strategy as described previously[49]. Table 1 summarizes the experimental information obtained for all protein spots together with the protein spot number in ref. 49.

The protein in spot M1[1] (designations labeled M listed in Table 1 refer to Fig. 1b spots for AMAP, with the number in brackets referring to the corresponding APAP spot in Fig. 1d) was identified as Life Tech mouse embryo 8 5dpc *Mus musculus* cDNA clone, based upon a PSD mass spectrum obtained from a tryptic peptide with mass value at m/z 992.13 employing MS-Tag to interrogate the dbEST database. The MALDI spectrum of the in-gel digest mixture from spot M1 was virtually identical to that from spot 1, except for the presence of a few additional minor peaks observed at m/z 1064.16, 1192.19, 1809.42, and 3265.44. These mass values correspond to those anticipated for four additional tryptic peptides from this unknown protein bearing C-terminal lysine residues.

Evidence of the presence of methionine adenosyltransferase was found in both spots M2 and M8[12]. The protein forms occurring in these two spots differ from each other both in their isoelectric points and apparent molecular weights. However, no sequence differences were found through analysis of their respective MALDI mass spectra, except that a few more peptide mass peaks were detected in spot M8 than in spot M2.

Table 1. Mouse liver proteins targeted by AMAP metabolites

Spot No.	Protein identified[*]	Subcellular location
M1/1	Life Tech mouse embryo 8 5 dpc *Mus musculus* cDNA clone	unknown
M2/M8/12	Methionine adenosyltransferase	Cytoplasmic
M3/M4/M5	Selenium-binding liver protein	Cytoplasmic
M13/9/10/14	AP56=acetaminophen binding protein	Cytoplasmic
30	Selenium binding protein	Cytoplasmic
M6/M6'	Endoplasmic reticulum transmembrane protein precursor	ER membrane
M6/M6'/M10	Protein disulfide isomerase/Thyroid hormone binding protein	ER lumen
M12	HMA CoA synthase and	Cytoplasmic
	Homologous to proteasome subunit C8 from human and rat	Cytoplasmic
M7	alpha-1 protease inhibitor 1	Cytoplasmic
	alpha-1 protease inhibitor 3	Cytoplasmic
	alpha-1 protease inhibitor 5	Cytoplasmic

[*]Proteins underlined are the manor targets for AMAP metabolites but are minor targets for APAP metabolites

Consequently, the calculated peptide sequence coverage for methionine adenosyltransferase in spot M8 is 9% higher than that in spot M2 (45% for spot 8 and 36% for spot M2). This could be due to the higher protein abundance in spot M8 than that in spot M2. However, in a subsequent experiment a larger number of plugs representing the same gel spot were excised corresponding to exposure 2, and pooled for in-gel digestion so as to have comparable intensities to those in the mass spectrum of in-gel digest from spot M8. Despite having more protein in this experiment, several mass values of higher abundance components present in spot 8 were still not observed (data not shown). These components include peptides with m/z values of 1068.32, 1082.32, 1118.36, 1285.59, and 1360.56. The sequences of the two most prominent peptides (m/z 1068.32 and 1285.59) were determined by MALDI-PSD analysis to be TACimYGHFGR (corresponding amino acid residues: 375-383) and NFDLRPGVIVR (corresponding amino acid residues: 353-363) (Table 1), respectively. Both of these tryptic peptides are located near the C-terminus of methionine adenosyltransferase (with total 396 amino acid residues). The other three mass values can also be assigned to sequences of tryptic peptides within this C-terminal region (between 353-396) (Table 1). These data clearly indicate that the form present in gel spot M2 must be a C-terminally truncated version of the methionine adenosyltransferase identified in gel spot M8, although determination of the exact site of cleavage would require further investigation. The origin of this truncated form remains to be established, but may have resulted from residual protease activity in the whole liver homogenate.

Spot M8 corresponds to spot 12, also one of the major exposures on the radiogram for APAP (Fig. 1d). Protein in the spot M2 location was not detected by either exposure or Coomassie blue stain for APAP.

Four isoforms of selenium binding protein were identified in spots M3[9], M4[10], M5[14], and M13[30]. These isoforms are highly homologous to each other[50]. Only a few amino acid residues are known to differ among their sequences. Still several subtle differences were detected in the case of acetaminophen-binding protein in spot M4[10], the most heavy exposure, compared with the other two selenium binding proteins (Table 1).

For example, peptide LAGQIFLGGSIVR having a mass value of m/z 1330.69 was found in spots M3/9 and M13/30, but not in spot M4/10; instead, the peptide specific for acetaminophen-binding protein, LTGQIFLGGSIVR (m/z 1360.80), is present in spot M4/10. The presence of acetaminophen binding protein in spot M4/10 is further supported by the much higher covering percentage of this isoform by tryptic peptides from this spot (55%) than those from the other three spots (Table 1). The sequences of both these variant peptides were established by PSD analysis. Other sequence variants detected are listed in Table 1. Assignments for the other three isomers cannot be made due to their similarity and low percentage coverage. Furthermore, the complete sequence of one of these four isomers is unknown (only three are listed in NCBInr protein database). Finally, EndoA cytokeratin was also identified in spot M5[14] simply by matching 28 mass values to theoretical masses of tryptic peptides from this protein. These 28 peptides cover 45% of its complete sequence.

Both protein disulfide isomerase/thyroid hormone binding protein and endoplasmic reticulum transmembrane protein precursor were present in spots M6 and M6'. These two exposures were partially overlapping, as was the case for their corresponding Coomassie blue stained spots. However, their individual identities can be easily discriminated by the number of tryptic peptides specific for each protein detected. Forty-one tryptic peptides from protein disulfide isomerase were detected in spot M6, while only sixteen masses from spot M6' can be matched to the sequences of tryptic peptides from this protein. On the other hand, only eight masses from spot M6, compared to twenty-two masses from spot M6', were specific for an endoplasmic reticulum transmembrane protein precursor. This contrast demonstrates that protein disulfide isomerase/thyroid hormone binding protein is accumulated more in spot M6 than M6', whereas the opposite is true for endoplasmic reticulum transmembrane protein precursor. These two proteins are major targets for reactive metabolites of AMAP based on the level of these two exposures (Fig. 1d), but they must be only minor targets for those of APAP, because corresponding spots for M6 and M6' are very light on the radiogram for APAP-treated mouse liver proteins (Fig. 1d). Protein disulfide isomerase/thyroid hormone binding protein is also found in spot M10, although only a few peptides were detected due to the limited amount of protein present in this spot.

The exposure at M7 is more diffuse compared with others, as is the corresponding stained gel spot. Several protease inhibitors were identified in this diffuse gel spot. Interpretation of a PSD spectrum of a mass with m/z at 947.65 establishes its sequence as LAQIHIPR, which is specific only for alpha-1 protease inhibitor 5 (Table 1). The identity of the other prominent mass peak was established by PSD analysis to be RLAQIHFPR, which is specific for both alpha-1 protease inhibitors 1 and 3, but not 5. The intensities of other mass peaks are too low to permit carrying out meaningful PSD experiments. However, many of them can be assigned to anticipated sequences of tryptic peptides from these protease inhibitors (Table 1). No corresponding exposure for M7 was detected in the case of APAP-treated mice.

One of the two proteins characterized in spot M12 is highly homologous to both human and rat proteasome subunit C8. It is also found in spot 11. This identification was easily made by the interpretation of the PSD spectrum corresponding to the conserved N-terminally acetylated peptide (AcetSSIGTGYDLSASTFSPDGR, Table 1). The other protein identified in spot M12 is HMA CoA synthase. PSD analysis of a peptide with mass at 1095.40 Da established its sequence as APLVLEQGLR, which is specific for HMA CoA synthase. This identification can be strengthened by attributing thirteen other masses to sequences of tryptic peptides arising from this protein. HMA CoA synthase is also present in spot 11 (Fig. 1c), but not found in our previous studies when mice were treated with 375 mg/kg APAP[49].

In summary, these studies have established the identity of eight different proteins occurring in eleven gel spots as main mouse liver target proteins for reactive metabolites of AMAP. Metabolites of both APAP and AMAP bind to four different proteins in nine gel spots at a similar level. Target proteins that are in common include Life Tech mouse embryo 8 5dpc *Mus musculus* cDNA clone in spot M1[1], acetaminophen binding protein in spot M4[10], and methionine adenosyltransferase in spots M2 and M8[12], which are three major cytosolic protein binding targets for both metabolites of APAP and AMAP. However, the major targets for reactive metabolites of AMAP, protein disulfide isomerase/thyroid hormone binding protein and endoplasmic reticulum transmembrane protein precursor in spot M6 and M6', are only slightly adducted by reactive metabolites of APAP (Figs. 1b and 1d).

DISCUSSION

In retrospect, it is not surprising to learn that certain proteins modified covalently by reactive metabolites of AMAP *in vivo* are also modified by those of APAP, because the reactive metabolites of both AMAP and APAP are electrophilic quinones that are known to conjugate preferentially with free thiol functions[44,51,52]. However, it was of prime importance to ascertain which particular suite of protein targets are implicated in common, because that information would suggest that covalent modifications to these common proteins do not contribute to the pathogenesis associated with APAP toxicity. Interestingly, modification of one such common target protein, protein disulfide isomerase, has been proposed to impair the posttranslational modification of plasma membrane proteins by APAP metabolites and thus cause a subsequent loss of membrane integrity that leads to death of hepatocytes[53,54]. Our results do not agree with this hypothesis because protein disulfide isomerase is one of the two major targets for AMAP metabolites (Fig. 1b), but it was only slightly modified by APAP metabolites (Fig. 1d). Of course, there is still the possibility that APAP metabolites bind to a lethal site(s) of these common target proteins while AMAP metabolites bind adventitiously elsewhere on the same proteins. This possibility appears unlikely when we consider the striking chemical similarity between AMAP metabolites and NAPQI, the major reactive metabolite of APAP. However, resolution of this possibility must await detailed structural elucidation and site localization of such adducts.

Two more of these major common binding targets, selenium binding proteins in four different spots (M3[9], M4[10], M5[14], and M13[30], Table 1) and Life Tech mouse embryo 8 5dpc *Mus musculus* cDNA clone in spot M1[1], are highly abundant in the liver (Figs. 1a and 1c) and their functions are unknown. Our discovery of these proteins as common targets for different electrophiles may indicate that they play a protective role in detoxifying electrophiles, particularly in the case of acetaminophen binding protein in spot M4[10], because it contains eight cysteine residues which could conjugate with electrophiles. Indirect evidence suggests that the acetaminophen binding proteins are also targeted by electrophilic metabolites of bromobenzene[55], 2,6-dimethyl APAP[56], and 3-methyl indole[57].

Despite there being many protein targets common for reactive metabolites of both APAP and AMAP, the differences between their covalent binding to proteins are marked. More than twice the number of radioactive gel spots were revealed on the 2D gel of whole liver homogenate from APAP treated mice (29 spots, Fig. 1a) than those from AMAP-treated mice (12 spots, Fig. 1c). Furthermore, the fact that twice the amount of time (one month vs. two weeks) was required to develop comparable images on the radiogram for 2D gels of protein from AMAP-treated mice indicates that significantly less radioactivity from ^{14}C-labeled AMAP modified protein has entered the gel than in the case of ^{14}C-labeled APAP, despite the fact that similar levels of total protein covalent binding were observed after administration of both APAP and AMAP. This same observation has been reported previously by Myers et al.[47] where it was attributed to possible instability of AMAP protein adducts during sample preparation or electrophoresis. Since AMAP metabolites bind mainly to microsomal proteins[48], these membrane proteins tend to precipitate at the entry of 2D gel electrophoresis[58]. If this is true, it is likely that many AMAP adducts are very hydrophobic and do not enter the gel as effectively as the majority of APAP adducts, which are more water-soluble.

The other main difference is the subcellular location of these target proteins. Except for the two major protein targets for reactive metabolites of AMAP, protein disulfide isomerase/thyroid hormone binding protein and endoplasmic reticulum (ER) transmembrane protein precursor in spot M6 and M6', that are located in endoplasmic reticulum (lumen and membrane), all other protein targets for AMAP are of cytosolic origin. In contrast, the major protein targets for APAP are not located in the ER. Instead, besides the common cytosolic protein targets, many of them are mitochondrial proteins including ATP synthetase α subunit, mitochondrial aldehyde dehydrogenase, glutathione peroxidase and glutathione transferase. These general results are consistent with previous findings that reactive metabolites of APAP bind more extensively to mitochondrial proteins than those of AMAP[19]. Because these metabolites are generated by cytochrome P450 in the ER, it has been suggested that AMAP metabolites are more reactive than APAP metabolites, so that they are trapped more rapidly by nearby protein thiols as well as other protein nucleophiles, whereas APAP metabolites diffuse throughout the cell, binding to other more critical protein targets[48].

Mitochondrial damage, i. e., inhibition of hepatic respiration and impaired oxidative energy metabolism, are early events after a hepatotoxic dose of acetaminophen[59-63]. There is also a solid correlation between mitochondrial GSH depletion and hepatic necrosis[64]; in other words, agents which deplete the mitochondrial glutathione pool are hepatotoxic while drugs which do not deplete the mitochondrial glutathione pool are not hepatotoxic. For example, acetaminophen[19,45,65], bromobenzene and iodobenzene[66], 1,2-dibromoethane[67,68], allyl alcohol[69,70], ethacrynic acid[71,72], hexathionine and heptathionine sulfoximine[73,74] are capable of depleting mitochondrial GSH substantially and are highly hepatotoxic. The opposite is true for 3'-hydroxyacetanilide[19,45], adriamycin[75], diethylmaleate[33,67,71], phorone[75], BCNU[33] and buthionine sulfoximine[76]. Interestingly, co-treatments of AMAP with buthionine sulfoxiamine (BSO)[33] and BCNU with adriamycin are both hepatotoxic by an unknown mechanism, but they do cause substantial mitochondrial GSH depletion[71]. Since most of these hepatotoxic agents and/or their metabolites can conjugate with reduced glutathione, they are likely to bind to proteins with free cysteine residues as well, like reactive metabolites of APAP. Given this strong correlation, the lack of mitochondrial protein covalent binding by nonhepatotoxic AMAP may suggest that mitochondrial proteins targeted by APAP metabolites are more important than others, if in fact covalent binding plays a crucial role in the initiation of APAP hepatotoxicity.

Among these mitochondrial protein targets for APAP identified in these and in our earlier studies[49], ATP synthetase α-subunit and glutathione peroxidase are of particular interest. It is conceivable that modification on ATP synthetase α-subunit may result in its impaired function, which would explain the marked decrease of hepatocellular ATP concentrations within 2 hr of administration of hepatotoxic doses of APAP to mice[46,77]. On the other hand, inhibition of mitochondrial glutathione peroxidase may be responsible for the development of oxidative stress in the second stage[15,19,78]. Metabolites and/or chemicals entering mitochondria may not only covalently bind to various proteins and thus compromise their normal functions, but also deplete the mitochondrial GSH pool which is essential to maintain the intracellular redox balance. Hence, the role of oxidative stress developed in the second stage of APAP toxicity certainly cannot be excluded. In fact, this may be the irreversible step during the processes leading to cell death, because many sulfhydryl antidotes such as N-acetylcysteine can protect against liver damage as late as 10 hrs after ingestion of APAP when maximum covalent binding has already occurred[79-81]. There is also evidence that both N-acetylcysteine and dithiol threitol can reverse otherwise lethal oxidative damage to hepatocytes after covalent binding has taken place[82]. In fact, hepatotoxicity of APAP has been proposed as a two-stage process with the second stage oxidative damage which continues after cessation of APAP metabolism and protein covalent binding[78]. Toxicity in liver slices proceeds after removal of APAP following an initial two hour incubation with 10mM APAP[78]. Events such as protein covalent binding occurring at the early time points may be responsible for the initiation of oxidative stress in the second stage.

ACKNOWLEDGEMENTS

Financial support was provided by the UC Systemwide Campus Laboratory Collaboration grant; NIH National Center for Research Resources RR 01614 (to ALB) and GM 36633 (to LZB), and Liver Center grant AM 27643 (to ALB).

REFERENCES

1. Miner, D. J. and Kissinger, P. T. (1979) *Biochem Pharmacol* **28**, 3285-3290.
2. Corcoran, G. B., Mitchell, J. R., Vaishnav, Y. N., and Horning, E. C. (1980) *Mol Pharmacol* **18**, 536-542.
3. Dahlin, D. C. and Nelson, S. D. (1982) *J Med Chem* **25**, 885-886.
4. Dahlin, D. C., Miwa, G. T., Lu, A. Y., and Nelson, S. D. (1984) *Proc Natl Acad Sci U S A* **81**, 1327-1331.
5. Harvison, P. J., Guengerich, F. P., Rashed, M. S., and Nelson, S. D. (1988) *Chem Res Toxicol* **1**, 47-52.
6. Holme, J. A., Dahlin, D. C., Nelson, S. D., and Dybing, E. (1984) *Biochem Pharmacol* **33**, 401-406.
7. Hinson, J. A., Pohl, L. R., Monks, T. J., and Gillette, J. R. (1981) *Life Sci* **29**, 107-116.
8. Jollow, D. J., Mitchell, J. R., Potter, W. Z., Davis, D. C., Gillette, J. R., and Brodie, B. B. (1973) *J Pharmacol Exp Ther* **187**, 195-202.
9. Potter, W. Z., Davis, D. C., Mitchell, J. R., Jollow, D. J., Gillette, J. R., and Brodie, B. B. (1973) *J Pharmacol Exp Ther* **187**, 203-210.
10. Potter, W. Z., Thorgeirsson, S. S., Jollow, D. J., and Mitchell, J. R. (1974) *Pharmacology* **12**, 129-143.
11. Mitchell, J. R., Jollow, D. J., Potter, W. Z., Gillette, J. R., and Brodie, B. B. (1973) *J Pharmacol Exp Ther* **187**, 211-217.
12. Mitchell, J. R., Thorgeirsson, S. S., Potter, W. Z., Jollow, D. J., and Keiser, H. (1974) *Clin Pharmacol Ther* **16**, 676-684.
13. Roberts, D. W., Bucci, T. J., Benson, R. W., Warbritton, A. R., McRae, T. A., Pumford, N. R., and Hinson, J. A. (1991) *Am J Pathol* **138**, 359-371.
14. Bartolone, J. B., Cohen, S. D., and Khairallah, E. A. (1989) *Fundam Appl Toxicol* **13**, 859-862.
15. Wendel, A., Feuerstein, S., and Konz, K. H. (1979) *Biochem Pharmacol* **28**, 2051-2055.
16. Gerson, R. J., Casini, A., Gilfor, D., Serroni, A., and Farber, J. L. (1985) *Biochem Biophys Res Commun* **126**, 1129-1137.
17. Albano, E., Rundgren, M., Harvison, P. J., Nelson, S. D., and Moldéus, P. (1985) *Mol Pharmacol* **28**, 306-311.
18. Albano, E., Poli, G., Chiarpotto, E., Biasi, F., and Dianzani, M. U. (1983) *Chem Biol Interact* **47**, 249-263.
19. Tirmenstein, M. A. and Nelson, S. D. (1989) *J Biol Chem* **264**, 9814-9819.
20. Younes, M., Sause, C., Siegers, C. P., and Lemoine, R. (1988) *J Appl Toxicol* **8**, 261-265.
21. Younes, M. and Siegers, C. P. (1985) *Chem Biol Interact* **55**, 327-334.

22. Younes, M., Cornelius, S., and Siegers, C. P. (1986) *Res Commun Chem Pathol Pharmacol* **51**, 89-99.

23. Mitchell, D. B., Acosta, D., and Bruckner, J. V. (1985) *Toxicology* **37**, 127-146.

24. Jollow, D. J., Mitchell, J. R., Zampaglione, N., and Gillette, J. R. (1974) *Pharmacology* **11**, 151-169.

25. Mitchell, J. R., Potter, W. Z., Hinson, J. A., and Jollow, D. J. (1974) *Nature* **251**, 508-511.

26. Timbrell, J. A., Mitchell, J. R., Snodgrass, W. R., and Nelson, S. D. (1980) *J Pharmacol Exp Ther* **213**, 364-369.

27. Mitchell, J. R., Thorgeirsson, U. P., Black, M., Timbrell, J. A., Snodgrass, W. R., Potter, W. Z., Jollow, H. R., and Keiser, H. R. (1975) *Clin Pharmacol Ther* **18**, 70-79.

28. Nelson, S. D., Mitchell, J. R., Snodgrass, W. R., and Timbrell, J. A. (1978) *J Pharmacol Exp Ther* **206**, 574-585.

29. Evans, M. A. (1983) *J Pharmacol Exp Ther* **224**, 73-79.

30. Bouis, P. and Boelsterli, U. A. (1990) *Toxicol Appl Pharmacol* **104**, 429-439.

31. Hinson, J. A., Pohl, L. R., and Gillette, J. R. (1979) *Life Sci* **24**, 2133-2138.

32. Hinson, J. A., Nelson, S. D., and Mitchell, J. R. (1977) *Mol Pharmacol* **13**, 625-633.

33. Tirmenstein, M. A. and Nelson, S. D. (1991) *Chem Res Toxicol* **4**, 214-217.

34. Baker, J. A., Hayden, J., Marshall, P. G., Palmer, C. H. R., and T. D. Whittet (1963) *J. Pharm. Pharmacol.* **15**, 97T-100T.

35. Nelson, E. B. (1981) *Chem. Abstr* **94**, 114719n.

36. Bulger, W. H., Temple, J. E., and Kupfer, D. (1983) *Toxicol Appl Pharmacol* **68**, 367-374.

37. Fouin-Fortunet, H., Lettéron, P., Tinel, M., Degott, C., Flejou, J. F., and Pessayre, D. (1984) *J Pharmacol Exp Ther* **229**, 851-858.

38. Hanzlik, R. P., Gillesse, T. J., and Wiley, R. A. (1981) *Adv Exp Med Biol* **136 Pt A**, 381-386.

39. Gottschall, D. W., Wiley, R. A., and Hanzlik, R. P. (1983) *Toxicol Appl Pharmacol* **69**, 55-65.

40. Wiley, R. A., Hanzlik, R. P., and Gillesse, T. (1979) *Toxicol Appl Pharmacol* **49**, 249-255.

41. Monks, T. J., Hinson, J. A., and Gillette, J. R. (1982) *Life Sci* **30**, 841-848.

42. Lau, S. S., Monks, T. J., Greene, K. E., and Gillette, J. R. (1984) *Toxicol Appl Pharmacol* **72**, 539-549.

43. Fieser, L. F. and Fieser, M., *Advanced Organic Chemistry*. 1961, New York: Reinhold publishing Corp. 845-851.

44. Rashed, M. S. and Nelson, S. D. (1989) *Chem Res Toxicol* **2**, 41-45.

45. Rashed, M. S., Myers, T. G., and Nelson, S. D. (1990) *Drug Metab Dispos* **18**, 765-770.

46. Tirmenstein, M. A. and Nelson, S. D. (1990) *J Biol Chem* **265**, 3059-3065.

47. Myers, T. G., Dietz, E. C., Anderson, N. L., Khairallah, E. A., Cohen, S. D., and Nelson, S. D. (1995) *Chem Res Toxicol* **8**, 403-413.

48. Matthews, A. M., Hinson, J. A., Roberts, D. W., and Pumford, N. R. (1997) *Toxicol Lett* **90**, 77-82.

49. Qiu, Y., Benet, L. Z., and Burlingame, A. L. (1998) *J. Biol. Chem.* **273**, 17940-17953.

50. Bansal, M. P., Mukhopadhyay, T., Scott, J., Cook, R. G., Mukhopadhyay, R., and Medina, D. (1990) *Carcinogenesis* **11**, 2071-2073.

51. Streeter, A. J., Bjorge, S. M., Axworthy, D. B., Nelson, S. D., and Baillie, T. A. (1984) *Drug Metab Dispos* **12**, 565-576.

52. Streeter, A. J., Dahlin, D. C., Nelson, S. D., and Baillie, T. A. (1984) *Chem Biol Interact* **48**, 349-366.

53. Zhou, L., McKenzie, B. A., Eccleston, E. D., Jr., Srivastava, S. P., Chen, N., Erickson, R. R., and Holtzman, J. L. (1996) *Chem Res Toxicol* **9**, 1176-1182.

54. Holtzman, J. L. (1997) *J Investig Med* **45**, 28-34.

55. Manautou, J. E., Khairallah, E. A., and Cohen, S. D. (1995) *J Toxicol Environ Health* **46**, 263-269.

56. Birge, R. B., Bartolone, J. B., McCann, D. J., Mangold, J. B., Cohen, S. D., and Khairallah, E. A. (1989) *Biochem Pharmacol* **38**, 4429-4438.

57. Kaster, J. K. and Yost, G. S. (1996) *Toxicologist* **30**, 283.

58. Chevallet, M., Santoni, V., Poinas, A., Rouquié, D., Fuchs, A., Kieffer, S., Rossignol, M., Lunardi, J., Garin, J., Rabilloud, T. (1998) *Electrophoresis* **19**, 1901-1909.

59. Esterline, R. L., Ray, S. D., and Ji, S. (1989) *Biochem Pharmacol* **38**, 2387-2390.

60. Katyare, S. S. and Satav, J. G. (1989) *Br J Pharmacol* **96**, 51-58.

61. Donnelly, P. J., Walker, R. M., and Racz, W. J. (1994) *Arch Toxicol* **68**, 110-118.

62. Burcham, P. C. and Harman, A. W. (1990) *Toxicol Lett* **50**, 37-48.

63. Strubelt, O. and Younes, M. (1992) *Biochem Pharmacol* **44**, 163-170.

64. Comporti, M. (1987) *Chem Phys Lipids* **45**, 143-169.

65. Uhlig, S. and Wendel, A. (1992) *Life Sci* **51**, 1083-1094.

66. Casini, A. F., Pompella, A., and Comporti, M. (1985) *Am J Pathol* **118**, 225-237.

67. Botti, B., Bini, A., Calligaro, A., Meletti, E., Tomasi, A., and Vannini, V. (1986) *Toxicol Appl Pharmacol* **83**, 494-505.

68. Broda, C., Nachtomi, E., and Alumot, E. (1976) *Gen Pharmacol* **7**, 345-348.

69. Penttil, K. E. (1988) *Chem Biol Interact* **65**, 107-121.

70. Comporti, M. (1989) *Chem Biol Interact* **72**, 1-56.
71. Meredith, M. J. and Reed, D. J. (1982) *J Biol Chem* **257**, 3747-3753.
72. Griffith, O. W. and Meister, A. (1985) *Proc Natl Acad Sci U S A* **82**, 4668-4672.
73. Griffith, O. W. (1982) *J Biol Chem* **257**, 13704-13712.
74. Meredith, M. J. and Reed, D. J. (1983) *Biochem Pharmacol* **32**, 1383-1388.
75. van Doorn, R., Leijdekkers, C. M., and Henderson, P. T. (1978) *Toxicology* **11**, 225-233.
76. Drew, R. and Miners, J. O. (1984) *Biochem Pharmacol* **33**, 2989-2994.
77. Jaeschke, H. (1990) *J Pharmacol Exp Ther* **255**, 935-941.
78. Mourelle, M., Beales, D., and McLean, A. E. (1990) *Biochem Pharmacol* **40**, 2023-2028.
79. Whitehouse, L. W., Wong, L. T., Paul, C. J., Pakuts, A., and Solomonraj, G. (1985) *Can J Physiol Pharmacol* **63**, 431-437.
80. Rumack, B. H., Peterson, R. C., Koch, G. G., and Amara, I. A. (1981) *Arch Intern Med* **141**, 380-385.
81. Bruno, M. K., Cohen, S. D., and Khairallah, E. A. (1988) *Biochem Pharmacol* **37**, 4319-4325.
82. Tee, L. B., Boobis, A. R., Huggett, A. C., and Davies, D. S. (1986) *Toxicol Appl Pharmacol* **83**, 294-314.

EXOCYCLIC DNA ADDUCTS AS SECONDARY MARKERS FOR OXIDATIVE STRESS: APPLICATIONS IN HUMAN CANCER ETIOLOGY AND RISK ASSESSMENT

Helmut Bartsch and Jagadeesan Nair

German Cancer Research Center (DKFZ)
Division of Toxicology and Cancer Risk Factors
Im Neuenheimer Feld 280
69120 Heidelberg
Germany

INTRODUCTION

Rationale for Using DNA Adduct Measurements in Cancer Etiology and Risk Assessment

DNA-bound carcinogen adducts reflect the amount of an exogenous chemical or its metabolite that covalently interacted with nucleic acid bases at the target site (biologically effective dose) or in surrogate tissues. DNA adducts are mechanistically more relevant to carcinogenesis than internal doses of genotoxins, since they take into account interindividual differences in metabolism and of DNA repair capacity. The rationale for measuring DNA adducts as relevant dosimeters of biological effects and predictor of cancer risk is derived from extensive experimental and human data, supporting their role in the initiation and probably the progression of cancer. Several hundreds of DNA adducts, many with miscoding properties, are known to be produced by some 20 classes of carcinogens and through endogenous processes including oxidized DNA bases (Hemminki et al., 1994). These lesions provide powerful tools for studying disease pathogenesis, etiology and for verifying preventive measures in human cancer or other chronic degenerative diseases.

Supportive evidence for the biological significance of DNA adducts in carcinogenesis includes: (i) over 80 % of identified (or suspected) human chemical carcinogens, often after metabolic activation, react with nucleic acids and proteins to form macromolecular products (adducts); (ii) carcinogen-DNA adducts represent the initiating events leading to mutations in oncogenes and tumor suppressor genes and malignant cell transformation; (iii) the carcinogenic potency of a large number of chemicals was shown to be proportional to their ability to bind to rodent liver DNA; (iv) human subjects who inherited syndromes with a defect in DNA repair are highly cancer-prone.

Biological Reactive Intermediates VI, Edited by Dansette *et al.*
Kluwer Academic / Plenum Publishers, 2001

Biological effect markers in the multistage carcinogenesis process are defined as indicators of irreversible damage resulting from toxic interactions at the target site; the resulting adverse effects should be pathologically linked to cancer development. The DNA adducts do not represent totally irreversible lesions, as they undergo repair (which may not be complete). In addition the adducted cell may cease to exist due to apoptososis or necrosis, before the adduct is converted into a mutation during cell replication. For all these reasons they were in the past considered not to be biological effect markers in the strict sense. However, as the carcinogen dose is linked to cancer outcome, DNA adducts as markers of the biologically effective dose, of which a part is converted into permanent mutations, must also be associated with cancer risk. This has been shown for several carcinogens and their adducts in experimental animals, when certain critical toxico-kinetic parameters are taken into account, e.g. steady-state concentration of the adduct, measurement of the critical adduct among other biologically less relevant ones, organ, cell and gene selectivity, and adduct persistence after cessation of exposure. Also more recently in a prospective study in humans, aromatic-DNA adducts were a predictor of individual lung cancer risk (Tang et al., 2000).

Not all types of DNA adducts are associated with the same cancer risk. A comparison of DNA adduct concentrations derived from alkylating agents, aflatoxins and aromatic amines in target tissues of experimental animals (that induced 50 % tumor incidence, TD_{50}) revealed an up to 40 fold range in the ability of DNA adducts to induce the same tumor incidence (Ottender and Lutz, 1999). This makes a prediction of tumor induction potential of unknown DNA adducts difficult. In the past, most available assays for DNA adduct determination provided information on the total amount of adducts in bulk genomic DNA, but new methods are capable of pinpointing critical targets in DNA at a gene or nucleotide sequence level. Because of the multistage and multifactorial nature of human carcinogenesis, carcinogen-macromolecular adducts *per se* have so far not been shown to be a precise and quantitative predictor of individual cancer risk. Therefore, risk estimation has so far been confined to a group level (Kriek et al., 1998). However, incorporation of DNA adduct measurements (and of other critical endpoints involved in carcinogenesis) will reduce (a) the enormous uncertainties associated today with high-to-low dose and species-to-species extrapolation and (b) yield information on inter-individual and inter-ethnic differences among human populations that can be integrated into risk assessment procedures (La and Swenberg, 1996; Wild and Pisani, 1997).

By using ultrasensitive analytical methods (e.g. immunoaffinity chromatography coupled with ^{32}P-postlabelling) DNA adducts have been detected in unexposed humans and untreated animals. Lipid peroxidation (LPO) products, often produced in response to oxidative stress, are increasingly associated with a variety of pathological conditions including cancer. As a consequence of physiological processes, LPO products such as hydroxyalkenals are formed in human tissues and react with DNA to yield background levels of a variety of adducts including etheno (ε) DNA adducts (Gupta and Lutz, 1999). These background adducts have been shown in general to increase with age, both in humans and in experimental animals, also great variations between human subjects were found in particular, when they were affected by risk factors producing oxidative stress, including chronic inflammatory processes, infections, nutritional imbalances, and metal storage disorders. In addition to DNA-reactive LPO products, oxidized DNA bases have been recognized as common lesions that occur more frequently in cells where antioxidant defense mechanisms are impaired (Will et al., 1999). The biological relevance of both types of DNA lesions is supported by the fact that many are miscoding lesions and are recognized by specific DNA-repair enzymes. Moreover, exogenous carcinogens can induce oxidative stress, and secondary oxidative DNA base damage is produced in addition to agent-specific DNA adducts. It is assumed that both types of lesions play a role in the initiation and promotion of the multistage carcinogenesis process, but at present their individual contribution (e.g., on the shape of the dose-response curve for a given carcinogen) cannot be discerned. Several questions remain to be answered: (i) What is the significance of endogenously formed adducts in human cancer, in particular in relation to spontaneous tumors? (ii) Have cancers arising from environmental agents *vs* those arising from endogenous processes been overestimated? (iii) Can one protect humans against endogenously derived DNA damage and degenerative diseases by administration of chemopreventive agents or dietary regimens using DNA adduct measurement to verify their preventive efficacy? This brief review summarizes recent results on the formation of endogenous sources, occurrence, detection and possible role of etheno- (ε) DNA adducts in carcinogenesis and risk prediction. Because of space limitations the literature citations are not exhaustive, and for a more detailed treatise the reader is referred to a recent compilation of this research area (Singer and Bartsch, 1999).

RESULTS AND DISCUSSION

Methods for Detection of Etheno Adducts

Etheno (ε) modified DNA bases 1,N^6-ethenodeoxyadenosine (εdA), 3,N^4-etheno-deoxycytidine (εdC), N^2,3-ethenodeoxyguanosine (εdG) are generated from the carcinogens vinyl chloride and urethane, but also by reactions of DNA with products derived from lipid peroxidation (LPO) and oxidative stress *via* endogenous pathways (El Ghissassi et al., 1995 reviewed in Chung et al., 1996; Bartsch, 1999). Fig. 1 depicts a mechanism for the formation of etheno adducts from DNA nucleosides and LPO products such as *trans*-4-hydroxy-2-nonenal (HNE) derived from PUFAs such as linoleic acid and of exogenous carcinogens.

Recently developed ultrasensitive methods allowed to detect these ε-adducts *in vivo* and to study their role in experimental and human carcinogenesis. εdA and εdC were analyzed in DNA of various human and animal tissues by immunoaffinity/^{32}P-postlabelling method (Nair et al., 1995). In brief 10 – 25 μg of DNA hydrolyzed to nucleotide-3'-monophosphates, purified by immunoaffinity columns prepared from specific monoclonal antibodies (MAb), labeled with ^{32}P-ATP and quantified after two-dimensional TLC by scintillation counting.

A new method has been developed to detect εdA at a cellular level by immuno-histochemistry (Yang et al., 2000) using an MAb raised against εdA in DNA (antibody clone EM-A-4). Semi-quantitative image analysis of relative pixel intensity showed ~1.5 times higher adduct levels ($p < 0.05$) in the livers of rats treated with vinyl chloride when compared to untreated controls.

εdA is detected in urine and could arise either by formation from HNE and deoxyadenosine in the nucleoside pool or as a consequence of DNA repair (Saparbaev et al., 1995; Hang et al., 1996). εdA can be analyzed according to a recently developed method (Nair, 1999). In spot urine samples from male and female volunteers levels ranged from 0.27 to 4.4 fmol εdA / μmol creatinine. Hereby urine samples were spiked with 1,N^6-ethenodeoxyadenosine-[^{3}H] (internal standard) and purified on an anion exchange-C18 column, followed by semi-preparatory HPLC on a C18 column and isocratic elution with ammonium acetate/methanol and an immunoaffinity column (prepared with MAb EM-A-1) and analyzed by an HPLC-fluorescence detection. These methods provided new tools to study the role of promutagenic etheno-DNA adducts as lead markers for oxidative stress and LPO-induced DNA damage that may drive cells towards malignancy, as briefly described below.

Background Levels in Unexposed Rodents and Humans and Diet-Related Effects

The use of the ultrasensitive immunopurification-^{32}P-postlabelling assay revealed the existence of background levels of εdA and εdC in tissues from unexposed rodents and humans (Nair et al. 1995, 1999). In adult and 12-day-old mice, levels of both adducts ranged from 0.3 to 1.4 x 10^{-8} in liver and lungs, showing some strain variations. Preliminary data suggested that these background levels in liver DNA were affected by the type of animal diet. Animals given AIN-76 diet exhibited higher etheno adduct levels than mice fed other diets, i.e. a high fat diet, Wayne's Breeder Blocks or a non-purified natural ingredient diet (NIH-07) (Fernando et al. 1996). AIN-76 as compared to NIH-07 diet has been associated with a higher spontaneous incidence of mouse liver neoplasias. This difference has been related to cancer protective antioxidants present in NIH-07 diet, which might also inhibit the formation of endogenous ε-adducts from LPO. More studies are warranted to elucidate the role of ε-adducts in spontaneous neoplasia of the liver. Liver DNA samples from humans with unknown exposures showed the presence of εdA and εdC residues in the range of $\leq$ 0.05 to 4 adducts per 10^8 parent bases. Comparable but variable εdA and εdC levels were detected in DNA of asymptomatic human esophageal, colonic and pancreatic tissues and in leukocytes from male and female volunteers on a low linoleic acid diet.

Etheno-DNA Adducts as Lead Markers for Secondary Oxidative DNA-Base Damage: Possible Risk Markers in the Promotion and Progression of Multistage Carcinogenesis

The highly variable background levels of εdA and εdC that were detected in asymptomatic human tissue DNA are apparently a result of LPO occurring under normal physiological conditions. However, etheno adduct levels were increased by several known pathological conditions predisposing to a higher cancer risk (summarized in Table 1). Elevated ε-adducts were found in hepatic DNA from patients and rodents with metal storage diseases, such as patients with Wilson's disease and primary hemachromatosis, after overproduction of nitric oxide (NO) by inducible NO-synthese (iNOS) in a mouse model, in colonic polyps of familial adenomatous polyposis (FAP) patients and in affected organs of patients with chronic inflammatory conditions, such as those with inflammatory bowel disease or chronic pancreatitis; all these are known as risk factors for many human cancers (Ohshima & Bartsch, 1994).

Table 1. Acquired or inherited cancer risk factors that enhanced significantly (2-40-fold) the formation of etheno-DNA adducts (εdA; εdC) in organs or white blood cells of humans and experimental animals. These adducts from endogenous sources are likely generated in cancer-prone organs as a consequence of oxidative stress *via* DNA-reactive LPO-derived hydroxyalkenals.

Predisposing pathological condition / risk factor	Affected organ/cell with increased etheno-DNA adducts	Reference
Human		
Wilson's disease (WD)	Liver	Nair et al., 1998*a*
Primary hemochromatosis	Liver	Nair et al., 1998*a*
Familial adenomatous polyposis (FAP)	Colonic polyps	Schmid et al., 2000
Inflammatory bowel disease	Colon	Nair et al., 2000*a*
Chronic pancreatitis	Pancreas	Bartsch and Nair, 1999
High linoleic acid diet in female volunteers	White blood cells	Nair et al., 1997
Experimental animals		
Long Evans Cinnamon rats (18 weeks old, model for WD)	Liver	Nair et al., 1996
SD rats treated by an iron overload	Liver	Nair et al., 1999
SJL mice with 50-fold overproduction of NO by iNOS	Spleen	Nair et al., 1998*b*

A high ω-6-polyunsaturated fatty (linoleic) acid diet increased ε-DNA adducts in white blood cells of female subjects (Nair et al., 1997). As evident from Table 1 adduct levels were thus found to be progressively elevated in preneoplasias of cancer-prone patients and rodents, suggesting that these promutagenic lesions, together with other oxidative damage can drive cells to malignancy (Nair et al., 1998*a*; Nair et al., 2000*b*). The notion that "oxidative stress" indeed plays a critical role in cancer development is supported by an increasing body of evidence, showing that antioxidants such as the vitamins A, C, and E, as well as plant-derived phenolic and thiol compounds can effectively prevent or delay cancer development in man and experimental animals (Tanaka, 1997).

The hypothesis implicating DNA damage by oxidative stress and lipid peroxidation as a cause of genetic instability, observed at later stages of tumor promotion was further investigated in NMRI mouse skin carcinogenesis induced by 7,12-dimethylbenz[a]anthracene (DMBA) and promoted by 12-*O*-tetradecanoylphorbol-13-acetate (TPA). In this model a clear-cut distinction of defined stages of carcinogenesis has been elaborated, being a prerequisite for attributing genetic alterations to distinct phenotypic changes (Marks and Fürstenberger, 1995; Di Giovanni, 1992). Cancer induction in mouse skin is initiated by single administration of a low dose of DMBA, thought to cause an activating mutation of the cellular proto-oncogene ras (Brown et al., 1994). Repeated applications of the tumor promoter (TPA) are thought to stimulate clonal expansion due to different susceptibilities of initiated cells *versus* normal cells. Biological endpoints of the promotion stage are papillomas which either regress or grow autonomously. The latter either spontaneously or under continuous promoter treatment develop into squamous cell carcinomas of the skin (Fürstenberger and Kopp-Schneider, 1995).

The process of malignant progression in mouse skin carcinogenesis is accompanied by an accumulation of additional genetic alterations, pointing to an increased genetic instability of papilloma cells (Portella et al., 1994), whereby increased "oxidative stress" is thought to contribute to genetic instability by generating a cascade of endogenous genotoxins. An abundant source of DNA-reactive oxygen species is provided by lipid hydroperoxides as intermediary reaction products generated along cyclooxygenase- (COX) and lipoxygenase- (LOX) catalyzed arachidonic acid (AA) oxygenation (Marnett, 1995). Experimentally induced skin tumors accumulate large amounts of prostaglandins and 8-and 12-hydroxyeicosatetraenoic acid (8- and 12-HETE) and constitutively overexpress the corresponding enzymes, i.e. COX-2, 8-LOX and the platelet-type 12-LOX (Müller-Decker et al., 1995; Krieg et al., 1995, Bürger et al., 1999), indicating a tumor-specific deregulation of these pathways. Inhibition of COX and LOX has been shown to significantly reduce tumor development in mouse skin (Müller-Decker et al., 1998).

We have now, for the first time, shown a close correlation between an upregulation of LOX-catalyzed AA metabolism and the formation of εdA and εdC adducts in DNA during tumor development by the initiation-promotion protocol of mouse skin carcino-genesis (Fig. 2). A close correlation between accumulation of 8-HETE and 12-HETE and formation of etheno-DNA adducts was observed at distinct stages of tumor development (Nair et al., 2000*b*). Reversible papillomas showed the highest contents of 8- and 12-HETE and at the same time the highest ε-adduct levels, when compared to age-matched control mice. DNA from irreversible papillomas and carcinomas showed decreasing contents of both HETE correlating with reduced ε-DNA adduct levels. Moreover, both increased levels of HETE and elevated numbers of DNA adducts were restricted to the preneoplastic tissue, whereas these parameters were only slightly increased in chronic hyperplastic epidermis obtained by TPA treatment alone (Fig. 2).

Reversible papillomas obtained after 20 weeks of TPA treatment had 15 to 68 times higher contents of HETEs, which was paralleled by 9 to 12 times increased amounts of etheno adducts. When compared to vehicle-treated control skin, these elevations were statistically significant. In irreversible papillomas harvested after 40 weeks of TPA treatment, the levels of HETEs and etheno-DNA adducts were found to be slightly reduced, as compared to reversible papillomas, but were still increased over control levels in age-matched mice. Comparison of mean group values by simple regression analysis showed highly significant positive correlations between HETEs and etheno-DNA adduct levels, indicating that the oxidative metabolism of AA and related polyun-saturated fatty acids may be involved in ε-adduct generation. Taken together our data suggest a mechanistic link between AA metabolism, persistent oxidative stress and LPO-

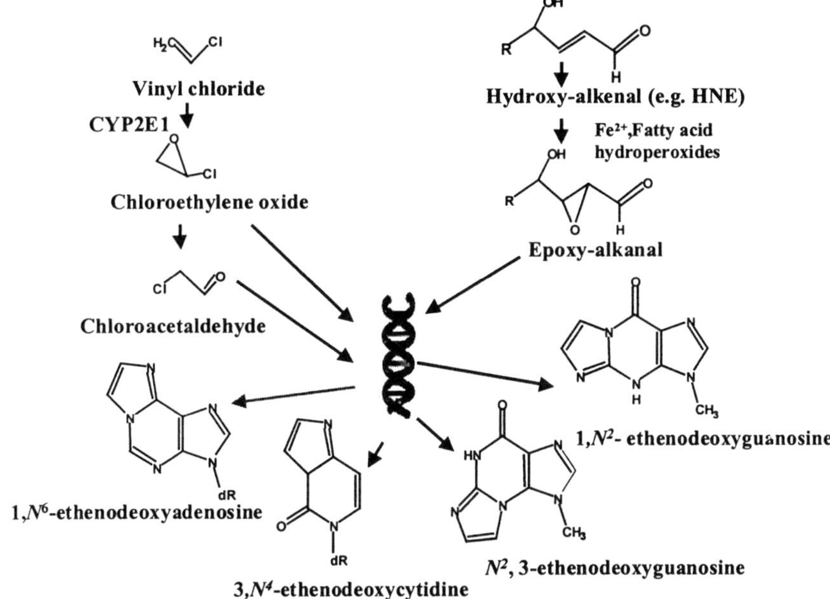

Figure 1. Mechanisms for the formation of etheno adducts from DNA nucleosides from exogenous and endogenous sources, as exemplified for vinyl chloride metabolites and hydroxyalkenals (HNE, *trans*-4-hydroxy-2-nonenal) derived by lipid peroxidation from polyunsaturated fatty acids.

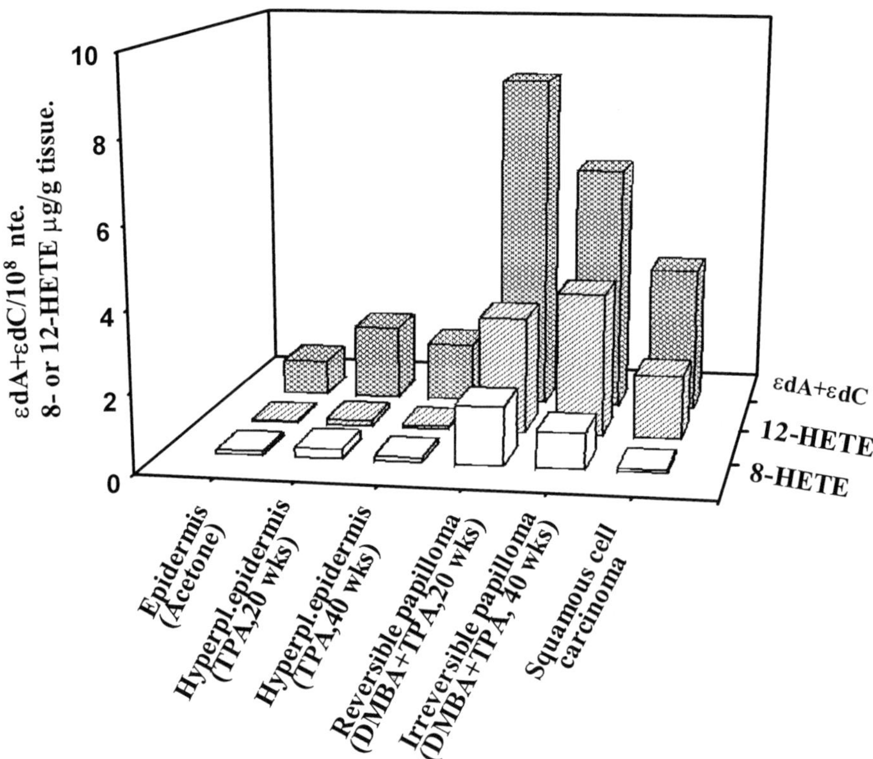

Figure 2. Formation of the lipoxygenase- (LOX) catalyzed metabolites of arachidonic acid 8- and 12-hydroxyeicosatetraenoic acid (8- and 12-HETE), and of the exocyclic DNA adducts, 1,N^6-ethenodeoxyadenosine (εdA) and 3,N^4-ethenodeoxycytidine (εdC) in the course of DMBA-TPA induced tumor development in mouse skin (extracted from Nair et al. 2000*b* and replotted).

induced DNA damage in multistage skin carcinogenesis. When constitutively overexpressed in mouse skin papillomas, we hypothesize that 8- and 12-LOX (or COX-2) catalyze AA oxygenation that is accompanied by generation of reactive oxygen species (ROS). These can trigger LPO of polyunsaturated fatty acids in an autocatalytic process. The reactive aldehydes generated can attack DNA to form exocyclic etheno-base adducts (Fig. 1). In rapidly dividing papilloma cells, the resulting mutational and cytogenetic damage may drive premalignant cells to genetic instability and subsequently to malignancy.

Increasing experimental data indicate the concurrent formation of oxidative DNA damage and/or of ε-DNA bases through LPO, either during or after exposure to exogenous genotoxins. This implies that agent-specific DNA adduct formation and toxicity may be accompanied, particularly at high exposure doses and/or GSH depletion, by secondary, oxidative stress-related DNA lesions, which cause genetic changes resulting in disease progression. If true, the extent of secondary DNA damage may be a rate limiting step in carcinogenesis and could serve as a better predictor of individual cancer risk, rather than the initial carcinogen adduct level alone. This question was addressed using the skin tumor model with outbred male NMRI mice. The latency time for the appearance of a papilloma was used as an indicator of the individual cancer risk following treatment of mice twice weekly with 7,12-dimethylbenz[a]anthracene (DMBA) applied to back skin (Fischer and Lutz, 1995).

Levels of DMBA-DNA adducts, of 8-hydroxy-2'-deoxyguanosine (8-oxodG) and various measures of the kinetics of cell division were determined in the epidermis of the treated skin area. The levels of 8-oxodG and the fraction of cells in DNA replication (labeling index after incorporation of 5-bromo-2'-deoxyuridine) were significantly higher in those mice that showed short latency times. On the other hand, the levels of DMBA-DNA adducts were lowest in animals with short latency times. The latter finding can be explained as a consequence of the inverse correlation seen for the labeling index: with each round of cell division, the adduct concentration is reduced to 50 % because the new DNA strand is free of DMBA adducts until the next treatment. In this animal model, therefore, oxygen radical-related genotoxicity and the rate of cell division, rather than levels of initial carcinogen-DNA adducts, were found to be of predictive value as indicators of an individual cancer risk. Given the strong increase of ε-adducts in skin DNA in the two stage DMBA-TPA skin carcinogenesis model (Fig. 2), one could assume that this secondary stable marker for oxidative stress could even be a more reliable parameter for individual risk prediction. This in view of the problems associated with the correct quantitation of 8-oxodG level and possible artifact formation (Collins et al., 1996).

For cancer risk assessment at low levels of DNA damage, exposure-related adducts must be discussed in relation to background DNA damage and its inter- and intra-individual variability as effected by many dietary and pathological factors (Gupta and Lutz, 1999). Therefore, identification of endogenous sources for DNA damage and quantification of the resulting oxidative modification of cellular components together with systematic DNA-biomarker application in human studies should provide a better basis for risk assessment of chronic degenerative diseases, perhaps even at an individual level.

CONCLUSIONS AND PERSPECTIVES

Persistent oxidative stress and LPO cause DNA damage and disturbance of cell signaling pathways, and they are implicated in human cancers and other degenerative diseases. Etheno (ε) modified DNA bases (εdA, εdC, εdG) are generated not only from vinyl chloride and urethane, but also by reactions of DNA with LPO products (4-hydroxyalkenals) derived from endogenous sources. Thus, these more stable secondary oxidation products are among the useful markers for oxidative stress-derived DNA damage. The recent development of ultrasensitive methods has made it possible to detect these ε-adducts *in vivo* and to study their formation and role in experimental and human carcinogenesis. Highly variable background levels of ε-adducts were detected in tissues from unexposed humans and rodents, suggesting an endogenous pathway of formation. The level of these DNA lesions in target organs was increased by several known cancer risk factors, e.g. high ω-6 PUFA diet, chronic inflammatory processes, NO overproduction, upregulation of LOX/COX-2. ε-Adducts were found to be significantly elevated in organs (liver, pancreas, colon, skin) of cancer-prone patients and/or rodents, suggesting that promutagenic ε-adducts when formed by persistent oxidative stress may play a role in the acquisition of genetic instability and malignancy.

Increasing experimental data indicate the possible concurrent formation of oxidative DNA base damage and/or of ε-DNA bases through LPO, either during or after exposure to exogenous genotoxins (Guichard et al., 1996). This implies that agent-specific DNA adduct formation and toxicity may be accompanied, particularly at high exposure doses and/or GSH depletion, by secondary, oxidative stress-related DNA lesions, which cause genetic changes resulting in disease progression. This could be one of the reasons why agent-specific genetic fingerprints of the primary carcinogen are often not found in cancer-relevant genes of the tumor but show a complex mutational profile, possibly caused by secondary DNA modifications. Measurement of DNA adduct levels related to many genotoxic carcinogens in target tissues has the potential to be not only an exposure marker but also a predictor of cancer risk at a group level. For risk prediction at an individual level, oxygen related genotocicity and the rate of cell division, rather than levels of the carcinogen DNA-adduct were found to predictor of skin cancer risk in a mouse tumor model with DMBA as carcinogen.

Furthermore, for cancer risk assessment at low levels of DNA damage, external exposure-related adducts must be considered in relation to background DNA damage (from both exogenous and endogenous sources) and its inter- and intra-individual variability, as effected by many dietary and pathological factors. Therefore, identification of endogenous sources for DNA damage and quantification of the resulting oxidative modification of cellular components will give new mechanistic insights into the carcinogenesis process, in particular for human cancers with an ill-defined etiopathogenesis. Systematic DNA-biomarker application in human studies should provide a better basis for risk assessment of chronic degenerative diseases.

Acknowledgement – The authors' research in this area was in part supported by EU contract ENV4-CT97-0505. The authors wish to acknowledge the work contributed by A. Barbin, K. Schmid, Y. Yang and the collaboration with H.G. Beger, G. Fürstenberger and G. Winde. Dr. P. Lorenz, University of Essen, Essen, Germany is thanked for providing the MAb used in these studies. C. Ditrich and I. Hofmann are thanked for skilled technical assistance and S. Fuladdjusch for excellent secretarial help.

References

Bartsch, H., and Nair, J., 1999, Persistent oxidative DNA damage as a driving force in pancreas and colon carcinogenesis, *Langenbeck's Archives of Surgery* 384-397.

Bartsch, H., 1999, Exocyclic adducts as new risk markers for DNA damage in man, in: *Singer, B., Bartsch, H.(eds) Exocyclic DNA Adducts in Mutagenesis and Carcinogenesis*, IARC Sci. Publ. No 150, IARC, Lyon 1-16.

Brown, K., Kemp, C., Burns, P., and Balmain, A., 1994, Importance of genetic alterations in tumour development, *Arch. Toxicol. Suppl.* 16:253-260.

Bürger, F., Krieg, P., Kinzig, A. Schurich, B., Marks, F., and Fürstenberger, G., 1999, Constitutive expression of 8-lipoxygenase in papillomas and clastogenic effect of lipoxygenase-derived arachidonic acid metabolites in keratinocytes, *Mol. Carcinogenesis* 24:108-118.

Chung, F.L., Chen, H.J.C., Nath, R.G., 1996, Lipid peroxidation as a potential endogenous source for the formation of exocyclic DNA adducts, *Carcinogenesis* 17:2105-2111.

Collins, A.R., Dusinska, M., Gedik, C.M., Stetina, R., 1996, Oxidative damage to DNA: do we have a reliable biomarker? *Environ Health Perspect* 104 Suppl 3:465-469

Di Giovanni, J., 1992, Multistage carcinogenesis in mouse skin, *Pharmacol. Ther.* 54: 63-128.

El Ghissassi, F.F., Barbin, A., Nair, J., Bartsch, H., 1995, Formation of $1,N^6$-ethenoadenine and $3,N^4$-ethenocytosine by lipid peroxidation products and nucleic acid bases, *Chem Res Toxicol* 8:278-283.

Fernando, R.C., Nair, J., Barbin, A., Miller, J.A., Bartsch H.,1996, Detection of $1,N^6$-ethenodeoxyadenosine and $3,N^4$-ethenodeoxycytidine by immunoaffinity/^{32}P-postlabelling in liver and lung DNA of mice treated with ethyl carbamate (urethane) or its metabolites, *Carcinogenesis* 17:1711-1718.

Fischer, W.H., and Lutz, W.K., 1995, Correlation of individual papilloma latency time with DNA adducts, 8-hydroxy-2'-deoxyguanosine, and the rate of DNA synthesis in the epidermis of mice treated with 7.12-dimethylbenz[a]anthracene, *Proc. Natl. Acad. Sci. USA* 92:5900-5904.

Fürstenberger, G., and Kopp-Schneider, A., 1995, Malignant progression of papillomas induced by the initiation-promotion protocol in NMRI mouse skin, *Carcinogenesis* 16:61-69.

Guichard, Y., El Ghissassi, F., Nair, J., Bartsch, H., Barbin, A., 1996, Formation and accumulation of DNA ethenobases in adult Sprague Dawley rats exposed to vinyl chloride, *Carcinogenesis* 17:1553-1559.

Gupta, R.C., and Lutz, W.K., 1999, From endogenous and unavoidable exogenous carcinogens – a basis for spontaneous cancer incidence? *Mutat Res* 424:1-8.

Hang, B., Chenna, A., Rao, S., Singer, B., 1996, $1,N^6$-Ethenoadenine and $3,N^4$-ethenocytosine are excised by separate human DNA glycosylases, *Carcinogenesis* 17:155-157.

Hemminki, K., Dipple, A., Shuker, D.E.G., Kadlubar, F.F., Segerbäck, D., Bartsch, H., (eds), 1994, DNA Adducts: Identification and Biological Significance, *IARC Sci Publ 125.* IARC, Lyon 1-478.

Krieg, P., Kinzig, A., Ress-Löschke, M., Vogel, S., Vanlandingham, B., Stephan M., Lehmann W. D., Marks, F., and Fürstenberger, G., 1995, 12-Lipoxygenase isoenzymes in mouse skin tumor development. *Mol. Carcinogenesis*, 14:118-129.

Kriek, E., Rojas, M., Alexandrov, K., Bartsch, H., 1998, Polycyclic aromatic hydrocarbon-DNA adducts in humans: relevance as biomarkers for exposure and cancer risk, *Mutat Res* 400:215-231.

La, D.K., Swenberg, J.A., 1996, DNA adducts: biological of exposure and potential applications to risk assessment, *Mutat Res* 365(1-3):129-146.

Marks, F., and Fürstenberger, G., 1995, in: *Chemical induction of cancer, Arcos, J.C., Argus, M.F. and Woo, Y.-T. (eds), Birkhäuser, Boston,* 125-160.

Marnett, L.J., 1995, Generation of mutagens during arachidonic acid metabolism, *Cancer Metastasis Rev.* 13:303-308.

Müller-Decker, K., Scholz, K., Marks, F., and Fürstenberger, G., 1995, Differential expression of prostaglandin H synthase isozymes during multistage carcinogenesis in mouse epidermis, *Mol. Carcinogenesis* 12:31-41.

Müller-Decker, K., Kopp-Schneider, A., Marks, F., Seibert, K., and Fürstenberger, G., 1998, Localization of prostaglandin H synthase isoenzymes in murine epidermal tumors: suppression of skin-tumor promotion by PGHS-2 inhibition, *Mol. Carcinogenesis* 23:36-44.

Nair, J., Barbin, A., Guichard, Y., Bartsch, H., 1995, $1,N^6$-ethenodeoxyadenosine and $3,N^4$-ethenodeoxy-cytidine in liver DNA from humans and untreated rodents detected by immunoaffinity/^{32}P-postlabelling, *Carcinogenesis* 16:613-617

Nair, J., Sone, H., Nagao, M., Barbin, A., Bartsch, H., 1996, Copper-dependent formation of miscoding ehteno-DNA adducts in the liver of Long Evans Cinnamon (LEC) rats developing hereditary hepatitis and hepatocellular carcinoma, *Cancer Res* 56:1267-1271.

Nair, J., Vaca, C.E., Velic, I., Mutanen, M., Valsta, L.M., Bartsch, H., 1997, High dietary ω-6 polyunsaturated fatty acids drastically increase the formation of etheno-DNA base adducts in white blood cells of female subjects. *Cancer Epidemiol Biomarkers Prev* 6:597-601.

Nair, J., Carmichael, P.L., Fernando, R.C., Phillips. D,H,, Strain, A.J., Bartsch, H., 1998*a*, Lipid peroxidation-induced etheno-DNA adducts in liver of patients with the genetic metal storage disorders Wilson's disease and primary hemochromatosis, *Cancer Epidemiol Biomarkers Prev* 7:435-440.

Nair, J., Gal, A., Tamir, S., Tannenbaum, S., Wogan, G., Bartsch, H., 1998*b*, Etheno adducts in spleen DNA of SJL mice stimulated to overproduce nitric oxide, *Carcinogenesis*, 19:2081-2084.

Nair, J., 1999, Lipid peroxid-induced etheno-DNA adducts in humas, in: *Singer, B., Bartsch, H.,(eds) Exocyclic DNA Adducts in Mutagenesis and Carcinogenesis, IARC Sci. Publ. No 150,* IARC, Lyon 55-62.

Nair, J., Barbin, A., Velic, I., Bartsch, H., 1999, Etheno DNA-base adducts from endogenous reactive species, *Mut Res* 424:59-69.

Nair, J., Schmid, K., Winde, G., Beger, H., Dolara, P., Bartsch, H., 2000*a*, Miscoding etheno-DNA adducts in colonic epithelium of patients suffering from Crohn's disease, ulcerative colitis and familial adenomatous polyposis, *Proc AACR* 41:594

Nair, J., Fürstenberger, G., Bürger, F., Marks, F., Bartsch, H., 2000*b*, Promutagenic etheno-DNA adducts in multistage mouse skin carcinogenesis: correlation with lipoxygenase-catalyzed arachidonic acid metabolism, *Chem Res Toxicol* in press

Ottender, M., and Lutz, W.K., 1999, Correlation of adduct levels with tumor incidence: Carcinogenic potency of DNA adducts, *Mutat Res* 424:237-247.

Ohshima, H., Bartsch, H., 1994, Chronic infections and inflammatory processes as cancer risk factors: possible role of nitric oxide in carcinogenesis, *Mutat Res* 305:253-264.

Portella, G., Liddell, J., Crombie, R., Haddow, S., Clarke, M., Stoler, A. B., and Balmain, A., 1994, Molecular mechanisms of invasion and metastasis during mouse skin tumor progression, *Invasion & Metastasis* 14:1-6.

Saparbaev, M., Kleibl, K., Laval, J., 1995, *Escherichia coli, Saccharomyces cerevisiae*, rat and human 3-methyladenine DNA glycosylases repair 1,N^6-ethenoadenine when present in DNA, *Nucleic Acids Res* 23:3750-3755.

Schmid, K., Nair, J., Winde, G., Velic, I., Bartsch, H., 2000, Increased levels of promutagenic etheno-DNA adducts in colonic polyps of FAP patients, *Int J Cancer* 87:1-4.

Singer, B., Bartsch, H., 1999, (eds) Exocyclic DNA adducts in mutagenesis and carcinogenesis, IARC, *Sci. Publ. No 150,* IARC, Lyon 1-36.

Tanaka, T., 1997, Effect of diet on human carcinogenesis, *Crit Rev Oncol Hematol* 25:73-95.

Tang, D., Mooney, L., Philipps, D., Hsu, Y., Stampfer, M., Perera, F., 2000, Aromatic DNA adducts as a predictor of lung cancer risk in a nested case-control study, *Proc. AACR* 41:221.

Wild, C.P., Pisani, P., 1997, Carcinogen-DNA and carcinogen-protein adducts in molecular epidemiology, in: *Toniolo P, Boffetta P, Shuker DEG, Rothman N, Hulka B, Peace N (eds) Application of Biomarkers in Cancer Epidemiology. IARC Sci Publ 142.* IARC, Lyon 143-158.

Will, O., Mahler, H.-C., Arrigo, A.-P., Epe, A., 1999, Influence of glutathione levels and heat-shock on the steady-state levels of oxidative DNA base modifications in mammalian cells, *Carcinogenesis* 20:333-337.

Yang, Y., Nair, J., Barbin, A., Bartsch, H., 2000, Immunohistochemical detection of 1,N^6-ethenodeoxyadenosine, a promutagenic DNA adduct, in liver of rats exposed to vinyl chloride or an iron overload, *Carcinogenesis* 21:777-781.

DOSE DEPENDENT INDUCTION OF DNA ADDUCTS, GENE MUTATIONS, AND CELL PROLIFERATION BY THE ANTIANDROGENIC DRUG CYPROTERONE ACETATE IN RAT LIVER

T. Wolff, J. Topinka, E. Deml , D. Oesterle, L. R. Schwarz

GSF-Forschungszentrum für Umwelt und Gesundheit, Institut für Toxikologie, D-85764 Neuherberg , Germany

1. INTRODUCTION

Formation of DNA adducts is regarded as the initial event in the multistep process of chemical carcinogenesis. The primary DNA damage can become permanent when the cells carrying the primary DNA damage divide and gene mutations arise. A wide variety of bacterial and mammalian in vitro systems have been developed to test for mutations and a plethora of chemicals have been already identified as mutagens in these systems. However, the in vitro tests have limitations. For example, the metabolic systems employed to activate promutagens may be not appropriate to activate every test compound. Moreover, in vitro systems are not suitable to study all factors affecting the mutagenicity of a compound in vivo, such as pharmakokinetics of the genotoxicant, cell proliferation and apoptosis. Therefore, in vivo studies are required to obtain realistic data on the mutagenicty of chemical agents which may improve the scientific basis for risk estimation.

In this presentation we report on in vivo studies carried out to assay the progesterone analogue, cyproterone acetate (CPA) for mutagenicity and to examine the role of DNA adduct formation and cell proliferation as modifying factors in mutagenesis. CPA is an antiandrogenic drug widely used in women for long term treatment of excess androgen levels leading to acne and hirsutism. The compound has turned out to be an unique example among liver tumorigens. It causes the formation of benign liver tumors in rats at high doses after long term treatment (Schuppler and Günzel, 1979). The tumorigenic activity was formerly attributed to a tumor promoting activity (Schulte-Hermann et al. 1983). Findings from our and other laboratories, however, indicated that CPA causes massive DNA damage. CPA induces DNA repair synthesis in hepatocytes of female rats (Neumann et al. 1992) and in human hepatocytes (Martelli et al., 1996). The drug induces the formation of specific DNA adducts (Topinka et al. 1993, 1996) extensively studied in our lab. A strong sex specificity was noted for the adduct pattern and the adduct levels. High adduct levels were found in hepatocytes and liver of female rats, whereas males showed about 40 times lower adduct levels (Topinka et al., 1993). The adducts formed showed a remarkably high persistence in the liver of fenale rats with a half life time of about 6-7 weeks. Accordingly, after repeated administration of low doses of CPA an almost linear accumulation of DNA adduct levels was observed (Werner et al., 1995). The pathway of metabolic activation of CPA was studied in cultured hepatocytes of female rats. As depicted in the reaction scheme in Fig. 1, the steroid is activated to DNA binding metabolites via O-sulfonylation of the 3-hydroxy-intermediate formed by reduction of the 3-keto group of CPA (Werner et al., 1996).

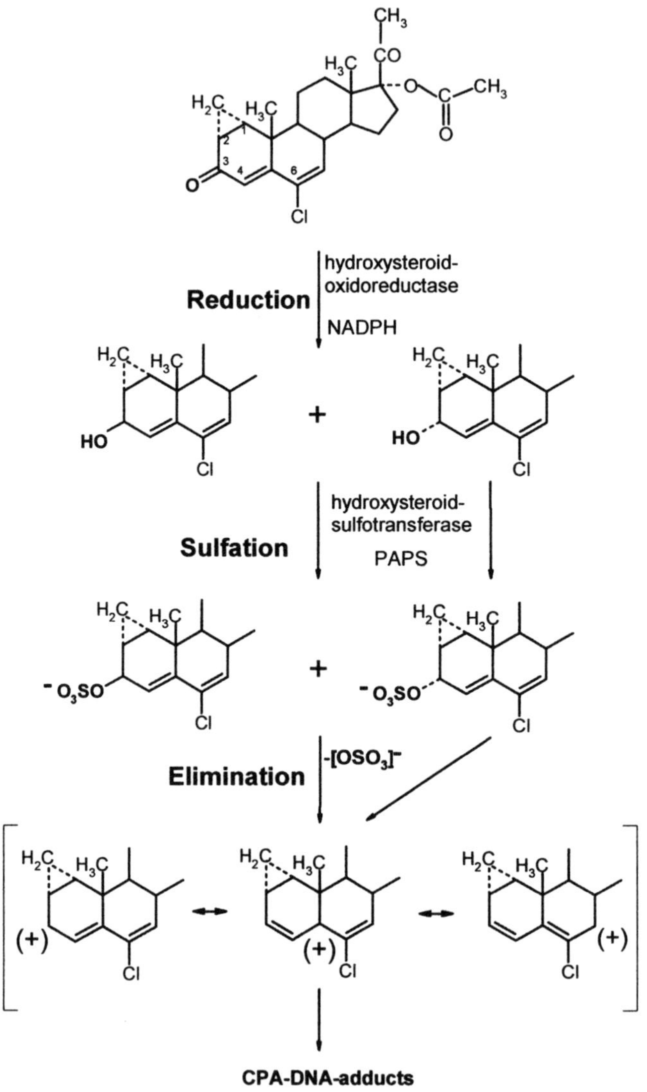

Figure 1. Proposed activation pathway of CPA (modified from Werner et al., 1996)

These findings strongly suggest that CPA is a mutagenic compound. Standard mutation assays, however, failed to support this notion. CPA was not mutagenic when tested with S. typhimurium bacteria and by the HPRT test with V79 Chinese hamster ovary cells in the presence and absence of the S9-fraction from rat liver (Lang and Reimann, 1993). Moreover, CPA did not induce micronuclei in bone marrow of rhesus monkeys (Schering, personal communication). The activation pathway (Fig. 1) provides an explanation for the failure of the assays to detect a mutagenic potential for CPA. Most probably, the activating enzyme hydroxysteroid sulfotransferase is neither present in the bacteria and mammalian cells used including the bone marrow cells of the monkeys. Therefore, we decided to assay for mutations in the liver, the target tissue of the genotoxic and tumorigenic activity of CPA. Since a few years, such studies can be carried out by using transgenic rodents, in which a bacterial reporter gene for mutagenic compounds is incorporated in a recoverable lambda vector. In the present studies we have used the commercially available Big Blue^{TM} rats.

2. METHODS

2.1. Animals and treatment

Mutations and DNA adducts were determined in the liver of homozygous female transgenic Fischer 344 Big Blue^{TM} rats purchased from Stratagene, La Jolla, USA, at the age of 3 months. Groups of 2-5 animals were treated by gavage with single doses of CPA as indicated in the figures as a microcristalline suspension. Control groups received the vehicle only.

2.2. Mutation assay

2.2.1. DNA isolation. After termination of the experiment, genomic DNA was isolated from liver tissue using the "Recoverease" isolation kit (Stratagene, La Jolla, USA). In brief, aliquots of gently homogenized tissue were sterile filtered and shortly centrifuged at low speed. The pellets were digested with RNAse and proteinase K and the samples dialysed with TE-buffer and stored at 4°C.

2.2.2. Determination of mutation frequency. The mutation frequency of the lacI transgene was determined according to the protocol of Dycaico et al. (1994). Briefly, the lambda LIZ shuttle vector containing the LacI target gene was recovered from genomic liver DNA and packed with envelope proteins to produce infectious lambda phages using theTranspack^{TM} packaging extract (Stratagene, La Jolla, USA) according to the manufacturers protocol. The packaged phages were adsorbed onto Escherichia coli SCS8 bacteria plated on X-Gal containing agar plates. After incubation for 20 hours. The plates were visually scored for blue mutant plaques. Mutant frequencies were determined by dividing the number of blue plaques by the total number of nonmutated clear plaques counted within represenative areas of the plates. Generally, a total number of 300.000 plaques per animal per dose was screened.

2.3. Determination of DNA adducts

DNA adducts were identified and quantified by the ^{32}P-postlabelling technique using the butanol extraction procedure (Gupta et al., 1984). Liver DNA was isolated and purified by treatment with RNAse/proteinase K/phenol.

2.4. Determination of cell replication

Animals were given a single i. p. dose of 30 mg bromodesoxyuridine (BrdU), 2 hours before sacrifice. BrdU incorporation into DNA indicating replicative DNA synthesis was demonstrated on paraffine sections of the liver by indirect immune peroxidase staining using a monoclonal mouse-anti BrdU antibody (Boehringer, Mannheim). About 3000-4000 cells per animal were scored. Mitotic figures were recorded in liver sections stained with hemalaun. About 2000 cells per animal were scored.

3. RESULTS AND DISCUSSION

3.1. Dose dependent induction of DNA adducts and mutations in rat liver after a manifestation time of 6 weeks

In a previous study we have shown that CPA in fact is mutagenic in the liver of transgenic Big Blue[TM] rats (Krebs et al., 1998). In that study female F344 Big Blue [TM] rats had been used to assay for mutations of the heterologous Lac I gene in liver DNA. Groups of 5 animals were treated with various single doses of CPA between 25 and 200 mg/kg b.w.. The experiment was terminated 6 weeks post exposure (Fig. 2). A dose dependent linear increase of mutation frequency was noted at doses of 75, 100 and 200 mg CPA/kg b.w.. Mutation frequencies recorded at doses of 25 and 50 mg CPA/kg, however, did not significantly differ from those determined in control animals, although high levels of DNA adducts were generated at these doses (Fig. 2). The highest non-effective dose of 50 mg CPA/kg represents a threshold for mutagenicity under the conditions employed in this experiment. The question arises whether the observed threshold of 50 mg CPA/kg is valid when mutations are recorded after a shorter observation period post exposure. It is conceivable that the manifestation time, i.e. the duration of exposure influences the response to a mutagenic agent and hence the shape of the dose-response relationship.

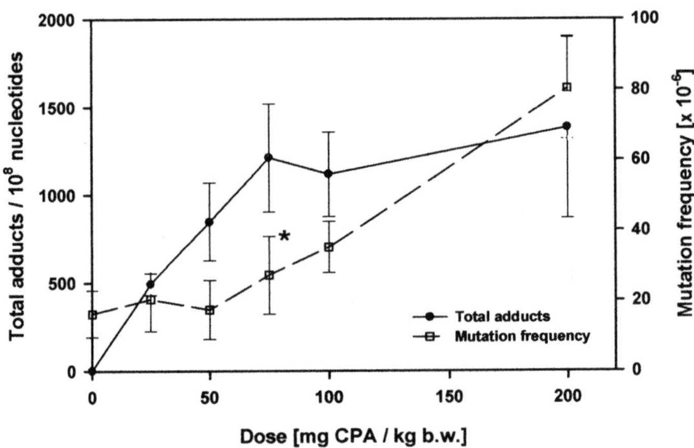

Figure 2. Dose dependence of DNA adduct levels and mutation frequencies, 6 weeks post exposure. Groups of 5 female Big Blue[TM] rats were treated with single oral doses of CPA as indicated in the figure (modified from Krebs et al., 1998). Mean values ± 95% confidence levels are shown. * Significantly differing from controls (p < 0,03).

3.2. Induction of DNA adducts and mutations as a function of the manifestation time

To examine the influence of the manifestation time on mutation frequency and to find conditions of maximum susceptibility, transgenic female rats were treated with a single dose of 100 mg CPA/kg b.w. and DNA adduct levels and mutation frequencies were determined at various time intervals post exposure within a period of 8 weeks.

DNA adduct levels strongly increased within the first 3 days and subsequently decreased to a level of about 40 % of the maximum value, 8 weeks post exposure (Fig. 3). From these data, a half life time for the persistence of DNA adducts of about 6 weeks can be estimated, which agrees well to the value of 6-7 weeks previously calculated for low dose exposure (Werner et al., 1995).

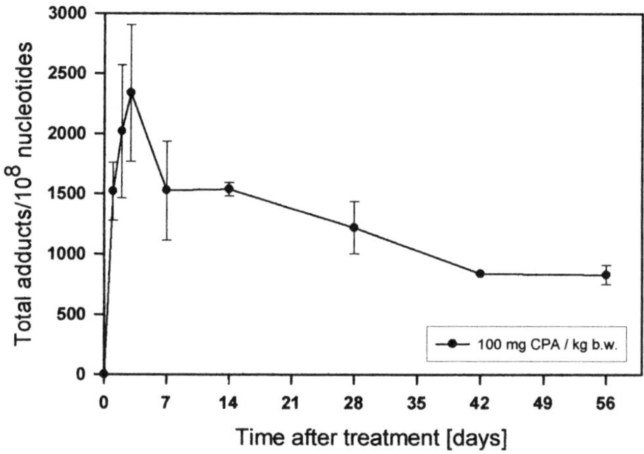

Figure 3. Time course of DNA adduct levels. 40 female Big Blue™ rats were treated with a single oral dose of 100 mg CPA/kg b.w. and the DNA analysed for adduct formation in groups of 5 animals at time intervals indicated in the figure. Mean values ± SD are shown.

The time course of mutation frequencies differed from that of DNA adduct levels. A strong rise within 2 days post treatment was followed by a plateau remaining until 2 weeks post exposure (Fig, 4). Thereafter, mutation frequencies strongly decreased by about 80 % within 2 weeks, whereas DNA adduct levels decreased by 13 % only within this period. We conclude from these data that the rapid reduction of mutation frequency is most likely due to the specific elimination of hepatocytes carrying mutated Lac I genes. The elimination process, i.e. the biological signals, which cause the liver to particularly eliminate mutated hepatocytes, is an open question. A hypothesis that could be offered is based on the assumption that mutations are predominantly expressed in a population of hepatocytes that are, after undergoing mitosis, particularly susceptible to apoptosis. Four weeks post exposure, the level of the mutation-carrying hepatocytes that had survived the elimation process had attained values which remained constant for another 4 weeks. This population of mutation-carrying hepatocytes may represent a population of cells resistant to apoptosis.

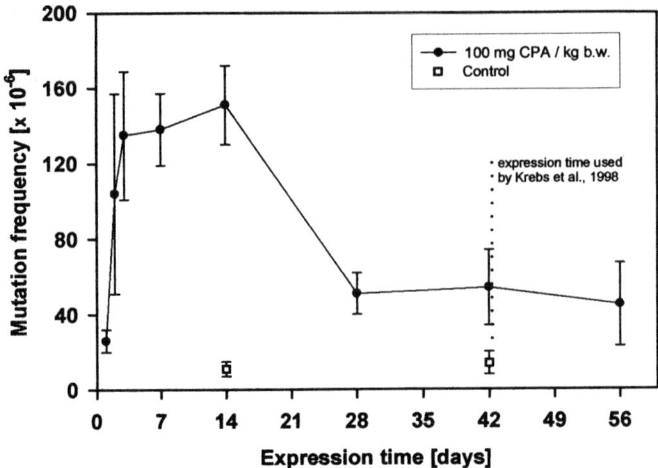

Figure 4. Time course of mutation frequencies. Groups of 10 female Big Blue™ rats were treated with a single dose of 100 mg CPA/kg b.w. and the mutation frequency determined in groups of 5 animals at the time intervals indicated in the figure. Groups of 5 animals that had received the vehicle only were analysed at the time indicated. Mean values ± SD are shown.

3.3. Dose dependent induction of DNA adducts and mutations at a manifestation time of 2 weeks

According to the time course-study, an exposure period of 1-2 weeks promised to provide maximum susceptibility for the expression of mutations. Therefore, the dose-response relationship was reexamined at an expression time of 2 weeks post exposure (Fig. 5). The DNA adduct levels almost linearly rose between the lowest doses of 5 mg and the highest dose of 160 mg CPA/kg, respectively.

The mutation frequencies started to rise at a dose of 10 mg CPA/kg. The maximum level of mutation frequency, which was 11-fold higher than in controls was attained at a dose of 160 mg CPA/kg. Compared to the previous study performed at a manifestation time of 6 weeks post exposure, mutation frequencies determined in this study were higher by a factor of 4. This factors agrees well with the factor of 4 between the mutagenic response at 2 weeks and at 6 weeks post exposure, which can be calculated from the time course study in Fig. 4.

Interestingly, the non-effective dose for mutations was 5 mg CPA/kg b.w., which is by a factor of 10 lower than that of 50 mg CPA/kg recorded in the previous dose dependence study (Krebs et al., 1998). Thus, the dose at which mutations can be detected obviously depends on the manifestation time at which mutations are recorded. Since the present experiment was performed under conditions of maximum susceptibility for mutations, our finding suggests that the lowest single dose of CPA, which is non-effective in the induction of mutations is 5 mg CPA/kg b.w. More data on mutation frequencies obtained at low doses are required to find out whether the 5 mg dose represents a threshold level.

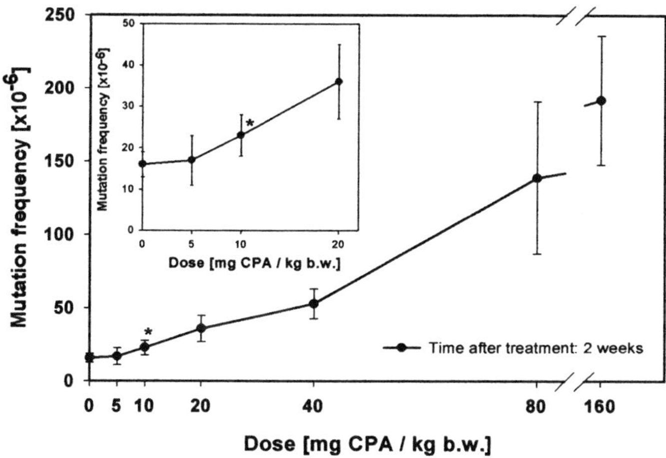

Figure 5. Dose dependence of mutation frequencies, 2 weeks post exposure. Groups of 5 female Big Blue[TM] rats were treated with single oral doses of CPA as indicated in the figure. Mean values ± SD are shown. *Significantly differing from controls (p < 0,027).

3.4. Time dependent induction of cell proliferation and mutations by CPA

In a further set of experiments we addressed the question whether the mutagenic activity of CPA may be due to the mitogenic activity of the compound, which has been shown to be operative at high doses of CPA (Schulte-Hermann et al. 1980) and is supposed to stimulate the conversion of CPA-DNA adducts into gene mutations. To test for this assumption we determined the emergence of S-phase cells and of mitotic figures in the liver of transgenic rats at various times intervals within 3 days post exposure, the time period when CPA-induced cell proliferation occurs.

A time and dose dependent induction was noted for both S-phase cells and mitoses (Fig. 6). Twenty four hours after administration of a high dose of 160 mg/kg, a maximum of almost 50 % of S-pase cells was reached. One day later, mitoses attained a maximum. A lower dose of 40 mg CPA/kg exhibited similar but less pronounced effects. Mutation frequencies were recorded at doses of 100 and 40 mg CPA/kg. At the higher dose, a 2-fold increase at day 1 and a 6-fold increase at day 2 post exposure was observed. The lower dose increased mutation frequency 3-fold, 2 days post exposure. These data suggest that the mitogenic activity of CPA triggers the expression of CPA-specific mutations.

The emergence of S-phase cells was recorded as a function of the dose of CPA within a dose range of 1 to 200 mg CPA/kg b.w. (Fig. 8). The relative portion of S-phase cells showed a linear dose dependent increase over the whole dose range. At the highest dose of 160 mg CPA/kg, 50 % of the cells were in the S-phase (Fig. 7).

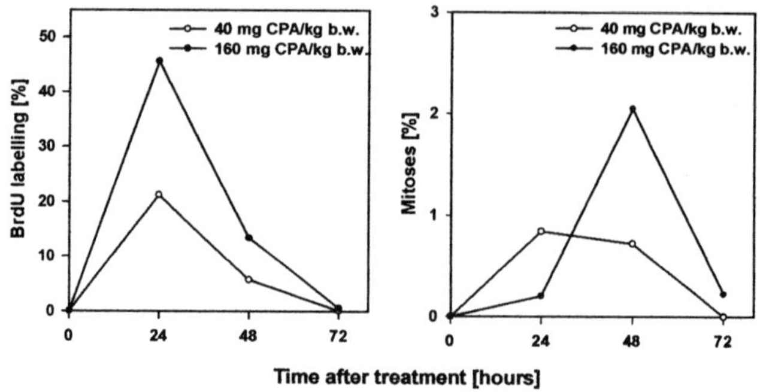

Figure 6. Dose and time dependent emergence of BrdU labelled (S-phase) cells and mitoses. Female Big Blue™ rats were treated with a single oral doses of CPA as indicated. Groups of 2 animals were analysed for cell proliferative activity at the time intervals indicated. Mean values are shown.

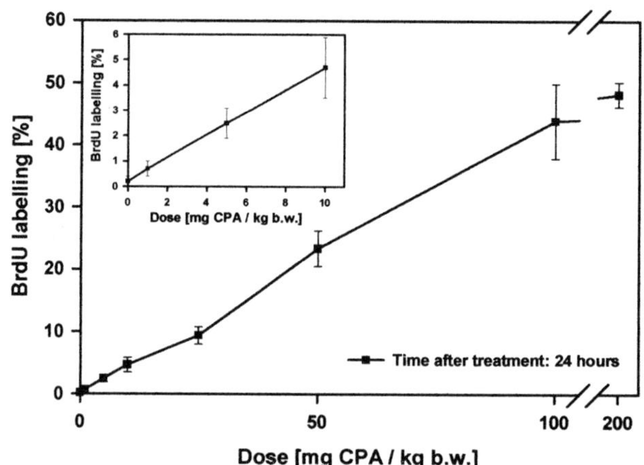

Figure 7. Dose dependent induction of S-phase cells, 24 hours post exposure. Groups of 3 female wild type Fischer F344 rats were treated with single oral doses of CPA as indicated in the figure. Mean values ± SD are shown.

4. SUMMARY AND CONCLUSIONS

1. CPA does not only induce the formation of DNA adducts but also of mutations in female rat liver.

2. The mutation frequency exhibited a characteristic time course. Within a period of 3 days post administration, a tremendous increase was noted, which remained at a high level until 2 weeks post exposure. Thereafter, most mutation-carrying cells were eliminated within a period of 2 weeks leaving a cell population remaining at a constant level for another 4 weeks. Thus, the length of the observation period post exposure, i. e. the manifestation time, seems to be a critical factor for the strength of the mutagenic response. The highest sensitivity was observed between 1 and 2 weeks post exposure.

Correspondingly, the dose response curve recorded 2 weeks post exposure showed a higher mutagenic response than the curve after 6 weeks of exposure recorded previously.

3. When CPA-induced mutations were recorded as a function of the dose, mutation frequencies at the lower dose range were found that did not differ from those of controls. The non-effective dose recorded 2 weeks post exposure was much lower than that recorded after 6 weeks of exposure indicating that it is a function of the manifestation time. Since DNA adducts were formed in high amounts at the non-effective doses, we assume that the mitogenic activity required for the conversion of DNA adducts into mutations was not sufficiently strong. The liver of adult animals exhibits a very low endogeneous proliferation rate, which is not likely to contribute significantly to the expression of mutations. We conclude that it is the mitogenic activity of CPA itself, which stimulates the expression of mutations.

REFERENCES

Dycaico, M. J., Provost, G. S., Kretz, P. L., Ransom, S.L., Moores, J.C., and Short, J.M., 1994, the use of shuttle vectors for mutation analysis in transgenic mice and rats, *Mutat. Res.* **307:** 461-478.

Gupta, R. C., 1985, Enhanced sensitivity of ^{32}P-postlabeling analysis of aromatic carcinogen adducts, *Cancer Res.* **45:** 5656-5662.

Krebs, O., Schäfer, B., Wolff, T., Oesterle, D., Deml, E., Sund, M., and Favor, J., The DNA damaging drug cyproterone acetate causes gene mutations and induces glutathione-S-transferase P in the liver of female Big Blue™ transgenic F344 rats, *Carcinogenesis* **19:** 241-245.

Lang, R., and Reimann, R., 1993, Studies for a genotoxic potential of some exogenous and endogenous sex steroids. 1. Communication: Examination for the induction of gene mutations using Ames Salmonella/microsome test and the HGPRT test in V79 cells, *Environm. Mol. Mutagenesis* **21:**272-304.

Martelli, A., Matioli, F., Ghia, M., Mereto, E., and Brambilla, G., 1996, Comparative study of DNA repair induced by cyproterone acetate in primary cultures of human and rat hepatocytes, *Carcinogenesis* **17:** 1153-1156.

Martelli, A., Brambilla-Campart G., Allavena, A., Mereto, E., Brambilla, G., 1996, Induction of micronuclei and initiation of enzyme-altered foci in the liver of female rats treated with cyproterone acetate, chlormadinone acetate, or megestrol acetate, *Carcinogenesis* **17:**551-554.

Neumann, I., Thierau, D., Andrae, U., Greim, H., and Schwarz, I. R., 1992, Cyproterone acetate induces DNA-damage in cultured rat hepatocytes and preferentially induces DNA synthesis in y-glutamyltranspepetidase positive cells. *Carcinogenesis,* **13:** 373-378.

Schulte-Hermann, R., Hoffmann, V., Parzefall, W., Kallenbach, M., Gerhard, A., and Schuppler, J., 1980, Adaptive response of rat liver to the gestagen and antiandrogen cyproterone acetate and other inducers. II. Induction of growth. *Chem. Biol. Interact.* **31:** 287-300.

Schulte-Hermann, R., Timmermann-Trosiener , I., and Schuppler, J., Promotion of
spontaneous preneoplastic cells in rat liver as a possible explanation of tumor
production by nonmutagenic compounds, *Cancer Res.* **43:** 839-844.

Schuppler, J., and Günzel, P., 1979, Liver tumors and steroid hormones in rats and mice,
Arch Toxicol. **2:**181-195.

Topinka, J., Andrae, U., Schwarz, L.R., and Wolff, T., 1993, Cyproterone acetate
generates DNA adducts in rat liver and primary rat hepatocytes, *Carcinogenesis*
14:423-427.

Topinka, J., Binkova, B., Schwarz, L.R., and Wolff, T., 1996, Cyproterone acetate is an
integral part of hepatic DNA adducts induced by this steroidal drug,
17:167-169.

Werner, S., Topinka, J., Wolff, T., and Schwarz, L, R., 1995, Accumulation and
persistence of DNA adducts of the synthetic steroid cyproterone acetate in rat liver,
Carcinogenesis **16:** 2369-2372.

Werner, S., Kunz, S., Wolff, T., and Schwarz, L.R., 1996, The steroidal drug cyproterone
acetate is activated to DNA-binding metabolites by sulfonation, *Cancer Res.*
56:4391-4397.

DOSE-DEPENDENT DIFFERENCES IN THE PROFILE OF MUTATIONS INDUCED BY CARCINOGENIC (R,S,S,R) BAY- AND FJORD-REGION DIOL EPOXIDES OF POLYCYCLIC AROMATIC HYDROCARBONS

Allan H. Conney[1], Richard L. Chang[1], Xiao Xing Cui[1], Maria Schiltz[1], Haruhiko Yagi[2], Donald M. Jerina[2], and Shu-Jing C. Wei[1]

[1]Laboratory for Cancer Research, Department of Chemical Biology, College of Pharmacy, Rutgers, The State University of New Jersey, Piscataway, NJ, 08854

[2]Section on Oxidation Mechanisms, Laboratory of Bioorganic Chemistry, National Institute of Diabetes and Digestive and Kidney Diseases, National Institutes of Health, Bethesda, MD, 20892

ABSTRACT

Chinese hamster V79 cells were exposed to a high or low concentration of the highly carcinogenic (R,S,S,R) or the less active (S,R,R,S) bay- or fjord-region diol epoxides of benzo[a]pyrene, benzo[c]phenanthrene or dibenz[c,h]acridine. Independent 8-azaguanine-resistant clones were isolated, and base substitutions at the hypoxanthine (guanine) phosphoribosyltransferase (hprt) locus were determined. For the three (R,S,S,R) diol epoxides studied, the proportion of mutations at AT base pairs increased as the concentration of diol epoxide decreased. Concentration-dependent differences in the mutational profile were not observed, however, for the three (S,R,R,S) diol epoxides. In studies with V-H1 cells (a DNA repair deficient variant of V79 cells), a concentration-dependent difference in the profile of mutations for the (R,S,S,R) diol epoxide of benzo[a]pyrene was not observed. These results suggest that concentration-dependent differences in the mutational profile are dependent on an intact DNA repair system. In additional studies, we initiated mouse skin with a high or low dose of benzo[a]pyrene and promoted the mice for 26 weeks with 12-O-tetradecanoylphorbol-13-acetate. Papillomas were examined for mutations in the c-Ha-*ras* proto-oncogene. Dose-dependent differences in the profile of c-Ha-*ras* mutations in the tumors were observed. In summary, (i) dose-dependent differences in mutational profiles at the *hprt* locus were observed in Chinese hamster V79 cells treated with several highly mutagenic and carcinogenic (R,S,S,R) bay- or fjord-region diol epoxides but not with their less active (S,R,R,S) diol epoxide enantiomers, (ii) a dose-dependent difference in the mutational profile was not observed for the (R,S,S,R) diol epoxide of benzo[a]pyrene in a DNA-repair defective V79 cell line, and (iii) a dose-dependent difference in the mutational profile in the c-Ha-*ras* proto-oncogene was observed in tumors from mice treated with a high or low dose of benzo[a]pyrene.

Biological Reactive Intermediates VI, Edited by Dansette *et al.*
Kluwer Academic / Plenum Publishers, 2001

INTRODUCTION

Although experimental carcinogenesis and mutagenesis studies are often done with high doses of chemicals, humans and other organisms are exposed to only small amounts of mutagens and carcinogens in their natural environment. Because of concerns about the significance or relevance for humans of studies on the mutational profiles of high concentrations of chemicals, we initiated comparative studies on the effects of high and low concentrations of some ultimate carcinogens on the profile of mutations in the hypoxanthine (guanine) phosphoribosyltransferase (*hprt*[1]) gene in Chinese hamster V79 cells.

In the present manuscript, we have described studies from our laboratory indicating dose-dependent differences in the profile of mutations induced by (*R,S,S,R*) bay- or fjord-region diol epoxides of benzo[*a*]pyrene, benzo[*c*]phenanthrene and the nitrogen heterocycle, dibenz[*c,h*]acridine at the *hprt* locus in Chinese hamster V79 cells (1-5). The structures and abbreviations for these compounds are described in Figure 1 and footnote 1. In additional studies, we found that the mutational profile in the c-Ha-*ras* proto-oncogene in skin tumors obtained by initiation of mouse skin with a low dose of benzo[*a*]pyrene was different from that obtained by initiation with a high dose of benzo[*a*]pyrene (6).

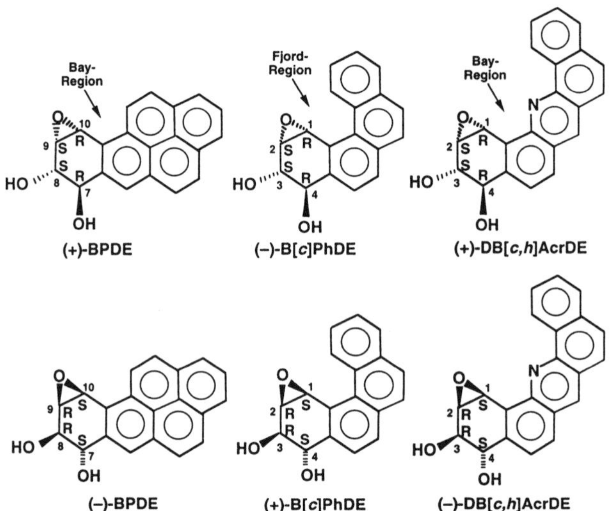

Figure 1. Structures of the optically active (*R,S,S,R*) and (*S,R,R,S*) bay- or fjord-region diol epoxides of benzo[*a*]pyrene (BP), benzo[*c*]phenanthrene (B[*c*]Ph) and dibenz[*c,h*]acridine (DB[*c,h*]Acr). The (*R,S,S,R*) enantiomers have high carcinogenic and mutagenic activity whereas the (*S,R,R,S*) enantiomers are less active. All diol epoxides shown are diasteriomers in which the benzylic hydroxyl group and epoxide oxygen are *trans*.

---

[1] The abbreviations used are: BP, benzo[a]pyrene; (+)-BPDE, (+)-(*7R,8S,9S,10R*)-7,8-dihydroxy-9,10-epoxy-7,8,9,10-tetrahydrobenzo[*a*]pyrene; (−)-BPDE, (−)-(*7S,8R,9R,10S*)-7,8-dihydroxy-9,10-epoxy-7,8,9,10-tetrahydrobenzo[*a*]pyrene; (−)-B[*c*]PhDE, (−)-(*1R,2S,3S,4R*)-3,4-dihydroxy-1,2-epoxy-1,2,3,4-tetrahydrobenzo[*c*]phenanthrene; (+)-B[*c*]PhDE, (+)-(*1S,2R3R,4S*)-3,4-dihydroxy-1,2-epoxy-1,2,3,4-tetrahydrobenzo[*c*]phenanthrene; (+)-DB[*c*]AcrDE, (+)-(*1R,2S,3S,4R*)-3,4-dihydroxy-1,2-epoxy-1,2,3,4-tetrahydrodibenz[*c,h*]acridine; (−)-DB[*c,h*]AcrDE, (−)-(*1S,2R,3R,4S*)-3,4-dihydroxy-1,2-epoxy-1,2,3,4-tetrahydrodibenz[*c,h*]acridine; *hprt*, hypoxanthine (guanine) phosphoribosyltransferase; TPA, 12-O-tetradecanoylphorbol-13-acetate.

DOSE-DEPENDENT DIFFERENCES IN (+)-BPDE-INDUCED BASE SUBSTITUTIONS IN THE CODING REGION OF THE *hprt* GENE IN V79 CELLS

Treatment of Chinese hamster V79 cells with a low dose (0.01-0.02 µM), an intermediate dose (0.04-0.10 µM), or a high dose (0.30-0.48 µM) of (+)-BPDE for 1 h resulted in 97, 100, and 32% cell survival, respectively, and the frequency of mutations (8-azaguanine resistant colonies/10^5 surviving cells) was increased 9-, 51-, and 513-fold, respectively, over the mutation frequency for control cells treated with DMSO. At all dose levels of (+)-BPDE studied, most mutations were at GC base pairs (Table 1, Fig. 1) which is consistent with studies indicating that guanine is the major target for the formation of (+)-BPDE adducts in DNA (7,8).

Decreasing the dose of (+)-BPDE decreased the proportion of base substitution mutations at GC base pairs and increased the proportion of base substitutions at AT base pairs (Table 1, ref. 2). At the high dose, 7 of 120 base substitutions occurred at AT base pairs (6%) and 113 at GC base pairs (94%). At the intermediate dose, 20 of 82 base substitutions occurred at AT base pairs (24%) and 62 at GC pairs (76%). At the low dose, 27 of 76 base substitutions were at AT base pairs (36%) and 49 were at GC base pairs (64%). As the dose of (+)-BPDE was decreased, there was a dose-dependent decrease in the proportion of GC→TA transversions (from 69% to 42% of the total base substitutions) and a dose-dependent increase in the proportion of AT→CG transversions (from 1% to 25% of the base substitutions). A smaller increase in the proportion of AT→GC transitions was also observed as the dose of (+)-BPDE was decreased. When additional base substitution data from aberrant splicing mutants were added to our results on base substitutions in the coding region of the *hprt* gene, similar results were observed (3).

Table 1. Effect of the dose of (+)-BPDE on the kinds of base substitutions in the coding region of the *hprt* gene in 8-azaguanine-resistant mutant clones.

Mutations	Number of mutations observed (%)							
	High dose (0.30-0.48 µM)		Intermediate dose (0.04-0.10 µM)		Low dose (0.01-0.02 µM)		DMSO control	
At G•C base pairs								
G•C → T•A	83	(69)	43	(52)	32	(42)	3	(11)
G•C → C•G	23	(19)	12	(15)	11	(14)	3	(11)
G•C → A•T	7	(6)	7	(9)	6	(8)	1	(4)
Total	113	(94)	62	(76)	49	(64)	7	(26)
At A•T base pairs								
A•T → G•C	4	(3)	8	(10)	5	(7)	16	(59)
A•T → T•A	2	(2)	6	(7)	3	(4)	0	(0)
A•T → C•G	1	(1)	6	(7)	19	(25)	4	(15)
Total	7	(6)	20	(24)	27	(36)	20	(74)

Chinese hamster V-79 cells were treated with a high (0.30-0.48 µM), intermediate (0.04-0.10 µM) or low (0.01-0.02 µM) dose of (+)-BPDE or with DMSO vehicle alone. Independent 8-azaguanine-resistant clones were examined for base substitution mutations (taken from ref. 2). Numbers in parentheses represent the percent of total base substitutions. Very few mutations were observed in DMSO-treated control cells, and data from these control cells were accumulated during the past 10 years (taken from refs. 2 and 9).

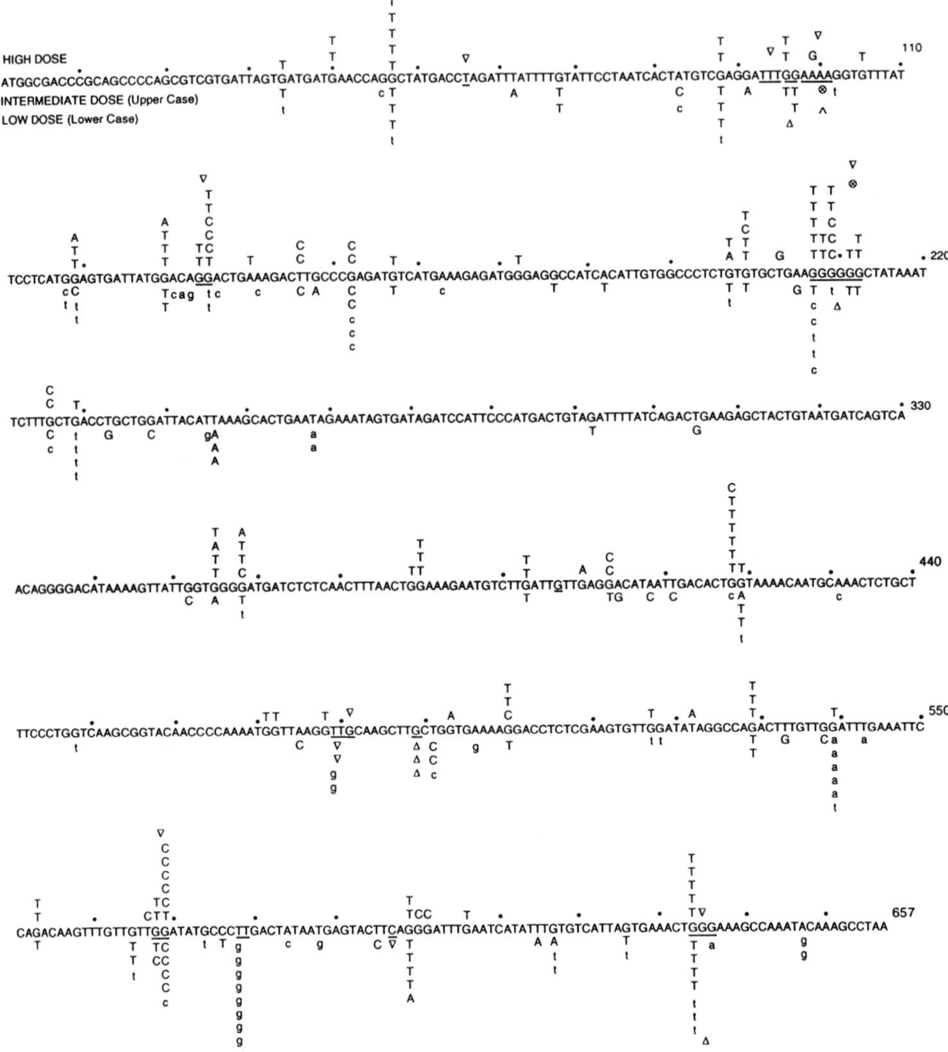

Figure 2. (+)-BPDE-induced mutations in the coding region of the *hprt* gene in V79 cells. Mutations obtained from the high dose of (+)-BPDE (0.30-0.48 μM) are shown above the wild type sequence, mutations from the intermediate dose of (+)-BPDE (0.04-0.10 μM) are shown as capital letters below the wild-type sequence, and mutations from the low dose of (+)-BPDE (0.01-0.02 μM) are shown as lower case letters below the wild-type sequence. The numbers indicate the nucleotide positions relative to the first base of the start codon. The letters used for the base substitutions indicate the new bases found in the *hprt* mutant clones. Taken from ref. 2.

DOSE-DEPENDENT DIFFERENCES IN HOT SPOTS FOR (+)-BPDE-INDUCED BASE SUBSTITUTIONS IN THE CODING REGION OF THE *hprt* GENE IN V79 CELLS

An evaluation of the frequency of mutations at specific bases in the coding region of the *hprt* gene in V79 cells indicated differences in the profile of hot spots induced by the different doses of (+)-BPDE (Fig. 2; ref. 2). Hot spots were defined as target bases that had at least 3.3% of the total base substitutions at the dose level studied (significantly different from random mutations; P<0.01). Eleven hot spots were observed for the high dose group, 7 hot spots were observed for the intermediate dose group, and 6 hot spots were observed for the low dose group. These results are consistent with more extensive damage of the *hprt* gene with a high dose of (+)-BPDE than with an intermediate or low dose. Hot spots found in the high dose group that were not seen in the low dose group occurred at G-47, G-130, G-135, G-199, G-209, G-355, G-358, G-418, and G-569. Hot spots that were found in the low dose group that were not seen in the high dose group occurred at G-152, G-229, G-539, and T-578. A hot spot at G-634 was observed in all three dose groups. In an additional study, it was found that the mutational spectra induced by benzo[*a*]pyrene in a human lymphoblastoid cell line with a cytochrome P-450 metabolic activation system was dependent upon the conditions of exposure (10). A low concentration of benzo[*a*]pyrene and a long exposure time resulted in a different profile of mutations than a high concentration of benzo[*a*]pyrene and a short exposure time (10).

LACK OF DOSE-DEPENDENT DIFFERENCES IN (–)-BPDE-INDUCED BASE SUBSTITUTIONS IN THE CODING REGION OF THE *hprt* GENE IN V79 CELLS

The kinds of base substitutions induced by the weakly tumorigenic (*S,R,R,S*) (–)-BPDE in the coding region of the *hprt* gene were determined in V79 cells (11). No statistically significant dose-dependent differences were observed. At a high cytotoxic dose of (–)-BPDE (2.0 μM; 29% cell survival), 5 of 36 base substitutions occurred at AT base pairs (14%) and 31 at GC base pairs (86%). At the low non-cytotoxic dose studied (0.5 μM; 100% cell survival) , 7 of 34 base substitutions were at AT base pairs (21%) and 27 were at GC base pairs (79%). Although the results suggested that as the dose of (–)-BPDE was decreased, there was a trend towards an increase in the proportion of mutations at AT base pairs, this trend was not statistically significant (p=0.54; Fisher's exact test). Five hot spots were observed for (–)-BPDE, and these were at G-46, G-134, T-170, G-199 and G-355 (11). This hot spot profile for (–)-BPDE is different from that observed for either the high or low dose of (+)-BPDE (Fig. 2).

DOSE-DEPENDENT DIFFERENCES IN (–)-B[*c*]PhDE-INDUCED BASE SUBSTITUTIONS IN THE CODING REGION OF THE *hprt* GENE IN V79 CELLS

At a high cytotoxic dose of the highly tumorigenic (*R,S,S,R*) (–)-B[*c*]PhDE (1.0-1.25 μM; 26% cell survival), 44 of 64 base substitutions occurred at AT base pairs (69%) and 20 occurred at GC base pairs (31%) (Table 2; ref.4). At a low non-cytotoxic dose of (–)-B[*c*]PhDE (0.01-0.1 μM; 97% cell survival), 49 of 55 base substitutions were at AT base pairs (89%), and only 6 were at GC base pairs (11%) (Table 2; ref. 4). The distribution of different kinds of base substitutions induced by (–)-B[*c*]PhDE in the coding region of the *hprt* gene are shown in Table 2. The results indicated that as the dose of (–)-B[*c*]PhDE was decreased, there was an increase in the proportion of mutations at AT base pairs and a decrease in the proportion of mutations at GC base pairs

(4). The largest dose-dependent effect was for GC→TA transversion, which decreased from 22% to 4% of the total base substitutions as the dose decreased. AT→TA transversions were the most common mutation observed for cells treated with either the high or low dose of (–)-B[c]PhDE. At all dose levels of (–)-B[c]PhDE studied, most mutations were at AT base pairs which is consistent with studies indicating that adenine is the major target for (–)-B[c]PhDE adduct formation in DNA (12).

The distribution of the kinds of (+)-B[c]PhDE-induced base substitutions in the coding region of the *hprt* cDNA is shown in Table 2 (4). At a high cytotoxic dose of (+)-B[c]PhDE (2.0-3.0 μM; 31% cell survival), 23 of 45 base substitutions occurred at GC base pairs (51%) and 22 at AT base pairs (49%). At the low non-cytotoxic dose of (+)-B[c]PhDE (0.12-0.50 μM; 95% cell survival), 28 of 56 base substitutions were at GC base pairs (50%) and 28 were at AT base pairs (50%). Our data indicated that there was no (+)-B[c]PhDE-induced dose-dependent difference in base substitutions targeted at GC and AT base pairs, and there was also no (+)-B[c]PhDE-induced dose-dependent difference in the distribution of different kinds of base substitutions.

Mutational hot spots for (–)- and (+)- B[c]PhDE differed from each other and from the mutational hot spots for (+)- and (–)- BPDE (2, 4, 11).

Table 2. Effect of the dose of (–)-B[c]PhDE and (+)-B[c]PhDE on the kinds of base substitutions in the coding region of the hprt gene in 8-azaguanine-resistant mutant clones.

Base substitution at:	(–)-B[c]PhDE				(+)-B[c]PhDE				DMSO control	
	High dose (1.0-1.25 μM)		Low dose (0.01-0.1 μM)		High dose (2.0-3.0 μM)		Low dose (0.12-0.5 μM)			
G•C base pairs										
G•C → T•A	14	(22)	2	(4)	18	(40)	19	(34)	3	(11)
G•C → C•G	2	(3)	3	(5)	4	(9)	2	(4)	3	(11)
G•C → A•T	4	(6)	1	(2)	1	(2)	7	(13)	1	(4)
Total	20	(31)	6	(11)	23	(51)	28	(50)	7	(26)
A•T base pairs										
A•T → G•C	6	(9)	11	(20)	3	(7)	8	(14)	16	(59)
A•T → T•A	35	(55)	34	(62)	17	(38)	19	(34)	0	(0)
A•T → C•G	3	(5)	4	(7)	2	(4)	1	(2)	4	(15)
Total	44	(69)	49	(89)	22	(49)	28	(50)	20	(74)

Chinese hamster V-79 cells were treated with high or low doses of (–)-B[c]PhDE or (+)-B[c]PhDE, or with DMSO vehicle alone. Independent 8-azaguanine-resistant clones were examined for base substitution mutations. Numbers in parentheses represent the percentage of total base substitutions. Taken from ref. 4.

DOSE-DEPENDENT DIFFERENCES IN (+)-DB[c,h]AcrDE-INDUCED BASE SUBSTITUTIONS IN THE CODING REGION OF THE *hprt* GENE IN V79 CELLS

Chinese hamster V79 cells were treated with a high cytotoxic (1.1-1.3 μM; 20% cell survival) or a low non-cytotoxic (0.1-0.3 μM; 98% cell survival) concentration of (+)-DB[c,h]AcrDE (5). Mutation frequencies for the low or high dose were 33- or 442-fold above background, respectively. At the high dose, 48 of 62 base substitutions occurred at GC base pairs (77%) and 14 at AT base pairs (23%). At the low dose, 43 of 72 base substitutions were at GC base pairs (60%) and 29 were at AT base pairs (40%). The data indicate that as the concentration of (+)-DB[c,h]AcrDE is decreased, an increase in the proportion of mutations targeted at AT base pairs is observed (p<0.05). In similar studies with (–)-DB[c,h]AcrDE, no dose-dependent difference in the mutational profile was observed.

LACK OF DOSE-DEPENDENT DIFFERENCES IN THE PROFILE OF (+)-BPDE-INDUCED BASE SUBSTITUTIONS IN THE *hprt* GENE OF REPAIR DEFICIENT V-H1 CELLS

A possible explanation for dose-dependent differences in mutational profiles induced by (+)-BPDE, (−)-B[*c*]PhDE, and (+)-DB[*c,h*]AcrDE might involve dose-dependent differences in DNA repair activities for the removal of diol epoxide-induced guanine and adenine adducts from DNA in the *hprt* gene. Although no dose dependence in the ratio of guanine to adenine adducts in total cellular DNA was observed after exposing V79 cells to (+)-BPDE (13), the ratio of guanine/adenine adducts in the *hprt* gene may be altered by the dose of (+)-BPDE.

In order to enhance our understanding of the role of DNA repair for dose-dependent mutagenesis by (+)-BPDE in Chinese hamster V79 cells, we examined the effect of the concentration of (+)-BPDE on its mutational profile at the *hprt* gene in a repair deficient Chinese hamster V79 cell mutant, V-H1, described by Zdzienicka and Simons (14). The V-H1 cell line belongs to the second complementation group of UV-sensitive mutant cell lines of rodent origin (15). This mutant is defective in the XPD/ERCC2 gene (16), which encodes for an ATP-dependent DNA helicase. The *ERCC2* protein is a subunit of the BTF2/TFIIH complex that is essential for transcription initiation and nucleotide excision repair (17). The V-H1 cell line is extremely sensitive to UV radiation and is deficient in cyclobutane pyrimidine dimer repair but shows intermediate levels of (6-4) photoproduct repair (18). In addition, an earlier mutational study of UV-induced *hprt* mutant clones from V-H1 cells suggested a deficiency in the preferential repair of UV-induced mutagenic lesions from the transcribed strand (19). In our studies, V-H1 cells removed (+)-BPDE adducts in DNA at ~ 50% the rate of V79 cells, and the V-H1 cells exhibited many (+)-BPDE-induced base substitutions at guanine on the transcribed strand of the *hprt* gene (32% of the mutations at guanine) (9). These results are in marked contrast to our earlier observations with V79 cells in which only 0.5% of the base substitutions targeted at guanine occurred on the transcribed strand of the *hprt* gene (2,3). These results indicate that V-H1 cells have a marked deficiency in the preferential removal of (+)-BPDE-guanine adducts from the transcribed strand of the *hprt* gene, suggesting a deficiency in transcription-coupled repair of (+)-BPDE adducts (9).

The spectra of mutations induced by high and low concentrations of (+)-BPDE in the coding region and flanking intron sequences of the *hprt* gene in V-H1 mutant clones revealed that the predominant base substitutions observed were GC→TA transversions (Table 3). An examination of the 111 base substitutions observed in the coding region and flanking intron sequences of the *hprt* gene in V-H1 mutant clones obtained after treating V-H1 cells with (+)-BPDE, revealed that at the high concentration of (+)-BPDE (40-48 nM; 31% cell survival), 52 of 61 (85%) of the base substitutions occurred at GC base pairs and 9 (15%) at AT base pairs. At the low concentration of (+)-BPDE (4-6 nM; 95% cell survival), 40 of 50 (80%) of the base substitutions occurred at GC base pairs and 10 (20%) at AT base pairs (Table 3). The results indicate that decreasing the dose of (+)-BPDE (i) did not cause a significant change in the proportion of base substitutions targeted at GC or AT base pairs in V-H1 cells, (ii) did not affect the pattern of nucleotide substitutions described in Table 3, and (iii) did not alter the global mutation spectra (type/site of mutation) (9). The results of our studies suggest that dose-dependent differences in the mutational profile of (+)-BPDE in the *hprt* gene of V79 cells may be explained by dose-dependent differences in the repair of (+)-BPDE DNA adducts. This concept is supported by a recent study indicating different mechanisms of repair of BPDE DNA adducts in human fibroblasts exposed to a low or high concentration of racemic BPDE (20).

Table 3. Kinds of base substitutions in the coding region and flanking intron sequences of the *hprt* gene in (+)-BPDE-induced 8-azaguanine-resistant mutant clones in Chinese hamster V-H1 and V79 cells

Base substitution target	V-H1[a] High dose[b] (40-48 nM)		Low dose[b] (4-6 nM)		V79[a] High dose[c] (300-480nM)		Intermediate dose[c] (40-100 nM)		Low dose[c] (10-20 nM)	
G•C base pairs										
G•C → T•A	31	(51)	25	(50)	101	(71)	50	(52)	44	(43)
G•C → C•G	11	(18)	9	(18)	24	(17)	16	(17)	13	(13)
G•C → A•T	10	(16)	6	(12)	9	(6)	9	(9)	7	(7)
Total	52	(85)	40	(80)	134	(94)	75	(78)	64	(63)
A•T base pairs										
A•T → G•C	3	(5)	1	(2)	5	(4)	8	(8)	15	(15)
A•T → T•A	5	(8)	7	(14)	2	(1)	7	(7)	3	(3)
A•T → C•G	1	(2)	2	(4)	1	(1)	7	(7)	20	(20)
Total	9	(15)	10	(20)	8	(6)	22	(22)	38	(38)

[a]Numbers in parenthesis represent the percentage of total base substitutions.
[b]Taken from Schiltz et al. (13)
[c]Taken from Wei et al. (2) and Hennig et al. (3)

DOSE-DEPENDENT DIFFERENCES IN THE PROFILE OF c-Ha-*ras* MUTATIONS IN SKIN TUMORS FROM MICE TREATED WITH BP AND OTHER CARCINOGENS

CD-1 female mice were initiated once with a single topical application of a low dose (20-50 nmol) or a high dose (800 nmol) of BP, and the mice were then treated twice weekly with 5 nmol of TPA for 26 weeks. Tumors were examined for mutations in the c-Ha-*ras* gene (Table 4; ref. 6). Eighty-five percent of the tumors from the low dose group had a mutation in the c-Ha-*ras* gene and 74% of the tumors from the high dose group had a mutation in the c-Ha-*ras* gene. The kinds of base substitutions observed in the c-Ha-*ras* gene for different tumors obtained from the low and high dose BP groups are shown in Table 4. The distribution of mutations as a percentage of total base substitutions for G→A (35), G→T (35), G→C (37), G→T (38), C→A (181), A→T (182), and A→G (182) in the low dose group was 5, 2, 11, 74, 0, 7 and 2%, respectively, and the distribution of these mutations in the high dose group was 3, 7, 13, 61, 15, 1 and 0%, respectively. The numbers in parentheses indicate the nucleotide position in the coding sequence of the c-Ha-*ras* gene. The one tumor observed in control mice (acetone/TPA treatment) had an A→G (182) transition mutation. Although only the mutation at C→A (181) was significantly different (0% vs. 15%) when individual base substitutions were compared between the two dose groups (p<0.001, Fisher's exact test), differences in the global mutation spectra (site and kind of all substitution mutations) in the c-Ha-*ras* gene between the high and low dose group tumors were also statistically significant (p<0.004, Fisher's exact test). The major differences observed between the high and low dose tumor groups were mutations at exon 2 of the c-Ha-*ras* gene (p<0.001, Fisher's exact test). The proportion of base substitution mutations at AT base pairs in tumors from the high dose group was 1%, whereas the proportion of mutations at AT base pairs in tumors from the low dose group was 9% (Table 4). This increase in the proportion of mutations at AT base pairs when the dose of BP is lowered is in agreement with increased (+)-BPDE-induced mutations at AT base pairs in V79 cells as the concentration of (+)-BPDE is decreased (Table 1).

There have been a few reports of dose-dependent differences in the mutational profile in chemically-induced tumors. When skin tumors induced by racemic B[*c*]PhDE-2 (epoxide <u>trans</u> to the benzylic hydroxyl group) in mice were analyzed for mutations at codon 61 of the c-Ha-*ras* gene, a dose-dependent difference in the frequency of CAA→CTA mutations was observed. Increased mutations at AT base pairs were observed with decreasing dose (21). These observations are in agreement with our studies indicating that decreasing the concentration of (−)-B[*c*]PhDE increased the proportion of base substitution mutations at AT base pairs and decreased the proportion of mutations at GC base pairs in the *hprt* gene of Chinese hamster V79 cells (Table 2). In other studies, mutations at the Ki-*ras* gene were examined from lung tumors induced by 4-(methylnitrosamino)-1-(3-pyridyl)-1-butanone in A/J mice, and a dose-dependent difference in the frequency of GGT→GAT mutations at codon 12 was reported (22). However, this conclusion was not very strong since it was based on the analysis of 46 low dose tumors (22) that were compared with mutation profile data from 10 high dose tumors obtained in a different laboratory (23). It is of interest that the dosing procedure has also been reported to affect the pattern of *ras* gene mutations in liver tumors induced by 4-aminoazobenzene, N-hydroxy-2-acetylaminofluorene, and N-nitrosodiethylamine in CD-1 mice. For each of these carcinogens, the multiple low dose tumor group (versus a single high dose group) had fewer Ki-*ras* and N-*ras* mutations and more Ha-*ras* codon 61 (C→A) mutations (23).

Our present findings and the reports of others indicate that the dose of a carcinogen can influence the kinds of mutations found in chemically-induced tumors. These results suggest a need for a greater emphasis on mutagenesis and carcinogenesis studies with low doses of carcinogens/mutagens that are more relevant for humans who are exposed to small amounts of environmental chemicals.

Table 4. Mutations in c-Ha-*ras* detected in tumors obtained from mice initiated with a high or low dose of BP

Base substitution (nucleotide number)	Amino acid change (codon number)	Number of mutations (%)			
		High dose		Low dose	
G→A (35)	Gly→Glu(12)	2[a,c]	(3)	3[a,c]	(5)
G→T (35)	Gly→Val(12)	5[a,c]	(7)	1[a,c]	(2)
G→C (37)	Gly→Arg(13)	10[a,c]	(13)	7[a,c]	(11)
G→T (38)	Gly→Val(13)	46[a,c]	(61)	45[a,c]	(74)
C→A (181)	Gln→Lys(61)	11[b,c,d]	(15)	0[b,c,d]	(0)
A→T(182)	Gln→Leu(61)	1[b,c]	(1)	4[b,c]	(7)
A→G (182)	Gln→Arg(61)	0[b,c]	(0)	1[b,c]	(2)

CD-1 mice were initiated with a single application of either a low (25-50 nmol) or high (800 nmol) dose of BP or acetone vehicle and promoted twice weekly with 5 nmol of TPA for 26 weeks. DNA from tumor samples were amplified by PCR and examined for c-Ha-*ras* mutations by SSCP analysis and DNA sequencing. Statistical analysis was evaluated for entries with the same superscript. Taken from ref. 6.

[a]When different base substitutions in exon 1 are compared between the high and low dose group tumors, p=0.395, Fisher's exact test.

[b]When different base substitutions in exon 2 are compared between the high and low dose group tumors, p=9.7 x 10^{-4}, Fisher's exact test.

[c]When the global mutation spectra are compared between the high and the low dose group tumors, p=3.35 x 10^{-3}, Fisher's exact test.

[d]When C→A (181) transversions were compared between the high and low dose group tumors, p=1.09 x 10^{-3}, Fisher's exact test.

ACKNOWLEDGEMENTS

We thank Drs. R. Lehr, S. Kumar and N. Shirai for samples of (+)- and (–)-DB[*c,h*]AcrDE, and we thank Ms. F. Florek and Ms. C. Burrows for their help in the preparation of this manuscript. A.H.C. is the William M. and Myrle W. Garbe Professor of Cancer and Leukemia Research. Supported in part by National Cancer Institute Grant CA-49756. Send reprint requests to Allan H. Conney, Laboratory for Cancer Research, Department of Chemical Biology, College of Pharmacy, 164 Frelinghuysen Road, Piscataway, NJ 08854-8020.

REFERENCES

1. S.-J.C. Wei, R.L. Chang, C.Q.,Wong, N. Bhachech, X.X. Cui, E. Hennig, H. Yagi, J.M. Sayer, D.M. Jerina, B.D. Preston, and A.H. Conney, Dose-dependent differences in the profile of mutations induced by an ultimate carcinogen from benzo[*a*]pyrene, *Proc. Natl Acad. Sci. USA*, 88: 11227-11230 (1991).

2. S.-J.C. Wei, R.L. Chang, N. Bhachech, X.X. Cui, K.A. Merkler, C.Q Wong, E. Hennig, H. Yagi, D.M. Jerina, and A.H. Conney, Dose-dependent differences in the profile of mutations induced by (+)-7R,8S-dihydroxy-9S,10R-epoxy-7,8,9,10-tetrahydrobenzo[*a*]pyrene in the coding region of the hypoxanthine (guanine) phosphoribosyltransferase gene in Chinese hamster V79 cells, *Cancer Res.*, 53: 3294-3301 (1993).

3   E. Hennig, A.H. Conney, and S.-J.C Wei, Characterization of *hprt* splicing mutations induced by the ultimate carcinogenic metabolite of benzo[*a*]pyrene in Chinese hamster V79 cells, *Cancer Res.*, 55: 1550-1558 (1995).

4. S.-J.C. Wei, R.L. Chang, X.X. Cui, K.A. Merkler, C.Q. Wong, H. Yagi, D.M. Jerina, and A.H. Conney, Dose-dependent differences in the mutational profiles of (-)-(1R,2S,3S,4R)-3,4-dihydroxy-1,2-epoxy-1,2,3,4-tetrahydrobenzo[*c*]phenanthrene and its less carcinogenic enantiomer, *Cancer Res.*, 56: 3695-3703 (1996).

5. X.X. Cui, R.L. Chang, M. D'Ayala, A. Kolly, S. Kumar, N. Shirai, D. Jerina, R. Lehr, A.H. Conney, and S.C. Wei, Dose dependent difference in mutation profile of (+)-1R,2S-epoxy-3S,4R-dihydroxy-1,2,3,4-tetrahydrodibenz[c,h]acridine ((+)-DB[c,h]AcrDE), *Proc. Am. Assoc. Cancer Res.*, 39: 186 (1998).

6. S-J.C. Wei, R.L. Chang, K.A. Merkler, M.Gwynne, X.X. Cui, B. Murthy, M-T. Huang, J-G. Xie, Y-P. Lu, Y-R. Lou, D.M. Jerina and A.H. Conney, Dose-dependent mutation profile in the c-Ha-ras proto-oncogene of skin tumors in mice initiated with benzo[a]pyrene, *Carcinogenesis*, 20: 1689-1696 (1999).

7. T. Meehan and K. Straub, Double-stranded DNA steroselectively binds benzo-(a)pyrene diol epoxide, *Nature (Lond.)*, 277: 410-412 (1977)

8. D.M. Jerina, A. Chadha, A.M. Cheh, M.E. Schurdak, A.W. Wood and J.M. Sayer, Covalent bonding of bay-region diol epoxides to nucleic acids. *In:* C.M. Witmer, R. Snyder, D.J. Jollow, G.F. Kalf, J.J. Kocsis and I.G. Sipes (eds), Biological Reactive Intermediates IV. Molecular and Cellular Effects and Their Impact on Human Health, pp. 533-553. New York: Plenum Publishing Corp. (1991).

9. M. Schiltz, X.X. Cui, Y-P. Lu, H. Yagi, D.M. Jerina, M.Z. Zdzienicka, R.L. Chang, A.H. Conney and S-J.C. Wei, Characterization of the mutational profile of (+)-7R,8S-dihydroxy-9S,10R-epoxy-7,8,9,10-tetrahydrobenzo[a]pyrene at the hypoxanthine (guanine) phosphoribosyltransferase gene in repair-deficient Chinese hamster V-H1 cells, *Carcinogenesis*, 20: 2279-2285 (1999).

10. J. Chen and W.G. Tilly, Mutational spectra vary with exposure conditions: benzo[a]pyrene in human cells, *Mutat. Res.*, 357: 209-217 (1996).

11. S-J.C. Wei, R.L. Chang, E. Hennig, X.X. Cui, K.A. Merkler, C-Q. Wong, H. Yagi, D.M. Jerina and A.H. Conney, Mutagenic selectivity at the *HPRT* locus in V-79 cells: comparison of mutations caused by bay-region benzo[a]pyrene 7,8-diol-9,10-epoxide enantiomers with high and low carcinogenic activity, *Carcinogenesis*, 15: 1729-1735 (1994).

12. S.K. Agarwal, J.M. Sayer, H.J.C. Yeh, L.K. Pannell, B.D. Hilton, M.A. Pigott, A. Dipple, H. Yagi and D.M. Jerina, Chemical characterization of DNA adducts derived from the configurationally isomeric benzo(c)phenanthrene-3,4-diol 1,2-epoxides, *J. Am. Chem. Soc.*, 109: 2497-2504 (1987).

13. S-J.C. Wei, R.L. Chang, C-Q. Wong, X.X. Cui, N. Dandamudi, Y-P. Lu, K.A. Merkler, J.M. Sayer, A.H. Conney and D.M. Jerina, The ratio of deoxyadenosine to deoxyguanosine adducts formed by (+)-(*7R,8S,9S,10R*)-7,8-dihydroxy-9,10-epoxy-7,8,9,10-tetrahydrobenzo[*a*]pyrene in purified calf thymus DNA and DNA in V-79 cells is independent of dose, *Int. J. Oncol.* 14: 509-513 (1999).

14. M.Z. Zdzienicka and J.W.I.M. Simons, Mutagen-sensitive cell lines are obtained with a high frequency in V79 Chinese hamster cells, *Mutat. Res.*, 178: 235-244 (1987).

15. A.R. Collins, Mutant rodent cell lines sensitive to ultraviolet light, ionizing radiation and cross-linking agents: a comprehensive survey of genetic and biochemical characteristics, *Mutat. Res., DNA repair,* 293: 99-118 (1993).

16. S. Kadkhodayan, E.P. Salazar, M.J. Ramsey, K. Takayama, M. Zdzienicka, J.D. Tucker, and C.A. Weber, Molecular analysis of *ERCC2* mutations in the repair deficient hamster mutants UVL-1 and V-H1, *Mutat. Res.*, 385: 47-57 (1997).

17. J.H.J. Hoeijmakers, J-M. Egly, and W. Vermeulen, TFIIH: a key component in multiple DNA transactions, *Curr. Opin. Genet. Dev.*, 6: 26-33 (1996).

18. D.L. Mitchell, M.Z. Zdzienicka, A.A. van Zeeland, and R.S. Nairn, Intermediate (6-4) photoproduct repair in Chinese hamster V79 mutant V-H1 correlates with intermediate levels of DNA incision and repair replication, *Mutat. Res.*, 226: 43-47 (1989).

19. H. Vrieling, M.L. Van Rooijen, N.A. Groen, M.Z. Zdzienicka, J.W.I.M. Simons, P.H.M. Lohman, and A.A. van Zeeland, DNA strand specificity for UV-induced mutations in mammalian cells. *Mol. Cell. Biol.*, 9: 1277-1283 (1989).

20. D.R. Lloyd and P.C. Hanawalt, p53-dependent global genomic repair of benzo[a]pyrene-7,8-diol-9,10-epoxide adducts in human cells, *Cancer Res.*, 60: 517-521 (2000).

21. Z.A. Ronai, S. Gradia, K. El-Bayoumy, S. Amin, S.S. Hecht, Contrasting incidence of *ras* mutations in rat mammary and mouse skin tumors induced by anti-benzo[c]phenanthrene-3,4-diol-1,2-epoxide, *Carcinogenesis*, 14: 2113-2116 (1994).

22. B. Chen, L. Liu, A. Castonguay, R.R. Maronpot, M.W. Anderson and M. You, Dose-dependent *ras* mutation spectra in N-nitrosodiethylamine induced mouse liver tumors and 4-(methylnitrosamino-1-(3-pyridyl)-1-butanone induced mouse lung tumors, *Carcinogenesis*, 14: 1603-1608 (1993).

23. S.A. Belinsky, T.R. Devereux, R.R. Maronpot, G.D. Stoner, and M.W. Anderson, Relationship between the formation of promutagenic adducts and the activation of the K ras protooncogene in lung tumors from A/J mice treated with nitrosamines. *Cancer Res.*, 49: 5305-5311 (1989).

24. S. Manam, G.A. Shinder, D.J. Joslyn, A.R. Kraynak, C.L. Hammermeister, K.R. Leander, B.J. Ledwith, S. Prahalada, M.J. van Zwieten, and W.W. Nichols, Dose-related changes in the profile of ras mutations in chemically induced CD-1 mouse liver tumors, *Carcinogenesis*, 16: 1113-1119 (1995).

NUCLEIC ACID MICROARRAY TECHNOLOGY FOR TOXICOLOGY: PROMISE AND PRACTICALITIES

Diane E. Heck[1], Amit Roy[2] and Jeffrey D. Laskin[2]
Departments of [1]Pharmacology and Toxicology, Rutgers University and [2]Environmental and Community Medicine, UMDNJ-Robert Wood Johnson Medical School, 170 Frelinghuysen Road, Piscataway, NJ 08854

ABSTRACT

Much ongoing research in toxicology focuses on a hypothesis-driven mechanism of action approach aimed at understanding the molecular events mediating the actions of the chemicals of interest. Using this approach, investigators develop hypotheses based on observations, which may be derived from a host of resources but most frequently have been made within their own laboratories, or uncovered by others and reported in the scientific literature. Although the bulk of current understanding of biochemical toxicology emerged using studies based on observations derived in this way, this process, which is essentially based on existing information, may often limit the expansion knowledge. More simply expressed, one only finds that which one seeks. Without a clear understanding of the processes targeted by a specific toxin the problem of making observations that globally and accurately reflect the events mediating pathology which have been induced by the toxic agent is challenging. Recently, the development of high-throughput technologies for biochemical analysis of gene expression has led to innovative approaches in addressing the problem of making broad-based observations that more accurately reflect the entire spectrum of molecular lesions induced by specific toxins. These strategies include the use of new techniques in analysis of gene expression to convey information on alterations in mRNA levels, one of the earliest cellular signs initiated in response to a potential toxin. Prior to this time studies on toxicant-induced altered gene expression were limited to single, or small numbers of identified genes chosen by an investigator who reasoned, based on an existing observations, that levels of the proteins encoded by these genes were likely to be altered during toxic injury. Now, using cDNA or oligonucleotide genome-wide arrays, toxin-induced alterations in gene expression of thousands of genes can be examined simultaneously. Using these tools, molecular toxicologists can for the first time employ reasoned strategies to make observations, and then formulate hypotheses based on these observations.

DISCOVERING HYPOTHESIS: MICROARRAYS AND MECHANISMS

The use of microarray techniques is already opening the door to numerous mechanistic insights by identifying many genes not previously recognized as important in cellular responses to specific toxins (1-3). Identifying these genes, and understanding the processes mediated by the proteins they code for, will fundamentally alter our understanding of many mechanisms regulating toxicity. For example, in one such study, we have used microarray technology to address a problem concerning the toxicity of chlorinated polycyclic hydrocarbons. Numerous excellent investigators have amassed significant information on the biological and biochemical effects of 2,3,7,8-tetrachlorodibenzo-*p*-dioxin (TCDD, 4) in the liver, however, it is not clear that the toxic effects of this environmental contaminant to humans are mediated by this tissue. Indeed epidemiological studies indicate that in adult humans TCDD-induced pathology is primarily manifested in alterations to epithelial tissues, primarily skin, as well as in cardiovascular defects (5). Although many of the biochemical events resulting from exposure to TCDD originally uncovered in liver studies may be important in the development of pathology in other tissues, numerous unrecognized interactions may also be important. It is recognized that TCDD is a rodent carcinogen and in laboratory animals, TCDD may induce profound developmental lesions (6,7). Although numerous studies indicate that, similar to rodents, the toxicity of TCDD in humans is largely dependent on transcriptional regulation by its intracellular receptor, the aryl hydrocarbon (Ah) receptor (8, 9), the overall effects of this environmental contaminant in humans are less clear-cut. However, two specific pathological results have been characterized which may occur following exposure of adult humans to TCDD, chloracne and cardiovascular disease (10). Both of these conditions are characterized by alterations to microvascular endothelial cells. These observations prompted us to treat human microvascular endothelial cells with TCDD. We reasoned that these treatments would be likely to perturb the cells and initiate responses that are mediated by changes in gene expression. Potentially, among the genes exhibiting altered expression are genes likely to be important to cellular processes perturbed by TCDD. Therefore, we treated human microvascular endothelial cells with TCDD and evaluated changes in the expression of 8556 unique genes using Incyte Pharmaceuticals human GEM V2 unigene based microarrays. We found that 3660 genes were expressed by the cells and that expression of 568 genes was potentially altered following exposure to TCDD. Although some of the genes whose mRNA levels changes were previously known to be regulated by TCDD, prior to this time, the vast majority of the genes whose level of expression was changed were not recognized as regulated by TCDD. Analysis of these studies indicates that several distinct cellular processes, also not previously associated with dioxin toxicity, were specifically targeted by the environmental contaminant (11).

Nucleic acid microarrays consist of nucleic acid probes immobilized on a surface and arranged in a matrix so that each probe can be identified, generally by its position within the matrix. Nucleic acid material, either RNA or cDNA derived from mRNA contained in a biological sample is then hybridized to the probes. Expression level for the hundreds to thousands of probes in the matrix is determined by quantifying the amount of specifically hybridized material (12-15). Microarray technologies available for the analysis of gene expression fall into several categories. Several corporate and nationally supported facilities currently conduct analysis for collaborators or clients using 'chips' which contain probes for thousands of genes. These chip arrays provide a mechanism for genome-wide screens of mRNA levels. The quality control and efficacy of these screening methods have been rigorously tested, and the strengths and weaknesses of each system have been well evaluated (see further below). However, the availability of this technology may be limited by the personnel and resources available. Alternatively, equipment for devising custom arrays is widely available and there are numerous shared resource facilities available for individual investigators. In these instances, issues such as array fabrication, sample handling, data analysis and resource management may directly affect the quality of the data derived. A third alternative is the use of commercial glass or membrane-based arrays. These media, although by far the most accessible, are limited to smaller specific gene sets determined as likely to be important for numerous applications by their manufacturers. All of these platforms use a common approach for evaluation of gene expression. In an array experiment, mRNA, isolated from samples, which may be derived from animals, isolated tissues or cells, is compared to mRNA harvested from a similar source that has undergone some experimental treatment. Whichever platform for conducting the array, glass microscope slides, plastic cassettes or woven membranes, has been chosen the end results, the relative amounts of hybridized material, reflects alterations in mRNA levels mediated by the experimental conditions (16, 17). At this point, it is important to note that the overall physiological state of the cells, tissues, and/or animals assayed is reflected by the results of array. Therefore, variables not necessarily altered by the experimental treatment such as diurnal or hormonal status may mediated some of the mRNA alterations observed. Because we have little information about global regulation of gene expression related to many of these events, careful consideration must be given to the overall physiological status of the test subjects when designing array experiments.

PITFALLS AND ERROR MANAGEMENT

Although understanding of error development in this emerging technology is far from complete, clearly the largest obstacles in deriving data, which accurately reflects the experimental condition, is the quality of the nucleic acid samples and the experimental design (as mentioned above). Because the optimal conditions for sample hybridization to the individual probes included in an array may vary widely, these individual conditions are seldom met by the optimal conditions for overall sample hybridization to the array. Therefore, high quality nucleic acid samples are of the utmost importance because any higher structure, concentration-dependent interactions or degradation of RNA can lead to erroneous information (18, 19). In addition, error at the labeling step may be generated by unbalanced incorporation of the detectable molecules into samples or inaccuracies in detection. In commercially available genome-wide arrays, these problems have been extensively addressed through use and construction of standards and protocol refinements. Without appropriate controls and careful evaluation, significant error may be introduced at these steps in experiments when using user-built arrays and commercial membrane arrays.

When using individually constructed arrays, the substrate supporting the probes, most commonly glass microscope slides, may introduce error from artifactual signal generation generated by dust or handling-induced surface damage as well as from inherent contamination of some glass. The detection devises may also add to mechanistic error through impaired spatial resolution of the hybridized spots, photobleaching of fluorescent probes or spectral overlap between detected channels in dual labeling experiments (20). Automated detection for large arrays involves identification and quantification of each spot without human guidance. This requires advanced computer vision algorithms and is often complicated by the requirement for identification of irregular spot shapes and uneven distribution of hybridized material within the spot. Hybridization detection for custom arrays and commercial membranes is generally semi-automatic, allowing the investigator to determine the position of the spots. However, subtle alterations in the dimensions and/or topography may not be adequately accounted for and/or cause difficulties in spot identification. In the case of user built arrays adequate technology to address this issue is often unavailable leading to significant variability between experiments. This may be complicated by detection software that does not include sufficiently sensitive mechanisms for detection and resolution of signals compromised by sub-optimal spot geometry (21, 22).

FUTURE VISIONS: USING MICROARRAYS TO PREDICT TOXICITY

Current understanding of the molecular events mediating toxicity is often confounded by difficulties in distinguishing adaptive and maladaptive responses. For example, in many instances inflammation is required for healing and tissue repair (for review see 23, 24), however, under some circumstances inflammatory processes may enhance tissue injury (25-27). If the sequella of events which facilitate tissue recovery, adaptive events, could be isolated from those maladaptive ones which mediate injury, sites for therapeutic intervention could be delineated. Additionally, inhibition of appropriate adaptive responses is a critical target for the molecular actions of toxins. Microarray experiments provide toxicologists with powerful tools for understanding both cellular adaptation, and its absence, toxicity. Those responses that initially appear to be 'non-specific', because they are induced by numerous toxins or classes of toxins, may be the most likely to be important in adaptation to insult. Culling through microarray results from unrelated experiments has already led investigators to begin to understand adaptation (28, 29). Analysis of this neglected 'non-specific' data by molecular toxicologists, who are in a unique position because their work focuses on maladaptive responses, is highly likely to add much to current knowledge of cellular adaptation to insults and stresses. Prior to the advent of microarrays, many toxicologists tended to view any biochemical alterations found following experimental treatment with a toxin as mediators of toxicity. The possibility that the alterations were adaptive responses, mediating events that required the reestablishment of homeostasis was seldom considered. Results from microarray experiments have led investigators to reconsider the importance of adaptive responses in an array of toxic processes including the effects of TCDD and acetaminophen on the liver, thalidomide on cultured cells, and nitric oxide on early skeletal development (11, 30, 31). Advances in defining and understanding the molecular events regulating adaptation may provide information for the identification of sensitive targets of impending toxicity, as well as new strategies for therapeutic intervention.

MICROARRAY ANALYSIS AND ENVIRONMENTAL TOXICOLOGY

Determining, in advance, if exposure to an environmental agent may result in tissue injury or human disease is the primary challenge confronting modern toxicology. Until the recent advent of technological developments to facilitate this process, identification of the maladaptive responses, which ultimately lead to tissue injury, has often been virtually impossible. For the past half-century rodent assays have been the backbones of toxicological assessment. Unfortunately, these studies are seldom done to predict the outcome of exposure to a newly devised potential product; sadly, they are generally initiated following tragic incidents resulting from exposure to untested agents. The focus of these studies is to identify and characterize toxic agent(s) responsible for the detrimental effects from an array of potential environmental contaminants. The endpoints evaluated are often not reflective of the agent's mechanism of toxic action, making it difficult to identify other chemicals likely to initiate the same effects, susceptible individuals, or effective strategies to limit toxic injury.

To address these issues there is a growing interest in the use of microarray techniques to screen large numbers of substances for potential toxicity at an early stage in product development, long before they are released into the environment. To effectively accomplish this aim it has been proposed that groups of genes important in maladaptive responses to toxins be used to screen the effects of agents on an assortment of cell types. Many investigators are using microarray techniques to identify molecular 'fingerprints', sequella of genes with altered expression following toxic exposure (32, 33). Determining which genes to analyze, a requirement for the development of these screens, is likely to benefit from pooling of the results of many individual studies. In addition, devising gene sets reflecting adaptive responses may also assist investigators in predicting toxicity. Years of studies on the syndromes comprising numerous genetic diseases and inborn errors in metabolism have clearly demonstrated the consequences of deficits in appropriate adaptive responses (34). Clearly, in the future, information accumulated from the results of microarray analysis will play a central role in understanding, intervening and predicting toxicity.

ACKNOWLEDGMENTS

Research in the investigators laboratories is supported by NIH grants ES06897 and ES05022.

REFERENCES

1.   A. Brazma and J. Vilo, Gene expression data analysis, *FEBS Lett.* 480: 17-24 (2000).
2.   J. Khan, M.L. Bittner, Y. Chen, P.S. Meltzer, J.M. Trent, DNA microarray technology: the anticipated impact on the study of human disease, *Biochim. Biophys. Acta.* 1423: M17-M28 (1999).
3.   K.M. Kurian, C.J. Watson and A.H. Wyllie, DNA chip technology, *J. Pathol.* 187: 267-271 (1999).
4.   P.C. Mann, Selected lesions of dioxin in laboratory rodents, *Toxicol. Pathol.* 25: 72-79 (1997).
5.   M.H. Sweeney and P. Mocarelli, Human health effects after exposure to 2,3,7,8-TCDD, *Food Addit. Contam.* 17: 303-316 (2000).

6.   D. Neubert, Reflections on the assessment of the toxicity of "dioxins" for humans, using data from experimental and epidemiological studies, *Teratog. Carcinog. Mutagen.* 17: 157-215 (1997-98).

7.   Y.P. Dragan and D. Schrenk, Animal studies addressing the carcinogenicity of TCDD (or related compounds) with an emphasis on tumour promotion. *Food Addit. Contam.* 17: 289-302 (2000).

8.   L. Poellinger, Mechanistic aspects--the dioxin (aryl hydrocarbon) receptor, *Food Addit. Contam.* 17: 261-266 (2000).

9.   S.T. Okino and J.P. Whitlock, The aromatic hydrocarbon receptor, transcription, and endocrine aspects of dioxin action, *Vitam. Horm.* 59:241-264 (2000).

10.   P.A. Bertazzi, I. Bernucci, G. Brambilla, D. Consonni and A.C. Pesatori, The Seveso studies on early and long-term effects of dioxin exposure: a review, *Environ. Health Perspect.* S2: 625-633 (1998).

11.   D.E. Heck, Nitric oxide biosynthesis and metabolism in keratinocytes: the functions and consequences of nitric oxide production in the skin, *In*: J.D. Laskin and D.L. Laskin (eds), Cellular and Molecular Biology of Nitric Oxide. pp. 293-308. New York: Marcel Dekker (1999).

12.   C.A. Harrington, C. Rosenow and J. Retief, Monitoring gene expression using DNA microarrays, *Curr. Opin. Microbiol.* 3: 285-291 (2000).

13.   G. Sherlock, Analysis of large-scale gene expression data, *Curr Opin Immunol.* 12: 201-205 (2000).

14.   C.B. Epstein and'R.A. Butow, Microarray technology - enhanced versatility, persistent challenge, *Curr. Opin. Biotechnol.* 11: 36-41 (2000).

15.   D.D. Bowtell, Options available--from start to finish--for obtaining expression data by microarray, *Nature Genet.* 21S: 25-32 (1999).

16.   V.G. Cheung, M. Morley, F. Aguilar, A. Massimi, R. Kucherlapati and G. Childs, Making and reading microarrays, *Nature Genet.* 21S: 15-19 (1999).

17.   D. Ghosh, High throughput and global approaches to gene expression, *Comb. Chem. High Throughput Screen,* 3: 411-420 (2000).

18.   J. Aach, W. Rindone and G.M. Church, Systematic management and analysis of yeast gene expression data, *Genome Res.* 10: 431-445 (2000).

19.   M.D. Kane, T.A. Jatkoe, C.R. Stumpf, J. Lu, J.D. Thomas and S.J. Madore, Assessment of the sensitivity and specificity of oligonucleotide (50mer) microarrays, *Nucleic Acids Res.* 28: 4552-4557 (2000).

20.   M. Schena, D. Shalon, R.W. Davis and P.O. Brown, Quantitative monitoring of gene expression patterns with a complementary DNA microarray, *Science.* 270: 467-470 (1995).

21.   J.J. Chen, R. Wu, P.C. Yang, J.Y. Huang, Y.P. Sher, M.H. Han, W.C. Kao, P.J. Lee, T.F. Chiu, F. Chang, Y.W. Chu, C.W. Wu and K. Peck, A proximal CCD imaging system for high-throughput detection of microarray-based assays, *IEEE Eng. Med Biol. Mag.* 18:120-122 (1999).

22.   J.A. Ferguson, T.C. Boles, C.P. Adams and D.R. Walt, A fiber-optic DNA biosensor microarray for the analysis of gene expression, *Nature Biotechnol.* 14: 1681-1684 (1996).

23.   P.K. Kim and C.S. Deutschman, Inflammatory responses and mediators, *Surg. Clin. North Am.* 80: 885-894 (2000).

24.   D.K. Podolsky, Review article: healing after inflammatory injury--coordination of a regulatory peptide network, *Pharmacol. Ther.* 14S: 87-93 (2000).

25.   J.D. Laskin, D.E. Heck and D.L. Laskin, Multifunctional role of nitric oxide in inflammation, *Trend. Endro. Metabol.* 5: 377-382 (1994).

26.   G.W. Sullivan, I.J. Sarembock and J. Linden, The role of inflammation in vascular diseases, *J. Leukoc. Biol.* 67: 591-602 (2000).

27.   G. Ramadori and B. Christ, Cytokines and the hepatic acute-phase response, *Semin. Liver Dis.* 19: 141-155 (1999).

28.   R. Carmaciu, The adaption to environmental factors: hazard or programme, *Rom. J. Physiol.* 35: 247-252 (1998).

29.   T. Yamaguchi, H. Nishimura, T. Watanabe, S. Saito, M. Yabuki, K. Shiba, N. Isobe, F. Kishida, M. Kumano, F. Shono, H. Adachi and M. Matsuo, A research to develop a predicting system of mammalian subacute toxicity. I. Prediction of subacute toxicity using the biological parameters of acute toxicities, *Chemosphere*. 32: 979-998 (1996).

30.   C.R. Gardner, D.E. Heck, C.S. Yang, P.E. Thomas, X.J. Zhang, G.L. DeGeorge, J.D. Laskin and D.L. Laskin, Role of acetaminophen-induced hepatotoxicity in the rat, *Hepatology*, 27: 748-754 (1998).

31.   D.E. Heck, L. Louis, M.A. Gallo and J.D. Laskin, Modulation of the development of pleutei by nitric oxide in the sea urchin *Arbacia punctulata*, *Biological Bulletin* 199: 195-197 (2000).

32.   C.A. Afshari, E.F. Nuwaysir and J.C. Barrett, Application of complementary DNA microarray technology to carcinogen identification, toxicology, and drug safety evaluation, *Cancer Res.* 59: 4759-4760 (1999).

33.   M.J. Cunningham, S. Liang, S. Fuhrman, J.J. Seilhamer and R. Somogyi, Gene expression microarray data analysis for toxicology profiling, *Ann N.Y. Acad. Sci.* 919: 52-67 (2000).

34.   J.V. Leonard and A.A. Morris, Inborn errors of metabolism around time of birth, *Lancet.* 356: 583-587 (2000).

USE OF COVALENT BINDING IN RISK ASSESSMENT

H. Greim

Institute of Toxicology and Environmental Hygiene
Technical University München, Hohenbachernstraße 15-17
D-85350 Freising-Weihenstephan, Germany

ABSTRACT

Risk characterization comprises hazard identification describing the intrinsic toxic potential of a chemical, toxicokinetics, as well as the toxic mechanisms, information about dose response and exposure assessment. Compounds that induce reversible effects, which are repaired during and after exposure, are considered thresholded and allow definition of a NOEL. If damage is not repaired, the effect persists and accumulates upon repeated exposure. In such cases a NOEL cannot be determined. Biological reactive intermediates of chemicals have the potential to bind covalently to cellular macromolecules like proteins and DNA. Such interaction is not repaired completely and may persist. Thus, data on covalent binding (CB) are of qualitative and quantitative significance in the risk assessment process.

Qualitatively, CB, especially with DNA and in correlation with this to proteins, is indicative for an irreversible and non-thresholded mutagenic and carcinogenic effect. Absence or presence of CB assists to differentiate between primarily genotoxic and thresholded non-genotoxic carcinogens. Quantitatively, CB is used to understand internal exposure and target dose, which is a prerequisite for species-species extrapolation, and to justify extrapolation from high dose to low dose. The reactive intermediates of ethylene, propylene and styrene have been determined in rodents and humans and modeled to predict dose responses of internal exposure. It is described in this communication that such information, together with other parameters like cell proliferation as a result of cytotoxicity, is the basis for quantitative risk assessment of human exposure to these compounds.

INTRODUCTION

Risk characterization comprises hazard identification describing the toxic potential of a chemical, toxicokinetics, as well as the toxic mechanisms, information about dose response and exposure assessment. Compounds that induce reversible effects, which are repaired during and after exposure, are considered thresholded and allow definition of the NOEL. If damage is not repaired, the effect persists and accumulates upon repeated exposure. In such cases no NOEL can be determined. Biological reactive intermediates of chemicals have the potential to bind covalently to cellular macromolecules like proteins and DNA. Such interaction is not repaired completely and may persist.

In the risk assessment process data on covalent binding (CB) are of qualitative and quantitative significance.

Qualitatively, CB, especially with DNA and in correlation with this to proteins, is indicative for an irreversible and non-thresholded mutagenic and carcinogenic effect. Absence or presence of CB assists to differentiate between primarily genotoxic and non-genotoxic carcinogens. Lung tumors in mice induced by 1,2-butadiene, coumarine, and possibly styrene may be the result of cytotoxic rather than genotoxic effects, because little or relatively low DNA-binding in the lungs has been observed as compared to the rat, which has no lung tumors. In such cases carcinogenic effects are considered thresholded.

Quantitatively, CB is used to evaluate carcinogenic potency as proposed by Lutz (1979). Later, this was extended by correlating DNA modifications at N7- and O6-alkylguanine with tumor incidence (Vogel *et al.,* 1990). Increasingly, CB became a tool for risk assessment, esp. to understand internal exposure and target dose, which is a prerequisite for species-species and high dose to low dose extrapolation. CB of ethylene oxide (Csanády *et al.,* 2000), propylene oxide (Segerbäck *et al.,* 1994) and styrene oxide (Filser *et al.,* 1993), the reactive intermediates of ethylene, propylene and styrene, have been determined in rodents and humans and modeled to predict dose responses of internal exposure. However, to understand dose response of tumor incidences in experimental animals, other parameters like cell proliferation as a result of cytotoxicity must also be considered.

Such information is the basis for quantitative risk assessment of human exposure to carcinogens and for a science-based classification of these chemicals as carcinogens.

So far, the mechanism of action and potency of a carcinogen have either not been taken into account for classification, or at best have been used as supporting arguments. The advancing knowledge of reaction mechanisms and the potency of carcinogens have initiated a reevaluation of the traditional concepts. The International Agency for Research of Cancer (IARC 1997), the OECD (1997a, b; Sanner *et al.,* 1997), and the European Commission (1996) are still discussing applying information on carcinogenic mechanisms and potency as criteria for a revised classification. The US Environmental Protection Agency (EPA 1996) and a committee of the Deutsche Forschungsgemeinschaft (Neumann *et al.,* 1997, 1998) have published modified concepts for further discussion. Both committees propose to use data on the carcinogenic mechanism and information whether carcinogenicity is not likely below a certain dose. The American Conference of Governmental Industrial Hygienists (ACGIH 1997) has been using a concept which considers carcinogenic potency for classification since 1995.

RISK ASSESSMENT OF STYRENE, ETHYLENE AND ETHYLENE OXIDE

Styrene (ST)

The epidemiological studies which have been carried out to date on workers in the styrene producing and processing industries (IARC, 1994; Kolstad *et al.,* 1995) have not yielded clear evidence of carcinogenic effects. Increased incidence of tumors of the lymphatic and hematopoietic systems was inconsistent and the observed tumor incidences did not correlate with the cumulative exposure to styrene. In addition, the subjects were also exposed to other substances as well as styrene (e.g., 1,3-butadiene, ethylbenzene, dyes, benzene) in most studies. However, Vodicka *et al.,* (1995, 1999) described higher DNA-adduct levels in peripheral lymphocytes of laminators than in factory controls and Somerovska *et al,* observed DNA-strand-breaks in mononuclear leukocytes of workers in the reinforced plastic industry.

Mutagenicity studies *in vitro* required the presence of metabolic activation systems to produce positive results. Cytogenetic studies with experimental animals and exposed workers have produced contradictory results (Henschler, 1987; IARC, 1994).

Increased tumor incidences were found in 3 of 11 long-term studies with rodents: two demonstrated an increase in the incidence of lung tumors in mice and one of mammary tumors in rats. The latter, however, was detected only for the medium dose range. Cruzan *et al.* (1999, 2000) confirmed these studies by reporting an increased incidence of bronchioalveolar adenomas in male and female CD-1 mice and lung carcinomas in females after 104 weeks exposure to 160 ppm ST by inhalation, but not in rats.

Styrene is metabolized in the organism to the epoxide styrene-7,8-oxide (SO), which alkylates macromolecules *in vitro* and *in vivo* (IARC, 1994; Osterman-Golkar *et al.*, 1995, see Figure 1).

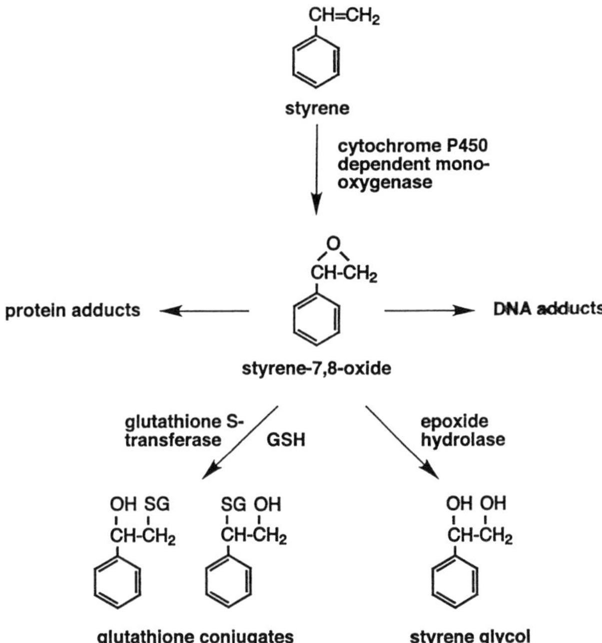

Figure 1. Metabolism of styrene and styrene-7,8-oxide.

The epoxide is mutagenic *in vitro* and carcinogenic in an animal study (IARC, 1994). SO is further metabolized by hydrolase and GSH-S-transferase activities. Species differences of these activities are given in Table 1.

Table 1. Species-specific enzyme activities in styrene and styrene oxide metabolism

Activity/mg protein	Mouse	Rat	Man
Epoxidation	++	+	(+)
GSH- S-Transferase	++	++	+
Hydrolase	+	++	+++

These data suggest that in humans body burden of SO after ST or SO exposure is lower than in rats or mice. Humans have a lower capacity to produce the epoxide than rodents, but higher hydrolase activity for further metabolism. This agrees with the findings of Ostermann-Golkar *et al.* (1995), who have investigated species differences of Hb-adduct formation at low dose ST and SO exposure (Table 2).

Table 2. Slopes of dose response curves for Hb-adduct formation (hydroxyphenethylvaline) at low doses of styrene oxide and styrene (Ostermann-Golkar *et al.*, 1995)

	Species	Relative slope
Styrene-7,8-oxide	Mouse	4
	Rat	2
Styrene	Mouse	1
	Rat	0.3
	Man	0.03

The relative slopes of the dose response curves were lowest in humans, higher in rats, highest in mice.

These species differences in the metabolic activation and inactivation of ST and SO might also explain the increased incidence of lung tumors in mice, which did not occur in rats (Cruzan *et al.*, 1999, 2000). Accordingly, Carlson (1999) observed profound species differences in CYP2E1 and CYP2F2 activities in Clara cells among mouse, rat and man (Table 3).

Table 3. Relative activities of styrene metabolism by CYP2E1 and CYP2F2 in the lungs of mice, rats, and humans (GP Carlson, 1999)

	CYP2E1			CYP2F2		
	Mouse	Rat	Human	Mouse	Rat	Human
Clara cells						
Large bronchioles	3	0	N	4	0	N
Term. bronchioles	2	1	0	3	2	0

No activities in other lung cells
N = not present, 0–4 increasing activities

Species differences in the activities of P-450, epoxide hydrolase and GST-S-transferases in the lungs have also been observed by Oberste-Frielinghaus *et al.* (1999), indicating that the lung burden of SO after ET inhalation is highest in the mouse, less in the rat and lowest in humans. Filser's group presented further data during this meeting (Oberste-Frielinghaus *et al.*), that at low ST concentrations when metabolism follows first order kinetics, the ratio of the rates of SO formation from ST to those of metabolic elimination of SO was almost 5 times higher in lung cells of mice than in those of rats. In vivo, 80 ppm ST resulted in a GSH loss of 30% in the lungs of mice and 60% at 300 ppm. In rats, 300 ppm ST reduced GSH content by 20 % only. Moreover, SO-lung burden in rats at 400 ppm without tumors has been calculated to be about twofold higher that in mice at 40 ppm, at which tumors have occurred (Filser 2000). Table 4 summarizes the information that allows the following conclusion: At exposure concentrations of about 100 ppm, SO lung burden is highest in the mouse, lower in rats and even lower in man. At that exposure GSH-depletion only occurs in the mouse, not in rats or humans.

Table 4. Species-specific lung toxicity after inhalation of about 100 ppm styrene (Oberste-Frielinghaus *et al.*, 2000)

	Mouse	Rat	Man
Tumor formation	+	–	?
SO lung burden	++	+	(+)
GSH-depletion	++	0	0
DNA-Adducts	(±)	(±)	?

Since DNA-adducts are near background levels and similar in mice and rats, the increased tumor incidence in the mouse lung does not correlate with the DNA adduct levels. Thus, lung tumors in mice most likely are the result of GSH-depletion rather than the consequence of DNA-adduct formation and primary genotoxicity. Furthermore, because there is little SO formation in humans, high activity of SO hydrolase and no GSH depletion at 100 ppm, lung tumors in mice after ST inhalation are of little relevance to man.

Estimations of the cancer risk associated with exposure to styrene have been carried out on the basis of the body burden of styrene-7,8-oxide or of that of its adducts with hemoglobin and DNA, and by taking into account the results of the long-term studies with experimental animals (Csanády et al., 1995; Filser et al., 1993). For 40 years of styrene exposure at work (styrene vapor concentration of 20 ml/m^3 air, 8 h/day, 5 days/week, 48 weeks/year) cancer risks were determined to range between 1.7 and 7.5 per 100 000 exposed persons. These risks are smaller than the unavoidable risk resulting from endogenously formed ethylene oxide of 1.2 per 10 000 persons (Greim 1993). This risk assessment considered a primary genotoxic mechanism for tumor induction. However, the more recent findings indicate that at least lung tumors in mice are the result of cytotoxicity rather than genotoxicity, suggesting a thresholded mechanism.

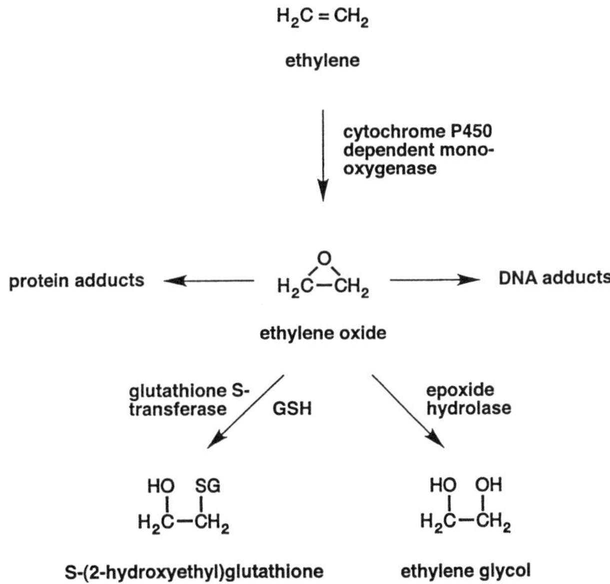

Figure 2. Metabolism of ethylene and ethylene oxide.

Ethylene (ET) and Ethylene Oxide (EO)

In the mammalian organism ET is metabolized to EO (Ehrenberg et al., 1977; Filser and Bolt 1984, see Figure 2).

From the rate of endogenous production of ET, its rate of metabolism to EO, and pharmacokinetic data of EO, the average concentration of endogenous EO in the human body has been calculated to be 0.17 nmol/kg body weight (Filser et al., 1992). EO is almost completely metabolized by epoxide hydrolase and glutathione S-transferase (Filser and Bolt, 1984). Following ET and EO exposure of animals and workers, hemoglobin- and

DNA-adducts of EO have been reported (see Csanády *et al.*, 2000). As summarized by IARC (1994) EO, not ET, was found to be mutagenic in vitro and in vivo and carcinogenic in experimental animals. To understand the discrepant results of the rodent studies with ET and EO and their relevance to humans several toxicokinetic models have been developed. They permit prediction of the body burden of EO and the formation of 2-hydroxyethyl adducts with hemoglobin and DNA in ET-exposed animals and humans (Csanády *c* 2000). As given in Table 5, DNA-adduct formation after exposure at 50 ppm ethylene or 1 ppm ethylene oxide is similar in mice, rats and humans, whereas formation of Hb-adducts differs. However, adjusting these data to the life-spans of the erythrocytes in the different species (mouse = 40 d, rat = 60 d, man = 126 d) the Hb-adduct levels agree.

Several conclusions can be drawn from these investigations:

1. Exposure of 50 ppm ET leads to similar amounts of EO-Hb and EO-DNA levels like exposure to 1 ppm EO (Table 5).
2. The different Hb-adduct levels among mice, rats and humans can be explained by the different life-spans of the erythrocytes (Table 4). Taking this into account, there are no species differences in the EO body burdens, at least up to ET exposures of 300 ppm.
3. Carcinogenicity studies with ET in rodents are negative because the body burden of the EO metabolically formed during ET.exposure is too low to result in increased tumor incidence.
4. The carcinogenic risk at the workplace exposure is high. For 40 years of exposure at work at concentrations of 50 ppm ET/m^3 air or 1 ppm EO/m^3, 8 h/day, 5 days/week, 48 weeks/year, a cancer risk of 1 per 100 has been determined.

Table 5. Hb- and DNA-adducts after ethylene and ethylene oxide exposure in different species (Csanady *et al.*, 2000)

Species	50 ppm ethylene		1 ppm ethylene oxide	
	nmol HEV/g Hb	nmol HEG/g DNA	nmol HEV/g Hb	nmol HEG/g DNA
Mice	0.65	0.44	0.62	0.40
Rats	1.4	0.64	1.3	0.58
Humans	2.6	0.56	2.4	0.52

HEV= N-(2-hydroxyethyl)valine), HEG=N7-(2-hydroxyethyl)guanine

CONCLUSION

Cancer research has made great progress and carcinogenic substances can now be better differentiated on the basis of their mechanisms of action and potency. Since risk assessment has to consider quantitative aspects, quantitative information of dose-dependence and toxicokinetics is required. For risk assessment of carcinogenic chemicals, covalent binding to proteins or DNA is a valid tool to investigate the body burden of biological reactive intermediates of carcinogenic chemicals. Furthermore, the biological relevance of reactive intermediates and covalent binding, especially in the low dose range, must be evaluated. Such information is essential for species-species extrapolation and definition of the carcinogenic potency of human exposure. International bodies increasingly use such information to evaluate carcinogenic compounds.

REFERENCES

ACGIH: American Conference of Governmental Industrial Hygienists: Threshold Limit Values (TLVs) and Biological Exposure Indices (BEIs). ACGIH, Technical Affairs Office, Cincinnati, OH 45240, 2000
Carlson GP: personal communication (2000)

Cruzan G, Cushman JR, Andrews LS, Granville GC, Johnson KA, Hardey CJ, Coombs DW, Mullins PA, Brown WR: Chronic toxicity/oncogenicity study of styrene in CD rats by inhalation exposure for 104 weeks. Toxicol Sci 46, 266–281, 1998

Cruzan G, Cushman JR, Andrews LS, Granville GC, Johnson KA, Bevan Ch, Hardey CJ, Coombs DW, Mullins PA, Brown WR: Chronic toxicity/oncogenicity study of styrene in CD-1 mice by inhalation exposure for 104 weeks. Toxicol Sci 2000, in press

Csanády GA, Kessler W, and Filser JG (1995). Carcinogenic risk estimates for inhaled styrene based on the body burden of its metabolite styrene-7,8-oxide. Naunyn-Schmiedeberg's Arch. Pharmacol., Suppl. 351, 122.

Csanády GA, Denk B, Pütz C, Kreuzer PE, Kessler W, Baur C, Gargas ML, Filser JG: A physiological toxicokinetic model for exogenous and endogenous ethylene and ethylene oxide in rat, mouse, and human: Formation of 2-hydroxyethyl adducts with hemoglobin and DNA. Toxicol Appl Pharmacol 165, 1–26, 2000

Ehrenberg L, Ostermann-Golkar S, Segerbäck D, Svenson K, Calleman CJ: Evaluation of genetic risks of alkylating agents. III. Alkylation of hemoglobin after metabolic conversion of ethene to ethene oxide in vivo. Mutat Res 45, 175–184, 1977

EPA: Environmental Protection Agency: Proposed Guidelines for Carcinogenic Risk Assessment. Office of Research and Development. U.S. Environmental Protection Agency. Washington, DC. EPA/600/P-92/003C, April 1996, Federal Register 61, 17960–18011, 1996

European Commission: Guidelines for Setting Specific Concentration Limits for Carcinogens in Annex I of Directive 67/548/EEC. Inclusion of Potency Considerations. Commission Working Group on the Classification and Labelling of Dangerous Substances. 16.07.1996

Filser J, Bolt HM: Inhalation pharmacokinetics based on gas uptake studies. VI. Comparative evaluation of ethylene oxide and butadiene monoxide as exhaled reactive metabolites of ethylene and 1,3-butadiene in rats. Arch Toxicol 55, 219–223, 1984

Filser JG, Denk B, Törnquist M, Kessler W, Ehrenberg L: Pharmacokinetics of ethylene in man: Body burden with ethylene oxide and hydroxyethylation of hemoglobin due to endogenous and environmental ethylene. Arch Toxicol 66, 157–163, 1992

Filser JG, Kessler W, and Csanády GA: Different approaches to estimate the carcinogenic risk of styrene based on animal studies. The SIRC Review 3 , 54, 1993.

Filser JG: personal communication 2000

Greim H (Ed.): Ethylene. Occupational Toxicants. Wiley-VCH, Weinheim, 1993.

Henschler D: (Ed.): Styrol. Toxikologisch-arbeitsmedizinische Begründungen von MAK-Werten, 13. Lieferung, VCH-Verlag, Weinheim, 1987

IARC: International Agency for Research of Cancer. IARC Monographs, Preamble, 12. Evaluation, Vol. 68, Lyon 1997

IARC: International Agency for Research on Cancer. Some industrial chemicals: Ethylene oxide, Styrene. IARC Monographs, Volume 60, Lyon 1994.

Kolstad HA, Juel K, Olsen J, and Lynge E: Exposure to styrene and chronic health effects: mortality and incidence of solid cancers in the Danish reinforced plastics industry. Occup. Environ. Med. 52, 320–327, 1995.

Lutz WR: In vivo covalent binding of organic chemicals to DNA as a quantitative indicator in the process of chemical carcinogenesis. Mutation Res 65, 289–356, 1979

Neumann HG, Thielmann HW, Filser JG, Gelbke HP, Greim H, Kappus H, Norpoth KH, Reuter U, Vamvakas S, Wardenbach P, Wichmann HE: Vorschläge zur Änderung der Einstufung krebserzeugender Arbeitsstoffe. Arbeitsmed. Sozialmed. Umweltmed. 32, 298–304, 1997

Neumann HG, Thielmann HW, Filser JG, Gelbke HP, Greim H, Kappus H, Norpoth KH, Reuter U, Vamvakas S, Wardenbach P, Wichmann HE: Changes in the classification of carcinogenic chemicals in the work area. J. Cancer Res. Clin. Oncol. 124, 661–669, 1998

NRC: National Research Council. Science and judgment in risk assessment. Committee on Risk Assessment of Hazardous Air Pollutants, Commission on Life Sciences, NRC. Washington, DC: National Academy Press 1994

Oberste-Frielinghaus M, Dhawan-Robl M, Pütz C, Csanády J, Filser JG: Styrene metabolism and glutathione depletion in lungs of rats and mice. Abstract, Sixth International Symposium on Biological Reactive Intermediates, Paris 2000

Oberste-Frielinghaus M, Dhawan-Robl M, Pütz C, Csanády J, Filser JG: Metabolism of styrene and racemic styrene-7,8-oxide in microsomes and cytosol of lung obtained from mouse, rat and human. Naunyn-Schmiedeberg's Arch Pharmacol R 157, 1999

OECD 1995a: Classification Systems on Carcinogens in OECD Countries—Similarities and Differences. Prepared by Norway and The Netherlands. Oslo and Bilthoven, Final Report, September 1995

OECD 1995b: Report from the OECD Working Group on Harmonization of Classification and Labeling of Carcinogens, Washington DC, October 17–18, 1995

Osterman-Golkar S, Christakopoulos A, Zorec V, Svensson K: Dosimetry of styrene 7,8-oxide in styrene- and styrene oxide-exposed mice and rats by quantification of haemoglobin adducts. Chem-Biol Interact 95, 79–87, 1995.

Sanner T, Dybing E, Kroese D, Roelfzema H, Hardeng S: Carcinogen Classification Systems: Similarities and Differences. Reg. Toxicol. Pharmacol 23, 128–138, 1996

Segerbäck D, Ostermann-Golkar S, Molholt B, Nilsson R: *In vivo* tissue dosimetry as a basis for cross-species extrapolation in cancer risk assessment of propylene oxide. Regulatory Toxicol 20, 1–14, 1994

Somerovská M, Jahnová E, Tulinská J, Zámecniková M, Sarmanová J, Terenová A, Vodicková L, Lisková A, Vallová B, Soucek P, Hemminki K, Norppa H, Hirvonen A, Tates AD, Fuortes L, Dusinská M, Vodická P, Biomonitoring of occupational exposure to styrene in a plastics lamination plant. Mutat Res 428, 255–269, 1999

Vodicka P, Bastlova T, Vodickova I, Peterkova K, Lambert B, Hemminki K: Biomarkers of styrene exposure in lamination workers: levels of O6-guanine DNA adducts, DNA strand breaks and mutant frequencies in the hypoxanthine guanine phosphoribosyltransferase gene in T-lymphocytes. Carcinogenesis 16, 1473–1481, 1995

Vodicka P, Tyrdi9k T, Ostermann-Goökar S, Vodickova L, Peterkova K, Soucek P, Sarmanova J, Farmer PB, Granath F, Lambert B, Hemminki K: An evaluation of styrene genotoxicity using several biomarkers in a 3-year follow-up study of hand-lamination workers. Mutat Res 445, 205–224, 1999

Vogel EW, Barbin A, Nivard MJM, Bartsch H: Nucleophilic selectivity of alkylating agents and their hypermutability in *Drosophila* as predictors of carcinogenic potency in rodents. Carcinogenesis 11, 2211–2217, 1990

AUTHOR INDEX